Springer Collected Works in Mathematics

More information about this series at http://www.springer.com/series/11104

Photo by B. S. Skobelzin

Igor R. Shafarevich 1983

Igor R. Shafarevich

Collected Mathematical Papers

Reprint of the 1989 Edition

Springer

Igor R. Shafarevich
Steklov Mathematical Institute
Moscow
Russia

ISSN 2194-9875
Springer Collected Works in Mathematics
ISBN 978-3-662-47153-1 (Softcover)
978-3-540-13618-7 (Hardcover)

Springer Heidelberg Dordrecht London New York

Library of Congress Control Number: 2012954381

Mathematics Subject Classification (2010): 14-XX, 01A70, 10.0X, 12.00

Printed on acid-free paper

Springer-Verlag GmbH Berlin Heidelberg is part of Springer Science+Business Media
(www.springer.com)

IGOR R. SHAFAREVICH

COLLECTED MATHEMATICAL PAPERS

Springer-Verlag
Berlin Heidelberg New York
London Paris Tokyo

Igor R. Shafarevich
Steklov Mathematical Institute
ul. Vavilova 42, 117966 Moscow, USSR

Mathematics Subject Classification (1980):
00, 10, 14, 20, 17, 22, 13

ISBN 978-3-540-13618-7 Springer-Verlag Berlin Heidelberg New York
ISBN 978-3-540-13618-7 Springer-Verlag New York Berlin Heidelberg

Library of Congress Cataloging-in-Publication Data
Shafarevich, I. R. (Igor Rostislavovich), 1923–
Collected mathematical papers.
Text in English, French, and German, translated from Russian.
Bibliography: p.
1. Mathematics–Collected works. 2. Shafarevich, I. R. (Igor Rostislavovich), 1923–
QA3.S542512 1989 510 87-36932

2141/3140-543210 – Printed on acid-free paper

Preface

This volume contains almost all of my mathematical papers published between 1943 and 1984. Not included are: lectures given at conferences, if their content is covered by other papers; notes of lecture courses; books.

All papers here are not in their original form, but are published in English translations (with two exceptions, where the translations are in French and German). Some translations were made especially for this edition. Most of them however were already published and are merely reproduced here. I am grateful to the publishing houses who permitted using their editions.

Some mistakes in translations are corrected in the text, but only if they made the text unintelligible. All other corrections and remarks about the subsequent development of the subjects considered in the papers are included in the "Notes" at the end of the volume. References to these are given in the margin.

My thanks are due to Professor H. G. Zimmer for translations of several papers and to Professors G. van der Geer and D. Zagier who checked the translations, corrected the English of my "Notes" and made other valuable remarks.

I am grateful to Springer-Verlag for the suggestion to publish this collection of my papers. To go over my old papers and to recollect the hopes and ideas connected with them turned out to be much more interesting and edifying to me than I could have imagined. I was especially struck by the following observation: how often the published paper was only a small part of far more ambitious plans! Some papers now appear to me as remainders of a shipwreck, although the reader probably will not notice it. I wonder whether this is so with other mathematicians. Can it be a subconscious device analogous to Machiavelli's "arcieri prudenti" who chose a much higher aim in order to hit a remote object?

Another strong feeling that I experienced while looking through my old papers was a feeling of deep gratitude to all those with whom I had the opportunity to cooperate in the field of mathematics: to my pupils, colleagues and to those from whom I have learned – either through personal contact or from their works.

Moscow, June 1988 I. R. Shafarevich

Table of contents

On the normalizability of topological fields

Dokl. Akad. Nauk SSSR **40,** 149–151 (1943)
[C.R. Acad. Sci. (Dokl.), Paris **40,** 133–135 (1943)]

(Communicated by I. M. Vinogradow, Member of the Academy, 19. 2. 1943)

1 In this paper necessary and sufficient conditions are given in order that a given field be normalizable. These conditions make it possible to reduce some questions of the theory of topological fields to the theory of normalized fields.

In particular, the whole theory of locally bicompact fields can be obtained as a particular case of the theory of normalized fields. Let us say that a topological field is normalizable if we can define a function $\varphi(x)$ of its elements possessing the following properties:

1) $\varphi(x) > 0$, *if* $x \neq 0$; $\varphi(0) = 0$;
2) $\varphi(x+y) \leqq \varphi(x) + \varphi(y)$;
3) $\varphi(xy) = \varphi(x)\varphi(y)$;
4) a system of sets $U_\varepsilon\{\varphi(x) < \varepsilon\}$ forms a full system of neighbourhoods of zero.

In order to formulate the conditions of normalizability, the following denotations are necessary. Let R be the set of elements p of the field K such that $p^n \to 0$ for $n \to \infty$, and $\tilde{R}$ be the sets of elements q such that $q^{-1} \notin R$. We shall say that a
2 set W of elements of K is bounded if for any given neighbourhood U of zero there exists a neighbourhood V of zero such that $WV \in U$.

Theorem 1. *In order that a topological field K be normalizable, it is necessary and sufficient that the set R be open and the set $\tilde{R}$ be bounded.*

The necessity of these conditions is obvious. We shall prove their sufficiency by constructing a norm $\varphi(x)$.

Let p be an element of the set R different from zero. Then the sets $p^n R$ $(-\infty < n < +\infty)$ form a full system of neighbourhoods of zero. They increase according as n decreases, and their totality covers K. Let x be any element of the field K different from zero. We denote by $n(x)$ the largest number n, for which we have $x \in p^n R$, and form the set $\alpha_k(x) = \dfrac{n(x^k)}{k}$. It is easy to prove that

$$\alpha_{k_l}(x) \geqq \alpha_k(x), \quad \alpha_{k+1}(x) \geqq \alpha_k(x) - \frac{1}{k+1}.$$

Hence it follows that for $n \to \infty$ $\alpha_k(x)$ tends to a certain limit which will be denoted by $\alpha(x)$.

Now we take an arbitrary number ϱ $(0 < \varrho < 1)$ and define a function $\psi_\varrho(x)$ in the following manner:

$$\psi_\varrho(0) = 0,$$
$$\psi_\varrho(x) = \varrho^{\alpha(x)} \quad \text{for} \quad x \neq 0.$$

It is easy to see that the function $\psi_\varrho(x)$ satisfies to the conditions 1), 3) and 4); however, it is not a norm, since the condition 2) does not hold for any ϱ.

A weaker condition

$$2')\quad \psi_\varrho(x+y) \leqslant a_\varrho\{\psi_\varrho(x)+\psi_\varrho(y)\}$$

holds instead, a_ϱ being fixed. Indeed, if 2′) were not fulfilled, the set of the numbers $\frac{\psi_\varrho(x+y)}{\psi_\varrho(x)+\psi_\varrho(y)}$ $(x, y \in K)$ would not be bounded. Then a set z_n would exist such that

$$\frac{\psi_\varrho(1+z_n)}{1+\psi_\varrho(z_n)} \to \infty \quad \text{for} \quad n \to \infty.$$

Hence it would follow that $\frac{1}{\psi_\varrho(1+z_n)} = \psi_\varrho\left(\frac{1}{1+z_n}\right) \to 0$ for $n \to \infty$. But since for $\psi_\varrho(x)$ 4) holds, we should have $\frac{1}{1+z_n} \to 0$ for $n \to \infty$. Hence, in virtue of $\frac{\psi_\varrho(1+z)}{1+\psi_\varrho(z)} = \frac{\psi_\varrho(1+z^{-1})}{1+\psi_\varrho(z^{-1})}$, it would follow that $\frac{1}{1+z_n^{-1}} \to 0$ for $n \to \infty$, which is impossible in virtue of $\frac{1}{1+z_n} + \frac{1}{1+z_n^{-1}} = 1$.

It remains now to choose such a ϱ that $a_\varrho \leqslant 1$. This inequality holds if we take $\varrho' = \varrho^{\frac{1}{a_\varrho+1}}$, and thus the proof of the theorem is accomplished.

From Theorem 1 we can obtain some corollaries relating to locally bicompact fields, since, as we can easily prove, the conditions formulated here are satisfied for any field of this kind. This enables us to study the structure of all locally bicompact fields by using the theorems of Ostrowsky [1] on the normalized fields.

In fact, let a locally bicompact field K be connected. In virtue of Theorem 1, it is normalizable; it is easy to see that its norm $\varphi(x)$ is Archimedean.

If K is not connected, then the norm $\varphi(x)$ is non-Archimedean. We even can prove that it is discrete. For, since K is non-Archimedean, the set $R\{\varphi(y)<1\}$ forms a ring, and the sets $R_\varepsilon\{\varphi(x)<\varepsilon<1\}$ are its ideals. But since K is locally bicompact, the quotient rings R/R_ε are finite. This contradicts to the hypothesis that between two given numbers ε_1 and ε_2 there exists an infinity of numbers ε_α such that $\varphi(x_a)=\varepsilon_a$ holds for certain elements x_a of K.

Thus, we have proved the following

Theorem 2. *Any locally bicompact field is normalizable. If it is connected, its norm is Archimedean; otherwise it is discrete.*

From Theorem 2 and the theorems by Ostrowski it follows at once that a connected locally bicompact topological field is a field of real or of complex numbers [2], and that a non-connected locally bicompact topological field is a field of p-adic or of z-adic numbers with a finite coefficient field [3].

As a last application of Theorem 1 we shall consider the relation between two norms $\varphi_1(x)$ and $\varphi_2(x)$ giving the same topology for a field K. If $\varphi(x)$ is the norm constructed with the aid of the process described in the proof of Theorem 1, and if it defines in K the same topology as $\varphi_1(x)$ and $\varphi_2(x)$, we can use an element $p \in R$ and the numbers $\varrho_1 < 1$, $\varrho_2 < 1$, and obtain

$$\varphi(x) = \varphi_1(x)^{\alpha_1}; \quad \varphi(x) = \varphi_2(x)^{\alpha_2},$$

where

$$\alpha_1 = \frac{\log \varrho_1}{\log \varphi_1(p)}; \quad \alpha_2 = \frac{\log \varrho_2}{\log \varphi_2(p)},$$

whence it follows

$$\varphi_1(x) = \varphi_2(x)^{\beta}, \quad \beta = \frac{\alpha_2}{\alpha_1}.$$

Thus, we come to

Theorem 3. *If the norms $\varphi_1(x)$ and $\varphi_2(x)$ define the same topology in a field K, they are connected by a relation $\varphi_1(x) = \varphi_2(x)^{\beta}$.*

V. A. Stekloff Mathematical Institute,
Academy of Sciences of the USSR

Received 19. 2. 1943

References

1. Ostrowsky, Acta mathem., *41* (1918).
2. L. Pontrjagin, Ann. of Math., II s., *33* (1932).
3. Van-Dantzig, Studien over topologische Algebra, Amsterdam, 1931.

On Galois groups of $\mathfrak{p}$-adic fields

Dokl. Akad. Nauk SSSR **53**, No. 1, 15–16 (1946)
[C.R. Acad. Sci. (Dokl.), Paris **53**, 15–16 (1946)]

(Communicated by I. M. Vinogradow, Member of the Academy, 18. II. 1946)

In this note the following problem is solved; let R be the field of $\mathfrak{p}$-adic numbers; k, a normal extension of R; and K, the Abelian extension of k defined by its group of norms H in k; to find the Galois group K/R (if it is normal), k and H being given.

Let us denote the Galois groups: K/R by $G = \{g, h, \dots\}$, k/R by $F = \{\sigma, \tau, \dots\}$ and K/k by $\mathfrak{A} = \{\alpha, \beta, \dots\}$.

Let $(F:1) = m$, $(\mathfrak{A}:1) = n$. By assumption, $\mathfrak{A}$ is an Abelian normal divisor of G with the factor-group F.

For the determination of G by $\mathfrak{A}$ and F it is sufficient to know in $\mathfrak{A}$ the automorphisms $\alpha \to \alpha^\sigma$ and the system of factors ([1], p. 93). Moreover, if $\bar{\sigma}$, $\bar{\tau}$ is a system of representatives of the classes of G by $\mathfrak{A}$ then

$$\bar{\sigma}\,\bar{\tau} = \alpha_{\sigma,\tau}\,\overline{\sigma\tau}\,. \tag{1}$$

Let A be a normal simple algebra over R of degree m with the invariant $1/m$ ([2], p. 112). It is known that A contains a maximal commutative subfield of degree m isomorphic to k ([2], p. 113). Then

$$A = (t_\sigma, k, a_{\sigma,\tau}) \qquad a_{\sigma,\tau} \in k\,.$$

Let, finally, $a\,H \to \alpha(a\,H)$ be an isomorphism between k^*/H and $\mathfrak{A}$ established by means of Hasse's symbol ([3], p. 149).

Theorem. *K/R is normal if and only if H is invariant with respect to $\sigma \in F$.*
1 *If K/R is normal, then the automorphisms and the system of factors in $\mathfrak{A}$ are determined thus*

$$\alpha^\sigma(a\,H) = \alpha(a^\sigma H) \tag{2}$$

$$\alpha_{\sigma,\tau} = \alpha(a_{\sigma,\tau} H) \tag{3}$$

The assertion (2) easily follows from the properties of Hasse's symbol ([3], p. 151). To prove (3), consider besides A another two simple algebrae B and C with centres R and k of degrees $m\,n$ and n and with invariants $1/m\,n$ and $1/n$:

$$B = (u_g, K/R, b_{g,h})\,; \qquad C = (v_\alpha, K/k, c_{\alpha,\beta})\,.$$

It is known [4] that by means of C we may in the following way compute $\alpha_{\sigma,\tau}$:

$$\alpha_{\sigma,\tau} = \alpha(V(v_{\alpha_{\sigma,\tau}})H) = \alpha\Big(\prod_{\beta \in \mathfrak{A}} c_{\beta,\alpha_{\sigma,\tau}} H\Big)\,. \tag{4}$$

The algebra B enables us to deduce (3) from (4). To this end observe that the subalgebra of B consisting of elements commutable with k has the form $(u_\alpha, K/k, b_{\alpha,\beta})$. Its invariant over k is $1/n$ [5] and, consequently, it is isomorphic to C. From (4) we obtain

$$\alpha_{\sigma,\tau} = \alpha\,(\prod_{\beta\in\mathfrak{A}} b_{\beta,\alpha_{\sigma,\tau}} H)\,. \tag{5}$$

In order to connect the algebra A with B, consider B^n. On the one hand, it is known [6] that

$$B^n \sim (k/R, N_{K/k}(b_{\sigma,\tau}) \prod_{\beta\in\mathfrak{A}} b_{\beta;\overline{\sigma\tau}}\, b^{-1}_{\beta,\bar{\sigma}\bar{\tau}})$$

and, on the other, the invariant of B^n is $1/m$, and, consequently, $B^n \sim A$, i.e.

$$a_{\sigma,\tau} \sim N_{K/k}(b_{\sigma,\tau}) \prod_{\beta\in\mathfrak{A}} b_{\beta,\overline{\sigma\tau}}\, b^{-1}_{\beta,\bar{\sigma}\bar{\tau}}\,. \tag{6}$$

Combining (5) and (6), we observe that (3) is equivalent to the following formula:

$$N_{K/k}(b_{\sigma,\tau}) \prod_{\beta\in\mathfrak{A}} b_{\beta,\overline{\sigma\tau}}\, b^{-1}_{\beta,\bar{\sigma}\bar{\tau}} \sim \prod_{\beta\in\mathfrak{A}} b_{\beta,\alpha_{\sigma\tau}} H\,. \tag{7}$$

To prove it multiply the conditions of associativeness for the $b_{g,h}$

$$b^{\beta}_{g,h}\, b_{\beta,gh} = b_{\beta,g}\, b_{\beta g,h}$$

over all $\beta\in\mathfrak{A}$. In the result we obtain

$$N_{K/k}(b_{g,h}) \prod_{\beta\in\mathfrak{A}} b_{\beta,gh} = \prod_{\beta\in\mathfrak{A}} b_{\beta,g}\, b_{\beta g,h}\,.$$

Substitute here $g = \alpha_{\sigma,\tau}$ and $h = \overline{\sigma\tau}$ and replace according to (1) (since $\alpha_{\sigma,\tau}\in\mathfrak{A}$, $\prod_\beta b_{\beta\alpha_{\sigma,\tau},\bar{\sigma}\bar{\tau}} = \prod_\beta b_{\beta,\bar{\sigma}\bar{\tau}}$)

$$N_{K/k}(b_{\alpha_{\sigma,\tau},\bar{\sigma}\bar{\tau}}) \prod_\beta b_{\beta,\bar{\sigma}\bar{\tau}} = \prod_\beta b_{\beta,\alpha_{\sigma,\tau}}\, b_{\beta,\bar{\sigma}\bar{\tau}}$$

i.e.

$$\prod_\beta b_{\beta,\alpha_{\sigma\tau}} H = N_{K/k}(b_{\alpha_{\sigma,\tau}\bar{\sigma}\bar{\tau}}) \prod_\beta b_{\beta,\overline{\sigma\tau}}\, b^{-1}_{\beta,\bar{\sigma}\bar{\tau}} H\,.$$

Since we may omit all factors of the form $N_{K/k}(b)\in H$, this formula proves (7).

References

1. Zassenchaus, H.: Lehrbuch der Gruppentheorie, 1937
2. Deuring, M.: Algebren, 1935
3. Chevalley: J. reine u. angew. Math. *169* (1932), 140
4. Nakayama, F.: Math. Ann. *112* (1935), 85
5. Köthe, G.: Math. Ann. *107* (1933), 761
6. Witt, E.: J. reine u. angew. Math. *173* (1936), 191

Received 18|II|1946

On p-extensions

Mat. Sb., Nov. Ser. **20** (62), 351 – 363 (1947). Zbl. **41,** 171
[Transl., II. Ser., Am. Math. Soc. **4,** 59 – 72 (1956)]

Introduction

In this article are considered finite, normal, separable extensions, whose degrees are powers of a prime p. In what follows such extensions will be called p-extensions.

The work is divided into two parts, depending on the field, whose extensions we consider.

In the first part (§ 1-2) we take for our fundamental field a field of p-adic numbers, not containing the p-th roots of unity. Here $\mathfrak{p}$ is a prime ideal, dividing p.

We denote the fundamental field by k, and its degree over the field of p-adic rationals by n_0. The fundamental result can then be formulated as follows:

Theorem 1. *The p-extensions of a field k are in one-to-one correspondence with the normal subgroups of a free group S with $n_0 + 1$ generators, whose indices are powers of p. Moreover, the correspondence is such that if a p-extension K corresponds to the normal subgroup N, then the Galois group of K is isomorphic to the quotient group S/N. If two p-extensions K and K_1 correspond to the normal subgroups N and N_1 respectively, then $K \supset K_1$ implies $N \subset N_1$ and conversely.*

1 This theorem can be regarded as a generalisation of the local class field theory.

The theorem enable one to reduce a number of questions about p-extensions of k to questions in group theory. By means of group-theoretic considerations we solve here the following problems:

1. The existence of an extension of k with a given Galois p-group.
2. The number of extensions of k with a given Galois p-group.
3. The possibility of p-extensions contained in p-extensions with a larger Galois group.

The answers are as follows:

1. A given finite p-group is realized as a Galois group of an extension of the field k if and only if the number d of its generators does not exceed $n_0 + 1$.

2. If $d \leq n_0 + 1$, if p^n is the order of the p-group and if α is the number of its automorphisms, then the number of extensions, whose Galois groups are isomorphic to this group is

$$\frac{1}{\alpha} p^{(n_0+1)(n-d)}(p^{(n_0+1)}-1)(p^{n_0+1}-p)\cdots(p^{n_0+1}-p^{d-1}).$$

3. Let G and $\overline{G}$ be two p-groups and let $\overline{G}$ be a homomorphic image of G

by a given fixed homomorphism. If the number of generators of G (and hence of $\overline{G}$) does not exceed $n_0 + 1$, then for every field with a Galois group $\overline{G}$ there exists an extension with a Galois group G (over k) such that the natural homomorphism of G onto $\overline{G}$ coincides with the given homomorphism.

§ 1 is devoted to the proof of Theorem 1, and § 2 to the derivation of the corollaries.

In the second part (§ 3) we take for the fundamental field an algebraic function field with an algebraically closed field of constants of characteristic $p \neq 0$ and consider only the unramified p-extensions.

It appears that completely analogous results hold in this case. The only change consists in the number of generators of the free group S by means of which we describe the unramified p-extensions. Instead of $n_0 + 1$ the number is γ, which has been determined for any algebraic function field of characteristic p by Hasse and Witt [5].

These results coincide in the special case of Abelian fields with the results of Witt and Schmid [6].

§ 1

We shall introduce the following notation:

k is a field of $\mathfrak{p}$-adic numbers, not containing the p-th roots of unity;

K is a p-extension of the field k;

K^* is the multiplicative group of K;

$(K^*)^p$ is the subgroup of K^*, generated by the p-th powers;

K_1 is the extension of k which is composed of all cyclic fields of degree p; if K_{i-1} is given, then $K_i = (K_{i-1})_1$;

n_0 is the degree of k over the field of p-adic rationals.

S is a free group with $n_0 + 1$ generators;

$G^{(1)}$ is the subgroup of a group G, generated by the commutators and p-th powers; if $G^{(i-1)}$ is given, then $G^{(i)} = (G^{(i-1)})^1$;

$\mathfrak{P}^{(i)} = S/S^{(i)}$.

We shall prove Theorem 1, formulated in the indroduction.

It is known that if G has a finite number of generators, then $(G:G^{(1)}) < \infty$ [1]; it can then be easily shown by induction on i that any $G^{(i)}$, being a subgroup with a finite index of a group with a finite number of generators, has also a finite number of generators and therefore $(G:G^{(i)}) < \infty$ for every i.

In particular, $\mathfrak{P}^{(i)}$ is always a finite group.

It is known from class field theory that K_{i+1}/K_i is a finite extension, and that its Galois group is isomorphic to $K_i^*/(K_i^*)^p$. It is clear that K_i is the maximal of all those p-extensions whose Galois group satisfies the condition $G^{(i)} = (e)$. From the fundamental theorem of Galois theory all such extensions are in one-to-one correspondence with the normal subgroups of the Galois group of K_i/k, and that this correspondence enjoys all the properties listed in the formulation of Theorem 1.

On the other hand all the normal subgroups of S, whose quotient groups satisfy the condition $G^{(i)} = (e)$, contain $S^{(i)}$, and conversely if a normal subgroup contains $S^{(i)}$, then its quotient group satisfies this condition. Hence the normal subgroups whose quotient groups satisfy the condition $G^{(i)} = (e)$ are in one-to-one correspondence with the normal subgroups of $\mathfrak{P}^{(i)}$. Since for every p-group $G^{(1)} \neq G$ [1], it follows that $G^{(i)} = (e)$ for some i.

Hence the theorem will follow if we can prove that the Galois group of K_i/k is isomorphic with $\mathfrak{P}^{(i)}$.

First of all we determine the number of generators of the Galois group of K_i/k. By Burnside's theorem [1] the number of generators of a p-group G is equal to d if

$$(G : G^{(1)}) = p^d.$$

If G is the Galois group of K_i/k, then $G^{(1)}$ belongs to K_1. Hence we must calculate $[K_1 : k]$. Since, by assumption, k does not contain a p-th root of unity, it follows that

$$[K_1 : k] = (k^* : (k^*)^p) = p^{n_0 + 1} \text{ [2]}.$$

Hence the number of generators of the Galois group of K_i/k is $n_0 + 1$. Hence this group is isomorphic with the quotient group of S with respect to a normal subgroup which we shall denote by N_i.

It now remains to prove that

$$N_i = S^{(i)}.$$

It is clear that $N_i \supseteq S^{(i)}$. Since the relation $G^{(i)} = (e)$ holds for the Galois group of K_i/k, it is sufficient to show that the indices of N_i and $S^{(i)}$ are the same, or, in other words that

$$(S : S^{(i)}) = [K_i : k]. \tag{1}$$

Formula (1) is proved by induction on i. For $i = 1$

$$(S : S^{(i)}) = [K_i : k] = p^{n_0 + 1}.$$

Suppose that we have already shown that

$$(S : S^{(i)}) = [K_i : k] \tag{2}$$

$$(S : S^{(i+1)}) = (S : S^{(i)})(S^{(i)} : S^{(i+1)}). \tag{3}$$

But since $S^{(i+1)} = (S^{(i)})^{(1)}$, it follows that

$$(S^{(i)} : S^{(i+1)}) = (S^{(i)} : (S^{(i)})^{(1)}) = p^{d_i}, \tag{4}$$

where d_i is the number of generators of $S^{(i)}$. By Schreier's theorem [3]

$$d_i = (S:S^{(i)})(n_0+1-1)+1 = (S:S^{(i)})n_0+1. \tag{5}$$

On the other hand

$$[K_{i+1}:k] = [K_{i+1}:K_i]\,[K_i:k]. \tag{6}$$

But since $K_{i+1} = (K_i)_1$ it follows that

$$[K_{i+1}:K_i] = [(K_i)_1:K_i] = p^{n_i+1}, \tag{7}$$

where n_i is the degree of K_i over the field of p-adic rationals.

Since

$$n_i = [K_i:k]\,n_0, \tag{8}$$

it follows from (3), (4) and (5) that

$$(S:S^{(i+1)}) = (S:S^{(i)})\,p^{(S:S^{(i)})n_0+1},$$

while, on the other hand, from (6), (7) and (8) we have

$$[K_{i+1}:k] = [K_i:k]\,p^{[K_i:k]\,n_0+1}.$$

Hence by (2), formula (1) is established and the proof of Theorem 1 follows.

§ 2

In this section we will derive some corollaries of Theorem 1 of the previous section.

Corollary 1. *A p-group G is a Galois group of some extension of the field k if and only if the number of its generators does not exceed n_0+1.*

The proof is obvious.

Corollary 2. *Let $(G:e) = p^n$, let d be the number of generators of G and a be the number of automorphisms of G. If $d \le n_0+1$, then the number of extensions of k, whose Galois groups are isomorphic to G is*

$$\frac{1}{a}\,p^{(n_0+1)(n-d)}(p^{n_0+1}-1)(p^{n_0+1}-p)\cdots(p^{n_0+1}-p^{d-1})^{*}. \tag{9}$$

By Theorem 1 the number of extensions having G for its Galois group is equal to the number of normal subgroups of S, whose quotient groups are isomorphic to G.

To every choice of n_0+1 generators of G corresponds a normal subgroup of S, whose quotient group is isomorphic with G and conversely. But to two different systems of generators $s_1, \cdots, s_{n_0+1}$ and $t_1, \cdots, t_{n_0+1}$ of G can correspond the same normal subgroup. This happens if and only if the two systems $s_1, \cdots, s_{n_0+1}$ and $t_1, \cdots, t_{n_0+1}$ satisfy the same relations. That is, if

$$f(s_1, \cdots, s_{n_0+1}) = e$$

implies

$$f(t_1, \cdots, t_{n_0+1}) = e$$

* An analogous formula is found in Witt [4], but for a quite different case.

and conversely. This happens if and only if the correspondence

$$s_i \rightarrow t_i$$

defines an automorphism of G.

Hence, we must divide all the systems of generators of $n_0 + 1$ elements of G into systems of intransitivity of the automorphism group. The number of such systems of intransitivity will then be the answer to our problem. But an automorphism is determined by its action on a system of generators and therefore the number of systems of generators in each system of intransitivity is the same and is equal to the number of automorphisms of G. Hence the number of extensions which we seek is

$$\frac{1}{\alpha}\phi_{n_0+1}(G),$$

where $\phi_{n_0+1}(G)$ is the number of systems of generators with $n_0 + 1$ elements of G.

By the basis theorem [1], a system of elements $s_1, \cdots, s_{n_0+1}$ forms a system of generators of G if and only if the totality of elements $s_1 G^{(1)}, \cdots, s_{n_0+1} G^{(1)}$ is a system of generators of $G/G^{(1)}$. Therefore we must first determine the number $\phi_{n_0+1}(G/G^{(1)})$.

But $G/G^{(1)}$ is an elementary Abelian group of order p^d. We know the number $\frac{1}{\alpha_1}\phi_{n_0+1}(G/G^{(1)})$ where α_1 is the number of automorphism of $G/G^{(1)}$. This number is equal to the number of subgroups of order p^{n_0+1-d} of an elementary Abelian group of order p^{n_0+1}, that is

$$\frac{1}{\alpha_1}\phi_{n_0+1}(G/G^{(1)}) = \frac{(p^{n_0+1}-1)(p^{n_0+1}-p)\cdots(p^{n_0+1}-p^{d-1})}{(p^d-1)(p^d-p)\cdots(p^d-p^{d-1})}\ [1].$$

But

$$\alpha_1 = (p^d-1)(p^d-p)\cdots(p^d-p^{d-1})\ [1].$$

Therefore

$$\phi_{n_0+1}(G/G^{(1)}) = (p^{n_0+1}-1)(p^{n_0+1}-p)\cdots(p^{n_0+1}-p^{d-1}). \tag{10}$$

Any system of generators of G composed on $n_0 + 1$ elements can be obtained from such a system of generators of $G/G^{(1)}$ by considering any system of representatives of the corresponding cosets. In this way

$$\phi_{n_0+1}(G) = \phi_{n_0+1}(G/G^{(1)})(G^{(1)}:e)^{n_0+1}. \tag{11}$$

Since the order of $G^{(1)}$ is p^{n-d}, equations (10) and (11) establish formula (9).

Corollary 3. *Let G and $\overline{G}$ be two p-groups with $\overline{G}$ a homomorphic image of G by some fixed homomorphism, such that the number of generators of G, and hence of $\overline{G}$ does not exceed $n_0 + 1$.*

Then for every extension $\overline{K}/k$ with a Galois group $\overline{G}$ there exists an extension K/k with a Galois group G such that $K \supset \overline{K}$ and that the given homomorphism *of G onto $\overline{G}$ is realized by a natural homomorphism of a Galois group of a field onto a Galois group of a subfield.*

Proof. Let G_0 be the normal subgroup of G, which is the center of the homomorphism of G onto $\overline{G}$.

In view of Theorem 1 we have to show that for any normal subgroup $\overline{N}$ of S for which $S/\overline{N} \cong \overline{G}$ there exists a smaller normal subgroup N for which $S/N \cong G$ and such that $\overline{N}$ corresponds to G_0 under this isomorphism. Since the number of generators of G does not exceed $n_0 + 1$, it follows that S must contain a normal subgroup N_1 such that $S/N_1 \cong G$, and such that to G_0 corresponds a normal subgroup $\overline{N}_1$, and $S/\overline{N}_1 \cong \overline{G}$.

We now select a sufficiently large i such that all the subgroups of S under consideration contain $S^{(i)}$. Then to all these subgroups of S will correspond subgroups in $\mathfrak{P}^{(i)}$, which we will distinguish by a dash. Our statement will be proved if we can show that there exists an automorphism in $\mathfrak{P}^{(i)}$ which transforms $\overline{N}_1'$ onto $\overline{N}'$. In fact this automorphism will transform N_1' onto the normal subgroup N', whose image in S will have all the necessary properties.

The quotient groups of $\mathfrak{P}^{(i)}$ with respect to $\overline{N}_1'$ and $\overline{N}'$ are isomorphic. Therefore it is sufficient to prove the following:

Two normal subgroups of $\mathfrak{P}^{(i)}$ go into each other by some automorphism if and only if the corresponding quotient groups are isomorphic.

The number of normal subgroups of $\mathfrak{P}^{(i)}$ having a given quotient group G is known, by Corollary 2. We now count the number of normal subgroups of $\mathfrak{P}^{(i)}$ obtained from a given one with a quotient group G by applying automorphisms $\mathfrak{P}^{(i)}$. If these two numbers turn out to be the same then our proposition is proved.

As a preliminary investigation we study the relations in the group $\mathfrak{P}^{(i)}$ and its automorphisms.

Let $s_1, s_2, \cdots, s_{n_0+1}$ be the generators of the group $\mathfrak{P}^{(i)}$, which are images of the free generators of S. The relations between the generators s_i can be easily described in the following way:

Consider the function

$$[x, y] = xyx^{-1}y^{p-1} \text{ in } S.$$

We shall call a form of the first degree any expression of the form

$$[x_1, y_1]^{a_1} [x_2, y_2]^{a_2} \dots [x_r, y_r]^{a_r}.$$

Assuming that the form of degree $k-1$ is already defined, we define a form of degree k as an expression of the form

$$[\xi_1, \eta_1]^{a_1} [\xi_2, \eta_2]^{a_2} \dots [\xi_r, \eta_r]^{a_r},$$

where ξ_i and η_i are any forms of degree $k-1$.

It is easily seen that all relations among the s_i in $\mathfrak{P}^{(i)}$ have the form

$$f(s_1, \cdots, s_{n_0+1}) = e, \tag{12}$$

where f is any form of degree not less than i.

We shall call the system of generators $t_1, \cdots, t_{n_0+1}$ in $\mathfrak{P}^{(i)}$ free if the relations among the t_i are the same as those among the s_i.

We need the following

Lemma. *Any system of $n_0 + 1$ generators in $\mathfrak{P}^{(i)}$ is free.*

In fact, let $t_1, \cdots, t_{n_0+1}$ be a new system of generators. If we express t_i in terms of s_i we see that any form of degree k in t_i is also a form of degree k in s_i. Therefore t_i satisfies all the relations of the form (12). This shows that any normal subgroup of S which is composed of relations among the t_i contains $S^{(i)}$, but it cannot exceed $S^{(i)}$ because it has of the same index as $S^{(i)}$. Therefore they coincide, which proves the lemma.

Corollary to the lemma. *The transformation of s_i into any system of $n_0 + 1$ generators determines an automorphism of $\mathfrak{P}^{(i)}$.*

Let $\mathfrak{R}$ be any normal subgroup of $\mathfrak{P}^{(i)}$ with a quotient group G. We designate by $\mathfrak{A}$ the group of all the automorphisms of $\mathfrak{P}^{(i)}$, and by $\mathfrak{B}$ the group of all those automorphisms which leave $\mathfrak{R}$ unaltered, and finally by $\mathfrak{C}$ the normal subgroup of $\mathfrak{B}$ composed of all the automorphism of $\mathfrak{P}^{(i)}$, which leave all the cosets with respect to $\mathfrak{R}$ unaltered.

The number of normal subgroups into which $\mathfrak{R}$ is transformed by applying to it the automorphism of $\mathfrak{P}^{(i)}$ is equal to $(\mathfrak{A}:\mathfrak{B})$. We have

$$(\mathfrak{A}:\mathfrak{B}) = \frac{(\mathfrak{A}:\mathfrak{C})}{(\mathfrak{B}:\mathfrak{C})} = \frac{(\mathfrak{A}:\mathfrak{C})}{\alpha'}, \tag{13}$$

where α' is the number of those automorphisms of G, which are induced by the automorphisms of $\mathfrak{P}^{(i)}$. It is clear that

$$\alpha' \leq \alpha.$$

We now determine $(\mathfrak{A}:\mathfrak{C})$. In what follows for brevity we shall designate the group $(\mathfrak{P}^{(i)})^{(1)}$ by $\overline{\mathfrak{P}}$. Consider the group $\mathfrak{B}'$ of automorphisms of $\mathfrak{P}^{(i)}$ which leave unaltered the cosets with respect to $\mathfrak{R}\overline{\mathfrak{P}}$. It is clear that

$$(\mathfrak{A}:\mathfrak{C}) = (\mathfrak{A}:\mathfrak{B}')(\mathfrak{B}':\mathfrak{C}). \tag{14}$$

In view of Burnside's basis theorem, $\mathfrak{R}\overline{\mathfrak{P}}$ is transformed under the automorphisms of $\mathfrak{P}^{(i)}$ into any subgroup of the same order, lying between $\mathfrak{P}^{(i)}$ and $\overline{\mathfrak{P}}$. The number of such subgroups is equal to

$$\frac{(p^{n_0+1}-1)(p^{n_0+1}-p)\cdots(p^{n_0+1}-p^{d-1})}{(p^d-1)(p^d-p)\cdots(p^d-p^{d-1})}. \tag{15}$$

Moreover, making again a change in generators, it is possible to induce in a quotient group with respect to any such subgroup

$$(p^d-1)(p^d-p)\cdots(p^d-p^{d-1}) \tag{16}$$

automorphisms. It follows from (15) and (16) that

$$(\mathfrak{A}:\mathfrak{B}')=(p^{n_0+1}-1)(p^{n_0+1}-p)\cdots(p^{n_0+1}-p^{d-1}). \tag{17}$$

We now find $(\mathfrak{B}':\mathfrak{C})$. We write each automorphism of $\mathfrak{P}^{(i)}$ in the form

$$s_k^{\sigma}=a_{k,\sigma}\cdot s_k.$$

This automorphism belongs to $\mathfrak{B}'$ is and only if all $a_{k,\sigma}\in\mathfrak{R}\overline{\mathfrak{P}}$: it belongs to $\mathfrak{C}$ if and only if $a_{k,\sigma}\in\mathfrak{R}$.

We consider the group $\mathfrak{P}$, which is the direct product of n_0+1 groups isomorphic to $\mathfrak{P}^{(i)}$. In $\mathfrak{P}$ to every automorphism corresponds a fixed element, but not every element of $\mathfrak{P}$ corresponds to an automorphism, and those elements which correspond to automorphisms do not form a subgroup. We shall denote their totality by $\mathfrak{L}$. We denote by $\widetilde{\overline{\mathfrak{P}}}$ the normal subgroup composed of those elements of $\mathfrak{P}$ whose components are in $\overline{\mathfrak{P}}$. Analogously we define $\widetilde{\mathfrak{R}}$. By $(M:e)$ we denote the number of elements of a finite set M. It can be easily seen that

$$(\mathfrak{B}':\mathfrak{C})=\frac{(\mathfrak{L}\cap\widetilde{\mathfrak{R}}\widetilde{\overline{\mathfrak{P}}}:e)}{(\mathfrak{L}\cap\widetilde{\mathfrak{R}}:e)}. \tag{18}$$

By the lemma, the system of elements $a_{k,\sigma}$ of $\mathfrak{P}^{(i)}$ defines an automorphism if and only if $a_{k,\sigma}s_k$ is a system of generators of $\mathfrak{P}^{(i)}$. But in view of the basis theorem this will take place if and only is $a_{k,\sigma}s_k\overline{\mathfrak{P}}$ is a system of generators for $\mathfrak{P}^{(i)}/\overline{\mathfrak{P}}$.

In this way, the property that an element λ of $\mathfrak{P}$ belongs to $\mathfrak{L}$ is determined by the class $\lambda\widetilde{\overline{\mathfrak{P}}}$, which corresponds to it. If we denote by x the number of classes $\lambda\widetilde{\overline{\mathfrak{P}}}\,(\lambda\in\mathfrak{L})$ in which the elements of $\widetilde{\mathfrak{R}}\widetilde{\overline{\mathfrak{P}}}$ lie, we will have

$$\left.\begin{aligned}(\mathfrak{L}\cap\widetilde{\mathfrak{R}}\widetilde{\overline{\mathfrak{P}}}:e)&=x(\widetilde{\overline{\mathfrak{P}}}:e),\\(\mathfrak{L}\cap\widetilde{\mathfrak{R}}:e)&=x(\widetilde{\overline{\mathfrak{P}}}\cap\widetilde{\mathfrak{R}}:e).\end{aligned}\right\} \tag{19}$$

But

$$\left.\begin{aligned}(\widetilde{\overline{\mathfrak{P}}}:e)&=(\overline{\mathfrak{P}}:e)^{n_0+1},\\(\widetilde{\overline{\mathfrak{P}}}\cap\widetilde{\mathfrak{R}}:e)&=(\overline{\mathfrak{P}}\cap\mathfrak{R}:e)^{n_0+1}.\end{aligned}\right\} \tag{20}$$

From (17), (18) and (19) we obtain

$$(\mathfrak{B}':\mathfrak{C})=(\overline{\mathfrak{P}}:\overline{\mathfrak{P}}\cap\mathfrak{R})^{n_0+1}=(\overline{\mathfrak{P}}\mathfrak{R}:\mathfrak{R})^{n_0+1}=p^{(n_0+1)(n-d)}. \tag{21}$$

Comparing equations (13), (14), (17) and (21) we obtain that the number of normal subgroups which we seek is equal to

$$\frac{1}{\alpha'}\,p^{(n_0+1)(n-d)}(p^{n_0+1}-1)(p^{n_0+1}-p)\cdots(p^{n_0+1}-p^{d-1}).$$

Here $\alpha'\leq\alpha$. On the other hand, by definition, this number cannot exceed the number defined by (9). From this it follows that these two numbers are equal, which proves Corollary 3.

As a side result we obtain that $\alpha = \alpha'$, that is that any automorphism of any quotient group of $\mathfrak{P}^{(i)}$ is induced by some automorphism of $\mathfrak{P}^{(i)}$.

§ 3

In this section we prove a theorem which is in many ways analogous to Theorem 1.

For the fundamental field we take an algebraic function field k with an algebraically closed field of constants of characteristic $p \neq 0$. We differ from § 1 by considering not all the p-extensions of k, but only the unramified ones. We shall denote by γ the birational invariant, which is designated by the same letter in Hasse and Witt [5], and by S a free group with γ generators.

Theorem 2. *The unramified p-extensions of k are in one-to-one correspondence with the normal subgroups of the group S, whose indices are powers of p. This correspondence has all the properties which are enumerated in the statement of Theorem 1.*

In what follows we shall consider only unramified p-extensions. Keeping in mind this limitation we can retain all the notation of § 1, (with the exception of n_0, which does not have any meaning in this case).

Almost all the arguments of § 1 hold in this case, so we will only note the few places where new arguments become necessary.

The existence for the field k of a finite extension K_1 which is a composite of all cyclic extensions of degree p has been shown for our case in the paper of Hasse and Witt [5]. We denote the degree of K_1/k by p^{γ}.

As in § 1, the proof depends on showing the equality of the two numbers $(S:S^{(i)})$ and $[K_i:k]$.

Arguing again by induction we see that the assertion of the theorem is equivalent to proving the formula

$$\gamma_i = [K_i:k]\,(\gamma - 1) + 1 \tag{22}$$

where γ_i is the number γ defined for the field K_i.

We shall devote this section to the proof of formula (22).

We shall establish a more general formula

$$\gamma' = [K:k]\,(\gamma - 1) + 1 \tag{23}$$

3 for any extension K/k (which need to be normal), which can be obtained by a chain of cyclic extensions of degree p, where by γ' we understand the number γ for the field K.

It is sufficient to prove formula (23) for an extension of degree p, since then we can easily prove it by induction for the degree K/k. In fact, if $K \supset \overline{K} \supseteq k$, and if $K/\overline{K}$ is a cyclic extension of degree p, and if the formula is proved for $\overline{K}/k$, then

$$\gamma' = [K:\overline{K}]\,(\overline{\gamma}-1) + 1,$$

but

$$\overline{\gamma} = [\overline{K}:k]\,(\gamma-1) + 1,$$

hence

$$\gamma' = [K:\overline{K}]\,[\overline{K}:k]\,(\gamma-1) + 1 = [K:k]\,(\gamma-1) + 1.$$

We will now establish formula (23) for a cyclic extension K/k of degree p. In this case $K = k(\theta)$, where

$$\wp(\theta) = v, \ \wp(x) = x^p - x, \ v \in k \quad [7].$$

All the elements of the field K have the form

$$\alpha = a_0 + a_1\theta + \cdots + a_{p-1}\theta^{p-1}, \ a_i \in k.$$

It is known from [5] that

$$p^{\gamma} = (b:\wp(a)),$$

where b goes over the additive group of all the elements of k for which the equation $\wp(x) = b$ is solvable in any $\mathfrak{p}$-adic closure $k_{\mathfrak{p}}$ of the field k (such elements will be called in the future unramified), while a goes over the additive group of all the elements of k. Similarly

$$p^{\gamma'} = (\beta:\wp(\alpha)),$$

where β and α have a similar meaning in K.

We denote by Γ_i the additive group of all unramified elements of K of the form

$$\beta = b_0 + b_1\theta + \cdots + b_{i-1}\theta^{i-1}.$$

In particular $\Gamma_0 = 0$.

Then

$$(\beta:\wp(\alpha)) =$$

$$= (\beta:\Gamma_{p-1}+\wp(\alpha))\,(\Gamma_{p-1}+\wp(\alpha):\Gamma_{p-2}+\wp(\alpha))\cdots(\Gamma_1+\wp(\alpha):\Gamma_0+\wp(\alpha)).$$

Formula (23) will be establish if we can show that

$$(\beta:\Gamma_{p-1}+\wp(\alpha)) = p^{\gamma}, \tag{24}$$

$$(\Gamma_i+\wp(\alpha):\Gamma_{i-1}+\wp(\alpha)) = p^{\gamma-1} \ \text{for } i = p-1, \cdots, 1 \tag{25}$$

since then

$$p^{\gamma'} = (\beta:\wp(\alpha)) = p^{\gamma}(p^{\gamma-1})^{p-1} = p^{p(\gamma-1)+1}.$$

In order to prove formulas (24) and (25) we need some preliminary investigations.

In the first place we must clarify what elements $b_0, b_1, \cdots, b_{p-1}$ will produce an element $b_0 + b_1\theta + \cdots + b_{p-1}\theta^{p-1}$ of the form $\wp(\alpha)$.

Let $\alpha = a_0 + a_1\theta + \cdots + a_{p-1}\theta^{p-1}$, then

$$\wp(\alpha) = \wp(a_0) + \wp(a_1\theta) + \cdots + \wp(a_{p-1}\,\theta^{p-1}).$$

Replacing everywhere θ^p by $\theta + v$, we get

$$\left.\begin{aligned}
&b_{p-1} = \wp(a_{p-1}),\\
&b_{p-2} = \wp(a_{p-2}) + a_{p-1}^p c_{p-1}^1 v,\\
&\cdots\cdots\cdots\cdots\cdots\cdots\cdots\cdots,\\
&b_{p-k} = \wp(a_{p-k}) + a_{p-(k-1)}^p c_{p-(k-1)}^1 v + \cdots + a_{p-1}^p c_{p-1}^{k-1} v^{k-1},\\
&\cdots\cdots\cdots\cdots\cdots\cdots\cdots\cdots\cdots\cdots\cdots\cdots.
\end{aligned}\right\} \quad (26)$$

In what follows we shall need the following lemma.

Lemma. *If the element $\alpha = a_0 + a_1\theta + \cdots + a_i\theta^i$ is unramified, then a_i is unramified. Conversely for every unramified element a_i, one can find elements a_0, $a_1, \cdots, a_{i-1}$ such that the element $a_0 + a_1\theta + \cdots + a_i\theta^i$ is unramified.*

Let us suppose that α is unramified. Let us denote the automorphism of K/k which transforms θ into $\theta + 1$ by σ. It is obvious that $\sigma^k\alpha$ is unramified.

We shall show that there exist elements $r_0, r_1, \cdots, r_i$ of a simple subfield R_p of the field k such that $r_0\alpha + r_1\sigma\alpha + \cdots + r_i\sigma^i\alpha = a_i$. Obviously a_i will be also an unramified element.

We select r_k, satisfying the conditions

$$\begin{aligned}
&r_0 + r_1 + \cdots + r_i = 0,\\
&0 \cdot r_0 + 1 \cdot r_1 + \cdots + ir_i = 0,\\
&0^2 \cdot r_0 + 1^2 \cdot r_1 + \cdots + i^2 \cdot r_i = 0,\\
&\cdots\cdots\cdots\cdots\cdots\cdots\cdots,\\
&0^{i-1} r_0 + 1^{i-1} r_1 + \cdots + i^{i-1} \cdot r_i = 0,\\
&0^i r_0 + 1^i r_1 + \cdots + i^i \cdot r_i = 1.
\end{aligned}$$

The determinant of this system is a Vandermonde determinant and is different from zero, therefore this system has a unique solution.

We shall now show that

$$\theta^k (r_0 + r_1\sigma + \cdots + r_i\sigma^i) = \delta_{ik}.$$

In fact

$$\begin{aligned}
\theta^k (r_0 + r_1\sigma + \cdots + r_i\sigma^i) &= r_0\theta^k + r_1(\theta+1)^k + \cdots + r_i(\theta+i)^k =\\
&= \theta^k(r_0 + r_1 + \cdots + r_i) + \theta^{k-1} c_k^1 (0 \cdot r_0 + 1 \cdot r_1 + \cdots + ir_i) + \cdots\\
&\cdots + (0^k r_0 + 1^k \cdot r_1 + \cdots + i^k r_i) = \delta_{ik}.
\end{aligned}$$

Our assertion is now proved for any α, since

$$(r_0 + r_1\sigma + \cdots + r_i\sigma^i)\,(a_0 + a_1\theta + \cdots + a_i\theta^i) = a_i.$$

Let now suppose that a_i is unramified. Then in order to select the elements $a_{i-1}, a_{i-2}, \cdots, a_0$ which we need, we will have to solve successively, as can be seen from equation (26), equations of the form

$$a_k + \wp(a_k^{\mathfrak{p}}) = c_k^{\mathfrak{p}}$$

for elements $a_k \in k$ and $a_k^{\mathfrak{p}} \in k_{\mathfrak{p}}$ for given $c_k^{\mathfrak{p}}$. The solvability of such equations is shown in paper [6].

We can now proceed to the proof of formulas (24) and (25).

We introduce a homomorphism χ_i, defined in Γ_i by

$$\chi_i(b_0 + b_1\theta + \cdots + b_{i-1}\theta^{i-1}) = b_{i-1}.$$

We note that $\chi_i(\alpha) = 0$ if and only if $\alpha \in \Gamma_{i-1}$. From the proof of the lemma it follows that

$$\chi(\alpha) = (r_0 + r_1\sigma + \cdots + r_i\sigma^i)\,\alpha. \tag{27}$$

We first prove formula (24). We have

$$(\beta : \Gamma_{p-2} + \wp(\alpha)) = (\chi_p(\beta) : \chi_p(\wp(\alpha)).$$

In view of equation (27)

$$\chi_i(\alpha^p) = (r_0 + r_1\sigma + \cdots + r_i\sigma^i)\,\alpha^p = \chi_i(\alpha)^p,$$

since r_i belongs to a simple field R_p. Hence

$$\chi_p(\wp(\alpha)) = \wp(\chi_p(\alpha)) = \wp(a_{p-1}).$$

On the other hand it follows from the lemma that the value of $\chi_p(\beta)$ ranges over the whole group of unramified elements of k. Hence

$$(\beta : \Gamma_{p-1} + \wp(\alpha)) = (b : \wp(\alpha)) = p^{\gamma}.$$

We now prove formula (25).

We apply homomorphism χ_i to the groups in the left hand side of formula (25). Then, arguing as in the first case, we obtain:

$$(\Gamma_i + \wp(\alpha) : \Gamma_{i-1} + \wp(\alpha)) = (b : \chi_i(\Gamma_i \cap \wp(\alpha)).$$

Now we must find out what elements are in the group $\chi_i(\Gamma_i \cap \wp(\alpha))$. In order to do this we make use of (26) letting

$$b_{p-1} = 0,\ b_{p-2} = 0,\ \cdots,\ b_{i+1} = 0.$$

We then obtain

$$\wp(a_{p-1}) = 0, \text{ that is } a_{p-1} = r_{p-1} \in R_p,$$
$$\wp(a_{p-2}) + r_{p-1}c_{p-1}^{1}v = 0,$$

that is if $r_{p-1} \neq 0$, then

$$v = \wp\left(-\frac{1}{r_{p-1}c_{p-1}^{1}}\,a_{p-2}\right)$$

which is impossible, that is

$$a_{p-1} = 0,\ \wp(a_{p-2}) = 0.$$

Proceeding further in this way, we obtain

$$a_{p-1} = 0,\ a_{p-2} = 0, \cdots, a_{i+1} = r \in R_p,\ b_i = \wp(a_i) + r(i-1)v.$$

Conversely if $b_i = \wp(a_i)$ then $\wp(a_i\theta^i)$ is an element of the form

$$b_0 + \cdots + b_i\theta^i,$$

and if $b_i = \wp(a_i) + rv$, $r \neq 0$, then $\wp(a_i\theta^i + \frac{1}{r(i-1)}\theta^{i+1})$ will be an element of the same form. In this way

$$\chi_i(\Gamma_i \cap \wp(\alpha)) = \wp(a) + rv,$$

$$(\Gamma_i + \wp(\alpha) : \Gamma_{i-1}\wp(\alpha)) = (b : \wp(\alpha) + rv) = p^{\gamma-1}.$$

This completes the proof of the theorem.

Obviously all the corollaries derived from Theorem 1 in § 2, hold in this case also.

BIBLIOGRAPHY

[1] H. Zassenhaus, *Lehrbuch der Gruppentheorie*, Bd. I, Teubner, Leipzig-Berlin, 1937.

[2] C. Chevalley, *La théorie du corps de classes*, Ann. of Math. (2) 41 (1940), 394-418.

[3] O. Schreier, *Die Untergruppen der freien Gruppen*, Abh. Math. Sem. Hamburg. Univ. 5 (1927), 161-183.

[4] E. Witt, *Konstruktion von galoisschen Körpern der Charakteristik* p *zu vorgegebener Gruppe der Ordnung* p^f, J. Reine Angew. Math. 174 (1936), 237-245.

[5] H. Hasse and E. Witt, *Zyklische unverzweigte Erweiterungskörper vom Primzahlgrade* p *über einem algebraischen Funktionenkörper der Charakteristik* p, Monatsh. Math. Phys. 43 (1936), 477-492.

[6] L. Schmid and E. Witt, *Unverzweigte abelsche Körper vom Exponenten* p^n *über einem algebraischen Funktionenkörper der Charakteristik* p, J. Reine Angew. Meth. 176 (1936), 168-173.

[7] A. Albert, *Modern higher algebra*, Univ. of Chicago Press, 1937.

ON p-EXTENSIONS

I. R. ŠAFAREVIČ

(Résumé)

The main results of this paper are two theorems.

Theorem 1. *Let* k *be a* $\mathfrak{p}$*-adic number field containing no* p*-th roots of the unit and let* n_0 *be its degree over the rational* p*-adic field. All the normal extensions of* k*, whose degrees are powers of* p *(*p*-extensions) are in one-to-one correspondence with the normal subgroups of the free group* $\mathfrak{G}$ *with* $n_0 + 1$ *gener-*

ators whose indices are powers of p. The correspondence has the following properties: 1) if a p-extension K corresponds to the normal subgroup N, then its Galois-group is isomorphic to the quotient group $\mathfrak{G}/N$; *2) if p-extensions* K_1 *and* K_2 *correspond to the normal subgroups* N_1 *and* N_2, *respectively, then* $K_1 \subset K_2$ *implies* $N_1 \supset N_2$ *and vice versa.*

Theorem 2. *Let k be an algebraic function field with algebraically closed constant field of characteristic p. All non-degenerate normal non-ramified extensions of k, whose degrees are powers of p, are in one-to-one correspondence with with normal subgroup of the free group* $\mathfrak{G}$ *with* γ *generators whose indices are powers of p. Here* γ *is the invariant of the field k introduced by Hasse and Witt* [5]. *The correspondence has the properties 1) and 2) formulated in Theorem* 1.

Theorem 1 can be regarded as generalization of the local class field theory.

On the particular case of Abelian p-extensions Theorem 2 reduces to well known results of Witt and Schmid [6].

Theorems 1 and 2 permit to prove some properties of p-extensions with pure group theoretic methods.

We shall formulate the following 3 consequences only for the case *k* being $\mathfrak{p}$-adic field not containing p-th root of unit. In the case *k* being an algebraic function field analogous consequences can be proved in quite the same way.

Consequence 1. *For a preassigned p-group G there exists an extension K/k whose Galois-group is isomorphic to G if and only if the minimal number of generators of the group G does not exceed* $n_0 + 1$.

Consequence 2. *Let d be the minimal number of generators of the p-group G of the order* p^n *and* α *its number of automorphisms. If* $d \leq n_0 + 1$, *the number of different extensions with Galois-group G is equal to*

$$\frac{1}{\alpha} p^{(n_0+1)(n-d)} (p^{n_0+1} - 1)(p^{n_0+1} - p) \cdots (p^{n_0+1} - p^{d-1}).$$

Consequence 3. *Let G be a p-group and* $\overline{G}$ *another p-group homomorphic to G with some preassigned homomorphism. For each extension K/k with Galois-group G there exists an extension* $\overline{K}/k$ *containing K and whose Galois-group is* $\overline{G}$ *while the preassigned homomorphism* $\overline{G}$ *on G is realised as the homomorphism of the Galois-group of the field on the Galois-group of the subfield.*

Translated by:
Emma Lehmer

A general reciprocity law

Mat. Sb., Nov. Ser. **26** (68), 113 – 146 (1950). Zbl. **36,** 159
[Transl., II. Ser., Am. Math. Soc. **4,** 73 – 106 (1956)]

Introduction

In this article we solve a problem which is connected with three questions in the theory of algebraic numbers, namely with an analogy between algebraic numbers and functions, with a general reciprocity law and with class field theory.

1. An analogy between algebraic numbers and functions. The idea of a deep analogy between algebraic number fields and algebraic function fields was inherent in the works of Gauss and Kummer, but apparently was first expressed by Kronecker. It was noted that simple ideals in the theory of algebraic numbers play the same part as points on the Riemann surface in algebraic function fields, and that the prime divisors of the discriminant correspond to branch points on the Riemann surface, etc. The problem arose of a systematic carrying over of results from the theory of algebraic functions which have been proved analytically, into the theory of algebraic numbers.

The theory of abelian integrals is the most erudite part of the theory of algebraic numbers. D. Hilbert ([8], p. 365, 483) began investigating the analog of this theory in algebraic number fields. Into every algebraic number field Ω, containing an n-th root of unity, a symbol $\left(\frac{\alpha,\beta}{\mathfrak{p}}\right)$ is introduced which puts into correspondence with every pair of numbers α, β and a prime ideal $\mathfrak{p}$ of the field Ω an n-th root of unity ([3], p. 53). For this symbol, which is called the norm residue symbol, it can be proved that ([3], p. 54)

$$\prod_{\mathfrak{p}} \left(\frac{\alpha,\beta}{\mathfrak{p}}\right) = 1, \tag{a}$$

where α and β are fixed numbers of the field Ω and $\mathfrak{p}$ runs over all prime ideals of Ω. As Hilbert pointed out, this relation plays the same role in the theory of algebraic numbers as does Cauchy's integral theorem in the theory of abelian integrals. The theory of the norm residue symbol which was developed by Hilbert for $n = 2$ has been generalized to an arbitrary n in the papers of N. G. Čebotarev, E. Artin and H. Hasse.

Further developments in the theory of abelian integrals, [10] and [11], as well as in the theory of algebraic numbers, show that relation (a) is more precisely an analog not of Cauchy's integral theorem, but of its corollary, namely that the sum of the residues of the abelian differential $\alpha d\beta$ in all the points of the Riemann surface is equal to zero. From this point of view the norm residue symbol $\left(\frac{\alpha,\beta}{\mathfrak{p}}\right)$ is analoguous to the residue of the abelian differential $\alpha d\beta$ at the point $\mathfrak{p}$.

In spite of the fact that the norm residue symbol plays the part of a residue, its definition has nothing in common with the definition of a residue of an abelian differential. Moreover, although the symbol $\left(\frac{\alpha,\beta}{\mathfrak{p}}\right)$, as well as the abelian differential $\alpha\, d\beta$, depends only on the behaviour of α and β in the point $\mathfrak{p}$, that is on the expansion of α and β in a $\mathfrak{p}$-adic series, nevertheless the definition of the symbol $\left(\frac{\alpha,\beta}{\mathfrak{p}}\right)$ depends materially on the properties of Ω as a whole, that is on the properties of all its prime ideals.

In this paper we give a construction of the symbol $\left(\frac{\alpha,\beta}{\mathfrak{p}}\right)$, which is exactly analogous to the definition of the residue of an abelian differential.

In doing this we prove that the expansion of an algebraic function in a series in a neighborhood of a point on the Riemann surface corresponds to the decomposition of an algebraic number into a certain product (§ 1, part 3). If we apply to the product into which the numbers α and β decompose the same operations which are applied to the series expansions of functions in calculating the residue, we get the value of the symbol (α, β), which is equal to $\left(\frac{\alpha,\beta}{\mathfrak{p}}\right)$. All the fundamental properties of abelian differentials are carried over to the field of algebraic numbers. In doing this, as in the theory of algebraic functions, the most difficult part is the proof that the symbol (α, β) is independent of the choice of the local uniformizing variable.

2. General reciprocity law. The general law of reciprocity expresses in an explicit form the relation between n-th power residues $\left(\frac{\alpha}{\beta}\right)\left(\frac{\beta}{\alpha}\right)^{-1}$ in terms of the numbers α and β. In case α and β are rational and $n = 2$, the reciprocity law was proved by Gauss using the theory of quadratic form. If α and β lie in the cyclotomic field of p-th roots of unity, $n = p$, the reciprocity law was proved by Kummer and applied to his work on Fermat's Last Theorem. Hilbert ([9], p.310) proposed a problem on the discovery of a general reciprocity law.

It can be easily derived from formula (a), that [3], p. 58)

$$\left(\frac{\alpha}{\beta}\right)\left(\frac{\beta}{\alpha}\right)^{-1} = \prod_{\mathfrak{p}|n}\left(\frac{\alpha,\beta}{\mathfrak{p}}\right). \tag{b}$$

This reduces the problem of finding a general reciprocity law to the problem of finding expression for the norm residue symbols $\left(\frac{\alpha,\beta}{\mathfrak{p}}\right)$ in terms of the numbers α, β.

The construction of the norm residue symbol given in this paper provides such an expression. In this way the present paper contains both the formulation and the proof of the general reciprocity law.

3. Class Field Theory. The class field theory studies arithmetic properties of the abelian extensions of algebraic number fields and of $\mathfrak{p}$-adic number fields. In spite of the great importance of the fundamental theorems of this theory, the

proofs of these theorems are for the most part artificial and not widely understood. The reason for this seems to be the following. As was pointed out by Hasse ([4], p. 150) a natural structure of the theory would be such that the $\mathfrak{p}$-adic class field theory would be first constructed, and the class field theory for algebraic fields would be made to follow from it. But the norm residue symbol $\left(\frac{\alpha, \beta}{\mathfrak{p}}\right)$, which plays such an important part in the $\mathfrak{p}$-adic class field theory, was first defined only through algebraic number fields (or by means of algebraic theories which introduced extraneous methods.)

The construction of the norm residue symbol given in this paper provides such a $\mathfrak{p}$-adic definition. In this way it opens up a possibility of a more natural construction of class field theory.

As in the case of function fields, the proof of the independence of the symbol (α, β) on the choice of the local uniformizing variable is quite formidable. In order that long calculations do not obscure the idea of the proof, the first part of this paper is devoted to the definitions and fundamental properties of the symbol (α, β) in general, and a proof of the invariance of the symbol is given for odd prime exponents. The following part contains the invariance proof for an arbitrary exponent.

4. **Fundamental definitions.** We will formulate here the fundamental definitions and introduce the notation which we shall use throughout this paper.

All our discussion will concern a fixed discreet field k of characteristic 0 with a valuation ([2], p. 304). We shall also suppose that the field of residue classes of the field k ([2], p. 304) has a characteristic $p \neq 0$ and is perfect.

We shall call the elements of the field k numbers. The exponent ([2], p. 304) of a number α of the field k will be denoted by $w(\alpha)$, and it will be assumed that it goes over all integer values as α goes over the numbers of the field k. A prime number, that is a number whose exponent is equal to 1, will be denoted by π. Any number α can be uniquely represented in the form

$$\alpha = \xi \pi^{a}$$

where a is an integer and ξ is a unit, that is $w(\xi) = 0$. Here $w(\alpha) = a$. If we shall need another prime number in k we shall denote it by τ:

$$\tau = \xi \pi$$

where ξ is a unit.

It is known that ([11], p. 132) in the ring of integer elements of k a unique system of representatives of classes with respect to an ideal (π) can be chosen, which is closed under multiplication. We shall call this system multiplicative and denote it by R.

Any number in k can be uniquely expressed in the form

$$\alpha = \sum_{i=n}^{\infty} \alpha_i \pi^i$$

where n is an integer and $\alpha_i \in R$.

We shall denote by ϵ and call principal those units which satisfy the congruence

$$\epsilon \equiv 1 \ (\pi).$$

For the characteristic p of the field of residue classes, $w(p) > 0$. We shall denote $w(p)$ by e and call it the ramification order of k. In the field k

$$p = \xi \pi^e.$$

If $e - 1$, then k is called an unramified field. Among all the subfields of k which are unramified there exist a maximal field. This field is called the field of inertia of k and will be denoted by $\mathfrak{K}$. The degree of k over $\mathfrak{K}$ is e. Since $\mathfrak{K} \supset R$, every number in $\mathfrak{K}$ can be represented in the form

$$\alpha = \sum_{i=n}^{\infty} \alpha_i p^i$$

where $\alpha_i \in R$.

Because the field of residue classes is perfect, the correspondence $\alpha \to \alpha^p$ in the field $\mathfrak{K}$ is defined by

$$\left(\sum_{i=n}^{\infty} \alpha_i p^i\right)^p = \sum_{i=n}^{\infty} \alpha_i^p p^i,$$

where $\alpha_i \in R$, is an automorphism. We can now introduce the function

$$\wp(\alpha) = \alpha^p - \alpha .$$

It possesses the property

$$\wp(\alpha + \beta) = \wp(\alpha) + \wp(\beta).$$

We shall always suppose that the field k contains a primitive p^n-th root of unity. Here n is a fixed number throughout the article.

We shall select one and for all a certain primitive p^n-th root of unity and denote it by ζ. It is known that

$$(\mathfrak{K}(\zeta) : \mathfrak{K}) = p^{n-1}(p-1).$$

Therefore

$$e \equiv 0 \ (p^{n-1}(p-1)).$$

We introduce the notation

$$\frac{e}{p-1} = e_1, \quad \frac{e}{p^{n-1}(p-1)} = \frac{e_1}{p^{n-1}} = e_0.$$

It is clear that $e + e_1 = pe_1$.

A finite extension K/k is called unramified if π remains a prime in K. A number in k is called p^n primary and will be denoted be ω if $k\left(\sqrt[p^n]{\omega}\right)/k$ is unramified.

In what follows we shall make use of a notation of Hasse in which groups and their elements are denoted by the same letter. So that for example the group of principal units will be denoted by ϵ the group of p^n-primary numbers by ω and so on.

The reader who is interested only in applications to the theory of algebraic numbers can think of the field k as the $\mathfrak{p}$-adic closure of an algebraic field with respect to a prime ideal $\mathfrak{p}$. In this case the multiplicative system R consists of 0 and of the roots of degree $N(\mathfrak{p})-1$ of unity, where $N(\mathfrak{p})$ is the absolute norm of the prime ideal $\mathfrak{p}$.

§1. Canonical Decompositions

The object of this section is to study the quotient groups of all the principal units of the field k with respect to subgroups of p^n-th powers. To do this we select from the group of principal units a particular fixed system of generators and study the representation of the principal units by means of these generators.

1. The function $E(\alpha, x)$. In what follows, an important part is played by functions introduced by Artin and Hasse, [1]. We shall define them here and enumerate some of their principal properties. The proofs can be found in the work of Hasse [5].

In the ring of formal power series in x with coefficients in $\mathfrak{K}$ we consider the series

$$L(\alpha, x) = \alpha x + p^{-1}\alpha^p x^p + p^{-2}\alpha^{p^2} x^{p^2} + \cdots,$$

where α is an integer in $\mathfrak{K}$.

Obviously the series $L(\alpha, x)$ has the following properties

$$L(\alpha_1 + \alpha_2, x) = L(\alpha_1, x) + L(\alpha_2, x),$$
$$L(a\alpha, x) = aL(\alpha, x),$$

where a is an integer rational p-adic number. From this it follows that the series

$$E(\alpha, x) = e^{L(\alpha, x)}$$

has the properties

$$E(\alpha_1 + \alpha_2, x) = E(\alpha_1, x)\, E(\alpha_2, x) \tag{1}$$
$$E(a\alpha, x) = E(\alpha, x)^a. \tag{2}$$

In case $\alpha \in R$ we have the following identity

$$E(\alpha, x) = \prod (1 - \alpha^m x^m)^{-\frac{\mu(m)}{m}}, \tag{3}$$

where $\mu(m)$ is the Möbius function and m goes over all the natural numbers prime to p. This identity shows that for $\alpha \in R$, $E(\alpha, x)$ has integral p-adic coefficients. In view of relation (1) and (2) this is true for any $\alpha \in \mathfrak{K}$. From this it follows that $E(\alpha, x)$ converges if we substitute for x any number of k, which is divisible by π. Moreover equation (3) shows that for $\alpha \in R$

$$E(\alpha, x) \equiv 1 + \alpha x \ (x^2). \tag{4}$$

The function $E(1, \xi)$ has integral p-adic coefficients and in view of (4) the equation

$$1 + \eta = E(1, \xi)$$

is solvable in the form of a power series $\xi = Q(\eta)$ in powers of η with integer p-adic coefficients. Moreover

$$\xi \equiv \eta \ (\eta^2).$$

Applying this to the case $1 + \eta = \zeta$, $\eta = \zeta - 1$ we see that for $\xi = Q(\zeta - 1)$ we will have

$$E(1, \xi) = \zeta, \tag{5}$$

$$\xi \equiv \zeta - 1 \ ((\zeta - 1)^2). \tag{6}$$

Let $\overline{\mathfrak{K}}$ be that discrete unramified field with a valuation, which is obtained from $\mathfrak{K}$ by the algebraic closure of the residue class field. For every integer α of $\mathfrak{K}$, there exists an integer A in $\overline{\mathfrak{K}}$ such that

$$\wp(A) = \alpha, \tag{7}$$

We introduce the notation

$$E(p^n A, \xi) = E(\alpha). \tag{8}$$

This notation is meaningful because it can be easily seen that different integers A satisfying relation (7) give the same $E(p^n A, \xi)$.

It can be proved that $E(\alpha)$ belongs to k and even to $\mathfrak{K}(\zeta)$. From this it follows that it is p^n-primary, since $E(\alpha)$ is a p^n-th power of the number $E(A, \xi)$ which lies in the unramified extension $\overline{\mathfrak{K}}/k$ of the field k.

It is clear that $E(\alpha)$ satisfies the conditions

$$E(\alpha + \beta) = E(\alpha) E(\beta),$$

$$E(a\alpha) = E(\alpha)^a,$$

where a is an integral rational p-adic number.

It can also be proved that every p^n-primary number ω of k is representable in the form

$$\omega = E(\alpha) \lambda^{p^n},$$

where α is an integer in $\mathfrak{K}$ and λ is any number in k.

2. **Hensel's Theorem.** K. Hensel found very broad conditions for a system of principal units to be the system of generators of the principal units group. Since in what follows we will need not only the formulation of Hensel's theorem, but also some formulas which are contained in the proof of the theorem, we will give here a short proof. A more detailed proof will be found in Hensel's paper [7].

Hensel's Theorem. *Let us select for every i such that $i \neq 0$ (p) for $1 \leq i < pe_1$, as well as for $i = pe_1$, and for every $\alpha \in R$ a principal unit $\epsilon_i(\alpha)$, satisfying the condition $\epsilon_i(\alpha) \equiv 1 + \alpha\pi^i \ (\pi^{i+1})$. Then every principal unit ϵ can be represented in the form $\epsilon = \prod_{i,r} \epsilon_i(\alpha_{i,r})^{p^r}$, where the product extends over all the above mentioned values of i and over all non-negative integer values of r.*

The proof is in three parts.

A. Suppose that we have a system of principal units $\epsilon_i(\alpha)$ depending on the subscript i and on the number $\alpha \in R$, satisfying the congruences

$$\epsilon_i(\alpha) \equiv 1 + \alpha\pi^i \ (\pi^{i+1}). \tag{9}$$

Then every principal unit ϵ can be represented in the form

$$\epsilon = \prod_{i=1}^{\infty} \epsilon_i(\alpha_i). \tag{10}$$

In fact equation (10) is equivalent to the congruences

$$\epsilon = \prod_{i=1}^{n} \epsilon_i(\alpha_i) \ (\pi^{n+1}) \tag{11}$$

for every n. Suppose that (11) has been proved for $n = r$. Let

$$\varepsilon\left(\prod_{i=1}^{r} \varepsilon_i(\alpha_i)\right)^{-1} \equiv 1 + \alpha_{r+1}\pi^{r+1} \ (\pi^{r+2}).$$

Then

$$\varepsilon\left(\prod_{i=1}^{r} \varepsilon_i(\alpha_i)\right)^{-1} \equiv \varepsilon_{r+1}(\alpha_{r+1}) \ (\pi^{r+2})$$

or

$$\varepsilon \equiv \prod_{i=1}^{r+1} \varepsilon_i(\alpha_i) \ (\pi^{r+2}),$$

that is equation (11) is proved for $n = r + 1$.

B. Let $\eta \in R$. We study the formula of raising to the p-th power the unit $1 + \eta\pi^i + \cdots$

$$(1 + \eta\pi^i + \cdots)^p = 1 + p\eta\pi^i + \cdots + \eta^p\pi^{pi} + \cdots,$$

where the dots represent terms divisible by a higher power of π, than one of those which have been written down. The term $p\eta\pi^i$ is divisible by the $(e+i)$-th power of π and $\eta^p\pi^{pi}$ by the pi-th power of π. Therefore we have the formulas

$$(1 + \eta\pi^i + \cdots)^p \equiv 1 + \eta^p\pi^{pi} \ (\pi^{pi+1}), \quad \text{if} \quad i < e_1, \tag{12_1}$$
$$(1 + \eta\pi^i + \cdots)^p \equiv 1 + \gamma_e\eta\pi^{e+i} \ (\pi^{e+i+1}), \quad \text{if} \quad i > e_1, \tag{12_2}$$
$$(1 + \eta\pi^{e_1} + \cdots)^p \equiv 1 + (\eta^p + \gamma_e\eta)\,\pi^{pe_1} \ (\pi^{pe_1+1}). \tag{12_3}$$

Here $\gamma_e \in R$ and is a coefficient in the formula

$$p = \gamma_e\pi^e + \gamma_{e+1}\pi^{e+1} + \cdots.$$

From (12_3) if follows that if $\epsilon \equiv 1 \ (\pi^{pe_1+1})$, then ϵ is a p-th power of some principal unit. This will be important in what follows.

C. It follows from (12_2) that if for some $i > pe_1$ we replace in the system $\epsilon_i(\alpha)$ of A all the $\epsilon_i(\alpha)$ by $\epsilon_{i-1}(\gamma_e^{-1}\alpha)^p$, we again obtain a system which satisfied condition (9). Analogously it follows from (12_1) that if for $i = ph < pe_1$ we replace $\epsilon_i(\alpha)$ by $\epsilon_h(\alpha^{p^{-1}})^p$, then we also obtain a system satisfying (9). It is possible to extract the p-th root of $\alpha \in R$ because the field of residue classes of k is perfect.

It we carry out the above replacements we will obtain a system $\epsilon_i(\alpha)^{p^r}$, where i goes over all the values less than pe_1, which are not divisible by p and also takes on the value pe_1, r goes over all the non-negative numbers and α goes over all the elements of R. This system will satisfy condition (9). Hence by A, every principal unit can be represented in the form

$$\epsilon = \prod \epsilon_i(\alpha_{i,r})^{p^r}.$$

This proves Hensel's Theorem.

3. Canonical decompositions. We now apply Hensel's theorem to the proof of the following theorem:

Theorem on canonical decomposition. *Every principal unit ϵ can be represented in the form $\epsilon = E(\alpha) \prod E(\alpha_i, \pi^i)$, where α and α_i are integer in $\mathfrak{R}$, and i goes over all the values between 1 and pe_1, which are not divisible by p.*

Proof. We shall prove that both $E(\alpha_i, \pi^i)$ and $E(\alpha)$, with α_i and $\alpha \in R$, satisfy the conditions of Hensel's theorem. For $E(\alpha_i, \pi^i)$ this is obvious in view of (4). It remains to show that

$$E(\alpha) \equiv 1 + \beta\pi^{pe_1} \; (\pi^{pe_1+1}),$$

where $\beta \in R$ goes over all the elements of R, as α goes over all the elements of R.

By its construction $E(\alpha)$ lies in $\mathfrak{R}(\zeta)$. By formula (8), $E(\alpha)$ is a p^n-th power of a number $E(A, \xi)$, which lies in $\overline{\mathfrak{R}}(\zeta)$. By (4) and (6)

$$E(A, \xi) \equiv 1 + A(\zeta-1) \; ((\zeta-1)^2).$$

We now raise $E(A, \xi)$ to the powers p, p^2, p^3, and so on. In raising it up to and including the p^{n-1}-st power we can make use of the forumla (12_1), since for the field $\mathfrak{R}(\zeta)$ we have $e = p^{n-1}(p-1)$, $pe_1 = p^n$. Therefore

$$E(A, \xi)^{p^{n-1}} \equiv 1 + A^{p^{n-1}} (\zeta-1)^{p^{n-1}} \; ((\zeta-1)^{p^{n-1}+1}).$$

In raising this congruence to the p-th power, however, we must use formula (12_3) and obtain

$$E(A, \xi)^{p^n} \equiv 1 + (A^{p^n} + \gamma_e A^{p^{n-1}})(\zeta-1)^{p^n} \; ((\zeta-1)^{p^n+1}). \tag{13}$$

We must now calculate γ_e. Denoting $\zeta - 1$ by λ we have

$$(\lambda+1)^{p^{n-1}(p-1)} + (\lambda+1)^{p^{n-1}(p-2)} + \cdots + (\lambda+1)^{p^{n-1}} + 1 = 0.$$

All the coefficients of this equation, except the first are divisible by p, therefore

$$\lambda^{p^{n-1}(p-1)} + p(1 + a_1\lambda + \cdots) = 0,$$

$$p = -\frac{\lambda^{p^{n-1}(p-1)}}{1 + a_1\lambda + \cdots} = -\lambda^{p^{n-1}(p-1)} + \cdots.$$

Hence $\gamma_e = -1$ and equation (13) becomes

$$E(p^n A, \xi) \equiv 1 + \wp(A)^{p^{n-1}} (\zeta - 1)^{p^n} \quad ((\zeta-1)^{p^n+1}).$$

Suppose that in k

$$\zeta - 1 = \gamma_{e_0} \pi^{e_0} + \gamma_{e_0+1} \pi^{e_0+1} + \cdots;$$

then since $\alpha = \wp(A)$ and $p^n_n e_0 = pe_1$,

$$E(\alpha) \equiv 1 + \alpha^{p^{n-1}} \gamma_{e_0}^{p^n} \pi^{pe_1} \quad (\pi^{pe_1+1}). \tag{14}$$

Since as α goes over all the elements of R, $\alpha^{p^{n-1}} \gamma_{e_0}^{p^n}$ does the same, the applicability of Hensel's theorem follows. But by this theorem every principal unit is representable in the form

$$\epsilon = \prod_r E(\alpha_r)^{p^r} \prod_{i,r} E(\alpha_{i,r}, \pi_i)^{p^r}.$$

In view of properties (1) and (2) of the function $E(\alpha, x)$ and of analogous properties of the function $E(\alpha)$

$$\prod_r E(\alpha_{i,r}, \pi^i)^{p^r} = E\Big(\sum_r \alpha_{i,r} p^r, \pi^i\Big),$$

$$\prod_r E(\alpha_r)^{p^r} = E\Big(\sum_r \alpha_r p^r\Big),$$

$$\varepsilon = E\Big(\sum_r \alpha_r p^r\Big) \prod_i E\Big(\sum_r \alpha_{i,r} p^r, \pi^i\Big).$$

This proves the theorem. In what follows we shall call the decomposition of the principal unit into the divisors $E(\alpha)$ and $E(\alpha_i, \pi^i)$, whose existence we have just established, a canonical decomposition. If in a product we write subscripts i, j or h we shall assume that they go over the system of numbers from 1 to pe_1, which are not divisible by p.

We note that from the above considerations follows a fact important in the sequel; namely that if $E(\alpha)$ is a p-th power, then $\alpha \equiv \wp(\beta)\,(p)$. In fact if $E(\alpha) = \epsilon^p$, the equations (14) and (12) show that

$$\varepsilon \equiv 1 + \eta\pi^{e_1} \quad (\pi^{e_1+1}),$$

$$\alpha^{p^{n-1}} \gamma_{e_0}^{p^n} \equiv \eta^p + \gamma_e \eta \quad (p).$$

But if

$$\zeta - 1 = \gamma_{e_0} \pi^{e_0} + \gamma_{e_0+1} \pi^{e_0+1} + \cdots,$$

$$p = -(\zeta - 1)^{p^{n-1}(p-1)} + \cdots,$$

then

$$p = -\gamma_{e_0}^{p^{n-1}(p-1)} \pi^{p^{n-1}(p-1)e_0} + \cdots,$$

that is

$$\gamma_e \equiv -\gamma_{e_0}^{p^{n-1}(p-1)} \ (p), \quad \eta^p + \gamma_e \eta \equiv \gamma_{e_0}^{p^n}((\eta\gamma_{e_0}^{-p^{n-1}})^p - (\eta\gamma_{e_0}^{-p^{n-1}})) \ (p).$$

Hence

$$\alpha^{p^{n-1}} \equiv \wp(\eta\gamma_{e_0}^{-p^{n-1}}) \ (p), \quad \alpha \equiv \wp(\eta^{p^{-(n-1)}} \gamma_{e_0}^{-1}) \ (p).$$

4. The uniqueness of the canonical decomposition. We study the uniqueness of the canonical decomposition of the principal units.

Lemma 1. *If* $E(\alpha) \prod E(\alpha_i, \pi^i) = E(\beta) \prod E(\beta_i, \pi^i)$, *then*

$$\alpha \equiv \beta + \wp(\eta) \ (p), \ \alpha_i \equiv \beta_i \ (p).$$

Proof. We denote $\alpha_i - \beta_i$ by γ_i and $\alpha - \beta$ by γ. From property (1), and the analogous property of $E(\alpha)$ we have:

$$E(\gamma) \prod E(\gamma_i, \pi^i) = 1.$$

We first show that all $\gamma_i \equiv 0$ (p). Suppose that this is not the case, and that γ_j is the first γ_i which is not divisible by p.

$$\prod_{i \geq j} E(\gamma_i, \pi^i) = E(\gamma)^{-1} \epsilon^p.$$

But this is impossible because of congruence (12_1), (12_2) and (12_3), since

$$\prod_{i \geq j} E(\gamma_i, \pi^i) \equiv 1 + \gamma_j \pi^j \ (\pi^{j+1}).$$

Hence all $\gamma_i \equiv 0$ (p). We shall now show that $\gamma \equiv \wp(\eta)$ (p). Since it is true that all $\gamma_i \equiv 0$ (p),

$$E(\gamma) = \epsilon^p,$$

from which it follows from the remark made in part 3 that $\gamma \equiv \wp(\eta)$ (p). This proves the lemma.

Lemma 2. *In any canonical decomposition*

$$\zeta_1 = E(\theta) \prod E(\theta_i, \pi^i) \tag{15}$$

of a p-th root of unity ζ_1, *the following congruences hold:*

$$\theta \equiv \wp(\eta) \ (p^{n-1}),$$
$$\theta_i \equiv 0 \ (p^{n-1}).$$

Proof. Let us suppose that any canonical decomposition of ζ_1 satisfies the conditions:

$$\theta \equiv \wp(\eta) \ (p^r),$$
$$\theta_i \equiv 0 \ (p^r) \tag{16}$$

for some $r \leq n-2$. We shall show that analogous conditions hold for $r+1$. By the definition of $E(\alpha)$

$$E(\wp(\eta)) = E(p^n\eta, \xi) = E(p^n\sigma) \prod E(p^n\sigma_i, \pi^i), \tag{17}$$

where

$$E(\eta, \xi) = E(\sigma)\, \Pi\, E(\sigma_i, \pi^i).$$

It follows from (15) and (17) that ζ_1 has a canonical decomposition

$$\zeta_1 = E(\bar{\theta})\, \Pi\, E(\bar{\theta}_i, \pi^i),$$

where

$$\begin{aligned} \bar{\theta} &= \theta - \wp(\eta) + p^n \sigma \equiv \theta - \wp(\eta)\ (p^n), \\ \bar{\theta}_i &= \theta_i + p^n \sigma_i \equiv \theta_i\ (p^n), \end{aligned} \tag{18}$$

where in view of (16) the following congruences hold:

$$\bar{\theta} \equiv 0\ (p^r)$$
$$\bar{\theta}_i \equiv 0\ (p^r).$$

Suppose that

$$\bar{\theta} p^{-r} = \lambda, \quad \bar{\theta}_i p^{-r} = \lambda_i,$$

and that

$$\zeta = E(\mu)\, \Pi\, E(\mu_i, \pi^i)$$

is a canonical decomposition of a p^n-th root of unity ζ.

By (15)

$$(E(\lambda)\, \Pi\, E(\lambda_i, \pi^i))^{p^{r+1}} = 1,$$

hence for some integer a

$$E(\lambda) \prod E(\lambda_i,\ \pi^i) = \zeta^{ap^{n-r-1}} = E(ap^{n-r-1}\mu)\, E(ap^{n-r-1}\mu_i, \pi^i).$$

By Lemma 1 we have

$$\lambda \equiv \wp(\eta_1) + ap^{n-r-1}\mu\ (p),$$
$$\lambda_i \equiv ap^{n-r-1}\mu_i\ (p),$$

and since $n \geq r+2$,

$$\lambda \equiv \wp(\eta_1)\ (p),$$
$$\lambda_i \equiv 0\ (p),$$

that is

$$\bar{\theta} - \wp(p^r \eta_1) \equiv 0\ (p^{r+1}).$$
$$\bar{\theta}_i \equiv 0\ (p^{r+1}).$$

From congruences (18) it follows that

$$\theta \equiv \wp(\eta + p^r \eta_1)\ (p^{r+1}),$$
$$\theta_i \equiv 0\ (p^{r+1}).$$

This proves the lemma.

Uniqueness Theorem. *If* $E(\alpha)\, \Pi\, E(\alpha_i, \pi^i) = E(\beta)\, \Pi\, E(\beta_i, \pi^i)$, *then* $\alpha \equiv \beta + \wp(\eta)\ (p^n)$, $\alpha_i \equiv \beta_i\ (p^n)$.

Proof. By Lemma 1, $\alpha \equiv \beta + \wp(\eta)\ (p)$, $\alpha_i \equiv \beta_i\ (p)$. By (17)

$$E(\alpha)\prod E(\alpha_i, \pi^i) = E(\overline{\alpha})\prod E(\overline{\alpha}_i, \pi^i),$$

$$\begin{aligned} \overline{\alpha} &\equiv \alpha - \wp(\eta)\ (p^n), \\ \overline{\alpha}_i &\equiv \alpha_i\ (p^n), \\ \overline{\alpha} &\equiv \beta\ (p). \end{aligned} \tag{19}$$

We denote $\frac{\overline{\alpha}-\beta}{p}$ by γ and $\frac{\overline{\alpha}_i-\beta_i}{p}$ by γ_i, then

$$E(p\gamma)\,\Pi\, E(p\gamma_i, \pi^i) = 1,$$

therefore

$$\zeta_1 = E(\gamma)\,\Pi\, E(\gamma_i, \pi^i)$$

is a canonical decomposition of a p-th root of unity ζ_1. By Lemma 2

$$\gamma \equiv \wp(\eta_1)\ (p^{n-1}),$$
$$\gamma_i \equiv 0\ (p^{n-1}),$$

that is

$$\overline{\alpha} \equiv \beta + \wp(\eta_1 p)\ (p^n),$$
$$\overline{\alpha}_i \equiv \beta_i\ (p^n).$$

In view of (19)

$$\alpha \equiv \beta + \wp(\eta + p\eta_1)\ (p^n),$$
$$\alpha_i \equiv \beta_i\ (p^n),$$

and the theorem is proved.

We note that from (17) and the formula

$$\left(E(\lambda)\prod E(\lambda_i, \pi^i)\right)^{p^n} = E(p^n\lambda)\prod E(p^n\lambda_i, \pi^i),$$

it follows that in the canonical decomposition

$$\epsilon = E(\alpha)\,\Pi\, E(\alpha_i, \pi^i)$$

we can replace the α_i by $p^n\beta_i$, and α by $\wp(\eta) + p^n\beta$, thus multiplying by the p^n power of some principal unit.

§ 2. The Symbol (λ, μ)

1 **1. The Function $\delta\epsilon$.** In § 1 we have proved that every principal unit can be represented in the form

$$\epsilon = E(\alpha)\,\Pi\, E(\alpha_i, \pi^i),$$

where if ϵ is considered up to a multiplier, which is a p^n-th power, while α_i (mod p^n), and α up to an additive constant of the form $\wp(\beta)$ (mod p^n), and that the correspondence is one to one.

In what follows we shall be interested only in equalities up to a multiplier, which is a p^n-th power. The fact that λ and μ differ by a multiplier which is a p^n-th power we shall express by: $\lambda \approx \mu$.

To every ϵ in its canonical decomposition corresponds the first factor $E(\alpha)$, determined up to a p^n-th power.

We introduce a function defined over the group of principal units

$$\delta\epsilon \approx E(\alpha).$$

It is clear that

$$\delta\epsilon_1\epsilon_2 \approx \delta\epsilon_1\,\delta\epsilon_2,$$
$$\delta\epsilon^{p^n} \approx 1.$$

Therefore the mapping $\epsilon \to \delta\epsilon$ is a homomorphism of the group ϵ/ϵ^{p^n} on the group $\omega/\omega \cap \epsilon^{p^n}$.

It should be noted that the factor $E(\alpha)$ in the canonical decomposition of ϵ depends not only on ϵ, but on the choice of the prime π by means of which we effect the canonical decomposition. If this fact needs to be emphasized, we shall write $\delta_\pi\epsilon$ for $\delta\epsilon$.

In case the field of residue of the field k is finite, the group $\omega/\omega \cap \epsilon^{p^n}$ is a cyclic group of order p^n. In fact in this case $\mathfrak{K}$ is a finite extension of the field $\mathfrak{K}_0$ of rational p-adic numbers. We shall understand by $S(\alpha)$ the trace in $\mathfrak{K}/\mathfrak{K}_0$. Then the mapping

$$\chi E(\alpha) = \zeta^{S(\alpha)}$$

is an isomorphism of the group $\omega/\omega \cap \epsilon^{p^n}$ and of the cyclic group generated by ζ.

That this mapping is multiplicative follows from the properties of the trace, and of the function $E(\alpha)$. That χ maps into a unit over the group $\omega/\omega \cap \epsilon^{p^n}$ follows from the equation

$$S(\alpha^P - \alpha) = S(\alpha^P) - S(\alpha) = 0.$$

That the group $\omega \cap \epsilon^{p^n}$ is actually the center of the homomorphism follows from the fact that for $S(\alpha) \equiv 0\ (p^n)$, $\alpha \equiv \wp(\beta)\ (p^n)$ ([11], p. 134). Finally that χ gives a mapping over the whole group follows from the fact that finite fields are separable.

The expression of a p^n- primary number ω in the form $E(\alpha)$ depends on the choice of ζ as well as on the function χ. But as Hasse ([5], p. 182) has shown the resulting mapping

$$\chi\omega = \zeta^{S(\alpha)}$$

does not depend on the choice of ζ.

Using the mapping χ we can go from the function $\delta\epsilon$, in case of a finite field of residue classes, to the function $\chi\delta\epsilon$, whose values are the p^n-th roots of unity. However, for the most part we will consider an arbitrary field of residue classes and will not make use of the mapping χ.

2. The Symbol (λ, μ). Every number λ of the field k can be written in the form

$$\lambda = \pi^a w \epsilon \qquad \text{([6], p. 189)}$$

where a is an integer, w an element of R and ϵ a principal unit. We are interested in equalities up to a multiple of a p^n-th power, and therefore we shall dispense with the multiplier w, since it is a p^n-th power, and we will consider the exponent a (mod p^n).

We can then represent all numbers as

$$\lambda \approx \pi^a E(\alpha) \prod E(\alpha_i, \pi^i). \tag{20}$$

Let another number μ be represented in the same form

$$\mu \approx \pi^b E(\beta) \prod E(\beta_j, \pi^i).$$

We introduce the function

$$(\lambda, \mu) \approx E(a\beta - b\alpha + \gamma), \tag{21}$$

where

$$E(\gamma) \approx \delta \prod_{i,j} E(i\alpha_i \beta_j, \pi^{i+j}). \tag{22}$$

The function (λ, μ) is a function of two variables over the group λ/λ^{p^n}, whose values belong to the group $\omega/\omega \cap \epsilon^{p^n}$. In fact, changing a, b, α_i, β_i by additive constants which are multiples of p^n does not alter $a\beta - b\alpha + \gamma$ (mod p^n) and therefore does not change $E(a\beta - b\alpha + \gamma)$ up to a multiple of a p^n-th power. Similarly one can change α and β by additive constants of the form $\wp(\eta)$ (mod p^n).

The symbol (λ, μ) possesses the following four properties, whose proof is the main object of this paper.

I. **Bilinearity**

$$(\lambda_1 \lambda_2, \mu) \approx (\lambda_1, \mu)(\lambda_2, \mu),$$
$$(\lambda, \mu_1 \mu_2) \approx (\lambda, \mu_1)(\lambda, \mu_2).$$

II. **Anti-symmetry**

$$(\mu, \lambda) \approx (\lambda, \mu)^{-1}.$$

III. **Separability.** If in the field of residue classes not every element is represented in the form $x^p - x$, then

from $(\lambda, \mu) \approx 1$ for all μ follows $\lambda \approx 1$,

from $(\lambda, \mu) \approx 1$ for all λ follows $\mu \approx 1$.

IV. **Invariability:** that is the independence of the values of (λ, μ) on the choice of the prime π by means of which (λ, μ) is defined.

Theorem 1. (Bilinearity).

$$(\lambda_1 \lambda_2, \mu) \approx (\lambda_1, \mu)(\lambda_2, \mu) \text{ and } (\lambda, \mu_1 \mu_2) \approx (\lambda, \mu_1)(\lambda, \mu_2).$$

Proof. Let

$$\lambda_1 \approx \pi^{a_1} E(\alpha^{(1)}) \prod E(\alpha_i^{(1)}, \pi^i),$$
$$\lambda_2 \approx \pi^{a_2} E(\alpha^{(2)}) \prod E(\alpha_i^{(2)}, \pi^i),$$

$$\mu \approx \pi^b E(\beta) \prod E(\beta_j, \pi^j).$$

Then

$$(\lambda_1, \mu) \approx E(a_1\beta - b\alpha^{(1)} + \gamma^{(1)}),$$

where

$$E(\gamma^{(1)}) \approx \delta \prod_{i,j} E(i\alpha_i^{(1)}\beta_j, \pi^{i+j}),$$

and

$$(\lambda_2, \mu) \approx E(a_2\beta - b\alpha^{(2)} + \gamma^{(2)}),$$

where

$$E(\gamma^{(2)}) \approx \delta \prod_{i,j} E(i\alpha_i^{(2)}\beta_j, \pi^{i+j});$$

$$\lambda_1\lambda_2 \approx \pi^{a_1+a_2} E(\alpha^{(1)} + \alpha^{(2)}) \prod E(\alpha_i^{(1)} + \alpha_i^{(2)}, \pi^i),$$

$$(\lambda_1\lambda_2, \mu) \approx E((a_1 + a_2)\beta - b(\alpha^{(1)} + \alpha^{(2)}) + \gamma),$$

where

$$E(\gamma) \approx \delta \prod_{i,j} E(i(\alpha_i^{(1)} + \alpha_i^{(2)})\beta_j, \pi^{i+j}) \approx$$

$$\approx \delta \prod_{i,j} E(i\alpha_i^{(1)}\beta_j, \pi^{i+j})\, \delta \prod_{i,j} E(i\alpha_i^{(2)}\beta_j, \pi^{i+j}) \approx E(\gamma^{(1)})\, E(\gamma^{(2)});$$

$$(\lambda_1\lambda_2, \mu) \approx E((a_1 + a_2)\beta - b(\alpha^{(1)} + \alpha^{(2)}))\, E(\gamma) \approx$$

$$\approx E(a_1\beta - b\alpha^{(1)}) E(\gamma^{(1)})\, E(a_2\beta - b\alpha^{(2)})\, E(\gamma^{(2)}) \approx$$

$$\approx E(a_1\beta - b\alpha^{(1)} + \gamma^{(1)})\, E(a_2\beta - b\alpha^{(2)} + \gamma^{(2)}) \approx (\lambda_1, \mu)(\lambda_2, \mu).$$

The second formula is proved in a similar manner.

3. Anti-symmetry.

Lemma. *Let M be some principal unit of the field $k(\sqrt[p^n]{\pi})$. Then $\delta N(M) \approx 1$, where the norm is taken in the field $k\left(\sqrt[p^n]{\pi}\right)/k$.*

Proof. The number $\sqrt[p^n]{\pi}$ is a prime in the field $k\left(\sqrt[p^n]{\pi}\right)$. We denote it by π_n. In this field we denote the numbers e, e_1 and e_0 by $e^{(n)}$, $e_1^{(n)}$ and $e_0^{(n)}$. It is clear that $e^{(n)} = p^n e$, $e_1^{(n)} = p^n e_1$, $e_0^{(n)} = p^n e_0 = pe_1$. In particular

$$\zeta - 1 = \gamma_{e_0} \pi_n^{pe_1} + \gamma_{e_0+1} \pi_n^{pe_1+1} + \cdots.$$

Therefore for $i \not\equiv 0\ (p)$

$$(1 - \alpha\pi_n^i)^{\sigma-1} = \frac{1 - \zeta^i\alpha\pi_n^i}{1 - \alpha\pi_n} = 1 - \alpha(\zeta^i - 1)\pi_n^i + \cdots \equiv 1 - \alpha\gamma_{e_0}\pi_n^{pe_1+i}\ (\pi^{pe_1+i+1}),$$

where by σ is denoted the automorphism of the field $k(\sqrt[p^n]{\pi})/k$ which maps π_n into $\zeta\pi_n$.

It therefore follows that the system of numbers composed of $E(\alpha, \pi_n^i)$ for $i < pe_1$

and $i \not\equiv 0\ (p)$ together with the numbers $(1-\alpha\pi_n^{i-pe_1})^{\sigma-1}$ for $i \not\equiv 0\ (p)$ and $pe_1 < i < pe_1^{(n)}$, and the numbers $E(\alpha)$ satisfies the conditions of Hensel's theorem in the field $k\left(\sqrt[p^n]{\pi}\right)$.

It follows from Hensel's theorem that M can be represented in the form

$$M \approx E(\alpha) \prod_{i<pe_1} E(\alpha_i, \pi_n^i)\, A^{\sigma-1},$$

where A is some principal unit in the field $k\left(\sqrt[p^n]{\pi}\right)$.

In view of the obvious identity

$$N(1-\alpha\pi_n^m) = 1-\alpha^{p^n}\pi^m$$

for $m \not\equiv 0\ (p)$ and $\alpha \in R$ formula (3) becomes

$$N(E(\alpha, \pi_n^i)) = E(\alpha^{p^n}\pi^i)$$

for $i \not\equiv 0\ (p)$. Realizing that $E(\alpha)$ lies not only in $k\left(\sqrt[p^n]{\pi}\right)$, but also in k we get

$$N(M) = E(p^n\alpha) \prod E(\alpha_i^{p^n}, \pi^i),$$

$$\delta N(M) \approx 1.$$

Theorem 2. Anti-symmetry.

$$(\mu, \lambda) \approx (\lambda, \mu)^{-1}.$$

Proof. From (21) and (22)

$$(\lambda, \mu)(\mu, \lambda) \approx \delta \prod_{i,j} E((i+j)\alpha_i\beta_j, \pi^{i+j}).$$

The theorem will be proved if we can show that for every l

$$\delta E(l\cdot\alpha, \pi^l) \approx 1. \tag{23}$$

It is also clear that it is sufficient to prove formula (23) for $\alpha \in R$. Let $l = p^r h$, $h \not\equiv 0\ (p)$. By (3) it is sufficient to show that

$$\delta(1-\alpha\pi^{p^r h})^{p^r} \approx 1.$$

By lemma this will be proved if we can show that

$$(1-\alpha\pi^{p^r h})^{p^r} = N(M)$$

for some $M \in k\left(\sqrt[p^n]{\pi}\right)$. That this is the case can be seen from

$$(1-\alpha\pi^{p^r h})^{p^r} = \prod_{a=1}^{p^r} (1-\zeta^{ap^{n-r}}\alpha^{p^{-r}}\pi^h)^{p^r} = \prod_{a=1}^{p^r} N(1-\zeta^a\alpha^{p^{-n}}\pi_n^{p^r h}) =$$

$$= N\left(\prod_{a=1}^{p^r} (1-\zeta^a\alpha^{p^{-n}}\pi_n^{p^r h})\right).$$

This proves the theorem.

4. Separability.

Theorem 3 (Separability). *If in a field of residue classes of the field k not every element is representable in the form $x^p - x$, then from $(\lambda, \mu) \approx 1$ for all μ it follows that $\lambda \approx 1$, and from $(\lambda, \mu) \approx 1$ for all λ it follows that $\mu \approx 1$.*

Proof. It is clear that the second assertion of the theorem follows from the first because of cosymmetry. Let

$$\lambda \approx \pi^a E(\alpha) \prod E(\alpha_i, \pi^i).$$

We must show that from $(\lambda, \mu) \approx 1$ for all μ follows that A) $a \equiv 0\ (p^n)$, B) $\alpha \equiv \wp(\beta)\ (p^n)$, C) $\alpha_i \equiv 0\ (p^n)$.

A) We select a $\beta \in R$ such that $\beta \not\equiv \wp(\eta)\ (p)$. This is possible because of the condition imposed on the field of residue classes. The relation $(\lambda, E(\beta)) \approx 1$ gives

$$E(a\beta) \approx 1, \quad a\beta \equiv \wp(\eta)\ (p^n),$$

therefore because of the choice of β it follows that $a \equiv 0\ (p^n)$.

B) From the relation $(\lambda, \mu) \approx 1$, it follows that

$$E(-\alpha) \approx 1, \quad E(\alpha) \approx 1,$$

therefore $\alpha = \wp(\beta)\ (p^n)$.

C) Let p^r be the highest power of p, which divides all the α_i. We shall show that the assumption that $r < n$ leads to a contradiction.

Let j be the smallest subscript for which α_j is not divisible by p^{r+1}. We introduce the notation

$$\alpha_i = p^r \bar{\alpha}_i.$$

By assumption $\bar{\alpha}_i \equiv 0\ (p)$ for $i < j$, but $\bar{\alpha}_j \not\equiv 0\ (p)$.

We make use of the relation $(\lambda, \mu) \approx 1$ for $\mu = E(\beta p^{n-r-1}, \pi^{pe_1 - j})$. It is clear that $0 < pe_1 - j < pe_1$ and $pe_1 - j \not\equiv 0\ (p)$, hence μ is already represented in the canonical form

$$(\lambda, \mu) \approx \delta \prod_i E(p^{n-1}\bar{\alpha}_i\beta, \pi^{pe_1+i-j}) \approx$$

$$\approx \delta \prod_{i<j} E(p^{n-1}\bar{\alpha}_i\beta, \pi^{pe_1+i-j})\, \delta E(p^{n-1}\bar{\alpha}_j\beta, \pi^{pe_1})\, \delta \prod_{i>j} E(p^{n-1}\bar{\alpha}_i\beta, \pi^{pe_1+i-j}) \approx 1.$$

We will show that the first and last factor of this product ≈ 1, and therefore the middle factor must also be ≈ 1. For the first factor this follows because $\bar{\alpha}_i \equiv 0\ (p)$ for $i < j$, and therefore this factor is a p^n-th power. For the last factor it is a consequence of

$$E(\bar{\alpha}_i\beta, \pi^{pe_1+i-j}) \equiv 1\ (\pi^{pe_1+1}),$$

for $i > j$, and therefore, according to the remark made in § 1, section 2 B, $E(\bar{\alpha}_i\beta, \pi^{pe_1+i-j})$ is a p-th power. In this way we obtain

$$\delta E(p^{n-1}\bar{\alpha}_j\beta, \pi^{pe_1}) \approx 1. \tag{24}$$

In view of the remark in § 1, section 2 B to which we have just referred and formula (14)

$$E(\bar{\alpha}_j\beta, \pi^{pe_1}) = E(\gamma_{e_0}^{-p}\alpha^{p-n+1})\,\varepsilon^p. \tag{25}$$

Therefore it follows from (24) that

$$E(p^{n-1}\gamma_{e_0}^{-p}(\bar{a}_j\beta)^{p-n+1}) \approx 1,$$

that is

$$\gamma_{e_0}^{-p}(\bar{\alpha}_j\beta)^{p-n+1} \equiv \wp(\eta)\ (p).$$

Because of the condition imposed on the field of residue classes and because $\bar{a}_j \not\equiv 0\ (p)$ we can select β in such a way that this relation could not hold for any η. This contradiction proves the theorem.

§3. Invariance

We used the canonical decomposition of the numbers λ and μ in defining the symbol (λ, μ). But a canonical decomposition depends materially on the prime π by means of which it is constructed. It does not follow from the foregoing that we would get the same value for the symbol (λ, μ) if we should calculate it using a different prime τ. The invariance of the symbol (λ, μ) as we go from one prime to another is its deepest property. In the present section we shall prove invariance for the case $n = 1$, $p \neq 2$. In view of the assumption $n = 1$, the symbol $\approx$ will be used in this section for equalities up to a p-th power.

Theorem 4. (Invariance). *In case $n = 1$, $p \neq 2$, the symbol (λ, μ) depends only on the numbers λ and μ, but does not depend on the prime π which enters into its definition.*

Because of bilinearity it is sufficient to show that the following five formulas hold whether the symbol (λ, μ) is calculated with the help of the prime π, or with respect to another prime τ.

$$(E(\alpha), \varepsilon) \approx (\varepsilon, E(\alpha)) \approx 1. \tag{26}$$

$$(\pi, E(\alpha)) \approx (E(\alpha), \pi)^{-1} \approx E(\alpha), \tag{27}$$

$$(\pi, \pi) \approx 1, \tag{28}$$

$$(\pi, E(\alpha, \pi^i)) \approx 1, \tag{I}$$

$$(E(\alpha, \pi^i), E(\beta, \pi^j)) \approx \delta_\pi E(i\alpha\beta, \pi^{i+j}). \tag{II}$$

Formula (26) does not depend on the choice of π. We next prove the invariance of (27). Let τ be another prime, and let $\Pi = \tau w \epsilon$. If we now calculate $(\pi, E(\alpha))$, using the prime τ, we get

$$(\pi, E(\alpha)) = (\tau\epsilon, E(\alpha)) \approx (\tau, E(\alpha))(\epsilon, E(\alpha)) \approx \delta_\tau E(\alpha) \approx E(\alpha).$$

Finally formula (28) is a special case of the formula

$$(\lambda, \lambda) \approx 1,$$

which is true for any prime and for any λ, because of anti-symmetry of (λ, μ) and because p is odd.

The difficulty lies in the proof of the invariance of formulas (I) and (II) to which we devote the present section.

1. Auxiliary formulas.

Lemma. *The following formulas hold:*

$$E(\alpha, \pi^{pu}) = E(p\alpha^{p^{-1}}, \pi^u)\, e^{-p\alpha^{p^{-1}}\pi^u} \approx e^{-p\alpha^{p^{-1}}\pi^u} \tag{29}$$

and

$$e^{\alpha\pi^a} \approx E(\alpha, \pi^a) \quad \textit{for} \quad a > e. \tag{30}$$

Proof of formula (29).

$$L(\alpha, \pi^{pu}) = \alpha\pi^{pu} + p^{-1}\alpha^p\pi^{p^2u} + p^{-2}\alpha^{p^2}\pi^{p^3u} + \cdots = pL(\alpha^{p^{-1}}, \pi^u) - p\alpha^{p^{-1}}\pi^u,$$

equation (29) follows from the fact that $E(\alpha, x) = e^{L(\alpha, x)}$.

Proof of formula (30).

$$E(\alpha, \pi^a) = e^{\alpha\pi^a} \cdot e^{p^{-1}\alpha^p\pi^{pa}} \cdot e^{p^{-2}\alpha^{p^2}\pi^{p^2a}} \cdot \ldots$$

We prove that for $a > e$, $m > 1$ and $p \neq 2$

$$e^{p^{-m}\alpha^{p^m}\pi^{p^m a}} \equiv 1 \; (\pi^{pe_1+1}). \tag{31}$$

From this it will follow that all $e^{p^{-m}\alpha^{p^m}\pi^{p^m a}}$ are p-th powers and therefore formula (30) holds.

$$w(p^{-m}\alpha^{p^m}\pi^{p^m a}) = p^m a - me > (p^m - m)e > pe_1 \quad \text{for} \quad p \neq 2.$$

But for $w(x) > e_1$, $w(e^x - 1) = w(x)$, which proves formula (31).

Using the above lemma we will derive a formula which generalizes the relation

$$(E(\alpha, \pi^i), E(\beta, \pi^j)) \approx \delta\, E(i\alpha\beta, \pi^{i+j})$$

to the case where $E(\alpha, \pi^i)$ and $E(\beta, \pi^j)$ are not canonical factors, that is when either i or $j > pe_1$, or $\equiv 0$ (p).

We shall show that for any r and s the following formula holds:

$$(E(\alpha, \pi^r), E(\beta, \pi^s)) \approx \delta E(r\alpha\beta, \pi^{r+s}) \prod_{m=1}^{\infty} E(r\alpha^{p^m}\beta, \pi^{p^m r+s})\, E(-s\alpha\beta^{p^m}, \pi^{r+p^m s}). \tag{32}$$

Here, as usual, it is assumed that δ applies to all that follows.

Proof of formula (32). If either r or $s \geq pe_1$, then formula (32) is obvious, since in that case either $E(\alpha, \pi^r)$ or $E(\beta, \pi^s)$ is a p-primary number and therefore the left side of formula (32) ≈ 1 by (27), and all the factors of the right are p-th powers. We shall assume now that both r and $s < pe_1$. If $r \not\equiv 0$ (p), $s \not\equiv 0$ (p), then $p^m r + s \not\equiv 0$ (p), and $r + p^m s \not\equiv 0$ (p), and formula (32) follows from the definition of the symbol (λ, μ).

It remains to consider two cases:

A. One of the numbers r and s is divisible by p, while the other is not.

B. Both r and s are divisible by p.

Case A. It is clear that because of anti-symmetry it is sufficient to consider the case $r \neq 0$ (p), $s = pu$. In this case $r + s \neq 0$ (p) and $r + p^m s \neq 0$ (p), therefore formula (32) is equivalent to

$$(E(\alpha, \pi^r), E(\beta, \pi^s)) \approx \delta \prod_{m=1}^{\infty} E(r\alpha^{p^m}\beta, \pi^{p^m r+s}). \tag{33}$$

Since formula (32) holds for $s \geq pe_1$, we can suppose that it holds for $s > pu$ and prove it for $s = pu$. By (29)

$$E(\beta, \pi^{pu}) \approx e^{-p\beta^{p-1}\pi^u},$$

$$(E(\alpha, \pi^r), E(\beta, \pi^{pu})) \approx (E(\alpha, \pi^r), e^{-p\beta^{p-1}\pi^u}). \tag{34}$$

Let

$$-p\beta^{p-1}\pi^u = \sum_a \lambda_a \pi^a. \tag{35}$$

Here all $a \geq e+u > e$, and $\lambda_a \in R$. From (35) and (30)

$$e^{-p\beta^{p-1}\pi^u} \approx \prod_a E(\lambda_a, \pi^a),$$

$$(E(\alpha, \pi^r), E(\beta, \pi^{pu})) \approx \prod_a (E(\alpha, \pi^r), E(\lambda_a, \pi^a)).$$

Since $pu < pe_1$, it follows that all $a \geq e+u > pu$, and therefore formula (32) is proved for $s = a$

$$(E(\alpha,\pi^r), E(\lambda_a, \pi^a)) \approx \delta E(r\alpha\lambda_a, \pi^{r+a}) \prod_{m=1}^{\infty} E(r\alpha^{p^m}\lambda_a, \pi^{p^m r+a}) E(-a\alpha\lambda_a^{p^m} \pi^{r+p^m a}).$$

But since $r \not\equiv 0\ (p)$, $p^m a + r \not\equiv 0\ (p)$ for $m \geq 1$, therefore

$$\delta E(-a\alpha\lambda_a^{p^m}, \pi^{r+p^m a}) \approx 1,$$

$$(E(\alpha, \pi^r), E(\lambda_a, \pi^a)) \approx \delta \prod_{m=0}^{\infty} E(r\alpha^{p^m}\lambda_a, \pi^{p^m r+a}).$$

By (30)

$$E(r\alpha^{p^m}\lambda_a, \pi^{p^m r+a})) \approx e^{r\alpha^{p^m}\lambda_a \pi^{p^m r+a}}$$

and by (35)

$$\sum_a r\alpha^{p^m}\lambda_a \pi^{p^m r+a} = -pr\alpha^{p^m}\beta^{p-1}\pi^{p^m r+u}.$$

Therefore

$$\prod_{a,m} E(r\alpha^{p^m}\lambda_a, \pi^{p^m r+a}) = \prod_{m=0}^{\infty} e^{-pr\alpha^{p^m}\beta^{p-1}\pi^{p^m r+u}}$$

By (29)

$$\prod_{a,m} E(r\alpha^{p^m}\lambda_a, \pi^{p^m r+a}) \approx \prod_{m=0}^{\infty} E(r\alpha^{p^{m+1}}\beta, \pi^{p^{m+1}r+pu}) =$$

$$= \prod_{m=1}^{\infty} E(r\alpha^{p^m}\beta, \pi^{p^m r+pu}).$$

Comparing with the previous formula we get

$$(E(\alpha, \pi^r), E(\beta, \pi^{pu})) \approx \delta \prod_{m=1}^{\infty} E(r\alpha^{p^m}\beta, \pi^{p^m r + pu}),$$

that is formula (33) for $s = pu$.

Case B. In this case all the factors on the right hand side of (32) are p-th powers. We must therefore show that for any u and v

$$(E(\alpha, \pi^{pu}), E(\beta, \pi^{pv})) \approx 1.$$

We apply relation (29)

$$E(\alpha, \pi^{pu}) \approx e^{-p\alpha^{p^{-1}}\pi^u},$$

$$E(\beta, \pi^{pv}) \approx e^{-p\beta^{p^{-1}}\pi^v},$$

$$(E(\alpha, \pi^{pu}), E(\beta, \pi^{pv})) \approx (e^{-p\alpha^{p^{-1}}\pi^u}, e^{-p\beta^{p^{-1}}\pi^v}).$$

It is clear that

$$e^{-p\alpha^{p^{-1}}\pi^u} \equiv e^{-p\beta^{p^{-1}}\pi^v} \equiv 1 \quad (\pi^{e+1}).$$

Our formula will be proved if we can show that for any ϵ_1 and ϵ_2 for which

$$\begin{gathered} \epsilon_1 \equiv \epsilon_2 \equiv 1 \ (\pi^{e+1}), \\ (\epsilon_1, \epsilon_2) \approx 1. \end{gathered} \tag{36}$$

In fact if we represent ϵ_1 and ϵ_2 in a canonical form

$$\epsilon_1 = E(\alpha) \prod E(\alpha_i, \pi^i),$$

then all i for which $\alpha_i \not\equiv 0$ (p) will be greater than e. Hence

$$\varepsilon_1 \approx E(\alpha) \prod_{i>e} E(\alpha_i, \pi^i),$$

$$\varepsilon_2 \approx E(\beta) \prod_{j>e} E(\beta_j, \pi^j),$$

$$(\varepsilon_1, \varepsilon_2) \approx \delta \prod_{i,j} E(i\alpha_i \beta_j, \pi^{i+j}).$$

But $i + j > 2e > pe_1$, and therefore $E(i\alpha_i\beta_j, \pi^{i+j})$ is a p-th power. This proves (36).

2. Change of Variable. We now consider the decomposition of the number $E(\alpha, \pi^i)$ into a product $\prod_r E(\alpha_r, \tau^r)$ by means of a new prime τ. We shall also suppose that $r < pe_1$ and $r \not\equiv 0$ (p), so that the decomposition will not be necessarily in a canonical form.

Let $\pi = \phi(\tau)$, where ϕ is a power series without a constant term with coefficients in $\mathfrak{R}$. Since $E(\alpha, \pi^i) = e^{L(\alpha, \pi^i)}$, it is sufficient to show how $L(\alpha, \phi(\tau)^i)$ can be represented as a sum $\sum_r L(\alpha_r, \tau^r)$.

We shall introduce the notation

$$L(1, x) = L(x).$$

Lemma. *The following identity holds*

$$L(x+y)=L(x)+L(y)+\sum_{r=1}^{\infty}\sum_{i=1}^{p^r-1}A_r^{(i)}L(x^i y^{p^r-i}), \tag{37}$$

where the $A_r^{(i)}$ *are determined from*

$$\frac{(x+y)^{p^r}-(x^p+y^p)^{p^{r-1}}}{p^r}=\sum A_r^{(i)}x^i y^{p^r-i}. \tag{38}$$

That the $A_r^{(i)}$ are rational integers follows from the congruence

$$(x+y)^{p^r}\equiv(x^p+y^p)^{p^{r-1}}\ (p^r),$$

which can be easily proved by induction on r.

Proof of Lemma. We must show that

$$\sum_{s=0}^{\infty}p^{-s}\sum_{r=1}^{\infty}\sum_{i=1}^{p^r-1}A_r^{(i)}(x^i y^{p^r-i})^{p^s}=L(x+y)-L(x)-L(y). \tag{39}$$

By (38)

$$\sum_{i=1}^{p^r-1}A_r^{(i)}(x^i y^{p^r-i})^{p^s}=\frac{(x^{p^s}+y^{p^s})^{p^r}-(x^{p^{s+1}}+y^{p^{s+1}})^{p^{r-1}}}{p^r}.$$

Therefore

$$\sum_{s=0}^{\infty}p^{-s}\sum_{r=1}^{\infty}\sum_{i=1}^{p^r-1}A_r^{(i)}(x^i y^{p^r-i})^{p^s}=$$

$$=\sum_{s=0}^{\infty}\sum_{r=1}^{\infty}p^{-(r+s)}\left((x^{p^s}+y^{p^s})^{p^r}-(x^{p^{s+1}}+y^{p^{s+1}})^{p^{r-1}}\right), \tag{40}$$

$$\sum_{s=0}^{\infty}\sum_{r=1}^{\infty}p^{-(r+s)}(x^{p^s}+y^{p^s})^{p^r}=\sum_{s=0}^{\infty}p^{-s}\sum_{r=1}^{\infty}p^{-r}(x^{p^s}+y^{p^s})^{p^r}=$$

$$=\sum_{s=0}^{\infty}p^{-s}\left(L(x^{p^s}+y^{p^s})-x^{p^s}-y^{p^s}\right), \tag{41}$$

$$\sum_{s=0}^{\infty}\sum_{r=1}^{\infty}p^{-(r+s)}(x^{p^{s+1}}+y^{p^{s+1}})^{p^{r-1}}=\sum_{s=0}^{\infty}p^{-(s+1)}\sum_{r=1}^{\infty}p^{-(r-1)}(x^{p^{s+1}}+y^{p^{s+1}})^{p^{r-1}}=$$

$$=\sum_{s=0}^{\infty}p^{-(s+1)}L(x^{p^{s+1}}+y^{p^{s+1}}). \tag{42}$$

Substituting (41) and (42) into (40) we obtain

$$\sum_{s=0}^{\infty}p^{-s}\sum_{r=1}^{\infty}\sum_{i=1}^{p^r-1}A_r^{(i)}(x^i y^{p^r-i})^{p^s}=$$

$$=L(x+y)-\sum_{s=0}^{\infty}p^{-s}(x^{p^s}+y^{p^s})=L(x+y)-L(x)-L(y),$$

that is formula (39). This proves the lemma.

It is clear that formula (37) can be extended to any number of summands, in other words we have

$$L\left(\sum x_i\right)=\sum L(x_i)+\sum_{r=1}^{\infty}\sum_{i_1+i_2+\ldots=p^r} A_r^{(i_1 i_2\ldots)} L(x_1^{i_1}x_2^{i_2}\ldots), \tag{43}$$

where the $A_r^{(i_1 i_2\ldots)}$ are determined from the formula

$$p^{-r}\left(\left(\sum x_i\right)^{p^r}-\left(\sum x_i^p\right)^{p^{r-1}}\right)=\sum_{i_1+i_2+\cdots=p^r} A^{(i_1 i_2\ldots)} x_1^{i_1}x_2^{i_2}\ldots$$

Now let $f(t)$ be a power series in t without a constant term and with coefficients which are integral in $\mathfrak{K}$:

$$f(t)=\sum_{n=1}^{\infty}\alpha_n t^n.$$

If every number in the inertia field of the form

$$\alpha_n \sum \alpha_n^{(i)} p^i,\ \alpha_i \in R,$$

be written as a sum of elements of R:

$$\alpha=\alpha_n^{(0)}+\underbrace{\alpha_n^{(1)}+\cdots+\alpha_n^{(1)}}_{p \text{ times}}+\underbrace{\alpha_n^{(2)}+\cdots+\alpha_n^{(2)}}_{p^2 \text{ times}}+\cdots,$$

then $f(t)$ can be written in the form $\sum x_i$, where every x_i is a product of an element of R by a power of t. It is clear that

$$\sum x_i=f(t),$$
$$\sum x_i^p=f^p(t),$$

where by $f^p(t)$ we understand the power series $\sum \alpha_n^p t^{pn}$. Applying formula (43) we obtain

$$L(f(t))=\sum L(\beta_h, t^h),$$

where

$$\sum \beta_h t^h=f(t)+p^{-1}(f(t)^p-f^p(t))+p^{-2}(f(t)^{p^2}-f^p(t)^p)+\cdots$$

or, if we introduce the notation

$$\Lambda_r f(t)=p^{-r}(f(t)^{p^r}-f^p(t)^{p^{r-1}}),$$

$$\sum \beta_h t^h=f+\sum_{r=1}^{\infty}\Lambda_r f. \tag{44}$$

If $\alpha=\alpha_0+\alpha_1 p+\cdots,\ \alpha_i\in R$, then

$$L(\alpha, x)=L(\alpha_0 x)+pL(\alpha_1 x)+\cdots.$$

Applying to this relation formula (44) we get

$$L(\alpha, f(t))=\sum L(\beta_k, t^k),$$

where

$$\sum \beta_h t^h=\alpha f+\sum_{r=1}^{\infty}\alpha^{P^r}\Lambda_r f.$$

We write the function $E(\alpha, f(t))$ as follows: if $f(t)=\sum \alpha_n t^n$ is a power series in t without a constant term with coefficients in $\mathfrak{K}$, and if π is a prime in

k, then

$$\mathscr{E}_\pi(f(t)) = \prod_n E(\alpha_n \pi^n).$$

It is clear that

$$\mathscr{E}_\pi(f_1(t)+f_2(t)) = \mathscr{E}_\pi(f_1(t))\mathscr{E}_\pi(f_2(t)). \tag{45}$$

The change of variable formula which we have obtained can now be written in the form

$$E(\alpha, \pi^i) = \mathscr{E}_\tau(\alpha\varphi^i + \sum_{r=1}^{\infty}\alpha^{p^r}\Lambda_r\varphi^i), \tag{46}$$

where $\pi = \phi(\tau)$.

3. Invariance of formula (I). Let π be expressed in terms of τ with coefficients in R as

$$\pi = \tau w \psi(\tau), \tag{47}$$

where $w \in R$, and $\psi(\tau)$ is a power series with constant term 1, and is a principal unit and therefore can be written in the form

$$\psi(\tau) = \prod_{s=1}^{\infty} E(\beta_s, \tau^s),\ \beta_s \in R.$$

This is not a canonical decomposition because s goes over all integer values. Such a representation is possible in view of section 2 A of § 1 and congruence (4).

$$(\pi,\ E(\alpha, \pi^i)) = (\tau,\ E(\alpha, \tau^i w^i \psi(\tau)^i))(\psi(\tau),\ E(\alpha, \tau^i w^i \psi(\tau)^i))$$

or by (46) and (47)

$$(\pi, E(\alpha, \pi^i)) = (\tau, \mathscr{E}_\tau(\alpha w^i \tau^i \psi^i + \sum \alpha^{p^r} w^{p^r i}\Lambda_r \tau^i \psi^i))\times$$
$$\times\left(\mathscr{E}_\tau\left(\sum \beta_s t^s\right), \mathscr{E}_\tau(\alpha w^i \tau^i \psi^i + \sum \alpha^{p^r} w^{p^r i}\Lambda_r \tau^i \psi^i)\right). \tag{48}$$

We introduce in the ring of power series in t with integer coefficients in $\mathfrak{K}$ a bilinear function fDg, satisfying the condition

$$\beta t^j D\alpha t^i = i\alpha\beta t^{i+j} + \sum_{m=1}^{\infty}(i\alpha^{p^m}\beta t^{p^m i+j} - j\alpha\beta^{p^m} t^{i+p^m j}).$$

Then in view of (32) we have the formula

$$(\mathscr{E}_\tau(f(t)),\ \mathscr{E}_\tau(g(t))) \approx \delta_\tau \mathscr{E}_\tau(gDf). \tag{49}$$

By means of (49) and the relation

$$(\tau, \epsilon) \approx \delta_\tau \epsilon$$

(48) can be written in the form

$$(\pi, E(\alpha, \pi^i)) \approx \delta_\tau \mathscr{E}_\tau(\Phi(t)),$$

where

$$\Phi(t) = \rho(t\psi)^i + \sum_{r=1}^{\infty}\rho^{p^r}\Lambda_r(t\psi)^i + (\rho(t\psi)^i + \sum\rho^{p^r}\Lambda_r(t\psi)^i)\, D \sum\beta_s t^s,$$

and where for brevity we have let $\alpha w^i = \rho$.

Besides the function D we shall introduce in the ring of power series two more bilinear functions d and δ satisfying the conditions

$$\beta t^j d\alpha t^i = i\alpha\beta t^{i+j},$$

$$\beta t^j \partial\alpha t^i = \sum_{m=1}^{\infty}(i\alpha^{p^m}\beta t^{p^m i+j} - j\alpha\beta^{p^m} t^{i+p^m j}).$$

It is clear that

$$fDg = fdg + f\delta g. \tag{50}$$

We shall write $l\,dg$ simply as dg. The function dg has the property

$$dfg = fdg + gdf. \tag{51}$$

A power series f for which a power series g with integer coefficients can be found such that $f = dg$ will be called homologous to 0 and we shall write $f \sim 0$. Relation (51) shows that

$$fdg \sim -gdf. \tag{52}$$

Since $f\delta g = -g\delta f$, it follows that

$$fDg \sim -gDf. \tag{53}$$

Formula (23) shows that if $f \sim 0$, then

$$\delta_\pi \mathfrak{E}_\pi(f(t)) = 1,$$

therefore if we can show that $\Phi \sim 0$, the invariance of formula (I) will then be established.

We first note the identity

$$f\partial g = f\sum_{m=1}^{\infty}(dg)^{p^m} - g\sum_{m=1}^{\infty}(df)^{p^m}. \tag{54}$$

Because of the bilinearity of both sides of this relation it is sufficient to check it for $f = \beta t^i$, $g = \alpha t^i$, in which case it is obvious.

By (50) and (54)

$$\Phi = \rho(t\psi)^i + \sum \rho^{p^r}\Lambda_r(t\psi)^i + (\rho(t\psi)^i + \sum\rho^{p^r}\Lambda_r(t\psi)^i)\,d\sum\beta_s t^s -$$
$$-\sum\beta_s t^s \cdot \sum_{m=1}^{\infty}(d(\rho(t\psi)^i + \sum\rho^{p^r}\Lambda_r(t\psi)^i)^{p^m} + (\rho(t\psi)^i +$$
$$+\sum\rho^{p^r}\Lambda_r(t\psi)^i)\sum_{m=1}^{\infty}(d\sum\beta_s t^s)^{p^m}.$$

We shall show that for any $l = 0, 1, \cdots$, the sum of all the terms of this expression which contain ρ^{p^l}, is homologous to 0. We must consider two cases.

A. $l = 0$. The coefficient of ρ is equal to

$$(t\psi)^i + (t\psi)^i d\sum\beta_s t^s + (t\psi)^i\sum_{m=1}^{\infty}(d\sum\beta_s t^s)^{p^m} = (t\psi)^i(1 + \sum_{m=0}^{\infty}(d\sum\beta_s t^s)^{p^m}).$$

We now calculate $\psi^{-1}d\psi$

$$\psi^{-1}d\psi = d\log\psi = \sum dL(\beta_s, t^s) = \sum_{m=0}^{\infty}(d\sum\beta_s t^s)^{p^m}. \tag{55}$$

Hence the coefficient of ρ is equal to

$$(t\psi)^i(1+\psi^{-1}d\psi).$$

But

$$d(t\psi)^i = \psi^i dt^i + t^i d\psi^i = it^i\psi^i + it^i\psi^{i-1}d\psi = it^i\psi^i(1+\psi^{-1}d\psi).$$

Since $i \not\equiv 0\ (p)$

$$(t\psi)^i(1+\psi^{-1}d\psi) = d\left(\frac{1}{i}(t\psi)^i\right) \sim 0.$$

B. $l \geq 1$. The coefficient of ρ^{p^l} is equal to

$$\Lambda_l(t\psi)^i + \Lambda_l(t\psi)^i d\sum\beta_s t^s + \Lambda_l(t\psi)^i \cdot \sum_{m=1}^{\infty}\left(d\sum\beta_s t^s\right)^{p^m} -$$

$$-\sum\beta_s t^s \cdot \left((d(t\psi)^i)^{p^l} + \sum_{m=1}^{l-1}(d\Lambda_m(t\psi)^i)^{p^{l-m}}\right).$$

The first three terms of this expression can be transformed as in case A and are equal to

$$\Lambda_l(t\psi)^i(1+\psi^{-1}d\psi).$$

In order to transform the fourth term we note that for any f

$$\begin{aligned} dp^{-m}f^{p^m} &= f^{p^m-1}df, \\ dp^{-m}f^{pp^{m-1}} &= f^{(p^{m-1}-1)p}(df)^p, \\ d\Lambda_m f &= f^{p^m-1}df - f^{(p^{m-1}-1)p}(df)^p \end{aligned} \tag{56}$$

and therefore

$$(df)^{p^l} + \sum_{m=1}^{l-1}(d\Lambda_m f)^{p^{l-m}} = (df)^{p^l} +$$

$$+\sum_{m=1}^{l-1}\left(f^{(p^m-1)p^{l-m}}(df)^{p^{l-m}} - f^{(p^{m-1}-1)p^{l-m+1}}(df)^{p^{l-m+1}}\right) = f^{(p^{l-1}-1)p}(df)^p.$$

We apply this relation to $f = (t\psi)^i$ and obtain for the fourth term

$$i(t\psi)^{Pp^{l-1}i}(1+(\psi^{-1}d\psi)^P).$$

Hence the coefficient of ρ^{p^l} is equal to

$$\Lambda_l(t\psi)^i(1+\psi^{-1}d\psi) - \sum\beta_s t^s \cdot i(t\psi)^{Pp^{l-1}i}(1+(\psi^{-1}d\psi)^P). \tag{57}$$

We now make use of the fact that $n = 1$, and that therefore we are interested in the power series only (mod p). Because of the identity

$$df^p g = f^p dg + pf^{p-1}g\,df \equiv f^p dg\ (p)$$

the p-th powers of the series play the role of constants and therefore do not influence the homology to 0. In particular we can eliminate in (57) the multiplier $(t\psi)^{Pp^{l-1}i}$ without changing the homologous to 0 character of the whole expression. This expression now becomes

$$\psi^{-Pp^{l-1}i}\Lambda_l\psi^i(1+\psi^{-1}d\psi)-i\sum\beta_s t^s(1+(\psi^{-1}d\psi)^P). \tag{58}$$

$$\psi^{-Pp^{l-1}i}\Lambda_l\psi^i=p^{-l}\left((\psi^p\psi^{-P})^{p^{l-1}i}-1\right).$$

Using formula (29) we obtain

$$\psi^p\psi^{-P}=e^{p\sum\beta_s t^s},$$

$$\psi^{-Pp^{l-1}i}\Lambda_l\psi^i=p^{-l}\left(e^{ip^l\sum\beta_s t^s}-1\right).$$

By assumption $p\neq 2$ and therefore

$$p^{-l}\left(e^{ip^l\sum\beta_s t^s}-1\right)\equiv i\sum\beta_s t^s\ \ (p^l).$$

Finally we can write (58) in the form,

$$i\sum\beta_s t^s(1+\psi^{-1}d\psi)-i\sum\beta_s t^s(1+(\psi^{-1}d\psi)^P)=i\sum\beta_s t^s(\psi^{-1}d\psi-(\psi^{-1}d\psi)^P). \tag{59}$$

But

$$\psi^{-1}d\psi=\sum_{m=0}^{\infty}\left(d\sum\beta_s t^s\right)^{pm},$$

$$(\psi^{-1}d\psi)^P=\sum_{m=1}^{\infty}\left(d\sum\beta_s t^s\right)^{pm},$$

$$\psi^{-1}d\psi-(\psi^{-1}d\psi)^P=d\sum\beta_s t^s$$

and (59) can be written

$$i\sum\beta_s t^s d\sum\beta_s t^s=d\frac{i}{2}\left(\sum\beta_s t^s\right)^2\sim 0$$

since $p\neq 2$, and therefore $i/2$ is an integer.

4. Transformation of Formula (II). We shall show that if $\pi=\phi(\tau)$ then

$$(E(\alpha,\pi^i),E(\beta,\pi^j))=\delta_\tau\mathscr{E}_\tau(i\alpha\beta\varphi^{i+j-1}\,d\varphi). \tag{60}$$

From (46) and (49)

$$(E(\alpha,\pi^i),\ E(\beta,\pi^j))=\delta_\tau\mathscr{E}_\tau(\Phi(t)),$$

where

$$\Phi(t)=\left(\beta\varphi^j+\sum_{r=1}^{\infty}\beta^{P^r}\Lambda_r\varphi^j\right)D\left(\alpha\varphi^i+\sum_{s=1}^{\infty}\alpha^{P^s}\Lambda_s\varphi^i\right).$$

Equation (60) will be proved, if we can show that

$$\Phi\sim i\alpha\beta\varphi^{i+j-1}d\varphi\ \ (\mathrm{mod}\,p).$$

We denote ϕ^i by f and ϕ^j by g. Then we get

$$\Phi=\left(\beta g+\sum_{r=1}^{\infty}\beta^{P^r}\Lambda_r g\right)D\left(\alpha f+\sum_{s=1}^{\infty}\alpha^{P^s}\Lambda_s f\right).$$

In view of (50) and (54)

$$\Phi=\left(\beta g+\sum\beta^{P^r}\Lambda_r g\right)d\left(\alpha f+\sum\alpha^{P^s}\Lambda_s f\right)+$$

$$+\left(\beta g+\sum\beta^{P^r}\Lambda_r g\right)\sum_{m=1}^{\infty}\left(d\left(\alpha f+\sum\alpha^{P^s}\Lambda_s f\right)\right)^{p^m}-$$

$$-\left(\alpha f+\sum\alpha^{P^s}\Lambda_s f\right)\sum_{m=1}^{\infty}\left(d\left(\beta g+\sum_r\beta^{P^r}\Lambda_r g\right)\right)^{p^m}$$

Because the function fdg is bilinear, we have

$$\Phi=\alpha\beta gdf+\sum_{r,s}\beta^{p^r}\Lambda_r g d\alpha^{p^s}\Lambda_s f+\beta g d\sum_s \alpha^{p^s}\Lambda_s f+$$

$$+\Big(\sum_r \beta^{p^r}\Lambda_r g\Big)d\alpha f+\beta g\sum_{m=1}^{\infty}(d\alpha f)^{p^m}+\beta g\sum_{m,s}(d\alpha^{p^s}\Lambda_s f)^{p^m}+$$

$$+\Big(\sum_r \beta^{p^r}\Lambda_r g\Big)\sum_m (d\alpha f)^{p^m}+\sum_{r,s,m}\beta^{p^r}\Lambda_r g\,(d\alpha^{p^s}\Lambda_s f)^{p^m}-\alpha f\sum_m (d\beta g)^{p^m}-$$

$$-\alpha f\sum_{m,r}(d\beta^{p^r}\Lambda_r g)^{p^m}-\Big(\sum_s \alpha^{p^s}\Lambda_s f\Big)\sum_m (d\beta g)^{p^m}-\sum_{r,s,m}\alpha^{p^s}\Lambda_s f\,(d\beta^{p^r}\Lambda_r g)^{p^m}.$$

We shall show that in this expression the sum of all the terms with the exception of the first term $\alpha\beta gdf$ is homologous to 0. In fact we shall show that the coefficient of $\alpha^{p^a}\beta^{p^b}$, where either a or $b>0$ is homologous to zero. We must separate two cases.

A. $b=0,\ a>0$. The coefficient of $\alpha^{p^a}\beta$ is equal to

$$gd\Lambda_a f+g\Big((df)^{p^a}+\sum_{s=1}^{a-1}(d\Lambda_s f)^{p^{a-s}}\Big)=g\Big((df)^{p^a}+\sum_{s=1}^{a}(d\Lambda_s f)^{p^{a-s}}\Big).$$

By (56) this equal to $gf^{p^a-1}df$. Remembering that $f=\phi^i$, $g=\phi^j$ we have

$$df^{p^a-1}df=i\phi^{p^ai+j-1}d\phi.$$

Since $j\not\equiv 0\ (p)$, $p^ai+j\not\equiv 0\ (p)$ and

$$gf^{p^a-1}df=d\left(\frac{1}{p^ai+j}\varphi^{p^ai+j}\right)\sim 0.$$

Analogously we can consider the case $a=0,\ b>0$.

B. $a>0,\ b>0$. The coefficient of $\alpha^{p^a}\beta^{p^b}$ is equal to

$$\Lambda_b g d\Lambda_a f+\Lambda_b g\Big((df)^{p^a}+\sum_{s=1}^{a-1}(d\Lambda_s f)^{p^{a-s}}\Big)-\Lambda_a f\Big((dg)^{p^b}+\sum_{r=1}^{b-1}(d\Lambda_r)^{p^{b-r}}\Big).$$

Applying (56) we find that this is equal to

$$\Lambda_b g\,(f^{p^a-1}df-f^{(p^{a-1}-1)P}(df)^P)+\Lambda_b g\cdot f^{(p^{a-1}-1)P}(df)^P-\Lambda_a f\cdot g^{(p^{b-1}-1)P}(dg)^P=$$

$$=\Lambda_b g\cdot f^{p^a-1}df-\Lambda_a f\cdot g^{(p^{a-1}-1)P}(dg)^P. \qquad (61)$$

We note that the following holds

$$h^{-1}\Lambda_1\varphi^h\equiv\varphi^{P(h-1)}\Lambda_1\varphi\ (p). \qquad (62)$$

In fact

$$h^{-1}\frac{\Lambda_1\varphi^h}{\Lambda_1\varphi}=h^{-1}\frac{\varphi^{ph}-\varphi^{Ph}}{\varphi^p-\varphi^P}=\varphi^{P(h-1)}h^{-1}\frac{(\varphi^p\varphi^{-P})^h-1}{(\varphi^p\varphi^{-P})-1}.$$

But since $\phi^p\equiv\phi^P\ (p)$, it follows that

$$h^{-1}\frac{(\varphi^p\varphi^{-P})^h-1}{(\varphi^p\varphi^{-P})-1}\equiv 1\ (p),$$

which proves (62).

If we now substitute ϕ^i for f and ϕ^j for g, then because of the identity

$\Lambda_a f = p^{-a+1}\Lambda_1 f^{p^{a-1}}$ and (62), the expression (61) will be congruent modulo p to

$$ij\Lambda_1\varphi\,(\varphi^{p^a i+p^b j-p-1}d\varphi - \varphi^{p^a i+p^b j-2p}(d\varphi)^p). \tag{63}$$

We replace ϕ^P by ϕ^p whenever possible. We also suppress $\phi^{p^a i+p^b j-2p}$ in (63), since we have shown in part 3 that this does not alter whether the expression is homologous to zero or not. Finally there remains

$$ij\Lambda_1\phi(\phi^{p-1}d\phi-(d\phi)^P) = ij\Lambda_1\phi d\Lambda_1\phi.$$

Since $p \neq 2$

$$ij\Lambda_1\phi d\Lambda_1\phi = d\frac{ij}{2}(\Lambda_1\phi)^2 \sim 0.$$

5. Invariance of Formula (II). In order to show the invariance of formula (II), we must show in view of (60) that

$$\delta_\pi E(i\alpha\beta,\pi^{i+j}) \approx \delta_\tau\mathscr{E}_\tau(i\alpha\beta\varphi^{i+j-1}d\varphi),$$

where $\pi = \phi(\tau)$ expresses the prime π in terms of the prime τ. In other words we must show that for any $m > 1$

$$\delta_\pi E(\alpha,\pi^m) \approx \delta_\tau\mathscr{E}_\tau(\alpha\varphi^{m-1}d\varphi). \tag{64}$$

For $m > pe_1$ formula (64) is obvious, since both sides are p-th powers. Therefore we shall prove the formula for a fixed value of m, assuming that it has already been proved for larger values of m.

If $m \not\equiv 0\ (p)$, then the formula is obvious, since the left hand side is equal to 1 by definition, and the right hand side because

$$\alpha\varphi^{m-1}d\varphi = d\left(\frac{\alpha}{m}\varphi^m\right) \sim 0.$$

We must again consider two cases.

A. $m = pe_1$. By (14) with $n = 1$

$$\delta_\pi E(\alpha,\pi^{pe_1}) = E(\gamma_{e_0}^{-p}\alpha). \tag{65}$$

Let

$$\pi = \phi(\tau) = \lambda_1\tau + \cdots.$$

Substituting the expression into

$$\zeta - 1 = \gamma_{e_0}\pi^{e_0} + \cdots,$$

we get

$$\zeta - 1 = \gamma_{e_0}\lambda_1^{e_0}\tau^{e_0} + \cdots,$$

that is

$$\gamma'_{e_0} = \gamma_{e_0}\lambda_1^{e_0}. \tag{66}$$

On the other hand

$$[\alpha\varphi^{m-1}d\varphi]_{t=\tau} = \alpha\lambda_1^{pe_1}\tau^{pe_1} + \cdots = \alpha\lambda_1^{pe_0}\tau^{pe_1} + \cdots,$$

since $e_0 = e_1$, because $n = 1$. Therefore

$$\delta_\tau \mathfrak{E}_\tau(\alpha\varphi^{m-1}d\varphi) \approx \delta_\tau E(\alpha\lambda_1^{pe_0}, \tau^{pe_1}) \approx E((\gamma'_{e_0})^{-p}\alpha\lambda_1^{pe_0}). \tag{67}$$

Substituting (66) into (67) and comparing it with (65) we get the desired formula (64) in this case.

B. $m = pu < pe_1$. By (29) we have

$$E(\alpha, \pi^{pu}) \approx e^{-p\alpha^{P-1}}\pi^u. \tag{68}$$

In order to make an analogous transformation of the left hand side of (64) we note that

$$\alpha\varphi^{pu-1}d\varphi - \alpha\varphi^{P(u-1)}(d\varphi)^P = d\alpha u^{-1}\Lambda_1\varphi^u \sim 0,$$

therefore

$$\alpha\varphi^{pu-1}d\varphi \sim \alpha\varphi^{P(u-1)}(d\varphi)^P,$$

$$\delta_\tau\mathfrak{E}_\tau(\alpha\varphi^{pu-1}d\varphi) \approx \delta_\tau\mathfrak{E}_\tau(\alpha\varphi^{P(u-1)}(d\varphi)^P). \tag{69}$$

From the definition of $\mathfrak{E}_\tau$ and by (29)

$$\mathfrak{E}_\tau(\alpha\varphi^{P(u-1)}(d\varphi)^P) \approx [e^{-p\alpha^{P-1}\varphi^{u-1}d\varphi}]_{t=\tau}. \tag{70}$$

Let

$$-p\alpha^{P-1}\pi^u = \sum_{v=e+u}^{\infty} \lambda_v\pi^v, \tag{71}$$

where $\lambda_v \in R$ and all $v \geq e + u > pu$, since $u < e_1$. As long as $\pi = \phi(\tau)$ we will have

$$-p\alpha^{P-1}\varphi(\tau)^u = \sum \lambda_v(\varphi(\tau))^v \tag{72}$$

or multiplying by $[\phi^{-1}d\phi]_{t=\tau}$,

$$-p\alpha^{P-1}[\varphi^{u-1}d\varphi]_{t=\tau} = \sum \lambda_v[\varphi^{v-1}d\varphi]_{t=\tau}. \tag{73}$$

It follows from (69), (70), (73) and (30) that

$$\delta_\tau\mathfrak{E}_\tau(\alpha\varphi^{pu-1}d\varphi) \approx \delta_\tau\mathfrak{E}_\tau\left(\sum \lambda_v\varphi^{v-1}d\varphi\right), \tag{74}$$

and from (68), (71) and (30) that

$$E(\alpha, \pi^{pu}) \approx \mathfrak{E}_\pi\left(\sum \lambda_v t^v\right). \tag{75}$$

But in formulas (74) and (75) $v > pu$, and therefore the formula is proved for $m = v$, that is

$$\delta_\pi\mathfrak{E}_\pi(\lambda_v t^v) \approx \delta_\tau\mathfrak{E}_\tau(\lambda_v\varphi^{v-1}d\varphi),$$

but from (45)

$$\delta_\pi\mathfrak{E}_\pi\left(\sum \lambda_v t^v\right) \approx \delta_\tau\mathfrak{E}_\tau\left(\sum \lambda_v\varphi^{v-1}d\varphi\right). \tag{76}$$

Comparing relations (74), (75) and (76) we obtain formula (64).

§4. Reciprocity Law

1. **Norm residue symbol.** We again suppose as in § 3 that $n = 1$ and $p \neq 2$.

Theorem 5. *If the field k is obtained as a closure of some field of algebraic* *numbers with respect to a prime ideal* $\mathfrak{p}$, *then the symbol* $\chi(\lambda, \mu)$ *coincides with the norm residue symbol* $\left(\frac{\lambda, \mu}{\mathfrak{p}}\right)$.

Proof. Let π be any prime. We consider four cases.

A. $\lambda = \pi$, $\mu = E(\alpha)$. By the definition of the symbol $\chi(\lambda, \mu)$

$$\chi(\pi, E(\alpha)) = \chi E(\alpha).$$

On the other hand it was shown by Hasse ([5], p. 182) that

$$\chi E(\alpha) = (\sqrt[p^n]{E(\alpha)})^{1-\sigma},$$

where σ in Artin's automorphism of the unramified extension $k(\sqrt[p^n]{E(\alpha)})/k$. From the definition of the norm residue symbol it follows that

$$\left(\frac{\pi, E(\alpha)}{\mathfrak{p}}\right) = \chi E(\alpha).$$

B. Let $\lambda = \pi$, and μ be any principal unit ϵ. Let

$$\epsilon = E(\alpha) \prod E(\alpha_i, \pi^i);$$

then

$$\chi(\pi, \epsilon) = \chi E(\alpha). \tag{77}$$

On the other hand

$$\left(\frac{\pi, \varepsilon}{\mathfrak{p}}\right) = \left(\frac{\pi, E(\alpha)}{\mathfrak{p}}\right) \prod \left(\frac{\pi, E(\alpha_i, \pi^i)}{\mathfrak{p}}\right). \tag{78}$$

Since $E(\alpha_i, \pi^i)$ are norms in $k(\sqrt[p^n]{\pi})/k$, it follows that

$$\left(\frac{\pi, E(\alpha_i, \pi^i)}{\mathfrak{p}}\right) = 1. \tag{79}$$

Comparing relations (77), (78) and (79) we obtain the required result.

C. Let $\lambda = \epsilon_1$, $\mu = \epsilon_2$ be any two principal units. Here we shall make use of the invariance of the symbol (λ, μ) which was proved in § 3.

We denote the prime $\pi\epsilon_1$ by τ. Then

$$\chi(\varepsilon_1, \varepsilon_2) = \chi(\pi\varepsilon_1, \varepsilon_2)\chi(\pi, \varepsilon_2)^{-1} = \chi(\tau, \varepsilon_2)\chi(\pi, \varepsilon_2)^{-1},$$

$$\left(\frac{\varepsilon_1, \varepsilon_2}{\mathfrak{p}}\right) = \left(\frac{\tau, \varepsilon_2}{\mathfrak{p}}\right)\left(\frac{\pi, \varepsilon_2}{\mathfrak{p}}\right)^{-1}.$$

Because of the invariance of the symbol (λ, μ) we can apply the result of case B to the prime τ as well as to the prime π. We obtain

$$\chi(\pi, \varepsilon_2) = \left(\frac{\pi, \varepsilon_2}{\mathfrak{p}}\right),$$

$$\chi(\tau, \varepsilon_2) = \left(\frac{\tau, \varepsilon_2}{\mathfrak{p}}\right),$$

which proves the theorem in this case.

D. The general case. Let

$$\lambda \approx \pi^a \varepsilon_1, \qquad \chi(\lambda, \mu) = \chi(\pi, \varepsilon_1)^a \chi(\pi, \varepsilon_2)^{-b} \chi(\varepsilon_1, \varepsilon_2),$$

$$\mu \approx \pi^b \varepsilon_2, \qquad \left(\frac{\pi^a \varepsilon_1, \pi^b \varepsilon_2}{\mathfrak{p}}\right) = \left(\frac{\pi, \varepsilon_1}{\mathfrak{p}}\right)^a \left(\frac{\pi, \varepsilon_2}{\mathfrak{p}}\right)^{-b} \left(\frac{\varepsilon_1, \varepsilon_2}{\mathfrak{p}}\right).$$

In view of B and C,

$$\chi(\pi, \varepsilon_1) = \left(\frac{\pi, \varepsilon_1}{\mathfrak{p}}\right),$$

$$\chi(\pi, \varepsilon_2) = \left(\frac{\pi, \varepsilon_2}{\mathfrak{p}}\right),$$

$$\chi(\varepsilon_1, \varepsilon_2) = \left(\frac{\varepsilon_1, \varepsilon_2}{\mathfrak{p}}\right),$$

which proves the theorem in general.

The theorem makes sense only for a finite field of residue classes. In the case of an arbitrary field k it is possible to give an analogous proof that the symbol (λ, μ) coincides with the symbol introduced by E. Witt ([12], p. 153) by means of algebraic theory.

2. An explicit form of the reciprocity law. For p-th power residues the following implicit form of the reciprocity law has been derived ([3], p. 58)

$$\left(\frac{\lambda}{\mu}\right)\left(\frac{\mu}{\lambda}\right)^{-1} = \prod_{\mathfrak{p} \mid p} \left(\frac{\lambda, \mu}{\mathfrak{p}}\right). \tag{80}$$

Applying Theorem 5, we obtain the following theorem:

Theorem 6 (explicit form of the reciprocity law). *If λ and μ are any numbers in the field of algebraic numbers Ω, containing a p-th root of unity for $p \neq 2$, then*

$$\left(\frac{\lambda}{\mu}\right)\left(\frac{\mu}{\lambda}\right)^{-1} = \prod_{\mathfrak{p} \mid p} (\lambda, \mu)_{\mathfrak{p}}, \tag{81}$$

where by $(\lambda, \mu)_{\mathfrak{p}}$ we denote the symbol $\chi(\lambda, \mu)$ in the field $\Omega_{\mathfrak{p}}$.

3. Kummer's reciprocity law. We shall now show that Kummer's law of reciprocity is a special case of the law of reciprocity derived in part 2 of this section.

Formulation of Kummer's law of reciprocity. *Let Ω be the field of p-th roots of unity, where p is an odd prime, and let λ and μ be two algebraic numbers satisfying the condition*

$$\lambda \equiv \mu \equiv 1 \ (\zeta - 1).$$

We represent λ in the form $f(\zeta)$, where $f(1) = 1$, and f is an integer valued polynomial of degree not greater than p. Let

$$\left.\frac{d^h \log f(e^v)}{dv^h}\right|_{v=0} = l_h(\lambda).$$

The symbol $l_h(\mu)$ has analogous meaning. Then we have the formula

$$\left(\frac{\lambda}{\mu}\right)\left(\frac{\mu}{\lambda}\right)^{-1} = \zeta^{l_1(\lambda)\, l_{p-1}(\mu) - l_2(\lambda)\, l_{p-2}(\mu) + \cdots - l_{p-1}(\lambda)\, l_1(\mu)} \tag{82}$$

Proof. Because of the identity

$$\sum_{h=1}^{p-1} (-1)^h \left.\frac{d^h F(x+1)}{dx^h} \frac{d^{p-h} G(x+1)}{dx^{p+h}}\right|_{x=0} \equiv \sum (-1)^h \left.\frac{d^h F(e^v)}{dv^h} \frac{d^{p-h} G(e^v)}{dv^{p-h}}\right|_{v=0} (p),$$

which holds for any function E and G, we can replace $l_h(\lambda)$ by $\bar{l}_h(\lambda)$

$$\bar{l}_h(\lambda) = \left.\frac{d^h \log f(x+1)}{dx^h}\right|_{x=0}.$$

We denote $\zeta - 1$ by π and note that

$$\lambda = f(1+\pi) = \phi(\pi),$$

where $\phi(x)$ is an integer valued polynomial with constant term 1. In order to calculate (λ, μ) we must know the coefficient α_i in the canonical decomposition

$$\lambda = \varphi(\pi) = E(\alpha) \prod_{i=1}^{p-1} E(\alpha_i, \pi^i).$$

These are determined from the identity $\phi(x) = \sum_{i=1}^{p-1} \alpha_i x^i (x^p)$, that is

$$\alpha_i \equiv \frac{1}{i!} \left.\frac{d^i \varphi \log(x)}{dx^i}\right|_{x=0} \equiv \frac{1}{i!} \bar{l}_i(\lambda) \ (p).$$

Therefore

$$\lambda \approx E(\alpha) \prod_{i=1}^{p-1} E\left(\frac{1}{i!} \bar{l}_i(\lambda), \pi^i\right), \quad \mu \approx E(\beta) \prod_{j=1}^{r-1} E\left(\frac{1}{j!} \bar{l}_j(\mu), \pi^j\right),$$

$$(\lambda, \mu) \approx \delta \prod_{i,j} E\left(\frac{i}{i!\,j!} \bar{l}_i(\lambda)\, \bar{l}_j(\mu), \pi^{i+j}\right).$$

But if $i + j > p$, or $i + j < p$, then $\delta E\left(\frac{i}{i!\,j!} \bar{l}_i(\lambda) \bar{l}_j(\mu), \pi^{i+j}\right) \approx 1$. For $i+j=p$

$$\frac{i}{i!\,j!} \equiv (-1)^{i-1} \ (p),$$

therefore, $\chi(\lambda, \mu) \approx \delta \prod_{i=1}^{p-1} E((-1)^{i-1} \bar{l}_i(\lambda)\, \bar{l}_{p-i}(\mu), \pi^p)$, that is

$$\chi(\lambda, \mu) = \zeta^{\sum (-1)^{i-1} \bar{l}_i(\lambda)\, \bar{l}_{p-i}(\mu)}.$$

If we now apply the reciprocity law derived in part 2 of this section noting that the right hand side depends on only one prime ideal $\mathfrak{p} = (\zeta - 1)$, we will obtain formula (82).

BIBLIOGRAPHY

[1] E. Artin and H. Hasse, *Die beiden Ergäzungssätze zum Reziprozitätsgesetz der l^n-ten Potenzreste im Körper der l^n-ten Einheitswurzeln*, Abh. Math. Sem. Hamburg. Univ. 6 (1928), 146-162.

[2] B. L. van der Waerden, *Moderne Algebra*, Tl. 1, 2. Aufl., Springer, Berlin, 1937.

[3] H. Hasse, *Bericht über neuere Untersuchungen und Probleme aus der Theorie der algebraischen Zahlkörper*, II, *Reziprozitätsgesetz*, Teubner, Leipzig-Berlin, 1930.

[4] H. Hasse, *Die Normenresttheorie relativ-Abelscher Zahlkörper als Klassenkörpertheorie im Kleinen*, J. Reine Angew. Math. 16 2 (1930), 145-168.

[5] H. Hasse, *Die Gruppe der p^n-primären Zahlen für einen Primteiler $\mathfrak{p}$ von p*, J. Reine Angew. Math. 176 (1936), 174-183.

[6] K. Hensel, *Theorie der algebraischen Zahlen*, Bd. I, Teubner, Leipzig, 1908.

[7] K. Hensel, *Die multiplikative Darstellung der algebraischen Zahlen für den Bereich eines beliebigen Primteilers*, J. Reine Angew. Math. 146 (1916), 189-215.

[8] D. Hilbert, *Gesammelte Abhandlungen*, Bd. I, Springer, Berlin, 1932.

[9] D. Hilbert, *Gesammelte Abhandlungen*, Bd. III, Springer, Berlin, 1935.

[10] H. L. Schmid, *Über das Reziprozitätsgesetz in relativ-zyklischen algebraischen Funktionenkörpern mit endlichem Konstantenkörper*, Math. Z. 40 (1935), 94-109.

[11] E. Witt, *Zyklische Körper und Algebren der Charakteristik p vom Grad p^n*, J. Reine Angew. Math. 176 (1936), 126-140.

[12] E. Witt, *Schiefkörper über diskret bewerteten Körpern*, J. Reine Angew. Math. 176 (1936), 153-156.

Translated by:
Emma Lehmer

A new proof of the Kronecker-Weber theorem

Tr. Mat. Inst. Steklova **38**, 382–387 (1951). Zbl. **53**, 355
[Translated for this volume into English]

The theorem of Kronecker-Weber, stating that every abelian extension of the rational number field is contained in some cyclotomic field, is a central theorem in the theory of abelian extensions of the rational number field. This theorem not only yields a classification of absolutely abelian fields, but also determines the decomposition laws in these fields and permits one to describe the structure of their discriminants and to obtain an explicit formula for the ideal class number.

Since Kronecker [7] enunciated this theorem in 1853, many proofs of it have been proposed on the basis of various principles [1, 2, 5, 6, 8].

The present article contains a new proof of the Kronecker-Weber theorem that differs in two respects from earlier known proofs: in the first place it is considerably simpler than they are and in the second it reveals to a greater extent the essence of the matter. Indeed, it turns out that the theorem of Kronecker-Weber is essentially a p-adic fact. Once we have proved the p-adic analogue of this theorem, the theorem itself is obtained automatically by applying Minkowski's theorem on the existence of critical prime numbers*.

For the convenience of the reader we shall make use only of the fundamentals of the theory of p-adic fields from the exposition of Weyl ([4], Chap. III, Sect. 10 and 12) and of the above-mentioned theorem of Minkowski ([4], Chap. IV, Sect. 6). Here we shall employ Minkowski's theorem in an equivalent form, called by N. G. Tchebotarev the arithmetic monodromy theorem. According to this theorem the join of the inertia groups of all critical prime ideals of a normal extension of the rational number field is its Galois group.

As algebraic tools we need only the fundamental theorem of Galois theory and the theorem of Lagrange about cyclic fields ([3], Chap. VII).

Since every abelian field is the composite of fields of prime power degree we may in the sequel confine ourselves to considering fields of degree l^n. We denote by $K(l^m; p_1, \ldots, p_r)$ the union of all cyclic fields K whose discriminants are divisible only by the prime numbers $p_1, \ldots, p_r$ and whose order divides l^m. Then $K(l^m; p_1, \ldots, p_r)$ is a finite extension because these fields K are only finite in number. Indeed, if k designates the l^m-th cyclotomic field, we have $Kk = k(\sqrt[l^m]{\alpha})$ where in α all prime ideals, except for the divisors of $p_1, \ldots, p_r$, occur to an exponent divisible by l^m. The set of all numbers α having this property is obtained from a finite subset upon multiplication by l^m-th powers of numbers, and hence the number of these fields is finite.

In view of the fact that every abelian field of degree l^n is contained in some field $K(l^m; p_1, \ldots, p_r)$, it suffices to show that each $K(l^m; p_1, \ldots, p_r)$ is a subfield of a

* i.e. primes dividing the discriminant (Translator's note)

cyclotomic field. This will be done, but first we shall investigate analogous extensions of p-adic number fields.

The field $K(l^m; p_1, \ldots, p_r)$ appears as an analogue of an algebraic function field whose Riemann surface has an abelian monodromy group and given points of ramification with ramification orders not exceeding a fixed number.

In the sequel we shall need the following three simple lemmas.

Lemma 1. *Let R be a field whose characteristic is not divisible by the prime number l, set $k = R(\zeta)$, where ζ is an l-th root of unity, and denote by s a generating automorphism of the Galois group of k such that $\zeta^s = \zeta^g$. Then the field $k(\sqrt[l]{\alpha})$ is abelian over R if and only if $\alpha^s = \alpha^g \alpha_0^l$ for some α_0.*

Proof. Since $k(\sqrt[l]{\alpha})$ is normal over R, we have $\alpha^s = \alpha^h \alpha_0^l$. Let $\bar{s}$ and σ denote the automorphisms of $k(\sqrt[l]{\alpha})$ defined by the formulas

$$\sqrt[l]{\alpha}^{\bar{s}} = \sqrt[l]{\alpha^h}\,\alpha_0\,; \qquad \xi^{\bar{s}} = \xi^s \quad \text{for } \xi \in k\,,$$
$$\sqrt[l]{\alpha}^{\sigma} = \zeta \sqrt[l]{\alpha}\,; \qquad \xi^{\sigma} = \xi \quad \text{for } \xi \in k\,,$$

respectively. If $k(\sqrt[l]{\alpha})$ is abelian we obtain $\bar{s}\sigma = \sigma\bar{s}$ or, via application to $k(\sqrt[l]{\alpha})$: $g = h$. On repeating the argument in reverse order one shows the sufficiency of the condition.

Lemma 2. *Let R be a field of characteristic different from 2. For the field $R(\sqrt{a})$ to be a subfield of a cyclic field of 4-th degree, it is necessary that a have the form $u^2 + v^2$ for $u, v \in R$.*

Proof. Let $k = R(\sqrt{a}, \sqrt{x + y\sqrt{a}})$. *Since* k is normal it follows that

$$x - y\sqrt{a} = (x + y\sqrt{a})(r + s\sqrt{a})^2\,.$$

The automorphism σ defined by the formulas

$$\sqrt{x + y\sqrt{a}}^{\,\sigma} = \sqrt{x + y\sqrt{a}}\,(r + s\sqrt{a})\,, \quad \sqrt{a}^{\,\sigma} = -\sqrt{a}\,,$$

is a generator of the Galois group of k. The condition $\sigma^2 \neq 1$ yields

$$-1 = (r + s\sqrt{a})(r - s\sqrt{a})$$

or

$$-1 = r^2 - s^2 a\,,$$

i.e.

$$a = \left(\frac{r}{s}\right)^2 + \left(\frac{1}{s}\right)^2.$$

Lemma 3. *Let R_l be the field of l-adic numbers, let ζ and k have the same meaning as in Lemma 1, and put $\lambda = 1 - \zeta$. For a unit $\varepsilon \in k$ congruent to 1 modulo λ to be an l-th power it is necessary and sufficient that $\varepsilon \equiv 1\,(\lambda^{l+1})$.*

Proof. λ is a prime element of k, and $l = \lambda^{l-1}\xi$, where

$$\xi = \frac{(1-\zeta)(1-\zeta^2)\ldots(1-\zeta^{l-1})}{(1-\zeta)^{l-1}} = (1+\zeta)(1+\zeta+\zeta^2)\ldots(1+\zeta+\ldots+\zeta^{l-2})$$

([4], Chap. III, Sect. 4). We remark that $\xi \equiv (l-1)! \equiv -1\,(l)$.

Suppose now that $\varepsilon = \varepsilon_0^l$ for $\varepsilon_0 = 1 + a\lambda + \beta\lambda^2$, where a is a rational integer and β an arbitrary number in k. Then we have

$$\varepsilon = \varepsilon_0^l \equiv 1 + l\,a\,\lambda + a^l\,\lambda^l \equiv 1 + a(\xi + 1)\,\lambda^l \equiv 1\,(\lambda^{l+1})\,.$$

Suppose conversely that $\varepsilon \equiv 1\,(\lambda^{l+1})$. We show that, for any $i > l$, there is an ε_i such that

$$\varepsilon_{i+1} \equiv \varepsilon_i\,(\lambda^{i-l+1})\,; \qquad \varepsilon_i^l \equiv \varepsilon\,(\lambda^i)\,.$$

For ε_{l+1} we take 1. Let ε_i already be chosen and

$$\varepsilon\,\varepsilon_i^{-1} \equiv 1 + a_i\,\lambda^i \quad (\lambda^{i+1})\,.$$

Consider $\eta_i = 1 - a_i\,\lambda^{l-i+1}$. Then we have

$$\eta_i^l \equiv 1 - a_i\,l\,\lambda^{l-i+1} \equiv 1 + a_i\,\lambda^i \equiv \varepsilon\,\varepsilon_i^{-1}\,(\lambda^{i+1})\,.$$

We may choose ε_{i+1} to be $\varepsilon_i\,\eta_i$. On putting $\lim \varepsilon_i = \varepsilon_0$, we obtain $\varepsilon = \varepsilon_0^l$.

Let us now turn to the p-adic analogue of the theorem of Kronecker-Weber. We shall consider abelian extensions of the p-adic number field R_p and assume the degree of these extensions to be a power of a prime number l. Designate by $K(l^m, p)$ the composite of all cyclic extensions of R_p whose order is a divisor of l^m. By $Z(l^m, p)$ we denote the 2^{m+2}-nd cyclotomic field in the case of $p = l = 2$, or the subfield of degree l^m of the l^{m+1}-st cyclotomic field in the case of $p = l \neq 2$, or the maximal subfield of exponent dividing l^m of the p-th cyclotomic field in the case of $p \neq l$, respectively. The degree of $Z(l^m, p)$ is 2^{m+1} for $p = l = 2$, l^m for $p = l \neq 2$, and l^h with $l^h = (l^m, p-1)$ for $p \neq l$, respectively.

Local analogue of the Kronecker-Weber theorem. *The field $K(l^m, p)$ is the composite of the unramified field $W(l^m)$ of degree l^m and the cyclotomic field $Z(l^m, p)$.*

Proof. It is clear that $K(l^m, p)$ contains both $W(l^m)$ and $Z(l^m, p)$ because these fields occur among those whose composite constitutes $K(l^m, p)$. Hence it is sufficient to show that the degree of $K(l^m, p)$ does not exceed the degree of $W(l^m)\,Z(l^m, p)$.

The proof of this assertion is divided into three parts.

1) $p \neq l$. The degree f of the prime ideal of the field $K(l^m, p)$ coincides with the degree of $W(l^m)$ since $W(l^m)$ is the inertia field of $K(l^m, p)$, and it suffices for us to show that the ramification index e does not exceed the degree of $Z(l^m, p)$, i.e. l^h. From the fact that e is the order of the cyclic (in view of $p \neq l$) inertia group, it follows that e divides l^m. It remains to verify that e divides $p-1$.

Let σ be the Frobenius automorphism, i.e. the mapping inducing the automorphism $A \to A^p$ on the residue class field. Denote by τ a generator of the inertia group and by Π a prime element of $K(l^m, p)$. Then

$$\Pi^\tau \equiv \alpha\,\Pi\,(\Pi^2)$$

and the order of α with respect to the modulus Π equals e. Since the field is abelian we have

$$\Pi^{\sigma\tau} \equiv \alpha\,\Pi^\sigma \equiv \Pi^{\tau\sigma} \equiv (\alpha\,\Pi)^\sigma \equiv \alpha^p\,\Pi^\sigma\,(\Pi^2)\,,$$

hence

$$\alpha^{p-1} \equiv 1\,(\Pi)\,,$$

which was to be shown.

2) $p=l\neq 2$. The Galois group of $W(l^m)\,Z(l^m,l)$ is the direct product of two cyclic groups of order l^m, i.e. a maximal group of exponent l^m with two generators. In view of this we need only prove that the Galois group of the field $K(l^m,l)$ has no more than two generators. Since the number of generators of an abelian l-group is equal to the number of generators of its maximal factor group of exponent l, everything amounts to showing that the degree of the field $K(l,l)$ does not exceed l^2.

By virtue of Lemma 1, this degree is equal to the index $(\alpha:\beta^l)$, where α and β are non-zero numbers of K and $\alpha^s=\alpha^g\alpha_0^l$. We write α in the form $\lambda^h\eta$ with a unit η. In view of the fact that we have $\lambda^s=\lambda\eta_0$ and $g\not\equiv 1\,(l)$, the integer h must be divisible by l. Let $\eta\equiv a(\lambda)$ for a rational integer a, hence $\eta\equiv a^l(\lambda)$ and $\eta\,a^{-1}=\varepsilon\equiv 1\,(\lambda)$. In this way, $\alpha=\varepsilon\gamma^l$, and for computing the index $(\alpha:\beta^l)$ we may assume that $\alpha\equiv\beta\equiv 1\,(\lambda)$. According to Lemma 3, the index $(\alpha:\beta^l)$ does not exceed the number of those residue classes (mod λ^{l+1}) for which $\alpha^s\equiv\alpha^g\,(\lambda^{l+1})$.

The quantity $\zeta=1-\lambda$, by definition of s and g, satisfies this relation. If $\alpha=1+a\,\lambda(\lambda^2)$, then $\alpha\,\zeta^a=\alpha_1\equiv 1\,(\lambda^2)$ satisfies the same relation. We suppose that $\alpha_1\equiv 1+a_i\,\lambda^i(\lambda^{i+1})$. From $\lambda=1-\zeta$ and $\zeta^s=\zeta^g$ it follows that $\lambda^s\equiv g\,\lambda(\lambda^2)$ and hence

$$\alpha_1^s\equiv 1+a_i\,g^i\,\lambda^i\equiv\alpha_1^g\equiv 1+a_1\,g\,\lambda^i(\lambda^{i+1}),\qquad g^{i-1}\equiv 1\,(l),\qquad i=l\,.$$

Thus $\alpha=\zeta^a(1+\lambda^l)^b\,\varepsilon_0^l$, i.e. the index does not exceed l^2.

3) $p=l=2$. The Galois group of $W(2^m)\,Z(2^m,2)$ has the invariants $(2,2^m,2^m)$, i.e. is the maximal 2-group on three generators having exponent 2^m and one invariant equal to 2. In view of this fact we need only show that the Galois group of the field $K(2^m,2)$ has: a) three generators and b) at least one invariant equal to 2.

a) As in 2) it suffices to show that the degree of $K(2,2)$ is equal to 8. This degree is the same as the index $(a:a^2)$ for a ranging over all non-zero numbers of R_2: $a=2^h c$ for $c\equiv 1\,(2)$, where in accordance with Lemma 3, $c=b^2$ if and only if $c\equiv 1\,(8)$. Consequently,

$$a=2^n(-1)^m\,5^n\,a_0^2\,;\qquad (a:a_0^2)=8\,.$$

b) If all invariants of the Galois group of $K(2^m,2)$ were at least equal to 4, the field $K(4,2)$ would have a Galois group with invariants $(4,4,4)$. From this it would follow that each quadratic subfield is contained in some cyclic field of 4-th degree. However, by Lemma 2 this cannot be true for $R_2(\sqrt{-1})$. Indeed, the equation $-1=u^2+v^2$ is unsolvable in R_2, even modulo 4.

The proof of the local theorem is finished.

Proof of the Kronecker-Weber theorem. The field $K(l^m;p_1,\dots,p_r)$ contains the fields $Z(l^m,p_i)$ for $i=1,\dots,r$ and hence also their composite $\bar{K}$. We show that it coincides with this composite. To this end we observe that the Galois group of $\bar{K}$ is the direct product of the groups of the $Z(l^m,p_i)$ because the intersection of the composite of all fields but the i-th with this i-th field is unramified and hence coincides with the field of rational numbers. The ramification index $\bar{e}_i$ of the prime number p_i in $\bar{K}$ is equal to the degree of $Z(l^m,p_i)$ (see [4], Chap. III, Sect. 12) and hence the degree of $\bar{K}$ is equal to $\prod_i \bar{e}_i$.

On the other hand, the field $K(l^m; p_1, \ldots, p_r)$, when p_i-adically completed, becomes a subfield of the composite $W(l^m)\,K(l^m, p_i)$ in accordance with the local theorem. Therefore, the ramification index e_i of the prime p_i in $K(l^m; p_1, \ldots, p_r)$ satisfies the inequality $e_i \leq \bar{e}_i$. According to the arithmetic monodromy theorem, the Galois group of $K(l^m; p_1, \ldots, p_r)$ is the join of the inertia groups and from this it follows that the degree of this field does not exceed $\prod e_i$ since it is an abelian field. By means of the inequality $\prod e_i \leq \prod \bar{e}_i$ involving the degree $\prod \bar{e}_i$ of $\bar{K}$ we conclude that $K(l^m; p_1, \ldots, p_r) = \bar{K}$. This proves the theorem.

References

1. Tchebotarev, N. G.: Proof of the theorem of Kronecker-Weber concerning abelian extensions. Mat. Sb. *31* (1923)
2. Delone, B. N. (Delaunay, B.): Zur Bestimmung algebraischer Zahlkörper durch Kongruenzen; eine Anwendung auf die abelschen Gleichungen. J. reine angew. Math. *152* (1923), 120–123
3. van der Waerden, B. L.: Algebra I. Springer-Verlag, Heidelberg 1955
4. Weyl, H.: Algebraic Theory of Numbers. Ann. of Math. Studies, Princeton Univ. Press 1940
5. Fueter, R.: Abelsche Gleichungen in quadratisch-imaginären Zahlkörpern. Math. Ann. *75* (1914), 177–255
6. Hilbert, D.: Ges. Abh., Bd. I. Berlin 1932, or Chelsea Reprint, New York 1965
7. Kronecker, L.: Über die algebraisch auflösbaren Gleichungen. Ber. Königl. Akad. Wiss., Berlin 1853
8. Speiser, A.: Die Zerlegungsgruppe. J. reine angew. Math. *149* (1919), 174–188
9. Weber, H.: Theorie der abelschen Zahlkörper. Acta Math. *8/9* (1886/1887), 193–263/105–130

Translated by H. G. Zimmer

On the imbedding problem for fields

Dokl. Akad. Nauk SSSR **95**, No. 3, 459–461 (1954). Zbl. **55**, 31
[Translated for this volume into English]

Suppose given a finite group G and the homomorphic image F of it with a fixed homomorphism from G to F. The imbedding problem consists in the search for conditions under which a given normal extension K/k with Galois group F can be imbedded into an extension $\bar{K}/k$ with Galois group G in such a way that the prescribed homomorphism of G to F coincides with the natural homomorphism of the Galois group of the field onto the Galois group of the subfield.

The extension of F to G is called *semi-direct* if the corresponding factor system is associated to 1. In this case G is determined by the given F and the kernel $\mathfrak{G}$ of the homomorphism from G to F as an operator group with operator domain F. We shall in this case write $G = F \times \mathfrak{G}$. Scholz [1] showed that in case $\mathfrak{G}$ is abelian and G a semi-direct extension, the imbedding problem is always solvable. We obtain the following generalization of this fact to the case of a nilpotent normal subgroup $\mathfrak{G}$:

Theorem 1. *If G is a semi-direct extension of F and the normal subgroup $\mathfrak{G}$ has order l^α and class c, and if $l \geq c$, then an arbitrary algebraic number field L/k with group F can be imbedded into a field $K \supset L$ with group G over k.*

The proof rests on the following arguments. Put $(F{:}1) = m$. We designate by $\mathfrak{G}_{d,F}^{(c,k)}$ the factor group of the free product of md cyclic groups of order l^k with generators $s_{\sigma,i}$ ($\sigma \in F$; $i = 1, \ldots, d$) modulo the c-th term of the lower central series. The rule

$$s_{\sigma,i}^{\tau} = s_{\sigma\tau,i}, \qquad \tau \in F,$$

turns F into an operator domain for $\mathfrak{G}_{d,F}^{(c,k)}$. Any group of order l^α with operator domain F is an operator-homomorphic image of some group $\mathfrak{G}_{d,F}^{(c,k)}$. It suffices therefore to prove the theorem for the case in which $\mathfrak{G} = \mathfrak{G}_{d,F}^{(c,k)}$.

There exists an abelian group N with operator domain F such that the center of the group $\bar{\mathfrak{G}} = N \cdot \mathfrak{G}_{d,F}^{(c,k)}$ decomposes into a direct product of admissible subgroups each of which in turn is a direct product of m cyclic subgroups of order l^k, these latter being conjugate with respect to F. This follows from the fact that the group of all automorphisms of the group $\mathfrak{G}_{d,F}^{(c,k)}$ induces on its center a Lie representation [2] of the full linear group modulo l^k. In view of the assumption $l \geq c$, the Lie representation appears as the first component of the tensor representation [3], where the tensor representation, being restricted to the group F, decomposes into a direct sum of regular representations.

Let L/k be a field with group F. Since an extension K/L with group $\mathfrak{G}_{d,F}^{(c,k)}$ has group $F \times \mathfrak{G}_{d,F}^{(c,k)}$ over k, we shall construct it as a Scholzian extension over L with respect to l^{k+1} (see the note [4], the notions and results of which we shall use in the sequel). In view of the above explanations the field K can be constructed in the following manner. Suppose a field K_0/k with group $F \times \mathfrak{G}_{d,F}^{(c-1,k)}$ has already been constructed. Then the field K is to be found among the subfields of the composite of fields $\bar{K}_i$ each of which in turn is the composite of m fields K_i^σ for $\sigma \in F$, where these latter are conjugate over k. The field K_i itself contains K_0, is normal over L and has, over it, a completely determined Galois group which is an extension of $\mathfrak{G}_{d,F}^{(c-1,k)}$ by a central normal subgroup of order l^k. Here the K_i^σ must be independent over K_0.

The conditions, under which K_0 can be imbedded into such fields $\bar{K}_i$ in such a way that their composite is a Scholzian field, consist in that its invariants $[\chi, X]$, $[X]_\Psi$ and $(X)_{l^{k+1}}$ introduced in [4] must be equal to 1. This is achieved as in the cited paper. In fact it is shown there that, for any number d, we have a number $D > d$ such that the following is satisfied: For an arbitrary Scholzian field $\bar{K}_0 \supset L \supset k$ with group $F \times \mathfrak{G}_{D,F}^{(c-1,k)}$ over k, there exists a subfield, normal over k and with group $F \times \mathfrak{G}_{d,F}^{(c-1,k)}$, whose invariants are all equal to 1.

From this we obtain the assertion of the theorem by induction on c (for an arbitrary d, and for given k and F).

Analogous arguments lead to the following result:

Theorem 2. *If $G = F \times \mathfrak{G}$, the orders of F and $\mathfrak{G}$ are relatively prime, and if $\mathfrak{G}$ is nilpotent, then an arbitrary algebraic number field L/k with group F can be imbedded into a field $K \supset L$ with group G over k.*

The invariants introduced in the paper [4] are closely related to the imbedding problem also in other cases, as becomes clear from the subsequent simplest example.

Let k denote an algebraic number field, let l be a prime number different from 2 and 3, and suppose that k contains the roots of unity of order l^2. We consider three numbers $\alpha_1, \alpha_2, \alpha_3$ of the field k such that the field $k(\sqrt[l]{\alpha_1}, \sqrt[l]{\alpha_2}, \sqrt[l]{\alpha_3})$ has degree l^3 and is a Scholzian field with respect to l^2. In the given case the requirement to be a Scholzian field amounts to the properties of $\alpha_1, \alpha_2, \alpha_3$ to be l-hyperprimary and to satisfy the relations

$$\left(\frac{\alpha_i}{\mathfrak{p}_j}\right) = 1\,, \qquad \mathfrak{p}_j | \alpha_j, \quad i \neq j\,, \qquad i,j = 1,2,3\,.$$

Let us consider the class X of l-invariant numbers of the field $k(\sqrt[l]{\alpha_1}, \sqrt[l]{\alpha_2})$ which corresponds to the imbedding of this field into a field with non-commutative Galois group of order l^3 whose elements are all of order l. We take χ to be the character of the Galois group of the field $k(\sqrt[l]{\alpha_1}, \sqrt[l]{\alpha_2}, \sqrt[l]{\alpha_3})$ for which $\alpha_\chi = \alpha_3$. In this case the invariant $[\chi, X]$ introduced in the paper [4] is defined (here all $l_i = l$, and $G = \mathfrak{G}_i$ is an elementary abelian group of order l^3). We shall denote it by $[\alpha_1, \alpha_2, \alpha_3]$. This symbol possesses the following properties:

$$[\alpha_1 \beta_1, \alpha_2, \alpha_3] = [\alpha_1, \alpha_2, \alpha_3][\beta_1, \alpha_2, \alpha_3]\,, \qquad [\alpha_i, \alpha_j, \alpha_k] = \pm [\alpha_1, \alpha_2\, \alpha_3]\,,$$

where i, j, k is a permutation of the numbers 1, 2, 3 and the + and − signs correspond to even and odd permutations, respectively.

We denote by G the factor group of the free product of three cyclic groups of order l modulo the third term of the lower central series. The factor group of G

modulo the commutator subgroup is isomorphic to the Galois group of the field $k(\sqrt[l]{\alpha_1}, \sqrt[l]{\alpha_2}, \sqrt[l]{\alpha_3})$, and the commutator subgroup is abelian (but not contained in the center). The link between the symbol $[\alpha_1, \alpha_2, \alpha_3]$ and the imbedding problem consists in the following:

Theorem 3. *A necessary and sufficient condition for the field $k(\sqrt[l]{\alpha_1}, \sqrt[l]{\alpha_2}, \sqrt[l]{\alpha_3})$ to be imbeddable into a field with Galois group G over k is that the symbol $[\alpha_1, \alpha_2, \alpha_3]$ be equal to one.*

D. K. Faddeev [5] and later on independently Hasse [6] found a necessary condition for a field with an arbitrary group F to be imbeddable into a field with group G which is an extension of F. Since we assumed the field $k(\sqrt[l]{\alpha_1}, \sqrt[l]{\alpha_2}, \sqrt[l]{\alpha_3})$ to be Scholzian this condition is fulfilled for it. On the other hand, with given α_1 and α_2, it is always possible to choose α_3 in such a way that the field $k(\sqrt[l]{\alpha_1}, \sqrt[l]{\alpha_2}, \sqrt[l]{\alpha_3})$ becomes Scholzian, while the symbol $[\alpha_1, \alpha_2, \alpha_3]$ assumes an arbitrary preassigned value. In view of theorem 3 this shows that the necessary imbedding condition, found by D. K. Faddeev and Hasse, is not sufficient, as this was conjectured by Hasse [6].

References

1. Scholz, A.: Über die Bildung algebraischer Zahlkörper mit auflösbarer Galoisscher Gruppe. Math. Z. *30* (1929), 332–356
2. Thrall, R. M.: A note on a theorem by Witt. Bull. Amer. Math. Soc. *47* (1941), 303–308
3. Brauer, R.: On algebras which are connected with the semisimple continuous groups. Ann. of Math. *38* (1937), 858–872
4. Shafarevich, I. R.: On extensions of fields of algebraic numbers solvable in radicals. Dokl. Akad. Nauk SSSR (N.S.) 95_2 (1954), 225–227
5. Delone, B. N., Faddeev, D. K.: Investigations in the geometry of the Galois theory. Mat. Sb. N.S. *15* (57) (1944), 243–284
6. Hasse, H.: Existenz and Mannigfaltigkeit abelscher Algebren mit vorgegebener Galoisgruppe über einem Teilkörper des Grundkörpers. I, II, III. Math. Nachr. *1* (1948), 40–61, 213–217, 277–283

Received 21|I|1954

Translated by H. G. Zimmer

On the construction of fields with a given Galois group of order l^a

Izv. Akad. Nauk SSSR, Ser. Mat. **18**, 261 – 296 (1954). Zbl. **56**, 33
[Transl., II. Ser., Am. Math. Soc. **4**, 107 – 142 (1956)]

This paper develops a new method of construction of fields with a given Galois group of order l^a, which gives a wider class of fields than those obtained earlier by the method of Scholz-Reichardt. In particular this method is applicable to the case $l = 2$.

Introduction

The existence of fields with a given Galois group of order l^a $(l>2)$ over a given field of algebraic numbers has been established in 1937 by A. Scholz [9].

A simpler proof based on the same fundamental ideas was given by H. Reichardt [8].

A solution of the same problem without the restriction $l > 2$ was announced by T. Tannaka [10]. It appears that it has not been noticed until now that Tannaka's proof contains an error. Part 3 of § 2 of the present paper contains an indication of the error and a construction of a counter example of the statement on which Tannaka bases his proof.

The class of fields with a given Galois group of order l^a which can be constructed by means of the Scholz-Reichardt method is limited in two respects.

In the first place the limitation $l > 2$ leaves open the question of the existence of fields with a given Galois group of order 2^a. This problem is of interest in itself, for example from the point of view of the theory of constructions with ruler and compass.

Secondly, extensions are obtained only over rational number fields with the Scholz-Reichardt method. It is true that from an extension K/R over a rational number field one can obtain an extension Kk/k over an arbitrary number field, which has the same Galois group only if $K \cap k = R$. In this way only a narrow class of fields is constructed. For instance all such fields are invariant under all the automorphisms of k/R.

It is easily seen that in order to construct fields with some very simple Galois groups it is necessary to learn how to construct fields with Galois groups of order l^a which cannot be constructed by the Scholz-Reichardt method.

Let G be, for example, a group of order l^a, let Γ be the direct product of n groups G_i, isomorphic with G, and let $\mathfrak{G}$ be a group having Γ for a normal subgroup and such that the elements of the quotient group $\mathfrak{G}/\Gamma$ induce in Γ an automorphism, which leads to a transposition in G_i. Let us suppose that there exists a field K over the rational number field R with the given Galois group $\mathfrak{G}$, and let k be a subfield of Γ. The field K/k is a composite of n fields K_i/k, which are conjugate and relatively prime to each other over k and have a Galois group G. As was mentioned above, such fields cannot be constructed by the method of Scholz-Reichardt. Analogously, the construction of fields with arbitrary extensions of Galois groups is closely connected with the construction of sufficiently general fields with Galois groups of order l^a.

The present paper contains a method of construction of fields with a given Galois group of order l^a without the restrictions imposed on the field by the Scholz-Reichardt method. In particular it is proved that there exists an extension over an arbitrary field of algebraic numbers with a given Galois group of order 2^a.

In this construction use is made of conditions found by Scholz under which a field can be centrally imbedded in a field with a given Galois group. Fields which satisfy this condition will be called Scholz fields. The fundamental problem is to find conditions under which a given Scholz field can be centrally imbedded in a Scholz field with a given Galois group. In order to do this two invariants of a Scholz field (χ, X) and $(X)_h$ are constructed, the invariant (χ, X) playing the most important role. These invariants are determined from the arithmetic properties of the fields. They represent a scalar product of a one-dimensional and a two-dimensional group cohomology, which resembles by its properties the index of intersection. The equality of the invariants to unity constitutes the necessary and sufficient condition of imbeddedness.

The problem is therefore reduced to the construction of Scholz fields with invariants equal to unity. This is done in § 3, where it is shown that a Scholz field, whose Galois group has a sufficient number of generators, must contain a subfield with a given Galois group and with invariants equal to unity. In order to do this the theory of homomorphisms of groups of order l^a is constructed, and the concept of their composition is introduced, which for abelian groups is reduced to multiplication of characters.

In this paper Čebotarev's density law is used repeatedly, so that it may appear that use is made of analytic considerations. However, in studying the corresponding arguments it is easy to see that it would have been possible to limit ourselves to the use of Frobenius theorem for groups of order l^a. As was pointed out by Deuring [6], Frobenius theorem can be proved non-analytically in this case.

The author is deeply grateful to D. K. Faddeev, who looked over the manuscript of this paper and made a number of valuable remarks.

§ 1. Invariants of central extensions

In this section we shall investigate properties of algebraic number fields whose Galois groups are of prime power order. We shall suppose that the order of all the groups which we will encounter is a power of the same prime l.

1. **Algebraic properties of central extensions.** The group G_1 will be called the central extension of the group G if G is a homomorphic image of G_1 and if the kernel of the homomorphism lies in the center of G_1 and has exponent l. In this way the homomorphism enters into the notion of an extension so that two different homomorphisms of G_1 on G will give different extensions. If the kernel of the homomorphism of G_1 on G is a cyclic group of order l, then the extension will be called simple and central.

Let k be an arbitrary field, Ω be its finite normal separable extension, K be an intermediaite subfield $k \subset K \subset \Omega$ and let K/k be normal, then the Galois group of Ω/k is an extension of the Galois group of K/k. If this extension is central, or simple and central, then we shall call Ω/k a central, or correspondingly, a simple central extension of the field K/k.

Let us suppose that k has a characteristic $\neq l$ and contains an l-th root of unity, and that Ω/k is a simple central extension of K/k. Then Ω/k is a cyclic field of degree l and therefore $\Omega = K(\sqrt[l]{\mu})$ for some $\mu \in K$ (see [1], p. 217).

It is known (see [8], p. 3) that the fact that Ω/k is a central extension of the field K/k is equivalent to μ satisfying the relation

$$\mu^{\sigma} \approx \mu, \tag{1}$$

where σ is any automorphism of K/k. Here, and in what follows, the notation

$$\alpha \approx \beta$$

signifies that $\alpha\beta^{-1}$ is an l-th power, and the relation itself will be called an l-equality. The numbers μ which satisfy (1) will be called l-invariants.

Let G_1 be the Galois group of Ω/k, G be the Galois group of K/k and Z be the Galois group of Ω/K. Since G_1 is a central extension of G, this defines a factor set $a(\sigma, \tau)$ on G with values in Z and with identity automorphisms (see [2], p. 334). The number μ which we have selected defines by means of the relation

$$\sqrt[l]{\mu}^{z} = X(z)\sqrt[l]{\mu}, \quad z \in Z,$$

a character X of the group Z. Applying X to $a(\sigma, \tau)$ we get a factor set

$$X(a(\sigma, \tau)) = \zeta(\sigma, \tau),$$

given on G with values in the group of l-th roots of unity.

It can be easily checked that two l-invariant numbers define the same factor set $\zeta(\sigma, \tau)$ if and only if their ratio is l-equal to an invariant number (a number in k). It is easily seen that a product of factor sets corresponds to a product of l-invariant numbers. We shall denote by $M(G)$ the group of all the classes of non-associated factor sets on G with values in the group of l-th roots of unity. It follows from what we have developed above that the quotient group of all the l-invariant numbers with respect to the subgroup of numbers which are l-equal to the invariant numbers is isomorphic with some subgroup M' of the group $M(G)$:

$$(\mu)/(mC^l) \cong M' \subseteq M(G). \tag{2}$$

The following representation of $M(G)$ will be found useful. Let $s_1, \cdots, s_d$ be some minimal system of generators of G. It defines a representation of G in the form S/N, where S is a free group with d generators and N is its normal subgroup. We denote by N' the subgroup of N generated by the commutators of elements of N with elements of S and by l-th powers of the elements of N.

Lemma 1. *$M(G)$ is isomorphic with the character group of the group N/N'.*

This lemma is a consequence of a well known theorem (see [2], p. 335) which states that for any group G, which is represented as a quotient group of a free group S with respect to the normal subgroup N, the group of all classes of factor sets with identity automorphisms and with values in an arbitrary abelian group A is isomorphic to the quotient group of all homomorphisms of $M/[S, N]$ in A with respect to those homomorphisms which are induced by the homomorphisms of S in A.

Here $[S, N]$ is the normal subgroup of S generated by the commutators of the elements in S with the elements in N.

In the case under consideration A is the group of l-th roots of unity. The homomorphisms of $N/[S, N]$ in A must therefore transform into 1 all the l-th powers of the elements of N, that is they coincide with the character of the group N/N'.

In order to derive our lemma from the above theorem it remains to show that any character of the group N/N' induced by a character of order l of the group S is equal to 1.

This is equivalent to saying that $N \subset S'$, where S' is a subgroup of S, generated by commutators and l-th powers. The same condition can be expressed in the form of a relation

$$S/S' \cong G/G',$$

where G' is a subgroup of G generated by l-th powers and commutators (a Frattini subgroup). The above relation follows from the fact that groups on both sides of this relation are isomorphic with simple abelian groups of order l^d. For S/S' this is obvious. For G/C' this follows from a theorem about Frattini subgroups (see [2], p. 402), which states that $s_1, \cdots, s_d$ is a minimal system of generators on G, if and only if it is such a system in G/G'.

2. Scholz fields. One of the fundamental problems in the theory of fields whose

Galois group is an l-group is the establishment of conditions under which the given normal extension K/k can be imbedded in a central extension Ω/k, whose Galois group is the given central extension of the group of K/k, Brauer (see [5], p. 56) found necessary and sufficient conditions in terms of decompositions of certain algebras. However, it is hard to make use of Brauer's conditions. In case k is an algebraic number field it is possible, as has been pointed out by Scholz [9], to strengthen slightly Brauer's conditions and obtain a condition which is sufficient to insure that K/k can be imbedded in the extension Ω/k, whose Galois group is any central extension of the Galois group of K/k:

1°. The prime divisors of l in k split completely in K.

2°. The real infinite primes of k remain real in k.

3°. All ramified prime ideals $\mathfrak{p}_i$ of the extension K/k have relative degree 1.

4°. The absolute norm of the ramified prime ideals satisfies the condition: $\mathfrak{N}(\mathfrak{p}_i)^m \equiv 1(l^h)$ for some m, relatively prime to l and a sufficiently large h (if $(K:k) = l^a$, then it is sufficient that $h > a$).

We note that if k contains an l-th root of unity, then we can let $m = 1$ in condition 4°, since in that case $\mathfrak{N}(\mathfrak{p}) \equiv 1(l)$.

We shall call the fields which satisfy conditions 1°-4°, Scholz fields. For Scholz fields the group M' in (2) coincides with $M(G)$, that is

$$(\mu)/(mC^l) \cong M(G).$$

Because of this and Lemma 1, we shall denote by the same letter X the following:

1) the class of l-invariant numbers with respect to the subgroup of numbers which are l-equal to the invariant numbers,

2) an element of $M(G)$

3) a character of the group N/N'.

Central extensions (even simple ones) of a Scholz field K/k are not, in general, Scholz fields. In fact even if $K(\sqrt[l]{\mu})/k$ is a Scholz field we can find a number $m \in k$, such that the field $K(\sqrt[l]{\mu m})/k$, having the same Galois group over k as $K(\sqrt[l]{\mu})$, is no longer a Scholz field.

In order to show the existence of a field with a given group of order l^a it is important to know when a Scholz field K/k possesses a Scholz extension having a Galois group which is a central extension of the Galois group of K/k. Scholz showed that such extensions always exist if k is a field of rational numbers and $l > 2$. This fact is the basis of his proof about the existence of extensions of a field of rational numbers with a given group of odd order l^a.

We shall now give the necessary and sufficient conditions for the existence of such an extension. In order to do this we must introduce some new invariants of fields whose Galois group is of order l^a, under the assumption that the base field contains an l-th root of unity.

3. **Arithmetical structure of l-invariant numbers.** We study first of all the factorization of an l-invariant number μ in the field K. We consider one of its prime divisors $\mathfrak{P}$. Suppose that $\mathfrak{P}^\sigma$ divides μ to the power $a(\sigma)$. Because μ is l-invariant it follows that

$$a(\sigma) \equiv a(1)\ (l).$$

Hence that factor of μ which consists of ideals conjugate with $\mathfrak{P}$ can be split into the product of all these distinct ideals to one and the same power, for instance $a(1)$, and into an l-th power. The first product is obviously an invariant ideal. Proceeding in the same manner with the other divisors of μ we find that the ideal (μ) splits into an l-th power and an invariant ideal. Since every invariant ideal is the product of ramified prime ideals by an ideal of the base field, we obtain for μ the decomposition

$$(\mu) = \mathfrak{C}^l \mathfrak{D}\, \mathfrak{m}, \tag{3}$$

where $\mathfrak{m}$ is an ideal of the field k, relatively prime to the discriminant of K/k, and $\mathfrak{D}$ is an invariant ideal, composed only of prime divisors of the discriminant of K/k, where neither $\mathfrak{D}$ nor $\mathfrak{m}$ contains l-th powers. It is clear that such a decomposition is unique.

The prime ideals $\mathfrak{P}$, which divide $\mathfrak{D}$ can also be characterized by the following property: the order of the generating automorphism of the inertia group of the ideal $\mathfrak{P}$ over K is one l-th as large as the order of any element of the Galois group $K(\sqrt[l]{\mu})/k$, which induces this isomorphism. In other words a generator of the inertia group of $\mathfrak{P}$ in K increases its order in the extension of K to $K(\sqrt[l]{\mu})$. We can speak of a generator of the inertia group of $\mathfrak{P}$ in K/k, since this group is cyclic. This follows from the fact that $(l, \mathfrak{P}) = 1$ by condition 1° in the definition of a Scholz field.

To prove our statement we note that because of the condition $(l, \mathfrak{P}) = 1$ the inertia group of any prime divisor of $\mathfrak{P}$ in the field $K(\sqrt[l]{\mu})$ is cyclic. The order of the group, or what is the same thing of its generator, is l times as large as the order of the inertia of $\mathfrak{P}$ in K, since the ramification order is increased l fold in going from K to $K(\sqrt[l]{\mu})$. In this way a generator of the inertia group increases its order l times. But if σ is any automorphism of $K(\sqrt[l]{\mu})/k$ which induces in K an automorphism of the inertia group of $\mathfrak{P}$, then it has the same order as the generator of an automorphism which generates the inertia group of a prime divisor of $\mathfrak{P}$ in $K(\sqrt[l]{\mu})$, since they differ only by a multiplier, belonging to the center G_1, and having order l.

We note that $\mathfrak{D}$ does not change by multiplication by a number of the form

mC^l and therefore is an invariant of class X. We shall therefore denote $\mathfrak{D}$ by $\mathfrak{D}(X)$. Obviously we have the relation

$$\mathfrak{D}(X_1X_2) \approx \mathfrak{D}(X_1)\mathfrak{D}(X_2).$$

4. Invariants (χ, X). We shall call two ideals $\mathfrak{a}$ and $\mathfrak{b}$ l-relatively prime to each other, if any prime divisor $\mathfrak{p}$ which they have in common, goes into one of them to a degree divisible by l. We recall that if $\mathfrak{a}$ and $\mathfrak{b}$ are l-relatively prime to each other, then Legendre symbol $(\frac{\mathfrak{a}}{\mathfrak{b}})$ is defined.

We shall say that an ideal $\mathfrak{a}$ in the field k is l-relatively prime to an invariant ideal $\mathfrak{B}$ of the field K, if every prime ideal $\mathfrak{p}$ which goes into $\mathfrak{a}$ with a power not divisible by l has a prime divisor $\mathfrak{P}$ in K, which goes into $\mathfrak{B}$ with a power which is divisible by l.

Let $\mathfrak{p}$ be an arbitrary prime ideal of the field k, l-relatively prime to μ, and let $\mathfrak{P}$ be an arbitrary prime divisor of $\mathfrak{p}$ in K. The l-th power Legendre symbol $(\frac{\mu}{\mathfrak{P}})$ does not depend on the choice of $\mathfrak{P}$ among the divisors of $\mathfrak{p}$. In fact if $\mathfrak{P}^\sigma$ is any other divisor, then

$$\left(\frac{\mu}{\mathfrak{P}^\sigma}\right)=\left(\frac{\mu^{\sigma^{-1}}}{\mathfrak{P}}\right)=\left(\frac{\mu}{\mathfrak{P}}\right),$$

since $\mu^{\sigma-1} \approx \mu$. Therefore we shall write the symbol $(\frac{\mu}{\mathfrak{P}})$ (which in fact depends only on $\mathfrak{p}$) as $\left[\frac{\mu}{\mathfrak{p}}\right]$. If $\mathfrak{p}$ is not l-relatively prime to μ, then $\left[\frac{\mu}{\mathfrak{p}}\right]$ is not defined. If we extend this symbol by multiplication over all ideals $\mathfrak{a}$ of the field k which are l-relatively prime to $\mathfrak{m}\mathfrak{D}$, we will obtain the symbol $\left[\frac{\mu}{\mathfrak{a}}\right]$. Obviously if $\mathfrak{a}$ consists of prime ideals of the first order in K, there exists $\mathfrak{A} \in K$, such that $\mathfrak{a} = N_{K/k}(\mathfrak{A})$ and

$$\left[\frac{\mu}{\mathfrak{a}}\right]=\left(\frac{\mu}{\mathfrak{A}}\right). \tag{4}$$

Let α be a number of the field k, which is l-relatively prime to $\mathfrak{D}(X)$ and satisfying the following conditions

1) α is l-hyperprimary

2) α is totally positive (this condition exists only for $l = 2$)

3) all the prime ideals, which appear in α to powers which are not multiples of l, split in K into prime divisors of the first degree.

If $(\alpha, \mathfrak{m}) \neq 1$, then we multiply $\mathfrak{m}$ by a number $m \in k$ such that $(\alpha, \mathfrak{m}m) = 1$. Then the symbol $\left[\frac{\mu'}{\alpha}\right]$ is defined for $\mu' = \mu m$.

We consider the expression

$$\left[\frac{\mu'}{\alpha}\right]\left(\frac{\alpha}{\mathfrak{m}'}\right)^{-1} \quad \text{with } \mathfrak{m}' = \mathfrak{m}m. \tag{5}$$

It is clear that this expression does not depend on the auxilliary number m. If we replace m by m_1, the expression (5) would be multiplied by

$$\left(\frac{mm_1^{-1}}{\alpha}\right)\left(\frac{\alpha}{mm_1^{-1}}\right)^{-1},$$

which is equal to 1, by the reciprocity law and because α is hyperprimary. Moreover the expression (5) does not depend on the choice of μ from the class X of l-invariant numbers, which can be proved in the same way. Therefore we designate the expression (5) by (α, X).

We now study a character χ of order l of the Galois group of K/k. As is known (see [11], p. 45), it corresponds uniquely up to an l-equality to a definite number α_χ of k such that

$$\sqrt[l]{\alpha_\chi} \in K, \quad \sqrt[l]{\alpha_\chi}^{\sigma} = \chi(\sigma)\sqrt[l]{\alpha_\chi}.$$

Since the field K/k is a Scholz field, it follows that if α_χ is l-relatively prime to $\mathfrak{D}(X)$, it must satisfy the conditions, which we have imposed on α, for which we have just defined the symbol (α, X). In this way we have defined a symbol (α_χ, X), which does not depend on the choice of the number α_χ, which corresponds to χ, and therefore can be denoted by (χ, X).

We shall next derive some of the properties of this symbol. First of all it is obvious that the symbol is multiplicative with respect to the argument χ:

$$(\chi_1\chi_2, X) = (\chi_1, X)(\chi_2, X), \tag{6}$$

since $\alpha_{\chi_1\chi_2} \approx \alpha_{\chi_1}\alpha_{\chi_2}$, and the symbol $\left[\frac{\mu}{\alpha}\right]$ is multiplicative by its definition, we have also with respect to the argument X:

$$(\chi, X_1, X_2) = (\chi, X_1)(\chi, X_2), \tag{7}$$

where in equations (6) and (7) the symbol on the left hand side of the equation is defined if the two symbols on the right hand side are defined.

Let us clarify the connection between the symbols (χ, X) in different fields. Let K/k be a normal subfield of a Scholz extension K_1/k, so that its Galois group G is a quotient group of the group G_1 of the field K_1/k. Then every factor set of G is at the same time a factor set of G_1, and therefore every element X of the group $M(G)$ can be regarded as an element X_1 of the group $M(G_1)$. In the same way, a character χ of order l of the group G can be regarded as a character χ_1 of the group G_1. We state that

$$(\chi, X) = (\chi_1, X_1). \tag{8}$$

In fact, in this case $\alpha_\chi \approx \alpha_{\chi_1}$. By condition 3) imposed on α,

$$(\alpha_\chi) = N_{K/k}(\mathfrak{A}_\chi), \quad \mathfrak{A}_\chi \in K,$$
$$(\alpha_{\chi_1}) = N_{K_1/k}(\mathfrak{A}_{\chi_1}), \ \mathfrak{A}_{\chi_1} \in K_1,$$

where $\mathfrak{A}_\chi$ and $\mathfrak{A}_{\chi_1}$ are selected in such a way that

$$\mathfrak{A}_\chi = N_{K_1/K}(\mathfrak{A}_{\chi_1}).$$

From the definitions of the symbols (χ, X) and $\left[\frac{\mu}{\mathfrak{A}}\right]$ and from formula (4) and the known properties of Legendre symbols, we have

$$(\chi_1, X_1) = \left(\frac{\mu}{\mathfrak{A}_{\chi_1}}\right)_{K_1} \left(\frac{\alpha_{\chi_1}}{\mathfrak{m}}\right)_k^{-1} = \left(\frac{\mu}{\mathfrak{A}_{\chi}}\right)_K \left(\frac{\alpha_\chi}{\mathfrak{m}}\right)^{-1} = (\chi, X).$$

5. The invariants $(X)_h$. In order to develop the construction of other invariants of a Scholz field, which together with the invariants (χ, X) determine whether it can be imbedded in central Scholz extensions, we need the following lemma.

Lemma 2. *Every class X of l-invariant numbers of a Scholz field K/k contains an l-hyperprimary number.*

Proof. Suppose that the field K/k satisfies only condition 1°, defining a Scholz field. By this condition any prime divisor $\mathfrak{l}_i$ of the number l in k splits completely in K and therefore the ring $K_{\mathfrak{l}_i}$, which is obtained as an extension of the base field k over K as its $\mathfrak{l}_i$-adic closure, is a direct sum of the fields $k_{\mathfrak{l}_i}$. In this direct sum the automorphisms of K/k induce, as can be easily seen, a regular transposition of the components. If we single out one of these components, then any other component will be defined by that automorphism σ which transforms it into the chosen component. Therefore

$$K_{\mathfrak{l}_i} = \Sigma\, k^{\sigma}_{\mathfrak{l}_i},$$

and we can identify $K_{\mathfrak{l}_i}$ with a ring of functions, given over a Galois group G of the field K/k with values in $k_{\mathfrak{l}_i}$, where an automorphism τ operates on a function $\alpha(\sigma)$ according to the following rule:

$$\alpha(\sigma)^\tau = \alpha(\sigma\tau).$$

If the function $\mu_i(\sigma)$ corresponds to an l-invariant number μ in the ring $K_{\mathfrak{l}_i}$, then we shall have

$$\mu_i(\sigma\tau) = \mu_i(\sigma)\,\gamma_i(\tau)^l \tag{9}$$

for some function $\gamma_i(\tau)$. Substituting $\sigma = 1$ in (9) we get

$$\mu_i(\tau) = \mu_i(1)\,\gamma_i(\tau)^l.$$

Let $\mathfrak{l}^{E_i}$ be the moduli of l-hyperprimarity, that is such exponents of $\mathfrak{l}_i$ that from

$$\alpha \equiv 1(\mathfrak{l}_i^{E_i}), \quad \alpha \in k_{\mathfrak{l}_i},$$

it follows that $\alpha \approx 1$ in $k_{\mathfrak{l}_i}$. We can find such numbers C in K and m in k that if $C_i(\sigma)$ are functions in $K_{\mathfrak{l}_i}$, corresponding to C, then

$$\left.\begin{aligned} C_i(\sigma) &\equiv \gamma_i(\sigma)\,(\mathfrak{l}_i^{E_i}), \\ m &\equiv \mu_i(1)\,(\mathfrak{l}_i^{E_i}) \quad \text{for all} \quad \mathfrak{l}_i \mid l. \end{aligned}\right\} \tag{10}$$

Then an l-invariant number $\mu m^{-1} C^{-l}$ will lie in the same class X as μ and will be l-hyperprimary in view of (10). The lemma is proved.

We note that all hyperprimary numbers belonging to the same class X can be obtained from each other by multiplication by a number of the form mC^l, where m is an l-hyperprimary number.

We now pass to the construction of the second set of invariants. Let k_h be the maximal extension of k, having an abelian group of exponent l, which is contained in $k(\zeta_h)$, where ζ_h is a primitive l^h-th root of unity. It follows from elementary properties of absolutely abelian extensions that k_h has the following structure. Let k contain an l^r-th root of unity, but not an l^{r+1}-st root of unity. If $h \leq r$, then $k_h = k$. If $h > r$, and $l \neq 2$, or if $l = 2$, and $r > 1$, then k_h is obtained by adjoining to k an l^{r+1}-st root of unity, and therefore $(k_h : k) = l$. If $l = 2$ and $r = 1$ then $k_2 = k(\sqrt{-1})$ and $k_h = k(\sqrt{-1}, \sqrt{2})$ for $h > 2$.

Let X be a class of l-invariant numbers, μ its l-hyperprimary representative, and $\mathfrak{m}$ a divisor of μ in the decomposition (3). We denote by $(X)_h$ a coset with respect to a subgroup, which according to class field theory (see [7]) correspond to the field k_h, and which contains $\mathfrak{m}$. The class $(X)_h$ does not depend on the choice of μ in X. In fact, the use of another representative implies the multiplication of $\mathfrak{m}$ by a l-hyperprimary number m, which must belong to the subgroup corresponding to k_h. This follows because the field which corresponds to the group of l-hyperprimary numbers contains the maximal abelian field of exponent l in which only the divisors of l ramify. Since this property holds in k_h, the subgroup corresponding to it must contain, according to class field theory, a group of l-hyperprimary numbers.

We can describe the class $(X)_h$ in greater detail as follows. We denote by Λ_h the group of rational numbers congruent to 1 modulo l^{r_1}, where $r_1 = r$, if $h \leq r$, $r_1 = r + 1$, , if $h > r$, and $l \neq 2$, or if $l = 2$ and $r > 1$, $r_1 = 2$, if $l = 2$, $r = 1$, $h = 2$ and $r_1 = 3$, if $l = 2$, $r = 1$, $h > 2$. We note that in all these cases

$$r_1 \geq \min(h, r). \tag{11}$$

The invariant $(X)_h$ can be characterised as the class $\mathfrak{N}(\mathfrak{m})\,\Lambda_h$, where $\mathfrak{N}(\mathfrak{m})$ denotes the absolute norm of $\mathfrak{m}$.

Since the field k contains the field of l^r-th root of unity

$$\mathfrak{N}(\mathfrak{m}) \equiv 1 \; (l^r).$$

Therefore $(X)_h$ can have only a finite number of values, namely l values for $l \neq 2$, and 2 or 4 values for $l = 2$. Similarly, it is obvious that relations analogous to (6) and (7) hold, namely

$$(X_1 X_2)_h = (X_1)_h (X_2)_h, \tag{12}$$

$$(X)_h = (X_1)_h \tag{13}$$

with the same notation which was adopted for the derivation of (7) and (8).

§2. Conditions for imbeddedness of Scholz extensions

In this section we will derive conditions under which a Scholz field K/k can be imbedded in a Scholz field with a given Galois group, which is a simple central extension of the group K/k. Since in the condition 4° in the definition of Scholz fields enters an exponent h, we shall suppose that this condition is satisfied for K/k for some h and will require that K/k be imbedded in a field for which this condition is satisfied for the same h. The conditions will therefore depend on h.

In order to simplify the discussion we shall suppose that the field K intersects with the field $\mathfrak{K}$, which is the maximal unramified extension of k with an abelian Galois group of exponent l of the base field k. We shall show that K/k can be imbedded in a field, having the same property.

1. The case of a fundamental field containing an l-th root of unity. We shall suppose that the fundamental field k contains an l-th root of unity. Then for any Scholz extension K/k with a Galois l-group, the invariants (χ, X) and $(X)_h$ introduced in the previous section are defined.

Theorem 1. *In order that the Scholz field K/k be imbedded in a Scholz field with a Galois group, which is a simple central extension of the Galois group of K/k, corresponding to the character X, it is necessary and sufficient that the invariants (χ, X) for all characters χ, and $(X)_h$ of the field K be equal to 1.*

We note that the equality to unity of these invariants is equivalent to the conditions

$$\left[\frac{\mu}{\alpha_\chi}\right] = \left(\frac{\alpha_\chi}{\mathfrak{m}}\right), \tag{14}$$

$$\mathfrak{N}(\mathfrak{m}) \in \Lambda_h \tag{15}$$

for l-invariant and l-hyperprimary numbers μ of class X and for all the characters χ. The condition $(\alpha_\chi, \mathfrak{m}) = 1$ is here always satisfied since $\alpha_\mathfrak{m}$ does not contain any divisor of the discriminant of K/k.

We shall prove the necessity of conditions (14) and (15). We know that it is sufficient to check them for a single representative of the class X of l-invariant numbers. On the other hand, because of the conditions of the theorem, this class contains at least one number μ for which $K(\sqrt[l]{\mu})$ is a Scholz field. Let $\mathfrak{P}$ be a ramified prime ideal of K/k. If $(\mathfrak{P}, \mathfrak{D}(X)) = 1$, then because $\mathfrak{P}$ is of the first degree in $K(\sqrt[l]{\mu})$, it follows that

$$\left[\frac{\mu}{\mathfrak{p}}\right] = \left(\frac{\mu}{\mathfrak{P}}\right) = 1.$$

Since α_χ is composed of ramified prime ideals and is l-relatively prime to $\mathfrak{D}(X)$, it follows that

$$\left[\frac{\mu}{\alpha_\chi}\right] = 1. \tag{16}$$

On the other hand, every prime ideal $\mathfrak{q}/\mathfrak{m}$ must be split completely in K, for otherwise it would be a ramified prime ideal in $K(\sqrt[l]{\mu})$, not of the first degree. Therefore $\mathfrak{q}$ splits completely in $k(\sqrt[l]{\alpha_\chi})$, that is

$$\left(\frac{\alpha_\chi}{\mathfrak{q}}\right) = 1 \quad \text{for} \quad \mathfrak{q}/\mathfrak{m}, \quad \text{i.e.,} \left(\frac{\alpha_\chi}{\mathfrak{m}}\right) = 1. \tag{17}$$

Equations (16) and (17) show that conditions (14) are satisfied.

Conditions (15) are also satisfied because for every prime ideal $\mathfrak{q}/\mathfrak{m}$

$$\mathfrak{N}(\mathfrak{q}) \equiv 1\ (l^h),$$

that is

$$\mathfrak{N}(\mathfrak{m}) \equiv 1\ (l^h)$$

and always

$$\mathfrak{N}(\mathfrak{m}) \equiv 1\ (l^r);$$

this shows in view of (11) that $\mathfrak{N}(\mathfrak{m}) \in \Lambda_h$.

We now show the sufficiency of conditions (14) and (15). Since K/k is a Scholz field, we can find in it an l-invariant number μ such that the field $K(\sqrt[l]{\mu})/k$ possesses a Galois group, which is a simple central extension of the Galois group of K/k corresponding to the character X. Without changing the Galois group of $K(\sqrt[l]{\mu})/k$ we can change μ arbitrarily in its class X.

Therefore we have to show that if conditions (14) and (15) are satisfied we can then find in the class X of l-invariant numbers, a number μ such that in the field $K(\sqrt[l]{\mu})$ the four conditions which define a Scholz field will hold.

That the first property holds has already been shown in Lemma 2, and we can therefore choose in X an l-invariant, l-hyperprimary number μ_0.

That the second condition holds can be shown in an analogous and even simpler way. This condition is only for $l = 2$. Let $\mathfrak{p}_{\infty,i}$ be the real infinite primes in k. Let $K_{\mathfrak{p}_{\infty,i}}$ be the direct sum of fields of real numbers in $k_{\mathfrak{p}_{\infty,i}}$. From arguments similar to those employed in the proof of Lemma 2, it follows that to a number μ_0 in each $K_{\mathfrak{p}_{\infty,i}}$ there corresponds a real function $\mu_{0,i}(\sigma)$ such that

$$\mu_{0,i}(\sigma\tau) = \mu_{0,i}(\sigma)\,\beta_i(\tau)^2.$$

From this it follows that all the $\mu_{0,i}(\sigma)$ have the same sign, which corresponds with the sign of $\mu_{0,i}(1)$. We select in k an l-hyperprimary number m_0, such that the sign of its components in each $k_{\mathfrak{p}_{\infty,i}}$ is the same as the sign of the corresponding $\mu_{0,i}(1)$. Then $\mu_0 m_0$ will also be l-hyperprimary and totaly positive. This new number we shall denote by μ'. Conditions 1° and 2° hold for μ'.

Let $\mathfrak{p}_i$ be all the ramified ideals in K/k which are relatively prime to the ideal $\mathfrak{D}(X)$. We shall find an l-hyperprimary totally-positive number $m' \in k$, such that

$$\left(\frac{m'}{\mathfrak{p}_i}\right) = \left[\frac{\mu'}{\mathfrak{p}_i}\right]^{-1}, \text{ i.e., } \left(\frac{\mu' m'}{\mathfrak{P}_i}\right) = 1 \quad \text{for all } \mathfrak{p}_i.$$

This condition reduces to a finite number of congruences with respect to the moduli $\mathfrak{p}_i$, relatively prime to each other and to l, and therefore such a number m' exists. We denote the number $\mu' m'$ by μ'' and get as before

$$(\mu'') = (\mathfrak{C}'')^l \mathfrak{D}\mathfrak{m}''.$$

We now consider the modulus $\mathfrak{F}$, which is a product of all the ramified prime ideals $\mathfrak{p}_i$ which are prime to $\mathfrak{D}(X)$, the real infinite primes $\mathfrak{p}_{\infty,i}$ and all $\mathfrak{l}_i^{E_i}$ (for a definition of E_i see the proof of Lemma 2). Let $S_{\mathfrak{F}}$ be the group of l-th powers of all classes mod $\mathfrak{F}$, let L/k be the class field of the group $S_{\mathfrak{F}}$ and let Z_h be the field obtained by adjoining to k an l^h-th root of unity. We denote by σ an automorphism of L/k, which is an Artin automorphism of the ideal $\mathfrak{m}''$. We shall prove that there exists an automorphism $\bar{\sigma}$ in the composite KZ_hL, which induces in K and Z_h an identity automorphism and in L the automorphism σ.

In fact $KZ_h \cap L$ is a field K_1k_h, where k_h is the field which appeared in the definition of the invariant $(X)_h$ and K_1 is obtained by adjoining to k all the $\sqrt[l]{\alpha_\chi}$ for which α_χ is l-relatively prime to $\mathfrak{D}(X)$ (see figure 1). We shall show that σ induces in $KZ_h \cap L$ an identity automorphism. From what was said above, σ induces in $KZ_h \cap L$ an automorphism, which is an Artin automorphism of the ideal m'' in K_1k_h.

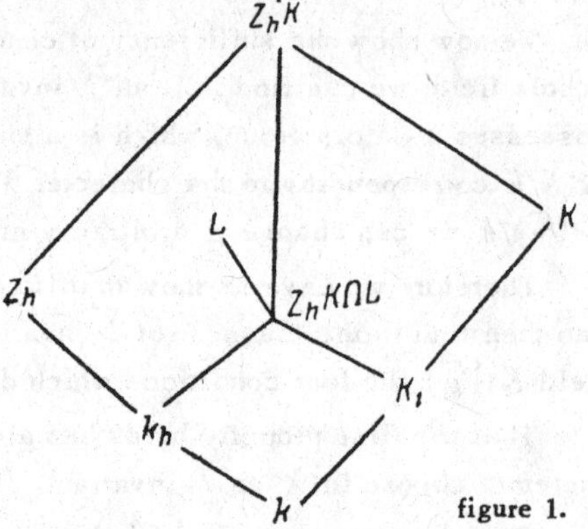

figure 1.

In k_h, m'' belongs to an identity automorphism, because of the condition $(X)_h = 1$.

For all $\mathfrak{p}_i$, relatively prime to $\mathfrak{D}(X)$, by construction of μ''

$$\left[\frac{\mu''}{\mathfrak{p}_i}\right] = 1, \text{ i.e., } \left[\frac{\mu''}{\alpha_\chi}\right] = 1$$

for α_χ, l-relatively prime to $\mathfrak{D}(X)$. From the condition $(\chi, X) = 1$, it follows that for such α_χ

$$\left(\frac{\alpha_\chi}{\mathfrak{m}''}\right) = 1.$$

This means that $\mathfrak{m}''$ belongs to an identity automorphism in K_1. Hence $\mathfrak{m}''$ belongs to K_1 and to k_h, and therefore to an identity automorphism in K_1k_h. As long as an automorphism σ in the field L induces an identity automorphism in $KZ_h \cap L$, it can be extended to KZ_hL in such a way as to induce in KZ_h an identity automorphism. This extension we shall denote by $\bar{\sigma}$.

By Čebotarev's density law (see [4], p. 148), there exists in k a prime ideal $\mathfrak{q}$ belonging in Z_hLK to the automorphism $\bar{\sigma}$. Since $\mathfrak{q}$ belongs in L to the same automorphism as $\mathfrak{m}''$, it means that $\mathfrak{q}$ is in the same class with respect to $S_{\mathfrak{F}}$ as $\mathfrak{m}''$ and therefore $\mathfrak{q} = \mathfrak{m}''(m)\mathfrak{i}^l$, where m is l-hyperprimary, totally positive, and congruent to 1 with respect to all the moduli $\mathfrak{p}_i$. We shall show that the number $\mu = \mu''m$ satisfies all four conditions, which we need. It is l-hyperprimary and totally positive, since so were μ'' and m. In the field $\Omega = K(\sqrt[l]{\mu})$ the ramified prime ideals of K are of degree 1, since the degree is unchanged for all the divisors of $\mathfrak{D}(X)$, and only the ramification order is increased l times. For $\mathfrak{p}_i$ relatively prime to $\mathfrak{D}(X)$ this follows from

$$\left(\frac{\mu}{\mathfrak{P}_i}\right)=\left(\frac{\mu'' m}{\mathfrak{P}_i}\right)=\left[\frac{\mu''}{\mathfrak{p}_i}\right]\left(\frac{m}{\mathfrak{p}_i}\right)=1.$$

The extension Ω/K will have for ramified prime ideals, in view of

$$(\mu)=(\mu'' m)=(\mathfrak{C}''^{i-1})^l\mathfrak{D}\,\mathfrak{q},$$

divisors of $\mathfrak{D}$, which, as has already been shown, are of degree 1, and divisors of $\mathfrak{q}$ which are also of degree 1, because, by construction, $\mathfrak{q}$ belongs to an identity automorphism in K/k. Finally, the fourth condition also holds. It was assumed for ramified prime ideals in K/k, and for $\mathfrak{q}$ it follows, because by construction $\mathfrak{q}$ belongs to an identity automorphism in Z_h, that is

$$\mathfrak{N}(\mathfrak{q})\equiv 1\ (l^h).$$

The theorem is proved.

In case the character $X \neq 1$, the maximal subfield of the field K, having an abelian group with exponent l does not change when K is extended to $K(\sqrt[l]{\mu})$, so that $K(\sqrt[l]{\mu})$ will also intersect with the unramified field $\mathfrak{K}$ over k. If $X=1$, then μ is l-equal to a number α in k. By construction it is divisible by the prime ideal $\mathfrak{q}$, which is relatively prime to all the α_X. The maximal subfield of the field $K(\sqrt[l]{\mu})$ with an abelian group of exponent l is obtained from such a subfield of the field K by adjoining $\sqrt[l]{\alpha}$, and be the above, will also intersect with $\mathfrak{K}$ over k.

We note that it follows from the proof of the theorem that we can select μ and its class X in such a way that the ideal $\mathfrak{m}$ in the decomposition (3) is relatively prime to any given ideal.

It follows from the theorem that in order that a Scholz field K/k can be imbedded in a Scholz extension, whose Galois group is a simple, central extension of the Galois group of K/k, corresponding to the classes $X_1, \cdots, X_r$, it is necessary and sufficient that all the invariants $(\chi, X_1), \cdots, (\chi, X_r)$ for all χ, and all the invariants $(X_1)_h, \cdots, (X_r)_h$ become 1.

Corollary. *It is sufficient that all the invariants* $(X)_h$ *and* (χ, X) *become unity for imbedding a Scholz extension* K/k *in a Scholz extension, whose Galois group is any central (not only simple) extension of the Galois group.*

Since K/k is a Scholz field, it contains l-invariant numbers $\mu_1, \cdots, \mu_m$, such that the field $\Omega = K(\sqrt[l]{\mu_1}, \cdots, \sqrt[l]{\mu_m})$ possesses the necessary Galois group. We can arbitrarily vary the numbers $\mu_1, \cdots, \mu_m$ within their classes X without changing in doing so the Galois group of Ω/k. We shall show that the $\mu_1, \cdots, \mu_m$ can be selected from their classes in such a way that Ω is a Scholz field. Let us suppose that among the numbers $\mu_1, \cdots, \mu_m$ the numbers $\mu_1, \cdots, \mu_r$ are multiplicatively independent with respect to the subgroup of numbers, which are l-equal to the invariant numbers, as modulus, and that the remaining numbers $\mu_{r+1}, \cdots, \mu_m$ are l-equal to some products of $\mu_1, \cdots, \mu_r$ by invariant numbers $\alpha_{r+1}, \cdots, \alpha_m$.

We shall adjoin to K in turn $\sqrt[l]{\mu_1}, \cdots, \sqrt[l]{\mu_r}$.

The number of generators of the Galois groups of the fields $K(\sqrt[l]{\mu_1}, \cdots, \sqrt[l]{\mu_i})$ obtained in this way will be the same number d, as that of the Galois group of the field K/k. In fact if the number of generators should increase, then the quotient group with respect to the Frattini subgroup would also increase, that is the field $K\sqrt[l]{\mu_1}, \cdots, \sqrt[l]{\mu_i})$ would contain a number $\sqrt[l]{\alpha}$, $\alpha \in k$, which is not contained in K. We would then have

$$\alpha \approx \mu_1^{x_1} \cdots \mu_i^{x_i} \text{ in } K, \text{ but not all } x_i \equiv 0\ (l),$$

which could contradict the choice of $\mu_1, \cdots, \mu_r$. In this way, the successive adjoining of $\sqrt[l]{\mu_1}, \cdots, \sqrt[l]{\mu_r}$ leaves the characters χ unaltered.

At each step we will adjoin $\sqrt[l]{\mu_i}$, corresponding to the class X_i, where μ_i lies not only in $K(\sqrt[l]{\mu_1}, \cdots, \sqrt[l]{\mu_{i-1}})$, but also in K. Because all the characters χ remain the same, we get by (8) and (13) that

$$(\chi, X_i) = 1, \quad (X_i)_h = 1,$$

and that shows that if we multiply the μ_i by numbers of k, the field $K(\sqrt[l]{\mu_1}, \cdots, \sqrt[l]{\mu_i})$ becomes a Scholz field.

Now it remains to adjoin to the Scholz field $K(\sqrt[l]{\mu_1}, \cdots, \sqrt[l]{\mu_r})$ the numbers $\sqrt[l]{\alpha_{r+1}}, \cdots, \sqrt[l]{\alpha_m}$, also one at the time. Since we adjoin here numbers of the unit class, all the invariants are equal to unity, and by Theorem 1, we can select the numbers $\alpha_{r+1}, \cdots, \alpha_m$ in such a way that all the fields become Scholz fields.

2. The case when the fundamental field does not contain an l-th root of unity. We suppose that the fundamental field does not contain an l-th root of unity Let K/k be a Scholz field. We adjoin to k and K an l-th root of unity ζ, and denote $k(\zeta)$ by $\bar{k}$, and $K(\zeta)$ by $\bar{K}$. We denote the degree $(\bar{k} : k)$ by m. Obviously $m \mid l-1$, and therefore $(m, l) = 1$.

We note that the conditions for a field to be a Scholz field are equivalent for the fields K/k and $\bar{K}/\bar{k}$. We check this for all four conditions.

Condition 2° applies only when $l = 2$, but in that case $\bar{k} = k$, and $\bar{K} = K$, therefore our assertion is obvious.

For conditions 1° and 3° the assertion follows from the same considerations. Let $\mathfrak{p}$ be a prime ideal in k, and $\bar{\mathfrak{p}}$, its prime divisor in $\bar{k}$. We shall show that $\mathfrak{p}$ splits completely in K (and correspondingly is of degree 1) if and only if $\mathfrak{p}$ splits completely (and is of degree 1) in $\bar{K}$. The ring $K_{\mathfrak{p}}$ is a direct sum of fields $K_{\mathfrak{P}}$, which are extensions of $k_{\mathfrak{p}}$ of degree $n_{\mathfrak{p}}$ (order $r_{\mathfrak{p}}$), and $\bar{k}_{\mathfrak{p}}$ is a direct sum of the fields $\bar{k}_{\bar{\mathfrak{p}}}$ of degree $m_{\mathfrak{p}}$ (order $s_{\mathfrak{p}}$) over $k_{\mathfrak{p}}$. Moreover $n_{\mathfrak{p}}(r_{\mathfrak{p}})$ is a power of l, and $m_{\mathfrak{p}}(s_{\mathfrak{p}})$ divide m, and therefore are relatively prime to l. The ring $\overline{K}_{\mathfrak{p}}$ is a direct sum of the fields $K_{\mathfrak{P}}\bar{k}_{\bar{\mathfrak{p}}}$ of degree $n_{\mathfrak{p}}m_{\mathfrak{p}}$ (order $r_{\mathfrak{p}}s_{\mathfrak{p}}$) over $k_{\mathfrak{p}}$ and $n_{\mathfrak{p}}(r_{\mathfrak{p}})$ over $\bar{k}_{\bar{\mathfrak{p}}}$, since $(n_{\mathfrak{p}}, m_{\mathfrak{p}}) = 1$. In this way, $\overline{K}_{\mathfrak{p}}$ is a sum of fields of degree $n_{\mathfrak{p}}$ (order $r_{\mathfrak{p}}$) over $\bar{k}_{\mathfrak{p}}$. In order that $\mathfrak{p}$ should split completely (have order 1) in K, it is necessary and sufficient that $n_{\mathfrak{p}}(r_{\mathfrak{p}})$ be equal to one. In order that $\bar{\mathfrak{p}}$ should split completely (have order 1) in $\overline{K}$ it is necessary and sufficient that $m_{\mathfrak{p}}(s_{\mathfrak{p}})$ equal 1.

It remains to prove the equivalence of condition 4° for the fields K/k and $\overline{K}/\bar{k}$. If $\mathfrak{p}$ is a divisor of the discriminant of K/k in k, $\bar{\mathfrak{p}}$ is its prime divisor in $\bar{k}$, then $\mathfrak{N}(\bar{\mathfrak{p}}) = \mathfrak{N}(\mathfrak{p})^f$, where f is the relative degree of $\bar{\mathfrak{p}}$ in $\bar{k}/k$, which is relatively prime to l. Our assertion follows from this in an obvious manner.

Hence if K/k is a Scholz field, then the invariants (χ, X) and $(X)_h$, defined in the previous section, are also defined for $\overline{K}/\bar{k}$.

We shall also show that the condition under which the field K intersects in k with the maximum nonramified extention $\mathfrak{K}$ of the field k, having an abelian group of exponent l, is satisfied if and only if it is satisfied for $\overline{K}/\bar{k}$. We denote the corresponding extension of the field $\bar{k}$ by $\bar{\mathfrak{K}}$ and show that

$$\overline{K} \cap \bar{\mathfrak{K}} = (K \cap \mathfrak{K})\bar{k}.$$

It is obvious that the right hand side is contained in the left hand side, and we must prove that also the left hand side is contained in the right hand side. To do this we note that since $\overline{K} = K\bar{k}$, $K \cap \bar{k} = k$, the Galois group of $\overline{K}/k$ is the direct product of the Galois group of K/k having order l^{a}, and the Galois group of $\bar{k}/k$ of order m, relatively prime to l. From this it follows that $\overline{K} \cap \mathfrak{K}$ is a composite of $\bar{k}$ and the subfield $\mathfrak{K}_1$ of the field K, where $\mathfrak{K}_1 \cap \bar{\mathfrak{K}}$. Because the degrees $(\mathfrak{K}_1 : k)$ and $(\bar{k} : k)$ are relatively prime, $\bar{\mathfrak{K}}_1$ must be nonramified over k. Since, moreover, the Galois group of $\mathfrak{K}_1/k$ is isomorphic with the Galois group of $\mathfrak{K}_1\bar{k}/\bar{k} \subset \bar{\mathfrak{K}}/\bar{k}$, it is abelian and of exponent l. Therefore $\mathfrak{K}_1 \subseteq \mathfrak{K}$, which proves our assertion.

Theorem 2. *In order that the Scholz field K/k be imbedded in a Scholz field with a Galois group, which is a simple central extension of the Galois group of K/k, corresponding to the characters X, it is necessary and sufficient that the invariants (χ, X) for all characters χ, and $(X)_h$ of the field $\overline{K}/\bar{k}$ be equal to 1.*

The proof of necessity reduces directly to Theorem 1. Let us suppose that Ω/k is a central simple Scholz extension of K/k with the required Galois group. The field $\overline{\Omega} = \Omega(\zeta)$ is also a Scholz extension of the field $\overline{K}/\bar{k}$, with the same Galois group. It follows from Theorem 1 that all the invariants of the field $\overline{K}/\bar{k}$ are equal to 1.

The proof of sufficiency also reduces to the corresponding statement of Theorem 1, only in a more complicated way. We shall make use of the following notation. We denote by t a generating automorphism of the field $\overline{k}/k$, so that $t^m = 1$. Let g be a number such that

$$\zeta^t = \zeta^g, \quad g^m \not\equiv 1 \ (l^2).$$

Since $\overline{k} \neq k$, we also have

$$g \not\equiv 1 \ (l).$$

We denote by s the expression $t^{m-1} + t^{m-2}g + \cdots + tg^{m-2} + g^{m-1}$. We make use of the following relation, which holds for numbers as well as for ideals of the field $\overline{k}$:

It follows from

$$\alpha^s \approx 1 \tag{18}$$

that $\alpha \approx \beta^{t-g}$ for some β and conversely any such α satisfies (18);

it follows from

$$\alpha^{t-g} \approx 1 \tag{19}$$

that $\alpha \approx \beta^s$ for some β, and conversely, any such α satisfies (19).

All the material necessary for the proof of (18) and (19) will be found in Reichardt [8].

We denote $1 - g^m$ by lx, where $(l, x) = 1$, and define y by the congruence $xy \equiv 1 \ (l)$. Then it can be shown by straight substitution that numbers of the form β^{t-g} satisfy condition (18), and β^s satisfy (19), if we make use of the fact that

$$s(t-g) = t^m - g^m = 1 - g^m \equiv 0 \ (l).$$

Let $\alpha^s = \gamma^l$. Raising both sides to the $t-g$ power we get

$$\alpha^{lx} = \gamma^{l(t-g)}, \quad \text{i.e.,} \quad \alpha^x = \zeta^r \gamma^{t-g}.$$

We raise both sides of this equation by the y-th power. We get

$$\alpha \approx \zeta^{ry} \gamma^{y(t-g)}.$$

We must now show that the multiplier $\zeta^{ry} \approx 1$. To do this we substitute the above expression for α in (18). Since $\zeta^t = \zeta^g$, then $\zeta^s = \zeta^{mg^{-1}}$. Therefore

$$1 \approx \alpha^s \approx \zeta^{rymg^{-1}} \gamma^{y(t-g)s} \approx \zeta^{rymg^{-1}},$$

and since $mg^{-1} \not\equiv 0 \ (l)$, it follows that $\zeta^{ry} \approx 1$.

Let $\alpha^{t-g} = \gamma^l$. Raising both sides to the s degree, we get, as before,

$$\alpha^{lx} = \gamma^{ls}, \quad \alpha^x = \zeta^r \gamma^{t-g}, \quad \alpha \approx \zeta^{ry} \gamma^{y(t-g)}.$$

In view of $\zeta^s = \zeta^{mg^{-1}}$, we have $\zeta^{ry} = \zeta^{rym^{-1}gs}$ and

$$\alpha \approx (\zeta^{rymg^{-1}} \gamma^y)^s.$$

Finally, we shall make use of the fact that condition (19) is necessary in order that the field $\overline{k}(\sqrt[l]{\alpha})$ be a composite of the field $\overline{k}$ and a cyclic field of degree l over k. The proof consists of verification.

First of all we construct, making use of Theorem 1, a field $\overline{K}(\sqrt[l]{\overline{\mu}})$, which is a Scholz extension of $\overline{K}/\overline{k}$ with the required Galois group. Next we make use of the results of Reichardt [8]. He showed that $\overline{k}$ contains a number $\overline{m}$, such that the l-invariant number $\mu_0 = \overline{\mu}\,\overline{m}$ satisfies the condition

$$\mu_0^{t-g} \approx 1, \tag{20}$$

and the field $\overline{K}(\sqrt[l]{\mu_0})$ is a direct composite of $\overline{k}$ and some field Ω, normal over k and having the same Galois group as $\overline{K}(\sqrt[l]{\overline{\mu}})/\overline{k}$. Our problem would be solved if we could show that Ω is a Scholz extension of k. This is equivalent to saying that $\overline{K}(\sqrt[l]{\mu_0})$ is a Scholz extension of $\overline{k}$. Without violating the property (20) of the number μ_0, and the properties of the field $\overline{K}(\sqrt[l]{\mu_0})$ given above, we can replace μ_0 by a number of the form $\mu_0 a^s$, where a is any number of $\overline{k}$. The theorem will be proved if we can show that in this way we obtain a Scholz field $\overline{K}(\sqrt[l]{\mu_0 a^s})/\overline{k}$.

We begin with condition 1°, defining a Scholz field. Let $\mathfrak{l}$ be any prime divisor of l in k. Since the field $\overline{K}(\sqrt[l]{\overline{\mu}})$ is a Scholz field,

$$\overline{\mu} \approx 1 \text{ in } \overline{K}_{\mathfrak{l}}.$$

For the number $\mu_0 = \overline{\mu}\,\overline{m}$ we get

$$\mu_0 \approx \overline{m} \text{ in } \overline{K}_{\mathfrak{l}},$$

and by (20)

$$\overline{m}^{t-g} \approx 1 \text{ in } \overline{K}_{\mathfrak{l}}.$$

The ring $K_{\mathfrak{l}}$ is decomposible, as has been shown in the proof of Theorem 2, into the direct sum of rings:

$$\overline{K}_{\mathfrak{l}} = \Sigma\, \overline{k}_{\mathfrak{l}}^{\sigma},$$

where σ goes over all the automorphisms of K/k. All the components of the number $\overline{m}^{t-g}$ in this decomposition are equal. Therefore

$$\overline{m}^{t-g} \approx 1 \text{ in } \overline{k}_{\mathfrak{l}}, \text{ that is } \overline{m} \approx b_1^s \text{ in } \overline{k}_{\mathfrak{l}}.$$

Selecting a number a_1 in $\overline{k}$ such that $a_1 b_1 \approx 1$ in $\overline{k}_{\mathfrak{l}}$ we obtain an l-hyperprimary number $\mu_0 a_1^s$. We let

$$\mu_0 a_1^s = \mu_1 = \overline{\mu}\,\overline{m}_1. \tag{21}$$

Condition 2° is satisfied in our case, since $l \neq 2$. Before going on to the consideration of condition 3°, we make the following remark. From the decomposition

$$(\mu_1) = \overline{\mathfrak{C}}^{l}\,\overline{\mathfrak{D}}\,\overline{\mathfrak{m}}_1$$

and from (20) and (21), it follows that

$$\overline{\mathfrak{m}}_1^{t-g} \approx 1, \text{ i.e., } \overline{\mathfrak{m}}_1 \approx \overline{\mathfrak{n}}_1^{s}, \quad \mathfrak{n}_1 \in \overline{k}.$$

Let $\overline{\mathfrak{p}}$ be a ramified prime ideal in $\overline{K}/\overline{k}$, relatively prime to $\mathfrak{D}(X)$. Because by construction $\overline{K}$ is normal over k, it follows that $\bar{\mathfrak{p}}^l$ is ramified in $\overline{K}/\bar{k}$, and relatively prime to $\mathfrak{D}(X)$. From (20) and (21) and because $\overline{K}(\sqrt[l]{\bar{\mu}})$ is a Scholz field, it follows that

$$\left(\frac{\bar{m}_1}{\mathfrak{p}^l}\right)_{\bar{k}} = \left(\frac{\bar{m}_1}{\mathfrak{P}^l}\right)_{\overline{K}} = \left(\frac{\bar{\mu}}{\mathfrak{P}^l}\right) = \left(\frac{\bar{\mu}_1^{l-1}}{\mathfrak{P}}\right)^g = \left(\frac{\bar{\mu}_1}{\mathfrak{P}}\right) = \left(\frac{\bar{m}_1}{\mathfrak{p}}\right), \tag{22}$$

where $\overline{\mathfrak{P}}$ is a prime divisor of $\bar{\mathfrak{p}}$ in $\overline{K}$, By (22), we shall denote $\left(\frac{\bar{m}_1}{\mathfrak{p}_1}\right)$ by $\zeta(\mathfrak{p})$, where $\mathfrak{p} = N_{\bar{k}/k}(\bar{\mathfrak{p}})$.

Because $\overline{K}$ is normal over k it follows that

$$\bar{\alpha}_\chi^{t-g} \approx 1, \tag{23}$$

where χ is any character of order l of the Galois group of $\overline{K}/\bar{k}$. (For the definition of the number $\bar{\alpha}_\chi$, see § 1, part 4.) From here, as in the derivation of (22) it follows that for any prime ideal $\bar{\mathfrak{q}} \in \bar{k}$

$$\left(\frac{\bar{\alpha}_\chi}{\mathfrak{q}^t}\right) = \left(\frac{\bar{\alpha}_\chi}{\mathfrak{q}}\right).$$

Moreover it follows from (23) that

$$(\bar{\alpha}_\chi) \approx \mathfrak{A}_\chi^s. \tag{24}$$

Finally we note that if $\bar{a}$ is any number in k, then

$$\left(\frac{\bar{a}^s}{\bar{\mathfrak{p}}}\right) = \prod_{r=0}^{m-1}\left(\frac{\bar{a}^{t^r g^{m-1-r}}}{\bar{\mathfrak{p}}}\right) = \prod_{r=0}^{m-1}\left(\frac{\bar{a}}{\bar{\mathfrak{p}}^{t^{-r}}}\right)^{g^{m-1}} = \left(\frac{\bar{a}}{N_{\bar{k}/k}(\mathfrak{p})}\right)^{g^{m-1}} = \left(\frac{\bar{a}}{\mathfrak{p}}\right)^{g^{-1}}. \tag{25}$$

We now consider condition 3° in the definition of a Scholz field. We determine two numbers $\bar{c}_1$ and $\bar{c}_2$ in $\bar{k}$ in such a way that

$$\left(\frac{\bar{c}_1}{\alpha_\chi}\right) = \left(\frac{\bar{\alpha}_\chi}{\mathfrak{n}_1}\right), \quad \left(\frac{\bar{c}_2}{\mathfrak{p}_i}\right)^{g^{-1}} = \zeta(\mathfrak{p}_i)^{-1}, \tag{26}$$

where χ goes over all the characters of order l of the Galois group $\overline{K}/\bar{k}$, for which $\bar{\alpha}_\chi$ is l-relatively prime to $\mathfrak{D}(X)$, and $\bar{\mathfrak{p}}_i$ are all the ramified prime ideals in $\overline{K}/\bar{k}$, which are relatively prime to $\mathfrak{D}(X)$.

The existence of the number $\bar{c}_2$ is obvious, because this number must satisfy only some congruences with relatively prime moduli $\mathfrak{p}_i$, which are not l-th powers.

In order to prove the existence of the number $\bar{c}_1$, satisfying conditions (26), we note that these conditions are multiplicative with respect to $\bar{\alpha}_\chi$. We select from the numbers $\bar{\alpha}_\chi$ those $\bar{\alpha}_{\chi_1}, \cdots, \bar{\alpha}_{\chi_d}$ which are multiplicatively independent with respect to a modulus which is an l-th power, then all $\bar{\alpha}_\chi$ are l-equal to their product. It is sufficient to find $\bar{c}_1$ for which (26) is satisfied for $\bar{\alpha}_{\chi_1}, \cdots, \bar{\alpha}_{\chi_d}$. We note that the ideals $(\bar{\alpha}_{\chi_1}), \cdots, (\bar{\alpha}_{\chi_d})$ are also multiplicatively independent with respect to an l-th power as a modulus in the group of ideals. In fact if some product were an l-th power, then that would mean that some $\overline{\alpha}_\chi$ is singular and hyperprimary, that is that $\overline{K}$ intersects with the unramified field $\overline{\mathfrak{K}}$ at least in $\overline{k}(\sqrt[l]{\overline{\alpha}_\chi})$, which contradicts the assumptions made. In this way the ideals $(\overline{\alpha}_{\chi_1}), \cdots, (\overline{\alpha}_{\chi_d})$ are multiplicatively independent with respect to an l-th power modulus, and we can find a number $\overline{c}_1$, which has preassigned values of the Legendre symbol.

By (24), the number $\overline{c}_1$ is defined by the values of the symbol $\left(\frac{\overline{c}_1}{\mathfrak{p}_i^s}\right)$. Since

$$\mathfrak{p}_i = \mathfrak{p}_i^{1+t+\cdots+t^{m-1}},$$

all the ideals $\mathfrak{p}_i$ and $\overline{\mathfrak{p}}_i^s$ are multiplicatively independent with respect to an l-th power as modulus. Moreover, they are all relatively prime to l. Therefore there exists an l-hyperprimary number $\overline{c}_0$, for which relations analogous to (26) hold:

$$\left(\frac{\overline{c}_0}{\overline{\alpha}_\chi}\right) = \left(\frac{\overline{\alpha}_\chi}{\overline{\pi}_1}\right)^{-1}, \quad \left(\frac{\overline{c}_0}{\mathfrak{p}_i}\right)^{\sigma^{-1}} = \zeta(\mathfrak{p}_i)^{-1}. \tag{27}$$

We denote by $\mathfrak{F}$ the product of all $\mathfrak{p}_i$ and the modulus of l-hyperprimarity; by $\overline{S}_{\mathfrak{F}}$, the group of all l-th power classes of ideals mod $\mathfrak{F}$, and by L, the class field of the group $\overline{S}_{\mathfrak{F}}$ over $\overline{k}$. Let σ_1 be the Artin automorphism of the ideal $\overline{\pi}\overline{c}_0$ in $\overline{L}$. The field $\overline{K} \cap \overline{L}$ is obtained from $\overline{k}$ by adjoining $\sqrt[l]{\overline{\alpha}_\chi}$ for all $\overline{\alpha}_\chi$, l-relatively prime to $\mathfrak{D}(X)$. By (27)

$$\left(\frac{\overline{\alpha}_\chi}{\overline{\pi}_1\overline{c}_0}\right) = \left(\frac{\overline{\alpha}_\chi}{\overline{\pi}_1}\right)\left(\frac{\overline{\alpha}_\chi}{\overline{c}_0}\right) = \left(\frac{\overline{\alpha}_\chi}{\overline{\pi}_1}\right)\left(\frac{\overline{c}_0}{\overline{\alpha}_\chi}\right) = 1,$$

so that σ_1 induces in $\overline{K} \cap \overline{L}$ the identity automorphism. From this it follows that there exists an automorphism σ in $\overline{K}\overline{L}$, which induces the identity automorphism in $\overline{K}$, and the automorphism σ_1 in $\overline{L}$. By Čebotarev's density law, $\overline{k}$ contains a prime ideal $\overline{\mathfrak{q}}_1$, belonging to σ in $\overline{K}\overline{L}$. Since $\overline{\mathfrak{q}}$ belongs to the same automorphism in $\overline{L}$ as $\overline{\pi}_1\overline{c}_0$, $\overline{\mathfrak{q}}_1$ lies in the same class as $\overline{\pi}_1\overline{c}_0$ with respect to $\overline{S}_{\mathfrak{F}}$, that is $\overline{\mathfrak{q}}_1 = \overline{\pi}_1\overline{c}_0(\overline{c})\,\mathfrak{i}^l$, where $\overline{c}$ is l-hyperprimary and congruent to 1, modulo all $\mathfrak{p}_i$. We can prove that conditions 1°, 2° and 3° hold for the number $\mu_2 = \mu_1(\overline{c}\,\overline{c}_0)$.

Condition 1° has not been altered since, by construction, $\overline{c}_0\overline{c}$ is hyperprimary. Condition 2° is automatically satisfied in our case. In connection with condition 3° we note that

$$(\mu_2) = \mathfrak{C}_2^l\,\overline{\mathfrak{D}}\,\overline{\mathfrak{q}}_1^s.$$

The ideal $\overline{\mathfrak{q}}_1$ is of degree 1 in $\overline{K}$, since it belongs by construction to the identity automorphism in $\overline{K}$. Finally, the ideals $\overline{\mathfrak{p}}_i$, are relatively prime to $\mathfrak{D}(X)$ and are of degree 1 in $\overline{K}(\sqrt[l]{\mu_2})$, since by (21), (25) and (27)

$$\left(\frac{\mu_2}{\mathfrak{P}_i}\right) = \left(\frac{\mu_1}{\mathfrak{P}_i}\right)\left(\frac{(c_0 c)}{\mathfrak{P}_i}\right) = \left(\frac{\overline{m}_1}{\overline{\mathfrak{p}}_i}\right)\left(\frac{\overline{c}_0^s}{\overline{\mathfrak{p}}_i}\right)\left(\frac{\overline{c}^s}{\overline{\mathfrak{p}}_i}\right) = 1,$$

and those ramified ideals in $\overline{k}$, which divide $\overline{\mathfrak{D}}$ acquire in $\overline{K}(\sqrt[l]{\mu_2})$ a ramification order which is l times as large, and therefore remain of degree 1.

It remains to discuss condition 4°. In order to do this we find in $\overline{K}$ a number ξ such that

$$\mathfrak{N}(\xi)^{1-g}\mathfrak{N}(\overline{\mathfrak{q}}_1) \equiv 1\ (l^h).$$

Since $g \not\equiv 1\ (l)$, and $\mathfrak{N}(\overline{\mathfrak{q}}_1) \equiv 1\ (l)$, since $\overline{k}$ contains an l-th root of unity, such a choice is possible.

We denote by H the group of ideal classes in $\overline{k}$ with respect to a modulus which is the least common multiple of l^h and the modulus of hyperprimarity, and by $\overline{\mathfrak{K}}_1$, its abelian field. All the ramified prime ideals of $\overline{\mathfrak{K}}_1$ are divisors of l, and its maximal unramified field, having a group of exponent l, coincides with $\overline{\mathfrak{K}}$.

Since $\overline{K}$ is a Scholz field, its ramified prime ideals are not divisors of l, and therefore $\overline{K} \cap \overline{\mathfrak{K}}_1$ is nonramified over $\overline{k}$. But, by assumption, $\overline{\mathfrak{K}} \cap \overline{K} = \overline{k}$, therefore $\overline{K} \cap \overline{\mathfrak{K}}_1 = \overline{k}$.

Because of this, and Čebotarev's density law we can find a number $\overline{\kappa}$ in $\overline{k}$, such that $\overline{\mathfrak{q}}_1(\overline{\kappa}) = \overline{\mathfrak{q}}$, where $\overline{\mathfrak{q}}$ is a prime ideal, belonging to the identity automorphism in $\overline{K}$: $\overline{\kappa}^{-1}\xi^{t-g} \in H$ and $\overline{\kappa} \equiv 1\ (\overline{\mathfrak{p}}_i)$, for all $\overline{\mathfrak{p}}_i$.

The field $\overline{K}(\sqrt[l]{\mu_2\overline{\kappa}^s})$ will not lose the properties 1°, 2° and 3°. It is obvious for properties 2° and 3°, and for 1° it follows from

$$\overline{\kappa}_2^s H = \xi^{(t-g)s} H = \xi^{lx} H,$$

that is $\overline{\kappa}_2^s$ is l-hyperprimary. The field $\overline{K}(\sqrt[l]{\mu_2\overline{\kappa}^s})$ will also possess property 4°.

In fact, $\overline{\mathfrak{q}}$ is the only prime divisor of the discriminant for which we have to check property 4°. For it we have

$$\mathfrak{N}(\overline{\mathfrak{q}}) = \mathfrak{N}(\overline{\mathfrak{q}}_1)\,\mathfrak{N}(\overline{\kappa}) \equiv \mathfrak{N}(\xi^{t-g})\,\mathfrak{N}(\overline{\mathfrak{q}}_1) \equiv \mathfrak{N}(\xi)^{1-g}\,\mathfrak{N}(\overline{\mathfrak{q}}_1) \equiv 1\ (l^h),$$

since $\kappa \equiv \xi^{t-g}\ (l^h)$. Theorem 2 is proved.

3. Example. In conclusion we shall give an example, which shows that the invariants which we have constructed are not always equal to 1, we will limit ourselves to the invariant (χ, X).

Let K_1/k be a Scholz field, and let μ be an l-invariant number in the field, which does not lie in the principal class X. Since the ideal $\mathfrak{m}$ in the decomposition (3) is relatively prime to the discriminant of K_1/k, the class field of the group which is composed of l-th power classes mod $\mathfrak{m}$ does not intersect with K_1/k. Therefore we can find, in view of Čebotarev's density theorem, a prime, l-hyperprimary (for $l = 2$, a totally positive) number $\pi \in k$, which splits completely in K_1/k, but is inert in $K_1(\sqrt[l]{\mu})$, and for which $(\frac{\pi}{\mathfrak{m}}) = 1$.

If we denote $K_1(\sqrt[l]{\pi})$ by K (see Figure 2), the class of l-invariant numbers to which μ belongs by X, and the characters for which $\pi = a_\chi$, by χ, then

$$(\chi, X)=\left(\frac{\mu}{\mathfrak{P}}\right)_{K_1}\left(\frac{\pi}{\mathfrak{m}}\right)^{-1}\neq 1, \quad \pi=N_{K_1/k}(\mathfrak{P}).$$

These considerations hold for $l = 2$ even if k is a field of rational numbers R. The simplest concrete example of this type is as follows.

Let $K = R(\sqrt{17}, \sqrt{281})$. K is a Scholz field, since

$$17>0, \quad 281>0, \quad 17\equiv 281\equiv 1\,(8) \quad \left(\frac{281}{17}\right)=1, \quad \left(\frac{17}{281}\right)=1.$$

Let χ be a character of the Galois group K for which $\alpha_\chi = 281$, and X be the class which corresponds to the imbedding of $R(\sqrt{17})$ in a cyclic field of the 4th degree. We shall show that $(\chi, X) = -1$.

We can choose for a representative of class X, the number μ;

$$\mu = 17 + 4\sqrt{17} = \sqrt{17}(4+\sqrt{17}),$$

since

$$\mu^\sigma = \mu(4+\sqrt{17})^2.$$

We have $\mathfrak{m} = 1$, $\mathfrak{D} = \sqrt{17}$, since $4 + \sqrt{17}$ is a unit in $R(\sqrt{17})$.

In the field $R(\sqrt{17})$

$281 = N(37-8\sqrt{17})$.

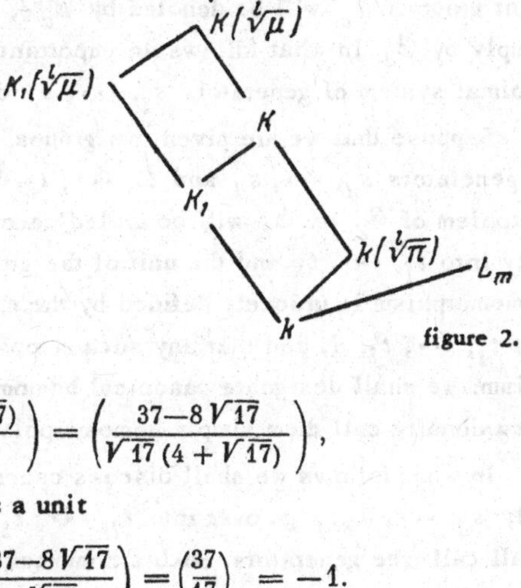

figure 2.

Obviously

$$\left(\frac{\alpha_\chi}{\mathfrak{m}}\right)=\left(\frac{281}{1}\right)=1.$$

We find $\left[\frac{\mu}{\alpha_\chi}\right]$. Since

$\alpha_\chi = 281 = N(37-8\sqrt{17})$,

then

$$\left[\frac{\mu}{\alpha_\chi}\right]=\left(\frac{\sqrt{17}\,(4+\sqrt{17})}{37-8\sqrt{17}}\right)=\left(\frac{37-8\sqrt{17}}{\sqrt{17}\,(4+\sqrt{17})}\right),$$

by the reciprocity law. Since $4 + \sqrt{17}$ is a unit

$$\left(\frac{37-8\sqrt{17}}{\sqrt{17}\,(4+\sqrt{17})}\right)=\left(\frac{37-8\sqrt{17}}{\sqrt{17}}\right)=\left(\frac{37}{17}\right)_R=-1.$$

Therefore

$$\left(\frac{\alpha_\chi}{\mathfrak{m}}\right)=1, \quad \left[\frac{\mu}{\alpha_\chi}\right]=-1, \quad (\chi, X)=\left[\frac{\mu}{\alpha_\chi}\right]\left(\frac{\alpha_\chi}{\mathfrak{m}}\right)^{-1}=-1.$$

Although the invariants (χ, X) are not in general equal to 1, their values cannot be arbitrary. For instance if χ is any character of a Galois group of the Scholz field $K = k(\sqrt[l]{\alpha}, \sqrt[l]{\beta})$, and if X corresponds to imbedding K in a field with a non-abelian group of order l^3 and exponent l, then $(\chi, X) = 1$.

In Tannaka's paper, which was discussed in the introduction, it was stated that if $k \ni \zeta$, and the field K is, in our terminology, a Scholz field, then it can be imbedded in a Scholz field with a preassigned Galois group, which is a central extension of the Galois group of K/k.

By Theorem 1 this is possible if and only if the invariants (χ, X) and $(X)_h$ are equal to 1. The last example shows that this is not always the case.

The mistake in Tannaka's reasoning consists in that he first multiplies an l-invariant number μ by m, in order that all the ramified $\mathfrak{p}_i$ in K/k would be of degree 1 in $K(\sqrt[l]{\mu})$, and then by m', in order that all the divisors of $\mu m m'$ be of the first degree in K. He did not notice, however, that he lost in the second multiplication what he gained in the first.

§3. The existence of extensions with a given Galois group of order l^a

1. Canonical homomorphisms. We consider a sequence of normal subgroups N_c of a free group S with d generators, defined inductively as follows: $N_0 = S$, N_{c+1} is a subgroup of N_c, generated by the commutators of the elements of N_c with the elements of S, and by the l-th powers of the elements of N_c. The quotient group S/N_c will be denoted by $\mathfrak{G}_d^{(c)}$, or in case it does not create confusion, simply by $\mathfrak{G}_d$. In what follows an important role will be played by a certain fixed minimal system of generators $s_1, \cdots, s_d$ of the group $\mathfrak{G}_d$.

Suppose that we are given two groups $\mathfrak{G}_d$ and $\mathfrak{G}_\delta$ with fixed minimal systems of generators $s_1, \cdots, s_d$ and $t_1, \cdots, t_\delta$. We shall suppose that $\delta \leq d$. The homomorphism of $\mathfrak{G}_d$ on $\mathfrak{G}_\delta$ will be called canonical if the generators $s_1, \cdots, s_d$ go only into $t_1, \cdots, t_\delta$ and the unit of the group $\mathfrak{G}_\delta$. It is clear that the canonical homomorphism is uniquely defined by the mapping of the set $s_1, \cdots, s_d$ into the set $t_1, \cdots, t_\delta, 1$, and that any such mapping defines some canonical homomorphism. We shall designate canonical homomorphisms by the letters S, T, etc. and occasionally call them simply homomorphisms.

In what follows we shall discuss canonical homomorphisms S under which only $s_1, \cdots, s_{\delta-1}$ go over into $t_1, \cdots, t_{\delta-1}$. The generators $s_\delta, \cdots, s_d$ we shall call free generators. Such a homomorphism S is uniquely defined by the set of these free generators, which go into t_δ. This set we shall denote by $[S]$. The number of elements of $[S]$ we will denote by $|S|$. Such homomorphisms shall be called of type $(s_1, \cdots, s_{\delta-1}; t_1, \cdots, t_{\delta-1} | t_\delta)$.

Two homomorphisms S and T will be called independent, if the sets $[S]$ and $[T]$ do not intersect. Then there exists a single homomorphism R for which

$$[R] = [S] \bigcup [T].$$

R will be called the composite of S and T and denoted by $S * T$. The operation of composing the homomorphisms is commutative and associative and is applicable to any number of independent in pairs homomorphisms. Obviously, we have $|S_1 * \cdots * S_m| = |S_1| + \cdots + |S_m|$.

2. Functions of homomorphisms. In what follows we shall call for short any functions $f(S)$, defined over the set of all homomorphisms and assuming values, which are elements of some Galois group of order g, a function of homomorphisms.

For functions of homomorphisms of the type $(s_1, \cdots, s_{\delta-1}; t_1, \cdots, t_{\delta-1} | t_\delta)$, a degree is defined with respect to the generator t_δ. We consider the function

$$\phi(S, T) = f(S) f(T) f(S * T)^{-1}, \tag{28}$$

defined for any pair of independent homomorphisms S and T and symmetric in S and T. We shall assign the degree 0 to a function which takes on the value 1 for every S. We shall say that $f(S)$ is of degree $\leq n$, if $\phi(S, T)$ is a function of degree $\leq n-1$ in T for every S. We note that we do not assign a degree to every function. For instance, the function which takes on only one value, different from 1, for all S, has no degree.

The definition of a degree can be formulated in a different but equivalent way. A function $f(S)$ is of degree $\leq n$, if for any set of $n+1$ independent in pairs homomorphisms $S_{i_1}, \cdots, S_{i_{n+1}}$, the following relation holds

$$\prod_{(j_1, \cdots, j_k)} f(S_{j_1} * \cdots * S_{j_k})^{(-1)^k} = 1, \tag{29}$$

where $(j_1, \cdots, j_k)$ go over all non-empty subsets of the set $(i_1, \cdots, i_{n+1})$.

We shall prove the equivalence of these two definitions. The proof will be by induction. We shall first show, that for every independent in pairs set $S_{i_1}, \cdots, S_{i_{n+1}}$, the following identity holds.

$$\prod_{(r_1, \dots, r_k)} \varphi(S_{i_1}, S_{r_1} * \cdots * S_{r_k})^{(-1)^k} = \prod_{(j_1, \dots, j_k)} f(S_{j_1} * \cdots * S_{j_k})^{(-1)^k}, \tag{30}$$

where on the left $(r_1, \cdots, r_k)$ go over all non-empty subsets of the set $(i_2, \cdots, i_{n+1})$, and on the right $(j_1, \cdots, j_k)$ are non-empty subsets of $(i_1, \cdots, i_{n+1})$. We replace ϕ in the left hand side of (30) by f according to (28). We get

$$\prod_{(r_1, \dots, r_k)} f(S_{i_1})^{(-1)^k} \prod_{(r_1, \dots, r_k)} f(S_{r_1} * \cdots * S_{r_k})^{(-1)^k} \cdot$$

$$\cdot \prod_{(r_1, \dots, r_k)} f(S_{i_1} * S_{r_1} * \cdots * S_{r_k})^{(-1)^{k+1}}. \tag{31}$$

We compare the separate factors in (31) and in the right hand side of (30). The first product is $f(S_{i_1})$ raised to the power

$$\sum_{(r_1, \dots, r_k)} (-1)^k = \sum_{k=1}^{n} (-1)^k C_n^k = -1,$$

that is it coincides with the factor $f(S_{i_1})^{-1}$ on the right side of (30). Every factor of the second and third product coincides in an obvious manner with a definite factor on the right of (30). In this way (30) is established.

If we suppose that $f(S)$ is of degree $\leq n$ according to the first definition, then it follows from the hypothesis of the induction that the left hand side of (30) becomes 1. Hence the right hand side is also 1 and that means that $f(S)$ is of degree $\leq n$, according to the second definition. If $f(S)$ is of degree $\leq n$ according to the second definition, then the right hand side, and consequently, the left hand side of (30) are equal to 1. But according to the hypothesis of induction, $\phi(S_{i_1}, T)$ is for arbitrary S_{i_1} of degree $\leq n-1$ with respect to T, that is $f(S)$ is of degree $\leq n$ according to the second definition.

3. Existence theorem. We shall prove a theorem which guarantees under certain conditions the existence of solutions of the equation $f_i(S) = 1$ where $f_i(S)$ are functions of homomorphisms having a definite degree. In doing this we shall consider canonical homomorphisms of the group $\mathfrak{G}_d$ with a varying number of generators d, but a fixed group $\mathfrak{G}_\delta$ with all the homomorphisms being of the same type.

Theorem 3. *For every pair of natural numbers k and n there exists a natural number $C(k, n)$ depending on them and having the following properties: for any k functions of homomorphisms $f_1(S), \cdots, f_k(S)$ of degree $\leq n$ over the group $\mathfrak{G}_d$ with $d \geq C(k, n)$, a homomorphism S of this group can always be found such that $f_i(S) = 1$ for $i = 1, \cdots, k$.*

The proof is by induction on the degree n of the functions $f_1, \cdots, f_k$. We select a homomorphism T_1 for which $|T_1| = 1$, and consider the functions

$$\phi_i(T_1, T) = f_i(T_1) f_i(T) f_i(T_1 * T)^{-1}, \qquad i = 1, \cdots, k,$$

as functions of T. They are defined for homomorphisms T independent of T_1, or in other words for homomorphisms of a group of $d-1$ generators, which can be obtained from the group $\mathfrak{G}_d$ by letting the generator $[T_1]$ become 1.

We shall suppose that $d \geq C(k, n-1) + 1$, and select from the free generators of the group $\mathfrak{G}_d$ a subset M_1 of $C(k, n-1)$ generators, not containing $[T_1]$. We consider the functions $\phi_i(T_1, T)$ only over the set of those homomorphisms T for which $[T] \subset M_1$, or what is the same thing, over the group with $\delta + C(k, n-1)$ generators. We can apply to them our hypothesis of induction and suppose that there exists a homomorphism T_2 with $[T_2] \subset M_1$ such that

$$\phi_i(T_1, T_2) = 1,$$

that is

$$f_i(T_1 * T_2) = f_i(T_1) f_i(T_2), \qquad i = 1, \cdots, k.$$

We now consider $3k$ functions

$$\phi_i(T_1, T), \quad \phi_i(T_2, T), \quad \phi_i(T_1 * T_2, T), \quad i = 1, \cdots, k. \tag{32}$$

These functions are defined over the homomorphisms T, independent of T_1 and T_2, or over a group with $d - 1 - C(k, n-1)$ generators, which would be obtained if we replaced by 1 all the generators of the group $\mathfrak{G}_d$ which belong to $[T_1] \cup M_1$. We will suppose that

$$d \geq 1 + C(k, n-1) + C(3k, n-1),$$

and select from the number of generators of the group $\mathfrak{G}_d$, the subset M_2 which does not intersect with $[T] \cup M_1$ and has $C(3k, n-1)$ generators. We consider the function (32) only over the set of those homomorphisms T for which $[T] \subset M_2$, that is over the group with $C(3k, n-1)$ generators. We can apply to them the hypothesis of induction and obtain that there exists a homomorphism T_3 with $[T_3] \subset M_2$, such that

$$\phi_i(T_1, T_3) = 1,$$
$$\phi_i(T_2, T_3) = 1,$$
$$\phi_i(T_1 * T_2, T_3) = 1,$$

that is

$$f_i(T_1 * T_3) = f_i(T_1) f_i(T_3),$$
$$f_i(T_2 * T_3) = f_i(T_2) f_i(T_3),$$
$$f_i(T_1 * T_2 * T_3) = f_i(T_1) f_i(T_2) f_i(T_3) \quad (i = 1, \cdots, k).$$

Continuing this process r steps, we will get, as can be easily seen, that if

$$d \geq 1 + C(k, n-1) + C(3k, n-1) + \cdots + C((2^{r-1} - 1)k, n-1),$$

then there exist independent in pairs homomorphisms $T_1, \cdots, T_r$, for which the following relation holds:

$$f_i(T_{j_1} * \cdots * T_{j_s}) = f_i(T_{j_1}) \cdots f_i(T_{j_s}) \quad (i = 1, \cdots, k), \tag{33}$$

if $(j_1, \cdots, j_s)$ is any subset of $(1, \cdots, r)$. We let $r = g^{k+1}$, and correspondingly

$$C(k, n) = 1 + C(k, n-1) + C(3k, n-1) + \cdots + C\left((2^{g^{k+1}-1} - 1)k, n-1\right).$$

As long as, by assumption of the theorem, $d \geq C(k, n)$, homomorphisms $T_1, \cdots, T_r$ exist having the desired properties. Since we have k functions $f_i(S)$ each of which can assume g values, and since the number of homomorphisms T in this set is g^{k+1}, there exists a system of homomorphisms $T_{j_1}, \cdots, T_{j_g}$ for which the following relations hold

$$f_i(T_{j_1}) = f_i(T_{j_2}) = \cdots = f_i(T_{j_g}) \quad (i = 1, \cdots, k). \tag{34}$$

We let

$$T_{j_1} * T_{j_2} * \cdots * T_{j_g} = S.$$

By (33) and (34) we have

$$f_i(S) = f_i(T_{j_1}) f_i(T_{j_2}) \cdots f_i(T_{j_g}) = f_i(T_{j_1})^g = 1,$$

since the values of the functions f_i belong to a group of order g. In this way, the homomorphism S satisfies all the conditions of the theorem.

We note that if the group $\mathfrak{G}_{d_1}^{(c)}$ with $d_1 \leq d$ is a homomorphic image of the group $\mathfrak{G}_d^{(c)}$ under some canonical homomorphism, then any function of a homomorphism over $\mathfrak{G}_{d_1}^{(c)}$ is also a function of a homomorphism over $\mathfrak{G}_d^{(c)}$, since any canonical homomorphism of $\mathfrak{G}_{d_1}^{(c)}$ is also a canonical homomorphism over $\mathfrak{G}_d^{(c)}$.

Corollary. *If under the conditions of Theorem 3, the group $\mathfrak{G}_d$ has m specified free generators $s_{i_1}, \cdots, s_{i_m}$ and if $d \geq C(k, n) + m$, then the group $\mathfrak{G}_d$ can be mapped by means of a canonical homomorphism on the group $\mathfrak{G}_{\delta+m}$ with generators $s'_1, \cdots, s'_{\delta+m}$, all of whose functions $f_1, \cdots, f_k$ become unity under the homomorphism $s'_1 \to t_1, \cdots, s'_\delta \to t_\delta, s'_{\delta+1} \to 1, \cdots, s'_{\delta+m} \to 1$.*

Proof. We map the group $\mathfrak{G}_d$ on $\mathfrak{G}_{d-m}$, letting

$$s_{i_1} \to 1, \cdots, s_{i_m} \to 1.$$

We can apply Theorem 3 to the group $\mathfrak{G}_{d-m}$ thus obtained and find a canonical homomorphism S for which

$$f_1(S) = 1, \cdots, f_k(S) = 1.$$

S will be a canonical homomorphism of $\mathfrak{G}_d$. We consider the canonical homomorphism of $\mathfrak{G}_d$ on the group $\mathfrak{G}_{\delta+m}$ with generators $s'_1, \cdots, s'_\delta, s'_{\delta+1}, \cdots, s'_{\delta+m}$, which is defined for all generators, except $s_{i_1}, \cdots, s_{i_m}$, in the same way as S, and maps $s_{i_1}, \cdots, s_{i_m}$ into $s'_{\delta+1}, \cdots, s'_{\delta+m}$. Obviously, $\mathfrak{G}_{\delta+m}$ will possess the properties described in the corollary.

4. Invariants (χ, X) and $(X)_h$ for fields having the group $\mathfrak{G}_d^{(c)}$. Suppose that we have a Scholz field K/k, whose Galois group is $\mathfrak{G}_d^{(c)}$, with a selected system of generators $s_1, \cdots, s_d$. We study the invariants (χ, X) and $(X)_h$ in this case. First of all we consider the basis $\chi_1, \cdots, \chi_d$ of the character group of order l, reciprocal to the system of generators $s_1, \cdots, s_d$, that is satisfying the relations

$$\chi_i(s_j) = 1, \ i \neq j,$$
$$\chi_i(s_i) \neq 1.$$

The numbers a_{χ_i} will be denoted simply by a_i. In what follows we shall suppose that the numbers are l-relatively prime in pairs.

We note that if we represent $\mathfrak{G}_d^{(c)}$ in a form of a quotient group N/S of a free group with d generators, as we have done in § 1, then N will coincide with N_c and N' of Lemma 1 with N_{c+1}. Obviously N/N' is nothing else but the kernel of the homomorphism of $\mathfrak{G}_d^{(c+1)}$ on $\mathfrak{G}_d^{(c)}$, which is the center of the group $\mathfrak{G}_d^{(c+1)}$.

Hence, in order to study the characters X, we must know the center Z of the
1 group $\mathfrak{G}_d^{(c+1)}$. This has been studied by A. I. Skopin [3] who showed that Z is decomposible into the direct sum

$$Z = Z_1 + Z_2 + \cdots + Z_{c+1}, \tag{35}$$

where Z_ν are isomorphic to a module of Lie polynomials of degree ν in d variables $x_1, \cdots, x_d$ over the field of residues modulo l. Moreover the variables x_i are connected with the generators s_i by the relation $s_i = 1 + x_i$. From this it follows that the homomorphism of the group $\mathfrak{G}_d^{(c+1)}$ under which some s_i go into 1, induces in the module of Lie polynomials a homomorphism under which the corresponding x_i go into 0.

It also follows from A. I. Skopin's construction that the generators x_i of the module Z_1 map into $s_i^{l^{c+1}}$ under an isomorphic mapping of Z_1 into the center of $\mathfrak{G}_d^{(c+1)}$. It follows from this that in extending the group $\mathfrak{G}_d^{(c)}$ corresponding to a character X of the group Z, which annihilates the subgroup Z_1, none of the elements of the form $s_i^r u$, where $(r, l) = 1$, and u belongs to the Frattini subgroup of the group $\mathfrak{G}_d^{(c)}$, increase their order.

In order to show this we prove that in the group S

$$(s_i^r u)^{l^c} \equiv s_i^{rl^c} \, (N_{c+1}), \text{ if } u \in N_1.$$

The easiest way to prove this is by induction on c. We have

$$(s_i^r u)^{l^{c-1}} \equiv s_i^{rl^{c-1}} \, (N_c),$$

that is

$$(s_i^r u)^{l^{c-1}} \equiv s_i^{rl^{c-1}} v, \ v \in N_c.$$

Raising both sides to the l-th power and making use of the fact that the elements of N_c commute with the elements of S with respect to the modulus N_{c+1}

$$(s_i^r u)^{l^c} \equiv (s_i^{rl^{c-1}} v)^l \equiv s_i^{rl^c} v^l \equiv s_i^{rl^c} (N_{c+1}),$$

since $v^l \in N_{c+1}$, if $v \in N_c$.

It follows from the above relation that every element of the form $s_i^r u$ is of order l^c in $\mathfrak{G}_d^{(c)}$ and of order l^{c+1} in the extension corresponding to X only if $\chi(s_i^{rl^c}) \neq 1$.

Corresponding to the decomposition (35) we can represent every character X of the group Z in the form

$$X = X_1 + X',$$

where X_1 annihilates all the subgroups Z_i, except Z_1, and X' annihilates the subgroup Z_1. For ideals $\mathfrak{D}(X)$, introduced in part 3 of § 1, we have

$$\mathfrak{D}(X) \approx \mathfrak{D}(X_1) \, \mathfrak{D}(X').$$

Since we have proved that $\mathfrak{D}(X)$ contains those and only those prime ideals, whose generating automorphisms of the inertia group increase their order in extensions corresponding to X, it follows from the above that

$$\mathfrak{D}(X') \approx 1, \quad \mathfrak{D}(X) \approx \mathfrak{D}(X_1).$$

We examine more closely the characters X_1. The group Z_1 is isomorphic with

$$\mathfrak{G}_d^{(1)} = \mathfrak{G}_d^{(c+1)} / \Phi(\mathfrak{G}_d^{(c+1)}),$$

where $\Phi(\mathfrak{G}_d^{(c+1)})$ denotes a Frattini subgroup of $\mathfrak{G}_d^{(c+1)}$, and where the homomorphism is given by the mapping $s \to s^{l^c}$. In fact the mapping

$$x \to x^{l^c} N_{c+1}$$

is a homomorphism because of the relation

3 $$(xl)^{lc} \equiv x^{lc} y^{lc} (N_{c+1}),$$

which can be proved independently by induction on c. We know that $u \in \Phi(\mathfrak{G}_d^{(c+1)})$ maps into 1 under this homomorphism, and the generators s_i into the generators Z_1, which proves our statement.

In this way the characters X_1 of the group Z_1 are in one to one correspondence with the character χ or order l of the group $\mathfrak{G}_d^{(c+1)}$. The character X_1 which corresponds to the character χ we shall denote by $\widetilde{\chi}$. We shall make clear for what pairs of characters χ_i, χ_j the invariant $(\chi_i, \widetilde{\chi}_j)$ is defined. In order that this invariant be defined it is necessary and sufficient that α_i be l-relatively prime to $\mathfrak{D}(\widetilde{\chi}_j)$. As we know, α_i will not be l-relatively prime to $\mathfrak{D}(\widetilde{\chi}_j)$ if it contains a prime divisor $\mathfrak{p}$ to a degree, which is not divisible by l, the generator of whose inertia group increases its order in the extension corresponding to $\widetilde{\chi}_j$. As can be easily seen in view of the fact that the numbers $\alpha_1, \cdots, \alpha_d$ are l-relatively prime in pairs, the generator of the inertia group of such a divisor will be an automorphism σ for which $\chi_i(\sigma) \neq 1$, $\chi_j(\sigma) = 1$ for $i \neq j$. Such an automorphism increases its order for extensions corresponding to the character $\widetilde{\chi}_j$, only if $i = j$.

In fact the element σ for which $\chi_i(\sigma) \neq 1$, $\chi_j(\sigma) = 1$ for $i \neq j$ has the form $s_i^r u$, where $(l, r) = 1$, and u is an element of the Frattini subgroup. Such an element increases its order in an extension, corresponding to $\widetilde{\chi}_j$ only if

$$\widetilde{\chi}_j((s_i^r u)^{lc}) \neq 1.$$

Since

$$(s_i^r u)^{l^c} = s_i^{rl^c}, \quad \widetilde{\chi}_j(s_i^{rl^c}) = \chi_j(s_i^r),$$

this happens only if $i = j$. Hence the invariants (χ_i, X') are defined for all characters χ_i and X', and the invariants $(\chi_i, \widetilde{\chi}_j)$ if and only if $i \neq j$.

5. The invariants (χ, X) and $(X)_h$ as functions of homomorphisms. As in part 4 we will consider a Scholz field K, whose Galois group is the group $\mathfrak{G}_d^{(c)}$ with a selected system of generators $s_1, \cdots, s_d$, and with numbers $\alpha_1, \cdots, \alpha_d$, l-relatively prime in pairs. We consider a canonical homomorphism of the group $\mathfrak{G}_d^{(c)}$ on some group $\mathfrak{G}_\delta^{(c)}$. Its kernel $\mathfrak{N}$ is a normal subgroup of the group $\mathfrak{G}_d^{(c)}$, so that to $\mathfrak{N}$ belongs a normal subfield of the field K. We denote this subfield by K^S. Obviously K^S has the Galois group $\mathfrak{G}_\delta^{(c)}$, having a system of generators $t_1, \cdots, t_\delta$, which are images of the generators $s_1, \cdots, s_d$. The numbers $\alpha_1, \cdots, \alpha_\delta$ corresponding to the subfield K^S are products of the numbers $\alpha_1, \cdots, \alpha_d$ which have no factors in common, so that $\alpha_1, \cdots, \alpha_\delta$ are l-relatively prime in pairs. This follows from the fact that if $[S] = (s_{i_1}, \cdots, s_{i_m})$ the numbers $\alpha_1, \cdots, \alpha_{\delta-1}$ corresponding to K^S coincide with such numbers corresponding to K, while the number α_δ of the field K^S has the form $\alpha_{i_1}, \cdots, \alpha_{i_m}$, where $\alpha_{i_1}, \cdots, \alpha_{i_m}$ are numbers corresponding to the field K. The last statement can be checked independently. In fact $\sqrt[l]{a_1}, \cdots, \sqrt[l]{a_{\delta-1}}$ and $\sqrt[l]{a_{i_1} \cdots a_{i_m}}$ are invariant with respect to the automorphisms of the field K which compose the kernel of the homomorphism S, and correspond to the characters $\chi_1, \cdots, \chi_d$.

We consider the characters $\chi_1, \cdots, \chi_\delta$ of the group $\mathfrak{G}_\delta^{(c)}$ and the characters $\widetilde{\chi}_1, \cdots, \widetilde{\chi}_\delta$ and X' of the center Z of the group $\mathfrak{G}_\delta^{(c+1)}$.

Let S be an arbitrary canonical homomorphism of the group $\mathfrak{G}_d^{(c)}$ on the group $\mathfrak{G}_\delta^{(c)}$. In the field K^S are defined the invariants

$$(\chi_i, X'), \quad (\chi_i, \widetilde{\chi}_j) \quad \text{for } i \neq j, \ (X')_h \text{ and } (\widetilde{\chi}_i)_h. \tag{36}$$

For a fixed field K each one of them is uniquely defined by a homomorphism S, and is therefore a function of this homomorphism. We shall denote by k the number of invariants in (36), or by k_δ if it becomes necessary to emphasize its dependence on the number of generators δ of the group $\mathfrak{G}_\delta^{(c)}$. In this way we have k functions of homomorphisms $f_1(S), \cdots, f_k(S)$. Our next problem is to study their degrees.

Since by (35) each character X is the sum of characters X_ν of the group Z_ν, which annihilate Z_μ if $\mu \neq \nu$, we shall confine ourselves to the consideration of the characters X_ν. Every such character is a character of the module of Lie polynomials of degree ν in d variables $x_1, \cdots, x_d$ over a field of residues modulo l. We shall say that a character is of degree $\leq n$ with respect to the generator x_i, if it annihilates all the Lie monomials in which x_i appears to a degree greater than n. In a manner analogous to what we have done for the group $\mathfrak{G}_d^{(c)}$, we define canonical homomorphisms for Lie modules of a given degree in d variables. If $\Lambda_\delta^{(\nu)}$ is a module of Lie polynomials of degree ν in the variables $y_1, \cdots, y_\delta$ while $\Lambda_d^{(\nu)}$ has d variables $x_1, \cdots, x_d$, and S is a canonical homomorphism of $\Lambda_d^{(\nu)}$ on $\Lambda_\delta^{(\nu)}$, then every character X_ν of the module $\Lambda_\delta^{(\nu)}$ will also be a character of the module $\Lambda_d^{(\nu)}$. Such a character we denote by X_ν^S.

Lemma 3. *If a character X_ν of the module $\Lambda_\delta^{(\nu)}$ is of degree $\leq n$ with respect to y_δ, and annihilates all the polynomials which depend only on $y_1, \cdots, y_{\delta-1}$, then*

$$\sum_{(i_1, \ldots, i_k)} (-1)^k X_\nu^{S_{i_1} * \cdots * S_{i_k}} = 0, \tag{37}$$

where $S_{i_1}, \cdots, S_{i_k}$ go over all non-empty subsets of the set $S_1, \cdots, S_{n+1}$ of $n+1$ relatively prime in pairs homomorphisms of the type $(x_1, \cdots, x_{\delta-1}; y_1, \cdots, y_{\delta-1} | y_\delta)$.

Proof. Let ϕ be an arbitrary Lie polynomial of the module $\Lambda_d^{(\nu)}$. We denote by p_i the operator, which replaces by y_δ all the variables of the set $[S_i]$. The operator which replaces by zero all the free generators of the set $[S_i]$ we denote by q_i, and the operator which replaces by zero all the free generators, except those which enter into $[S_1] \cup \cdots \cup [S_{n+1}]$ we denote by q_0. Then condition (37) can be written as follows

$$X_\nu\Big(\sum(-1)^k p_{i_1}\dots p_{i_k} q_{j_1}\dots q_{j_{n+1-k}} q_0\varphi\Big)=0, \tag{38}$$

where the sum goes over all the separations $(i_1, \cdots, i_k, j_1, \cdots, j_{n+1-k})$ of the set $(1, \cdots, n+1)$ into two parts, in which the first part is not empty.

We note that the operators p_i and p_j commute with each other as they operate on different generators. This also holds for operators q_i, q_j and for p_i, q_j for $i \neq j$, while q_0 commutes with all the operators. Therefore we have

$$\sum(-1)^k p_{i_1}\dots p_{i_k} q_{j_1}\dots q_{j_{n+1-k}} q_0 = \\ = q_0(q_1-p_1)\dots(q_{n+1}-p_{n+1})-q_0q_1\dots q_{n+1}. \tag{39}$$

We apply the operator (39) to an arbitrary Lie polynomial ϕ. If for some i, the polynomial ϕ does not contain a generator of the set $[S_i]$, then $p_i\phi = q_i\phi = \phi$ and therefore

$$q_0(q_1-p_1)\cdots(q_{n+1}-p_{n+1}) \tag{40}$$

transforms ϕ into 0. If ϕ contains at least one generator of each $[S_i]$ the operator (40) will transform ϕ into a polynomial of degree at least $n+1$ in γ_δ. Finally the operator $q_0q_1\cdots q_{n+1}$ transforms ϕ into a polynomial depending only on $\gamma_1, \cdots, \gamma_{\delta-1}$. In this way the operator (39) transforms any Lie polynomials into the sum of a polynomial of degree not less than $n+1$ in γ_δ and a polynomial depending only on $\gamma_1, \cdots, \gamma_{\delta-1}$. For such polynomials the character X_ν becomes 0, which proves (38) and therefore (37).

Corollary. *Under the assumptions of Lemma 3, for any homomorphism S_0, independent of $S_1, \cdots, S_{n+1}$*

$$\sum_{(i_1,\dots,i_k)}(-1)^k X_\nu^{S_0*S_{i_1}*\dots*S_{i_k}}=0. \tag{41}$$

To prove this it is sufficient to apply first a canonical homomorphism which acts like S_0 on the generators belonging to $[S_0]$ and does not alter the other generators, and then to apply the lemma to the resulting Lie module.

In what follows we shall use the multiplicative notation, so that the formulas (37) and (41) take the form

$$\prod_{(i_1,\dots,i_k)}(X_\nu^{S_{i_1}*\dots*S_{i_k}})^{(-1)^k}=1 \tag{42}$$

and

$$\prod_{(i_1,\dots,i_k)}(X_\nu^{S_0*S_{i_1}*\dots*S_{i_k}})^{(-1)^k}=1. \tag{43}$$

Theorem 4. *If the characters X_ν is of degree $\leq n$ with respect to γ_δ does not annihilate any monomial which depend on γ_δ, but does annihilate all the monomials depending on $\gamma_1, \cdots, \gamma_{\delta-1}$, then the invariants (χ_i, X_ν) and $(X_\nu)_h$ have, as functions of homomorphisms of the type $(s_1, \cdots, s_{\delta-1}; t_1, \cdots, t_{\delta-1}|t_\delta)$*

of the Galois group $\mathfrak{G}_d^{(c)}$ *of the Scholz field K the following degrees:*

$$(X_\nu)_h \text{ is of degree } \leq n,$$
$$(\chi_i, X_\nu) \text{ for } i \neq \delta \text{ is of degree } \leq n,$$
$$(\chi_\delta, X_\nu) \text{ is of degree } \leq n+1.$$

Hence all the invariants are of degree $\leq c+2$.

Proof. We make use of the second of the two definitions of degree given in part 2, and will give a separate proof for each one of the three cases formulated in the theorem.

We denote the invariant $(X_\nu)_h$ of the subfield K^S by $f(S)$ and prove that it satisfies relation (29). The character X_ν connected with the Galois group of the field K^S defines a character X_ν^S connected with the Galois group of the field K, where by formula (13) of § 1

$$f(S) = (X_\nu)_h = (X_\nu^S)_h.$$

Formula (29) thus reduces to the formula

$$\prod_{(i_1, \ldots, i_k)} (X_\nu^{S_{i_1} * \cdots * S_{i_k}})_h^{(-1)^k} = 1.$$

In view of (12) of § 1 we have

$$\prod_{(i_1, \ldots, i_k)} (X_\nu^{S_{i_1} * \cdots * S_{i_k}})_h^{(-1)^k} = \Big(\prod_{(i_1, \ldots, i_k)} (X^{S_{i_1} * \cdots * S_{i_k}})^{(-1)^k} \Big)_h.$$

The right hand side of this formula is equal to *1* by (42).

We now denote by $f(S)$ the invariant (χ_i, X_ν) of the subfield K^S of the field K, remembering that it is defined if X_ν is one of the characters X', that is $\nu > 1$, or is the character $\widetilde{\chi}_j$ with $j \neq i$. The character χ of the Galois group of the field K^S is the character χ^S of the Galois group of the field K, while the character X_ν connected with the Galois group of the field K^S defines a character X_ν^S connected with the Galois group of the field K. In the case under consideration all the characters χ_i^S coincide, since the character χ_i does not depend on the generator t_δ. We shall denote all of them by $\overline{\chi}_i$. We note that the invariants $(\overline{\chi}_i, X_\nu^S)$ are defined when the invariants (χ_i, X_ν) are defined. By (7) we have

$$f(S) = (\chi_i, X_\nu) = (\overline{\chi}_i, X_\nu^S),$$

and formula (29) reduces in our case to the equality

$$\prod_{(i_1, \ldots, i_k)} (\overline{\chi}_i, X_\nu^{S_{i_1} * \cdots * S_{i_k}})^{(-1)^k} = 1.$$

By (6) we have

$$\prod_{(i_1, \ldots, i_k)} (\overline{\chi}_i, X_\nu^{S_{i_1} * \cdots * S_{i_k}})^{(-1)^k} = \Big(\overline{\chi}_i, \prod_{(i_1, \ldots, i_k)} (X_\nu^{S_{i_1} * \cdots * S_{i_k}})^{(-1)^k}\Big).$$

The right hand side of the formula is equal to *1* by (42).

It remains to consider the last case. As in the previous cases we have

$$f(S) = (\chi_\delta, X_\nu) = (\chi_\delta^S, X_\nu^S).$$

The left hand side of (19) reduces in our case to the expression

$$\prod_{(i_1,\dots,i_k)} (\chi_\delta^{S_{i_1}*\cdots*S_{i_k}}, X_\nu^{S_{i_1}*\cdots*S_{i_k}})^{(-1)^k}, \tag{44}$$

where $(S_{i_1}, \cdots, S_{i_k})$ goes over non-empty subsets $(S_0, S_1, \cdots, S_{n+1})$ of $n+2$ independent homomorphisms. We apply Lemma 3 to the character χ_δ. We obtain

$$\chi_\delta^{S_{i_1}*\cdots*S_{i_k}} = \chi_\delta^{S_{i_1}} \cdots \chi_\delta^{S_{i_k}}.$$

We note that, as can be independently checked, the invariants $(\chi_\delta^{S_i}, X_\nu^{S_i})$ are defined, when we define the invariants (χ_δ, X_ν) so that we can write (44) in the form

$$\prod_{(i_1\dots i_k)} \prod_{r=1}^{k} (\chi_\delta^{S_{i_r}}, X_\nu^{S_{i_1}*\cdots*S_{i_k}})^{(-1)^k}.$$

If in the above product we collect all the terms in which the first place is occupied by $\chi_\delta^{S_j}$, we get the expression

$$\Pi\,(\chi_\delta^{S_j}, X_\nu^{S_j*S_{i_1}*\cdots*S_{i_k}})^{(-1)^{k+1}},$$

where $S_{i_1}, \cdots, S_{i_k}$ go over all the subsets of the set $S_0, \cdots, S_{n+1}$ which do not contain S_j. We can rewrite this expression as follows

$$(\chi_\delta^{S_j}, \prod_{(i_1,\dots,i_k)} (X_\nu^{S_{i_1}*\cdots*S_{i_k}})^{(-1)^k})^{-1},$$

which is equal to 1 in view of (43). Hence the theorem is proved.

We note that if in the assumption of Theorem 4, X annihilates all the polynomials depending on y_δ, then for $i \neq \delta$, the invariant (χ_i, X) has no degree, and the invariant (χ_δ, X) is of degree 1.

6. The existence of extensions with a given group of order l^a.

Theorem 5. *For every natural number δ there exists a natural number $C(\delta)$ having the following properties: in any Scholz field K with a Galois group $\mathfrak{G}_d^{(c)}$, $d > C(\delta)$, and with a selected system of generators $s_1, \cdots, s_d$, and l-relatively prime in pairs $\alpha_1, \cdots, \alpha_d$, there exists a subfield K^S with a Galois group $\mathfrak{G}_\delta^{(c)}$ and with all the invariants (χ, X) and $(X)_h$ equal to 1.*

Proof. We select a basis in the module of Lie polynomials of degree $c+1$ consisting of Lie monomials, and consider the basis of the character group reciprocal to this basis.

If a basis Lie monomial ϕ_i contains the variables $x_1, \cdots, x_d$ to degrees $m_1, \cdots, m_d$, then the corresponding basis character X_i will annihilates all the Lie monomials in which the variables enter to different powers. If moreover, $m_{i_1} \neq 0, \cdots, m_{i_r} \neq 0$ and the remaining degrees $m_j = 0$, then we shall say that

X_i depends only on $x_{i_1}, \cdots, x_{i_r}$. Since every character is composed from basis characters and since the invariants (χ, X) and $(X)_h$ are multiplicative, we shall endeavor to make equal to unity only those invariants which contain basis characters.

We shall rewrite all the invariants (χ, X) and $(X)_h$ with basis characters χ and X in a definite order. First we shall write those invariants which contain the characters X depending only on x_1 and x_2 and the characters χ_1 and χ_2. Moreover it is easily seen that if the character χ is χ_1, then X depends on x_2, and if χ is χ_2, then X depends on x_1. We shall denote the numbers of invariants in this first group by k_2. Next we write down the invariants in which χ is either χ_1, χ_2 or χ_3 and X depends only on x_1, x_2, x_3, where either χ is χ_3 or X depends on x_3. The number of invariants of this group we shall denote by k_3. Analogously k_i will designate the number of invariants for which χ is one of the characters $\chi_1, \cdots, \chi_i$, and X depends only on $x_1, \cdots, x_i$, and either $\chi = \chi_i$ or X depends on x_i. Obviously the number of such groups is δ, and every invariant which contains only basis characters will be in one of these groups. Here $C(k, n)$ are the numbers whose existence was established in Theorem 3. In that theorem the numbers of generators δ of a group which maps the group $\mathfrak{G}_d^{(c)}$ by a canonical homomorphism was fixed. As this number will now change, we shall denote the corresponding number $C(k, n)$ by $C(k, n, \delta)$. We shall prove that the number $C(\delta)$, whose existence was established by the theorem, can be defined by the formula

$$C(\delta) = 1 + C(k_2, c+2, 2) + C(k_3, c+2, 3) + \cdots + C(k_\delta, c+2, \delta).$$

In fact let $f_1^{(2)}(S), \cdots, f_{k_\delta}^{(2)}(S); \cdots; f_1^{(\delta)}(S), \cdots, f_{k_\delta}^{(\delta)}(S)$ be the groups of invariants which we have just described, where we consider the invariants as functions of the homomorphisms of a Galois group given over a Scholz field K. We consider a homomorphism of the type $(s_1; t_1 \mid t_2)$. Under these homomorphisms the invariants of the first group have, by Theorem 4, degrees $\leq c + 2$. We apply the corollary of Theorem 3 and obtain that the field K contains a subfield K^{S1}, whose Galois group has

$$2 + C(k_3, c+2, 3) + \cdots + C(k_\delta, c+2, \delta)$$

generators, and which in its turn has a subfield K^{T1}, whose Galois group has two generators, and for which all the k_2 invariants of the first group become unity.

In what follows we shall consider only the field K^{S1} and homomorphisms of the type $(s_1, s_2; t_1, t_2 \mid t_3)$ over its Galois group. Under these homomorphisms the invariants of the first group do not change, that is remain 1, and the invariants of the second group have, by Theorem 4, degree $\leq c + 2$. We apply again the corollary of Theorem 3 and obtain subfields K^{S2} and K^{T2}. The Galois group of the field K^{S2} has $3 + C(k_4, c+1, 4) + \cdots + C(k_\delta, c+1, \delta)$ generators. The Galois group of the field K^{T_2} has three generators. All the invariants of the first and second group are equal to 1 in the field K^{T_2}. Proceeding in this manner for δ steps, we will arrive at a field $K^{S_\delta} = K^{T_\delta}$, whose Galois group is the group $\mathfrak{G}_\delta^{(c)}$ all of whose invariants are equal to 1. The theorem is proved.

Theorem 6. *There exists an extension over every field of algebraic numbers with a preassigned Galois group of order* l^a.

Proof. Since every group of class c with d generators is a quotient group of the group $\mathfrak{G}_d^{(c)}$, it is sufficient to show the existence of extensions with the group $\mathfrak{G}_d^{(c)}$. We consider separately two cases.

1. k contains an l-th root of unity. We can even show the existence of a Scholz extension with the group $\mathfrak{G}_d^{(c)}$ with l-relatively prime numbers $\alpha_1, \cdots, \alpha_d$. The proof will be for any number of generators d by induction on the class c. We must prove the existence of a field with the given properties and with the group $\mathfrak{G}_\delta^{(c)}$, where we can assume that there exists a field with the same properties and with the group $\mathfrak{G}_d^{(c-1)}$ for any d. We let $d = C(\delta)$, where $C(\delta)$ is a number whose existence has been established in Theorem 5. Let K be the Scholz field with the group $\mathfrak{G}_d^{(c-1)}$, whose existence is our hypothesis of induction. By Theorem 5, it contains a subfield K^S with the Galois group $\mathfrak{G}_\delta^{(c-1)}$, all of whose invariants are equal to 1. As a subfield of a Scholz field, the field K^S is a Scholz field. It has been noted in part 5 of this section that all the numbers $\alpha_1, \cdots, \alpha_\delta$ of this field are l-relatively prime in pairs. By the corollary to Theorem 1, the field K^S can be imbedded in the Scholz field K_1 with the Galois group $\mathfrak{G}_\delta^{(c)}$. It remains to show that the numbers $\alpha_1, \cdots, \alpha_\delta$ are l-relatively prime in pairs in the field K_1. If $c > 1$, then this is obvious because they are the same as in the field K. Therefore we have to study in greater detail only the first step in the induction. In this case $K_1 = k(\sqrt[l]{\alpha_1}, \cdots, \sqrt[l]{\alpha_\delta})$ and by the proof of the corollary of Theorem 1 this field is constructed by adjoining consecutively the numbers $\sqrt[l]{\alpha_i}$ to $k(\sqrt[l]{\alpha_1}, \cdots, \sqrt[l]{\alpha_{i-1}})$ by Theorem 1. It remains to recall that in the proof of Theorem 1 the ideal $\mathfrak{m}$ (in our case this is (α_i)) can be made relatively prime to any ideal given in advance.

2. k does not contain an l-th root of unity. We shall use the same notation as in the proof of Theorem 2. As in the first case we shall show that we can construct the field K_1 to be a Scholz field, and such that the numbers $\overline{\alpha}_1, \cdots, \overline{\alpha}_\delta$ be l-relatively prime in pairs in the field $\overline{K}_1$. The proof is again by induction on c and we begin by considering the field K having all the required properties and having the Galois group $\mathfrak{G}_d^{(c-1)}$ with $d = C(\delta)$. By Theorem 5, the field $\overline{K}$ contains a subfield $\overline{K}^S$, having the required properties, all of whose invariants are equal to 1. It is easily seen, that the field K contains a subfield $K_0 = K^S$ such that $\overline{K}_0 = \overline{K}^S$. By Theorem 2, K_0 can be imbedded in the Scholz field K_1 with the group $\mathfrak{G}_d^{(c)}$. The fact that the numbers $\alpha_1, \cdots, \alpha_\delta$ of K can be taken l-rela-

tively prime in pairs has been proved in the consideration of the first case. Hence the theorem follows.

BIBLIOGRAPHY

[1] B. L. Van der Waerden, *Moderne Algebra,* Tl. 1, 2. Aufl., Springer, Berlin, 1937.

[2] A. G. Kurosh, *The theory of groups,* 2nd ed., Gostehizdat, Moscow, 1953. (Russian) English translation published by Chelsea, New York, 1955, 1956.

[3] A. I. Skopin, *Factor groups of an upper central series of free groups,* Dokl. Akad. Nauk SSSR (N.S.) 74(1950), 425-428. (Russian)

[4] N. G. Čebotarev, *Fundamentals of Galois theory,* v. II, Gostehizdat, Leningrad-Moscow, 1937. (Russian)

[5] R. Brauer, *Über die Konstruktion der Schiefkörper, die von endlichem Rang in Bezug auf ein gegebenes Zentrum sind,* J. Reine Angew. Math. 168 (1932), 44-64.

[6] M. Deuring, *Neuer Beweis des Bauerschen Satzes,* J. Reine Angew. Math. 173 (1935), 1-4.

[7] H. Hasse, *Bericht über neuere Untersuchungen und Probleme aus der Theorie der algebraischen Zahlkörper,* I, Ia, II, Jber. Deutsch. Math. Verein. 35 (1926), Abt. I, 1-55; 36 (1927), Abt. I, 233-311; Ergänzungsband 6(1930).

[8] H. Reichardt, *Konstruktion von Zahlkörpern mit gegebener Galoisgruppe von Primzahlpotenzordnung,* J. Reine Angew. Math. 177 (1937), 1-5.

[9] A. Scholz, *Konstruktion algebraischer Zahlkörper mit beliebiger Gruppe von Primzahlpotenzordnung,* I, Math. Z. 42 (1936), 161-188.

[10] T. Tannaka, *Über die Konstruktion der galoisschen Körper mit vorgegebener p-Gruppe,* Tôhoku Math. J. 43 (1937), 252-260.

[11] E. Witt, *Der Existenzsatz für abelsche Funktionenkörper,* J. Reine Angew. Math. 173 (1935), 43-51.

Translated by:
Emma Lehmer

On an existence theorem in the theory of algebraic numbers

Izv. Akad. Nauk SSSR, Ser. Mat. **18**, 327–334 (1954). Zbl. **56**, 34
[Transl., II. Ser., Am. Math. Soc. **4**, 143–150 (1956)]

In this paper is proved the existence of algebraic numbers for which the Legendre symbols with respect to the ideals conjugate to the prime divisors of these numbers have given values, and which satisfy certain other conditions.

In some problems of Galois theory an important part is played by algebraic numbers, which are connected by some arithmetic relations with their conjugates, for example for which the Legendre symbols with respect to their conjugates or with respect to the prime factors of their conjugates have definite values.

A problem of this type was first considered by N. G. Čebotarev (see [1], p. 164). In fact, he proposed the problem of the existence of primary principal ideals in a normal field K/k for which the Legendre symbol $(\frac{\pi^{\Sigma u}}{\pi})$ has preassigned values. Here the sum in the exponent goes over all automorphisms u different from 1 of the field K/k, the Legendre symbol is an l-th power character where l is a prime and k contains l-th roots of unity.

N. G. Čebotarev showed (see [1], p. 226-234) that if K is the field of Gauss numbers and $l = 2$, the Legendre symbol $(\frac{\pi^u}{\pi})$ is equal to 1 for a primary number π, and therefore the problem proposed by him does not always have a positive solution. On the other hand he showed that for many quadratic fields this problem is solvable. In what follows we shall use considerations which were used by N. G. Čebotarev (see the proof of formula (6)).

The problem which will concern us differs from the problem of N. G. Čebotarev. We consider the question of the existence of algebraic numbers α for which the Legendre symbols $(\frac{\alpha}{\mathfrak{p}^u})$ with respect to the ideals $\mathfrak{p}^u$, conjugate to the prime divisors $\mathfrak{p}$ of α have preassigned values. We shall also require that α satisfy certain other conditions. The solution of this problem is necessary to the construction of fields with a solvable Galois group.

In the future the following notation will be used: K/k is a normal field with a Galois group F, l is a prime number such that a primitive l-th root of unity ζ is contained in K, where $\zeta^u = \zeta^{g(u)}$, $u \in F$, $\mathfrak{q}_1, \cdots, \mathfrak{q}_r$ are ideals in K, relatively prime to each other and to l.

Theorem. *Let $\xi_1, \cdots, \xi_r$ be any given l-th roots of unity, and let $\zeta(u)$ be a function over F with values in the group of l-th roots of unity, satisfying the condition $\zeta(u^{-1}) = \zeta(u)^{g(u)}$, then there exist infinitely many numbers α in K, satisfying the conditions*

$$\left(\frac{\alpha}{\mathfrak{q}_i}\right)=\xi_i, \quad i=1,\ldots,r, \tag{1}$$

$$\left(\frac{\alpha}{\mathfrak{p}^u}\right)=\zeta(u) \textit{ for all prime } \mathfrak{p}\mid\alpha \textit{ with } u\in F,\ u\neq 1. \tag{2}$$

Condition (2) assumes that if $\mathfrak{p}\mid\alpha$ and $u\neq 1$, then $\mathfrak{p}^u$ does not divide α.

For $l=2$ we must impose the following additional conditions:

1′. $\mathfrak{q}_1,\cdots,\mathfrak{q}_r$ are relatively prime to the discriminant of K/k.

2′. All the prime divisors of 2 in k split completely in K.

3′. The real infinite primes in k remain real in K.

4′. If u_i are involutory automorphisms ($u_i^2=1$), K_{u_i} are the subfields belonging to them, and if $\mathfrak{D}_{u_i}$ are the differents of K/K_{u_i}, then it follows from the fact that $\mathfrak{D}_{u_i}\cdots\mathfrak{D}_{u_m}$ is a square of an ideal of K, that

$$\zeta(u_1)\cdots\zeta(u_m)=1.$$

The proof is in two parts, depending on whether l is odd, or $l=2$.

Case I. Let $l\neq 2$. We shall show that α can be chosen in the form of a product of two principal prime ideals of absolutely the first degree. In order to do this we construct in K a sequence of l-hyperprimary prime ideals $\pi_1,\cdots,\pi_t,\cdots$, having absolute degree 1, which are not divisible by l and which satisfy the condition

$$\left(\frac{\pi_t}{\mathfrak{q}_i}\right)=\xi_i^{\frac{1}{2}}. \tag{3}$$

The exponent $1/2$ has a meaning because $l\neq 2$.

We shall suppose that the above conditions hold for all the terms of the series, which we construct by recursion, selecting π_t after $\pi_1,\cdots,\pi_{t-1}$ have been constructed in such a way that

$$\left(\frac{\pi_t}{\pi_s^u}\right)=\left(\frac{\pi_s}{\pi_s^u}\right)^{-1}\zeta(u), \quad s=1,2,\ldots,t-1. \tag{4}$$

Conditions (3) and (4) imposed on π_t, assign values of the Legendre symbols of the number π_t with respect to the prime ideals $\mathfrak{q}_1,\cdots,\mathfrak{q}_r,(\pi_1),\cdots,(\pi_{t-1})$ (it is assumed in condition (3) that $(\pi_t,\mathfrak{q}_i)=1$). In view of the generalized theorem on arithmetical progressions there exists a principal prime ideal π_t satisfying the conditions imposed by us. Therefore we can continue the sequence $\pi_1,\cdots,\pi_t$ as far as we like.

Since for any number π the symbol $(\frac{\pi}{\pi^u})$ can assume only a finite number of values (not more than l^n, $n=(K:k)$), for a sufficiently large number of terms in the sequence two numbers π_s and π_t $(s<t)$ can be found such that

$$\left(\frac{\pi_s}{\pi_s^u}\right)=\left(\frac{\pi_t}{\pi_t^u}\right) \quad \text{for all} \quad u\in F, \quad u\neq 1. \tag{5}$$

We shall show that the number $\alpha=\pi_s\pi_t$ satisfies the conditions of the theorem. It is obvious that condition (1) for α follows from condition (3) for π_s and π_t. Condition (2) for α now has the form

$$\left(\frac{\pi_s \pi_t}{\pi_s^u}\right) = \zeta(u), \quad \left(\frac{\pi_s \pi_t}{\pi_t^u}\right) = \zeta(u).$$

The first of these is satisfied because of condition (4) imposed on π_t. It remains to show that the second condition is satisfied. In order to do this we note that

$$\left(\frac{\pi_t}{\pi_s^u}\right) = \left(\frac{\pi^{u^{-1}}}{\pi_s}\right)^{g(u^{-1})} = \left(\frac{\pi_s}{\pi_t^{u^{-1}}}\right)^{g(u^{-1})}$$

and therefore by (5)

$$\left(\frac{\pi_s}{\pi_s^u}\right) = \left(\frac{\pi_s^{u^{-1}}}{\pi_s}\right)^{g(u^{-1})} = \left(\frac{\pi_s}{\pi_s^{u^{-1}}}\right)^{g(u^{-1})} = \left(\frac{\pi_t}{\pi_t^{u^{-1}}}\right)^{g(u^{-1})}.$$

Hence

$$\left(\frac{\pi_s \pi_t}{\pi_s^u}\right) = \left(\frac{\pi_s}{\pi_t^{u^{-1}}}\right)^{g(u^{-1})} \left(\frac{\pi_t}{\pi_t^{u^{-1}}}\right)^{g(u^{-1})} = \left(\frac{\pi_s \pi_t}{\pi_t^{u^{-1}}}\right)^{g(u^{-1})} = \zeta(u^{-1})^{g(u^{-1})} = \zeta(u).$$

Therefore the theorem is proved in this case.

Case II. Let $l = 2$. Unfortunately in this case the proof is much more cumbersome because we cannot impose condition (3) (since the exponent $1/2$ loses its meaning), and therefore we cannot suppose that α is a product of two factors. We will show however that we can find a desired number α consisting of three factors. We will first suppose that u is an unvolutory automorphism: $u^2 = 1$. In this case we can find a prime hyperprimary number π such that

$$\left(\frac{\pi}{\pi^u}\right) = \zeta(u).$$

In fact let K_u be a subfield of elements invariant with respect to u, $K = K_u(\sqrt{\Delta_u})$ and $\mathfrak{D}_u$ is a different of K/K_u. We shall prove that for any hyperprimary totally positive ξ the following relation holds:

$$\left(\frac{\xi}{\xi^u}\right) = \left(\frac{\xi}{\mathfrak{D}^u}\right), \quad \text{if} \quad (\xi, \xi^u) = 1. \tag{6}$$

Let $\xi = \alpha + \beta\sqrt{\Delta_u}$. Since $\xi + \xi^u = 2\alpha$ and ξ^u is hyperprimary then

$$\left(\frac{\xi}{\xi^u}\right) = \left(\frac{\xi + \xi^u}{\xi^u}\right) = \left(\frac{2\alpha}{\xi^u}\right) = \left(\frac{\alpha}{\xi^u}\right) = \left(\frac{\xi^u}{\alpha}\right).$$

It can happen that α has a common divisor with two. In case the numerator in the Legendre symbol is hyperprimary and the denominator has a factor in common with two, we shall calculate the Legendre symbol by means of the relation

$$\left(\frac{\nu}{\mathfrak{l}}\right) = 1, \text{ if } \nu \text{ is hyperprimary}, \ \mathfrak{l} \mid 2.$$

As can be easily seen the law of reciprocity remains in force.

We have moreover

$$\left(\frac{\xi^u}{\alpha}\right)=\left(\frac{\alpha-\beta\sqrt{\Delta_u}}{\alpha}\right)=\left(\frac{-\beta\sqrt{\Delta_u}}{\alpha}\right)=\left(\frac{-\beta}{\alpha}\right)\left(\frac{\sqrt{\Delta_u}}{\alpha}\right)=\left(\frac{\sqrt{\Delta_u}}{\alpha}\right).$$

In view of these relations we must first of all note that α is relatively prime to ξ^u since $2\alpha = \xi + \xi^u$ and $(\xi, \xi^u) = 1$. Secondly, since $\alpha \in K_u$, the symbol $\left(\frac{-\beta}{\alpha}\right)$ in the field K is equal to the symbol $\left(\frac{\beta^2}{\alpha}\right)$ in the field K_u and therefore is equal to unity. Analogously we have

$$\left(\frac{\sqrt{\Delta_u}}{\alpha}\right)=\left(\frac{\Delta_u}{\alpha}\right),$$

where the left hand symbol is calculated in K, and the right hand one in K_u.

By condition 2′, Δ_u must be hyperprimary and totally positive. Hence

$$\left(\frac{\Delta_u}{\alpha}\right)=\left(\frac{\alpha}{\Delta_u}\right)=\left(\frac{\alpha}{\sqrt{\Delta_u}}\right),$$

where the last symbol is again calculated in K.

If $\mathfrak{D}_u$ is the different in K/K_u, then

$$(\sqrt{\Delta_u}) = \mathfrak{D}_u\mathfrak{C}, \quad \mathfrak{C} \in K_u,$$

$$\left(\frac{\alpha}{\sqrt{\Delta_u}}\right)=\left(\frac{\alpha}{\mathfrak{C}}\right)\left(\frac{\alpha}{\mathfrak{D}_u}\right)=\left(\frac{\alpha}{\mathfrak{D}_u}\right),$$

since again

$$\left(\frac{\alpha}{\mathfrak{C}}\right)_K=\left(\frac{\alpha^2}{\mathfrak{C}}\right)_{K_u}=1.$$

In this way

$$\left(\frac{\xi}{\xi^u}\right)=\left(\frac{\alpha}{\mathfrak{D}_u}\right).$$

Analogously we check that

$$\left(\frac{\xi}{\mathfrak{D}_u}\right)=\left(\frac{\alpha+\beta\sqrt{\Delta_u}}{\mathfrak{D}_u}\right)=\left(\frac{\alpha}{\mathfrak{D}_u}\right),$$

which proves (6).

Condition 4′ together with the generalized theorem about arithmetical progressions guarantees the existence of prime hyperprimary totally positive π such that

$$\left(\frac{\pi}{\mathfrak{D}_u}\right)=\zeta(u)$$

for every involutory automorphism. By formula (6) we have

$$\left(\frac{\pi}{\pi^u}\right)=\zeta(u).$$

We now consider also the non-involutory automorphism. Let $v^2 \neq 1$, that is $v \neq v^{-1}$. Suppose that we have found three primes, hyperprimary totally positive numbers π, κ, ρ satisfying the following conditions:

$$\left.\begin{aligned}
&\left(\frac{\pi}{\varkappa^{v}}\right)=1, &&\left(\frac{\pi}{\varkappa^{v^{-1}}}\right)=\left(\frac{\varkappa}{\varkappa^{v}}\right)\zeta(v),\\
&\left(\frac{\pi}{\rho^{v}}\right)=\left(\frac{\rho}{\rho^{v}}\right)\zeta(v), &&\left(\frac{\pi}{\rho^{v^{-1}}}\right)=1,\\
&\left(\frac{\varkappa}{\rho^{v}}\right)=1, &&\left(\frac{\varkappa}{\rho^{v^{-1}}}\right)=\left(\frac{\rho}{\rho^{v}}\right)\zeta(v),\\
&\left(\frac{\pi}{\pi^{v}}\right)=\left(\frac{\varkappa}{\varkappa^{v}}\right)=\left(\frac{\rho}{\rho^{v}}\right).
\end{aligned}\right\} \tag{7}$$

We shall prove that the number $\alpha=\pi\kappa\rho$ satisfies the conditions

$$\left(\frac{\alpha}{\pi^{v}}\right)=\left(\frac{\alpha}{\varkappa^{v}}\right)=\left(\frac{\alpha}{\rho^{v}}\right)=\zeta(v). \tag{8}$$

In fact

$$\left(\frac{\alpha}{\pi^{v}}\right)=\left(\frac{\pi\varkappa\rho}{\pi^{v}}\right)=\left(\frac{\pi}{\pi^{v}}\right)\left(\frac{\varkappa}{\pi^{v}}\right)\left(\frac{\rho}{\pi^{v}}\right).$$

Since

$$\left(\frac{\varkappa}{\pi^{v}}\right)=\left(\frac{\varkappa^{v^{-1}}}{\pi}\right)=\left(\frac{\pi}{\varkappa^{v^{-1}}}\right)=\left(\frac{\varkappa}{\varkappa^{v}}\right)\zeta(v),$$

$$\left(\frac{\rho}{\pi^{v}}\right)=\left(\frac{\rho^{v^{-1}}}{\pi}\right)=\left(\frac{\pi}{\rho^{v^{-1}}}\right)=1,$$

$$\left(\frac{\pi}{\pi^{v}}\right)=\left(\frac{\varkappa}{\varkappa^{v}}\right),$$

therefore

$$\left(\frac{\alpha}{\pi^{v}}\right)=\left(\frac{\varkappa}{\varkappa^{v}}\right)\left(\frac{\varkappa}{\varkappa^{v}}\right)\zeta(v)=\zeta(v).$$

Analogously

$$\left(\frac{\alpha}{\varkappa^{v}}\right)=\left(\frac{\pi}{\varkappa^{v}}\right)\left(\frac{\varkappa}{\varkappa^{v}}\right)\left(\frac{\rho}{\varkappa^{v}}\right)=\left(\frac{\varkappa}{\varkappa^{v}}\right)\left(\frac{\rho}{\rho^{v}}\right)\zeta(v)=\zeta(v),$$

$$\left(\frac{\alpha}{\rho^{v}}\right)=\left(\frac{\pi}{\rho^{v}}\right)\left(\frac{\varkappa}{\rho^{v}}\right)\left(\frac{\rho}{\rho^{v}}\right)=\left(\frac{\rho}{\rho^{v}}\right)^{2}\zeta(v)=\zeta(v).$$

Similarly we can check that

$$\left(\frac{\alpha}{\pi^{v^{-1}}}\right)=\zeta(v),\quad \left(\frac{\alpha}{\varkappa^{v^{-1}}}\right)=\zeta(v),\quad \left(\frac{\alpha}{\rho^{v^{-1}}}\right)=\zeta(v). \tag{9}$$

We check only the first equality

$$\left(\frac{\alpha}{\pi^{v^{-1}}}\right)=\left(\frac{\pi}{\pi^{v^{-1}}}\right)\left(\frac{\varkappa}{\pi^{v^{-1}}}\right)\left(\frac{\rho}{\pi^{v^{-1}}}\right).$$

Since

$$\left(\frac{\pi}{\pi^{v^{-1}}}\right)=\left(\frac{\pi^{v}}{\pi}\right)=\left(\frac{\pi}{\pi^{v}}\right)=\left(\frac{\rho}{\rho^{v}}\right),$$

$$\left(\frac{\varkappa}{\pi^{v^{-1}}}\right)=\left(\frac{\varkappa^{v}}{\pi}\right)=\left(\frac{\pi}{\varkappa^{v}}\right)=1,$$

$$\left(\frac{\rho}{\pi^{v^{-1}}}\right)=\left(\frac{\rho^{v}}{\pi}\right)=\left(\frac{\pi}{\rho^{v}}\right)=\left(\frac{\rho}{\rho^{v}}\right)\zeta(v),$$

it follows that

$$\left(\frac{\alpha}{\pi^{v^{-1}}}\right)=\left(\frac{\rho}{\rho^{v}}\right)^{2}\zeta(v)=1.$$

We now separate the totality of all automorphisms of K/k different from one into three parts. In the first part, which we shall denote by U we will assemble all the involutory automorphisms. The non-involutory automorphisms we will divide into two parts, separating each pair of reciprocal automorphisms v and v^{-1} by putting arbitrarily one into one part and the other into the other part. These two parts we will denote by V and V^{-1}. In this way

$$F = 1 + U + V + V^{-1}.$$

Now, as we have done for an odd l, we construct a sequence of prime hyperprimary totally positive numbers $\pi_1, \cdots, \pi_t, \cdots$. We shall suppose that all the numbers π_t satisfy the conditions

$$\left(\frac{\pi_t}{\mathfrak{q}_i}\right)=\xi_i, \quad i=1, \ldots, r,$$

$$\left(\frac{\pi_t}{\mathfrak{D}_u}\right)=\zeta(u), \quad u \in U.$$

Otherwise we shall generate our sequence recursively by the following rather complicated rule. Suppose that the numbers $\pi_1, \cdots, \pi_{t-1}$ have already been constructed. We shall divide this sequence into two parts which we shall call first and second. Just how this division into two parts is accomplished will be shown later. Now we shall select π_t in such a way that besides the above conditions it also satisfies the following conditions:

$$\left(\frac{\pi_t}{\pi_s^{u}}\right)=1, \quad u \in U,$$

$$\left(\frac{\pi_t}{\pi_s^{v}}\right)=1, \quad \left(\frac{\pi_t}{\pi_s^{v^{-1}}}\right)=\left(\frac{\pi_s}{\pi_s^{v}}\right)\zeta(v),$$

if $v \in V$ and π_s belongs to the first part, and

$$\left(\frac{\pi_t}{\pi_s^{v}}\right)=\left(\frac{\pi_s}{\pi_s^{v}}\right)\zeta(v), \quad \left(\frac{\pi_t}{\pi_s^{v^{-1}}}\right)=1,$$

if $v \in V$ and π_s belongs to the second part.

We shall assume that we have defined the division into the first and second parts in such a way that the following condition is satisfied: for every natural number N there exists an M such that for any division of the sequence $\pi_1, \cdots, \pi_M$ into N parts, at least one part will contain three numbers π_r, π_s, π_t, $(r < s < t)$ having the following properties: when the sequence $\pi_1, \cdots, \pi_{t-1}$ is divided into the first and second parts π_r and π_s go into different parts, but when the sequence $\pi_1, \cdots, \pi_{s-1}$ is divided into two parts, π_r does not go into the same part into which it went in the division of the series $\pi_1, \cdots, \pi_{t-1}$. In this case our problem is solved.

In fact let us denote by ν the number of elements in the set v and let N be

2^ν. For a division into N parts we take a division which corresponds to the values of the symbol $\left(\frac{\pi_i}{\pi_i^v}\right)$, $v \in V$, putting into the same part all those π_i for which the symbols $\left(\frac{\pi_i}{\pi_i^v}\right)$ are equal. Let π_r, π_s, π_t, be three numbers satisfying all our conditions. If we take these numbers for π, κ and ρ then condition (7) will be satisfied. In fact if π_s belongs to the first part and π_r to the second part when we divide the sequence $\pi_1, \cdots, \pi_{t-1}$ into two parts, then condition (7) will be satisfied if we let $\pi_t = \pi$, $\pi_s = \kappa$, $\pi_r = \rho$. If on the other hand π_s belongs to the second part and π_r to the first, then we must let $\pi_t = \pi$, $\pi_s = \rho$, $\pi_r = \kappa$.

It remains to show how to perform the division into the first and second parts. The possibility of such a division is contained in the following combinatorial lemma.

Lemma. *For every natural number t, a division of the numbers $1, 2, \cdots, t-1$ into two parts, the first part and the second part, can be affected, which is such that for every natural number N there exists an M such that if we divide the numbers $1, 2, \cdots, M$ into N parts at least one part must contain three numbers r, s, and t $(r < s < t)$, such that on division of $1, 2, \cdots, t-1$ into two parts r and s go into different parts, and on division of $1, 2, \cdots, s-1$ into two parts, r goes into a different part from the one into which it went when $1, 2 \cdots, t-1$ was divided into two parts.*

The solution of this problem is as follows. In dividing the numbers $1, 2, \cdots, t-1$ into two parts we put in the first part those numbers k for which $t-k$ is divisible by an odd power of 2, and the other numbers go into the second part.

To show that in doing so the required property is achieved we make use of a theorem of van der Waerden (see [2]) for a particular case of arithmetical progression of length three. By this theorem there exists for every N an M such that for any division of the numbers $1, 2, \cdots, M$ into N parts, at least one part contains three numbers in arithmetical progression. In order to derive our result from this theorem we need to check that in the above division into the first and second parts any arbitrary numbers r, s and t, which are in arithmetical progression will satisfy our conditions. This is almost obvious.

By assumption $t-s = s-r$. Let $t-s$ be divisible by an odd power of 2. Then on division of the numbers $1, 2, \cdots, t-1$ into two parts s goes into the first part. Since

$$t-r = (t-s) + (s-r) = 2(t-s),$$

then $t-r$ is divisible by an even power of 2 and therefore r belongs to the second part. In the division of $1, 2, \cdots, s-1$ into two parts r goes obviously into the first part. If $t-s$ is divisible by an even power of 2 then we must repeat the argument. The lemma is proved.

Returning to the proof of the theorem we see that if we let

$$\pi = \pi_t, \quad \kappa = \pi_s, \quad \rho = \pi_r, \quad \alpha = \pi\kappa\rho,$$

then conditions (8) and (9) for α will be satisfied for every $v \in V$, that is, (8) will be satisfied for any non-involutory automorphism v.

Analogous conditions

$$\left(\frac{\alpha}{\pi^u}\right) = \left(\frac{\alpha}{\varkappa^u}\right) = \left(\frac{\alpha}{\rho^u}\right) = \zeta(u) \tag{10}$$

will also be satisfied for any involutory automorphism u. In fact by construction

$$\left(\frac{\pi_i}{\pi_j^u}\right) = 1 \quad \text{for} \quad i \neq j \quad \text{and} \quad \left(\frac{\pi_i}{\mathfrak{D}_u}\right) = \zeta(u),$$

that is by (6), $(\pi_i/\pi_i^u) = \zeta(u)$. Hence

$$\left(\frac{\alpha}{\pi^u}\right) = \left(\frac{\pi}{\pi^u}\right)\left(\frac{\varkappa}{\pi^u}\right)\left(\frac{\rho}{\pi^u}\right) = \left(\frac{\pi}{\pi^u}\right) = \zeta(u).$$

Analogously the other two relations in (10) may be verified. Hence conditions (2) are satisfied for α. But conditions (1) are also satisfied because by construction

$$\left(\frac{\pi}{\mathfrak{q}_i}\right) = \left(\frac{\varkappa}{\mathfrak{q}_i}\right) = \left(\frac{\rho}{\mathfrak{q}_i}\right) = \xi_i \text{ and } \left(\frac{\alpha}{\mathfrak{q}_i}\right) = \xi_i^3 = \xi_i.$$

Therefore we have proved the existence of a number α satisfying the conditions of the theorem. Since, in this construction, we can select all the π_i, and hence also α, relatively prime to any preassigned modulus, there are infinitely many choices for α. The theorem is proved.

We note, that in the construction of the sequence $\pi_1, \cdots, \pi_t, \cdots$ there is a great deal of arbitrariness which we can make use of if needed. For example if we are given a normal field Ω/K we can select π_t so that they belong in Ω to a given class of automorphisms σ:

$$\left[\frac{\Omega}{\pi_t}\right] = \sigma.$$

In doing this it is necessary only that there exist infinitely many prime primary numbers π belonging to class σ in Ω and satisfying the conditions:

$$\left(\frac{\pi}{\mathfrak{q}_i}\right) = \xi_i^{\frac{1}{2}} \quad \text{for} \quad l \neq 2, \quad \text{and} \quad \left(\frac{\pi}{\mathfrak{q}_i}\right) = \xi_i \quad \text{for} \quad l = 2.$$

BIBLIOGRAPHY

[1] N. G. Čebotarev, *Collected papers*, v. 1, Izdat. Akad. Nauk SSSR, Moscow, 1949.

[2] A. Ya. Khinchin, *Three pearls of number theory*, 2d ed., OGIZ, Moscow, 1948; English translation: Graylock Press, Rochester, N.Y., 1952.

Translated by:
Emma Lehmer

On the problem of imbedding fields

Izv. Akad. Nauk SSSR, Ser. Mat. **18,** 389–418 (1954). Zbl. **57,** 33
[Transl., II. Ser., Am. Math. Soc. **4,** 151–183 (1956)]

In this paper is discussed the problem of imbedding a normal field of algebraic numbers in a larger field in case the relative Galois group is or order l^a, while the absolute Galois group is a decomposible extension.

Introduction

The problem of imbedding fields is as follows. Given a normal extension k/Ω with a Galois group F and a group G, having a normal subgroup $\mathfrak{G}$, with $G/\mathfrak{G} \cong F$. It is required to find when there exists a field $K \supset k$ normal over Ω such that its Galois group over Ω is G, and such that the natural homomorphism of G on F coincides with the natural homomorphism of a Galois group of a field on a Galois group of a subfield.

The simplest case of this problem arises when G is a decomposible extension of F with respect to $\mathfrak{G}$. In this case G is defined by the automorphisms which induce the elements of F in $\mathfrak{G}$, that is by defining $\mathfrak{G}$ as an operator group with the group of operators F. Conversely for every group $\mathfrak{G}$ with a group of operators F there exists a group G, which is a decomposible extension of F with respect to $\mathfrak{G}$. This extension is called a semi-direct product of F and $\mathfrak{G}$. We shall denote it by $F \cdot \mathfrak{G}$ to distinguish it from the direct product, which for the two groups A and B will be denoted by $A \times B$.

In the present paper the imbedding problem is discussed in case G is a semi-direct product $F \cdot \mathfrak{G}$.

In case the group $\mathfrak{G}$ is commutative the imbedding problem for semi-direct products was solved completely by Scholz [7], who showed that in this case the desired field K always exists.

We shall discuss in the present paper the next case in difficulty, namely when $\mathfrak{G}$ is of order l^a. It is shown that here as well as for a commutative group $\mathfrak{G}$ the problem has a positive solution in either of the following two cases.

1) The group $\mathfrak{G}$ is of class c with $l > c$,

2) the orders of the groups F and $\mathfrak{G}$ are relatively prime.

The case discussed by Scholz of the commutative group $\mathfrak{G}$ is obviously contained in case 1), because for an abelian group $c = 1$.

The results which are obtained as well as the methods of proof make it very likely that the imbedding problem can be solved positively for any semi-direct product and for any nilpotent normal subgroup $\mathfrak{G}$.

In most of the present paper (§ 2 and § 3) it is assumed that the reader if familiar with the paper [4]. In particular the concepts developed in that paper will be made use of here without further references.

§1. Extensions with a nilpotent disposition group

1. Nilpotent disposition groups. We consider a finite group G with a finite group of operators F. If G is abelian, then, as has been shown by Scholz [7], G is an operator-homomorphic image of some relatively simply constructed operator group, called a disposition group. We shall show that the concept of a disposition group can be generalized to the case when G is an arbitrary group of order l^a.

We denote by $\mathfrak{G}_d^{(c)}$ the group with generators $s_1, \cdots, s_d$ satisfying the conditions:

1) For some, once and for all, fixed k

$$s_i^{lk} = 1, \quad i = 1, \cdots, d;$$

2) any commutator of length c composed of generators $s_1, \cdots, s_d$ is equal to 1.

In other words $\mathfrak{G}_d^{(c)}$ is a quotient group of a free product of d cyclic groups of order l^k with respect to the c-th term of the lower central series. It follows readily from Schreier's theorem about subgroups of free groups that $\mathfrak{G}_d^{(c)}$ is a finite group. The order of $\mathfrak{G}_d^{(c)}$ is some power of l, the number of generators is d and the class is c.

Let F be an arbitrary finite group of order m. We consider the group $\mathfrak{G}_{md}^{(c)}$ with md generators, which we will denote by $s_{u,i}$, $u \in F$, $i = 1, \cdots, d$. The mapping which is defined for the generators $s_{u,i}$ by the condition

$$s_{u,i}^v = s_{uv,i}$$

goes over the whole group $\mathfrak{G}_{md}^{(c)}$ and is an automorphism, since all the relations imposed on the generators $s_{u,i}$ in $\mathfrak{G}_{md}^{(c)}$ are invariant with respect to the transposition of generators. The automorphisms thus defined form a group isomorphic with F*. We shall denote by $\mathfrak{G}_{d,F}^{(c)}$ the group $\mathfrak{G}_{md}^{(c)}$ with the operator group isomorphic to F and shall call it a nilpotent disposition group. In case $c = 1$, we obtain the abelian disposition groups of Scholz.

The generalization of the fundamental result of Scholz about abelian disposition groups to nilpotent disposition groups is contained in the following theorem.

Theorem 1. *An arbitrary group G of order l^a and class c with an operator group F is the operator-homomorphic image of some nilpotent disposition group $\mathfrak{G}_{d,F}^{(c)}$.*

Proof. Let the group G have d generators $s_1, \cdots, s_d$ and let the maximum order of its elements be l^k. We shall show that G is the operator-homomorphic

* We shall use the definition of multiplication of operators in the same order as is given in [3], page 97.

image of the group $\mathfrak{G}^{(c)}_{d,F}$. In order to do this we extend the mapping

$$\phi(s_{u,i}) = s_i^u$$

of the generators of $\mathfrak{G}^{(c)}_{d,F}$ onto the generators of G to all the elements of $\mathfrak{G}^{(c)}_{d,F}$, because of multiplicativity. Since all the relations taking place in $\mathfrak{G}^{(c)}_{d,F}$ between the generators $s_{u,i}$ hold for all the elements of G, it follows that ϕ is a homomorphism of $\mathfrak{G}^{(c)}_{d,F}$ onto G. It remains to show that this homomorphism is operative, that is that

$$\phi(\sigma^v) = \phi(\sigma)^v, \quad \sigma \in \mathfrak{G}^{(c)}_{d,F}, \quad v \in F. \tag{1}$$

Since ϕ is a homomorphism, it is sufficient to check this relation for $\sigma = s_{u,i}$. In this case we have

$$\phi(s^v_{u,i}) = \phi(s_{uv,i}) = s_i^{uv},$$
$$\phi(s_{u,i})^v = (s_i^u)^v = s_i^{uv},$$

which proves relation (1) and therefore proves the theorem.

2. **Central disposition groups.** We shall call the group G of order l^a with an operator group F a central disposition group if its center is a disposition group. Hence the center of the group G must have the following structure: It must be decomposible into the direct product of admissible groups, each of which must be the direct product of m conjugate with respect to F cyclic groups. Obviously the notion of a central disposition group and a nilpotent disposition group coincide for abelian groups, but as we shall see later, non-abelian nilpotent disposition groups are not in general central disposition groups.

Theorem 2. *Any group G of order l^a and class c with an operator group F is the operator-homomorphic image of some central disposition group for $l > c$.*

Proof. In view of Theorem 1 it is sufficient to prove the theorem in case $G = \mathfrak{G}^{(c)}_{d,F}$ and $l > c$. The center of the group $\mathfrak{G}^{(c)}_{d,F}$ in case $l > c$ has been determined by Thrall [8]. It follows from his considerations that the center of the group $\mathfrak{G}^{(c)}_d$ for $l > c$ is isomorphic (in the additive notation) to the module of Lie polynomials in d variables $x_1, \cdots, x_d$ with coefficients in the ring of residue classes modulo l^k. The variables $x_1, \cdots, x_d$ are in one to one correspondence with the generators of $s_1, \cdots, s_d$ of the group $\mathfrak{G}^{(c)}_d$ and the automorphism $\mathfrak{G}^{(c)}_d$ which reduces to the transposition of the generators $s_1, \cdots, s_d$ induces in the center the same transposition of the variables $x_1, \cdots, x_d$.

For the group $\mathfrak{G}^{(c)}_{d,F}$ we have a module of Lie polynomials of degree c with variables $x_{u,i}$ and coefficients in the ring of residue classes modulo l^k. The transformation of the variables

$$x^v_{u,i} = x_{uv,i} \tag{2}$$

defines an automorphism of the module of Lie polynomials. We define the module of Lie polynomials of degree c with a group of operators defined by formula (2) and isomorphic to F by $\Lambda(F)$.

We introduce into the consideration the module of multilinear forms with c md-dimensional vectors $\xi^{(j)}$, $j = 1, \cdots, c$, where $\xi^{(j)}$ is a vector with coordinates $x_{u,i}^{(j)}$, $u \in F$, $i = 1, \cdots, d$.

We define the action of the elements of F on the elements of the module by the formulas

$$x_{u,i}^{(j)v} = x_{uv,i}^{(j)}.$$

We shall denote this operator module by $M(F)$.

The proof of Theorem 2 follows from the following two considerations:

A) $\Lambda(F)$ is an operator direct summand of $M(F)$, that is there exists in $M(F)$ an F-invariant submodule $\Lambda'(F)$ which is operator-isomorphic to $\Lambda(F)$ and an F-invariant submodule $N(F)$ such that

$$M(F) = \Lambda'(F) + N(F). \tag{3}$$

B) $M(F)$ is a disposition group.

We shall show first of all that the statement of Theorem 2 follows from assumptions A) and B). We consider the direct product of the group $\mathfrak{G}_{d,F}^{(c)}$ and the group $\mathfrak{N}$, which is operator-isomorphic with $N(F)$. The center of the group $\mathfrak{G}_{d,F}^{(c)} \times \mathfrak{N}$ is operator-isomorphic with $M(F)$ by assumption A) and is therefore a disposition group by assumption B). Hence $\mathfrak{G}_{d,F}^{(c)} \times \mathfrak{N}$ is a central disposition group and since $\mathfrak{G}_{d,F}^{(c)}$ is its operator-homomorphic image, Theorem 2 follows.

We proceed to prove propositions A) and B).

Proof of proposition A). We shall make use of considerations applied by Thrall [8] to the study of the group $\mathfrak{G}_d^{(c)}$. If instead of transformations (2) we consider any reversible linear transformations of md variables $x_{u,i}$ with coefficients in the ring of residue classes modulo l^k, then the module Λ of Lie polynomials as well as the module M of multilinear forms becomes an operator module with respect to the group $\mathfrak{A}$ of all reversible linear transformations module l^k. Formulas (2) define an isomorphic image of F in $\mathfrak{A}$. Hence the transition of F to $\mathfrak{A}$ gives an enlargement of the group of operators, so that it is sufficient to show the existence of the decomposition (3) for modules Λ and M with respect to the domain of operators $\mathfrak{A}$. We shall therefore suppose that Λ has n generators (in our application $n = md$) $x_1, \cdots, x_n$ and that M consists of forms in n-dimensional vectors $x^{(j)}$ with coordinates $x_1^{(j)}, \cdots, x_n^{(j)}$.

Since Λ and M are modules over rings of residue classes modulo l^k, the operators $\mathfrak{A}$ define representations of $\mathfrak{A}$ modulo l^k. The representation defined by the module Λ is called the c-th Lie representation, and the representation defined by the module M is called the c-th tensor representation.

First of all it is clear that Λ is an operator-homomorphic image of the module M. The homomorphism is given by the mapping

$$x_{i_1}^{(1)} x_{i_2}^{(2)} \cdots x_{i_c}^{(c)} \to [x_{i_1} \; x_{i_2} \cdots x_{i_c}],$$

where $[x_{i_1} \cdots x_{i_c}]$ denotes the product $(\cdots(x_{i_1} x_{i_2}) \cdots x_{i_c})$ in a Lie ring.

A tensor representation of a full linear group is completely decomposible over a field of characteristic 0, (see [2], p. 177) so that the Lie representation is contained in it as a direct summand. As Brauer [6] pointed out, formulas which give a decomposition of the c-th tensor representation of a linear group over a field of characteristic 0 into a direct sum of indecomposables, contain in the denominator only numbers $\leq c$, and hence can be applied for an analogous decomposition over a ring of residue classes modulo l^k if $l > c$. Therefore it follows that for $l > c$, the c-th Lie representation is contained as a direct summand in the c-th tensor representation, that is that the module Λ is a direct summand of the module M, which proves proposition A).

Proof of proposition B). The monomials $x_{u_1, i_1}^{(1)} \; x_{u_2, i_2}^{(2)} \cdots x_{u_c, i_c}^{(c)}$ form a basis of the module M, where $u_1, \cdots, u_c$ are elements of the group F and $i_1, \cdots, i_c$ are any numbers from 1 to d. Let $w_2, \cdots, w_c$ be any elements of F. We consider the module $T(w_2, \cdots, w_c; i_1, \cdots, i_c)$ which is generated by all the elements $x_{v, i_1}^{(1)} \; x_{w_2 v, i_2}^{(2)} \cdots x_{w_c v, i_c}^{(c)}$ where v is any element of F. It is obvious that M is the direct sum of the modules $T(w_2, \cdots, w_c; i_1, \cdots, i_c)$, where each of these modules is invariant with respect to F and is a sum of cyclic submodules, generated by the elements $x_{v, i_1}^{(1)} \; x_{w_2 v, i_2}^{(2)} \cdots x_{w_c v, i_c}^{(c)}$ and conjugate with respect to F. This proves proposition B). The theorem is proved.

Using the considerations which were used in the proof of Theorem 2 we shall derive another result, which we will need in what follows. We denote by Z the center of the group $\mathfrak{G}_{d,F}^{(c)}$. It is the direct product of cyclic groups of order l^k. We denote by Z_r the totality of elements of Z, which are l^r-th powers of elements of Z. Obviously Z_r is a characteristic subgroup of $\mathfrak{G}_{d,F}^{(c)}$. We denote the group $\mathfrak{G}_{d,F}^{(c)}/Z_r$ by $\mathfrak{G}_{d,F}^{(c,r)}$ and Z_z/Z_{r+1} by $Z^{(r)}$. In doing so

$$\mathfrak{G}_{d,F}^{(c,0)} = \mathfrak{G}_{d,F}^{(c-1)}, \quad \mathfrak{G}_{d,F}^{(c,k)} = \mathfrak{G}_{d,F}^{(c)}.$$

Corollary. *For every $r \leq k-1$ there exists a group $\mathfrak{G}$ with a domain of operators F and with the following properties: $\mathfrak{G}$ contains an elementary admissible central subgroup A such that*

1) $\mathfrak{G}/A \cong \mathfrak{G}_{d,F}^{(c,r)}$;

2) *A is the direct product of admissible subgroups, each of which is a direct product of m conjugate with respect to F cyclic groups.*

3) *There exists a homomorphism of $\mathfrak{G}$ on $\mathfrak{G}_{d,F}^{(c,r+1)}$ such that its kernel is contained in A.*

To prove this we must note that the admissible subgroup $Z^{(r)}$ lies in the center of $\mathfrak{G}_{d,F}^{(c,r+1)}$ and is operator-isomorphic with respect to the module of Lie polynomials of degree c with dm variables $x_{u,i}$. We now apply propositions A) and B), which we used in the proof of Theorem 2 and obtain that there exists an elementary abelian group $\mathfrak{N}$ with a group of operators F such that $A = Z^{(r)} \times \mathfrak{N}$ possesses property 2). Therefore we can take $\mathfrak{G}_{d,F}^{(c,r+1)} \times \mathfrak{N}$ for $\mathfrak{G}$.

In what follows we will need the following remark. The normal subgroup A can be represented as an intersection of subgroups B_i such that $(A : B_i) = l$. The groups $\mathfrak{G}/B_i$ are simple central extensions of $\mathfrak{G}_{d,F}^{(c,r)}$. Similarly Z_r can be represented as an intersection of subgroups C_i, such that $(Z_r : C_i) = l$. The groups $\mathfrak{G}_{d,F}^{(c,r+1)}/C_i$ are also simple central extensions of $\mathfrak{G}_{d,F}^{(c,r)}$. Obviously the factor sets which correspond to the second extensions are contained in the factor sets corresponding to the first extensions. We must note that the first extensions do not produce any new factor sets. This follows from the fact that $\mathfrak{G} = \mathfrak{G}_{d,F}^{(c,r+1)} \times \mathfrak{N}$.

3. Fields with a central disposition group. Suppose that we are given a central disposition group $\mathfrak{G}$ with a group of operators F, where in view of the corollary to Theorem 2 we can limit ourselves to the case, where the center of $\mathfrak{G}$ is the direct product of m cyclic groups of order l, conjugate with respect to F. We denote this center by Z:

$$Z = \{z\} \times \{z^{u_1}\} \times \cdots \times \{z^{u_{m-1}}\}. \tag{4}$$

We denote by $\mathfrak{G}_1$ the quotient group of $\mathfrak{G}$ with respect to the subgroup $Z_1 = \{z^{u_1}\} \times \cdots \times \{z^{u_{m-1}}\}$, and $\mathfrak{G}/Z$ by $\overline{\mathfrak{G}}$. Obviously $\overline{\mathfrak{G}}$ has F for an operator group, but $\mathfrak{G}_1$ is not an operator group. We denote the semi-direct product of $\overline{\mathfrak{G}}$ and F by $\overline{G}$.

Suppose that we have a chain of fields $\Omega \subset k \subset \overline{K} \subset K_1$ for which $\overline{K}/\Omega$, K_1/k and k/Ω are normal, and let the Galois group of k/Ω be F, $\overline{K}/\Omega$ be $\overline{G}$, $\overline{K}/k$ be $\overline{\mathfrak{G}}$, and K_1/k be $\mathfrak{G}_1$.

Theorem 3. *If the fields K_1^u are independent over $\overline{K}$ for all $u \in F$, then the composite of all K_1^u has for its Galois group over Ω the semi-direct product of $\mathfrak{G}$ and F.*

We call here the extensions $K_1, \cdots, K_r$ of the field k independent over k if the intersection of the composite $K_1 \cdots K_{i-1} K_{i+1} \cdots K_r$ with K_i if k for every $i = 1, \cdots, r$, so that the Galois group of the composite $K_1 \cdots K_r$ over k is isomorphic to the direct product of the Galois groups of K_i/k.

Proof. We denote the composite of the fields K_1^u by K and calculate the Galois group of K/Ω. First we show that the Galois group of K/k coincides with $\mathfrak{G}$. To do this we study the group $\mathfrak{G}$ in more detail. The group $\mathfrak{G}$ is an extension of the group $\overline{\mathfrak{G}}$ with a central normal subgroup Z and therefore can be given by its factor set with values in Z. Because Z has the decomposition (4), this factor set can be written in the form

$$\alpha'(\sigma,\tau) = z^{\sum_u a_u(\sigma,\tau)u}, \quad \sigma,\tau \in \overline{\mathfrak{G}}.$$

We now pass to an associated factor set by selecting a more convenient system of representatives. Let the factor set $\alpha'(\sigma,\tau)$ correspond to a selection of representatives for which the class σ is represented by $[\sigma]$. If we denote $a_e(\sigma,\tau)$ simply by $a(\sigma,\tau)$, then the group $\mathfrak{G}_1$ will be a central extension of the group $\overline{\mathfrak{G}}$ with a factor set $z^{a(\sigma,\tau)}$. We consider the representatives $[\sigma^{u^{-1}}]^u$, lying of course in the same cosets as $[\sigma]$. We consider at the same time the group $\mathfrak{G}_u = \mathfrak{G}/Z_1^u$. This group has for its center $\{z^u\}$, and is a central extension of $\overline{\mathfrak{G}}$ with a factor set $z^{ua(\sigma^{u^{-1}},\tau^{u^{-1}})}$ (for the above choice of representatives $[\sigma^{u^{-1}}]^u$). From this it follows that we can select representatives in cosets of $\mathfrak{G}$ with respect to Z in order to obtain instead of $\alpha'(\sigma,\tau)$, an associated factor set

$$\alpha(\sigma,\tau) = z^{\sum ua(\sigma^{u^{-1}},\tau^{u^{-1}})}$$

The Galois group of K/k is obviously an extension of $\overline{\mathfrak{G}}$ with the kernel of homomorphism isomorphic to Z, so that in order to show that this group is isomorphic to $\mathfrak{G}$ we must only establish that it has the same factor set as $\mathfrak{G}$. The proof repeats almost word for word the above discussion. The quotient group of the Galois group of K/k with respect to the subgroup $Z_1 = \{z^{u_1}\} \times \cdots \times \{z^{u_{m-1}}\}$ is isomorphic to the Galois group of K_1/k, that is to $\mathfrak{G}_1$. The center of $\mathfrak{G}_1$ is generated by the element z, the quotient group with respect to the center is $\overline{\mathfrak{G}}$, and with a proper choice of representatives of the cosets with respect to the center the factor set will be $z^{a(\sigma,\tau)}$. We denote our selected representatives by $[\sigma]$. The automorphism $[\sigma]$ is some extension of the automorphism σ from the field $\overline{K}$ onto the field K_1. We now consider the field K_1^u, belonging to the subgroup Z_1^u and extend the automorphism σ to this field. In order to do this we note that every number of the field K_1^u is represented in the form α^u, where $\alpha \in K_1$, and that for primitive elements of K_1^u such a representation is unique. We define the automorphism $[\sigma]_u$ in K_1^u as follows:

$$(\alpha^u)^{[\sigma]_u} = (\alpha^{[\sigma^{u^{-1}}]})^u.$$

It can be easily seen that we have actually obtained an automorphism of the field K_1^u, which extends the automorphism σ of K. The Galois group of K_1^u is an extension of $\overline{\mathfrak{G}}$ with the kernel of homomorphism $\{z^u\}$, where for a given choice of representatives of cosets with respect to $\{z^u\}$, the factor set can be easily seen to be $z^{a(\sigma^{u^{-1}},\tau^{u^{-1}})u}$. If we extend in this way the automorphism σ to the field K_1^u for every u, then we get a certain extension of this automorphism in the field K, that is a certain fixed choice of representatives of the cosets of the Galois group of K/k with respect to Z. Therefore to this choice of representatives corresponds the factor set

$$z^{\sum\limits_{u} a(\sigma^{u^{-1}}, \tau^{u^{-1}})u},$$

which proves that the Galois group of K/k is isomorphic with $\mathfrak{G}$.

We now show that the isomorphism which we have established between the Galois group of K/k and the group $\mathfrak{G}$ is an operator isomorphism with respect to the group of operators F. We have established an isomorphism between our two groups by an isomorphic mapping into each other of their centers and of certain systems of representatives of residue classes with respect to the center, which we have constructed. The isomorphism of the centers is obviously an operator isomorphism, since both centers are direct products of groups of the form $\{z^u\}$, $u \in F$. It remains to check that we can apply the operators of F to the constructed correspondence between representatives of cosets with respect to the centers. We denote by $\langle\sigma\rangle$ the representative of the system which we have just constructed. This representative is uniquely defined by the choice of the representative $[\sigma]$ in the coset with respect to the center of the group $\mathfrak{G}_1 = \mathfrak{G}/Z_1$ and by the fact that for $u \in F$

$$\langle\sigma\rangle Z_1^u = \langle\sigma^{u^{-1}}\rangle^u Z_1^u = ([\sigma^{u^{-1}}] Z_1)^u. \tag{5}$$

But this defines uniquely the action of the operators of F on the representative $\langle\sigma\rangle$. In fact applying to (5) the automorphism $v \in F$, we get

$$\langle\sigma\rangle^v Z_1^{uv} = \langle\sigma^{u^{-1}}\rangle^{uv} Z_1^{uv}$$

or, denoting uv by w

$$\langle\sigma\rangle^v Z_1^w = \langle\sigma^{vw^{-1}}\rangle^w Z_1^w = ([\sigma^{vw^{-1}}] Z_1)^w \quad \text{for all} \quad w \in F,$$

which defines the elements $\langle\sigma\rangle^v$ uniquely.

The Galois group of K/Ω is an extension of F with the kernel of the homomorphism isomorphic with $\mathfrak{G}$, where, as we have shown, the kernel of the homomorphism is even operator-isomorphic with $\mathfrak{G}$. It remains to show that the extension is semi-direct, that is that with a proper choice of representatives $\overline{u}$ in the coset u with respect to $\mathfrak{G}$, the representatives $\overline{u}$ will form a group.

By the condition of the theorem the automorphisms u of the field k can be extended to automorphisms in the field $\overline{K}$, which we shall also denote by u, in such a way that the new automorphisms will as before form a group. In order to show the extension of the automorphism v from the field $\overline{K}$ to the field K it is sufficient to point out its action on the elements of the fields K_1^u, since their composite gives all K. We make use of the representation of the elements of the field K_1^u in the form α^u, $\alpha \in K_1$, $u \in F$, and suppose that

$$(\alpha^u)^{\overline{v}} = \alpha^{uv}.$$

It is obvious that this automorphism actually extends the automorphism v of the field $\overline{K}$ and that

$$\overline{v_1 v_2} = v_1 v_2.$$

Hence Theorem 3 is proved.

§2. The condition for imbeddedness for Scholz extensions

In order to construct the extension K/k having the group $\mathfrak{G}_{d,F}^{(c,k)}$ for the Galois group over k, and the semi-direct product $F \cdot \mathfrak{G}_{d,F}^{(c,k)}$ for the Galois group over Ω, we shall make use on the whole of the same considerations which will be found in paper [4]. In fact we will construct K/k as a Scholz field using the induction on c.

In this section we shall consider the following question. Let Ω, k and $\overline{K}$ have the same meaning as in Theorem 3, $K_1 \supset \overline{K}$ and $(K_1 : \overline{K}) = l$, where K_1 is normal over k and K is the composite of fields conjugate with K_1 over Ω. We shall suppose that $\overline{K}/k$ is a Scholz field and make clear when it is possible to construct a field K_1 with a given Galois group over k so that K is also a Scholz field over k. We shall suppose that the prime divisors of the discriminant of the field $\overline{K}/k$ are of degree 1 in k/Ω. That this can also be done for K/k is clear at each step of the discussion below.

We shall make use of the notation already defined without repeating the definitions.

1. **Construction of the invariants** $[\chi, X]$. We shall suppose in this section that k contains an l-th root of unity. Let $K_1 = \overline{K}(\sqrt[l]{\mu})$, where μ is an l-invariant number of the field $\overline{K}/k$. We denote the class of l-invariant numbers which contains μ by X. We consider any character χ of order l of the Galois group of $\overline{K}/k$ and the corresponding to its number α_χ of k (see [4], p. 268). This number will have the following decomposition into primes:

$$(\alpha_\chi) = \prod_{\mathfrak{p}} \mathfrak{p}^{\Sigma \varphi_{\mathfrak{p}}(u) u},$$

where $\mathfrak{p}$ goes over prime ideals in k conjugate to each other with respect to Ω, and the sum in the exponent goes over all the automorphisms u of the field k/Ω.

Let $\mathfrak{D}$ be that invariant ideal in $\overline{K}/k$, which is composed of ramified prime ideals and enters into the decomposition

$$(\mu) = \mathfrak{C}^l \mathfrak{D} \mathfrak{m} \tag{6}$$

(see [4], p. 266). We consider the function $g(u)$ defined by the fact that for any l-th root of unity

$$\zeta^u = \zeta^{g(u)},$$

and suppose that $\mathfrak{D}^{\Sigma \phi_{\mathfrak{p}}(u^{-1}) g(u) u}$ is relatively prime to the prime divisors $\mathfrak{P}$ of the ideal $\mathfrak{p}$ in $\overline{K}$ for every $\mathfrak{p}$. It can be easily checked that this condition does not depend on the choice of $\mathfrak{P}$ among the divisors of $\mathfrak{p}$ in $\overline{K}$. We shall therefore write it in the form

$$(\mathfrak{D}^{\Sigma \phi_{\mathfrak{p}}(u^{-1}) g(u) u}, \mathfrak{p}) = 1. \tag{7}$$

If condition (7) is satisfied, we shall define the symbol $[\chi, X]$ as follows:

$$[\chi, X] = \prod_{\mathfrak{p}} \left[\frac{\mu^{\Sigma \varphi_{\mathfrak{p}}(u^{-1}) g(u) u}}{\mathfrak{p}} \right] \cdot \left(\frac{\alpha_\chi}{m} \right)^{-1}. \tag{8}$$

Expression (7) guarantees that the symbol in the right hand side of the equality (8) is defined.

The form of relation (7) by means of which the symbol $[\chi, X]$ is defined, as well as the definition itself, depends not only on the numbers χ and X but also on the choice of α_χ, corresponding to χ, the number μ in the class X and on the prime ideals $\mathfrak{p}$ among the totality of all conjugate to each other ideals. We shall prove that the conditions under which this symbol is defined as well as the symbol itself depend in fact only on χ and X.

a) The independence of the symbol on the choice of α_χ is obvious, for if another α'_χ corresponding to the same χ were selected, then $\alpha_\chi \approx \alpha'_\chi$, and therefore $\phi_{\mathfrak{p}}(u) \equiv \phi'_{\mathfrak{p}}(u)\ (l)$ for any $\mathfrak{p}$ and u.

b) We prove the independence of the choice of μ in the class X. Any other l-invariant number of the class X can be obtained from μ by multiplying by an l-th power a number in $\overline{K}$ and by a number in k. The multiplication of μ by an l-th power of a number in $\overline{K}$ obviously does not alter either condition (7), or the symbol $[\chi, X]$. Let us multiply μ by a number $m \in k$, relatively prime to the discriminant of $\overline{K}/k$. Obviously condition (7) does not change, since we have not changed $\mathfrak{D}$. The right hand side of formula (8) is multiplied by

$$\prod_{\mathfrak{p}} \left(\frac{m^{\Sigma \varphi_{\mathfrak{p}}(u^{-1}) g(u) u}}{\mathfrak{p}} \right) \cdot \left(\frac{\alpha_\chi}{m} \right)^{-1}. \tag{9}$$

Since

$$\left(\frac{m^{\varphi_{\mathfrak{p}}(u^{-1}) g(u) u}}{\mathfrak{p}} \right) = \left(\frac{m}{\mathfrak{p}^{\Sigma \varphi_{\mathfrak{p}}(u^{-1}) u^{-1}}} \right),$$

we have

$$\prod_{\mathfrak{p}} \left(\frac{m^{\Sigma \varphi_{\mathfrak{p}}(u^{-1}) g(u) u}}{\mathfrak{p}} \right) = \prod_{\mathfrak{p}} \left(\frac{m}{\mathfrak{p}^{\Sigma \varphi_{\mathfrak{p}}(u^{-1}) u^{-1}}} \right) = \left(\frac{m}{\alpha_\chi} \right).$$

Since the field $\overline{K}$ is a Scholz field by assumption, and since α_χ is hyperprimary, we have

$$\left(\frac{m}{\alpha_\chi} \right) = \left(\frac{\alpha_\chi}{m} \right),$$

which shows that expression (9) is equal to 1.

c) To show the independence on the choice of $\mathfrak{p}$ we suppose that $\mathfrak{p}^v$ is any prime ideal conjugate with $\mathfrak{p}$. Then

$$\mathfrak{p}^{\Sigma \varphi_{\mathfrak{p}}(u) u} = \mathfrak{p}^{v \Sigma \varphi_{\mathfrak{p}}(u) v^{-1} u} = \mathfrak{p}^{v \Sigma \varphi_{\mathfrak{p}}(vu) u}.$$

In this way replacing $\mathfrak{p}$ by $\mathfrak{p}^v$ leads to the replacement of the function $\phi_{\mathfrak{p}}(u)$ by the function $\phi_{\mathfrak{p}}(vu)$. Condition (7) takes on the form

$$(\mathfrak{D}^{\sum \varphi_p(vu^{-1})g(u)u}, \mathfrak{p}^v) \overset{\bullet}{\approx} 1. \tag{10}$$

To prove this condition we apply to (7) the automorphism v. We obtain

$$(\mathfrak{D}^{\sum \varphi_p(u^{-1})g(u)uv}, \mathfrak{p}^v) \approx 1.$$

Replacing in the sum in the exponent the variable uv by u we obtain

$$(\mathfrak{D}^{\sum \varphi_p(vu^{-1})g(uv^{-1})u}, \mathfrak{p}^v) \approx 1.$$

Since $g(uv^{-1}) \equiv g(u)g(v^{-1}(l)$, we have

$$\mathfrak{D}^{\sum \varphi_p(vu^{-1})g(uv^{-1})u} \approx (\mathfrak{p}^{\sum \varphi_p(vu^{-1})g(u)u})^{g(v^{-1})},$$

which proves the truth of relation (10).

Therefore the condition which defines $[\chi, X]$ does not depend on the choice of $\mathfrak{p}$ among the conjugates. We show that the value of the symbol $[\chi, X]$ also does not depend on the choice of $\mathfrak{p}$.

If we replace $\mathfrak{p}$ by $\mathfrak{p}^v$ the multiplier

$$\left[\frac{\mu^{\sum \varphi_p(u^{-1})g(u)u}}{\mathfrak{p}}\right] \tag{11}$$

in the right hand side of (8) will be replaced by

$$\left[\frac{\mu^{\sum \varphi_p(vu^{-1})g(u)u}}{\mathfrak{p}^v}\right]. \tag{12}$$

Making use of the relation

$$\left(\frac{\alpha}{\mathfrak{b}^v}\right) = \left(\frac{\alpha^{g(v^{-1})v^{-1}}}{\mathfrak{b}}\right),$$

we see that the expression (12) becomes equal to

$$\left[\frac{\mu^{\sum \varphi_p(vu^{-1})g(u)g(v^{-1})uv^{-1}}}{\mathfrak{p}}\right].$$

If we now replace uv^{-1} by u we obtain (11).

If $k = \Omega$, the symbol $[\chi, X]$ becomes the symbol (χ, X), defined in paper [4]. Obviously the symbol $[\chi, X]$ as well as the symbol (χ, X) possess the properties

$$[\chi_1\chi_2, X] = [\chi_1, X][\chi_2, X], \tag{13}$$

$$[\chi, X_1X_2] = [\chi, X_1][\chi, X_2], \tag{14}$$

$$[\chi, X] = [\chi', X']. \tag{15}$$

The last equality means that if $\overline{K}$ is a subfield of a normal field Ω of the field $\overline{K}'$, then the symbol $[\chi, X]$ can be calculated in both fields and has the same values in both fields.

2. The invariants $[X]_\psi$. These invariants are defined only for classes X which contain l-invariant representatives μ for which the field $\overline{K}(\sqrt[l]{\mu})$ is a Scholz field. In other words these are the classes for which all the invariants (χ, X) and

$(X)_h$ are equal to one. In what follows we shall suppose that the representative μ of the class X is selected so that the field $\overline{K}(\sqrt[l]{\mu})$ is a Scholz field.

We consider the class of l-invariant numbers X and the function $\psi(u)$ defined over the group F with values in the group of residues modulo l, satisfying the conditions

1. $\psi(1) = 0,\ \psi(u^{-1}) = -\psi(u)\,g(u),\ u \in F,$
2. $\mu^{\Sigma\psi(u)u} \approx \alpha_\psi \in k.$

If $l = 2$ we shall impose an additional condition. Let u be an involutory automorphism in $\overline{K}/k : u^2 = 1$, $\overline{K}_u$ be a subfield of $\overline{K}$ composed of numbers invariant with respect to u, let $\mathfrak{D}_u$ be the different of $\overline{K}_u/\overline{K}_u$. We shall require that the function $\psi(u)$ satisfy the condition

3. $\prod_u \mathfrak{D}_u^{\psi(u)} \approx 1$ in $\overline{K}$,

where the product goes over all the involutory automorphisms of F.

It is obvious that all these conditions refer only to the function ψ and to the class of l-invariant numbers X, but do not depend on the choice of the representative μ in X. If conditions 1 and 2 are satisfied, and for $l = 2$ condition 3 as well, then we define the symbol $[X]_\psi$ as follows: we select from the class X an l-hyperprimary representative μ such that $\overline{K}(\sqrt[l]{\mu})/k$ is a Scholz field and

$$(\mathfrak{m}, \mathfrak{m}^u) = 1, \text{ for all } u \neq 1, \tag{16}$$

where $\mathfrak{m}$ is the ideal in the decomposition (6) and obtain

$$\mathfrak{m}_\psi = \mathfrak{m}^{\Sigma\psi(u)u}, \quad [X]_\psi = \left[\frac{\mu}{\mathfrak{m}_\psi}\right]. \tag{17}$$

In view of condition (16), the right hand side of (17) is defined.

We show next that $[X]_\psi$ does not depend on the choice of μ in the class X. It is obvious that if we multiply μ by an l-th power of a number in $\overline{K}$, the symbol $[X]_\psi$ remains unaltered. We show that neither does it change if we multiply μ by a number m in k, such that the field $\overline{K}(\sqrt[l]{\mu m})$ is a Scholz field. The following difficulty may appear in the proof: if

$$(\mu) = \mathfrak{C}^l\mathfrak{D}\mathfrak{m}, \quad (\mu') = \mathfrak{C}'^l\mathfrak{D}\mathfrak{m}',$$

where $\mathfrak{m}$ and $\mathfrak{m}'$ satisfy condition (16) and $\mu' = \mu m$, then it does not necessarily follow that m satisfies a condition analogous to (16):

$$(m, m^u) = 1, \quad u \in F, \quad u \neq 1.$$

In order not to run into this difficulty in what follows we introduce a μ'' for which

$$\mu'' = \mu\alpha, \quad \mu' = \mu''\beta, \quad (\mu'') = \mathfrak{C}''\mathfrak{D}\mathfrak{m}'', \tag{18}$$

and suppose that the following conditions are satisfied:

$$\begin{gathered}(\mathfrak{m}'', \mathfrak{m}''^u) = 1, \quad (\mathfrak{m}'', \mathfrak{m}^u) = 1, \quad (\mathfrak{m}'', \mathfrak{m}'^u) = 1,\\ (\alpha, \alpha^u) = 1, \quad (\beta, \beta^u) = 1.\end{gathered} \tag{19}$$

In order to do this it is sufficient to select a prime ideal $\mathfrak{p}$ of absolutely the first degree, relatively prime to all the $\mathfrak{m}^u$ and $\mathfrak{m}'^u$ and equivalent to $\mathfrak{m}$. An independent check shows that conditions (18) and (19) are satisfied if we let

$$(\alpha) = \mathfrak{p}\mathfrak{m}^{-1}, \quad (\beta) = \mathfrak{m}'\mathfrak{p}^{-1}.$$

Hence all we have to show is that

$$\left[\frac{\mu}{\mathfrak{m}_\psi}\right] = \left[\frac{\mu'}{\mathfrak{m}'_\psi}\right], \tag{20}$$

if

$$\mu = \mu' m, \quad (\mathfrak{m}, \mathfrak{m}^u) = (\mathfrak{m}', \mathfrak{m}'^u) = (m, m^u) = 1.$$

In doing this we suppose that μ, μ' and m are l-hyperprimary. In this case

$$\mathfrak{m}'_\psi = \mathfrak{m}_\psi m^{\Sigma\psi(u)u}$$

and

$$\left[\frac{\mu'}{\mathfrak{m}'_\psi}\right] = \left[\frac{\mu}{\mathfrak{m}_\psi}\right]\left(\frac{m}{\mathfrak{m}_\psi}\right)\left[\frac{\mu}{m^{\Sigma\psi(u)u}}\right]\left(\frac{m}{m^{\Sigma\psi(u)u}}\right). \tag{21}$$

Equation (20) will be established if we can show that the product of the second and third factors as well as the fourth factor in the right hand side of (21) is equal to 1. In fact

$$\left[\frac{\mu}{m^{\Sigma\psi(u)u}}\right] = \left[\frac{\mu^{\Sigma\psi(u)g(u)u^{-1}}}{m}\right] = \left[\frac{\mu^{-\Sigma\psi(u^{-1})u^{-1}}}{m}\right] = \left(\frac{\alpha_\psi}{m}\right)^{-1},$$

by conditions 1 and 2. Since m is l-hyperprimary and is an l-th power residue with respect to the ramified ideals it follows from the law of reciprocity that

$$\left[\frac{\mu}{m^{\Sigma\psi(u)u}}\right]\left(\frac{m}{\mathfrak{m}_\psi}\right) = \left(\frac{\alpha_\psi}{m}\right)^{-1}\left(\frac{m}{\alpha_\psi}\right) = 1.$$

The proof of the fact that

$$\left[\frac{m}{m^{\Sigma\psi(u)u}}\right] = 1, \tag{22}$$

is different in the two cases $l \neq 2$ and $l = 2$.

For $l \neq 2$ we have

$$\left(\frac{m}{m^{\Sigma\psi(u)u}}\right) = \left(\frac{m^{\Sigma\psi(u)g(u)u^{-1}}}{m}\right) = \left(\frac{m}{m^{\Sigma\psi(u)g(u)u^{-1}}}\right) = \left(\frac{m}{m^{\Sigma\psi(u)u}}\right)^{-1}.$$

From this it follows that

$$\left(\frac{m}{m^{\Sigma\psi(u)u}}\right)^2 = 1, \text{ i.e., } \left(\frac{m}{m^{\Sigma\psi(u)u}}\right) = 1.$$

For $l = 2$, it is obvious that $g(u) = 1$ for all u. If u is a non-involutory automorphism, then $u \neq u^{-1}$ and

$$\left(\frac{m}{m^{\psi(u)u+\psi(u^{-1})u^{-1}}}\right) = \left(\left(\frac{m}{m^u}\right)\left(\frac{m}{m^{u^{-1}}}\right)^{-1}\right)^{\psi(u)}. \tag{23}$$

Since

$$\left(\frac{m}{m^{u^{-1}}}\right)=\left(\frac{m^u}{m}\right)=\left(\frac{m}{m^u}\right),$$

it follows that this expression is equal to 1.

We now consider the product

$$\left(\frac{m}{m^{\Sigma\psi(u)u}}\right),$$

where the sum goes only over involutory automorphisms of F different from 1. It was shown in [5] that if u is an involutory automorphism of the field $\overline{K}$ in which the prime divisors of 2 and the real infinite primes split completely, then for the Legendre symbol of the second degree the following relation holds:

$$\left(\frac{m}{m^u}\right)=\left(\frac{m}{\mathfrak{D}_u}\right),$$

where $\mathfrak{D}_u$ is the different of $\overline{K}/\overline{K}_u$, and $\overline{K}_u$ is a subfield belonging to u. Applying this relation to our product and making use of condition 3 we have

$$\left(\frac{m}{m^{\Sigma\psi(u)u}}\right)=\prod\left(\frac{m}{m^u}\right)^{\psi(u)}=\left(\frac{m}{\Pi\mathfrak{D}_u^{\psi(u)}}\right)=1.$$

This equation together with (23) gives (22), which proves the independence of the symbol $[X]_\psi$ on the choice of μ from the class X.

For the symbol $[X]_\psi$ we have the equality

$$[X]_\psi=[X']_\psi,$$

analogous to (15), but we have no relation analogous to (14).

3. **Conditions for imbeddedness.** Besides the two invariants $[\chi, X]$ and $[X]_\psi$, which we have just introduced, we shall also consider the invariants $(X)_h$ which we have introduced in paper [4], for a class of l-invariant numbers of the field $\overline{K}/k$. With their help we can formulate the following necessary and sufficient condition of imbeddedness in a Scholz field.

Let $\overline{K}\supset k\supset\Omega$, $\overline{K}/k$ be a Scholz field with the Galois group of degree l^a, $\overline{K}/\Omega$ be normal, X be the class of l-invariant numbers in $\overline{K}/k$, K_1 be any simple central extension of $\overline{K}$, corresponding to X, and K be the composite of all the fields conjugate with K_1 over Ω.

Theorem 4. *In order that there should exist a field K_1 for which K is a Scholz field over k, it is necessary and sufficient that all the invariants $(X)_h$, $[X]_\psi$ and $[\chi, X]$ be equal to 1 for every ψ and χ for which they are defined.*

Proof. The necessity of the theorem can be checked independently. The fact that $(X)_h$ is equal to unity follows from the corresponding assertion in Theorem 1 of paper [4], since from the fact that K/k is a Scholz field follows the fact that K_1/k is also a Scholz field. In formula (17), defining $[X]_\psi$, $\mathfrak{m}_\psi$ consists only of prime divisors of the ideals $\mathfrak{m}^u$. Since K/k is a Scholz field

$$\left[\frac{\mu^u}{\mathfrak{p}}\right]=1, \text{ i.e., } \left[\frac{\mu}{\mathfrak{p}^u}\right]=1 \quad \text{for all} \quad u \neq 1, \quad u \in F, \quad \mathfrak{p} \mid \mathfrak{m},$$

from which it follows that $[X]_\psi = 1$.

In formula (8), which defines $[\chi, X]$, for every prime $\mathfrak{q} \mid \mathfrak{m}$, it follows since K_1/k is a Scholz field that

$$\left(\frac{\alpha_\chi}{\mathfrak{q}}\right)=1, \text{ i.e., } \left(\frac{\alpha_\chi}{\mathfrak{m}}\right)=1.$$

On the other hand since K/k is a Scholz field and by (7),

$$\left[\frac{\mu^{\Sigma \psi_\mathfrak{p}(u^{-1})\, g(u)\, u}}{\mathfrak{p}}\right]=1, \quad \mathfrak{p} \mid \alpha_\chi,$$

from which it follows that $[\chi, X] = 1$.

We now prove the sufficiency of the conditions of the theorem. To do this we note that the invariants (χ, X) defined in paper [4] are contained among the invariants $[\chi, X]$. In fact in order that the invariant (χ, X) be defined it is necessary that all the prime ideals contained in α_χ to a degree not divisible by l be relatively prime to $\mathfrak{D}$. It is obvious from this that condition (7) is satisfied, that is the symbol $[\chi, X]$ is also defined. We note that

$$[\chi, X]=\prod_{\mathfrak{p}}\left[\frac{\mu^{\Sigma \varphi_\mathfrak{p}(u^{-1})\, g(u)\, u}}{\mathfrak{p}}\right]\cdot\left(\frac{\alpha_\chi}{\mathfrak{m}}\right)^{-1}=$$

$$=\prod_{\mathfrak{p}}\left[\frac{\mu}{\mathfrak{p}^{\Sigma \varphi_\mathfrak{p}(u^{-1})\, u^{-1}}}\right]\cdot\left(\frac{\alpha_\chi}{\mathfrak{m}}\right)^{-1}=\left[\frac{\mu}{\alpha_\chi}\right]\left(\frac{\alpha_\chi}{\mathfrak{m}}\right)^{-1}=(\chi, X).$$

In this way from the conditions of the theorem follows the equality to unity of the invariants (χ, X) and $(X)_h$. We now apply Theorem 1 of paper [4] and see that K_1 can be taken as a Scholz field over k. In what follows we will always suppose that K_1/k is a Scholz field.

It is obvious that the first, second and fourth conditions for a Scholz field, being satisfied by the field K_1, are also satisfied by its conjugates K_1^u, and therefore by their composite K. In this way it remains to show that if $K_1 = \overline{K}(\sqrt[l]{\mu})$, then μ may be multiplied by a number $m \in k$ such that $\overline{K}(\sqrt[l]{\mu m})$ will be as before a Scholtz field over k, and such that the third condition for a Scholz field will be satisfied by the composite of the fields conjugate with $\overline{K}(\sqrt[l]{\mu m})$ with respect to Ω. The multiplication of μ by m will not disturb the Scholz character of K_1 if m is totally positive (for $l = 2$), if it is divisible only by prime ideals which split completely in $\overline{K}$ and if it is congruent to 1 with respect to a sufficiently high power of l. This power we will denote by l^g.

We will attain a Scholz field K in two steps. First we multiply μ by a number m_1 in order that K_1 remain a Scholz field, and that in K all the ramified prime ideals of $\overline{K}$ be of degree 1. Next we multiply μm_1 by a number m_2 so that these properties remain unchanged and moreover all the ramified prime ideals in K be of degree 1.

Let $\mathfrak{p}$ be a ramified prime ideal of $\overline{K}$. In order that the field K, obtained by adjoining to $\overline{K}$ all the $\sqrt[l]{\mu^u}$ have all the prime divisors of $\mathfrak{p}$ of the first degree, it is necessary and sufficient that any function $f_{\mathfrak{p}}(u)$ for which

$$(\mathfrak{D}^{\Sigma f_{\mathfrak{p}}(u)u}, \mathfrak{P}) \approx 1, \quad \mathfrak{P} \mid \mathfrak{p},$$

satisfies the condition

$$\left[\frac{\mu^{\Sigma f_{\mathfrak{p}}(u)u}}{\mathfrak{p}}\right] = 1.$$

Let

$$\xi_{f\mathfrak{p}} = \left[\frac{\mu^{\Sigma f_{\mathfrak{p}}(u)u}}{\mathfrak{p}}\right]$$

and show that there exists a totally positive number m_1, consisting of prime ideals which split completely in $\overline{K}$ and which is congruent to 1 modulo l^g for which

$$\left(\frac{m_1^{\Sigma f_{\mathfrak{p}}(u)u}}{\mathfrak{p}}\right) = \xi_{f_{\mathfrak{p}}}^{-1} \tag{24}$$

for any $\mathfrak{p}$ and $f_{\mathfrak{p}}$. Obviously then μm_1 will satisfy the conditions which we need.

We take an arbitrary primitive l-th root of unity ζ and let

$$\left(\frac{m_1}{\mathfrak{p}^u}\right) = \zeta^{x_{\mathfrak{p},u}}, \qquad \xi_{f_{\mathfrak{p}}}^{-1} = \zeta^{b_{f\mathfrak{p}}}. \tag{25}$$

Equation (24) can be written

$$\left(\frac{m_1}{\mathfrak{p}^{\Sigma f_{\mathfrak{p}}(u)g(u)u^{-1}}}\right) = \xi_{f_{\mathfrak{p}}}^{-1}$$

or

$$\sum_u f_{\mathfrak{p}}(u^{-1})\, g(u^{-1})\, x_{\mathfrak{p},u} = b_{f\mathfrak{p}} \quad \text{for all } \mathfrak{p} \text{ in } f_{\mathfrak{p}}, \tag{26}$$

when the equation is considered in the field of residues modulo l. We next find under what conditions there exists a totally-positive number m_1 satisfying conditions (25) and consisting of prime ideals which split in $\overline{K}$ and which are congruent to 1 modulo l^g.

We denote by $\mathfrak{F}$ the product of all $\mathfrak{p}^u$, of l^g and of the infinite primes, by H the group of all l-th power classes of ideals modulo $\mathfrak{F}$, and by L the class field of H. We show that equation (25) and the condition that m_1 be congruent to 1 modulo l^g and is totally positive are equivalent to the ideal (m_1) belonging to a definite automorphism in L. In fact if the numbers m_1 and m_1' satisfy these conditions then $(m_1')H = (m_1)H$, and conversely if an ideal $\mathfrak{a}$ is such that $\mathfrak{a}H = (m_1)H$, then $\mathfrak{a} = (\alpha)\mathfrak{b}^l$, where α satisfies all the required conditions. We denote by σ an an automorphism of the field L such that if the ideal (m_1) belongs to this automorphism, then m_1 satisfies all the conditions which we have imposed on it.

In order to find a number m_1, consisting of prime divisors which split completely in $\overline{K}$ and which belongs to the automorphism σ in L, it is necessary and sufficient that σ should induce an identity automorphism in the intersection $\overline{K} \cap L$. Obviously $\overline{K} \cap L$ is the field obtained by adjoining to k all the $\sqrt[l]{\alpha_\chi}$. In this way we must find the conditions under which σ induces in each of the fields $k(\sqrt[l]{\alpha_\chi})$ an identity automorphism.

We find the automorphism induced by σ in $k(\sqrt[l]{\alpha_\chi})$. In order to do this we take an arbitrary divisor $\mathfrak{a}$, belonging to σ in L. We have seen that $\mathfrak{a} = (\alpha)\mathfrak{b}^l$ and therefore (α) belongs to σ in L. The automorphism induced by σ in $k(\sqrt[l]{\alpha_\chi})$ can be reduced to the multiplication of $\sqrt[l]{\alpha_\chi}$ by $\left(\frac{\alpha_\chi}{\mathfrak{a}}\right) = \left(\frac{\alpha_\chi}{\alpha}\right)$. Since α_χ is l-hyperprimary and totally positive we have

$$\left(\frac{\alpha_\chi}{\alpha}\right) = \left(\frac{\alpha}{\alpha_\chi}\right) = \prod_{\mathfrak{p}, u} \left(\frac{\alpha}{\mathfrak{p}^u}\right)^{\varphi_{\mathfrak{p},\chi}(u)} = \zeta^{\sum\limits_{\mathfrak{p},u} x_{\mathfrak{p},u}\varphi_{\mathfrak{p},\chi}(u)},$$

if

$$(\alpha_\chi) = \prod_{\mathfrak{p}} \mathfrak{p}^{\Sigma \varphi_{\mathfrak{p},\chi}(u)}.$$

Hence in order that there should exist a totally-positive number m_1, composed of completely decomposible in $\overline{K}$ prime ideals and congruent to 1 modulo l^g and satisfying condition (25), the following relation must hold

$$\sum_{\mathfrak{p},u} x_{\mathfrak{p},u}\varphi_{\mathfrak{p},\chi}(u) = 0 \quad \text{for all } \chi. \tag{27}$$

It remains to show that there exists an $x_{\mathfrak{p},u}$, satisfying equations (26) and (27). We shall write down the conditions for compatability of these two equations and show that this condition is that the invariant $[\chi, X]$ be equal to unity.

If the system of equations composed of (26) and (27) is incompatible then there exists a linear combination of these equations, all of whose coefficients of the variables $x_{\mathfrak{p},u}$ are equal to zero, and the constant term is not equal to zero. Let

$$\sum_\chi c_\chi \sum_{\mathfrak{p},u} x_{\mathfrak{p},u}\varphi_{\mathfrak{p},\chi}(u) + \sum_{f_\mathfrak{p},\mathfrak{p}} d_{f_\mathfrak{p},\mathfrak{p}} \sum_u f_\mathfrak{p}(u^{-1})\, g(u^{-1})\, x_{\mathfrak{p},u} = \sum_{f_\mathfrak{p},\mathfrak{p}} d_{f_\mathfrak{p},\mathfrak{p}}\, b_{f_\mathfrak{p}} \tag{28}$$

be such a linear combination.

Letting

$$\prod_\chi \chi^{c_\chi} = \chi_0, \quad \sum_{f_\mathfrak{p}} d_{f_\mathfrak{p},\mathfrak{p}} f_\mathfrak{p}(u) = F_\mathfrak{p}(u),$$

we get

$$\sum_{\chi} c_\chi \varphi_{\mathfrak{p},\chi}(u) = \varphi_{\mathfrak{p},\chi_0}(u),$$

$$\sum_{f_\mathfrak{p}} d_{f_\mathfrak{p},\mathfrak{p}} b_{f_\mathfrak{p}} = b_{F_\mathfrak{p}}.$$

Taking into account that in (28) the coefficients of all the $x_{\mathfrak{P},u}$ are equal to zero, we get

$$\varphi_{\mathfrak{p},\chi_0}(u) = -F_\mathfrak{p}(u^{-1})g(u^{-1}) \text{ for all } \mathfrak{P} \text{ and } u.$$

From this we conclude that

$$(\mathfrak{D}^{\sum \varphi_{\mathfrak{p},\chi_0}(u)g(u)u}, \mathfrak{p}) \approx 1$$

and that

$$\prod_{\mathfrak{p}} \left[\frac{\mu^{\sum \varphi_{\mathfrak{p},\chi_0}(u)g(u)u}}{\mathfrak{p}} \right] = \zeta^{\sum_\mathfrak{p} b_{F_\mathfrak{p}}}. \tag{29}$$

In order that the system (28) be compatible it is necessary and sufficient that

$$\sum_{f_\mathfrak{p},\mathfrak{p}} d_{f_\mathfrak{p},\mathfrak{p}} b_{f_\mathfrak{p}} = \sum b_{F_\mathfrak{p}} = 0,$$

that is that the right hand side of (29) be equal to 1. But the left hand side of this equation is $[\chi_0, X]$, since the field K_1 is a Scholz field by assumption and $\left(\frac{\alpha_\chi}{\mathfrak{m}}\right) = 1$. Hence the system of equations (26) and (27) is compatible and hence the number m_1 with the desired properties exists.

We denote the number μm_1 by μ_1. We have shown that the field $\overline{K}(\sqrt[l]{\mu_1})$ is a Scholtz field, and that in the composite of all its conjugates the ramified ideals of $\overline{K}$ are of degree 1. We must find a number m_2, such that all the properties which we have found for μ_1 would hold true for $\mu_1 m_2$, and such that all the ramified prime ideals in the composite of the fields conjugate with $\overline{K}(\sqrt[l]{\mu_1 m_2})$ be of the first degree.

We select m_2 to be totally positive, and consisting of prime ideals which split completely in $\overline{K}$ satisfying the conditions

$$\left.\begin{aligned} &\left(\frac{m_2}{\mathfrak{p}}\right) = 1 \quad \text{for all ramified } \mathfrak{P} \text{ in } \overline{K}/k, \\ &m_2 \equiv 1\ (l^\rho), \quad (m_2, m_2^u) = 1 \quad u \in F, \quad u \neq 1. \end{aligned}\right\} \tag{30}$$

Since $\overline{K}/\Omega$ is normal, $\mathfrak{P}^u$ will be ramified in $\overline{K}/k$ as well as $\mathfrak{P}$. From this it follows that for this choice of m_2, the properties which we have found for μ_1 will hold for $\mu_1 m_2$. We now denote by K the composite of conjugates of $\overline{K}(\sqrt[l]{\mu_1 m_2})$ over Ω. It remains to choose m_2 so that it satisfies conditions (30), and such that in the field K/k the ramified ideals of $K/\overline{K}$ be of degree 1. The ramified ideals of $K/\overline{K}$ which are not ramified in $\overline{K}$ will be divisors of $(\mathfrak{m}_1 m_2)^u$, only if

$$(\mu_1) = \mathfrak{C}_1^l \mathfrak{D} \mathfrak{m}_1.$$

We will suppose that

$$(\mathfrak{m}_1, \mathfrak{m}_1^u) = 1,$$

as this can be easily accomplished by multiplying μ_1 by a number m_2, satisfying conditions (30).

In order that the divisors of $\mathfrak{m}_1 m_2$ be of degree 1 in K it is necessary and sufficient that the following conditions be satisfied:

$$\left[\frac{\mu_1 m_2}{\mathfrak{q}^u}\right] = 1 \quad \text{for all} \quad \mathfrak{q} \mid \mathfrak{m}_1 m_2.$$

These conditions can be considered in two parts:

$$\left[\frac{\mu_1 m_2}{\mathfrak{q}^u}\right] = 1, \text{ i.e., } \left(\frac{m_2}{\mathfrak{q}^u}\right) = \left[\frac{\mu_1}{\mathfrak{q}^u}\right]^{-1} \quad \text{for } \mathfrak{q} \mid \mathfrak{m}_1$$

and

$$\left(\frac{m_2}{\mathfrak{r}^u}\right) = \left[\frac{\mu_1}{\mathfrak{r}^u}\right]^{-1} \quad \text{for } \mathfrak{r} \mid m_2.$$

We shall choose m_2 so that all the prime ideals $\mathfrak{r}$ dividing it satisfy the condition

$$\left[\frac{\mu_1}{\mathfrak{r}^u}\right] = \zeta(u), \tag{31}$$

where $\zeta(u)$ is some system of l-th roots of unity. The number m_2 itself must satisfy the conditions

$$\left(\frac{m_2}{\mathfrak{q}^u}\right) = \left[\frac{\mu_1}{\mathfrak{q}^u}\right]^{-1}, \quad \mathfrak{q} \mid \mathfrak{m}_1,$$

$$\left(\frac{m_2}{\mathfrak{r}^u}\right) = \zeta(u)^{-1}, \quad u \in F, \quad u \neq 1, \quad \mathfrak{r} \mid m_2. \tag{32}$$

The question of the existence of such numbers was considered in paper [5]. In order to make use of the theorem which was proved in that paper, we will suppose that the l-th roots of unity $\zeta(u)$ satisfy the condition

$$\zeta(u) = \zeta(u^{-1})^{g(u^{-1})} \tag{33}$$

and in case $l = 2$, we shall require that for any involutory automorphisms $u_1, \cdots, u_r$ in F, the condition

$$\zeta(u_1) \cdots \zeta(u_r) = 1 \text{ for every } \mathfrak{D}_{u_1} \cdots \mathfrak{D}_{u_r} \approx 1, \tag{34}$$

is satisfied for the differents $\mathfrak{D}_{u_i}$ of the fields $\overline{K}/\overline{K}_{u_i}$, where $\overline{K}_{u_i}$ is a field belonging to the automorphism u_i.

It follows from the theorem of paper [5], that there exist infinitely many numbers m_2 satisfying conditions (32) for any $\zeta(u)$, satisfying conditions (33) and (34). It remains to clarify whether at the same time we can satisfy conditions (30) and (31). Here we must consider separately the cases $l \neq 2$ and $l = 2$.

Suppose that $l \neq 2$. In this case by the theorem of paper [5], m_2 can be found as the product of two primes $\pi_1 \pi_2$ satisfying the conditions

$$\left(\frac{\pi_i}{\mathfrak{q}^u}\right) = \left[\frac{\mu_1}{\mathfrak{q}^u}\right]^{-\frac{1}{2}}, \quad i = 1, 2,$$

where conditions (30) and (31) will be satisfied for m_2 if we can show the existence of infinitely many primes π, which split completely in $\overline{K}$ and satisfy conditions (33) and the following:

$$\left[\frac{\mu_1}{\pi^u}\right] = \zeta(u); \qquad (35)$$

$$\left(\frac{\pi}{\mathfrak{p}}\right) = 1 \text{ for } \mathfrak{p} \text{ ramified in } \overline{K}/k; \qquad (36)$$

$$\pi \equiv 1 \ (l^g); \qquad (37)$$

$$(\pi) \neq (\pi^u), \ u \in F, \ u \neq 1. \qquad (38)$$

We will clarify under what conditions there exist primes π satisfying these conditions. As in the first part of the proof we will construct a modulus $\mathfrak{F}$ composed of the product of all $\mathfrak{p}$, $\mathfrak{q}^u$, l^g and the infinite primes. We denote by H the group of l-th power residue classes modulo $\mathfrak{F}$ and by L the class field of H, and we construct an automorphism σ in L such that if an ideal $\mathfrak{a}$ belongs to σ, then $\mathfrak{a} = (\alpha)\mathfrak{b}^l$, where α is totally-positive and

$$\left(\frac{\alpha}{\mathfrak{q}^u}\right) = \left[\frac{\mu_1}{\mathfrak{q}^u}\right]^{-\frac{1}{2}}, \quad \left(\frac{\alpha}{\mathfrak{p}}\right) = 1, \quad \alpha \equiv 1 \ (l^g).$$

Condition (35) that the prime ideal $\mathfrak{r}$ split completely in $\overline{K}$ and satisfies the conditions

$$\left(\frac{\mu_1}{\mathfrak{r}^u}\right) = \zeta(u)$$

or

$$\left(\frac{\mu_1^u}{\mathfrak{r}}\right) = \zeta(u^{-1})^{g(u)},$$

is equivalent to $\mathfrak{r}$ belonging to a definite automorphism τ of the field K' obtained by adjoining to $\overline{K}$ all the $\sqrt[l]{\mu_1^u}$, $u \in F$, $u \neq 1$. Hence we must find out when there exists a prime ideal of the field k, which belongs in L to the automorphism σ, and in K' to the automorphism τ.

For this it is necessary and sufficient that σ and τ should induce the same automorphism in $K' \cap L$. Let us see when this can happen. The field $K' \cap L$ is obtained by adjoining to k all the $\sqrt[l]{\alpha_\chi}$ and the $\sqrt[l]{\alpha_\phi}$ where ϕ is a function over the group F with values in a group of residues modulo l, such that

$$\mu^{\Sigma\varphi(u)u} \approx \alpha_\varphi \in k, \quad \varphi(1) = 0.$$

Hence we must find out when σ and τ induce the same automotphism in the fields $k(\sqrt[l]{\alpha_\chi})$ and $k(\sqrt[l]{\alpha_\phi})$. It is obvious that σ and τ are identity automorphisms in $k(\sqrt[l]{\alpha_\chi})$. The automorphism which σ induced in $k(\sqrt[l]{\alpha_\phi})$ can be found by exactly the same method as was used in the first part of the proof. This automorphism reduces to multiplying $\sqrt[l]{\alpha_\phi}$ by $\left[\frac{\mu_1}{\mathfrak{m}_\varphi}\right]$, where $\mathfrak{m}_\phi = \mathfrak{m}_1^{\Sigma\phi(u)u}$. Analogously we can also find the automorphism which is induced by τ in $k(\sqrt[l]{\alpha_\phi})$. This automorphism is reduced to multiplying $\sqrt[l]{\alpha_\phi}$ by $\prod_u \zeta(u^{-1})^{g(u)\phi(u)}$. Hence the coincidence of the two automorphisms σ and τ in $K' \cap L$ is equivalent to the conditions:

$$\prod_u \zeta(u^{-1})^{g(u)\varphi(u)} = \left[\frac{\mu_1}{\mathfrak{m}_\varphi}\right]^{-\frac{1}{2}} \quad \text{for all } \phi. \tag{39}$$

We must find out when there exist root of unity $\zeta(u)$ satisfying conditions (33) and (39).

We take an arbitrary primitive l-th root of unity, letting

$$\zeta(u) = \zeta^{x_u}, \quad \left[\frac{\mu_1}{\mathfrak{m}_\varphi}\right]^{-\frac{1}{2}} = \zeta^{b_\varphi}.$$

Then conditions (33) and (39) for x_u can be written in the form of equations in a residue field (mod l)

$$\begin{gathered} x_u = x_{u^{-1}}\, g(u^{-1}), \\ \sum_u x_{u^{-1}}\, g(u)\, \varphi(u) = b_\varphi. \end{gathered} \tag{40}$$

We put down the conditions for this system to be compatible and show that it consists in the invariants $[X]_\psi$ being equal to unity.

If the system (40) is not compatible then there exists a linear combination of the equations in which all the coefficients of the variables x_u are equal to 0, while the constant term $\neq 0$.

Let

$$\sum_u c_u (x_u - x_{u^{-1}}\, g(u^{-1})) + \sum_\varphi d_\varphi \sum_u x_{u^{-1}}\, g(u)\, \varphi(u) = \sum_\varphi d_\varphi\, b_\varphi \tag{41}$$

be such a linear combination.

Let

$$\sum_\varphi d_\varphi \varphi(u) = \psi(u).$$

Then obviously

$$\sum_\varphi d_\varphi\, b_\varphi = b_\psi.$$

Setting all the coefficients in (41) equal to zero we get

$$g(u^{-1})\, \psi(u^{-1}) = c_{u^{-1}} - c_u\, g(u^{-1}),$$

from which it follows that

$$\psi(u^{-1}) = -\psi(u)\, g(u), \quad \psi(1) = 0.$$

Hence the function ψ satisfies the conditions in the definition of the symbol $[X]_\psi$. From this we can conclude that

$$\zeta^{b_\psi} = \left[\frac{\mu_1}{\mathfrak{m}_\psi}\right]^{-\frac{1}{2}} = 1, \text{ i.e., } b_\psi = 0,$$

and the compatibility of the system (40) follows.

We can proceed analogously in case $l = 2$. From the proof of the theorem of the paper [5] the number m_2 is found in this case to be the product of three prime factors $\pi_1 \pi_2 \pi_3$, satisfying the conditions

$$\left(\frac{\pi_i}{q^u}\right) = \left[\frac{\mu_1}{q^u}\right]^{-1} \quad (i = 1, 2, 3),$$

where m_2 will satisfy conditions (36) and (37) if we can prove the existence of infinitely many primes π, which split completely in K and satisfy the conditions

$$\left[\frac{\mu_1}{\pi^u}\right] = \zeta(u)$$

together with the conditions (33), (34), (36), (37) and (38).

As in the case $l \neq 2$ it is necessary and sufficient for the existence of such numbers that the following system of congruences

$$\left.\begin{aligned} &x_u = x_{u^{-1}}, \\ &x_{u_1} + \cdots + x_{u_r} = 0, \quad \text{if} \quad \mathfrak{D}_{u_1} \cdots \mathfrak{D}_{u_r} \approx 1, \\ &\sum_u x_{u^{-1}} \varphi(u) = b_\varphi \end{aligned}\right\} \tag{42}$$

has a simultaneous solution. The assumption that this system is incompatible leads again to a linear combination of these equations in which the coefficients of all the variables are equal to zero, while the constant term is not equal to zero. Let this linear combination be

$$\sum_u c_u (x_u - x_{u^{-1}}) + \sum_\varphi d_\varphi \sum_u x_{u^{-1}} \varphi(u) +$$

$$+ \sum_{(u_1, \ldots, u_r)} e_{u_1, \ldots, u_r} (x_{u_1} + \cdots + x_{u_r}) = \sum d_\varphi b_\varphi,$$

where the third sum on the left hand side goes over all totalities of involutory automorphisms $(u_1, \cdots, u_r)$ for which

$$\mathfrak{D}_{u_1} \cdots \mathfrak{D}_{u_r} \approx 1.$$

We suppose again that

$$\sum_\phi d_\phi \phi(u) = \psi(u)$$

and get, as in the case $l \neq 2$

$$\psi(u^{-1}) c_{u^{-1}} - c_u + \sum_{u \in (u_1, \ldots, u_r)} e_{u_1, \ldots, u_r} = 0.$$

From this we see that for $u \neq u^{-1}$

$$\psi(u^{-1}) \neq -\psi(u),$$

and also that in the set of involutory automorphisms

$$\psi(u) = \sum_{(u_1, \ldots, u_r)} \varepsilon_{u_1, \ldots, u_r}(u) \cdot e_{u_1, \ldots, u_r}, \tag{43}$$

where $\epsilon_{u_1, \ldots, u_r}(u)$ is a characteristic function of the set $(u_1, \cdots, u_r)$. Since by the definition of the sets $(u_1, \cdots, u_r)$,

$$\prod_u \mathfrak{D}_u^{\varepsilon_{u_1, \ldots, u_r}(u)} \approx 1,$$

where the product goes over all involutory automorphisms u, it follows from (34) that

$$\prod_u \mathfrak{D}_u^{\psi(u)} \approx 1.$$

Hence the function $\psi(u)$ satisfies all the conditions entering into the definition of the invariant $[X]_\psi$ and hence

$$\zeta^{b_\psi} = \left[\frac{\mu_1}{m_\psi}\right] = [X]_\psi = 1,$$

guarantees the compatability of the system (42). This proves the theorem.

We will apply Theorem 4 to find conditions under which a Scholz field $\overline{K}$ with the Galois group $F \cdot \mathfrak{G}_{d,F}^{(c,r-1)}$ over Ω can be imbedded in a field K with the Galois group $F \cdot \mathfrak{G}_{d,F}^{(c,r)}$ over Ω. We shall denote the subfield of the field $\overline{K}$ belonging to $\mathfrak{G}_{d,F}^{(c,r-1)}$ as before by k.

Theorem 5. *If in the field $\overline{K}$ with the Galois group $F \cdot \mathfrak{G}_{d,F}^{(c,r-1)}$ over Ω all the invariants $[\chi, X]$, $(X)_h$ and $[X]_\psi$ are equal to 1, for all X, corresponding to the extension of $\mathfrak{G}_{d,F}^{(c,r-1)}$ to $\mathfrak{G}_{d,F}^{(c,r)}$, then $\overline{K}$ can be imbedded in the normal Scholz field K over Ω, having the Galois group $F \cdot \mathfrak{G}_{d,F}^{(c,r)}$ over Ω.*

Proof. Applying the corollary to Theorem 3 we see that the operator group $A = Z^{(r)} \times \mathfrak{N}$ can be represented in the form

$$A = A_1 \times \cdots \times A_m,$$

where every group A_i in its turn can be represented as

$$A_i = \prod_{u \in F} Z_i^u,$$

where Z_i is a cyclic group of order l. We introduce in A a chain of groups B_i such that

$$B_i = \mathfrak{N} A_{i+1} \cdots A_m, \quad B_0 = A, \quad B_m = \mathfrak{N}.$$

Obviously

$$B_{i-1}/B_i \cong A_i / A_i \cap B_i. \tag{44}$$

We shall construct a chain of Scholz fields K_i containing $\overline{K}$ and having Galois group $F \cdot \mathfrak{G}_{d,F}^{(c,r)} \times \mathfrak{N}/B_i$ over Ω. Since K_m has the Galois group $\mathfrak{G}_{d,F}^{(c,r)}$ it can be taken for the field K. Let K_{i-1} be already constructed. We denote by X_i the character, which becomes 1 in all the cyclic groups Z_j^u with $j \neq 1$ or $u \neq 1$, and is different from 1 in Z_i. We now apply Theorem 4 to the field K_{i-1} as $\overline{K}$ and show that K_{i-1} can be imbedded in a cyclic field of degree l over K_{i-1} and normal over k, and such that the Galois group of this field is an extension of the Galois group of K_{i-1}/k, corresponding to X_i, and that the composite of all the fields conjugate to this field over Ω be the Scholz field K_i. In order to do this we must check that all the conditions of Theorem 4 are satisfied.

We distinguish two cases:

1) $c = 1$, $r = 1$ and

2) either c or r is different from 1.

We show that in both cases the invariants $[\chi, X_i]$, $(X_i)_h$ and $[X_i]_\psi$ are equal to 1. In case 1) this is obvious, since $X_i = 1$. In case 2) we use the fact that X_i becomes 1 on all A_j, $j < i$, and therefore corresponds to some extension of the Galois group $\mathfrak{G}_{d,F}^{(c,r-1)}$ of the field $\overline{K}$. On the other hand the quotient group with respect to the Frattini subgroup has not changed in going from the Galois group of the field $\overline{K}$ to the Galois group of the field K_{i-1}, since the characters χ are the characters of $\mathfrak{G}_{d,F}^{(c,r-1)}$. Hence equation (15) and its analogues for $(X)_h$ and $[X]_\psi$ show that the invariants $[\chi, X_i]$, $(X_i)_h$ and $[X_i]_\psi$ in the field K_{i-1} coincide with some invariants of the field $\overline{K}$. Since by assumption the invariants of the field K are equal to 1, this also holds for the invariants of the field K_{i-1} under consideration.

We have therefore shown that the field $\underset{\sim}{K}_{i-1}$ can be imbedded in the Scholz field $\widetilde{K}_i$. Relations (44) show that the field $\widetilde{K}_i$ contains a subfield of the field K_i with the Galois group $F \cdot \mathfrak{G}_{d,F}^{(c,r)}/B_i$. As a subfield of a Scholz field, the field K_i is a Scholz field. This proves the theorem.

§3. The construction of fields, satisfying the conditions of the imbedding problem

1. **Operator canonical homomorphisms.** The contents of this section are closely related to § 3 of paper [4]. Therefore we shall leave out some of the proofs, in case they repeat word for word proofs given in that paper, and will confine ourselves to references to the corresponding places of that paper.

Instead of the groups $\mathfrak{G}_d^{(c)}$ of paper [4], we shall consider the operator nilpotent, disposition groups $\mathfrak{G}_{d,F}^{(c)}$ defined above. Since for the most part the letters c, k and F will remain unchanged throughout the discussion, we will denote these groups simply by $\mathfrak{G}_d$.

In all future discussions we will use a fixed system of generators $s_{u,i}$ of the group $\mathfrak{G}_d$. All the generators $s_{u,i}$ for a fixed i and all possible $u \in F$, we shall call a class of generators. The group $\mathfrak{G}_d$ has in this way d classes of generators.

Suppose that we are given the groups $\mathfrak{G}_d$ and $\mathfrak{G}_\delta$ $(\delta \leq d)$ with the generators $s_{u,i}$ $(i = 1, \cdots, d)$ and $t_{u,i}$ $(i = 1, \cdots, \delta)$. The homomorphism of $\mathfrak{G}_d$ on $\mathfrak{G}_\delta$ is called canonical if the generators $s_{u,i}$, $u \in F$, $i = 1, \cdots, \delta - 1$ go under this homomorphism into the generators $t_{u,i}$ for the same u and i, and the remaining generators $s_{u,i}$ go for some i into $t_{u,\delta}$ with the same u and for the other values of i into 1. It follows from this definition that every canonical homomorphism is an operator homomorphism with respect to the group F. We will denote canonical homomorphisms by the letters S, T, etc. The above homomorphism may be called a homomorphism of type $(s_{u,1}, \cdots, s_{u,\delta-1}; t_{u,1}, \cdots, t_{u,\delta-1} | t_{u,\delta})$. We shall denote by $[S]$ the set of generators of the group $\mathfrak{G}_d$ for which under the homomorphism S we have $s_{u,i} \to s_{u,\delta}$. It is obvious that the homomorphism S of the above type is uniquely determined by the set $[S]$. In what follows we shall consider only canonical homomorphisms and therefore we shall call them simply homomorphisms.

Two homomorphisms S and T will be called independent if $[S]$ and $[T]$ do not intersect. Then there exists a unique homomorphism R for which

$$[R] = [S] \cup [T].$$

We shall call R the composite of S and T and denote it by $S * T$.

2. Functions of homomorphisms. We consider the functions $f(S)$ given over the set of all homomorphisms with values in some finite abelian group of order ν. A degree with respect to the class of generators is defined for functions of homophisms of the same type $(s_{u,1}, \cdots, s_{u,\delta-1}; t_{u,1}, \cdots, t_{u,\delta-1} | t_{u,\delta})$. The function which is identically equal to 1 is of degree 0. The function $f(S)$ is of degree $\leq n$ if

$$\varphi(S, T) = f(S)^{-1} f(T)^{-1} f(S * T)$$

is of degree $\leq n-1$ as a function of T for every S.

The function $f(S)$ is of degree $\leq n$ if and only if the following relation

$$\prod_{(i_1, \ldots, i_k)} f(S_{i_1} * \ldots * S_{i_k})^{(-1)^k} = 1, \tag{45}$$

holds for any set of $n+1$ independent in pairs homomorphisms $S_1, \cdots, S_{n+1}$, where $(i_1, \cdots, i_k)$ go over all non-empty subsets of the set $(1, \cdots, n+1)$. The proof of this fact repeats word for word the proof of the analogous assertion of section 2 of § 3 of our paper [4] and will be omitted here.

Theorem 6. *For any pair of natural numbers k and n there exists a number $C(k, n)$ depending on them and having the following property: given any k functions $f_1(S), \cdots, f_k(S)$ of homomorphisms of degree $\leq n$ over the group $\mathfrak{G}_d$ with $d \geq C(k, n)$, a homomorphism S can always be found such that $f_i(S) = 1, i = 1, \cdots, k$.*

Corollary. *If under the assumptions of Theorem 6 the group $\mathfrak{G}_d$ has m classes of generators $s_{u,i_1}, \cdots, s_{u,i_m}$, $i_1, \cdots, i_m \geq \delta - 1$ and if $d \geq C(k, n+m)$, then the group $\mathfrak{G}_d$ can be mapped by means of a canonical homomorphism onto the group $\mathfrak{G}_{\delta+m}$ with generators $s'_{u,1}, \cdots, s'_{u,\delta+m}$, for which all the functions $f_1, \cdots, f_k$ go into unity under the homomorphism $s'_{u,1} \to t_{u,1}, \cdots, s'_{u,\delta} \to t_{u,\delta}, s'_{u,\delta+1} \to 1, \cdots, s'_{u,\delta+m} \to 1$.*

The proofs of Theorem 6 and of the corollary repeat word for word the proofs of Theorem 3 and its corollary in section 3 of § 3 of paper [4]. They will be omitted.

In what follows we shall apply the theory of canonical homomorphisms not to the groups $\mathfrak{G}^{(c)}_{d,F}$ but to the quotient groups $\mathfrak{G}^{(c,r)}_{d,F}$ for some fixed value of r. It is easily seen that every canonical homomorphism of the group $\mathfrak{G}^{(c)}_{d,F}$ on the group $\mathfrak{G}^{(c)}_{\delta,F}$ defines some homomorphism of $\mathfrak{G}^{(c,r)}_{d,F}$ onto $\mathfrak{G}^{(c,r)}_{\delta,F}$ for any $r \leq k$. These homomorphisms will also be called canonical and they will concern us in what follows. We shall denote by $Z^{(r-1)}_{\delta}$ and $Z^{(r-1)}_{d}$ respectively the group $Z^{(r-1)}$ of central elements of order l of the groups $\mathfrak{G}^{(c,r)}_{\delta,F}$ and $\mathfrak{G}^{(c,r)}_{d,F}$.

Suppose that we are given a chain of fields $\Omega \subset k \subset \overline{K}$, where k/Ω is normal and has the Galois group F, $\overline{K}/\Omega$ is also normal and $\mathfrak{G}^{(c,r)}_{d,F}$ is the Galois group of $\overline{K}/k$, while that of $\overline{K}/\Omega$ is the semi-direct product of $\mathfrak{G}^{(c,r)}_{d,F}$ by F. We study the invariants $[\chi, X]$ and $[X]_{\psi}$ for the classes of l-invariant numbers which correspond to simple central extensions of $\mathfrak{G}^{(c,r)}_{d,F}$ to various quotient groups $\mathfrak{G}^{(c,r+1)}_{d,F}$. All such extensions are in one to one correspondence with the characters of order l of the group $Z^{(r)}_{d}$. The character which corresponds in this way to the class X will also be denoted by X.

We consider the basis $\chi_{u,i}$ $(i = 1, \cdots, d)$ of the character group of order l, reciprocal to the system of generators $s_{u,i}$. The numbers $\alpha_{\chi_{u_i}} \in k$ will be denote for brevity by $\alpha_{u,i}$. If we transfer the operators of F from the group $\mathfrak{G}^{(c,r)}_{d,F}$ to its character group by means of the rule

$$\chi^u(\sigma^u) = \chi(\sigma),$$

we will get

$$\chi_{u,i} = \chi^u_{1,i}, \quad \alpha_{u,i} = \alpha^u_{1,i}.$$

Therefore we can denote $\alpha_{1,i}$ by α_i and we will suppose that

$$\alpha_{u,i} = \alpha^u_i.$$

Beginning at this point we can always suppose that α_i and α^u_j are relatively prime for all i, j and u. This condition will be satisfied if for instance α_i are divisible only by prime ideals of degree 1 over Ω, where if $\mathfrak{p} \mid \alpha_i$, then $\mathfrak{p}^u \nmid \alpha_j$ for all i, j and u.

It can be easily checked, as was done in paper [4], p. 290, that the generating automorphism of the group of ideals $\mathfrak{p}$ dividing α^u_i is an automorphism σ satisfying the conditions:

$$\chi_{u,i}(\sigma) \neq 1, \quad \chi_{v,j}(\sigma) = 1 \quad \text{for } v \neq u \quad \text{or } i \neq j.$$

Such an automorphism is obviously of order l^k in the quotient group of the group $\mathfrak{G}^{(c,r)}_{d,F}$ with respect to its commutator. Hence if $c \neq 1$, then the generating automorphism of the inertia group of the divisors of α^u_i does not increase its order in going from the group $\mathfrak{G}^{(c,r)}_{d,F}$ to the extension of this group, corresponding to the class X. From this it follows that the ideal $\mathfrak{D}(X)$ is l-relatively prime to α^u_i, and therefore that all the invariants $[\chi, X]$ are defined.

If $c = 1$, then the classes X are in one to one correspondence with the characters χ. We shall denote by $\widetilde{\chi}$ the class corresponding to the character X. The invariant $[\chi_{u,i}, \widetilde{\chi}_{v,j}]$ is always defined with the exception $u = v$, $i = j$. The proof of these assertions repeats verbatim the proofs of the corresponding assertions in section 4 of § 3 of paper [4].

We now consider a Scholz field $\overline{K} \supset k \supset \Omega$ with the Galois group $\mathfrak{G}^{(c,r)}_{d,F}$ over k and $F \cdot \mathfrak{G}^{(c,r)}_{d,F}$ over Ω. We consider the canonical operator homomorphism S of $\mathfrak{G}^{(c,r)}_{d,F}$ on $\mathfrak{G}^{(c,r)}_{\delta,F}$ for some $\delta \leq d$. Its kernel $\mathfrak{N}$ is a normal subgroup of the group $\mathfrak{G}^{(c,r)}_{d,F}$ so that $\mathfrak{N}$ contains a normal subfield of the field $\overline{K}$ which we will denote by $\overline{K}^S$. The field $\overline{K}^S$ has the Galois group $\mathfrak{G}^{(c,r)}_{\delta,F}$ over k, and the group $F \cdot \mathfrak{G}^{(c,r)}_{\delta,F}$ over Ω. Moreover the numbers α^u_i which correspond to the field $\overline{K}^S$ are l-relatively prime to each other, as has been shown in section 5 of § 3 of paper [4]. Each of the invariants $[\chi, X]$, $(X)_h$ and $[X]_\psi$ defined for the field $\overline{K}^S$ is for a fixed $\overline{K}$ a function of the homomorphism S.

We can identify X with a character of the group $Z^{(r)}$. On the other hand the group $Z^{(r)}$ is operator-isomorphic with an additively written module of Lie polynomials of degree c in the variables $x_{u,i}$ $(i = 1, \cdots, \delta)$. All the variables $x_{u,i}$ for a fixed i and any $u \in F$ will be called a class of variables.

Theorem 7. *If the character X of degree $\leq n$ with respect to the totality of the variables of the class $x_{u,\delta}$ does not annihilate a single polynomial depending on at least one of the variables $x_{u,\delta}$, but annihilates all the polynomials which depend only on $x_{u,i}$ $(i = 1, \cdots, \delta - 1)$ then the invariants $[\chi_{u,i}, X]$ and $(X)_h$ as functions of the homomorphisms of the type $(s_{u,1}, \cdots, s_{u,\delta-1}; t_{u,1}, \cdots, t_{u,\delta-1} \mid t_{u,\delta})$ of the Galois group $\mathfrak{G}^{(c,r)}_{d,F}$ of the Scholz field $\overline{K}$ have the following degrees: $(X)_h$ is of degree $\leq n$, $[\chi_{u,i}, X]$ for $i \neq \delta$ is of degree $\leq n$, $[\chi_{u,\delta}, X]$ is of degree $\leq n + 1$. Hence all these invariants are of degree $\leq c + 1$.*

The proof of this theorem repeats verbatim the proof of Theorem 4 of section 5 of § 3 of our paper [4], where the lemma which precedes this theorem must also be proved. We omit the proof.

It remains to study the behaviour of the invariant $[X]_\psi$ as a function of homomorphisms.

We confine ourselves to the case when all the invariants $[\chi, X]$ and $(X)_h$ are equal to 1 in the field $\overline{K}/k$ with the Galois group $\mathfrak{G}^{(c,r)}_{d,F}$.

Theorem 8. *If in the field $\overline{K}$ all the invariants $[\chi, X]$ and $(X)_h$ are equal to 1, and if the character X of the module $Z^{(r)}_d$ of degree $\leq n$ with respect to the totality of the variables of the class $x_{u,\delta}$ does not annihilate a single polynomial in at least one of the $x_{u,\delta}$, but annihilates all the polynomials which depend only on $x_{u,i}$ $(i = 1, \cdots, \delta - 1)$ then the invariant $[X]_\psi$ as a function of the automorphisms of the type $(s_{u,1}, \cdots, s_{u,\delta-1}; t_{u,1}, \cdots, t_{u,\delta-1} \mid t_{u,\delta})$ of the Galois group of the field $\overline{K}$ is of degree $\leq 2n$.*

Proof. We note first of all that in the case under consideration the question for what values of ψ and X the invariant $[X]_\psi$ is defined is solved if ψ is known as a function of F, X as a factor set with respect to the group $\mathfrak{G}^{(c,r)}_{d,F}$, and the field k independently of the properties of the field $\overline{K}$. In fact conditions 1 and 3 in the definition of the symbol $[X]_\psi$ have to do only with the properties of the function ψ and of the field k, while condition 2 is equivalent to

$$X^{\sum\psi(u)u} = 1. \tag{46}$$

Hence X^u denotes the image of X under the automorphism which u induces in the character group of the group $Z_d^{(r)}$. It is obvious that condition (46) does not depend on the field $\overline{K}$.

Let X_1, X_2 and X_3 be three characters for which the invariant $[X]_\psi$ is defined. We prove that they satisfy the relation

$$[X_1X_2X_3]_\psi [X_1X_2]_\psi^{-1} [X_1X_3]_\psi^{-1} [X_2X_3]_\psi^{-1} [X_1]_\psi [X_2]_\psi [X_3]_\psi = 1, \tag{47}$$

analogous to a special case of (45) with $n = 2$.

Relation (47) is proved by verification. To do this we select from the classes of l-invariant numbers corresponding to X_1, X_2 and X_3 representatives μ_1, μ_2 and μ_3 such that the corresponding ideals $\mathfrak{m}_1$, $\mathfrak{m}_2$ and $\mathfrak{m}_3$ satisfy the relations: $(\mathfrak{m}_i, \mathfrak{m}_j^u) = 1$, $i, j = 1, 2, 3$, except in the case $i = j$, $u = 1$. Let $\mathfrak{m}_\psi^{(i)} = \mathfrak{m}_i^{\sum\psi(u)u}$, $i = 1, 2, 3$. Then (47) can be written

$$\left[\frac{\mu_1\mu_2\mu_3}{\mathfrak{m}_\psi^{(1)}\mathfrak{m}_\psi^{(2)}\mathfrak{m}_\psi^{(3)}}\right]\left[\frac{\mu_1\mu_2}{\mathfrak{m}_\psi^{(1)}\mathfrak{m}_\psi^{(2)}}\right]^{-1}\left[\frac{\mu_1\mu_3}{\mathfrak{m}_\psi^{(1)}\mathfrak{m}_\psi^{(3)}}\right]^{-1}\left[\frac{\mu_2\mu_3}{\mathfrak{m}_\psi^{(2)}\mathfrak{m}_\psi^{(3)}}\right]^{-1}\left[\frac{\mu_1}{\mathfrak{m}_\psi^{(1)}}\right]\left[\frac{\mu_2}{\mathfrak{m}_\psi^{(2)}}\right]\left[\frac{\mu_3}{\mathfrak{m}_\psi^{(3)}}\right] = 1. \tag{48}$$

In its turn equation (48) can be verified independently if we make use of the multiplicative property of the symbol $[\frac{\mu}{\mathfrak{a}}]$.

We now consider all the characters X for which the invariant $[X]_\psi$ is defined for a fixed function ψ. As is seen from (46) these characters form a group. Let $X_1, \cdots, X_m$ be the basis of this group and let

$$X = X_1^{c_1} \cdots X_m^{c_m} \tag{49}$$

be the representation of any character in terms of the basis. For a fixed field $\overline{K}$ the invariant $[X]_\psi$ can be written in the form $\zeta^{f(c_1, \cdots, c_m)}$, where ζ is some l-th root of unity, $f(c_1, \cdots, c_m)$ is a function of m variables, where both the arguments and the functions go over values of the field of residue classes modulo l.

We shall prove that the following representation holds

$$\begin{aligned} f(c_1, \ldots, c_m) &= \sum_{i,k=1}^{m} a_{i,k}c_ic_k + \sum_{i=1}^{m} b_ic_i \quad (a_{i,k} = a_{k,i}) \quad \text{for } l \neq 2, \\ f(c_1, \ldots, c_m) &= \sum_{1\leqslant i<k\leqslant m} a_{i,k}c_ic_k + \sum_{i=1}^{m} b_ic_i \quad \text{for } l = 2. \end{aligned} \tag{50}$$

In order to do this we denote by $\overline{c}$ the vector with coordinates $(c_1, \cdots, c_m)$ in the field of residue classes modulo l. Then (47) can be written

$$f(c+d+e)-f(c+d)-f(c+e)-f(d+e)+f(c)+f(d)+f(e)=0.$$

From this it follows that the function ϕ, symmetric in c and d,

$$\phi(c,d)=f(c+d)-f(c)-f(d) \tag{51}$$

is linear in the argument c as well as in d.

For $l\neq 2$, we can write $\phi(c,d)$ in the form

$$\varphi(c,d)=\sum_{i,k=1}^{m} a_{i,k}c_i d_k \quad (a_{i,k}=a_{k,i}).$$

If we let

$$f_1(c_1,\ldots,c_m)=\sum_{i,k=1}^{m}\frac{1}{2}a_{i,k}c_i c_k, \tag{52}$$

then f_1 can be easily checked to satisfy the same equation (51) as f. From this it follows that the function $f(c)-f_1(c)$ is linear in c, that is

$$f=f_1+\sum_{i=1}^{m} b_i c_i,$$

which together with (52) gives (50).

For $l=2$, we can write ϕ in the form

$$\varphi=\sum_{1\leqslant i<k\leqslant m} a_{i,k}(c_i d_k+d_i c_k)+\sum_{i=1}^{m}\alpha_i c_i d_i.$$

Let

$$f_1(c_1,\ldots,c_m)=\sum_{1\leqslant i<k\leqslant m} a_{i,k}c_i c_k.$$

Then it can be easily verified that

$$f_1(c+d)-f_1(c)-f_1(d)=\sum_{1\leqslant i<k\leqslant m} a_{i,k}(c_i d_k+d_i c_k).$$

Hence if we let $f_2=f-f_1$ we will get

$$f_2(c+d)-f_2(c)-f_2(d)=\sum_{i=1}^{m}\alpha_i c_i d_i. \tag{53}$$

Let $c=d$ in (53). In the left hand side we get

$$f_2(c+c)-f_2(c)-f_2(c)=0,$$

and therefore

$$\sum_{i=1}^{m}\alpha_i c_i^2=\sum_{i=1}^{m}\alpha_i c_i=0.$$

Since this holds for all c_i it follows that $\alpha_1=\cdots=\alpha_m=0$. Therefore the function $f(c)-f_1(c)$ is linear in c, from which (50) follows as before.

We now consider any character X of the group $Z_\delta^{(r)}$ for which the symbol $[X]_\psi$ is defined. For any canonical homomorphism S to X corresponds a character X^S of the group $Z_d^{(r)}$ for which the invariant $[X^S]_\psi$ is also defined. If we write X^S in the form (49) then c_i will be a function $c_i(S)$ of the homomorphism S. From (50) we have

$$f(S) = \sum a_{i,k} c_i(S) c_k(S) + \sum b_i c_i(S).$$

On the other hand formula (37) of paper [4] shows that every function $c_i(S)$ is of degree $\leq n$ with respect to S. The statement of the theorem will be proved if we can show that the product of two functions of degree $\leq n$ is of degree $\leq 2n$.

We will show that in general if $f(S)$ and $\phi(S)$ are of degree $\leq n$ and $\leq m$ respectively, then $f(S)\phi(S)$ is of degree $\leq n+m$. We make use of induction on m and n. We must show that

$$f(S*T)\phi(S*T) - f(S)\phi(S) - f(T)\phi(T) \tag{54}$$

is of degree $\leq m+n-1$ with respect to both S and T.

To simplify the notation we let $f(S*T) - f(S) - f(T)$ be Δf. By comparing corresponding terms we can check the identity

$$\Delta(f\varphi) = \Delta f \Delta\varphi + f(S)\Delta\varphi + f(T)\Delta\varphi + \varphi(S)\Delta f + \varphi(T)\Delta f + \\ + f(S)\varphi(T) + \varphi(S) f(T). \tag{55}$$

If m or n is equal to 0, then obviously $f\phi$ is of degree $m+n$, so that in this case our assertion is proved. We now consider the case in which neither m nor n is zero and find the degree of $\Delta(f\phi)$ with respect to S (in view of symmetry of S and T the degree with respect to T will be the same). The degree of $\Delta f \leq m-1$ and of $\Delta\phi$ is $\leq n-1$. We assume as a hyposethis of the induction that the degree of $\Delta f \Delta\phi$ is $\leq m+n-2$, the degree of $f(S)\Delta\phi$ is $\leq m+n-1$ and the degree of $\phi(T)\Delta f$ is $\leq m+n-1$. Since we are interested in the degree with respect to S, $f(T)$ and $\phi(T)$ are constants. Obviously the degree is unchanged by multiplying the function by a constant. Hence the degree of $f(T)\Delta\phi$ is $\leq n-1$, the degree of $\phi(T)\Delta f$ is $\leq m-1$, the degree of $f(T)\phi(S)$ is $\leq n$ and the degree of $f(S)\phi(T)$ is $\leq m$. Since $m \geq 1$ and $n \geq 1$, $m \leq m+n-1$ and $n \leq m+n-1$. We have shown therefore that every summand in (55) is of degree $\leq m+n-1$, and therefore the degree of $\Delta(f\phi)$ is $\leq m+n-1$. The theorem is proved.

We can now proceed with the proof of the statement that every field of algebraic numbers k/Ω with the Galois group F can be imbedded for $l > c$ in a field K, normal over Ω and having over k the Galois group $\mathfrak{G}^{(c)}_{d,F}$ (as an operator group).

The proof is very closely related to the contents of section 6 of § 3 of the paper [4].

Theorem 9. *For every natural number δ there exists a natural number $C(\delta)$ with the following properties: in every Scholz field $\overline{K}$ having the Galois group $\mathfrak{G}^{(c,r)}_{d,F}$ over k and the group $F \cdot \mathfrak{G}^{(c,r)}_{d,F}$ over Ω with $d > C(\delta)$ and with l-relatively prime in pairs numbers α^u_i, there exists a subfield $\overline{K}^S$, whose Galois group over k is $\mathfrak{G}^{(c,r)}_{\delta,F}$, and over Ω is $F \cdot \mathfrak{G}^{(c,r)}_{d,F}$ and all the invariants $[\chi, X]$, $(X)_h$ and $[X]_\psi$*

are equal to 1. Here by $\overline{K}^S$ *we denote a subfield corresponding to some operator canonical homomorphism S.*

Proof. First we must prove the existence of the number $C_1(\delta)$, which possesses the same properties as the number $C(\delta)$ in the formulation of Theorem 9, with the exception that in the field $\overline{K}^S$ the invariant $[X]_\psi$ need not necessarily be equal to 1. The proof repeats word for word the proof of Theorem 5 of paper [4]. It is omitted. Next we assume that in the field $\overline{K}$ the invariants $[\chi, X]$ and $[X]_\psi$ are already equal to 1. Then this will be true for any subfield, corresponding to some operator canonical homomorphism of its Galois group. We show that for that field there exists a number $C_2(\delta)$, which possesses all the properties of the number $C(\delta)$ formulated in Theorem 9. The proof again repeats the arguments of Theorem 5 of paper [4]. Obviously if we take for $C(\delta)$ the number $C_1(C_2(\delta))$, then $C(\delta)$ will satisfy all the requirements of Theorem 9.

Theorem 10. *Let* k/Ω *be an arbitrary normal field of algebraic numbers with the Galois group* F*, where* F *is the operator group of* $\mathfrak{G}$ *of order* l^a *and class* c*, with* $l > c$*. Then there exists a field* K *containing* k*, normal over* Ω*, and having over* k *the Galois group* $\mathfrak{G}$ *and over* Ω *the group* $F \cdot \mathfrak{G}$*.*

Proof. We consider separately two cases.

1. k **contains an** l**-th root of unity.** In this case we can apply all the preceding considerations. We will also show that the field $\overline{K}$ can be constructed to be a Scholz field and such that the numbers α_i^u which correspond to it be l-relatively prime in pairs.

Since every group c is an operator homomorphic image of some group $\mathfrak{G}^{(c)}_{d,F}$ it is sufficient to discuss the case when $\mathfrak{G} = \mathfrak{G}^{(c)}_{d,F}$. We will consider a more general case, $\mathfrak{G} = \mathfrak{G}^{(c,r)}_{d,F}$. The proof will be by induction on c, and for a fixed c by induction on r. Since $\mathfrak{G}^{(c,0)}_{d,F} = \mathfrak{G}^{(c-1)}_{d,F}$ and $\mathfrak{G}^{(c,k)}_{d,F} = \mathfrak{G}^{(c+1)}_{d,F}$ it is sufficient to consider only the induction on r for a fixed c.

Suppose that we have to construct a field K with the Galois group $F \cdot \mathfrak{G}^{(c,r)}_{\delta,F}$ over Ω. We shall suppose that the field $\overline{K}$ with the group $\mathfrak{G}^{(c,r-1)}_{d,F}$ has already be constructed for any d. We take d for the number $C(\delta)$ whose existence is guaranteed by Theorem 9. Let $\overline{K}$ be the field with the Galois group $F \cdot \mathfrak{G}^{(c,r-1)}_{d,F}$ over Ω, whose existence follows from the hypotheses of the induction. By Theorem 9 it contains a subfield $\overline{K}^S$ with the Galois group $F \cdot \mathfrak{G}^{(c,r-1)}_{\delta,F}$ over Ω and with all the invariants $[\chi, X]$, $(X)_h$ and $[X]_\psi$ equal to 1. As a subfield of a Scholz field $\overline{K}$, $\overline{K}^S$ is a Scholz field and all the numbers α_i^u which correspond to it are l-relatively prime in pairs.

From Theorem 5 of the present paper it follows that the field $\overline{K}^S$ can be imbedded in the Scholz field K with the Galois group $F \cdot \mathfrak{G}^{(c,r)}_{\delta,F}$ over Ω. By repeating the argument used in the proof of Theorem 6 of paper [4] it can be shown that the numbers α_i^u of the field K are l-relatively prime in pairs.

2. k does not contain an l-th root of unity. Suppose that by adjoining to k such a root of unity we obtain a field $\bar{k}$ with the Galois group $\bar{F}$ over Ω. The group $\bar{F}$ has a normal subgroup H, which is the Galois group of $\bar{k}/k$. We consider $\mathfrak{G}$ as an operator group with a group of operators $\bar{F}$, assuming that to the elements of H corresponds the identity operator, and that to every $\bar{u} \in \bar{F}$ corresponds the same operator as to $u \in F$ provided $u = \bar{u}H$. Since $\bar{k}$ contains an l-th root of unity we can apply to it the result established in case 1 and construct a field $\bar{K}$ with the Galois group $\bar{F} \cdot \mathfrak{G}$ over Ω. Then H is a normal subgroup of the group $\bar{F} \cdot \mathfrak{G}$. To it belongs the normal subfield $K \supset k$, which obviously has the Galois group $\bar{F} \cdot \mathfrak{G}/H \cong F \cdot \mathfrak{G}$, over Ω. The theorem is proved.

We note in conclusion that the condition $l > c$, which was imposed by us throughout the paper was used only once in the proof of Theorem 2. This condition was essential to the proof that the Lie representation is the direct sum of a tensor representation. As is known (see [1], p. 211) if $(m, l) = 1$ then any representation of the group F of order m in the field of characteristic l is completely reducible. From this it follows that the Lie representation, which is a factor of the tensor representation, is contained in it as a direct summand. Hence all the preceding arguments remain in force if we replace the condition $l > c$ by the condition $(m, l) = 1$. We get the following result.

Theorem 11. *Let k/Ω be a normal algebraic number field with the Galois group F of order m and let G be any group which has a normal subgroup $\mathfrak{G}$ of order l^a, such that $G/\mathfrak{G} \cong F$, $(m, l) = 1$. Then there exists a field $K \supset k$, normal over Ω and having the Galois group G over it.*

In the formulation of the theorem the condition that G be a semi-direct product of F and $\mathfrak{G}$ was not included, since, because of the condition $(m, l) = 1$, this follows from a theorem of Schur (see [3], p. 386).

BIBLIOGRAPHY

[1] van der Waerden, *Moderne Algebra*, Bd. 2, 2. Aufl., Springer, Berlin, 1940.

[2] H. Weyl, *Classical groups, their invariants and representations*, Princeton, 1939.

[3] A. G. Kurosh, *The theory of groups*, 2nd ed., Gostehizdat, Moscow, 1953; English translation published in 2 vols. by Chelsea, New York, 1955, 1956.

[4] I. R. Šafarevič, *On the construction of fields with a given Galois group of order l^a*, Izv. Akad. Nauk SSSR. Ser. Mat. 18 (1954), 261-296. (Russian) = Amer. Math. Soc. Transl. (2) 4 (1956), 107-142.

[5] I. R. Šafarevič, *On an existence theorem in the theory of algebraic numbers*, Izv. Akad. Nauk SSSR. Ser. Mat. **18** (1954), 327-334. (Russian) = Amer. Math. Soc. Transl. (2) 4 (1956), 143-150.

[6] R. Brauer, *On algebras which are connected with the semisimple continuous groups*, Ann. of Math. (2) 38 (1937), 857-872.

[7] A. Scholz, *Über die Bildung algebraischer Zahlkörper mit auflösbarer Galoisscher Gruppe*, Math. Z. **30** (1929), 332-356.

[8] R. M. Thrall, *A note on a theorem by Witt*, Bull. Amer. Math. Soc. 47 (1941), 303-308.

Translated by:
Emma Lehmer

Construction of fields of algebraic numbers with given solvable Galois group

Izv. Akad. Nauk SSSR, Ser. Mat. **18**, 525–578 (1954). Zbl. **57**, 274
[Transl., II. Ser., Am. Math. Soc. **4**, 185–237 (1956)]

It is proved in this paper that over any field of albegraic numbers there exist infinitely many extensions with a given in advance solvable Galois group.

Introduction

In this paper is developed a method of constructing extensions of a given field of algebraic numbers, having a given in advance solvable Galois group. In particular a proof is given of the existence of such extensions.

The fundamental tools of which we shall make use are the invariants of the fields, which we have introduced in paper [6], whose Galois group is of order l^a, and the theory of composition of homomorphisms of groups of order l^a and of functions of homomorphisms, which were developed in the same paper. These considerations relating to the fields with Galois groups of order l^a are applicable here because the problem of constructing fields with a given solvable Galois group is very closely connected with some questions in the theory of Galois groups of order l^a. This connection is given by a group-theoretical theorem proved independently by D. K. Faddeev (see [4] and [5]) and Gaschütz [1]. If the Faddeev-Gaschütz theorem is stated not in terms of abstract groups, but about Galois groups it can be formulated as follows (see Kochendörffer [2]). Let $\mathfrak{G}$ be a given group having an abelian normal subgroup A of order l^a and let $\Omega \subset L \subset k \subset K \subset K^*$ be a sequence of fields having the following properties: k/Ω is normal and has the Galois group $G \cong \mathfrak{G}/A$, L belongs in k to a Sylow l-subgroup H of the group G, K/L is normal and its Galois group is isomorphic with the Sylow subgroup $\mathfrak{H}$ of the group $\mathfrak{G}$, where the fields conjugate with K/Ω are independent over k and their composite is K^*. Then the field K^*/k has a subfield with its Galois group isomorphic with $\mathfrak{G}$. Because of this theorem, for an already constructed field k/Ω with the Galois group G, the construction of a field with the Galois group $\mathfrak{G}$ over Ω is reduced to the construction of the field K/L containing k with the given Galois group $\mathfrak{H}$ of order l^a, such that the fields conjugate with K/Ω are independent over k. In order to construct a field with a given solvable Galois group this method has to be applied several times in such a way that at each step the construction of the required field with the given Galois group of order l^a be actually accomplished. The solution of this problem is given in the present paper.

We suppose that all the fields k/Ω under consideration are Scholz fields with respect to a certain modulus M (defined in § 1). This guarantees that in the

field k/L (L belongs to a Sylow l-subgroup of the Galois group of k/Ω) any factor set consisting of l-th roots of unity, is decomposible and therefore the field K with the above properties can be constructed by means of a lower central series in the group $\mathfrak{H}$ through a normal subgroup A.

The main difficulty lies in the fact that we require that the newly constructed field K^* be a Scholz field. We introduce some invariants of the field k/Ω, whose values being unity is a necessary and sufficient condition that the field K^* can be constructed as a Scholz field. Finally, making all the invariants of k/Ω equal to unity can be achieved as follows. We take for G a certain "reduced free" group, such that every solvable group is a quotient group of a certain "reduced free" group. Instead of constructing the field k with the Galois group G we construct a field $\widetilde{k}$ with the group $\widetilde{G}$, which is a "reduced free" group of the same type as G, but with a greater number of generators. By making use of the theory of composition of homomorphisms developed in paper [6] we can show that $\widetilde{k}$ contains the subfield k with the Galois group G, all of whose invariants are equal to 1.

We will assume that the reader is familiar with the papers [6], [7] and [8]. A short account of the results here obtained has been published earlier (see [9]). In comparison with the exposition in [9] we have achieved some simplifications in the present paper. The most significant of these is as follows. As is known (see Ore [3]) every solvable group contains a semi-complete nilpotent normal subgroup. Therefore every solvable group is a quotient group of some semidirect product of groups of order l^a. The above construction is much simpler for these groups than for arbitrary solvable groups. The idea that Ore's result, mentioned above, could be applied to the construction of fields with solvable Galois groups was suggested to me by D. K. Faddeev with whose kind permission I make use of it here.

In this paper we shall refer often to results of D. K. Faddeev. However, I am indebted to him considerably more than could be inferred from the references to his work. I should like to express my deep gratitude for the many discussions and advice given to me by Dmitry Konstantinovič.

§1. Invariants of relative abelian fields

1. Formulation of the problem. We will give here a definition of a class of fields for which it is known that the imbedding problem has a positive solution in a number of the most important cases.

Let Ω be a field of algebraic numbers, k/Ω be a normal extension and M a natural number, divisible by all the prime divisors of the degree of k/Ω. The extension k/Ω will be called a Scholz extension with respect to M if the following conditions are satisfied:

1° All the prime ideals of k, which are ramified in k/Ω have relative degree 1 over Ω.

2° For every ramified prime ideal $\mathfrak{P}$ of the field k/Ω

$$\mathfrak{N}(\mathfrak{P}) \equiv 1 (M),$$

where $\mathfrak{N}(\mathfrak{P})$ denotes the absolute norm of $\mathfrak{P}$.

3° All the prime divisors of M in Ω split completely in k.

4° All the real infinite primes in Ω split in k only into real primes.

If λ is any prime divisor of M, we shall denote the highest power of λ dividing M by M_λ.

If $l \mid M$, $k \ni \zeta$, where ζ is an l-th root of unity, and if $(k:\Omega)$ and M are powers of l, then the notion of a Scholz field coincides with the one introduced for fields of degree l^α in paper [6].

If a number M and its prime divisor l are given, then the normal extension k/Ω containing an l-th root of unity ζ, will be called a Scholz extension with respect to M and l if $k/\Omega(\zeta)$ is Scholz with respect to M and if for every prime divisor $\mathfrak{p}$ not dividing l of the discriminant of the field k/Ω

$$\mathfrak{N}(\mathfrak{p}) \equiv 1 (l). \tag{1}$$

It is obvious that if $\Omega \ni \zeta$ then the notion of a Scholz field with respect to M, and with respect to M and l coincide. In general these two notions are connected by the following two statements.

Lemma 1. *If k/Ω is a Scholz field with respect to M and if $l \mid M$, then $k(\zeta)/\Omega$ is a Scholz field with respect to M and l.*

First we verify all the requirements for a Scholz field with respect to M for the field $k(\zeta)/\Omega(\zeta)$.

1) Let $\overline{\mathfrak{P}}$ be a ramified prime ideal of $k(\zeta)/\Omega(\zeta)$, let $\mathfrak{P}$ be its divisor in k, $\overline{\mathfrak{p}}$ in $\Omega(\zeta)$ and $\mathfrak{p}$ in Ω. Obviously $\mathfrak{P}$ is either ramified in k/Ω or else it divides l, since $k(\zeta) = k \cdot \Omega(\zeta)$. The second case cannot take place. In fact if $\overline{\mathfrak{P}} \mid l$, then $\mathfrak{p}$ splits completely in k and the resulting $\mathfrak{p}$-adic closure of the algebra $k_\mathfrak{p}$ is a direct sum of the fields $\Omega_\mathfrak{p}$. The algebra $\Omega(\zeta)_\mathfrak{p}$ is the direct sum of the fields $\Omega(\zeta)_{\overline{\mathfrak{p}}}$ and therefore the algebra $k(\zeta)_\mathfrak{p} = k_\mathfrak{p}\Omega(\zeta)_\mathfrak{p}$ is also the direct sum of the fields $\Omega(\zeta)_{\overline{\mathfrak{p}}}$. It follows therefore that the order and ramification index of $\overline{\mathfrak{P}}$ in $k(\zeta)/\Omega$ are the same as for $\overline{\mathfrak{p}}$ in $\Omega(\zeta)$. For the multiplicative properties of these numbers in successive extensions and because they are equal to 1 in k/Ω, it follows that they are equal to 1 in $k(\zeta)/\Omega(\zeta)$. Hence $(\overline{\mathfrak{P}}, l) = 1$ and the prime divisors of l in $\Omega(\zeta)$ split completely in $k(\zeta)$.

We see that $\mathfrak{P}$ is ramified in k/Ω. Since it is there of degree 1, any number in k is congruent to a number in Ω mod $\mathfrak{P}$. From this it follows that any number in $k(\zeta)$ is congruent to a number in $\Omega(\zeta)$ mod $\overline{\mathfrak{P}}$, and this shows that $\overline{\mathfrak{P}}$ is of degree 1.

2) From the preceding discussion it follows that if $\overline{\mathfrak{P}}$ is ramified in

$k(\zeta)/\Omega(\zeta)$, then its multiple $\mathfrak{P}$ is ramified in k/Ω. Since

$$\mathfrak{N}(\overline{\mathfrak{P}}) = \mathfrak{N}(\mathfrak{P})^f, \quad \mathfrak{N}(\mathfrak{P}) \equiv 1\,(M),$$

then

$$\mathfrak{N}(\overline{\mathfrak{P}}) \equiv 1\,(M).$$

3) Property 3° is already proved for the prime divisors of l. If $\lambda \mid M$, $\lambda \neq l$, then the proof for the prime divisors λ repeats verbatim the proof of the property 3°.

Finally the property 4° is trivial for $k(\zeta)/\Omega(\zeta)$ since for $l = 2$, $\zeta \in \Omega$, and for $l > 2$, $\Omega(\zeta)$ is purely imaginary.

To check congruence (1) we note that the prime divisors of the discriminant of $k(\zeta)/\Omega$ are relatively prime to l, and are, as we have shown, prime divisors of the discriminant of k/Ω for which equation (1) follows from conditions 1° and 2° and from the definition of a Scholz field, since $l \mid M$.

The inverse of Lemma 1 is in general not true, however, the following slightly weaker statement holds.

Lemma 2. *If k/Ω is Scholz with respect to M, k' is a normal extension of k and $(k':k)$ is a power of l, where l is a prime divisor of M, and if $k'(\zeta)/\Omega$ is Scholz with respect to M and l, then k'/Ω is Scholz with respect to M.*

We check all the requirements for k'/Ω to be a Scholz field with respect to M.

1) We denote $k'(\zeta)$ by $\overline{k'}$, $k(\zeta)$ by $\overline{k}$, $\Omega(\zeta)$ by $\overline{\Omega}$. Let $\overline{\mathfrak{P}}'$ be a prime ideal in $\overline{k}'$. We denote the prime ideals which are divisible by it in k' by $\mathfrak{P}'$, in $\overline{k}$ by $\overline{\mathfrak{P}}$, in k by $\mathfrak{P}$, in $\overline{\Omega}$ by $\overline{\mathfrak{p}}$ and in Ω by $\mathfrak{p}$. We denote the field of residue classes of the ideal $\mathfrak{p}$ in Ω by $\Omega[\mathfrak{p}]$.

Let $\mathfrak{P}'$ be a ramified ideal in k'/Ω. By condition (1)

$$\overline{k}[\overline{\mathfrak{P}}] = k[\mathfrak{P}].$$

On the other hand

$$\overline{k}'[\overline{\mathfrak{P}}'] = \overline{k}[\overline{\mathfrak{P}}]\,k'[\mathfrak{P}'] = k'[\mathfrak{P}'],$$

since

$$\overline{k}[\overline{\mathfrak{P}}] = k[\mathfrak{P}] \subset k'[\mathfrak{P}'].$$

By our condition

$$\overline{k}'[\overline{\mathfrak{P}}'] = \overline{k}[\overline{\mathfrak{P}}].$$

From this it follows that

$$k'[\mathfrak{P}'] = k[\mathfrak{P}].$$

Since by our condition

$$k[\mathfrak{P}] = \Omega[\mathfrak{p}],$$

then

$$k'[\mathfrak{P}'] = \Omega[\mathfrak{p}],$$

which shows that $\mathfrak{P}'$ is of degree 1 in k'/Ω.

2) Since we have shown that

$$\bar{k}'[\bar{\mathfrak{P}}'] = k'[\mathfrak{P}'] = k[\mathfrak{P}],$$

it follows that

$$\mathfrak{N}(\bar{\mathfrak{P}}') = \mathfrak{N}(\mathfrak{P}').$$

Hence

$$\mathfrak{N}(\mathfrak{P}') = \mathfrak{N}(\bar{\mathfrak{P}}') \equiv 1\,(M).$$

3) Let now $\bar{\mathfrak{R}}'$ be a divisor of M in $\bar{k}'$. By our condition we have

$$k_{\mathfrak{R}} = \Omega_{\mathfrak{r}}, \quad \bar{k}'_{\bar{\mathfrak{R}}'} = \bar{k}_{\bar{\mathfrak{R}}}.$$

From the relations

$$\bar{k}'_{\bar{\mathfrak{R}}'} = \bar{k}_{\bar{\mathfrak{R}}} k'_{\mathfrak{R}'} = \bar{\Omega}_{\bar{\mathfrak{r}}} k'_{\mathfrak{R}'}, \quad \bar{k}_{\bar{\mathfrak{R}}} = \bar{\Omega}_{\bar{\mathfrak{r}}} k_{\mathfrak{R}} = \bar{\Omega}_{\bar{\mathfrak{r}}}$$

we can derive that

$$\bar{k}_{\bar{\mathfrak{R}}} = \bar{k}_{\bar{\mathfrak{R}}} k'_{\mathfrak{R}'}. \tag{2}$$

Since by our condition $(k':k)$, and therefore $(k'_{\mathfrak{R}'} : k_{\mathfrak{R}})$ is a power of l, and $(\bar{k}:k)$, and therefore $(\bar{k}_{\bar{\mathfrak{R}}} : k_{\mathfrak{R}})$ divide $l-1$, and therefore are relatively prime to l, we can conclude from (2) that $k'_{\mathfrak{R}'} = k_{\mathfrak{R}} = \Omega_{\mathfrak{r}}$, which shows that the prime divisors of M split completely in k'/Ω.

Finally condition 4° is obvious since for $l = 2$, $\zeta \in \Omega$ and for $l > 2$, k'/k is of odd degree and is therefore purely imaginary.

The lemma is proved.

We next consider the sequences of fields which correspond, according to the Galois theory, to the sequence of groups investigated in the Faddeev-Gaschütz theorem.

We now consider a sequence of algebraic number fields

$$\Omega \subset L \subset k \subset K \subset K^*, \tag{3}$$

having the following properties: k/Ω is the normal extension with the Galois group G, L belong to the Sylow l-subgroup H, K/L is normal and is a simple central extension of k/L, K/Ω is not normal and the fields conjugate with K with respect to Ω are independent over k and their composite is K^*.

In this and the following section we shall consider the following problem: to find conditions which must be satisfied by the field k/Ω which is a Scholz field with respect to M and l so that there exists a field K, whose Galois group is a given simple central extension of the Galois group of k/L and is such that K^*/Ω is a Scholz field with respect to M and l. The answer will come out in terms of some invariants of the field k/Ω to whose determination we next proceed. In what follows we shall always suppose that k/Ω is a Scholz field with respect to M and l. We shall therefore suppose that k contains an l-th root of unity ζ and we will let

$$\zeta^{\sigma} = \zeta^{g(\sigma)}, \quad \sigma \in G.$$

Obviously

$$g(\sigma\tau) \equiv g(\sigma)\, g(\tau) \quad (l). \tag{4}$$

2. Invariants $[\chi, X]$. We take an arbitrary character χ of degree l of the group H and a number α_χ of L which corresponds to it and such that

$$\sqrt[l]{\alpha_\chi} \in k, \quad \sqrt[l]{\alpha_\chi}^{\,h} = \chi(h) \sqrt[l]{\alpha_\chi}, \quad h \in H.$$

We consider the decomposition of α_χ into prime divisors in k. In the first place we have

$$(\alpha_\chi) = \mathfrak{a}_\chi \mathfrak{b}^l,$$

where $\mathfrak{a}_\chi$ and $\mathfrak{b}$ are ideals in L, and $\mathfrak{a}_\chi$ consists only of ramified prime ideals in k/L. Since α_χ is determined uniquely by the character χ up to an l-equality, $\mathfrak{a}_\chi$ depends only on χ and not on the choice of the α_χ which corresponds to it.

If $\mathfrak{p}$ is a prime divisor of $\mathfrak{a}_\chi$ in L, then it follows from 1° in the definition of a Scholz field that

$$\mathfrak{p} = \mathfrak{P}^{\Sigma h},$$

where $\mathfrak{P}$ is a prime ideal of k and the sum in the exponent goes over all $h \in H$. If $\sigma \in G$ and $\tau \in G$, then it is easily seen that

$$\mathfrak{P}^{\sigma \Sigma h} = \mathfrak{P}^{\tau \Sigma h},$$

if and only if the classes with respect to the double moduli $\mathfrak{Z}_{\mathfrak{P}}\, \sigma H$ and $\mathfrak{Z}_{\mathfrak{P}}\, \tau H$ coincide. Here $\mathfrak{Z}_{\mathfrak{P}}$ denotes the decomposition group of $\mathfrak{P}$.

Hence the decomposition of α_χ into its prime divisors in k has the form

$$(\alpha_\chi) = \mathfrak{b}^l \prod_{\mathfrak{P}} \mathfrak{P}^{\Sigma \varphi_{\mathfrak{P},\chi}(\mathfrak{S}) \sigma \Sigma h}$$

where $\mathfrak{P}$ goes over all conjugate with each other ramified ideals of k/L, $\mathfrak{S}$ are cosets of G with respect to the double modulus $(\mathfrak{Z}_{\mathfrak{P}}, H)$ and σ is an arbitrary representative of $\mathfrak{S}$.

We now consider the class X of l-invariant numbers of the field k/L. Let $\mathfrak{D}(X)$ be an ideal which corresponds to the class X (for its definition see [6], p. 266) and let μ be an arbitrary representative of X. We suppose that for any ramified ideal $\mathfrak{P}$

$$(\mathfrak{D}(X)^{\Sigma \varphi^*_{\mathfrak{P},\chi}(\mathfrak{T})\tau}, \mathfrak{P}) \approx 1. \tag{5}$$

Here $\phi^*_{\mathfrak{P},\chi}(\mathfrak{T})$ denotes a function over cosets with respect to the double modulus $(H, \mathfrak{Z}_{\mathfrak{P}})$, defined by the relation

$$\varphi^*_{\mathfrak{P},\chi}(\mathfrak{T}) \equiv \varphi_{\mathfrak{P},\chi}(\mathfrak{T}^{-1})\, g(\tau) \quad (l),$$

where τ, as before, is an arbitrary representative of $\mathfrak{T}$.

We note that the function $g(\tau)$ satisfies the conditions

$$g(z\tau) = g(\tau z) = g(\tau h) = g(h\tau) = g(\tau), \quad h \in H, \quad z \in \mathfrak{Z}_{\mathfrak{P}}, \tag{6}$$

so that $\phi^*(\mathfrak{T})$ is a function over cosets with respect to the modulus $(H, \mathfrak{Z}_{\mathfrak{P}})$.

To prove conditions (6) it is sufficient to show that

$$g(h) \equiv 1 \ (l), \quad h \in H, \tag{7}$$

$$g(z) \equiv 1 \ (l), \quad z \in \mathfrak{Z}_{\mathfrak{P}}. \tag{8}$$

Since the order of H is a power of l, (7) follows from (4) and from the fact that

$$g(h)^{l-1} \equiv 1 \ (l).$$

It is obvious that the automorphisms $z \in \mathfrak{Z}_{\mathfrak{P}}$ induce in the field $\Omega(\zeta)/\Omega$ automorphisms of the decomposition group of a prime ideal in this field, which is divisible by $\mathfrak{P}$. By (1) and by assumption that k/Ω is a Scholz field with respect to M and l, the ideal $\mathfrak{p}$ of Ω splits completely in $\Omega(\zeta)$, since the decomposition group of its prime divisor in this field is equal to 1. From this it follows that $\zeta^z = \zeta$ for $z \in \mathfrak{Z}_{\mathfrak{P}}$, in other words, condition (8).

In order that condition (5) have a definite meaning we must show that this condition does not depend on the choice of the representatives τ in the classes $\mathfrak{T}$ with respect to the modulus $(H, \mathfrak{Z}_{\mathfrak{P}})$.

It is obvious that

$$\mathfrak{D}(X)^{\tau} = \mathfrak{D}(X)^{h\tau}$$

for $h \in H$, since $\mathfrak{D}(X)$ is an invariant ideal with respect to h. We denote by $[\mathfrak{D}(X)]_{\mathfrak{P}}$ that power of $\mathfrak{P}$, which goes into $\mathfrak{D}(X)$. From the obvious identities

$$\begin{gathered} [\mathfrak{D}(X)^{\tau z}]_{\mathfrak{P}} = ([\mathfrak{D}(X)^{\tau}]_{\mathfrak{P}})^{z} = [\mathfrak{D}(X)^{\tau}]_{\mathfrak{P}}, \\ \mathfrak{P}^{z} = \mathfrak{P}, \quad z \in \mathfrak{Z}_{\mathfrak{P}}, \end{gathered} \tag{9}$$

we obtain

$$[\mathfrak{D}(X)^{\tau z}]_{\mathfrak{P}} = [\mathfrak{D}(X)^{\tau}]_{\mathfrak{P}},$$

which proves our assertion.

If condition (5) is not satisfied, we define the symbol $[\chi, X]$ by

$$[\chi, X] = \prod_{\mathfrak{P}} \left(\frac{\mu^{\Sigma \varphi^*_{\mathfrak{P},\chi}(\mathfrak{T})\tau}}{\mathfrak{P}} \right)_k \left(\frac{\alpha_\chi}{\mathfrak{m}} \right)_L^{-1}, \tag{10}$$

where $\mathfrak{P}$ goes over all conjugate ramified prime ideals of k/Ω, and $\mathfrak{m}$ is the ideal in the decomposition

$$\mu = \mathfrak{C}^{l} \mathfrak{D}(X) \mathfrak{m} \tag{11}$$

(see [6], p. 266). If condition (5) is not satisfied the symbol $[\chi, X]$ is not defined.

We next prove that the symbol $[\chi, X]$ does not depend on the choice of

a) the representative τ of the class $\mathfrak{T}$,

b) the ideal $\mathfrak{P}$ among the totality of conjugate ideals,

c) the choice of μ in the class X.

To prove a) we first show that if $z \in \mathfrak{Z}_{\mathfrak{P}}$, then for any A for which

$$(A^{l-z}, \mathfrak{P}) = 1,$$

we have

$$\left(\frac{A^{l-z}}{\mathfrak{P}}\right) = 1, \quad z \in \mathfrak{Z}_{\mathfrak{P}}. \tag{12}$$

It is obvious that the residue class A^{l-z} mod $\mathfrak{P}$ has in the multiplicative group of residue class an order, which is divisible by the order $e_{\mathfrak{P}}$ of the decomposition group of $\mathfrak{P}$. Since the order of the group of residue classes is $\mathfrak{N}(\mathfrak{P}) - 1$, from condition (12) follows the congruence

$$\mathfrak{N}(\mathfrak{P}) \equiv 1 \ (le_{\mathfrak{P}}). \tag{13}$$

In fact it follows from (13), in view of what has been said above, that the index of the subgroup generated by the class A^{l-z} mod $\mathfrak{P}$ in the group of residue classes is divisible by l and therefore this class is an l-th power of some other class, which is equivalent to (12).

To prove (13) we make use of the arbitrary choice of M, with respect to which we suppose that k/Ω is a Scholz field. We will require now that any prime divisor l of degree $(k:\Omega)$ goes into M to a power at least by a unit larger than the power to which it appears in the exponent (the greatest common divisor of the order of the elements) of the group G. Under this condition equation (13) is obviously satisfied, since on one hand, because $(\mathfrak{P}, l) = 1$,

$$\mathfrak{N}(\mathfrak{P}) \equiv 1 \ (e_{\mathfrak{P}}),$$

and on the other hand, $\mathfrak{N}(\mathfrak{P}) - 1$ is divisible by a power of l which is at least one larger that $e_{\mathfrak{P}}$, which follows from the condition which we introduced that $e_{\mathfrak{P}}$ is the order of a cyclic (because k is a Scholz field) group $\mathfrak{Z}_{\mathfrak{P}}$.

Statement a) can now be proved very simply. We must prove that expression (10) is unaltered if τ is replaced by another representative $h\tau z$, $h \in H$, $z \in \mathfrak{Z}_{\mathfrak{P}}$, of the same coset with respect to the double modulus $(H, \mathfrak{Z}_{\mathfrak{P}})$. The invariance with respect to multiplication by h follows from the l-invariance of μ in k/L, and with respect to z by (12).

We now prove statement b). Let $\mathfrak{P}^{\theta}$ be an ideal conjugate with $\mathfrak{P}$. Then

$$\mathfrak{P}^{\sigma} = \mathfrak{P}^{\theta\theta^{-1}\sigma} \text{and} \, (\alpha_{\chi}) = \mathfrak{b}^{l} \prod \mathfrak{P}^{\theta\Sigma\varphi_{\mathfrak{P},\chi}(\mathfrak{S})\theta^{-1}\Sigma h}.$$

Here σ goes over a complete system of representatives of cosets with respect to the modulus $(\mathfrak{Z}_{\mathfrak{P}}, H)$, while $\theta^{-1}\sigma$ also goes over a complete system of representatives of cosets with respect to the modulus $(\theta^{-1}\mathfrak{Z}_{\mathfrak{P}}\theta, H) = (\mathfrak{Z}_{\mathfrak{P}}\theta, H)$ since from

$$G = \Sigma \, \mathfrak{Z}_{\mathfrak{P}} \sigma_i H$$

follows

$$G = \Sigma\, \theta^{-1}\, \mathfrak{Z}_{\mathfrak{P}}\, \theta\theta^{-1}\, \sigma_i H.$$

If $\mathfrak{S} = \mathfrak{Z}_{\mathfrak{P}}\, \sigma H$, then we shall denote $\theta^{-1}\mathfrak{Z}_{\mathfrak{P}}\,\theta\theta^{-1}\sigma_i H$ by $\theta^{-1}\mathfrak{S}$. In this way

$$(\alpha_\chi) = \mathfrak{b}^l \prod \mathfrak{P}^{\theta\Sigma\varphi_{\mathfrak{P}^\theta,\chi}(\mathfrak{R})\rho\Sigma h},$$

where $\mathfrak{R}$ goes over all the classes with respect to the modulus $(\mathfrak{Z}_{\mathfrak{P}}\theta, H)$, ρ is a representative of $\mathfrak{R}$ and

$$\phi_{\mathfrak{P}^\theta,\chi}(\mathfrak{R}) = \phi_{\mathfrak{P},\chi}(\theta\mathfrak{R}).$$

It is now obvious that

$$(\mathfrak{D}(X)^{\Sigma\varphi^*_{\mathfrak{P}^\theta,\chi}(\mathfrak{U})\nu}, \mathfrak{P}^\theta) = (\mathfrak{D}(X)^{\Sigma\varphi^*_{\mathfrak{P},\chi}(\mathfrak{U}\theta^{-1})\nu\theta^{-1}\theta}, \mathfrak{P}^\theta)^{\sigma(\theta^{-1})},$$

where $\mathfrak{U}$ goes over all the cosets with respect to the modulus $(H, \mathfrak{Z}_{\mathfrak{P}}\theta)$, $\nu \in \mathfrak{U}$. If we let

$$\mathfrak{U}\,\theta^{-1} = \mathfrak{T}, \quad \nu\,\theta^{-1} = \tau,$$

then $\mathfrak{T}$ will go over all the classes modulo $(H, \mathfrak{Z}_{\mathfrak{P}})$, $\tau \in \mathfrak{T}$. Therefore

$$(\mathfrak{D}(X)^{\Sigma\varphi^*_{\mathfrak{P}^\theta,\chi}(\mathfrak{U})\nu}, \mathfrak{P}^\theta) = (\mathfrak{D}(X)^{\Sigma\varphi^*_{\mathfrak{P},\chi}(\mathfrak{T})\tau}, \mathfrak{P})^{\theta\sigma(\theta^{-1})}.$$

Therefore condition (5) which defines the symbol $[\chi, X]$ does not depend on the choice of $\mathfrak{P}$ among conjugate ideals.

Analogously the symbol itself

$$\left(\frac{\mu^{\Sigma\varphi^*_{\mathfrak{P}^\theta,\chi}(\mathfrak{U})\nu}}{\mathfrak{P}^\theta}\right) = \left(\frac{\mu^{\Sigma\varphi^*_{\mathfrak{P},\chi}(\mathfrak{T})\tau\theta}}{\mathfrak{P}^\theta}\right)^{\sigma(\theta^{-1})} = \left(\frac{\mu^{\Sigma\varphi^*_{\mathfrak{P},\chi}(\mathfrak{T})\tau}}{\mathfrak{P}}\right),$$

which shows the independence of the symbol on the choice of $\mathfrak{P}$ among conjugate ideals.

It remains to prove proposition c). Obviously it is sufficient to check the invariance of the expression for $[\chi, X]$ by multiplying μ by an l-th power and by a number $m \in L$, relatively prime to the discriminant of k/L. The first is obvious, the second does not alter $\mathfrak{D}(X)$ and therefore the invariance of condition (5) is obvious. We consider the symbol $[\chi, X]$ itself

$$\prod_{\mathfrak{P}}\left(\frac{(\mu m)^{\Sigma\varphi^*_{\mathfrak{P},\chi}(\mathfrak{T})\tau}}{\mathfrak{P}}\right)\left(\frac{\alpha_\chi}{mm}\right)^{-1} =$$

$$= \prod_{\mathfrak{P}}\left(\frac{\mu^{\Sigma\varphi^*_{\mathfrak{P},\chi}(\mathfrak{T})\tau}}{\mathfrak{P}}\right)\left(\frac{\alpha_\chi}{m}\right)^{-1}\prod_{\mathfrak{P}}\left(\frac{m^{\Sigma\varphi^*_{\mathfrak{P},\chi}(\mathfrak{T})\tau}}{\mathfrak{P}}\right)\left(\frac{\alpha_\chi}{m}\right)^{-1}.$$

We must show that

$$\prod_{\mathfrak{P}}\left(\frac{m^{\Sigma\varphi^*_{\mathfrak{P},\chi}(\mathfrak{T})\tau}}{\mathfrak{P}}\right)_k\left(\frac{\alpha_\chi}{m}\right)_L^{-1} = 1. \tag{14}$$

In fact

$$\prod_{\mathfrak{P}}\left(\frac{m^{\Sigma\overset{*}{\varphi}_{\mathfrak{P},\chi}(\mathfrak{T})\tau}}{\mathfrak{P}}\right)_k=\prod_{\mathfrak{P}}\left(\frac{m}{\mathfrak{P}^{\Sigma\overset{*}{\varphi}_{\mathfrak{P},\chi}(\mathfrak{T})\tau^{-1}g(\tau^{-1})}}\right)_k=$$

$$=\prod_{\mathfrak{P}}\left(\frac{m}{\mathfrak{P}^{\Sigma\varphi_{\mathfrak{P}}(\mathfrak{S})\sigma}}\right)_k=\prod_{\mathfrak{P}}\left(\frac{m}{\mathfrak{P}^{\Sigma\varphi_{\mathfrak{P},\chi}(\mathfrak{S})\sigma\Sigma h}}\right)_L=\left(\frac{m}{\alpha_\chi}\right)_L. \tag{15}$$

Now (14) follows from (15) and from the reciprocity law since α_χ is l-hyperprimary, in view of condition 3° in the definition of a Scholz field.

The invariants $[\chi, X]$ have the following properties:

$$[\chi_1\chi_2, X]=[\chi_1, X][\chi_2, X], \tag{16}$$

$$[\chi, X_1X_2]=[\chi, X_1][\chi, X_2], \tag{17}$$

$$[\chi_1, X_1]=[\chi, X], \tag{18}$$

where it is supposed in (16) and (17) that both symbols on the right hand side are defined, and the relations state that the left hand side is also defined. It is supposed in (18) that $k\subset k_1$, and that the degree (k_1, k) is a power of l. The character χ and the class X are considered in k, while the character χ_1 and class X_1 correspond to them in k_1.

We note that if k/Ω is a Scholz field, then k/L is also a Scholz field, and if the invariant (χ, X) is defined (see [6], p. 248), then the invariant $[\chi, X]$ is also defined and they coincide.

3. The invariants $(X)_M$. We denote by L_M the maximal extension of L having an abelian group of exponent l and containing $L(\zeta_M)$, where ζ_M is a primitive M-th root of unity. We select in the class X a representative μ such that all the prime divisors of M in $\Omega(\zeta)$ split completely in $k(\sqrt[l]{\mu})/\Omega$. In case the divisors of M are also divisors of l the existence of such a representative was proved in paper [6], p. 269. If $q\mid M$, $(q, l)=1$, $q\in L$, then because k/Ω is a Scholz field with respect to M, q is unramified and is of degree 1 in k/Ω. In order that q should possess these same properties in $k(\sqrt[l]{\mu})/\Omega$, it is necessary and sufficient that the following condition is satisfied

$$\left[\frac{\mu}{q}\right]=1. \tag{19}$$

Obviously if we can find m satisfying the condition

$$\left(\frac{m}{q}\right)=\left[\frac{\mu}{q}\right]^{-1}, \tag{20}$$

then condition (19) will be satisfied for μm. It is also obvious that condition (20) for various $q\mid M$, $(q, l)=1$ and also corresponding conditions for $q\mid l$ are compatible.

We define the invariant $(X)_M$ as the class $\mathfrak{m}\Lambda_M$ with respect to the subgroup Λ_M, corresponding to the field L_M/L in the class field theory sense to which belongs the ideal $\mathfrak{m}$ in the decomposition (11) of the representative μ of the class X selected in this way.

We shall show that $(X)_M$ does not depend on the choice of μ in X only if this choice satisfies the above conditions. In fact another representative must be of the form μm, where m is l-hyperprimary and

$$\left(\frac{m}{\mathfrak{q}}\right) = 1 \text{ for } \mathfrak{q} \mid M, \quad (\mathfrak{q}, l) = 1.$$

In this way m lies in the group $S_{\mathfrak{F}}$, consisting of l-th power ray classes modulo $\mathfrak{F}$, which is the product of the modulus of l-hyperprimarity and of all the $\mathfrak{q}$ which divide M, but do not divide l. The field which corresponds to $S_{\mathfrak{F}}$ is a maximal extension of the field L, having an abelian group of exponent l and a discriminant consisting only of divisors of M. Since L_M also possesses these properties with the exception of being maximal, it is contained in this field and therefore $\Lambda_M \supset S_{\mathfrak{F}}$, and from $m \in S_{\mathfrak{F}}$ follows that $m \in \Lambda_M$.

The invariant $(X)_M$ has the following properties

$$(X_1 X_2)_M = (X_1)_M (X_2)_M,$$

$$(X_1)_M = (X)_M,$$

analogous to (17) and (18).

It is obvious that from $(X)_M = 1$ follows that $(X)_{M_1} = 1$ for $M_1 \mid M$. It is also obvious that $(X)_M$ can assume only a finite number of values, in fact not more than M.

4. The invariants $[X]_\psi$. These invariants are defined only for those classes X of l-invariant numbers of the field k/L for which all these invariants $[\chi, X]$, which are defined are equal to 1, and also $(X)_M = 1$.

It will be shown in § 2 that a representative μ can be selected in each of these classes such that in the field K^*, obtained by adjoining to the field k all the $\sqrt[l]{\mu^\sigma}$, all the ramified prime ideals of k split into ideals of the first degree. This is equivalent to

$$\left(\frac{\mu^{\Sigma f_{\mathfrak{P}}(S)\sigma}}{\mathfrak{P}}\right) = 1 \tag{21}$$

for all ramified ideals $\mathfrak{P}$ of the field k/Ω and for all functions $f_{\mathfrak{P}}(S)$ over the right cosets of G with respect to H for which

$$(\mathfrak{D}(X)^{\Sigma f_{\mathfrak{P}}(S)\sigma}, \mathfrak{P}) = 1.$$

As long as Theorem 1 of § 2 has not yet been proved we shall assume that the invariant $[X]_\psi$ is defined for all classes X for which a representative μ exists satisfying condition (21).

In what follows we shall suppose that the choice of the representative μ of X is affected in just such a way.

We consider the function $\psi(\mathfrak{S})$ over the cosets $\mathfrak{S}$ of the group G with respect to the double modulus (H, H) with values in the field of residue classes mod l, possessing the following properties:

1°. $\psi(1) \equiv 0$, $\psi^*(\mathfrak{S}) \equiv -\psi(\mathfrak{S})(l)$.

2°. $\mu^{\Sigma\psi(S)\sigma} \approx m_\psi \in L$ for $\mu \in X$.

Here S goes over the right cosets of G with respect to H, σ is any representative of S, and $\psi(S)$ denotes $\psi(\mathfrak{S})$, where $\mathfrak{S}$ is the coset with respect to the double modulus (H, H) containing S.

3°. If $m_\psi = \mathfrak{d}\mathfrak{m}_\psi$ is a decomposition of m_ψ into a divisor $\mathfrak{d}$ consisting only of prime divisors of the discriminant of k/L, and $\mathfrak{m}_\psi$ is relatively prime to the discriminant, then

$$\mathfrak{d} = (\alpha_\chi) \prod_{\mathfrak{P}} \mathfrak{P}^{\Sigma f_{\mathfrak{P}}(S)\sigma\Sigma h}, \quad \left(\mathfrak{D}(X)^{\Sigma f^* \mathfrak{P}(\mathfrak{T})\tau}, \mathfrak{P}\right) \approx 1, \tag{22}$$

where χ is one of the characters of order l of the group H.

We note that by multiplying m_ψ by $\alpha_\chi^{-l} = (\sqrt[l]{\alpha_\chi})^{-l}$ we can eliminate α_χ from formula (22). Then $\mathfrak{d}$ will be uniquely defined up to such a multiple of α_χ that the invariant $[\chi, X]$ is defined.

For $l = 2$ we impose another additional condition. Let $\mathfrak{S}$ be an involutory class with respect to a double modulus, that is a class $\mathfrak{S}$ which coincides with its inverse: $\mathfrak{S} = \mathfrak{S}^{-1}$. We put into correspondence with each such class an ideal $\mathfrak{C}_{\mathfrak{S}}$ of the field L by the following rule.

Let σ be a representative of the class $\mathfrak{S}$ satisfying the condition

$$\sigma^2 \in H. \tag{23}$$

The existence of such a representative follows from the condition $\mathfrak{S} = \mathfrak{S}^{-1}$ as is shown in the appendix.

We denote by LL^σ the composite of the fields L and L^σ belonging to the subgroup $H \cap H^\sigma$, and by Λ_σ the subfield belonging to the subgroup $\{\sigma, H \cap H^\sigma\}$ generated by σ and $H \cap H^\sigma$. By condition (22)

$$\{\sigma, H \cap H^\sigma\} = H \cap H^\sigma + \sigma H \cap H^\sigma$$

and by (23)

$$(LL^\sigma : \Lambda_\sigma) = 2.$$

Denoting by $\mathfrak{D}_\sigma$ the different of the field LL^σ/Λ_σ, we let

$$\mathfrak{C}_{\mathfrak{S}} = N_{LL^\sigma/L}(\mathfrak{D}_\sigma).$$

The condition which we have imposed on the function ψ can be written in the following form:

4°. $\prod_{\mathfrak{S}} \mathfrak{C}_{\mathfrak{S}}^{\psi(\mathfrak{S})} \approx (\alpha_\chi)$ in L, $[\chi, X]$ is defined,

where the product goes over all the involutory classes $\mathfrak{S}$ with respect to the double modulus (H, H).

It is obvious that all these conditions depend only on the function ψ and on the class X, but not on the choice of μ in X. If conditions 1°, 2° and 3° are satisfied and for $l = 2$ also 4°, then we can define the symbol $[X]_\psi$ as follows. We select in X such a representative μ for which

$$(\mathfrak{m}, \mathfrak{m}^\sigma) = 1 \text{ for } \sigma \notin H \tag{24}$$

and let

$$[X]_\psi = \left[\frac{\mu}{\mathfrak{m}_\psi}\right], \tag{25}$$

where $\mathfrak{m}_\psi = \mathfrak{m}^{\Sigma\psi(S)\sigma}$, and $\mathfrak{m}$ is the ideal in the decomposition (11). The symbol $\left[\frac{A}{\mathfrak{b}}\right]$ is defined in paper [6], p. 267.

In view of condition $\psi(1) \equiv 0\ (l)$ and condition (24), the right hand side of (25) is actually defined.

We now prove that the symbol $[X]_\psi$ does not depend on the choice of the representative μ of the class X. We remind the reader that here we talk of a choice of a representative which satisfies all the conditions set forth in the beginning of this section.

Letting

$$\mu' = \mu m, \quad \mathfrak{m}' = \mathfrak{m}(m), \quad m \in L,$$

where m is composed of prime ideals of L of the first degree in L/Ω, which split completely in k.

By repeating word for word the argument on page 400 of paper [8] it can be checked that the following conditions are satisfied:

$$(\mathfrak{m}, \mathfrak{m}^\sigma) = 1, \quad (m, m^\sigma) = 1, \quad (\mathfrak{m}', \mathfrak{m}'^\sigma) = 1, \quad \sigma \notin H,$$

where μ, m and μ' are l-hyperprimary.

We must show that

$$\left[\frac{\mu}{\mathfrak{m}_\psi}\right] = \left[\frac{\mu'}{\mathfrak{m}'_\psi}\right], \tag{26}$$

where

$$\mathfrak{m}'_\psi = \mathfrak{m}_\psi m^{\Sigma\psi(S)\sigma}.$$

We note that $m^{\Sigma\psi(S)\sigma}$ is a number of L. In fact it is proved in the appendix that if S goes over all the right cosets with respect to H, which are connected in one two sided class with respect to (H, H) and if σ is any representative of S, then $\alpha^{\Sigma\sigma} \in L$, if $\alpha \in L$. It follows from the fact that ψ is constant over cosets with respect to the double modulus (H, H), that $m^{\Sigma\psi(S)\sigma}$ is a product of such numbers and therefore belongs to L.

In order to prove (25) we note that

$$\left[\frac{\mu'}{\mathfrak{m}'_\psi}\right] = \left[\frac{\mu}{\mathfrak{m}_\psi}\right]\left(\frac{m}{\mathfrak{m}_\psi}\right)\left[\frac{\mu}{m^{\Sigma\psi(S)\sigma}}\right]\left(\frac{m}{m^{\Sigma\psi(S)\sigma}}\right). \tag{27}$$

Equation (26) will be proved if we can show that the product of the second and third factors, as well as the fourth factor of expression (27) is equal to 1.

In fact

$$\left[\frac{\mu}{m^{\Sigma\psi(S)\sigma}}\right]=\left[\frac{\mu^{\Sigma\psi(S)\sigma^{-1}g(\sigma^{-1})}}{m}\right],$$

where σ^{-1} goes over representatives of left cosets S^{-1} of the group G with respect to H. Since $S\in\mathfrak{S}$ it follows that $S^{-1}\in\mathfrak{S}^{-1}$, and by conditions 1° and 2° we have

$$\left[\frac{\mu^{\Sigma\psi(S)\sigma^{-1}g(\sigma^{-1})}}{m}\right]=\left[\frac{\mu^{\Sigma\psi^*(S)\sigma}}{m}\right]=\left(\frac{m_\psi}{m}\right)^{-1}, \tag{28}$$

that is

$$\left(\frac{m}{\mathfrak{m}_\psi}\right)\left[\frac{\mu}{m^{\Sigma\psi(S)\sigma}}\right]=\left(\frac{m}{\mathfrak{m}_\psi}\right)\left(\frac{m_\psi}{m}\right)^{-1}. \tag{29}$$

We will prove that

$$\left(\frac{m}{\mathfrak{m}_\psi}\right)=\left(\frac{m}{m_\psi}\right) \tag{30}$$

and then it will follow from (29) and the reciprocity law that

$$\left(\frac{m}{\mathfrak{m}_\psi}\right)\left[\frac{\mu}{m^{\Sigma\psi(S)\sigma}}\right]=1.$$

To prove (30) we note that

$$(m_\psi)=\mathfrak{b}\,\mathfrak{m}_\psi,$$

where $\mathfrak{b}$ satisfies condition (22). In view of that condition

$$\left(\frac{m}{\mathfrak{b}}\right)_L\prod_{\mathfrak{P}}\left(\frac{m}{\mathfrak{P}^{\Sigma f_{\mathfrak{P}}(S)\sigma}}\right)_k=\prod\left(\frac{m^{\Sigma f^*_{\mathfrak{P}}(T)\tau}}{\mathfrak{P}}\right)_k, \tag{30'}$$

where

$$\left(\mathfrak{D}^{\Sigma f^*_{\mathfrak{P}}(T)\tau},\ \mathfrak{P}\right)\approx 1. \tag{31}$$

Since μ satisfies condition (21), it follows from (22) that

$$\left(\frac{\mu^{\Sigma f^*_{\mathfrak{P}}(T)\tau}}{\mathfrak{P}}\right)=1,$$

and since this must hold for μm instead of μ we have

$$\left(\frac{m^{\Sigma f^*_{\mathfrak{P}}(T)\tau}}{\mathfrak{P}}\right)=1.$$

From this and (30) it follows that

$$\left(\frac{m}{\mathfrak{m}_\psi}\right)=\left(\frac{m}{m_\psi}\right)\left(\frac{m}{\mathfrak{b}}\right)^{-1}=\left(\frac{m}{m_\psi}\right).$$

The proof that

$$\left(\frac{m}{m^{\Sigma\psi(S)\sigma}}\right)=1, \tag{32}$$

is different for $l\neq 2$ and for $l=2$.

For $l\neq 2$ we have

$$\left(\frac{m}{m^{\Sigma\psi(S)\sigma}}\right)=\left(\frac{m^{\Sigma\psi(S)\sigma^{-1}g(\sigma^{-1})}}{m}\right),$$

since $\psi(S)$ is constant over classes with respect to the double modulus (H,H) and in view of formula (104) of the appendix. By condition 1° we have as in the proof of (29) that

$$\left(\frac{m^{\Sigma\psi(S)\sigma^{-1}g(\sigma^{-1})}}{m}\right)=\left(\frac{m^{\Sigma\psi(S)\sigma}}{m}\right)^{-1}.$$

Since m is l-hyperprimary, it follows that

$$\left(\frac{m}{m^{\Sigma\psi(S)\sigma}}\right)^2=1,$$

which is equivalent to (32) for $l\neq 2$.

For $l=2$ we consider first the non-involutory class $\mathfrak{S}$. Then from condition 1° and from formula (104) of the appendix

$$\left(\frac{m}{m^{\Sigma\psi(S)\sigma+\psi(S^{-1})\sigma^{-1}}}\right)=\left(\frac{m}{m^{\Sigma\sigma}}\right)^{\psi(\mathfrak{S})}\left(\frac{m}{m^{\Sigma\sigma^{-1}}}\right)^{\psi(\mathfrak{S}^{-1})}=\left(\frac{m}{m^{\Sigma\sigma}}\right)^{\psi(\mathfrak{S})+\psi(\mathfrak{S}^{-1})}=1,$$

where S goes over all the right cosets with respect to H, which are contained in $\mathfrak{S}$.

We consider the expression

$$\left(\frac{m}{m^{\Sigma\psi(S)\sigma}}\right),$$

in which the sum goes over only those classes S, which are different from 1, and and which are contained in involutory two sided classes $\mathfrak{S}$. We have

$$\left(\frac{m}{m^{\Sigma\psi(S)\sigma}}\right)=\prod_{\mathfrak{S}}\left(\frac{m}{m^{\mathfrak{S}}}\right)^{\psi(\mathfrak{S})},$$

where the product goes over all involutory classes $\mathfrak{S}$, and σ goes over representatives of the right cosets with respect to H, which are contained in $\mathfrak{S}$.

It is proved in the appendix (formula (126)) that

$$\left(\frac{m}{m^{\mathfrak{S}}}\right)=\left(\frac{m}{\mathfrak{C}_{\mathfrak{S}}}\right),$$

therefore we have

$$\left(\frac{m}{m^{\Sigma\psi(S)\sigma}}\right)=\left(\frac{m}{\prod_{\mathfrak{S}}\mathfrak{C}_{\mathfrak{S}}^{\psi(\mathfrak{S})}}\right)=\left(\frac{m}{\alpha_\chi}\right)=1,$$

by condition 4° and relation (21).

This proves the independence of the symbol $[X]_\psi$ on the choice of the representative μ of the class X.

For the symbol $[X]_\psi$ we have the equality

$$[X]_\psi=[X_l]_\psi,$$

analogous to (18), but there is no equality analogous to (17).

§2. Conditions for imbeddedness for Scholz extensions

In this section we will give the solution of the problem which was formulated in part 1 of the previous section. According to that formulation we are given in the chain (3) the fields Ω, L and k, as well as the Galois group of K/L as a simple central extension of the Galois group H of the field k/L. Because k/Ω is a Scholz field, and therefore also k/L, to this extension of the group H corresponds a class of l-invariant numbers of the field k/L, which we will denote by X.

Theorem 1. *In order that the field k/L can be imbedded in a field K, corresponding to the class X and such that K^*/Ω is a Scholz field, it is necessary and sufficient that all the invariants $(X)_M$, $[\chi, X]$ and $[X]_\psi$ be equal to 1 for all χ and X for which they are defined.*

Proof. The necessity of the invariant to be equal to 1 can be checked independently. The equality to unity of $(X)_M$ follows from the fact that the class X contains a number μ such that all the prime divisors of M in $\Omega(\zeta)$ split completely in $k(\sqrt[l]{\mu})$, while for any prime divisor $\mathfrak{p}$ of the ideal $\mathfrak{m}$ the following relation holds: $\mathfrak{N}(\mathfrak{p}) \equiv 1\,(M)$. In formula (10) defining $[\chi, X]$ for any prime $\mathfrak{q} \mid \mathfrak{m}$, it follows from the fact that K/L is a Scholz field that

$$\left(\frac{\alpha_\chi}{\mathfrak{q}}\right) = 1, \text{ i.e., } \left(\frac{\alpha_\chi}{\mathfrak{m}}\right) = 1.$$

On the other hand, since K^*/L is a Scholz field and in view of (5)

$$\left(\frac{\mu^{\Sigma\varphi^*_{\mathfrak{P}}(T)\tau}}{\mathfrak{P}}\right) = 1,$$

from which it follows that $[\chi, X] = 1$.

Since K^*/L is a Scholz field it follows that

$$\left[\frac{\mu^{\Sigma\sigma^{-1}}}{\mathfrak{p}}\right] = \left[\frac{\mu}{\mathfrak{p}^{\Sigma\sigma}}\right] = 1, \quad \mathfrak{p} \mid \mathfrak{m},$$

where the sum goes over all the representatives σ of all the right cosets of G with respect to H, which are contained in a different from 1 class with respect to (H, H). Since $\mathfrak{m}_\psi$ consists of factors $\mathfrak{p}^{\Sigma\sigma}$, $\mathfrak{p} \mid \mathfrak{m}$, it follows that $[X]_\psi = 1$.

We now prove the sufficiency. In order to do this we note that from the equality to unity of the invariants $(X)_{M_l}$ and $[\chi, X]$ follows the equality to unity of the invariant (χ, X) of paper [6] for the field k/L. Applying Theorem 1 of paper [6] we find that we can select such a representative μ_0 of the class X that the field $k(\sqrt[l]{\mu_0})/L$ is a Scholz field with respect to M_l.

We first show that we can select a representative μ_1 in X such that $k(\sqrt[l]{\mu_1})/L$ is as before a Scholz field and that the third condition for a Scholz field is satisfied in $k(\sqrt[l]{\mu_1})/\Omega$. To do this we find a number $m_0 \in L$ consisting of prime ideals of the first degree, which split completely in k, which is l-hyperprimary and for $l = 2$ totally positive and which satisfies the conditions:

$$\mathfrak{N}(q) \equiv 1 \ (M_l), \quad q \mid m_0,$$
$$\left(\frac{m_0}{\mathfrak{r}}\right) = \left[\frac{\mu_0}{\mathfrak{r}}\right]^{-1}, \quad \left(\frac{m_0}{\mathfrak{p}}\right) = 1 \tag{33}$$

for all prime divisors $\mathfrak{r}$ of the number M/M_l in L and for prime divisors $\mathfrak{p}$ of the discriminant of the field k/L. Such a choice is possible because $\mathfrak{r}$ is relatively prime to l and to the discriminant of k/L. It is obvious that for $\mu_1 = \mu_0 m_0$, the field $k(\sqrt[l]{\mu_1})/L$ will be as before a Scholz field with respect to M_l and that all the prime divisors of M/M_l in $\Omega(\zeta)$ will split completely in $k(\sqrt[l]{\mu_1})$ in view of (33).

We now show that there exists a representative μ_2 in X such that the second and third conditions for a Scholz field with respect to M and l are satisfied in $k(\sqrt[l]{\mu_2})/\Omega$ and that $k(\sqrt[l]{\mu_2})/L$ is as before a Scholz field. To do this we let

$$\mu_1 = \mathfrak{C}_1^l \mathfrak{D}(X) \mathfrak{m}_1$$

and denote by $\mathfrak{F}$ the product of the modulus of l-hyperprimarity in L by all the ramified prime ideals of k/L and M, and by $S_{\mathfrak{F}}$ the principal union modulo $\mathfrak{F}$. We find in the class $\mathfrak{m}_1 S_{\mathfrak{F}}$ a prime ideal q of the first degree over Ω which splits completely in k and satisfies the condition

$$\mathfrak{N}(q) \equiv 1 \ (M). \tag{34}$$

Such an ideal q actually exists. To show this we denote by $\mathfrak{F}'$ the product of all ramified prime ideals k/L by the modulus of hyperprimarity and by M_l, so that

$$\mathfrak{F} = \mathfrak{F}' M/M_l.$$

We denote by Σ' the class field of $S_{\mathfrak{F}'}$ and by Σ'' the class field of S_{M/M_l}. It is obvious that

$$\Sigma = \Sigma' \Sigma''.$$

We must find a prime ideal q of the first degree in Ω which lies in the classes $\mathfrak{m}_1 S_{\mathfrak{F}'}$, $\mathfrak{m}_1 S_{M/M_l}$, splits completely in k and satisfies condition (34). In other words it must belong to the same automorphism as $\mathfrak{m}_1$ in the fields Σ' and Σ'', and to the identity automorphism in the fields k and $L(\zeta_M)$. In order to show the existence of such a prime ideal, it is sufficient to show that the above automorphisms can be extended to a single automorphism of the field $\Sigma' \cdot \Sigma'' \cdot k \cdot L(\zeta_M)$.

That there exists an automorphism in the field $\Sigma' \cdot k \cdot L(\zeta_{M_l})$, which induces the above automorphisms in Σ', k and $L(\zeta_{M_l})$ follows from the fact that $\mathfrak{m}_1$ consists of prime ideals which split completely in k, and that the invariant $(X)_{M_l}$ is equal to 1. The proof will be found on p. 273 of paper [6].

We now consider the field $\Sigma'' L(\zeta_{M/M_l})$. It is obvious that the field $\Sigma'' \cap L(\zeta_{M/M_l})$ is in fact the field L_{M/M_l}, in the definition of the invariant $(X)_{M/M_l}$. Since $(X)_{M/M_l} = 1$, $\mathfrak{m}_1$ belongs in L_{M/M_l} to the identity automorphism. From this follows the existence of the required automorphism.

We represent $\Sigma'\Sigma'' k\,L(\zeta_M)$ as the composite of the fields $\Sigma'\cdot k\cdot L(\zeta_{M_l})$ and $\Sigma''\cdot L(\zeta_{M/M_l})$. The discriminants of these fields are relatively prime and therefore their intersection coincides with the intersection of Σ' and Σ'' and is the maximal abelian non-ramified extension Σ_0/L with exponent l. The automorphisms which we have constructed induce in Σ_0/L, the same automorphism to which $\mathfrak{m}_1$ belongs. Therefore the whole field $\Sigma'\Sigma'' k\,L(\zeta_M)$ contains an automorphism with the required properties.

If $\mathfrak{q}$ is a prime ideal of the first degree in Ω belonging to this automorphism then

$$\mathfrak{q} = \mathfrak{m}_1(m_1)\,\mathfrak{c}^l.$$

Letting $\mu_2 = \mu_1 m_1$ we obtain, as can be easily seen, a number with the required properties.

Finally we show that there exists a representative μ_3 in X such that the first, second and third conditions for a Scholz field are satisfied in $k(\sqrt[l]{\mu_3})/\Omega$. Since $k(\sqrt[l]{\mu_2})/L$ is already a Scholz field, we must make sure that the ramified prime ideals of L, which are not ramified in $k(\sqrt[l]{\mu_2})/L$ split completely in that field, without disturbing its Scholz character or the second and third conditions for the field $k(\sqrt[l]{\mu_2})/\Omega$ to be a Scholz field with respect to M and l. In order to do this we find an l-hyperprimary and totally positive number m_2 which is composed of prime ideals of the first degree, which split completely in k and satisfy the conditions:

$$\left(\frac{m_2}{\mathfrak{p}}\right) = 1 \text{ for all } \mathfrak{p}, \text{ ramified in } k/L;$$

$$\left(\frac{m_2}{\mathfrak{r}}\right) = 1,\ \mathfrak{r}\,|\,M/M_l;$$

$$\mathfrak{N}(\mathfrak{q}) \equiv 1\ (M),\ \ \mathfrak{q}\,|\,m_2;$$

$$\left(\frac{m_2}{\mathfrak{p}'}\right) = \left[\frac{\mu_2}{\mathfrak{p}'}\right]^{-1}, \text{ for } \mathfrak{p}' \text{ ramified in } L \text{ but not in } k/L.$$

Since all the ideals $\mathfrak{p}'$ in the last condition are relatively prime to all the ideals $\mathfrak{q}$, $\mathfrak{r}$, and M in the preceding conditions and to the discriminant of k/L, all these conditions are compatible. It is obvious that $\mu_3 = \mu_2 m_2$ possesses all the required properties.

It is easily seen that the 4-th condition for a Scholz field is satisfied in $k(\sqrt[l]{\mu})/\Omega$ if it is satisfied in k/Ω and $k(\sqrt[l]{\mu})/L$ (this needs to be verified only for $l = 2$). Hence letting $\mu' = \mu_3$ we can say that the field $K = k(\sqrt[l]{\mu'})/\Omega$ is a Scholz field with respect to M and l. We next see which of the conditions for a Scholz field are satisfied in K^*. It is obvious that the third and fourth conditions are satisfied and that the second condition is also satisfied provided the first condition is satisfied. Hence we must multiply μ' by a number $m' \in L$ such that for

$K = k(\sqrt[l]{\mu' m'})$ the first condition for a Scholz field will be satisfied in K^*/Ω, without disturbing the second, third and fourth conditions in the field $k(\sqrt[l]{\mu'})/\Omega$. We now divide condition 1° into two parts:

A. The requirement that the ramified ideals of k split in K^* into ideals of the first degree.

B. The requirement that the ramified prime ideals in K^*, relatively prime to the discriminant of k/Ω, be of degree 1 with respect to Ω.

We consider first requirement A. The multiplication of μ' by m' will not disturb the second, third and fourth conditions for $k(\sqrt[l]{\mu'})/\Omega$ to be a Scholz field with respect to M and l, provided m' is a principal prime ideal of the first degree in L/Ω, which splits completely in k, is totally positive, hyperprimary and satisfies the conditions

$$\left(\frac{m'}{\mathfrak{r}}\right) = 1, \quad \mathfrak{r} \mid M/M_l,$$
$$\mathfrak{N}(m') \equiv 1 \ (M).$$

All these conditions will be in the future designated by conditions (*).

Let $\mathfrak{P}$ be a ramified prime ideal of k/Ω. In order that the prime divisors of $\mathfrak{P}$ be of degree 1 in K^*, which is obtained by adjoining all the $\sqrt[l]{\mu^\sigma}$ to k, it is necessary and sufficient that for any function $f_{\mathfrak{P}}(T)$ over the right cosets T of the group G with respect to H for which

$$(\mathfrak{D}(X)^{\Sigma f_{\mathfrak{P}}(T)r}, \mathfrak{P}) \approx 1, \tag{35}$$

the following condition is satisfied

$$\left(\frac{\mu^{\Sigma f_{\mathfrak{P}}(T)\tau}}{\mathfrak{P}}\right) = 1. \tag{36}$$

Let

$$\xi_{f_{\mathfrak{P}}} = \left(\frac{\mu'^{\Sigma f_{\mathfrak{P}}(T)\tau}}{\mathfrak{P}}\right).$$

We will show that there exists a number m' satisfying conditions (*) and such that

$$\left(\frac{m'^{\Sigma f_{\mathfrak{P}}(T)\tau}}{\mathfrak{P}}\right) = \xi_{f_{\mathfrak{P}}}^{-1} \tag{37}$$

for any $\mathfrak{P}$ and $f_{\mathfrak{P}}$ satisfying (35). It is obvious that in that case $\mu' m'$ will satisfy condition A.

Because the equality

$$\left(\frac{m'}{\mathfrak{P}^{\sigma h}}\right) = \left(\frac{m'}{\mathfrak{P}^{\sigma}}\right), \quad \left(\frac{m'}{\mathfrak{P}^{z\sigma}}\right) = \left(\frac{m'}{\mathfrak{P}^{\sigma}}\right), \qquad h \in H, \quad z \in \mathfrak{Z}_{\mathfrak{P}},$$

we can denote $\left(\frac{m'}{\mathfrak{P}^{\sigma}}\right)$ by $\zeta^{x_{\mathfrak{P}}, \mathfrak{S}}$, where $\mathfrak{S}$ is the coset with respect to the double modulus $(\mathfrak{Z}_{\mathfrak{P}}, H)$ containing σ. Moreover we introduce the notation.

$$\xi_{f_{\mathfrak{P}}}^{-1} = \zeta^{b_{f_{\mathfrak{P}}}}.$$

Condition (37) can now be written in the form

$$\sum_S f_{\mathfrak{P}}^{*}(S)\, x_{\mathfrak{P},\mathfrak{S}} \equiv b_{f_{\mathfrak{P}}}(l), \tag{38}$$

where the sum in (38) goes over all the left cosets S of the group G with respect to H, and $\mathfrak{S}$ is the coset modulo $(\mathfrak{Z}_{\mathfrak{P}}, H)$ containing S.

We clarify under what conditions the number m' satisfying conditions (*) and the relations

$$\left(\frac{m'}{\mathfrak{P}^{\sigma}}\right)_k = \left(\frac{m'}{N_{k/L}\mathfrak{P}^{\sigma}}\right)_L = \zeta^{x_{\mathfrak{P},\mathfrak{S}}} \tag{39}$$

actually exists for given $x_{\mathfrak{P},\mathfrak{S}}$ (mod l).

In order to do this we denote by $\mathfrak{F}$ the product of all the ideals L ramified in k/L or L/Ω, by the modulus of hyperprimarity, by M_l and by the infinite primes, and by $S_{\mathfrak{F}}$ the group unions of ideal classes modulo $\mathfrak{F}$, and by $\Sigma_{\mathfrak{F}}$ the class field of $S_{\mathfrak{F}}$. Repeating word for word the arguments on p. 404 of paper [8] we obtain that conditions (*) and (39) are equivalent to the requirement that the ideal (m_1) should belong to a definite automorphism in $\Sigma_{\mathfrak{F}}$ and to the identity automorphisms $L(\zeta_{M/M_l})$, k/L and the class field Σ' of the group of unions with respect to the modulus which is the product of all the prime divisors of M/M_l. Since the discriminants of $\Sigma_{\mathfrak{F}}\cdot k$ and $\Sigma' L(\zeta_{M/M_l})$ are relatively prime, the compatibility of the conditions (*) and (39) is equivalent to the statement that the above automorphisms of the field $\Sigma_{\mathfrak{F}}$ induce an identity automorphism in $\Sigma_{\mathfrak{F}} \cap k$.

Repeating again the corresponding considerations on p. 404 of paper [8] we see that $\Sigma_{\mathfrak{F}} \cap k$ is obtained by adjoining to L all the $\sqrt[l]{\alpha_\chi}$ and that the automorphism of the field $\Sigma_{\mathfrak{F}}$ under discussion reduces in $L(\sqrt[l]{\alpha_\chi})$ to multiplication of $\sqrt[l]{\alpha_\chi}$ by

$$\zeta^{\sum_{\mathfrak{P},\mathfrak{S}} x_{\mathfrak{P},\mathfrak{S}}\varphi_{\mathfrak{P},\chi}(\mathfrak{S})},$$

if

$$(\alpha_\chi) = \prod_{\mathfrak{P}} \mathfrak{P}^{\sum \varphi_{\mathfrak{P},\chi}(S)\,\sigma \sum h}$$

is the decomposition of α_χ into prime divisors in k.

In this way the compatibility of conditions (*) and (39) is equivalent to the compatibility of the system composed of congruences (38) and

$$\sum_{\mathfrak{P},\mathfrak{S}} x_{\mathfrak{P},\mathfrak{S}}\varphi_{\mathfrak{P},\chi}(\mathfrak{S}) \equiv 0\,(l). \tag{40}$$

We will write down the conditions for the compatibility of this system, and show that they coincide with the condition that all the invariants $[\chi, X]$ be equal to 1.

If the system consisting of congruences (38) and (40) is incompatible, then there exists a linear combination of these congruences in which the coefficients of all the variables are equal to zero, and the constant term is different from zero.

Let

$$\sum_{\chi} c_\chi \sum_{\mathfrak{P},\mathfrak{S}} x_{\mathfrak{P},\mathfrak{S}} \varphi_{\mathfrak{P},\chi}(\mathfrak{S}) + \sum_{\mathfrak{P},f_{\mathfrak{P}}} d_{\mathfrak{P},f_{\mathfrak{P}}} \sum_{S} f^*_{\mathfrak{P}}(S)\, x_{\mathfrak{P},\mathfrak{S}} \equiv \sum_{\mathfrak{P},f_{\mathfrak{P}}} d_{\mathfrak{P},f_{\mathfrak{P}}} b_{f_{\mathfrak{P}}} \ (l) \quad (41)$$

be such a linear combination. Let

$$\prod_{\chi} \chi^{c_\chi} = \chi_0, \qquad \sum_{f_{\mathfrak{P}}} d_{\mathfrak{P},f_{\mathfrak{P}}} f(T) = F_{\mathfrak{P}}(T).$$

Then obviously

$$\sum_{\chi} c_\chi \varphi_{\mathfrak{P},\chi}(\mathfrak{S}) \equiv \varphi_{\mathfrak{P},\chi_0}(\mathfrak{S}) \ (l),$$

$$\sum_{f_{\mathfrak{P}}} d_{\mathfrak{P},f_{\mathfrak{P}}} b_{f_{\mathfrak{P}}} \equiv b_{f_{\mathfrak{P}}} \ (l).$$

Taking into consideration that in (41) the coefficients of all $x_{\mathfrak{P},\mathfrak{S}}$ are equal to zero, we get

$$\varphi_{\mathfrak{P},\chi_0}(\mathfrak{S}) \equiv -\sum_{S\in\mathfrak{S}} F^*_{\mathfrak{P}}(S) \ (l)$$

or

$$\varphi^*_{\mathfrak{P},\chi_0}(\mathfrak{T}) \equiv -\sum_{T\in\mathfrak{T}} F_{\mathfrak{P}}(T) \ (l) \quad (42)$$

for all $\mathfrak{P}$ and $\mathfrak{T}$.

Since all the $f_{\mathfrak{P}}$, and therefore also $F_{\mathfrak{P}}$, satisfy condition (35) it follows from (42) that

$$\left(\mathfrak{D}(X)^{\sum\varphi^*_{\mathfrak{P},\chi_0}(T)\tau}, \mathfrak{P}\right) \approx 1. \quad (43)$$

Moreover it follows from (42) that

$$\prod_{\mathfrak{P}} \left(\frac{\mu'^{\sum\varphi^*_{\mathfrak{P},\chi_0}(T)\tau}}{\mathfrak{P}}\right) = \zeta^{\sum_{\mathfrak{P}} b_{F_{\mathfrak{P}}}}. \quad (44)$$

For the compatibility of our system it is necessary and sufficient that

$$\sum d_{\mathfrak{P},f_{\mathfrak{P}}} b_{f_{\mathfrak{P}}} \equiv \sum b_{F_{\mathfrak{P}}} \equiv 0(l),$$

that is that the right hand side be equal to 1. The left hand side of the equality is exactly $[\chi_0, X]$, since by assumption the field $k(\sqrt[l]{\mu'})$ is a Scholz field it follows that

$$\left(\frac{\alpha_{\chi_0}}{\mathfrak{m}}\right) = 1.$$

Hence the symbol $[\chi_0, X]$ is defined in view of (43). Therefore the system consisting of equations (38) and (40) is compatible and therefore the number $\bar{\mu} = \mu' m'$, satisfying condition A exists.

Starting with the number $\bar{\mu}$ we must find an $\bar{m} \in L$, such that the number $\mu = \bar{\mu}\bar{m}$ should satisfy condition A as before, and moreover, also condition B as well.

We choose $\overline{m}$ to be l-hyperprimary and (for $l=2$) totally positive and satisfying the condition

$$\left(\frac{\overline{m}}{\mathfrak{P}^{\Sigma f^*(S)\sigma}}\right)=1, \tag{45}$$

if $\mathfrak{P}$ is ramified in k/Ω,

$$(\mathfrak{D}(X)^{\Sigma f(T)r}, \mathfrak{P}) \approx 1$$

and if

$$\left(\frac{\overline{m}}{\mathfrak{r}}\right)=1, \quad \mathfrak{r}\,|\,M/M_l,$$

$$\mathfrak{N}(q)\equiv 1\ (M), \quad q\,|\,\overline{m}.$$

We let

$$\left(\frac{\overline{m}}{\mathfrak{P}^{\sigma}}\right)=\zeta^{y_{\mathfrak{P},\mathfrak{S}}}, \tag{46}$$

where $\mathfrak{S}$ is a coset with respect to the double modulus $(\mathfrak{Z}_{\mathfrak{P}}, H)$. Then condition (45) is equivalent to (46) and to

$$\Sigma f^*(S)\, y_{\mathfrak{P},\mathfrak{S}}\equiv 0\ (l), \quad \text{if} \quad (\mathfrak{D}(X)^{\Sigma f(T)r}, \mathfrak{P})\approx 1. \tag{47}$$

All these conditions we will call in the future conditions (**). It is obvious that if they are satisfied for $\overline{m}$ then $\overline{\mu}\overline{m}$ will as before satisfy condition A.

We will examine under what conditions $\overline{\mu}\overline{m}$ satisfies requirement B. The ramified prime ideals in K^*, which are not ramified in k, will be for $K=k(\sqrt[l]{\overline{\mu}\overline{m}})$ prime divisors of $(\overline{\mathfrak{m}}\, m)^{\sigma}$, in case

$$\overline{\mu}=\overline{\mathfrak{C}}^{\,l}\,\mathfrak{D}(X)\,\overline{\mathfrak{m}}.$$

We suppose that

$$(\overline{\mathfrak{m}}, \overline{\mathfrak{m}}^{\sigma})=1 \quad \text{for} \quad \sigma\notin H,$$

as this can be achieved by multiplying $\overline{\mu}$ for a number satisfying conditions (**). In order that $\overline{\mu}\overline{m}$ should satisfy condition B it is necessary and sufficient that the condition

$$\left(\frac{\overline{\mu}\overline{m}}{\mathfrak{Q}^{\sigma}}\right)=1, \tag{48}$$

be satisfied, where the Legendre symbol is evaluated in k, and $\mathfrak{Q}$ is any prime divisor of $\overline{\mathfrak{m}}\overline{m}$ and $\sigma\notin H$.

The conditions can be naturally divided into two groups:

$$\left(\frac{\overline{\mu}\overline{m}}{\mathfrak{B}^{\sigma}}\right)=1, \text{ i.e., } \left(\frac{\overline{m}}{\mathfrak{B}^{\sigma}}\right)=\left(\frac{\overline{\mu}}{\mathfrak{B}^{\sigma}}\right)^{-1} \quad \text{for} \quad \mathfrak{B}\,|\,\overline{\mathfrak{m}}$$

and

$$\left(\frac{\overline{m}}{\mathfrak{Q}^{\sigma}}\right)=\left(\frac{\overline{\mu}}{\mathfrak{Q}^{\sigma}}\right)^{-1} \quad \text{for} \quad \mathfrak{Q}\,|\,\overline{m}.$$

We select $\overline{m}$ in such a way that all the prime ideals which divide it satisfy the condition

$$\left(\frac{\overline{\mu}}{\mathfrak{Q}^{\sigma}}\right)=\zeta(\sigma),$$

where $\zeta(\sigma)$ is some set of l-th root of unity. The number $\overline{m}$ itself must satisfy the conditions

$$\left(\frac{\overline{m}}{\mathfrak{B}^{\sigma}}\right)=\left(\frac{\overline{\mu}}{\mathfrak{B}}\right)^{-1}, \quad \mathfrak{B}\,|\,\overline{m}, \tag{49}$$

$$\left(\frac{\overline{m}}{\mathfrak{Q}^{\sigma}}\right)=\zeta(\sigma)^{-1}, \quad \sigma \notin H, \quad \mathfrak{Q}\,|\,\overline{m}. \tag{50}$$

We shall consider the set $\zeta(\sigma)$ of roots of unity constant over the left cosets S of the group G with respect to H and we will denote $\zeta(\sigma)$ by $\zeta(S)$ if $\sigma \in S$. If $\mathfrak{S}$ is a coset of G with respect to the double modulus (H, H), then by $\zeta(\mathfrak{S})$ we will understand

$$\zeta(S_1)\cdots\zeta(S_r),$$

where $S_1, \cdots, S_r$ are distinct left cosets of G with respect to H, which are contained in $\mathfrak{S}$. We will suppose that the roots of unity $\zeta(S)$ so defined satisfy the conditions

$$\zeta(\mathfrak{S})^{g(\mathfrak{S})} = \zeta(\mathfrak{S}^{-1}), \tag{51}$$

and in case $l = 2$ we shall also require that for any involutory classes $\mathfrak{S}_1, \cdots, \mathfrak{S}_u$ the condition

$$\zeta(\mathfrak{S}_1)\cdots\zeta(\mathfrak{S}_u) = 1 \tag{52}$$

is satisfied whenever

$$\mathfrak{E}_{\mathfrak{S}_1}\cdots\mathfrak{E}_{\mathfrak{S}_u} \approx (\alpha_\chi) \text{ in } L.$$

In the appendix to this paper we have shown the existence of infinitely many numbers $\overline{m}$ satisfying the conditions (50) for arbitrary roots of unity $\zeta(S)$, satisfying the conditions (51) and (52). It remains to show that it is possible to make m satisfy conditions (**). Here we must consider separately the two cases $l \neq 2$ and $l = 2$.

We first suppose that $l \neq 2$. In this case $\overline{m}$, according to the theorem proved in the appendix, can be obtained as the product of two prime satisfying the conditions

$$\left(\frac{\varkappa_i}{\mathfrak{B}^{\sigma}}\right)_k=\left(\frac{\varkappa_i}{N_{k/L}\mathfrak{B}^{\sigma}}\right)_L=\left(\frac{\overline{\mu}}{\mathfrak{B}^{\sigma}}\right)^{-\frac{1}{2}}, \quad \mathfrak{B}\,|\,\overline{m}, \quad i=1,\,2, \tag{53}$$

where we can satisfy conditions (**) for $\overline{m}$, if we can prove the existence of an infinitely many primes κ of the first degree satisfying the conditions (**), (53) and the relation

$$\left(\frac{\overline{\mu}}{\mathfrak{Q}^{\sigma}}\right)_k=\zeta(S). \tag{54}$$

We shall clarify under what conditions prime numbers exist satisfying all these conditions.

First we shall suppose that the numbers $\gamma_{\mathfrak{P},\mathfrak{S}}$ are given in conditions (46) and (47) and find out when there exists a κ satisfying (46) and (47), the remaining conditions (**) and also (53) and (54). In order to do this we denote by $\mathfrak{F}$ the product $N_{k/L}\mathfrak{P}$ of all ramified prime ideals $\mathfrak{P}$ in k/Ω, by all $N_{k/L}\mathfrak{B}^{\sigma}$ for $\mathfrak{B}\,|\,\overline{\mathfrak{m}}$, the prime divisors $\mathfrak{B}$ of the number M/M_l, the modulus of hyperprimarity, and the real infinite primes L, and by $S_{\mathfrak{F}}$ the group of unions of ideal classes modulo $\mathfrak{F}$, and by $\Sigma_{\mathfrak{F}}$ the class field of $S_{\mathfrak{F}}$. Repeating the arguments on p. 404 of paper [8] we get that condition (46) and the remaining conditions (**) as well as the equality

$$\left(\frac{\overline{m}}{\mathfrak{P}^{\sigma}}\right)=\left(\frac{\overline{\mu}}{\mathfrak{P}^{\sigma}}\right)^{-\frac{1}{2}} \tag{55}$$

are equivalent to the fact that the ideal $(\overline{m})$ belongs to a completely defined automorphism s in $\Sigma_{\mathfrak{F}}$ and to the identity automorphism in the field $L(\zeta_M)$, and that the conditions

$$\left(\frac{\overline{\mu}}{\mathfrak{Q}^{\sigma}}\right)=\zeta(S),\quad \mathfrak{Q}\,|\,\varkappa,\quad \sigma\notin H,$$

that is

$$\left(\frac{\overline{\mu}^{\tau}}{\mathfrak{Q}}\right)=\zeta(T^{-1})^{\sigma(\tau)},$$

where S is the left, and T is the right coset of G with respect to H, are equivalent to the fact that (κ) belongs to a completely defined automorphism t in the field K^*, which is obtained by adjoining to k all the $\sqrt[l]{\overline{\mu}^{\sigma}}$, $\sigma\in H$.

As in the first part of the proof we see that these conditions are compatible only if the automorphisms s and t coincide in $K^*\cap\Sigma_{\mathfrak{F}}$.

The field $K^*\cap\Sigma_{\mathfrak{F}}$ is obtained by adjoining to L all the $\sqrt[l]{\alpha_\chi}$ and $\sqrt[l]{m_\Phi}$, where $\Phi(T)$ is such a function of the right cosets T of the group G with respect to H that

$$\mu^{\Sigma\Phi(T)\tau}\approx m_\Phi\in L,\quad \Phi(1)=0. \tag{56}$$

We note that from $m_\Phi\in L$, that is $m_\Phi^h=m_\Phi$, it follows, since $\mathfrak{m}$ is divisible by prime ideals of the first degree, that identically

$$\left(\sum\Phi(T)\tau\right)h\equiv\sum\Phi(T)\tau\ \ (l).$$

This shows that $\Phi(T)$ is a constant over cosets with respect to the modulus (H,H).

We must find out when s and t coincide first in $L(\sqrt[l]{m_\Phi})$ and secondly in $L(\sqrt[l]{\alpha_\chi})$. Repeating the arguments on page 404 of paper [8], we see that in $L(\sqrt[l]{\alpha_\chi})$ the identity automorphism is induced by t, while s induces an automorphism which reduces to multiplication of $\sqrt[l]{\alpha_\chi}$ by

$$\zeta^{\sum\limits_{\mathfrak{P},\mathfrak{S}}\varphi_{\mathfrak{P},\chi}(\mathfrak{S})\,y_{\mathfrak{P},\mathfrak{S}}}$$

if

$$(\alpha_\chi)=\prod_{\mathfrak{P}}\mathfrak{P}^{\Sigma\varphi_{\mathfrak{P},\chi}(\mathfrak{S})\,\sigma\Sigma h}.$$

In the field $L(\sqrt[l]{m_\Phi})$, however, s induces an automorphism which can be reduced to multiplication of $\sqrt[l]{m_\Phi}$ by

$$\left[\frac{\bar{\mu}}{\mathfrak{m}_\Phi}\right]^{-\frac{1}{2}} \prod_{\mathfrak{P}} \zeta^{\sum_{\mathfrak{S}} F_{\mathfrak{P},\Phi}(\mathfrak{S})\, y_{\mathfrak{P},\mathfrak{S}}},$$

if

$$\mathfrak{d}_\Phi = \prod_{\mathfrak{P}} \mathfrak{P}^{\sum_{\mathfrak{S}} F_{\mathfrak{P},\Phi}(\mathfrak{S})\, \sigma\Sigma h}, \qquad m_\Phi = \mathfrak{d}_\Phi \mathfrak{m}_\Phi.$$

The automorphism t induces in the field $L(\sqrt[l]{m_\Phi})$ an automorphism which reduces to the multiplication of $\sqrt[l]{m_\Phi}$ by

$$\prod_T \zeta(T^{-1})^{g(T^{-1})\Phi(T)},$$

where the product goes over all the right cosets of the group G with respect to H.

We denote $\left[\frac{\bar{\mu}}{m_\Phi}\right]^{-1/2}$ by ζ^{a_Φ}, $\prod_T \zeta(T^{-1})^{g(T^{-1})\Phi(T)}$ by ζ^{b_Φ}. Then we must discuss the compatibility of the system of congruences

$$\sum f_{\mathfrak{P}}(S)\, y_{\mathfrak{P},\mathfrak{S}} \equiv 0\ (l), \tag{57}$$

if $f_{\mathfrak{P}}(S)$ satisfies (35),

$$\sum \phi_{\mathfrak{P},\chi}(\mathfrak{S})\, y_{\mathfrak{P},\mathfrak{S}} \equiv 0\ (l), \tag{58}$$

$$\sum F_{\mathfrak{P},\Phi}(\mathfrak{S})\, y_{\mathfrak{P},\mathfrak{S}} + a_\Phi \equiv b_\Phi(l). \tag{59}$$

If this system were incompatible, there would exist a linear combination of these congruences such that all the coefficients of $y_{\mathfrak{P},S}$ would be equal to 0, while the constant term would be $\neq 0$. We note that a linear combination of the various congruences (57) is again one of the congruences only with a different $f_{\mathfrak{P}}(S)$, and that the same remark applies to the congruences (58) and (59). In this way if our system were incompatible, there would exist an $f_{\mathfrak{P}}, \chi$ and Φ such that in the congruence

$$\sum f_{\mathfrak{P}}(S)\, y_{\mathfrak{P},\mathfrak{S}} + \sum \varphi_{\mathfrak{P},\chi}(\mathfrak{S})\, y_{\mathfrak{P},\mathfrak{S}} + \sum F_{\mathfrak{P},\Phi}(\mathfrak{S})\, y_{\mathfrak{P},\mathfrak{S}} + a_\Phi \equiv b_\Phi\ (l)$$

all the coefficients of $y_{\mathfrak{P},\mathfrak{S}}$ would be equal to zero and the constant term $b_\Phi - a_\Phi$ be different from zero. From the above assumption it follows that

$$F_{\mathfrak{P},\Phi}(\mathfrak{S}) \equiv - \sum_{S\in\mathfrak{S}} f_{\mathfrak{P}}(S) - \varphi_{\mathfrak{P},\chi}(\mathfrak{S})\ (l),$$

that is

$$\mathfrak{d}_\Phi = (\alpha_\chi)^{-1} \prod_{\mathfrak{P}} \mathfrak{P}^{-\Sigma f_{\mathfrak{P}}(S)\sigma\Sigma h}, \tag{60}$$

where $\Sigma_{S\in\mathfrak{S}} f_{\mathfrak{P}}(S)$ satisfies condition (35). The condition for compatibility of our system can now be written in the form $a_\Phi \equiv b_\Phi\ (l)$, that is

$$\prod \zeta(T^{-1})^{g(T^{-1})\Phi(T)} = \left[\frac{\bar{\mu}}{\mathfrak{m}_\Phi}\right]^{-\frac{1}{2}} \tag{61}$$

for any Φ which is constant over cosets modulo (H, H), satisfies condition (56) and for which the ideal $\mathfrak{b}_\Phi$ in the decomposition of m_Φ is of the form (22).

It remains to find out when there exist roots of unity $\zeta(S)$ satisfying conditions (51), (52) and (61).

We let

$$\zeta(S)=\zeta^{x_S}, \quad \left[\frac{\bar{\mu}}{\mathfrak{m}_\Phi}\right]^{-\frac{1}{2}}=\zeta^{b_\Phi};$$

then conditions for x_S can be written in terms of the congruences:

$$\begin{gathered} g(\mathfrak{S}) \sum_{S\in\mathfrak{S}} x_S \equiv \sum_{S'\in\mathfrak{S}^{-1}} x_{S'}\ (l), \\ \sum_T x_{T^{-1}}\Phi(T)\, g(T^{-1}) \equiv b_\Phi\ (l). \end{gathered} \tag{62}$$

We find the conditions for the compatibility of this system. If the system is not compatible, then there exists a linear combination of these congruences in which all the coefficients of the variables x_S are equal to zero and the constant term is $\neq 0$. Let

$$\sum_{\mathfrak{S}} C_{\mathfrak{S}}\, g(\mathfrak{S})\Big(\sum_{S\in\mathfrak{S}} x_S - \sum_{S'\in\mathfrak{S}^{-1}} x_{S'}\Big) + \sum_\Phi d_\Phi \sum_T x_{T^{-1}}\Phi(T)\, g(T^{-1}) \equiv \sum_\Phi d_\Phi b_\Phi\ (l) \tag{63}$$

be such a combination. Let

$$\sum_\Phi d_\Phi\, \Phi(T) = \psi(T).$$

Then, obviously

$$\sum_\Phi d_\Phi b_\Phi \equiv b_\psi\ (l).$$

Taking into account that in (63) the coefficients of all x_S are equal to 0, we get

$$g(T)\psi(T^{-1}) \equiv C_{\mathfrak{T}^{-1}} - C_{\mathfrak{T}}\, g(T)\ (l)$$

that is

$$\psi(T) \equiv g(\mathfrak{T})\, C_{\mathfrak{T}} - C_{\mathfrak{T}^{-1}}\ (l). \tag{64}$$

We have already seen that $\psi(T)$ is a constant over cosets modulo (H,H). This follows once more from (64), moreover, if we set

$$\psi(\mathfrak{S}) = \psi(S), \quad S\in\mathfrak{S},$$

then it follows from (64) that

$$g(\mathfrak{S})\,\psi(\mathfrak{S}^{-1}) \equiv -\psi(\mathfrak{S})\ (l).$$

This shows together with (60) that $\psi(\mathfrak{S})$ satisfies all the conditions in the definition of the number $[X]_\psi$. From this we conclude that

$$\zeta^{b_\psi} = \left[\frac{\bar{\mu}}{\mathfrak{m}_\phi}\right]^{-1/2} = 1,$$

that is $b_\psi \equiv 0\ (l)$, from which follows the compatibility of the system (62).

We proceed in an analogous way in case $l=2$. By the proof of the theorem given in the appendix the number $\bar{m}$ can in this case be found as a product

$\kappa_1\kappa_2\kappa_3$ of three primes satisfying the conditions

$$\left(\frac{\kappa_i}{\mathfrak{P}^\sigma}\right)=\left(\frac{\overline{\mu}}{\mathfrak{P}^\sigma}\right), \quad \sigma \notin H, \quad i=1,2,3, \tag{65}$$

where we will satisfy conditions (**) and (48) for $\overline{m}$ if we can prove that there exist infinitely many primes κ satisfying conditions (**), (54) and (65).

As in the case of $l \neq 2$ it is necessary and sufficient for the existence of such numbers, that the following system be compatible:

$$\sum_{S\in\mathfrak{S}} x_S \equiv \sum_{S'\in\mathfrak{S}^{-1}} x_{S'}\ (l), \tag{66}$$

$$\sum_{S\in\mathfrak{S}_1+\cdots+\mathfrak{S}_u} x_S \equiv 0\ (l), \quad \text{if} \quad \mathfrak{C}_{\mathfrak{S}_1}\cdots\mathfrak{C}_{\mathfrak{S}_u} \approx (\alpha_\chi) \text{ and } [\chi, X] \text{ is defined,}$$

$$\sum_T x_{T^{-1}}\Phi(T) \equiv b_\Phi\ (l)$$

for every Φ satisfying condition (56).

The assumption as to the incompatibility of this system leads again to a linear combination of the congruences in which the coefficients of all the x_S are zero, while the constant tem is different from zero. Let this linear combination be

$$\sum_{\mathfrak{S}} C_{\mathfrak{S}}\Big(\sum_{S\in\mathfrak{S}} x_S - \sum_{S'\in\mathfrak{S}^{-1}} x_{S'}\Big) + \sum_{\Phi} d_\Phi \sum_T x_S \Phi(S^{-1}) +$$

$$+ \sum_{(\mathfrak{S}_1,\ldots,\mathfrak{S}_u)} e_{\mathfrak{S}_1,\ldots,\mathfrak{S}_u} \sum_{S\in\mathfrak{S}_1+\ldots+\mathfrak{S}_u} x_{\mathfrak{S}} \equiv \sum_\Phi d_\Phi b_\Phi\ (l),$$

where the outside sum in the third summand goes over the totality of all involutory classes $(\mathfrak{S}_1, \cdots, \mathfrak{S}_u)$ modulo (H, H) for which

$$\mathfrak{C}_{\mathfrak{S}_1}\cdots\mathfrak{C}_{\mathfrak{S}_u} \approx (\alpha_\chi) \text{ and } [\chi, X] \text{ is defined.} \tag{67}$$

We suppose once more that

$$\sum_\Phi d_\Phi \Phi(T) = \psi(T)$$

and obtain as in the case $l \neq 2$ that

$$\psi(\mathfrak{S}^{-1}) \equiv C_{\mathfrak{S}^{-1}} + C_{\mathfrak{S}} + \sum_{(\mathfrak{S}_1,\ldots,\mathfrak{S}_u)} e_{\mathfrak{S}_1,\ldots,\mathfrak{S}_u} \varepsilon_{\mathfrak{S}_1,\ldots,\mathfrak{S}_u}(\mathfrak{S})\ (2), \tag{68}$$

where $e_{\mathfrak{S}_1,\cdots,\mathfrak{S}_u}(\mathfrak{S})$ is a characteristic function of the set $\mathfrak{S}_1, \cdots, \mathfrak{S}_u$ of involutory automorphisms, if condition (67) is satisfied for the set, and is zero otherwise. From this we conclude that

$$\psi(\mathfrak{S}^{-1}) \equiv \psi(\mathfrak{S})\ (2).$$

Since by definition

$$\prod_{\mathfrak{S}} \mathfrak{C}_{\mathfrak{S}}^{e_{\mathfrak{S}_1,\ldots,\mathfrak{S}_u}(\mathfrak{S})} \approx (\alpha_\chi), \quad [\chi, X] \quad \text{is defined.}$$

where the product goes over all the involutory automorphisms, it follows from (68) that

$$\prod_{\mathfrak{S}} \mathfrak{E}_{\mathfrak{S}}^{\psi(\mathfrak{S})} \approx (\alpha_\chi), \quad [\chi, X] \text{ is defined.}$$

Here the function $\psi(\mathfrak{S})$ satisfies all the conditions which enter into the definition of $[X]_\psi$ and hence

$$(-1)^{b_\psi} = \left[\frac{\mu}{m_\psi}\right] = [X]_\psi = 1,$$

which shows the compatibility of the system (66). The theorem is proved.

§3. A general condition of imbeddedness for Scholz fields

In what follows we shall need a generalization of Theorem 1 to a more general case where the field K/L in the chain (3) is a central, but not a simple central extension of the field k/L. To do this it will be necessary to introduce another set of invariants of the field k.

We denote the Galois group of the field K/L by $\mathfrak{G}$, and of the field K/k by $\mathfrak{Z}$. It is obvious that $\mathfrak{G}/\mathfrak{Z} \cong G$ and that $\mathfrak{Z}$ lies in the center of $\mathfrak{G}$. In what follows we shall suppose that $\mathfrak{Z}$ is of index l. Any character X of the group $\mathfrak{Z}$ defines a class of l-invariant numbers X in k. It is obvious that these classes form a group which is a homomorphic image of the character group $\mathfrak{X}$ of the group $\mathfrak{Z}$. We select in $\mathfrak{X}$ some fixed basis $X_1, \cdots, X_r$ which we shall use in what follows. We will always suppose that the invariants $[\chi, X_i]$, $(X_i)_M$ and $[X_i]_M$ are equal to 1 for $i = 1, \cdots, r$.

Let $\phi_1(S), \cdots, \phi_s(S)$ be functions over right cosets of G with respect to H, which are constant over the cosets of G modulo (H, H). We shall understand by A^ϕ the expression $A^{\Sigma\phi(S)\sigma}$, where S in the exponent goes over all right cosets of the group G with respect to H, and $\sigma \in S$. Since ϕ is constant over the cosets modulo (H, H) it follows from $\alpha \in L$, that $\alpha^\phi \in L$, and from the l-invariance of μ in k/L follows the l-invariance of μ^ϕ. Therefore X^ϕ is a definite class of l-invariant numbers if X is a class of l-invariant numbers.

We suppose that the functions $\phi_1, \cdots, \phi_{s-1}$ are such that

$$X_1^{\varphi_1} \ldots X_{s-1}^{\varphi_{s-1}} = 1, \tag{69}$$

$$X_s^{\overset{*}{\varphi}_1} = \ldots = X_s^{\overset{*}{\varphi}_{s-1}} = 1. \tag{70}$$

We select in the classes $X_1, \cdots, X_{s-1}$ some l-hyperprimary representatives $\mu_1, \cdots, \mu_{s-1}$. In view of (69) we have

$$\mu_1^{\varphi_1} \ldots \mu_{s-1}^{\varphi_{s-1}} \approx m_{\varphi_1, \ldots, \varphi_{s-1}} \in L. \tag{71}$$

We let

$$[X_1, \ldots, X_{s-1}; X_s]_{\varphi_1, \ldots, \varphi_{s-1}} = [m_{\varphi_1, \ldots, \varphi_{s-1}}, X_s],$$

where the symbol $[\alpha, X]$ is defined by formulas (4′) and (10) by replacing α_χ by α.

We prove that the symbol $[X_1, \cdots, X_{s-1}; X_s]_{\phi_1, \cdots, \phi_{s-1}}$ does not depend on the choice of the l-hyperprimary representatives $\mu_1, \dots, \mu_{s-1}$ in the classes $X_1, \dots, X_{s-1}$. The change to another system of representatives $\mu_1 m_1, \dots, \mu_{s-1} m_{s-1}$, $m_i \in L$, leads to multiplication of $m\phi_1, \dots, \phi_{s-1}$ by $m_1^{\phi_1} \dots m_{s-1}^{\phi_{s-1}}$. Because the symbol $[\alpha, X]$ is obviously multiplicative in α, the symbol $[X_1, \dots, X_{s-1}; X_s]_{\phi_1, \dots, \phi_{s-1}}$ is multiplied by

$$[m_1^{\phi_1} \dots m_{s-1}^{\phi_{s-1}}, X_s].$$

Because of the obvious identity

$$[\alpha^{\phi}, X] = [\alpha, X^{\phi^*}],$$

we have

$$[m_1^{\varphi_1} \dots m_{s-1}^{\varphi_{s-1}}, X_s] = [m_1^{\varphi_1}, X_s] \dots [m_{s-1}^{\varphi_{s-1}}, X_s] =$$
$$= \left[m_1, X_s^{\overset{*}{\varphi_1}}\right] \dots \left[m_{s-1}, X_s^{\overset{*}{\varphi_{s-1}}}\right]. \tag{72}$$

In view of (70)

$$X_s^{\phi_i^*} = 1, \quad i = 1, \dots, s-1,$$

and since m_i is l-hyperprimary it follows that

$$\left[m_i, X^{\phi_i^*}\right] = 1.$$

In conjunction with (72) this shows the independence of the symbol

$$[X_1, \dots, X_{s-1}; X_s]_{\phi_1, \dots, \phi_{s-1}}$$

on the choice of $\mu_i \in X_i$.

We note that although the $m\phi_1, \dots, \phi_{s-1}$ are not defined uniquely in L up to an l-equality by formula (71), they are defined uniquely up to a multiplier α_χ. Since we have supposed that $[\chi, X_S] = 1$, the symbol $[X_1, \dots, X_{s-1}; X_s]_{\phi_1, \dots, \phi_{s-1}}$ remains unaltered by multiplying $m\phi_1, \dots, \phi_{s-1}$ by α_χ.

Theorem 2. *In order that the Scholz field k with invariants $[\chi, X]$, $(X)_M$ and $[X]_\psi$ equal to unity can be imbedded in a field K with the Galois group $\mathfrak{G}$ over L such that K^*/Ω be a Scholz field, it is necessary and sufficient that all the invariants $[X_1, \dots, X_{s-1}; X_s]_{\phi_1, \dots, \phi_{s-1}}$, for $s = 2, \dots, r$ be equal to one.*

Proof. In order to show the necessity we suppose that the field k is imbedded in the field K with the desired properties. This means that we can select in the classes $X_1, \dots, X_r$ the representatives $\mu_1, \dots, \mu_r$, such that the field which is obtained by adjoining to k all the $\sqrt[l]{\mu_i^\sigma}$, $i = 1, \dots, r$, $\sigma \in G$, is a Scholz field. Moreover the field K_S obtained by adjoining to k all the $\sqrt[l]{\mu_i^\sigma}$, $i = 1, \dots, s$, is also a Scholz field. Obviously K_S is obtained by adjoining to K_{s-1} all the $\sqrt[l]{\mu_s^\sigma}$, $\sigma \in G$.

By Theorem 1, in order that K_S be a Scholz field it is necessary that $[\chi, X_s] = 1$ in K_{s-1}. The number $m\phi_1, \dots, \phi_{s-1}$ is such a number of L that $\sqrt[l]{m\phi_1, \dots, \phi_{s-1}} \in K_{s-1}$, that is one of the numbers α_χ. Hence $[m\phi_1, \dots, \phi_{s-1}, X_s]$ must be equal to 1, which proves the necessity of Theorem 2.

The sufficiency of the conditions of Theorem 2 will follow if we can show that the field k can be imbedded for any $s \leq r$ in a field K_s for which K_s^* is a Scholz field. Theorem 2 will then follow from this assertion since $K_r = K$. The assertion itself will be proved by induction on s. We can suppose that we have selected in the classes $X_1, \cdots, X_{s-1}$ the numbers $\mu_1, \cdots, \mu_{s-1}$ such that the field K_{s-1}^*, which is obtained by adjoining to k all the $\sqrt[l]{\mu_i^\sigma}$, $i = 1, \cdots, s-1, \sigma \in G$, is a Scholz field. We shall also suppose that the ideals $\mathfrak{m}_i \neq 1$, which correspond to them satisfy the conditions:

$$(\mathfrak{m}_i, \mathfrak{m}_j^\sigma) = 1 \quad \text{for} \quad i \neq j, \quad \sigma \notin H. \tag{73}$$

By Theorem 1 it is necessary and sufficient for the existence of a field K_s with the required properties that all the invariants $[\chi, X_s]$, $(X_s)_M$ and $[X_s]_\psi$ be equal to 1 in K_{s-1}^*. Since X_s contains l-invariant numbers in k, the invariants $(X_s)_M$ and $[X_s]_\psi$ can be calculated in the field k, where they are equal to 1 by assumption. For the same reason the invariants $[\chi, X_s]$ are equal to 1 for the characters χ of the Galois group of the field k/L.

All the numbers α_χ for any character χ of the Galois group of the field K_{s-1}^*/L can be obtained from the numbers α_χ corresponding to the characters of the Galois group of k/L, by multiplication by the numbers $m_{\phi_1, \cdots, \phi_{s-1}}$, where $\phi_1, \cdots, \phi_{s-1}$ are functions satisfying conditions (69).

In fact we must find numbers $m \in L$ such that $\sqrt[l]{m} \in K_{s-1}$. Since, moreover, $m \in k$, then $m \approx \mu_1^{\phi_1} \cdots \mu_{s-1}^{\phi_{s-1}}$ in k.

We now consider the ideals $\mathfrak{m}_i$ and note that in view of (73) we can suppose that

$$\mathfrak{m}_1^{\phi_1} \in L, \cdots, \mathfrak{m}_{s-1}^{\phi_{s-1}} \in L.$$

From this it follows as on page 208 that the $\phi_1, \cdots, \phi_{s-1}$ are constant over the cosets of G modulo (H, H) and that $m \approx m_{\phi_1, \cdots, \phi_{s-1}}$ in k and therefore $m \approx m_{\phi_1, \cdots, \phi_{s-1}} \alpha_\chi$ in L.

Therefore we must achieve that the equality

$$[m_{\phi_1, \cdots, \phi_{s-1}}, X_s] = 1 \tag{74}$$

be satisfied for all $\phi_1, \cdots, \phi_{s-1}$ satisfying conditions (69).

We can achieve the equality (74) by multiplying every μ_i by $m_i \in L$, $i = 1, \cdots, s-1$, if in doing so we do not alter the Scholz character of the field K_{s-1}^*. Under such a multiplication $[m_{\phi_1, \cdots, \phi_{s-1}}, X_s]$ is multiplied by $[m_1^{\phi_1} \cdots m_{s-1}^{\phi_{s-1}}, X_s]$. Therefore we must select the m_i such that

$$\left[m_1^{\varphi_1} \ldots m_{s-1}^{\varphi_{s-1}}, X_s\right] = [m_{\varphi_1, \ldots, \varphi_{s-1}}, X_s]^{-1}$$

or

$$\left[m_1, X_s^{\overset{*}{\varphi}_1}\right] \ldots \left[m_{s-1}, X_s^{\overset{*}{\varphi}_{s-1}}\right] = [m_{\varphi_1, \ldots, \varphi_{s-1}}, X_s]^{-1}. \tag{75}$$

We denote $[m_i, X_s^{\phi_i^*}]$ by $\zeta_i(\phi_i)$, and $[m\phi_1, \dots, \phi_{s-1}, X_s]$ by $\eta(\phi_1, \dots, \phi_{s-1})$. Equation (75) for m_i can then be written

$$\left[m_i, X_s^{\phi_i^*}\right] = \zeta_i(\phi_i)^{-1}, \tag{76}$$

where

$$\zeta_1(\phi_1) \cdots \zeta_{s-1}(\phi_{s-1}) = \eta(\phi_1, \dots, \phi_{s-1}). \tag{77}$$

It is obvious that

$$\zeta_i(\varphi_i + \varphi_i') = \zeta_i(\varphi_i)\zeta_i(\varphi_i'), \tag{77'}$$

$$\eta(\varphi_1 + \varphi_1', \dots, \varphi_{s-1} + \varphi_{s-1}') = \eta(\varphi_1, \dots, \varphi_{s-1})\,\eta(\varphi_1', \dots, \varphi_{s-1}'). \tag{77''}$$

We must next see what conditions must be imposed on m_i in order that replacing μ_i by $m_i\mu_i$ would not alter the Scholz character of the field K_{s-1}^*. We denote by $\mathfrak{P}$ an arbitrary prime divisor of the discriminant of the field $k/\Omega(\zeta)$, by $\mathfrak{Q}_i$ an arbitrary prime divisor of m_i, and by $\mathfrak{B}_i$ an arbitrary prime divisor of $\mathfrak{m}_i$. The following conditions guarantee that the multiplication of μ_i by m_i does not alter the Scholz character of the field K_{s-1}^*:

$$\left(\frac{m_i}{\mathfrak{P}}\right) = 1, \quad \left(\frac{m_i}{\mathfrak{Q}_j^\sigma}\right) = 1, \quad \left(\frac{\mu_i}{\mathfrak{Q}_j^\sigma}\right) = 1, \quad \left(\frac{m_i}{\mathfrak{B}_j^\sigma}\right) = 1. \tag{78}$$

$$\mathfrak{N}(\mathfrak{Q}_i) \equiv 1\ (M), \quad \left(\frac{m_i}{\mathfrak{N}}\right) = 1, \quad \mathfrak{N} \mid M. \tag{79}$$

In order to satisfy condition (71) we let

$$\left(\frac{\mu_s^{\varphi_i^*}}{\mathfrak{Q}_i}\right) = 1, \quad \left(\frac{m_i}{m_s^{\varphi_i^*}}\right) = \zeta_i(\varphi_i). \tag{80}$$

Conditions (77) contain the condition

$$\left(\frac{m_i}{\mathfrak{Q}_i^\sigma}\right) = 1, \quad \mathfrak{Q}_i \mid m_i.$$

In order to satisfy these conditions we must apply the theorem which is proved in the appendix. According to the proof of that theorem we must look for a number m_i of the form $\kappa_i\kappa_i'$ for $l \neq 2$, and of the form $\kappa_i\kappa_i'\kappa_i''$ for $l = 2$, where κ_i, κ_i', κ_i'' are principal prime ideals. We will put down the conditions which κ_i, κ_i' and κ_i'' must satisfy in order that m_i should satisfy conditions (76) and (78). In doing this we will put down only the conditions for κ_i, assuming that they must also hold for κ_i', and for $l = 2$ also for κ_i'':

$$\left(\frac{\kappa_i}{\mathfrak{P}}\right) = 1, \tag{81}$$

$$\left(\frac{\kappa_i}{\mathfrak{Q}_j^\sigma}\right) = 1, \quad i \neq j, \tag{82}$$

$$\left(\frac{\mu_j}{\mathfrak{Q}_i^\sigma}\right) = 1, \quad \left(\frac{\kappa_i}{\mathfrak{B}_j^\sigma}\right) = 1, \tag{83}$$

$$\left(\frac{\mu_s^{\varphi_i^*}}{\mathfrak{Q}_i}\right) = 1, \qquad \left(\frac{\varkappa_i}{\mathfrak{m}_s^{\varphi_i^*}}\right) = \zeta_i(\varphi_i)^\epsilon, \tag{84}$$

$$\varkappa_i \equiv 1 \ (M), \tag{85}$$

where $\epsilon = \frac{1}{2}$ for $l \neq 2$ and $\epsilon = 1$ for $l = 2$. The numbers κ_i will be chosen successively for $i = 1, \cdots, s-1$, making sure that condition (83) is satisfied for any κ_t, for $i = t$ and for all $j = 1, \cdots, s-1$, and condition (82) for $i = t$, $j < t$ and for $i < t$, $j = t$, while condition (81) for $i = t$.

We note that conditions (81), (82), (83) and (84) are compatible since they are equivalent to the fact that κ_t belong to the identity automorphism in some field Σ. In order to clarify the compatibility of these conditions with condition (84) we must find the intersection of Σ with a field Σ' such that if κ belongs to a certain automorphism in Σ' this is equivalent to condition (84). It is obvious that the discriminant of Σ' consists only of divisors $\mathfrak{m}_s$. From this it follows easily that the intersection of Σ with Σ' is a composite of the fields $L(\sqrt[l]{m_{\phi^*}})$, where $\mu_s^{\phi^*} \approx m_{\phi^*} \in L$ for all functions ϕ, for which this equality holds. Hence conditions (78), (79) and (80) are compatible provided

$$\zeta_i(\phi) = 1 \tag{85'}$$

for every ϕ for which

$$\mu_s^{\phi^*} \approx m_{\phi^*} \in L.$$

It remains to show that there exists a function $\zeta_i(\phi)$ satisfying (69) and (70). In order to do this we consider the direct product $\Phi^{(s-1)}$ of the $(s-1)$st iterate of the group Φ of functions $\phi(S)$ constructed over cosets of G with respect to (H,H) and determined mod l. Formula (77″) shows that $\eta(\phi_1, \cdots, \phi_{s-1})$ is the character of the subgroup Φ_0 of the group $\Phi^{(s-1)}$, consisting of those choices of the functions $(\phi_1, \cdots, \phi_{s-1})$ for which

$$X_1^{\phi_1} \cdots X_{s-1}^{\phi_{s-1}} = 1.$$

On the other hand $\zeta_1(\phi_1) \cdots \zeta_{s-1}(\phi_{s-1})$, where $\zeta_i(\phi_i)$ satisfies (77′) is the most general character over $\Phi^{(s-1)}$. Conditions (85′) state that this character must be 1 with respect to the subgroup Φ_1 of the group $\Phi^{(s-1)}$ composed of a choice of functions $(\phi_1, \cdots, \phi_{s-1})$, for which

$$X_s^{\phi_1^*} = \cdots = X_s^{\phi_{s-1}^*} = 1.$$

Our problem consists in showing the existence of a character of the group $\Phi^{(s-1)}$, which coincides with $\eta(\phi_1, \cdots, \phi_{s-1})$ in Φ_0, and with the unit character in Φ_1. It is obvious that for the existence of such a character it is necessary that the character η should coincide with the unit character of $\Phi_0 \cap \Phi_1$. But this is assured by the equality to unity of the invariant $[X_1, \cdots, X_{s-1}; X_s]_{\phi_1, \cdots, \phi_{s-1}}$.

Hence the theorem is proved.,

We note that for the invariant $[X_1,\cdots,X_{s-1};X_s]_{\phi_1,\cdots,\phi_{s-1}}$ the following relations hold

$$[X_1X_1',\ldots,X_{s-1}X_{s-1}';X_s]_{\varphi_1,\ldots,\varphi_{s-1}}= \\ =[X_1,\ldots,X_{s-1};X_s]_{\varphi_1,\ldots,\varphi_{s-1}}[X_1',\ldots,X_{s-1}';X_s]_{\varphi_1,\ldots,\varphi_{s-1}}, \tag{86}$$

$$[X_1,\ldots,X_{s-1};X_sX_s']_{\varphi_1,\ldots,\varphi_{s-1}}= \\ =[X_1,\ldots,X_{s-1};X_s]_{\varphi_1,\ldots,\varphi_{s-1}}[X_1,\ldots,X_{s-1};X_s']_{\varphi_1,\ldots,\varphi_{s-1}} \tag{87}$$

as well as relations analogous to (18) for the invariants $[\chi, X]$.

§4. Construction of field with solvable Galois groups

1. Canonical homomorphisms. We consider an arbitrary finite group F or order m and a free group S_d with md generators, which we will denote by $s_{u,i}$, $u \in F$, $i=1,\cdots,d$. We shall consider normal subgroups N_c of S_d defined as follows: $N_0=S_d$, N_{c+1} is a normal subgroup of N_c generated by the l^k-th powers of the elements of S_d and by the commutators of the elements of S_d with the elements of N_c. Hence the prime l and the index k are fixed once and for all. We denote S_d/N_c by $\mathfrak{G}_d^{(c)}$. Obviously this quotient group depends on c, d, l, k and m, but since l, k and the group F remain fixed throughout the discussion, the notation involves only c and d. To every $v \in F$ corresponds a transformation of the variables according to the rule:

$$s_{u,i}^{v}=s_{uv,i}, \quad i=1,\cdots,d, \quad u \in F. \tag{88}$$

Such a transformation defines an automorphism in S_d, which generates a group isomorphic with F. Since, as can be easily seen, N_c is a characteristic subgroup of S_d, we have a group of automorphisms in $\mathfrak{G}_d^{(c)}$, which is isomorphic to F. The elements $x \in \mathfrak{G}_d^{(c)}$ which belong to the center of $\mathfrak{G}_d^{(c)}$ and satisfy the equation $x^{l^{k-r}}=1$ form a normal subgroup of $\mathfrak{G}_d^{(c)}$. We denote the quotient group of $\mathfrak{G}_d^{(c)}$ with respect to this normal subgroup by $\mathfrak{G}_d^{(c,r)}$. Obviously

$$\mathfrak{G}_d^{(c,0)}=\mathfrak{G}_d^{(c-1)},\quad \mathfrak{G}_d^{(c,k)}=\mathfrak{G}_d^{(c)}$$

and $\mathfrak{G}_d^{(c,r)}$ is homomorphic with $\mathfrak{G}_d^{(c,r-1)}$. We shall denote the kernel of this homomorphism by Z_r.

It is clear that the groups $\mathfrak{G}_d^{(c,r)}$ which we have defined are quotient groups of $\mathfrak{G}_d^{(c)}$ with respect to normal subgroups which are admissable relative to the operators of F defined by (88). In this way all these groups are operator groups in the domain of operators F.

We consider in F a fixed Sylow l-subgroup $\mathfrak{H}$. We define in $\mathfrak{G}_d^{(c,r)}$ a series of subgroups $Z_{r,s}$ by the following rule: $Z_{r,0}=Z_{r-1}$, $Z_{r,s}=[\mathfrak{H},Z_{r,s-1}]$ is a subgroup generated by the elements z^{1-h}, where $z \in Z_{r,s-1}$, $h \in \mathfrak{H}$. We denote $\mathfrak{G}_d^{(c,r)}/Z_{r,s}$ by $\mathfrak{G}_d^{(c,r,s)}$. It is obvious that for some m depending only on the group F,

$$\mathfrak{G}_d^{(c,r,m)} = \mathfrak{G}_d^{(c,r)}.$$

Suppose that in the chain of fields $\Omega \subset L \subset k \subset K$ the Galois group of K/Ω is $F \cdot \mathfrak{G}_d^{(c,r)}$, where F is the Galois group of k/Ω and $\mathfrak{G}_d^{(c,r)}$ is the Galois group of K/k. We consider a field $\overline{K} \supset K$ such that the Galois group of $\overline{K}/k$ is $\mathfrak{G}_d^{(c,r+1,s)}$, and that of $\overline{K}/L$ is $\mathfrak{H} \cdot \mathfrak{G}_d^{(c,r+1,s)}$. If the fields conjugate with $\overline{K}/\Omega$ are independent over k, then their composite will be denoted by $\overline{K}^*$.

The Galois group of the field $\overline{K}^*/\Omega$ has the following structure (this can be easily obtained by repeating the discussion on p. 395 of paper [8]).

We denote $Z_r/Z_{r+1,s}$ by $\mathfrak{A}$, and $Z_r/Z^u_{r+1,s}$, $u \in F$ by $\mathfrak{A}^u$. There exists a natural isomorphism of $\mathfrak{A}$ onto $\mathfrak{A}^u$. Its action on $\alpha \in \mathfrak{A}$ we shall denote by α^u. Moreover different from u, belonging to the same right coset U of the group F with respect to $\mathfrak{H}$ have the same effect on α. Therefore we can denote α^u by α^U, and $\mathfrak{A}^u$ by $\mathfrak{A}^U$. Let $a(\sigma, \tau)$ be a factor set of $\mathfrak{G}_d^{(c,r)}$ corresponding to the extension $\mathfrak{G}_d^{(c,r+1,s)}$. We consider the extension $\mathfrak{G}_d^{(c,r+1,s)*}$ of the group $\mathfrak{G}_d^{(c,r)}$ with a normal subgroup which is the direct product of all $\mathfrak{A}^U$ and with a factor set

$$A(\sigma, \tau) = (\ldots, \alpha(\sigma u^{-1}, \tau u^{-1})^U, \ldots).$$

The elements of F define in a natural way automorphisms in the group $\mathfrak{G}_d^{(c,r+1,s)*}$ which we will now consider as an F-operator group. The Galois group of the field $\overline{K}^*/\Omega$ is a semidirect product of $F \cdot \mathfrak{G}_d^{(c,r+1,s)*}$.

In what follows we shall consider canonical operator homomorphisms of the group $\mathfrak{G}_d^{(c)}$ on the group $\mathfrak{G}_\delta^{(c)}$, defined in paper [8]. It is easily seen that these homomorphisms define the homomorphisms $\mathfrak{G}_d^{(c,r)}$ on $\mathfrak{G}_\delta^{(c,r)}$ and $\mathfrak{G}_d^{(c,r,s)*}$ on $\mathfrak{G}_\delta^{(c,r,s)*}$. It is obvious that these homomorphisms are F-operator. We shall apply to these homomorphisms the notions of composition of automorphisms, of functions of homomorphisms and their powers, which were introduced in papers [6] and [8].

We shall apply these notions to class field theory using the following considerations. Suppose that we are given a chain of fields $\Omega \subset L \subset k \subset K$, where k/Ω is normal with the Galois group F, L belongs to the Sylow subgroup $\mathfrak{H}$, and K/k has for its Galois group one of the groups $\mathfrak{G}_d^{(c)}$, $\mathfrak{G}_d^{(c,r)}$, $\mathfrak{G}_d^{(c,r,s)*}$. Moreover K/Ω is normal and has respectively the Galois group $F \cdot \mathfrak{G}_d^{(c)}$, $F \cdot \mathfrak{G}_d^{(c,r)}$ or $F \cdot \mathfrak{G}_d^{(c,r,s)*}$. We suppose that K/Ω is a Scholz field with respect to M and l. To every homomorphism S of the Galois group of K/k corresponds a subfield $K(S)$ which is a Scholz field with respect to M and l. In this field the invariants $[\chi, X]$ and $(X)_M$ are defined, as well as the invariant $[X]_\psi$, in case $[\chi, X] = (X)_M = 1$, and the invariants $[X_1, \ldots, X_{s-1}; X_s]_{\phi_1, \ldots, \phi_{s-1}}$, in case $[\chi, X] = (X)_M = [X]_\psi = 1$, so that each one of these invariants is a function of the homomorphism S.

2. Powers of invariants as functions of homomorphisms. In papers [6] and [8] considerable difficulty arose from the fact that the invariants $[\chi, X]$ and $[X]_\psi$

of the field k were defined not for all χ and X and fields k, and therefore the variant of discussing these invariants as functions of homomorphisms was not defined for all homomorphisms. In our more general case these difficulties are mounting. However, they can be circumvented by means of the following proposition.

Lemma 3. *For every Scholz field k with respect to M and l, and for all its subfields and for all χ and X, a symbol $\{\chi, X\}$ can de defined which coincides with $[\chi, X]$, whenever that symbol is defined and which satisfies the conditions:*

$$\{\chi_1\chi_2, X\} = \{\chi_1, X\}\{\chi_2, X\}, \tag{89}$$

$$\{\chi, X_1X_2\} = \{\chi, X_1\}\{\chi, X_2\}, \tag{90}$$

$$\{\chi_1, X_1\} = \{\chi, X\}, \tag{91}$$

analogous to (16), (17) *and* (18).

Proof. It is sufficient to define the symbol $\{\chi, X\}$ satisfying conditions (89) and (90) for any χ and X in the field k itself. Then the symbol will be defined in any subfield by means of relation (91), and all the conditions of Lemma 3 will be satisfied, as can be easily seen.

The problem of defining the symbol $[\chi, X]$ ahead of the symbol $\{\chi, X\}$ is essentially the problem of extending the function $\phi(a, b)$ given over some vector pairs of two vector spaces A and B to a bilinear function given over any pair of vectors. In general it can be easily clarified when such an extension is possible. In order to do this we consider the Kronecker product $A \circ B$ of the spaces A and B and put into correspondence with every pair of vectors a, b a vector $a \circ b \in A \circ B$. The bilinearity of the function $\phi(a, b)$ is equivalent to the fact that the function

$$f(a \circ b) = \phi(a, b),$$

given in $A \circ B$ can be linearly extended over all $A \circ B$. Hence in order that the function $\phi(a, b)$ can be bilinearly extended over all $a \in A$ and $b \in B$ it is necessary and sufficient that the following condition be satisfied: if $\phi(a_1, b_1)$, $\phi(a_2, b_2)$ and $\phi(a_3, b_3)$ are defined and if

$$a_1 \circ b_1 + a_2 \circ b_2 \equiv a_3 \circ b_3,$$

then

$$\phi(a_1, b_1) + \phi(a_2, b_2) = \phi(a_3, b_3), \tag{92}$$

where the notation $\equiv$ implies the equality of the vectors in $A \circ B$.

We now return to our given function $[\chi, X]$ remembering that we must replace the additive notation by the multiplicative. We select in the group of classes X a basis $X_1, \cdots, X_t$ and in each class X_i a representative μ_i. We let

$$\mu_X = \mu_{X_1}^{r_1} \cdots \mu_{X_t}^{r_t},$$

if

$$X = X_1^{r_1} \cdots X_t^{r_t}.$$

Obviously $X \to \mu_X$ is an isomorphism of the group X onto the group $k^\times/k^{\times l}$. We take an arbitrary prime ideal $\mathfrak{P}$, and define a function

$$\alpha_{\mathfrak{P}}(\chi, X) \approx \mu_X^{\Sigma\varphi^*_{\mathfrak{P},\chi}(\mathfrak{T})\tau}$$

using notation of § 1. Obviously

$$\alpha_{\mathfrak{P}}(\chi_1\chi_2, X) \approx \alpha_{\mathfrak{P}}(\chi_1, X)\alpha_{\mathfrak{P}}(\chi_2, X),$$
$$\alpha_{\mathfrak{P}}(\chi, X_1X_2) \approx \alpha_{\mathfrak{P}}(\chi, X_1)\alpha_{\mathfrak{P}}(\chi, X_2),$$

from which it follows that the mapping $\chi \circ X \to \alpha_{\mathfrak{P}}(\chi, X)$ defines a homomorphism of a Kronecker product of the character group χ and the class group X into the group $k^\times/k^{\times l}$. In particular if

$$(\chi_1 \circ X_1)(\chi_2 \circ X_2) \equiv (\chi_3 \circ X_3),$$

then

$$\alpha_{\mathfrak{P}}(\chi_1, X_1)\alpha_{\mathfrak{P}}(\chi_2, X_2) \approx \alpha_{\mathfrak{P}}(\chi_3, X_3). \tag{93}$$

Since

$$[\chi, X] = \prod_{\mathfrak{P}} \left(\frac{\alpha_{\mathfrak{P}}(\chi, X)}{\mathfrak{P}}\right) \tag{94}$$

is defined only if the symbols $\left(\frac{\alpha_{\mathfrak{P}}(\chi, X)}{\mathfrak{P}}\right)$ are defined it follows from (94) that (92) is satisfied for the function $[\chi, X]$. Therefore the lemma is proved.

Exactly the same considerations are applicable to the invariants $[X_1, \cdots, X_{s-1}; X_s]_{\phi_1, \cdots, \phi_{s-1}}$ and it can be shown that it is possible to define in the field k and in all its subfields the invariant $\{X_1, \cdots, X_{s-1}; X_s\}_{\phi_1, \cdots, \phi_{s-1}}$ which coincides with $[X_1, \cdots, X_{s-1}; X_s]_{\phi_1, \cdots, \phi_{s-1}}$, for any $X_1, \cdots, X_{s-1}; X_s$, when the latter invariant is defined, and which possesses the properties

$$\{X_1X_1', \ldots, X_{s-1}X_{s-1}'; X_s\}_{\varphi_1, \ldots, \varphi_{s-1}} =$$
$$= \{X_1, \ldots, X_{s-1}; X_s\}_{\varphi_1, \ldots, \varphi_{s-1}} \{X_1', \ldots, X_{s-1}'; X_s'\}_{\varphi_1, \ldots, \varphi_{s-1}}, \tag{95}$$

$$\{X_1, \ldots, X_{s-1}; X_s X_s'\}_{\varphi_1, \ldots, \varphi_{s-1}} =$$
$$= \{X_1, \ldots, X_{s-1}; X_s\}_{\varphi_1, \ldots, \varphi_{s-1}} \{X_1, \ldots, X_{s-1}; X_s'\}_{\varphi_1, \ldots, \varphi_{s-1}}, \tag{96}$$

$$\{X_1', \ldots, X_{s-1}'; X_s'\}_{\varphi_1, \ldots, \varphi_{s-1}} = \{X_1, \ldots, X_{s-1}; X_s\}_{\varphi_1, \ldots, \varphi_{s-1}}, \tag{97}$$

analogous to (89), (90) and (91).

Because of (95) we can decompose any invariant

$$\{X_1, \cdots, X_{s-1}; X_s\}_{\phi_1, \cdots, \phi_{s-1}}$$

into a product of invariants

$$\{1, \cdots, X_i, \cdots, 1; X_s\}_{\phi_1, \cdots, \phi_{s-1}}.$$

In what follows we shall make use of the notation

$$\{1, \cdots, X_i, \cdots, 1; X_s\}_{\phi_1, \cdots, \phi_{s-1}} = \{X_i, X_s\}^{(i)}_{\phi_1, \cdots, \phi_{s-1}}$$

and will confine ourselves to the investigation of the invariants $\{X, X'\}^{(i)}_{\phi_1, \ldots, \phi_{s-1}}$.

Lemma 4. *Given a Scholz field k with respect to M and l in which $[\chi, X] = (X)_M = 1$ for all χ and X. The symbol $\{X\}_\psi$ can be defined for the field k and for all its subfields, which coincides with $[X]_\psi$ whenever this symbol is defined and which satisfies the conditions*

$$\{XX'X''\}_\psi \{X'X''\}_\psi^{-1} \{XX''\}_\psi^{-1} \{XX'\}_\psi^{-1} \{X\}_\psi \{X'\}_\psi \{X''\}_\psi = 1, \tag{98}$$

$$\{X_1\}_\psi = \{X\}_\psi, \tag{99}$$

where (99) is analogous to (91). We also suppose that the function ψ satisfies condition 1° in the definition of the symbol $[X]_\psi$.

Proof. We select in the group of classes X a basis $X_1, \cdots, X_t$ and in each class X_i an arbitrary representative μ_{X_i}. We make this selection so that if

$$\mu_{X_i} = \mathfrak{C}_i^l \mathfrak{D}_i \mathfrak{m}_i,$$

then

$$(\mathfrak{m}_i, \mathfrak{m}_l^\sigma) = 1, \quad \sigma \notin H. \tag{100}$$

We let

$$\mu_X = \mu_{X_1}^{r_1} \cdots \mu_{X_t}^{r_t},$$

if

$$X = X_1^{r_1} \cdots X_t^{r_t}.$$

We denote the ideal $\mathfrak{m}$ in the decomposition (11) for μ_X by $\mathfrak{m}_X$. It is obvious that

$$\mu_{XX'} = \mu_X \mu_{X'}, \quad \mathfrak{m}_{XX'} = \mathfrak{m}_X \mathfrak{m}_{X'}. \tag{101}$$

We let

$$\{X\}_\psi = \left[\frac{\mu_X}{\mathfrak{m}_X^\psi}\right].$$

In this way the invariant $\{X\}_\psi$ which we have defined coincides with $[X]_\psi$ when $[X]_\psi$ is defined, that is when conditions 2° and 3° are satisfied and for $l = 2$ also condition 4° in the definition of $[X]_\psi$, since in this case $\{X\}_\psi$ differs from $[X]_\psi$ only in the choice of the representative μ_X in the class X, and as we have proved in § 1, $[X]_\psi$ does not depend on this choice.

The symbol $\{X\}_\psi$ also satisfies conditions (98). In fact the left hand side of (98) is equal to

$$\left[\frac{\mu_{XX'X''}}{\mathfrak{m}^\psi_{XX'X''}}\right]\left[\frac{\mu_{X'X''}}{\mathfrak{m}^\psi_{X'X''}}\right]^{-1}\left[\frac{\mu_{XX''}}{\mathfrak{m}^\psi_{XX''}}\right]^{-1}\left[\frac{\mu_{XX'}}{\mathfrak{m}^\psi_{XX'}}\right]^{-1}\left[\frac{\mu_X}{\mathfrak{m}^\psi_X}\right]\left[\frac{\mu_{X'}}{\mathfrak{m}^{\psi'}_{X'}}\right]\left[\frac{\mu_{X''}}{\mathfrak{m}^\psi_{X''}}\right].$$

Making use of relations (100) and (101) and of the multiplicative property of the symbol $\left[\frac{\mu}{\mathfrak{a}}\right]$, we obtain after factoring that this expression is equal to one.

We take expression (99), as in the proof of Lemma 3, for the definition of the symbol $\{X\}_\psi$ in the subfields of the field k and obtain in this way the symbol $\{X\}_\psi$, satisfying all the conditions formulated in the conditions of the lemma.

We consider an arbitrary l-group G of class c with a given system of generators $s_1, \cdots, s_d$. Let $G = S_d/N$ be a representation of G in the form of a quotient group of a free group S_d corresponding to this choice of generators. We denote by N' the subgroup of N generated by the commutators of the elements of N with the elements of S_d and by the l-th powers of the elements of N. The classes X of the factor sets over G with values in the group of l-th roots of unity are in one-to-one correspondence with the characters of the group N/N' (see [6], Lemma 1). On the other hand the group S_d/N' is of class $\leq c+1$ and therefore all its elements, and in particular all the elements of N/N' can be represented as commutator forms (see [10]) of degree $\leq c+1$ in the generators $s_1, \cdots, s_d$. Therefore N/N' is a homomorphic image of the module of Lie forms of degree $\leq c+1$ in d variables $x_1, \cdots, x_d$. The variables $x_1, \cdots, x_d$ are in one-to-one correspondence with the generators $s_1, \cdots, s_d$. If a group of automorphisms is given in G, which consists of transformations of generators, then the same transformations of $x_1, \cdots, x_d$ define in the Lie module a group of automorphisms with respect to which the above homomorphism is an operator homomorphism. Because of the above properties we shall discuss the classes X as characters of the Lie module.

We return to the chain of fields $\Omega \subset L \subset k \subset K$, considered previously, where the Galois group of K/k is one of the groups $\mathfrak{G}_d^{(c)}$, $\mathfrak{G}_d^{(c,r)}$, $\mathfrak{G}_d^{(c,r,s)*}$ and where K/Ω is a Scholz field. We define in K the symbols $\{\chi, X\}$, $\{X', X''\}^{(i)}_{\phi_1, \cdots, \phi_{s-1}}$ and $\{X\}_\psi$ by Lemmas 3 and 4. As was pointed out, considered for subfields $K(S)$, they are functions of homomorphisms.

Theorem 3. *If the characters X, X' and X'' are of degrees $< n$, n' and n'' respectively in the totality of variables $x_{u,\delta}$ and if X and at least one of X' or X'' does not annihilate at least one Lie polynomial, depending on at least one $x_{u,\delta}$ and if X, X' and X'' annihilate all polynomials depending only on $x_{u,i}$, $i < \delta - 1$, while $\chi(s_{u,i}) = 1$, $i \leq \delta - 1$, then the invariants $\{\chi, X\}$, $(X)_M$, $\{X\}_\psi$ and $\{X', X''\}^{(i)}_{\phi_1, \cdots, \phi_{s-1}}$ have as functions of homomorphisms of the type $(s_{u,1}, \cdots, s_{u,\delta-1}; t_{u,1}, \cdots, t_{u,\delta-1} \mid t_{u,\delta})$ of the Scholz field K with respect to M and l the following degrees:*

$(X)_M$ is of degree $\leq n$,

$\{\chi, X\}$ is of degree $\begin{cases} \leq n, & \text{if } \chi(s_{u,\delta}) = 1 \text{ for all } u, \\ \leq n+1, & \text{if } \chi(s_{u,\delta}) \neq 1 \text{ for some } u, \end{cases}$

$\{X\}_\psi$ is of degree $\leq 2n$,

$\{X', X''\}^{(i)}_{\phi_1, \cdots, \phi_{s-1}}$ is of degree $\leq n' + n''$.

The proof of this theorem concerning the invariants $(X)_M$ and $\{\chi, X\}$ coincides with Theorem 4 of paper [6], and the part concerning the invariant $\{X\}_\psi$ follows the proof of Theorem 7 of paper [8]. It remains to prove the assertion of Theorem 3 relating to the degree of the invariant $\{X', X''\}^{(i)}_{\phi_1, \cdots, \phi_{s-1}}$.

Our considerations will be very close to those employed in Theorem 7 of paper [8].

We select a basis $X_1, \cdots, X_t$ in the group of characters X and write X' and X'' in terms of this basis:

$$X' = X_1^{c_1'} \cdots X_t^{c_t'}, \quad X'' = X_1^{c_1''} \cdots X_t^{c_t''}.$$

For a fixed field K the invariant $\{X', X''\}^{(i)}_{\phi_1, \cdots, \phi_{s-1}}$ can be written in the form

$$\zeta^{f(c_1', \cdots, c_t'; c_1'', \cdots, c_t'')} = \zeta^{f(c', c'')},$$

where c' and c'' are vectors with coordinates $(c_1', \cdots, c_t')$, $(c_1'', \cdots, c_t'')$ in a t-dimensional space over the field of residue classes mod l.

The relations

$$\{X_1'X_2', X'''\}^{(i)}_{\varphi_1, \dots, \varphi_{s-1}} = \{X_1', X'''\}^{(i)}_{\varphi_1, \dots, \varphi_{s-1}} \{X_2', X'''\}^{(i)}_{\varphi_1, \dots, \varphi_{s-1}},$$

$$\{X', X_1'''X_2''\}^{(i)}_{\varphi_1, \dots, \varphi_{s-1}} = \{X', X_1''\}^{(i)}_{\varphi_1, \dots, \varphi_{s-1}} \{X', X_2'''\}^{(i)}_{\varphi_1, \dots, \varphi_{s-1}}$$

give us

$$f(c_1' + c_2', c'') \equiv f(c_1', c'') + f(c_2', c'') \ (l),$$

$$f(c', c_1'' + c_2'') \equiv f(c', c_1'') + f(c', c_2'') \ (l),$$

which show that the function $f(c', c'')$ is bilinear. From this it follows that there exists a representation

$$f(c', c'') \equiv \sum_{i,k=1} a_{i,k} c_i' c_k'' \ (l).$$

If we consider the invariant $\{X', X''\}^{(i)}_{\phi_1, \cdots, \phi_{s-1}}$ as a function of the canonical homomorphism S of the field K, then $c_i'(S)$ and $c_k''(S)$ are functions of this homomorphism, where

$$f(S) \equiv \sum a_{i,k} c_i'(S) c_k''(S) \ (l).$$

Relations (37) of paper [6] show that $c_i'(S)$ is of degree $\leq n'$, while $c_i''(S)$ is of degree $\leq n''$ with respect to S. On the other hand it was shown in paper [8], p. 416 that the degree of a product of two functions does not exceed the sum of their degrees. Therefore the degree of $f(S)$ is $\leq n' + n''$, which proves the theorem.

3. The construction of a field with a given solvable Galois group. At this point we shall make use of the following notation. If in the chain of fields $\Omega \subset L \subset k \subset K$, the field K/k has the Galois group $\mathfrak{G}_d^{(c)}$, $\mathfrak{G}_d^{(c,r)}$ or $\mathfrak{G}_d^{(c,r,s)*}$ we will denote it respectively by $K_d^{(c)}$, $K_d^{(c,r)}$ or $K_d^{(c,r,s)*}$.

Theorem 4. *For any set of natural numbers δ, r and s there exists a natural number $c_1(\delta, r, s)$ with the following properties: in any Scholz field $K_d^{(c,r,s)*}/\Omega$ with respect to M and l with the Galois group $F \cdot \mathfrak{G}_d^{(c,r,s)*}$ with $d > c_1(\delta, r, s)$ there exists a subfield $K_\delta^{(c,r,s)*} = K_d^{(c,r,s)*}(S)$ with the Galois group $F \cdot \mathfrak{G}_\delta^{(c,r,s)*}$ and with all the invariants $\{\chi, X\}$, $(X)_M$, $\{X\}_\psi$ and $\{X', X''\}^{(i)}_{\phi_1, \cdots, \phi_{s-1}}$ equal to unity.*

The proof of this theorem repeats verbatim the proof of Theorem 5 of paper [6], if we change the reference to Theorem 4 of paper [6] to a reference to Theorem 3 of the present paper.

Lemma 5. *For any set of numbers* δ, r *and* s *there exists a number* $c(\delta, r, s)$ *with the following properties: in a Scholz field* $K_d^{(c,r)}/\Omega$ *with respect to* M *and* l *and with the Galois group* $F \cdot \mathfrak{G}_d^{(c,r)}$ *over* Ω *for* $d > c(\delta, r, s)$, *there exists a subfield* $K_\delta^{(c,r)} = K_d^{(c,r)}(S)$ *with the Galois group* $F \cdot \mathfrak{G}_\delta^{(c,r)}$ *over* Ω, *which can be imbedded in the Scholz field* $K_\delta^{(c,r+1,s)*}$ *with the Galois group* $F \cdot \mathfrak{G}_\delta^{(c,r+1,s)*}$ *over* Ω.

We will prove the lemma by induction on s. Suppose that for every δ we find the number $c(\delta, r, s-1)$. We prove that we can let

$$c(\delta, r, s) = c(c_1(\delta, r, s), r, s-1),$$

where the number $c_1(\delta, r, s)$ is defined in Theorem 4. In fact let $d > c(\delta, r, s)$ and $K_d^{(c,r)} \supset k$ be the given Scholz field with the Galois group $F \cdot \mathfrak{G}_d^{(c,r)}$ over Ω. By the hypothesis of the induction in the field $K_d^{(c,r)}$ there exists a subfield

$$K_{d_1}^{(c,r)} = K_d^{(c,r)}(S_1),$$

which is imbedded in the Scholz field $K_{d_1}^{(c,r+1,s-1)*}$ with the Galois group $F \cdot \mathfrak{G}_{d_1}^{(c,r+1,s-1)*}$ over Ω with $d_1 > c_1(\delta, r, s)$. By Theorem 4 the field $K_{d_1}^{(c,r+1,s-1)*}$ contains a subfield

$$K_\delta^{(c,r+1,s-1)*} = K_{d_1}^{(c,r+1,s-1)*}(S)$$

all of whose invariants are equal to 1. Applying Theorem 2 we see that $K_\delta^{(c,r+1,s-1)*}$ can be imbedded in the Scholz field K^*, where obviously $K_\delta^{(c,r+1,s)*} \subset K^*$ by the definition of the field K. Hence the field $K_\delta^{(c,r+1,s)*}$ is a Scholz field and the lemma is proved.

For some natural number m

$$\mathfrak{G}_\delta^{(c,r+1,m)} = \mathfrak{G}_\delta^{(c,r+1)}.$$

We shall denote $\mathfrak{G}_\delta^{(c,r+1,m)*}$ by $\mathfrak{G}_\delta^{(c,r+1)*}$. It follows easily from Lemma 5 that

Corollary. *For any* δ, c *and* r *there exists a number* $c(\delta, c, r)$ *with the following properties: in any Scholz field* $K_d^{(c,r)}$ *with* $d > c(\delta, c, r)$ *there exists a subfield* $K_\delta^{(c,r)} = K_d^{(c,r)}(S)$ *which can be imbedded in the Scholz field* $K_\delta^{(c,r+1)*} \supset k$ *with the Galois group* $F \cdot \mathfrak{G}_\delta^{(c,r+1)*}$ *over* Ω.

For proof it is sufficient to put $s = m$ in Lemma 5.

Theorem 5. *Every Scholz field* k/Ω *with respect to* M *and* l *with the Galois group* F *can be imbedded in a Scholz field* $K \supset k$ *with respect to* M *and* l *with the Galois group* G *over* k *and* $F \cdot G$ *over* Ω *for every* F*-operator group* G *of order* l^a. *In doing so* M *must be divisible by* l *to a degree at least one larger than the product of the exponents of* G *and the Sylow* l*-subgroup of the group* F.

Proof. Since every F-operator group is an operator-homomorphic image of the group $\mathfrak{G}_d^{(c)}$, having the same exponent, it is sufficient to prove the theorem for $G = \mathfrak{G}_d^{(c)}$. We shall prove it for a more general case of $G = \mathfrak{G}_\delta^{(c,r)}$. Since

$$\mathfrak{G}_\delta^{(c,k)} = \mathfrak{G}_\delta^{(c)}, \quad \mathfrak{G}_\delta^{(c,0)} = \mathfrak{G}_\delta^{(c-1)},$$

we can prove the theorem by induction on r for an arbitrary fixed δ.

Suppose that the theorem is proved for $\mathfrak{G}_\delta^{(c,r-1)}$. Then the field k can be imbedded in a Scholz field $K_d^{(c,r-1)}$ with the Galois group $F \cdot \mathfrak{G}_d^{(c,r-1)}$ over Ω for any d. We take, in particular, $d > c(\delta, r-1)$, where $c(\delta, r-1)$ is the number whose existence was proved in the corollary to Lemma 5. By that corollary

$$K_d^{(c,r-1)} \supset K_\delta^{(c,r-1)} = K_d^{(c,r-1)} (S),$$

where $K_\delta^{(c,r-1)} \supset k$ and can be imbedded in a Scholz field $K_\delta^{(c,r)*}$. The fields

$$\Omega \subset L \subset K_\delta^{(c,r-1)} \subset K_\delta^{(c,r,m)} \subset K_\delta^{(c,r)*}$$

form a chain of fields to which the Faddeev-Gaschütz theorem is applicable. According to that theorem there exists a field $K_\delta^{(c,r)}$ with the Galois group $F \cdot \mathfrak{G}_\delta^{(c,r)}$ over Ω between the fields $K_\delta^{(c,r-1)}$ and $K_\delta^{(c,r)*}$. Since $K_\delta^{(c,r)}$ is a subfield of the Scholz field $K_\delta^{(c,r)*}$ it is itself a Scholz field. Hence the theorem is proved.

We note that the condition which we have imposed on M in the formulation of Theorem 5 was used in the derivation of the properties of the symbol $[\chi, X]$ on page 532 and in applying Scholz's theorem, which states that for central extensions the imbedding problem always has a solution for fields of degree l^α.

Theorem 6. *The conclusions of Theorem 5 hold if the requirements that the field be Scholz with respect to M and l, be replaced by it being Scholz with respect to l alone.*

Proof. If Ω contains an l-th root of unity ζ, then the notions of a field being Scholz with respect to M and l, or only with respect to M coincide. If Ω does not contain an l-th root of unity, then we denote $k(\zeta)$ by $\bar{k}$, its group by $\bar{F}$, the natural homomorphism of $\bar{F}$ on F by $\phi(\bar{u}) = u$, and the kernel of the homomorphism by C. We consider the semi-direct product $\bar{F} \cdot G$, defining the action of the operators of $\bar{F}$ on G by the formula

$$\sigma^{\bar{u}} = \sigma^{\phi(\bar{u})}, \quad \sigma \in G, \quad \bar{u} \in \bar{F}, \quad \phi(\bar{u}) \in F.$$

By Lemma 1, the field $\bar{k}$ is a Scholz field with respect to M and l and by Theorem 5 it can be imbedded in a Scholz field $\bar{K}$ with respect to M and l with the Galois group $\bar{F} \cdot G$ over Ω. In the field $\bar{K}$, to the subgroup C of its Galois group corresponds a subfield K with the Galois group $F \cdot G$ over Ω. Obviously, $K \supset k$ gives a solution of our imbedding problem. Since $K(\zeta) = \bar{K}$ is a Scholz field with respect to M and l, we can conclude from Lemma 2 that K is a Scholz field with respect to M. The theorem is proved.

Corollary. *Theorem 6 holds for any nilpotent normal subgroup of G, if the* *condition on M is imposed on every Sylow subgroup of G.*

The proof is obvious.

Theorem 7. *For every field of algebraic numbers there exists an extension with a given in advance solvable Galois group.*

Proof. Let G be the given solvable group. It has been shown by Ore [3] that the maximal nilpotent normal subgroup N is semi-complete, that is, that there exists a proper subgroup $G_1 \subset G$ such that $G = G_1 \cdot N$. From this it follows that G is a homomorphic image of the semi-direct product $G_1 \cdot N$. Applying the same reasoning to G_1 we can continue the argument until we reach the identity subgroup. As a result we obtain the group $\widetilde{G}$, whose homomorphic image is the group G and which possesses the following series of normal subgroups:

$$\widetilde{G} \supset N_1 \supset N_2 \supset \cdots \supset N_r = (l)$$

with the following properties: $\widetilde{G}/N_i$ is the semi-direct product of the subgroup F_i, isomorphic to $\widetilde{G}/N_{i-1}$ and a nilpotent normal subgroup N_{i-1}/N_i. We choose for M a number containing any prime divisor l of the order of $\widetilde{G}$ to a degree higher than it enters into the products of the exponents of the Sylow l-subgroups of the groups N_{i-1}/N_i. Applying r times the corollary of Theorem 6 we obtain the field $\widetilde{K}/\Omega$ with the Galois group $\widetilde{G}$. Since G is a homomorphic image of $\widetilde{G}$, the field $\widetilde{K}$ contains the subfield K/Ω with the Galois group G.

The theorem is proved.

We note that in the proof of Theorem 7 it is possible to arrange that the discriminants of all the fields under consideration be prime to some ideal $\mathfrak{M}$ given in advance. From this it follows that the number of extensions of the field Ω, having a given solvable Galois group is infinite.

APPENDIX

We will prove here an existence theorem, which generalizes to the case of non-normal fields the principal result of paper [7].

Let k/Ω be a normal field of algebraic numbers with the Galois group G. We shall suppose that k contains an l-th root of unity ζ and we will let

$$\zeta^\sigma = \zeta^{g(\sigma)}, \quad \sigma \in G.$$

Let H be a subgroup of G, whose order is a power of l, and let L be the subfield of k which corresponds to it. It is obvious that

$$g(\sigma_1 \sigma_2) \equiv g(\sigma_1) g(\sigma_2), \quad g(h) \equiv 1 \ (l), \quad h \in H,$$

and therefore $g(\sigma)$ is a constant in both right and left cosets of G with respect to H, and also in the cosets of G with respect to the double modulus (H, H). We shall write $g(S)$, where S is one of these classes instead of $g(\sigma)$, $\sigma \in S$.

Let the function $\zeta(S)$ be defined over the left cosets S of the group G with respect to H with value in the group of l-th roots of unity. If $\mathfrak{S}$ is a coset of G with respect to the modulus (H, H), composed of classes $S_1, \cdots, S_r$, we will let

$$\zeta(\mathfrak{S}) = \zeta(S_1) \cdots \zeta(S_r).$$

We will suppose that the function $\zeta(S)$ is such that for every class $\mathfrak{S}$ modulo (H,H)

1. $\zeta(\mathfrak{S}) = \zeta(\mathfrak{S}^{-1})^{g(\mathfrak{S})}$.

In case $l = 2$, we must impose some additional conditions.

Let $\mathfrak{S}$ be an involutory class, that is, $\mathfrak{S} = \mathfrak{S}^{-1}$. We shall prove that there exists a representative σ in $\mathfrak{S}$ such that $\sigma^2 \in H$. In fact if σ is any representative of $\mathfrak{S}$, it follows from the fact that $\mathfrak{S}$ is involutory that

$$\sigma_1^{-1} = h_1 \sigma_1 h_2, \quad h_1, h_2 \in H.$$

From this we get

$$(\sigma_1 h_1)(\sigma_1 h_1) h_1^{-1} h_2 = 1,$$

that is,

$$(\sigma_1 h_1)^2 = h_2^{-1} h_1 \in H.$$

Hence it is sufficient to take $\sigma_1 h_1$ for σ.

It is obvious that σ lies in the normalizer of the subgroup $H \cap H^\sigma$. On the other hand from

$$\sigma^2 \in H, \quad \sigma^2 = \sigma^{-1} \sigma^2 \sigma \in \sigma^{-1} H \sigma = H^\sigma$$

it follows that $\sigma^2 \in H \cap H^\sigma$. Therefore the subgroup $\{\sigma, H \cap H^\sigma\}$, generated by σ and $H \cap H^\sigma$ contains $H \cap H^\sigma$ as a normal subgroup of index 2. We denote by LL^σ the composite of the fields L and L^σ, belonging to the subgroup $H \cap H^\sigma$ and by Λ_σ the subfield belonging to $\{\sigma, H \cap H^\sigma\}$. It is obvious that

$$LL^\sigma \supset \Lambda_\sigma, \quad (LL^\sigma : \Lambda_\sigma) = 2.$$

We denote by $\mathfrak{D}_\sigma$ the different of the field LL^σ/Λ_σ and set

$$\mathfrak{E}_{\mathfrak{S}} = N_{LL^\sigma/L}(\mathfrak{D}_\sigma).$$

It is not hard to show that $\mathfrak{E}_{\mathfrak{S}}$ does not depend on the choice of σ in $\mathfrak{S}$.

We can now formulate the additional conditions which are needed in the case of $l = 2$.

1′. All the prime divisors of 2 split completely in k/Ω.

2′. The real infinite primes of Ω split into reals in k.

3′. For any involutory classes $\mathfrak{S}_1, \cdots, \mathfrak{S}_r$ for which $\mathfrak{E}_{\mathfrak{S}_1} \cdots \mathfrak{E}_{\mathfrak{S}_r} \approx \alpha_\chi$ in L,

$$\zeta(\mathfrak{S}_1) \cdots \zeta(\mathfrak{S}_r) = 1.$$

Here α_χ is such a number in L that $\sqrt[l]{\alpha_\chi} \in k$.

Theorem. *Given a field k/Ω and the function $\zeta(S)$ satisfying condition* 1 *and for $l = 2$, conditions* 1′, 2′ *and* 3′, *as well as the ideals of the first degree $\mathfrak{D}_1, \cdots, \mathfrak{D}_m$ of the field L/Ω which are non-ramified and not conjugate to each* *other in k/L, and the l-th roots of unity $\xi_1, \cdots, \xi_m$. Then there exist infinitely many numbers $\alpha \in L$ satisfying the conditions:*

$$\left(\frac{\alpha}{\mathfrak{Q}_i}\right) = \xi_i, \quad i = 1, \cdots, m, \tag{102}$$

$$\left(\frac{\alpha}{\mathfrak{P}^\sigma}\right) = \zeta(S) \text{ for all } \mathfrak{P} \mid \alpha,\ \sigma \in G,\ \sigma \notin H. \tag{103}$$

Condition (103) presupposes that if $\mathfrak{P} \mid \alpha$ and $\sigma \notin H$, then $(\mathfrak{P}^\sigma, \alpha) = 1$.

In the proof we make use of the following notation. Let $\mathfrak{T}$ be a coset of G modulo (H, H) and let $\mathfrak{T} = T_1 + \cdots + T_r$, where T_i are right cosets of G with respect to (H, H). If α is a number of L, then we will understand by $\alpha^{\mathfrak{T}}$ the product $\alpha^{\tau_1 \cdots \tau_r}$, where $\tau_i \in T_i$. The expression $\mathfrak{a}^{\mathfrak{T}}$ has an analogous meaning, where $\mathfrak{a}$ is an ideal of L.

We shall need the following lemma.

Lemma. *If α and $\mathfrak{b}$ are a number and an ideal of L, then $\alpha^{\mathfrak{T}}$ and $\mathfrak{b}^{\mathfrak{T}}$ also lie in L, and*

$$\left(\frac{\alpha}{\mathfrak{b}^{\mathfrak{T}}}\right) = \left(\frac{\alpha^{\mathfrak{T}^{-1}}}{\mathfrak{b}}\right)^{\rho(\mathfrak{T})}. \tag{104}$$

Proof. It follows from the fact that $\mathfrak{T}$ is a coset with respect to the double modulus (H, H) that $T_i = T_1 h_i$, $h_i \in H$, where h_i are the representatives of the cosets H with respect to $H \cap H^{\tau_1}$. Therefore

$$\alpha^{\mathfrak{T}} = \alpha^{\tau_1(h_1 + \cdots + h_r)}$$

and correspondingly

$$\mathfrak{b}^{\mathfrak{T}} = \mathfrak{b}^{\tau_1(h_1 + \cdots + h_r)}.$$

The number α^{τ_1} and the ideal $\mathfrak{b}^{\tau_1}$ belong to the field LL^{τ_1}. The isomorphisms of the extension LL^{τ_1}/L are induced, as can be easily seen, by the automorphisms $h_1, \cdots, h_r$ of the field k/L. In this way we have

$$\alpha^{\mathfrak{T}} = N_{LL^{\tau_1}/L}(\alpha^{\tau_1}), \quad \mathfrak{b}^{\mathfrak{T}} = N_{LL^{\tau_1}/L}(\mathfrak{b}^{\tau_1}),$$

from which it follows that $\alpha^{\mathfrak{T}}$ and $\mathfrak{b}^{\mathfrak{T}}$ belong to L.

Applying these considerations to the Legendre symbol $\left(\frac{\alpha}{\mathfrak{b}^{\mathfrak{T}}}\right)$, we get

$$\left(\frac{\alpha}{\mathfrak{b}^{\mathfrak{T}}}\right)_L = \left(\frac{\alpha}{N_{LL^{\tau_1}/L}(\mathfrak{b}^{\tau_1})}\right)_L = \left(\frac{\alpha}{\mathfrak{b}^{\tau_1}}\right)_{LL^{\tau_1}}. \tag{105}$$

We now apply the automorphism τ_1^{-1} to α, $\mathfrak{b}^{\tau_1}$ and LL^{τ_1} and obtain

$$\left(\frac{\alpha^{\tau_1^{-1}}}{\mathfrak{b}}\right)_{LL^{\tau_1^{-1}}} = \left(\frac{\alpha}{\mathfrak{b}^{\tau_1}}\right)_{LL^{\tau_1}}^{\rho(\mathfrak{T}^{-1})} \quad \text{or} \quad \left(\frac{\alpha}{\mathfrak{b}^{\tau_1}}\right)_{LL^{\tau_1}} = \left(\frac{\alpha^{\tau_1^{-1}}}{\mathfrak{b}}\right)_{LL^{\tau_1^{-1}}}^{\rho(\mathfrak{T})}. \tag{106}$$

Finally, analogously to (105), we have

$$\left(\frac{\alpha^{\tau_1^{-1}}}{\mathfrak{b}}\right)_{LL^{\tau_1^{-1}}} = \left(\frac{N_{LL^{\tau_1^{-1}}/L}(\alpha^{\tau_1^{-1}})}{\mathfrak{b}}\right)_L = \left(\frac{\alpha^{\mathfrak{T}^{-1}}}{\mathfrak{b}}\right)_L. \tag{107}$$

Combining (105), (106) and (107) we obtain

$$\left(\frac{\alpha}{\mathfrak{b}\mathfrak{T}}\right)_L = \left(\frac{\alpha^{\mathfrak{T}^{-1}}}{\mathfrak{b}}\right)_L^{g(\mathfrak{T})}.$$

This proves the lemma*.

Proof of Theorem. We first consider the case $l \neq 2$. We construct as in paper [7] a sequence of prime l-hyperprimary principal ideals $\pi_t \in L$, having degree 1 in L/Ω which split completely in k/L, are relatively prime to l and satisfying the conditions

$$\left(\frac{\pi_t}{\mathfrak{Q}_i}\right) = \xi_i^{1/2}, \quad i = 1, \cdots, m. \tag{108}$$

These conditions must be satisfied for all the members of the sequence which is constructed recursively, where after the construction of $\pi_1, \cdots, \pi_{t-1}$, the number π_t is constructed so that the following conditions are satisfied:

$$\left(\frac{\pi_t}{\mathfrak{P}_s^\sigma}\right) = \left(\frac{\pi_s}{\mathfrak{P}_s^\sigma}\right)^{-1} \zeta(S), \quad s = 1, \ldots, t-1, \quad \sigma \notin H, \tag{109}$$

$$\left(\frac{\pi_s}{\mathfrak{P}_t^\sigma}\right) = \left(\frac{\pi_s}{\mathfrak{P}_s^\sigma}\right)^{-1} \zeta(S), \quad s = 1, \ldots, t-1, \quad \sigma \notin H, \tag{110}$$

where $\mathfrak{P}_s$ is a fixed prime divisor of π_s in k.

We next show that such a choice of π_t is possible. We denote by $\mathfrak{f}$ the product of the modulus of hyperprimarity of all the $N_{k/L}(\mathfrak{P}_s^\sigma)$ and $N_{k/L}(\mathfrak{Q}_i)$, by $S_\mathfrak{f}$ the group of classes modulo $\mathfrak{f}$ in L, composed of l-th powers of the principal ideals, by $\Sigma_\mathfrak{f}$ the field of classes of $S_\mathfrak{f}$. That the prime ideal $\mathfrak{p} \in L$ is a principal ideal ($\mathfrak{p} = (\pi)$, where π is l-hyperprimary) and satisfies conditions (108) and (109) is equivalent to the fact that $\mathfrak{p}$ belongs to a completely defined automorphism σ in $\Sigma_\mathfrak{f}$. That $\mathfrak{p}$ splits completely in k/L and satisfies condition (110) is equivalent to the fact that $\mathfrak{p}$ belongs to a definite automorphism τ in the field k', which is obtained by adjoining to k all the $\sqrt[l]{\pi_i^\sigma}$, $i = 1, \cdots, t-1$, $\sigma \notin H$. In order that a prime ideal $\mathfrak{p}$ exist satisfying all these conditions it it necessary and sufficient by Čebotarev's density theorem, that there exist an automorphism in the composite $\Sigma_\mathfrak{f} \cdot k'/L$, which induces σ in $\Sigma_\mathfrak{f}$ and τ in k'. For this it is in turn necessary and sufficient that σ and τ should induce the same automorphism in $\Sigma_\mathfrak{f} \cap k'$.

We examine the properties of the field $\Sigma_\mathfrak{f} \cap k'$. Since the discriminant of k/L and all π_i^σ $(i = 1, \cdots, t-1)$ are relatively prime to $\mathfrak{Q}_i$, the discriminant of k'

* Analogously a more general relation can be proved, namely

$$\left(\frac{\alpha}{\mathfrak{b}\mathfrak{T}}\right)_{L_1} = \left(\frac{\alpha^{\mathfrak{T}^{-1}}}{\mathfrak{b}}\right)_{L_2}^{g(\mathfrak{T})},$$

where α is a number of the field L_1, belonging to the subgroup H_1, $\mathfrak{b}$ is an ideal of the field L_2, belonging to the subgroup H_2, $\zeta \in L_1 \cap L_2$, and $\mathfrak{T}$ is a coset of G with respect to the double modulus (H_1, H_2).

must also be prime to $\mathfrak{D}_i$ and therefore the discriminant of $\Sigma_{\mathfrak{f}} \cap k'$ does not contain the ideals $\mathfrak{D}_i$. Similarly the ramified prime ideals of k/L do not enter into the discriminant of $\Sigma_{\mathfrak{f}} \cap k'$, which is therefore composed only of the prime divisors of π_i^σ and l. We denote by k_0 the maximal abelian subfield of the field k, whose discriminant is composed only of the prime divisors of l and whose Galois group over the maximal unramified subfield is of exponent l. It is obvious that $\Sigma_{\mathfrak{f}} \cap k'$ contains k_0.

On the other hand $\Sigma_{\mathfrak{f}} \cap k'$ is abelian over L and has over k_0 a Galois group of exponent l, that is it is obtained by adjoining to k_0 the radicals $\sqrt[l]{\eta_i}$, $\eta_i \in k_0$.

Let η be any number of the field k_0 such that $\sqrt[l]{\eta} \in \Sigma_{\mathfrak{f}} \cap k'$. Then

$$\sqrt[l]{\eta} \in k', \quad \eta \in k.$$

Therefore we have for η the representation

$$\eta = \prod_i \pi_i^{\Sigma f_i(T)\tau} \cdot \beta^l, \tag{111}$$

where $f_i(T)$ is a function over right cosets of G with respect to H, and $\beta \in k$. We denote by H_0 the subgroup of the group G to which k_0 belongs. Obviously $H_0 \subset H$. Since $\eta \in k_0$, $\eta^{h_0} = \eta$ for $h_0 \in H_0$. We substitute into this relation the expression (111) for η. Since π_i is an ideal of the first degree, the exponents of π_i^τ on the left and right hand sides of the equation must be congruent mod l. We get that

$$(\Sigma f_i(T)\tau) h_0 \equiv \Sigma f_i(T)\tau \ (l),$$
$$f_i(Th_0) \equiv f_i(T) \ (l) \text{ for } h_0 \in H_0.$$

Therefore $f_i(T)$ can be considered as constants in the cosets with respect to the double modulus (H, H_0). From this it follows that

$$\pi_i^{\Sigma f_i(T)\tau} \in k_0,$$

and since $\eta \in k_0$, we have also that $\beta^l \in k_0$. Obviously

$$\sqrt[l]{\pi_i^{\Sigma f_i(T)\tau}} \in \Sigma_{\mathfrak{f}} \cap k',$$

and since $\sqrt[l]{\eta} \in \Sigma_{\mathfrak{f}} \cap k'$, therefore also $\beta \in \Sigma_{\mathfrak{f}} \cap k'$. We see that

$$k_0(\beta) \subset \Sigma_{\mathfrak{f}} \cap k', \quad \beta^l \in k_0.$$

The above discussion shows that the discriminant of $k_0(\beta)/L$ can be divisible only by divisors of l. It is easy to see that the Galois group of $k_0(\beta)$ over its maximal unramified subfield is of exponent l, and is abelian over L. Hence $k_0(\beta)$ possesses all the properties by means of which we have defined k_0, and since k_0 is maximal, they must coincide, so that $\beta \in k_0$. Therefore $\Sigma_{\mathfrak{f}} \cap k'$ is contained in a field which can be obtained by adjoining to k_0 all the

$$\sqrt[l]{\pi_i^{\Sigma f_i(T)\tau}},$$

where $f_i(T)$ are constant over the cosets of G with respect to the modulus (H, H_0). We will suppose that

$$k_0(\sqrt[l]{\eta}) \subset \Sigma_{\mathfrak{f}} \cap k'.$$

As we have shown

$$\eta = \prod_i \pi_i^{\Sigma f_i(T)\tau}.$$

From the fact that $\Sigma_{\mathfrak{f}} \cap k'/L$ is abelian it follows that $\eta^h = \eta$, $h \in H$. Arguing as before we see that $f_i(T)$ must be constant over the cosets of G with respect to the double modulus (H, H). Conversely, it is obvious that all the numbers

$$\pi_i^{\Sigma f_i(T)\tau} \in L,$$

so that

$$\sqrt[l]{\pi_i^{\Sigma f_i(T)\tau}} \in \Sigma_{\mathfrak{f}} \cap k',$$

if $f_i(T)$ is constant in the cosets with respect to the modulus (H, H). Hence we have shown that $\Sigma_{\mathfrak{f}} \cap k'$ is a composite of the fields k_0 and $L(\sqrt[l]{\pi_i^{\mathfrak{S}}})$ where $\mathfrak{S}$ are the cosets of G with respect to the modulus (H, H).

We now find the automorphisms which σ and τ induce in each of these fields. In k_0, σ and τ induce, as can be easily seen, identity automorphisms. We now take any field $L(\sqrt[l]{\pi_i^{\mathfrak{S}}})$ and a prime ideal $\mathfrak{p}$ belonging to the automorphism σ in $\Sigma_{\mathfrak{f}}$. By construction this means that $\mathfrak{p}$ is a principal ideal $(\mathfrak{p} = (\pi))$ and

$$\left(\frac{\pi}{\mathfrak{P}_s^{\sigma}}\right) = \left(\frac{\pi_s}{\mathfrak{P}_s^{\sigma}}\right)^{-1} \zeta(S) \text{ for any } s = 1, \cdots, t-1.$$

From this it follows that

$$\left(\frac{\pi_i^{\mathfrak{S}}}{\pi}\right) = \left(\frac{\pi}{\pi_i^{\mathfrak{S}}}\right) = \left(\frac{\pi_i}{\pi_i^{\mathfrak{S}}}\right)^{-1} \zeta(\mathfrak{S}).$$

Hence π belongs in $L(\sqrt[l]{\pi_i^{\mathfrak{S}}})$ to an automorphism which can be reduced to multiplication of $\sqrt[l]{\pi_i^{\mathfrak{S}}}$ by

$$\left(\frac{\pi_i}{\pi_i^{\mathfrak{S}}}\right)^{-1} \zeta(\mathfrak{S}),$$

and therefore this is the automorphism which σ induces in this field.

We now consider a prime ideal $\mathfrak{p}$, which belongs to τ in k'. Obviously then $\mathfrak{p} = N_{k/L}(\mathfrak{P})$, where

$$\left(\frac{\pi_i^{\sigma}}{\mathfrak{P}}\right) = \left(\frac{\pi_i^{\sigma}}{\mathfrak{P}_i}\right)^{-1} \zeta(S^{-1})^{g(\sigma)}.$$

Arguing as before we see that τ can be reduced in $L(\sqrt[l]{\pi_i^{\mathfrak{S}}})$ to the multiplication of $\sqrt[l]{\pi_i^{\mathfrak{S}}}$ by

$$\left(\frac{\pi_i^{\mathfrak{S}}}{\pi_i}\right) \zeta(\mathfrak{S}^{-1})^{g(\mathfrak{S})}.$$

Since π_i is l-hyperprimary, we have

$$\left(\frac{\pi_i^{\mathfrak{S}}}{\pi_i}\right) = \left(\frac{\pi_i}{\pi_i^{\mathfrak{S}}}\right),$$

and because of condition 1°, which is imposed on $\zeta(S)$ we have

$$\zeta(\mathfrak{S}^{-1})^{g(\mathfrak{S})} = \zeta(\mathfrak{S}).$$

Hence σ and τ can be reduced to the multiplication of $\sqrt[l]{\pi_i^{\mathfrak{S}}}$ by one and the same number, and therefore they coincide in $\Sigma_{\mathfrak{f}} \cap k'$. This proves the possibility of choosing π_t at each step so that it satisfies conditions (108), (109) and (110).

We now argue exactly as in paper [7]. The sequence $\pi_1, \pi_2, \cdots$ must contain two numbers π_i and π_j such that

$$\left(\frac{\pi_i}{\mathfrak{P}_i^{\sigma}}\right) = \left(\frac{\pi_j}{\mathfrak{P}_j^{\sigma}}\right)$$

for every $\sigma \in G$, $\sigma \notin H$. An independent check shows that $\alpha = \pi_i \pi_j$ satisfies the conditions of the theorem.

We now consider the case $l = 2$. Obviously now $g(\sigma) = 1$. We divide all the classes, which are different from unity, with respect to the double modulus (H, H) into three groups. In the first group we put all the involutory classes. The non-involutory classes we divide into two parts putting arbitrarily for each pair of mutually inverse classes $\mathfrak{S}$ and $\mathfrak{S}^{-1}$ one into one part and the other into the other part. These three parts we will denote by U, V and V^{-1} and we will consider them in the future not only as sets of classes $\mathfrak{S}$, but also as sets of automorphisms which are contained in them.

As for an odd l, we construct a sequence of prime principal ideals $\pi_1, \cdots, \pi_t, \cdots$ of the field L, which are l-hyperprimary, totally positive, which split completely in k/L and satisfy the conditions

$$\left(\frac{\pi_t}{\mathfrak{Q}_i}\right) = \xi_i, \quad i = 1, \ldots, m, \quad \left(\frac{\pi_t}{\mathfrak{E}_{\mathfrak{S}}}\right) = \zeta(\mathfrak{S}), \quad \mathfrak{S} \in U. \tag{112}$$

We now construct the sequence recursively by the following rule. Let $\pi_1, \cdots, \pi_{t-1}$ be constructed. We divide them into two parts, the first and the second by a method which will be explained later. We denote these two parts by Π_1 and Π_2. We select π_t such that it satisfies the conditions:

if $\pi_s \in \Pi_1$, $\sigma \in U$, then

$$\left(\frac{\pi_t}{\mathfrak{P}_s^{\sigma}}\right) = 1, \tag{113}$$

$$\left(\frac{\pi_s}{\mathfrak{P}_t^{\sigma}}\right) = \left(\frac{\pi_s}{\mathfrak{P}_s^{\sigma}}\right)\zeta(S); \tag{114}$$

if $\pi_s \in \Pi_2$, $\sigma \in U$, then

$$\left(\frac{\pi_t}{\mathfrak{P}_s^{\sigma}}\right) = \left(\frac{\pi_s}{\mathfrak{P}_s^{\sigma}}\right)\zeta(S), \tag{115}$$

$$\left(\frac{\pi_s}{\mathfrak{P}_t^{\sigma}}\right) = 1; \tag{116}$$

if $\pi_s \in \Pi_1$, $\sigma \in V$, then

$$\left(\frac{\pi_t}{\mathfrak{P}_s^\sigma}\right) = 1, \tag{117}$$

$$\left(\frac{\pi_s}{\mathfrak{P}_t^\sigma}\right) = \left(\frac{\pi_s}{\mathfrak{P}_s^\sigma}\right)\zeta(S); \tag{118}$$

if $\pi_s \in \Pi_2$, $\sigma \in V$, then

$$\left(\frac{\pi_t}{\mathfrak{P}_s^\sigma}\right) = \left(\frac{\pi_s}{\mathfrak{P}_s^\sigma}\right)\zeta(S), \tag{119}$$

$$\left(\frac{\pi_s}{\mathfrak{P}_t^\sigma}\right) = 1; \tag{120}$$

if $\pi_s \in \Pi_1$, $\sigma \in V^{-1}$, then

$$\left(\frac{\pi_t}{\mathfrak{P}_s^\sigma}\right) = \left(\frac{\pi_s}{\mathfrak{P}_s^\sigma}\right)\zeta(S), \tag{121}$$

$$\left(\frac{\pi_s}{\mathfrak{P}_t^\sigma}\right) = 1; \tag{122}$$

if $\pi_s \in \Pi_2$, $\sigma \in V^{-1}$, then

$$\left(\frac{\pi_t}{\mathfrak{P}_s^\sigma}\right) = 1, \tag{123}$$

$$\left(\frac{\pi_s}{\mathfrak{P}_t^\sigma}\right) = \left(\frac{\pi_s}{\mathfrak{P}_s^\sigma}\right)\zeta(S). \tag{124}$$

Here $\mathfrak{P}_s$ is a prime divisor of π_s in k, selected arbitrarily, but once and for for all for every s.

As in the first of the proof (for $l \neq 2$) we must check that such a choice is indeed possible. This verification is done exactly as in the first part.

We denote by $\mathfrak{f}$ the product of the prime ideals of L, ramified in k/L or in L/Ω, by the modulus of l-hyperprimarity, and by all the $N_{k/L}(\mathfrak{P}_s^\sigma)$ and $N_{k/L}(\mathfrak{O}_i)$ and construct as in the first part of the proof the fields $\Sigma_\mathfrak{f}$ and k'. Let to every prime ideal $\mathfrak{r}$ in L, ramified in k/L or L/Ω, correspond a square root of unity $(-1)^{x_\mathfrak{r}}$. Then the fact that the prime ideal $\mathfrak{p} \in L$ is a principal ideal ($\mathfrak{p} = (\pi)$, where π is 2-hyperprimary) and that it satisfies conditions (113), (115), (117), (119), (121), (123) and

$$\left(\frac{\pi}{\mathfrak{r}}\right) = (-1)^{x_\mathfrak{r}}, \tag{125}$$

is equivalent to the fact that $\mathfrak{p}$ belongs in $\Sigma_\mathfrak{f}$ to a perfectly definite automorphism ω_1. In the same way the fact that $\mathfrak{p}$ splits completely in k and satisfies the conditions (116), (118), (120), (122), (124) and (126) is equivalent to $\mathfrak{p}$ belonging in k' to a very definite automorphism ω_2. Our problem is now reduced to showing that for some choice of $x_\mathfrak{r}$ the automorphisms ω_1 and ω_2 will induce in $\Sigma_\mathfrak{f} \cap k'$ one and the same automorphism and that the following conditions will be satisfied:

$$\sum f(\mathfrak{r})\, x_\mathfrak{r} \equiv b(\mathfrak{S}) \ (2), \quad \text{if} \quad \mathfrak{E}_\mathfrak{S} \approx \prod_\mathfrak{r} \mathfrak{p}^{f(\mathfrak{r})}, \quad \zeta(\mathfrak{S}) = (-1)^{b(\mathfrak{S})}.$$

Arguing as in the first part of the proof we see that $\Sigma_{\mathfrak{f}} \cap k'$ is a composite of the fields k_0, $L(\sqrt{a_\chi})$ and $L(\sqrt{\pi_s^{\mathfrak{S}}})$, where a_χ are such numbers of L that $\sqrt{a_\chi} \in k$, and $\mathfrak{S}$ are any cosets with respect to the modulus (H, H). It can be easily seen that in k_0, ω_1 and ω_2 induce the same automorphism, which is the identity automorphism. We consider the field $L(\sqrt[l]{\pi_s^{\mathfrak{S}}})$ where $\pi_s \in \Pi_1$, and $\mathfrak{S} \in U$. Arguing as in the case $l \neq 2$ we see that ω_1 induces in this field an identity automorphism and ω_2 an automorphism which reduces to multiplying $\sqrt{\pi_s^{\mathfrak{S}}}$ by

$$\left(\frac{\pi_s}{\pi_s^{\mathfrak{S}}}\right)\zeta(\mathfrak{S}).$$

We must therefore show that

$$\left(\frac{\pi_s}{\pi_s^{\mathfrak{S}}}\right) = \zeta(\mathfrak{S}).$$

Since by condition

$$\left(\frac{\pi_s}{\mathfrak{E}_{\mathfrak{S}}}\right) = \zeta(\mathfrak{S}),$$

it is sufficient to show that

$$\left(\frac{\pi_s}{\pi_s^{\mathfrak{S}}}\right) = \left(\frac{\pi_s}{\mathfrak{E}_{\mathfrak{S}}}\right).$$

We will prove that

$$\left(\frac{\alpha}{\alpha^{\mathfrak{S}}}\right) = \left(\frac{\alpha}{\mathfrak{E}_{\mathfrak{S}}}\right), \quad \mathfrak{S} \in U, \tag{126}$$

for any 2-hyperprimary number α.

In view of formula (105) we have

$$\left(\frac{\alpha}{\alpha^{\mathfrak{S}}}\right)_L = \left(\frac{\alpha}{\alpha^{\tau_1}}\right)_{LL^{\tau_1}}.$$

We select for τ_1 that representative of $\mathfrak{S}$, whose square is contained in H. Then τ_1 induces in the field $LL^{\tau_1}/\Lambda_{\tau_1}$ an automorphism of the second order. We can apply to that field formula (6) of paper [7] and obtain

$$\left(\frac{\alpha}{\alpha^{\tau_1}}\right)_{LL^{\tau_1}} = \left(\frac{\alpha}{\mathfrak{D}_{\tau_1}}\right)_{LL^{\tau_1}} = \left(\frac{\alpha}{N_{LL^{\tau_1}/L}(\mathfrak{D}_{\tau_1})}\right)_L = \left(\frac{\alpha}{\mathfrak{E}_{\mathfrak{S}}}\right)_L.$$

The case of the field $k(\sqrt{\pi_s^{\mathfrak{S}}})$, where $\pi_s \in \Pi_2$, and $\mathfrak{S} \in U$ can be considered in an analogous way.

We next consider the field $k(\sqrt{\pi_s^{\mathfrak{S}}})$, where $\pi_s \in \Pi_1$, $\mathfrak{S} \in V$. Equations (117) show that ω_1 induces in this field an identity automorphism. We take any prime ideal $\mathfrak{p}$, belonging to ω_2 in k'. Equations (122) show that

$$\left(\frac{\pi_s}{\mathfrak{P}^{\sigma}}\right)_k = 1, \text{ if } \sigma \in V^{-1}, \pi_s \in \Pi_1, N_{k/L}(\mathfrak{P}) = \mathfrak{p},$$

that is,

$$\left(\frac{\pi_s^{\sigma}}{\mathfrak{P}}\right)_k = 1, \text{ if } \sigma \in V, \pi_s \in \Pi_1.$$

From these equations we see that

$$\left(\frac{\pi_s^{\mathfrak{S}}}{\mathfrak{P}}\right)_k=\left(\frac{\pi_s^{\mathfrak{S}}}{\mathfrak{p}}\right)_L=1,$$

since by the lemma $\pi_s^{\mathfrak{S}} \in L$. Hence ω_2 also induces in $k(\sqrt{\pi_s^{\mathfrak{S}}})$ an identity automorphism.

We consider the field $k(\sqrt{\pi_s^{\mathfrak{S}}})$, where $\pi_s \in \Pi_2$ and $\mathfrak{S} \in V$. Equations (119) show that ω_1 induces in this field an automorphism which reduces to the multiplication of $\sqrt{\pi_s^{\mathfrak{S}}}$ by

$$\left(\frac{\pi_s}{\pi_s^{\mathfrak{S}}}\right)\zeta(\mathfrak{S}).$$

Equations (124) show that for $\sigma^{-1} \in V^{-1}$, that is, $\sigma \in V$

$$\left(\frac{\pi_s^{\mathfrak{S}}}{\mathfrak{p}}\right)=\left(\frac{\pi_s^{\mathfrak{S}}}{\pi_s}\right)\zeta(\mathfrak{S}^{-1}),$$

if $\mathfrak{p}$ belongs to ω_2 in k'. Hence ω_2 induces in $L(\sqrt{\pi_s^{\mathfrak{S}}})$ an automorphism which reduces to the multiplication of $\sqrt{\pi_s^{\mathfrak{S}}}$ by

$$\left(\frac{\pi_s^{\mathfrak{S}}}{\pi_s}\right)\zeta(\mathfrak{S}^{-1}).$$

Since

$$\left(\frac{\pi_s}{\pi_s^{\mathfrak{S}}}\right)=\left(\frac{\pi_s^{\mathfrak{S}}}{\pi_s}\right),$$

in view of 2-hyperprimarity of π_s, and $\zeta(\mathfrak{S}) = \zeta(\mathfrak{S}^{-1})$ by assumption, then ω_1 and ω_2 coincide in $L(\sqrt{\pi_s^{\mathfrak{S}}})$. The consideration of the case $\mathfrak{S} \in V^{-1}$ is carried out analogously.

Finally we consider the field $L(\sqrt{\alpha_\chi})$. The automorphism which ω_1 induces in this field can be reduced to the multiplication of $\sqrt{\alpha_\chi}$ by

$$(-1)^{\Sigma\varphi_\chi(\mathfrak{r})x_{\mathfrak{r}}}, \quad \text{if} \quad \alpha_\chi \approx \prod \mathfrak{r}^{\varphi_\chi(\mathfrak{r})},$$

while ω_2 induces in it an identity automorphism. Hence we must select $x_{\mathfrak{r}}$ so that the following congruences hold:

$$\Sigma f_{\mathfrak{S}}(\mathfrak{r})x_{\mathfrak{r}} \equiv b(\mathfrak{S})\ (2), \text{ if } \mathfrak{C}_{\mathfrak{S}} = \Pi\, \mathfrak{r}^{f_{\mathfrak{S}}(\mathfrak{r})},$$
$$\mathfrak{S} \in U,\ \zeta(\mathfrak{S}) = (-1)^b(\mathfrak{S}),$$
$$\Sigma \phi_\chi(\mathfrak{r})x_{\mathfrak{r}} \equiv 0\ (2), \text{ if } (\alpha_\chi) = \prod_{\mathfrak{r}} \mathfrak{r}^{\phi_\chi(\mathfrak{r})}.$$

If this system were compatible, there would exist a linear combination of the left hand sides with all the coefficients of $x_{\mathfrak{r}}$ equal to zero and with the constant term different from zero.

Let

$$\sum_{\mathfrak{S}\in U} c(\mathfrak{S})\sum_{\mathfrak{r}} f_{\mathfrak{S}}(\mathfrak{r})x_{\mathfrak{r}} + \sum_{\chi} d_\chi \sum \varphi_\chi(\mathfrak{r})x_{\mathfrak{r}} \equiv \sum_{\mathfrak{S}} c(\mathfrak{S})\,b(\mathfrak{S})\ (2) \qquad (127)$$

be such a combination. We denote $\Pi\,\chi^{d_\chi}$ by χ_0. Taking into consideration that in (127) all the coefficients of $x_{\mathfrak{r}}$ are 0, we get

$$\phi_{\chi_0}(\mathfrak{r}) \equiv \sum_{\mathfrak{S} \in U} c(\mathfrak{S}) f_{\mathfrak{S}}(\mathfrak{r}) \ (2)$$

or

$$(\alpha_{\chi_0}) = \prod_{\mathfrak{S} \in U} \mathfrak{F}_{\mathfrak{S}}^{c}(\mathfrak{S}).$$

But in this case by the condition of the theorem

$$\prod_{\mathfrak{S}} \zeta(\mathfrak{S})^{c(\mathfrak{S})} = 1, \text{ that is, } \sum_{\mathfrak{S}} b(\mathfrak{S}) c(\mathfrak{S}) \equiv 0 \ (2),$$

which shows that the constant term of (127) becomes 0.

Hence a selection of the required number π_t is possible at each step no matter how we divide the sequence $\pi_1, \cdots, \pi_{t-1}$ into the sets Π_1 and Π_2. We shall perform this division in such a way that for every number n there exists a $C(n)$ with the following properties: no matter how we divide the sequence $\pi_1, \cdots, \pi_{C(n)}$ into n subsequences, at least one of them will contain three numbers π_r, π_s and π_t $(r < s < t)$ such that if we divide $\pi_1, \cdots, \pi_{t-1}$ into Π_1 and Π_2, then π_r and π_s will go into different parts, while in dividing $\pi_1, \cdots, \pi_{s-1}$ into Π_1 and Π_2, the number π_r will go into a different part from the one into which it went on the division of $\pi_1, \cdots, \pi_{t-1}$. The possibility of performing such a division at each step has been shown in a lemma contained in paper [7].

We shall divide $\pi_1, \cdots, \pi_{C(n)}$ into subsequences putting into one subsequence all the π_i with the same value of $\left(\frac{\pi_i}{\mathfrak{P}_i^{\sigma}}\right)$, $\sigma \notin H$. Hence $n = 2^{(G:H)-1}$. Because of the above property of division into Π_1 and Π_2 we can find three numbers π_r, π_s and π_t satisfying the conditions

$$\left(\frac{\pi_r}{\mathfrak{P}_r^{\sigma}}\right) = \left(\frac{\pi_s}{\mathfrak{P}_s^{\sigma}}\right) = \left(\frac{\pi_t}{\mathfrak{P}_t^{\sigma}}\right).$$

An independent check, analogous to the one carried out in paper [7] shows that the number $\alpha = \pi_r \pi_s \pi_t$ satisfies the conditions of the theorem. Hence the theorem is proved.

We note that it follows from the proof of the theorem that π_i can be selected as belonging to a given automorphism in any extension K/L of the field k, if this automorphism induces in k an identity automorphism and if there exist principal prime ideals which belong in K to this automorphism and satisfy conditions (108) and (112).

BIBLIOGRAPHY

[1] W. Gaschütz, *Zur Erweiterungstheorie der endlichen Gruppen*, J. Reine Angew. Math. **190** (1952), 93-107.

[2] R. Kochendörffer, *Zwei Reduktionssätze zum Einbettungsproblem für abelsche Algebren*, Math. Nachr. **10** (1953), 75-84.

[3] O. Ore, *Contributions to the theory of groups of finite order,* Duke Math. J. 5 (1939), 431-460.

[4] D. K. Faddeev, *On the theory of homology in groups,* Izv. Akad. Nauk SSSR. Ser. Mat. **16** (1952), 17-22. (Russian)

[5] D. K. Faddeev, *On a theorem in the theory of homologies in groups,* Dokl. Akad. Nauk SSSR (N.S.) 92 (1953), 703-705. (Russian)

[6] I. R. Šafarevič, *On the construction of fields with a given Galois group of order* l^a, Izv. Akad. Nauk SSSR. Ser. Mat. **18** (1954), 261-296. (Russian) = Amer. Math. Soc. Transl. (2) 4 (1956), 107-142.

[7] I. R. Šafarevič, *On an existence theorem in the theory of algebraic numbers,* Izv. Akad. Nauk SSSR. Ser. Mat. **18** (1954), 327-334. (Russian) = Amer. Math. Soc. Transl. (2) 4 (1956), 143-150.

[8] I. R. Šafarevič, *On the problem of imbedding fields,* Izv. Akad. Nauk SSSR. Ser. Mat. **18** (1954), 389-418. (Russian) = Amer. Math. Soc. Transl. (2) 4 (1956), 151-183.

[9] I. R. Šafarevič, *On extensions of fields of algebraic numbers solvable in radicals,* Dokl. Akad. Nauk SSSR (N.S.) 95 (1954), 225-227. (Russian)

[10] H. Zassenhaus, *Lehrbuch der Gruppentheorie,* Bd. I, Teubner, Leipzig-Berlin, 1937.

Translated by:
Emma Lehmer

On birational equivalence of elliptic curves

Dokl. Akad. Nauk SSSR **114**, No. 2, 267–270 (1957). Zbl. **81**, 153
[Translated for this volume into English]

(Presented by Academician I. M. Vinogradov on 13|XI|1956)

1. We consider elliptic curves (curves of genus 1)* over an arbitrary field k about which we only assume that its characteristic is different from 2 and 3. The classification of such curves from the viewpoint of birational equivalence is well-known in the case of an algebraically closed field k [1]. In this case every elliptic curve γ is birationally equivalent to a curve ω having Weierstrass normal form

$$y^2 = x^3 + a\,x + b\,; \qquad a, b \in k\,. \tag{1}$$

Two curves having Weierstrass form in turn are birationally equivalent if and only if their absolute invariants defined by the formula

$$j = \frac{4\,a^3}{4\,a^3 + 27\,b^2}$$

coincide.

In case the field k is not algebraically closed an elliptic curve given over k may be transformed into Weierstrass form over some finite extension of the field k, and its absolute invariant still lies in k [2].

Since among the curves with given absolute invariant there is also one having Weierstrass form over k, we obtain all elliptic curves over k with given value of the absolute invariant by taking the curve ω with equation (1) and considering all elliptic curves over k which are birationally equivalent to it over a finite extension of the field k.

2. We consider a certain finite normal extension K/k of the field k and investigate the elliptic curves γ given over k and birationally equivalent to ω over K. We denote by M a general point of γ over k and by $k(M)$ and $\mathrm{K(M)}$ the fields of rational functions on γ with coefficients in k and K, respectively. According to the hypotheses, the field $K(M)$ is isomorphic to $K(x, y)$, where (x, y) is a general point of ω.

Any automorphism σ of the field K/k may be extended to an automorphism $\varphi_\gamma(\sigma)$ of the field $K(M)$ by putting $M^\sigma = M$ and shifting to $K(x, y)$ via the isomorphism between $K(M)$ and $K(x, y)$. Obviously,

$$\varphi_\gamma(\sigma\tau) = \varphi_\gamma(\sigma)\,\varphi_\gamma(\tau)\,. \tag{2}$$

Conversely, if we are given an isomorphism $\sigma \mapsto \varphi(\sigma)$ of the Galois group of K/k into the group of automorphisms of the field $K(x, y)$ such that $\sigma = \varphi(\sigma)$ on K, then the subfield of functions in $K(x, y)$ invariant under all $\varphi(\sigma)$ defines a

---

* It is not being assumed that the curve has a given rational point over k. (Translator's note)

certain elliptic curve over k which is birationally equivalent to ω over K. It is easy to verify that two curves γ_1 and γ_2 of the type under consideration are birationally equivalent over k if and only if

$$\varphi_{\gamma_1}(\sigma) = t\,\varphi_{\gamma_2} t^{-1}, \tag{3}$$

where t is some automorphism of the field $K(x, y)$ leaving fixed the elements of the field K.

The automorphism $s_\gamma(\sigma)$ of the field $K(x, y)$ defined by the equation

$$\varphi_\gamma(\sigma) = s_\gamma(\sigma)\,\varphi_\omega(\sigma)$$

obviously does not change the elements of the field K and hence is an automorphism of $K(x, y)/K$. For the relation (2) to be fulfilled it is necessary and sufficient that $s_\gamma(\sigma)$ satisfy the condition

$$s_\gamma(\sigma\tau) = s_\gamma(\sigma)\,s_\gamma(\tau)^{\varphi_\omega(\sigma)},$$

where

$$s_\gamma(\tau)^{\varphi_\omega(\sigma)} = \varphi_\omega(\sigma)\,s_\gamma(\tau)\,\varphi_\omega(\sigma)^{-1},$$

and the condition (3) may be rewritten in the form

$$s_{\gamma_1}(\sigma) = t\,s_{\gamma_2}(\sigma)\,t^{-\varphi_\omega(\sigma)}.$$

The automorphisms s of the field $K(x, y)/K$ are well-known [3]. They have the form

$$s(x, y) = \varepsilon(x, y) + P, \tag{4}$$

where P is a point on ω with coordinates in K, the addition being understood in the sense of addition on ω, and if $j \neq 0, 1$, then

$$\varepsilon(x, y) = (x, \pm y),$$

while if $j = 0$ or 1, then ε can be one of six or four automorphisms leaving fixed the point at infinity of ω. On writing down the automorphisms $s_\gamma(\sigma)$ in the form (4) we arrive at the following result:

Theorem 1. *Every elliptic curve γ over k which over K is birationally equivalent to the curve ω with equation (1) defines a system of automorphisms $\varepsilon_\gamma(\sigma)$ of the field $K(x, y)$ and of points $P_\gamma(\sigma)$ on the curve ω over K. The relations*

$$\begin{aligned} \varepsilon_\gamma(\sigma\tau) &= \varepsilon_\gamma(\sigma)\,\varepsilon_\gamma(\tau)^\sigma, \\ P_\gamma(\sigma\tau) &= P_\gamma(\sigma)\,\varepsilon_\gamma(\tau)^\sigma + P_\gamma(\tau)^\sigma \end{aligned} \tag{5}$$

hold.

Any system of automorphisms and points satisfying these conditions is defined by some curve γ. The curves γ_1 and γ_2 are birationally equivalent over k if and only if there exist an automorphism ε and a point P on ω such that

$$\begin{aligned} \varepsilon_{\gamma_1}(\sigma) &= e_{\gamma_2}(\sigma)\,\varepsilon^{1-\sigma}, \\ P_{\gamma_1}(\sigma) &= \varepsilon^{-\sigma}(P_{\gamma_2}(\sigma) + \varepsilon_{\gamma_2}(\sigma)\,P - P^\sigma). \end{aligned} \tag{6}$$

3. In the set of curves we are investigating there may be several curves having Weierstrass form. Any of them may be taken as ω. One can show that a curve γ

can be transformed over k into Weierstrass form if and only if γ is birationally equivalent over k to a curve γ' for which $P_{\gamma'}(\sigma)=0$ (zero of the group of points on ω). Thus in the collection of all curves with one system of automorphisms $\varepsilon_\gamma(\sigma)$ there is, up to birational equivalence, exactly one curve having Weierstrass form over k. If we took it as ω, then one could show that to all curves, to which the system of automorphisms $\varepsilon_\gamma(\sigma)$ corresponded in our first description, the identity system of automorphisms would now correspond. In view of this fact we confine ourselves to considering curves for which $\varepsilon_\gamma(\sigma)=1$. The equations (5) and (6), the latter with $\varepsilon=1$, then define a crossed homomorphism and a principal crossed homomorphism, respectively [4].

If $\varepsilon_\gamma(\sigma)=1$ for a curve γ, then the curve ω is its Jacobian variety [5].

In fact, since $K(x,y)=K(M)$, we have

$$(x,y)=\Phi(M),$$

where Φ is a birational mapping of γ to ω over K. For a divisor $\mathfrak{A}=\mathfrak{P}_1^{a_1}\dots\mathfrak{P}_r^{a_r}$ on γ we put

$$\Phi(\mathfrak{A})=a_1\,\Phi(\mathfrak{P}_1)+\dots+a_r\,\Phi(\mathfrak{P}_r).$$

As is well-known, if $d(\mathfrak{A})=d(\mathfrak{B})=0$, then $\mathfrak{A}\sim\mathfrak{B}$ on γ if and only if $\Phi(\mathfrak{A})=\Phi(\mathfrak{B})$. It remains for us only to verify that in case $d(\mathfrak{A})=0$ and $\mathfrak{A}$ is defined over some field $k'\supset k$, then $\Phi(\mathfrak{A})$ is also defined over k'. This follows easily from (4): if σ is an isomorphism of Kk'/k', then

$$(\Phi(\mathfrak{A}))^\sigma=\Phi^\sigma(\mathfrak{A}^\sigma)=\Phi(\mathfrak{A}^\sigma)+(\Sigma\,a_r)\,P_\sigma=\Phi(\mathfrak{A})+d(\mathfrak{A})\,P_\sigma=\Phi(\mathfrak{A}).$$

Thus the set of elliptic curves over k having ω as their Jacobian variety and birationally equivalent to ω over K is mapped onto the set of crossed homomorphisms of the Galois group G of the field K/k into the group $\mathfrak{A}_K$ of points on ω with coordinates in K. The formulas (6) show that two curves of the type under consideration are birationally equivalent over k only if: either 1) their associated crossed homomorphisms differ only by a principal crossed homomorphism, or 2) one of them is obtained from the other by application of an automorphism ε of the curve ω, regarded as an algebraic group. If we include in the concept of an elliptic curve with given Jacobian variety ω also the canonical function Φ which maps the group of divisor classes of degree zero on γ to the curve ω, then the transformations of the second type with $\varepsilon\neq1$ are eliminated, and the birational classes (in the new sense) of curves γ are in one-to-one correspondence with the elements of the one-dimensional cohomology group [4] $\mathrm{H}^1(G,\mathfrak{A}_K)$. The group operation defined on $H^1(G,\mathfrak{A}_K)$ carries over in a natural way to these classes. One obtains all curves with Jacobian variety ω by considering the union of all fields K and the corresponding inductive limit of the groups $H^1(G,\mathfrak{A}_K)$. Thus we arrive at the following result:

Theorem 2. *The birational classes of elliptic curves over k with given Jacobian variety ω form a group isomorphic to $H^1(\mathfrak{G},\mathfrak{A}_{\bar{k}})$, where $\mathfrak{G}$ is the Galois group of the separable algebraic closure $\bar{k}$ of the field k and $\mathfrak{A}_{\bar{k}}$ the group of points on ω with coordinates in $\bar{k}$. Here the crossed homomorphisms are to be taken as continuous, where the groups $\mathfrak{G}$ and $\mathfrak{A}_{\bar{k}}$ carry the Krull topology* [6] *and the discrete topology, respectively.*

As a union of cohomology groups of finite groups, the group $H^1(\mathfrak{G}, \mathfrak{A}_{\bar{k}})$ is a torsion group. One can show that the order of an elliptic curve γ in this group is divisible by the least order of a divisor class on γ.

4. If k is an algebraic number field, then by the Mordell-Weil theorem [7], the group $\mathfrak{A}_K$ has a finite number of generators. By the same token this is true also for $H^1(G, \mathfrak{A}_K)$; but since this group is a torsion group it is finite. Together with Theorems 1 and 2 this leads to the following result:

Theorem 3. *There is only a finite number of birationally non-equivalent curves over an algebraic number field k having a given value of their absolute invariant and a prime divisor of degree one in a fixed finite extension K/k.*

By the theorem of Lutz [8], this is also true in case k is a $\mathfrak{p}$-adic number field. However, as there is only a finite number of extensions of fixed degree over a $\mathfrak{p}$-adic number field, an even stronger assertion holds:

Theorem 4. *All elliptic curves of a fixed degree and with a fixed value of their absolute invariant j fall into a finite number of birational classes over a $\mathfrak{p}$-adic number field k.*

An analogous theorem does not hold for the rational numbers, even if one restricts oneself to curves with given Jacobian variety. Indeed, the curves with equation (in homogeneous coordinates)

$$a_0 x_0^3 + a_1 x_1^3 + a_2 x_2^3 = 0$$

have the curve

$$a_0 a_1 a_2 x_0^3 + x_1^3 + x_2^3 = 0$$

as their Jacobian variety. Hence, with $a_i = p^i$ $(i = 0, 1, 2)$ for an arbitrary prime p, we obtain infinitely many curves with common Jacobian variety and mutually birationally non-equivalent, for they are non-equivalent even over the field of p-adic numbers.

5. In his paper [9] which appeared recently, Weil defines the group of principal homogeneous algebraic spaces with a given abelian operator group. Since every elliptic curve appears as an homogeneous space with respect to its Jacobian variety, the result of Weil applies to our case. One can show that the group defined by him is isomorphic to the one described in Theorem 2.

Exactly the same reasoning as the one used in deriving this theorem applies also to the case considered by Weil, so that the following theorem holds.

Theorem 5. *The group of principal homogeneous algebraic varieties over k with a given abelian variety G is isomorphic to the first cohomology group $H^1(\mathfrak{G}, \mathfrak{A}_{\bar{k}})$, where $\mathfrak{G}$ is the Galois group of the separable algebraic closure $\bar{k}/k$ and $\mathfrak{A}_{\bar{k}}$ the group of points on G with coordinates in $\bar{k}$.*

References

1. Walker, R. J.: Algebraic Curves. Dover, New York 1950
2. Deuring, M.: Invarianten und Normalformen elliptischer Funktionenkörper. Math. Z. *47* (1942), 47–56

3. Hasse, H.: Zur Theorie der abstrakten elliptischen Funktionenkörper. I, II, III. J. reine angew. Math. *175* (1936), 55–62, 69–88, 193–208
4. Kurosch, A. G.: The Theory of Groups I, II. Chelsea, New York 1955/56
5. Chow, W. L.: The Jacobian variety of an algebraic curve. Amer. J. Math. *76* (1954), 453–476
6. Krull, W.: Galoissche Theorie der unendlichen algebraischen Erweiterungen. Math. Ann. *100* (1928), 687–698
7. Weil, A.: Sur un théorème de Mordell. Bull. Sci. Math. (2) *54* (1930), 182–191
8. Lutz, E.: Sur l'équation $y^2 = x^3 - Ax - B$ dans les corps p-adiques. J. reine angew. Math. *177* (1937), 238–247
9. Weil, A.: On algebraic groups and homogeneous spaces. Amer. J. Math. *77* (1955), 493–512

Received 12|XI|1956

Translated by H. G. Zimmer

Exponents of elliptic curves

Dokl. Akad. Nauk SSSR **114**, No. 4, 714–716 (1957). Zbl. **81**, 154
[Translated for this volume into English]

(Presented by Academician I. M. Vinogradov on 26|XII|1956)

Besides the genus g one can assign several other integral invariants to an algebraic curve γ defined over a field k: the least positive degree f of a divisor on γ; the least degree d of a curve birationally equivalent to γ over k; the least degree v of a prime divisor on γ. The relationship between these invariants has been little studied in the case of arbitrary curves, but for elliptic curves ($g=1$) they can all be expressed in terms of one. Indeed, it follows easily from the theorem of Riemann-Roch that for an elliptic curve, we have $v=f$, $d=f$ for $f>2$; $d=4$ for $f=2$; and $d=3$ for $f=1$. The invariant f we shall call the *exponent* of the curve γ.

For a curve of genus g, we have $f|2g-2$ and hence, if $g \neq 1$, the exponent can assume only a finite number of values. As was already noted (see, for instance, [1, 2]), for elliptic curves it is unknown which values the exponent can assume. In this note it will be proved that, over the rational number field R, there exist elliptic curves γ having an arbitrarily large exponent; and in addition, one can even arbitrarily prescribe the Jacobian curve ω of γ.

For the proof we shall use the group $H(\omega)$ formed by the k-birational equivalence classes of curves with given Jacobian curve ω. This group was for the first time defined by A. Weil [3]. Further we shall make use of the construction of the group $H(\omega)$ which was given in our paper [4]. As was shown in that paper, the exponent of the curve γ is a multiple of the order of γ as an element of the group $H(\omega)$. It was also proved there that the subgroup $H(K,\omega)$ of the group $H(\omega)$ consisting of all curves in $H(\omega)$ having a prime divisor of first degree in a given normal extension K of the field R is isomorphic to the group $H^1(G,\mathfrak{A}_K)$, where G is the Galois group of K/R, $\mathfrak{A}_K$ the group of points on ω with coordinates in K, and $H^1(G,\mathfrak{A}_K)$ the group of crossed homomorphisms of G to $\mathfrak{A}_K$.

If $\mathfrak{p}$ is a prime divisor of K, p the prime number divided by it, $G_\mathfrak{p}$ the decomposition group of $\mathfrak{p}$, and $K_\mathfrak{p}$ and R_p the corresponding local fields, then we have a natural embedding homomorphism

$$\varphi_\mathfrak{p}\colon H^1(G,\mathfrak{A}_K) \to H^1(G_\mathfrak{p},\mathfrak{A}_{K_\mathfrak{p}}). \qquad (1)$$

We first consider the group $H^1(G_\mathfrak{p},\mathfrak{A}_{K_\mathfrak{p}})$.

Suppose the equation of ω has the form

$$y^2 = x^3 + a\,x + b\,, \quad \Delta = 4\,a^3 + 27\,b^2 \neq 0\,,$$

where a and b may be taken to be integers. We designate by H_p the subgroup of those elements of a group H whose order is relatively prime to p.

For the study of the group $H^1(G_\mathfrak{p},\mathfrak{A}_{K_\mathfrak{p}})_p$ one may assume that the field $K_\mathfrak{p}$ has no higher ramification, since in the contrary case one could pass to some subfield. We denote by $\mathfrak{A}^r_{K_\mathfrak{p}}$ the subgroup of $\mathfrak{A}_{K_\mathfrak{p}}$ considered by Lutz [5] and consisting of the points (x, y) on ω for which $x\,\mathfrak{p}^{2(r-1)}$ and $y\,\mathfrak{p}^{3(r-1)}$ are non-integral. It is easy to show that for $r \geq 1$ the factor group $\mathfrak{A}^r_{K_\mathfrak{p}}/\mathfrak{A}^{r+1}_{K_\mathfrak{p}}$ is a p-group. From this one easily infers that the group $H^1(G_\mathfrak{p}, \mathfrak{A}_{K_\mathfrak{p}})_p$ is isomorphic to the group $H^1(G_\mathfrak{p}, \mathfrak{A}_{K_\mathfrak{p}}/\mathfrak{A}^1_{K_\mathfrak{p}})_p$.

We suppose moreover that $p \nmid \Delta$. Then, as was shown in [5], the factor group $\mathfrak{A}_{K_\mathfrak{p}}/\mathfrak{A}^1_{K_\mathfrak{p}}$ is isomorphic to $\mathfrak{A}_{\mathfrak{K}_\mathfrak{p}}$, where $\mathfrak{K}_\mathfrak{p}$ is the residue class field of $K_\mathfrak{p}$ with respect to $\mathfrak{p}$ and $\mathfrak{A}_{\mathfrak{K}_\mathfrak{p}}$ is the group of points on the curve ω, considered modulo $\mathfrak{p}$. Let us denote by $F_\mathfrak{p}$ the inertia group of $\mathfrak{p}$ in $K_\mathfrak{p}$. Then $G_\mathfrak{p}/F_\mathfrak{p}$ is the Galois group of the field $\mathfrak{K}_\mathfrak{p}$ and $G_\mathfrak{p}$ appears as an operator group for $\mathfrak{A}_{\mathfrak{K}_\mathfrak{p}}$, where $F_\mathfrak{p}$ acts trivially. As pointed out already,

$$H^1(G_\mathfrak{p}, \mathfrak{A}_{K_\mathfrak{p}})_p \cong H^1(G_\mathfrak{p}, \mathfrak{A}_{\mathfrak{K}_\mathfrak{p}})_p .$$

Let us consider the restriction homomorphism

$$H^1(G_\mathfrak{p}, \mathfrak{A}_{\mathfrak{K}_\mathfrak{p}})_p \rightarrow H^1(F_\mathfrak{p}, \mathfrak{A}_{\mathfrak{K}_\mathfrak{p}})_p . \tag{2}$$

The kernel of this map consists of those homomorphisms which become 0 on $F_\mathfrak{p}$ and hence are homomorphisms from $G_\mathfrak{p}/F_\mathfrak{p}$ to $\mathfrak{A}_{\mathfrak{K}_\mathfrak{p}}$. Since for curves over a finite field, as is well-known [6], $f = 1$, we have

$$H^1(G_\mathfrak{p}/F_\mathfrak{p}, \mathfrak{A}_{\mathfrak{K}_\mathfrak{p}}) = 0$$

and therefore, the homomorphism (2) is a monomorphism. It is easy to see that the image of this homomorphism coincides with the group $\mathrm{Hom}_{G_\mathfrak{p}/F_\mathfrak{p}}(F_\mathfrak{p}, \mathfrak{A}_{\mathfrak{K}_\mathfrak{p}})$ of $G_\mathfrak{p}/F_\mathfrak{p}$-operator homomorphisms of $F_\mathfrak{p}$ to $\mathfrak{A}_{\mathfrak{K}_\mathfrak{p}}$. We arrive at the following result:

Theorem 1. *If $p \nmid \Delta$ and the field $K_\mathfrak{p}$ does not have higher ramification, then*

$$H^1(G_\mathfrak{p}, \mathfrak{A}_{K_\mathfrak{p}})_p \cong \mathrm{Hom}_{G_\mathfrak{p}/F_\mathfrak{p}}(F_\mathfrak{p}, \mathfrak{A}_{\mathfrak{K}_\mathfrak{p}}) .$$

Now we construct a field K in which there is a crossed homomorphism $f \in H^1(G, \mathfrak{A}_K)$ of given order m. To this end we consider all algebraic points P on ω for which $mP = 0$. They form a group $\mathfrak{A}_m$ which is a direct sum of two cyclic groups of order m. By T_m we designate the field obtained by adjunction to R of the coordinates of all $P \in \mathfrak{A}_m$. This field is normal. We denote its Galois group by $\bar{G}$. It is obvious that $P^\sigma \in \mathfrak{A}_m$ for $\sigma \in \bar{G}$ and $P \in \mathfrak{A}_m$ and that σ defines an automorphism of the group $\mathfrak{A}_m$. We consider a group G which contains a normal subgroup A isomorphic to $\mathfrak{A}_m$, and which is the semi-direct product of $\bar{G}$ and A with automorphisms

$$\alpha^\sigma = \varphi^{-1}(\varphi(\alpha)^\sigma), \quad \alpha \in A , \tag{3}$$

where φ is some fixed isomorphism of A onto $\mathfrak{A}_m$.

We construct a field K which contains T_m, is normal over R and has over R the Galois group G. The existence of such a field follows from results of Scholz [7] and Delone and Faddeev [8]. The construction of the field K by any of these methods can be carried out in such a way that an arbitrary prime divisor $\mathfrak{p}$ of the field T_m has in K ramification index m. We choose $\mathfrak{p}$ relatively prime to the discriminant of T_m/R, to m, Δ and 2.

Let use define a mapping f of the group G to $\mathfrak{A}_K$ in the following way:

$$f(\sigma\alpha) = \varphi(\alpha), \quad \sigma \in \bar{G}, \ \alpha \in A .$$

It can be easily derived from (3) that f is a crossed homomorphism, i.e. $f \in H^1(G, \mathfrak{A}_K)$. Obviously, $mf=0$. We show that f has in $H^1(G, \mathfrak{A}_K)$ the exact order m. For this purpose it suffices to prove that its image $\varphi_{\mathfrak{p}} f$ under the homomorphism $\varphi_{\mathfrak{p}}$ defined by (1) has order m. By virtue of Theorem 1, this will be proved if we show that $\varphi_{\mathfrak{p}} f$, as an operator homomorphism of $F_{\mathfrak{p}}$ to $\mathfrak{A}_{\mathfrak{K}_{\mathfrak{p}}}$, has order m. The latter assertion is a consequence of the following observation. By the choice of $\mathfrak{p}$, the group $F_{\mathfrak{p}}$ is cyclic of order m. The homomorphism f maps a generator of $F_{\mathfrak{p}}$ onto an element P of $\mathfrak{A}_m$ which has order m, and since $\mathfrak{p}$ is not critical* in T_m, the image of P in $\mathfrak{A}_{\mathfrak{K}_{\mathfrak{p}}}$ has the same order.

We have come to our final result:

Theorem 2. *In the group of elliptic curves having over the field of rational numbers a given Jacobian curve, there exist elements of arbitrary order.*

Corollary. *Among all elliptic curves having over the field of rational numbers a given Jacobian curve, there exist curves whose exponent is divisible by any given number.*

All the proofs remain valid, with considerable simplifications, also for the field of rational functions over the complex number field. The fact itself was noted in this case without detailed proofs by Enriques [9].

We remark that for the field of p-adic numbers the analogous theorem is not true. Indeed, Theorem 1 shows that, if the discriminant of ω is not divisible by p, the exponents of curves having ω as their Jacobian curve can be divisible only by p and by prime divisors of the number of points on ω over the residue field modulo p.

References

1. Hasse, H.: Number Theory. Akademie-Verlag, Berlin 1979, resp. Springer-Verlag, Berlin-Heidelberg-New York 1980
2. Weil, A.: Remarques sur un mémoire d'Hermite. Arch. Math. *5* (1954), 197–202
3. Weil, A.: On algebraic groups and homogeneous spaces. Amer. J. Math. *77* (1955), 493–512
4. Shafarevich, I. R.: On birational equivalence of elliptic curves. Dokl. Akad. Nauk 114_2 (1957), 267–270
5. Lutz, E.: Sur l'équation $y^2 = x^3 - Ax - B$ dans les corps $\mathfrak{p}$-adiques. J. reine angew. Math. *177* (1937), 238–247
6. Schmidt, F. K.: Analytische Zahlentheorie in Körpern der Charakteristik p. Math. Z. *33* (1931), 1–32
7. Scholz, A.: Über die Bildung algebraischer Zahlkörper mit auflösbarer Galoisscher Gruppe. Math. Z. *30* (1929), 332–356
8. Delone, B. N., Faddeev, D. K.: Investigations in the geometry of the Galois theory. Mat. Sb. N.S. *15* (57) (1944), 243–284 (Russian)
9. Enriques, F.: Sur les problèmes qui se rapportent à la résolution des équations algébriques renfermant plusieurs inconnues. Math. Ann. *51* (1899), 134–153

Received 20|XII|1956

Translated by H. G. Zimmer

* i.e. not ramified in T_m/R. (Translator's note)

Cohomology groups of nilpotent algebras

(with A. I. Kostrikin)

Dokl. Akad. Nauk SSSR **115**, No. 6, 1066–1069 (1957). Zbl. **83**, 30
[Translated for this volume into English]

(Presented by Academician I. M. Vinogradov on 26|III|1957)

In this note we study cohomology groups $H^n(N, k)$ for a nilpotent associative algebra N of finite rank over an arbitrary field k, where, as N-module of coefficients, we take the field k itself with N acting trivially. All properties of cohomology groups needed in what follows are to be found in the book [1].

It is known that the group $H^n(N, k)$ does not change if the algebra N is replaced by an algebra A obtained from N by formal adjunction of a unit element. In this way one obtains every algebra whose quotient algebra by its radical is isomorphic to the ground field. This property is shared in particular by the group algebra A of a finite p-group G over the (prime) field of p elements. In this case, the group $H^n(A, k)$ coincides with the cohomology group $H^n(G, Z_p)$ of the group G with coefficients in the cyclic group Z_p of order p.

The group $H^n(N, k)$ is a vector space of finite dimension over the field k. The dimension of this space is denoted by b_n and will be called the *n-th Betti number* of the algebra N. An analogous notation will be adopted for the groups $H^n(G, Z_p)$. The investigation of Betti numbers is the main goal of the present paper. The formulation of the results obtained in this direction becomes more natural if one considers the Poincaré function $R_N(t) = \sum_{n=0}^{\infty} b_n t^n$ associated with the algebra N; for a finite p-group G, the function $R_G(t)$ is defined in an analogous manner.

We have the following results:

Theorem 1. *Let $N = N_1 + \ldots + N_m$ be the direct sum of m nilpotent algebras; denote by $R_N(t) = R(t)$ and $R_{N_i}(t) = R_i(t)$ respectively their corresponding Poincaré functions. Then*

1
$$\frac{1}{R(t)} - 1 = \sum_{i=1}^{m} \left(\frac{1}{R_i(t)} - 1 \right).$$

Theorem 2. *The Betti numbers of a nilpotent algebra N are related by the inequalities*

$$b_n - b_{n-1} + \ldots + (-1)^n b_0 \geq \frac{1 + (-1)^n}{2}, \qquad n = 1, 2, \ldots, \tag{1}$$

which can be rewritten in the form

$$\frac{1}{1+t} R_N(t) \gg \frac{1}{1-t^2}. \tag{2}$$

Here the relation $F(t) \gg G(t)$ signifies that each coefficient of the power series $G(t)$ does not exceed the corresponding coefficient of $F(t)$.

In particular, for $n=2$ the relation (1) yields $b_2 \geq b_1$. In the case of a p-group this shows that the number of non-equivalent extensions of the cyclic group Z_p by a p-group G is not less than p^d, where d is the number of generators of G.

On multiplying the relation (2) by $1+t$, we obtain

$$R_N(t) \gg \frac{1}{1-t},$$

from which we derive Theorem 3.

Theorem 3. *The Betti numbers of a nilpotent algebra and of a finite p-group are positive.*

An upper estimate for the Betti numbers is given in Theorem 4.

Theorem 4. *For the function $R_N(t)$ corresponding to a nilpotent algebra N of rank r, the inequality*

$$R_N(t) \ll \frac{1}{1-rt}$$

holds.

For the function $R_G(t)$ corresponding to a group G of order p^ν, the inequality

$$R_G(t) \ll \frac{1}{(1-t)^\nu} \tag{3}$$

holds.

Corollary. *The radius of convergence ϱ of the series $R_N(t)$ is contained within the bounds*

$$\frac{1}{r} \leq \varrho \leq 1.$$

The radius of convergence of the series $R_G(t)$ is equal to 1.

We list some examples.

1. For a cyclic group G,

$$R_G(t) = \frac{1}{1-t}.$$

Since for the direct product $G_1 \times G_2$ of two groups G_1 and G_2 we have

$$R_{G_1 \times G_2}(t) = R_{G_1}(t) \cdot R_{G_2}(t)$$

in accordance with the well-known Künneth formulas, we conclude that for an abelian group G with d generators,

$$R_G(t) = \frac{1}{(1-t)^d}.$$

2. If a group G is the semi-direct extension of a cyclic group by a cyclic group:

$$G = \{x, y\}, \quad x^{p^\alpha} = y^{p^\beta} = 1, \quad x^{-1} y x = y^{1+u}, \quad (1+u)^{p^\alpha} - 1 = m p^\beta;$$

we obtain in the case of
a) $m \equiv 0 \pmod{p}$:

$$R_G(t) = \frac{1}{(1-t)^2}$$

and in the case of
b) $(m,p)=1$:

$$R_G(t) = \frac{1}{(1-t)^2\,(1+t^2+t^4+\ldots+t^{2(p-1)})} = \frac{1+t}{(1-t)\,(1-t^{2p})}\,.$$

3. If N is the quotient algebra of the polynomial algebra in d variables having no constant term by the ideal of polynomials of degree ≥ 3, then

$$R_N(t) = \frac{(1+t)^d}{1-\sum_1^d \binom{i+1}{2}\binom{d+2}{i+2} t^{i+1}}\,.$$

4. If N is the quotient algebra of the polynomial algebra in two variables having no constant term by the ideal of polynomials of degree $\geq r$, then

$$R_N(t) = \frac{1+t}{1-t-r\,t^2}\,.$$

5. Let N be the reduced free non-commutative algebra on d generators with the relation $N^r=0$. Then

$$R_N(t) = \frac{1+d\,t}{1-d^r\,t^2}\,.$$

These and many other examples we have studied allow us to state the conjec-
2 ture that for an arbitrary nilpotent algebra N of finite rank, the series $R_N(t)$ is a
rational function of t. Then, in the case of a finite p-group G, all poles of the
function $R_G(t)$ would be roots of unity.

The following theorem also supports this conjecture.

Theorem 5. *If the Betti numbers of a nilpotent algebra N over a finite field k are bounded, then $R_N(t)$ is a rational function.*

The following example shows that there exist infinitely many algebras with bounded Betti numbers.

Let N have two generators e_1 and e_2; suppose that $N^5=0$; let k be an arbitrary field of characteristic $\neq 2$ and assume that

$$e_2 e_1 = -e_1 e_2 + a e_1^3 + b e_1^2 e_2 + c e_1^4; \quad e_2^2 = f e_1^2 + g e_1^2 e_2 + h e_1^4;$$

$$e_1^3 e_2 = 0; \quad b \cdot f \left(f + \left(\frac{g-a}{b} \right)^2 \right) \neq 0;$$

then

$$R_N(t) = \frac{1+2t+2t^2+t^3}{1-t^4} = \frac{1+t+t^2}{1-t+t^2-t^3}\,.$$

As is well-known, the same Poincaré function belongs to the generalized quaternion group.

The proofs of the asserted theorems follow easily from a consideration of the special complete resolution

$$0 \xleftarrow{\partial_{-1}} X_{-1} \xleftarrow{\partial_0} X_0 \leftarrow \dots \leftarrow X_{n-1} \xleftarrow{\partial_n} X_n \leftarrow \dots, \tag{4}$$

in which

$$X_{-1} = k, \quad X_0 = A = \alpha e + N, \quad \alpha \in k, \quad e^2 = e, \quad e v = v,$$
$$\partial_0 (\alpha e + v) = \alpha \quad (v \in N).$$

Set

$$\mathrm{Ker}(\partial_n) = J_n$$

and denote by b_{n+1} the minimal number of generators of J_n as an A-module. It is known that b_{n+1} is equal to the dimension of the linear space $J_n - NJ_n$.

We set

$$X_{n+1} = E^{b_{n+1}} \otimes A,$$

where $E^{b_{n+1}}$ is the linear space of dimension b_{n+1} over k and $\otimes$ designates the tensor product. It is easy to see that the numbers b_n coincide with the Betti numbers of the algebra N. For the proof of Theorem 1 one may, of course, confine oneself to the case $m=2$. Let $N = N_1 + N_2$, $R_1(t) = \sum a_n t^n$, $R_2(t) = \sum b_n t^n$, $R(t) = \sum c_n t^n$; and denote by A_n, B_n and C_n the kernels of ∂_n in the corresponding exact sequences of type (4). Obviously,

$$C_0 = A_0 + B_0,$$
$$C_1 = A_1 + B_1 + E^{a_1} \otimes B_0 + E^{b_1} \otimes A_0 = A_1 + B_1 + E^{a_1} \otimes C_0 + E^{b_1 - a_1} \otimes A_0.$$

From this, by induction, we easily deduce the relations

$$C_{n-1} = A_{n-1} + B_{n-1} + \sum_1^{n-1} E^{d_i} \otimes C_{n-1-i} + \sum_1^{n-1} E^{b_i - d_i} \otimes A_{n-1-i}, \tag{5}$$

where

$$d_n = \sum_0^{n-1} (b_i - d_i)\, a_{n-i} \quad (d_0 = 0). \tag{6}$$

From (5) and (6) it follows that

$$c_n = b_n + \sum_0^n d_{n-i} c_i, \quad b_n + \sum_0^n d_i a_{n-i} - \sum_0^n b_i a_{n-i} = 0,$$

i.e.

$$R(t) = R_2(t) + R(t)\, D(t), \tag{7}$$

$$R_2(t) + R_1(t)\, D(t) - R_1(t)\, R_2(t) = 0, \tag{8}$$

where

$$D(t) = \sum d_n t^n.$$

Theorem 1 easily follows from (7) and (8).

Theorem 2 follows from the fact that, in the sequence (4), the ranks of the modules X_n and J_n over the field k are connected by the relations

$$r(X_n) = b_n(1 + r(N)) = r(J_n) + r(J_{n-1}).$$

From this we obtain by induction that

$$r(J_{n-1}) = (b_{n-1} - b_{n-2} + \ldots + (-1)^{n-1} b_0)(1 + r(N)) + (-1)^n,$$

from which, in view of the inequality

$$b_n \geq \frac{r(J_{n-1})}{1 + r(N)},$$

Theorem 2 is follows easily.

Theorems 3, 4 and 5 follow similarly by examining the sequence (4). The only exception is the inequality (3) in Theorem 4. It is derived from the fact that the Betti numbers of a central extension are majorized by the Betti numbers of the direct product. This fact is deduced in the familiar manner by studying the Hochschild-Serre spectral sequence.

Note added in proof. After giving the manuscript in print the authors learned that an analogous problem was considered by B. B. Venkow. His results concerning p-groups partly coincide with ours.

References

Cartan, H., Eilenberg, E.: Homological Algebra. Princeton University Press, Princeton 1973

Received 21|III|1957

Translated by H. G. Zimmer

The imbedding problem for split extensions

Dokl. Akad. Nauk SSSR **120**, No. 6, 1217–1219 (1958). Zbl. **85**, 26
[Translated for this volume into English]

(Presented by Academician I. M. Vinogradov on 18|II|1958)

The imbedding problem concerns a given normal extension k/Ω with Galois group F, a group G and an epimorphism $\varphi: G \to F$. It is required to find conditions under which there exists a normal extension K/Ω with Galois group G such that $K \supset k$ and the epimorphism φ coincides with the canonical homomorphism of the Galois group of the field onto the Galois group of the subfield.

The group G is said to be a *split extension of its image F under the homomorphism φ* if it contains a subgroup which is isomorphic to F under φ. There may be several such subgroups. In what follows we shall fix one of them and denote it by F. If the kernel of φ is N, then $G = F \cdot N$. We shall say then that G is a *split extension of the group F with kernel N*.

The aim of the present note is to report the following result:

Theorem 1. *The imbedding problem is solvable for any algebraic number field k if G is a split extension with nilpotent kernel.*

This theorem contains as a special case a series of results obtained earlier about the imbedding problem and the task of constructing fields with given Galois group.

If $F=1$, Theorem 1 shows the existence of an extension with an arbitrary nilpotent Galois group N. This was proven by Scholz [1] and Reichardt [2] for nilpotent groups N of odd order and by the author [3] for arbitrary nilpotent groups N. In the case of an abelian kernel N, Theorem 1 was proven by Scholz [4] and, by more direct methods, by Delone and Faddeev [5]. In the cases in which N is a p-group of class $c \le p$ or in which the orders of the groups G and N are relatively prime, Theorem 1 was established by the author [6]. Since every solvable group $\mathfrak{G}$ occurs as the factor group of a group G resulting from a chain of split extensions with nilpotent kernels [7], it follows from Theorem 1 that there exist algebraic number fields with an arbitrary solvable Galois group. This fact was shown by the author [8] on the basis of the solution of a certain more artificial imbedding problem.

The proof of Theorem 1 rests on arguments similar to those exploited in the papers [3, 6, 8] of the author. The main difference is that the concept of a Scholzian field used in those papers is now replaced by the concept of a relatively Scholzian field. We may obviously confine ourselves to the case in which N is a group of order l^α, where l denotes a prime number. We first consider the case in which G too is an l-group. The subfield of K belonging to the subgroup F will be designated by L.

The field K will be called relatively Scholzian (with respect to k) if the following conditions are satisfied for it:

1. Every prime divisor of the discriminant of the field k/Ω, considered as a divisor of L, contains a prime factor of first degree to the first power.

2. The prime divisors of the discriminant of K/k decompose in K into prime divisors of first degree but are not critical* in k/Ω and have degree 1 in k/Ω.

3. The absolute norms of the prime divisors of the discriminant of K/k satisfy the conditions

$$\mathfrak{N}(\mathfrak{p}) \equiv 1\,(l^h),$$

where h is sufficiently large.

4. The prime divisors of l split completely in K/k.

5. The real infinite divisors of k stay real in K.

The connection of this concept with the imbedding problem is based on the following theorem. Let Z be a normal subgroup of order l in N. We denote by $\bar{N}$, $\bar{G}$ and $\bar{K}$ the groups N/Z, G/Z and the field having Galois group $\bar{G}$, respectively. The homomorphism $\varphi: G \to \bar{G}$ maps F isomorphically. On identifying φF with F we may write $\bar{G} = \bar{N} \cdot F$. The subfield of $\bar{K}$ corresponding to $\bar{N}$ will be denoted by k.

Theorem 2. *The imbedding problem for the field $\bar{K}$ and the homomorphism $G \to \bar{G}$ is solvable if the field $\bar{K}$ is relatively Scholzian (with respect to k).*

Now let the order of G be arbitrary. We designate by H the l-Sylow subgroup of G. Let A be a minimal abelian normal subgroup of G lying in N. We put $G/A = \bar{G}$, $H/A = \bar{H}$, $N/A = \bar{N}$ and denote by φ the homomorphism from G onto $\bar{G}$. The field with Galois group $\bar{G}$ over a field $\varkappa$ will be designated by $\bar{K}/\varkappa$, and its subfields corresponding to $\bar{H}$ and $\bar{N}$ by Ω and k, respectively. Applying the reduction theorem of Faddeev [9] and Kochendörffer [10] one derives from Theorem 2 the following result.

Theorem 3. *The imbedding problem for the field $\bar{K}/\varkappa$ and the homomorphism $\varphi: G \to \bar{G}$ is solvable if the field $\bar{K}/\Omega$ is relatively Scholzian (with respect to k).*

The field K appearing as a solution of the imbedding problem stated in Theorem 3 will in general not be relatively Scholzian. The conditions under which it may be chosen relatively Scholzian are to be found analogously to the manner in which this was carried out in the papers [3, 6, 8]. They consist in the requirement that certain invariants $\chi_i(\bar{K})$ of the field $\bar{K}$ whose values are l-th roots of unity be equal to one.

We suppose that the group N has d generators $s_1, \ldots, s_d$ and that the group F has order m. Let us consider the reduced free group of class c, $\mathfrak{N}_d^c$ on md generators $\sigma_{f,i}$, $i = 1, \ldots, d$, $f \in F$, whose factor group is N. Defining the commutation of f_1 with $\sigma_{f,i}$ by the rule

$$f_1^{-1} \sigma_{f,i} f_1 = \sigma_{ff_1,i},$$

we obtain a group $F \cdot \mathfrak{N}_d^c$ which we denote by $\mathfrak{G}_d^c$. The mapping

$$f \mapsto f, \quad \sigma_{f,i} \mapsto s_i^f$$

---

* i.e., not ramified. (Translator's note)

defines a homomorphism of $\mathfrak{G}_d^c$ to G which maps F isomorphically onto F. From this it follows that the solvability of the imbedding problem for the field k and the homomorphism $\mathfrak{G}_d^c \to F$ implies the solvability of the initial imbedding problem.

Suppose that, for any d, we have established the existence of a field $K_d^{c-1} \supset k$ possessing Galois group $\mathfrak{G}_d^{c-1}$. For the possibility of imbedding this field into the field K_d^c it is necessary that the invariants $\chi_i(K_d^{c-1})$ mentioned above equal 1. On employing the apparatus developed in the papers [3, 6, 8] one ends up with the following assertion.

Theorem 4. *For any positive integer δ, there exists a number $d(\delta)$ such that in an arbitrary field K_d^{c-1}, containing k and being relatively Scholzian, there is a subfield $\bar{K}_\delta^{c-1}$, also containing k, whose invariants $\chi_i(\bar{K}_\delta^{c-1})$ are equal to 1.*

Such a field $\bar{K}_\delta^{c-1}$ can therefore be imbedded into a relatively Scholzian field K_δ^c (over k). Successive application of this process also yields the proof of Theorem 1.

References

1. Scholz, A.: Konstruktion algebraischer Zahlkörper mit beliebiger Gruppe von Primzahlpotenzordnung I. Math. Z. *42* (1937), 161–188
2. Reichardt, H.: Konstruktion von Zahlkörpern mit gegebener Galoisgruppe von Primzahlpotenzordnung. J. reine angew. Math. *177* (1937), 1–5
3. Shafarevich, I. R.: On the construction of fields with given Galois group of order l^α. Izv. Akad. Nauk SSSR *18*$_3$ (1954), 261–296. Amer. Math. Soc. Transl. Ser (2) *4* (56), 107–142
4. Scholz, A.: Über die Bildung algebraischer Zahlkörper mit auflösbarer Galoisscher Gruppe. Math. Z. *30* (1929), 332–356
5. Delone, B. N., Faddeev, D. K.: Investigations in the geometry of the Galois theory. Mat. Sb. N.S. *15* (57) (1944), 243–284 (Russian)
6. Shafarevich, I. R.: On the problem of imbedding fields. Izv. Akad. Nauk SSSR *18*$_5$ (1954), 389–418. Amer. Math. Soc. Transl. Ser. (2) *4* (56), 151–183
7. Ore, O.: Contributions to the theory of groups of finite order. Duke Math. J. *5* (1939), 431–460
8. Shafarevich, I. R.: Construction of fields of algebraic numbers with given solvable Galois group. Izv. Akad. Nauk SSSR *18*$_6$ (1954), 525–578. Amer. Math. Soc. Transl. (2) *4* (56), 185–237
9. Faddeev, D. K.: On a theorem of the theory of homologies in groups. Dokl. Akad. Nauk SSSR N.S. *92* (1953), 703–705 (Russian)
10. Kochendörffer, R.: Zwei Reduktionssätze zum Einbettungsproblem für abelsche Algebren. Math. Nachr. *10* (1953), 75–84

Received 17|II|1958

Translated by H. G. Zimmer

The group of algebraic principal homogeneous varieties

Dokl. Akad. Nauk SSSR **124**, No. 1, 42–43 (1959). Zbl. **115**, 389
[Translated for this volume into English]

In this note we investigate the group $\mathfrak{H}(\alpha, k)$ of algebraic principal homogeneous varieties on which an abelian variety α operates, all varieties being regarded as defined over a fixed field k. The definition of this group is contained in the paper [1] of Weil. In the case in which α is an elliptic curve, the group $\mathfrak{H}(\alpha, k)$ coincides with the group of elliptic curves having α as their Jacobian curve.

We introduce the following notations: α_K is the group of points on α which are rational over an extension K/k. $G(K/k)$ is the Galois group of K/k. We assume K/k to be a normal separable algebraic extension (which may be infinite). $\varkappa$ is the separable algebraic closure of k; $\mathfrak{G} = G(\varkappa/k)$, $\mathfrak{a} = \alpha_\varkappa$.

For a positive integer n and an abelian group A, A_n stands for the subgroup of elements of period n in A, and $A^{(n)} = A/nA$. In the sequel the number n will be assumed relatively prime to the characteristic of k (and arbitrary in case k is of characteristic zero). k_n designates the smallest extension of k over which all points of $\mathfrak{a}_n$ are rational, $\mathfrak{N}$ the corresponding normal subgroup of $\mathfrak{G}$, and $F = \mathfrak{G}/\mathfrak{N} = G(k_n/k)$. K^* denotes the multiplicative group of the field K.

1. *Arbitrary field k.* As in known [2], $\mathfrak{H}(\alpha, k) \cong H^1(\mathfrak{G}, \mathfrak{a})$. From the exactness of the sequence

$$0 \to \mathfrak{a}_n \to \mathfrak{a} \overset{v}{\to} \mathfrak{a} \to 0,$$

in which v is the homomorphism of multiplication by n, one easily deduces the exact sequence

$$0 \to \alpha_k^{(n)} \to H^1(\mathfrak{G}, \mathfrak{a}_n) \to H^1(\mathfrak{G}, \mathfrak{a})_n \to 0.$$

Hence it suffices for the determination of the group $H^1(\mathfrak{G}, \mathfrak{a})_n$ to find the group $H^1(\mathfrak{G}, \mathfrak{a}_n)$.

We consider finite separable extensions K/k_n which are normal over k and for which the groups $G(K/k)$ have the following structure:

I. $G(K/k_n)$ is isomorphic to a subgroup of the group $\mathfrak{a}_n$. We fix some monomorphism

$$i\colon G(K/k_n) \to \mathfrak{a}_n.$$

II. i represents an F-operator homomorphism. Obviously, $\mathfrak{a}_n$ and $G(K/k_n)$ are F-operator groups.

III. If $\xi \in H^2(F, G(K/k_n))$ is the cohomology class defining the extension $G(K/k)/G(K/k_n) = F$ in accordance with the theory of Schreier, then $i\xi = 0$.

We fix a cocycle $\xi_0 \in \xi$ and a cochain η for which

$$\delta\eta = \xi_0.$$

The set of triples (K, i, η) constitutes an abelian group with respect to the natural group operation. We denote this group by $\mathfrak{A}_n$.

Theorem 1. *The groups $H^1(\mathfrak{G}, \mathfrak{a}_n)$ and $\mathfrak{A}_n$ are isomorphic.*

2. *Algebraic number field k.* In this case the existence of an infinite number of fields K/k_n satisfying conditions I–III even with $\xi = 0$ has been proved (see [3, 4]). Since, according to the theorem of Mordell-Weil, $\alpha_k^{(n)}$ is finite one easily deduces from this the following result:

Theorem 2. *The group $H^1(\mathfrak{G}, \mathfrak{a})_n$ contains infinitely many elements of order n.*

This result was obtained by me earlier [5] by more special considerations.

We denote by $\mathfrak{h}(\alpha, k)_n$ the subgroup of the group $\mathfrak{H}(\alpha, k)_n$ consisting of homogeneous spaces that contain a rational point in every $\mathfrak{p}$-adic completion $k_\mathfrak{p}$ of the field k.

The following assertion is a simple consequence of Theorem 1:

Theorem 3. *The group $\mathfrak{h}(\alpha, k)_n$ is finite.*

The fact that the group $\mathfrak{h}(\alpha, k)_n$ may be non-trivial was first discovered by Selmer [6] for $n = 3$ in the case in which α is an elliptic curve. In the case of elliptic curves Theorem 3 has the following significance.

We unite in one class all elliptic curves of given degree over k which are birationally equivalent over k, and in one genus all curves which are birationally equivalent over every $k_\mathfrak{p}$. Then every genus consists of a finite number of classes.

As Lang and Tate communicated to me, they have independently proved Theorem 2.

3. *$\mathfrak{p}$-adic number field k.* We confine ourselves to the case in which α is an elliptic curve. As D. K. Faddeev [7] showed, there exists over an arbitrary field a scalar product $\alpha_k \otimes H^1(G, \alpha_K) \to B_k$, where $G = G(K/k)$ is the Galois group of K/k and B_k the Brauer group of k. Over a $\mathfrak{p}$-adic number field the Brauer group is isomorphic to the group W of all roots of unity, and we obtain a scalar product

$$\varphi: \alpha_k \otimes \mathfrak{H}(\alpha, k) \to W.$$

Theorem 4. *The scalar product φ is non-degenerate, so that the groups α_k and $\mathfrak{H}(\alpha, k)$ are the character groups of one another.*

As Tate [8] communicated to me, he has independently proved Theorem 4 and also obtained its generalization to the case in which α is an arbitrary abelian variety.

References

1. Weil, A.: On algebraic groups and homogeneous spaces. Amer. J. Math. *77* (1955), 493–512
2. Shafarevich, I. R.: On birational equivalence of elliptic curves. Dokl. Akad. Nauk SSSR *114*$_2$ (1957), 267–270
3. Scholz, A.: Über die Bildung algebraischer Zahlkörper mit auflösbarer Galoisscher Gruppe. Math. Z. *30* (1929), 332–356
4. Delone, B. N., Faddeev, D. K.: Investigations in the geometry of the Galois theory. Mat. Sb. *15* (57) (1944), 243–284 (Russian)
5. Shafarevich, I. R.: Exponents of elliptic curves. Dokl. Akad. Nauk SSSR *114*$_4$ (1957), 714–716

6. Selmer, E. S.: The diophantine equation $a x^3 + b y^3 + c z^3 = 0$. Acta Math. *85* (1951), 203–362
7. Faddeev, D. K.: To the theory of algebras over the field of algebraic functions of one variable. Vestn. Leningrad. Univ. Ser. Mat. Mech. Astr. *12*$_7$ (1957), 45–51 (Russian; Engl. summary)
8. Tate, J.: *WC*-groups over p-adic fields. Sém. Bourbaki *10*e année 1957, 156/1–13

Received 13|IX|1958

Translated by H. G. Zimmer

The imbedding problem for local fields
(with S. P. Demushkin)

Izv. Akad. Nauk SSSR, Ser. Mat. **23**, No. 6, 823–840 (1959). Zbl. **93**, 44
[Transl., II. Ser., Am. Math. Soc. **27**, 267–288 (1963)]

This article considers the problem of imbedding a finite field extension in a larger extension, with given absolute Galois group. We show that in the case of local fields the condition of consistency of D. K. Faddeev is necessary and sufficient for the solution of this problem.

Introduction

In this article we give conditions under which any given finite normal extension k/Ω with Galois group F may be imbedded in a larger extension K/Ω with Galois group G by a given epimorphism $\phi: G \longrightarrow F$ which is the realization of some homomorphism of the Galois group of the field onto the Galois group of the subfield. We consider such conditions only when the kernel A of the epimorphism ϕ is abelian, i.e., when the Galois group of K/k is abelian.

The problem consists in finding invariants of the field k, the group G and the epimorphism ϕ, on which the solution of the imbedding problem depends. We call such invariants obstructions. A series of conditions which are necessary for the solution of the imbedding problem were found by D. K. Faddeev [3] and H. Hasse [4], somewhat later. Such conditions are what we call the first obstruction. The definition of the first obstruction is given in §1. One considers elements of $H^2(F_x, k^*)$, where $x \in \mathrm{Hom}(A, k^*)$ and F_x is the subgroup which leaves x invariant (if we consider $\mathrm{Hom}(A, k^*)$ as an F-operator group). If the first obstruction vanishes, k/Ω is said to be consistent with the solution to the imbedding problem. In general, the condition that k/Ω is consistent with the solution is not sufficient for the solution.

Furthermore, we make use of the reduction theorem of Kochendörffer [5], according to which conditions for the imbedding problem are reduced to the case when G is a p-group. We may, therefore, assume that A contains a normal subgroup A_1 of the group G such that $(A: A_1) = p$. Then A/A_1 lies in the center of G/A_1. If the field k/Ω is consistent with the initial imbedding problem, then the imbedding problem becomes, as is easy to see, the problem for a field with Galois group G/A_1. Its solution is a field $k(\sqrt[p]{\mu})$ in which the number μ is determined precisely by a substitution $\mu \longrightarrow \mu m$, $m \in \Omega$. We show when this field $k(\sqrt[p]{\mu})$ may be chosen so that it is consistent with the solution of the imbedding problem for a field with group G.

It follows from this that, in this case, the first obstruction lies in a set of cyclic algebras $(a_{x_1}, b_{x_1})_{k_{x_1}}$ of index p, where $x_1 \in \mathrm{Hom}(A_1, k^*)$. Here, k_{x_1} is the subfield left stationary by the subgroup F_{x_1} of the homomorphism x_1, $a_{x_1} \in k_{x_1}$ is defined so that $k_{x_1}(\sqrt[p]{a_{x_1}}) = k_x$, k_x being the invariant field of F_x, in which x extends x_1 from A_1 to A, and b_{x_1} is an element of k_{x_1}, which is defined by means of μ. Under the substitution μ to μm all b_{x_1} are multiplied by m. This result will follow as a corollary when we derive the properties of the first obstruction.

Having obtained these results, we apply them to the case of local fields (finite extensions of p-adic number fields). Using properties of the cohomology groups of local fields, we obtain the principal result of this work: for local fields, the triviality of the first obstruction (i.e., the complete condition of Faddeev-Hasse) is necessary and sufficient for the solution to the imbedding problem. It follows that in the case of an algebraic number field k/Ω, the condition of Faddeev-Hasse for the solution to the imbedding problem is equivalent to the solution to all corresponding p-adic problems for the fields $k_{\mathfrak{p}}/\Omega_p$ and all prime divisors $\mathfrak{p}$.

§1. The first obstruction to the imbedding problem. The problem, which we are considering, may be formulated in the following form. Given a finite group G, an epimorphism $\phi: G \longrightarrow F$ and a normal separable extension k_0/Ω with Galois group F. We require conditions on k_0, G and ϕ so that there is a field $K \supset k_0$, normal over Ω, whose Galois group is isomorphic to G, and such that the homomorphism ϕ coincides with the homomorphism of the Galois group of the field K onto that of the field k_0. This problem is called the problem of imbedding. We designate this problem by

$$(k_0/\Omega, G, \phi).$$

It is of interest to state the imbedding problem in such a way that the solution K can be an admissible and regular algebra. Regular (see [4]) means a semisimple commutative algebra K, with a given group of automorphisms G, and also a regular representation of G in the vector space K/Ω. On the other hand, for induction, it will be necessary to place the problem of imbedding over certain other imbedding problems, for which we may obtain, as solutions, certain regular algebras. Therefore, from the beginning, we consider the possibility that the solutions to imbedding problems may lie over regular algebras k. Here, it is important to note that all regular algebras which we meet in this work will always be normal extensions, over Ω, of the field k_0, over which we place our primary problem. We will maintain the field k_0 as the quality of a subalgebra. k_0 is called the kernel or the regular algebra. Thus, we consider the imbedding problem $(k/\Omega, G, \phi)$, where k is a regular algebra, normal over Ω, and as a possible solution, k_0 an admissible and regular algebra K.

Denote by A the kernel of the homomorphism ϕ. We then have an exact sequence

$$1 \longrightarrow A \xrightarrow{i} G \xrightarrow{\phi} F \longrightarrow 1. \tag{1}$$

We call the problem $(k/\Omega, G, \phi)$, the problem associated with the sequence (1).

We shall be concerned with the imbedding problem only when the following conditions are satisfied:

1) The group A is abelian.

2) The characteristic of the algebra k is relatively prime to the order of the group A.

3) The field Ω has an infinite number of elements.

4) The kernel k_0 of the algebra k contains all mth roots of unity, where m is the period of A (the least common multiple of the orders of all the elements). Besides, we suppose that the kernel k_0 of the algebra k corresponds to a subgroup $\mathfrak{A}_0$, for which $\phi^{-1}(\mathfrak{A}_0)$ is abelian. This condition is included because we are considering the imbedding problem with abelian kernel A, and all regular algebras which occur will be extensions of k_0.

R. Brauer [1] settled the imbedding problem in a case which plays a basic role here. In order to formulate Brauer's result, we remark that the inclusion i of A in G (as A is abelian) gives A the structure of an F-operator group. We shall consider this F-operator structure. Furthermore, the exact sequence (1) defines, according to the theorem of Schreier, a cohomology class $a \in H^2(F, A)$ which is called the fundamental class of the exact sequence.

Definition 1. An F-operator group A is called k-elementary, if there is an F-operator monomorphism

$$x\colon\ A \rightarrowtail k_0^*$$

of this group into the multiplicative group of the kernel k_0 of the algebra k.

It is clear that a k-elementary group is always cyclic. If the mth roots of unity lie in Ω, then the condition that A be k-elementary coincides with the requirement that A is a trivial operator group.

We say that the imbedding problem $(k/\Omega, G, \phi)$ is elementary, if in the exact sequence (1) the group A is a k-elementary F-operator group.

Theorem of Brauer. *The elementary imbedding problem $(k/\Omega, G, \phi)$ can be solved if and only if the cohomology class $x^*(a)$, where $x^*\colon H^2(F, A) \longrightarrow H^2(F, k^*)$ is the induced homomorphism, splits in k^*.*

In other words, the fundamental class a must lie in the kernel of the homomorphism

$$x^*: H^2(F, A) \mapsto H^2(F, k^*)$$

defined by the homomorphism x.

We introduce, in the set of all imbedding problems, a partial ordering. For this, we introduce the two following operations on the imbedding problems:

I. *Extension of the basic field.* If $(k/\Omega, G, \phi)$ is any imbedding problem, and k' is any subfield of the algebra k, we suppose that $G' = \phi^{-1}(F')$, where F' is the Galois group of k/k' and take ϕ' to be the restriction of ϕ to G'. Then we obtain an imbedding problem $(k/k', G', \phi')$.

II. *Replacement of the group G with a factor-group by a normal subgroup lying in A.* Let $A' \subset A$ be a normal subgroup of the group G. Put

$$G' = G/A'; \ \phi' = \phi\psi'^{-1},$$

where $\psi': G \longrightarrow G'$ is the projection map. It is evident that $(k/\Omega, G', \phi')$ is also an imbedding problem.

Definition 2. We say that the imbedding problem $(k/k', G', \phi')$ is concomitant with the imbedding problem $(k/\Omega, G, \phi)$, if one can be obtained from the other after a sequence of operations of the form I or II above.

It is easy to see that the relation "concomitant" can be used to introduce a partial ordering in the set of imbedding problems.

From the solution to a given imbedding problem $(k/\Omega, G, \phi)$ we may obtain the solutions to all imbedding problems which are concomitant to it. It is sufficient to show this for an imbedding problem which is obtained by applying either operation I or II. If K/Ω is a solution to the imbedding problem $(k/\Omega, G, \phi)$, then K/k' is a solution to the problem $(k/k', G', \phi')$, obtained from operation I, and if the algebra $\overline{K} \subset K$ corresponds to a normal subgroup A', then $\overline{K}/\Omega$ is a solution to $(k/\Omega, G', \phi')$, obtained from operation II.

It follows from this that each condition which is necessary for the solution to some imbedding problem, which is concomitant to a given imbedding problem $(k/\Omega, G, \phi)$, will also be necessary for the solution to the problem $(k/\Omega, G, \phi)$.

We derive, at once, necessary and sufficient conditions for the solution of an elementary imbedding problem, concomitant to the problem $(k/\Omega, G, \phi)$, and thus obtain necessary conditions for the problem $(k/\Omega, G, \phi)$ (as is well known in view of Brauer's theorem).

Let $x: A \longrightarrow k_0^*$ be any homomorphism of $A = \operatorname{Ker}\phi$ into the multiplicative group k_0^* of the kernel k_0 of the algebra k. Denote by F_x the subgroup consisting of those elements $f \in F$ for which

$$x(a^f) = x(a)^f, \quad a \in A.$$

Because $x(a) \in k_0$, for $f \in \mathfrak{A}_0$, we have

$$x(a)^f = x(a).$$

On the other hand, we suppose that $\phi^{-1}(\mathfrak{A}_0)$ is an abelian group, as $a^f = a$, $f \in \mathfrak{A}_0$. Hence, it follows that $F_x \supset \mathfrak{A}_0$ and thus F_x corresponds to a k_x which is contained in the kernel k_0, that is, it is a field.

Set

$$A_x = \mathrm{Ker}(x), \quad \overline{G}_x = \phi^{-1}(F_x), \quad G_x = \overline{G}_x / A_x$$

and denote by ϕ_x the natural homomorphism of G_x onto F_x. We see that the imbedding problem $(k/k_x, G_x, \phi_x)$ is concomitant with the problem $(k/\Omega, G, \phi)$. For this, the kernel of ϕ_x i.e., the group A/A_x, is an elementary F_x-operator group. In fact

$$x\colon\ A/A_x \longrightarrow k_0^*$$

is an F_x-monomorphism. Therefore, the imbedding problem $(k/k_x, G_x, \phi_x)$ is elementary.

It is easy to see that any elementary imbedding problem, which is concomitant to the problem $(k/\Omega, G, \phi)$, is of necessity concomitant to a problem of the form $(k/k_x, G_x, \phi_x)$. Therefore, conditions for the solution to the problem $(k/k_x, G_x, \phi_x)$ are simultaneously conditions for the solution to the problem $(k/\Omega, G, \phi)$. Conditions for the solution to $(k/k_x, G_x, \phi_x)$ follow from Brauer's theorem. One concludes, therefore, that the cohomology class

$$\rho a \in H^2(F_x, A)$$

is sent, under the homomorphism

$$x^*\colon\ H^2(F_x, A) \longrightarrow H^2(F_x, k^*),$$

into the identity. Here, we denote by ρ the restriction map from cohomology classes of F to F_x, and by a the fundamental class of the exact sequence (1). The class $x^*\rho a$ will be denoted C_x.

Definition 3. The cohomology class

$$C_x = x^*\rho a \in H^2(F_x, k^*), \quad x \in \mathrm{Hom}(A, k_0^*),$$

is called the first obstruction for the imbedding problem $(k/\Omega, G, \phi)$. If the first obstruction is trivial, we say that k/Ω is consistent with the imbedding problem.

§2. **The second obstruction.** In this section we consider the case when the group G is a p-group. As Kochendörffer's results [5] show, the imbedding problem may be solved in this case. If G is a p-group, then A contains a subgroup

A_1, which is normal in G and has index p in A. Set $G/A_1 = F^1$, and let ψ be the natural homomorphism of F^1 onto F. It is clear that the imbedding problem $(k/\Omega, F^1, \psi)$ is concomitant with the initial problem $(k/\Omega, G, \phi)$.

By assumption, the pth roots of unity lie in the kernel k_0 of the algebra k, and as $(k: \Omega)$ is a power of p, they are contained in Ω. The group A/A_1 is a normal subgroup of order p in the p-group F^1, and hence, lies in its center. It follows that the imbedding problem $(k/\Omega, F^1, \psi)$ is elementary. If we suppose that the algebra k/Ω is consistent with the imbedding problem $(k/\Omega, G, \phi)$, then the imbedding problem $(k/\Omega, F^1, \psi)$ may be solved. Denote by k^1 one of the solutions of this imbedding problem. We suppose initially that the original imbedding problem is solved, with K a solution. The subalgebra k^1 of K, belonging to A_1, is on the one hand a solution to $(k/\Omega, F^1, \psi)$; on the other hand, the imbedding of k^1 in K is the solution of some new imbedding problem which is connected with an exact sequence

$$1 \to A_1 \to G \xrightarrow{\phi_1} F^1 \to 1. \tag{2}$$

In particular, the solution k^1 is consistent with this imbedding problem.

In the light of all this, we introduce in this paragraph the following imbedding problem: when is it that the solution to the imbedding problem $(k/\Omega, F^1, \psi)$, which we write k^1, can be chosen so that it is consistent with the imbedding problem connected with the exact sequence (2)? We will thus be able to obtain necessary conditions for the solution to the initial imbedding problem $(k/\Omega, G, \phi)$. We will call this the second obstruction. We do not compute the second obstruction here, but we will derive some of its properties, which will make it possible for us to show that, for local fields Ω it vanishes.

Furthermore, we investigate the situation in somewhat more generality. Given an imbedding problem $(k/\Omega, G, \phi)$, corresponding to the exact sequence (1) and a group A with subgroup A_1 of index p which is normal in G, so that the problem $(k/\Omega, F^1, \psi)$, corresponding to the exact sequence

$$1 \to A/A_1 \to F^1 \to F \to 1,$$

is elementary, central and solvable. Let k^1 be one of its solutions. We investigate the connection between the first obstruction for the problem $(k/\Omega, G, \phi)$ and the first obstruction for the problem $(k^1/\Omega, G, \phi_1)$, corresponding to the exact sequence (2).

Let $x_1 \in \operatorname{Hom}(A_1, k_0^*)$ and let x be the extension of x_1 on A. Denote by G_x and G_{x_1} the centralizers of x and x_1 in G. It is clear that

$$G \supset G_{x_1} \supset G_x \supset A \supset A_1.$$

The corresponding chain of algebras is

$$\Omega \subset k^1_{x_1} \subset k_x \subset k \subset k^1.$$

In light of an earlier remark, $k^1_{x_1}$ and k_x are fields.

Denote

$$G_x/A = F_x, \quad G_{x_1}/A = F_{x_1},$$
$$G_x/A_1 = F^1, \quad G_{x_1}/A_1 = F^1_{x_1}.$$

The first obstruction to the problem $(k/\Omega, G, \phi)$ has the form

$$C_x = x^* \rho a,$$

and to the problem $(k^1/\Omega, G, \phi_1)$,

$$C_{x_1} = x_1^* \rho_1 a_1,$$

where a and a_1 are the fundamental classes of the sequences (1) and (2), and ρ is the restriction homomorphism from F to F_x, and ρ_1 is the restriction homomorphism from F^1 to $F^1_{x_1}$.

Theorem 1. $C_{x_1} \otimes k_x \sim C_x$.

Proof. As is known

$$C_{x_1} \otimes k_x \sim (k^1/k_{x_1}, r x_1 \rho_1 a_1)$$

where r is the restriction homomorphism from $F^1_{x_1}$ to F^1_x.

On the other hand

$$C_x \sim (k^1/k_x, \lambda x \rho a),$$

where λ is the lift homomorphism of F_x to F^1_x. Consequently, the theorem asserts that

$$r x_1^* \rho_1 a_1 = \lambda x^* \rho a.$$

It is clear that $r x_1^* = x_1^* r$. The homomorphism $r\rho_1$ is denoted r_1 and is the restriction from F^1 to F^1_x. Then, we must show that

$$x_1^* r_1 a_1 = \lambda x^* \rho a.$$

The cocycle on the left-hand side belongs to $H^2(F^1_x, x_1(A_1))$, while the cocycle on the right belongs to $H^2(F^1_x, x(A))$.

Consider these cocycles in $H^2(F^1_x, k^{1*})$. We show that they coincide in $H^2(F^1_x, x(A))$. For this purpose we use Lemma 1, which will be proved in the next section, in the case where we take for each a the element $\rho^F_{F_x} a$. Then, it is clear that

$$(\rho^F_{F_x} a)_1 = \rho^{F^1}_{F_x} a_1$$

and Lemma 1 gives

$$i^* \rho^{F^1}_{F^1_x} a_1 = \lambda^{F^1_x}_{F_x} \rho^F_{F_x} a.$$

Applying x^*, we have,

$$x^*_1 \rho^{F^1}_{F^1_x} a_1 = \lambda^{F^1_x}_{F_x} x^* \rho^F_{F_x} a$$

(x commutes with $\lambda^{F^1_x}_{F_x}$).

Thus, Theorem 1 is proved.

Theorem 1 shows that in order for the condition of consistency with the solution to the imbedding problem $(k/\Omega, G, \phi)$ to be satisfied, i.e., $C_x \sim 1$, the first obstruction, after the first step, C_{x_1} must be split in k_x, i.e.,

$$C_{x_1} \otimes k_x \sim C_x \sim 1.$$

This means that the algebra C_{x_1} is a cyclic algebra. Therefore, it follows that it is sufficient to show that $(k_x : k^1_{x_1}) = p$ or 1. To prove this last assertion, we introduce a function defined on F_{x_1}:

$$f(\sigma) = \frac{x(\bar{z}^{\sigma})}{x(\bar{z})^{\sigma}},$$

where z is any element of A/A_1, $\bar{z}$ being a representative in A. The function f does not depend on the choice of the representative $\bar{z}$ in A or on the extension of x_1 on A. The equality $f(\sigma) = 1$ signifies $\sigma \in F_x$. In addition, a straightforward computation shows that

$$f(\sigma\tau) = f(\sigma)^{\tau} f(\tau),$$

i.e., f is a cocycle. It is clearly the case that the value of f is a pth root of unity, as

$$f(\sigma)^p = \frac{x((\bar{z}^p)^\sigma)}{x(\bar{z}^p)^\sigma} = \frac{x_1((\bar{z}^p)^\sigma)}{x_1(\bar{z}^p)^\sigma} = 1.$$

Therefore, we must have the equality

$$f(\sigma\tau) = f(\sigma)f(\tau).$$

Hence, in particular, we find that F_x is a normal subgroup of F_{x_1}, as

$$f(\tau\sigma_x\tau^{-1}) = f(\tau)f(\sigma_x)f(\tau^{-1}) = f(\sigma_x) = 1,$$

for $\sigma_x \in F_x$.

Now, we show that $(F_{x_1} : F_x) = p$ or 1. It is sufficient to consider the case $(F_{x_1} : F_x) \neq 1$ and show that

$$(F_{x_1} : F_x) = p.$$

Let σ be an element in F_{x_1}, so that $f(\sigma) = \zeta_p$. Then $\sigma \in F_x$, $\sigma^p \in F_x$ and

$$F_{x_1} = \sum_{i=0}^{p-1} F_x\sigma^i,$$

i.e.,

$$(F_{x_1} : F_x) = p.$$

The cocycle f is constant on residue classes with respect to F_x, and it follows that it may be obtained from a cocycle on F_{x_1}/F_x. Denote this again by f.

Consider the cocycle f in k_x^*. As a 1-dimensional cocycle it splits, that is in k_x there is a number α_{x_1}, depending only on x_1, so that

$$f(\sigma) = \alpha_{x_1}^{\sigma-1}.$$

If in this equality we raise both sides to the pth power, we get

$$(\alpha_{x_1}^p)^{\sigma-1} = f(\sigma)^p = 1,$$

i.e., $a_{x_1} = \alpha_{x_1}^p \in k_{x_1}^1$. Hence, we see that

$$k_x = k_{x_1}^1(\sqrt[p]{a_{x_1}}).$$

Thus, if $C_x \sim 1$, then $C_{x_1} \sim (a_{x_1}, b_{x_1})$ and is a cyclic algebra.

We proceed to formulate Theorem 2. Let there be given three exact sequences:

$$1 \to A \to G \to F \to 1 \quad \text{(with class } a\text{)},$$
$$1 \to A \to \bar{G} \to F \to 1 \quad \text{(with class } b\text{)},$$
$$1 \to A \to \tilde{G} \to F \to 1 \quad \text{(with class } a \cdot b\text{)}.$$

Let A_1 be a normal subgroup of G, contained in A. Then A_1 is normal in $\bar{G}$ and $\tilde{G}$. Consequently, we may consider the imbedding problems

$$(k/\Omega, F^1, \psi), \ (k/\Omega, \bar{F}^1, \bar{\psi}), \ (k/\Omega, \tilde{F}^1, \tilde{\psi}),$$

where

$$F^1 = G/A_1, \quad \bar{F}^1 = \bar{G}/A_1, \quad \tilde{F}^1 = \tilde{G}/A_1.$$

We suppose that the problems $(k/\Omega, F^1, \psi)$ and $(k/\Omega, \bar{F}^1, \bar{\psi})$ are solved, with k^1 and $\bar{k}^1$ as solutions. Then, consider the algebra $K = k^1 \otimes_k \bar{k}^1$. Denote the group of K over Ω by Γ. It is the direct product of F^1 and $\bar{F}^1$ relative to the factor-group F.* It is in the same way an extension of the direct product $A/A_1 \times \bar{A}/A_1$ over the group F with cocycle $aA_1 \times \overline{bA_1}$. In the group $A/A_1 \times \bar{A}/A_1$ we single out a subgroup $\tilde{A}$, which consists of elements $(a, \bar{a}^{-1})$. Let the algebra $\tilde{k}^1$, by definition, correspond to the subgroup $\tilde{A}$, in K. The algebra $\tilde{k}^1$ has, over Ω, the group $\Gamma/\tilde{A}$, which is isomorphic to the group $\tilde{F}^1$. Consequently, $\tilde{k}^1$ is a solution to the imbedding problem $(k/\Omega, \tilde{F}^1, \tilde{\psi})$.

Let $x_1 \in \mathrm{Hom}(A_1, k_0^*)$. Consider, corresponding to x_1, the first obstructions for the imbedding problems

$$(k^1/\Omega, G, \phi_1), \ (\bar{k}^1/\Omega, \bar{G}, \bar{\phi}_1), \ (\tilde{k}^1/\Omega, \tilde{G}, \tilde{\phi}_1),$$

where $\phi_1, \bar{\phi}_1, \tilde{\phi}_1$ are the natural homomorphisms from $G, \bar{G}, \tilde{G}$ onto $F^1, \bar{F}^1, \tilde{F}^1$.

We form

$$C_{x_1} = (k^1/k^1_{x_1}, F^1_{x_1}, x_1 \rho a_1),$$
$$\bar{C}_{x_1} = (\bar{k}^1/\bar{k}^1_{x_1}, \bar{F}^1_{x_1}, x_1 \bar{\rho} b_1),$$
$$\tilde{C}_{x_1} = (\tilde{k}^1/\tilde{k}^1_{x_1}, \tilde{F}^1_{x_1}, x_1 \tilde{\rho} (ab)_1),$$

where $\rho, \bar{\rho}$ and $\tilde{\rho}$ are restrictions respectively of F^1 to $F^1_{x_1}$, $\bar{F}^1$ to $\bar{F}^1_{x_1}$ and $\tilde{F}^1$ to $\tilde{F}^1_{x_1}$. But, as it is easy to see, the fields $k^1_{x_1}, \bar{k}^1_{x_1}, \tilde{k}^1_{x_1}$ coincide (we denote them by $k^1_{x_1}$). Therefore, the algebras $C_{x_1}, \bar{C}_{x_1}, \tilde{C}_{x_1}$ are defined over

*Translator's note: A precise definition occurs in the next section.

the same field.

Let Γ_{x_1} be the subgroup of Γ to which $k^1_{x_1}$ corresponds in K.

Theorem 2. $C_{x_1} \otimes \overline{C}_{x_1} \sim \widetilde{C}_{x_1}$.

Proof. As is well known,

$$C_{x_1} \sim (K/k^1_{x_1}, \Gamma_{x_1}, \lambda x_1 \rho a_1),$$
$$\overline{C}_{x_1} \sim (K/k^1_{x_1}, \Gamma_{x_1}, \overline{\lambda} x_1 \overline{\rho} b_1),$$
$$\widetilde{C}_{x_1} \sim (K/k^1_{x_1}, \Gamma_{x_1}, \widetilde{\lambda} x_1 \widetilde{\rho} (ab)_1),$$

where λ lifts from $F^1_{x_1}$ to Γ_{x_1}, $\overline{\lambda}$ from $\overline{F}^1_{x_1}$ to Γ_{x_1}, and $\widetilde{\lambda}$ from $\widetilde{F}^1_{x_1}$ to Γ_{x_1}.

Theorem 2 asserts that the classes in which the cycles $\lambda x_1 \rho a_1 \cdot \overline{\lambda} x_1 \overline{\rho} b_1$ and $\widetilde{\lambda} x_1 \widetilde{\rho} (ab)_1$ lie coincide. But, it is evident that x_1 commutes with all the homomorphisms λ, $\overline{\lambda}$ and $\widetilde{\lambda}$, and, therefore, it is sufficient to show that the classes $\lambda \rho a_1 \cdot \overline{\lambda} \overline{\rho} b_1$ and $\widetilde{\lambda} \widetilde{\rho} (ab)_1$ coincide.

In order to simplify the last assertion, consider the commutative diagrams

$$\begin{array}{ccc} F^1 & \xleftarrow{\Pi} & \Gamma \\ i\uparrow & & \uparrow I \\ F^1_{x_1} & \xleftarrow{\pi} & \Gamma_{x_1} \end{array} \qquad \begin{array}{ccc} \overline{F}^1 & \xleftarrow{\overline{\Pi}} & \Gamma \\ \overline{i}\uparrow & & \uparrow I \\ \overline{F}^1_{x_1} & \xleftarrow{\overline{\pi}} & \Gamma_{x_1} \end{array} \qquad \begin{array}{ccc} \widetilde{F}^1 & \xleftarrow{\widetilde{\Pi}} & \Gamma \\ \widetilde{i}\uparrow & & \uparrow I \\ \widetilde{F}^1_{x_1} & \xleftarrow{\widetilde{\pi}} & \Gamma_{x_1}. \end{array}$$

Here, the homomorphisms denoted by π and Π are projections, while the homomorphisms denoted by i and I are injections.

Then the following diagrams of cohomology groups are commutative (we write them only in dimension 2):

$$\begin{array}{ccc} H^2(F^1, A_1) & \xrightarrow{\Lambda} & H^2(\Gamma, A_1) \\ \rho\downarrow & & \downarrow \mathrm{P} \\ H^2(F^1_{x_1}, A_1) & \xrightarrow{\lambda} & H^2(\Gamma_{x_1}, A_1) \end{array}$$

$$\begin{array}{ccc} H^2(\overline{F}^1, A_1) & \xrightarrow{\overline{\Lambda}} & H^2(\Gamma, A_1) \\ \overline{\rho}\downarrow & & \downarrow \mathrm{P} \\ H^2(\overline{F}^1_{x_1}, A_1) & \xrightarrow{\overline{\lambda}} & H^2(\Gamma_{x_1}, A_1) \end{array} \qquad \begin{array}{ccc} H^2(\widetilde{F}^1, A_1) & \xrightarrow{\widetilde{\Lambda}} & H^2(\Gamma, A_1) \\ \widetilde{\rho}\downarrow & & \downarrow \mathrm{P} \\ H^2(\widetilde{F}^1_{x_1}, A_1) & \xrightarrow{\widetilde{\lambda}} & H^2(\Gamma_{x_1}, A_1). \end{array}$$

Hence, we obtain

$$\lambda\rho = \mathrm{P}\Lambda,$$
$$\overline{\lambda}\overline{\rho} = \mathrm{P}\overline{\Lambda},$$
$$\widetilde{\lambda}\widetilde{\rho} = \mathrm{P}\widetilde{\Lambda}.$$

The equality, which we want to prove, reduces to the following.

$$\mathrm{P}\Lambda a_1 \cdot \mathrm{P}\bar{\Lambda} b_1 = \mathrm{P}\widetilde{\Lambda}(ab)_1,$$

or

$$\Lambda a_1 \cdot \bar{\Lambda} b_1 = \widetilde{\Lambda}(ab)_1.$$

But this last equality follows from Lemma 2 which we prove in the next section.

Let us apply Theorem 2, in the case when $b = 1$. The solution k^1 of the imbedding problem $(k/\Omega, F^1, \psi)$ has the form

$$k^1 = k(\sqrt[p]{\mu_0}).$$

The group $\bar{F}^1$ can be written as a direct product of F and A/A_1, and the solution to $(k/\Omega, \bar{F}^1, \bar{\psi})$ is

$$\bar{k}^1 = k(\sqrt[p]{m}), \quad m \in \Omega.$$

By the same construction,

$$\tilde{k}^1 = k(\sqrt[p]{\mu_0 m})$$

and therefore is a solution to the problem $(k/\Omega, F^1, \psi)$.

Theorem 2 gives, in this case, for $x_1 \in \mathrm{Hom}(A_1, k_0^*)$,

$$C_{x_1}(\mu_0) \otimes \bar{C}_{x_1}(m) \sim \widetilde{C}_{x_1}(\mu_0 m).$$

As we are considering the case where $ab = a$ and $\widetilde{G} = G$,

$$\widetilde{C}_{x_1}(\mu_0 m) = C_{x_1}(\mu_0 m)$$

and Theorem 2 shows that in the transition from the solution $k(\sqrt[p]{\mu_0})$ to the problem $(k/\Omega, F^1, \psi)$ to the solution $k(\sqrt[p]{\mu_0 m})$, the algebra $C_{x_1}(\mu_0)$ is multiplied by some algebra $\bar{C}_{x_1}(m)$.

A direct computation shows that

$$\bar{C}_{x_1}(m) \sim (a_{x_1}, m).$$

From these remarks, and the remarks on Theorem 1, we obtain the following assertion:

Theorem 3. *If for the imbedding problem* $(k/\Omega, G, \phi)$ *the condition of consistence is satisfied, then after the first step (the construction of the field* $k^1 = k(\sqrt[p]{\mu_0})$) *the first obstruction for the problem* $(k^1/\Omega, G, \phi_1)$ *is a cyclic algebra of the form*

$$(a_{x_1}, b_{x_1})_{k_{x_1}}, \quad x_1 \in \mathrm{Hom}(A_1, k_0^*),$$

and moreover, the transitions from a solution $k^1 = k(\sqrt[p]{\mu_0})$ *to a solution* $k(\sqrt[p]{\mu_0 m})$ *always consist of multiplication by a cyclic algebra* $(a_{x_1}, m)_{k_{x_1}}$.

Thus, the second obstruction consists in the impossibility of finding $m \in \Omega$ so that

$$(a_{x_1}, m) \sim (a_{x_1}, b_{x_1})^{-1}$$

for all $x_1 \in \mathrm{Hom}(A_1, k_0^*)$.

The imbedding problem decomposes, as follows, into two steps:

1) Is it possible, for given $x_1 \in \mathrm{Hom}(A_1, k_0^*)$, to find $m_{x_1} \in \Omega$ so that $(a_{x_1}, m_{x_1}) \sim (a_{x_1}, b_{x_1})^{-1}$?

2) If for each x_1 we can find such an $m_{x_1} \in \Omega$, then under what conditions can the choice be made consistent?

In what remains, we will need some properties of the transfer homomorphism t, which is defined as follows [2]: If $f \in \mathrm{Hom}_H(A, B)$, H being a subgroup of G, and A and B being G-modules, and if $G = \sum_i Hs_i$ is a given coset decomposition of G, then

$$(tf)(a) = \sum_i f(a^{s_i^{-1}})^{s_i}.$$

With this definition tf is a G-homomorphism from A to B. If a cohomology class is taken as a homomorphism of a suitable term in a resolution, then one can define the transfer for cohomology classes.

The transfer has the following properties:

1. Invariants of the algebras C_{x_1} and tC_{x_1} coincide (in the local case).

2. $tC_{x_1} = C_{tx_1}$.

The first assertion follows from the theory of algebras over local fields.

Proof of the second assertion. The cohomology class a_1 is some F^1-homomorphism of the second member X_2 of a resolution for F^1 into A_1; $a_1 \in \mathrm{Hom}_{F^1}(X_2, A_1)$; ρa_1 is the same homomorphism, only considered as a $F^1_{x_1}$-homomorphism. We apply x_1 to this last cocycle. Then

$$x_1 \rho a_1 \in \mathrm{Hom}_{F^1_{x_1}}(X_2, k_0^*).$$

Let ξ be some element in X_2. Then

$$(t(x_1 \rho a_1))(\xi) = \prod_i [(x_1 \rho a_1)(\xi^{s_i^{-1}})]^{s_i}.$$

But $a_1 \in \mathrm{Hom}_{F^1}(X_2, A_1)$ and thus

$$(t(x_1\rho a_1))(\xi) = \prod_i [x_1(a_1(\xi)^{s_i^{-1}})]^{s_i} = \prod_i x_1^{s_i}(a_1(\xi))$$

$$= (\prod_i x_1^{s_i})(a_1(\xi)) = (tx_1)(a_1(\xi)),$$

i.e.,

$$tC_{x_1} = C_{tx_1}.$$

§3. **Two lemmas.** In this section, the two lemmas which were used in the proofs of Theorems 1 and 2 are proved.

Suppose that the group G is imbedded in the exact sequence

$$1 \to A \to G \to F \to 1$$

whose fundamental class we denote by a, let A_1 be a normal subgroup of G which lies in A. Then the group G can be considered as an extension of A_1. This corresponds to an exact sequence

$$1 \to A_1 \to G \to F^1 \to 1,$$

whose fundamental class we denote by a_1.

Suppose that we are given three exact sequences:

$$1 \to A \to G \to F \to 1 \text{ (fundamental class } a), \tag{1}$$

$$1 \to A \to \overline{G} \to F \to 1 \text{ (fundamental class } b), \tag{$\overline{1}$}$$

$$1 \to A \to \widetilde{G} \to F \to 1 \text{ (fundamental class } a \cdot b).$$

If G_1 and G_2 have normal subgroups A_1 and A_2, and $\epsilon\colon G_1/A_1 \to G_2/A_2 \doteq E$ is an isomorphism, then the elements $(g_1, g_2) \in G_1 \times G_2$, for which

$$g_2 A_2 = \epsilon(g_1 A_1)$$

form a subgroup G' called the direct product of G_1 and G_2 relative to the factor group E. We denote this subgroup by $G_1 \times {}^E G_2$. This group contains a subgroup isomorphic to $A_1 \times A_2$ and

$$G'/A_1 \times A_2 \cong E.$$

The groups A_1 and A_2 are both normal subgroups of G' and

$$G'/A_1 \cong G_1, \quad G'/A_2 \cong G_2.$$

If ν: $A_1 \to A_2 = A$ is an isomorphism, then the elements $(a_1, a_2) \in A_1 \times A_2$, for which $a_2\nu(a_1) = 1$, form a normal subgroup of G' which we denote by U. The factor group G'/U is denoted $G_1 \times {}^E_A G_2$.

Let $\Gamma = F^1 \times {}^F \bar{F}^1$ and i: $A_1 \to A$ be the imbedding of a primitive, normal subgroup in the group G.

Then we have the following two assertions:

Lemma 1. $i^* a_1 = \lambda_F^{F^1} a$.

Lemma 2. $\lambda^\Gamma_{F^1} a_1 \cdot \lambda^\Gamma_{\bar{F}^1} b_1 = \lambda^\Gamma_{\bar{F}^1} (ab)_1$.

Here, λ is a lift homomorphism. We shall prove these relations between two-dimensional cohomology classes, by interpreting them as relations between the corresponding group extensions. This is convenient because we shall pass from a to a_1 defined in group-theoretic terms. But for this, we need to recall the group-theoretic properties of multiplication of cohomology classes and of the lift map λ (which are well known).

1°. **Multiplication.** If a and b are fundamental cocycles of the sequences (1) and ($\bar{1}$), then $a \cdot b$ is the fundamental class of the sequence

$$1 \to A \to G \times {}^F_A \bar{G} \to F \to 1.$$

2°. **Lift.** If a is the fundamental class of the sequence (1), and

$$\Phi \to F \tag{2}$$

is an epimorphism with kernel N, and moreover, N is trivial on A, then $\lambda_F^\Phi a$ is the fundamental class of the sequence

$$1 \to A \to G \times {}^F \Phi \xrightarrow{P} \Phi \to 1.$$

Conversely, if we are given an epimorphism (2), whose kernel N acts trivially on A, and in the sequence

$$1 \to A \to \mathfrak{G} \xrightarrow{P} \Phi \to 1$$

$P^{-1}(N) = A \times N'$, where $N' \cong N$ and N' is normal in $\mathfrak{G}$, then the fundamental class $\mathfrak{A}$ of this sequence is represented in the form

$$\mathfrak{A} = \lambda_F^{\Phi} a,$$

where a is the fundamental class of the sequence

$$1 \longrightarrow A \longrightarrow \mathfrak{G}/N' \longrightarrow F \longrightarrow 1.$$

Proof of Lemma 1. The class a_1 is the fundamental class of the sequence

$$1 \longrightarrow A_1 \longrightarrow G \longrightarrow F^1 \longrightarrow 1.$$

In the inclusion of A_1 in A, the group G passes over into the group $\mathfrak{G}$, which is obtained as a simple extension of the normal subgroup A_1 in A. In the group $\mathfrak{G}$, therefore, there are two normal subgroups, which are isomorphic to A, and moreover, their intersection is A_1. In order to distinguish them, we denote the second of them, which is a formal extension of A_1, by the symbol $\bar{A}$. Since $A \cap \bar{A} = A_1$, $\bar{A}_1 = A_1$.

We introduce in $\mathfrak{G}$ the subgroup consisting of elements of the form $a \cdot \bar{a}^{-1}$. This subgroup is denoted by D and the mapping $a \longrightarrow a \cdot \bar{a}^{-1}$ is written

$$d(a) = a\bar{a}^{-1}.$$

This homomorphism has kernel A_1. Thus, d: $A/A_1 \longrightarrow D$ is an isomorphism. The subgroup D, as is easily seen, is normal in $\mathfrak{G}$, even a direct factor, and is isomorphic to A/A_1. The subgroup A does not intersect D, and therefore, in $\mathfrak{G}$ is contained in a normal subgroup $A \times D$. Under the projection p of $\mathfrak{G}$ on F^1, this normal subgroup goes to

$$p(A \times D) = p(D) = A/A_1,$$

and hence, p is an isomorphism of D onto A/A_1. Thus,

$$p^{-1}(A/A_1) = A \times D$$

and

$$D \cong A/A_1.$$

According to 2°, the cocycle $i^* a_1$ is constant on the cosets of $\mathfrak{G}/D$, and is obtained by lifting to F^1 the fundamental cocycle of the sequence

$$1 \longrightarrow A \longrightarrow \mathfrak{G}/D \longrightarrow F \longrightarrow 1.$$

But $\mathfrak{G}/D$ is isomorphic to G, and thus, we have finally

$$i^* a_1 = \lambda_F^{F^1} a.$$

Lemma 1 is thus proved.

Proof of Lemma 2. According to 2°, $\lambda_{F^1}^{\Gamma} a_1$ is the fundamental class of the sequence

$$1 \to A_1 \to \mathfrak{G} \to \Gamma \to 1$$

where $\mathfrak{G} = G \times^{F^1} \Gamma$. The normal subgroup A/A_1 of the group F^1 is denoted by B, and similarly, that of $\overline{A/A_1}$ is denoted $\underline{B}$. Then $\Gamma \supset B \times \underline{B}$ and

$$\Gamma/\underline{B} \cong F^1, \quad \Gamma/B \cong \bar{F}^1,$$
$$\Gamma/B \times \underline{B} \cong F.$$

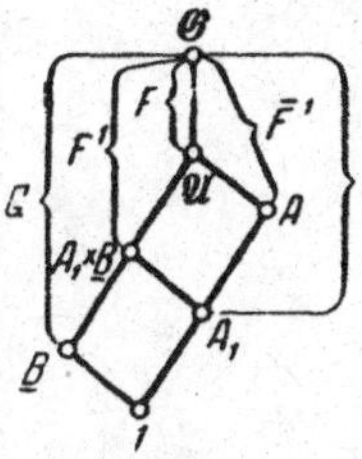

Figure 1

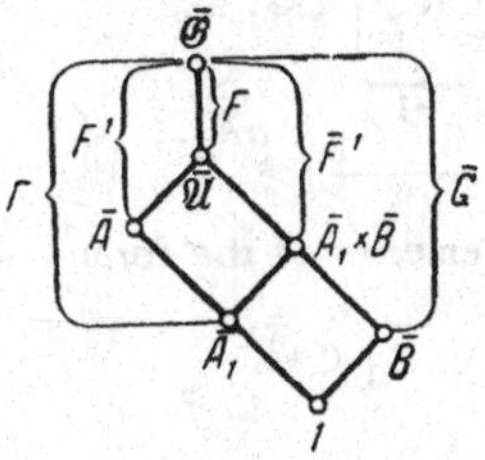

Figure 2

By the definition of the direct product relative to the factor group, $\mathfrak{G}$ contains a subgroup isomorphic to $A_1 \times \underline{B}$.

In the group $\mathfrak{G}$, there is a subgroup generated by the elements (a, aA_1). This subgroup is isomorphic to A, and we denote it also by A. Denote $A \times \underline{B} \subset \mathfrak{G}$ by $\mathfrak{A}$. Then

$$\mathfrak{G}/A \times \underline{B} = F.$$

In other words, we see that

$$\mathfrak{G} = G \times^{F} \bar{F}^1.$$

With the analogous notation for b, we obtain

$$\bar{\mathfrak{G}} = \bar{G} \times^{F} F^1.$$

In order to get an expression for $\lambda^{\Gamma}_{F^1} a_1 \cdot \lambda^{\Gamma}_{\bar{F}^1} b_1$, we must, according to 1°, consider the direct product $\mathfrak{G} \times \bar{\mathfrak{G}}$.

Put $\mathfrak{G} \times^{\Gamma} \bar{\mathfrak{G}} = \bar{\bar{\mathfrak{G}}}$. To learn more about this group, we introduce the set of elements $(g, \bar{h}) \in \mathfrak{G} \times \bar{\mathfrak{G}}$ for which $\epsilon(gA_1) = \overline{hA_1}$, where ϵ is the isomorphism

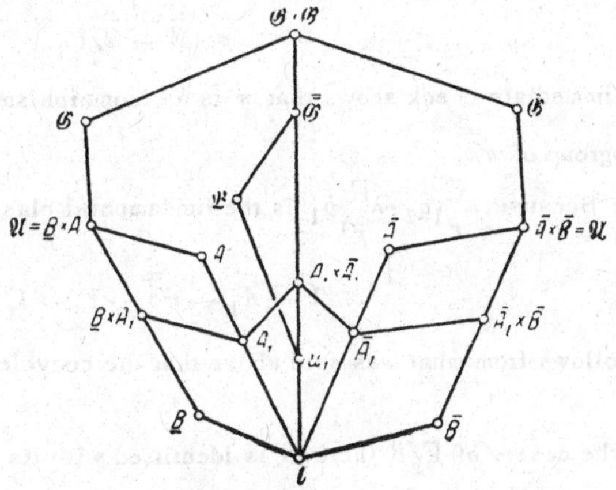

Figure 3

$$\epsilon: \mathfrak{G}/A_1 \to \bar{\mathfrak{G}}/\bar{A}_1 = \Gamma.$$

Notice that

$$\epsilon B = \epsilon A/A_1 = \bar{B}, \quad \epsilon \underline{B} = \underline{B}.$$

In the group $\underline{B} \times A \times \bar{A} \times \bar{B} \subset \mathfrak{G} \times \bar{\mathfrak{G}}$, we consider the subgroup consisting of elements of the form

$$\underline{a^{-1}A_1} \times a \times \bar{a}^{-1} \times aA_1.$$

This subgroup is denoted $\mathfrak{B}$. It is clear that $\mathfrak{B} \subset \bar{\bar{\mathfrak{G}}}$, as

$$\epsilon(\underline{a^{-1}A_1} \times aA_1) = \overline{\underline{a^{-1}A_1}} \times \overline{aA_1}.$$

Finally, denote by U_1 the subgroup of elements of the form

$$(a_1, \bar{a}_1^{-1}) \in A_1 \times \bar{A}_1 \subset \bar{\bar{\mathfrak{G}}}.$$

By definition,

$$\bar{\bar{\mathfrak{G}}}/U_1 = \mathfrak{G} \times^{\Gamma}_{A_1} \bar{\mathfrak{G}}.$$

It is clear that $U_1 \subset \mathfrak{B}$ and $\mathfrak{B}/U_1 \cong A/A_1$. The subgroup $\mathfrak{B}/U_1$ of the group $\bar{\bar{\mathfrak{G}}}/U_1$ is denoted by $\widetilde{B}$.

Consider the homomorphism

$$\pi\colon\ \bar{\bar{\mathfrak{G}}} \longrightarrow \bar{\bar{\mathfrak{G}}}/U_1.$$

An immediate check shows that π is an isomorphism on A and that $\widetilde{B}$ is a normal subgroup of $\pi\bar{\bar{\mathfrak{G}}}$.

Because $\lambda^{\Gamma}_{F^1}a_1 \cdot \lambda^{\Gamma}_{\bar{F}^1} b_1$ is the fundamental class of the sequence

$$1 \longrightarrow A_1 \longrightarrow \pi\bar{\bar{\mathfrak{G}}} \longrightarrow \Gamma \longrightarrow 1,$$

it follows from what was said above that the cocycle $\lambda^{\Gamma}_{F^1}a_1 \cdot \lambda^{\Gamma}_{\bar{F}^1} b_1$ is constant on the cosets of $\Gamma/\widetilde{B}$ (here $\widetilde{B}$ is identified with its image in $\pi\bar{\bar{\mathfrak{G}}}/\pi A_1 \cong \Gamma$). In the group Γ, the subgroup $\widetilde{B}$ consists of elements $b_1 \times \bar{b}_2 \in B \times \bar{B}$ for which $b_1 b_2 = 1$. Thus, $\Gamma/\widetilde{B} \cong \bar{F}^1$ and we see that

$$\lambda^{\Gamma}_{F^1}a_1 \cdot \lambda^{\Gamma}_{\bar{F}^1} b_1 = \lambda^{\Gamma}_{\widetilde{F}^1} c,$$

where c is some class in $H^2(\widetilde{F}^1, A_1)$. In order to find that class, we must, according to 2°, consider $\pi\bar{\bar{\mathfrak{G}}}/\widetilde{B} = \hat{G}$. The class c is the fundamental class of the sequence

$$1 \longrightarrow A_1 \longrightarrow \hat{G} \longrightarrow \hat{F}^1 \longrightarrow 1.$$

Thus, $\widetilde{B} = \pi\mathfrak{B}$, and $\hat{G} = \pi\bar{\bar{\mathfrak{G}}}/\pi\mathfrak{B} \cong \bar{\bar{\mathfrak{G}}}/\mathfrak{B}$, and it is sufficient to show that

$$\hat{G} \cong \bar{\bar{\mathfrak{G}}}/\mathfrak{B} \cong \widetilde{G}.$$

For this, notice that

$$\mathfrak{G}/\underline{B} \cong G, \quad \bar{\mathfrak{G}}/\bar{B} \cong \bar{G}$$

and that there is an isomorphism

$$\eta\colon\ \mathfrak{G} \times \bar{\mathfrak{G}}/\underline{B} \times \bar{B} \longrightarrow G \times \bar{G}.$$

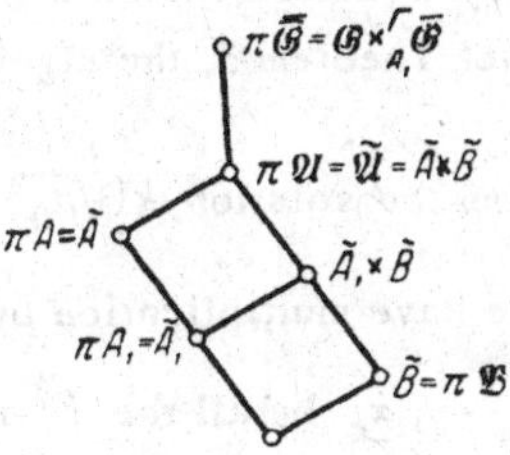

Figure 4

An immediate verification shows that

$$\eta\bar{\bar{\mathfrak{G}}} \cong G \times {}^{F}\bar{G}, \quad \eta\mathfrak{B} \cong U$$

where $U \subset A \times \bar{A} \subset G \times \bar{G}$ is the subgroup consisting of elements of the form $(a, \bar{a}^{-1})$.

Thus

$$\hat{G} \cong \bar{\bar{\mathfrak{G}}}/\mathfrak{B} \cong \eta\bar{\bar{\mathfrak{G}}}/\eta\mathfrak{B} \cong G \times {}^{F}\bar{G}/U = G \times^{F}_{A} G = \widetilde{G},$$

as was required.

§4. The imbedding problem for local fields. Suppose now that Ω is a local field. In this condition we have

Theorem 4. *If the condition of consistency is satisfied, the imbedding problem $(k/\Omega, G, \phi)$ may be solved.*

Proof. As we have already noted, it is sufficient to prove this theorem in the case when G is a p-group. Then, there is contained in A a normal subgroup A_1 of G, and $(A\colon A_1) = p$. The imbedding problem, which corresponds to the exact sequence

$$1 \longrightarrow A/A_1 \longrightarrow F^1 \longrightarrow F \longrightarrow 1,$$

is concomitant with an initial problem, and hence must be solvable.

Let one of its solutions be $k^1 = k(\sqrt[p]{\mu_0})$, $\mu_0 \in k$. The element μ_0 is defined precisely in terms of multiplication by m in the field Ω. The problem is the selection of the multiple m so that the algebra $k(\sqrt[p]{\mu_0 m})$, which is a solution to the imbedding problem $(k/\Omega, F^1, \psi)$, is consistent with the group G. Let us show that such a choice of the multiple is possible. For this, notice that because tC_{x_1} has the same invariants as does C_{x_1}, it is sufficient to consider the algebra tC_{x_1}.

But $tC_{x_1} = C_{tx_1}$, and the character tx_1 is invariant with respect to the entire group F^1. Therefore, it is sufficient to consider the obstruction only for the character $x_1 \in \mathrm{Hom}_{F^1}(A_1, k_0^*)$.

Denote by x the extension of x_1 on the entire group A.

In light of Theorem 3, the algebra $C_{x_1}(\mu_0)$ has the form (a_{x_1}, b_{x_1}) and for the transition from the solution $k(\sqrt[p]{\mu_0})$ of the problem $(k/\Omega, F^1, \psi)$ to the solution $k(\sqrt[p]{\mu_0 m})$, we have multiplication by (a_{x_1}, m).

Let $x_1, \cdots, x_n$ be all the F^1-invariant characters of A_1. We need to select a multiple m so that the algebra (a_{x_i}, m) takes in advance the given value $C_{x_i}^{-1}(\mu_0)$. The expression for (a_{x_i}, m) is the character for determining m, and the system of equations

$$(a_{x_i}, m) = C_{x_i}^{-1}(\mu_0)$$

may be solved, if for any relation

$$\prod_i (a_{x_i}, m)^{\beta_i} = 1$$

(for all m), it follows that

$$\prod_i (C_{x_i}^{-1}(\mu_0))^{\beta_i} = 1.$$

We show that this is in fact the case. The equality

$$\prod_i (a_{x_i}, m)^{\beta_i} = 1$$

means that $\prod_i a_i^{\beta_i}$ is a pth power in Ω, i.e.,

$$\prod_i \alpha_{x_i}^{\beta_i} \in \Omega.$$

This in turn means that

$$\prod_i f_i^{\beta_i} = 1,$$

where f_i is a cocycle constructed according to the formula

$$f_i(\sigma) = \frac{x_i'(\bar{z}^\sigma)}{x_i'(\bar{z})^\sigma}$$

(x_i' is the extension of x_i to A). It follows immediately that

$$(C_{\prod_i x_i^{\beta_i}}(\mu_0))^{-1} \sim 1.$$

It remains to notice that for F^1-invariant characters x_1, x_2,

$$C_{x_1 x_2}(\mu_0) \sim C_{x_1}(\mu_0) \otimes C_{x_2}(\mu_0).$$

From $C_{\prod_i x_i^{\beta_i}}(\mu_0)^{-1} \sim 1$, it follows that

$$\prod_i (C_{x_i}^{-1}(\mu_0))^{\beta_i} \sim 1.$$

The proof of the theorem is thus complete.

Remark 1. The condition of consistence for the imbedding problem $(k/\Omega, G, \phi)$ follows from the condition that all algebras $C_x = x\rho a$, where ρ is the restriction from F to F_x, split. But, as we have seen, tC_x and C_x have the same invariants, in the local case, $tC_x = C_{tx}$, and tx is already invariant under the whole group F. Therefore, in the local case, it is sufficient to verify the condition of consistence for just the character $x \in \mathrm{Hom}_F(A, k_0^*)$.

We are stopped by the question of when the solution to the imbedding problem $(k/\Omega, G, \phi)$ (k is a local field) is a field. In the case where G is a p-group, this problem is solved.

Proposition. *If the number of generators of the group G is equal to the number of generators of the group F, then all solutions to the imbedding problem $(k/\Omega, G, \phi)$ will be fields.*

For the proof of the proposition, we maintain that it is necessary to solve the imbedding problem successively for steps which are central and of power p (this is possible because G is a p-group). Then, because of the condition of the equality of the number of generators of G and F, for each step, the number of generators will not increase. It is thus sufficient to check the condition in the case when $(G: F) = p$. The solutions of the problem $K = k(\sqrt[p]{\mu})$ are fields, if and only if $\mu \neq \mu'^p$, $\mu' \in k$.

If $\mu = \mu'^p$ in k, then G is a direct product of F and A (as is easily verified). From this we obtain a proof of the proposition.

BIBLIOGRAPHY

[1] R. Brauer, *Über die Konstruktion der Schiefkörper, die von endlichem Rang in Bezug auf ein gegebenes Zentrum sind,* J. Reine Angew. Math. **168** (1932), 44-64.

[2] H. Cartan and S. Eilenberg, *Homological algebra,* Princeton Univ. Press, Princeton, N. J., 1956.

[3] B. N. Delone and D. K. Faddeev, *Investigations in the geometry of the Galois theory,* Mat. Sb. (N.S.) **15**(57) (1944), 243-284. (Russian)

[4] H. Hasse, *Existenz und Mannigfaltigkeit abelscher Algebren mit vorgegebener Galoisgruppe über einem Teilkörper des Grundkörpers.* I, Math. Nachr. **1** (1948), 40-61.

[5] R. Kochendörffer, *Zwei Reduktionssätze zum Einbettungsproblem für abelsche Algebren,* Math. Nachr. **10** (1953), 75-84.

Translated by:
D. W. Kahn

Impressions from the International Mathematical Congress in Edinburgh

Usp. Mat. Nauk **14**, No. 2, 243–246 (1958)
[Translated for this volume into English]

For me the most interesting aspect of the congress was that it confirmed and intensified a feeling which had already arisen earlier under the influence of the mathematical literature – namely, that during the last 10 years there had been fundamental changes in mathematics which did not simply push it forward in the same direction as it had developed in earlier, but which changed the very direction of this development. In various fields of mathematics progress was achieved based on ideas which are very similar to one another, and one has begun to work out a language common to all these fields.

The fields I have in mind are algebraic number theory, algebraic geometry, the theory of analytic manifolds (in particular, the theory of analytic functions of several variables), the theory of differentiable manifolds, and topology. These fields have many points in common with analytic number theory, Lie groups, and functional analysis.

It seems to me particularly important that, in spite of the diversity of the concrete material encompassed here, one can yet speak of a single direction in mathematics based on a small number of new ideas and concepts. The basic idea which unites this new direction is homological algebra. Concepts like cohomology groups of modules and groups, spectral sequences and cohomology groups of sheaves, which did not exist 10–15 years ago, have now become standard apparatus.

Obviously these new ideas have had an indirect influence also on areas of research in which they are not directly applied (for example, in the papers of Bott and Milnor on the orders of homotopy groups of spheres, the relationship to Bernoulli numbers is established with the help of techniques worked out by Hirzebruch in connection with his researches on the homology theory of sheaves and the theorem of Riemann-Roch).

In the sequel I will report on those results presented to the meeting which appeared to me particularly interesting. Of course they are by no means all directly connected with homological algebra, but it seems to me that the phenomenon particularly noticeable during the congress, namely that both the number and the significance of the papers in these fields has been growing during the last years, is linked to the emergence of the idea of a homological algebra.

I suppose that cohomological methods are at present applied most fruitfully in algebraic geometry. At the conference a survey lecture by Grothendieck was devoted to the cohomological theory of algebraic varieties. In his lecture Grothendieck reported that he had succeeded in defining in a "reasonable manner" the concept of Betti groups for algebraic varieties defined over an arbitrary field. The "reasonable manner" is to be understood in the sense that the usual theorems on the behaviour of Betti groups under continuous (in the given case rational) mappings, especially the Lefschetz fixed-point theorem, are preserved. Obviously this theory should imply results for arbitrary algebraic varieties analogous to those obtained by Weil for algebraic curves. This would lead to very interesting consequences in the theory of congruences in many variables.

Another interesting result of Grothendieck was set forth in the survey lecture by Hirzebruch, namely a generalization of the theorem of Riemann-Roch. The generalization consists in the fact that the Todd genus satisfies a transformation law under rational mappings. The original Riemann-Roch theorem in Hirzebruch's formulation is obtained from this result by applying it to the mapping of a variety to a point. The proof of this generalization turns out to be considerably simpler than the proof of the Riemann-Roch theorem found by Hirzebruch. In particular, it is algebraic, so that it applies to varieties over arbitrary fields, and it does not make use of the cobordism theory of Thom, on which Hirzebruch's proof was based.

Of course, at the conference one could learn about many results which were not lectured about. Among these was a new foundation of intersection theory of algebraic varieties carried out by Serre by means of homology theory of local rings.

The methods of homological algebra in topology were the theme of a survey lecture by Eilenberg. He reported on a very beautiful method of inductively computing cohomology groups of spaces $K(\Pi, n)$. It turns out that the cohomology groups of the space $K(\Pi, n)$ can be expressed in terms of homology theory of rings (the so-called relative homology theory), where for the ring one takes the cohomology ring of the space $K(\Pi, n-1)$.

A very great impression resulted from the lecture of the young topologist Adams. Building on the theory of secondary cohomological operations, he showed that mappings with Hopf invariant 1 can exist only for spheres of dimensions 2, 4 and 8. From this a proof of a famous conjecture results according to which over the field of real numbers there exist division algebras only in dimensions 1, 2, 4 and 8. It is interesting that the latter result was independently obtained and also reported on at the congress by Kervaire, who obtained it by starting off from a weaker topological result – he established the non-parallelizability of spheres of dimension different from 1, 3 and 7.

A sensation was the report of Morrey that any compact analytic real n-dimensional manifold can be analytically embedded into a $(2n+1)$-dimensional Euclidean space. Up to now it was not even known whether there exists on such a manifold a non-constant analytic function. Hirsch reported on very strong embedding theorems of differential manifolds. Thus he proved that a three-dimensional compact manifold can be embedded into a four-dimensional Euclidean space and a six-dimensional projective space into a seven-dimensional Euclidean space.

From the field of topology I already will, in conclusion, mention two results of Milnor. The first result was reported on in the survey lecture by Hirzebruch. It consists in the description of necessary and sufficient conditions for a given collection of integers to occur as the set of Chern numbers of a compact almost complex manifold*. The significance of the theorem is clear in view of the role Chern numbers play in algebraic geometry. The second result concerns the orders of homotopy groups of spheres. He shows that the order of a certain subgroup of this group (the image under the Whitehead homomorphism) is divisible by an integer which I shall not write out and which is defined by means of Bernoulli numbers. It is even likely that the order of the subgroup mentioned is equal to this number. Therefore there seems to be some hope that at least the orders of the homotopy groups of spheres can be explicitly determined.

The theory of functions of several complex variables was presented in survey lectures by Cartan and Grauert. The lecture of Cartan was concerned with the generalization of the concept of complex analytic manifold corresponding to the passage from an algebraic variety without singular points to a variety with singularities. In this way one arrives at so-called analytic spaces which are locally equivalent, not to a region of n-dimensional complex space, but to a subset of this space defined by a finite number of analytic equations. Analytic spaces often occur as quotient spaces of analytic manifolds with respect to discrete groups, where the transformations of these groups have fixed points. Cartan mentioned a series of results related to the difficulties entailed by the passage from analytical manifolds to analytic spaces, e.g. in connection with the integration of differential forms (in view of the "singularities" the integrals diverge) and with the embedding of one space into another one.

The lecture of Grauert was devoted to the theory of analytic sheaves and analytic fibre bundles with a complex Lie group as structure group. It is interesting that one obtains as a very special case of his theory a result of Birkhoff-Lappo-Danilevsky on the existence of systems of differential equations with given monodromy group.

An intermediate position between the theories of functions in several variables and in one variable was occupied by the lecture of Bers. This lecture dealt with such a basic and at the same time uninvestigated concept as the space of all compact Riemann surfaces of given genus p. After Riemann's results showing that this space has dimension $6p-6$ (for $p>1$), almost no fundamental results had been obtained in this field, but it is now clearly reviving. Thus Bers showed how to introduce in this space the structure of an analytic and even Kählerian manifold.

Going over to algebra it is natural to mention in the first place the sensational result of Nagata – the negative solution of Hilbert's XIV-th problem. Namely Nagata constructed an example of a group G of linear transformations of 32 variables $x_1, \ldots, x_{32}$ such that in the polynomial ring of $x_1, \ldots, x_{32}$, those polynomials which are invariant with respect to G form a subring with an infinite number of generators.

In the lecture of Chevalley a survey was given of a field which is being much studied at present, namely the theory of algebraic groups. By this one means groups which at the same time are algebraic varieties, where the group operations

* The word "almost" is missing in the original text. The problem of characterizing Chern numbers of complex manifolds is unresolved even in dimension 2 ("geography of surfaces"). (Translator's note)

are given by rational functions in coordinates of variable points. According to the theorem of Chevalley-Barsotti-Rosenlicht there exists in every such group a maximal normal subgroup which is an algebraic subgroup of matrices (linear group), the factor group being an abelian variety (which, in the case of the complex ground field, means that it is compact). Abelian varieties were deeply studied during the last 15 years by Weil and his school. In the seminar which Chevalley conducted last year simple linear algebraic groups were studied. It was shown that if the ground field is "sufficiently big" one obtains groups exactly corresponding to the classical simple groups. This result can be applied to the investigation of groups over finite fields which lead to finite simple groups. It was shown by Hertzig that in this way one obtains all simple finite groups known up to now and two new ones.

From the field of number theory I would like to mention the theory of indeterminate equations having a sufficiently large number of variables relative to their degree. Davenport's survey lecture was devoted to this question. A large number of conversations in the hallways gave the impression that many people are now interested in this question. Here there are three conjectures of increasing difficulty: I. The equation $f(x_1, \ldots, x_m) = 0$, where f is an integral form of degree n and $m > n^2$, has a non-zero solution as a congruence to any modulus. II. If the form f is indefinite, then, under the same hypothesis, the equation $f = 0$ has a non-zero integral solution. III. If the coefficients of the form f are real and the other assumptions remain the same as in II, then the inequality $|f| < \varepsilon$ for $\varepsilon > 0$ has a non-zero integral solution. Conjecture I was proved for $n = 2$ and 3, Conjecture II only for $n = 2$, and Conjecture III is not known for any n. Davenport established II for any n, but with the condition $m > n^2$ replaced by the condition "m sufficiently large compared to n", and obtained the analogous result to III for $n = 2$.

In conclusion I wish to emphasize that these necessarily fragmentary impressions can in no way give a complete picture of the work of the congress even in those fields which I touched upon. This is due, quite apart from the inevitable subjectivity of the point of view, simply to the impossibility of following sufficiently attentively the activities of a great number of parallelly working sections. Also I have not referred to the lectures of Soviet mathematicians because their work is well-known to us anyway.

Translated by H. G. Zimmer

Principal homogeneous spaces defined over a function field

Tr. Mat. Inst. Steklova **64**, 316–346 (1961). Zbl. **129**, 128
[Transl., II. Ser., Am. Math. Soc. **37**, 85–114 (1964)]

Introduction

In this paper we investigate principal homogeneous spaces over abelian varieties. Roughly speaking, we shall be dealing with algebraic varieties, defined over some field k, which are biregularly equivalent over the algebraic closure of k to abelian varieties. Thus in the particular case of algebraic curves we shall be dealing with elliptic curves, i.e., curves of genus 1.

We shall be concerned here with the case when k is the field of algebraic functions of one variable ** over an algebraically closed field k_0. (In the last section of the paper we discuss also more general k.)

In some particular cases this question was investigated a long time ago in connection with the theory of algebraic surfaces. If the homogeneous space is one-dimensional, then it can be interpreted as an algebraic surface over k_0 containing a pencil of elliptic curves. Many papers have been devoted to such surfaces, or, more precisely, to the pencils on them. Apparently Painlevé [13] and then Enriques [3] were the first to consider them. There is a survey of later work, due mainly to Enriques and Severi, in the book [4].

The point of view in the present paper is the usual one in the theory of numbers, arguing from the local to the global. At first we look at the local question; that is, the ground field is taken to be the completion of k with respect to a prime divisor. Theorem 1 describes the invariants belonging to the principal homogeneous spaces in this case. On going over to the case when the ground field is a field of algebraic functions we meet the two following fundamental problems:

1) Which systems of local invariants can be realized by a global principal homogeneous space?

2) For what sort of principal homogeneous space do all the local invariants vanish?

---

* Translator's note: The author usually identifies a cocycle with the corresponding element of the cohomology group. He also usually identifies a principal homogeneous space with its isomorphism class. In the translation both of these identifications have been retained, except in one or two places where it seemed particularly important to make a distinction.

** Translator's note: I.e., k/k_0 has transcendence degree 1.

1 These questions are investigated in §2.

In the rest of the introduction we describe our notation and some well-known results.

For any commutative group A we shall denote by A_n the subgroup of elements of A of order n and by $A^{(n)}$ the quotient group A/nA. We denote the number of elements of any finite group G by $[G]$.

We shall denote by A an abelian variety of dimension d defined over the field k. All the general facts about an abelian variety which we need can be found in the book of Lang [9].

We shall say that A acts on the algebraic variety V as a group of regular transformations if there is a regular map $(a, v) \to av$ defined over k of $A \times V$ into V such that $e \cdot v = v$ (e is the unit of A) and $a_1(a_2v) = (a_1a_2)v$.

Following Weil [18] we shall say that a variety V on which A acts as a group of regular transformations is a principal homogeneous space if it satisfies the following conditions:

1) For any $v, v' \in V$ the equation

$$av = v' \tag{1}$$

has precisely one solution $a \in A$.

2) The map of $V \times V$ into A which maps (v, v') onto a is regular.

An isomorphism of two principal homogeneous spaces V, V' is defined to be a biregular map $V \to V'$ which commutes with the action of the elements $a \in A$.

A principal homogeneous space having a point v_0 defined over the ground field is isomorphic to A (the isomorphism being given by $a \to av_0$). Any principal homogeneous space has a point defined over some finite separable extension of the ground field. Using these two facts one can give the following description of the set of all principal homogeneous spaces (cf. [11] and [20]).

The isomorphism classes of principal homogeneous spaces are in one-one correspondence with the elements of a one-dimensional cohomology group $H^1(\mathfrak{G}, \mathfrak{A})$. Here $\mathfrak{G}$ is the galois group of the maximal separable extension $\mathfrak{K}$ of k and $\mathfrak{A}$ is the group of points of A which are defined over $\mathfrak{K}$. The group $H^1(\mathfrak{G}, \mathfrak{A})$ consists of continuous cocycles when $\mathfrak{G}$ is topologized in the usual way and $\mathfrak{A}$ is taken to be discrete.

This result allows us to transfer the group operation from $H^1(\mathfrak{G}, \mathfrak{A})$ to the set of isomorphism classes of principal homogeneous spaces. We thus get a group which will be denoted by $\mathfrak{H}(A, k)$. From the general properties of cohomology groups it follows that $\mathfrak{H}(A, k)$ is a torsion group.

Let us suppose that the natural number n does not divide the characteristic of k (and is arbitrary if k is of characteristic 0). Then we have the exact sequence

$$0 \to \alpha^{(n)} \xrightarrow{\delta} H^1(\mathfrak{G}, \mathfrak{A}_n) \xrightarrow{\iota} H^1(\mathfrak{G}, \mathfrak{A})_n \to 0 \tag{2}$$

in which α denotes the group of points of A defined over k, the homomorphism ι is induced by the injection of $\mathfrak{A}_n$ in $\mathfrak{A}$ and the homomorphism δ is defined as follows: let $a \in \alpha$; then there certainly exists a $b \in \mathfrak{A}$ such that $nb = a$ and δa is given by the cocycle

$$f(\sigma) = b - \sigma b, \tag{3}$$

where σ runs through $\mathfrak{G}$ and it is easily verified that $f(\sigma) \in \mathfrak{A}_n$.

§1. A discretely normed field of constants

1. The structure of the group of rational points. In what follows we shall use some facts about the structure of the groups of rational points on an abelian variety defined over a discretely normed field, the detailed proofs of which will be given elsewhere.

Let k be a discretely normed field with k_0 as its residue class field and let A be an abelian variety of dimension d over k. It is well known (cf. [22] and [23]) that the group α is topologizable and that it contains open subgroups γ_i, isomorphic to the direct sum of d copies of the ring O of integral elements of k. The subgroups γ_i form a complete system of neighborhoods of the identity for the group α and α is naturally isomorphic to the projective limit of the α/γ_i:

$$\alpha = \varprojlim \alpha/\gamma_i. \tag{4}$$

Then the following statements are true:

A. The groups α/γ_i can be given the structure of an algebraic group (or, more precisely, of a quasi-algebraic group) in such a way that the natural homomorphism $\alpha/\gamma_j \to \alpha/\gamma_i$ (for $\gamma_i \supset \gamma_j$) is a homomorphism of algebraic groups. In order words (4) determines on α the structure of a proalgebraic group in the sense of Serre [15].

B. The group $\pi_0(\alpha) = \alpha/\alpha^0$, where α^0 is the connected component of the identity in α, is finite. ($\pi_0(\alpha)$ is the zero-dimensional homotopy group of α in the terminology of Serre.) Hence

$$\alpha \cong \pi_0(\alpha) \times \alpha^0, \tag{5}$$

if k_0 is algebraically closed, where α^0 consists of precisely those elements of α which are divisible by every natural number relatively prime to the characteristic p of the ground field k.

C. The group α contains a maximal proalgebraic subgroup γ such that (i) γ is unipotent (i.e., is the projective limit of algebraic groups for which all the terms in the composition series are isomorphic to the additive group of the field k_0) and (ii) the quotient group $\Gamma = \alpha/\gamma$ is an algebraic group of dimension $\leq d$. The group Γ is an abelian variety of dimension d when and only when A has a good reduction module the unique prime divisor of k.

2. Computation of $H^1(\mathfrak{G}, \mathfrak{A}_n)$. In what follows we shall always suppose that the residue class field k_0 is algebraically closed. Then $\mathfrak{G}$ has a normal divisor $\mathfrak{N}$ which is a topological p-group such that the quotient group $\mathfrak{T} = \mathfrak{G}/\mathfrak{N}$ is the projective limit of cyclic groups whose order is prime to p:

$$\mathfrak{T} = \varprojlim Z_m, \ (m, p) = 1, \tag{6}$$

with the natural system of mappings Z_m into $Z_{m'}$ when m is divisible by m'. In other words $\mathfrak{T}$ is the completion of the infinite cyclic group Z for the topology in which the groups Z_m with $(m, p) = 1$ form a basic system of neighborhoods of 0. We denote by σ a generator of Z. Then σ is a topological generator of $\mathfrak{T}$.

We shall denote by $\mathfrak{A}_n^{\mathfrak{N}}$ the subgroup of $\mathfrak{A}_n$ consisting of elements invariant under $\mathfrak{N}$. It consists of those points of order n on A which are defined over an extension of k_0 without higher ramification. (In contrast to the simplest case discussed by Igusa [7] it is not generally true that all the points of $\mathfrak{A}_n$ have this property, as is shown by simple examples.) This group $\mathfrak{A}_n^{\mathfrak{N}}$ is an $\mathfrak{T}$-module.

Lemma 1. *The group $H^1(\mathfrak{G}, \mathfrak{A}_n)$ is isomorphic to*

$$\frac{\mathfrak{A}_n^{\mathfrak{N}}}{(1-\sigma)\,\mathfrak{A}_n^{\mathfrak{N}}}.$$

The proof of Lemma 1 depends on the following simple fact:

Lemma 2. *Let X be a $\mathfrak{G}$-module for some group $\mathfrak{G}$ and let $\mathfrak{N}$ be a normal subgroup of $\mathfrak{G}$ such that $H^1(\mathfrak{N}, X) = 0$. Then $H^1(\mathfrak{G}, X) \cong H^1(\mathfrak{T}, X^{\mathfrak{N}})$ where $\mathfrak{T} = \mathfrak{G}/\mathfrak{N}$ and $X^{\mathfrak{N}}$ denotes the submodule of X consisting of elements invariant under $\mathfrak{N}$.*

For the proof we consider the homomorphism

$$\lambda\colon H^1(\mathfrak{T}, X^{\mathfrak{N}}) \to H^1(\mathfrak{G}, X),$$

for which each cocycle for $\mathfrak{T}$ is regarded as a cocycle for $\mathfrak{G}$. We verify that λ is an isomorphism under the conditions of Lemma 2.

Suppose, first, that $\phi \in H^1(\mathfrak{T}, X^{\mathfrak{N}})$ is such that $\lambda\phi = 0$. Then there exists an $x \in X$ such that

$$(\lambda\phi)(g) = x - gx \tag{7}$$

for all $g \in \mathfrak{G}$. But now

$$(\lambda\phi)(e_{\mathfrak{T}}) = \phi(h) = \phi(e_{\mathfrak{N}}) = 0,$$

for $h \in \mathfrak{N}$, where $e_{\mathfrak{T}}$ is the identity of $\mathfrak{T}$, and so $x = hx$, i.e., $x \in X^{\mathfrak{N}}$. If $\overline{g}$ denotes the image in $\mathfrak{T}$ of an element $g \in \mathfrak{G}$, then (7) takes the form

$$\phi(\overline{g}) = x - \overline{g}x,$$

that is, $\phi = 0$. Hence the kernel of λ is 0. Now we show that λ is surjective. Let $\psi \in H^1(\mathfrak{G}, X)$. By the hypothesis of the lemma, the restriction of ψ to $\mathfrak{N}$ vanishes, i.e., there exists an $x \in X$ such that

$$\psi(h) = x - hx$$

for all $h \in \mathfrak{N}$. Put

$$\psi_1(g) = \psi(g) - x + gx.$$

Clearly $\psi_1(h) = 0$ for $h \in \mathfrak{N}$ and hence

$$\psi_1(gh) = \psi_1(g) + g\psi_1(h) = \psi_1(g),$$

i.e., ψ_1 is constant on cosets of $\mathfrak{G}$ modulo $\mathfrak{N}$ and so determines a function ϕ (say) on $\mathfrak{T}$. Since

$$\psi_1(hg) = \psi_1(h) + h\psi_1(g) = h\psi_1(g)$$

and

$$\psi_1(hg) = \psi_1(g),$$

the values of ψ_1 (and consequently of ϕ) lie in $X^{\mathfrak{N}}$. Clearly $\lambda\phi = \psi$. Hence λ is an isomorphism. We remark that Lemma 2 also follows from an exact sequence obtainable from the spectral sequence of Hochschild-Serre [6].

It is now easy to prove Lemma 1. Since $\mathfrak{N}$ is a topological p-group but $(n, p) = 1$ by hypothesis the conditions of Lemma 2 are satisfied and

$$H^1(\mathfrak{G}, \mathfrak{A}_n) = H^1(\mathfrak{T}, \mathfrak{A}_n^{\mathfrak{N}}).$$

Let σ_m denote the image of the topological generator σ of $\mathfrak{T}$ in Z_m in the representation (6) of $\mathfrak{T}$. Then

$$H^1(Z_m, \mathfrak{A}_n^{\mathfrak{N}}) = \frac{\operatorname{Ker}(1 + \sigma_m + \cdots + \sigma_m^{m-1})}{\operatorname{Im}(1 - \sigma_m)},$$

where Ker and Im denote the kernel and image respectively of the corresponding homomorphisms of $\mathfrak{A}_n^{\mathfrak{N}}$.

Since $\mathfrak{A}_n$ is finite the group $\mathfrak{T}$ contains a subgroup $\mathfrak{T}_0$ of finite index which acts on $\mathfrak{A}_n$ trivially. Put

$$(\mathfrak{T} : \mathfrak{T}_0) = r.$$

Clearly $(r, p) = 1$. We shall show that $m \equiv 0\ (rn)$ implies

$$\operatorname{Im}(1-\sigma_m) = (1-\sigma)\mathfrak{A}_n, \tag{8}$$

$$\operatorname{Ker}(1+\sigma_m+\cdots+\sigma_m^{m-1}) = \mathfrak{A}_n. \tag{9}$$

The equation (8) is clear because the kernel of the homomorphism of $\mathfrak{T}$ onto Z_m contains $\mathfrak{T}_0$ (as $r \mid m$) and σ coincides with σ_m on $\mathfrak{A}_n$. Further $\sigma_m^r = 1$ on $\mathfrak{A}_n$ (also as $r \mid m$) and hence

$$1+\sigma_m+\cdots+\sigma_m^{m-1} = (1+\sigma_m+\cdots+\sigma_m^{r-1})\frac{m}{r}.$$

This homomorphism annihilates the whole of $\mathfrak{A}_n$ because $n \mid m/r$. Clearly

$$\mathfrak{T} = \varprojlim Z_m,\quad m \equiv 0\ (rn),\ (m, p) = 1$$

and so

$$H^1(\mathfrak{G}, \mathfrak{A}_n) = H^1(\mathfrak{T}, \mathfrak{A}_n^{\mathfrak{N}}) = \varprojlim H^1(Z_m, \mathfrak{A}_n^{\mathfrak{N}}) = \frac{\mathfrak{A}_n}{(1-\sigma)\mathfrak{A}_n}.$$

This concludes the proof of Lemma 1.

Corollary. *The order of* $H^1(\mathfrak{G}, \mathfrak{A}_n)$ *is equal to that of* $\mathfrak{a}_n$.

For from Lemma 1 and the exact sequence

$$0 \longrightarrow \mathfrak{a}_n \xrightarrow{1-\sigma} \mathfrak{A}_n^{\mathfrak{N}} \longrightarrow (1-\sigma)\mathfrak{A}_n^{\mathfrak{N}} \longrightarrow 0$$

we have

$$[H^1(\mathfrak{G}, \mathfrak{A}^{\mathfrak{N}})] = \frac{[\mathfrak{A}_n^{\mathfrak{N}}]}{[(1-\sigma)\mathfrak{A}_n^{\mathfrak{N}}]}$$

and

$$[(1-\sigma)\mathfrak{A}_n^{\mathfrak{N}}] = \frac{[\mathfrak{A}_n^{\mathfrak{N}}]}{[\mathfrak{a}_n]},$$

respectively. On substituting one formula in the other we obtain

$$[H^1(\mathfrak{G}, \mathfrak{A}_n)] = [\mathfrak{a}_n].$$

3. **Multiplication in cohomology groups.** In what follows we shall use the Weil product. This product is a function $e_n(x, \xi)$ which is defined for $x \in \mathfrak{A}_n$, $\xi \in \hat{\mathfrak{A}}_n$ (where $\hat{\mathfrak{A}}$ denotes the group of points on the abelian variety $\hat{A}$ due to A) and takes its values in the group $\mathfrak{z}_n$ of the nth roots of 1.

The function e_n is bilinear; i.e.,

$$e_n(x+x', \xi) = e_n(x, \xi)\, e_n(x', \xi),$$

$$e_n(x, \xi+\xi') = e_n(x, \xi)\, e_n(x, \xi');$$

and also nondegenerate; i.e., $e_n(x_0, \xi) = 1$ for all $\xi \in \hat{\mathfrak{A}}_n$ implies $x_0 = 0$ and $e_n(x, \xi_0) = 1$ for all $x \in \mathfrak{A}_n$ implies $\xi_0 = 0$ (cf. Lang [9], p. 189).

As is known (Cartan-Eilenberg [2], p. 246) the existence of a product on two $\mathfrak{G}$-modules A and B with values in a module C, permits us to define a multiplication of elements of the cohomology groups $\hat{H}^r(\mathfrak{G}, A)$ and $\hat{H}^s(\mathfrak{G}, B)$ with values in $\hat{H}^{r+s}(\mathfrak{G}, C)$ (r, s any integers). This multiplication is bilinear. In particular, the Weil product gives a product:

$$\hat{H}^r(\mathfrak{G}, \mathfrak{A}_n) \otimes \hat{H}^s(\mathfrak{G}, \hat{\mathfrak{A}}_n) \to \hat{H}^{r+s}(\mathfrak{G}, \mathfrak{Z}_n).$$

We shall use this product in the case $r + s = 1$, since then there is a simple interpretation of the group $\hat{H}^1(\mathfrak{G}, \mathfrak{Z}_n)$: it is isomorphic to k^*/k^{*n} where k^* denotes the multiplicative group of the field k. The proof of this last isomorphism follows from the exact cohomology sequence

$$H^0(\mathfrak{G}, \mathfrak{K}^*) \xrightarrow{n} H^0(\mathfrak{G}, \mathfrak{K}^*) \xrightarrow{\delta} H^1(\mathfrak{G}, \mathfrak{Z}_n) \to H^1(\mathfrak{G}, \mathfrak{K}^*)$$

belonging to the exact sequence

$$1 \to \mathfrak{Z}_n \to \mathfrak{K}^* \xrightarrow{n} \mathfrak{K}^* \to 1,$$

where $\mathfrak{K}$ is the separable algebraic closure of k and $\xrightarrow{n}$ denotes raising each element to the nth power. Clearly

$$H^0(\mathfrak{G}, \mathfrak{K}^*) = k^*$$

and

$$H^1(\mathfrak{G}, \mathfrak{K}^*) = 1$$

because of the Hilbert-Speiser theorem. Hence

$$H^1(\mathfrak{G}, \mathfrak{Z}_n) \cong k^*/k^{*n}.$$

Thus we have defined a product

$$\hat{H}^r(\mathfrak{G}, \mathfrak{A}_n) \otimes \hat{H}^{1-r}(\mathfrak{G}, \hat{\mathfrak{A}}_n) \to k^*/k^{*n}.$$

Hitherto we have not used any special properties of k. We shall now make use of the facts that k is discretely normed and that n is relatively prime to the characteristic of the residue class field k_0. Then

$$k^* = \{\pi\} \times U,$$

where π is a prime element of k and U is the group of units. Since k_0 is algebraically closed we have $U = U^n$. Hence

$$k^*/k^{*n} = \{\pi\}/\{\pi^n\} = Z_n$$

and we have an isomorphism

$$s: H^1(\mathfrak{G}, \mathfrak{z}_n) \to Z_n.$$

Thus we have a product

$$\hat{H}^r(\mathfrak{G}, \mathfrak{A}_n) \otimes \hat{H}^{1-r}(\mathfrak{G}, \hat{\mathfrak{A}}_n) \to Z_n.$$

Let us now consider the case $r = 1$. We shall show that

$$\hat{H}^{1-r}(\mathfrak{G}, \hat{\mathfrak{A}}_n) = \hat{H}^0(\mathfrak{G}, \hat{\mathfrak{A}}_n) = \hat{\mathfrak{a}}_n.$$

For

$$\mathfrak{G} = \varprojlim \mathfrak{g}_i$$

is the projective limit of the finite galois groups $\mathfrak{g}_i$ of finite extensions K_i of k such that $\mathfrak{K} = \bigcup K_i$. Hence

$$\hat{H}^0(\mathfrak{G}, \hat{\mathfrak{A}}_n) = \varprojlim \hat{H}^0(\mathfrak{g}_i, \hat{\mathfrak{A}}_n),$$

where

$$\hat{H}^0(\mathfrak{g}_i, \hat{\mathfrak{A}}_n) = \hat{\mathfrak{a}}_n / N_i \hat{\mathfrak{A}}_n$$

and

$$N_i = \sum_{\sigma \in \mathfrak{g}_i} \sigma.$$

We now argue as in the proof of Lemma 1. The groups $\mathfrak{g}_i$ can be chosen so that the Z_m in the proof of Lemma 1 are the corresponding quotient groups. Then (9) implies that $N_i \hat{\mathfrak{A}}_n = 0$ and hence $\hat{H}^0(\mathfrak{g}_i, \hat{\mathfrak{A}}_n) = \hat{\mathfrak{a}}_n$.

Thus finally we have defined a product

$$H^1(\mathfrak{G}, \mathfrak{A}_n) \otimes \hat{\mathfrak{a}}_n \to Z_n,$$

which we shall denote by the same symbol e_n as the Weil product.

Lemma 3. *$e_n(x, \xi) = 1$ for $x \in \mathfrak{a}\delta$, $\xi \in \hat{\mathfrak{a}}_n^0$. Here δ denotes the homomorphism in the exact sequence* (2) *and $\hat{\mathfrak{a}}^0$ is the component of the identity in $\hat{\mathfrak{a}}$.*

If $x = \delta a$, $a \in \mathfrak{a}$, then by (3)

$$x(\sigma) = b - \sigma b,$$

where $nb = a$. We shall use the decomposition (5) of the group $\mathfrak{a}$. If $a \in \mathfrak{a}^0$, then by statement B of section 1 we may choose b in $\mathfrak{a}^0$ and hence $x(\sigma) = 0$. If $a \in \pi_0(\mathfrak{a})$ and its order is a power of the characteristic p of k_0, then $e_n(\delta a, \xi) = 1$ because the order of $e_n(\delta a, \xi)$ is both a power of p and a divisor of n, which is prime to p. There remains thus only the case when $a \in \pi_0(\mathfrak{a})$ and its order m is prime to p. Clearly then $mnb = 0$.

We use now the fact that a point $\xi \in \hat{\mathfrak{a}}_n^0$ can be divided in $\hat{\mathfrak{a}}^0$ by any natural number m prime to the characteristic p of k_0 (statement B of section 1) and put $\xi = m\eta$, $\eta \in \hat{\alpha}^0_{mn}$.

Then

$$e_n(x, \xi) = e_n(b - \sigma b, \xi) = e_n(b - \sigma b, m\eta). \tag{10}$$

By well-known properties of e_n (cf. Lang [9], p. 190) we have

$$e_n(b - \sigma b, m\eta) = e_{mn}(b - \sigma b, \eta) \tag{11}$$

and

$$e_{mn}(b - \sigma b, \eta) = e_{mn}(b, \eta)\, e_{mn}(\sigma b, \eta)^{-1}. \tag{12}$$

Trivially

$$e_{mn}(\sigma b, \eta) = e_{mn}(\sigma b, \sigma\eta), \tag{13}$$

since $\eta \in \hat{\alpha}$.

By the definition of the function e_n we have

$$e_{mn}(\sigma b, \sigma\eta) = e_{mn}(b, \eta)^{\sigma^{-1}}. \tag{14}$$

But now the mnth roots of 1 are in k_0, because k_0 is algebraically closed, and thus in k by Hensel's lemma. Hence $e_{mn}(b, \eta)^{\sigma^{-1}} = e_{mn}(b, \eta)$. Together with (10) – (14) this gives

$$e_n(\delta a, \xi) = e_n(x, \xi) = 1,$$

which completes the proof of the lemma.

The exact sequence (2) together with Lemma 3 allows us to define a product

$$H^1(\mathfrak{G}, \mathfrak{A})_n \otimes \hat{\alpha}^0_n \to Z_n,$$

which we shall denote by $\epsilon_n(u, v)$, with $u \in H^1(\mathfrak{G}, \mathfrak{A})_n$, $v \in \hat{\alpha}^0_n$. Thus

$$\epsilon_n(u, v) = s e_n(\iota^{-1} u, v)$$

by definition, where Lemma 3 guarantees that this expression does not depend on the choice of the representative $\iota^{-1} u$ of $u \in H^1(\mathfrak{G}, \mathfrak{A})_n$ in $H^1(\mathfrak{G}, \mathfrak{A}_n)$. Clearly the product ϵ_n is bilinear.

4. The theorem of duality.

Theorem 1. *The product $\epsilon_n(u, v)$ is nondegenerate and hence represents each of the groups $H^1(\mathfrak{G}, \mathfrak{A})_n$, $\hat{\alpha}^0_n$ as the character group of the other.*

Proof. We must prove the two following assertions:

$$v_0 = 0 \text{ if } \epsilon_n(u, v_0) = 1 \text{ for all } u \in H^1(\mathfrak{G}, \mathfrak{A})_n, \tag{15}$$

$$u_0 = 0 \text{ if } \epsilon_n(u_0, v) = 1 \text{ for all } v \in \hat{\alpha}^0_n. \tag{16}$$

We shall prove (15). Let $f = \iota^{-1} u$. Then $e_n(f, v_0) = 1$ for all $f \in H^1(\mathfrak{G}, \mathfrak{A}_n)$.

The equation $e_n(f, v_0) = 1$ means that the cocycle

$$\psi(\sigma) = e_n(f(\sigma), v_0)$$

is cohomologous to 0 in $\mathfrak{z}_n$. As in the proof of Lemma 3, we have $\mathfrak{z}_n \subset k$, and so $\mathfrak{G}$ acts trivially on $\mathfrak{z}_n$. Thus the fact that the cocycle $\psi(\sigma)$ cobounds means simply that

$$e_n(f(\sigma), v_0) = 1.$$

We shall show now that all elements of $\mathfrak{A}_n$ can be represented as the sum of elements $f(\sigma)$, where f runs through $H^1(\mathfrak{G}, \mathfrak{A}_n)$ and σ runs through $\mathfrak{G}$. Then because of the nondegeneracy of e_n it will follow that $v_0 = 0$.

Clearly every element of the type $x - \sigma x$, $x \in \mathfrak{A}_n$, $\sigma \in \mathfrak{G}$, is of the type $f(\sigma)$ (for we can simply put $f(\sigma) = x - \sigma x$). On the other hand, every $y \in \mathfrak{A}_n^{\mathfrak{N}}$ is also of the type $f(\sigma)$, for we can put $g(\sigma) = y$ in the notation of Lemma 1 to obtain a g in $H^1(\mathfrak{T}, \mathfrak{A}_n^{\mathfrak{N}})$ and then put $f = \lambda g$.

We shall show that in fact every element of $\mathfrak{A}_n$ is the sum of elements of the type $x - \sigma x$ and elements y from $\mathfrak{A}_n^{\mathfrak{N}}$.

Since $\mathfrak{A}_n$ is finite, there is a normal subgroup of $\mathfrak{N}$ of finite index which acts trivially on $\mathfrak{A}_n$. The order of the quotient group Π (say), being a power of p, is relatively prime to n. Hence our assertion will follow from the following simple group-theoretic lemma.

Lemma 4. *Let X be a Π-module, where Π is a finite group, and suppose that the elements of X can be divided by the order P of Π. Then every element of X is the sum of elements of the type $a - ha$, $h \in \Pi$ and a Π-invariant element.*

For since

$$1 = \frac{1}{P} \sum_{h \in \Pi, h \neq 1} (1 - h) + \frac{1}{P} \sum_{h \in \Pi} h,$$

we have

$$x = \sum_{h \in \Pi, h \neq 1} \left(\frac{x}{P} - h \frac{x}{P} \right) + \sum_{h \in \Pi} h \frac{x}{P},$$

for any $x \in X$, and the second sum is clearly Π-invariant.

This concludes the proof of (15). In other words we have shown that the group $\hat{\alpha}_n^0$ is isomorphically injected into the character group of $H^1(\mathfrak{G}, \mathfrak{A})_n$. We shall show that these groups are isomorphic, which will imply (16). It is enough to show that the groups have the same order.

The exact sequence (2) implies that

$$\frac{[H^1(\mathfrak{G}, \mathfrak{A}_n)]}{[\alpha^{(n)}]} = [H^1(\mathfrak{G}, \mathfrak{A})_n]$$

and so

$$[H^1(\mathfrak{G}, \mathfrak{A})_n] = \frac{[\alpha_n]}{[\alpha^{(n)}]} \tag{17}$$

by the corollary to Lemma 1.

From (5) and the infinitely divisible property of the elements of α^0 we obtain

$$\alpha^{(n)} = \pi_0(\alpha)^{(n)}. \tag{18}$$

Since

$$\alpha_n = \pi_0(\alpha)_n \times \alpha_n^0$$

and

$$\pi_0(\alpha)_n = \pi_0(\alpha)^{(n)}$$

(because of the finiteness of $\pi_0(\alpha)$), the formulas (17) and (18) give

$$[H^1(\mathfrak{G}, \mathfrak{A})_n] = [\alpha_n^0].$$

It thus remains only to show that the orders of α_n^0 and $\hat{\alpha}_n^0$ are equal. We note that

$$\alpha_n^0 = \Gamma_n^0,$$

where Γ^0 is the component of identity of the algebraic group Γ defined in statement C of section 1, and thus it is enough to show that

$$[\Gamma_n^0] = [\hat{\Gamma}_n^0], \tag{19}$$

where $\hat{\Gamma}$ is the group corresponding to $\hat{A}$. As is well known (cf. Lang [9], p. 117), the varieties A and $\hat{A}$ are isogenous, and hence so also are the groups Γ and $\hat{\Gamma}$. From general theorems about the structure of commutative algebraic groups (cf. Serre [14], p. 50) it follows that the groups of elements of order n of two isogeneous connected algebraic groups are isomorphic. This gives (19) and therefore (16).

This concludes the proof of Theorem 1.

From general theorems about the structure of commutative algebraic groups it follows that Γ^0 has a subgroup isomorphic to the direct product of d' copies of the multiplicative group of k^0, the quotient group being isomorphic to an abelian variety of dimension d'', where $d' + d'' \leq d$.

Corollary. *$H^1(\mathfrak{G}, \mathfrak{A})_n$ is isomorphic to $d' + 2d''$ copies of the group Z_n.*

We remark that Theorem 1 is susceptible of another formulation. For if we denote by ${}_pA$ the product of the l-components of the abelian group A for all primes $l \neq p$ (where A can either be a discrete torsion group or a compact zero-dimensional group), then Theorem 1 is equivalent to the following statement:

The group ${}_p\mathfrak{H}(A, k)$ *is dual to* ${}_p\pi_1(\alpha)$. *Here* $\pi_1(\alpha)$ *is the fundamental group of the proalgebraic group* α *in the sense of Serre* [15].

It would be very interesting to know whether there is a similar duality between $\mathfrak{H}(A, k)$ and $\pi_1(\alpha)$, at least when the characteristic of the ground field
2 k is 0.

5. **Exponents of principal homogeneous spaces.** The exponent $m(V)$ of a principal homogeneous space $V \in \mathfrak{H}(A, k)$ is by definition the greatest common divisor of the degrees of the separable extensions K/k such that V has a point defined over K. It is known (see [11], for example) that $m(V)$ is divisible by the order n of V in $\mathfrak{H}(A, k)$ and also that for an arbitrary field k it need not coincide with it. We shall show that $m(V)$ need not coincide with n, even for a discretely normed field k.

By Theorem 1 an element $V \in \mathfrak{H}(A, k)$ can be regarded as an element of the character group $\operatorname{Char} \hat{\alpha}_n^0$ of $\hat{\alpha}_n^0$. Let K/k be a finite separable extension corresponding to the subgroup $\mathfrak{G}_0$ of $\mathfrak{G}$ of finite index. If V is regarded as an element of $\mathfrak{H}(A, K)$ it corresponds similarly to an element of the character group $\operatorname{Char} \hat{\beta}_n^0$, where $\hat{\beta}$ is the group of points of $\hat{A}$ defined over K. This gives us a homomorphism

$$\operatorname{Char} \hat{\alpha}_n^0 \rightarrow \operatorname{Char} \hat{\beta}_n^0. \tag{20}$$

Let us denote by N the norm homomorphism

$$N: \hat{\beta} \rightarrow \hat{\alpha}$$

defined in the natural way on $\hat{\beta}$. We shall denote the restriction of N to $\hat{\beta}^0$ and $\hat{\beta}_n^0$ by the same letter N.

Lemma 5. *The homomorphism* (20) *is conjugate to the homomorphism* N.

Proof. We consider a cocycle $\psi \in H^1(\mathfrak{G}, \mathfrak{A})_n$ belonging to the principal homogeneous space V. From the way the correspondence between cocycles and principal homogeneous spaces was set up it follows that the same space considered over the new ground field K corresponds to the cocycle $\rho\psi \in H^1(\mathfrak{G}_0, \mathfrak{A})$ where ρ is the restriction map from $\mathfrak{G}$ to $\mathfrak{G}_0$. Lemma 5 thus asserts that

$$\epsilon_n(\rho x, \xi) = \epsilon_n(x, N\xi)$$

for all $x \in H^1(\mathfrak{G}, \mathfrak{A})_n$. This identity is readily verified from the definition of ϵ_n

We choose now a prime $l \neq p$ and put $n = l^\nu$. If $m(V) = n$ for some $V \in \mathfrak{H}(A, k)_n$, then there exists an extension K_1/k, the degree of which is divisible by l precisely to the power ν, such that V has a point defined over K_1. From the known structure of the group $\mathfrak{G}$ it is readily deduced that the least normal extension of k which contains K_1 also has these properties; so we may suppose without loss of generality that K_1 is normal. The field K_1 contains a subfield K of degree l^ν over k such that $[K_1 : K] \not\equiv 0\ (l)$. It is readily deduced from Lemma 2 that V has a point already defined over K. On the other hand, K is uniquely defined by its degree: $K = k(\sqrt[l^\nu]{\pi})$. Hence the equation $m(V) = n$ is equivalent to the statement that V has a point defined over the fixed field K.

We shall now prove that there exists a V for which this is not the case. If all $V \in \mathfrak{H}(A, k)_n$ for some A had a point defined over K, then by Lemma 5 we would have

$$N\hat{\beta}_n^0 = 0.$$

We shall produce an example of an A for which this is not so. It is isomorphic over K to Ω^n, the direct product of n copies of an elliptic curve Ω defined over k. By the theorem about the descent of the ground field (cf., for example, Serre [14], Ch. 5, §4, p. 108) an A is given by a homomorphism ϕ of the galois group G of K/k into the group of automorphisms of Ω^n defined over k. We recollect that the points of A are characterized by the fact that they are invariant under the transformation $\sigma \times \phi(\sigma)$, $\sigma \in G$, of Ω^n. Hence the application of the transformation σ to the coordinates of points of A is equivalent to the application of $\phi^{-1}(\sigma)$ to Ω^n.

We take for ϕ the regular representation of G as a transposition group on the factors of Ω^n. From what was said above, the homomorphism N reduces to the addition of the components of the points of Ω^n defined over K. Of course, this homomorphism is not zero on α_n^0 if only we choose the curve Ω so that $\omega_n^0 \neq 0$, where ω denotes the group of points on Ω defined over k.

The example constructed above is for any number n of the type l^ν. From it we can easily construct a similar example for any $n \not\equiv 0\ (p)$.

We note that the variety A just constructed enjoys the following property, as is readily verified: the group α of points of A defined over k is isomorphic to the group of points of Ω defined over K. Ju. I. Manin has shown the author that for any variety X over a field k and for any finite extension K/k the set of points on X defined over K may be naturally identified with the points defined over k on a second variety $\overline{X}$. This was remarked independently by A. Weil [19].

The examples we have constructed for which $m(V)$ and n do not coincide are all abelian varieties of dimension greater than 1. For varieties of dimension 1, i.e., for elliptic curves, we have $m(V) = n$. In other words the exponent of an elliptic curve V is equal to its order in $\mathfrak{H}(A, k)$. This may be verified as follows (supposing for simplicity that $p \neq 2$):

We write V in Weierstrass normal form

$$y^2 = x^3 + ax + b, \quad \Delta = 4a^3 + 27b^2 \neq 0.$$

The elements a, b of k can be taken to be integral and not simultaneously divisible by the fourth and sixth power respectively of the prime element π of k. The group $\mathfrak{a}$ can easily be investigated by the methods of Lutz [12]. We omit details and give only the final results in which it is necessary to distinguish three cases.

I. $\Delta \not\equiv 0$ (π). In this case the group Γ is connected and is an elliptic curve over k_0. All the points of $\mathfrak{A}_n$ are defined over k by Hensel's lemma. Hence $N = n \cdot 1$ and

$$N\beta_n^0 = n \cdot \beta_n^0 = 0$$

for any field K of degree n over k.

II. $\Delta \equiv 0$ (π) but a and b are not simultaneously divisible by π. In this case the group Γ is not connected, as is easy to see. Its connected component Γ^0 is isomorphic to the multiplicative group of the field k_0, and so $\mathfrak{a}_n^0 = Z_n$. This is true also for any finite extension K/k of k. Hence again the points of β_n^0 are all defined over k and

$$N \cdot \beta_n^0 = 0.$$

III. $a \equiv b \equiv 0$ (π). In this case $\Gamma^0 = 0$, as is easy to see, so $\mathfrak{a}_n^0 = 0$ and hence always

$$N \cdot \beta_n^0 = 0.$$

In this way we have shown that N takes β_n^0 into 0 for any extension K/k of degree n (admittedly, the extension is in fact unique). Hence any curve V with order n in $\mathfrak{H}(A, k)$ has a point defined over K.

§2. A function field as a field of constants

In this section we consider the case when the field of constants is an extension of transcendence degree 1 of an algebraically closed field k_0.

1. **Introductory remarks.** In the case when k_0 is the field of complex numbers (or, more generally, has characteristic 0) the galois group $\mathfrak{G}$ of the algebraic closure of k is well known. This is fundamental for our future investigations. For the case when k_0 has a characteristic other than 0 an analogous result was obtained by Grothendieck [5], but only for extensions of k without higher ramification. Consequently we must introduce the following restriction on the abelian varieties A under discussion when the field k has characteristic p:

All points of finite order $n \not\equiv 0$ (p) on A are defined over an extension of k without higher ramification.

It will be assumed that this restriction is satisfied for all abelian varieties considered in this section. Apparently, this does not eliminate too many cases. For example, it is satisfied in the case treated by Igusa [7], not to mention the case when the characteristic of k_0 is 0.

We shall denote by κ the maximal extension of k without higher ramification and by $\mathfrak{g}$ its galois group over k. The group of points on A defined over κ is denoted by $\mathfrak{a}$.

It is easy to see that

$$\mathfrak{H}(A, k)_n \cong H^1(\mathfrak{g}, \mathfrak{a})_n. \tag{21}$$

The proof is the same as that of

$$\mathfrak{H}(A, k)_n \cong H^1(\mathfrak{G}, \mathfrak{A})_n,$$

it being only necessary to verify that every principal homogeneous space $V \in \mathfrak{H}(A, k)_n$ has a point defined over κ. Indeed, as was shown in [21], V has a point defined over a field K which is normal over $k(\mathfrak{A}_n)$ and has its galois group isomorphic to $\mathfrak{A}_n$ (here $k(\mathfrak{A}_n)$ is obtained from k by adjoining the coordinates of the points of $\mathfrak{A}_n$). Then K clearly has no higher ramification, because $k(\mathfrak{A}_n)/k$ has none by our hypothesis and $K/k(\mathfrak{A}_n)$ has none because $(n, p) = 1$.

We note too that the analogue

$$0 \to \mathfrak{a}^{(n)} \to H^1(\mathfrak{g}, \mathfrak{a}_n) \to H^1(\mathfrak{g}, \mathfrak{a})_n \to 0 \tag{22}$$

of the sequence (2) is also exact, the proof being almost the same. The only point that requires attention is the proof that $\mathfrak{a} \xrightarrow{n} \mathfrak{a}$ is an epimorphism. In other words, we must show that if $x \in \mathfrak{a}$ and $ny = x$, then $y \in \mathfrak{a}$. This follows from the fact that $\kappa(y)/\kappa(x)$ is a normal extension whose galois group is a subgroup of $\mathfrak{a}_n$. Hence $\kappa(y)$ does not have higher ramification, i.e., $y \in \mathfrak{a}$.

In what follows we shall need some relations which are analogous to (21) and (22) and which can be regarded in a sense as approximations to them. We shall denote by S a finite set of prime divisors of k which includes all the prime divisors for which A does not have a good reduction (this latter number is finite; cf. Shimura [16]). It is easy to see that S contains all ramification points of $k(\mathfrak{a}_n)/k$. Let us denote by κ^S the maximal subfield of κ in which at most the $\mathfrak{p} \in S$ ramify, by $\mathfrak{g}^S$ the galois group of κ^S/k, by $\mathfrak{a}^S$ the group of points of A defined over κ^S, and by $\mathfrak{H}(A, k)^S$ the group of those $V \in \mathfrak{H}(A, k)$ which have a point defined over each local completion $k^{\mathfrak{q}}$ $(\mathfrak{q} \notin S)$.

Finally we denote by $k^{\mathfrak{p}}$ the completion of the field k with respect to its prime divisor $\mathfrak{p}$, by $\mathfrak{G}^{\mathfrak{p}}$ the galois group of its algebraic closure, and by $\mathfrak{a}_{\mathfrak{p}}$ the group of points on A defined over $k^{\mathfrak{p}}$.

Lemma 6.

$$\mathfrak{H}(A,k)_n^S \cong H^1(\mathfrak{g}^S, \mathfrak{a}^S)_n.$$

The sequence

$$0 \to \mathfrak{a}^{(n)} \to H^1(\mathfrak{g}^S, \mathfrak{a}_n) \to H^1(\mathfrak{g}^S, \mathfrak{a}^S)_n \to 0 \tag{23}$$

is exact.

To prove the first statement we must show that each $V \in \mathfrak{H}(A,k)_n^S$ has a point defined over κ^S. Without loss of generality we may suppose that the points of $\mathfrak{a}_n$ are defined over the ground field k, because by our choice of the set S we have $k(\mathfrak{a}_n) \subset \kappa^S$. In this case V belongs to an element of $\mathrm{Hom}(\mathfrak{g}, \mathfrak{a}_n)$ and it has a point defined over the field K belonging to the subgroup $\mathrm{Ker}\, f$ of $\mathfrak{G}$. The fact that V has a point defined over $k^{\mathfrak{q}}$ means that $f = 0$ on $\mathfrak{g}^{\mathfrak{q}}$, where $\mathfrak{g}^{\mathfrak{q}}$ is the splitting group of the prime divisor $\mathfrak{g}$. Hence $\mathfrak{g}^{\mathfrak{q}} \subset \mathrm{Ker}\, f$ for all $\mathfrak{g} \notin S$, and so $K \subset \kappa^S$, as required.

To show the exactness of (23), it is necessary only to check that division by n is possible in the group $\mathfrak{a}^S$. Suppose that $ny = x$, $y \in \mathfrak{a}^S$. We must show that $y \in \mathfrak{a}^S$. But this is a purely local statement which follows at once from the fact that division by n is possible in $\mathfrak{a}_{\mathfrak{q}}$ if A has a good reduction modulo $\mathfrak{q}$.

2. Relations between the local invariants. We now go over to the main subject of this section. We shall consider a principal homogeneous space $V \in \mathfrak{H}(A,k)$ over the different local completions $k^{\mathfrak{p}}$ of k. We obtain a homomorphism

$$\phi_{\mathfrak{p}}: \mathfrak{H}(A,k) \to \mathfrak{H}(A, k^{\mathfrak{p}})$$

for each $\mathfrak{p}$ and so a homomorphism

$$\phi: \mathfrak{H}(A,k) \to \prod_{\mathfrak{p}} \mathfrak{H}(A, k^{\mathfrak{p}}).$$

We shall investigate the image of this homomorphism.

Lemma 7. *For given $V \in \mathfrak{H}(A,k)$ the image $\phi_{\mathfrak{p}}(V)$ is 0 for all except a finite number of $\mathfrak{p}$.*

This statement is equivalent to saying that V has a point defined over all except a finite number of $k^{\mathfrak{p}}$. By Hensel's lemma V has a point defined over $k^{\mathfrak{p}}$ if A has a good reduction modulo $\mathfrak{p}$; and so the lemma follows from the theorem of Shimura [16] already quoted, which states that A has a good reduction for all except a finite number of $\mathfrak{p}$.

Hence ϕ takes $\mathfrak{H}(A,k)$ into the direct sum of the $\mathfrak{H}(A, k^{\mathfrak{p}})$. We shall denote this homomorphism also by the same letter:

$$\phi: \mathfrak{H}(A,k) \to \sum_{\mathfrak{p}} \mathfrak{H}(A, k^{\mathfrak{p}}).$$

We denote by ϕ_n the restriction of ϕ to elements of order n:

$$\phi_n : \mathfrak{H}(A, k)_n \to \sum_{\mathfrak{p}} \mathfrak{H}(A, k^{\mathfrak{p}})_n.$$

But now by Theorem 1 we have

$$\mathfrak{H}(A, k^{\mathfrak{p}})_n = \mathrm{Char}\,(\hat{\mathfrak{a}}_{\mathfrak{p}}^{\mathfrak{g}})_n, \tag{24}$$

where $\hat{\mathfrak{a}}_{\mathfrak{p}}$ denotes the group of points of the dual variety $\hat{A}$ defined over $k^{\mathfrak{p}}$. Hence we have the homomorphism

$$\phi_n : \mathfrak{H}(A, k)_n \to \sum_{\mathfrak{p}} \mathrm{Char}\,(\hat{\mathfrak{a}}_{\mathfrak{p}}^{\mathfrak{g}})_n = \mathrm{Char} \prod_{\mathfrak{p}} (\hat{\mathfrak{a}}_{\mathfrak{p}}^{\mathfrak{g}})_n. \tag{25}$$

In (25) the character group and the product Π are understood in the sense of the theory of topological groups.

But now the groups $(\hat{\mathfrak{a}}_{\mathfrak{p}}^{\mathfrak{g}})_n$ have a common subgroup which we must consider together with its character group. We shall denote by $\mathfrak{t}$ the subgroup of $\hat{\mathfrak{a}}$ which is mapped into $\hat{\mathfrak{a}}_{\mathfrak{p}}^{\mathfrak{g}}$ for each $\mathfrak{p}$ under the injection $\hat{\mathfrak{a}} \to \hat{\mathfrak{a}}_{\mathfrak{p}}$. Clearly for each $\mathfrak{p}$ we have an injection

$$\psi_n^{\mathfrak{p}} : \mathfrak{t}_n \to (\hat{\mathfrak{a}}_{\mathfrak{p}}^{\mathfrak{g}})_n$$

and so an injection

$$\psi_n : \mathfrak{t}_n \to \prod_{\mathfrak{p}} (\hat{\mathfrak{a}}_{\mathfrak{p}}^{\mathfrak{g}})_n$$

and its conjugate homomorphism

$$\psi_n^* : \mathrm{Char} \prod_{\mathfrak{p}} (\hat{\mathfrak{a}}_{\mathfrak{p}}^{\mathfrak{g}})_n \to \mathrm{Char}\,\mathfrak{t} \tag{26}$$

for the character groups. Writing (24), (25), (26) together we obtain the following sequence of homomorphisms:

$$\mathfrak{H}(A, k)_n \xrightarrow{\phi_n} \sum_{\mathfrak{p}} \mathfrak{H}(A, k^{\mathfrak{p}})_n \xrightarrow{\psi_n^*} \mathrm{Char}\,\mathfrak{t} \to 0. \tag{27}$$

Lemma 8.

$$\psi_n^* \phi_n = 0.$$

Proof. Let $V \in \mathfrak{H}(A, k)_n$ and consider an element $f = \iota^{-1} V$ of $H^1(\mathfrak{g}, \mathfrak{a}_n)$. For each $\mathfrak{p}$ let $\mathfrak{P}$ denote an extension to κ and let $\rho_{\mathfrak{P}}$ be the restriction homomorphism from $\mathfrak{g}$ to the splitting group $\mathfrak{g}_{\mathfrak{P}}$ of $\mathfrak{P}$. Then ϕ_n can be interpreted as the sum of all the homomorphisms $\rho_{\mathfrak{P}}$. The assertion of the lemma will thus follow from the verification of the identity

$$\sum_{\mathfrak{p}} \epsilon_n (\rho_{\mathfrak{P}} f, t) \equiv 0 \ (n) \tag{28}$$

for all $f \in H^1(\mathfrak{g}, \mathfrak{a}_n)$ and $t \in \mathfrak{t}_n$.

For this we require an explicit formula for $\epsilon_n(\rho_{\mathfrak{P}} f, t)$. Let ζ be a fixed primitive nth root of 1 and let $\sigma_{\mathfrak{P}}$ be a generator of $\mathfrak{g}_{\mathfrak{P}}$ for which

$$\sqrt[n]{\pi}^{\,\sigma_{\mathfrak{P}}-1} = \zeta \tag{29}$$

where π is a prime element at $\mathfrak{P}$. The interpretation of the homomorphism δ in the exact sequence (2) shows that

$$\epsilon_n(\rho_{\mathfrak{P}} f, t) \equiv a \pmod n$$

if and only if

$$e_n(\rho_{\mathfrak{P}} f, t)(\sigma_{\mathfrak{P}}) = \zeta^a.$$

Hence the identity (28) is equivalent to

$$\prod_{\mathfrak{p}} e_n(\rho_{\mathfrak{P}} f, t)\, \sigma_{\mathfrak{P}} = 1. \tag{30}$$

To prove this we note that $e_n(f, t)$ can be regarded as an element f_t of the group

$$H^1(\mathfrak{g}, \mathfrak{z}_n) = \operatorname{Hom}(\mathfrak{g}, \mathfrak{z}_n),$$

because $t \in \hat{\mathfrak{a}}$. Hence (30) can be written as

$$\prod_{\mathfrak{p}} f_t(\sigma_{\mathfrak{P}}) = 1$$

or

$$f_t\Big(\prod_{\mathfrak{p}} \sigma_{\mathfrak{P}}\Big) = 1. \tag{31}$$

But now the homomorphism $f_t = \operatorname{Hom}(\mathfrak{g}, \mathfrak{z}_n)$ corresponds to a $\mu \in k$ such that

$$f_t(\sigma) = \sqrt[n]{\mu}^{\,\sigma-1}.$$

The relation (31) now follows from the definition (29) of the $\sigma_{\mathfrak{P}}$ together with the fact that the principal divisor (μ) has degree 0.

Theorem 2. *The sequence* (27) *is exact.*

Clearly all that is needed is to prove exactness in (27) at the point $\Sigma\, \mathfrak{H}(A, k^{\mathfrak{p}})_n$.

We shall have shown that (27) is exact if we can show that the sequence

$$\mathfrak{H}(A, k)_n^S \xrightarrow{\phi_n^S} \sum_{\mathfrak{p} \in S} \mathfrak{H}(A, k^{\mathfrak{p}})_n \xrightarrow{(\psi_n^S)^*} \operatorname{Char} \mathfrak{t}_n \to 0 \tag{32}$$

is exact for all sufficiently large S. In particular, we may choose S so large that the image of $(\psi_n^S)^*$ is the whole of $\operatorname{Char} \mathfrak{t}_n$ and that the results of section 1 apply.

By Lemma 8 we have $(\psi_n^S)^* \phi_n^S = 0$, that is, $\mathrm{Im}\,\phi_n^S \subset \mathrm{Ker}\,(\psi_n^S)^*$. In other words, Char $\mathfrak{t}_n$ is a homomorphic image of $\mathrm{Coker}\,\phi_n^S$. We must show that this homomorphism is an isomorphism, and this will follow from the equality of the orders of the groups,

$$[\mathrm{Coker}\,\phi_n^S] = [\mathrm{Char}\,\mathfrak{t}_n]. \tag{33}$$

We shall now prove (33). By Lemma 6 we have

$$[\mathrm{Coker}\,\phi_n^S] = [\Sigma\, H^1(\mathfrak{G}^{\mathfrak{p}}, \mathfrak{A}_n)/\{\phi_n^S H^1(\mathfrak{G}^S, \mathfrak{A}_n) + \sum_{\mathfrak{p}\in S} \delta a_{\mathfrak{p}}^{(n)}\}].$$

We shall now use the known structure of the groups $\mathfrak{g}^{\mathfrak{p}}$ and $\mathfrak{g}^S$ and of their one-dimensional cohomology groups. The group $H^1(\mathfrak{g}^{\mathfrak{p}}, \mathfrak{a}_n)$ was determined in Lemma 1. The group $\mathfrak{g}^S$ may be determined on the basis of the theory of Grothendieck [5]; namely, we can choose generators σ_i in the splitting groups of the divisors $\mathfrak{p}_i \in S$ $(1 \le i \le m)$ so that $\mathfrak{g}^S$ is generated by $\sigma_1, \cdots, \sigma_m$ and $2g$ generators $\tau_1, \cdots, \tau_{2g}$ (g = the genus of k), which are connected by the sole relation

$$\sigma_1 \cdots \sigma_m (\tau_1, \tau_2)(\tau_3, \tau_4) \cdots (\tau_{2g-1}, \tau_{2g}) = 1.$$

Here (τ_i, τ_j) denotes the commutator $\tau_i \tau_j \tau_i^{-1} \tau_j^{-1}$.

An element $f \in H^1(\mathfrak{g}^S, \mathfrak{a}_n)$ is uniquely determined by giving $a_i = f(\sigma_i)$ $(i = 1, \cdots, m)$ and $c_j = f(\tau_j)$ $(j = 1, \cdots, 2g)$ such that

$$\sum_{k=1}^{m} \sigma_1 \cdots \sigma_{k-1} a_k + \sigma_1 \cdots \sigma_m \sum_{k=1}^{g} (\tau_1, \tau_2) \cdots (\tau_{2k-3}, \tau_{2k-2})$$

$$\times ((1 - \tau_{2k-1}\tau_{2k}\tau_{2k-1}^{-1}) c_{2k-1} + \tau_{2k-1}(1 - \tau_{2k}\tau_{2k-1}^{-1}\tau_{2k}) c_{2k}) = 0. \tag{34}$$

Going over to a representation with standard cochains we may write

$$[\mathrm{Coker}\,\phi_n^S] = [\mathrm{Map}(S, \mathfrak{a}_n)/\overline{\mathrm{Map}(S, \mathfrak{a}_n)}],$$

where $\mathrm{Map}(S, \mathfrak{a}_n)$ denotes the group of mappings of S into $\mathfrak{a}_n$ and $\overline{\mathrm{Map}(S, \mathfrak{a}_n)}$ is its subgroup generated by mappings of the three following types:

$$f_0^{\mathfrak{p}}(\mathfrak{q}) = \begin{cases} (1 - \sigma_{\mathfrak{q}}) x & \text{for } \mathfrak{q} = \mathfrak{p} \text{ and some } x \in \mathfrak{a}_n, \\ 0 & \text{for } \mathfrak{q} \ne \mathfrak{p}, \end{cases}$$

$$f_1(\mathfrak{q}) = (\delta a)(\sigma_{\mathfrak{q}}), \text{ where } a \in \mathfrak{a}^{(n)},$$

$$f_2(\mathfrak{q}_i) = a_i,$$

where the $a_i \in \mathfrak{a}_n$ satisfy (34) for some c_j.

We denote the subgroups generated by the functions of these three types by

$\mathfrak{b}_0$, $\mathfrak{b}_1$, $\mathfrak{b}_2$, respectively, so that

$$\overline{\mathrm{Map}(S, \mathfrak{a}_n)} = \mathfrak{b}_0 + \mathfrak{b}_1 + \mathfrak{b}_2.$$

Let us consider the map γ (say) of $\mathrm{Map}(S, \mathfrak{a}_n)$ into $\mathfrak{a}_n$,

$$\gamma: f \to f(\mathfrak{p}_1) + \sigma_1 f(\mathfrak{p}_2) + \cdots + \sigma_1 \cdots \sigma_{m-1} f(\mathfrak{p}_m).$$

Clearly $\overline{\mathrm{Map}(S, \mathfrak{a}_n)} \supset \mathrm{Ker}\,\gamma$ since $\mathrm{Ker}\,\gamma$ consists of functions of the type f_2 (where indeed the c_j all vanish). Hence

$$[\mathrm{Coker}\,\phi_n^S] = [\mathrm{Map}(S, \mathfrak{a}_n) / \overline{\mathrm{Map}(S, \mathfrak{a}_n)}] = [\mathfrak{a}_n / \overline{\mathfrak{a}}_n],$$

where $\overline{\mathfrak{a}}_n$ denotes the image of $\overline{\mathrm{Map}(S, \mathfrak{a}_n)}$ under γ.

It remains to compute the index $(\mathfrak{a}_n : \overline{\mathfrak{a}}_n)$. We do this by means of the Weil product $e_n(x, \xi)$ for $x \in \mathfrak{a}_n$, $\xi \in \hat{\mathfrak{a}}_n$, where, as before, $\hat{\mathfrak{a}}_n$ is the set of points of order n on the dual variety $\hat{A}$. Clearly the index $(\mathfrak{a}_n : \overline{\mathfrak{a}}_n)$ is the order of the annihilator $\mathfrak{a}_n'$ (say) of $\overline{\mathfrak{a}}_n$ in $\hat{\mathfrak{a}}_n$. The group $\overline{\mathfrak{a}}_n$ is the sum of the groups

$$\overline{\mathfrak{b}}_i = \gamma\,\mathfrak{b}_i, \quad i = 0, 1, 2,$$

and so $\mathfrak{a}_n'$ is the intersection of the respective annihilators $\mathfrak{b}_n' \subset \hat{\mathfrak{a}}_n$ $(i = 0, 1, 2)$.

We shall now determine the annihilators of $\overline{\mathfrak{b}}_0$. Consider first the functions $f_0^{\mathfrak{p}_1}$. For these

$$\gamma f_0^{\mathfrak{p}_1} = (1 - \sigma_1)\, x$$

and so a $\xi \in \hat{\mathfrak{a}}_n$ is orthogonal to all of them when and only when $\xi^{\sigma_1} = \xi$. We shall consider this condition satisfied in what follows.

Consider secondly the functions $f_0^{\mathfrak{p}_2}$. For these we have

$$\gamma f_0^{\mathfrak{p}_2} = \sigma_1 (1 - \sigma_2)\, x.$$

Since $\xi^{\sigma_1} = \xi$, the condition that ξ be orthogonal to all the $\gamma f_0^{\mathfrak{p}_2}$ implies that $\xi^{\sigma_2} = \xi$. By induction it follows that ξ must be invariant under all the σ_i.

We now use the orthogonality of ξ to functions of the type f_2. Clearly the images of functions of type f_2 under γ are precisely the elements of the form

$$(\sigma_1 \cdots \sigma_m) \sum_{k=1}^{g} (\tau_1, \tau_2) \cdots (\tau_{2k-2}, \tau_{2k-1}) \left((1 - \tau_{2k-1} \tau_{2k} \tau_{2k-1}^{-1})\, c_{2k-1} \right.$$
$$\left. + \tau_{2k-1} (1 - \tau_{2k} \tau_{2k-1}^{-1} \tau_{2k})\, c_{2k}\right) \tag{35}$$

for arbitrary $c_1, \cdots, c_{2g} \in \mathfrak{a}_n$. Put $c_2 = c_3 = \cdots = c_{2g} = 0$. Since $\xi^{\sigma_i} = \xi$ $(i = 1, \cdots, m)$ the orthogonality of ξ to such elements is equivalent to its orthogonality to the elements $(1 - \tau_1 \tau_2 \tau_1^{-1})\, c_1$ and so to the equation $\xi^{\rho_2} = \xi$

where $\rho_2 = \tau_1 \tau_2 \tau_1^{-1}$.

Now let us put

$$c_1 = c_3 = c_4 = \cdots = c_{2g} = 0$$

in (35). We obtain the element

$$(\tau_1 - \tau_1 \tau_2 \tau_1^{-1} \tau_2^{-1}) c_2 = (1 - \tau_1 \tau_2 \tau_1^{-1} \tau_2^{-1} \tau_1^{-1}) \tau_1 c_2 = (1 - \rho_1) c_2',$$

where $\rho_1 = \tau_1 \tau_2 \tau_1^{-1} \tau_2^{-1} \tau_1^{-1}$ and c_2' is an arbitrary element of $\mathfrak{a}_n$. The orthogonality of ξ to the $(1 - \rho_1) c_2'$ is thus equivalent to

$$\xi^{\rho_1} = \xi.$$

Hence $\xi^{\rho_1} = \xi^{\rho_2} = \xi$. Since

$$\tau_1 = \rho_2^{-1} \rho_1^{-1} \rho_2, \quad \tau_2 = \rho_2^{-1} \rho_1 \rho_2 \rho_1^{-1} \rho_2,$$

we have $\xi^{\tau_1} = \xi^{\tau_2} = \xi$. By induction we obtain similarly $\xi^{\tau_i} = \xi$ for all τ_i. Hence ξ is invariant under all the generators of $\mathfrak{g}^S$, i.e., ξ is defined over k, i.e., $\xi \in \hat{\mathfrak{a}}_n$.

It remains to use the condition that ξ is orthogonal to the γf_1. We saw in §1 that for each $\mathfrak{p}$ this is equivalent to the condition that ξ be in the connected component $\hat{\mathfrak{a}}^0_{\mathfrak{p}}$ of $\hat{\mathfrak{a}}_{\mathfrak{p}}$. Hence the condition that ξ be orthogonal to all three groups $\overline{\mathfrak{b}}_1, \overline{\mathfrak{b}}_2, \overline{\mathfrak{b}}_3$ is equivalent to $\xi \in \mathfrak{t}_n$. This shows that the annihilator of $\overline{\mathfrak{a}}_n$ is precisely $\mathfrak{t}_n$, which gives (33). This concludes the proof of Theorem 2.

3. Locally trivial homogeneous spaces. The theorem just proved allows us to say something about the kernel of ϕ. This kernel, which we shall denote by $\mathfrak{h}$ is the group of principal homogeneous spaces having a point defined over each $k^{\mathfrak{p}}$. We shall call such homogeneous spaces locally trivial. The structure of $\mathfrak{h}$ depends essentially on the structure of $\mathfrak{t}$. We shall say something below about the interpretation of this latter group, but for the moment let us see what can be said under the hypothesis $\mathfrak{t} = 0$.

Theorem 3. *If* $\mathfrak{t} = 0$, *the group* $\mathfrak{h}$ *is infinitely divisible and satisfies the isomorphism*

$$\mathfrak{h} \cong {}_pZ^e.$$

Here ${}_pZ$ *is the group of all roots of* 1 *of order prime to* p *and* ${}_pZ^e$ *is the direct sum of* e *copies of* ${}_pZ$, *where*

$$e = 2d(m + 2g - 2) - \Sigma\, e_{\mathfrak{p}} - r,$$

$e_{\mathfrak{p}}$ *is the rank of the group* $\pi_1(\mathfrak{a}_{\mathfrak{p}})$ *for those prime divisors* $\mathfrak{p}$ *for which* A *does not have a good reduction,* m *is the number of such prime divisors, and* r *is the*

3 *rank of A over k.*

Proof. We have the exact sequences

$$0 \to \mathfrak{h}_n \to \mathfrak{S}(A, k)_n \to \Sigma\, \mathfrak{S}(A, k^{\mathfrak{p}})_n \to 0$$

and

$$0 \to \mathfrak{h}_n \to \mathfrak{S}(A, k)_n^S \to \sum_{\mathfrak{p} \in S} \mathfrak{S}(A, k^{\mathfrak{p}})_n \to 0, \tag{36}$$

where for S we may take the set of all prime divisors to which A does not have a good reduction.

Clearly the theorem is equivalent to

$$[\mathfrak{h}_n] = n^e \tag{37}$$

for any $n \not\equiv 0\ (p)$. Since by Theorem 1 we have

$$[\mathfrak{S}(A, k^{\mathfrak{p}})_n] = n^{e_{\mathfrak{p}}},$$

and so

$$[\sum_{\mathfrak{p} \in S} \mathfrak{S}(A, k^{\mathfrak{p}})_n] = n^{\Sigma e_{\mathfrak{p}}},$$

it remains only to compute the order of $\mathfrak{S}(A, k)_n^S$. For this we use Lemma 6.

We have

$$[\mathfrak{S}(A, k)_n^S] / [\alpha^{(n)}] = [H^1(\mathfrak{g}^S, \mathfrak{a}_n^S)]. \tag{38}$$

Let us compute the order of $H^1(\mathfrak{g}^S, \mathfrak{a}_n)$, using the arguments and the notation that were used in the proof of Theorem 2. Clearly a cocycle $f \in Z^1(\mathfrak{g}^S, \mathfrak{a}_n)$ is given by the $m + 2g$ elements $a_1, \cdots, a_m, c_1, \cdots, c_{2g}$ connected by (34), and we may choose $a_2, \cdots, a_m, c_1, \cdots, c_{2g}$, arbitrarily, leaving a_1 to be uniquely determined by (34). Hence

$$Z^1(\mathfrak{g}^S, \mathfrak{a}_n) \cong \mathfrak{a}_n^{m+2g-1}.$$

Now consider the coboundaries, i.e., elements of $B^1(\mathfrak{g}^S, \mathfrak{a}_n)$. They are given by the sets $a_1, \cdots, a_m, c_1, \cdots, c_{2g}$, which can be put in the form

$$a_i = (1 - \sigma_i)\, x,$$

$$c_j = (1 - \tau_j)\, x \tag{39}$$

for some $x \in \mathfrak{a}_n$. The formula (39) can be regarded as a map

$$\psi : \mathfrak{a}_n \to \mathfrak{a}_n^{m+2g}$$

whose image is just $B^1(\mathfrak{g}^S, \mathfrak{a}_n)$. The kernel of ψ consists of those elements x of $\mathfrak{a}_n$ for which

$$(1 - \sigma_i)\, x = (1 - \tau_j)\, x = 0, \quad i = 1, \cdots, m, \quad j = 1, \cdots, 2g.$$

This means that $\operatorname{Ker}\psi = \mathfrak{a}_n$ and we have the exact sequence

$$0 \longrightarrow \mathfrak{a}_n \longrightarrow \mathfrak{a}_n \longrightarrow B^1(\mathfrak{g}^S, \mathfrak{a}_n) \longrightarrow 0,$$

from which

$$[B^1(\mathfrak{g}^S, \mathfrak{a}_n)] = [\mathfrak{a}_n]/[\mathfrak{a}_n]$$

follows and hence

$$[H^1(\mathfrak{g}^S, \mathfrak{a}_n)] = \frac{[Z^1(\mathfrak{g}^S, \mathfrak{a}_n)]}{[B^1(\mathfrak{g}^S, \mathfrak{a}_n)]} = [\mathfrak{a}_n]^{m+2g-2}[\mathfrak{a}_n]. \tag{40}$$

By the theory of the rank of an abelian variety (cf. [10]) we have

$$[\mathfrak{a}^{(n)}] = n^r[\mathfrak{a}_n], \tag{41}$$

where r is the rank of A over k. On inserting (40) and (41) in (38) we obtain

$$[\mathfrak{H}(A, k)_n^S] = [\mathfrak{a}_n]^{m+2g-2} n^{-r} = n^{2d(m+2g-2)-r}.$$

The exact sequence (23) now gives us (37), as required.

4. The unramified trace of an abelian variety. We shall now consider the group $\mathfrak{t}$ in more detail and, in particular, we shall try to find when it vanishes. For this we require the following definitions:

An abelian variety is called unramified if it has a good reduction to every prime divisor of k.

An abelian subvariety A_0 of the abelian variety A is called the unramified trace of A if it is a maximal unramified subvariety.

The existence of the unramified trace can be proved by the same argument which gives the existence of the k/k_0-trace (cf. [9], p. 212). Clearly $A_0 \supset A^*$, where A^* is the k/k_0-trace.

Theorem 4. *The group $\mathfrak{t}$ is isomorphic to the group of points on A_0 defined*
4 *over k which have a finite order prime to p.*

Proof. It is clear that the points on A_0 defined over k with finite order prime to p are in $\mathfrak{t}$. We must prove the converse.

Let L be the maximal unramified extension of k. By definition the points of $\mathfrak{t}$ are infinitely divisible over every field $k^{\mathfrak{P}}$ and so over L. Hence we have an abelian variety A defined over k and a group $\mathfrak{t}$ of its points of finite order which are infinitely divisible over L. Denote by θ the group of all points of A of finite order which are infinitely divisible over L. We shall show that θ is the set of points of finite order not divisible by p on some subvariety A_0.

Let us consider A over the ground field L. The points of θ are then defined over the new ground field.

Let ϕ be the l-adic representation of the galois group $\mathfrak{G}$ of the algebraic closure of L acting on the points of finite order of A, where l is any prime other than p (cf. [9], p. 182). The group θ will correspond to a submodule, of dimension ν (say), of the complete representation module, which has dimension $2d$, where $\dim A = d$. Over the field of l-adic numbers the representation ϕ will contain the unit representation precisely ν times. Hence the endomorphism ring of A contains an endomorphism which maps A onto an abelian subvariety A_0 of the required kind. In particular, ν must be even: $\nu = 2d_0$.

We shall show now that A_0 is defined not merely over L but over k. Let g be any automorphism of L/k. The abelian variety A_0^g has dimension $2d_0$ and again contains θ, so A_0^g and A_0 have precisely the same points of finite order not divisible by p. Hence $A_0^g = A_0$, and thus A_0 is defined over k.

All the points of finite order not divisible by p on A_0 are defined over L and so over the completion $k^{\mathfrak{p}}$ of k with respect to any prime divisor $\mathfrak{p}$ of k. Hence for any $\mathfrak{p}$ and any $n \not\equiv 0\ (p)$ the group $(\alpha_{\mathfrak{p}})_n$ consists of n^{2d_0} points. In the notation of section 1 of §1 this means that Γ is an abelian variety of dimension $2d_0$ over k_0 and hence by statement C of section 1 of §1 there is a good reduction to modulus $\mathfrak{p}$. In other words A_0 is unramified.

We note that by the theorem on rank (cf. [10]) the group of points defined over k on the variety A/A^* has a finite number of generators. A fortiori this will be true for A_0/A^*.

Let ${}_p(\alpha_0)$, ${}_p(\alpha^*)$, ${}_p(\bar{\alpha})$ denote the groups of points defined over k of finite order not divisible by p on the varieties A_0, A^*, $\bar{A} = A_0/A^*$, respectively. Then

$${}_p(\alpha_0) \cong {}_p(\alpha^*) \times {}_p(\bar{\alpha}),$$

as is readily verified, since ${}_p(\alpha^*)$ is infinitely divisible.

Corollary. $\mathfrak{t} = 0$ *when and only when the unramified trace of* A *has no points of finite order not divisible by* p *which are defined over* k. *The group* $\mathfrak{t}$ *and hence also* $\mathfrak{H}$ *is infinitely divisible when the torsion group of the points on* A_0/A^* *defined over* k *is a* p*-group.*

5. Remark on unramified varieties.

Theorem 5. *If* k *is a field of rational functions* $k = k_0(t)$, *then every unramified variety over* k *is defined over* k_0.

Proof. Let A^* be the k/k_0-trace of A. We must show that $A = A^*$. By the theorem on rank (cf. [10]) the group of points on A/A^* defined over k has a finite number of generators; in particular, its torsion group will have a finite order t (say). Hence

$$[\alpha_n] = n^{2d^*}$$

for any integer n prime to p and t, where d^* is the dimension of A^*. On the other hand A has a good reduction to every prime divisor $\mathfrak{p}$ by hypothesis. Hence the points of $\mathfrak{A}_n$ are all defined over every $k^{\mathfrak{p}}$ and so their least field of definition over k is unramified over k. But $k = k_0(t)$ has no unramified extensions except itself, so all the points of $\mathfrak{A}_n$ are defined over k. Hence

$$[\alpha_n] = n^{2d}, \tag{43}$$

where d is the dimension of A. On comparing (42) and (43) we obtain $d = d^*$, so $A = A^*$, as required.

Corollary. *For an abelian variety defined over a field $k = k_0(t)$ of rational functions the group ${}_p\mathfrak{H}$ of locally trivial principal homogeneous spaces of finite order prime to p is infinitely divisible.*

When k is not a field of rational functions over k_0 the position is more complicated. We shall now describe the construction of a class of abelian varieties over k which have a good reduction to every prime divisor $\mathfrak{p}$ but which are not defined over k_0.

Let $\bar{k}/k$ be a finite normal unramified extension with galois group G. If we interpret k and $\bar{k}$ as the fields of functions of the nonsingular complete curves X, $\bar{X}$, respectively, then the injection of k in $\bar{k}$ gives a map

$$\pi: \bar{X} \longrightarrow X,$$

which in our case is an unramified covering with galois group G. We may regard $\bar{X}$ as a principal fibre space over X with structure group G. We shall suppose that G is isomorphic to a group of automorphisms of an abelian variety A^* defined over k_0. Then we can consider the associated fibre space A over X with fibre A^*. In other words A is defined as the factor space of $\bar{X} \times A^*$ with respect to the group G, where G acts according to the rule

$$(\bar{x}, a)\, g = (g\bar{x}, ag^{-1}), \ (\bar{x}, a) \in \bar{X} \times A^*, \ g \in G.$$

It is easy to see that A is an abelian variety defined over k and that it has a good reduction to every prime divisor $\mathfrak{p}$ of k. The reduced variety coincides with A^* for every $\mathfrak{p}$. From the point of view of the theory of the variation of structures of Kodaira and Spencer [8], A is a fibre space with basis X while an arbitrary unramified abelian variety over k defines a family of structures. We shall say that an unramified abelian variety A obtained by the above method is of fibre-space type.

Theorem 6. *Every unramified one-dimensional abelian variety is of fibre-space type. For any dimension $d > 1$ there exist unramified abelian varieties which are not of fibre-space type.*

Proof. Let A be an unramified elliptic curve over the field k of algebraic functions. We write A in Weierstrass normal form,

$$y^2 = x^3 + ax + b, \quad a, b \in k.$$

The absolute invariant j given by

$$j = \frac{4a^3}{4a^3 + 27b^2}$$

belongs to k. Since A has a good reduction to every prime divisor $\mathfrak{p}$ of k the value $j(\mathfrak{p})$ is finite for every $\mathfrak{p}$. Hence j has no poles; i.e., it is a constant: $j \in k_0$. Hence over some finite normal extension K of k the elliptic curve A is equivalent to a curve A^* defined over k_0. We may assume that K is normal over k.

We shall show that A is already equivalent to A^* over some subfield $\bar{k}$ of K which is unramified over k. For this we use the theory of the descent of the field of definition (cf. [14]). By this theory A is defined by a representation ϕ of the galois group G of K/k in the group of automorphisms of A^*. We shall show that $\phi(\sigma) = 1$ for any automorphism σ in the inertia group of any prime divisor of K. Hence the kernel H of the representation ϕ will belong to an unramified extension $\bar{k}$ of k, so that A is already equivalent to A^* over $\bar{k}$, which is what we wanted to prove.

Let then σ belong to the inertia group $\mathfrak{T}$ of the prime divisor $\mathfrak{P}$ of K. Let us denote by $\mathfrak{a}_{\mathfrak{p}}$ the group of points of A defined over $k^{\mathfrak{p}}$ and by $\mathfrak{a}_{\mathfrak{P}}$ the group of points of A^* defined over $K_{\mathfrak{P}}$. It is readily verified that $\mathfrak{a}_{\mathfrak{p}}$ coincides with the subgroup of $\mathfrak{a}_{\mathfrak{P}}$ consisting of those elements which are invariant under all $\sigma \cdot \phi(\sigma)$ for $\sigma \in \mathfrak{T}$. Let $\Gamma_{\mathfrak{p}}$ and $\Gamma_{\mathfrak{P}}$ be the algebraic groups corresponding to the proalgebraic groups $\mathfrak{a}_{\mathfrak{p}}$ and $\mathfrak{a}_{\mathfrak{P}}$ (cf. section 1 of §1). By hypothesis A has a good reduction to modulus $\mathfrak{p}$ and the reduced variety coincides with A^*. Hence $\Gamma_{\mathfrak{p}} = \Gamma_{\mathfrak{P}} = A^*$ and $\Gamma_{\mathfrak{p}}$ is the subgroup of $\Gamma_{\mathfrak{P}}$ invariant under all the $\phi(\sigma)$, $\sigma \in \mathfrak{T}$. This is possible only if $\phi(\sigma) = 1$ for all $\sigma \in \mathfrak{T}$; thus the proof is complete.

The assertion of Theorem 6 which we have just proved is for dimension 1. We note that the hypothesis $d = 1$ was used only at the very beginning of the proof when we used the invariant j. The point of that argument is this. The variety M^1 of moduli of elliptic curves over k_0 is just the affine straight line k_0^1 (with coordinate j). If the field k corresponds to the complete curve X defined over k_0 (as above), then any unramified abelian variety A defined over k gives a regular map $X \longrightarrow M^1 = k_0^1$, and this can only be the map into a single point. What is essential here is not the explicit form of M^1 but the fact that it is impossible to map a complete curve nontrivially into it. This is not the case for $d > 1$, which completes the proof of the existence of unramified abelian varieties not of fibre-space type.

We shall consider the case when k_0 is the field of complex numbers and restrict attention to the d-dimensional abelian varieties belonging to a single principal matrix. It is known (see [1]) that for these the variety of moduli M^d is an open subset of a normal projective variety $\overline{M}$ of dimension $d(d+1)/2$, where $\overline{M} - M^d$ is an algebraic subvariety of dimension $d(d-1)/2$. If $d > 1$ then $\overline{M} - M^d$ has codimension greater than 1 in $\overline{M}$ and so there exists a complete nonsingular algebraic curve X in M^d. Let i be the injection map

$$i : X \longrightarrow M^d.$$

The inverse image under i of the family of abelian varieties over M^d defines a family of varieties over X, i.e., an abelian variety over $k_0(X) = k$. Clearly this variety has a good reduction to every prime divisor $\mathfrak{p}$ of k, but it is not defined over k_0 because the reduced varieties to the various $\mathfrak{p}$ are not equivalent. This concludes the proof of the theorem.

We remark that similarly one can ask whether there exist unramified curves of genus g which are not of fibre-space type. The answer to this is apparently unknown. The question is related to the construction of the variety of moduli of curves of genus g and, in particular, to the question whether this variety is affine. For $g = 2$ it is easy to show that every unramified curve is of fibre-space type.

6. **Constant abelian varieties.** The investigations above on the group $\mathfrak{H}(A, k)$ of principal homogeneous spaces can be made much more simple and precise when the abelian variety A is defined over the field of constants k_0. We shall call such abelian varieties constant. It is very easy to determine what becomes of the results of this section when A is constant. We shall not do this because in such a case the whole theory can be deduced under the much wider hypothesis that $k = k_0(X)$ is the field of functions on a nonsingular variety X of arbitrary dimension defined over k_0.

The concept of a principal homogeneous space is in this case related to that of a principal fibre-space defined over k_0 and having A as its structure group (cf. [17]). Every principal fibre-space P over X defined over k_0 and having A as its structure group defines a principal homogeneous space V over A defined over k. Here V is the fibre P over a generic point of X. The converse statement, namely, that not every principal homogeneous space over A defined over k can be obtained from a principal fibre-space with structure group A, is false.

Serre [17] considers the property of a principal fibre-space P of having a formal section at each point of X ([17], p. 12). It is easy to see that this property is equivalent to the property of the corresponding principal homogeneous space of having a point defined over the field of quotients $k^{\mathfrak{p}}$ of the complete local ring of each point $\mathfrak{p}$ of X. As before, we shall then say that V is locally trivial and denote the group of locally trivial principal homogeneous spaces by $\mathfrak{H}$.

We shall consider the exact sequence (2) for the fields k and $k^{\mathfrak{p}}$. Since A is a constant variety the points of $\mathfrak{A}_n$ are defined over k_0 and a fortiori over k. Hence

$$H^1(\mathfrak{G}, \mathfrak{A}_n) = \mathrm{Hom}(\mathfrak{G}, \mathfrak{A}_n),$$

$$H^1(\mathfrak{G}^{\mathfrak{p}}, \mathfrak{A}_n) = \mathrm{Hom}(\mathfrak{G}^{\mathfrak{p}}, \mathfrak{A}_n).$$

From considerations similar to those in §1 it is easy to deduce that $\alpha_{\mathfrak{p}}$ is infinitely divisible in our case. Hence (2) takes the form

$$\mathfrak{H}(A, k^{\mathfrak{p}}) = \mathrm{Hom}(\mathfrak{G}^{\mathfrak{p}}, \mathfrak{A}_n)$$

for $k^{\mathfrak{p}}$. The exact sequence (2) for k is

$$0 \longrightarrow \alpha^{(n)} \xrightarrow{\delta} \mathrm{Hom}(\mathfrak{G}, \mathfrak{A}_n) \xrightarrow{\iota} \mathfrak{H}(A, k)_n \longrightarrow 0. \tag{44}$$

By the theory of Kummer extensions every element $f \in \mathrm{Hom}(\mathfrak{G}, \mathfrak{A}_n)$ determines an abelian extension $K_f/k = k(\sqrt[n]{\alpha_1}, \cdots, \sqrt[n]{\alpha_m})$ whose galois group is isomorphic to a subgroup of $\mathfrak{A}_n$. Let us suppose that $f \in \iota^{-1}\mathfrak{H}$. Then each function $\alpha_1, \cdots, \alpha_m$ must be an nth power in each $k^{\mathfrak{p}}$, so that K_f/k determines an unramified covering of X. More precisely, the normalization X_f of the variety X in the field K_f is an unramified covering of X. It is clear that the principal fibre-space becomes trivial over X_f, which is just the definition of an isotrivial fibre-space in [17]. We have thus arrived at the following result:

If a principal fibre-space with nonsingular base X and structure group A has a formal section at every point of X, then it is isotrivial.

This gives an answer to a problem posed by Serre ([17], p. 12), although only in the very simple case when the structure group is an abelian variety.

Now let us look at $\mathfrak{H}$. We have shown that then in (2) we may replace $\mathfrak{G}$ by the galois group of the maximal unramified covering of X, and indeed of the maximal abelian covering, since we are interested only in $\mathrm{Hom}(\mathfrak{G}, \mathfrak{A}_n)$.

Let us suppose now that the Néron-Severi group of X has no torsion. It is known that in this case the abelian unramified coverings of X correspond to isogenies of its Albanese variety B (say). Hence (44) gives

$$0 \longrightarrow \alpha^{(n)} \xrightarrow{\delta} \mathrm{Hom}(\mathfrak{B}_n, \mathfrak{A}_n) \longrightarrow \mathfrak{H}_n \longrightarrow 0, \tag{45}$$

where $\mathfrak{B}_n$ is the group of points on B of order n.

It only remains to give an explicit formula for $\mathfrak{a}^{(n)}$, but this is easy. The elements of $\mathfrak{a}$ are points of A with coordinates in $k_0(X)$, i.e., rational maps of X into A. It is known that such maps are given by elements of Hom(B, A) combined with translations of the variety A; indeed, this is the definition of the Albanese variety ([9], p. 41). Since the group of points on A defined over k_0 is infinitely divisible, we have

$$\mathfrak{a}^{(n)} = \mathrm{Hom}(B, A)^{(n)}.$$

The standard interpretation of δ in (45) gives us finally

$$\mathfrak{H}_n \cong \mathrm{Hom}(\mathfrak{B}_n, \mathfrak{A}_n)/\rho_n \mathrm{Hom}(B, A), \tag{46}$$

where ρ_n is the restriction homomorphism from Hom(B, A) to $\mathfrak{B}_n$.

We conclude with a simple application of (46); namely, to find an expression for ${}_p\mathrm{Ext}(B, A)$ in the category of abelian varieties defined over k_0. For this we note that it follows from the proof of Poincaré's theorem ([9], p. 28) that to every $C \in \mathrm{Ext}(B, A)$ there corresponds a principal homogeneous space V_C over the group A defined over the field $k_0(B)$. The map $C \longrightarrow V_C$ defines a monomorphism ϕ of Ext(B, A) into $\mathfrak{H}(A, k_0(B))$. In particular, Ext(B, A) is a torsion group. The image of the monomorphism is $\mathfrak{H}(A, k_0(B))$, as follows either from Poincaré's theorem or from the results of Serre ([17], pp. 12, 28). On applying (46) for any $n \not\equiv 0$ (p) we get

$$\mathrm{Ext}(B, A)_n \cong \frac{\mathrm{Hom}(\mathfrak{B}_n, \mathfrak{A}_n)}{\rho_n \mathrm{Hom}(B, A)}.$$

BIBLIOGRAPHY

[1] H. Cartan, *Sur la compactification de Satake*, Séminaire H. Cartan 1957/58, Exposés 12 bis and 14–23, Secretariat Mathématique, 11 Rue Pierre Curie, Paris, 1958. MR 21 #2750.

[2] H. Cartan and S. Eilenberg, *Homological algebra*, Princeton Univ. Press, Princeton, N. J., 1956. MR 17, 1040.

[3] F. Enriques, *Sur les problèmes qui se rapportent à la résolution des équations algébriques renfermant plusieurs inconnues*, Math. Ann. 51 (1899), 134–153.

[4] ———, *Le superficie algebriche*, Nicola Zanichelli, Bologna, 1949. MR 11, 202.

[5] A. Grothendieck, *Géometrie formelle et géometrie algébrique*, Séminaire Bourbaki 1958/59, Exposé 182, Paris.

[6] G. Hochschild and J.-P. Serre, *Cohomology of group extensions*, Trans. Amer. Math. Soc. 74 (1953), 110–134. MR 14, 619.

[7] J.-I. Igusa, *Abstract vanishing cycle theory*, Proc. Japan Acad. 34 (1958), 589–594. MR 21 #50.

[8] K. Kodaira and D. C. Spencer, *On deformations of complex analytic structures*. I, Ann. of Math. (2) 67 (1958), 328–401. MR 22 #3009.

[9] S. Lang, *Abelian varieties*, Interscience, New York, 1959. MR 21 #4959.

[10] S. Lang and A. Néron, *Rational points of abelian varieties over function fields*, Amer. J. Math. 81 (1959), 95–118. MR 21 #1311.

[11] S. Lang and J. Tate, *Principal homogeneous spaces over abelian varieties*, ibid. 80 (1958), 659–684. MR 21 #4960.

[12] E. Lutz, *Sur l'équation* $y^2 = x^3 - Ax - B$ *dans les corps* $\mathfrak{p}$*-adiques*, J. Reine Angew. Math. 177 (1937), 238–247.

[13] P. Painlevé, *Leçons sur la théorie analytique des équations differentielles*, Stockholm, 1895.

[14] J.-P. Serre, *Groupes algébriques et corps de classes*, Actualités Sci. Ind. no. 1264 = Publ. Inst. Math. Univ. Nancago. VII, Hermann, Paris, 1959. MR 21 #1973.

[15] ———, *Groupes proalgébriques*, Inst. Hautes Études Sci. Publ. Math. No. 7 (1960), 1–67. MR 22 #9493.

[16] G. Shimura, *Reduction of algebraic varieties with respect to a discrete valuation of the basic field*, Amer. J. Math. 77 (1955), 134–176. MR 16, 1616.

[17] J.-P. Serre, *Espaces fibrés algébriques*, Séminaire C. Chevalley de l'Ecole Norm. Sup. 1958/59, Exposé 1.

[18] A. Weil, *Algebraic groups and homogeneous spaces*, Amer. J. Math. 77 (1955), 493–512. MR 17, 533.

[19] ———, *Adèles et groupes algébriques*, Séminaire Bourbaki 1958/59, Exposé 186.

[20] I. R. Šafarevič, *Birational equivalence of elliptic curves*, Dokl. Akad. Nauk SSSR 114 (1957), 267–270. (Russian) MR 20 #867.

[21] ———, *The group of principal homogeneous algebraic manifolds*, ibid. 124 (1959), 42–43. (Russian) MR 21 #4961.

[22] A. Mattuck, *Abelian varieties over p-adic ground fields*, Ann. of Math. (2) 62 (1955), 92–119. MR 17, 87.

[23] J.-I. Igusa, *Analytic groups over complete fields*, Proc. Nat. Acad. Sci. U.S.A. 42 (1956), 540–541. MR 18, 935.

Translated by:
J. W. S. Cassels

The second obstruction for the imbedding problem of algebraic number fields

(with S. P. Demushkin)

Izv. Akad. Nauk SSSR, Ser. Mat. **26**, No. 6, 911–924 (1962). Zbl. **115**, 37
[Transl., II. Ser., Am. Math. Soc. **58**, 245–260 (1966)]

In this paper we investigate the problem of imbedding of algebraic number fields into fields with larger Galois group. The second obstruction for the solvability of the imbedding problem will be computed.

We consider the problem of imbedding of the field k, normal over Ω, with Galois group F, into a larger field K with group G over Ω, where the field K should be such that the natural homomorphism

$$G = \mathfrak{G}(K/\Omega) \longrightarrow \mathfrak{G}(k/\Omega) = F$$

of the Galois group of the field on the Galois group of the subfield coincides with a preassigned homomorphism $\phi: G \longrightarrow F$. Such an imbedding problem will be written in the form $(k/\Omega, G, \phi)$. As in the paper [1] we shall suppose that the kernel A of the homomorphism $\phi: G \longrightarrow F$ is an abelian group and that the mth roots of unity (m = period of the group A) are contained in k. Furthermore, since the problem of immersion is reduced to the problem of imbedding with p-groups (see [3]), we will throughout assume that G is a p-group.

Some necessary conditions for the solvability of the problem of imbedding – the conditions of compatability – were discovered by D. K. Faddeev.

In a form convenient for us these have been given in the paper [1]. We call these conditions the first obstruction. Further we shall throughout assume that the first obstruction for the problem $(k/\Omega, G, \phi)$ vanishes.

We shall solve the problem of imbedding by successive steps. In A we choose an F-invariant series

$$A \supset A_1 \supset \cdots \supset A_i \supset \cdots \supset 1$$

with $[A_i : A_{i+1}] = p$ (since G is a p-group it is possible to choose such a series). Then we solve the problem $(k/\Omega, G/A_1, \overline{\phi})$ which is solvable since it is of the same form as the problem $(k/\Omega, G, \phi)$. Let k^1 be the solution of the problem $(k/\Omega, G/A_1, \phi)$. We ask: when is it possible to choose the field k^1 so that, for the problem $(k^1/\Omega, G, \phi_1)$ the first obstruction vanishes? We shall call the invariants, that yield the answer for this question, the second obstruction for the problem of imbedding. In an implicit form the second obstruction is given in the paper [1]. In the present paper the second obstruction is found for the case when k is an algebraic number field.

§1. The compatibility problem for cyclic algebras; construction of invariants

Since all the results, of which we shall be speaking in what follows, refer to the case when the field k to be imbedded is an algebraic number field we will suppose from the very beginning, that k is such a field. We shall be interested in the second obstruction for the imbedding problem, i.e. invariants permitting us to answer the question as to when there will exist a solution k^1 of the imbedding problem $(k/\Omega, G/A_1, \bar{\phi})$ such that the second obstruction vanishes for the problem $(k^1/\Omega, G, \phi_1)$. As is known from the paper [1], the second obstruction is determined by a set of cyclic algebras over subfields of the field k given by characters of the group A_1. For us this dependence of the cyclic algebras on the characters of the group A_1 will not be important, therefore in denoting these algebras we omit their dependence on characters.

Thus, let there be given a finite number of cyclic algebras

$$C_i = (\alpha_i, \beta_i)_{k_i} \qquad (k_i \subset k).$$

We ask: when will there exist in the field Ω such a number m that all the algebras C_i can be represented in the form $C_i \approx (\alpha_i, m)_{k_i}$ for all i? If such a number exists the second obstruction vanishes; if such a number m does not exist the second obstruction does not vanish.

We see that the problem of finding the second obstruction for the imbedding problem is reduced to a certain problem on cyclic algebras. It is with this last problem that we will deal now. Let us formulate it.

Problem of compatibility of cyclic algebras. Given a certain number of algebraic p-extensions $\{k_i\}$ of the algebraic number field Ω, and over each field k_i is given a cyclic algebra $C_i = (\alpha_i, \beta_i)_{k_i}$ of degree p. What are the conditions in order that all the algebras C_i can simultaneously be represented in the form

$$C_i \approx (\alpha_i, m)_{k_i}, \quad m \in \Omega?$$

It is easy to point out necessary conditions for the problem of compatibility of cyclic algebras. These are local conditions: for every prime divisor $\mathfrak{p}$ of the field Ω there must exist such a number $m_\mathfrak{p} \in \Omega_\mathfrak{p}$ (where $\Omega_\mathfrak{p}$ is the closure of Ω at $\mathfrak{p}$) that

$$\left(\frac{\alpha_i, \beta_i}{\mathfrak{P}_i}\right)_{k_i} = \left(\frac{\alpha_i, m_\mathfrak{p}}{\mathfrak{P}_i}\right)_{k_i}$$

for all i and for all prime factors $\mathfrak{P}_i$ of the divisor $\mathfrak{p}$ in the field k_i. Here, as in what follows,

$$\left(\frac{\alpha_i, \beta_i}{\mathfrak{P}_i}\right)_{k_i}$$

will denote the norm residue symbol at the prime $\mathfrak{P}_i$.

Lemma 1. *For cyclic algebras $\{C_i\}$, arising from the imbedding problem, the local conditions for their compatibility are fulfilled.*

Proof. Let $(k/\Omega, G, \phi)$ be the problem of imbedding from which arise the algebras $\{C_i\}$, $\mathfrak{p}$ – a prime divisor of the field Ω, $\mathfrak{P}_i$ – its factors in the field k_i. We need to prove the existence of such a number $m_\mathfrak{p} \in \Omega_\mathfrak{p}$ that

$$\left(\frac{\alpha_i, \beta_i}{\mathfrak{P}_i}\right)_{k_i} = \left(\frac{\alpha_i, m_\mathfrak{p}}{\mathfrak{P}_i}\right)_{k_i}$$

for all i and $\mathfrak{P}_i \mid \mathfrak{p}$, if the algebras $\{C_i\}$ arose from the imbedding problem. Let us first explain what the local conditions for the algebras $\{C_i\}$ signify from the point of view of the imbedding problem. We assert that the local conditions are equivalent to the assertion that for the imbedding problem $P_\mathfrak{p} = (k_\mathfrak{p}/\Omega_\mathfrak{p}, G, \phi)$, where $k_\mathfrak{p}$ is the algebra obtained from k/Ω by extension of Ω to $\Omega_\mathfrak{p}$, the second obstruction vanishes.

In fact, from the vanishing of the first obstruction for the imbedding problem $P = (k/\Omega, G, \phi)$ follows its vanishing for the problem $P_\mathfrak{p}$. The algebras $C_i(P_\mathfrak{p})$ are equivalent to the algebras $(\alpha_i, \beta_i)_{k_i\mathfrak{p}}$, since the passage from the problem P to the problem $P_\mathfrak{p}$ is obtained by extension of Ω to $\Omega_\mathfrak{p}$. If the second obstruction vanishes for the problem $P_\mathfrak{p}$ there exists such a number $m_\mathfrak{p} \in \Omega_\mathfrak{p}$ that all the algebras $C_i(P_\mathfrak{p})$ can be represented in the form

$$C_i(P_\mathfrak{p}) \approx (\alpha_i, m_\mathfrak{p})_{k_i\mathfrak{p}}.$$

It remains to notice that the decomposition of the algebra over the center, which is a direct sum of subalgebras (in our case subfields), is equivalent to the assertion that the components of the algebra decompose over the corresponding components of the center. Since $k_{i\mathfrak{p}}$ is the direct sum of the fields $k_{i\mathfrak{P}_i}$, the representation

$$(\alpha_i, \beta_i)_{k_i\mathfrak{p}} \approx (\alpha_i, m_\mathfrak{p})_{k_i\mathfrak{p}}$$

is equivalent to the fact that

$$\left(\frac{\alpha_i, \beta_i}{\mathfrak{P}_i}\right)_{k_i} = \left(\frac{\alpha_i, m_\mathfrak{p}}{\mathfrak{P}_i}\right)_{k_i}$$

for all i and $\mathfrak{P}_i \mid \mathfrak{p}$.

Consequently, for the proof of the lemma it is sufficient to show that the second obstruction vanishes for the imbedding problem $P_\mathfrak{p}$. We show, in fact, that the embedding problem for $P_{\mathfrak{p}}$ is solvable. In this manner, it is a question of the proof of the following assertion: for the solvability of the imbedding problem $(k/\Omega, G, \phi)$, where Ω is a local field and k an algebra, the vanishing of the first obstruction is sufficient. Here it is not possible to apply directly Theorem 4 of [1], since there the whole argument was for the case when k is a field.

Let

$$k = \sum k e^{\sigma_j}$$

be a resolution of the algebra k into a direct sum of fields ke^{σ_j} and let H be the subgroup of F acting in the normal field $\kappa = ke$. Then we obtain the resolution

$$F = \sum H\sigma_j.$$

Consider the imbedding problem $(\kappa/\Omega, \phi^{-1}H, \phi)$. For such a problem the first obstruction vanishes, since it is solvable owing to the main theorem of [1]. Let $\mathcal{K}$ be the solution of this problem. Consider the algebra

$$K = \sum \mathcal{K} e^{\bar{\sigma}_j},$$

which is a direct sum of algebras isomorphic to $\mathcal{K}$. Here $\bar{\sigma}_j$ are the inverse images of the elements σ_j in G. The group G can be represented in the form

$$G = \sum \varphi^{-1} H \bar{\sigma}_j.$$

Let us define the action of the elements of the group G in K by the formula:

$$(\alpha e^{\bar{\sigma}_j})^{h\bar{\sigma}_i} = \alpha^{h_1} e^{\bar{\sigma}_l},$$

if

$$\bar{\sigma}_j h \bar{\sigma}_i = h_1 \bar{\sigma}_l.$$

It is easy to see that the obtained representation will be regular and that the algebra K will be a solution of the imbedding problem $(k/\Omega, G, \phi)$.

The lemma is proved.

In order to find the second obstruction for the imbedding problem it is sufficient by Lemma 1 to answer the following question: Suppose that a system C_i of cyclic algebras satisfies the local conditions for compatibility; when can the system of algebras C_i be made compatible? Let us now consider that question.

Consider in the field Ω the aggregate of numbers b possessing the property: for every prime divisor $\mathfrak{p}$ in Ω the number b can be represented in $\Omega_{\mathfrak{p}}$ in the form

$$b = \prod_{i,\, \mathfrak{P}_i | \mathfrak{p}} (N_{\mathfrak{P}_i} \alpha_i)^{b_{\mathfrak{P}_i}} \lambda^p, \tag{1}$$

where $N_{\mathfrak{P}_i}$ is the norm from the field $k_{i\mathfrak{P}_i}$ into the field $\Omega_{\mathfrak{P}}$, $b_{\mathfrak{P}_i}$ are certain integers and $\lambda \in \Omega_{\mathfrak{p}}$.

The numbers b with such a property will be called root numbers. The product of two root numbers will again be a root number.

Let us remark that, for example, all norms $N_{k_i|\Omega}\,\alpha_i$ will be root numbers, since such a norm can always be decomposed into the product of local norms.

Let $\{b\}$ be the subgroup of the multiplicative group of the field Ω generated by the root numbers b.

Lemma 2. *The factor-group* $\{b\}/\Omega^{*p}$ *is finite.*

Proof. Consider the normal closure of the fields $k_i(\sqrt[p]{\alpha_i})$. Let this be the field k'. It will be a finite extension of the field Ω since the number of algebras C_i is finite.

For the proof of the lemma it is sufficient to show that

$$\Omega(\sqrt[p]{b}) \subset k'.$$

Next, to prove this last inclusion it is sufficient to establish that every prime divisor $\mathfrak{p}$ in Ω, completely decomposing in k', completely decomposes in $\Omega(\sqrt[p]{b})$. We show that for such a divisor $\mathfrak{p}$ the number b is a pth power in $\Omega_{\mathfrak{p}}$. In fact, α_i are pth powers in $k_{i\mathfrak{P}_i}$:

$$\alpha_i = A_i^p, \quad A_i \in k_{i\mathfrak{P}_i},$$

therefore

$$b = \prod_{i,\,\mathfrak{P}_i|\mathfrak{p}} (N_{\mathfrak{P}_i}\alpha_i)^{b_{\mathfrak{P}_i}} \lambda^p = \Big[\prod_{i,\,\mathfrak{P}_i|\mathfrak{p}} (N_{\mathfrak{P}_i} A_i)^{b_{\mathfrak{P}_i}} \lambda \Big]^p.$$

The lemma is proved.

With each root number b we associate

$$I(b) = \prod_{\mathfrak{p},\,i,\,\mathfrak{P}_i|\mathfrak{p}} \left(\frac{\alpha_i,\,\beta_i}{\mathfrak{P}_i}\right)_{k_i}^{b_{\mathfrak{P}_i}},$$

where $b_{\mathfrak{P}_i}$ are obtained from the representation (1).

We will call $I(b)$ root functions for the problem of compatibility of cyclic algebras $\{C_i\}$.

Lemma 3. *If the cyclic algebras* $\{C_i\}$ *satisfy the local conditions for compatibility, then the root functions* $I(b)$ *do not depend on the representation of the number* b *in the form* (1).

Proof. Let there be two representations of the number b:

$$b = \prod_{i,\,\mathfrak{P}_i|\mathfrak{p}} (N_{\mathfrak{P}_i}\alpha_i)^{b_{\mathfrak{P}_i}} \lambda_{\mathfrak{p}}^p$$

and

$$b = \prod_{i,\, \mathfrak{P}_i | \mathfrak{p}} (N_{\mathfrak{P}_i} \alpha_i)^{b'_{\mathfrak{P}_i}} \lambda'^{p}_{\mathfrak{p}}.$$

Then

$$1 = \prod_{i,\, \mathfrak{P}_i | \mathfrak{p}} (N_{\mathfrak{P}_i} \alpha_i)^{b_{\mathfrak{P}_i} - b'_{\mathfrak{P}_i}} (\lambda_a / \lambda'_{\mathfrak{p}})^p$$

and it is sufficient to show that $I(1) = 1$ for every representation of the identity. Let

$$1 = \prod_{i,\, \mathfrak{P}_i | \mathfrak{p}} (N_{\mathfrak{P}_i} \alpha_i)^{n_{\mathfrak{P}_i}} \nu_{\mathfrak{p}}^{p}.$$

Then, utilizing the basic properties of the norm residue symbol and the fact that the algebras $\{C_i\}$ satisfy the local conditions for their compatibility we obtain

$$I(1) = \prod_{\mathfrak{p},\, i \mathfrak{P}_i | \mathfrak{p}} \left(\frac{\alpha_i, \beta_i}{\mathfrak{P}_i} \right)_{k_i}^{n_{\mathfrak{P}_i}} = \prod_{\mathfrak{p},\, i,\, \mathfrak{P}_i | \mathfrak{p}} \left(\frac{\alpha_i, m_{\mathfrak{p}}}{\mathfrak{P}_i} \right)_{k_i}^{n_{\mathfrak{P}_i}}$$

$$= \prod_{\mathfrak{p},\, i,\, \mathfrak{P}_i | \mathfrak{p}} \left(\frac{N_{\mathfrak{P}_i} \alpha_i, m_{\mathfrak{p}}}{\mathfrak{p}} \right)_{\Omega}^{n_{\mathfrak{P}_i}} = \left(\frac{\prod_{i,\, \mathfrak{P}_i | \mathfrak{p}} (N_{\mathfrak{P}_i} \alpha_i)^{n_{\mathfrak{P}_i}}, m_{\mathfrak{p}}}{\mathfrak{p}} \right)_{\Omega} = \prod_{\mathfrak{p}} \left(\frac{\nu_{\mathfrak{p}}^{-p}, m_{\mathfrak{p}}}{\mathfrak{p}} \right)_{\Omega} = 1.$$

The lemma is proved.

Since the value $I(b)$ depends only on the class of b in $\{b\}/\Omega^{*p}$ there are only a finite number of root functions. It is easy to verify that root functions behave multiplicatively with respect to root numbers, i.e.,

$$I(b_1 b_2) = I(b_1)\, I(b_2).$$

§2. The main theorems

Theorem 1. *For the solvability of the problem of compatibility of cyclic algebras, satisfying the local conditions, it is necessary and sufficient that all root*
1 *functions are* 1.

Proof. Let the problem of compatibility of cyclic algebras be solvable, i.e.

$$\left(\frac{\alpha_i, \beta_i}{\mathfrak{P}_i} \right)_{k_i} = \left(\frac{\alpha_i, m}{\mathfrak{P}_i} \right)_{k_i}, \quad m \in \Omega,$$

for all i and $\mathfrak{P}_i \mid \mathfrak{p}$ in k_i, and let $I(b)$ be the root function

$$I(b) = \prod_{\mathfrak{p}, i, \mathfrak{P}_i \mid \mathfrak{p}} \left(\frac{\alpha_i, \beta_i}{\mathfrak{P}_i}\right)_{k_i}^{b_{\mathfrak{P}_i}},$$

$$b = \prod_{i, \mathfrak{P}_i \mid \mathfrak{p}} (N_{\mathfrak{P}_i} \alpha_i)^{b_{\mathfrak{P}_i}} \lambda_{\mathfrak{p}}^{p}.$$

Then, using the basic properties of the norm residue symbol, we obtain

$$I(b) = \prod_{\mathfrak{p}, i, \mathfrak{P}_i \mid \mathfrak{p}} \left(\frac{\alpha_i, \beta_i}{\mathfrak{P}_i}\right)_{k_i}^{b_{\mathfrak{P}_i}} = \prod_{\mathfrak{p}, i, \mathfrak{P}_i \mid \mathfrak{p}} \left(\frac{\alpha_i, m}{\mathfrak{P}_i}\right)_{k_i}^{b_{\mathfrak{P}_i}}$$

$$= \prod_{\mathfrak{p}, i, \mathfrak{P}_i \mid \mathfrak{p}} \left(\frac{N_{\mathfrak{P}_i} \alpha_i, m}{\mathfrak{p}}\right)_{\Omega}^{b_{\mathfrak{P}_i}} = \prod_{\mathfrak{p}} \left(\frac{\prod_{i, \mathfrak{P}_i \mid \mathfrak{p}} (N_{\mathfrak{P}_i} \alpha_i)^{b_{\mathfrak{P}_i}}, m}{\mathfrak{p}}\right)_{\Omega}$$

$$= \prod_{\mathfrak{p}} \left(\frac{b \lambda_{\mathfrak{p}}^{-p}, m}{\mathfrak{p}}\right) = \prod_{\mathfrak{p}} \left(\frac{b, m}{\mathfrak{p}}\right)_{\Omega} = 1.$$

In this manner the necessity of the condition of the theorem is proved. Let us proceed to the proof of its sufficiency.

Let $\mathfrak{M}$ be the subgroup of idèles $m = \{m_{\mathfrak{p}}\}$ for which

$$\left(\frac{\alpha_i, m_{\mathfrak{p}}}{\mathfrak{P}_i}\right)_{k_i} = 1$$

for all i and $\mathfrak{P}_i \mid \mathfrak{p}$. Consider also the subgroup $\mathfrak{N} = \mathfrak{M} m$, $m \in \Omega$. According to the condition of the theorem there exists an idèle

$$\overline{m}_0 = \{m_{0\mathfrak{p}}\},$$

for which

$$\left(\frac{\alpha_i, \beta_i}{\mathfrak{P}_i}\right)_{k_i} = \left(\frac{\alpha_i, m_{0\mathfrak{p}}}{\mathfrak{P}_i}\right)_{k_i}$$

for all i and $\mathfrak{P}_i \mid \mathfrak{p}$. Consequently, for the proof of the theorem it is necessary to show that $\overline{m}_0 \in \mathfrak{N}$.

Let us remark that the subgroup $\mathfrak{N}$ is an admissible subgroup (see [2]). In order to show this let us denote by S the finite set of divisors in Ω which are either factors of p, or are such that their factors in k_i are ramified in $k_i(\sqrt[p]{\alpha_i})$. Then $\mathfrak{N}$ will contain the set of ideal elements A_S (the group A_S is defined as the set of idèles $\overline{n} = \{n_{\mathfrak{p}}\}$, for which $n_{\mathfrak{p}} = 1$ if $\mathfrak{p} \in S$ and $n_{\mathfrak{p}}$ is a $\mathfrak{p}$-adic unit if $\mathfrak{p} \notin S$).

An admissible subgroup is characterized by the condition that all its elements are orthogonal to a certain finite set of numbers in Ω (see [2]). Let $b_1, b_2, \cdots, b$ be such numbers for $\mathfrak{N}$. We show that these are root numbers. For this we choose as $\overline{m} = \{m_{\mathfrak{p}}\}$ the idèle for which $m_{\mathfrak{p}}$ is arbitrary for fixed $\mathfrak{p}$ and $m_{\mathfrak{q}} = 1$ for $\mathfrak{q} \neq \mathfrak{p}$. Then the condition

$$\left(\frac{\alpha_i, m_{\mathfrak{p}}}{\mathfrak{P}_i}\right)_{k_i} = 1$$

for all i and $\mathfrak{P}_i \mid \mathfrak{p}$ will signify that

$$(N_{\mathfrak{P}_i}, \alpha_i m_{\mathfrak{p}})_{\Omega_{\mathfrak{p}}} \approx 1,$$

i.e. $m_{\mathfrak{p}}$ is orthogonal to a finite set of numbers in $\Omega_{\mathfrak{p}}$. For the numbers b_i themselves we must have

$$[b_i, \overline{m}] = 1,$$

i.e.

$$(b_i, m_{\mathfrak{p}})_{\Omega_{\mathfrak{p}}} \approx 1.$$

Since the orthogonal complement to the orthogonal complement coincides with the original group, then

$$b_i = \prod_{i,\, \mathfrak{P}_i \mid \mathfrak{p}} (N_{\mathfrak{P}_i} \alpha_i)^{b_{\mathfrak{P}_i}} \lambda_{\mathfrak{p}}^{p}.$$

Therefore $b_1, b_2, \cdots, b_r$ are root numbers.

In order to complete the proof of the theorem it is sufficient to show that the element $\overline{m}_0 = \{m_{0\mathfrak{p}}\}$ is orthogonal to all root numbers. Let

$$b = \prod_{i,\, \mathfrak{P}_i \mid \mathfrak{p}} (N_{\mathfrak{P}_i} \alpha_i)^{b_{\mathfrak{P}_i}} \lambda_{\mathfrak{p}}^{p}, \quad \lambda_{\mathfrak{p}} \in \Omega_{\mathfrak{p}}.$$

Then we have:

$$[b, \overline{m}_0] = \prod_{\mathfrak{p}} \left(\frac{b, m_{0\mathfrak{p}}}{p}\right)_{\Omega} = \prod_{\mathfrak{p}} \left(\frac{\prod\limits_{i,\, \mathfrak{P}_i \mid \mathfrak{p}} (N_{\mathfrak{P}_i} \alpha_i)^{b_{\mathfrak{P}_i}} \lambda_{\mathfrak{p}}^{p},\ m_{0\mathfrak{p}}}{p}\right)$$

$$= \prod_{\mathfrak{p},\, i,\, \mathfrak{P}_i \mid \mathfrak{p}} \left(\frac{N_{\mathfrak{P}_i} \alpha_i,\ m_{0\mathfrak{p}}}{p}\right)_{\Omega}^{b_{\mathfrak{P}_i}} = \prod_{\mathfrak{p},\, i,\, \mathfrak{P}_i \mid \mathfrak{p}} \left(\frac{\alpha_i, m_{0\mathfrak{p}}}{\mathfrak{P}_i}\right)_{k_i}^{b_{\mathfrak{P}_i}}.$$

$$= \prod_{\mathfrak{p},\, i,\, \mathfrak{P}_i \mid \mathfrak{p}} \left(\frac{\alpha_i, \beta_i}{\mathfrak{P}_i}\right)_{k_i}^{b_{\mathfrak{P}_i}} = I(b) = 1$$

(according to the condition). The theorem is proved.

Let, as before, k' denote the normal closure over Ω of the fields $k_i(\sqrt[p]{\alpha_i})$. Denote by $g(\mathfrak{p})$, where $\mathfrak{p}$ is a prime divisor in Ω, the number of prime factors of $\mathfrak{p}$ in the field k'.

Theorem 2. *If the highest common factor of the numbers* $g(\mathfrak{p})$, $\mathfrak{p} \in \Omega$, *is equal to* 1, *then the cyclic algebras* $\{C_i\}$, *satisfying the local conditions, can always be made compatible.*

For the proof of the theorem it is sufficient to show that all the root expressions in the above case are equal to 1. Since we are considering the problem of compatibility of cyclic algebras C_i over fields k_i, which are p-extensions of the field Ω, then k' also will be a p-extension of Ω. Therefore from the fact that the greatest common factor of the numbers $g(\mathfrak{p})$ is equal to unity it follows that there exists a prime divisor $\mathfrak{p}_0$ of the field Ω for which $g(\mathfrak{p}_0) = 1$, i.e. $\mathfrak{p}_0$ has only one prime divisor $\mathfrak{P}_0$ in k'.

Let b be a root number. We prove that $l(b) = 1$. For this we consider the equality (1), defining the number b at the prime $\mathfrak{p}_0$:

$$b = \prod_{i,\, \mathfrak{P}_{0i} | \mathfrak{p}_0} (N_{\mathfrak{P}_{0i}} \alpha_i)^{b_{\mathfrak{P}_{0i}}} \lambda_0^p, \quad \lambda_0 \in \Omega_{\mathfrak{p}_0}.$$

Since $\mathfrak{p}_0$ has only one factor in k' there will also be one factor $\mathfrak{P}_{0i}$ in the field k_i. Furthermore, the decomposition group of the divisor $\mathfrak{P}_0$ will coincide with the entire group $\mathfrak{G}(k'/\Omega)$ and local norms $N_{\mathfrak{P}_{0i}}$ will coincide with the usual norms from the fields k_i in Ω. Therefore at the prime $\mathfrak{p}_0$ we obtain the equality:

$$b = \prod_i (N_{k_i|\Omega} \alpha_i)^{n_i} \lambda_0^p$$

(the numbers $b_{\mathfrak{P}_{0i}}$ are here denoted by n_i). Consider the number b' in the basic field:

$$b' = b \Big/ \prod_i (N_{k_i|\Omega} \alpha_i)^{n_i}.$$

It is a root number since all the numbers $N_{k_i|\Omega}\, \alpha_i$ are root numbers. The number b' is a pth power in $\Omega_{\mathfrak{p}_0}$:

$$b' = \lambda_0^p.$$

Let us show that it is a pth power in Ω. In fact, suppose $b' \neq u^p$, $u \in \Omega$. Then the field $\Omega(\sqrt[p]{b'})$ will be intermediate between Ω and k' (from the proof of Lemma 2 it follows that $\Omega(\sqrt[p]{b'}$ is a subfield of the field k'):

$$\Omega \in \Omega(\sqrt[p]{b'}) \subset k'.$$

Passing to the completion at $\mathfrak{p}_0$ we obtain

$$\Omega_{\mathfrak{p}_0} = \Omega(\sqrt[p]{b'})_{\mathfrak{p}_0} \subset k'_{\mathfrak{P}_0},$$

which contradicts the fact that

$$[k'_{\mathfrak{P}_0} : \Omega_{\mathfrak{p}_0}] = [k' : \Omega]$$

when $g(\mathfrak{p}_0) = 1$. Consequently

$$b' = u^p, \qquad u \in \Omega.$$

But u^p is a root number for which $I(u^p) = 1$. Therefore

$$I(b) = \prod_i I(N_{k_i|\Omega}\alpha_i)^{n_i},$$

and it is sufficient to verify that

$$I(N_{k_i|\Omega}\alpha_i) = 1.$$

For this purpose let us consider the decomposition of $N_{k_i|\Omega}\alpha_i$ as a product of local norms:

$$N_{k_i|\Omega}\alpha_i = \prod_{\mathfrak{P}_i|\mathfrak{p}} N_{\mathfrak{P}_i}\alpha_i.$$

Hence $b_{\mathfrak{P}_i} = 1$, $b_{\mathfrak{P}_j} = 0$ for $j \neq i$. In such a case

$$I(N_{k_i|\Omega}\alpha_i) = \prod_{\mathfrak{p},\, j,\, \mathfrak{P}_j|\mathfrak{p}} \left(\frac{\alpha_j, \beta_j}{\mathfrak{P}_j}\right)_{k_j}^{b_{\mathfrak{P}_j}} = \prod_{\mathfrak{p},\, \mathfrak{P}_i|\mathfrak{p}} \left(\frac{\alpha_i, \beta_i}{\mathfrak{P}_i}\right)_{k_i} = 1$$

(by the law of reciprocity). The theorem is proved.

Corollary 1. *If the 3-dimensional cohomology group*

$$H^3(\mathfrak{G}(k'/\Omega), k'^*) = 1,$$

then the algebras $\{C_i\}$, satisfying the local conditions, can always be made compatible.

For the proof it is sufficient to remark that for algebraic number fields the three-dimensional group of cohomology is cyclic and its order is equal to the greatest common divisor of the numbers $g(\mathfrak{p})$, $\mathfrak{p} \in \Omega$.

Corollary 2. *If the field k' is cyclic over Ω, the algebras $\{C_i\}$, satisfying the local conditions, are always compatible.*

For the proof it is sufficient to remark that in the case when the field k' is cyclic over Ω, the three-dimensional cohomology group $H^3(\mathfrak{G}(k'/\Omega), k'^*)$ will be the identity or alternatively that, owing to "Law of Density" of Čebotarev, there exists in Ω a prime divisor $\mathfrak{p}$, belonging to a generating element of the group $\mathfrak{G}(k'/\Omega)$, and then for such a $\mathfrak{p}$ we have:

$$g(\mathfrak{p}) = 1.$$

Let us derive the results, relating to the problem of imbedding $(k/\Omega, G, \phi)$ of algebraic number fields k, that follow immediately from Theorem 2 and its corollaries.

Let there be given the character-group $\mathfrak{A}$ of the normed subgroup A in the imbedding problem. Then the factor-group F is canonically mapped into the group of automorphisms of the group $\mathfrak{A}$. Denote by F' the kernel of this mapping, by k' the subfield belonging to F' in k, and by $g(\mathfrak{p})$ the number of factors in k' of the prime divisor $\mathfrak{p}$ in Ω.

Theorem 3. *If the greatest common divisor of the numbers* $g(\mathfrak{p})$, $\mathfrak{p} \in \Omega$, *is equal to* 1, *then the vanishing of the first obstruction to the imbedding problem* $(k/\Omega, G, \phi)$ *of the algebraic number field* k *is sufficient for its solvability.*

For the proof it is sufficient to notice that, when passing over to the imbedding problem $(k^1/\Omega, G, \phi_1)$, the condition of the theorem is preserved.

Corollary 1. *If the three-dimensional group of cohomology*

$$H^3(F/F', k'^*) = 1,$$

then the vanishing of the first obstruction to the imbedding problem $(k/\Omega, G, \phi)$ *of an algebraic number field* k *is sufficient for its solvability.*

Corollary 2. *If the factor-group* F *induces in the character group of the normal subgroup a cyclic group of automorphisms, then the vanishing of the first obstruction to the imbedding problem* $(k/\Omega, G, \phi)$ *of an algebraic number field* k *is sufficient for its solvability.*

By means of the results of [3] it is easy to show that Theorem 3 and its corollaries will also be valid for the case when G is not a p-group.

Remark. Even though the proofs of the Corollaries 1 and 2 are arithmetical their formulation is purely algebraical. It would be interesting to elucidate whether these corollaries are valid or not for an arbitrary field k. The conjecture that it is so is verified in the case when k is a local field. In this case there are no three-dimensional homologies and, in fact, the vanishing of the first obstruction is sufficient for imbedding.

§3. An example

Consider an example for the computation of the second obstruction. (This example is taken from [5]).

Let the normal subgroup A in the imbedding problem $(k/\Omega, G, \phi)$ be a cyclic group of order 8: $A = \{\alpha\}$. We will assume that every divisor $\mathfrak{p}$ in Ω has in the field k', which was defined in the previous paragraph, either two or four divisors (see Theorem 3). This will imply in particular, that the factor-group F induces in the character-group of the group A a non-cyclic group of automorphisms. But the group of automorphisms of the group A is the product of two cyclic groups of second order. Consequently, the group F should generate the entire group of automorphisms and therefore the field k' has the form:

$$k' = \Omega(\sqrt{a}, \sqrt{b}).$$

Let us carry out the first step in the imbedding problem $(k/\Omega, G, \phi)$. After the first step the normal subgroup will be

$$A_1 = \{\alpha^2\}.$$

It has two characters χ_1 and χ_2 that are essential for us:

$$\chi_1(\alpha^2) = -1,$$
$$\chi_2(\alpha^2) = i.$$

Let x_1 and x_2 be the extensions of the characters χ_1 and χ_2 to A. It is clear that always

$$F_{x_2} \subset F_{x_2},$$
$$F_{x_1} \subset F_{\chi_1} = F,$$
$$F_{\chi_2} \subset F_{\chi_1} = F, \quad F_{x_2} \subset F_{x_1}$$

and that the indices of all the written subgroups are either equal to 1 or to 2. It is not difficult to see that the indices are actually all equal to 2. We define

$$k_{x_1} = \Omega(\sqrt{a}) \quad [k_{x_2} = \Omega(\sqrt{a}, \sqrt{b})],$$
$$k_{\chi_2} = \Omega(\sqrt{a}) \quad [k_{\chi_1} = \Omega].$$

Therefore

$$a_{x_2} = b,$$
$$a_{x_1} = a.$$

The number a will be a root number, but it cannot yield the second obstruction since it is the norm of $a_{x_1} = a$.

We shall prove that the number b will also be a root number. For this it is necessary that

$$b \approx a^{i_\mathfrak{p}} \prod_{\mathfrak{P}_i | \mathfrak{p}} (N_{\mathfrak{P}_i} b)^{i_{\mathfrak{P}_i}} \text{ in } \Omega_\mathfrak{p}. \tag{2}$$

The sign $\approx$ signifies 2-equality, i.e. equality to within squares. Since the norm $N_{\mathfrak{P}_i} b$ will always be equal to b or b^2 independently of the factor $\mathfrak{P}_i$ of the divisor $\mathfrak{p}$ we will take in the 2-equality (2) only one of the factors of the divisor $\mathfrak{p}$:

$$b \approx a^{i_\mathfrak{p}} N_\mathfrak{p} b^{j_\mathfrak{p}} \tag{2'}$$

($N_\mathfrak{p} b$ denotes here the norm $N_{\mathfrak{P}_i} b$ for one of the factors $\mathfrak{P}_i$ of the divisor $\mathfrak{p}$). If $a \approx 1$ in $\Omega_\mathfrak{p}$, then

$$N_\mathfrak{p} b = b$$

and therefore in the 2-equality (2′) it is necessary to take $i_{\mathfrak{p}} = 0$, $j_{\mathfrak{p}} = 1$. If $a \not\approx 1$ in $\Omega_{\mathfrak{p}}$, then

$$N_{\mathfrak{p}} b = b^2 \approx 1.$$

Therefore when $b \approx 1$ one must take $i_{\mathfrak{p}} = 0$, and when $b \neq 1$ one must have $b \approx a$ $(i_{\mathfrak{p}} = 1)$, i.e. $ab \approx 1$. This last relation is fulfilled since, in the contrary case, the degree $[\Omega_{\mathfrak{p}}(\sqrt{a}, \sqrt{b}): \Omega_{\mathfrak{p}}]$ would be equal to 4, and this would signify that in the field k' there will only be one factor for $\mathfrak{p}$.

Consequently, we obtain a representation of the number b in the form (2′):

1) $b \approx 1$ in $\Omega_{\mathfrak{p}}$; $i_{\mathfrak{p}} = 0$, $j_{\mathfrak{p}} = 0$;

2) $b \not\approx 1$, but $a \approx 1$ in $\Omega_{\mathfrak{p}}$; $i_{\mathfrak{p}} = 0$, $j_{\mathfrak{p}} = 1$;

3) $b \not\approx 1$ and $a \not\approx 1$ in $\Omega_{\mathfrak{p}}$; $i_{\mathfrak{p}} = 1$, $j_{\mathfrak{p}} = 0$.

Since we assume that the first obstruction to our problem vanishes the algebras C_{χ_1} and C_{χ_2} will have the form:

$$C_{\chi_1} \approx (a, \beta_{\chi_1})_{\Omega},$$
$$C_{\chi_2} \approx (b, \beta_{\chi_2})_{\Omega(\sqrt{a})}.$$

Hence the root function $I(b)$ takes the following form:

$$I(b) = \prod_{\mathfrak{p}} \left(\frac{a, \beta_{\chi_1}}{\mathfrak{p}}\right)^{i_{\mathfrak{p}}} \prod_{\mathfrak{p}} \left(\frac{b, \beta_{\chi_2}}{\mathfrak{p}}\right)^{j_{\mathfrak{p}}}_{\Omega(\sqrt{a})}.$$

Here by

$$\left(\frac{b, \beta_{\chi_2}}{\mathfrak{p}}\right)_{\Omega(\sqrt{a})}$$

we denote the norm residue symbol

$$\left(\frac{b, \beta_{\chi_2}}{\mathfrak{P}_i}\right)_{\Omega(\sqrt{a})}$$

for one of factors $\mathfrak{P}_i$ of the divisor $\mathfrak{p}$.

Consider now a numerical example. Let the field to be imbedded be

$$k = R(\sqrt{c}, \sqrt{2}, i),$$

where R is the field of rational numbers. The Galois group F of the field k/R is the direct product of the groups

$$F_1 = \mathfrak{G}(k_1/R), \quad k_1 = R(\sqrt{c}),$$

and

$$F_2 = \mathfrak{G}(k_2/R), \quad k_2 = R(\sqrt{2}, i).$$

Let $G = G_1 \times F_2$, where

$$G_1 = \{\alpha, \beta;\ \alpha^8 = 1, \beta^2 = \alpha^4,\ \beta^{-1}\alpha\beta = \alpha^7\}$$

and let $\phi : G \longrightarrow \mathfrak{G}(k/R)$ be the homomorphic mapping of the group G on $\mathfrak{G}(k/R)$ given in the following manner: $\phi(\alpha) = 1$, $\phi(\beta)$ is the generating element of the group F_1 and ϕ is an isomorphism on F_2. Then we shall have the imbedding problem $(k/R, G, \phi)$ with the cyclic normal subgroup, of 8th order, generated by the element α. The field k contains 8th roots of unity since $\sqrt{2} \in k$ and $i \in k$.

The necessary and sufficient condition for the vanishing of the first obstruction for such a problem is the decomposibility of the cyclic algebra

$$C_0 = (-1, -1)_{R(\sqrt{-c}, \sqrt{2})}\,.$$

The decomposibility of the algebra C_0 signifies that in the field $R(\sqrt{-c}, \sqrt{2})$ the equation

$$x^2 + y^2 + z^2 = 0$$

is solvable.

The field k' will in the present case be equal to $R(\sqrt{-c}, \sqrt{2})$ wherein

$$k_{\chi_1} = R(\sqrt{-c}).$$

In order that in the field $R(\sqrt{-c}, \sqrt{2})$ a three-dimensional cocycle may not vanish (so that 2 may be a root number) the following condition should be fulfilled: if $-c \not\approx 1$ in R_p, then either $2 \approx 1$ or $-2c \approx 1$ in R_p. For the problem which we are now considering the algebras C_{χ_1} and C_{χ_2} obtained after the first step (construction of the field $k^1 = R(\sqrt{\mu}, \sqrt{c}, \sqrt{2}, i)$, where μ is an arbitrary number in R) will have the form

$$C_{\chi_1} \approx (\mu, -c)_R;$$

$$C_{\chi_2} \approx (2, \mu)_{R\sqrt{-c}} \otimes (-1, -1)_{R(\sqrt{-c})}\,.$$

Hence the root function $I(2)$ will be expressed in the following form:

$$I(2) = \prod_p \left(\frac{\mu, -c}{p}\right)_R^{i_p} \prod_p \left(\frac{2, \mu}{p}\right)_{R(\sqrt{-c})}^{j_p} \prod_p \left(\frac{-1, -1}{p}\right)_{R\sqrt{-c})}^{j_p}$$

We know that the root function does not depend on the number μ, therefore in the expression $I(2)$ we may take $\mu = 1$. Furthermore

$$\left(\frac{-1, -1}{p}\right)_{R(\sqrt{-c})} = 1$$

for all $p \neq 2$. Therefore

$$I(2) = \prod_p \left(\frac{1, -c}{p}\right)_R^{i_p} \prod_p \left(\frac{2,1}{p}\right)_{R(\sqrt{-c})}^{j_p} \left(\frac{-1, -1}{2}\right)_{R(\sqrt{-c})}^{j_2} = \left(\frac{-1, -1}{2}\right)_{R(\sqrt{-c})}^{j_2}$$

In order that the second obstruction not vanish it is necessary and sufficient that the following conditions should be fulfilled:

$$\left(\frac{-1,-1}{2}\right)_{R(\sqrt{-c})} = -1 \text{ and } j_2 = 1.$$

We will show that if for c we take any product of prime numbers of the form $16n+7$ all the above mentioned conditions will be fulfilled for the problem $(k/R, G, \phi)$.

For such a problem the first obstruction vanishes since $-c \equiv 1 \pmod 8$, $-c$ is a square in R_2, and the equation $x^2+y^2+z^2=0$ is solvable in $R_2(\sqrt{2})$. A solution of the last equation will be, for example,

$$x = -1+\sqrt{-c}+(1+(\sqrt{-c})\sqrt{2},$$
$$y = -1-\sqrt{-c}+(-1+\sqrt{-c})\sqrt{2},$$
$$z = \sqrt{6(c-1)}.$$

It is easy to see that the root of the number $6(c-1)$ can be extracted.

Furthermore, the following condition will also be fulfilled, namely, that the degree

$$[R_p(\sqrt{-c}, \sqrt{2}) : R_p] < 4$$

for every prime number p. In fact, if $p=2$, then $-c \approx 1$ in R_2; if $p \mid c$, then $(2/p) = 1$ and $2 \approx 1$ in R_p; if $p \neq 2$ does not divide c and

$$\left(\frac{2}{p}\right) = -1, \quad \left(\frac{-c}{p}\right) = -1,$$

then

$$\left(\frac{-2c}{p}\right) = \left(\frac{2}{p}\right)\left(\frac{-c}{p}\right) = 1$$

and $-2c \approx 1$ in R_p.

At the same time the equation $x^2+y^2+z^2=0$ is not solvable in R_2. This means that

$$\left(\frac{-1,-1}{2}\right)_{R_2} = -1.$$

Furthermore, $j_2 = 1$, since $2 \not\approx 1$ in R_2 and $-c \approx 1$ in R_2.

In this manner, for the imbedding problem $(R(\sqrt{c}, \sqrt{2}, i)/R, G, \phi)$ under the condition that c has only prime factors of the form $16n+7$ the second obstruction does not vanish. Consequently this imbedding problem is not solvable.

BIBLIOGRAPHY

[1] S. P. Demuškin and I. R. Šafarevič, *The imbedding problem for local fields*, Izv. Akad. Nauk SSSR Ser. Mat. 23 (1959), 823–840; English transl., Amer. Math. Soc. Transl. (2) 27 (1963), 267–288. MR 22 #1566.

[2] H. Weyl, *Algebraic theory of numbers*, Annals. of Math. Studies, no. 1, Princeton Univ. Press, Princeton, N.J., 1940; Russian transl., Moscow, 1947. MR 2, 37.

[3] R. Kochendörffer, *Zwei Reduktionssätze zum Einbettungsproblem für abelsche Algebren*, Math. Nachr. 10 (1953), 75–84. MR 15, 282.

[4] B. N. Delone and D. K. Faddeev, *Investigations of the geometry of the Galois theory*, Mat. Sb. 15 (57) (1944), 243–284. MR 6, 200.

[5] D. K. Faddeev, *On a hypothesis of Hasse*, Dokl. Akad. Nauk SSSR 94 (1954), 1013–1016. MR 15, 938.

Translated by:

T. S. Bhanu Murthy

Algebraic number fields

Proc. Int. Congr. Math., Stockholm 1962, Inst. Mittag-Leffler, Djursholm, 163–176 (1963).
Zbl. **126**, 69
[Transl., II. Ser., Am. Math. Soc. **31**, 25–39 (1963)]

Introduction

Algebraic number theory has two aspects; one aspect concerns the construction of a general theory of algebraic number fields, its object being to obtain complete classification of these fields and the description of their arithmetics; the other aspect is connected with applications of this general theory to such concrete questions as, for example, diophantine equations, complex multiplication of elliptic and Abelian functions, and integration of algebraic differentials.

The present paper is concerned with some problems covered by the first aspect of algebraic number theory.

A model for all theories developed here is the class field theory which exhaustively describes all Abelian (namely, with commutative Galois group) extensions of a fixed algebraic number field. Our knowledge of the domain of such extensions with non-Abelian Galois groups is far from complete. Often it is not even clear how questions should be stated, and in what terms their solutions should be found. Sometimes it is easier to formulate questions if we consider this domain from a more general point of view.

Algebraic number fields and rings of algebraic integers belonging to them are the simplest examples; at the same time, they are starting points for much more general theories. Such connections are of special importance, since they help us to find new formulations of questions by comparing the latter theory with similar ones in which we can use analytical apparatus or geometrical intuition. This explains why we shall often be concerned later with questions which, strictly speaking, do not belong to the theory of algebraic number fields.

§ 1. Fields of algebraic functions

An essential part of the development of the theory of algebraic numbers has been played by its connections and similarity with the theory of algebraic functions of one variable. Let k be a field of algebraic functions of one variable with field of constants k_0. We are interested in the problem of classification of finite extensions of the field k. If $k_0 = C$ is the field of complex numbers, the solution of this problem is well known. In this case the field k may be associated with its Riemann surface $\mathfrak{R}$. If k' is a finite extension of k, then its Riemann surface $\mathfrak{R}'$ is continuously mapped in the natural way onto $\mathfrak{R}$: $\mathfrak{R}' \to \mathfrak{R}$. Thus, surface $\mathfrak{R}'$ is a covering of surface $\mathfrak{R}$. This covering is of the same degree as the extension k'/k and has a finite number of ramification points. It follows easily from the Riemann existence theorem that in this way we may obtain a one-to-one correspondence between finite extensions of the field k and the ramified coverings of finite degree of its Riemannian surface $\mathfrak{R}$. Thus the problem is reduced to a topological question, the answer to which is very well known. Let us fix a finite set $S = \{\mathfrak{p}_1, \ldots, \mathfrak{p}_m\}$ of points of surface $\mathfrak{R}$ and consider the coverings of a finite degree $\mathfrak{R}' \to \mathfrak{R}$ with ramification points only at the points of the above set. They correspond to nonramified coverings of surface $\mathfrak{R} - S$, which in turn are in a one-to-one correspondence with subgroups of a finite index of the fundamental group $\pi_1(\mathfrak{R} - S)$ of this surface. Finally, group $\pi_1(\mathfrak{R} - S)$ is well known; it has $m + 2g$ generators $u_1, \ldots, u_m, s_1, t_1, \ldots, s_g, t_g$ (where g is the genus of surface $\mathfrak{R}$) connected by one relation

$$u_1 \ldots u_m s_1 t_1 s_1^{-1} t_1^{-1} \ldots s_g t_g s_g^{-1} t_g^{-1} = 1. \tag{1}$$

We may note that although the source of these results is analytical, they may be given an algebraic interpretation. In fact, the points of the Riemann surface $\mathfrak{R}$ correspond to the prime divisors of field k and, in particular, ramification points of the covering $\mathfrak{R}' \to \mathfrak{R}$ may be characterised in terms of extensions k'/k as critical prime divisors (divisors of the discriminant). The group $\pi_1(\mathfrak{R} - S)$ may also be described in algebraic terms. For this purpose, let us consider the field $\mathfrak{K}_S$ which is the union of all extensions k'/k whose critical prime divisors belong to a set S. This extension is infinite and its Galois group $\mathfrak{G}_S$ is a topological group. Therefore $\mathfrak{G}_S$ may be defined for an arbitrary field of algebraic functions. Knowledge of it makes possible the enumeration of all extensions k'/k with the above property since, by the principal Galois theorem, they correspond to subgroups of a finite index. If $k_0 = C$, then $\mathfrak{G}_S$ is a complement of the group $\pi_1(\mathfrak{R} - S)$ in the topology defined by all subgroups of finite index. Thus relation (1) defines $\mathfrak{G}_S$ as a topological group.

It is known that those fields of algebraic functions whose field k_0 is finite are nearest to algebraic number fields. Therefore, from the point of view of algebraic number fields, it would be interesting to obtain for this class of fields a result similar to the one formulated above. The solution, however, is known only for the intermediate case, when the field of constants k_0 is algebraically closed.

If the characteristic of field k_0 is zero, we may apply the well-known Lefschetz principle. The problem is then easily reduced to the case where the field of coefficients k_0 has a finite transcendence degree over the field of rational numbers, and by imbedding such a field into the field of complex numbers, we reduce it to the case $k_0 = C$ considered above. Consequently, for any field of algebraic functions with an algebraically closed field of constants k_0 of characteristic 0, $\mathfrak{G}_S$ is the same as in the case where $k_0 = C$.

For an algebraically closed field k_0 of characteristic $p > 0$, a similar result is obtained by A. Grothendieck [1] for extensions with degree relatively prime to p. In order to describe this result, let us construct the field $\mathfrak{L}_S$ in the same way as we constructed the field $\mathfrak{K}_S$ above, but restricting ourselves to normal fields whose degrees are relatively prime to p. We find that the Galois group $\mathfrak{H}_S$ of the field $\mathfrak{L}_S$ is again the same as in the case $k_0 = C$. The proof is based on reduction of these fields modulo p and consequently on such a deep analytical fact as the Riemann existence theorem.

Now let us return to the case of a finite field of constants k_0 which is of special interest to us. Denote by $\bar{k}_0$ the algebraic closure of the field k_0 and by $\bar{k}$ the composite of k and $\bar{k}_0$. Clearly, $k \subset \bar{k} \subset \mathfrak{L}_S$. The Galois group $\bar{\mathfrak{H}}_S$ of the field $\mathfrak{L}_S/\bar{k}$ is described by Grothendieck's theorem, and the well-known Galois group $\mathfrak{F}$ of the field $\bar{k}/k$ is the free topological group with one generator. Since $\mathfrak{H}_S/\bar{\mathfrak{H}}_S = \mathfrak{F}$, it follows from the definition of $\mathfrak{H}_S$ that this is the problem of group extension. Since the group $\mathfrak{F}$ is free, the corresponding extension splits and consequently the structure of group $\mathfrak{H}_S$ is defined by the way in which the generator of $\mathfrak{F}$ acts on group $\bar{\mathfrak{H}}_S$ (the so-called Frobenius automorphism). Unfortunately, nothing is known about this, even for the simplest cases (e.g., in the case of a p-extension).

§ 2. Local fields

Before we investigate algebraic number fields, it is natural to consider local fields, i.e., completions of algebraic number fields with respect to their prime divisors.

Let k be a local field containing the field of p-adic numbers R_p. By the ramification theory of Hilbert-Dedekind, every normal extension K/k contains a maximal tamely ramified subfield k_1/k. The degree of the field K/k_1 is a power of p. Thus, every extension of a local field is in a sense constructed from a tamely ramified field and a p-extension. The theory of tamely ramified fields is quite simple, since it follows from Dedekind-Hilbert theory. Therefore we may assume that the main difficulty in the classification of local fields lies in the classification of p-extensions.

At present, there exists reasonably complete information on p-extensions of a local field. In order to describe them, let us consider, as for the above case of the field of algebraic functions, a field $\mathfrak{K}$ which is the union of all finite p-extensions of a local field k. Denote by $\mathfrak{G}$ the Galois group of the extension $\mathfrak{K}/k$. The group $\mathfrak{G}$ is a topological p-group and contains reasonably complete information on p-extensions of the field k, since they are in one-to-one correspondence with the subgroups of finite index of group $\mathfrak{G}$. Thus, the main efforts have been directed to the computation of this group.

Denote by v the degree of $[k\colon R_p]$. The structure of the group $\mathfrak{G}$ depends on this number, and on whether the field k contains a primitive pth root of unity ζ.

The following results hold:

1. *If $\zeta \notin k$, then the group $\mathfrak{G}$ is a free topological p-group with $v+1$ generators* [2].

2. *If $\zeta \in k$, then the group $\mathfrak{G}$ has $v+2$ generators and is defined (as a topological p-group) by one defining relation* ([3] *and* [4]).

In these results the statement on the number of generators easily follows from Burnside's theorem [5].

The proof of the statement on the number of relations is based on arguments which we shall consider here for a more general situation.

Let k be an arbitrary field, and let $\mathfrak{K}$ be its p-extension (finite or not) with Galois group $\mathfrak{G}$. Clearly, the number r of independent relations connecting the minimal system of generators of group $\mathfrak{G}$ is equal to the number of generators of its two-dimensional homology group

$$r = \dim_{Z_p} H^2(\mathfrak{G}, Z_p),$$

where Z_p is a group of order p.

Let K_α be a family of subfields of $\mathfrak{K}$, finite over k, and let G_α be their Galois groups. The following homomorphisms are defined:

$$j_\alpha\colon\ H^2(G_\alpha, Z_p) \to H^2(\mathfrak{G}, Z_p)$$

and $H^2(\mathfrak{G}, Z_p)$ is a union of groups $j_\alpha H^2(G_\alpha, Z_p)$. Every element $\xi \in H^2(G_\alpha, Z_p)$ defines an extension G_α^ξ of the group G_α by means of Z_p. Clearly, $j_\alpha \xi = 0$ if and only if K_α can be imbedded into the field $K' \subset \mathfrak{K}$ which has Galois group G_α^ξ. This enables us to define the group $H^2(\mathfrak{G}, Z_p)$ (and consequently also the number of relations r in group $\mathfrak{G}$) if we can solve the corresponding problem of imbedding. We may roughly say that r is equal to the number of conditions which must be satisfied in order for the imbedding problem to be solvable. In particular, Result 2 follows immediately from these arguments, since the condition for solubility of the imbedding problem is, by the well-known Brauer criterion [6], that some element of group $B(k)_p$ is equal to 1, where $B(k)$ is the Brauer group of the field k, and $B(k)_p$ is its subgroup consisting of elements of period p. It is known that for every n group $B(k)_n$ has one generator. Therefore, $r = 1$.

Result 1 describes the group $\mathfrak{G}$ completely. In Result 2 a relation remains to be found which will define the group $\mathfrak{G}$. This has been achieved by S. P. Demuškin [7]. It turns out that if μ is the greatest number such that the field k contains a primitive p^μth root of unity, the relation (for $p \neq 2$, or $p = 2, \mu \neq 1$) is

$$s_1^{p^\mu}(s_1, s_2)(s_3, s_4)\ldots(s_{\nu+1}, s_{\nu+2}) = 1, \qquad (2)$$

where (s, t) denotes $s t s^{-1} t^{-1}$. The case where $p = 2$, $\mu = 1$ (which occurs, for example, if $k = R_2$) has, as yet, not been investigated.

This result has an essentially group-theoretical character and is connected with an interesting question, namely, the study of topological p-groups defined by one relation. A relation f defining a p-group G is called complete if it cannot be transformed by a suitable change of generators into a relation which does not contain one of the generators. This requirement is equivalent to the requirement that group G cannot be decomposed into a free product. A relation f is called m-complete if it also holds for the group $G/G^{(m)}$, where $G^{(m)}$ is the mth term of the central chain of G.

For the group $\mathfrak{G}$ (the Galois group of maximal p-extension of a local field k) it is not difficult to evaluate the group $\mathfrak{G}/\mathfrak{G}^{(2)}$ and to verify that the relation defining $\mathfrak{G}$ is 2-complete. Demuškin has proved for an arbitrary p-group G, defined by one relation f (restricting ourselves, however, to groups with an even number of generators), that G is uniquely defined by $G/G^{(2)}$ if f is 2-complete. From this, the classification of all relations of such a type follows easily, since they can always be reduced to type (2) with some μ. The fact that the number μ has the above meaning for $\mathfrak{G}$ follows from the study of $G/G^{(1)}$.

It would be very interesting to continue these studies and discover how these results may be generalised to the case of m-complete relations for an arbitrary m.

Let us point out one group-theoretical corollary of this theorem on local fields, namely, that a subgroup of finite index of a p-group defined by relation (2) is itself defined by a similar relation (possibly, with different μ and ν). Naturally the question arises whether this fact can be generalised for every p-group with one relation, and whether in this case we can give some meaning to the words "relations of the same type".

It seems natural to expect that the study of any finite extensions of a local field may be reduced to the study of tamely ramified extensions and p-extensions. This problem is equivalent to computation of the Galois group of an algebraic closure of a local field. This problem is as yet unsolved, although H. Koch (see [8] and his dissertation submitted to the Humboldt University, Berlin) has found a number
2 of properties of this group and has proved that it is uniquely defined by simple invariants of field k.

§ 3. Algebraic number fields

In the theory of algebraic number fields it is natural to attempt to use the same approach as the one we have already used in fields of algebraic functions and in local fields. Let k be an algebraic number field which is a finite extension of the field R of rational numbers. Denote by S a finite set of prime divisors of k and consider all the finite extensions k'/k whose critical prime divisors (i.e., divisors of the discriminant) are only the prime divisors of S. Denote the union of all such fields by $\mathfrak{K}_S$ and the Galois group of $\mathfrak{K}_S/k$ by $\mathfrak{G}_S$. Obviously, one of the basic problems of the theory of algebraic numbers is the study of the groups $\mathfrak{G}_S$.

In the same way as for algebraic functional fields the following two questions arise:

Does $\mathfrak{G}_S$ always have a finite number of generators?

If this is so, can we then (for a fixed field k) find a bound for the number of generators, depending only on the number of divisors in set S?

The answers to these questions are not known. A positive answer to the second question would mean that for any finite normal extension K/k the number of generators of its Galois group is bounded, depending on the number of critical prime divisors. This question is interesting even for very simple extensions. For example, let k coincide with the field of rational numbers R. Denote by L/R the quadratic extension with prime discriminant p and let K be the Hilbert class field of L (i.e., its maximal nonramified extension). Clearly, K/R has only one critical prime number p. Since the Galois group of K/L coincides with the group of classes of divisors of L, and the number of generators of the Galois group of K/R can differ from the number of generators of the Galois group of K/L by not more than 1, the question which interests us is equivalent to the following:

Is the number of generators of a group of classes of divisors of a quadratic field with prime discriminant bounded?

Questions of this sort had already interested Gauss, who called a quadratic field regular if its group of classes of divisors was cyclic [9].

In his usual way, Gauss constructed tables [9] from which it follows that for $p < 3000$ and for $9000 < p < 10\,000$ the numbers of generators is equal to 1 or 2, where in the second case one of the generators is of order 3.

This simple example already shows that the study of groups $\mathfrak{G}_S$ leads to great difficulties. It seems that at present there are no general results in this direction. We can say much more if we restrict ourselves to the study of p-extensions. Let a prime p be fixed and denote by $\mathfrak{P}_S$ the Galois group of the p-extensions of k such that their prime divisors are contained in set S. Clearly, $\mathfrak{P}_S$ is a maximal p-quotient group of $\mathfrak{G}_S$.

It is easy to see that both questions, posed for groups $\mathfrak{G}_S$, have positive answers for groups $\mathfrak{P}_S$.

Denote the minimal number of generators of $\mathfrak{P}_S$ by $d(S)$. This number is a rather fine invariant of the field k and the prime divisors forming the set S.

The order of magnitude of $d(S)$ is close to the number of elements of S. More precisely, there exist two constants c_1 and c_2 independent of k such that

$$c_1 \leq d(S) - m \leq c_2 \,,$$

where m denotes the number of prime divisors contained in S for which $\zeta \in k$.

Of especial interest is the case when S is empty. Then we can denote $\mathfrak{P}_S$ simply by $\mathfrak{P}$. This is clearly the Galois group of maximal nonramified p-extension of k. It follows from class field theory that all the factors of a series of commutator-subgroups of $\mathfrak{P}$ are finite. Whether this series stops, i.e., whether the group $\mathfrak{P}$ is finite, is not known. This question constitutes the p-field tower problem.

A closely related case is the one when all the prime divisors of S are relatively prime to p. Here the factors of a series of commutator subgroups are finite. The question of termination of the series is similar to the same question for group $\mathfrak{P}$.

As in the theory of algebraic functional fields, it is natural to attempt to find relations defining the groups $\mathfrak{P}_S$. It is possible to find an upper bound for their number, for example, by using the general method of determining the number of relations in the p-group which is described in § 2. The determination of the number of relations is connected, as we have seen, with the solution of an imbedding problem. In our case it is a problem of the imbedding of a field whose all critical prime divisors are contained in S, into a field of the same type. If we examine this problem we obtain the following result:

The number of relations $r(\mathfrak{P}_S)$ defining the group $\mathfrak{P}_S$ satisfies the inequality

$$r(\mathfrak{P}_S) \leq \gamma + m + c \,,$$

where γ denotes the number of generators of a p-component of the group of classes of k, while m is the number of prime divisors of S for which $\zeta \in k$, and c is a constant depending only on the degree of $[k : R]$.

For example, for group $\mathfrak{P}$, we have

$$r(\mathfrak{P}) \leq \gamma + c \,.$$

Clearly, $d(\mathfrak{P}) \geq \gamma$, so that

3 $$r(\mathfrak{P}) \leq d(\mathfrak{P}) + c \,. \tag{3}$$

From rather rough group-theoretic considerations it follows that if for some family of p-groups G, we have $r(G) \leq d(G) + c$, where c is a constant, then the class of groups G tends to infinity together with $d(G)$. This yields the following result:

For a field k of fixed degree, the class of group $\mathfrak{P}$ grows indefinitely together with the number of generators of a p-component of the group of classes of k.

Scholz [10] has constructed some examples of fields k in which the groups $\mathfrak{P}$ have an arbitrary large class. It follows from the preceding result that this is, in fact, a general law, as was conjectured (in a slightly more general form) by Fröhlich [11].

However, one feels that the number of relations satisfying inequalities (3) for $d(\mathfrak{P})$ sufficiently large comperatively to c is too small to define a finite p-group. In fact, the simplest group, namely, the direct product d of groups Z_p, is defined by $d(d+1)/2$ relations

$$\sigma_i^p = 1, \quad i = 1, \ldots, d; \quad \sigma_i \sigma_j = \sigma_j \sigma_i \quad (1 \leq i < j \leq d).$$

In this connection, I think the following question is of fundamental importance and I would like to draw it to the attention of experts in the theory of groups:

What is the smallest number of relations necessary to define a finite p-group with d generators?

Let us state the question precisely. Define the function $\varrho(d)$ as follows:

$$\varrho(d) = \min_{d(G)=d} r(G),$$

where the minimum is taken over all finite groups with minimal number of generators equal to d. We speak here of finding a lower bound for the function $\varrho(d)$ as d increases. On considering the quotient group by the commutator subgroup, we find that $\varrho(d) \geq d$. If we can show that for $d \to \infty$,

$$\varrho(d) - d \to \infty,$$

which to me seems quite likely, then the negative solution of the problem of the p-field tower follows, with all its consequences: existence of fields which cannot be imbedded into fields with unique factorisation, estimation of the true order of growth of the minimal discriminant of an algebraic number field, etc. Since $r(G) = \dim_{Z_p} H^2(G, Z_p)$, $d(G) = \dim_{Z_p} H^1(G, Z_p)$, this question is equivalent to the problem of estimation of the order of the second homology group of a finite p-group with a fixed first homology group.

In conclusion, let us note that everything said above also applies to groups $\mathfrak{P}_S$ if all the prime divisors of S are relatively prime to p.

§ 4. Critical prime divisors of algebraic varieties over algebraic number fields

Questions connected with critical prime divisors of algebraic number fields may also be posed for algebraic varieties defined over algebraic number fields; this leads to interesting problems and to some results.

Up to now we have not defined a critical prime divisor of an extension k'/k. To formulate this definition let us consider a prime divisor $\mathfrak{p}$ of a field k, its local rings $\sigma_\mathfrak{p}$ and the integral closure $\sigma'_\mathfrak{p}$ of the ring $\sigma_\mathfrak{p}$ in k'. The ring $\sigma'_\mathfrak{p}/\mathfrak{p}$ is an algebra over the field $k(\mathfrak{p}) = \sigma_\mathfrak{p}/\mathfrak{p}$. If this algebra is separable, then $\mathfrak{p}$ is called a noncritical prime divisor, otherwise it is called a critical prime divisor.

Consider now a complete nonsingular algebraic variety X, defined over an algebraic number field k, and a prime divisor $\mathfrak{p}$ of k. The reduction $X/\mathfrak{p}$ of the variety X modulo $\mathfrak{p}$ (whose definition is given, for example, in [12]) is an algebraic set over $k(\mathfrak{p})$. It may happen that $X/\mathfrak{p}$ already has singularities. This means that the local rings of some of its points are not regular. In the case of the ring $\sigma'_{\mathfrak{p}}/\mathfrak{p}$ the condition of regularity of the local rings of its prime ideals coincides with the condition of separability. This leads to the following definition: a prime divisor $\mathfrak{p}$ of the field k is called noncritical for a complete nonsingular variety X defined over k if $X/\mathfrak{p}$ is a nonsingular variety over $k(\mathfrak{p})$; otherwise $\mathfrak{p}$ is called critical. In [12] it is proved that X has only a finite number of critical prime divisors.

For convenience, we associate with the concept of a prime divisor a birational invariant form by connecting it with a field of algebraic functions instead of the variety which is the X-model of this field. Let K be the field of rational functions over a complete nonsingular algebraic variety X defined over k. A prime divisor $\mathfrak{p}$ of k is called noncritical for K if there exists at least one nonsingular complete variety X' for which K is a field of functions, and $\mathfrak{p}$ is a noncritical prime divisor in the sense of the preceding definition.

One of the main theorems on algebraic numbers connected with the concept of discriminant is Hermite's theorem, which states that the number of extensions k'/k of given degree and given discriminant is finite. This theorem may be formulated as follows: *the number of extensions k'/k of given degree whose critical prime divisors belong to a given finite set S is finite.*

In such a form this statement may be formulated, as a question at least, either for algebraic varieties of any dimension n, or for corresponding fields of algebraic functions. Hermite's theorem corresponds to the case $n=0$. We now consider the case $n=1$ which is next in complexity, i.e., the case of algebraic curves or fields of analytic functions K of one variable. We shall assume that K is defined over the algebraic number field k and that k is algebraically closed in K. With these assumptions the following result, similar to Hermite's theorem, holds:

There exists only a finite number of fields of algebraic functions K/k of a given
5 *genus $g \neq 1$, the critical prime divisors of which belong to a given finite set S.*

This result also holds, with suitable modifications, for fields of genus $g=1$. For this we must, in addition to the genus, consider another invariant of the fields of algebraic functions of genus 1, namely, the greatest common divisor d of powers of all divisors of the field, which is called the exponent of the field. It is known [13] that the exponent may take arbitrarily large values. If we restrict ourselves to the fields of genus 1 with given exponent d, then the above formulated theorem of finiteness is true. For example, for $k=R$ and the simplest case $d=1$ (when K is a field of functions on an elliptic curve) there are 24 fields K with a unique critical prime 2. Corresponding elliptic curves are given by the equations

$$y^2 = x(x^2-a), \quad a = \pm 1, \pm 2, \pm 4, \pm 8,$$
$$y^2 = x(x-a\alpha)(x-a\alpha'), \quad a = \pm 1, \pm 2,$$
$$\alpha = 1+\sqrt{-1}, \quad 1+\sqrt{2}, \quad (1+\sqrt{2})^2, \quad 2+\sqrt{2}.$$

Can we expect to obtain a sufficiently complete classification of fields of algebraic functions of one variable by their sets of critical prime divisors?

The simplest case, when the genus g of K is equal to 0, speaks in favor of such a proposition. In this case, it is well known that the field of functions has as a model a curve given by equation

$$ax^2 + by^2 = 1 .$$

This model is uniquely defined by the field K up to a projective transformation. Classification of fields coincides, consequently, with classification of ternary quadratic forms up to a linear equivalence. It follows from the theory of quadratic forms that for $k = R$ a field of genus 0 is uniquely defined by the set of its critical primes. The set of all fields whose critical primes belong to a given set $S = \{p_1, \ldots, p_m\}$ forms an Abelian group, isomorphic to the direct product of m groups Z_2.

The second fundamental theorem of the theory of algebraic numbers connected with critical prime numbers is Minkowski's theorem, according to which every finite extension K/R of the field of rational numbers R has at least one critical prime. It would be interesting to find out to what extent this result may be generalised to arbitrary algebraic varieties. The question, in the case of fields of algebraic functions with one variable, is the following one:

6 Does there exist a field of algebraic functions of one variable with field of constants R which has no critical primes and is distinct from the field of rational functions $R(x)$?

The answer to this question is not known to me. For extensions of an arbitrary algebraic number field k the answer is negative. Moreover, there exist fields K of algebraic functions of one variable with the field of constants R such that for some algebraic number field k the field $K \cdot k$ has no critical prime divisors over k. An example is the field $R(x, y)$, where x and y are connected by the relation

$$x^3 + y^3 = 1 ,$$

and the field k is

$$R(\sqrt[3]{2}, \sqrt[4]{-3}) .$$

We can obtain finer invariants than the set of critical prime divisors if we turn from considering the fields to considering particular models. For a description of this situation it is convenient to use the language of "schemes" [14]. Let K/k be an extension of finite type of an algebraic number field k, and let X be a scheme over the ring of integers σ of k. Suppose that the scheme X is integral, flat and proper over σ, that K is its field of rational functions, and that the scheme $X \otimes k$ (which is an algebraic variety over k) is nonsingular. The concept of a critical prime divisor $\mathfrak{p}$ of the scheme X is defined, as before, by considering the scheme $X/\mathfrak{p}$. But in this case, we ascribe multiplicities r_i to prime critical divisors $\mathfrak{p}_i$, and define the discriminant of the system $D(X) = \Pi\, \mathfrak{p}_i^{r_i}$. Without pausing over the definitions of r_i, I need only mention that this definition is based on the concept of the different, introduced by E. Kähler [15] (which in these terms is a closed subscheme of the scheme X); similarly the definition of the discriminant of the algebraic number field is based on its different.

If field K/k has transcendence degree 1, i.e., if $K \otimes k$ is an algebraic curve, then the genus of the field K/k is called the genus of the scheme X. The following result holds:

7 *There exists only a finite number of schemes X of a given genus $g \neq 1$ (or of genus 1 and a given exponent d) with a given discriminant.*

The concept of the discriminant of the scheme is connected, seemingly, with other questions of the arithmetics of algebraic varieties. For example, A. Néron [16] has shown that for $g=1$, among all schemes X corresponding to the same field K, there exists one which is minimal in a certain sense. It can be shown that this scheme has a minimal discriminant. If, as we might expect, the theorem of existence of the minimal model holds for the case of an arbitrary genus $g>0$, this minimal model also has a minimal discriminant.

§ 5. Comparison with algebraic geometry

In the same way as for the theory of algebraic numbers, problems on algebraic curves over an algebraic number field k may be compared with similar problems for the case when k is a field of algebraic functions of one variable.

Let k be a field of algebraic functions of one variable with algebraically closed field of coefficients k_0, which is the field of rational functions over a complete algebraic curve T. As in § 4, let us consider a field K of algebraic functions of one variable with field of coefficients k.

Since K has transcendence degree 2 over k_0, it is a field of rational functions on the algebraic surface X, where the surface may be assumed to be nonsingular. Imbedding the field k into K defines a rational mapping $f\colon X \to T$. Using well-known methods of the theory of algebraic surfaces, we may reduce the problem to the case when f is regular. Thus, an analogue of the situation discussed in § 4 is a regular mapping of a nonsingular algebraic surface X on a nonsingular algebraic curve T. For all the points t of the curve T, apart possibly from a finite number, the inverse-image $f^{-1}(t)$ is a nonsingular algebraic curve of the same genus g, equal to the genus of the field K/k. Those points whose inverse images are curves with singularities are analogues of critical prime divisors.

We shall call them critical points.

In other words, the surface X is a fibre bundle of algebraic curves $f^{-1}(t)$. Critical points correspond to degenerate fibres which are curves with singular points.

Consider now fields without critical points.

To such fields correspond fibre bundles without degenerate fibres. In our case such fibre bundles really exist. An example is the direct product $X = F \times T$ of the basic curve T and an arbitrary curve F. A more general class of examples provides "constant" fibre bundles in which all the fibres are biregularly equivalent to each other. These fibre bundles are algebraic fibre spaces, locally trivial for $k_0 = C$ and locally isotrivial [17] for an arbitrary k_0. The problem concerning fields without critical points now takes the following form:

Does every fibre bundle without degenerate fibres, whose base is a projective straight line, coincide with the "constant" fibre bundle?

This question may be put into a different form if we consider the variety of modules M_g of algebraic curves of genus g. By associating with the fibre $f^{-1}(t)$ $(t \in T)$ the point which is the mapping of this curve into the variety M_g we define a regular mapping of the curve T into M_g. The question consists in finding out whether there are such mappings other than mappings into a point. A priori there are no such mappings if the variety M_g is affine, as is the case when $g = 2$.*

Let us consider the fibre bundles containing degenerate fibres. Here it is convenient (and possible without loss of generality) to restrict ourselves to minimal fibre bundles, i.e., fibre bundles $X \xrightarrow{f} T$ for which no fibre bundle $X' \xrightarrow{f'} T$ exist such that there is a regular birational mapping $g: X \to X'$ for which the triangle

$$\begin{array}{ccc} X & \xrightarrow{g} & X' \\ & {\scriptstyle f}\searrow \quad \swarrow{\scriptstyle f'} & \\ & T & \end{array}$$

is commutative.

Problems on fields with preassigned critical points may be given the following form:

8 Is the number of fibre bundles of curves of a given genus $g \neq 1$ finite if the basic curve T and the set S of all critical points of the fibre bundle is fixed?

In the number of particular cases (for example, if all the fibres are hyperelliptic) it can be shown that this question has a positive answer. It seems that the proof of the general case must be much more difficult than proving a similar fact for curves over algebraic number fields, just as the fact that the finiteness of the number of extensions with given ramification points of a field of algebraic functions requires a much more complicated proof [18] than Hermite's theorem in the theory of algebraic numbers. In this connection, it would be of interest to consider this question in the case $k_0 = C$ by using topological methods.

In §1 we have seen that for $k_0 = C$ the classification of finite extensions of a field of algebraic functions may be reduced to the topological classification of coverings.

Connections which exist between the theory of regular mappings of algebraic surfaces onto algebraic curves and a corresponding topological problem, namely, the theory of differentiable mappings of four-dimensional manifolds onto two-dimensional ones have seemingly not been investigated at all. Here three questions arise:

What is the classification of the differentiable mappings of four-dimensional manifolds onto those that are two-dimensional?

What differentiable mappings of a four-dimensional manifold onto an algebraic curve are realised as regular mappings of algebraic surfaces onto this curve?

Can two distinct regular mappings of an algebraic surface onto an algebraic curve be diffeomorphic?

The last question is similar to one solved positively by H. Grauert [19] for fibre bundles with vector fibres over Stein manifolds.

---

* D. Mumford has informed me that K. Kodaira has constructed an example of a "nonconstant" fibre bundle without degenerate fibres with base distinct from the projective line.

Conclusion

The present report is by no means aimed at giving a complete review of recent results in the theory of algebraic numbers. Here I have considered only one set of ideas in the theory of algebraic number fields and have assembled certain closely related results and problems in the theory of algebraic numbers and adjoining domains.

Bibliography

1. Grothendieck, A.: Géométrie formelle et géométrie algébrique. Séminaire Bourbaki (May, 1959) *182*, 1–28
2. Šafarevič, I. R.: On *p*-extensions. Mat. Sb. (N.S.) *20* (62) (1947), 351–363 (Russian)
3. Kawada, J.: On the structure of the Galois group of some infinite extensions. I, J. Fac. Sci. Univ. Tokyo Sect I *7* (1954), 1–18
4. Faddeev, D. K., Skopin, A. I.: Proof of a theorem of Kawada. Dokl. Akad. Nauk SSSR *127* (1959), 529–530 (Russian)
5. Zassenhaus, H.: Lehrbuch der Gruppentheorie. Teubner, Leipzig 1937
6. Brauer, R.: Über die Konstruktion der Schiefkörper, die von endlichem Rang in Bezug auf ein gegebenes Zentrum sind. J. Reine Angew. Math. *168* (1932), 44–64
7. Demuškin, S. P.: Group of maximal *p*-extension of a local field. Izv. Akad. Nauk SSSR Ser. Mat. *25* (1961), 329–346 (Russian)
8. Koch, H.: Über galoissche Erweiterungen *p*-adischer Zahlkörper. J. Reine Angew. Math. *209* (1962), 8–11
9. Gauss, K. F.: Works on the theory of numbers. Izdat. Akad. Nauk SSSR, Moscow 1959 (Russian)
10. Scholz, A.: Zwei Bemerkungen zum Klassenkörpertum. J. Reine Angew. Math. *161* (1929), 201–208
11. Fröhlich, A.: A note on the class field tower. Quart. J. Math. Oxford Ser. (2) *5* (1954), 141–144
12. Shimura, G.: Reduction of algebraic varieties with respect to a discrete valuation of the basic field. Amer. J. Math. *77* (1955), 134–176
13. Šafarevič, I. R.: Exponents of elliptic curves. Dokl. Akad. Nauk SSSR *114* (1957), 714–716 (Russian)
14. Grothendieck, A., Dieudonné, J.: Eléments de géométrie algébrique. I, Publ. Math. (I.H.E.S.), Paris, 8
15. Kähler, E.: Geometria arithmetica. Ann. Mat. Pura Appl. (4) *45* (1958)
16. Néron, A.: Modèles *p*-minimaux des variétés abéliennes. Séminaire Bourbaki (December, 1961) *227*, 1–16
17. Serre, J.-P.: Espaces fibrés algébriques. Séminaire C. Chevalley, École Norm. Sup. Exp. *1* (1958), 1–37
18. Lang, S., Serre, J.-P.: Sur les revêtements non ramifiés des variétés algébriques. Amer. J. Math. *79* (1957), 319–330
19. Grauert, H.: Analytische Faserungen über holomorph-vollständigen Räumen. Math. Ann. (2) *135* (1958), 263–273

Translated by Helen Alderson

Extensions with given ramification points

Publ. Math., Inst. Haut. Etud. Sci. **18**, 295–319 (1964). Zbl. **118**, 275
[Transl., II. Ser., Am. Math. Soc. **59**, 128–149 (1966)]

In this paper we investigate algebraic extensions K/k of a field k of algebraic numbers with given points of ramification. Such a formulation of the problem emphasizes the analogy with the theory of Riemann surfaces. Extensions with commutative galois group are considered in class field theory. Here we shall consider extensions whose galois groups are l-groups, i.e. have order of the form l^a where l is some fixed prime.

We shall fix a finite set S of prime divisors of k and consider the maximal l-extension K_S which is ramified only at the prime divisors $\mathfrak{p} \in S$. The galois group of K_S/k will be denoted by $\mathfrak{G}_S$. It is either a finite or a topological l-group. In §1 we determine the minimal number of generators of $\mathfrak{G}_S$. Our main aim is the determination of the mimimum number of relations between the generators. In §2 it is shown that this question is bound up with the conditions for the solvability of certain embedding problems of an arithmetical character. In §4 we give an estimate for the number of relations by considering arithmetical invariants which are introduced in §3. On the basis of these estimates we succeed in some cases in finding explicitly the form of $\mathfrak{G}_S$.

If none of the prime divisors of S divides l, then it is shown that the difference between the number of relations and the number of generators of $\mathfrak{G}_S$ does not exceed the number of generators of the unit group of k. In any case, the difference is bounded, if one considers only extensions of bounded degree over the field R of rational numbers.

The question whether $\mathfrak{G}_S$ is finite (if none of the prime divisors of S divides l) turns out to be very interesting. It follows from class field theory that the factors of the series of commutator groups of $\mathfrak{G}_S$ are finite. Hence the question of finiteness is equivalent to the question whether this series breaks off. If the set S is empty, this question is the same as the well-known problem of the class field tower, and in the general case is a natural generalization of it. In connection with the results which were discussed above, this question leads to new questions about finite l-groups. These questions are discussed in §6. In any case it will be shown that the length of the chain of commutator groups of $\mathfrak{G}_S$ grows indefinitely if the number of generators tends to infinity. In particular, the number of stories of the class field tower of a field k grows indefinitely with the number of classes of k, if we consider fields of bounded degree over R.

In conclusion, I should like to express my thanks to the referee, who read the manuscript of this paper very attentively and made a number of useful remarks.

§1. Number of generators

In this section we shall deduce a formula for the number of generators $d(S)$ of the galois group $\mathfrak{G}_S$ of K_S/k, the maximal l-extension of the field k of algebraic numbers which is ramified only at prime divisors of the set $S = \{\mathfrak{p}_1, \cdots, \mathfrak{p}_s\}$.

We shall suppose that S does not contain any complex infinite divisors. This is not an essential restriction since any complex prime is unramified.

By Burnside's theorem [1], $d(S)$ is the minimum number of generators of $\mathfrak{G}_S/\mathfrak{G}_S^{(1)}$, where $\mathfrak{G}_S^{(1)} = \mathfrak{G}_S^l\,(\mathfrak{G}_S, \mathfrak{G}_S)$. The subgroup $\mathfrak{G}_S^{(1)}$ corresponds to the subfield $K_S^{(1)} \subset K_S$ which is, clearly, the maximal abelian extension with period l of the field k, which is ramified only at prime divisors $\mathfrak{p} \in S$. Hence the determination of $d(S)$ reduces to a question about abelian extensions. This question can be solved by classical arguments using class field theory.

We shall use the formulation of class field theory in terms of ideles [2].

The following is the basic notation which we shall use (chosen, as far as possible, to coincide with [2]).

J (or J_k) is the group of ideles of k.

$\alpha_\mathfrak{p}$ is the $\mathfrak{p}$-component of the idele α.

k is the group of principal ideles and also the multiplicative group of k.

$\mathfrak{U}_\mathfrak{p}$ is the group of units of the $\mathfrak{p}$-adic completion $k_\mathfrak{p}$ of k.

$$\mathfrak{U}_S = \{\alpha \mid \alpha_q \in \mathfrak{U}_q,\ \alpha_\mathfrak{p} = 1 \text{ for } \mathfrak{p} \in S\}.$$

If S is empty, then $\mathfrak{U}_S$ is denoted by $\mathfrak{U}$: it is the group of unit ideles.

$$V_S = \{\alpha \in k \mid (\alpha) = \mathfrak{a}^l,\ \alpha \in k_\mathfrak{p}^l \text{ for } \mathfrak{p} \in S\}.$$

(α) is the principal divisor of the number α.

Clearly $V_S = \mathfrak{U}_S J^l \cap k$.

If S is empty, then V_S is denoted by V: it is the group of singular numbers [3].

Clearly $V_S \supset k^l$ and V_S/k^l is a finite group of period l, i.e. a finite-dimensional vector space over the field Z_l of l elements. We shall write

$$\sigma(S) = \dim_{Z_l}(V_S/k^l).$$

r is the number of generators of infinite order of the unit group of k:

$r = r_1 + r_2 - 1$, where r_1 is the number of real and r_2 the number of complex infinite divisors of k.

γ is the rank of the l-component of the divisor class group of k.

ζ is a primitive lth root of unity.

Theorem 1. *The number of generators* $d(S)$ *of* $\mathfrak{G}_S$ *is given by*

$$d(S) = t(S) + \lambda(S) + \sigma(S) - r - \delta, \tag{1}$$

where $t(S)$ *is the number of* $\mathfrak{p} \in S$ *for which* $\zeta \in k_{\mathfrak{p}}$,

$$\lambda(S) = \sum_{\mathfrak{p}|l, \mathfrak{p} \in S} n(\mathfrak{p}), \quad n(\mathfrak{p}) = [k_{\mathfrak{p}} : R_p], \quad \mathfrak{p}|p$$

and δ *is equal to* 0 *if* $\zeta \notin k$ *and to* 1 *if* $\zeta \in k$.

Proof. Let K be an abelian extension of k and B the corresponding subgroup of J. A prime divisor $\mathfrak{p}$ is unramified in K precisely when $\mathfrak{U}_{\mathfrak{p}} \subset B$ ([2], Theorem 3, Chapter 8). It follows easily that $K_S^{(1)}$ corresponds to $\mathfrak{U}_S J^l k$ and

$$d(S) = \dim_{Z_l} J/\mathfrak{U}_S \cdot J^l \cdot k.$$

We shall write

$$H_S = J/\mathfrak{U}_S \cdot J^l \cdot k$$

and if S is empty denote H_S by H.

Let us consider the sequence

$$(1) \longrightarrow V_S/k^l \xrightarrow{f_4} V/k^l \xrightarrow{f_3} \mathfrak{U}/\mathfrak{U}_S \mathfrak{U}^l \xrightarrow{f_2} H_S \xrightarrow{f_1} H \longrightarrow (1) \tag{2}$$

of groups where the homomorphisms f_i $(i = 1, 2, 3, 4)$ are defined as follows:

f_1 is the natural map of the group onto the factorgroup $H = H_S/(\mathfrak{U} \cdot J^l \cdot k/\mathfrak{U}_S J^l k)$.

$f_2(\mathfrak{a}) = \mathfrak{a}\mathfrak{U}_S J^l k$, $\mathfrak{a} \in \mathfrak{U}$.

f_4 is the natural injection, since clearly $V_S \subset V$.

To define f_3 we note that the number α belongs to V_S precisely when there is an idele $\mathfrak{a}$ such that $\alpha \mathfrak{a}^{-l} \in \mathfrak{U}$. Put

$$f_3(\alpha) = \alpha \mathfrak{a}^{-l} \mathfrak{U}_S \mathfrak{U}^l$$

for an arbitrary choice of such an $\mathfrak{a}$. Then $f_3(\alpha)$ does not depend on this choice since if $\alpha \mathfrak{a}_1^{-l} \in \mathfrak{U}$, then $\alpha \mathfrak{a}_1^{-l} = \alpha \mathfrak{a}^{-l} \mathfrak{i}$, where $\mathfrak{i} = (\mathfrak{a}_1/\mathfrak{a})^l \in \mathfrak{U}^l$.

We shall check now that the sequence (2) is exact.

For first $\operatorname{Im} f_1 = H$ and then, as $H = J/J\mathfrak{U}^l k$, we have

$$\operatorname{Ker} f_1 = \mathfrak{U} \cdot J^l \cdot k/\mathfrak{U}_S \cdot J^l \cdot k = \operatorname{Im} f_2$$

$$\operatorname{Ker} f_2 = \{\mathfrak{i}\mathfrak{U}_S\mathfrak{U}^l | \mathfrak{i} \in \mathfrak{U},\ \mathfrak{i} = \mathfrak{a}^l \mathfrak{b}_S \alpha,\ \mathfrak{a} \in J,\ \mathfrak{b}_S \in \mathfrak{U}_S,\ \alpha \in k\}.$$

As it follows from this that $\alpha \in V$, we have

$$\operatorname{Ker} f_2 = \operatorname{Im} f_3.$$

Finally

$$\operatorname{Ker} f_3 = \{\alpha | \alpha \in V,\ \alpha \in J^l \mathfrak{U}_S\} = \operatorname{Im} f_4$$

and

$$\operatorname{Ker} f_4 = 1.$$

The exactness of (2) implies

$$\dim_{Z_l} H - \dim_{Z_l} H_S + \dim_{Z_l} \mathfrak{U}/\mathfrak{U}_S\mathfrak{U}^l - \dim_{Z_l} V/k^l + \dim_{Z_l} V_S/k^l = 0, \tag{3}$$

where by definition

$$\dim_{Z_l} H = \gamma, \ \dim_{Z_l} H_S = d(S), \ \dim_{Z_l} V_S/k^l = \sigma(S). \tag{4}$$

Further

$$\dim_{Z_l} \mathfrak{U}/\mathfrak{U}_S \mathfrak{U}^l = \dim_{Z_l} \prod_{\mathfrak{p}\in S} \mathfrak{U}_\mathfrak{p}/\mathfrak{U}_\mathfrak{p}^l = \sum_{\mathfrak{p}\in S} \dim_{Z_l} \mathfrak{U}_\mathfrak{p}/\mathfrak{U}_\mathfrak{p}^l,$$

where it is known (cf. e.g. [4], Theorem 11.3) that

$$\dim_{Z_l} \mathfrak{U}_\mathfrak{p}/\mathfrak{U}_\mathfrak{p}^l = \begin{cases} 0, & \text{if } \ \mathfrak{p}\nmid l, \zeta\notin k_\mathfrak{p}, \\ 1, & \text{if } \ \mathfrak{p}\nmid l, \zeta\in k_\mathfrak{p}, \\ n(\mathfrak{p}), & \text{if } \ \mathfrak{p}\mid l, \zeta\notin k_\mathfrak{p}, \\ n(\mathfrak{p})+1, & \text{if } \ \mathfrak{p}\mid l, \zeta\in k_\mathfrak{p}. \end{cases} \tag{5}$$

Hence

$$\sum_{\mathfrak{p}\in S} \dim_{Z_l} \mathfrak{U}_\mathfrak{p}/\mathfrak{U}_\mathfrak{p}^l = t(S) + \lambda(S).$$

Finally, we shall show that

$$\dim_{Z_l} V/k^l = \gamma + r + \delta. \tag{6}$$

For when $\alpha \in V$, so $(\alpha) = \mathfrak{a}^l$, the map ϕ_1: $\alpha \longrightarrow \mathfrak{a}$ gives an isomorphism of V onto the group H_1 of elements of period l in the group of divisor classes of k. Clearly $\dim_{Z_l} H_1 = \dim_{Z_l} H$. We have an exact sequence

$$(1) \longrightarrow E/E^l \xrightarrow{\phi_2} V/k^l \xrightarrow{\phi_1} H \longrightarrow (1),$$

where E is the group of units of k and $\phi_2(\epsilon) = \epsilon k^l$ for $(\epsilon) \in E$.

Since $\dim_{Z_l} E/E^l = r + \delta$, this implies (6).

On substituting (4)–(6) in (3) we get

$$d(S) = \gamma + t(S) + \lambda(S) - (\gamma + r + \delta) + \sigma(S) = t(S) + \lambda(S) + \sigma(S) - r - \delta.$$

This proves Theorem 1.

§2. The number of relations and the embedding problem

In this section we shall consider an arbitrary field k and an l-extension K of it. Thus K/k is separable and normal and its galois group $\mathfrak{G}$ is an l-group, either a finite or a topological one. We shall expound a general method for computing the minimum number of relations in the group $\mathfrak{G}$ and reduce this problem to a certain embedding problem. In the particular case when k is the field of p-adic numbers, $\zeta \in k$ and K is the maximal l-extension of k an analogous method was applied in [5]. The arguments of this section are a generalization of the arguments there.

Let $\mathfrak{G}$ be a finite or topological zero-dimensional compact l-group. Then $\mathfrak{G}$ is the projective limit of finite l-groups:

$$\mathfrak{G} = \varprojlim G_\alpha. \tag{7}$$

We shall denote by $\mathfrak{F}$ a free topological l-group of which $\mathfrak{G}$ is a homomorphic image:

$$\mathfrak{G} = \mathfrak{F}/\mathfrak{N}. \tag{8}$$

We shall suppose that a minimal system of generators of $\mathfrak{F}$ is mapped into a minimal system for $\mathfrak{G}$. By the theorem of Burnside already quoted this means that the homomorphism $\mathfrak{F} \longrightarrow \mathfrak{G}$ derived from (8) determines an isomorphism of the groups $\mathfrak{F}/\mathfrak{F}_1$ and $\mathfrak{G}/\mathfrak{G}_1$ where $\mathfrak{F}_1 = \mathfrak{F}^l(\mathfrak{F}, \mathfrak{F})$, $\mathfrak{G}_1 = \mathfrak{G}^l(\mathfrak{G}, \mathfrak{G})$. In other words, $\mathfrak{N} \subset \mathfrak{F}_1$.

The minimal number of relations of $\mathfrak{G}$ is the minimal number of generators of the normal subgroup $\mathfrak{N}$ as an $\mathfrak{F}$-operational group.* This is equal to the minimum number of generators of the abelian (nonoperational) group $\mathfrak{N}/(\mathfrak{N}, \mathfrak{F})\mathfrak{N}^l$; the proof of this follows the proof of Burnside's theorem word for word. Hence the minimal number $r(\mathfrak{G})$ (say) of a relation in $\mathfrak{G}$ is given by

$$r(\mathfrak{G}) = \dim_{Z_l} \mathfrak{N}/(\mathfrak{N}, \mathfrak{F})\mathfrak{N}^l.$$

Here by dimension we mean the cardinal of a base of the topological vector space, but in all cases of interest to us this number will be finite.

We shall denote by $E(\mathfrak{G})$ the group of all extensions of $\mathfrak{G}$ by Z_l (the group which is more particularly denoted by $\mathrm{Ext}(\mathfrak{G}, Z_l)$, cf. e.g. [6], Chapter XI). Of course, $E(\mathfrak{G}) = H^2(\mathfrak{G}, Z_l)$ but in the future we shall meet this group only as a group of extensions and not as a homology group. Let us consider the group

$$\widehat{\mathfrak{N}/(\mathfrak{N}, \mathfrak{F})\mathfrak{N}^l} = \mathrm{Hom}\,(\mathfrak{N}/(\mathfrak{N}, \mathfrak{F})\mathfrak{N}^l, Z_l).$$

Clearly the sequence

$$(1) \longrightarrow \mathfrak{N}/(\mathfrak{N}; \mathfrak{F})\mathfrak{N}^l \longrightarrow \mathfrak{F}/(\mathfrak{N}, \mathfrak{F})\mathfrak{N}^l \longrightarrow \mathfrak{G} \longrightarrow (1)$$

defines some extension Θ of $\mathfrak{G}$ by $\mathfrak{N}/(\mathfrak{N}, \mathfrak{F})\mathfrak{N}^l$, i.e. an element

$$\Theta \in \mathrm{Ext}\,(\mathfrak{G}, \mathfrak{N}/(\mathfrak{N}, \mathfrak{F})\mathfrak{N}^l).$$

---

* Translator's note. What is apparently meant is that $\mathfrak{F}$ acts on $\mathfrak{N}$ by inner automorphisms. By an operational set of generators is meant a set of elements of $\mathfrak{N}$ which with their conjugates generate $\mathfrak{N}$ (if $\mathfrak{N}$ is finite) or an everywhere dense subgroup (if $\mathfrak{N}$ is a topological l-group).

A homomorphism $\phi \in \mathrm{Hom}(\mathfrak{N}/(\mathfrak{N}, \mathfrak{F})\mathfrak{N}^l, Z_l)$ takes this into

$$\phi\Theta \in E(\mathfrak{G}).$$

In this way we have defined a map

$$\xi: \mathfrak{N}/(\mathfrak{N}, \mathfrak{F})\mathfrak{N}^l \longrightarrow E(\mathfrak{G}), \; \xi(\phi) = \phi\Theta. \tag{9}$$

It is easy to verify that this map is an isomorphism (in the language of group homology this follows from the "second reduction theorem" [7]). Hence

$$r(\mathfrak{G}) = \dim_{Z_l} E(\mathfrak{G}).$$

We shall use the representation (7) of $\mathfrak{G}$. To each homomorphism

$$\phi_{\alpha,\beta}: G_\alpha \longrightarrow G_\beta$$

there corresponds a homomorphism

$$E(\phi_{\alpha,\beta}): E(G_\beta) \longrightarrow E(G_\alpha)$$

and then

$$E(\mathfrak{G}) = \varinjlim E(G_\alpha).$$

Each group G_α is a homomorphic image of $\mathfrak{G}$ and to the homomorphism

$$\phi_\alpha: \mathfrak{G} \longrightarrow G_\alpha$$

there corresponds a homomorphism

$$\psi_\alpha = E(\phi_\alpha): E(G_\alpha) \longrightarrow E(\mathfrak{G}).$$

The group $E(\mathfrak{G})$ is the union of the images of its subgroups $\psi_\alpha E(G_\alpha)$ and so

$$r(\mathfrak{G}) = \sup_\alpha \dim_{Z_l} \psi_\alpha E(G_\alpha).$$

On putting

$$\mathrm{Ker}\, \psi_\alpha = E(G_\alpha)^0,$$

we obtain

$$r(\mathfrak{G}) = \sup_\alpha \dim_{Z_l} E(G_\alpha)/E(G_\alpha)^0.$$

We shall now explain what it means for a $\theta \in E(G_\alpha)$ to lie in $E(G_\alpha)^0$. For this we note that the sequence

$$(1) \longrightarrow \mathfrak{N}_\alpha/(\mathfrak{N}_\alpha, \mathfrak{G})\mathfrak{N}^l_\alpha \longrightarrow \mathfrak{G}/(\mathfrak{N}_\alpha, \mathfrak{G})\mathfrak{N}^l_\alpha \longrightarrow G_\alpha \longrightarrow (1),$$

where $\mathfrak{N}_\alpha = \mathrm{Ker}\, \phi_\alpha$, determines a $\Theta \in \mathrm{Ext}(G_\alpha, \mathfrak{N}_\alpha/(\mathfrak{N}_\alpha, \mathfrak{G})\mathfrak{N}^l_\alpha)$. Every

$$\phi \in \mathfrak{N}_\alpha/(\mathfrak{N}_\alpha, \mathfrak{G})\mathfrak{N}^l_\alpha = \mathrm{Hom}(\mathfrak{N}_\alpha/(\mathfrak{N}_\alpha, \mathfrak{G})\mathfrak{N}^l_\alpha, Z_l)$$

determines a

$$\phi\Theta \in \mathrm{Ext}(G_\alpha, Z_l) = E(G_\alpha).$$

This is in $E(G_\alpha)^0$, as is easy to verify either by explicitly constructing it or by using the exact cohomology sequence with noncommutative coefficients. Further, it is also easily verified that

$$E(G_\alpha)^0 = \operatorname{Im} \xi, \quad \xi(\phi) = \phi\Theta$$

(this assertion being an immediate generalization of the assertion that ξ in (9) is an isomorphism, since $E(G_\alpha)^0 = 0$ there because $E(\mathfrak{F}) = 0$).

Let us consider the subfield K_α of K belonging to the subgroup G_α of $\mathfrak{G}$. If θ is in $E(G_\alpha)$, then by what was said above we have

$$\theta = \xi(\phi), \quad \phi \in \mathfrak{N}_\alpha/(\mathfrak{N}_\alpha, \mathfrak{G})\mathfrak{N}^l_\alpha.$$

We shall denote by $\mathfrak{N}_\theta$ the inverse image of Ker ϕ under

$$\mathfrak{N}_\alpha \longrightarrow \mathfrak{N}_\alpha/(\mathfrak{N}_\alpha, \mathfrak{G})\mathfrak{N}^l_\alpha$$

and by K_θ the corresponding subfield of K. Clearly

$$k \subset K_\alpha \subset K_\theta \subset K.$$

If G_θ is the galois group of K_θ/k and $\mathfrak{z}$ that of K_θ/K_α, we have an exact sequence

$$(1) \longrightarrow \mathfrak{z} \longrightarrow G_\theta \longrightarrow G_\alpha \longrightarrow (1)$$

and ϕ defines an injection of $\mathfrak{z}$ into Z_l. Hence G_θ defines an element of $E(G_\alpha)$ and ϕ takes it into θ.

More generally, let $\theta \in E(\overline{G})$, where $\overline{G}$ is the galois group of some normal extension $\overline{K}/k$.

By the embedding problem corresponding to θ we mean the question whether there exists a normal extension $\overline{\overline{K}}$:

$$k \subset \overline{K} \subset \overline{\overline{K}}$$

and an embedding ϕ of the galois group $G(\overline{\overline{K}}/\overline{K})$ into Z_l' such that

$$\theta = \phi\overline{\theta},$$

where $\overline{\theta}$ is the element of $E(\overline{G})$ which corresponds to the natural exact sequence

$$(1) \longrightarrow G(\overline{\overline{K}}/\overline{K}) \longrightarrow G(\overline{\overline{K}}/k) \longrightarrow \overline{G} \longrightarrow (1).$$

We shall speak of a K-embedding problem if there is a fixed extension K/k and it is required to find $\overline{\overline{K}}$ as a subfield of K.

All the preceding arguments lead to the following:

Theorem 2. *If $\mathfrak{G}$ is the galois group of some l-extension* K *of* k, *then*

$$r(\mathfrak{G}) = \operatorname*{Sup}_\alpha \dim_{Z_l} E(G_\alpha)/E(G_\alpha)^0,$$

where G_α runs through all finite factor groups of $\mathfrak{G}$ and $E(G_\alpha)^0$ consists of the elements of $E(G_\alpha)$ for which the K*-embedding problem is solvable.*

Clearly $\dim_{Z_l} E(G_\alpha)/E(G_\alpha)^0$ can be interpreted as the number of conditions that have to be fulfilled for the embedding problem to be solvable.

§3. Some arithmetical invariants of extensions

We shall now explain the nature of the conditions of solvability of the embedding problem formulated at the end of the last section for the case when k is a field of algebraic numbers.

Let us consider all possible extensions $\overline{K}$ of an arbitrary field K which have over K a galois group isomorphic either to Z_l or to 1. In the term "extension" we shall include a somewhat richer structure than is usual and will mean the pair consisting of the field $\overline{K}$ and an isomorphic injection of its galois group over K into Z_l. In other words we may suppose that we have fixed the field $\overline{K}$ and an element of $\mathrm{Hom}(G(\overline{K}/K), Z_l)$. This element can also be considered as an element of $\mathrm{Hom}(\mathfrak{a}_K, Z_l)$, where $\mathfrak{a}_K$ is the galois group of the maximal abelian extension of K. Conversely, each element χ of $\hat{\mathfrak{a}}_K = \mathrm{Hom}(\mathfrak{a}_K, Z_l)$ defines a field, namely that corresponding to the subgroup $\mathrm{Ker}\,\chi$ of $\mathfrak{a}_K$, and a homomorphism of its galois group into Z_l. Hence "extension" understood with this new structure is precisely an element of $\hat{\mathfrak{a}}_K = \mathrm{Hom}(\mathfrak{a}_K, Z_l)$. This treatment can already be found, for example, in [8]. What is important for us now is that the group structure thus carries over from the group $\hat{\mathfrak{a}}_K = \mathrm{Hom}(\mathfrak{a}_K, Z_l)$ to the set of extensions of K with group isomorphic to Z_l or 1. We shall denote the result of applying this operation to extensions $\overline{K}$ and $\overline{\overline{K}}$ by $\overline{K} \circ \overline{\overline{K}}$.

Let us suppose now that K is normal over some field k and has galois group G over it.

The galois group of each extension $\overline{K}$ of K which is normal over k defines a $\xi \in E(G)$. We shall now discuss in the case of number fields the question of when, conversely, to a $\xi \in E(G)$ there exists a corresponding extension $\overline{K}$ or, in other words, when the embedding problem corresponding to ξ and K is solvable. In what follows we shall call this the embedding problem (K, ξ).

For any prime divisor $\mathfrak{p}$ of k we shall denote by $K_\mathfrak{p}$ the algebra $K \otimes k_\mathfrak{p} = \prod_{\mathfrak{P} | \mathfrak{p}} K_\mathfrak{P}$. It has galois group G over $k_\mathfrak{p}$. Then the following statement holds:

The embedding problem (K, ξ) *is solvable when and only when the embedding problem* ($K_\mathfrak{p}$, ξ) *is solvable for any prime divisor* $\mathfrak{p}$ *of* k. *Further, if* $\zeta \in k$, *the solvability of the embedding problem for all* $\mathfrak{p}$ *except one implies its solvability also for that* $\mathfrak{p}$ (*cf.* [9] *or* [10]).

On the subject of the solvability of the embedding problem for the algebras $K_\mathfrak{p}$ the following results are known:

1. If the algebra $K_\mathfrak{p}$ is unramified over $k_\mathfrak{p}$ (i.e. all the fields $K_\mathfrak{P}/k_\mathfrak{p}$ for

$\mathfrak{P} | \mathfrak{p}$ are unramified), then the embedding problem is always solvable, and indeed there is always a solution $\Sigma_{\mathfrak{p}}$ which is an unramified algebra [9].

2. If $\zeta \notin k_{\mathfrak{p}}$ the embedding problem is always solvable. (As in §§1–2, G is considered as an l-group.) For $\mathfrak{p} \nmid l$ this assertion follows from the fact that under the given hypotheses $K_{\mathfrak{p}}$ is unramified. For $\mathfrak{p} | l$ it was proved in [11].

3. If $\zeta \in k_{\mathfrak{p}}$ there exists an element χ of Hom $(E(G), Z_l)$ such that $\chi(\xi) = 1$ is a necessary and sufficient condition for the solvability of the embedding problem ([12] and [4]).

We shall now suppose that K/k is ramified only at the prime divisors of a fixed set S. We shall denote by $\widetilde{E(G)}$ the subgroup consisting of all the elements of $E(G)$ for which the corresponding embedding problem is solvable. From the results enunciated above we have

Theorem 3. *There exists* $m = t(S) - \delta$ *elements* $\chi_1, \cdots, \chi_m$ *of* Hom $(E(G), Z_l)$ *such that* $\xi \in \widetilde{E(G)}$ *if and only if* $\chi_1(\xi) = 1, \cdots, \chi_m(\xi) = 1$. *Here, as in* §1, $t(S)$ *denotes the number of* $\mathfrak{p} \in S$ *such that* $\zeta \in k_{\mathfrak{p}}$ *and* $\delta = 0$ *if* $\zeta \notin k$ *but* $\delta = 1$ *if* $\zeta \in k$. *If* $\delta = 1$ *and* $S = \emptyset$ *then put* $m = 0$.

We shall now explain when a solution of the embedding problem can be chosen as an extension $\overline{K}/k$ ramified only at the prime divisors of S.

For this we recall that all solutions of the embedding problem can be expressed in terms of one of them. Let us denote by $\overline{k}$ an arbitrary extension of k with galois group isomorphic either to Z_l or to 1. We shall call the extension $K_0 = K \cdot \overline{k}$ trivial. The following result holds:

4. If $\overline{K}$ is one solution of the embedding problem, then any other has the form $\overline{\overline{K}} = \overline{K} \circ K_0$, where K_0 is a trivial extension ([9] and [10]).

Let $\mathfrak{p}$ be a prime divisor such that the algebra $K_{\mathfrak{p}}$ is unramified. Then the embedding problem $(K_{\mathfrak{p}}, \xi)$ has an unramified solution (Assertion 1). Let us denote it by $\Sigma_{\mathfrak{p}}$. Let $\overline{K}$ be some solution of the embedding problem (K, ξ). Then $\overline{K}_{\mathfrak{p}} = \Sigma_{\mathfrak{p}} \circ K_0$, where $K_0 = K_{\mathfrak{p}} \cdot \overline{k}_{\mathfrak{p}}$ is a trivial solution, and $\overline{k}_{\mathfrak{p}}/k_{\mathfrak{p}}$ is an extension with galois group Z_l or 1. (Assertion 4.) Like every extension of $k_{\mathfrak{p}}$, the field $\overline{k}_{\mathfrak{p}}$ corresponds to an element ϕ of Hom $(\mathfrak{a}_{K_{\mathfrak{p}}}, Z_l)$. By local class field theory, Hom $(\mathfrak{a}_{K_{\mathfrak{p}}}, Z_l) = $ Hom $(k_{\mathfrak{p}}, Z_l) = \hat{k}_{\mathfrak{p}}$. Hence we may consider that $\phi \in \hat{k}_{\mathfrak{p}}$. On the other hand, the unramified extension itself is determined uniquely up to a substitution (Assertion 4) $\Sigma_{\mathfrak{p}} \longrightarrow \Sigma_{\mathfrak{p}} \circ \Sigma_0$, where $\Sigma_0 = K_{\mathfrak{p}} \cdot \kappa$ and $\kappa/k_{\mathfrak{p}}$ is an unramified extension of $k_{\mathfrak{p}}$ with galois group isomorphic to Z_l or 1. To such an extension corresponds a $\phi_0 \in \hat{k}_{\mathfrak{p}}$ with the property $\phi_0 | \mathfrak{U}_{\mathfrak{p}} = 1$.

We see that ϕ is defined up to a factor of this type. In other words, a character $\psi \in$ Hom $(\mathfrak{U}_{\mathfrak{p}}, Z_l) = \hat{\mathfrak{U}}_{\mathfrak{p}}$ has been uniquely determined. We have the following result:

Lemma. *To each solution* $\overline{K}$ *of the embedding problem* (K, ξ) *and to a prime divisor* $\mathfrak{p}$ *of* k *unramified in* K *there corresponds an element* ψ *of* $\hat{\mathfrak{U}}_{\mathfrak{p}}$. *The equation* $\psi = 1$ *is necessary and sufficient for* $\mathfrak{p}$ *to be unramified in* $\overline{K}$.

Suppose that K is ramified only at the prime divisors of S, that $\mathfrak{p} \notin S$ and that $\overline{K}$ is some solution of the embedding problem. To it there corresponds a $\psi \in \hat{\mathfrak{U}}_{\mathfrak{p}}$ which we shall denote by $\psi_{\mathfrak{p}}$. Since $\overline{K}$ has only a finite number of points of ramification, we have $\psi_{\mathfrak{p}} = 1$ for almost all $\mathfrak{p}$. The set of all $\psi_{\mathfrak{p}}$ can, consequently, be regarded as an element of $\Pi_{\mathfrak{p} \notin S}\, \hat{\mathfrak{U}}_{\mathfrak{p}} = \hat{\mathfrak{U}}_S$. Any other solution of the same embedding problem has the shape $\overline{K} \circ K_0$, where $K_0 = K \cdot k$ and $\overline{k}/k$ is a cyclic extension of k, i.e. it corresponds by class field theory of an element of $\mathrm{Hom}\,(J/k, Z_l) = \widehat{J/k}$. If $\overline{\psi} \in \widehat{J/k}$, then the corresponding $\overline{\psi}_{\mathfrak{p}} \in \hat{\mathfrak{U}}_{\mathfrak{p}}$ is obtained, clearly, from the injection $\mathfrak{U}_{\mathfrak{p}} \longrightarrow J/k$. Thus we have the standard homomorphisms

$$\mathfrak{U}_S \xrightarrow{\phi} J/k, \quad \widehat{J/k} \xrightarrow{\hat{\phi}} \hat{\mathfrak{U}}_S.$$

To each solution $\overline{K}$ of the embedding problem corresponds a $\psi \in \hat{\mathfrak{U}}_S$ and one can find a solution ramified only at the prime divisors of S precisely when $\psi \in \mathrm{Im}\,\hat{\phi}$. It is clear that the element $\psi \cdot \mathrm{Im}\,\hat{\phi}$ of the group $\hat{\mathfrak{U}}_S/\mathrm{Im}\,\hat{\phi} = \mathrm{Coker}\,\hat{\phi}$ depends only on the embedding problem and not on the choice of solution $\overline{K}$.

Thus we have

Theorem 4. *To each element* ξ *of* $\widetilde{E(G)}$ *there corresponds a* $\psi(\xi) \in \hat{\mathfrak{U}}_S/\mathrm{Im}\,\hat{\phi} = \mathrm{Coker}\ \hat{\phi}$, *which is* 1 *precisely when there exists a solution of the embedding problem ramified only at the prime divisors of* S. *This function* ψ *is defined by the injection of* $\widetilde{E(G)}/(E)^0$ *into* $\mathrm{Coker}\,\hat{\phi}$.

§4. The number of relations

We shall again use the notation introduced in §1.

Theorem 5. *The minimum number of relations* $r(S)$ *for the group* $\mathfrak{G}_S$ *satisfies*

$$r(S) \leq t(S) + \sigma(S) - \delta,$$

if $S \neq \emptyset$ *or* $\zeta \notin k$, *and*

$$r(S) \leq \sigma(S)$$

if $S = \emptyset$, $\zeta \in k$.

Proof. By Theorem 2, we have

$$r(S) = \sup_{a} \dim_{Z_l}\, E(G_a)/E(G_a)^0.$$

Let us consider the group $\widetilde{E(G_a)}$ introduced in §3. Since

$$E(G_a)^0 \subset \widetilde{E(G_a)} \subset E(G_a),$$

we have

$$\dim_{Z_l} E(G_\alpha)/E(G_\alpha)^0 = \dim_{Z_l} E(G_\alpha)/\widetilde{E(G_\alpha)} + \dim_{Z_l} \widetilde{E(G_\alpha)}/E(G_\alpha)^0. \tag{11}$$

By Theorem 3 there exist $t(S) - \delta$ homomorphisms of $E(G_\alpha)$ into Z_l the intersection of whose kernels is just $\widetilde{E(G_\alpha)}$. It follows that

$$\dim_{Z_l} E(G_\alpha)/\widetilde{E(G_\alpha)} \leq t(S) - \delta, \tag{12}$$

if $S \neq \emptyset$ or $\zeta \notin k$. If $S = \emptyset$, $\zeta \in k$, then

$$E(G_\alpha) = \widetilde{E(G_\alpha)}. \tag{12'}$$

It follows from Theorem 4 that there is an injection of $\widetilde{E(G_\alpha)}/E(G_\alpha)^0$ into Coker $\hat{\phi}$. Hence

$$\dim_{Z_l} \widetilde{E(G_\alpha)}/E(G_\alpha)^0 \leq \dim_{Z_l} \operatorname{Coker} \hat{\phi}. \tag{13}$$

It remains to compute $\dim_{Z_l} \operatorname{Coker} \hat{\phi}$. The homomorphism $\hat{\phi}$ is conjugate, in the sense of pairings with values in Z_l with the homomorphism

$$\phi\colon \mathfrak{U}_S \longrightarrow J/k.$$

It is also conjugate to the homomorphism

$$\phi_l\colon \mathfrak{U}_S/\mathfrak{U}_S^l \longrightarrow J/J^l k,$$

so that $\hat{\phi} = \hat{\phi}_l$. But ϕ_l and $\hat{\phi}_l$ are already homomorphisms of vector spaces over Z_l. From the usual duality argument it follows that

$$\dim_{Z_l} \operatorname{Coker} \hat{\phi}_l = \dim_{Z_l} \operatorname{Ker} \phi_l. \tag{14}$$

Let us consider the sequence

$$(1) \longrightarrow V_S/k^l \xrightarrow{f} \mathfrak{U}_S/\mathfrak{U}_S^l \xrightarrow{\phi_l} J/J^l k, \tag{15}$$

in which f is determined by the formulas

$$f(\alpha) = i\mathfrak{U}_S^l, \quad \text{where} \quad \alpha = i\mathfrak{a}^l, \ i \in \mathfrak{U}_S, \ \mathfrak{a} \in J.$$

The existence of such a decomposition for $\alpha \in V_S$ follows from the fact that $V_S = \mathfrak{U}_S J^l \cap k$. It is easy to check that $f(\alpha)$ is independent of the choice of the ideal $\mathfrak{a}$ in this decomposition.

Let us check that the sequence (15) is exact. If $i \in \mathfrak{U}_S$ and $\phi_l(i) = 1$, then $i \in \mathfrak{U}_S \cap J^l k$, i.e. $i = \mathfrak{a}^l \alpha$, $\mathfrak{a} \in J$, $\alpha \in k$. It follows that $\alpha \in V_S$ and $i = f(\alpha)$. If $f(\alpha) = 1$, then in $\alpha = i\mathfrak{a}^l$ we have $i \in \mathfrak{U}_S^l$ i.e. $\alpha \in J^l$ and so $\alpha \in k^l$.

From the fact that the sequence (15) is exact, we get

$$\dim_{Z_l} \operatorname{Ker} \phi_l = \dim_{Z_l} V_S/k^l = \sigma(S). \tag{16}$$

The formulas (10)–(16) show that

$$r(S) \leq t(S) - \delta + \sigma(S),$$

if $S \neq \emptyset$ or $\zeta \notin k$ and

$$r(S) \leq \sigma(S)$$

if $S = \emptyset$, $\zeta \in k$.

This concludes the proof of the theorem.

Let us consider some particular cases.

1. Let us suppose that $\zeta \in k$, that the number of divisor classes of k is not divisible by $l \neq 2$, and that l is divisible in k by only one prime divisor: $l = \mathfrak{l}^e$. An example of such a field is the cyclotomic field belonging to the division of the circle into l^n parts if l is a regular prime. We shall take for S the set consisting of the single prime divisor $\mathfrak{l}$.

In this case, $t(S) = 1$, $\delta = 1$. We shall show that $\sigma(S) = 0$. For if $\alpha \in V_S$, then $k(\sqrt[l]{\alpha})$ is unramified over k, which is possible, in view of the restriction on the class number of k, only when $\alpha \in k^l$.

Hence in this case Theorem 5 shows that

$$r(S) = 0,$$

i.e. $\mathfrak{G}_S$ is a free group. For the number $d(S)$ of generators we have, by Theorem 1,

$$d(S) = [k: R] - r.$$

2. Let us suppose that $l = 2$, that the number of divisor classes is odd, that 2 is divisible in k only by one prime divisor: $2 = \mathfrak{l}^e$, and that k has only one infinite real prime divisor, which we shall denote by $\mathfrak{p}_\infty$. An example of such a field is the rational field R. We shall take for S the set consisting of the two prime divisors $\mathfrak{l}$ and $\mathfrak{p}_\infty$.

In this case $t(S) = 2$, $\delta = 1$ and also, as in Case 1, we have $\sigma(S) = 0$. Now it follows from Theorem 5 that

$$r(S) \leq 1.$$

On the other hand, a simple examination of the maximal abelian subfield of K_S shows that already the group $\mathfrak{G}_S/(\mathfrak{G}_S, \mathfrak{G}_S)$ is not a free abelian group. Hence $r(S) \neq 1$, i.e.

$$r(S) = 1,$$

i.e. the group $\mathfrak{G}_S$ is determined by a single relation.

In order to find it, let us denote by σ the isomorphism of k into the field of real numbers corresponding to the divisor $\mathfrak{p}_\infty$. We shall denote by the same letter σ an isomorphism of the field of all algebraic numbers into the complex field C, which extends this isomorphism of k. Finally, denote by τ the automorphism of the field C which takes any number into its complex conjugate. Since K_S is normal, the field K_S^σ is invariant with respect to σ. Hence $g = \sigma^{-1}\tau\sigma$ is an automorphism of K_S. Then $g^2 = 1$ since $\tau^2 = 1$.

From Burnside's theorem it is easy to deduce that g can be included in some system of generators of $\mathfrak{G}_S$: $g, g_1, \cdots, g_{d-1}$. Hence for this system of generators we have the relation

$$g^2 = 1. \tag{17}$$

It is easy to verify that this relation can be included in a minimal system of relations. More precisely, if $\mathfrak{G}_S = \mathfrak{F}/\mathfrak{N}$ is a representation of $\mathfrak{G}_S$ as a factorgroup of a free group with d generators, then $g^2 \in \mathfrak{N}$, $g^2 \notin (\mathfrak{N}, \mathfrak{F})\mathfrak{N}^2$. This follows from the fact that $g^2 \notin \mathfrak{N}(\mathfrak{F}, \mathfrak{F})$; since $g^2 = e$ is, as is easy to see, the defining relation of $\mathfrak{G}_S/(\mathfrak{G}_S, \mathfrak{G}_S)$. Since $r(S) = 1$, we can deduce now that $\mathfrak{G}_S$ is determined by the single relation (17). For the number of generators, by Theorem 1 we have

$$d(S) = [k: R] + 1 - r.$$

Hence the group $\mathfrak{G}_S$ is the 2-adic completion of the free product of the group Z_2 and the free group on $[k: R] - r$ generators. In particular

$$\mathfrak{G}_S = \overline{Z_2 * Z}$$

for $k = R$.

1 This result was obtained earlier in a different way by G. N. Markšaitis [13].

3. Let us suppose that all the prime divisors of S are coprime with l. We shall denote this by: $(S, l) = 1$.

Theorem 6. *If* $(S, l) = 1$, *then*

$$r(S) \leq d(S) + r, \tag{18}$$

for $S \neq \emptyset$ *or* $\zeta \notin k$ *and*

$$r(S) \leq d(S) + 1, \tag{18'}$$

for $S = \emptyset$ *and* $\zeta \in k$.

For then $\lambda(S) = 0$ in formula (1) and

$$r(S) - d(S) \leq \iota(S) + \sigma(S) - \delta - (\iota(S) + \sigma(S) - r - \delta) = r,$$

if $S \neq \emptyset$ or $\zeta \notin k$. Formula (18′) is proved similarly.

Hence for all fields of given degree over R and for all sets S on them with $(S, l) = 1$ the difference $r(S) - d(S)$ is bounded above.

In particular, if k is R or an imaginary quadratic field it follows from Theorem 6 that

$$r(S) \leq d(S),$$

except when $S = \emptyset$, $l = 2$.

On the other hand as $(S, l) = 1$ it follows from class field theory that $\mathfrak{G}_S/(\mathfrak{G}_S, \mathfrak{G}_S)$ is finite. Hence

$$r(S) = d(S)\ ^{1)} \tag{19}$$

since otherwise the abelian group $\mathfrak{G}_S/(\mathfrak{G}_S, \mathfrak{G}_S)$ would be defined by fewer relations than its number of generators, and consequently could not be finite.

For any group G let us construct the decreasing sequence of commutator groups

$$G^{(i)}: G^{(1)} = G,\ G^{(m+1)} = (G^{(m)}, G^{(m)}).$$

We shall denote by $h(G)$ the solvability class of G, i.e. the least value of m such that $G^{(m+1)} = e$ (and $h(G) = \infty$ if $G^{(m)} \neq e$ for all m). From Theorem 6 and a group-theoretic lemma whose proof we defer to the next section we have

Theorem 7. *For fields of fixed degree over* R *and for sets* S *with* $(S, l) = 1$ *we have*

$$h(\mathfrak{G}_S) \longrightarrow \infty, \ \textit{if}\ d(S) \longrightarrow \infty.$$

Proof. Let us suppose that it is not so and that for some sequence of fields k_i and sets S_i on them we have $h(\mathfrak{G}_{S_i}) < h$. We have already seen that the groups $\mathfrak{G}_S/\mathfrak{G}_S^{(2)}$ are finite. It is proved similarly that $\mathfrak{G}_S^{(m)}/\mathfrak{G}_S^{(m+1)}$ and hence also $\mathfrak{G}_{S_i}$ is finite (assuming $h(\mathfrak{G}_{S_i}) < h$). In the next section it is shown that there is a constant $\alpha > 0$ such that

$$r(G) > \alpha d(G)^2$$

for all l-groups G for which $h(G) \leq h$. Hence

$$r(S_i) > \alpha d(S_i)^2,$$

which, however, contradicts (18) for $d(S_i) \longrightarrow \infty$.

In particular, we can choose the set S to be empty. Then $d(S) = \gamma$ and K_S is the maximal unramified l-extension of k. If $\gamma \longrightarrow \infty$ for fields k of given degree over R, then $h(\mathfrak{G}_S) \longrightarrow \infty$ and hence the solubility class of the maximal solvable unramified extension of k tends to infinity. In other words, if we denote by k_{m+1} the Hilbert class field over k_m, with $k_1 = k$, then $k_{m+1} \neq k_m$ for all $m < h(\mathfrak{G}_S)$. The sequence of fields k_m is called the class field tower and the individual fields its stories: the number of distinct fields being the number of stories. Hence we have proved the

Corollary. *If for a sequence of fields of bounded degree over* R *the number of generators of the class group tends to infinity, then the number of stories of the class field tower also tends to infinity.*

We note that examples of fields of bounded degree for which γ is arbitrarily

---

1) If k is an imaginary quadratic field, $S = \emptyset$ and $l = 2$ it follows from the preceding arguments that $r(S) = d(S) + 1$ or $r(S) = d(S)$. Both cases actually occur, the first for $k = R(\sqrt{-21})$ ($\mathfrak{G}_S = Z_2 \times Z_2$) and the second for $k = R(\sqrt{-65})$ ($\mathfrak{G}_S$ is the group given by the relations (31) with $l = 2$, $n = 2$, $k = m = 1$).

large are very easy to construct. For example, when $l = 2$ we can take the fields $R(\sqrt{\mathfrak{D}})$ where $\mathfrak{D}$ is square-free and divisible by an increasing number of primes (by the fundamental theorem about genera). For arbitrary l we take t primes $p_1, \cdots, p_t$ such that $p_i \equiv l \pmod{l}$ and denote by k_i the subfield of degree l of the cyclotomic field for p_i. Then we take for k a subfield of $k_1, \cdots, k_t$ not contained in the compositum of any lesser number of the fields $k_1, \cdots, k_t$ and such that $[k: R] = l$. It is easy to see that $\gamma \geq t - 1$ for k, as follows from the fact that all the $k_i k/k$ are unramified.

§5. A group lemma

Lemma. *For any integer h there is a constant $\alpha > 0$ such that for all finite l-groups G for which $h(G) \leq h$ we have*

$$r(G) \geq \alpha d(G)^2 \tag{20}$$

for all sufficiently large $d(G)$ (i.e. $d(G) > \mathfrak{D}$, where $\mathfrak{D}$ depends only on h).

Proof. Let G be a group satisfying the conditions of the Lemma. Write $d(G) = d$, $r(G) = r$. Let F be a free group with d generators and

$$G = F/N$$

a representation of G as a factorgroup of F. By hypotheses N has r operational generators $\sigma_1, \cdots, \sigma_r$ if we regard N as an F-operational group.

For any group G we write $G_{i+1} = (G_i, G)G_i^l$, $G_1 = G$.

The fact that d is the minimal number of generators implies that $N \subset F_2$. For otherwise the homomorphism $F/F_2 \longrightarrow G/G_2$ would not be an isomorphism. This means that the images of the generators $x_1, \cdots, x_d$ of F in G would not be independent modulo G_2 and then it would follow by Burnside's theorem that one could find a system of fewer than d generators for G. It follows from $G^{(h)} = e$ that $N \supset F^{(h)}$. Put $c = 2^h$. Then a fortiori $NF_{c+1} \supset F^{(h)}$. Write

$$F/F_{c+1} = \mathfrak{F}, \quad NF_{c+1}/F_{c+1} = \mathfrak{N}.$$

The length of the l-central series of $\mathfrak{F}$ is c, i.e. $\mathfrak{F}_{c+1} = e$. Here

$$\mathfrak{N} \supset \mathfrak{F}^{(h)} \tag{21}$$

and, as before, $\mathfrak{N}$ has a system of r operational generators (as an $\mathfrak{F}$-operational group), which we shall denote by $S_1, \cdots, S_r$.

The group $\mathfrak{F}_{c+1}$ has period l^c. For $\mathfrak{F}_m^{l^r} \subset \mathfrak{F}_{m+r}$, since $\mathfrak{F}_m^l \subset \mathfrak{F}_{m+1}$, and in particular $\mathfrak{F}^{l^c} = \mathfrak{F}_1^{l^c} \subset \mathfrak{F}_{c+1} = e$.

The proof of (20) depends on estimating on the one hand the order $[\mathfrak{N}]$ of $\mathfrak{N}$ and on the other hand the order $[\mathfrak{F}^{(h)}]$ of $\mathfrak{F}^{(h)}$. Then one uses the inequality

$$[\mathfrak{N}] \geq [\mathfrak{F}^{(h)}] \tag{22}$$

which follows from (21).

1. The estimate of $[\mathfrak{N}]$. We define the following sequence of subgroups ${}^{(k)}\mathfrak{N}$ of $\mathfrak{N}$:

$$ {}^{(1)}\mathfrak{N} = \mathfrak{N}, \quad {}^{(k+1)}\mathfrak{N} = ({}^{(k)}\mathfrak{N}, \mathfrak{F}) $$

and will show that they enjoy the following properties

A. ${}^{(k)}\mathfrak{N} \subset \mathfrak{F}_{k+1}$.

B. ${}^{(k)}\mathfrak{N}/{}^{(k+1)}\mathfrak{N}$ is an abelian group with $\leq rd^{k-1}$ generators.

A is proved by induction. For $k = 1$ the assertion follows from the fact that $\mathfrak{N} \subset \mathfrak{F}_2$ since $N \subset F_2$. If one has already shown that ${}^{(k)}\mathfrak{N} \subset \mathfrak{F}_{k+1}$, then

$$ {}^{(k+1)}\mathfrak{N} = ({}^{(k)}\mathfrak{N}, \mathfrak{F}) \subset (\mathfrak{F}_{k+1}, \mathfrak{F}) \subset \mathfrak{F}_{k+2}. $$

To prove B we first verify that the commutators

$$ S_{i,j_1,\cdots,j_{k-1}} = (S_i, x_{j_1}, \cdots, x_{j_{k-1}}), \; i = 1, \cdots, r; \; j_s = 1, \cdots, d $$

form a system of operational generators of ${}^{(k)}\mathfrak{N}$ regarded as an $\mathfrak{F}$-operational group. This property is again verified by induction. For $k = 1$ it is part of the definition of the system $S_1, \cdots, S_r$. Suppose that it has been proved for k. It is then clear that a system of operational generators for $k + 1$ is given by the commutators

$$ (S_{i,j_1,\cdots,j_{k-1}}, x_{j_k}), $$

and that is just the $S_{i,j_1,\cdots,j_k}$.

It is now clear that the $S_{i,j_1,\cdots,j_k}\,{}^{(k+1)}\mathfrak{N}$ are a system of operational generators for ${}^{(k)}\mathfrak{N}/{}^{(k+1)}\mathfrak{N}$. This is a particular case of the well known fact that the elements of a normal subgroup $\mathfrak{H}$ of some l-group $\mathfrak{F}$ generate $\mathfrak{H}/(\mathfrak{H}, \mathfrak{F})$ as an l-operational group precisely when $t_1(\mathfrak{H}, \mathfrak{F}), \cdots, t_m(\mathfrak{H}, \mathfrak{F})$ generate $\mathfrak{H}/(\mathfrak{H}, \mathfrak{F})$ (generalized Burnside theorem). Since the number of $S_{i,j_1,\cdots,j_{k-1}}$ is rd^{k-1}, this concludes the proof of B.

From the fact that the period of $\mathfrak{F}$ is l^c it follows that that of ${}^{(k)}\mathfrak{N}$ cannot be greater. It follows from B that

$$ ({}^{(k)}\mathfrak{N} : {}^{(k+1)}\mathfrak{N}) \leq l^{crd^{k-1}}. \tag{23} $$

From A (with $k = c$) we have ${}^{(c)}\mathfrak{N} = e$. Hence

$$ [\mathfrak{N}] = \prod_{k=1}^{c-1} ({}^{(k)}\mathfrak{N} : {}^{(k+1)}\mathfrak{N}). $$

From this and from (23) follows

$$ [\mathfrak{N}] \leq l^{cr(1+d+\cdots+d^{c-2})}. \tag{24} $$

II. The estimate of $[\mathfrak{F}^{(h)}]$. We shall show that

$$[\mathfrak{F}^{(h)}] \geq l^{\binom{d}{c}}. \tag{25}$$

For this we consider a system $1 \leq i_1 < i_2 < \cdots < i_m \leq d$ $(m = 2^\rho,\ \rho \leq h)$ of indices and make a commutator $\xi_{i_1 \cdots i_m}$ on putting by induction

$$\xi_{i_1} = x_{i_1} \text{ (for } m = 1)$$

$$\xi_{i_1 \cdots i_{2m}} = (\xi_{i_1 \cdots i_m},\ \xi_{i_{m+1} \cdots i_{2m}}).$$

Thus

$$\xi_{i_1 i_2} = (x_{i_1},\ x_{i_2}),\ \xi_{i_1 i_2 i_3 i_4} = ((x_{i_1},\ x_{i_2}),\ (x_{i_3},\ x_{i_4})) \text{ etc.}$$

We shall show that the $\xi_{i_1 \cdots i_c}$ are in $\mathfrak{F}^{(h)}$ and that all $\binom{d}{c}$ such elements corresponding to all the $\binom{d}{c}$ sequences $1 \leq i_1 < i_2 < \cdots < i_c \leq d$ are independent in the group. From this (25) obviously follows.

For the proof we shall use Magnus' representation of the commutators in a free Lie ring [14]. Let us consider the ring of noncommutative power series in d variables $y_1, \cdots, y_d$ with integral coefficients: $Z\{y_1, \cdots, y_d\}$ and the following map μ of F into $Z\{y_1, \cdots, y_d\}$:

$$\mu x_i = 1 + y_i,\ \mu x_i^{-1} = 1 - y_i + y_i^2 - y_i^3 + \cdots.$$

It is not difficult to show that for the $\mu\xi$, $\xi \in \mathfrak{F}_c$ all the coefficients of terms of degree k, $0 < k \leq c$ with respect to $y_1, \cdots, y_d$ are divisible by l (see [15], for example). Hence

$$\mu\mathfrak{F}_{c+1} \equiv 1\ (l,\ (y_1, \cdots, y_d)^{c+1}).$$

On the other hand, it is easy to see that

$$\mu\xi_{i_1 \cdots i_c} \equiv 1 + \eta_{i_1 \cdots i_c}\ ((y_1, \cdots, y_d)^{c+1}),$$

where the commutators $\eta_{i_1 \cdots i_m}$, $m = 2^\rho$ are defined by the same formulas

$$\eta_{i_1} = y_{i_1}$$

$$\eta_{i_1 \cdots i_{2m}} = [\eta_{i_1 \cdots i_m},\ \eta_{i_{m+1} \cdots i_{2m}}],$$

as the $\xi_{i_1 \cdots i_m}$, but only in the ring $Z\{y_1, \cdots, y_d\}$.

Thus if there were a relation between the $\xi_{i_1 \cdots i_c}$ modulo $\mathfrak{F}_{c+1}$:

$$\Pi \xi_{i_1 \cdots i_c}^{a_{i_1 \cdots i_c}} \equiv 1\ (\mathfrak{F}_{c+1}),$$

then there would exist a relation with the same coefficients

$$\Sigma a_{i_1 \cdots i_c} \eta_{i_1 \cdots i_c} \equiv 0 \pmod{l} \tag{26}$$

between the homogeneous polynomials $\eta_{i_1 \cdots i_c}$ of degree c in the variables

$y_1, \cdots, y_d$.

Such a relation is impossible as all the forms $\eta_{i_1 \cdots i_c}$ depend on different systems of variables. More precisely, on putting all the $y_i = 0$ in the identity (26) except $y_{j_1}, \cdots, y_{j_c}$, we should take to 0 all the terms except $\eta_{j_1 \cdots j_c}$ and would deduce that $a_{j_1 \cdots j_c} \equiv 0 \pmod{l}$.

Thus we have proved (25). On combining it with (24) and (22) we get

$$\sum_{k=0}^{c-2} crd^k \geq \binom{d}{c} = \frac{d(d-1)\cdots(d-c+1)}{c!},$$

i.e.

$$r \geq \frac{d(d-1)\cdots(d-c+1)}{c(c!)\,(1+d+\cdots+d^{c-2})}.$$

Clearly for any $\alpha < 1/c(c!)$ the expression on the right hand side is $> \alpha d^2$ for sufficiently large d. This proves the Lemma and shows that we can take for α any number satisfying

$$\alpha < \frac{1}{2^h (2^h)!}$$

§6. Some remarks

At the present time it is unknown whether the group $\mathfrak{G}_S$ is finite or not when $(S, l) = 1$. In the case when S is empty this problem is the same as the so-called problem of l-class field towers. If the group $\mathfrak{G}_S$ were finite, then by Theorem 6 of §4 there would exist a sequence of finite l-groups G with a number $d(G_i)$ of generators tending to infinity for which the difference $r(G_i) - d(G_i)$ would be bounded. In this connection we come to the following problem:

1. Let us define $\rho(d)$ by

$$\rho(d) = \min_{d(G)=d} r(G),$$

where $r(G)$ is the minimal number of relations for a finite l-group G, the minimum being taken over all finite l-groups G with minimum number of generators d. Is it true that

2 $$\rho(d) - d \longrightarrow \infty \text{ for } d \longrightarrow \infty\,? \tag{27}$$

The simplest known l-groups have a very large number of relations $r(G)$ compared with the number d of generators. Thus

$$r(G) = \frac{d(d+1)}{2}$$

for commutative groups.

For the groups F/F_m (in the notation of §5) or for the l-Sylow subgroups of the symmetric groups the number $r(G)$ of relations is yet larger. If the assertion (27) were true, then it would follow that for any algebraic number field for which the number of generators of the divisor class group was sufficiently large in comparison with its degree over R the class field tower would not break off.[1)]

In connection with (19) of §4, which is valid when k is R or is a purely imaginary quadratic field, it is interesting to investigate the finite l-groups for which

$$r(G) = d(G). \tag{28}$$

Groups of a similar type were considered by I. Schur in connection with his investigation of "Multiplikatoren," i.e. the groups $H^2(G, T)$ where $T = R/Z$ is the group of all roots of unity. The connection with (28) is as follows. From the exact sequence

$$(1) \longrightarrow Z_l \longrightarrow T \xrightarrow{l} T \longrightarrow (1)$$

it is easy to deduce the exact sequence

$$0 \longrightarrow H'(G, T)/lH'(G, T) \longrightarrow H^2(G, Z_l) \longrightarrow H^2(G, T)_l \longrightarrow 0,$$

where $H^2(G, T)_l$ is the group of elements of order l in $H^2(G, T)$.

It is easy to see that $\dim_{Z_l} H^2(G, T)_l$ is just the number of generators $m(G)$ of the Multiplikator $H^2(G, T)$ and

$$\dim_{Z_l} H^1(G, T)/lH^1(G, T) = d(G).$$

Thus

$$r(G) - d(G) = m(G),$$

since

$$r(G) = \dim_{Z_l} H^2(G, Z_l).$$

In particular, the groups with (28) are precisely the groups whose Multiplikator is 0. Schur calls such groups closed.

In [16] Schur gives some series of closed groups. These groups have only one or two generators (and are described below). In [17][2)] there is an example of a closed group with three generators. In this connection there arises the following question:

II. Do there exist closed groups with arbitrarily many generators? In particular, do there exist closed groups with more than three generators?

---

1) Another proof of the connection between the conjecture (27) and the problem of the class field tower depends on the results of [23]. See [22].

2) J.-P. Serre drew the attention of the author to this paper.

Examples of closed groups are given, of course, by the cyclic groups

$$Z_{l^n},\ n \geq 1$$

the generalized quaternion groups

$$G = \{a,\ b\}$$

$$b^2 = a^{2^{n-1}},\ b^{-1}ab = a^{-1},\ n \geq 2$$

and the groups

$$G = \{a,\ b\}$$

$$b^2 = a^{2^{n-1}},\ b^{-1}ab = a^{-1+2^{n-1}},\ n > 2$$

(which last are defined by these relations only as 2-groups, i.e. as factorgroups of a free 2-group. Since the relations imply $a^{2^n(1-2^{n-2})} = 1$, whence it follows only in a 2-group that $a^{2^n} = 1$. This group can be defined by two relations also as a discrete group, cf. [24] [1]).

To find examples of closed l-groups with $d(G) = 2$ and $l > 2$ let us write the relations in the shape

$$a^{l^m}(a,\ b)^{\alpha}(a,\ b,\ b)^{\beta}(a,\ b,\ a)^{\gamma}\cdots = 1 \tag{29}$$

$$b^{l^m}(a,\ b)^{\alpha'}(a,\ b,\ b)^{\beta'}(a,\ b,\ a)^{\gamma'}\cdots = 1.$$

Let us suppose that one of the numbers α and α', say α, is not divisible by l. Then by a theorem of S. P. Demuškin [18] by a change of generators one can reduce the first relation to the shape

$$a^{l^m}(a,\ b) = 1. \tag{30}$$

In the group determined by the relation (30) every element can be written in the shape $a^x b^y$. In particular, the second relation can be written in this shape. After obvious transformations we will be able to write the system of relations in the form:

$$bab^{-1} = a^{1+l^m},\ b^{l^n} = a^{l^k}, \tag{31}$$

$$m \leq k \leq n.$$

Different triples (m, n, k) with $m \leq n \leq k$ give different groups since for groups defined by (31) one has

$$(G,\ G) \cong Z_{l^k},\ G/(G,\ G) \cong Z_{l^m} \times Z_{l^n}.$$

Some other examples of closed groups are known (cf. e.g. [19] [2]). It would be interesting to know what are the closed groups with two generators or, in other words,

1) This paper was mentioned to the author by J. Browkin.

2) This paper was mentioned to the author by A. I. Kostrikin.

III. When will an l-group defined by the relations (29) with $\alpha \equiv \alpha' \equiv 0 \pmod{l}$ be finite?

We remark that the situation of the last question can be realized for groups $\mathfrak{G}_S$. According to Fröhlich [20] this will be so if $k = R$, $l > 2$, $S = \{p_1, p_2\}$ and

$$p_1 - 1 = lm_1, \; p_2 - 1 = lm_2, \; (l, m_1 m_2) = 1,$$

where p_1 and p_2 are lth power residues one of the other, i.e. the congruences

$$p_1 \equiv X_1^l \pmod{p_2}$$

$$p_2 \equiv X_2^l \pmod{p_1}$$

are solvable. An example is $l = 3$, $p_1 = 79$, $p_2 = 97$. Apparently this is the simplest case when it is not known whether K_S is finite.

In the case when S is empty, that is in the case of l-class field towers, only very few examples are known in which the finiteness of the tower has been established. In all the cases for which $l > 2$, the groups are of one of the types given by (31). There is always finiteness, of course, when the divisor class group is cyclic. Besides this Scholtz and Taussky in [19] showed the finiteness of the 3-class field tower for the fields $R(\sqrt{-4027})$ and $R(\sqrt{-3299})$. The groups $\mathfrak{G}_S$ (where S is the empty set) are given here by (31) with $m = n = k = 1$ in the first case and $m = k = 1$, $n = 2$ in the second.

BIBLIOGRAPHY

[1] H. Zassenhaus, *Lehrbuch der Gruppentheorie*, Leipzig, 1937.

[2] E. Artin and J. Tate, *Class field theory*, Princeton Univ. Press, Princeton, N. J., 1961.

[3] H. Hasse, *Bericht über neuere Untersuchungen und Probleme aus der Theorie der algebraischen Zahlkörper*, Jber. Deutsch. Math.-Verein. 35 (1926).

[4] C. Chevalley, *Class field theory*, Nagoya Univ., Nagoya, 1954. MR 16, 678.

[5] D. K. Faddeev and A. I. Skopin, *Proof of a theorem of Kawada*, Dokl. Akad. Nauk SSSR 127 (1959), 529–530. MR 21 #5625.

[6] H. Cartan and S. Eilenberg, *Homological algebra*, Princeton Univ. Press, Princeton, N. J., 1956. MR 17, 1040.

[7] S. MacLane, *Cohomology theory in abstract groups. III; Operator homomorphisms of kernels*, Ann. of Math. (2) 50 (1949), 736–761. MR 11, 415.

[8] C. Chevalley, *La théorie du corps de classes*, Ann. of Math. (2) 41 (1940), 394–418. MR 2, 38.

[9] A. Scholtz, *Konstruktion algebraischer Zahlkörper mit beliebiger Gruppe von Primzahlpotenzordnung*, I, Math. Z. 42 (1936), 161–188.

[10] H. Reichardt, *Konstruktion von Zahlkörpern mit gegebener Galoisgruppe von Primzahlpotenzordnung*, J. Reine Angew. Math. 177 (1937), 1–5.

[11] I. R. Šafarevič, *On p-extensions*, Mat. Sb. 20(62) (1947), 351–363; English transl., Amer. Math. Soc. Transl. (2) 4 (1956), 59–72. MR 8, 560; MR 12, 1001, errata.

[12] R. Brauer, *Über die Konstruktion der Schiefkörper, die von endlichem Rang in Bezug auf gegebenes Zentrum sind*, J. Reine Angew. Math. 168 (1932), 44–64.

[13] G. N. Markšaitis, *On p-extensions with one critical number*, Izv. Akad. Nauk SSSR Ser. Mat. 27 (1963), 463–466. MR 27 #1437.

[14] W. Magnus, *Beziehungen zwischen Gruppen und Idealen in einem speziellen Ring*, Math. Ann. 111 (1935), 259–284.

[15] A. I. Skopin, *The factor groups of an upper central series of free groups*, Dokl. Akad. Nauk SSSR 74 (1950), 425–428. MR 12, 240.

[16] J. Schur, *Untersuchungen über die Darstellung der endlichen Gruppen durch gebrochene lineare Substitutionen*, J. Reine und Angew Math. 132 (1907), 85–137.

[17] J. Mennicke, *Einige endliche Gruppen mit drei Erzeugenden und drei Relationen*, Arch. Math. 10 (1959), 409. MR 22 #4777.

[18] S. P. Demuškin, *The group of a maximal p-extension of a local field*, Izv. Akad. Nauk SSSR Ser. Mat. 25 (1961), 329–346. MR 23 #A890.

[19] I. D. Macdonald, *On a class of finitely presented groups*, Canad. J. Math. 14 (1962), 602–614. MR 25 #3992.

[20] A. Fröhlich, *On fields of class two*, Proc. London Math. Soc. (3) 4(1954), 235–256. MR 16, 116.

[21] A. Scholtz and O. Taussky, *Die Hauptideale der kubischen Klassenkörper imaginär-quadratischer Zahlkörper*, J. Reine Angew. Math. 171 (1934), 19–42.

[22] J.-P. Serre, *Lecons sur la cohomologie galoisinen*, Cour au Collège de France, 1962–1963, 2nd. ed., Springer, Berlin, 1964. MR 31 #4785.

[23] K. Iwasawa, *A note on the group of units of an algebraic number field*, J. Math. Pures Appl. (9) 35 (1956), 189–192. MR 17, 946.

[24] B. Neumann, *On some finite groups with trivial multiplicator*, Publ. Math. Debrecen 4 (1956), 190–194. MR 18, 12.

Translated by:
J. W. S. Cassels

On class field towers
(with E. S. Golod)

Izv. Akad. Nauk SSR, Ser. Mat. **28,** No. 2, 261 – 272 (1964). Zbl. **136,** 26
[Transl., II. Ser., Am. Math. Soc. **48,** 91 – 102 (1965)]

In the paper we establish the existence of infinite unramified extensions of an algebraic number field, and of fields that cannot be embedded in fields of class number 1.

Introduction

One of the authors has shown that the class field tower problem has a negative solution if the following conjecture is true:

$$r(G) - d(G) \to \infty,$$

where $d(G)$ and $r(G)$ are the minimum numbers of generators and of defining relations for the group G, and where the limit is taken with respect to all finite p-groups (see [3] and [4], for another exposition see [2]). Here the number of relations is understood in the sense of the theory of pro-p-groups (see §1).

In the present paper we prove this conjecture; more precisely, we establish for an arbitrary finite p-group G the inequality

$$r(G) > \left(\frac{d(G)-1}{2}\right)^2.$$

The proof of this statement is reduced to the proof of an analogous statement on non-commutative local rings. This reduction is carried out in §1.

§2 contains the proof of the theorem on p-groups.

Finally, in §3 we give a summary of consequences relating to the theory of algebraic numbers.

Lemma 1 and the construction of the complex M in §3.2 are due to Šafarevič, Lemmas 2 and 3 to Golod.

The authors are deeply indebted to S. P. Demuškin for extremely stimulating conversations on this paper.

§1. Reduction to local rings

To begin with we recall some definitions and simple facts. A detailed exposition can be found in [2].

A pro-p-group is a group which is the projective limit of finite p-groups.

The group ring of a pro-p-group is defined as the projective limit of the group rings of its finite factor groups.

The group ring of a pro-p-group $\mathfrak{G}$ over the field Z_p of p elements (only this case will be considered in what follows) is denoted by $Z_p[\mathfrak{G}]$. This ring is zero-dimensional and compact. There is a canonical embedding $\mathfrak{G} \to Z_p[\mathfrak{G}]$ which we shall use to identify the elements of $\mathfrak{G}$ with the corresponding elements of $Z_p[\mathfrak{G}]$.

The ring $Z_p[\mathfrak{G}]$ has a unique maximal ideal, generated by the elements of a form $1 - \sigma$, $\sigma \in \mathfrak{G}$. Hence it is a local ring.

The association of a pro-p-group $\mathfrak{G}$ with its group ring $Z_p[\mathfrak{G}]$ is a functor from the category of pro-p-groups into the category of compact local rings.

A free pro-p-group with d generators (where d is a natural number) is the group $\mathfrak{F}_d$ which is the projective limit of the p-factor groups of a free group with d generators. Every pro-p-group $\mathfrak{G}$ with d generators can be represented in the form $\mathfrak{F}_d/\mathfrak{N}$, where $\mathfrak{N}$ is a closed normal subgroup of $\mathfrak{F}_d$. If we choose a representation $\mathfrak{G} = \mathfrak{F}_d/\mathfrak{N}$ in which the generators $\sigma_1, \cdots, \sigma_d$ of $\mathfrak{F}_d$ go over into a minimal system of generators of $\mathfrak{G}$, then

$$\mathfrak{N} \subset \mathfrak{F}_d^p(\mathfrak{F}_d, \mathfrak{F}_d),$$

where $(\mathfrak{F}_d, \mathfrak{F}_d)$ is the derived group of $\mathfrak{F}_d$ and $\mathfrak{F}_d^p$ is the subgroup of $\mathfrak{F}_d$ generated by the pth powers of its elements.

We denote by O_d the ring of formal power series in d non-commuting variables $x_1, \cdots, x_d$ with coefficients in Z_p. This is a local ring with the maximal ideal $I = (x_1, \cdots, x_d)$. The set $1 + I$ is a group which we denote by O_d^+.

The mapping

$$\mu: \ \mathfrak{F}_d \to O_d,$$

given by the conditions

$$\mu\sigma_i = 1 + x_i,$$

defines a homomorphism of $\mathfrak{F}_d$ into O_d^+. This homomorphism is an embedding and even defines an isomorphism

$$\bar{\mu}: \ Z_p[\mathfrak{F}_d] \to O_d.$$

The proof follows from the fact that side by side with the naturally defined homomorphism $\bar{\mu}: Z_p[\mathfrak{F}_d] \to O_d$ it is easy to construct the inverse homomorphism $\nu: O_d \to Z_p[\mathfrak{F}_d]$ which is defined by the fact that

$$\nu(x_i) = \sigma_i - 1.$$

For $\sigma \in \mathfrak{F}_d$ we obviously have $\mu\sigma \in O_d^+$.

Apart from these well-known facts we shall require the following simple results.

Lemma 1. *Suppose that a pro-p-group $\mathfrak{G}$ is represented in the form*

$$\mathfrak{G} = \mathfrak{F}_d/\mathfrak{N}, \tag{1}$$

where the normal subgroup $\mathfrak{N}$ *is generated by the elements* $\phi_1, \cdots, \phi_r \in \mathfrak{F}_d$.

We set

$$\mu\,\phi_i = 1 + f_i, \quad (f_1, \cdots, f_r) = \mathfrak{A}, \tag{2}$$

$$O_d/\mathfrak{A} = A. \tag{3}$$

Then the algebras A *and* $Z_p[\mathfrak{G}]$ *are isomorphic.*

Proof. We denote by π the homomorphism $\mathfrak{F}_d \to \mathfrak{G}$ corresponding to the representation (1) and by λ the homomorphism $O_d \to A$ corresponding to (3), and we shall show that we can construct a homomorphism $\tau\colon \mathfrak{G} \to A^*$ (the group of invertible elements of A) such that the diagram

$$\begin{array}{ccc} & & \mathfrak{F}_d \\ & \swarrow^{\pi} & \big\downarrow \\ \mathfrak{G} & & \lambda \circ \mu \\ & \searrow_{\tau} & \big\downarrow \\ & & A \end{array} \tag{4}$$

is commutative.

For this purpose it is sufficient to verify that

$$(\lambda \circ \mu)\,\mathfrak{N} = 1.$$

But this is obvious, because all the elements of $\mathfrak{N}$ are limits of products of elements of the form $\psi^{-1}\,\phi_i^{\pm 1}\,\psi$, $\psi \in \mathfrak{F}_d$, and $\lambda \circ \mu$ is a continuous homomorphism with $(\lambda \circ \mu)\,\phi_i = 1$ by (2).

The homomorphism τ defines a ring homomorphism:

$$\bar{\tau}\colon\ Z_p\,[\mathfrak{G}] \to A,$$

which, as we shall show, is in fact an isomorphism. The homomorphism π defines a ring homomorphism:

$$\bar{\pi}\colon\ Z_p\,[\mathfrak{F}_d] \to Z_p\,[\mathfrak{G}],$$

moreover the diagram

$$\begin{array}{ccc} & & Z_p\,[\mathfrak{F}_d] \\ & \swarrow^{\bar{\pi}} & \big\downarrow \\ Z_p\,[\mathfrak{G}] & & \lambda \circ \bar{\mu} \\ & \searrow_{\bar{\tau}} & \big\downarrow \\ & & A \end{array}$$

is commutative. Now $\bar{\tau}$ is an epimorphism, since $\lambda \circ \bar{\mu}$ is. In order to see that $\bar{\tau}$ is an isomorphism it is sufficient to verify that

$$\operatorname{Ker}\,(\lambda \circ \bar{\mu}) \subset \operatorname{Ker}\,\bar{\pi}.$$

But the ideal $\operatorname{Ker}\,(\lambda \circ \bar{\mu})$ is generated by the elements $\nu f_i = \phi_i - 1$, and

$$\bar{\pi}\,(\varphi_i - 1) = \bar{\pi}\,(\varphi_i) - 1 = 0.$$

This proves the lemma.

Deviating somewhat from the accepted terminology we shall call an algebra A with unit element nilpotent if it is of finite rank over the ground field k and $A/\mathfrak{m} = k$, where $\mathfrak{m}$ is the radical of A. The minimal number d of generators of the algebra A coincides with $\dim_k \mathfrak{m}/\mathfrak{m}^2$. Obviously there exists a representation $A = O_d/A$. The number of generators of the ideal A will be called the number of relations of A. As is known from the theory of local rings, the generators $\bar{x}_i = x_i \bmod \mathfrak{G}$ form a minimal system of generators of the algebra A if and only if $\mathfrak{A} \subset I^2$.

Corollary. *If there exist a finite p-group G with a minimal system of d generators, defined by r relations, then there exists a nilpotent algebra A over Z_p having a minimal system of d generators and defined by r relations.*

We choose for A the algebra $Z_p[G]$. If G has a minimal system of d generators $\sigma_1, \cdots, \sigma_d$ and is defined by r relations, then A has the system of d generators $1 - \sigma_i$. Since $\mathfrak{N} \subset \mathfrak{F}_d^p(\mathfrak{F}_d, \mathfrak{F}_d)$, we have $\mathfrak{A} \subset I^2$, and so the system of generators $1 - \sigma_i$ is minimal. By Lemma 1 the algebra A is defined by r relations.

§2. On the number of generators and relations of a p-group

Our object is the proof of the following results.

Theorem 1. *If a finite p-group G with a minimal number of d generators is defined by r relations, then*

$$r > \left(\frac{d-1}{2}\right)^2.$$

Here we say that a group G is defined by r relations if it has a representation

$$G = \mathfrak{F}_d/\mathfrak{N}, \tag{5}$$

where $\mathfrak{N}$ is a normal subgroup of the free pro-p-group $\mathfrak{F}_d$, generated (topologically) by r elements. Obviously, if we have a representation $G = F_d/N$, where F_d is a discrete free group and a normal subgroup N is generated by r elements, then by going over to completions we obtain a representation (5), so that Theorem 1 is a fortiori true for the ordinary (non-topological) way of defining a group by means of relations.

Theorem 2. *If a nilpotent algebra A over a field k has a minimal system of*
2 *d generators and is defined by r relations, then*

$$r > \left(\frac{d-1}{2}\right)^2.$$

Obviously, in view of the corollary to Lemma 1, Theorem 1 is an immediate consequence of Theorem 2. The proof of Theorem 2 follows from a number of auxiliary propositions.

1. We consider a representation of A in the form $O_d/\mathfrak{A}$. The filtration in O_d defined by the powers of the ideal $I = (x_1, \cdots, x_d)$ defines a filtration $\mathfrak{m}^n$ in A. Let

$$\hat{A} = \sum_{n=0}^{\infty} \mathfrak{m}^n/\mathfrak{m}^{n+1}$$

be the adjoint algebra. As a graded algebra it is the residue class ring of the ring of polynomials R_d in the non-commuting variables $x_1, \cdots, x_d$ with respect to the ideal $\hat{\mathfrak{A}}$ generated by the initial forms of the elements of $\mathfrak{A}$. $\hat{\mathfrak{A}}$ is contained in the ideal $\bar{\mathfrak{A}}$ generated by all the homogeneous components of the generators of $\mathfrak{A}$. For the proof of the Theorem 2 it is obviously sufficient to show that for

$$r \leqslant \left(\frac{d-1}{2}\right)^2$$

the graded algebra $\bar{A} = R_d/\bar{\mathfrak{A}}$ is infinite-dimensional. If the algebra A is generated by the r relations $f_1, \cdots, f_r \in I^2 \subset O_d$, then a basis of $\bar{\mathfrak{A}}$ consists of r forms of degree 2, r forms of degree 3, etc. (Some of these may be equal to zero.) The fact that the factor algebra $R_d/\bar{\mathfrak{A}}$ with respect to such an ideal $\bar{\mathfrak{A}}$ is always infinite-dimensional will be proved in Lemma 5.

2. Let A be a graded algebra over a field k and $A = \Sigma_{n=0}^{\infty} A_n$ its decomposition into homogeneous components. We assume that the numbers,

$$b_n = \dim_k A_n, \quad n \geqslant 0,$$

are all finite. The function

$$P_A(t) = \sum_{n=0}^{\infty} b_n t^n$$

is called the Poincaré function of the algebra A.

3. The resolution of a graded algebra. The subsequent arguments use the construction of a certain auxiliary complex M, which is a non-commutative analogue of the Koszul complex in a local algebra.

We assume that in R_d a finite or infinite sequence of homogeneous polynomials $f_1, f_2, \cdots$ of degree ≥ 2 is given and that this sequence contains only a finite number of polynomials of a given degree i. We denote this number by r_i $(i \geq 2)$. Here some of the polynomials f_i may be equal to zero, but all the same a definite degree must be assigned to them. Thus, 0 may occur in our sequence an infinite number of times, but a given degree is assigned to it only at a finite number of places.

We denote by $\mathfrak{A}$ the ideal of R_d generated by all the forms f_i $(i = 1, 2, \cdots)$ and set $A = R_d/\mathfrak{A}$. Obviously A is a graded algebra. We consider a sequence of variables $u_1, u_2, \cdots$ in one-to-one correspondence with the forms $f_1, f_2, \cdots$ and denote by M the algebra of non-commuting polynomials in $x_1, \cdots, x_d, u_1, u_2, \cdots$ with the coefficients in k.

Thus a basis of the algebra M over k consists of expressions of the form

$$a_0 u_{i_1} a_1 u_{i_2} \ldots a_{k-1} u_{i_k} a_k, \tag{6}$$

where $a_0, \cdots, a_k$ range over all monomials in $x_1, \cdots, x_d$.

The dimension of a monomial in M is defined as the number of variables u_i occurring in it. Thus (6) is the general form of a monomial of dimension k. We denote the linear space of all homogeneous polynomials of dimension k by $M^{(k)}$. Obviously $M = \Sigma_{k=0}^{\infty} M^{(k)}$ and in this way M is turned into a graded algebra.

In M we define a derivation ∂ by setting $\partial x_i = 0$, $\partial u_i = f_i$ and extending it so that the usual formula holds:

$$\partial\,(v \cdot w) = \partial v \cdot w + (-1)^m v \cdot \partial w,$$

where m is the dimension of v. In explicit form this derivation is given as follows:

$$\partial\,(a_0 u_{i_1} \ldots a_{k-1} u_{i_k} a_k) = \sum_{l=1}^{k} (-1)^l a_0 u_{i_1} \ldots a_{l-1} f_{i_l} a_l \ldots u_{i_k} a_k. \tag{7}$$

It is easy to verify that $\partial^2 = 0$ and that ∂ decreases the dimension by 1.

Obviously $H_0(M) = A$ and ∂ defines a sequence of homomorphisms of R_d-modules

$$0 \leftarrow A \leftarrow M^{(0)} \overset{\partial}{\leftarrow} M^{(1)} \overset{\partial}{\leftarrow} M^{(2)} \leftarrow \ldots,$$

which is exact at the terms A and $M^{(0)}$, but not exact, in general, at the other terms, that is, this resolution of A is not, in general, acyclic.

We introduce the concept of degree of an element M, by assigning to the x_i the degree 1 and to u_i the degree n_i, where n_i is the degree of the form f_i. By virtue of this condition the derivation does not change degrees.

We denote by M_n the set of all homogeneous polynomials of degree n. Obviously M_n is a vector space over k whose dimension is finite owing to the fact that there exist only a finite number of forms f_i of given degree. It is clear that $M = \Sigma_{n=0}^{\infty} M_n$ and that we have so defined another grading in M.

We set $M_n^{(k)} = M_n \cap M^{(k)}$. Obviously $M_n = \Sigma M_n^{(k)}$ and M_n is a sub-complex in the complex M. Finally, the grading $A = \Sigma_{n=0}^{\infty} A_n$ coincides with the grading

$$H_0\,(M) = \sum_{n=0}^{\infty} H_0\,(M_n),$$

i.e.

$$\dim_k H_0\,(M_n) = \dim_k A_n = b_n.$$

4. **Lemma 2.** *Under the assumptions of* §3

$$P_A(t)\left(1-dt+\sum_{i=2}^{\infty} r_i t^i\right) \geqslant 1,$$

where inequality between power series is understood coefficient-wise.

Proof. If $n \geq 1$, then M_n contains the sub-complex $\Sigma_{i=1}^d M_{n-1} x_i$ which is isomorphic (as a complex) to M_{n-1}^d, the sum of d copies of M_{n-1}. We set

$$M_n \Big/ \sum_{i=1}^{d} M_{n-1} x_i = \widetilde{M}_n.$$

Every monomial of M_n has as its extreme right factor either an x_i or a u_j. Therefore we can identify $\widetilde{M}_n$ as a vector space with $\Sigma_{n_i \leq n} M_{n-n_i} u_i$, where by the condition on the degrees of the forms f_i the sum is finite.

Let us find the action of ∂ in $\widetilde{M}_n$. If in the term (6) $a_k = 1$, $u_{i_k} = u_i$, then this term occurs in $M_{n-n_i} u_i$. In the formula (7) the last kth term is contained in $\Sigma_{i=1}^d M_{n-1} x_i$ and is therefore equal to 0 in $\widetilde{M}_n$. The remaining part of (7) coincides with the formula for the derivative of the term $a_0 u_{i_1} a_1 \cdots u_{i_{k-1}} a_{k-1}$ of dimensions $k-1$. Thus we have the isomorphism of complexes

$$\widetilde{M}_n^{(k)} \simeq \sum_{n_i \leqslant n} M_{n-n_i}^{(k-1)}.$$

In particular

$$H_0(\widetilde{M}_n) = 0, \quad H_q(\widetilde{M}_n) = \sum_{n_i \leqslant n} H_{q-1}(M_{n-n_i}). \tag{8}$$

From the exactness of the sequence of complexes

$$0 \to \sum_{i=1}^{d} M_{n-1} x_i \to M_n \to \widetilde{M}_n \to 0$$

there follows the exactness of the sequence of cohomologies of which we give the first terms:

$$0 \leftarrow H_0(\widetilde{M}_n) \leftarrow H_0(M_n) \leftarrow H_0\left(\sum_{i=1}^{d} M_{n-1} x_i\right) \leftarrow H_1(\widetilde{M}_n),$$

or by (8)

$$0 \leftarrow H_0(M_n) \leftarrow H_0(M_{n-1}^d) \leftarrow \sum_{n_i \leqslant n} H_0(M_{n-n_i}). \tag{9}$$

For the dimensions of the spaces occuring in this sequence we obtain the inequality:

$$b_n \geqslant d \cdot b_{n-1} - \sum_{n_i \leqslant n} b_{n-n_i} \quad (n \geqslant 1).$$

Multiplying this inequality by t^n and adding up for all $n \geq 1$ we obtain an inequality for the series:

$$\sum_{n=1}^{\infty} b_n t^n \geqslant \sum_{n=1}^{\infty} d \cdot b_{n-1} \cdot t^n - \sum_{n_i \leqslant n} t^n b_{n-n_i}. \tag{10}$$

When we set in the last sum $n = n_i + m$, we see that it is equal to

$$\sum_{n_i, m} t^{n_i+m} \cdot b_m = \sum_i t^{n_i} \sum_{m=0}^{\infty} b_m t^m = \left(\sum_{i=2}^{\infty} r_i t^i\right) P_A(t).$$

On the other hand,

$$\sum_{n=1}^{\infty} b_n t^n = P_A(t) - 1$$

(since $b_0 = 1$), and

$$\sum_{n=1}^{\infty} d \cdot b_{n-1} \cdot t^n = dt P_A(t).$$

Therefore the inequality (10) yields:

$$P_A(t) - 1 \geqslant dt P_A(t) - \left(\sum_{i=2}^{\infty} r_i t^i\right) P_A(t);$$

hence,

$$P_A(t) \left(1 - dt + \sum_{i=2}^{\infty} r_i t^i\right) \geqslant 1. \tag{11}$$

This proves Lemma 2.

Obviously there is no difficulty in interpreting the exact sequence (9) in terms of the algebra A, without resorting to the construction of the complex M.

Lemma 3. *If the coefficients of the power series*

$$\left(1 - dt + \sum_{i=2}^{\infty} r_i t^i\right)^{-1}$$

are non-negative, then

$$P_A(t) \geqslant \left(1 - dt + \sum_{i=2}^{\infty} r_i t^i\right)^{-1} \tag{12}$$

and the algebra A is infinite.

The inequality (12) is obtained from (11) by multiplying both sides by the power series

$$F(t) = \left(1 - dt + \sum_{i=2}^{\infty} r_i t^i\right)^{-1}, \tag{13}$$

which by assumption has non-negative coefficients. It remains to show that the algebra A is infinite. For this purpose it is sufficient to show that $b_n > 0$ for an infinite number of values of n, and this follows from (13) if we can show that the power series $F(t)$ is not a polynomial in t.

We set

$$1+\sum_{i=2}^{\infty} r_i t^i = U(t).$$

By (13)

$$F(t)\,(U(t)-dt)=1,$$

i.e.

$$F(t)\,U(t)=1+dtF(t).$$

Since the power series $U(t)$ has non-negative coefficients, this equation is only possible if $F(t)$ is an infinite power series.

Note. If the numbers r_i satisfy the inequalities $r_i \le s_i$ and all the coefficients of the power series $(1-dt+\Sigma_{i=2}^{\infty} s_i t^i)^{-1}$ are non-negative, then Lemma 3 is also applicable. This follows from the fact that if F and G are two power series with the free term 1 and if $F \le G$ and $G^{-1} \ge 0$, then $F^{-1} \ge G^{-1}$. For $F = G - U$, where U is a power series without a free term and with non-negative coefficients. We have then:

$$F = G(1-UG^{-1}),$$

from which we find:

$$F^{-1} = G^{-1}(1-UG^{-1})^{-1}.$$

Since $U \ge 0$ and $G^{-1} \ge 0$, we have $(1-UG^{-1})^{-1} \ge 1$. Multiplying both sides of this inequality by the non-negative power series G^{-1} we obtain $F^{-1} \ge G^{-1}$.

Lemma 4. *If $b \le (a/2)^2$, $c \le a/2$, then all the coefficients of the series*

$$\frac{1-ct}{1-at+bt^2}$$

are positive.

For if $b \le (a/2)^2$, then $1-at+bt^2 = (1-\alpha t)(1-\beta t)$, where α and β are real, $\alpha \ge \beta \ge 0$, $\alpha+\beta = a$. If $\alpha > \beta$, then by splitting into partial fractions we obtain:

$$\frac{1-ct}{1-at+bt^2} = \frac{\alpha-c}{\alpha-\beta}\cdot\frac{1}{1-\alpha t}+\frac{c-\beta}{\alpha-\beta}\cdot\frac{1}{1-\beta t},$$

hence:

$$\frac{1-ct}{1-at+bt^2} = \sum_{n=0}^{\infty}\frac{(\alpha-c)\,\alpha^n+(c-\beta)\,\beta^n}{\alpha-\beta}\cdot t^n.$$

Since $\alpha > c$, all the coefficients are positive. If $\alpha = \beta = a/2$, then the lemma is obvious.

Lemma 5. *If the ideal $\overline{\mathfrak{A}} \subset R_d$ is generated by r forms of degree 2, r forms of degree 3, etc. and if $r \le (d-1/2)^2$, then the algebra $A = R_d/\overline{\mathfrak{A}}$ is infinite.*

For in the notation of Lemma 2 we have:

$$r_i = r, \quad i = 2, 3, \ldots.$$

We consider the series

$$\frac{1}{1-dt+r\sum_{i=2}^{\infty}t^i}=\frac{1-t}{1-(d+1)t+(d+r)t^2}. \tag{14}$$

By Lemma 4, if $r+d \leq (d+1/2)^2$, i.e. if $r \leq (d-1/2)^2$, then all the coefficients of the series (14) are positive, so that by Lemma 3 the algebra A is infinite.

§3. Applications.

As was shown in [4], if k is a field of algebraic numbers and $\mathfrak{G}$ the Galois group of its maximal unramified p-extension, then the minimal number of generators d of $\mathfrak{G}$ and its number of relations r (as a pro-p-group) satisfy the condition $r \leq d+\rho$, where ρ is the number of generators of the group of units of k.

We recall that d is equal to the number γ of generators of the p-component of the group of divisor classes of k.

From Theorem 1 it follows that the group $\mathfrak{G}$ is necessarily infinite if

$$\left(\frac{\gamma-1}{2}\right)^2 \geqslant \gamma+\rho.$$

The inequality holds if $\gamma \geq 3+2\sqrt{\rho+2}$. Hence we have

Theorem 3. *Over a field of algebraic numbers k there exist infinite unramified extensions if the number of generators γ of its group of divisor classes and the number of generators ρ of its group of units are connected by the relation*

$$\gamma \geqslant 3+2\sqrt{\rho+2}. \tag{15}$$

In the case of a quadratic imaginary field $\rho = 1$ and the inequality (15) is satisfied when $\gamma \geq 7$. The simplest example is the field

$$k = R(\sqrt{-3\cdot5\cdot7\cdot11\cdot13\cdot17\cdot19}) = R(\sqrt{-4849845}),$$

in which the 2-component of the group of divisor classes has 7 generators. This follows from the Gauss Theorem on genera, and it can also be verified in a elementary way that $\gamma \geq 7$. For it is easily seen the fields

$$k(\sqrt{-3}),\quad k(\sqrt{5}),\quad k(\sqrt{-7}),\quad k(\sqrt{-11}),\quad k(\sqrt{13}),\quad k(\sqrt{17}),\quad k(\sqrt{-19})$$

are unramified independent quadratic extensions of k. Thus over this field there exists an infinite unramified 2-extension.

We recall that a class field tower of a field k is a sequence $k = K_0 \subset K_1 \subset K_2 \subset \cdots$ in which K_i is the maximal abelian unramified extension of K_{i-1}. From Theorem 3 it follows that for a field of the type indicated the class field tower is infinite. In particular, this is true for the field

$R(\sqrt{-3\cdot5\cdot7\cdot11\cdot13\cdot17\cdot19})$.

It is known that if the class field tower of a field k is infinite, then k cannot be embedded in a field of algebraic numbers of class number 1. For if $K \supset k$ were a finite extension of class number 1, then the field K_1K/K would be an abelian unramified extension of K, and since K is of class number 1, K_1K would have to coincide with K, according to the theory of class fields. Similarly, all the $K_i \subset K$, and this would mean that the class field tower of k is finite.

Thus from Theorem 3 follows:

Theorem 4. *If for a field k condition* (15) *is satisfied, then k is not contained in any field of algebraic numbers of class number* 1.

In particular, such a field is

$$k = R(\sqrt{-3\cdot5\cdot7\cdot11\cdot13\cdot17\cdot19}).$$

Finally, as E. Artin remarked, the class field tower problem is connected with the problem of the growth of the minimal discriminant of a field of algebraic numbers of degree n. We denote by D_n the minimum of the absolute values of the discriminants of fields of algebraic numbers of degree n. By Minkowski's Lemma there exists a constant $c > 1$ such that

$$\sqrt[n]{D_n} > c.$$

For example, by a result of Rogers [1] we may set $c = 32$ for all sufficiently large n.

On the other hand, if over a field k there exists an infinite sequence of unramified extensions K_i, and $[K_i : K] = n_i$, $[K : R] = n_0$, then

$$|D(K_i/R)| = |D_0|^{n_i}, \quad [K_i : R] = n_i n_0,$$

where $D(K_i/R)$ is the discriminant of the field K_i/R and D_0 the discriminant of K/R. Therefore

$$\sqrt[n_0 n_i]{D_{n_0 n_i}} \leqslant \sqrt[n_0]{|D_0|}.$$

Thus from Theorem 3 follows:

Theorem 5. *There exists an infinite sequence of natural numbers* $n_1 < m_2 < \cdots$ *and appropriate constants* c, c' *such that*

$$c' > \sqrt[n_i]{D_{n_i}} > c.$$

In particular, Roger's result and the example of the field

$$R(\sqrt{-3\cdot5\cdot7\cdot11\cdot13\cdot17\cdot19})$$

show that

$$4405 > \sqrt[n]{D_n} > 32$$

4 for $n = 2^k$ and all sufficiently large n.

BIBLIOGRAPHY

[1] C. A. Rogers, *The product of n real homogeneous linear forms*, Acta Math. 82 (1950), 185–208. MR 11, 501.

[2] J.-P. Serre, *Leçons sur la cohomologie Galoisienne*, Collège de France, 1962–63.

[3] I. R. Šafarevič, *Fields of algebraic numbers*, Proceedings of the International Congress of Mathematicians, Stockholm, 1962, pp. 163–176. (Russian)

[4] ———, *Extensions with given ramification points*, Publications Mathématiques de l'Institut des Hautes Études Scientifiques, No. 18, 1963. (Russian)

Translated by
K. A. Hirsch

Algebraic surfaces
(with B. G. Averbukh et al.)

Tr. Mat. Inst. Steklova **75** (1965). Zbl. **154**, 210
[Proc. Steklov Inst. Math. **75**, III – VI, 52 – 84, 162 – 182 (1965)]

Preface

Aeschylus said that his tragedies were fragments from the great banquets of Homer;
Athenaeus VIII, 39, 347e.

The basic facts in the theory of algebraic surfaces were discovered, in the latter half of the nineteenth century and the beginning of the twentieth, by Max Noether, Picard, Poincaré and in particular by the members of the classical Italian school of algebraic geometry, Castelnuovo, Enriques and Severi. Their results were the starting point for the next stage in the development of algebraic geometry, which was based on the application of topological, analytic and algebraic methods.

During this stage it became clear that the results making up the "classical" theory of algebraic surfaces fall into two fundamentally different classes.

Some of them are special cases of general theorems about algebraic varieties (or schemata) of arbitrary dimensionality. The clearest examples are provided by the theory of Picard varieties or of abelian varieties or by the Riemann-Roch theorem. It is interesting to note that almost all the results in the classical survey of Zariski [16] are exactly of this kind. At the present time there are in existence many excellent expositions of the theories relating to this part of the subject.

The results of the second class deal specifically with algebraic surfaces. Here belong such basic features of the subject as the criterion for whether an algebraic surface is rational or ruled, the solution of the problem of Luröth, the theory of minimal models, and the great complex of results which are grouped together by the Italian algebraic geometers under the heading of "classification of algebraic surfaces".

It seems that none of these results can be extended to varieties of higher dimension without the most essential changes, and at the present time even the very nature of such changes remains entirely unknown. Some attempts have been made to provide these results with proofs that are rigorous from a modern point of view and are based on present-day techniques, and also to extend the results as far as possible. The first (and basic) publications in this direction are the proof presented by Kodaira [48] for the criterion of rationality and the articles and book of Zariski [19] on the problem of minimal models. The purpose of the present book is to give a connected account of this whole range of questions. Below we give a short description of the contents of the book.

The entire theory is based on the connection between rational mappings of a variety into a projective space and the classes of divisors on the variety (see, for example, [30]). For algebraic curves it is well known that "almost everywhere" (for nonhyperelliptic curves) the mapping that corresponds to the canonical class is birational and defines the so-called canonical model of the curve uniquely up to a projective transformation. Thus the problem of birational classification is reduced to questions of the projective classification of curves in space.

A mapping corresponding to a canonical class of multiplicity two or three plays the same role for all curves other than rational and elliptic ones. For those curves a mapping corresponding to a canonical class of any multiplicity will not be birational. This method is in general inapplicable for their description, which, however, is easily obtained from other considerations.

The question is whether one obtains an analogous situation for algebraic surfaces. One first considers the class of those surfaces for which a canonical class of any multiplicity gives a birational mapping. It turns out that it is always sufficient to take a canonical class with a multiplicity not larger than nine (but a multiplicity of three, which plays the same role in the theory of curves, may be insufficient, as is shown by examples). The surfaces of this kind may be characterized in a simple manner: they are those nonrational surfaces for which the index of selfintersection of the canonical class is positive.

The rest of our task is the description of the remaining surfaces, those for which no canonical class of any multiplicity defines a birational mapping. These surfaces are analogous, from this point of view to the rational and elliptic curves. Their constructive description is also to a great extent analogous to the description of those curves. Namely, the surfaces we are considering fall into the following five groups: (1) rational surfaces; (2) two-dimensional abelian varieties; (3) ruled surfaces, i.e., surfaces made of families of rational curves; (4) surfaces made of families of elliptic curves; and (5) certain surfaces that are similar to abelian varieties in that their canonical class is zero, but which, unlike abelian varieties, have their first Betti number equal to zero.

In order to examine all these algebraic surfaces, we divide them into four groups on the basis of the value of an important invariant, which we denote by κ. The symbol κ stands for the maximal dimension of the image of the surface under rational mappings corresponding to different multiplicities of the canonical class. It is clear that κ is always less than or equal to two. If the linear systems corresponding to all the multiplicities of the canonical class are empty, then we set $\kappa = -1$. Thus κ may take on the four values -1, 0, 1 or 2. The goal of the classification is to give the character of the surfaces with a given value of κ with the aid of the so-called numerical invariants (the index of the selfintersection of a canonical class, plurigenera defined by the formula (11), and the irregularity defined by the formula (12)) and to give a constructive description of them. The results of the classification are given in the table at the end of the introduction.

The book also contains results outside the above mentioned theory, which are, however, related to it.

We shall discuss in detail the following: the theory of birational transformations of surfaces, the theory of minimal models, and Noether's theorem.

The theory of birational transformations of surfaces is based on the concept of the σ-process. This is a birational transformation

$$f\colon\ V \longrightarrow V'$$

of nonsingular surfaces V and V', which is biregular everywhere except at a point $P \in V$ and a curve $C \subset V'$, where, moreover, f^{-1} is regular and $f^{-1}(C) = P$. The following are basic results:

(1) if ϕ is a birational mapping

$$\phi\colon\ V \longrightarrow V'$$

of nonsingular surfaces such that ϕ^{-1} maps V' regularly onto V, then ϕ is the product of a finite number of σ-processes;

(2) any birational transformation of a nonsingular surface onto a nonsingular surface is the product of a finite number of σ-processes and a finite number of transformations inverse to σ-processes.

The theory of minimal models studies those surfaces V (called minimal) which are such that any regular birational transformation $f\colon\ V \longrightarrow V'$ is biregular. Every surface is birationally equivalent to a minimal one, from which it is obtained, consequently, by a finite number of σ-processes. The basic theorem says that in the class of surfaces birationally equivalent to each other, there is only one minimal one if the surfaces are not ruled.

The minimal models of ruled (in particular, rational) surfaces are all described.

Finally, Noether's theorem relates to the structure of the group of all birational transformations of a projective plane (or, what is the same, of the group of the automorphisms of the field of rational functions $k(x, y)$ of two variables). It shows that this group is generated by the so-called quadratic transformations:

$$x' = \frac{a_1x + b_1y + c_1}{a_2x + b_2y + c_2}, \qquad y' = \frac{a_3x + b_3y + c_3}{a_4x + b_4y + c_4}$$

The classical results presented in this book may almost all be found in the survey of Enriques [59]. The present work is very closely connected with Enriques' book. The proofs of a large part of the theorems are based on ideas of Enriques. At the same time, it would hardly be possible to carry out the details of Enriques' proofs, for example, following the customs of the time and his school, he frequently limited himself to the consideration of a "general" case, not choosing the most unpleasant cases that might be examined. On the other hand, for certain questions we can supplement the classical results with new ones. This is true, for example, of certain results in Chapters V, VII, and IX. Finally, there are a few divergences from assertions of Enriques.

We do not aim for the greatest possible generality in the conditions imposed on the base field. All results are true if this field coincides with the field of complex numbers. The majority of arguments, however, retain their validity if the
1 base field is algebraically closed and has characteristic 0, and some arguments remain valid for any algebraically closed field. These considerations are discussed in more detail in each chapter.

The present book is based on reports on seminars in the theory of algebraic surfaces held in 1961–1962 and 1962–1963 under the leadership of I. R. Šafarevič. The texts of the reports were then worked over, and certain parts were rewritten. The individual chapters were written by the following authors: Chapters I, II and III by A. B. Žižčenko; Chapters IV and VII by I. R. Šafarevič; Chapter V, §§1 and 2, by Ju. I. Manin §§3–6 by Ju. R. Vaĭnberg and Ju. I. Manin, §7 by A. N. Tjurin; Chapter VI by B. G. Moĭšezon; Chapters VIII and X by B. G. Averbuh; Chapter IX by G. N. Tjurin.*

---

* Translator's note: In a more recent article, *On special types of Kummer and Enriques surfaces*, Izv. Akad. Nauk SSSR 29 (1965), 1095–1118, Averbuh fills out some gaps in the classification which he began in Chapters VIII and X. This article has been translated as the Appendix to the present volume.

Ruled surfaces

CHAPTER IV

Ruled surfaces will be studied in this chapter, and it will be proved, in particular, that they are characterized by the condition $P_{12}=0$. If they are nonrational, then $q>0$. It will be shown that if $q>1$, then the condition $p_g=0$ is sufficient for the surface to be ruled. For $q=1$ this condition is insufficient. All the surfaces with $p_g=0$, $q=1$ will be found, and it will be verified that those among them for which $P_{12}=0$ are ruled. Thus, we give in this chapter a classification of all the surfaces with $p_g=0$ and $q>0$, not just a classification of the ruled nonrational surfaces. We will assume that the base field k is the field of complex numbers. Nevertheless, almost all the arguments remain valid when k is an algebraically closed field of characteristic 0. We will make special note of the places where the assumption $k=\mathbf{C}$ is essential.

§1. Elementary properties

Definition. *The surface V is said to be ruled if it is birationally equivalent to the direct product of an algebraic curve with a projective line.*

Theorem 1. *If the surface V is ruled, i.e., if V is birationally equivalent to $B\times P^1$, where B is an algebraic curve, then $P_n(V)=0$, $n\geq 1$ and the irregularity q of the surface V coincides with the (geometric) genus of the curve B.*

Proof. In view of the birational invariance of the numbers P_n and q we can assume that

$$V=B\times P^1, \tag{1}$$

where B is a nonsingular curve. In general, let $X=B\times C$ where B and C are nonsingular curves and let α and β be differential forms of degree 1 on B and C respectively, and let (α) and (β) be their divisors. If π_1 and π_2 are projections of V onto B and C, then the form

$$\omega=\pi_1^*(\alpha)\wedge\pi_2^*(\beta)$$

is a two-dimensional differential form on V. It is clear that

$$(\pi_1^*(\alpha))=\pi_1^{-1}((\alpha))=(\alpha)\times C,$$
$$(\pi_2^*(\beta))=\pi_2^{-1}((\beta))=B\times(\beta),$$

$$K(B\times C)=K(B)\times C+B\times K(C), \tag{2}$$
$$(\omega)=(\alpha)\times C+B\times(\beta). \tag{3}$$

From this it follows that, if $V = B \times P^1$, then

$$K \cdot (b \times P^1) = -2((B \times a) \cdot (b \times P^1)) = -2 < 0, \quad b \in B, \ a \in P^1$$

and consequently, $(nK \cdot (b \times P^1)) < 0$ for any $n > 0$. If we had $P_n > 0$ for at least one $n > 0$, there would exist a divisor D, $D > 0$, $D \sim nK$. Then necessarily $(D \cdot (b \times P^1)) < 0$. But this is impossible: if $D = \Sigma n_i c_i + \Sigma m_j (b_j \times P^1)$, then $(D \cdot (b \times P^1)) = \Sigma n_i (C_i \cdot (b \times P^1)) \geq 0$.

For the proof of the assertion about irregularity, we use the fact that, for any varieties B and C, the Albanese variety possesses the property

$$A(B \times C) = A(B) \times A(C).$$

In particular, from (1) we obtain

$$A(V) = A(B) \times A(P^1) = A(B),$$

and since

$$q = \dim A(V), \quad g = \dim A(B),$$

where g is the genus of the curve B, this proves our assertion.

The basic problem of this chapter consists in proving the converse assertion in its stricter form: a surface is ruled if $P_{12} = 0$. Here we shall begin from the theorem of Noether (Chapter I, §3). We reformulate that theorem geometrically: a rational mapping

$$\pi\colon V \to B$$

corresponds to the imbedding $K_1 = k(B) \to K = k(V)$, where if ξ is a generic point of the curve B, then the field K/K_1 is a field of functions on the curve $\pi^{-1}(\xi)$. Thus the theorem of Noether can be given the following formulation:

Theorem 2. *If there exists a rational mapping*

$$\pi\colon V \to B$$

of a surface V onto a curve B, such that the inverse image $\pi^{-1}(\xi)$ of a generic point ξ of the curve B is an irreducible curve of genus 0, then V is a ruled surface.

We shall use this theorem for the particular case in which π is a regular mapping. Then it determines a fibering of V into nonintersecting fibers $F_b = \pi^{-1}(b)$, $b \in B$. We shall frequently call π a fibering and F_b its fibers. The manner of the construction of such a fibering is based on the consideration of the Albanese mapping.

§2. The Albanese mapping for $p_g = 0$, $q > 0$

Theorem 3. *If $\alpha_V: V \longrightarrow A(V)$ is the Albanese mapping of a surface V for which $p_g = 0$, $q > 0$, then:* 1) *$\alpha_V(V)$ is an algebraic curve;* 2) *$\alpha_V(V)$ is nonsingular; and* 3) *the genus of $\alpha_V(V)$ is equal to q.*

Proof of 1). Since the variety $\alpha_V(V)$ generates all of $A(V)$, and $\dim A(V) = q > 0$, it is also true that $\dim \alpha_V(V) > 0$, i.e., $\dim \alpha_V(V) = 1$ or 2.

We shall show that $\dim \alpha_V(V) \neq 2$. Let us assume that $\dim \alpha_V(V) = 2$, and let a be a nonsingular point of the surface $\alpha_V(V)$. Since the mapping α_V is defined up to a translation of the variety $A(V)$, we can assume that $a = 0$. Let σ be a bivector, in the tangent space to $A(V)$ at the point 0, corresponding to the plane tangent to $\alpha_V(V)$ at this point; and let s be a two-dimensional element of the Grassmann algebra such that $(s, \sigma) \neq 0$. We denote by ω the invariant differential form corresponding to s. It is clear that ω is a differential form of the first kind on $A(V)$. From the fact that $(s, \sigma) \neq 0$, it follows that ω is not identically equal to zero at the point 0. From this it follows that the differential form of the first kind $\alpha_V^*(\omega)$ on V does not vanish identically, and this contradicts the assumption $p_g = 0$.

Proof of 2). Let $\alpha_V(V) = B$. Let B_N be a normalization of B and let $N: B_N \longrightarrow B$ be the canonical mapping. Since N is a birational equivalence, there exists a rational mapping $\nu: B \longrightarrow B_N$, $\nu = N^{-1}$. Let $\phi = \nu \cdot \alpha_V$, $\phi: V \longrightarrow B_N$. We have a commutative diagram

$$\begin{array}{ccc} A(V) & \xrightarrow{\alpha(\varphi)} & A(B_N) \\ \psi_2 \uparrow & & \uparrow \psi_1 \\ B & \xleftarrow{N} & B_N. \end{array} \qquad (4)$$

Here ψ_2 is the imbedding of B in $A(V)$. Since the (geometric) genus of the curve B, and thus also that of B_N, is distinct from 0 (cf. [31], Chapter II), the canonical mapping ψ_1 of the curve B_N into $A(B_N)$ is also an imbedding, which we can assume to be the identity mapping. The mapping $\alpha(\phi)$ is regular, and consequently the mapping $N': B \longrightarrow B_N$, $N' = (\alpha(\phi)|_B) \cdot \psi_2$ is also regular. From the diagram (4) it follows that $NN' = 1$, $N'N = 1$, i.e., N is a biregular equivalence of B and B_N, so that $B = B_N$ does not have singular points.

Proof of 3). Since $B = B_N$, (4) reduces to the diagram

$$\begin{array}{ccc} A(V) & \xrightarrow{\alpha(\varphi)} & A(B) \\ & {\scriptstyle\psi_2}\nwarrow \quad \nearrow{\scriptstyle\psi_1} & \\ & B & \end{array}$$

According to the universal mapping property of an Albanese variety there exists a mapping $\eta: A(B) \longrightarrow A(V)$, giving the commutative diagram

$$\begin{array}{ccc} A(V) & \overset{\eta}{\leftarrow} & A(B) \\ & {}^{\psi_2}\nwarrow \quad \nearrow^{\psi_1} & \\ & B & \end{array}$$

From this it follows that $\psi_2 = \eta\psi_1$ and $\psi_1 = \alpha(\phi)\,\psi_2$, i.e.,

$$\psi_1 = \alpha(\phi)\,\eta\psi_1, \quad \psi_2 = \eta\alpha(\phi)\,\psi_2.$$

Since $\mathrm{Im}\,\psi_1$ and $\mathrm{Im}\,\psi_2$ generate $A(B)$ and $A(V)$, it follows that $\alpha(\phi)\,\eta = 1$, $\eta\alpha(\phi) = 1$, i.e., $A(V)$ is isomorphic to $A(B)$. This proves 3) and Theorem 3.

We denote α_V by π and obtain a regular mapping (or fibering)

$$\pi: V \longrightarrow B$$

onto a nonsingular curve B of genus q.

Theorem 4. *A generic fiber of the fibering π is irreducible.*

Proof. We saw that the assertion of Theorem 4 is equivalent to the fact that the field $k(B)$ is algebraically closed in the field $k(V)$. If this is not so, let K' be the algebraic closure of $k(B)$ in $k(V)$ and let B' be a nonsingular model of the field K'. The inclusion $k(B') = K' \subset k(V)$ determines a rational mapping $V \longrightarrow B'$ and thus an epimorphism $A(V) \longrightarrow A(B')$.

From this it follows that the genus g of the curve B' is not greater than q. But B' is a covering of B (since $k(B) \subset k(B')$) and consequently $g \geq q$. We see that $g = q$. According to the formula of Hurwitz for the genus of a covering, the equality $g = q$ is possible only when $g = q = 1$. Thus, $A(V) = B$, $A(B') = B'$, and we have the mappings

$$A(V) \overset{\varphi}{\rightarrow} B', \quad B' \overset{\psi}{\rightarrow} B, \ \psi\varphi = \alpha_V.$$

According to the universality property of the Albanese variety, there exists a mapping $\chi: B \longrightarrow B'$, giving a commutative diagram

$$\begin{array}{ccc} A(V) & \overset{\alpha_V}{\rightarrow} & B \\ & {}^{\varphi}\searrow \quad \swarrow^{\chi} & \\ & B' & \end{array}$$

From this, as in the proof of Theorem 3, it is easy to obtain that χ and ψ are isomorphisms, i.e. that $B' = B$ and $K' = k(B)$. The theorem is proved.

§3. The case $q > 1$

In this section we will always assume that for the surface V

$$p_g = 0, \quad q > 1. \tag{5}$$

Lemma 1. *For a surface satisfying the conditions* (5),

$$(K^2) < 0.$$

Proof. According to formula (4) of the Introduction

$$\frac{(K^2) + \chi}{12} = 1 - q + p = 1 - q,$$

so that

$$(K^2) = -\chi + 12(1 - q).$$

Since $\chi = 2 - 4q + b_2$, we have

$$(K^2) = -2 + 4q - b_2 + 12(1 - q),$$

$$(K^2) = 8(1 - q) + 2 - b_2.$$

Since $q \geq 2$, $b_2 \geq 1$, it follows that $(K^2) \leq -7$.

Corollary. *If E is a hyperplane section, then $l(E + mK) = 0$ for sufficiently large m. If $l(E + nK) > 0$ and $l(E + mK) = 0$, for $m > n$, then*

$$E + nK \sim D > 0, \quad D \neq 0 \text{ and } p_a(C) \leq q \tag{6}$$

for any cycle C for which $0 < C < D$.

All the assertions except $D \neq 0$ follow from Lemma 3 (Chapter III, §1). The assertion $D \neq 0$ follows from the fact that otherwise we would have $E \sim -nK$ while $(E^2) > 0$, $(K^2) < 0$.

Lemma 2. *If C is an irreducible curve on V, π is the fibering introduced in the preceding section, and F_b is one of its fibers, then either $C = F_b$ or*

$$p_a(C) \geq (C \cdot F)(q - 1) + 1. \tag{7}$$

Proof. If $C \neq F_b$, then the mapping $f = \pi/C$ defines C as a covering of B. The degree of this covering is equal to $C \cdot F$. In fact by definition this degree is equal to the degree of the divisor $f^{-1}(b)$. Considering π on V locally as a function and using the equation

$$\deg(\pi/C)_0 = ((\pi_0) \cdot C)$$

($(\pi)_0$ and $(\pi/C)_0$ are zero divisors of the functions π and π/C on V and C respectively), we obtain that $\deg f^{-1}(b) = (F_b \cdot C)$. Applying now the formula of Hurwitz for the genus of a covering, we obtain the inequality $\gamma \geq (F \cdot C)(q - 1) + 1$ for the geometric genus γ of the curve C, and since $p_a(C) \geq \gamma$, (7) follows from this.

Lemma 3. *If the genus g of a generic fiber F of a fibering π is not 0, then $(F \cdot K) \geq 0$.*

By assumption

$$\frac{(F \cdot (F + K))}{2} + 1 \geqslant 1,$$

and since $(F^2) = 0$, we have $(F \cdot K) \geq 0$.

Corollary. *Under the conditions given for the divisor D in (6), $(D \cdot F) \geq 3$.*

Since $D \sim E + nK$, $(D \cdot F) = (E \cdot F) + n(K \cdot F) \geq (E \cdot F)$. It is clear that $(E \cdot F)$ is the degree of the curve F. If this curve had degree 1 or 2, it would be rational, which would contradict the assumption that $g > 0$.

Theorem 5. *An algebraic surface V with the invariants $p_g = 0$ and $q > 1$ is ruled.*

In the proof we may assume that V is a relatively minimal model and, consequently, does not contain exceptional curves of the first kind.

We consider the mapping $\pi \colon V \to B$ constructed in the preceding section. It satisfies all the assumptions of Theorem 2 except that we still do not know that the genus g of a generic fiber is equal to 0. If we can prove this, Theorem 5 will follow from Theorem 2.

We assume that $g \geq 1$. Let the divisor D whose existence is proven in the corollary of Lemma 1 have the form $D = \Sigma_1^m n_i C_i$, where the C_i are distinct irreducible curves. According to the corollary of Lemma 3, $(D \cdot F) \geq 3$.

We consider separately three cases, which together include all the possibilities:

1) for some C_i, for instance for $i = 1$, $(C_i \cdot F) \geq 2$,

2) for all C_i, $(C_i \cdot F) \leq 1$ but $m \geq 2$,

3) $D = nC$, $(C \cdot F) = 1$, $n \geq 3$.

In case 1) we can calculate $p_a(C_1)$ in different ways on the basis of Lemma 2 and formula (6). We obtain the contradiction

$$p_a(C_1) \leqslant q, \quad p_a(C_1) \geqslant 2(q-1) + 1 = 2q - 1.$$

In case 2), for at least one of the C_i, for instance for $i = 1$, $(C_i \cdot F) = 1$, since $(D \cdot F) > 0$. By assumption there exists another curve C_2. We apply formula (6) to $C = C_1 + C_2$:

$$p_a(C_1 + C_2) \leq q.$$

Since $p_a(C) = (C \cdot (C + K))/2 + 1$,

$$p_a(C_1 + C_2) = p_a(C_1) + p_a(C_2) + (C_1 \cdot C_2) - 1.$$

According to Lemma 2,

$$p_a(C_1) \geq q,$$

so that

$$p_a(C) \geq q + p_a(C_2) + (C_1 \cdot C_2) - 1.$$

If also $(C_2 \cdot F) = 1$, then $p_a(C_2) \geq q$, and since $(C_1 \cdot C_2) \geq 0$, we obtain a contradiction with (6):

$$p_a(C) \geq 2q - 1.$$

If $(C_2 \cdot F) = 0$, then $C_2 = F_b$, $(b \in B)$, $p_a(C_2) \geq 1$ and $(C_1 \cdot C_2) = (C_1 \cdot F) = 1$, so that we again obtain a contradiction with (6):

$$p_a(C) \geq q + 1.$$

In case 3) we argue in exactly the same way applying (6) and Lemma 2 to the curve $2C$. We get that

$$p_a(2C) \leqslant q,\ p_a(2C) \geqslant 2q + (C^2) - 1.$$

We obtain a contradiction if we show that $(C^2) \geq 0$. But $(C^2) < 0$ gives $(C \cdot D) < 0$, i.e. $(C \cdot (E + nK)) < 0$, and thus $(C \cdot K) < 0$, since $(C \cdot E) > 0$. Therefore $(C^2) + (C \cdot K) = 2p_a(C) - 2 < 0$, which is possible only if $(C^2) = -1$, $p_a(C) = 0$, i.e. if C is an exceptional curve of the first kind. Since we assumed that there were no such curves on V, this proves the theorem.

§4. Regular mappings of algebraic surfaces onto curves

We consider an arbitrary regular mapping $\pi: V \longrightarrow B$ of an algebraic surface V onto a nonsingular algebraic curve B with an irreducible generic fiber F. Let q and g be the genera of B and F. We will assume the following properties of such fiber spaces to be known (cf. for example, [25]).

The fiber $F_b = \pi^{-1}(b)$ is connected for all $b \in B$. For all points $b \in B$, except, perhaps, a finite number, F_b is an irreducible nonsingular algebraic curve with genus g. The set of points $\{b_1, \cdots, b_s\}$ for which this is not true will be denoted by S, and the corresponding fibers F_{b_i} will be said to be degenerate or singular.

The fiber space, $\pi: V - \pi^{-1}(S) \longrightarrow B - S$ is (if $k = \mathbf{C}$, the field of complex numbers) differentiably locally trivial.

We will denote by $\chi(L)$ the Euler characteristic of a topological space L. In particular, if

$$L = F_b = \sum_1^m n_i C_i,\ n_i > 0, \text{ then } \chi(F_b) = \chi(F'),\ F' = \sum_1^m C_i.$$

Theorem 6. *When* $k = \mathbf{C}$ *we have*

$$\chi(V) = \chi(F)\,\chi(B) + \sum_1^s (\chi(F_{b_i}) - \chi(F)). \tag{8}$$

Proof. (Proposed by A. B. Žižčenko.) We let $\widetilde{V} = V - \pi^{-1}(S)$. From the exact cohomology sequence determined by the space, the closed subspace, and its complement, it follows that

$$\chi(V) = \chi(\widetilde{V}) + \chi(\pi^{-1}(S)). \tag{9}$$

It is clear that

$$\chi(\pi^{-1}(S)) = \sum_{1}^{s} \chi(F_{b_i}). \tag{10}$$

Since $V \to B - S$ is locally trivial (as a differentiable fiber space), it follows from the spectral sequence of Leray for this fiber space that

$$\chi(\widetilde{V}) = \chi(F)\,\chi(B - S). \tag{11}$$

Finally,

$$\chi(B) = \chi(B - S) + s \tag{12}$$

(s is the number of points in S), as, for example, follows from the exact sequence analogous to the one considered in the derivation of (9). Comparing (9), (10), (11) and (12), we obtain (8).

Lemma 4. *If C is a connected curve (perhaps reducible) on a surface V, then*

$$\chi(C) \geqslant -(C\cdot(C + K)), \tag{13}$$

where equality holds only when C is an irreducible nonsingular curve.

Proof. Let $C = \Sigma C_i$, let $\overline{C}$ be a normalization of C, i.e. the unconnected sum of normalizations $\overline{C}_i$ of the curves C_i, and let

$$\phi\colon \overline{C} \to C$$

be the canonical regular mapping. At all points except a finite number, ϕ is a biregular equivalence. Therefore the usual manner of calculation of the Euler characteristic gives

$$\chi(C) = \chi(\overline{C}) - \delta,$$

where

$$\delta = \sum_{c\in C} (\deg(\varphi^{-1}(c)) - 1). \tag{14}$$

If the genus of the curve $\overline{C}_i$ is equal to g_i, then

$$\chi(\overline{C}) = \sum \chi(\overline{C}_i) = 2\sum(1 - g_i)$$
$$g_i = p_a(C_i) - \delta_i$$

(in view of formula (5) of the Introduction).

Thus,

$$\begin{aligned}\chi(C) &= 2\sum(1 - p_a(C_i)) + 2\sum\delta_i - \delta\\ &= -\sum(C_i\cdot(C_i + K)) + 2\sum\delta_i - \delta.\end{aligned}$$

Let n_i points of the curve C_i be taken into the point $c \in C$ under the mapping ϕ. Then the corresponding term in (14) is equal to

$$\left(\sum_1^m n_i\right) - 1 = \sum_1^l (n_i - 1) + l - 1,$$

if $n_i > 1$ for $i = 1, \cdots, l$ and $n_i = 1$ for $i = l+1, \cdots, m$. Here the term corresponding to the point c in the expression for δ_i is not smaller than $n_i - 1$, and the number $l - 1$ is not smaller than the multiplicity of the point c in the divisor $\Sigma_{i<j}(C_i \cdot C_j)$. From this it follows that

$$\delta \leqslant \sum_i \delta_i + \sum_{i<j} (C_i \cdot C_j). \tag{15}$$

Thus

$$2\delta_i - \delta \geqslant -2 \sum_{i<j} (C_i \cdot C_j), \tag{16}$$

so that

$$\chi(C) \geqslant -\sum (C_i (C_i + K)) - 2 \sum_{i<j} (C_i \cdot C_j)$$

$$= -((\sum C_i) \cdot (\sum C_i + K)) = -(C \cdot (C + K)).$$

In view of (16), equality holds in (13) only when all the $\delta_i = 0$ and $(C_i \cdot C_j) = 0$ for $i \neq j$. The first means that the curves are nonsingular, and the second that they are mutually nonintersecting. Since we are assuming a connected curve, this means that it consists of one component. The lemma is proved.

Lemma 5. *If a fiber F_b of the mapping π has the form $\Sigma n_i C_i$, then for* $C = \Sigma m_i C_i$, $m_i \geq 0$, we have $(C^2) \leq 0$.

We use the equality

$$(C \cdot F) = 0, \tag{17}$$

which is clear if the fiber F is taken different from F_b.

If it were true that $(C^2) > 0$, then for a hyperplane section E and a sufficiently large n

$$l(nC - E) > 0,$$

as follows directly from the Riemann-Roch inequality.

Let $nC - E \sim D > 0$, i.e.

$$nC \sim E + D.$$

Since $(D \cdot F) \geq 0$ and $(E \cdot F) > 0$, this implies that $(C \cdot F) > 0$, in contradiction with (17).

Theorem 7 (semicontinuity of the Euler characteristic). *If F is nonsingular and F_0 is a singular fiber of a mapping π and the surface V is a relatively minimal model, then*

$$\chi(F_0) \geq \chi(F), \tag{18}$$

where equality holds only when the genus of F is equal to 1 and F_0 is a nonsingular curve of genus 1 taken with some multiplicity.

Proof. It is clear that

$$\chi(F) = 2 - 2g = -(F\cdot(F+K)) = -(F\cdot K),$$

since $(F^2) = 0$.

Let $F_0 = \Sigma n_i C_i$, $n_i \geq 1$; we set $F' = \Sigma C_i$.

It follows from Lemma 4 that

$$\chi(F_0) - \chi(F) = \chi(F') - \chi(F) \geqslant -(F'^2) + ((F - F')\cdot K).$$

According to Lemma 5, $(F'^2) \leq 0$. It remains for us to show that $(F - F')\cdot K \geq 0$.

If for some C_i it were true that $(C_i \cdot K) < 0$, we would have

$$(C_i \cdot (C_i + K)) < (C_i^2),$$

and hence

$$2p_a(C_i) - 2 < (C_i^2).$$

Since $(C_i^2) \leq 0$ by Lemma 5, $p_a(C_i) - 1 < 0$, i.e. $p_a(C_i) = 0$, and $-2 < (C_i^2) \leq 0$. For C_i^2 there are consequently two values: -1 and 0. The first case would mean that C_i was an exceptional curve of the first kind, which would contradict the fact that V is a relatively minimal model. Let $(C_i^2) = 0$. From the condition $(C_i \cdot F_0) = 0$ we obtain

$$n_i(C_i^2) = -\sum_{j \neq i} n_j (C_i \cdot C_j) = 0.$$

Since $(C_i \cdot C_j) \geq 0$ for $i \neq j$, it follows from this that $(C_i \cdot C_j) = 0$ for all $i \neq j$, and this contradicts the connectedness of the fiber F_b if there exists a curve $C_j \neq C_i$. Thus, $F_0 = nC_i$. The inequality $p_a(F) \geq 0$ and the formula for p_a give us

$$0 \leqslant p_a(F) = p_a(nC_i) = np_a(C_i) + (n^2 - n)(C_i^2) + 1 - n = 1 - n,$$

i.e. $0 \leq 1 - n$, which if possible only for $n = 1$. Thus, all the $n_i = 1$. But in this case $F_0 = F'$, and $((F - F')\cdot K) = ((F_0 - F')\cdot K) = 0$. The inequality (18) is proved

We now explain when equality may hold. For it to hold, all the inequalities met along the way must be equalities. In particular, this refers to the inequality $\chi(F') \geq -(F'\cdot(F' + K))$, which, according to Lemma 4, is an equality only if $F_0 = nC$, where C is an irreducible nonsingular curve and $n \geq 2$. The inequality

$$-(F')^2 + ((F - F')\cdot K) \geqslant 0,$$

which we proved must also be an equality. Since then every member of the left side is nonnegative, then $((F - F')\cdot K) = (n-1)(C\cdot K) = 0$, and hence $(C\cdot K) = 0$ and $(F\cdot K) = 0$. From the fact that $(F^2) = 0$ it follows that, since $F_0 = nC$, $(C^2) = 0$. Hence

$$p_a(F) = \frac{(F^2)+(F\cdot K)}{2}+1=1,$$

$$p_a(C) = \frac{(C^2)+(C\cdot K)}{2}+1=1,$$

which is what was asserted.

§5. The case $q = 1$

Lemma 6. *For a surface* V *with the invariants* $p_g = 0$ *and* $q = 1$, *we have* $(K^2) \leq 0$.

In this case, $p_a(V) = 1 - q + p_g = 0$.

From formula (4) of the Introduction it follows that $(K^2) + \chi = 12p_a(V) = 0$, i.e.

$$(K^2) = -\chi.$$

We consider the projection $\pi: V \to B$ determined by the Albanese mapping. Since $\chi(B) = 0$, from Theorems 6 and 7 we obtain $\chi \geq 0$, which means that $(K^2) \leq 0$.

Remark. Another proof can be given by starting with formula (4) of the Introduction, which in our case gives

$$(K^2) = 2 - b_2.$$

It is sufficient for us to show that $b_2 \geq 2$, and for this to find two independent homology classes on V. The classes determined by the cycles E (a hyperplane section) and F (one of the fibers of the projection π), for example, will be such classes. They are independent, since $(E^2) > 0$, $(F^2) = 0$.

We will now examine separately the cases $(K^2) < 0$ and $(K^2) = 0$.

§6. The case $(K^2) < 0$

Theorem 8. *A surface with the invariants* $p_g = 0$, $q = 1$, *and* $(K^2) < 0$ *is ruled.*

Proof. We consider the same projection $\pi: V \to B$, which now coincides with the Albanese mapping. If the genus g of a generic fiber F is equal to 0, then the surface V is ruled. If $g = 1$, then X possesses a mapping onto elliptical curves. In Chapter VII, Theorem 3 it will be proved that for such a surface $(K^2) = 0$, which contradicts the assumption $(K^2) < 0$.

There remains to be considered the case $g > 1$.

The plan of the proof of Theorem 8 is the following. For some unramified covering C of a curve B with a projection $\phi: C \to B$ we consider the inverse image $V' = V \times_B C$ of the mapping π on C, i.e. the subvariety $C \times V$ consisting of the points (c, v) for which $\phi(c) = \pi(v)$. The projection $C \times V \to C$ determines on V a projection $\pi': V' \to C$ and a fiber space whose fibers are isomorphic to the fibers of π. The surface V' is itself an unramified covering of V. We will show, assuming that $g > 1$, that for a properly chosen covering C the surface V' will be the direct product $C \times F$. From this it follows, by formula (2), that on V', $K' \sim C \times K(F)$, where K' and $K(F)$ are canonical classes of V' and F. Therefore $(K'^2) = 0$. But if $f\colon V' \to V$ is an unramified covering of degree n, then, as it is easy to verify, $K' = f^*(K)$, and therefore

$$K' \cdot K' = f^*(K)\, f^*(K) = f^*(K \cdot K)$$

(we are considering $K' \cdot K'$ and $K \cdot K$ as cycles here, and not as numbers). Therefore $(K'^2) = n(K^2)$ and hence $(K^2) = 0$ in contradiction with the assumption of the theorem. This proves Theorem 8.

Applying the corollary of Lemma 1 (in which one may clearly replace E by $E + F_b - F_0$), we obtain that for any point $b \in B$ there exists an $n > 0$ and a divisor $D_b > 0$ such that

$$E + F_b - F_0 + nK \sim D_b, \quad D_b \neq 0, \quad p_a(C) \leqslant 1 \tag{19}$$

for any cycle C for which $0 < C < D_b$. From this it follows that the cycle D_b cannot contain as a component a fiber F, since for it $p_a(F_b) = g > 1$. Therefore any irreducible component C of the cycle D_b is mapped by the projection π onto B, and, consequently, is not a rational curve. Hence, for it $p_a(C) = 1$, while it has no singular points and has a genus of 1.

Let β be a generic point of the curve B. We denote by C_β some irreducible (over $k(\beta)$) component of the cycle D_β. For any $b \in B$ we will designate by C_b a specialization of the curve C_β. More exactly, we will designate by ξ_β a generic point of the curve C_β and will consider in $B \times V$ an irreducible subvariety Γ with a generic point (β, ξ_β). We set

$$C_b = pr_V((b \times V) \cdot \Gamma).$$

The cycle C_b consists of components of dimension 1. Otherwise its carrier would coincide with V, i.e. Γ would contain the component $b \times V$, which would contradict its irreducibility. Let $C_\beta = \Sigma C_\beta^{(i)}$ be a decomposition into absolutely irreducible components. Similarly, we define their specializations $C_b^{(i)}$. We choose for C_β a component of the cycle D_β not defined over k (in other words, the curve C_b does not remain constant under a change in $b \in B$). In order to prove that such a component exists, it is sufficient to show that the field of definition of the cycle D_β is transcendental over k. For this we denote by β' another generic point of the curve B that is independent from β, and we show that $D_\beta \neq D_{\beta'}$. The equality $D_\beta = D_{\beta'}$ would imply the relation $F_\beta \sim F_{\beta'}$. Let $(f) = F_\beta - F_{\beta'}$. The function f is the image of some function g on B: $f\colon \pi^*(g)$. In fact, for any constant c we have $((f - c)_0 \cdot F) = 0$. From this it follows that $(f - c)_0$ consists of fibers of the mapping π, i.e. f is constant on the fibers. This means that $f = \pi^*(g)$. But then obviously $(g) = \beta - \beta'$ and this is impossible, since the genus of the curve B is equal to 1. The field determined by the curve $C_\beta^{(i)}$ is also transcendental over k. We denote by U a nonsingular model of this field. Correspondingly, we will denote the curves $C_b^{(i)}$ by $\mathfrak{L}_u$, $u \in U$. They form a one-dimensional family of elliptic curves on V.

Let $\mathfrak{L}$ be the curve $\mathfrak{L}_u$ for some fixed $u \in U$, where $\mathfrak{L}$ is irreducible, does not have singular points, and $(\mathfrak{L} \cdot F) = n$ $(n > 0$, since $\mathfrak{L} \neq F)$. The mapping π determines on the curve $\mathfrak{L}$ the structure of a covering of the curve B. Since both curves are elliptic, this covering is unramified. As is known, there do not exist continuous systems of unramified coverings (cf. [32]. If $k = C$, this follows from the finiteness of the number of coverings of a given degree). Therefore we have on a generic curve $\mathfrak{L}_\xi$ (ξ is a generic point of U) a covering isomorphic to the one defined on $\mathfrak{L}$. This means that there exists an isomorphism $\phi: \mathfrak{L} \longrightarrow \mathfrak{L}_\xi$ such that for any $x \in \mathfrak{L}$

$$\phi(x) = \pi(\phi(x)).$$

We now consider the inverse image $\mathfrak{L} \times_B V$ of the mapping π on the covering $\mathfrak{L}$, i.e. the subvariety $V' \subset \mathfrak{L} \times V$ consisting of the pairs (y, v), $y \in \mathfrak{L}$, $v \in V$ for which $\pi(y) = \pi(v)$. The projection $\mathfrak{L} \times V \longrightarrow \mathfrak{L}$ defines on V' a fiber space $\pi': V' \longrightarrow \mathfrak{L}$ with base $\mathfrak{L}$, whose fibers are isomorphic to the fibers of π. We denote by $\mathfrak{L}'_\xi$ the curve on V' consisting of the points (y, v), $v \in \mathfrak{L}_\xi$, $x = \phi(y)$, $y \in \mathfrak{L}$, $v \in V$.

We note that it is possible that the mapping ϕ is defined over the larger field $k(\xi)$. The same is true of the curve $\mathfrak{L}'_\xi$. We denote by W a nonsingular curve such that the field of functions over it coincides with the field of definition of the curve $\mathfrak{L}'_\xi$. We denote by $\mathfrak{L}'_\eta$ the curve corresponding to a generic point η of the curve W, and denote its specialization for any point $w \in W$ by $\mathfrak{L}'_w$. It is easy to verify that $(\mathfrak{L}'_w \cdot F) = 1$, where F is any fiber of the mapping π'.

For any nonsingular fiber F_y, $y \in \mathfrak{L}$ of the projection π' the mapping

$$\psi(w) = \mathfrak{L}'_w \cdot F_y$$

determines a curve W as a covering of the curve F_y, or, in other words, defines $k(F_y)$ as a subfield of the field $k(W)$. As is known, there do not exist unconnected systems of subfields of order $g > 1$ (cf., for example, [15]). It follows from this that the relation of equivalence defined on W by the condition $w \sim w'$ if $\mathfrak{L}'_w \cdot F_y = \mathfrak{L}'_{w'} \cdot F_y$ does not depend on the choice of the fiber F_y. In other words, if the curves $\mathfrak{L}'_w$ and $\mathfrak{L}'_{w'}$ intersect in one point (lying on the fiber F_{y_0}), then they must have an infinite number of points of intersection (lying on all the fibers F_y), i.e. they must coincide. Since W parameterizes the family of curves $\mathfrak{L}'_w$, we obtain that $\mathfrak{L}'_w$ and $\mathfrak{L}'_{w'}$ do not intersect if $w' \neq w$.

Therefore the mapping $W \times \mathfrak{L} \to V'$ defined by the formula

$$(w, y) \to \mathfrak{L}'_w \cdot F_y, \quad w \in W, \; y \in \mathfrak{L},$$

is an isomorphism between $W \times \mathfrak{L}$ and V'. It is obvious that here F is mapped isomorphically onto W. Thus V' is isomorphic to $F \times \mathfrak{L}$. As we said in the beginning of the proof, the assertion of Theorem 8 follows from this.

§7. The case $(K^2) = 0$

The surfaces with the invariants $p_g = 0$, $q = 1$, $(K^2) = 0$ are the only surfaces (among those for which $p_g = 0$, $q > 0$) which can be not ruled. Their complete classification will be given below. The basic tool in investigating them will be the Albanese mapping $\pi: V \to B$. If the genus g of a generic fiber of the fibering π is equal to 0, then the surface is ruled. We will later consider the case $g > 0$.

We note that for surfaces of the type under consideration, $p_a(X) = 1 - q + p_g = 0$, and since $(K^2) = 0$, it is also true that $\chi(V) = 0$. The basic result which we obtain (Theorem 9) will be valid for any surfaces for which $p_a(V) = \chi(V) = 0$.

Theorem 9. *Let V be an algebraic surface such that $p_a(V) = \chi(V) = 0$, and let $\pi: V \to B$ be a regular mapping of V onto a nonsingular elliptic curve. If the genus g of a nondegenerate fiber F of the fibering π is greater than* 1, *then there exists an unramified covering $\overline{B} \to B$ such that the inverse image $\overline{V} = V \times_B \overline{B}$ of the mapping π on $\overline{B}$ is a direct product: $\overline{V} \simeq \overline{B} \times F$.*

Before proving Theorem 9 we give some useful propositions about mappings of surfaces onto curves.

Lemma 7. *If $\pi: V \to B$ is a regular mapping of the surface V onto an elliptic curve, a generic fiber of which is irreducible, and if $\chi(V) = 0$, then all the fibers of the mapping π are nonsingular or, if the genus of a generic fiber is equal to* 1, *all fibers are multiples of a nonsingular curve of genus* 1.

The lemma follows directly from Theorems 6 and 7, since in (8) $\chi(V) = \chi(B) = 0$ and thus $\chi(F_{b_i}) = \chi(F)$.

We now introduce several helpful concepts relating to fiber spaces $\pi: V \to B$ without degenerate fibers.

If f is a function on V belonging to the local ring $\mathfrak{D}_{F_b}$ of some fiber F_b, then its restriction to the fiber F_b yields a regular function on F_b. Thus the homomorphism

$$\rho_b: \mathfrak{D}_{F_b} \to k(F_b)$$

is defined.

In an analogous manner one can associate with any divisor D on the surface a divisor $D \cdot F_b$ on the curve F_b.

The intersection $D \cdot F_b$ is defined if the divisor D does not contain F_b as a component. We complete the definition for all divisors by assuming $F_b \cdot F_b = 0$ (not only as a number, but also as a divisor). Thus we obtain a homomorphism of the group of divisors on V onto the group of divisors on F_b. We designate it by ρ_b.

We finally define the concept of a differential on V over B. By this we will mean a differential in the field $k(V)/k(B)$. This field has a degree of transcendence 1, so that all its differentials have the form gDf, g, $f \in k(V)$, where by Df is meant the complete differential of the function f in the field $k(V)/k(B)$. The differentials on V over B form a module $D_{k(B)}(k(V))$ over $k(V)$. A divisor of the differential gDf is defined in the usual manner: for any divisor C not coinciding with a fiber of the mapping π, we choose a function $T \in k(V)$, such that $\nu_C(T) = 1$, we write gDf in the form $gDf = hDT$ and set $\nu_C(h) = m_C$,

$$(gDf) = \Sigma m_C C.$$

By definition, the fibers F_b do not enter into the divisor (gDf).

This definition may be given a more geometric character by considering the one-dimensional vector fiber space Θ on V, a fiber of which at the point v is a subspace of the tangent space at the point v which consists of the vectors tangent to the fiber $F_{\pi(v)}$ that passes through the point v. We denote by θ the fiber space dual to Θ. Then the differentials are the rational sections of this one-dimensional fiber space, and a divisor of a differential is a divisor of a rational section.

A differential on V over B is said to be regular at the fiber F_b if it can be written in the form gDf, g, $f \in \mathfrak{D}_{F_b}$. The set of all such differentials forms a module $D_{\mathfrak{D}_b}(\mathfrak{D}_{F_b})$ over the ring $\mathfrak{D}_{F_b}$. The homomorphism ρ_b extends to a homomorphism of the modules

$$D_{\mathfrak{D}_b}(\mathfrak{D}_{F_b}) \to D_k(k(F_b)),$$

which we will designate by ρ.

The homomorphisms introduced possess the following properties of commutativity:

$$\rho_b(f) = (\rho_b \cdot f), \quad f \in \mathfrak{D}_{F_b}, \tag{20}$$

$$\rho_b \cdot (gDf) = (\rho_b \cdot gd\rho_b f), \quad g, f \in D_{\mathfrak{D}_b}(\mathfrak{D}_{F_b}). \tag{21}$$

The proof of these relationships is by direct verification.

Let β be a generic point of B. The mapping ρ_β is an epimorphism and has as kernel the group consisting of the linear combinations of fibers.

The epimorphic character of the mapping ρ_β follows from the facts that all the principal divisors of the field $k(F_\beta)/k(\beta)$ are contained in $\operatorname{Im} \rho_\beta$, that the greatest common divisor of $\rho_\beta(C)$ and $\rho_\beta(C')$ is ρ_β of the greatest common divisor of C and C', and that all the divisors can be obtained as greatest common divisors of principal divisors.

Since mutually prime effective divisors on V and those not containing fibers intersect with a generic fiber in mutually prime divisors, we obtain the assertion about the kernel of the mapping ρ_β.

From the conjunction of these assertions it follows that if $C \cdot F_\beta \sim DF_\beta$ on F_β, then $C \sim D + \Sigma m_i F_{b_i}$ on V, where the F_{b_i} are certain fibers.

We now turn to the formulation of the result that lies at the basis of the proof of Theorem 9.

Definition. *A divisor* $D = \Sigma n_i C_i$ *(where the* C_i *are mutually distinct irreducible curves) is said to be unramified if the distinct curves* C_i *do not intersect and for each of them the projection* $\pi\colon C_i \to B$ *defines* C_i *as an unramified covering of the base* B.

Lemma 8. *Let* $\pi\colon V \to B$ *be a regular mapping of a surface* V *onto a curve* B *that does not have degenerate fibers. If there exists a function* $f \in k(V)$*, a divisor of which is the sum of a nonzero unramified divisor and some linear combination of fibers, then for some unramified covering of the base* $\overline{B} \to B$ *of the fiber bundle,* $\overline{V} = V \times_B \overline{B}$ *is the direct sum* $\overline{V} \simeq \overline{B} \times F$.

Proof. Let

$$(f) = \Sigma n_i C_i - \Sigma n_j' C_j' + \Sigma m_k F_{b_k},$$

where $n_i > 0$, $n_j' > 0$ and $\Sigma n_i C_i - \Sigma n_j' C_j'$ is an unramified divisor. We consider a fiber bundle L over B, a fiber of which is an affine line and which corresponds to the divisor $\Sigma m_k b_k$ on B. After adding to each fiber of L an infinite point, we will consider L as a fiber bundle, a fiber of which is a projective line.

We will show that the surface V can be considered as a ramified covering of the surface L. For this we suppose that B is covered by open sets W_i such that $b_i \in W_i$, $b_j \notin W_i$ for $i \neq j$, and that there exist functions τ_i such that $\tau_i(b_i) = 0$ and $\tau_i - \tau_i(b)$ is a local parameter at any point $b \in W_i$, where $\tau_i(b) \neq 0$ for $b \in W_i$, $b \neq b_i$. Then L can be given as the union of the open sets U_i, where $U_i \simeq W_i \times P^1$, where P^1 is a projective line and the points U_i and U_j are identified by the rule

$$b \times z \sim b' \times z' \ (b \in W_i,\ b' \in W_j,\ z,\ z' \in P^1),$$

if

$$b = b' \in W_i \cap W_j;\ z' = z \cdot \frac{\tau_i^{m_i}}{\tau_j^{m_j}}(b). \tag{22}$$

We define the mapping $\phi_i\colon \pi^{-1}(W_i) \to U_i$, setting

$$\varphi_i(v) = (\pi(v) \times \rho_{\pi(v)}(f\pi^*(\tau_i)^{-m_i})(v)),\ v \in \pi^{-1}(W_i).$$

We note that $f\pi^*(\tau_i)^{-m_i} \in O_{F_v}$, for $v \in \pi^{-1}(W_i)$, according to the choice of τ_i, and thus the function $\rho_{\pi(v)}(f\pi^*(\tau_i)^{-m_i})$ is defined. It is clear that the ϕ_i are regular mappings.

We will show that $\phi_i = \phi_j$ on $\pi^{-1}(W_i) \cap \pi^{-1}(W_j)$, and, thus, that the collection of mappings ϕ_i determines a unique regular mapping ϕ of the surface V onto L. In fact, if $x \in \pi^{-1}(W_i) \cap \pi^{-1}(W_j)$, then

$$\varphi_j(v) = (\pi(v) \times (\rho_{\pi(v)}(f\pi^*(\tau_j)^{-m_j}))(v))$$

and

$$\rho_{\pi(v)}(f\pi^*(\tau_j)^{-m_j})(v) = \rho_{\pi(v)}(f\pi^*(\tau_i)^{-m_i})(v) \cdot \frac{\tau_i^{m_i}}{\tau_j^{m_j}}(\pi(v))$$

in accordance with (22).

Thus, under the mapping ϕ each fiber F_b is mapped onto a projective line that is a fiber of L, where for $b \in W_i$ the mapping is effected by the function $\rho_b(f\pi^*(\tau_i)^{-m_i})$. We note that here none of the fibers F_b is mapped into a point. In other words, the function $\rho_b(f\pi^*(\tau_i)^{-m_i})$ is not constant on F_b. This follows from the fact that we can even indicate its divisor – it is equal to $\Sigma n_i C_i \cdot F_b - \Sigma n_j' C_j' \cdot F_b$ and is not equal to 0, since by assumption the curves C_i and C_j' do not intersect. Consequently, the mapping ϕ determines on each fiber a mapping of a covering of the projective line, where the degree of this covering is the same for all the fibers; it is equal to $\Sigma n_i (C_i \cdot F) = \Sigma n_j' (C_j' \cdot F)$.

We denote by W_b the divisor of the ramification of the covering, which the mapping ϕ determines on the fiber F_b. Since

$$(\rho_b(f\pi^*(\tau_i)^{-m_i})) = \Sigma n_i \rho_b C_i - \Sigma n_j' \rho_b C_j',$$

it follows that

$$W_b = (d\rho_b(f\pi^*(\tau_i)^{-m_i}))_0 + \Sigma (n_j' - 1)\, \rho_b C_j',$$

where $(\omega)_0$ designates a zero divisor of the differential ω. By (21)

$$(d\rho_b(f\pi^*(\tau_i)^{-m_i})) = \rho_b(Df),$$

and therefore $W_b = \rho_b W$, where W is a divisor on V,

$$W = (Df)_0 + \Sigma (n_j' - 1)\, C_j'.$$

From the definition of the divisor (Df) it follows that

$$W = \Sigma (n_i - 1)\, C_i + \Sigma (n_j' - 1)\, C_j' + \overline{W},\ \overline{W} > 0.$$

We will now demonstrate the central part of the proof of Lemma 8 – that the divisor $\overline{W}$ does not intersect the curves C_i and C_j'. In fact, since $(f_0) = \Sigma n_i C_i + \Sigma l_s F_s$, it follows from (20) that

$$(\rho_b (f\pi^* (\tau_i)^{-m_i}))_0 = \Sigma n_i \rho_b C_i = \Sigma n_i C_i \cdot F_b.$$

By the assumption of the theorem the covering $\pi: C_i \to B$ is unramified; therefore, if $(C_i \cdot F) = m_i$,

$$C_i \cdot F_b = Q_{b,i}^{(1)} + \ldots + Q_{b,i}^{(m_i)}, \quad Q_{b,i}^{(r)} \neq Q_{b,i}^{(s)} \text{ for } r \neq s.$$

From this it follows that

$$(\rho_b (f\pi^* (\tau_i)^{-m_i}))_0 = \Sigma n_i (Q_{b,i}^{(1)} + \ldots + Q_{b,i}^{(m_i)})$$

and thus that each of the points $Q_{b,i}^{(r)}$ occurs in the divisor $(\rho_b (f\pi^* (\tau_i)^{-m_i}))_0$ with multiplicity n_i. Therefore this point occurs in the divisor $(d(\rho_b (f\pi^* (\tau_i)^{-m_i})))$ with multiplicity $n_i - 1$. But it occurs in the divisor $(n_i - 1)\rho_b C_i$ with the same multiplicity. Hence none of these points occur in the divisor $\rho_b \overline{W}$. This means that the divisors $\overline{W}$ and C_i do not have common points on any fiber F_b; thus they do not intersect in general. The divisors C_j' are considered similarly.

Under the mapping ϕ the divisors $\Sigma n_i C_i$ and $\Sigma n_j' C_j'$ are preimages of zero and the infinite section of fiber bundle L:

$$\Sigma n_i C_i = \varphi^* (S_0), \quad \Sigma n_j' C_j' = \varphi^* (S_\infty).$$

The divisor V is the preimage of some divisor S on L: $V = \phi^* (S)$. From what was just proven it follows that S does not intersect either S_0 or S_∞.

Starting from this, we now show that for some unramified covering $B' \to B$ the preimage of L is trivial: $L' = L \times_B B' \simeq B' \times P^1$ and in L' the preimage S' of the divisor S is "constant": $S' = B' \times \Delta$, where Δ is an effective divisor on P^1. For this we denote by C an arbitrary irreducible component of $\overline{W}$. The projection defined in L, $L \to B$, determines C as a covering of B. We consider the fiber bundle $M = C \times_B L$. In it the points (c, c), $c \in C$, form a section that intersects neither zero nor the infinite section. A one-dimensional vector fiber bundle possessing such a section must necessarily be trivial. This means that L becomes trivial on some covering $C \to B$. For the divisor $\Sigma m_k b_k$ corresponding to L, this means that it becomes principal on the covering C. Such a divisor determines a class of divisors of finite order and then, as is known, becomes principal on some unramified covering $B' \to B$. Thus, $L' = L \times_B B' \simeq B' \times P^1$. For the preimage S' of the divisor S, the condition that S' does not intersect the zero section means that $(S' \cdot (B' \times 0)) = 0$, and since the divisor S' is effective, it follows from this that $S' = B' \times \Delta$.

We consider the preimage V' of the fibering π on B': $V' = V \times_B B'$. We have a mapping ϕ': $V' \to L'$ which determines on each fiber $F_{b'}$ a mapping of a covering over P^1 that is ramified at the points of a fixed (not depending on b') divisor Δ. From this it already follows that the covering $V' \to L'$ is a factor of the covering $V'' \to L'$, where $V'' = \overline{B} \times N$ for $\overline{B}$ an unramified covering of B' and N a covering of P^1 which is ramified only at points of the divisor Δ. This result follows from the theorem about unramified coverings of a direct product and from a remark of Abhyankar (cf. [1]).

The remark of Abhyankar is that it is possible to find a covering $H_1 \to P^1$ that is ramified only at points of Δ and is such that any covering $F_b \times_{P^1} H_1 \to H_1$ will be unramified. For this it is sufficient to take H_1 so that its indices of ramification at the points of Δ are divisible by the corresponding indices of ramification of the coverings $F_b \to P^1$ (for this it may be necessary to extend Δ by one point). One may already apply the theorem about unramified coverings of a direct product to the covering V: $X \times_{L'} (B' \times H_1) \to B' \times H_1$ (for $k = C$ it follows from the equality $\pi_1(X \times Y) = \pi_1(X) \times \pi_1(Y)$). We obtain that it is a factor of the covering $V'' = \overline{B} \times H \to B' \times H_1$, where $\overline{B}$ and H are unramified coverings of the curves B' and H_1. Since X' in turn is a factor of the covering $V \times_{L'} (B' \times H_1) \to B' \times H_1$, V' is also a factor of the covering V''.

We denote by $\overline{V}$ the surface $V' \times_{B'} \overline{B}$ that has the projection $\overline{\pi}$: $\overline{V} \to \overline{B}$, and we will show that $\overline{V} \simeq \overline{B} \times F_0$, where F_0 is an arbitrary fiber of the mapping $\overline{\pi}$. From what was said above it follows that $\overline{V}$ is a factor of V'', where, as it is easy to verify, the mapping u: $V'' \to \overline{V}$ commutes with the projections of both the surfaces onto $\overline{B}$.

Going if necessary from H to a larger covering, we can assume that H is a normal covering of P^1 with the Galois group G. From the construction it follows that we have a sequence of coverings

$$\overline{B} \times H \xrightarrow{u} \overline{V} \xrightarrow{v} \overline{B} \times P^1,$$

where the mappings u and v commute with the projections of all the members of this sequence onto $\overline{B}$. The covering $\overline{V} \to \overline{B} \times P^1$ belongs to some subgroup G_1 of the Galois group G of the covering $\overline{B} \times H \to \overline{B} \times P^1$. Since the automorphisms $g \in G$ act according to the law

$$g(\overline{b} \times h) = \overline{b} \times g(h),$$

the automorphisms $g_1 \in G_1$ also act in the same way. In view of this, $\overline{V} \simeq (\overline{B} \times H)/G_1 \simeq \overline{B} \times H/G_1$. Lemma 8 is proved.

In order to apply Lemma 8 to the proof of Theorem 9, we need to construct on a surface V satisfying the assumptions of the theorem, a function, a divisor of which consists of an unramified divisor and a linear combination of fibers. The following lemmas are connected with the construction of such a function.

We will designate by $C \approx D$ the algebraic, and by $C \overset{\approx}{\approx} D$ the numerical equivalence of divisors on a surface (if $k = \mathbf{C}$ one is concerned with homologousness and weak homologousness).

Lemma 9. *Let V be an algebraic surface with $(K^2) = 0$, $p_a(V) = 0$, and let $\pi: V \longrightarrow B$ be its mapping onto a curve B, where all the fibers of the mapping are not degenerate and have a genus $g > 1$. For a sufficiently large n (for example $n \geq 3$), there exists for any divisor D for which $D \overset{\approx}{\approx} nK$ a divisor C such that $C > 0$, $C \approx D$.*

Proof. It is sufficient for us to find a divisor C such that $C \approx D, l(C) > 0$. We will look for C in the form $D + F_b - F_0$, $b \in B$. We will show that $l(D + F_b - F_0) > 0$ for some $b \in B$, which is what we need.

Let $l(D + F_b - F_0) = 0$ for all $b \in B$. From the exact sequence

$$0 \to \mathcal{F}(D + F_b - F_0) \to \mathcal{F}(D + F_b) \to \mathcal{F}_{F_0}(D + F_b)\cdot F_0) \to 0$$

follows the exactness of the sequence

$$0 \to \mathcal{L}(D + F_b - F_0) \to \mathcal{L}(D + F_b) \to \mathcal{L}_{F_0}((D + F_b)\cdot F_0). \tag{23}$$

From the assumption made it follows that the restricted homomorphism

$$\rho_0 : \mathcal{L}(D + F_b) \to \mathcal{L}_{F_0}((D + F_b)\cdot F_0) \tag{24}$$

is an injection for any $b \in B$.

We calculate the dimensions of both of the spaces in (24). Since

$$((D + F_b)\cdot F_0) = ((nK + F_b - F_0)\cdot F_0) = 2n(g - 1),$$

the number $l((D + F_b)\cdot F_0)$ can be found by the Riemann-Roch theorem applied to the curve F_0:

$$l((D + F_b) F_0) = (2n - 1)(g - 1). \tag{25}$$

On the other hand, the Riemann-Roch theorem on V gives

$$l(D + F_b) \geqslant \frac{((D + F_b)\cdot(D + F_b - K))}{2} = \frac{(2n - 1)(K\cdot F_b)}{2} = (2n - 1)(g - 1). \tag{26}$$

From (25), (26), and the fact that the ρ_0 in (24) is an injection it follows that both spaces in (24) have the same dimension, and thus that ρ_0 is an isomorphism for any $b \in B$.

Thus, for any effective divisor Δ contained in the complete linear system

$(D + F_b) \cdot F_0|$ on F_0, there exists exactly one effective divisor $C_b \in |D + F_b|$ ch that $C_b \cdot F_0 = \Delta$.

We will show that at least one curve C_b passes through each point $v \in V$. For his we consider in $V \times B$ the subset Γ consisting of the points (v, b) for which $v \in C_b$. It is clear that Γ is an algebraic subvariety in $V \times B$. Under the mapping $\Gamma \longrightarrow B$ induced by the projection $V \times B \longrightarrow B$, each point $b \in B$ has for a preimage a curve C_b, from which it follows that $\dim \Gamma = 2$. The assertion that we wish to prove is that $pr_V \Gamma = V$. In view of the completeness of all the varieties considered, $pr_V \Gamma$ is an algebraic subvariety of V, and, consequently, coincides with V if $\dim pr_V \Gamma = 2$. If $pr_V \Gamma = U$, $\dim U = 1$, then $\Gamma = U \times B$, and this in turn means that $C_b = U$ for all $b \in B$. This cannot be true, since $C_b \neq C_{b'}$ for $b \neq b'$. In fact, it would follow from the equality $C_b = C_{b'}$ that $F_b \sim F_{b'}$. We saw in the proof of Theorem 8 that this leads to a contradiction: if $(f) = F_b - F_{b'}$, then $f = \pi^*(g)$, $g \in k(B)$, $(g) = (b) - (b')$, i.e. $b \sim b'$, and this contradicts the fact that the genus of the curve B is equal to 1.

We choose for the point v, in particular, an arbitrary point of the curve F_0 that is not contained in the divisor Δ. By assumption there exists a point $b \in B$ such that $C_b \ni v$. But then C_b intersects F_0 in more than $(C_b \cdot F_0)$ points, and this is possible only when C_b contains F_0 as a component. If $C_b = H + F_0$, $H > 0$, then $D + F_b \sim H + F_0 > 0$, i.e. $D + F_b - F_0 \sim H > 0$, which contradicts the assumption $l(D + F_b - F_0) = 0$. The lemma is proven.

Lemma 10. *If V is a surface satisfying the conditions of Lemma 9, then the divisor C, whose existence is established in that lemma, is distinct from 0 and is unramified.*

Let $C \approx nK$, $C > 0$. Since $(K \cdot F) > 0$, $nK \not\approx 0$, and thus $C \neq 0$. Let us assume that

$$C = \Sigma n_i C_i,\ C_i \neq C_j \ \text{ for } \ i \neq j,\ n_i > 0,$$

where the C_i are irreducible curves.

Since every irreducible curve lying on the surface V is either a covering of the base B or coincides with a fiber F, where the genus of B is equal to 1 and the genus of F is greater than 1, an irreducible rational curve cannot lie on V. Therefore

$$(C_i^2) + (C_i \cdot K) \geq 0. \tag{27}$$

We will show that $(C_i \cdot K) \geq 0$. From $(C_i \cdot K) < 0$ it would follow by (27) that $(C_i^2) > 0$. Since $C \approx nK$, then $(C_i \cdot K) < 0$ would give

$$(C_i \cdot \Sigma n_j C_j) = n_i (C_i^2) + \sum_{j \neq i} n_j (C_i \cdot C_j) < 0,$$

which is impossible since $(C_i \cdot C_j) \geq 0$ for $i \neq j$, and $(C_i^2) > 0$.

In view of the fact that $C \approx nK$, the equality $(K^2) = 0$ can be rewritten as

$$\sum n_i (C_i \cdot K) = 0 \tag{28}$$

or

$$(\sum n_i C_i)^2 = \sum n_i^2 (C_i^2) + \sum_{i \neq j} n_i n_j (C_i C_j) = 0. \tag{29}$$

From (28) it now follows that $(C_i \cdot K) = 0$, and from (27) and (29) that $(C_i^2) = 0$ and $(C_i \cdot C_j) = 0$ for $i \neq j$. Therefore

$$p_a(C_i) = \frac{(C_i^2) + (C_i \cdot K)}{2} + 1 = 1,$$

and since C_i is a nonrational curve, it is a nonsingular curve of genus 1. It cannot coincide with a fiber because the genus of a fiber is greater than 1, and is thus a covering of the base. Since the genus of each curve, both C_i and B, is equal to 1, this covering is unramified. Finally, the relation already shown, $(C_i \cdot C_j) = 0$ for $i \neq j$, completes the proof that the divisor C is unramified.

Lemma 11. *Let $\pi: V \longrightarrow B$ be a fiber space without degenerate fibers, let β be a generic point of the curve B and let F_β be a generic fiber. Every class of divisors of finite order on the curve F_β is defined over some unramified extension of the field $k(\beta)$.*

Proof. Let J_β be the Jacobian variety of the curve F_β. It is defined over the field $k(\beta)$ and for any $b \in B$ has as a specialization the Jacobian variety J_b of the curve F_b. We denote by $\mathfrak{a}_n(\beta)$ the cycle consisting of the points of order n on J_β. As is known, this cycle consists of n^{2g} points with multiplicity of one. Under the specialization of β into b the cycle $\mathfrak{a}_n(\beta)$ is specialized into the cycle $\mathfrak{a}_n(b)$ consisting of the points of order n on J_b. Since the fiber J_b is nondegenerate, the cycle $\mathfrak{a}_n(b)$ consists of n^{2g} distinct points. By the generalized lemma of Hensel (cf. [34]) it then follows that the cycle $\mathfrak{a}_n(\beta)$ is rational over the complete local ring of the point b. Since this is true for any point b, it follows from this that the field of definition of the cycle $\mathfrak{a}_n(\beta)$ is unramified over the field $k(\beta)$. The lemma is proved.

Remark. One may give this argument the following geometric form. We denote by J_b the Jacobian variety of the fiber F_b. In the set of all the varieties J_b, $b \in B$ it is possible to introduce the structure of an algebraic variety J equipped with a projection $\phi: J \longrightarrow B$. For each $b \in B$ the variety J_b contains n^{2g} points of order n. The collection of all these points for all the $b \in B$ forms, as it is easy to verify, a one-dimensional effective divisor $\mathfrak{a}_n \subset J$. The projection ϕ determines on $\mathfrak{a}_n$ the structure of a covering of B. Since for any $b \in B$ the cycle

$\mathfrak{a}_n \cdot J_b$ consists of n^{2g} dsitinct points, this covering is unramified. This is the statement of the lemma.

Lemma 12. *Let $K = k(B)$ be the field of functions over an algebraic curve B, U be an algebraic curve over K, and C be the class of divisors on U to which there corresponds a point, rational over K of the Jacobian variety of the curve U, where $l(C) > 0$. Then there exists in C an effective divisor defined over K.*

This assertion essentially coincides with the so-called criterion of rationality of Cartier.

Let L/K be a normal extension of the field K over which the class C is already defined, let G be the Galois group L/K, and let $\mathfrak{D}_L$, P_L, $\mathfrak{C}_L$, be the groups of divisors, of principal divisors, and of classes of divisors on U over L. All these groups are G-modules. By assumption $C^\sigma = C$ for $\sigma \in G$, i.e. $C \in H^0(G, \mathfrak{C}_L)$. We will show that there exists a $D' \in \mathfrak{D}_K$, $D' \in C$. For this we consider the exact sequence

$$H^0(G, \mathfrak{D}_L) \to H^0(G, \mathfrak{C}_L) \to H^1(G, P_L),$$

which is obtained from the exact sequence

$$(1) \to P_L \to \mathfrak{D}_L \to \mathfrak{C}_L \to (1).$$

It is clear that our assertion will be proved if we show that $H^1(G, P_L) = (1)$. For this we consider the exact sequence

$$(1) \to L^* \to L(U)^* \to P_L \to (1),$$

where L^* and $L(U)^*$ are the multiplicative groups of the fields L and $L(U)$, and the exact sequence

$$H^1(G, L(U)^*) \to H^1(G, P_L) \to H^2(G, L^*).$$

Since G is the Galois group $L(U)/K(U)$, $H^1(G, L(U)^*) = (1)$ by a well-known algebraic fact. On the other hand, $H^2(G, L^*) = 0$ by a theorem of Tsen, for $K = k(B)$ and the field k is algebraically closed. From this it follows that $H^1(G, P_L) = (1)$.

It remains to be shown that there exists an effective divisor D defined over K and equivalent to the divisor D'. But this follows from the fact that $l(C) > 0$ and that the dimension of a divisor is not reduced under a separable extension of the field of constants (cf. for example, [58], Chapter V, §6, Corollary 1 to Theorem 4).

Proof of Theorem 9. We choose a temporarily arbitrary number m and consider on the curve F_β the class of divisors $\mathfrak{E}$ whose order is equal to m. By Lemma 11, this class is defined over some unramified extension field $k(B)$. If $\overline{B}$ is a nonsingular model of this unramified extension, then we have an unramified covering $\phi: \overline{B} \to B$, and on the surface $\overline{V} = V \times_B \overline{B}$ the class of divisors $\mathfrak{E}$ of the curve $F_{\overline{\beta}}$ is defined already over the field $k(\overline{B})$. By Lemma 12, there exists an effective divisor $\mathfrak{D}$ of the curve $F_{\overline{\beta}}$ belonging to the class $K(F_{\overline{\beta}})+\mathfrak{E}$ and also defined over $k(\overline{B})$. To this divisor there corresponds an effective divisor D on $\overline{V}$ such that

$$D\cdot F_{\overline{\beta}}=\mathfrak{D}\in K(F_{\beta})+\mathscr{E}.$$

Then

$$(D+(n-1)K)\cdot F_{\overline{\beta}}\in nK(F_{\overline{\beta}})+\mathscr{E}, \tag{30}$$

$$m(D+(n-1)K)\cdot F_{\overline{\beta}}\in mnK(F_{\overline{\beta}}),$$

and hence on $F_{\overline{\beta}}$

$$m(D+(n-1)K)\cdot F_{\overline{\beta}}\sim mnK\cdot F_{\overline{\beta}}. \tag{31}$$

From this it follows that on V

$$m(D+(n-1)K)\sim mnK+\Sigma m_i F_{b_i},$$

and since

$$\Sigma m_i F_{b_i}\approx lF_b,$$

where $l=\Sigma m_i$, and b is any point on B,

$$m(D+(n-1)K)\approx mnK+lF_b. \tag{31a}$$

We will show that for a suitable choice of m, l must be divisible by m. For this it is sufficient to consider the indices of intersection of the separate parts (31a) with K. We obtain

$$m(D\cdot K)=2l(g-1),$$

and if m is relatively prime to $2(g-1)$, then it divides l. Until now m was arbitrary. Therefore we can choose it relatively prime to $2(g-1)$ and assume that $l=l'm$. Then

$$m(D+(n-1)K-l'F_b)\approx mnK,$$

and hence

$$D+(n-1)K-l'F_b\approxeq nK.$$

We can apply Lemma 9 to the divisor $D+(n-1)K-l'F_b$. We obtain a diviso $C'>0$ such that

$$C'\sim D+(n-1)K-l'F_b+F_{b_1}-F_0.$$

On the other hand, by the same Lemma 9 there exists a divisor $C>0$ such that

$$C\sim nK+F_{b_2}-F_0.$$

From this it follows from (31) that

$$mC'\sim mC+\Sigma s_i F_{b_i},$$

i.e. there exists a function f on X for which

$$(f) = -mC' + mC + \Sigma s_i F_{b_i}.$$

By Lemma 10 the divisor $-mC' + mC$ is not ramified. It is not equal to 0, since $C' \neq C$. In fact, by (30), it is even true that

$$C' \cdot F_{\overline{\beta}} \not\sim C \cdot F_{\beta}.$$

Applying Lemma 8, we obtain the assertion of Theorem 9.

We now consider the case when the genus g of a generic fiber of the Albanese mapping is equal to 1. As we will see later, in this case the exact analogue of Theorem 9 is not true. The following weaker statement, however, will follow from Lemma 7 and Corollary 3 of Theorem 7, Chapter VII.

Theorem 10. *If $\pi: V \to B$ is a regular mapping, a generic fiber of which is an elliptic curve, and $\chi(V) = 0$, then there exists a (perhaps ramified) covering $C \to B$ such that the inverse image of π on C is a direct product*

$$V \times_B C \simeq C \times F.$$

§8. Surfaces with $p_g = 0$, $q > 0$

(Classification and Theorem of Enriques)

In the preceding sections it was proven that a surface with the invariants $p_g = 0$, $q > 0$ is ruled, except perhaps for the case $q = 1$, $(K^2) = 0$ (Theorems 5 and 8). The basic analysis of this last case is Theorem 9. Starting from this theorem we now give a complete classification of those surfaces and verify that those of them which satisfy the condition $P_{12} = 0$ are ruled (the theorem of Enriques).

We consider separately the cases when the genus g of a generic fiber of the mapping π is greater than 1 and is equal to 1. We start with the case $g > 1$.

According to Theorem 9, in this case the surface V with invariants $p_g = 0$, $q = 1$, $(K^2) = 0$ have as an unramified covering the surface $\overline{V} = \overline{B} \times F$. We consider how it is possible to obtain the surface V from $\overline{V}$.

Theorem 9 shows that $\overline{V} = V \times_B \overline{B}$, where B is the Albanese variety of the surface V (in the given case, an elliptic curve), and $\overline{B} \to B$ is an unramified covering. We will assume without loss of generality that $\overline{B} \to B$ is a normal covering. We denote its Galois group by G. It is clear that $X = \overline{X}/G$. Under the isomorphism

$$\overline{V} \simeq \overline{B} \times F, \tag{32}$$

to the automorphism $\sigma \in G$ of the covering $\overline{B} \to B$ there corresponds the automorphism

$$\sigma(\bar{b}, f) = (\sigma\bar{b}, f\sigma^{-1}), \tag{33}$$

where $f \to f\sigma^{-1}$ is some automorphism of the curve F. This follows from the fact operation of the automorphism $\sigma \in G$ on $\overline{V}$, like that of the isomorphism of (32), commutes with the projection $\overline{V} \to \overline{B}$.

Thus, we have a homomorphism

$$\phi : G \to \mathrm{Aut}(F)$$

of the group G into the group of automorphisms of the curve F. Conversely, the existence of such a representation of the group G (which is itself given as the Galois group of the covering $\overline{B} \to B$) determines its operation as the group of automorphisms of the surface $\overline{V}$:

$$\sigma(\overline{b}, f) = (\sigma\overline{b}, f\varphi(\sigma)^{-1}).$$

It is immediately clear that we thus obtain a group operating on $\overline{V}$ without fixed points, and that the surface $V = \overline{V}/G$ does not have singular points. We see that we can obtain in this way all the surfaces interesting to us, but perhaps also many others. It is thus necessary for us to give the conditions which the covering $\overline{B} \to B$ the curve F, and the representation ϕ must satisfy in order that the surface V will have the invariants $p_g = 0$, $q = 1$. In order to formulate the result, we designate by L the curve $F/\phi(G)$.

Lemma 13. *For the surface* $V = \overline{V}/G$

$$q = 1 + \gamma, \tag{34}$$

$$p_g = \gamma, \tag{35}$$

where γ *is the genus of the curve* L.

Proof. The projection $\psi : \overline{V} \to V$ determines the mapping

$$\psi_i^* : \Omega^i(V) \to \Omega^i(\overline{V})$$

of the i-dimensional differential forms of first order. It is clear that ψ_i^* is an imbedding and that $\psi_i^*(\Omega^i(V)) = \Omega^i(\overline{V})^G$. Therefore it is necessary for us first to find the differential forms of first order on $\overline{V}$ and then to choose among them those that are invariant with respect to the operation of the automorphism from G.

We begin with the one-dimensional forms. It is obvious that

$$\Omega'(\overline{V}) = p_1^*(\Omega'(\overline{B})) \oplus p_2^*(\Omega'(F)), \tag{36}$$

where p_1 and p_2 are projections of $\overline{V}$ onto $\overline{B}$ and F. We now tell how an automorphism $\sigma \in G$ acts on a form from $\Omega'(\overline{V})$. By the definition (33) of the operation of σ on $\overline{V}$ the spaces $p_1^*\Omega'(\overline{B})$ and $p_2^*\Omega'(F)$ remain invariant under the operation of σ and are transformed in the same way as the space $\Omega'(\overline{B})$ under the operation of σ and the space $\Omega'(F)$ under the operation of $\phi(\sigma)^{-1}$. We now note that σ operates on $\Omega'(\overline{B})$ trivially, so that $\Omega'(\overline{B})^G = \Omega'(\overline{B})$. On the other hand, as is easily seen, $\Omega'(F)^{\phi(G)} \simeq \Omega'(L)$.

From this it follows that

$$\Omega'(V)^G \simeq \Omega'(\bar{B}) \oplus \Omega'(L),$$

which yields (34).

We now consider the forms of second degree. According to (2)

$$K(\bar{V}) = K(\bar{B}) \times F + \bar{B} \times K(F),$$

and since $K(\bar{B}) = 0$,

$$K(\bar{V}) = \bar{B} \times K(F). \tag{37}$$

Therefore

$$\Omega^2(\bar{V}) \simeq \Omega'(\bar{B}) \otimes \Omega'(F) \simeq \Omega'(F) \tag{38}$$

and

$$\Omega^2(\bar{V})^G \simeq \Omega'(F)^{\varphi(G)} \simeq \Omega'(L), \tag{39}$$

from which (35) follows.

Corollary. *The surface* $V = \bar{V}/G$ *has the invariants* $q = 1$, $p_g = 0$ *if and only if* $g = 0$, *i.e. the curve* $F/\phi(G)$ *is rational.*

We now turn to the construction of the surfaces with the invariants $p_g = 0$, $q = 1$ (again assuming that $g > 1$, i.e. that these surfaces are described by Theorem 9). We first note that the covering $\bar{B} \to B$, as an unramified covering of an elliptic curve, has an abelian Galois group with one or two generators:

$$G = \{\sigma_1\} \times \{\sigma_2\},\ \sigma_1^{m_1} = 1,\ \sigma_2^{m_2} = 1.$$

The field $k(F)/k(L)$ has as a Galois group the group $\phi(G)$. Consequently this Galois group has two generators; $\phi(\sigma_1)$ and $\phi(\sigma_2)$, the orders of which are divisors of the numbers m_1 and m_2. Since the curve L must be rational, the field $k(L)$ is isomorphic with the field of rational functions $k(L) = k(x)$, and the field $k(F)/k(L)$ can be obtained in the form $k(x, \sqrt[m_1]{P_1(x)}, \sqrt[m_2]{P_2(x)})$, (this easily follows from Galois theory), where $P_1(x)$ and $P_2(x)$ are any polynomials in x. Conversely, it is possible to define automorphisms s_1 and s_2 in the field $k(x, \sqrt[m_1]{P_1(x)}, \sqrt[m_2]{P_2(x)})$ by setting

$$s_1(\sqrt[m_1]{P_1(x)}) = \varepsilon_1 \sqrt[m_1]{P_1(x)},$$

$$s_1(\sqrt[m_2]{P_2(x)}) = \sqrt[m_2]{P_2(x)},$$

$$s_2(\sqrt[m_2]{P_2(x)}) = \varepsilon_2 \sqrt[m_2]{P_2(x)},$$

where ϵ_1 and ϵ_2 are primitive roots of 1 whose degrees are equal to the degrees of the fields $k(x, \sqrt[m_1]{P_1})/k(x)$ and $k(x, \sqrt[m_1]{P_1}, \sqrt[m_2]{P_2})/k(x, \sqrt[m_1]{P_1})$. Taking for F

a nonsingular model of the field $k(x, \sqrt[m_1]{P_1}, \sqrt[m_2]{P_2})$, we can define a homomorphism ϕ of the group G into the group $\operatorname{Aut} F$ by setting $\phi(\sigma_1) = s_1$, $\phi(\sigma_2) = s_2$, and on the surface $\overline{V} = \overline{B} \times F$ giving the operation of G by the formula (33). Since the curve $F/\phi(G) = L$ is rational, the surface $V = \overline{V}/G$ will have the invariants $p_g = 0$, $q = 1$. We have thus proven the following assertion.

Theorem 11. *For the case $g > 1$ all the surfaces V with the invariants $p_g = 0$, $q = 1$ can be obtained in the form $V = \overline{V}/G$, $\overline{V} = \overline{B} \times F$, where $\overline{B} \to B$ is an unramified covering of an elliptic curve, F is a nonsingular model of the field $k(x, \sqrt[m_1]{P_1(x)}, \sqrt[m_1]{P_2(x)})$, G is isomorphic to the Galois group of the covering $\overline{B} \to B$ and operates on $\overline{X}$ according to the rule $\sigma(\overline{b}, f) = (\sigma\overline{b}, f\phi(\sigma)^{-1})$, $\sigma \in G$, and $\phi(\sigma)$ is a homomorphism of the Galois group of the covering $\overline{B} \to B$ onto the Galois group of the field $k(x, \sqrt[m_1]{P_1}, \sqrt[m_2]{P_2})/k(x)$.*

We now consider the case when the genus of a generic fiber of the fibering π is equal to 1.

According to Theorem 10, there exists a covering $C \to B$ of the base such that

$$C \times_B V \simeq C \times F,$$

where F is a nonsingular curve of genus 1. We can of course assume that C is a normal covering with a Galois group G. It is clear that on $C \times F$ the group G operates according to the rule

$$\sigma(c \times f) = \sigma(c) \times f\varphi(\sigma)^{-1}, \ \sigma \in G, \tag{40}$$

where

$$\phi\colon G \to \operatorname{Aut} F$$

is a homomorphism of G into the group of biregular automorphisms of the curve F. It is obvious that

$$V = (C \times_B V)/G = (C \times F)/G.$$

Here we can limit ourselves to the case when ϕ is a monomorphism. In fact, if N is the kernel of ϕ, $G_1 = G/N$, $C_1 = C/N$, then it is easy to see that

$$V \times_B C_1 \simeq C_1 \times F, \quad V = (C_1 \times F)/G_1.$$

We now indicate which fixed points the automorphisms $\sigma \in G$ have.

It is clear that the point $\sigma = (c \times f)$ is fixed for an automorphism if and only if $\sigma c = c$, $f\phi(\sigma) = f$. Since for $\sigma \neq 1$ it is also true that $\phi(\sigma) \neq 1$, both σ and $\phi(\sigma)$ have a finite number of fixed points on C and F respectively, and consequently σ has only a finite number of fixed points on $C \times F$. Thus the covering $C \times F \to V$ can have only isolated branch points. From this and from the fact that neither $C \times F$ nor V have singular points, it follows that this covering does not generally have branch points. This follows, for example, from the theorem of Zariski about varieties of ramification [17], according to which the variety of ramification of any covering of a nonsingular variety by a nonsingular one has codimension 1. It is also possible to verify this directly, considering the subring $\mathfrak{D}_x^H$ consisting of the H-invariant elements of the local ring $\mathfrak{D}_x$ of a point $x \in C \times F$, where H is a stationary subgroup of the point x, and showing that this ring is nonregular (the ring $\mathfrak{D}_x^H$ is isomorphic, as it is easy to see, to the local ring of the image of the point x on V).

Conversely we take a normal covering $C \to B$ with a Galois group G and a monomorphism $\phi: G \to \operatorname{Aut} E$ of the group G into the group of automorphisms of the curve F of genus 1. We define the operation of G on $C \times F$ by formula (40) and explain when G operates on $C \times F$ without fixed points and when $(C \times F)/G$ has the invariants $p_g = 0$, $q = 1$.

The group $\phi(G)$ is a group of the automorphisms of a nonsingular curve F of genus 1. We introduce in F the structure of a one-dimensional abelian variety. Then, as is known, the finite group of the biregular transformations of F is a semi-direct product

$$\phi(G) = H \cdot \mathfrak{A}, \tag{41}$$

where $\mathfrak{A}$ consists of translations and H of the automorphisms of the abelian variety F, while $\mathfrak{A}$ has one or two generators and H is a cyclic group of order 1, 2, 3, 4, or 6. Here, naturally, $\mathfrak{A}$ is an H-operator group.

It is easy to verify that the elements of $\mathfrak{A}$ and only they do not have fixed points on F. Consequently, for the group G acting on $C \times F$ according to (33) not to have fixed points it is necessary and sufficient that in the group G of automorphisms of the covering $C \to B$ only the automorphisms of the form $\phi^{-1}(a)$, $a \in \mathfrak{A}$ have fixed points. This in turn means that the covering $C_1 \to B$, belonging to the subgroup $\phi^{-1}(\mathfrak{A})$ by Galois theory, is unramified.

We assume that $F/\phi(G) = L$ and we designate the genus of the curve L by γ. From formula (36) it follows that if $\overline{V} = C \times F$, $V = (C \times F)/G$,

$$\Omega'(V) \simeq \Omega'(\overline{V})^G \simeq \Omega'(C)^G \oplus \Omega'(F)^G \simeq \Omega'(B) \oplus \Omega'(L),$$

and therefore for V

$$q = 1 + \gamma,$$

so that $q = 1$, if and only if the curve L is rational. This in turn is equivalent to the fact that in (41) $H \neq 1$.

It follows from formula (38) that

$$\Omega^2(\overline{V}) \simeq \Omega'(C) \otimes \Omega'(F).$$

In order to calculate $\Omega^2(\overline{V})^G$, we first define $\Omega^2(\overline{V})^{\mathfrak{A}}$. Since the elements $a \in \mathfrak{A}$ are translations of the abelian variety F, $\Omega'(F)^{\mathfrak{A}} = \Omega'(F)$ and

$$\Omega^2(\overline{V})^{\mathfrak{A}} \simeq \Omega'(C)^{\mathfrak{A}} \otimes \Omega'(F) \simeq \Omega'(C_1) \otimes \Omega'(F),$$

where C_1 is a covering $C_1 \rightarrow B$ belonging to the subgroup $\phi^{-1}(\mathfrak{A})$ of the Galois group of the covering $C \rightarrow B$. The covering $C_1 \rightarrow B$ is by assumption unramified and we can apply formula (39) to it. Therefore

$$\Omega^2(\overline{V})^G = (\Omega^2(\overline{V})^{\mathfrak{A}})^H \simeq \Omega'(L),$$

and, consequently, $p_g = 0$ if and only if the curve L is rational, and hence $H \neq 1$. We have proved the following result.

Theorem 12. *For the case $g = 1$ all the surfaces V with the invariants $p_g = 0$, $q = 1$ can be obtained in the form $V = \overline{V}/G$, $\overline{V} = C \times F$, for $C \rightarrow B$ a covering of an elliptic curve B with a Galois group G of the type (41) with $H \neq 1$, where G is isomorphic to the group of biregular transformations of the curve F of genus 1, and in this isomorphism the elements of $\mathfrak{A}$ correspond to translations, and H to the automorphisms of the one-dimensional abelian variety F. Here the covering $C_1 \rightarrow B$ belonging in the sense of Galois theory to the subgroup $\mathfrak{A}$ of the Galois group G of the covering $C \rightarrow B$, must be unramified.*

It would have been possible to give an explicit construction of the extensions of the field $k(B)$ that have a Galois group of type (41) and satisfy all the conditions of Theorem 12, and at the same time give an explicit construction for the surfaces with the invariants $p_g = 0$, $q = 1$ for the case $g = 1$ also. A more elegant classification of such surfaces, however, will be obtained in the theory of surfaces with a pencil of elliptic curves (cf. Theorem 12, Chapter VII).

Theorem 13 (Criterion of Enriques). *A surface V is ruled if and only if $P_{12} = 0$ for it.*

Proof. First of all we verify that for a ruled surface $P_{12} = 0$. In view of the birational invariance of the number P_{12} it is sufficient to establish the equality $P_{12} = 0$ for the surface $V = B \times P^1$, where P^1 is a projective line. For this surface $P_n = 0$ for all $n > 0$. In fact, it follows from (2) that on V

$$(K \cdot (b \times P^1)) = -2 < 0.$$

Therefore if $P_n(V)$ were positive for some $n > 0$, we would have $nK \sim D > 0$,

$$(D \cdot (b \times P^1)) = (nK \cdot (b \times P^1)) = -2n < 0,$$

which is impossible.

We will show that it follows from the equation $P_{12} = 0$ that the surface V is ruled. First of all we note that because of this condition $p_g = 0$ and $p_2 = 0$.

Therefore, if $q = 0$, then the surface V is rational by the criterion of Castelnuovo and moreover is ruled. In the same way, if $q > 1$, the surface is ruled according to Theorem 5. It remains for us to consider the case $q = 1$.

In this case we can use Theorems 11 and 12 and represent V in the form $\overline{V}/G$, where $\overline{V} = C \times F$, $C \to B$ is a covering, perhaps ramified, and the group G defined in Theorems 11 and 12 also operates on $\overline{V}$ without fixed points. We have to determine $P_n(X)$, i.e. the dimension of the space of differentials of degree n and of first order on X.

If x and y are two functions that are algebraically independent on X, then they will be independent also on $\overline{X}$, and any differential of nth degree on $\overline{X}$ can be written in the form $f(dx \wedge dy)^n$, $f \in k(\overline{X})$. The invariance of this differential with respect to the operation of the automorphisms of G is equivalent to the invariance of the function f, and this in turn means that $f \in k(X)$, and hence the differential itself is also an image of a differential on X under the mapping of the differentials that is induced by the mapping $\pi\colon \overline{X} \to X$.

We have thus shown that the differentials of nth degree on $\overline{X}$ that are invariant with respect to the automorphisms of G coincide with the differentials of the form $\pi^*(\omega)$, where ω is a differential on X.

We now note that since the covering $\overline{X} \to X$ is unramified a differential ω is of first order on X if and only if $\pi^*(\omega)$ is of first order on $\overline{X}$. Thus,

$$\Omega_n(X) \simeq \Omega_n(\overline{X})^G,$$

where $\Omega_n(X)$ and $\Omega_n(\overline{X})$ are the spaces of the differentials of nth degree and first order on X and $\overline{X}$ respectively. It is clear that

$$\Omega_n(\overline{V})^G \supset \Omega_n(C)^G \otimes \Omega_n(F)^G \supset \Omega_n(B) \otimes \Omega_n(F)^G.$$

Since B is an elliptic curve, $\Omega_n(B) \simeq k$. Hence, in view of what we proved earlier,

$$P_n(V) \geqslant \dim_k \Omega_n(F)^{\varphi(G)}.$$

It remains to find the dimension of the space $\Omega_n(F)^{\phi(G)}$. For this we recall that $\phi(G)$ is the Galois group of the extension $k(F)/k(L)$ and that L is a rational curve. Since the differentials of nth degree that are invariant with respect to the automorphisms of $\phi(G)$, as we saw, have the form $\psi^*(\omega)$ where ψ is a covering $F \to L$ and ω is a differential on L, the space $\Omega_n(F)^{\phi(G)}$ coincides with the space of all differentials of nth degree on L for which $\psi^*(\omega)$ is of first order.

The surface V possesses a regular mapping $V \to B$ whose fibers are isomorphic to F (this mapping is induced by the mapping $\overline{V} \to \overline{B}$). Therefore if we show that the genus of the curve F is equal to 0, we will have proved at the same time that the surface V is ruled.

We see that the assertion of Theorem 11 reduces to the following lemma about algebraic curves.

Lemma 14. *Let $\psi: F \to L$ be a normal covering of a rational curve. If any differential ω of 12th degree on L is such that $\psi^*(\omega)$ of first order on F is equal to 0, then F is a rational curve.*

Proof. We choose a coordinate t on the projective line L so that all the branch points of the covering ψ are finite. Let $\omega = f(dt)^n$ be a differential of nth degree on L. We give the conditions to which the function f must be subjected in order for the differential $\psi^*(\omega)$ to be of first order.

Let a point $P_i \in F$, $\psi(P_i) = Q_i \neq \infty$ and let P_i be a branch point of multiplicity e_i for the covering ψ. Then $t = \tau_i^{e_i}$ on F, where τ_i is a local parameter at the point P_i, while $\nu_{P_i}(u) = 0$, and

$$\psi^*(\omega) = \psi^*(f)(\tau_i^{e_i-1}v)^n, \quad \nu_{P_i}(v) = 0.$$

From this it follows that ω is regular in P_i if and only if

$$\nu_{P_i}(\psi^*(f)) \geqslant -n(e_i - 1),$$

and since $\nu_{P_i}(\psi^*(f)) = e_i \nu_{Q_i}(f)$, the same condition can be written in the form

$$\nu_{Q_i}(f) \geqslant -n\left(1 - \frac{1}{e_i}\right),$$

or

$$\nu_{Q_i}(f) \geqslant -\left[n\left(1 - \frac{1}{e_i}\right)\right].$$

An analogous consideration at the infinite point (∞) yields the condition

$$\nu_\infty(f) \leqslant -2n.$$

We thus see that the differential $\psi^*(\omega)$ is a differential of first order if and only if

$$f \in \mathcal{L}(D), D = \Sigma\,[n(1 - 1/e_i)]\, Q_i - 2n(\infty).$$

We have to explain when there exists a differential $(\omega) \neq 0$ such that $\psi^*(\omega)$ is of first order on F. In other words, we must explain when

$$l(D) > 0.$$

Since on a curve of order 0 the dimension $l(D) > 0$ if and only if $\deg D \geq 0$, the case of interest to us takes the form

$$\Sigma\,[n(1 - 1/e_i)] \geqslant 2n. \tag{42}$$

We now assume that the genus of the curve F is different from 0, and we will show that the relationship (42) can be satisfied for $n = 2, 3, 4$, or 6. This then gives us the existence of a differential ω of degree $n = 2, 3, 4$, or 6 such that $\psi^*(\omega)$ is of first order of F. The differential $\omega^{12/n}$ will then satisfy all the conditions of the lemma.

Let the covering ψ: $F \to L$ have degree m. Then m/e_i branch points of multiplicity e_i lie over each of the points Q_i (because of the normality of the covering). Since the genus g of the curve F is by assumption ≥ 1,

$$\frac{1}{2}\Sigma\,(e_i - 1)\frac{m}{e_i} - m + 1 \geqslant 0,$$

i.e.

$$\Sigma\,(1 - 1/e_i) \geqslant 2. \tag{43}$$

Let $e_1 \leq e_2 \leq \cdots \leq e_k$. Since $e_i \geq 2$, if $k \geq 4$ condition (42) is satisfied already for $n = 2$:

$$2\,(1 - 1/e_i) = 2 - 2/e_i \geqslant 1,$$

$$\sum_1^k [2\,(1 - 1/e_i)] \geqslant k \geqslant 4.$$

If $3 \leq e_1 \leq e_2 \leq e_3$, then the relationship (42) is satisfied with $n = 3$:

$$3\,(1 - 1/e_i) \geqslant 2, \quad i = 1, 2, 3,$$

$$\sum_1^3 [3\,(1 - 1/e_i)] \geqslant 6.$$

If $e_1 = 2$, $4 \leq e_2 \leq e_3$, then (42) is satisfied for $n = 4$:

$$[4\,(1 - {}^1/_2)] + [4\,(1 - 1/e_2)] + [4\,(1 - 1/e_3)] \geqslant 2 + 3 + 3 = 8.$$

If $e_1 = 2$, $e_2 = 3$, $e_3 \geq 6$, then (42) is satisfied for $n = 6$:

$$[6\,(1 - {}^1/_2)] + [6\,(1 - {}^1/_3)] + [6\,(1 - 1/e_3)] \geqslant 3 + 4 + 5 = 12.$$

Thus there remain the unexamined cases $e_1 = 2$, $e_2 = 3$, $e_3 = 3$, 4, or 5. In these cases (corresponding to the groups of right polyhedra) the relationship of (43) is not satisfied:

$$1 - {}^1/_2 + 1 - {}^1/_3 + 1 - 1/e_3 \leqslant 1\frac{29}{30} < 2 \quad \text{for} \quad e_3 \leqslant 5.$$

Lemma 14, and with it the theorem of Enriques, is proved.

Remark. If the genus of the curve F is greater than 1, then for a sufficiently large n the dimension of the space of the differentials ω of nth degree on L such that $\psi^*(\omega)$ is of first order on F takes as large a value as is desired. In other words, if $V = (F \times B)/G$, then $\max P_n(V) = \infty$.

In fact, in this case instead of (43) we have the inequality

$$\theta = \Sigma\,(1 - 1/e_i) - 2 > 0,$$

and if $n \equiv 0 \pmod{e_i}$ for all e_i, then

$$\deg D = \Sigma\,[n\,(1 - 1/e_i)] - 2n = n\theta$$

and hence $l(D)$ grows indefinitely large along with n.

CHAPTER VII

This chapter studies surfaces with a pencil of elliptic curves. A description is given of the connection between this class of surfaces with other classes of algebraic surfaces (surfaces with $(K^2) = 0$, surfaces with $\kappa = 1$). Finally, there is presented a classification of surfaces with a pencil of elliptic curves, or, more precisely, a classification of the pencils themselves. This last question was considered in the works [25, 42, 57, 40]. We present the results of these papers without proofs, proving only those assertions which are not contained in them.

The base field k is assumed to be the field of complex numbers. The majority of the arguments remain valid when k is an algebraically closed field of characteristic 0 (or with even weaker restrictions on the characteristic). We shall note the places where the assumption $k = \mathbf{C}$ is essential in an argument.

§1. Basic concepts

Definition. *A surface with a pencil of elliptic curves is a triple* (V, B, π) *consisting of a nonsingular surface* V*, a nonsingular curve* B*, and a regular mapping* $\pi\colon V \longrightarrow B$ *such that a generic fiber of the fibering* π *is a nonsingular curve of genus* 1.

Two surfaces with a pencil of elliptic curves (V, B, π) *and* (V', B, π') *are said to be biregularly (respectively, birationally) equivalent if there exists a biregular (respectively, birational) mapping* $f\colon V \longrightarrow V'$ *such that* $\pi' f = \pi$.

A surface with an elliptic pencil (V, B, π) *is said to be a minimal model, if the fibers of the fibering* π *do not contain exceptional curves of the first kind, i.e (by the theorem of Castelnuovo, Chapter* II, §4) *it does not contain nonsingular rational curves* C *with* $(C^2) = -1$.

Every surface with an elliptic pencil is birationally equivalent to a minimal model – a transformation contracting an exceptional curve of the first kind contained in a fiber into a point clearly commutes with a projection.

Remark. A surface (V, B, π) with a pencil of elliptic curves can be a minimal model in the sense of the definition of this section, while at the same time the surface V is not a minimal model in the sense of Chapter II. As an example we consider two plane nonsingular cubic curves $G = 0$ and $H = 0$ that have nine distinct points of intersection $P_1, \cdots, P_9$. We denote by V the surface obtained by an application of a σ-process to the points $P_1, \cdots, P_9$ of the plane. The total preimages on V of the plane curves $\lambda G + \mu H = 0$ determine on V a pencil of elliptic curves. The surface obtained is a minimal model in the sense of the definition of this section. This can be easily verified directly if one assumes that no three of the P_i lie on the same line, and that no six of them lie on the same second order curve; one may also deduce it from the description of the possible types of degenerate fibers given in §6. The surface, however, is of course not a minimal model in the sense of the definition of Chapter II.

The concept of a minimal model of a surface with an elliptic pencil is analogous to the concept of a relatively minimal model of an arbitrary surface. It also plays the role of an absolute minimal model, however, as the following result shows.

Theorem 1. *If (V, B, π) and (V', B, π') are two surfaces with a pencil of elliptic curves, $f\colon V' \longrightarrow V$ is a birational mapping of V' onto V, and if V is a minimal model, then the mapping f is regular.*

Proof. Let F_β and F'_β be generic fibers of the fiberings π and π'. We denote by ξ and ξ' generic points of the curves F_β and F'_β, and by o_ξ, $o_{\xi'}$, O_ξ, and $O_{\xi'}$, their local rings on the curves F_β and $F_{\beta'}$ and on the surfaces V and V' respectively. It is clear that the mapping f induces a biregular isomorphism between F_β and F'_β. Thus

$$f(\xi') \in o_\xi,\ f(0') \in o_\xi \otimes k(\beta).$$

It follows easily from this that exceptional curves of the mappings f and f^{-1} are contained in fibers of the fiberings π and π'. If f were not regular, then there would exist on V exceptional curves of the first or second kind. Exceptional curves of the first kind cannot exist on V by definition of a minimal model. We shall show that there also do not exist any exceptional curves of the second kind.

For this we note that if there exists an exceptional curve of the second kind, then there also exists an irreducible exceptional curve of the second kind. The proof of this fact given in Chapter II (Theorem 1, §5) remains valid in the present case (we cannot directly apply this theorem because of the different sense of the term "minimal model").

If there were on V an irreducible exceptional curve of the second kind C contained in a fiber F_0, then, by Theorem 1, Chapter II, §1, we would have

$$(C^2) \geq 0.$$

Let

$$F_0 = nC + \sum n_i C_i,\quad n > 0.$$

It is clear that $(F_0 \cdot C) = 0$, since in the calculation of $(F_0 \cdot C)$ we can replace F_0 by any other fiber. Therefore

$$(C^2) = -\frac{1}{n}\sum n_i (C_i \cdot C) \leqslant 0.$$

Consequently it must be true that $(C^2) = 0$, which is possible only if all the $(C_i \cdot C) = 0$. This means that C coincides with a connected component of F_0, which, because of the connectedness of F_0, is possible only for $F_0 = nC$. But then we have $p_a(C) = 1$, at the same time that $p_a(C) = 0$ (by Theorem 1 of §1, Chapter II). The theorem is proved.

Corollary. *Two birationally equivalent minimal models are biregularly equivalent. A birational automorphism of a minimal model is biregular.*

We consider an example of the application of Theorem 1. Let (V, B, π) be the surface with a pencil described in the remark. It is clear that $B = \mathbf{P}^1$ (a point on B is a ratio $(\lambda : \mu)$). The lines $L_i = \sigma(P_i)$, $i = 1, \cdots, 9$ are mapped by the projection π biregularly onto B. Thus it is possible to define biregular mappings $s_i : B \to L_i$ such that $\pi s_i = 1$. Taking the point $F_\beta \cdot L_1$ for 0, we can define on the generic fiber F_β the structure of a one-dimensional abelian variety. The mappings ϕ_i, $i = 1, \cdots, 9$, defined by

$$\varphi_i(v) = v + s_i\pi(v)$$

are biregular transformations, if $v \in F_\beta$ are contractions onto the points $s_i(\beta)$. They are thus birational, and by the corollary are also biregular automorphisms of V. The group formed by these automorphisms is, as it is easy to show, the free abelian group with eight generators $\phi_2, \cdots, \phi_9$. We thus obtain an example of a surface that has an infinite group of automorphisms, but does not have, as is easily verified, an algebraic group (or, for $k = \mathbf{C}$, a Lie group) of automorphisms.

On the other hand, there exist on V curves with negative squares (for example, L_i). Using the automorphisms constructed, we can obtain an infinite number of such curves. Thus, we obtain an example of a surface containing an infinite number of curves with a negative square (and even exceptional curves of the first kind).

§2. The structure of fibers

Lemma 1. *Let V be a surface, π its regular mapping onto a curve B, F_0 one of the fibers of the fibering π, and $F_0 = \Sigma n_i C_i$, $n_i > 0$, where the C_i are irreducible curves. If $D = \Sigma m_i C_i$, then $(D^2) \leq 0$.*

Proof. Assume $(D^2) > 0$. It then follows from the Riemann-Roch theorem that for any E and for a sufficiently large n, $l(nD - E) > 0$. We choose for E a hyperplane section of the surface V and set

$$nD - E \sim D_1 > 0.$$

Then, if F is any fiber of the fibering π,

$$n(D \cdot F) = (E \cdot F) + (D_1 \cdot F). \tag{1}$$

Since all the components of the cycle D are contained in F_0, $(D \cdot F) = 0$. On the other hand, $(E \cdot F) > 0$ and $(D_1 \cdot F) \geq 0$, since we can choose an irreducible fiber for F. Thus we have a contradiction and the lemma is proved.

The assertion of Lemma 1 can be expressed differently. For this we consider the space X (over the field of rational numbers) a basis of which is formed by the irreducible components $C_1, \cdots, C_k$ of the fiber F_0. The index of intersection $(C \cdot D)$ determines in X a scalar product and the quadratic form

$$\psi(C) = (C^2). \tag{2}$$

The lemma asserts that this form is not positive.

Theorem 2 (cf. Zariski [22]). *If, in the notation of Lemma 1, the fiber F_0 is connected and $(D^2) = 0$, then $D = rF_0$, where r is a rational number.*

Since $(F_0 \cdot C_i) = 0$ for all the $C_1, \cdots, C_k$, the quadratic form ψ defined by (2) has rank $\leq k - 1$. The assertion of Theorem 2 is equivalent to saying that this rank is equal to $k - 1$. Assume that the rank of ψ is equal to $l \leq k - 2$. In the space X we denote by Y the subspace consisting of all D for which $(D \cdot C) = 0$ for all $C \in X$. It is clear that in the factor space X/Y the form ψ is negative definite. We can choose $C_1, \cdots, C_l$ such that they form a basis in X/Y, and then

$$(x_1 C_1 + \cdots + x_l C_l)^2 < 0 \tag{3}$$

for all $x_1, \cdots, x_l$ not simultaneously equal to zero.

We set

$$C_{l+1} = L + D,$$

where $D \in Y$ and L is a linear combination of $C_1, \cdots, C_l$. Let

$$L = L_1 - L_2,$$

where $L_1 > 0$, $L_2 > 0$, and L_1 and L_2 do not have common components. Then, on the one hand, $(C_{l+1} \cdot L_1) \geq 0$, since C_{l+1} and L_1 do not have common components, and on the other hand,

$$(C_{l+1} \cdot L_1) = (L_1^2) - (L_1 \cdot L_2) \leq 0.$$

Hence $(L_1^2) = 0$ and $(L_1 \cdot L_2) = 0$. But from (3) it follows that $L_1 = 0$,

$$C_{l+1} + L_2 \in Y,$$

and, in particular, $((C_{l+1} + L_2) \cdot C_i) = 0$. This means that the cycles C_i that are components of $C_{l+1} + L_2$ do not intersect with other cycles, and this, in turn, means that two sets of cycles constitute two connected components of the fiber F_0, which contradicts the assumption.

§3. The canonical class

Lemma 2. *On a surface with a pencil of elliptic curves, the canonical class contains a divisor consisting of components of fibers.*

Proof. Let F_β be a generic fiber of the fibering $\pi: V \longrightarrow B$. Since the canonical class of the curve F_β is equal to 0, $K \cdot F_\beta \sim 0$. If $D \in K$ and $D \cdot F_\beta = (f)$, then $\overline{D} = D - (f) \in K$ and $(\overline{D} \cdot F_\beta) = 0$. As we saw in Chapter IV, §7, it follows from this that D consists of components of fibers.

Remark. It follows from the proof of the lemma that any effective divisor consists of components of fibers.

Theorem 3. *On a minimal model of a surface with a pencil of elliptic curves* $(K^2) = 0$.

We have to show the impossibility of a) $(K^2) > 0$, b) $(K^2) < 0$.

a) It follows from the Riemann-Roch theorem that for any E and for some $n > 0$, $l(nK - E) > 0$. We take for E a hyperplane section of the surface V and let

$$nK - E \sim D > 0.$$

Since a fiber F is an elliptic curve and $(F^2) = 0$, we have $(K \cdot F) = 0$. Thus

$$0 = n\,(K \cdot F) = (E \cdot F) + (D \cdot F).$$

But $(E \cdot F) > 0$, and $(D \cdot F) \geq 0$, which leads to a contradiction.

b) According to Lemma 2, there exists in the canonical class a representative K of the form

$$K = \sum_0^m K_i,$$

where the K_i consist of components of fibers F_i.

Since $(K_i \cdot K_j) = 0$ for $i \neq j$,

$$(K^2) = \sum (K_i^2)$$

and it is sufficient for us to show that all the $(K_i^2) \geq 0$.

Let $(K_0^2) < 0$, and let C_0 be an irreducible component of the fiber F_0 contained in K_0. If $F_0 = nC_0$, then $(C_0^2) = 0$, and thus $(K_0^2) = 0$. If F_0 contains still other components $F_0 = nC_0 + \Sigma_1^k n_i C_i$, then

$$0 = (F_0 \cdot C_0) = n\,(C_0^2) + \sum n_i\,(C_i \cdot C_0).$$

Here $(C_i \cdot C_0) \geq 0$ and for at least one i, $(C_i \cdot C_0) > 0$; otherwise C_0 would be a connected component of the fiber F_0, which contradicts its connectedness. Therefore, $(C_0^2) < 0$. For at least one component C_0 of the cycle F_0 we have $(C_0 \cdot K_0) < 0$. For, if $(C_i \cdot K_0) \geq 0$ for $i = 0, \cdots, k$, then $(C_i \cdot K_0) = 0$ for all i – otherwise for some C_j it would be true that $(C_j \cdot K_0) > 0$, and since for all j, $(C_j \cdot K_0) \geq 0$, then $(F_0 \cdot K_0) > 0$ at the same time as $(F_0 \cdot K_0) = 0$. This proves that if for all i, $(C_i \cdot K_0) \geq 0$, then $(C_i \cdot K_0) = 0$. But then $(K_0^2) = 0$, which contradicts our assumption.

We have proved the existence of an irreducible component C_0 of the divisor K_0 for which $(C_0 \cdot K_0) < 0$. But then the inequality already established

$$(C_0 \cdot K) = (C_0 \cdot K_0) < 0, \quad (C_0^2) < 0$$

shows that

$$p_a(C_0) = \frac{(C_0 \cdot K) + (C_0^2)}{2} + 1 \leqslant 0,$$

which is possible only for $p_a(C_0) = 0$, $(C_0^2) = -1$. This means that C_0 is an exceptional curve of the first kind that is a component of a fiber, and this contradicts the minimality of V.

Theorem 4. *On a minimal model of a surface with a pencil of elliptic curves, the canonical class contains a divisor that is a rational combination of fibers.*

Proof. Let K be a divisor of the canonical class that consists of components of fibers, $K = \sum K_i$. Since by Theorem 3 $(K_i^2) = 0$, it follows from Theorem 1 that $K_i = r_i F_i$, where r_i is some rational number. This is the assertion of the theorem.

Remark. It follows from the remark after Lemma 2 that any effective divisor of the canonical class is a rational combination of fibers.

Definition. A fiber F_0 is said to be multiple if

$$F_0 = \sum n_i C_i, \quad n_i > 1.$$

Corollary to Theorem 4. *If the fibering $\pi: V \longrightarrow B$ does not contain multiple fibers, the canonical class contains a divisor that is an integral linear combination of fibers.*

§4. Surfaces with an elliptic pencil and surfaces with $(K^2) = 0$

Theorem 5. *Assume that the surface V is neither rational nor ruled, but is a minimal model and for it $(K^2) = 0$. Then either $12K = 0$ or there exists an m such that, for all sufficiently large n, the linear system $|mnK|$ has neither fixed curves nor base points and is composed of a pencil of elliptic curves. The property given* *uniquely determines this pencil.*

Proof. We consider two cases: A) All the $P_n \leq 1$, and B) some $P_n \geq 2$.

A) We consider separately the cases $p = 0$ and $p = 1$. If $p = 0$ and $q = 0$, then, for $P_2 = 0$, by the theorem of Castelnuovo (Chapter III, §2), the surface is rational, and for $P_2 = 1$, by the remark at the end of §1, Chapter VIII, we have an Enriques surface, for which $2K = 0$, and thus $12K = 0$. If $p = 0$ and $q > 1$, then by Theorem 5 (§3, Chapter IV) V is a ruled surface. Finally, for $p = 0$, $q = 1$, by Theorems 11 and 12 (§8, Chapter IV) V can be represented in the form $(B \times C)/G$, where B is an elliptic curve, C is a curve of arbitrary genus g, and G is the finite group of automorphisms without fixed points of the surface $B \times C$. If $g = 0$, the surface V is ruled. If $g > 1$, then by the remark after Lemma 14 (Chapter IV), $P_n(V)$ takes values as large as desired, and we have case B). It remains to consider the case when $g = 1$, i.e., when B and C are elliptic curves.

By the theorem of Enriques (Theorem 13, Chapter IV), if V is not a ruled surface, then $P_{12}(V) > 0$, i.e., $12K \sim D > 0$. We shall show that $D = 0$. Thus, if $f\colon B \times C \to V$ is a projection, then the canonical class $\overline{K}$ of the surface $B \times C$ has the form $f^*(K)$. Therefore, if $D > 0$, $D \neq 0$, then

$$12\overline{K} = f^*(12K) = f^*(D)$$

and it is clear that $f^*(D) > 0$, $f^*(D) \neq 0$. This, however, contradicts the fact that $B \times C$ is an abelian variety, so $\overline{K} = 0$ and $12\overline{K} = 0$.

We now consider the case $p = 1$ and thus $P_2 = 1$. Then

$$p_a(V) = 1 - q + p = 2 - q = \frac{\chi}{12} = \frac{2 - 4q + 2p + h^{1,1}}{12} = \frac{4 - 4q + h^{1,1}}{12}.$$

Since $h^{1,1} > 0$, it follows from this that $q \leq 2$. If $q = 0$, then by the Riemann-Roch theorem

$$l(-K) + l(2K) \geq 2,$$

i.e., $l(-K) \geq 1$. Since $l(K) \geq 1$, it follows from this that $K = 0$.

The case $q = 1$ is impossible by Theorem 1 (Chapter VIII). Finally, for $q = 2$, V is an abelian variety according to Theorem 3 (Chapter VIII), and thus again $K = 0$.

B) Let $P_\nu \geq 2$ for some ν, i.e. $l(\nu K) \geq 2$.

By Bertini's theorem the system $|\nu K|$ is composed of a pencil C_λ. If D is the fixed part of this system, then

$$\nu K \sim D + \Sigma C_i, \tag{4}$$

where the C_i are curves of the pencil C_λ and $(C_\lambda^2) \geq 0$. On the other hand, $(C_i \cdot K) \geq 0$ and $(D \cdot K) \geq 0$, for otherwise, by the lemma of Chapter II, §4, the surface V would be ruled or would not be a minimal model. Since $(K^2) = 0$, it follows from (4) by the multiplication of both the parts by K that $(C_i \cdot K) = 0$. Now multiplying both parts of (4) by C_i, we obtain from this (and from the fact that $(C_i^2) \geq 0$) that $(C_i^2) = 0$ and $(C_i \cdot C_j) = 0$. Thus, $p_a(C_i) = 1$, and distinct curves C_i do not intersect. It follows from Bertini's theorem that C_λ is a pencil of elliptic curves, where, since $(C_\lambda^2) = 0$, it does not have fundamental points. It determines a regular mapping of V onto some curve B.

It remains to prove the assertion about the fixed components of the system $|nK|$. By the remark after Theorem 4, the divisor $D + \Sigma C_i$ is a rational linear combination of fibers. Let ν' be the common denominator of the coefficients of this combination. We set $m = \nu \cdot \nu'$. Then $|mK|$ contains an integral linear combination of fibers:

$$mK \sim \Sigma n_i F_{b_i}, \quad n_i > 0.$$

Since for a sufficiently large n the class $n(\Sigma n_i b_i)$ on the curve B does not contain fixed components, the class mnK on V also does not contain fixed components.

Corollary. *If a surface V is a minimal model and is not ruled, then $\kappa = 1$ for it if and only if $(K^2) = 0$, $12K \neq 0$.*

§5. The Jacobian fibering

In this and the following sections we shall present the results of the works [25, 42, 40, 57] on the classification of surfaces with a pencil of elliptic curves. We shall not give proofs of the majority of the results set forth. The reader can find them in the works indicated.

Let $\pi: V \to B$ be a fibering, whose fibers determine a pencil of elliptic curves on the surface V. If β is a generic point of the curve B, then the fiber F_β is a curve of genus 1 defined over the field $k(B)$. We can thus apply to the analysis of the surfaces V the theory of curves of genus 1. This is the point of view of the works [42, 56, 57].

It is in general impossible to introduce on the curve F_β the structure of a one-dimensional abelian variety over the field $k(B)$; for this it is necessary that it have a rational point over this field. The existence of a rational point on the curve F_β over the field $k(B)$ is equivalent to the existence of a rational (and thus regular) mapping $\sigma: B \to V$ such that $\pi\sigma = 1$. The image $\sigma B = C$ is characterized by the fact that it is an irreducible curve on V and $(C \cdot F) = 1$, where F is any fiber of the fibering π. We shall call such a curve C a section of the fibering π. With each fibering π one may associate another fibering having the same base and already possessing a section. For this it is necessary to consider the Jacobian curve A_β of the curve F_β. The curve A_β possesses the property that there exists a birational mapping ϕ of the curve F_β onto A_β, defined over some finite extension of the field $k(B)$ and establishing an isomorphism between the group of the classes of divisors of degree zero of the curve F_β defined over some field $K \supset k(B)$ and the group of points on the curve A_β defined over the same field. It will be convenient later to assume that the curves F_β (and the corresponding surfaces V with a pencil of elliptic curves) are distinct if the mappings ϕ are distinct, i.e., do not differ by an automorphism of the one-dimensional abelian variety A_β. A generic point of the curve A_β determines some algebraic surface, which, in view of the inclusion $k(B) \subset k(A_\beta)$, has a pencil of elliptic curves with the base B.

We denote by J a nonsingular minimal model of this surface with a pencil of elliptic curves. We shall call J a Jacobian fibering of the fibering $\pi: V \to B$. Since the curve A_β has a rational point over the field $k(B)$, the zero point, the fibering J has a section σ, which we shall call the zero section. For each fibering $\pi: V \to B$ there exists a covering $C \to B$ such that the fibering $V \times_B C$ over C has a section. Every fibering with a section is isomorphic to its Jacobian fibering.

The method of classification to be used is to classify first all the fiberings having a section, and then all the fiberings having a given Jacobian fibering. An arbitrary fibering is associated with its Jacobian fibering in the following way. Let $C \to B$ be some normal covering with a Galois group G and let $\widetilde{J}_C$ be a nonsingular minimal model of the fibering $J \times_B C$ over C. The group G, naturally, operates on $\widetilde{J}_C$. The sections $\sigma: C \to \widetilde{J}_C$ form a group, which is clearly a G-operator group. We denote this group by $\mathfrak{A}_J(C)$. Then all the fiberings $\pi: V \to B$ for which the fibering $V \times_B C$ over C has a section are in one-to-one correspondence with the elements of the group $H^1(G, \mathfrak{A}_j(C))$.

The group structure can be carried over with the help of this correspondence onto the set of all the fiberings of V having J as a Jacobian fibering and for which $V \times_B C$ has a section. For any two fiberings V_1 and V_2 over B with the same Jacobian fibering it is possible to find a covering $C \to B$ such that $V_1 \times_B C$ and $V_2 \times_B C$ have a section. Using this, it is possible to introduce a group operation into the whole set of fiberings on elliptic curves $\pi: V \to B$ having a given Jacobian fibering J. The group obtained is denoted by $\mathfrak{H}(B, J)$. It is a torsion group.

Let $\pi: V \to B$ be a fibering on elliptic curves having the Jacobian fibering J. Let $C \to B$ be a covering such that $V \times_B C$ has a section over C and

$u \in H^1(G, \mathfrak{A}_J(C))$ is the element corresponding to V. Thus u is a one-dimensional cocycle, i.e., has the form $u_g \in \mathfrak{A}_J(C)$, $g \in G$. Since u_g is a section of $\widetilde{J}_C$ over C, the transformation $x \longrightarrow g(x) + u_g(\pi(x))$ is (since $\widetilde{J}_C$ is a minimal model) a biregular automorphism of $\widetilde{J}_C$. We thus have a mapping $\phi\colon G \longrightarrow \mathrm{Aut}(\widetilde{J}_C)$ which is a a monomorphism. Now V is defined in terms of $\widetilde{J}_C$ and u_g:

$$V \simeq \widetilde{J}_C/\phi(G). \tag{5}$$

§6. Fibers of a Jacobian fibering

The works [25] and [40] describe all the types of degenerate fibers that can be met in Jacobian fiberings. In order to present this description, we recall that a generic fiber F_β is birationally equivalent over $k(B)$ to the curve given by the equation

$$y^2 = x^3 + p\cdot x + q, \quad p, q \in k(B).$$

Let $b \in B$, let t be a local parameter at the point b, and let $\nu(f)$ be the index of the function $f \in k(B)$ at the point b. One may assume that in the above equation $\nu(p) \geq 0$, $\nu(q) \geq 0$, and $\min(\nu(p) - 4, \nu(q) - 6) < 0$. We set

$$\Delta = 4p^3 + 27q^2.$$

The fiber F_b is degenerate if and only if $\nu(\Delta) > 0$. We consider two cases.

A. For $k = 0$ or $k = 1$, one has the representation

$$p = t^{2k}a, \quad q = t^{3k}b, \quad \nu(a) = \nu(b) = 0, \quad n = \nu(4a^3 + 27b^2) > 0.$$

B. Such a representation does not exist.

We shall denote case A for $k = 0$ by A'_n depending on n, and for $k = 1$, by A''_n. The number n can take any positive integral value.

In case B the number $\nu(\Delta)$ can take the values 2, 3, 4, 6, 8, 9, 10. We shall denote these cases by B_n, $n = 2, 3, 4, 6, 8, 9, 10$.

The description of the degenerate fibers in all these cases is given below. Here the Θ_i are rational curves without singular points, except for the case A'_1, when Θ has one double point with different tangents, and for the case B_2, when Θ has one double point with a double tangent. The intersections of the curves Θ_i are shown in Figures 1–4. All the curves are transversal at the points of intersection, except for case B_3, when Θ_1 and Θ_2 have a tangent point of first order.

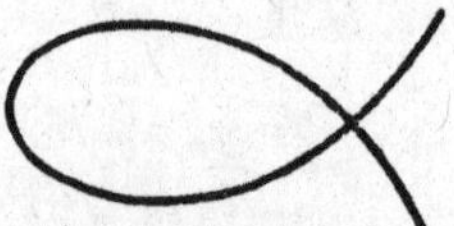

A'_1; $A_b = A_b^0$

Figure 1

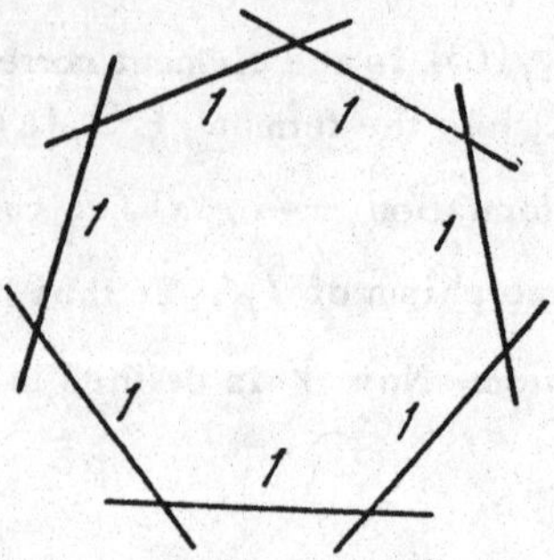

A'_n, $n > 1$; $A_b / A_b^0 = \mathbf{Z}_n$

Figure 2

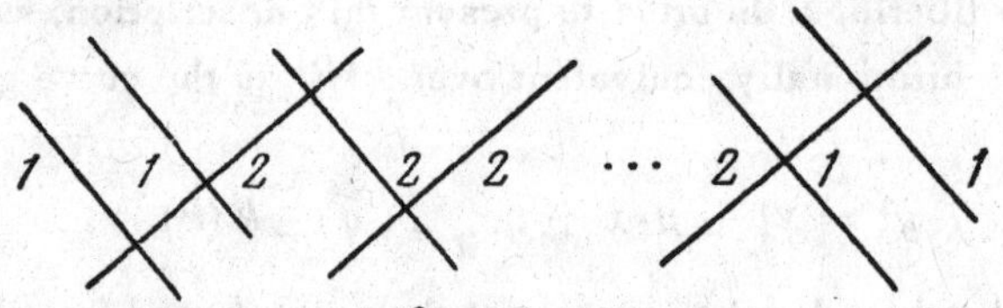

A''_n; for $n \equiv 0\,(2)$, $A_b / A_b^0 = \mathbf{Z}_2 \times \mathbf{Z}$; for $n \not\equiv 0\,(2)$, $A_b / A_b^0 = \mathbf{Z}_4$

Figure 3

B_2; $A_b = A_b^0$

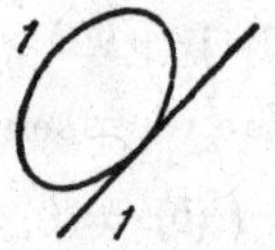

B_3; $A_b / A_b^0 = \mathbf{Z}_2$

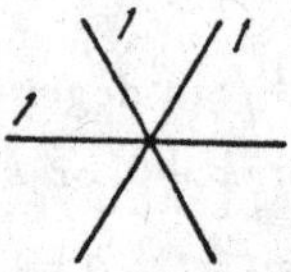

B_4; $A_b / A_b^0 = \mathbf{Z}_3$

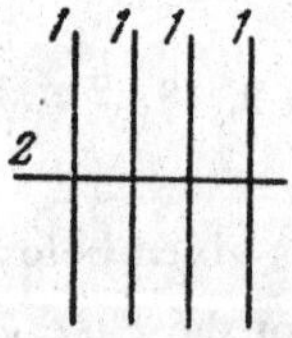

B_6; $A_b / A_b^0 = \mathbf{Z}_2 \times \mathbf{Z}_2$

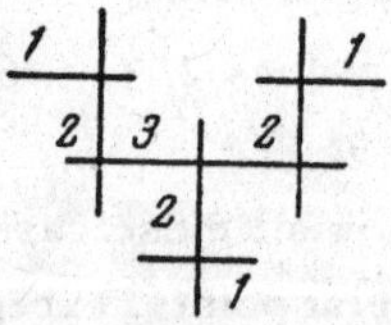

B_8; $A_b / A_b^0 = \mathbf{Z}_3$

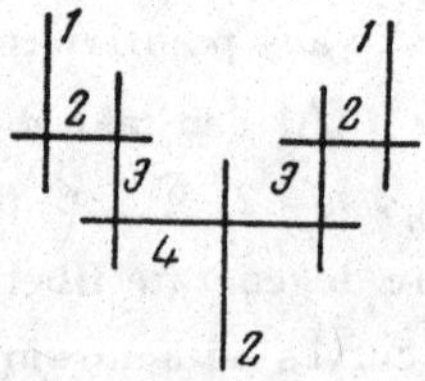

B_9; $A_b / A_b^0 = \mathbf{Z}_2$

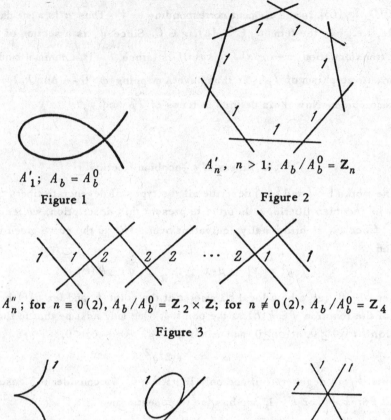

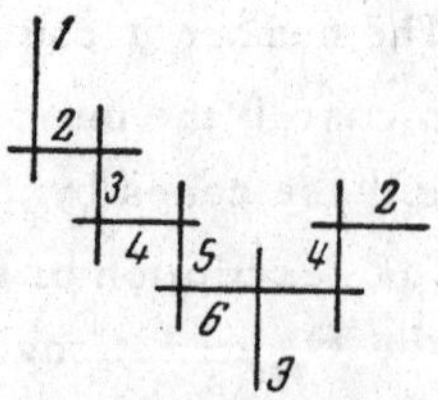

B_{10}; $A_b = A_b^0$

Figure 4

The fibers have the following structure (Figures 1–4):

$$A_1': \quad F_b = \Theta,$$

$$A_n',\ n > 1: F_b = \Theta_1 + \ldots + \Theta_n,$$

$$A_n'': \quad F_b = \Theta_0 + \Theta_1 + 2\Theta_2 + \ldots + 2\Theta_{n+2} + \Theta_{n+3} + \Theta_{n+4},$$

$$B_2: F_b = \Theta,$$

$$B_3: F_b = \Theta_1 + \Theta_2,$$

$$B_4: F_b = \Theta_1 + \Theta_2 + \Theta_3,$$

$$B_6: \quad F_b = 2\Theta_0 + \Theta_1 + \Theta_2 + \Theta_3 + \Theta_4,$$

$$B_8: \quad F_b = \Theta_1 + \Theta_2 + \Theta_3 + 2\Theta_4 + 2\Theta_5 + 2\Theta_6 + 3\Theta_7,$$

$$B_9: F_b = \Theta_1 + \Theta_2 + 2\Theta_3 + 2\Theta_4 + 2\Theta_5 + 3\Theta_6 + 7\Theta_7 + 4\Theta_8,$$

$$B_{10}: F_b = \Theta_1 + 2\Theta_2 + 2\Theta_3 + 3\Theta_4 + 3\Theta_5 + 4\Theta_6 + 4\Theta_7 + 5\Theta_8 + 6\Theta_9.$$

It is evident from the consideration of the separate cases that it is always true that $\nu(\Delta) = \chi(F_b)$.

More detailed properties of the degenerate fibers come from a consideration of the group structure on them. The group operation on a generic fiber F_β of the fibering J determines a regular mapping $F_\beta \times F_\beta \to F_\beta$. This mapping can be considered as a rational mapping $J \times_B J \to J$. It turns out that if J is a minimal model, this mapping is regular and defines the structure of an algebraic group on the set of nonsingular points of each fiber F_b. We denote this group by A_b. In particular, the component that intersects the zero section forms a subgroup A_b^0, a connected component of the unit of the group A_b. The group $A_b = A_b^0$ is an elliptic curve if the fiber F_b is nonsingular, is a multiplicative group in the case of A_n', and an additive group in all the remaining cases. The group A_b/A_b^0 is shown in Figures 1–4.

Finally, we shall indicate the form of Jacobian fiberings that do not have degenerate fibers.

Definition. Let A be an elliptic curve (with a fixed group structure) over the field k, let B be an arbitrary curve, let $\overline{B} \to B$ be a normal nonramified covering whose Galois group is isomorphic to some subgroup of the group of automorphisms of the curve A (as an abelian variety), and let $\phi: G \to \operatorname{Aut} A$ be the corresponding automorphism. We define the operation of G on $\overline{B} \times A$ according to the rule

$$\sigma(\bar{b} \times a) = \sigma\bar{b} \times a\varphi(\sigma)^{-1}.$$

The fibering $(B \times A)/G \to B$ does not have degenerate fibers. Such a fibering is said to be a *fiber bundle*. We have

Theorem 6. *Every Jacobian fibering without degenerate fibers is a fiber bundle.*

Corollary. *If B is a rational curve, then a Jacobian fibering without degenerate fibers over B has the form $B \times F$, where F is an elliptic curve.*

§7. Local classification

In order for a fibering of V over B by elliptic curves to be isomorphic to its Jacobian fibering, it is necessary and sufficient that it have a section.

Every section s over B gives a local section at any point $b \in B$. By this is meant the mapping $s_b : B \longrightarrow V$ given by a formal power series in the powers of the local parameter t at the point b, and such that $\pi s_b = 1$. In other words, a local section is a rational point on the curve F_β over the field of power series $k\{t\}$. An important necessary condition for the existence of a section is the existence of a local section at some point $b \in B$. In connection with this we introduce also the concept of a local isomorphism of fiberings, i.e. an isomorphism given by a formal power series in the powers of the local parameter t at the point b. In other words, the fiberings $\pi: V \longrightarrow B$ and $\pi': V' \longrightarrow B$ are locally isomorphic at the point $b \in B$ if the curves F_β and F'_β are isomorphic over the field of power series $k\{t\}$.

We now give the classification of a fibering up to a local isomorphism at a given point $b \in B$. For any fibering $V \longrightarrow B$ there exists a covering $C \longrightarrow B$ having one branch point over the point $b \in B$ such that $V \times_B C$ is formally isomorphic to $J \times_B C$, where J is the Jacobian fibering of V.

We denote by $\widetilde{V} \longrightarrow C$ a minimal nonsingular model of the fibering $V \times_B C$. Then the fiberings $\widetilde{V} \longrightarrow C$ and $\widetilde{J} \longrightarrow C$ are isomorphic. Since, moreover, there is a unique projection $\widetilde{J} \longrightarrow J$, we obtain a mapping $\widetilde{V} \longrightarrow J$. Let U_b and $\widetilde{F}_c$ be fibers of the fiberings $\widetilde{J}$ and $\widetilde{V}$ lying over the point b and its preimage $c \in C$ respectively. The mapping $\widetilde{V} \longrightarrow J$ determines a mapping $\widetilde{F}_c \longrightarrow U_b$, which, as can be easily seen, is an unramified cyclic covering. The Galois group H of this covering, naturally, is isomorphic to the subgroup of the Galois group G of the covering $C \longrightarrow B$ at the point b. The group G has a distinguished character ψ_0. In fact, the corresponding extension is obtained by the addition of the element $\sqrt[n]{t}$ to $k\{t\}$. The character ψ_0 is determined by $\psi_0(\sigma) = \sqrt[n]{t}^{1-\sigma}$. A restriction of the character ψ_0 determines some character on the Galois group H of the covering $\widetilde{F}_c \longrightarrow U_b$. Thus we have some unramified covering $\widetilde{F}_c \longrightarrow U_b$ and some character of the Galois group H of this covering. Since the group H (when $k = \mathbf{C}$) is a factor group of the group $H_1(U_b)$, we thus have some character ψ of the group $H_1(U_b)$:

$$\psi \in \operatorname{Char} H_1(U_b),$$

where Char denotes the group of characters of finite order, and $H_1(U_b)$ is the one-dimensional Betti group with integer coefficients.

The character ψ is a basic invariant of a fibering from the point of view of a formal isomorphism. It is convenient to replace it with another invariant, however. For this we consider the open set $A_b^0 \subset U_b$ and the natural mapping $j: H_1(U_b) \to \overline{H}_1(A_b^0)$, where $\overline{H}_1$ is the homology group with arbitrary carriers (if the fiber U_b is singular, A_b^0 is not compact).

It is easy to verify for all types of fibers that j is an isomorphism, and consequently determines an isomorphism j^* of the groups $\operatorname{Char} H_1(U_b)$ and $\operatorname{Char} \overline{H}_1(A_b^0)$.

Since A_b^0 is a variety, the group $\operatorname{Char} \overline{H}_1(A_b^0)$ is isomorphic to $H_1(A_b^0, \mathbf{Q}/\mathbf{Z})$, where $\mathbf{Q}$ is an additive group of rational numbers. As a result we have an isomorphism of the groups $\operatorname{Char} H_1(U_b)$ and $H_1(A_b^0, \mathbf{Q}/\mathbf{Z})$. We denote by $h_b(V)$ the element of the group $H_1(A_b^0, \mathbf{Q}/\mathbf{Z})$ corresponding under this isomorphism to the character ψ which we associated with the fibering of V.

The following is a basic result of the local classification.

Theorem 7. *For $k = \mathbf{C}$ the fiberings $\pi: V \to B$ and $\pi': V' \to B$ are locally isomorphic at the point $b \in B$ if and only if $h_b(V) = h_b(V')$.*

For a somewhat more general class of fields k an analogous result is obtained in the works [42] and [57].

Corollary 1. *The invariant $h_b(V)$ can be different from zero only if the fiber U_b is nonsingular or has the type A_n'.*

In fact, in the remaining cases A_b^0 is an additive group which is simply connected.

Corollary 2. *For an arbitrary fibering $\pi: V \to B$ a singular fiber F_b is either isomorphic to the corresponding fiber U_b of the Jacobian fibering or (when U_b is nonsingular or of type A_n') is multiple and is obtained from U_b by multiplication by some integer.*

If the fiber U_b is singular, but not of type A_n', the assertion follows from Corollary 1. The remaining cases are easily verified by a direct construction of the fiber F_b according to formula (5).

Corollary 3. *If the fibering $V \to B$ has no degenerate fibers other than multiple nonsingular fibers, there exists a covering $C \to B$ such that the fibering $V \times_B C$ over C is isomorphic to $C \times F$, where F is an elliptic curve.*

For the proof it is sufficient to choose C such that $V \times_B C \simeq J \times_B C$ and to apply Corollary 2 and Theorem 6.

Corollary 4. *The fibering $V \to B$ has a local section at a point $b \in B$ if and only if the fiber F_b is not multiple.*

The proof is obvious.

Corollary 5. *The fibering $V \to B$ has a local section at a point $b \in B$ if and only if it has a differentiable section in some neighborhood of b.*

In fact, if V does not have a local section, then the fiber F_b has the form mD, $m > 1$. If the fibering had a differentiable section s, then we would have $(s \cdot F_b) = 1$ at the same time that $(s \cdot F_b) = m(s \cdot D) > 1$.

§8. Classification of fiberings

For a fibering $\pi: V \to B$ and any point $b \in B$ we have defined the invariant $h_b(V) \in H^1(A_b^0, \mathbf{Q}/\mathbf{Z})$. It is easy to see that $h_b(V) = 0$ if F_b is a nondegenerate fiber of the fibering π. Thus for a given V, $h_b(V) \neq 0$ only for a finite number of points $b \in B$. Therefore the correspondence

$$V \to \{h_b(\nu),\ \ b \in B\}$$

determines the homomorphism

$$\varphi:\ \mathfrak{H}(B, J) \to \sum_{b \in B} H_1(A_b^0, \mathbf{Q}/\mathbf{Z}).$$

Our first goal is the description of the kernel and cokernel of the homomorphism ϕ. We begin with the description of the cokernel.

Theorem 8. *If J is not a direct product $B \times A$, where A is an elliptic curve, then the homomorphism ϕ is epimorphic. If $J \simeq B \times A$, then the cokernel of the homomorphism ϕ is isomorphic to the group of points of finite order of the variety A.*

The kernel of the homomorphism ϕ consists of those fiberings which, by Theorem 7, have a local section at any point $b \in B$. In other words, they are locally isomorphic to J at any point of the base. We shall call such fiberings locally trivial, and we shall denote by $\mathfrak{H}(B, J)$ the group consisting of all such fiberings.

Now let us assume that $k = \mathbf{C}$. The analysis of the group $\mathfrak{H}(B, J)$ can be conducted according to the classical example of the analysis of one-dimensional vector fiberings. Namely, it is based on the comparison of the algebraic and differential structure of the fiberings $V \in \mathfrak{H}(B, J)$.

We denote by $\mathfrak{H}_0(B, J)$ the subgroup of the group $\mathfrak{H}(B, J)$ consisting of those fiberings which are isomorphic to J as differentiable fiberings. These fiberings are also characterized by the fact that they possess a differentiable section.

Theorem 9. *The group $\mathfrak{D}(B, J) = \mathfrak{H}(B, J)/\mathfrak{H}_0(B, J)$ is finite.*

We first describe the group $\mathfrak{H}_0$ and then the group $\mathfrak{D}$ and we shall then show that $\mathfrak{H}$ is their direct sum.

Let $V \in \mathfrak{H}_0(B, J)$ and let $u_\sigma \in H^1(G, \mathfrak{A}_J(C))$ be a cocycle. It easily follows from the condition $V \in \mathfrak{H}_0(B, J)$ that

$$u_\sigma = s - \sigma s, \quad \sigma \in G, \tag{6}$$

where $s: C \to \widetilde{J}$ is a differentiable section of the fibering $\widetilde{J}$, and the section σs is defined by the formula

$$(\sigma s)(c) = s(\sigma^{-1}c).$$

The cocycle u_σ, as well as s and σs, are two-dimensional cycles of the variety $\widetilde{J}$. We denote by $\hat{u}_\sigma$, $\hat{s}$ and $\hat{\sigma s}$ the two-dimensional cycles corresponding to them — elements of the group $H^2(\widetilde{J}, \mathbf{Z})$. Let $P^{2,0}$ be the operator associating with an element of $H^2(\widetilde{J}, \mathbf{Z})$ its component of type $(2, 0)$. Since u_σ is an algebraic cycle, by a theorem of Lefschetz, $P^{2,0}\hat{u}_\sigma = 0$. Therefore, we obtain from (6) that

$$P^{2,0}\hat{s} = \sigma P^{2,0}\hat{s}. \tag{7}$$

The mapping $f: \widetilde{J} \to J$ determines the imbedding

$$f^*: H^2(J, \mathbf{C}) \to H^2(\widetilde{J}, \mathbf{C}).$$

As is known (and easily verified), here elements of $H^{2,0}(J, \mathbf{C})$ are mapped into elements of $H^{2,0}(\widetilde{J}, \mathbf{C})$ that are invariant with respect to the operation of the group G. Equation (7) shows then that there exists an element $x \in H^{2,0}(J, \mathbf{C})$ such that

$$f^*x = P^{2,0}\hat{s}.$$

Direct verification shows that the element x is determined by a given fibering of V uniquely up to a term of the type $P^{2,0}J$, $y \in H^2(J, \mathbf{Z})$.

The coset in the group $H^{2,0}(J, \mathbf{C})/P^{2,0}H^2(I, \mathbf{Z})$ determined by the element x is denoted by $\gamma(V)$.

Theorem 10. *The mapping*

$$\gamma: \mathfrak{H}_0(B, J) \to H^{2,0}(J, \mathbf{C})/P^{2,0}H^2(J, \mathbf{Z})$$

determines an isomorphism of the group $\mathfrak{H}_0(B, J)$ *and a torsion part of the group* $H^{2,0}(J, \mathbf{C})/P^{2,0}H^2(J, \mathbf{Z})$.

The homomorphism

$$P^{2,0}: H^2(J, \mathbf{Z}) \to H^{2,0}(J, \mathbf{C})$$

has a kernel, according to a theorem of Lefschetz, the group $H_a^2(J)$ consisting of algebraic cycles. The factor group

$$H_t^2(J) = H^2(J, \mathbf{Z})/H_a^2(J)$$

is said to be a group of transcendental cycles. This group is imbedded in the group $H^{2,0}(J, \mathbf{C})$ with the aid of the homomorphism $P^{2,0}$. On the other hand, if Y is a subgroup of an infinitely divisible group X, then the torsion part of the group X/Y is isomorphic, as can easily be seen, to the group $Y \otimes \mathbf{Q}/\mathbf{Z}$. In conjunction with Theorem 9 this gives us the following result.

Theorem 11. *The group* $\mathfrak{H}_0(B, J)$ *is isomorphic to* $H_t^2(J) \otimes \mathbf{Q}/\mathbf{Z}$.

It remains for us to describe the structure of the group $\mathfrak{D}(B, J)$.

Theorem 12. *If the fibering* J *has any degenerate fibers, or if it has the form* $B \times A$, *where* A *is an elliptic curve, then* $\mathfrak{D}(B, J) = 0$. *If* J *does not have degenerate fibers and* $J \not\simeq B \times A$, *then the group* $\mathfrak{D}(B, J)$ *is isomorphic, depending on the type of the group of automorphisms of a fiber, to a group of order* 1, 2, 3, *or* 4 *(in the last case to* $\mathbf{Z}_2 \oplus \mathbf{Z}_2$*).*

The proof is based on results found in [57]. Let us assume that J has a degenerate fiber. It follows from the above that the group $\mathfrak{D}(B, J)$ is isomorphic to the factor group of the group $\mathfrak{H}(B, J)$ over the subgroup of the infinitely divisible elements. According to [57] this factor group is isomorphic to the group of those sections s of the fibering J that have finite order in the group $\mathfrak{A}_J(B)$ of all sections and for any $b \in B$ intersect the fiber F_b in some point belonging to the subgroup A_b^0. We have to show that such a section s is equal to the zero section o.

We note first that if $s \neq o$, then s does not intersect the zero section. For, in the terminology of [57], s determines for each $b \in B$ a nonzero point of finite order on the curve F_β that is rational over the field $k\{t\}$ of power series in the powers of the local parameter t at the point b. This point belongs to a connected component α_b^0 of the group of all points α_b of the curve F_β that are rational over $k\{t\}$. On the other hand, under the specialization $\beta \to b$ the points of finite order in the group α_b^0 are mapped isomorphically onto the points of finite order of the group A_b^0. This means that $s \cdot A_b^0 \neq 0$, and since this is true for any $b \in B$, it follows from this that s does not intersect the zero section.

Let $ms = o$. Then the divisor $ms \cdot F_\beta - mo \cdot F_\beta$ is equivalent to 0 on F_β and consequently, determines some function f on F_β and thus on J. We want to apply Lemma 8 of Chapter IV to this function. This cannot be done directly, since the lemma applies to fiberings without degenerate fibers, while J has degenerate fibers. In order to be able to apply the result of Lemma 8, we remove from J in each degenerate fiber F_b all the components different from A_b^0. We denote the open set left by J^0. It is easy to verify that the proof of Lemma 8 remains valid without any changes for the surface J^0 if one considers in the formulation of the lemma only divisors C_i on J that intersect with singular fibers F_b only at points of the sets A_b^0. s and o are such divisors. The divisor $ms - mo$, as we have seen, is not ramified, and we can now apply Lemma 8. It gives us, in particular, that all the fibers of the fibering J^0 are isomorphic, which is possible only if the fibering J does not have degenerate fibers. Thus, if J has a degenerate fiber, then $s = o$.

The case $J = B \times A$ is analyzed on the basis of the description of the group $\mathfrak{D}(B, J)$ given in [25]. In this case $\mathfrak{D}(B, J)$ is a periodic subgroup of the group $\mathbf{Z} \oplus \mathbf{Z}$, and is consequently equal to zero.

Now let J not have degenerate fibers, and consequently be a fiber bundle, but let it not be a direct product. Then J has the form $(C \times A)/G$, where $f: C \to B$ is an unramified covering with a Galois group G, A is an elliptic curve, ϕ: $G \to \operatorname{Aut} A$ is a homomorphism, and G operates on $C \times A$ according to the rule

$$\sigma (c \times a) = \sigma (c) \times a\varphi (\sigma)^{-1}. \tag{8}$$

According to [57], the group $\mathfrak{D}(B, J)$ is isomorphic to the group of sections of finite order of the fibering J. We now indicate what these sections are like. Let $s: B \to J$ be a periodic section. It determines the section

$$s^*: \quad C \to J \times_B C, \quad s^*(c) = (sf)(c) \times c.$$

The section s^* possesses the property

$$\sigma s^*(c) = s^*(\sigma c),$$

where σ operates on $J \times_B C$ according to the rule

$$\sigma (x \times c) = x \times \sigma (c). \tag{9}$$

In our case, $J \times_B C \simeq A \times C$, and under this isomorphism the operation (8) of the automorphism σ goes into (9).

In view of this s^* can be written in the form

$$s^*(c) = u(c) \times c, \tag{10}$$

where $u: C \to A$ is a regular mapping. Rule (10) gives

$$u (\sigma (c)) = u (c) \varphi (\sigma)^{-1}. \tag{11}$$

Since s is a section of finite order, for some $m > 1$

$$ms^*(c) = 0, \quad mu(c) = 0.$$

Thus $u(c)$ is an element of period m in A. The number of such elements is finite, and then it follows from the regularity of the mapping u that this mapping is constant:

$$u(c) = u_0 \in A.$$

The condition (11) shows that

$$\phi(\sigma) u_0 = u_0. \tag{12}$$

As is known, σ can be an automorphism of order 2, 3, 4, or 6. It is easy to verify that the solutions of equation (12) form in these cases groups respectively of orders 4 (isomorphic to $Z_2 \oplus Z_2$), 3, 2, and 1.

§9. One particular case

We apply the obtained classification of fiberings on elliptic curves to the study of one type of surfaces which we have already met earlier (Chapter IV, §7). Namely, we consider the surfaces V with the invariants $p = 0$, $q = 1$ for which the Albanese mapping $\pi\colon V \to B$ has as fibers elliptic curves. We saw in Chapter IV, §7 that in this case the fibers can be only nonsingular or multiples of nonsingular curves. It follows from Corollary 2 of Theorem 7 that the Jacobian fibering J of a fibering of V does not have degenerate fibers and is, consequently, a fiber bundle

$$J \simeq (C \times A)/G.$$

We first establish that V has the invariants we need if and only if J is not a direct product (i.e., $\phi(G) \neq 1$). For this we have to give an argument very close to that which is contained in Chapter IV, §8. We note that by formula (5) of §5 the fibering of V is a factor of the fibering $\widetilde{J} \simeq J \times_B \overline{C}$ for some covering $\overline{C} \to B$ with a Galois group $\overline{G}$. We can choose $\overline{C}$ such that the covering $C \to B$ will be a factor of it. This gives a homomorphism $\psi\colon \overline{G} \to G$. Then, as it is easy to see, $\overline{J} \simeq \overline{C} \times A$. It is not difficult to interpret the operation of the group $\widetilde{G}$ on $\widetilde{J}$. A simple calculation shows that

$$\overline{\sigma}\,(\overline{c} \times a) = \overline{\sigma}\,(\overline{c}) \times a \cdot u\,(\overline{\sigma})^{-1},$$

where $u(\overline{\sigma})$ is an automorphism of the curve A (but not of the corresponding abelian variety), and $u\colon \overline{G} \to \operatorname{Aut} A$ is a homomorphism of the group $\overline{G}$ into the group of automorphisms on the curve A. Here $u(\overline{\sigma})$ is a combination of an automorphism and a translation of the abelian variety A

$$u\,(\overline{\sigma})\,(a) = \varphi\psi\,(\overline{\sigma})\,(a) + v\,(\overline{\sigma}),$$

where $v(\overline{\sigma})$ is a point of A.

Repeating the argument given in the proof of Lemma 13, §8, Chapter IV, we see that

$$\Omega^i\,(V) \simeq \Omega^i\,(\overline{C} \times A)^{\overline{G}}, \quad i = 1, 2.$$

It follows from this that

$$\Omega^1\,(V) \simeq \Omega^1\,(B) \oplus \Omega^1\,(A)^{\varphi(G)}$$

and thus $q = 1$ if and only if $\phi(G) \neq 1$, i.e. when J is not a direct product.

Analogously,

$$\Omega^2(V) \simeq (\Omega^1(\overline{C}) \oplus \Omega^1(A))^{\overline{G}}.$$

Let H be the kernel of the homomorphism $\psi\colon \overline{G} \to G$. Since

$$(\Omega^1(\overline{C}) \otimes \Omega^1(A))^{\overline{G}} = ((\Omega^1(\overline{C}) \otimes \Omega^1(A))^H)^G,$$

$$\Omega^1(\overline{C})^H \simeq \Omega^1(C),\ \Omega^1(A)^H = \Omega^1(A),\ \Omega^1(C)^G \simeq \Omega^1(B),$$

we have

$$\Omega^2(V) \simeq \Omega^1(B) \otimes \Omega^1(A)^{\varphi(G)},$$

from which it follows that $p = 0$ if and only if $\phi(G) \neq 1$.

Now applying the classification developed in the previous sections, we arrive at the following result.

Theorem 13. *The surfaces V with the invariants $p = 0$, $q = 1$, for which the fibers of the Albanese fibering $\pi\colon V \to B$ have genus 1 are classified in the following way. The Jacobian fibering of the fibering of V has the form*

$$J \simeq (C \times A)/G,$$

where $f\colon C \to B$ is an unramified covering with Galois group G that is not the identity mapping, A is an elliptic curve, $\phi\colon G \to \operatorname{Aut} A$ is an imbedding of G into the group of automorphisms of A (as an abelian variety), and G operates on $C \times A$ according to the rule $\sigma(c \times a) = \sigma(c) \times a\phi(\sigma)^{-1}$. The structure of the group $\mathfrak{H}(B, J)$ of all the fiberings of V with a given Jacobian fibering is determined from the exact sequence (F_b is a fiber of J over the point $b \in B$):

$$0 \to \mathfrak{h}(B, J) \to \mathfrak{H}(B, J) \to \sum_{b \in B} H_1(F_b, \mathbf{Q}/\mathbf{Z}) \to 0$$

and the relationship

$$\mathfrak{h}(B, J) = \mathfrak{D}(B, J), \tag{13}$$

where $\mathfrak{D}(B, J)$ has order 1, 2, 3, or 4 depending on whether the group $\phi(G)$ has order 6, 4, 3, or 2.

We need to verify only the relationship (13). It follows from the fact that $\mathfrak{D}(B, J) = \mathfrak{H}(B, J)/\mathfrak{H}_0(B, J)$ and from $\mathfrak{H}_0(B, J) = 0$. The last assertion follows from the fact that $H_t^2(J) = 0$. For, from the fact that $\chi = 0$, $q = 1$ it follows that $b_2 = 2$. On the other hand, we have immediately two nonhomologous algebraic cycles in J, for example a fiber and a section. Thus the rank of the group $H_a^2(J)$ is also equal to two. Since, according to the criterion of Lefschetz, the group $H_t^2(J) = H^2(J, \mathbf{Z})/H_a^2(J)$ is torsion-free, it must be equal to zero.

Remark. In his consideration of our class of surfaces, Enriques ([59], Chapter X, §11) asserts that V has a pencil of elliptic curves that are transversal to the fibers of the fibering $\pi: V \to B$. It is easy to see, however, that there is an elliptic curve on V distinct from the fibers of the fibering π only if V does not have multiple fibers. In fact, let L be such a curve. By assumption, $(L \cdot F_b) = r > 0$. Since the base B is, like L, an elliptic curve, the projection π determines on L the structure of an unramified covering of B. This means that $L \cdot F_b$ consists of r distinct points for any $b \in B$. But if F_b is a multiple fiber, $F_b = m \cdot U$, $m > 1$, then it is clear that $L \cdot F_b$ consists of the points of $L \cdot U$ taken m times.

Thus the classification of Enriques evidently only applies to fiberings without multiple fibers. As we saw, these are everywhere locally trivial. They form a finite subgroup $\mathfrak{D}(B, J) = \mathfrak{h}(B, J)$ of the group $\mathfrak{H}(B, J)$. The fiberings corresponding to the elements of the infinite factor group $\Sigma H_1(F_b, Q/Z)$ of this group are clearly omitted in Enriques' classification.

Galois theory of transcendental extensions and uniformization

(with I. I. Piatetskij-Shapiro)

Izv. Akad. Nauk SSSR, Ser. Mat. **30**, No. 3, 671–704 (1966). Zbl. **218**, 14024
[Transl., II. Ser., Am. Math. Soc. **69**, 111–145 (1968)]

In this paper we construct an algebraic analogue to the theory of uniformization of algebraic varieties by automorphic functions. The theory we construct is applicable to a certain class of algebraic varieties over an arbitrary field. In particular, over the field of complex numbers it is applicable to varieties that are uniformizable by arithmetic groups. In this case it is equivalent to the theory of Hecke operators.

Introduction

The classical theory of uniformization of algebraic varieties by means of automorphic functions of one or several complex variables is not an algebraic theory.

In the present paper we make an attempt to construct an algebraic theory that is applicable to algebraic varieties defined over arbitrary fields and is at the same time analogous to the classical theory of uniformization of complex algebraic varieties. On the other hand, this theory can be regarded as an algebraic analogue to the theory of Hecke operators.

Uniformizability of a complex manifold X means that X has a covering $\tilde{X} \to X$ which is unramified (or ramified only over a submanifold of X of smaller dimension) and is a homogeneous complex analytic manifold. For an algebraic variety X defined over an arbitrary field it is natural to consider algebraic coverings of X, that is, projective limits of finite coverings of it. One can raise the problem of those algebraic varieties X that have a covering $\tilde{X}$ (in the algebraic sense), which is unramified or ramified over a homogeneous subvariety of X of smaller dimension, and homogeneous. However, such varieties are a very rare exception; for example, on a curve of genus greater than 1 such homogeneous coverings (in the algebraic sense) do not exist.

The observation on which the paper is based consists in the following: for one class of algebraic varieties there exist unramified coverings $\tilde{X}$ (in the algebraic sense), which although not homogeneous are at least quasi-homogeneous, that is, the orbit of each point under the full group of automorphisms is everywhere dense in the Zariski topology. Among them are, for example, the manifolds X that are uniformizable by arithmetic groups with a compact fundamental domain.

Algebraic varieties X that are uniformizable by arithmetic groups with non-compact fundamental domain have quasi-homogeneous coverings that are ramified over a subvariety of X of smaller dimension.

A typical example, which plays a fundamental role in this paper, is the following.

Let H be the upper half-plane $\operatorname{Im} z > 0$, Γ the ordinary modular group, and Δ a subgroup of finite index of Γ. We denote by K_Δ the field of functions that are automorphic relative to Δ, and by K the field that is the union of all fields K_Δ, when Δ ranges over all subgroups of Γ of finite index. Let X_Δ be a nonsingular algebraic curve whose field of functions coincides with K_Δ. Obviously if $\Delta_1 \subset \Delta_2$ then $K_{\Delta_2} \subset K_{\Delta_1}$, and hence there exists a morphism of X_{Δ_1} into X_{Δ_2}. Let X be the projective limit of the algebraic varieties X_Δ. Clearly X is a covering in the algebraic sense of the curve X_Δ that is ramified only at a finite number of points. We show that the covering X is quasi-homogeneous, that is, the orbit of almost every point of X (not belonging to a certain subvariety of smaller dimension) is everywhere dense in X. We set $Y_\Delta = \Delta \backslash H$. Clearly $Y_\Delta \subset X_\Delta$ and $X_\Delta - Y_\Delta$ consists of a finite number of points. Let Y be the projective limit of the Y_Δ. It is clear that $Y \subset X$ and that $X - Y$ is a subvariety of X of dimension zero. An elementary argument shows that the orbit of every point of Y is everywhere dense in Y.

A similar proposition holds in the general case. Specifically: let D be a symmetric domain, and Γ an arithmetic group of analytic automorphisms of D. Let X be an algebraic variety uniformizable by functions that are automorphic relative to Γ. There always exists a quasi-homogeneous covering $\widetilde{X}$ of X that is ramified only over a subvariety of smaller dimension. This quasi-homogeneous covering can be constructed just as above, in the form of the projective limit of
1 algebraic varieties uniformizable by subgroups of Γ of finite index (see §7).

We also mention that a quasi-homogeneous covering can be obtained by using not all subgroups of finite index but only some of them. For example, we may restrict ourselves to congruence subgroups or even to congruence subgroups modulo p^m, where p is a given prime number and m arbitrary.

The examples of quasi-homogeneous proalgebraic varieties to be constructed in the present paper are defined over the field of complex numbers. Similar examples can be constructed over a field of positive characteristic. In an implicit form one such example is contained in the paper [15] by Igusa on the theory of elliptic curves. More general examples can be constructed on the basis of papers by Mumford [16].

We now return to the case when the ground field is the field of complex numbers. We do not know examples of quasi-homogeneous algebraic variety of general type, other than the ones obtained from arithmetic groups. In this context it seems to us a plausible and very interesting proposition that all quasi-homogeneous proalgebraic varieties of general type are connected with arithmetic groups acting in a symmetric domain. In other words, this means that if a finite-dimensional algebraic variety X defined over the field of complex numbers has a quasi-homogeneous proalgebraic covering of general type, then X can be uniformized by an arithmetic group acting in a certain symmetric domain D.

In the paper we also investigate the structure of the group of automorphisms of the field $k(\widetilde{X})$ of all rational functions over a given quasi-homogeneous variety $\widetilde{X}$. The field $k(\widetilde{X})$ is of finite transcendence degree, but is not, as a rule, generated by adjoining a finite number of elements. Among such fields $k(\widetilde{X})$ a specially important role is played by the fields that have a finite set F of generators (a set F is called the set of generators if every subfield K' of K that is invariant under all automorphisms of $k(\widetilde{X})$ and contains F coincides with $k(\widetilde{X})$ (see §3)).

In all the examples we discuss the groups of automorphisms of such fields turn out to be p-adic Lie groups or products of them of the type considered in the theory of idèles. It seems to us very interesting to clarify whether this is always the case.

§1. Fields of automorphic functions

Let D be a bounded symmetric domain in n-dimensional complex space, and Γ a discrete group of analytic automorphisms of D. Throughout this section we assume either that the factor space D/Γ is compact, or that Γ is an arithmetic group, or that D is the unit disc $|z| < 1$ and D/Γ has finite volume. In each of these cases the field K_Γ of functions automorphic relative to Γ is a field of algebraic functions of n unknowns, where n is the complex dimension of D (see [1], [2], [3], [12]).

We denote by $K(\Gamma)$ the union of all fields K_Δ, when Δ ranges over all subgroups of Γ of finite index. The fields $K(\Gamma)$ and some of their subfields are the main object of the present paper.

In this section we discuss the problem of the structure of the group $\mathfrak{G}$ of all automorphisms of $K(\Gamma)$. The following method of introducing a topology in $\mathfrak{G}$ is well known.

Let K be a subfield of $K(\Gamma)$, finitely generated over $\mathbf{C}$. We denote by $\mathfrak{G}(K)$ the collection of all automorphisms of $K(\Gamma)$ that leave every element of K fixed. Now we introduce a topology in $\mathfrak{G}$ by taking the set $\mathfrak{G}(K)$ as a basis of neighborhoods of the identity. It is easy to verify that all the axioms of a topological group are satisfied. $\mathfrak{G}$ is a locally compact zero-dimensional group.

We make the following definition. An analytic automorphism $z \to gz$ of D is called Γ-rational if the group $\Gamma \cap g^{-1}\Gamma g$ is of finite index in Γ. It can be shown (see [3]) that the Γ-rational elements form a group Γ' and that the set of Γ-rational elements for Γ does not change when Γ is replaced by a commensurable subgroup. With every $g \in \Gamma'$ we can associate the following automorphism of $K(\Gamma)$:

$$f(z) \longrightarrow f^g(z) = f(gz). \tag{1}$$

To show that the mapping (1) yields an automorphism $K(\Gamma)$ we need only verify that it carries $K(\Gamma)$ into itself. Let $f \in K(\Gamma)$; then there exists a subgroup Δ of finite index in Γ such that $f \in K_\Delta$. We set

$$\Delta_1 = \Delta \cap g^{-1}\Delta g.$$

From the fact that g is a Γ-rational element it follows that Δ_1 is a subgroup of finite index in Γ. An immediate verification shows that if $f \in K_\Delta$, then $f^g \in K_{\Delta_1}$.

We also mention the following easily verified proposition. If the mapping $f(z) \longrightarrow f(gz)$, where g is an arbitrary automorphism of D, is an automorphism of $K(\Gamma)$, then g is a Γ-rational element of the group of analytic automorphisms of D.

Let us show that an automorphism of $K(\Gamma)$ corresponding to any $\gamma \in \Gamma'$ $(\gamma \neq 1)$ is always nontrivial. For if $f(\gamma z) = f(z)$ for every $f \in K(\Gamma)$, then $\gamma \in \bigcap \Delta$, where Δ ranges over the set of all subgroups of finite index in Γ. Consequently it remains to show that the intersection of all subgroups of finite index of Γ is trivial. It is well known that this is the case when D is the unit disc. Therefore we need here only discuss the case when $\dim D > 1$ and the factor space D/Γ is compact or Γ is an arithmetic group.

If $\dim D > 1$ and D/Γ is compact, then D may be decomposed into the direct product of domains $D = D_0 \times D_1 \times \cdots \times D_r$, and we can find discrete groups $\Gamma_0, \cdots, \Gamma_r$ in $D_0, \cdots, D_r$ such that

$$\dim D_1 = \cdots = \dim D_r = 1, \quad \dim D_0 > 1,$$

Γ is commensurable with $\Gamma_0 \times \cdots \times \Gamma_r$ and Γ_0 is finitely generated and has a matrix representation with elements from an algebraic number field. This follows from results in [4] and [17]. Obviously it is sufficient for us to show that the intersection of the subgroups of finite index of Γ_0 is 1.

Thus the proof reduces to the case when Γ is finitely generated and has a matrix representation with elements from an algebraic number field; these properties also hold for arithmetic groups.

Let $\gamma_1, \gamma_2, \cdots, \gamma_t$ be generators of Γ. We consider a prime ideal $\mathfrak{P}$ of k that is not contained in the denominators of the matrix elements of $\gamma_1, \cdots, \gamma_t$.

We denote by Γ_n the subgroup of Γ consisting of all γ congruent to the unit matrix modulo $\mathfrak{P}^n$. It is not hard to see that Γ_n is of finite index and that the intersection of all subgroups Γ_n is trivial.

In Γ' we introduce the topology in which a basis of neighborhoods of the identity is formed by all subgroups of finite index in Γ. We denote by $\mathfrak{G}'$ the factor group of the completion of Γ' relative to this topology by the normal subgroup Δ, the intersection of all subgroups of finite index of Γ. Note that, as we have shown above, when D is a symmetric domain the group Δ is trivial.

To every element g of $\mathfrak{G}'$ there corresponds an automorphism $f \to f^g$ of $K(\Gamma)$ defined as follows:

$$f^g(z) = f(\gamma z), \tag{2}$$

where γ denotes an element of Γ' sufficiently close to g. It is not hard to see that this is well defined, that is, that for every function $f \in K(\Gamma)$ there exists a neighborhood U of the identity of $\mathfrak{G}'$ such that if $\gamma \in Ug$, the function $f(\gamma z)$ does not depend on γ. For U we may take the closure in $\mathfrak{G}'$ of the subgroup Δ which consists of all $\gamma \in \Gamma$ leaving f fixed.

To every element of $\mathfrak{G}'$ there corresponds a nontrivial automorphism of $K(\Gamma)$. Therefore $\mathfrak{G}'$ can be embedded in the group $\mathfrak{G}$ of all automorphisms of $K(\Gamma)$. It is easy to verify that the topology in $\mathfrak{G}'$ induced by the embedding coincides with the topology given in $\mathfrak{G}$.

The following circumstance is useful in the sequel. Let Δ be a subgroup of finite index in Γ. Then the field $K(\Gamma)$ is a normal algebraic extension of K_Δ, and its Galois group is the closure $\overline{\Delta}$ of Δ in $\mathfrak{G}$.

Proposition 1. *Let Γ be a discrete subgroup of analytic automorphisms of a bounded domain D. The group $\mathfrak{G}$ of all automorphisms of $K(\Gamma)$ coincides with the completion $\mathfrak{G}'$ of Γ' in the topology induced by the subgroups of Γ of finite index, provided one of the following three conditions hold:* 1) *the factor space D/Γ is compact;* 2) *D is the unit disk and the factor space D/Γ has finite*
3 *volume;* 3) *D is a symmetric domain and Γ an arithmetic group for which condition* A *is satisfied* (*see* §8, p. 27).

Proof. To show that $\mathfrak{G} = \mathfrak{G}'$ it is sufficient to verify that Γ' is everywhere dense in $\mathfrak{G}$.

Let σ be an arbitrary automorphism of $K(\Gamma)$. We show that for any neighborhood $U \subset \mathfrak{G}$ of the identity there exists a $\gamma \in \Gamma'$ such that $\sigma\gamma^{-1} \subset U$. From the definition of the topology in $\mathfrak{G}$ it is clear that without loss of generality we may assume that U coincides with the closure $\overline{\Delta}$ of some subgroup of finite index of Γ. Replacing, if necessary, Δ by a subgroup of finite index we may assume that Δ acts fixedpoint-free.

Consider the field $K_\Delta^{\sigma^{-1}}$. Let Δ_1 be the set of all $g \in \Delta$ that do not change the functions of $K_\Delta^{\sigma^{-1}}$. The group Δ_1 obviously is of finite index in Δ. From

$K_{\Delta}^{\sigma-1} \subset K_{\Delta_1}$ it follows that $K_{\Delta} \subset K_{\Delta_1}^{\sigma}$. As we have mentioned above, $K(\Gamma)$ is a normal algebraic extension of K_{Δ} with the Galois group $\overline{\Delta}$. Therefore to every subfield of $K(\Gamma)$ containing K_{Δ} there corresponds a subgroup of $\overline{\Delta}$. We denote by h the subgroup corresponding to $K_{\Delta_1}^{\sigma}$ and set $\Delta_2 = h \cap \overline{\Delta}$. From the fact that σ is an automorphism of $K(\Gamma)$ it follows that $K_{\Delta_1}^{\sigma}$ is a finite extension of K_{Δ}. Therefore the index of h in $\overline{\Delta}$ is finite. As is easy to verify, it is equal to the index of Δ_2 in Δ. Consequently

$$K_{\Delta_1}^{\sigma} = K_{\Delta_2}.$$

It remains to show that there exists an analytic automorphism g of D with the following properties:

1) $\Delta_2 = g^{-1}\Delta_1 g$;

2) the automorphism $f(z) \to f(gz)$ for $f \in K_{\Delta_1}$ coincides with our given automorphism σ.

For it follows from 1) that $g \in \Gamma'$, and from 2) that $\sigma g^{-1} \in \overline{\Delta} \subset U$. First of all we consider the case when D is a bounded domain and the factor space D/Γ is compact. As was shown in [5], in this case D/Γ is an absolutely minimal model of K_{Γ}. The same applies to factor spaces D/Δ_1 and D/Δ_2. From this it follows that every isomorphism of K_{Δ_1} to K_{Δ_2} induces an analytic mapping ϕ of D/Δ_1 onto D/Δ_2, which in turn induces an analytic mapping of the universal covering manifold of D/Δ_1 onto that of D/Δ_2 (D/Δ_1 and D/Δ_2 are manifolds, because we have assumed that Δ acts in D fixedpoint-free). Since D/Δ_1 and D/Δ_2 have one and the same universal covering, namely D, the mapping ϕ induces an analytic automorphism g of D onto itself, with the property that $\Delta_2 = g^{-1}\Delta_1 g$ and that the automorphism $f(z) \to f(gz)$, where $g \in K_{\Delta_1}$, coincides with our given automorphism σ.

Thus when D is a bounded domain and the factor space D/Γ is compact our proposition is proved.

It remains to consider the case when D/Γ is noncompact, but has finite volume. In this case, according to our assumption, either 1) D is the unit disk or 2) D is a symmetric domain and Γ an arithmetic group satisfying condition A. Both these cases will be treated in §8. There it is shown that every automorphism σ of $K(\Gamma)$ carrying K_{Δ_1} into K_{Δ_2} induces a one-to-one analytic mapping of D/Δ_1 onto D/Δ_2. Obviously from this it follows that it is induced by some analytic mapping g of D, and this mapping clearly has all the properties required.

This completes the proof.

In conclusion of this section we indicate a generalization of the construction field $K(\Gamma)$. This generalization is connected with the discussion not of all subgroups of finite index, but only some of them.

We consider a system $\mathfrak{M}$ of subgroups of finite index in Γ, and by taking it as a basis of neighborhoods of the identity we can give in Γ a Hausdorff topology. The conditions on the systems of subgroups $\mathfrak{M}$ that guarantee the possibility of giving a Hausdorff topology in Γ by means of it consist in the following:

1) $\bigcap_{\Delta \in \mathfrak{M}} \Delta = 1$;

2) if $\Delta \in \mathfrak{M}$ and $\gamma \in \Gamma$, then there always exists a Δ_1 such that $\Delta_1 \subset \gamma\Delta\gamma^{-1}$;

3) for every finite set of subgroups Δ_k, $k = 1, \cdots, N$, and suitable $\mathfrak{M}$ there exists a subgroup Δ such that $\Delta \subset \Delta_k$, $k = 1, \cdots, N$.

We denote by $K(\mathfrak{M}, \Gamma)$ the union of all fields K_Δ, with $\Delta \subset \mathfrak{M}$.

Now we indicate the construction of the group $\mathfrak{G}(\mathfrak{M})$ of all automorphisms of $K(\mathfrak{M}, \Gamma)$. Let $\Gamma'(\mathfrak{M})$ be the set of all $\gamma \in \Gamma'$ having the following property: for every subgroup $\Delta \in \mathfrak{M}$ and every $\gamma \in \Gamma'$ there exists a subgroup $\Delta_0 = \Delta_0(\Delta, \Gamma) \subset \mathfrak{M}$ such that $\Delta_0 \subset \gamma\Delta\gamma^{-1}$. It is easy to verify that transformation by $\gamma \in \Gamma$ induces an automorphism $K(\mathfrak{M}, \Gamma)$ if and only if $\gamma \in \Gamma'(\mathfrak{M})$.

In $\Gamma'(\mathfrak{M})$ we introduce a topology by taking all subgroups of the system $\mathfrak{M}$ as a basis of neighborhoods of the identity. We denote by $\mathfrak{G}'(\mathfrak{M})$ the completion of $\Gamma'(\mathfrak{M})$ in this topology. Then the following proposition holds.

Proposition 2. *Suppose that the conditions of Proposition 1 hold. Then the group $\mathfrak{G}'(\mathfrak{M})$ is isomorphic to the group of all automorphisms*[1] *of the field $K(\mathfrak{M}, \Gamma)$.*

The proof is entirely analogous to that of Proposition 1, and therefore we omit it.

In the next section we have to consider the more general situation when the group Γ acts in a complex space D which is possibly not a domain, but the disconnected union of a finite number of copies of a domain. Clearly all the results of this section carry over without any modification to this more general case.

§2. Arithmetic groups

Let G be a semisimple algebraic linear group defined over the field Q of rational numbers. As usual, we denote by G_k, where k is a ring, the set of all matrices with elements from k, satisfying the equations of the group, whose determinant is a unit of k. In particular, G_Z denotes the set of all integral matrices with determinant ± 1, and G_R the set of all nonsingular matrices. We recall a result in the well-known paper [6] by Borel and Harish-Chandra: the factor space G_R/G_Z always has finite volume. We denote by U a maximal compact subgroup of G_R, and by U^0 and G_R^0 the connected components of these groups.

1) We assume that in the group of all automorphisms of $K(\mathfrak{M}, \Gamma)$ a topology is introduced in the way just described.

G_R^0/U^0 is known to be a symmetric space. Under some additional conditions G_R^0/U^0 may be realized in the form of a bounded homogeneous domain D^0 in $\mathbf{C}^n$. In what follows we confine ourselves to this case. Then $D = G_R/U^0$ is the disconnected union of a finite number of copies of D^0. We use the term domain for D also. The group G_Z, realized as the group of analytic automorphisms of D is called an arithmetic group. The same name is applied to any group commensurable with G_Z.

We denote by $K(G)$ the union of all fields K_Γ, when Γ ranges over all subgroups of finite index of G_Z. Thus $K(G) = K(G_Z)$ in the notation of §1.

In this section we investigate the structure of the group $\mathfrak{G}$ of automorphisms of $K(G)$. Throughout the whole section we assume that the conditions of Proposition 1 in §1 are satisfied, which guarantee that the group $\mathfrak{G}$ of automorphisms of $K(G)$ coincides with the completion $\mathfrak{G}'$ of the group Γ' of rational elements of $\Gamma = G_Z$ in the topology induced by the subgroups of Γ of finite index.

As was shown in [3], the group Γ' of all rational elements of $\Gamma = G_Z$ coincides with the group of all automorphisms over Q of the Lie algebra of G. Therefore we assume henceforth that the group G coincides with the group of all automorphisms of its Lie algebra. As is not hard to verify, this assumption does not change the class of arithmetic groups under discussion. Under this assumption the group G_Q coincides with the group of all rational elements of G_Z, hence with any group Γ commensurable with G_Z.

We denote by $G_p = G_{Q_p}$ the collection of all p-adic points of G, and by $U_p = G_{Z_p}$ that of all integral p-adic points of G.

Let G_a denote the direct product of the groups G_p (p ranges only over non-Archimedean valuations) with the distinguished subgroups U_p, that is, the totality of sequences $\{g_p\}$, $g_p \in G_p$, in each of which only a finite number of g_p do not belong to U_p. The topology in G_a is introduced in the group of idèles, as usual.

We consider the diagonal monomorphism $\phi: G_Q \to G_a$ under which every element $g \in G_Q$ is associated with the sequence $\{g_p\}$, where $g_p = g$ for all p. In §1 we have shown that $G_Q \subset \mathfrak{G}$, that G_Q is everywhere dense in $\mathfrak{G}$, and that the topology induced in G_Q by the natural topology in $\mathfrak{G}$ coincides with the one that is defined by all subgroups of finite index of G_Z.

Let $\Gamma = G_Z$ and let m be an integer. We denote by $\Gamma(m)$ the subgroup of Γ consisting of all $\gamma \in \Gamma$ that are congruent to the unit matrix modulo m. Every subgroup of Γ containing $\Gamma(m)$ for some m is called a congruence subgroup. Obviously the index in Γ of every congruence subgroup is finite. In our notation congruence subgroups have the form $\phi^{-1}(\phi(G_Z) \cap U)$, where U is some open subgroup of the group $\Pi_p U_p$. We verify that the monomorphism ϕ is continuous in the topology given by all subgroups of finite index of G_Z. It then follows that it can be extended to a homomorphism $\overline{\phi}: \mathfrak{G} \to G_a$. The continuity of ϕ follows at once from the fact that if U is a sufficiently small open subgroup of G_a, then $\phi(G_Q) \cap U = \phi(G_Z) \cap U$ is a congruence subgroup in $\phi(G_Z)$, and all the more a subgroup of finite index.

Next we identify the elements of G_Q with the elements of G_a corresponding to them under the mapping ϕ.

Lemma 1. *The kernel of the homomorphism* $\overline{\phi}: \mathfrak{G} \to G_a$ *coincides with the intersection of the closures in* $\mathfrak{G}$ *of all congruence subgroups of* G_Q.

As we have just seen, there exists a basis U_K of neighborhoods of the identity in G_a for which $\phi(U_K) \cap \phi(G_Q)$ contains a certain congruence subgroup $\Gamma(K)$. Since the chosen open subsets U_K are compact we have $\overline{\phi}(\overline{\Gamma}(K)) \subset U_K$. If an element γ of $\mathfrak{G}$ is contained in the intersection of all $\overline{\Gamma}(K)$, as $\overline{\Gamma}(K)$ ranges over all congruence subgroups, then $\overline{\phi}(\gamma) \in U_K$ for all U_K, and hence $\overline{\phi}(\gamma) = 1$, that is, γ belongs to the kernel.

Suppose, conversely, that $\overline{\phi}(\gamma) = 1$. Let Γ be a congruence group and U an open subgroup of G_a for which

$$\phi^{-1}(U \cap \phi(G_Z)) = \Gamma.$$

We take any open subgroup $V \subset \overline{\phi}^{-1}(U)$ and an element $g \in G_Q$ such that $\gamma g^{-1} \in V$. Then $\overline{\phi}(g) \in U$, that is, $g \in \Gamma$. Consequently for every neighborhood V of the identity there exists an element $g \in \Gamma$ such that $\gamma g^{-1} \in V$. But this means that $\gamma \in \overline{\Gamma}$, and the lemma is proved.

Corollary. *The homomorphism* $\overline{\phi}$ *is proper.*

This follows from the fact that its kernel is compact.

We can now determine the image of the homomorphism $\overline{\phi}$. Since this is a proper homomorphism and the groups $\mathfrak{G}$ and G_a are locally compact, $\overline{\phi}$ is closed. Consequently $\overline{\phi}(\overline{G}_a) = \overline{\overline{\phi}(G_Q)}$. Thus we have proved the following proposition.

Proposition 1. *There exists a homomorphism of the Galois group* $\mathfrak{G}$ *of* $K(G)$ *into the group* G_a. *The image of this homomorphism coincides with* $\overline{G_Q}$ *(the closure of* G_Q *in* G_a*).*

Let G be any semisimple algebraic group defined over the field Q of rational numbers. We now investigate the structure of the closure of G_Q in G_a. For this purpose we have to discuss the universal covering (in the sense of the theory of algebraic groups) of G. We denote it by $\widetilde{G}$, its center by $\mathfrak{Z}$. Since by assumption G is given by the adjoint representation, its center is trivial and so we have the exact sequence

$$(1) \to \mathfrak{Z} \to \tilde{G} \xrightarrow{\varphi} G \to (1). \tag{1}$$

Let k be an arbitrary field and k' its separable algebraic closure. By $H^i(k, G)$, $i = 0, 1$, we denote the sets $H^i(\mathfrak{G}, G_{k'})$, where $\mathfrak{G} = \mathfrak{G}(k'/k)$ is the Galois group of k' over k. From (1) it follows that the sequence

$$(1) \to \mathfrak{Z}_{k'} \to \tilde{G}_{k'} \xrightarrow{\varphi_{k'}} G_{k'} \to (1)$$

is also exact, and that we have the exact sequence

$$(1) \to \mathfrak{Z}_k \to \tilde{G}_k \xrightarrow{\varphi_k} G_k \to H'(k, \mathfrak{Z}) \to H'(k, \tilde{G}). \tag{2}$$

To begin with we consider the case when $k = Q_p$ is the field of p-adic numbers. As Kneser [7] has shown, $H^1(Q_p, \tilde{G}) = 1$ for a simply connected group $\tilde{G}$. Therefore

$$G_{Q_p} / \varphi(\tilde{G}_p) \cong H^1(Q_p, \mathfrak{Z}). \tag{3}$$

Apart from the sequence (1) we consider the sequence

$$(1) \to \mathfrak{Z}_{\mathfrak{D}_p} \to \tilde{G}_{\mathfrak{D}_p} \to G_{\mathfrak{D}_p} \to (1), \tag{4}$$

where $\mathfrak{D}_p$ is the ring of integral elements in the maximal unramified extension k_1 of k. The exactness of the sequence (4) for almost all p follows from standard arguments. By analogy with (2) we have the exact sequence

$$(1) \to \mathfrak{Z}_{\mathfrak{D}_p} \to \tilde{G}_{Z_p} \to \tilde{G}_{Z_p} \to H^1(Z_p, \mathfrak{Z}) \to H^1(Z_p, \tilde{G}),$$

where $H^1(Z_p, G)$ denotes the set

$$H^1(\mathfrak{G}_1, G_{\mathfrak{D}_p}), \quad \mathfrak{G}_1 = \mathfrak{G}(k_1 / k).$$

From a theorem of Lang [8] and standard arguments it follows that in this case again $H^1(Z_p, G) = 1$ for every connected group G and almost all p. For such p we have an analogue to (3):

$$G_{Z_p} / \varphi_{Z_p}(\tilde{G}_{Z_p}) \simeq H^1(Z_p, \mathfrak{Z}). \tag{5}$$

We now consider the action of the homomorphism $\phi_a: \tilde{G} \to G_a$. Combining (3) and (5) we obtain

$$G_a / \varphi_a(\tilde{G}_a) = J(\mathfrak{Z}).$$

Here $J(\mathfrak{Z})$ denotes the group of idèles of the module $\mathfrak{Z}$, which is defined as the restricted direct product $\Pi H^1(Q_p, \mathfrak{Z})$ when in each group $H^1(Q_p, \mathfrak{Z})$ the subgroup $H^1(Z_p, \mathfrak{Z})$ is distinguished (see [9], p. 290).

As is known (see [9], p. 290), the group $H^1(Q, \mathfrak{Z})$ is mapped by localization into $J(\mathfrak{Z})$, that is, there exists a homomorphism

$$\alpha: H^1(Q, \mathfrak{Z}) \to J(\mathfrak{Z}).$$

The subgroup $P(\mathfrak{Z}) = \alpha H^1(Q, \mathfrak{Z})$ of $J(\mathfrak{Z})$ is called the group of principal idèles, and

$$\operatorname{Coker} \alpha = J(\mathfrak{Z})/P(\mathfrak{Z}) = \mathfrak{A}(\mathfrak{Z})$$

is called the group of idèle classes modulo $\mathfrak{Z}$.

Thus there exists a homomorphism

$$\delta: G_a \to \mathfrak{A}(\mathfrak{Z}),$$

which we call a determinant. For example, if $G = PL(2, Q)$, then $\widetilde{G} = SL(2, Q)$ and $\mathfrak{Z}$ is a group of order two. In this case, if

$$g \in G_a, \quad g = \{g_p\}, \quad g_p \in PL(2, Q_p),$$

then the determinants $\{\det g_p\}$ define an element of the group J/J^2 where J is the ordinary group of idèles. The element $\delta(g)$ coincides with the image of the element $\{\det g_p\}$ of $\mathfrak{A}/\mathfrak{A}^2$, where $\mathfrak{A}$ is the group of idèle classes of Q.

After these preliminary remarks we can pass on to the computation of the subgroup $\overline{G}_Q$ of G_a.

By the strong approximation theorem of Kneser [7] we have $\widetilde{G}_Q = \widetilde{G}_a$. Hence $\overline{G}_Q \supset \phi_a(\widetilde{G}_a)$, and it suffices to determine the image of $\overline{G}_Q$ in $J(\mathfrak{Z})$. For this purpose we write the sequence (2) for $k = Q$ and after localization consider the mapping of all its terms into the corresponding terms of the analogous sequence for G_a. So we obtain a commutative diagram

$$\begin{array}{ccc}
\widetilde{G}_a & \xleftarrow{\tilde{\alpha}} & \widetilde{G}_Q \\
\downarrow & \alpha & \downarrow \\
G_a & \longleftarrow & G_Q \\
\downarrow & & \downarrow \\
J(\mathfrak{Z}) & \longleftarrow & H^1(Q, \mathfrak{Z}) \\
\downarrow & & \downarrow \\
(1) & & (1).
\end{array}$$

From the commutativity of this diagram it follows that the image $\overline{G}_Q$ in $J(\mathfrak{Z})$ coincides with the closure $\overline{\alpha_{\mathfrak{Z}} H^1(Q, \mathfrak{Z})}$ of the image of $H^1(Q, \mathfrak{Z})$. However, the group $\alpha_{\mathfrak{Z}} H^1(Q, \mathfrak{Z})$, as is well known (see [9], p. 290), is discrete in $J(\mathfrak{Z})$. So we have proved the following proposition.

Proposition 2. *The group $\overline{G}_Q$ coincides with the subgroup of G_a consisting of the elements of unit determinant.*

We denote this subgroup by G_a^0. By working through all types of simple groups and their universal coverings we can show that if G is absolutely simple, and is a form of the Chevalley group corresponding to the cocycle with values in the group of outer automorphisms; then the module $\mathfrak{Z}$ is operator isomorphic to the group of all nth roots of unity, where n depends on G.

An exception is the group of type D_{2n} for which $\mathfrak{Z} = Z/2Z + Z/2Z$ with trivial action of the Galois group of the algebraic closure on it.

Consequently for all groups except D_{2n} the group $J(\mathfrak{Z})$ coincides with the group J_Q/J_Q^n where J_Q is the group of idèles of Q, and $\mathfrak{A}(\mathfrak{Z})$ coincides with the group $\mathfrak{A}(Q)/\mathfrak{A}(Q)^n$, where $\mathfrak{A}(A)$ is the group of idèle classes of Q. For groups of type D_{2n} we have

$$J(\mathfrak{Z}) = J_Q/J_Q^2 + J_Q/J_Q^2$$

and

$$\mathfrak{A}(\mathfrak{Z}) = \mathfrak{A}(Q)/\mathfrak{A}(Q)^2 + \mathfrak{A}(Q)/\mathfrak{A}(Q)^2.$$

Similarly it can be shown that if G is Q-simple but not absolutely simple, and k is the minimal field over which G splits into the product of absolutely simple groups, then

$$J(\mathfrak{Z}) = J_k/J_k^n, \quad \mathfrak{A}(\mathfrak{Z}) = \mathfrak{A}(k)/\mathfrak{A}(k)^n$$

with a natural modification for groups of type D_{2n}.

Proposition 3. *The Galois group $\mathfrak{G}$ of $K(G)$ is isomorphic to G_a^0 if and only if each subgroup of finite index in G_Z contains a congruence subgroup.*

The proof is easy to obtain by using the explicit construction of the homomorphism of $\mathfrak{G}$ into G_a that was indicated in the course of the proof of Proposition 1 of this section. We leave the details to the reader.

Now we proceed to the investigation of the subfields of $K(G)$ that are unions of fields K_Δ, where Δ ranges over a set $\mathfrak{M}$ of subgroups of finite index in G_Z. We assume that $\mathfrak{M}$ has the following property: if we take the groups belonging to it as a basis of neighborhoods in Γ, then a Hausdorff topology is introduced in Γ (see §1, p. 7).

For arithmetic groups there is a specific method of constructing such sets, which is based on the concept of a congruence subgroup. It consists in the following. Let S be an arbitrary finite or infinite set of prime numbers. We denote by $\mathfrak{M}(S, G)$ the set of all congruence subgroups $\Gamma(m)$ of $\Gamma = G_Z$ for which all the prime divisors of m belong to S.

We denote by $K(S) = K(S, G)$ the field that is the union of all fields K_Δ, where Δ ranges over the set of all subgroups belonging to $\mathfrak{M}(S, G)$. Our task consists in describing the Galois group $\mathfrak{G}(S)$ of $K(S, G)$.

We denote by G_S the restricted direct product of the groups G_p $(p \in S)$ with the distinguished subgroups U_p. In the special case when S is finite, G_S clearly coincides with ΠG_p. If S is the set of all prime numbers then G_S coincides with G_a. Obviously $G_a = G_S \times G_{\overline{S}}$, where $\overline{S}$ denotes the complementary set.

Similarly we may introduce the group $J_S(\mathfrak{Z})$ as the restricted direct product of the groups $H^1(Q_p, \mathfrak{Z})$ with the distinguished subgroups $H^1(Z_p, \mathfrak{Z})$, where p ranges over S. We denote by $K_S(\mathfrak{Z})$ the direct product $\Pi_{p \in S} H^1(Z_p, \mathfrak{Z})$ and by $H^1_S(Q, \mathfrak{Z})$ the subgroup of $H^1(Q, \mathfrak{Z})$ whose image under the natural homomorphism

$$\alpha : H^1(Q, \mathfrak{Z}) \to J(\mathfrak{Z})$$

is contained in $J_S(\mathfrak{Z}) \times K_{\overline{S}}(\mathfrak{Z})$, where $\overline{S}$ denotes the complementary set of prime numbers to S.

Let α_S be the natural homomorphism

$$H^1(Q, \mathfrak{Z}) \to J_S(\mathfrak{Z}).$$

Then we may define the group G^0_S as the subgroup of G_S whose image under the homomorphism

$$G_S \to J_S(\mathfrak{Z})$$

coincides with $\alpha_S(H^1_S(Q, \mathfrak{Z}))$.

The following proposition holds.

Proposition 4. *The Galois group $\mathfrak{G}(S)$ of $K(S, G)$ contains the group G^0_S as a subgroup of finite index.*

Proof. As was shown in Proposition 2 of §1, the group $\mathfrak{G}(S)$ can be found in the following way. We denote by $G_Q(S)$ the set of all $\gamma \in G_Q$ that transform the system of subgroups $\mathfrak{M}(S, G)$ into an equivalent one, that is, giving the same topology. The group $\mathfrak{G}(S)$ turns out to be isomorphic to the completion of $G_Q(S)$ in the topology in which the subgroups belonging to $\mathfrak{M}(S, G)$ form a basis of neighborhoods.

Thus to find $\mathfrak{G}(S)$ we have to classify in the first instance the subgroups of $G_Q(S)$. We denote by V the intersection of all subgroups of the form $\overline{\Delta}$, where $\Delta \subset \mathfrak{M}(S, G)$ and the bar denotes the closure in G_a.

Let $\gamma \in G_Q$. It is not difficult to see that $\gamma \in G_Q(S)$ if and only if γ belongs to the normalizer of V. From the fact that $G_a = G_S \times G_{\overline{S}}$ it follows that the normalizer of V is $G_S \times N$, where N is the normalizer of V in $G_{\overline{S}}$. Let us show that N is compact. Since the centralizer of V in $G_{\overline{S}}$ is trivial, it is sufficient to show that the group of all its automorphisms is compact.

Below we shall prove a lemma which states that the group of all automorphisms of a zero-dimensional compact finitely-generated topological group is compact. From this lemma it follows that N is compact.

The group $G_Q(S)$ is obviously the intersection of G_Q and $G_S \times N$. It is easy to show that the closure $\overline{G}_Q(S)$ contains $G_S^0 \times V$ and is contained in $G_S^0 \times N$.

From the definition of $\mathfrak{G}(S)$ it follows easily that it is isomorphic to the factor group of $\overline{G}_Q(S)$ with respect to V. To complete the proof of the theorem it remains to observe that N/V is compact.

Now we prove the lemma.

Lemma 2. *The group of automorphisms Φ of a zero-dimensional compact topological group K with a finite number of topological generators is compact.*

Proof. Consider all subgroups of finite index m of K. As is not hard to see, there is only a finite number of such subgroups. Their intersection is a fully invariant subgroup K_m of K. Consequently every automorphism of K induces an automorphism of the finite group $\Delta_m = K/K_m$. Let Φ_m be the group of automorphisms of Δ_m that are induced by automorphisms of K. Then the group of automorphisms Φ is the projective limit of the finite groups Φ_m; hence it is compact, and Lemma 2 is proved.

§3. Fields with a finite number of generators

Let K be an extension of finite transcendence degree of the field k, and $\mathfrak{G} = \mathfrak{G}(K/k)$ the group of all automorphisms of K that leave every element of k fixed. We say that K has a finite number of generators over k if K contains a finite set of elements F such that every subfield of K that contains k and F and is invariant under all automorphisms of K that leave k fixed coincides with K.

The main object of the present section is to clarify which of the fields considered in §2 have a finite number of generators. As a preliminary we make some general remarks.

Let K be an algebraic extension of k. It is not hard to verify that K has a finite number of generators over k if and only if it is a finite extension. In other words, an algebraic extension has a finite number of generators only when it is an extension of finite type, that is, obtained by adjoining to the ground field a finite number of elements. For transcendental extensions, as we shall show below, the situation is different: they may have a finite number of generators without being fields of finite type.

Let U be a subgroup of $\mathfrak{G}(K/k)$. We denote by K_U the subfield belonging to the group U, that is, consisting of all elements that are fixed under the action of U.

Lemma 1. *Let $\mathfrak{R}$ be a compact subgroup of $\mathfrak{G}$; then K is an algebraic normal* *extension of* $K_{\mathfrak{R}}$, *and the Galois group of this extension is isomorphic to* $\mathfrak{R}$.

Proof. First we verify that K is an algebraic extension of $K_{\mathfrak{R}}$. Let $\phi \in K$ and let U_ϕ be the set of all $g \in \mathfrak{G}$ left fixed by ϕ. The subgroup U_ϕ is open by the definition of the topology in $\mathfrak{G}$. Consequently

$$\mathfrak{R}_\phi = \mathfrak{R} \cap U_\phi$$

is a subgroup of finite index in $\mathfrak{R}$. Let $\widetilde{\mathfrak{R}}_\phi$ be a normal subgroup of finite index of $\mathfrak{R}$ contained in $\mathfrak{R}_\phi$, and $\Gamma = \mathfrak{R}/\widetilde{\mathfrak{R}}_\phi$. We consider the field $\widetilde{K} = K_{\mathfrak{R}\phi}$. The field K is invariant relative to Γ. Obviously if $\phi \in \widetilde{K}$ is fixed under the action of Γ then $\phi \in K_{\mathfrak{R}}$. From this it follows that $\widetilde{K}$ is a finite algebraic normal extension of $K_{\mathfrak{R}}$, and that its Galois group is isomorphic to $\Gamma = \mathfrak{R}/\mathfrak{R}_\phi$. At the same time we have shown that K is the union of finite normal algebraic extensions of $K_{\mathfrak{R}}$ and hence is itself a normal algebraic extension of $K_{\mathfrak{R}}$. It is not hard to verify, further, that its Galois group is isomorphic to $\Gamma = \mathfrak{R}/\widetilde{\mathfrak{R}}_\phi$.

Now we make a definition. We call a field K *normal* over k if $\mathfrak{G}$ contains an open compact subgroup U for which the field K_U is of finite type over k.

It is easy to show that if K is normal, then for every open compact subgroup U the field K_U is of finite type.

Note that all fields discussed in the preceding section are normal in the sense of the definition above.

Lemma 2. *A normal field K has a finite number of generators over k if and only if in a sufficiently small neighborhood of the identity of $\mathfrak{G}(K/k)$ there are no nontrivial normal subgroups.*

Proof. Let U be a neighborhood of the identity in $\mathfrak{G}$ having the required property. The field K is normal; therefore K_U is generated by a finite set Φ of elements of K. We show that Φ is a set of generators of K. For this it is sufficient to verify that K is the union of the fields $K_U(\Phi^g)$, $g \in \mathfrak{G}$, where Φ^g denotes the result of applying the automorphism g to the elements of Φ. The field K is normal over K_U. Therefore to every subfield of K containing K_U there corresponds a subgroup of U. In particular, to the field $K_U(\Phi^g)$ there corresponds $U \cap g^{-1}Ug$. To the union of the fields $K_U(\Phi^g)$ there corresponds $\bigcap g^{-1}Ug$, which by assumption is trivial. So we have shown that K contains a finite set of generators over k, provided there are no nontrivial normal subgroups in a sufficiently small neighborhood of the identity of $\mathfrak{G}$.

Now we show the converse: if every neighborhood of the identity of $\mathfrak{G}$ contains nontrivial normal subgroups, then K does not have a finite set of generators. For let Φ be a finite set of generators of K. We denote by U the subgroup consisting of all automorphisms of K that leave Φ fixed. Obviously U is open.

From the fact that Φ is a set of generators it follows that $\bigcap g^{-1}Ug = 1$, and hence U has no normal subgroups. The proof of the lemma is now complete.

Proposition 1. *The field $K(S, G)$ has a finite number of generators if and only if for all $p \in S$ every simple component of G_p is noncompact.*

Proof. We use the notation of Proposition 4 of §2. First of all we show that the condition stated in the proposition is necessary. For suppose that for some $p \in S$ the group G_p is compact; then, as is easy to see, the group $G'_p = \mathfrak{G}(S) \cap G_p$ is a compact normal subgroup of $\mathfrak{G}(S)$. Furthermore, if A is any normal subgroup of G_p, then $A' = \mathfrak{G}(S) \cap A$ is also a normal subgroup of $\mathfrak{G}(S)$. From this it follows that if for some $p \in S$ the group G_p is compact, then $\mathfrak{G}(S)$ contains arbitrarily small normal subgroups and hence, by Lemma 2, the field $K(S, G)$ does not have a finite number of generators.

Now we show that condition is sufficient. First of all we consider the case when S consists of a single prime number p. We show that the group G_p^0 has no compact normal subgroups. For suppose that this is not the case, and let $\mathfrak{R}$ be a compact normal subgroup of it. Since $\mathfrak{R}$ is closed, $\mathfrak{R}$ is also a p-adic Lie group. The Lie algebra of $\mathfrak{R}$ is therefore normal in the Lie algebra of G_p^0.

If G_p^0 is simple then it follows from what was shown above that the Lie algebra of $\mathfrak{R}$ must coincide with the Lie algebra of G_p^0. Consequently $\mathfrak{R}$ is an open normal subgroup of G_p^0. Now we use the fact that G is given by its adjoint representation. From the fact that the Lie algebra of $\mathfrak{R}$ is simple it follows that the centralizer of $\mathfrak{R}$ in G_p^0 is trivial. Hence the homomorphism of G_p^0 into the group of automorphisms of $\mathfrak{R}$ is a monomorphism. But this is impossible, because the group of automorphisms of a compact profinite group with a finite number of generators is compact, according to the lemma of §2. The cases when G_p is not simple and when S contains more than one prime number are treated similarly.

§4. Finite-dimensional proalgebraic varieties

In this section we describe a class of schemes that are models[1)] of fields of finite transcendence degree over a certain field k, but as a rule are not generated by a finite number of elements. Such fields are obviously infinite or finite algebraic extensions of the field of rational functions of a finite number of variables.

Definition 1. By an affine finite-dimensional proalgebraic variety we mean a spectrum X of an algebra A over a field k whose elements are all integral over a certain subalgebra of finite type over k.

[1)] A model of a field P is a scheme X for which the field of rational functions coincides with P.

Henceforth we always assume that A contains no nilpotent elements and that A is separable over some subalgebra of finite type.

An algebra A having this property can obviously always be represented as the union of a direct family of subalgebras $A^{(\alpha)}$ of finite type, and if $A^{(\alpha)} \subset A^{(\beta)}$ then all elements of $A^{(\beta)}$ are integral over $A^{(\alpha)}$. In other words, if $X^{(\alpha)} = \operatorname{spec} A^{(\alpha)}$, then

$$X = \varprojlim X^{(\alpha)},$$

where the $X^{(\alpha)}$ are finite-dimensional affine varieties over k and the morphisms $X^{(\beta)} \to X^{(\alpha)}$ are integral.

Definition 2. By a finite-dimensional proalgebraic variety over a field k we mean a scheme (X, O) over k (reduced) that is the union of a finite number of open sets U_i each of which is isomorphic to an affine finite-dimensional proalgebraic variety over k.

It is easy to see that if X is bicompact then it is sufficient to require that every point of X has a neighborhood isomorphic to a finite-dimensional affine proalgebraic variety.

Definition 3. By a projective finite-dimensional proalgebraic variety we mean the projective spectrum of a graded algebra $A = \Sigma_n A^n$ whose elements are all integral over a certain homogeneous subalgebra of finite type.

Clearly from this definition it follows that a projective finite-dimensional proalgebraic variety is a finite-dimensional proalgebraic variety in the sense of Definition 2.

Finite-dimensional proalgebraic varieties will be assumed below to be irreducible and normal; then its ring of rational functions $K = k(X)$ is a field. This field has finite transcendence degree. It can be shown that for each field K of finite transcendence degree over k there exists a certain model in the form of a finite-dimensional proalgebraic variety. A model can even be constructed in the form of a projective finite-dimensional proalgebraic variety.

Properties of the model are called biregular invariants, and properties of the field K, birational invariants.

Proposition 1. *A scheme X is a finite-dimensional proalgebraic variety if and only if it is integral over some algebraic variety.*

If the scheme X is integral over an algebraic variety X^0, then by definition there exists a morphism

$$f: X \to X^0$$

and a finite covering $X^0 = \bigcup U_i^0$ by affine open sets such that the sets $U_i = f^{-1}(U_i^0)$ are also affine and integral over U_i^0. This shows that the U_i are affine finite-dimensional proalgebraic varieties, and X a finite-dimensional proalgebraic variety.

4 Suppose, conversely, that X is a finite-dimensional proalgebraic variety and $X = \bigcup U_i$ a finite affine covering of it. For every open set U_i there exists, by assumption, an affine variety V_i (finite-dimensional) and an integral morphism $f_i: U_i \to V_i$. Obviously we may assume that $f(U_i) = V_i$. We denote by K the ring of rational functions on X (it need not be a field, because X may be reducible), by L_i the ring of rational functions on V_i, and by L the compositum of the rings L_i in K. We denote the coordinate rings of the affine schemes U_i and V_i by A_i and B_i. Then

$$B_i \subset A_i \subset K, \quad B_i \subset L_i \subset L \subset K.$$

Each of the rings L_i is the product of a finite number of fields of finite type over k. Since K has no nilpotent elements, neither has L. We construct an algebraic variety X^0 on which the ring of rational functions is contained in L, and an integral morphism $f: X \to X^0$. For this purpose we establish an equivalence relation between the points $x \in X$ by setting $x_1 \sim x_2$ if $\mathfrak{D}_{x_1} \cap L = \mathfrak{D}_{x_2} \cap L$, where $\mathfrak{D}_x$ is the local ring of x. We denote the class of equivalent points by x_0. To it there corresponds the local ring $\mathfrak{D}_{x_0} = \mathfrak{D}_x \cap L$, where $x \in x_0$. Thus the factor X_0 of the set X under this equivalence relation is equipped with a sheaf of local rings.

Let us show that X_0 is an algebraic variety with the ring of rational functions L. For this purpose we construct a covering of X_0 by affine open sets. We denote by A_i' the intersection $A_i \cap L$. Obviously the A_i' are algebras of finite type over k. For A_i' is integral over B_i and hence is contained in the interval closure $\overline{B}_i$ of B_i in L. $\overline{B}_i$ is known to have finite basis over B_i, and hence the same is true of A_i'.

We show that the affine schemes $U_i' = \operatorname{spec} A_i'$ form a covering of X_0. They are open in X_i by definition of the topology of the factor space. Every point $x_0 \in X_0$ arises from the local ring $\mathfrak{D}_x$, $x \in X$. The point x is contained in some affine set U_i and corresponds to a maximal ideal $\mathfrak{P}$ in A_i; and then x_0 corresponds to the maximal ideal $q = B_i' \cap \mathfrak{P}$ and $\mathfrak{D}_{x_0} = \mathfrak{D}$. Finally, the mapping $x \to x_0$ defined so that

$$\mathfrak{D}_x \cap L = \mathfrak{D}_{x_0}$$

is a regular and integral mapping of X onto X_0. Hence X_0 is a scheme and so an algebraic variety.

Proposition 2. *A scheme X is a finite-dimensional proalgebraic variety if and only if it may be represented in the form of a projective limit $X = \varprojlim X_\alpha$ of a set of algebraic varieties X_α, where the morphisms $f_{\alpha,\beta}: X_\beta \to X_\alpha$ are integral.*

We use the notation introduced in the proof of Proposition 1. If we have the representation $X = \varprojlim X_\alpha$, then the natural mapping $X \to X_\alpha$ satisfies the condition of Proposition 1, and hence X is a finite-dimensional proalgebraic variety. Let us show the converse. Let X be a finite-dimensional proalgebraic variety. We consider the algebra $L \subset K$ and all possible subalgebras L_α, $L \subset L_\alpha \subset K$, having a finite basis over L. As in the proof of Proposition 1, for each such algebra L_α we can construct an algebraic variety X_α, with L_α being the ring of rational functions on X_α, and if $L_\alpha \subset L_\beta$ then there exists an integral morphism $f_{\alpha,\beta}: X_\beta \to X_\alpha$ and X is integral over all the X_α.

On the other hand, $\overline{X} = \varprojlim X_\alpha$ is under our conditions a scheme that is integral over X_0. We have the natural morphism $X \to \overline{X}$, which commutes with the integral morphisms $X \to X_0$ and $\overline{X} \to X_0$. To verify that U is an isomorphism it is sufficient that U determines an isomorphism of the inverse images of an arbitrary affine covering of the set $U_0 \subset X_0$ in X and in $\overline{X}$. Thus the verification reduces to the affine case where it is obvious.

Let X be an irreducible proalgebraic variety. We denote by $\mathfrak{O}_x$ the local ring of the point $x \in X$, and by m_x the maximal ideal of x. We shall assume that the field k is perfect and that there exists a representation $X = \varprojlim X_\alpha$ such that P does not divide $[k(X_\alpha): k(X_\beta)]$ for $\alpha > \beta$, where P is the characteristic of the field k.

Definition 4. A point $x \in X$ is called interior if the ideal m_x has a finite number of generators, and a boundary point if it does not have a finite number of generators. We also call a point $x_\alpha \in X_\alpha$ interior if the point $x \in X$ lying over it is interior, and a boundary point if x is one.

In the examples to be constructed in §7 of proalgebraic varieties that are connected with automorphic functions the points lying over interior points of the domain of existence are also interior in the sense of Definition 4, and the points lying over boundary points of the domain of existence are also boundary points in the sense of Definition 4. We also mention that if X is the proalgebraic variety corresponding to the field of all algebraic functions of the given number of unknowns, then all points of X are boundary points.

Proposition 3. *Let $X = \varprojlim X_\alpha$ be a representation satisfying the conditions of Proposition 2. The point $x \in X$ is interior if and only if there exists a γ such that all morphisms $X_\alpha \to X_\gamma$ are unramified at x_γ (x_γ is the projection of x onto X_γ).*

Proof. Suppose that there exists an X_γ with the properties listed above; then the maximal ideal of the ring $\mathfrak{D}_{x_\alpha}$ $(\alpha > \gamma)$ is generated by the ideal m_{x_γ}, that is,

$$m_{x_\alpha} = \mathfrak{D}_{x_\alpha} m_{x_\gamma}.$$

Consequently,

$$m_x = \mathfrak{D}_x m_{x_\gamma},$$

from which it follows that m_x has the same finite basis.

Conversely, suppose that m_x has a finite basis. The ideal m_x is the union of all m_{x_α}; therefore its basis is contained in some m_{x_γ}. This means that

$$m_{x_\alpha} = \mathfrak{D}_{x_\alpha} m_{x_\gamma}$$

for every $\alpha > \gamma$, that is, the morphism $X_\alpha \to X_\gamma$ is unramified at x_γ. The proposition is proved.

A finite-dimensional proalgebraic variety is said to be regular if it has a representation $X = \varprojlim X_\alpha$ such that the points X_α over which the coverings X_β, $\beta > \alpha$, are ramified form a subvariety of lower dimension.

§5. The algebra of differentials

Let K be a field of finite transcendence degree n over the field of constants k. An expression of the form

$$\omega = f_0(df_1 \wedge \ldots \wedge df_n)^m, \tag{1}$$

where $f_0, f_1, \cdots, f_n$ are functions from K, is called a regular differential of degree m if ω is a regular differential in some subfield of K of finite type. Note that this subfield, in general, is larger than that generated by the functions f_0, $f_1, \cdots, f_n$.

We denote by $D^m = D^m(K)$ the set of all regular differentials of K of degree m. Obviously D^m is a vector space over k and $D^m D^s \subset D^{m+s}$. Let $D = \Sigma D^m$ be the graded algebra of differentials.

Definition 1. Let the algebra D satisfy the conditions of Definition 3 of §4; hence its projective spectrum is a projective proalgebraic variety. We denote it by X and the field of rational functions on it by $K(X)$. Obviously, always $K(X) \subset K$. A field K is said to be of general type if $K = K(X)$.

For example, the field of all algebraic functions of a given transcendence
5 degree is a field of general type.

Examples of fields that are not of general type are easy to obtain by considering fields of rational functions and fields of abelian functions. The fields discussed in §1 and §2 are apparently always fields of general type, although we have not succeeded in proving this.

Let K' be a subfield of K of finite type and of maximal transcendence degree over k. We denote by $D^m_{K'}$ the set of differentials $\omega \in D^m$ that have a representation of the form (1) with $f_0, f_1, \cdots, f_n \in K'$. Obviously $D^m_{K'}$ is a vector space of k. Every regular differential ω of K' is contained in $D^m_{K'}$, but the converse is not true in general. For example, if K is the field of all algebraic functions over k of a given transcendence degree n, then every differential ω of the form (1) is regular in a certain extension of K', and hence in this case $D^m_{K'}$ is infinite-dimensional.

We now make the following definition.

Definition 2. A field K is said to be locally finite if for every subfield K' of finite type, having maximal transcendence degree, and every $m > 0$ the space $D^m_{K'}$ is a finite-dimensional vector space over k.

For example, let X_0 be a complete nonsingular algebraic variety, and X a proalgebraic variety which is an unramified covering of it, of course not necessarily finite. The field $K = K(X)$ of all rational functions on X is locally finite. For let K' be some subfield of K of finite type having maximal transcendence degree. Without loss of generality we may assume that the field K_0 of rational functions on X_0 is contained in K'. Then X is an unramified covering of X_1, a nonsingular model of K_1.

In this case every differential $\omega \in D^m_{K'}$ is a regular differential in K', and hence

$$\dim D^m_{K'} < \infty.$$

Theorem 1. *Let k be a field of characteristic zero, and let X be a regular proalgebraic variety of dimension* 1. *Assume that in a certain representation $X = \varprojlim X_\alpha$ not all the X_α are of genus zero. Then the field $k(X)$ is locally finite if and only if the set Y of boundary points of the variety X is zero-dimensional.*

Proof. Let us assume that Y is zero-dimensional. Let $K(X) = \bigcup K_\alpha$, where $K_\alpha = K(X_\alpha)$ and the X_α are nonsingular complete algebraic curves. By assumption we may take it that all the X_α are coverings of a curve X_0 that are ramified only at the points of a finite set $S \subset X_0$. For every subfield K_0 of finite type there exists a γ such that $K_0 \subset K_\gamma$. Consequently it is sufficient to prove the finiteness of the dimension of the space $D^m_\gamma = D^m_{K_\gamma}$ of differentials of the form $\omega = f(dg)^m$, where $f, g \in K_\gamma$, for each of which there exists an α such that ω is regular on X_α. Without loss of generality we may assume that X_γ is chosen so that the morphism $X_\alpha \to X_\gamma$ is not ramified at all interior points of X_γ.

Now we find conditions on a differential ω given on X_γ under which this differential becomes regular on X_α for some $\alpha > \gamma$. At the interior points of X_γ such a differential ω must be regular. Let ξ be a boundary point of X_γ and t the local coordinate at this point. If the order of ramification of X_α at ξ is k

and τ is the local coordinate at a point $\eta \in X_a$ lying over ξ, then on X_γ

$$\nu_\xi(\omega) = \nu_\xi\left(\frac{\omega}{(dt)^m}\right),$$

and on X_a

$$\nu_\eta(\omega) = \nu_\eta\left(\frac{\omega}{(d\tau)^m}\right) = \nu_\eta\left(\frac{\omega}{(dt)^m}\right) + \nu_\eta\left(\left(\frac{dt}{d\tau}\right)^m\right) = \nu_\xi(\omega)\cdot k + m(k-1). \quad (2)$$

The condition $\nu_\eta(\omega) > 0$ gives

$$\nu_\xi(\omega) > -\left(1-\frac{1}{k}\right)m > -m.$$

The set of differentials ω satisfying these conditions at a finite number of points ξ is finite-dimensional, that is, the field $K(X)$ is locally finite.

Let $K(X)$ be locally finite. We wish to show that the set of points $S \subset X_0$ over which the covering $X_a \to X_0$ is unboundedly ramified is finite. Suppose that this is not the case and that $S' \subset S$ is a finite subset. We denote by $B^m(S')$ the set of differentials ω of degree m on Z_0 which are regular at all points that do not occur in S' and have at the points $\xi \in S'$ poles of order not exceeding $m-1$. Here m is any integer, greater than 1.

Every differential $\omega \in B^m(S')$ is regular on X, that is, on some X_a. For we can find an X_a such that the order of ramification at any point $\eta \in X_a$ lying over $\xi \in S'$ is greater than a preassigned number k. From (2) and the condition $\nu_\xi(\omega) \geq 1-m$ it follows that

$$\nu_\eta(\omega) \geqslant (1-m)k + m(k-1) = k-m,$$

and hence $\nu_\eta(\omega) \geq 0$ for $k \geq m$.

From the Riemann-Roch Theorem it follows that $\dim B^m(S') \to \infty$ when the number of points of S' increases unboundedly. Therefore if S is infinite the field $K(X)$ cannot be locally finite.

§6. Quasi-homogeneous proalgebraic varieties

Let X be an irreducible finite-dimensional proalgebraic variety over the field k and let $K(X)$ be the field of functions on X. The variety X is called quasi-homogeneous if:

1) the set Y of boundary points of X is different from X and is closed;

2) the orbit of every interior point relative to the group $G(X)$ of all automorphisms of X is everywhere dense in X.

Proposition 1. *If X is a quasi-homogeneous variety, then each element of $K(X)$ that is invariant under all automorphisms of $G(X)$ is algebraic over k. In particular, if k is algebraically closed in $K(X)$, then there are no invariant elements other than the elements of k.*

Proof. Let us assume that the element $u \in K(X)$ is invariant under $G(X)$ and not algebraic over k. Representing X in the form

$$X = \varprojlim X_\alpha,$$

where the X_α are algebraic varieties, we can find an X_γ such that $u \in K(X_\gamma)$. Then u may be regarded as a function on all the X_α, $\alpha > \gamma$.

By condition 1) in the definition of a quasi-homogeneous variety we can find a point $x_0 \in X_\gamma$ into which no point of the boundary is projected and at which the function u is regular. Let $u(x_0) = \alpha_0$. Then all the points of the subvariety $Y \subset X$ at which $u(x) = \alpha_0$ form a closed subset invariant under $G(X)$, different from X and containing interior points. But this contradicts condition 2) of the definition of a quasi-homogeneous variety.

We call an interior point x of an n-dimensional proalgebraic variety X simple if the ideal m_x of its local ring $\mathfrak{D}_x$ can be generated by n elements. It is easy to see that this condition is equivalent to the following: in the representation $X = \varprojlim X_\alpha$ there exists an X_γ such that for all $\alpha > \gamma$ the point $x_\alpha \in X_\alpha$ corresponding to x on X_α is simple on X_α. In this case the space m_x/m_x^2 has dimension n over k. The dual space is called the tangent space to X at the point x.

We prove some properties of quasi-homogeneous finite-dimensional proalgebraic varieties that are analogous to well-known properties of finite-dimensional varieties (see [10]).

Proposition 2. *All interior points of a quasi-homogeneous variety are simple.*

For let $U \subset X$ be the set of interior points of the quasi-homogeneous variety and $S \subset U$ the set of singular points. By definition the set S is invariant under $G(X)$, closed, and different from U. This contradicts condition 2) in the definition of a quasi-homogeneous variety.

Finally, we discuss the problem of birational equivalence of two quasi-homogeneous varieties. Obviously if X is a quasi-homogeneous variety then its group of automorphisms $G(X)$ is a subgroup of the group of all automorphisms of $k(X)$. Let Y be another quasi-homogeneous variety. An isomorphism of extensions of k

$$\phi: K(X) \rightarrow K(Y)$$

is called a birational isomorphism of the quasi-homogeneous varieties X and Y if it commutes with the action of the groups $G(X)$ and $G(Y)$ in the fields $K(X)$ and $K(Y)$.

Proposition 3. *Every birational isomorphism of quasi-homogeneous varieties X and Y determines a biregular isomorphism of the sets of their interior points.*

Proof. Let ϕ be a birational isomorphism of the quasi-homogeneous varieties X and Y, and let $X = \varprojlim X_\alpha$, $Y = \varprojlim Y_\alpha$ be representations of X and Y in the form of projective limits of irreducible algebraic varieties. Clearly the birational isomorphism ϕ determines a family of rational mappings ϕ_λ: $x_\lambda \to X_{\phi(\lambda)}$ that are permutable with the projections

$$f_{\lambda_1,\lambda_2}: X_{\lambda_1} \to X_{\lambda_2} \text{ and } g_{\mu_1,\mu_2}: Y_{\mu_1} \to Y_{\mu_2} \; (\lambda_1 > \lambda_2, \mu_1 > \mu_2).$$

The set of points at which ϕ_λ is not regular is an algebraic subvariety $W_\lambda \subset X_\lambda$ of codimension not less than 1. It is evident that if $\lambda_1 > \lambda_2$ then $f_{\lambda_1,\lambda_2} W_{\lambda_1} \subset W_{\lambda_2}$, and hence all the W_λ determine a closed subset W of the finite-dimensional proalgebraic variety X, of codimension not less than 1 in X. Since by hypothesis the isomorphism ϕ commutes with the action of $G(X)$ and $G(Y)$, the set W is invariant under $G(X)$. Now if W contained at least one interior point we would obtain that it is everywhere dense in X, but this contradicts our relation $\operatorname{codim}_X W \geq 1$. This shows that ϕ is regular at all interior points of X. Hence ϕ carries interior points of X into interior points of Y. When we now repeat the same arguments for the isomorphism ϕ^{-1}, we find that ϕ determines a biregular isomorphism of the sets of interior points of the quasi-homogeneous varieties X and Y.

Proposition 3 is now proved. It shows that the set of interior points of the quasi-homogeneous variety X is birationally invariant, that is, is determined only by the field of functions $K(X)$ and the action of the group $G(X)$ on it.

We denote by $\mathfrak{G}$ the group of all automorphisms of K. In general we cannot construct a model X of K on which all the automorphisms of K act regularly. This is impossible, for example, if K is purely a transcendental extension of k of degree n. Furthermore, if K has a quasi-homogeneous model X then it by no means follows that the groups $G(X)$ and $\mathfrak{G}$ coincide. Proposition 3 only implies that the normalizer of $G(X)$ in $\mathfrak{G}$ coincides with $G(X)$.

There is, however, an important type of field, the so-called fields of general type, for which such a model exists. For the algebra of differentials, obviously, is a birational invariant. Therefore its projective spectrum is a variety on which all the automorphisms of K act biregularly.

§7. Models of fields of automorphic functions

In this section we give an explicit description of the proalgebraic varieties that are models of the fields classified in §§1 and 2 of this paper. We also clarify the conditions under which these varieties are quasi-homogeneous proalgebraic varieties.

We begin with a construction.

Let D be a bounded domain in $\mathbf{C}^n$, and Γ a discrete group of analytic automorphisms of D with a compact factor space. We use the notation of §1. Let K

denote the closure of Γ in $\mathfrak{G}'$. The direct product $H = \mathfrak{K} \times D$ consists of the pairs

$$(k, z), \text{ where } k \in \mathfrak{K}, \ z \in D. \tag{1}$$

We determine the action of Γ in H in the following way:

$$\gamma: (k, z) \longrightarrow (k\gamma, \gamma^{-1}(z)).$$

It is not hard to see that Γ acts in H discretely.

Proposition 1. *The factor space* $X(\Gamma) = H/\Gamma$ *is a model of the field* $K(\Gamma)$ *in the form of a proalgebraic variety.*

Proof. We set $X_\Delta = D/\Delta$, where Δ is any subgroup of Γ of finite index. It is clear that X_Δ is an algebraic, and even a projective, variety. An embedding $\Delta_1 \subset \Delta_2$ induces an integral mapping $X_{\Delta_1} \longrightarrow X_{\Delta_2}$. It is not hard to see that X is the projective limit of the algebraic varieties X_Δ and hence a proalgebraic variety. The field K of rational functions on X is, by definition, the union of the fields $K_\Delta = K(X_\Delta)$, and consequently coincides with $K(\Gamma)$. Our proposition is proved.

Similarly one proves

Proposition 2. *Let* $\mathfrak{M}$ *be a system of subgroups of* Γ *satisfying conditions* 1), 2) *and* 3) *of* §1*(page* 7*); then the factor space* $X(\mathfrak{M}, \Gamma) = H(\mathfrak{M})/\Gamma$ *is a model of the field* $K(\mathfrak{M}, \Gamma)$, *where* $H(\mathfrak{M}) = \mathfrak{K}(\mathfrak{M}) \times D$, *and* $\mathfrak{K}(\mathfrak{M})$ *is the closure of* Γ *in the group* $\mathfrak{G}'(\mathfrak{M})$.

It is also easy to prove that all points of $X(\mathfrak{M})$ are interior in the sense of Definition 4 of §4.

Propositions 1 and 2 are concerned with the case when the factor space D/Γ is compact. Now we investigate what happens when the factor space D/Γ is not compact. In this case, if $\dim D > 1$, we may assume that Γ is an arithmetic group, but when $\dim D = 1$ we only assume that the factor space D/Γ has finite area. We recall that the theorem according to which K_Γ is a field of algebraic functions has only been proved in the cases listed above. Just as in the case when D/Γ is compact, we consider the factor space $X = H/\Gamma$, where H is defined as above. Obviously X is not compact, because the X_Δ are not compact. However, each X_Δ is known (see [2] and [3]) to be an open subset (in the sense of the Zariski topology) of some projective algebraic variety. X is the projective limit of the algebraic varieties X_Δ. As is not hard to verify, the morphisms $X_{\Delta_1} \longrightarrow X_{\Delta_2}$ are integral. Therefore X is a proalgebraic variety. All the points of X are obviously interior.

Let us show that there exists a projective proalgebraic variety Y whose set of interior points coincides with X.

With this aim we recall the construction (see [11], [2] and [3]) of the compactification of the factor space D/Γ. This well-known construction consists in the following. Certain points of the boundary of D are called Γ-rational. In the special case when D is the unit circle the rational points are the parabolic points. The general definition of rational points of the boundary is the following. Let z_0 be a point of the boundary of D, $G(z_0)$ its stability group, and $N(z_0)$ the maximal unipotent normal subgroup of $G(z_0)$. The group $N(z_0)$ is never trivial.

Then z_0 is called a Γ-rational point if the factor space $N(z_0)/\Gamma(z_0)$ is compact. It is not hard to see that the set of rational points of the boundary is one and the same for all commensurable arithmetic groups. We denote by $\widetilde{D}$ the union of D and all rational points of the boundary.

In the papers [2] and [3] it is shown that the factor space $\widetilde{D}/\Gamma$ can be endowed with a complex structure and a topology under which it becomes a compact analytic normal space, and in addition a projective algebraic variety. It is also shown that D/Γ is an open algebraic variety in the sense of the Zariski topology. We denote by $Y(\Gamma)$ the projective limit of $\widetilde{D}/\Delta$, where Δ ranges over the set of all subgroups of Γ of finite index. Then obviously $Y(\Gamma)$ is a projective proalgebraic variety. It is equally obvious that $X(\Gamma)$ can be embedded in $Y(\Gamma)$ in the form of an open subset in the Zariski topology. We show that $Y(\Gamma) - X(\Gamma)$ consists of boundary points. Let $y_0 \in Y_\Delta - X_\Delta$, where Δ is a subgroup of Γ of finite index. Then we have to show that there exists a subgroup $\Delta_1 \subset \Delta$ such that the covering $Y_{\Delta_1} \to Y_\Delta$ is ramified at the point $y_0 \in Y_\Delta$. For this purpose it is sufficient to take Δ_1 so that the index of $N(z_0) \cap \Delta_1$ in $N(z_0) \cap \Delta$ is greater than 1, where z_0 denotes any inverse image in $\widetilde{D}$ of the point y_0, and the subgroup $N(z_0)$ is defined above. The existence of such a subgroup Δ_1 clearly follows from the fact that the intersection of all subgroups of finite index of Γ is 1. Thus we have proved the following proposition.

Proposition 3. *Let D be a bounded symmetric domain, and Γ a discrete group of analytic automorphisms of D. Then there exists a projective proalgebraic variety $Y = Y(\Gamma)$ that is a model of the field $K(\Gamma)$ and has the following properties:*

1) *if the factor space D/Γ is compact then all points of Y are interior in the sense of Definition 4 of §4;*

2) *if D is the unit circle and D/Γ is not compact, but has finite area, or if D is any symmetric domain and Γ an arithmetic group, then all the points of Y lying over interior points of D are interior, and the points of Y lying over boundary points of D are boundary points in the sense of Definition 4 of §4.*

A similar proposition holds for the field $K(\mathfrak{M}, \Gamma)$, where $\mathfrak{M}$ is a system of subgroups of Γ satisfying the conditions of §1.

Proposition 4. *Let Γ be an arithmetic group acting in a symmetric domain. Then $Y = Y(\Gamma)$ is quasi-homogeneous.*

Proof. As was shown above, the set X of interior points of the model Y is the factor space H/Γ, where H is the direct product of $\mathcal{K}$ and D. To every element $g \in G_Q$ there corresponds a transformation in H defined in the following way:

$$g\colon\ (k, z) \to (gk\gamma^{-1}g^{-1},\ g\gamma(z)), \tag{2}$$

where γ is chosen relative to k and g such that $gk\gamma^{-1}g^{-1} \in \mathcal{K}$. Obviously such a choice of γ is always possible.

In H there is a natural Hausdorff topology arising from the representation of H as a direct product. Since the group Γ acts in H discretely, this topology can be extended to the factor space $X = H/\Gamma$. It is not hard to see that it is stronger than the Zariski topology. Therefore it is sufficient to show that the complete inverse image in H of the orbit of any preassigned point of X is everywhere dense in H. It is easy to see that this set has the form

$$(gk_0\alpha^{-1}g^{-1}\beta^{-1},\ \beta g\alpha(z_0)), \tag{3}$$

where k_0 and z_0 are fixed, g ranges over G_Q, $\alpha, \beta \in \Gamma$, and $g_0k_0\alpha^{-1}\beta^{-1}g_0 \in \mathcal{K}$. We set $r = \alpha^{-1}g^{-1}\beta^{-1}$; then (3) can be rewritten as

$$(\beta^{-1}r^{-1}\alpha^{-1}k_0r,\ r^{-1}(z)), \tag{4}$$

where $r \in G_Q$, $\alpha, \beta \in \Gamma$ subject to the condition

$$\beta^{-1}r^{-1}\alpha^{-1}k_0r \in \mathcal{K}.$$

Using the fact that the group G_Q, by Kneser's Theorem [7], is everywhere dense in the group G_∞ of all analytic automorphisms of D, it is not hard to show that the set (3) is everywhere dense in H. This completes the proof. Similarly one proves the following proposition.

Proposition 5. *Let S be a set of prime numbers. Denote by $Y = Y(S, G)$ the projective limit Y_Δ, where Δ ranges over the set of all congruence subgroups $\Gamma(m)$ of the group $\Gamma = G_Z$ for which all the prime divisors of m belong to S. If for at least one $p \in S$ the group G_p is not compact, then $Y(S, G)$ is a quasi-homogeneous proalgebraic variety.*

§8. The algebra of automorphic forms

Let Γ be a discrete group of analytic automorphisms of the bounded symmetric domain $D \subset C^n$ such that the factor space D/Γ is compact. Let Δ be a subgroup of Γ of finite index. We denote by A_Δ^m the set of all automorphic forms of weight m relative to Δ.

Let $A^m(\Gamma)$ be the union of all spaces A_Δ^m, where Δ ranges over the set of all subgroups of Γ of finite index. We denote by

$$A(\Gamma)=\sum_{m=0}^{\infty} A^m(\Gamma)$$

the graded algebra of automorphic forms, and by

$$D(\Gamma)=\sum D^m(\Gamma)$$

the graded algebra of regular differentials (see §5) of $K(\Gamma)$.

Proposition 1. *The algebras $A(\Gamma)$ and $D(\Gamma)$ are isomorphic. The projective spectrum of $D(\Gamma)$ coincides with $X(\Gamma)$ (see §7, Proposition 1).*

Proof. First of all we establish a one-to-one correspondence between $A(\Gamma)$ and $D(\Gamma)$. Let $h(z)\in A_\Delta^m$; then the expression

$$\omega=h(z)(dz_1\wedge\ldots\wedge dz_n)^m \tag{1}$$

is invariant under the transformations $z\longrightarrow\delta z$, $\delta\in\Delta$. Let $f_1,\cdots,f_n$ be any n analytically independent functions in K_Δ. Clearly the function

$$\varphi(z)=\frac{\omega}{(df_1\wedge\ldots\wedge df_n)^m} \tag{2}$$

is invariant under the transformations $z\longrightarrow\delta z$, where $\delta\in\Delta$, and hence belongs to K_Δ. Consequently ω is a differential of degree m of K_Δ. It remains to verify that ω is a regular differential. Without loss of generality we may assume that Δ has no transformations with fixed points, except the identity transformation (see [4]). In that case it is obvious that ω is a regular differential K_Δ. So we have shown that to every automorphic form of weight m there corresponds a regular differential of weight m. Clearly this correspondence is a homomorphism of $A(\Gamma)$ into $D(\Gamma)$ with trivial kernel. To show that the algebras $A(\Gamma)$ and $D(\Gamma)$ are isomorphic it is sufficient to check that the correspondence indicated above between $A(\Gamma)$ and $D(\Gamma)$ is an epimorphism. In other words, we have to show that every regular differential of $K(\Gamma)$ is representable in the form (1).

Let

$$\omega=f_0(df_1\wedge\ldots\wedge df_n)^m$$

be a regular differential of $K(\Gamma)$. Then there exists a subgroup Δ of finite index in Γ such that ω is a regular differential in K_Δ and $f_0,f_1,\cdots,f_n\in K_\Delta$. We set

$$df_1\wedge\ldots\wedge df_n=j(z)\,dz_1\wedge\ldots\wedge dz_n. \tag{3}$$

Then

$$\omega=f_0(z)\,(j(z))^m(dz_1\wedge\ldots\wedge dz_n)^m. \tag{4}$$

From the fact that ω is a regular differential of K_Δ it follows that the function $h(z)=f_0(z)\,j^m(z)$ is regular in D. From the fact that ω is a differential of K_Δ it follows that $h(z)$ is an automorphic phase of weight m.

Thus we have shown that $A(\Gamma)$ and $D(\Gamma)$ are isomorphic. Now we show that the spectrum of $A(\Gamma)$ coincides with $X(\Gamma)$. Clearly the spectrum of the algebra $A(\Gamma)$ is the same as the projective limit of the spectra of the algebras $A_\Delta = \Sigma A^m_\Delta$. It is well known that the spectrum of A_Δ is $X_\Delta = D/\Delta$. It remains to take into account that the projective limit of the X_Δ is $X(\Gamma)$. This completes the proof.

Let $\mathfrak{M}$ be a system of subgroups of Γ satisfying the conditions 1), 2) and 3) of §1(page 7). Similarly to the preceding we can define the algebra of automorphic forms and the algebra of differentials:

$$A(\mathfrak{M},\Gamma) = \sum_{\Delta\subset\mathfrak{M}} A_\Delta, \quad D(\mathfrak{M},\Gamma) = \sum_{\Delta\subset\mathfrak{M}} D_\Delta.$$

The following proposition is proved just as Proposition 1.

Proposition 2. *The algebras* $A(\mathfrak{M},\Gamma)$ *and* $D(\mathfrak{M},\Gamma)$ *are isomorphic. The projective spectrum of* $A(\mathfrak{M},\Gamma)$ *coincides with* $X(\mathfrak{M},\Gamma)$ *(see* §7).

We now proceed to the discussion of discrete groups for which the factor space D/Γ is not compact but has finite volume. First we consider the case when D is the unit disc $|z| < 1$. We denote by A^m_Δ the set of all automorphic forms $h(z)$ of weight m for which

$$(h, h) = \int_{D/\Delta} |h(z)|^2(1-|z|^2)^{2m-2}\,dx\,dy < \infty. \tag{5}$$

It is well known that $h \in A^m_\Delta$ if and only if $h(z)(1-|z|^2)^m$ is bounded in the unit disc. Therefore $A^{m_1}_\Delta A^{m_2}_\Delta \subset A^{m_1+m_2}_\Delta$. Consequently $A_\Delta = \Sigma A^m_\Delta$ is a graded algebra. A similar proposition is true for $A(\Gamma) = \Sigma_\Delta A_\Delta$.

Proposition 3. *The algebras* $A(\Gamma)$ *and* $D(\Gamma)$ *are isomorphic. The projective spectrum* $Y(\Gamma)$ *of* $A(\Gamma)$ *coincides with the set of all points of the field* $K(\Gamma)$.

Proof. A correspondence between the algebras $A(\Gamma)$ and $D(\Gamma)$ is established just as in the proof of Proposition 1. Here we only have to verify that if $h(z) \in A^m_\Delta$, then $\omega = h(z)(dz)^m$ is a regular differential of $K(\Gamma)$, that is, there exists a $\Delta_1 \subset \Delta$ such that ω is a regular differential of the field K_{Δ_1}. As we know, the set of all points of K_{Δ_1} is obtained by adjoining to D/Δ_1 all inequivalent parabolic vertices. Without loss of generality we may assume that Δ_1 contains no elliptic elements. In this case for every $\Delta_1 \subset \Delta$ the differential ω is regular at every point of D/Δ. Therefore it remains to consider the parabolic points. Let z_0 be a parabolic vertex of Δ. We map the unit disc D onto the upper half-plane $\operatorname{Im} z > 0$ so that the point z_0 is carried into the point at infinity, and the parabolic subgroup of Δ into the set of transformations of the form $z \longrightarrow z + n$, where n is an integer. From the fact that $h(z) \in A^m_\Delta$ it follows that $h(z)$ has the form

$$h(z) = \sum_{k=0}^{\infty} a_k e^{2\pi kzi}, \tag{6}$$

from which we obtain

$$\omega = h(z)\,(dz)^m \sim t^{1-m}(dt)^m,$$

where $t = e^{2\pi zi}$. Consequently ω has at $t = 0$ a pole of order at most $m - 1$. Let ν be the order of ramification at this point for the covering connected with the group Δ_1. Then $t = \tau^\nu$, where τ is a local parameter on the covering. We have

$$\omega \sim \lambda\tau^{(1-m)\nu+m(\nu-1)}(d\tau)^m = \tau^{\nu-m}(d\tau)^m, \tag{7}$$

where λ is a constant. Thus for a sufficiently large ν, ω is a regular differential in K_{Δ_1}.

By similar computations the converse proposition can be verified. Namely, if ω is a regular differential of $K(\Gamma)$, then ω can be represented in the form

$$h(z)(dz)^m, \tag{8}$$

where $h(z) \in A^m_\Delta$.

The statement on the structure of the projective spectrum of $A(\Gamma)$ follows from the fact that this spectrum is the projective limit of the spectra of the algebras $A_\Delta = \Sigma A^m_\Delta$. As we know, the spectrum of A_Δ coincides with the set of points of K_Δ. Proposition 3 is now proved.

We proceed to the case when D is any symmetric domain and Γ an arithmetic group acting in D. Let G be the linear algebraic group defined over Q from which the given arithmetic group Γ arises. We assume that the dimensions of the maximal tori of G splitting over Q and over $\mathfrak{R}$ are the same. We say that the arithmetic group Γ satisfies condition A) if it arises from an algebraic group G for which the above relationship holds.

Let D be a symmetric domain in $\mathbf{C}^n$, and Γ an arithmetic group of analytic automorphisms acting in D. An element of the invariant volume can always be written in the form

$$dv = \rho\, dz \wedge d\bar{z}, \tag{9}$$

where $dz = dz_1 \wedge \cdots \wedge dz_n$.

We denote by A^m_Γ the set of all Γ-automorphic forms of weight m for which

$$\int_{D/\Gamma} |f|^2 \rho^{-m}\, dv < \infty. \tag{10}$$

It is known (see [11]) that $f \in A^m_\Gamma$ if and only if the function $f\rho^{-m/2}$ is bounded in D. Consequently $A^{m_1}_\Gamma A^{m_2}_\Gamma \subset A^{m_1+m_2}_\Gamma$, and hence $A_\Gamma = \Sigma_m A^m_\Gamma$ is naturally endowed with the structure of a graded algebra. We set

$$A^m(\Gamma) = \sum A^m_\Delta,$$

where Δ ranges over the set of all subgroups of finite index in Γ, and we denote by

$$A(\Gamma) = \sum A^m_\Gamma$$

the algebra of automorphic forms.

Proposition 4. *Let Γ be an arithmetic group satisfying condition* A. *Then the algebras $A(\Gamma)$ and $D(\Gamma)$ are isomorphic. The projective spectrum of $A(\Gamma)$ is a quasi-homogeneous proalgebraic variety $W(\Gamma)$ for which the set of interior points coincides with $X(\Gamma)$, the projective limit of the factor spaces $X_\Delta = D/\Delta$, where Δ ranges over the set of all subgroups of finite index in Γ.*

Note. A similar proposition is valid when we discuss not all subgroups of finite index but only a certain set $\mathfrak{M}$ of subgroups of finite index satisfying the conditions of §1 (p. 7).

Proof of Proposition 4. To begin with we show that the algebras $A(\Gamma)$ and $D(\Gamma)$ are isomorphic. An isomorphism is established by the same method as in the proof of Propositions 1, 2 and 3. We need only verify that if $h(z) \in A^m_\Delta$, then

$$\omega = h(z)\,(dz_1 \wedge \ldots \wedge dz_n)^m \tag{11}$$

is a regular differential of K_{Δ_1}, where Δ_1 is a subgroup of finite index in Γ, and that every regular differential ω of $K(\Gamma)$ is representable in the form (11) for a certain form $h \in A^m_\Delta$. For this purpose we make use of the process of desingularization described in the papers [3], [13] and [14]. This process enables us to construct, for every group Γ satisfying condition A and not having elliptic elements, a nonsingular model W_Γ of K_Γ. Here it turns out that W_Γ has the following properties.

1) There exists a regular mapping π: $W_\Gamma \to Y_\Gamma$, where Y_Γ is a normal model of K_Γ defined in §7, and on $X_\Gamma = D/\Gamma$ this relationship is one-to-one.

2) Let Δ be a subgroup of finite index in Γ; then there exists a mapping $f_{\Delta,\Gamma}$: $W_\Delta \to W_\Gamma$ that is unramified at the points $\pi^{-1}(X_\Gamma)$ and ramified in $V_\Gamma = W_\Gamma - \pi^{-1}(X_\Gamma)$; moreover, at every pair of points $v' \in V_\Delta$, $v = f_{\Delta,\Gamma}(v') \in V_\Gamma$ there exist coordinate systems (t', u') and (t, u), where $t', t \in \mathbf{C}^1$; $u', u \in \mathbf{C}^{n-1}$ such that the mapping $f_{\Delta,\Gamma}$ has the form

$$f_{\Delta,\Gamma}(t', u') = (t'^n, u'),$$

i.e. it is ramified only at $t = 0$.

We also mention that the order of ramification n depends only on the subgroup Δ and can be made arbitrarily large by means of a suitable choice of subgroups. Using these facts it is not hard to verify by means of calculations similar to those in Proposition 3 that a differential ω of degree m of K_Δ is a regular differential of $K(\Gamma)$ if and only if ω is regular at each point of X_Γ and has at every point of V_Γ a pole with respect to t of order at most $m-1$. Using the explicit description of the process of desingularization it is easy to derive that the condition of the vanishing of the free term in the expansion of the so-called Fourier-Jacobi series of the automorphic form $h(z)$ is necessary and sufficient for the differential $\omega = h(z)(dz)^m$ to be a regular differential of $K(\Gamma)$.

It remains to observe that, as was shown in [11], the free term in the expansion of the automorphic form $h(z)$ in a Fourier-Jacobi series vanishes if and only if $h(z) \in A_\Delta^m$. Proposition 4 is proved.

The variety $W(\Gamma)$ is the projective spectrum of an algebra of differentials. Therefore every automorphism of $K(\Gamma)$ carries $W(\Gamma)$ into itself. The set $X(\Gamma)$ of interior points of $W(\Gamma)$ is carried by the automorphism g into itself. This leads immediately to the following proposition, which we have used in §1.

Every automorphism g of $K(\Gamma)$ carrying the field K_{Δ_1} into K_{Δ_2} carries X_{Δ_1} into X_{Δ_2}.

A similar proposition holds for the case when D is the unit disc $|z| < 1$.

Addendum

Let Γ be the modular group, $K = \mathbf{C}(j)$ where j is the modular invariant. We denote by S a set of prime numbers and by $P(S)$ the union of the fields of modular functions relative to all congruence subgroups $\Gamma(m)$, where m contains only prime divisors from S.

We know (see §2) that the Galois group of $P(S)$ can be described in the following way. Consider sequences $(g_{p_1}, g_{p_2}, \cdots)$, where $p_i \in S$, $i = 1, 2, \cdots$, all the g_{p_i} apart from a finite number are integers, and $\det g_{p_i} = \rho$, where ρ is a rational number for which the prime divisors of the numerator and denominator belong to S. Such sequences form a group $G(S)$. The factor group of $G(S)$ by the center is also the Galois group $\mathfrak{G}$ of $P(S)$.

Theorem. *Every subfield of $P(S)$ invariant under $\mathfrak{G}$, coincides with $P(S)$.*

Proof. We denote by T_m the operator

$$T_m f = f(mz) + \sum_{a=0}^{m-1} f\left(\frac{z+a}{m}\right).$$

Let P' be a subfield of $P(S)$ invariant under $\mathfrak{G}$. We set $P = K \cap P'$. Then

$$T_m P \subseteq P,$$

provided all prime divisors of m belong to S. For $T_m P' \subset P'$, and $T_m K \subset K$.

Next, it is easy to see that P is not empty, in fact larger than C. P, as a subfield of rational functions, is itself a field of rational functions of a certain $f_0 \in P$:

$$P = \mathbf{C}(f_0),$$

and we may assume that f_0 has a pole at ∞.

The assertion of the theorem reduces to

$$P = K.$$

Suppose that this is not the case. We denote by n_0 the order of the pole of f_0 at ∞ and consider first the case $n_0 = 1$. In this case the function f_0 necessarily has poles at finite points. We denote these points by $z_1, \cdots, z_\nu$ and their multiplicities by $n_1, \cdots, n_\nu$. We consider the function $f_m = T_m f_0$. It is not hard to see that it has a pole of order m at infinity. Its remaining poles are situated at the points

$$mz_i, \quad \frac{z_i + a}{m}, \quad a = 0, 1, \ldots, m-1,$$

with the multiplicities n_i.

From the fact that f_m has a pole of order m at ∞ and is a rational function of f_0 it follows that at $z_1, z_2, \cdots$ it must have poles of orders not less than $mn_1, mn_2, \cdots$, respectively. But this is impossible, because for sufficiently large m the points

$$mz_1, \ldots, mz_\nu, \quad \frac{z_1 + a}{m}, \ldots, \frac{z_\nu + a}{m}, \quad |a| \leqslant \lambda \sqrt{m},$$

where λ is a sufficiently small constant, are not equivalent to $z_1, \cdots, z_\nu$ (throughout the argument it was assumed that $z_1, \cdots, z_\nu$ are points of the fundamental domain).

We now consider the case $n_0 > 1$. First we show that f_0 has the form

$$f_0(z) = e^{-2\pi n_0 z i} + \sum_{k=0}^{\infty} a_k e^{2\pi k z i}. \tag{1}$$

Suppose that this is not so. We denote by n_1 the order of the pole at ∞ of the function $f_0 - e^{-2\pi n_0 z i}$; in other words, we assume that f_0 has the form

$$f_0(z) = e^{-2\pi n_0 z i} + a_1 e^{-2\pi n_1 z i} + \ldots. \tag{2}$$

We consider the function

$$\varphi = T_m f_0 - f_0^m.$$

This function has at ∞ a pole of order $(m-1)n_0 + n_1$, i.e. not divisible by n_0. This contradicts the fact that ϕ is a rational function of f_0. So we have shown that f_0 has the form (1).

Now we show that if $a_k \neq 0$ in (1) then $k \equiv 0 (\mathrm{mod}\, n_0)$. We denote by n_1 the natural number of least absolute value, not divisible by n_0, such that $a_{n_0} \neq 0$.

We set $r = [n_1/n_0] + 1$ and consider the function

$$\varphi = T_m f_0 - \sum_{k=0}^{r} \lambda_k f_0^{m-k}$$

It is not hard to see that for sufficiently large m and a suitable choice of the coefficients λ_k the function ϕ has at ∞ a pole of order

$$(m-1)n_0 - n_1,$$

but this is impossible because ϕ is a rational function of f_0 and hence the order of its pole is a multiple of n_0. So we have shown the index of every nonzero coefficient of f_0 is a multiple of n_0.

Let us show that K contains no functions whose nonzero coefficients are all divisible by some $n_0 > 1$. Suppose that such a function exists and denote it by f_0. Then f_0 is invariant under the modular group and the transformation $z \to z + 1/n_0$. It is not hard to see that the group generated by these transformations is always dense. Consequently such a function f_0 must be a constant. So we have shown that if $P = \mathbf{C}(f_0)$ then f_0 has a single pole of the first order at ∞. Obviously from this it follows that the modular invariant j is a rational function of f_0. We have shown that

$$P = \mathbf{C}(j) = K.$$

It is very probable that a similar proposition holds for arbitrary arithmetic groups; however, we have not succeeded in proving this.

BIBLIOGRAPHY

[1] C. L. Siegel, *Automorphic functions of several complex variables*, Russian transl., Moscow, 1954.

[2] W. L. Baily, Jr. and A. Borel, *On the compactification of arithmetically defined quotients of bounded symmetric domains*, Bull. Amer. Math. Soc. 70 (1964), 588–593. MR 29 #6058.

[3] I. I. Pjateckiĭ-Šapiro, *Arithmetic groups in complex domains*, Uspehi Mat. Nauk 19 (1964), no. 6 (120), 93–121 = Russian Math. Surveys 19 (1964), no. 6, 83–109. MR 32 #7790.

[4] A. Weil, *On discrete subgroups of Lie groups*. II, Ann. of Math. (2) 75 (1962), 578–602; Russian transl., Matematika 7 (1963), no. 1, 17–40. MR 25 #1242.

[5] T. Shioda, *On algebraic varieties uniformizable by bounded domains*, Proc. Japan Acad. 39 (1963), 617–619. MR 28 #5448.

[6] A. Borel and Harish-Chandra, *Arithmetic subgroups of algebraic groups*, Ann. of Math. (2) 75 (1962), 485–535. MR 26 #5081.

[7] M. Kneser, *Strong approximation*, Proc. Sympos. Pure Math., Vol. IX, pp. 187–196, Amer. Math. Soc., Providence, R.I., 1966.

[8] S. Lang, *Algebraic groups over finite fields*, Amer. J. Math. 78 (1956), 555–563. MR 19, 174.

[9] J. Tate, *Duality theorems in Galois cohomology over number fields*, Proc. Internat. Congress Math., Stockholm (1962), Inst. Mittag-Leffler, Djursholm, 1963, pp. 288–295. MR 31 #168.

[10] S. Lang, *Abelian varieties*, Interscience Tracts in Pure and Appl. Math., no. 7, Interscience, New York and London, 1959. MR 21 #4959.

[11] I. I. Pjateckiĭ-Šapiro, *Geometry of classical domains and the theory of automorphic functions*, Fizmatgiz, Moscow, 1961; French transl., Travaux et Recherches Mathématique, no. 12, Dunod, Paris, 1966. MR 25 #231; MR 33 #5949.

[12] W. L. Baily, Jr., *On the compactification of orbit spaces of arithmetic discontinuous groups acting on bounded symmetric domains*, Proc. Sympos. Pure Math., Vol. IX, pp. 281–295, Amer. Math. Soc., Providence, R.I., 1966.

[13] S. G. Gindikin and I. I. Pjateckiĭ-Šapiro, *On the algebraic structure of the field of Segal modular functions*, Dokl. Akad. Nauk SSSR 162 (1965), 1226–1229 = Soviet Math. Dokl. 6 (1965), 831–835. MR 31 #2218.

[14] Jun-ichi Igusa, *On the desingularization of Satake compactifications*, Proc. Sympos. Pure Math., Vol. IX, pp. 301–305, Amer. Math. Soc., Providence, R.I., 1966.

[15] ———, *Fibre systems of Jacobian varieties. III; Fibre systems of elliptic curves*, Amer. J. Math. 81 (1959), 453–457. MR 21 #3422.

[16] D. Mumford, *Projective invariants of projective structures and applications*, Proc. Internat. Congress Math., Stockholm (1962), Inst. Mittag-Leffler, Djursholm, 1963, pp. 526–530. MR 31 #175.

[17] A. Selberg, *On discontinuous groups in higher dimensional symmetric spaces*, Contributions to function theory (Internat. Colloq. Function Theory, Bombay, 1960), Tata Inst. of Fundamental Research, Bombay, 1960, pp. 147–164; Russian transl., Matematika 6 (1962), no. 3, 3–16. MR 24 #A188.

Translated by:
K. A. Hirsch

Cartan pseudogroups and Lie p-algebras

(with A. I. Kostrikin)

Dokl. Akad. Nauk SSSR **168,** No. 4, 740–742 (1966). Zbl. **158,** 38
[Sov. Math., Dokl. **7,** 715–718 (1966)]

1. In 1909 Cartan [1] classified all simple transitive pseudogroups of transformations. A modern presentation of part of the Cartan results can be found in [2–4]. In this presentation the group-theoretic problem is replaced by an equivalent problem concerning Lie algebras $\mathfrak{L}$ satisfying the following conditions:

a) $\mathfrak{L}$ is a topological Lie algebra with the linear topology determined by subspaces of finite codimension; $\mathfrak{L}$ is complete in this topology.

b) In $\mathfrak{L}$ there exists an (open) subalgebra $\mathfrak{L}_0$ such that the system of subalgebras $\mathfrak{L}_k$, defined by the conditions

$$\mathscr{L}_{-1} = \mathscr{L}; \qquad \mathscr{L}_{k+1} = \{x \in \mathscr{L}_k \mid [\mathscr{L}, x] \subset \mathscr{L}_k\}, \quad k \geqslant 0,$$

forms a complete system of neighborhoods of zero.

We shall call algebras with conditions a) and b) *infinite Lie algebras*. From the definition it follows that $[\mathfrak{L}_i, \mathfrak{L}_j] \subset \mathfrak{L}_{i+j}$; $\mathfrak{L}_k$ is an ideal in $\mathfrak{L}_0$, $k > 0$. In view of this, we have defined a representation Γ of the finite-dimensional algebra $L_0 = \mathfrak{L}_0/\mathfrak{L}_1$ into the finite-dimensional module $V = \mathfrak{L}_{-1}/\mathfrak{L}_0$.

In the papers mentioned it is proved that if $\mathfrak{L}$ is a simple Lie algebra over an algebraically closed field of characteristic zero and the representation Γ is irreducible, then $\mathfrak{L}$ belongs to one of the collections indicated below. Algebras of all the collections are realized in the ring of differential operators $\mathfrak{D}$ of the ring $K = k[[x_1, \dots, x_n]]$ of formal power series over the field k,

$$\mathscr{D} = \sum_1^n f_i \frac{\partial}{\partial x_i}, \quad f_i \in K. \tag{1}$$

The collections are:

1) $\mathfrak{L}$ consists of all operators of the form (1) (general algebra).

2) $\mathfrak{L}$ consists of all operators which leave invariant the differential form $dx_1 \wedge dx_2 \wedge \dots \wedge dx_n$ (special algebra).

3) $n = 2m$; $\mathfrak{L}$ consists of all operators which leave invariant the differential form $\omega = \sum_1^m dx_i \wedge dx_{m+i}$ (Hamiltonian algebra).

Cartan proved that all simple infinite Lie algebras are contained in collections 1)–3) and the collection

4) $n = 2m + 1$; $\mathfrak{L}_m$ consists of all operators which multiply the form $\omega = dx_n + \sum_{i=1}^m (x_i dx_{m+i} - x_{m+i} dx_i)$ by an element of the ring K (contact algebra).

In the present note we consider analogs of Cartan algebras over an algebraically closed field k of characteristic $p > 0$. The definition of each of the above collections retains meaning, but none of them remains simple. Namely, in $\mathfrak{L}$ there exists an ideal J, consisting of all operators of the form (1), for which $f_i \in K^{p\cdot} = K \cdot (x_1^p, \cdots, x_n^p)$. The factor-algebra $\mathfrak{L}/J$ has finite dimension over k for all four collections, and each is close to prime. More precisely, in case 1) the algebra $\mathfrak{L}/J = W_n$ is simple and dim $W_n = np^n$, $(n, p) \neq (1, 2)$; in case 2) the commutator group S_n of the algebra $\mathfrak{L}/J$ is simple and dim $S_n = (n-1)(p^n - 1), n > 2$; in case 3) the ideal H_n, of codimensional 1 in the commutator-group of the algebra $\mathfrak{L}/J$, is simple, and dim $H_n = p^n - 2$ $(p > 2$ or $n > 2)$. Finally, in case 4) the algebra K_n is simple, where $K_n = \mathfrak{L}/J$, if $n + 3 \not\equiv 0(p)$ and $K_n = (\mathfrak{L}/J, \mathfrak{L}/J]$, if $n + 3 \equiv 0(p), p > 2$. The corresponding dimension of K_n is p^n or $p^n - 1$.

The algebras W_n, S_n, H_n and K_n, upon which we again fix the titles general, special, Hamiltonian and contact, are Lie p-algebras. They have been described earlier (cf. [5–9]) in another way. We shall call algebras of these collections *algebras of Cartan type*, and simple p-algebras, obtained by reduction mod p from simple Lie algebras over a field of characteristic zero, we shall call *classical*. As mentioned above, all algebras of Cartan type are realized as "algebras of differentiation" of the truncated ring of polynomials $k[x_1, \cdots, x_n], x_i^p = 0$.

2. It seems likely to us that *algebras of Cartan type together with classical algebras exhaust all simple Lie p-algebras* (p > 5). In each case, no examples of other types are known.

Our basic result proves the conjecture only under some additional restrictions.

Theorem 1. *Let $\mathfrak{L}$ be a simple Lie p-algebra over an algebraically closed field k of characteristic $p > 7$, possessing a proper subalgebra $\mathfrak{L}_0$ such that* dim $V < p$, dim $L_0 < p$ *and the representation Γ is irreducible. Then $\mathfrak{L}$ is either a classical algebra or an algebra of Cartan type of one of the collections 1)–3). (The definition of V, L_0 and Γ was given in* §1.)

We shall elaborate on the basic steps in the proof of Theorem 1. It runs parallel to the proof of the corresponding theorem for infinite Lie algebras over a field of characteristic zero (cf. [2, 3]), but the various difficulties involved by the stipulation of a finite characteristic must be overcome. The elegant proof appearing in [3] does not apply in our case, since it utilizes the infiniteness of the algebra.

A filtration $\{\mathfrak{L}_k\}$ is constructed relative to the subalgebra $\mathfrak{L}_0$, as indicated in 1. From the irreducibility of the representation Γ, it follows that all the terms $\mathfrak{L}_k$ of the filtration are p-subalgebras. We distinguish between two cases.

1) $\mathfrak{L}_k = 0$ for $k \geq 2$;

2) $\mathfrak{L}_2 \neq 0$.

In case 1) it is proved that $\mathfrak{L}$ is an algebra of classical type. This follows from the results obtained earlier by one of the authors [10, 11]. In case 2) it is proved that $\mathfrak{L}$ is of Cartan type. Moreover, it is established that L_0 is a direct sum of algebras of classical type and, possibly, have one-dimensional center. This is clear over a field of characteristic zero, but is nontrivial in our case. The result obtained can be applied to an investigation of the representation Γ in the Curtis theory [12] of p-representations of algebras of classical type.

Proposition. *The difference of any two of the representations Γ can be expressed in the form of the difference of certain of the roots of the algebra L_0.*

This assertion is a strengthening of Lemma 5.5 of [4]. It can be proved that the conditions of the proposition are satisfied only by the algebras A_n, $A_n + k$, C_n, $C_n + k$ and their standard representations. The arguments already put forth enable us to define, relative to L_0, a univalent graduated algebra $L = V + L_0 + L_1 + \cdots$, associated with the filtration $\{\mathfrak{L}_k\}$. The coinciding of $\mathfrak{L}$ and L is proved the same way as in [2], since there are no restrictions on the characteristic, in this connection.

3. There is a basic difference between simple Lie p-algebras of Cartan type and classical algebras. Namely, in all of the former there exist proper subalgebras, defined in an absolutely invariant manner; i.e., invariant relative to all automorphisms of the algebra. We shall call such algebras invariant. In an algebra over a field of characteristic zero an invariant subalgebra must be an ideal, and therefore simple algebras cannot have proper invariant subalgebras. In exactly the same way a simple algebra of classical type over a field of characteristic $p > 0$ cannot have proper invariant subalgebras.

One can make the following conjecture: *if in a simple Lie p-algebra* $\mathfrak{L}$ $(p > 5)$ *there are no proper invariant subalgebras, then* $\mathfrak{L}$ *is of classical type.* It is clear that this conjecture follows from the classification conjecture formulated at the beginning of §2.

Let $\mathfrak{L}$ be a simple Cartan Lie p-algebra. There exists an important invariant subalgebra of $\mathfrak{L}$ which we shall denote by $\mathfrak{C}$. The subalgebra $\mathfrak{C}$ is generated by all elements $c \in \mathfrak{L}$ for which $(adc)^2 = 0$. The invariance of $\mathfrak{C}$ is clear from the definition. For the general properties of the algebra $\mathfrak{C}$, see [10]. A direct verification shows that, in algebras of Cartan type, $\mathfrak{C}$ is a proper subalgebra. Moreover, for an arbitrary algebra $\mathfrak{L}$ the subalgebra $\mathfrak{C}$ differs from zero if and only if there is a subalgebra $\mathfrak{L}_0$ in $\mathfrak{L}$ such that, in the filtration $\{\mathfrak{L}_k\}$ defined by it (cf. part 1) the term $\mathfrak{L}_2 \neq 0$.

Theorem 2. *In a Lie algebra* $\mathfrak{L}$ *of Cartan type, there exists a unique proper maximal invariant subalgebra* $\mathfrak{L}_0$, *coinciding with the normalizer* $N_{\mathfrak{L}}(\mathfrak{C})$ *of the subalgebra* $\mathfrak{C}$. *Under the realization of* $\mathfrak{L}$ *in the form of an algebra of differential operators of a truncated ring of polynomials,* $\mathfrak{L}_0$ *consists of all operators of the form* (1) *which have* $f_i \in (x_1, \cdots, x_n)$. *In the filtration* $\{\mathfrak{L}_k\}$ *determined by the subalgebra* $\mathfrak{L}_0$, *the term* $\mathfrak{L}_k$ *consists of operators of the form* (1) *which have* $f_i \in (x_1, \cdots, x_n)^{k+1}$.

In attempting to prove the classification conjecture, it would be especially useful to have the following assertion: *in any simple Lie p-algebra* $\mathfrak{L}$, *distinct from a classical algebra, there exists a unique proper maximal invariant subalgebra. It coincides with the normalizer* $N_{\mathfrak{L}}(\mathfrak{C})$ *of the subalgebra* $\mathfrak{C}$.

The validity of this conjecture would make it possible to select the most natural filtration in an arbitrary simple Lie p-algebra. As a matter of fact, the investigation is complicated by the presence of a collection of pathological filtrations, differing from the filtrations described in Theorem 2. For example, in the Hamiltonian algebra H_2 there exists a filtration $\{\mathfrak{L}_k'\}$ with last term $\mathfrak{L}'_{p-2} \neq 0$, in which the algebra $\mathfrak{L}_0'$ is a general algebra W_1, and the representation Γ is an irreducible representation of this algebra of degree $p - 1$.

V. A. Steklov Mathematical Institute
Academy of Sciences of the USSR

Received 4/MAR/66

BIBLIOGRAPHY

[1] E. Cartan, Ann. Sci. Ecole Norm. 26 (1909), 93.

[2] V. W. Guillemin and S. Sternberg, Bull. Amer. Math. Soc. 70 (1964), 16. MR **30** #533.

[3] Sh. Kobayashi and T. Nagano, J. Math. Mech. 14 (1965), no. 4, 679.

[4] J. M. Singer and S. Sternberg, J. Analyse Math. 15 (1965), 1.

[5] N. Jacobson, Duke Math. J. 10 (1943), 107. MR 4, 187.

[6] M. S. Frank, Proc. Nat. Acad. Sci. U. S. A. 40 (1954), 713. MR 16, 562.

7] A. A. Albert and M. S. Frank, Univ. e Politec. Torino Rend. Sem. Mat. 14 (1954–1955), 117. MR 18, 52.

8] R. Block, Trans. Amer. Math. Soc. 89 (1958), 421. MR 20 #6446.

9] M. S. Frank, Trans. Amer. Math. Soc. 112 (1964), 456.

10] A. I. Kostrikin, Trudy Mat. Inst. Steklov. 64 (1961), 79; English transl., Amer. Math. Soc. Transl. (2) 55 (1966), 195. MR 24 #A1933.

11] ______, Dokl. Akad. Nauk SSSR 150 (1963), 248 = Soviet Math. Dokl. 4 (1963), 637. MR 26 #6224.

12] C. W. Curtis, J. Math. Mech. 9 (1960), 307. MR 22 #1634.

Translated by:
J. J. Sember

The rank of elliptic curves
(with J. T. Tate)

Dokl. Akad. Nauk SSSR **175,** No. 4, 770–773 (1967). Zbl. **168,** 422
[Sov. Math., Dokl. **8,** No. 4, 917–920 (1967)]

Let $K = k(t)$ where k is a finite field, and let A be an elliptic curve defined over K. The object of this note is to show that the rank of the group A_K of rational points can be arbitrarily large for a suitable choice of A and for the fixed ground-field K. Analogous examples have been constructed by A. I. Lapin [1] over the field $k(t)$ when k is an algebraically closed field of characteristic zero.

We shall begin by the construction of simpler examples in which the field K is not fixed. We denote by k_f the field of p^f elements and by $\bar{k}$ the algebraic closure of any one of them. Let C be a geometrically irreducible complete curve and let A be an elliptic curve, both defined over k. Regarding A as a curve over the field $L = k(C)$ we denote by r the rank of the group A_L. Then r is known to be the rank of $\mathrm{Hom}_k(J(C), A)$, where $J(C)$ is the jacobian variety of C. Denote the numerators of the ζ-functions $Z_C(U)$ $(Z_A(U))$ of C (U) by $P_C(U)$ $(P_A(U))$. By Theorem 1 of [2] we have

$$r = 2h, \tag{1}$$

if $P_C = P_A^h \cdot G$ and P_A is irreducible over Q or $P_C = P_A^{h/2} \cdot G$ and $P_A = F^2$. Here $(P_A, G) = 1$.

We shall thus begin by the explicit computation of $P_C(U)$ for some C.

Theorem 1. *Let C be a complete nonsingular model of the curve defined over k_1 by*

$$y^e = \gamma x^f + \delta.$$

Suppose that $\gamma, \delta! \in k_1^*$, $p \nmid ef$, $2 \le e \le f$ and that $m = \text{l.c.m.}\ (e, f)$ divides $p^n + 1$ for some n. Let G be the direct product of the cyclic groups $\{\xi\}$ and $\{\eta\}$ of order e and f and for $\phi \in \mathrm{Hom}(G, \bar{k}^*)$ write $k_\phi = k_1(\phi(\xi), \phi(\eta))$; $d_\phi = [k_\phi : k_1]$.

If

$$\phi(\xi) \neq 1, \quad \phi(\eta) \neq 1, \quad \phi(\xi\eta) \neq 1, \tag{2}$$

the number d_ϕ is even, say $d_\phi = 2c_\phi$, and

$$P_C(U) = \prod (1 + p^{c_\phi} U^{d_\phi}), \tag{3}$$

the product being taken over all ϕ satisfying (2) but taking only one representative of each class of homomorphisms conjugate under the action of the galois group of the extension k_ϕ/k_1.

We shall use results of [3], a part of which we reproduce for the convenience of the reader.

We use the following notation:

ζ is a primitive mth root of unity in $\bar{k}$.

$\phi(\xi) = \zeta^{maf^{-1}}$; $\phi(\eta) = \zeta^{mbf^{-1}}$.

m_ϕ is the order of ϕ.

$a_0 = m_\phi a f^{-1}$ and $b_0 = m_\phi b f^{-1}$.

w is a generator of k^*_ϕ chosen so that

$$\zeta^{mm_\varphi^{-1}} = w^{(p^{d_\varphi}-1)m_\varphi^{-1}}.$$

χ is a character of the group k^*_ϕ for which $\chi(w) = e^{2\pi i m_\phi^{-1}}$.

In this notation the formulas on p. 493 of [3] give

$$P_C(U) = \prod L_\varphi(U),$$

where ϕ runs through the same values as in (3) and

$$L_\varphi(U) = 1 + \chi((\gamma^{-1}\delta)^{a_0}(-\delta)^{b_0})\, jU^{d_\varphi}; \tag{4}$$

$$j = \sum_{x+y+1=0} \chi(x)^{a_0}\chi(y)^{b_0}, \quad x, y \in k^*_\varphi. \tag{5}$$

Since m_ϕ is the l.c.m. of the orders of $\phi(\xi)$ and $\phi(\eta)$ in k^*_ϕ, we have $k_\phi = k_1(\zeta^{mm_\phi^{-1}})$ and d_ϕ is the least integer for which $m_\phi | (p^{d_\phi} - 1)$. Hence $p + m_\phi Z$ has order d_ϕ in $(Z/m_\phi Z)^*$ and $p^n + m_\phi Z$ has order 2 since $m_\phi | m$, $m | (p^n + 1)$ by hypothesis, and $m_\phi > 2$ by (2). It follows easily that $d_\phi = 2(n, d_p)$ and, in particular, that d_ϕ is even. We also see that

$$p^{c_\varphi} \equiv p^n \equiv -1\ (m_\varphi), \quad c_\varphi = d_\varphi/2. \tag{6}$$

We show now that $\chi = 1$ on $k^*_{c\phi}$. In fact $k^*_{c\phi} = \{w^{p^{c\phi}+1}\}$ and

$$\chi(w^{p^{c_\varphi}+1}) = e^{(2\pi i m_\varphi^{-1})(p^{c_\varphi}+1)} = 1$$

by (6).

We shall show now that

$$j = p^{c\phi}, \tag{7}$$

where j is the Jacobi sum defined by (5). The proof requires the identity

$$j = p^{-d_\varphi} g(a_0)\, g(b_0)\, g(-a_0 - b_0) \tag{8}$$

(formula (7) of [3]), where $g(r)$ is the gaussian sum

$$g(r) = \sum_{x \in k^*_\varphi} \chi(x)^r \psi(x),$$

and ψ the standard character of the additive group of k_ϕ. We shall show that

$$g(r) = \chi(c)^r p^{c\phi}, \tag{9}$$

when r is a_0, b_0 or $-a_0, -b_0$ and when $c \in k^*_{d\phi}$ has trace 0 relative to $k_{c\phi}$. Clearly (7) follows from (8) and (9). Formula (9) follows from the following lemma with $k = k_{d\phi}$, $k_0 = k_{c\phi}$, since $m \nmid r$ by (2) and so χ^r is not trivial on $k^*_{d\phi}$.

Lemma. *Let k be a quadratic extension of the finite field k_0 of q elements. Denote by θ a nontrivial character of k^* which is trivial on k^*_0 and by ψ the standard additive character of k. Then*

$$\sum_{x \in k^*} \theta(x)\psi(x) = \theta(c)\, q,$$

for some $c \in k^*$ which satisfies $\mathrm{Tr}_{k/k_0}(c) = 0$.

Let

$$k^* = \bigcup a_i k_0^*$$

be the decomposition of k^* into cosets relative to k_0^*. Since $\theta = 1$ on k_0^* we have

$$\sum_{x\in k^*} \theta(x)\psi(x) = \sum_i \theta(a_i) \sum_{y\in k_0^*} \psi(a_i y).$$

But

$$\sum_{y\in k_0^*} \psi(a_i y) = \sum_{y\in k_0} \psi(a_i y) - 1 = \begin{cases} -1, & \text{if } \psi \neq 1 \text{ on } a_i k_0, \\ q-1, & \text{if } \psi = 1 \text{ on } a_i k_0. \end{cases}$$

On noting also that $\Sigma_i \theta(a_i) = 0$ since $\theta \neq 1$ on k^*/k_0^*, we deduce that

$$\sum_{x\in k^*} \theta(x)\psi(x) = \left(\sum_i \theta(a_i)\right) q, \tag{10}$$

where the sum is taken over those i for which $\psi = 1$ on $a_i k_0$. In particular c is such an a_i and

$$\psi(cy) = \psi_1(\mathrm{Tr}_{k/k_1}(cy)),$$

by the definition of ψ, where ψ_1 is a character of k_1 and

$$\mathrm{Tr}_{k/k_1}(cy) = \mathrm{Tr}_{k_0/k_1}(\mathrm{Tr}_{k/k_0}(cy)), \quad \mathrm{Tr}_{k/k_0}(cy) = 0$$

by the choice of c.

On the other hand, the sum (10) cannot contain two distinct summands a_i and a_j, since otherwise we should have ψ trivial on $a_i k_0 + a_j k_0$, contrary to hypothesis. This completes the proof of (9). Since, further, $\chi = 1$ on k_1^* and $\gamma, \delta \in k_1^*$, it follows that $L_\phi = 1 + p^{c_\phi} U^{d_\phi}$: which proves (3).

Let us suppose, now, that $m \mid (p^n + 1)$ for an odd n. Then $d_\phi = 2(n, d_\phi)$ implies that c_ϕ is odd and that each factor $L_\phi(U)$ is divisible by $1 + pU^2$. When A is a hypersingular elliptic curve it is known that $1 + pU^2 = P_A(U)$. Hence in this case the number h in (1) is just the number of ϕ satisfying (2). In particular if $e = 2$ this is the number of divisors of the polynomial $x^f - 1$ irreducible over k_1 and distinct from $x - 1$ and also from $x + 1$ if $2 \mid f$. Thus we have the

Corollary. *Let A be a hypersingular elliptic curve defined over k_1 and let C be defined by*

$$y^2 = \gamma x^f + \delta. \tag{11}$$

Suppose that $f \mid (p^n + 1)$, $2 \nmid n$, and $p \neq 2$. Then the rank of A over $k_1(C)$ is $2h$, where h is the number of divisors of $x^f - 1$ irreducible over k_1 distinct from $x - 1$ and also from $x + 1$ if $2 \mid f$.

We note that there always exists a hypersingular* curve A defined over k_1. It is enough, by the results of [4] to show that p is the norm of an integer of a quaternion algebra $\mathfrak{A}$ over Q ramified only at p and ∞. But $\mathfrak{A} = (-p, -1)$ if $p \equiv -1 \pmod 4$ and $\mathfrak{A} = (-p, 1q)$ if $p \equiv +1 \pmod 4$ where q is a prime with $q \equiv -1 \pmod 4$ and $(q/p) = -1$. In both cases $\mathfrak{A}$ contains an element u with $u^2 + p = 0$.

We now go over to the construction of a curve defined over $k_1(x)$. On the curve C defined by (11) consider the automorphism s: $s(x) = x$, $s(y) = -y$ and on the surface $C \times A$ consider the automorphism $\sigma(c, a) = (s(c), -a)$, $c \in C$, $a \in A$. The projection $C \times A \to C$ defines a map $C \times A/\sigma \to C/s = \mathbf{P}^1$. Hence we may consider $\mathcal{K} = k_1(V)$, $V = (C \times A)/\sigma$ as a field of transcendence degree 1 over $k_1(\mathbf{P}^1) = k_1(x)$.

It is easy to see that the genus of $\mathcal{K}/k_1(x)$ is 1.

If A is given by $v^2 = u^3 + au^2 + bu + c$, the field $\mathcal{K}/k_1(x)$ is the field of rational functions on

* Translator's note. Or supersingular. Deuring [4] says: supersingulär.

the following curve:

$$v^2 = u^3 + a(\gamma x^f + \delta)u^2 + b(\gamma x^f + \delta)^2 u + c(\gamma x^f + \delta)^3. \tag{12}$$

It is easy to see that the rank of (12) over $k_1(x)$ is the rank of the group of maps $f\colon C \to A$ defined over k_1 for which

$$f(c^s) = -f(c). \tag{13}$$

For any map $f\colon C \to A$ the map $c \to f(c) + f(c^s)$ is constant, since it is the composition of $C \to \mathrm{P}^1$ and some map $\mathrm{P}^1 \to A$. From the fact that the group A_{k_1} is finite it follows that the group of maps satisfying (13) is of finite index in the group of points of A rational over $k_1(C)$. Hence the rank of (12) over $k_1(x)$ is the same as the rank of A over $k_1(C)$, which is determined by Theorem 1 and the Corollary. We have thus proved:

Theorem 2. *The rank of* (12) *over* $k_1(x)$ *when* $f \mid (p^n + 1)$, $2 \nmid n$, $p \neq 2$ *is equal to* $2h$, *where* h *is the number of divisors of* $x^f - 1$ *irreducible over* k_1 *distinct from* $x - 1$ *and also from* $x + 1$ *if* $2 \mid f$.

Let us suppose, in particular, that $f = p^n + 1$ and that n is prime. Then the number of irreducible divisors of $x^f - 1$ distinct from $x + 1$ is easily seen to be $(p^n - p)/2n + (p - 1)/2$. Thus the rank of (12) over $k_1(x)$ for $f = p^n + 1$, $p \neq 2$ with n prime is equal to $(p^n - p)/n + p - 1$.

In particular, the rank may take arbitrarily large values.

Institut des Hautes Etudes
Paris, France
Harvard University
U.S.A.
Steklov Mathematical Institute
Academy of Sciences of the USSR

Received 3/MAY/67

BIBLIOGRAPHY

[1] A. I. Lapin, *Subfields of hyperelliptic fields*, Izv. Akad. Nauk SSSR Ser. Mat. 28 (1964), 953–988. (Russian) MR 33 #7337.

[2] J. Tate, *Endomorphisms of abelian varieties over finite fields*, Inventiones Math. 2 (1966), 134–144.

[3] A. Weil, *Jacobi sums as "Grössencharaktere"*, Trans. Amer. Math. Soc. 73 (1952), 487–495. MR 14, 452.

[4] M. Deuring, *Die Typen der Multiplikatorenringe elliptischer Funktionenkörper*, Abh. Math. Sem. Hamburg 14 (1941), 197–272. MR 3, 104.

Translated by:
J. W. S. Cassels

On some infinitedimensional groups

In: Simposio Internazionale di Geometria Algebrica. Roma: Edizioni Cremonese, 208–212 (1967)
[Rend. Mat. Appl., V. Ser. **25**, 208–212 (1966)]

Several questions of algebraic geometry lead to consideration of groups that have some features of algebraic groups but at the same time are not finite dimensional algebraic varieties. In this lecture a definition of such an object is given and some of the simplest examples are investigated.

1. Infinitedimensional algebraic groups and their Lie algebras.

Let X be a set with fixed sequence of subsets X_n, each of which has a structure of a finitedimensional algebraic variety. We call X an infinitedimensional algebraic variety if the following conditions are satisfied:

1). $X = \bigcup_n X_n$

2). $X_n \subset X_{n+1}$ and is a closed algebraic subvariety.

A morphism of infinitedimensional algebraic varieties $X = \bigcup_n X_n$ and $Y = \bigcup_m Y_m$ is a map $f: X \to Y$ such, that for each n, $f(X_n) \subset Y_m$ for certain $m = m(n)$ and the restriction $f: X_n \to Y_m$ is a morphism of finitedimensional varieties.

An infinitedimensional algebraic variety $X = \bigcup X_n$ is called affine if all finitedimensional varietes X_n are affine. In this case the topological ring

$$k[X] = \varprojlim k[X_n]$$

is called the ring of regular functions on X.

An infinitedimensional algebraic group is a group in the category of infinitedimensional algebraic varieties. The group is called affine if the underlying variety is affine.

Example. The group $\mathrm{Aut}(\mathfrak{A}^n)$ of automorphisms of the n-dimensional affine space $\mathfrak{A}^n$ over the field k of characteristic 0 can be supplied by a structure of an infinitedimensional affine algebraic group.

Indeed, one can prove that a mapping

$$x_i' = f_i(x_i \dots x_n), \quad i = 1, \dots, n, \qquad f_i \in k[x_i \dots x_n] \ (*)$$

is an automorphism of $\mathfrak{A}^n$ if and only if the jacobian $\det\left(\frac{\partial f_i}{\partial x_i}\right)$ is
1 a constant. Take for X_m the subvariety in the space of all n-tuples $(f_1, \dots, f_n)$ $\deg f_i \leq m$, defined by the condition $\det\left(\frac{\partial f_i}{\partial x_j}\right) \in k$. Clearly X_m define in the group $\mathrm{Aut}(\mathfrak{A}^n)$ a structure of infinitedimensional affine algebraic group (*).

The group $\mathrm{Aut}(\mathfrak{A}^n)$ is not finitedimensional except when $n=1$. One can prove that the group is connected.

As A. Biyalinicki-Birula kindly informed me, one can attach to any group $\mathrm{Aut}(X)$ (X — *a* finitedimensional affine variety) a structure of infinitedimensional affine group, that is in some sense universal.

We define the local ring O_x of a point x of an infinitedimensional algebraic variety as a projective limit of local ringe $O_x(X_n)$ for all $X_n \ni x$. The point x is called simple if O_x is a regular topological local ring.

The union of the tangent spaces of X_n at x is called the tangent space of X at x Till the end of this paragraph we shall assume, that the ground field k has characteristic 0.

Theorem I. All points of an infinitedimensional algebraic group are simple.

The local ring of the point e (the unit element) of a group G has in a natural way the structure of a Hopf algebra. The tangent space Θ_e of G at e can be identified with the set of primitive elements of this Hopf algebra. This makes Θ_e a Lie algebra, called the Lie algebra of the group G and designated $L(G)$.

The following weak analogon of the Lie theorem is true:

(*) One can also introduce in this group the structure of an infinitedimensional algebraic group in the case of arbitrary field k by a direct and more elementary way without using the above theorem.

Theorem 2. For every closed subgroup H of an infinitedimensional algebraic group G, $L(H)$ is a Lie subalgebra of $L(G)$. If $L(H) = L(G)$ and G is connected, then $H = G$.

Now suppose that an infinitedimensional algebraic group G acts in a regular way on a finite dimensional affine variety X. This action defines an inclusion of the algebra $L(G)$ into the algebra $\mathcal{D}[X]$ of all differentiations of the algebra $k[X]$.

Theorem 3. If $G = \mathrm{Aut}(\mathfrak{A}^n)$, the image of $L(G)$ in $\mathcal{D}[\mathfrak{A}^n]$ consists of all operators of the form

$$\mathcal{D} = \sum_1^n F_i \frac{\partial}{\partial x_i}, \quad F_i \in k[x_1 \dots x_n]$$

with the property

$$\sum_1^n \frac{\partial F_i}{\partial x_i} \in k.$$

Applying theorems 2 and 3 one can obtain some information about the structure of the group $G = \mathrm{Aut}(\mathfrak{A}^n)$. Let B be the group of all automorphisms of $\mathfrak{A}^n$ of the form

$$x_i' = a_i x_i + f_i(x_1, \dots, x_{i-1}), \; f_i \in k[x_1, \dots, x_{i-1}], \quad a_i \neq 0$$

and $\mathcal{L}$-the group of all linear (not necessary homogeneous) automorphisms.

Theorem 4. The group G is generated by subgroups B and $\mathcal{L}$ as an algebraic group.

Theorem 5. The group G has a normal subgroup G_0, consisting of such automorphisms (*) that $\det\left(\frac{\partial f_i}{\partial x_j}\right) = 1$. The group G_0
2 is simple as an infinitedimensional algebraic group.

2. The group of automorphisms of an affine plane.

The structure of the group $\mathrm{Aut}(\mathfrak{A}^n)$ can be investigated in more details in the first nontrivial case $n = 2$. We suppose that the groundfield is an algebraically closed field of arbitrary characteristic.

Theorem 6. The group $G = \mathrm{Aut}(\mathfrak{A}^2)$ is generated by subgroups B and $\mathcal{L}$ as an abstract group.

The proof depends on the possibility of decomposition of any birational isomorphism of projective surfaces as a product of dilatations [1].

In the case of a ground field of characteristic 0 theorem 6 was proved by Engel [2].

Theorem 7. Let $B(\mathcal{L}) = \mathcal{L} \cap B$. The group G is isomorphic to the free product of groups B and $\mathcal{L}$, with amalgamated subgroup $B(\mathcal{L})$.

Theorem 8. Every algebraic finitedimensional subgroup of G is conjugated either to a subgroup of B or to a subgroup of $\mathcal{L}$.

Consequence 1. $\mathcal{L}$ is maximal algebraic finitedimensional subgroup of G. Every two such subgroups are conjugated.

Consequence 2. Each action of the multiplicative group on a plane can be reduced to a diagonal form.

This was earlier proved in another way by Gutwirth [3].

Consequence 3. Each action of the additive group on a plane over the field of characteristic can be reduced to a form:

$$x' = x$$

$$y' = y + tf(x), \quad f(x) \in k[x].$$

This was proved in another way by Ebey [4].

Let k be a not necessary algebraically closed field, K — a finite Galois extension of k with Galois group $\mathfrak{g}$ and G_K — the group of points of G with coordinates in K. The same reassoning that is used in the proof of theorem 8 shows, that from theorem 7 one can deduce: $H^1(\mathfrak{g}, G_K) = (1)$. This can be interpreted in the following way. A manifold X is called a from of Y (X and Y are defined over k) if X and Y are isomorphic over some separable extension K of k.

Theorem 9. Affine plane has no forms except itself.

Another reformulation: if A is such an algebra of the finite type over k that $A \otimes_k K \simeq K[X, Y]$, then $A \simeq k[X, Y]$.

3. Some questions.

The above results rise several interesting questions.

Can one introduce a universal structure of an infinitedimensional group in the group of all automorphisms (resp. all birational automorphisms) of arbitrary algebraic variety ?

Can one generalize theorems 8 and 9 to the case of an arbitrary $n \geq 2$?

From the theorem 8 it follows that the group G_0. mentioned in theorem 5, is for $n = 2$ a group « of rank I ». What are all the simple groups of rank I, acting on finitedimensional affine varieties ?

There are obvious generalizations of these question. It is clearly connected with Cartan's theory of infinite simple pseudo groups.

What are homogeneous affine surfaces on which infinitedimen-
3 sional algebraic groups are acting primitively ?

REFERENCES

[1] I. R. Shafarevich « *Algebraic surfaces* » (in Russian). Trudy Matematiceskogo Instituta im. Steklova, vol. 75, Moscowa 1965.

[2] W. Engel « *Ganze Cremona-Transformationen von Primzahlgrad in der Ebene* ». Math. Ann. 136 (1958), 319-325.

[3] A. Gutwirth « *The action of an algebraic torus on an affine plane* ». Trans. Amer. Math. Soc. 105 (1962) N. 3, 407-414.

[4] W. Ebey « *The operation of the fundamental domain on a plane* ». Proc. Amer. Math. Soc. 13 (1962) N. 5, 722-726.

Irreducible representations of a simple three-dimensional Lie algebra over a field of finite characteristic

(with A. N. Rudakov)

Mat. Zametki **2**, No. 5, 439 – 454 (1967). Zbl. **184**, 60
[Math. Notes **2**, Nos. 1, 2, 760 – 767 (1968)]

The irreducible representations of a simple three-dimensional Lie algebra over a field of finite characteristic are enumerated. The dimensionalities of all representations do not exceed the characteristics p of the base field. For any dimensionality < p there exists a unique representation of this dimensionality. The representations of dimensionality p form a three-dimensional algebraic set. Six literature references are cited.

The theory of irreducible representations of simple Lie algebras over an algebraically closed field f characteristic 0 was constructed by E. Cartan more than fifty years ago and became one of the most opular branches of algebra. Analogous problems over fields of finite characteristic have, almost com-letely, not been investigated. An exception is the p-representations of simple Lie algebras [1] which, as s evident in particular from the results of this work, comprise a very small part of the totality of all rep-esentations. Apparently, up until now all the irreducible representations had not been determined for a ingle simple Lie algebra over a field of finite characteristic.

In this note we give a description of all irreducible representations of a simple algebra of rank 3 (an $_1$ algebra in Cartan's notation), defined over an algebraically closed field k of characteristic $p \neq 2$. It is ot difficult to obtain formulas giving the irreducible representations by using the connection between this roblem and the properties of a universal covering algebra $\mathfrak{A}$ of the algebra A_1 and of the quotient field of he algebra $\mathfrak{A}$ [2]. In this connection it turns out that the irreducible representations may be of degree $\leq$ p, for any degree < p there exists a unique irreducible representation, and the formulas for an arbitrary -dimensional representation depend on the coordinates of a point x of some three-dimensional algebraic nanifold V.

In view of this result, it is desirable to assign a precise definition to these, which means that the set f representations is parametrized by an algebraic manifold, and afterwards to determine this manifold. 'ollowing classical patterns [3], we introduce the concept of a universal family of representations of the Lie lgebra L as the algebraic family of representations such that any other family is obtained from it as the in-erse image associated with some uniquely defined mapping of its basis into the basis of the universal amily. As often happens, in our case the universal family does not exist; however, one can prove its xistence if one provides the objects being classified with an additional "rigidity." In our case the rigidity s determined by the assignment in representation space of some weight vector (this replaces the choice of ighest weight in the classical theory, since in our case all weights are equivalent). A proof of the existence nd an explicit description of the universal family of rigid representations are the fundamental results of his work. In this connection, the basis of the universal family has the form X – S where X is a three-dim-nsional affine manifold, and S is a finite set of its singular points.

Later we investigate the degeneracy of the irreducible representations, i.e., those sets of representa-ions in which the common fiber is irreducible, but certain fibers are reducible. Under one additional re-triction on the type of degeneracy, it turns out that all families, admitting degeneracies of the type under con-ideration, are also described by a universal family whose basis is obtained from X by resolution of the ingular points $s \in S$.

In accordance with general principles holding in this field, finite-dimensional representations in inite characteristic are analogous in characteristic 0 to all representations, both finite dimensional and nfinite dimensional. Since the situation in finite characteristic is even easier (the representations are

---

M. V. Lomonosov Moscow State University. V. A. Steklov Mathematical Institute, Academy of ciences of the USSR. Translated from Matematicheskie Zametki, Vol. 2, No. 5, pp. 439-454, November, 967. Original article submitted June 15, 1967.

finite dimensional, in view of which analytic difficulties do not arise), then further investigation of this question may perhaps help analysis of the theory of finite-dimensional simple Lie algebras in characteristic 0.

1. Center of the Universal Covering Algebra

Let us consider the algebra A_1 over an algebraically closed field k of characteristic $p > 0$, $p \neq 2$, and let us denote by e, f, and h the elements of its standard basis.

Thus,

$$[e, h] = 2e, \quad [f, h] = -2f, \quad [e, f] = h. \tag{1}$$

The elements e, f, and h generate a universal covering algebra $\mathfrak{A}$ of the algebra A_1. We denote the center of the algebra $\mathfrak{A}$ by $\mathfrak{Z}$.

PROPOSITION 1. The algebra $\mathfrak{Z}$ is generated by the elements $x = e^p$, $y = f^p$, $z = h^p - h$, $t = (h+1)^2 - 4ef$. These elements are connected by the relation

$$z^2 - \prod_{k=0}^{p-1} (t - k^2) = 4xy, \tag{2}$$

which defines the algebra $\mathfrak{Z}$.

Let us verify that the elements x, y, z, $t \in \mathfrak{Z}$. For x, y, and z this follows from the existence of a p-algebra structure in the A_1 algebra and from the relations $e^{[p]} = 0$, $f^{[p]} = 0$, $h^{[p]} = h$, which one can easily derive, for example, by using a two-dimensional representation of the A_1 algebra. In order to establish that $t \in \mathfrak{Z}$, we regard e, f, h, and t as elements of the quotient field D of the algebra $\mathfrak{A}$ (compare with [4], Chapter V, § 3). From formulas (1) it follows that in D

$$e^{-1}he = h - 2, \quad f^{-1}hf = h + 2. \tag{3}$$

From here it follows that $[ef, h] = 0$ and hence, $[t, h] = 0$. Similarly, from (1) it follows that $e^{-1}(ef)e = ef - h$ and $e^{-1}te = (h-1)^2 - 4(ef-h) = t$, that is, $[e, t] = 0$. Also it can be shown that $[f, t] = 0$. From here it now follows that $t \in \mathfrak{Z}$.

Let us denote the quotient field of the rings $\mathfrak{Z}$ by K, and we assume $k(x, y, z) = K^* \subset K$ and $k[x, y, z] = \mathfrak{Z}^* \subset \mathfrak{Z}$. From the results of article [2] it follows that $\dim_{K^*} D = p^3$, and since $\dim_K D$ is a perfect square, then $[K : K^*] = p$, $\dim_K D = p^2$. It is easy to verify that $t \notin K^*$. Actually, from the equality $t = P(x, y, z) \cdot Q(x, y, z)^{-1}$ it follows that $Qt = P$, which leads to a contradiction if one compares terms of highest degree with respect to standard graduation of the algebra $\mathfrak{A}$. Thus, $K = K^*(t) = k(x, y, z, t)$.

In order to derive relations (2), we apply the principal norm N in the algebra D/K (see [5], p. 214). From the fact that $4ef = (h+1)^2 - t$, we obtain

$$N(4ef) = N((h+1)^2 - t), \text{ or } 4xy = \prod_{k=0}^{p-1} ((h+k)^2 - t), \tag{4}$$

since the element h generates a normal field K(h) with automorphisms $\sigma_k h = h + k$, $k = 0, \ldots, p-1$. In order to calculate the polynomial on the right hand side of (4), we introduce a new variable s, having set $t = s^2$. Then

$$\prod ((h+k)^2 - t) = \prod ((h+k)^2 - s^2) = \prod (h+k-s) \prod (h+k+s)$$
$$= ((h+s)^p - (h+s))((h-s)^p - (h-s)) = (h^p - h + s^p - s)(h^p - h - (s^p - s))$$
$$= (h^p - h)^2 - (s^p - s)^2 = z^2 - s^2(s^{p-1} - 1)^2 = z^2 - t(t^{(p-1)/2} - 1)^2.$$

Thus

$$N((h+1)^2 - t) = z^2 - t(t^{(p-1)/2} - 1)^2, \tag{5}$$

from which relation (2) follows.

Now we note that $k[x, y, z, t] = \mathfrak{Z}^*[t] \subset \mathfrak{Z}$ and both rings have one and the same quotient field of fractions K. In addition, $\mathfrak{Z}$ is a modulus of finite type over $\mathfrak{Z}^*[t]$ since $\mathfrak{Z} \subset \mathfrak{A}$ and $\mathfrak{A}$ are of finite type over $\mathfrak{Z}$ and the more so over $\mathfrak{Z}^*[t]$. As one can easily verify, $\mathfrak{Z}^*[t]$ is completely closed, from which it follows that $\mathfrak{Z} = \mathfrak{Z}^*[t]$. Proposition 1 is proved.

. Irreducible Representations

Knowing the structure of the algebra $\mathfrak{Z}$, it is not difficult to give a "set theoretical" enumeration of rreducible representations of the algebra A_1. In fact, every irreducible representation φ defines a maxi-al ideal $\mathfrak{m}_\varphi \in \operatorname{Spec} \mathfrak{Z}$. All representations with $\mathfrak{m}_\varphi = \mathfrak{m}$: these are irreducible representations of the nite-dimensional algebra $\mathfrak{A} \otimes \mathfrak{z}^k = \mathfrak{A}/\mathfrak{m}$. The degree of all irreducible representations does not exceed , and the maximal ideal $\mathfrak{m} \in \operatorname{Spec} \mathfrak{Z}$ then and only then corresponds to an irreducible representation of egree p (automatically unique), when $\mathfrak{A}/\mathfrak{m} \simeq k_p$, the algebra of matrices of order p over k ([2]).

According to Proposition 1, the ideal $\mathfrak{m} \in \operatorname{Spec} \mathfrak{Z}$ is given by the three-dimensional affine anifold Z which is defined by Eq. (2). Let us denote the homomorphism $\mathfrak{A} \to \mathfrak{A}/\mathfrak{m}$ by ρ. Let us assume aat the point $P = (\xi, \eta, \zeta, \tau)$ corresponding to $\mathfrak{m}$ is such that $\xi \neq 0$. Then in $\mathfrak{A}/\mathfrak{m}$ the element $\rho(e)$ has an iverse since $\rho(e)^p = \xi \neq 0$. It follows from formula (3) that the elements $\rho(e)$ and $\rho(h)$ generate a cross roduct which is isomorphic to k_p. From the fact that the element $\rho(e)$ has an inverse and the relation $ef = (h+1)^2 - t$ it follows that $\rho(f)$ can be expressed in terms of $\rho(e)$ and $\rho(h)$, and therefore $\mathfrak{A}/\mathfrak{m} \simeq k_p$, o that the point P corresponds exactly to one irreducible p-dimensional representation. One can carry out ae same discussion also in the case when $\eta \neq 0$.

Now let us assume that $\xi = \eta = 0$. Let us set $f^{p-1} + e = g$, and let us evaluate the element $\rho(g)^p$. Let s start with transformations in the algebra D. In it

$$g = e + yf^{-1} = (ef + y)f^{-1} = \left(\frac{(h+1)^2 - t}{4} + y\right)f^{-1},$$

$$g^p = \left(\left(\frac{(h+1)^2 - t}{4} + y\right)f^{-1}\right)^p = N\left(\frac{(h+1)^2 - t}{4} + y\right)y^{-1} = N\left(\frac{h^2 - t}{4} + y\right)y^{-1}.$$

Iaving substituted $t - 4y$ in place of t in relation (5), which is valid for any element $t \in \mathfrak{Z}$, we obtain the re-ult that $g^p = \frac{1}{4}(z^2 - (t-4y)((t-4y)^{(p-1)/2} - 1)^2)y^{-1}$. In view of relation (2), in the ring $\mathfrak{Z}$, $g^p \equiv x + 1 - t^{(p-1)/2}$ mod y), from which it follows that $\rho(x) = \rho(y) = 0$

$$\rho(g)^p = 1 - \rho(t)^{(p-1)/2}. \qquad (6)$$

First let us consider the case $\xi = \eta = 0$, $\tau^{(p-1)/2} \neq 1$, that is, $\tau \neq k^2$, $k \neq 0$. Then in view of (6), $(g)^p = 1 - \tau^{(p-1)/2} \neq 0$, so that the element $\rho(g)$ has an inverse in the algebra $\rho(\mathfrak{A})$. The previous discus-ion shows that $\rho(g)$ and $\rho(h)$ generate the algebra k_p. Since $4\rho(g)\,\rho(f) = \rho((h+1)^2 - t)$, then $\rho(f)$ and $\rho(e)$ an be expressed in terms of $\rho(g)$ and $\rho(h)$, and hence $\rho(\mathfrak{A}) = k_p$. Thus, even in this case the point P cor-esponds to a unique p-dimensional representation.

The case $\xi = \eta = \tau^{(p-1)/2} - 1 = 0$ remains to be considered. In view of (2), then $\zeta = 0$. Therefore, he irreducible representation corresponding to such a point is a p-representation. The classification of -representations of any classical, simple algebra is contained in article [1]. In the case of an A_1 algebra t leads to the result that there are p irreducible p-representations, where they are obtained by reductions vith respect to modulus p of irreducible representations of degree k, $1 \leq k \leq p$ of the A_1 algebra in charac-eristic 0. A simple test shows that the p-representation of degree k corresponds to the value $\tau = k^2$, and, herefore, any point $(0, 0, 0, k^2)$, $k \neq 0$, corresponds to two irreducible representations of degree k and p-k.

We obtain the following results.

PROPOSITION 2. The point $P = (\xi, \eta, \zeta, \tau)$ of the manifold Z, defined by Eq. (2), corresponds to a inique irreducible p-dimensional representation provided $P \neq (0, 0, 0, k^2)$, $k \neq 0$, and the points $P = (0, 0,$ $, k^2)$, $k \neq 0$, correspond to two irreducible representations of degree k and p-k.

We note that one can rewrite Eq. (2) in the form $z^2 - t\prod_{k=1}^{(p-1)/2}(t - k^2)^2 = 4xy$, from which one can asily verify that $(p-1)/2$ of the points $(0, 0, 0, k^2)$, $k \neq 0$, are singular points of the manifold Z. We denote he set of singular points of the manifold Z by S.

. Families of Representations

Let Y be a scheme over a field k. A family of representations of a k-algebra of Lie L with basis Y s defined as a vector bundle of E over Y and a representation L in the algebra End E, φ: $L \to \operatorname{End} E$. A epresentation φ_y: $L \to \operatorname{End} E_y$ is then defined for any $y \in Y$. The dimensionality of the family is called the dimensionality of the fibers of the fibering E. A family is called irreducible if the representation φ_y is irreducible for any $y \in Y$.

If f: $X' \to X$ is a morphism and $E \to X$ is a family of representations, then the inverse image $f^* E \to X'$ of the fibering E naturally furnishes the structure of the family of representations. We denote the totality of families of irreducible representations of dimension n with basis X (to within an isomorphism) by M_X.

The family of representations $U \to B$ is called universal if the mapping consisting of any morphism f: $X \to B$ of the family f^* U over X defines a one-to-one correspondence α: Hom (X, B)$\to M_X$. In particular, since one can regard a representation as a fibering with a basis which is a point, then each representation is encountered exactly once as a fiber of the universal family.

The best solution of the problem of description of irreducible representations of a given dimensionality n of a Lie algebra L would be construction of the family of its representations, universal in the set of all irreducible representations of dimension n. We shall show, however, that for $L = A_1$ and $n = p$ a universal family does not exist.

Let us consider an arbitrary family of representations $E \to Y$. The mapping φ: $L \to$ End E defines the homomorphism ψ: $\mathfrak{A} \to$ End E. In this connection, for any $y \in Y$ the homomorphism ψ_y: $\mathfrak{A} \to$ End E_y is, in view of Burnside's theorem, an epimorphism, and hence $\psi_y(\mathfrak{Z})$ consists of homothetic transformations of the space E_y. Therefore, there exists an homomorphism λ: $\mathfrak{Z} \to \Gamma(Y, \mathcal{O})$, such that $\psi_y(u)m = \lambda(u)m$ for $u \in \mathfrak{Z}$, $m \in E_y$. Since such an homomorphism is also defined for any open set of the manifold Y, then Y is a scheme over $\mathfrak{Z}$ and defines the morphism φ_Y: $Y \to \mathrm{Spec}\ \mathfrak{Z} = Z$.

For any point $y \in Y$, the point $\varphi_Y(y) \in Z$ corresponds, in the sense of Proposition 2, to the representation φ_y: $L \to E_y$. Since it is assumed that φ_y is irreducible, then $\varphi_Y(Y) \subset Z - S$ (we recall that $n = p$).

Let us assume that $E \to K$ is a universal family. Then the morphism φ_X: $X \to Z - S$ is one-to-one (on a set of closed points). This means that the manifold X is irreducible, the field k(X) contains the field k(Z) = K and the extension k(X)/K is purely nonseparable. On the other hand, a family $\overline{E} \to \overline{Z}$ will be constructed below, such that the morphism $\varphi_{\overline{Z}}$: $\overline{Z} \to Z$ is not ramified, from which it follows that $k(\overline{Z})/K$ is inseparable. The presence of a morphism φ: $\overline{Z} \to X$ such that $\varphi_{\overline{Z}} = \varphi_X \cdot \varphi$ follows from the premise that X is the basis of a universal set. We obtain the result that $k(\overline{Z}) \supset k(X) \supset K$, where $k(\overline{Z})/K$ is separable, and k(X)/K is purely nonseparable. This is possible only in the case k(X) = K.

Let us consider the homomorphism ψ_X: $\mathfrak{A} \to E_X$, corresponding to a generic point x of the base X of the family $E \to X$. Since $E_X = K^p$, then we obtain the embedding $D = \mathfrak{A} \otimes K \to \mathrm{End}\ K^p$ (in view of the fact that D is an algebra with division). However, D is an algebra with division over K, $\dim_K D = p^2$, and therefore D does not have representations by matrices of order p over K.

Now we change the concepts of representation and family of representations like so in order that for the new concepts it would be possible to prove the existence of a universal family.

A representation φ: $L \to$ End E and the choice in E of a weight vector a, i.e., an eigenvector for the algebra $\varphi(H)$, where H is a subalgebra of the Cartan algebra L, is called a rigid representation of the algebra L. A morphism of rigid representations is defined in a natural way. It is obvious that change of the vector a does not proportionally change the structure of a rigid representation.

We note that in all of this theory, we fix once and forever a certain Cartan subalgebra $H \subset L$.

In the case when $L = A_1$, $H = \{\alpha h\}$, and if a is a weight vector of the p-dimensional representation φ, then $\varphi(h)a = \lambda a$, $\lambda \in k$, from which it follows that $\varphi(h)\varphi(e)^i a = (\lambda - 2i)\varphi(e)^i a, \varphi(h)\varphi(f)^j a = (\lambda + 2j)\varphi(f)^j a$. Using these results, one can show that an irreducible p-dimensional representation φ possesses p different structures of a rigid representation.

A family of rigid representations is the name for a family $E \to X$ in which a section s: $X \to E$ is fixed such that for any $x \in X$ the vector $s(r) \neq 0$ and is a weight vector of the representation φ_X: $L \to \mathrm{End}\ E_X$ A morphism of families of rigid representations and the universal family are defined in a natural way.

We turn to the construction of a family of rigid p-dimensional representations, for which its universality will be proved. We denote by $\overline{\mathfrak{Z}}$ the subalgebra $\mathfrak{Z}$ [h] of the algebra $\mathfrak{A}$, and by $\overline{S}$ we denote the inverse image of the set S under the natural mapping $\overline{Z} = \mathrm{Spec}\ \overline{\mathfrak{Z}} \to Z$. It is obvious that $\overline{S}$ consists of $p(p-1)/2$ points. We denote the algebraic manifold $Z - S$ by B. It will be the base of the universal family. It is obvious at the morphism $\bar{Z} \to Z$ is not ramified, and therefore S coincides with the set of singular points of the ld $\bar{Z}$.

We start with a description of the ring $\bar{\mathfrak{Z}}$. By definition

$$\bar{\mathfrak{Z}} = \mathfrak{Z}[h] = k[x, y, z, t, h].$$

Since $t = (h+1)^2 - 4ef$, then $\bar{\mathfrak{Z}} = k[x, y, h, u]$, where $u = ef$. We determine the relation between x, y, and u, defining the algebra $\bar{\mathfrak{Z}}$. It is obtained from (2) by the substitution

$$z = h^p - h, \; t = (h+1)^2 - 4u$$

nd by reduction of similar terms. However, it is easier to determine this relation, having evaluated N(u) y two methods. On the one hand,

$$N(u) = N(ef) = xy.$$

But, on the other hand,

$$N(u) = \prod_{k=0}^{p-1} e^{-k}ue^{k} = \prod_{k=0}^{p-1} L_k, \quad L_k = e^{-k}ue^{k}.$$

From relations (3) we obtain

$$L_k = e^{-k}ue^{k} = e^{-(k-1)}fee^{k-1} = e^{-(k-1)}(u-h)e^{k-1} = L_{k-1} - h + 2(k-1),$$

rom where

$$L_k = u - kh + k(k-1).$$

Finally, we discover the relation

$$xy = \prod_{k=0}^{p-1} L_k(u, h) = \prod_{k=0}^{p-1} (u - kh + k(k-1)). \tag{7}$$

t is obvious that this agrees with the relation which is obtained from (2) by the substitution indicated bove. Thus, the algebra $\bar{\mathfrak{Z}}$ is determined by relation (7). It is easy to verify that the set $\bar{S}$ consists of oints for which $x = 0$, $y = 0$, $L_i(u, h) = 0$, $L_j(u, h) = 0$ for certain pairs of indices $i \neq j$.

We consider $\mathfrak{A}$ as a right $\bar{\mathfrak{Z}}$-module. It defines on the manifold $\bar{Z}$ a coherent algebraic bundle $\mathfrak{F}$ see [6], Sec. 41). We denote the restriction of this bundle on the open set B by $\mathfrak{F}$.

PROPOSITION 3. The sheaf $\mathfrak{F}$ is isomorphic to the direct sum of invertible sheaves $\mathfrak{F}_{D_i}$, $i = 0, \ldots,$ -1, corresponding to divisors D_i: $\mathfrak{F} = \mathfrak{F}_{D_0} \oplus \cdots \oplus \mathfrak{F}_{D_{p-1}}$. The divisors D_i are determined by the conitions $L_1 \ldots L_i = 0$, $x = 0$, $i \geq 1$; $D_0 = 0$.

Proof. The module $\mathfrak{A}$ over $\bar{\mathfrak{Z}}$ is obviously generated by p^2 elements $e^i f^j$, $0 \leq i, j < p$, but in view of he relation $ef = u \in \bar{\mathfrak{Z}}$ it is also generated by $2p-1$ elements e^i, f^j, $0 \leq i, j < p$. The following relations xist between these elements:

$$e^k y = f^{p-k} P_k, \quad f^{p-k} x = e^k Q_k, \quad P_k = L_1 \ldots L_k, \; Q_k = L_{k+1} \ldots L_p, \tag{8}$$

which one can easily prove by induction, using (6) and the fact that

$$x = e^p, \quad y = f^p.$$

Let us denote by $\bar{K}$ the quotient field of ring $\bar{\mathfrak{Z}}$ and let us assume $M_i = \mathfrak{A} \cap e^i\bar{K}$, $i = 0, 1, \ldots, p-1$. From the preceding remarks, it follows that $\mathfrak{A} = \oplus M_i$ (as a $\bar{\mathfrak{Z}}$-module) and M_i is generated by the elenents e^i and f^{p-i}. One can easily verify, considering each set separately

$$P_k \neq 0, \; Q_k \neq 0,$$
$$x \neq 0, \; y \neq 0,$$

hat the sheaf corresponding to the module M_k, is determined by relations (8) over the set B. On the other and, these same relations also determine the sheaf corresponding to the ideal (P_k, x).

Since the sheaf $\mathfrak{F}$ is locally free, then it determines the vector bundle $E \to B$. Considering $\mathfrak{A}$ as a left L-module, we define in this bundle the structure of the set of representations.

We calculate this structure. For this purpose, we note that the endomorphism α of the sheaf $\oplus_i \mathfrak{F}_{D_i}$ is given by matrices α_{ij}, where $\alpha_{ij} \in \mathrm{Hom}(\mathfrak{F}_{D_i}, \mathfrak{F}_{D_j})$, $0 \le i, j < p$. Since $\mathrm{Hom}(\mathfrak{F}_{D_i}, \mathfrak{F}_{D_j}) \simeq \Gamma(\mathfrak{F}_{D_j - D_i})$, then every endomorphism of the bundle E is determined by a system of elements $\alpha_{ij} \in \overline{\mathfrak{Z}}$, $\alpha_{ij} \equiv 0\ (D_j - D_i)$ on B.

PROPOSITION 4. The set of representations E → B is determined by the homomorphism φ: L → End E, where

$$\varphi(e)_{i,\,i+1} = 1, \quad i = 1, \ldots, p-1, \quad \varphi(e)_{p-1,\,p} = x,$$
$$\varphi(f)_{i,\,i-1} = L_i, \quad i = 0, \ldots, p-1, \quad \varphi(h)_{i,\,i} = h + 2i, \quad i = 1, \ldots, p, \tag{9}$$

and all components not written out explicitly are equal to 0.

Proof. If an element $a \in L$ is such that the endomorphism $\varphi(a)$ transforms a subbundle F_{D_i} in F_{D_j}, then an element $\alpha_{ij} = \varphi(a)_{ij} \in \overline{K}$ is determined from the conditions $am = m'\alpha_{ij}$ for all $m \in M_i$, $m' \in M_j$. In order to · etermine the elements α_{ij}, it is sufficient to set $m = e^i$. The first and third groups of formulas (9) then become obvious. The second group of formulas follows from the identities $fe^i = e^{i-1}L_i$, which are simple consequences of (8).

We introduce in E a structure of the set of rigid representations with the aid of a section corresponding to the element 1 of the algebra $\mathfrak{A}$.

THEOREM 1. The constructed set of representations E is universal for all sets of rigid irreducible p-dimensional representations of the algebra A_1.

Let X be an arbitrary k-scheme. In order to prove the theorem it is sufficient to construct the mapping β: $M_X \to \mathrm{Hom}(X, B)$, which is the inverse of the mapping α. Let ξ: E' → X be a set of rigid representations with rigidity defined by the section s. Then a homomorphism $\mathfrak{A} \to$ End E' is defined, where for an element $u \in \overline{\mathfrak{Z}}$ the sections $u \cdot s$ and u are linearly independent at any point—for $u \in \mathfrak{Z}$ this follows, as already indicated, from the irreducibility of the representation, and for u = h it follows from the definition of rigidity. Therefore, $us = \lambda_u s$, $\lambda_u \in \Gamma(X, \mathcal{O})$. We obtain a homomorphism $\overline{\mathfrak{Z}}$ in $\Gamma(X, \mathcal{O})$ and by the same token a morphism f: X → $\overline{Z}$. Let us consider a coherent sheaf $\overline{\mathfrak{F}}$, which $\mathfrak{A}$ determines over $\overline{Z}$. We construct a homomorphism of the sheaf $f^*\overline{\mathfrak{F}}$ over X into the sheaf $\mathcal{E}'$, corresponding to the bundle E'. First let us construct this homomorphism over affine open sets in X. Let X_0 be an affine open set in X, X_0 = Spec A, M' is the module over A, corresponding to E', and $\sigma \in M'$ corresponds to s. From the fact that E' is a set of representations, it follows that M' is an $\mathfrak{A}$-module and the operation commutes with the operation A.

The morphism f determines the homomorphism φ: $\overline{\mathfrak{Z}} \to A$. The sheaf $\overline{\mathfrak{F}}$ over $\overline{Z}$ corresponds to $\mathfrak{A}$, as a $\mathfrak{Z}$-module, but $f^*\overline{\mathfrak{F}}$ is the module $M'' = \mathfrak{A} \otimes_{\overline{\mathfrak{Z}}} A$. We construct the mapping h: $\mathfrak{A} \to M'$, having set $h(\nu) = \nu\sigma$. It is obvious that h is a dihomomorphism of the module, compatible with φ. Therefore, it determines the homomorphism $\overline{h}$: M" → M', which generates the homomorphism h^*: $f^*\overline{\mathfrak{F}} \to \mathcal{E}'$ over X_0. From the construction it is clear that a unique homomorphism over X is thus obtained.

We denote by M_x the maximal ideal of the local ring $\mathcal{O}_x$ of points x. The homomorphism h^* determines homomorphisms of the vector spaces over k:

$$h^*_x : \mathfrak{F}_{f(x)}/\mathfrak{m}_{f(x)}\mathfrak{F}_{f(x)} \to E'_x,$$

which are, as one can easily see, nonzero homomorphisms of L-module.

In view of the irreducibility of E^{ξ}_x, from here it follows that all h^*_x are epimorphisms.

If $f(x) \in \overline{S}$, then according to Conjecture 2, all composite factors of the L-module $\mathfrak{F}_{f(x)} / \mathfrak{m}_{f(x)}\mathfrak{F}_{f(x)}$ have dimensionality less than p, from which it follows that $f(X) \subset B$. We shall then consider f as a morphism of X into B. If $f(x) \overline{\in} \overline{S}$, then the module $\mathfrak{F}_{f(x)} / \mathfrak{m}_{f(x)}\mathfrak{F}_{f(x)}$ is isomorphic to E^{ξ}_x, and therefore h^*_x must be isomorphic.

We may now determine the mapping β, having set $\beta(\xi) = f$. It remains to be shown that $\beta\alpha = 1$ and $\alpha\beta = 1$. The first assertion, as one can easily see, is a tautology. The second means that if ξ: E' → X and f: X → B are corresponding morphisms, then f^*E and E' are isomorphic. Above we constructed a homorphism h^* of the bundle f^* E into E' and proved that on any fibre h^* it induces an isomorphism. From here, it follows that h is an isomorphism, and this also proves the theorem.

The sheaf $\mathfrak{F}$ is not locally free at points $\bar{s} \in S$ of the manifold $\bar{Z}$. We construct a manifold $\mathfrak{B}$ bi-ationally isomorphic to $\bar{Z}$, on which $\mathfrak{F}$ is locally free. Let us take a sufficiently fine covering U_α of the ıanifold $\bar{Z}$, which in each U_α contains not more than one point of the set $\bar{S}$. We denote by D(F) the open et consisting of the points where $F(x) \neq 0$. Let

$$F_{ij} = \Pi_{k \neq i,\, k \neq j} L_k, \quad i < j.$$

hen the equation of the manifold $\bar{Z}$ takes the form

$$L_i L_j F_{ij} = xy.$$

The point s_{ij}: $L_i = 0$, $L_j = 0$, $x = 0$, $y = 0$ is singular on this manifold. We resolve it, having carried ut a generalized σ-process in the embedding space (h, u, x, y) along the subspaces $L_i = 0$, $y = 0$. This rocess induces an isomorphism on B. Let us consider the transformation which the σ-process induces on ıe U_α, which contains the point s_{ij}, $i < j$, but which induces the identity in all remaining U_α. Combining ıem, we obtain the manifold $\mathfrak{B}$ and the morphism π: $\mathfrak{B} \to \bar{Z}$, which is an isomorphism on $\mathfrak{B} - \pi^{-1}(\bar{S})$ nd resolves each singular points s_{ij} into a projective line (the latter is obtained by a simple calculation).

The coherent sheaf $\mathfrak{F}$ was defined on the manifold $\bar{Z}$. We assume $\mathfrak{G} = \pi^* \mathfrak{F}$ and we prove that $\mathfrak{G}$ ı locally free. By construction in $\pi^{-1}(U_\alpha)$, $U_\alpha \ni s_{ij}$, $\mathfrak{B}$ is given by the equations

$$L_i L_j F_{ij} = xy, \quad L_i t_0 = y t_1 \tag{10}$$

ı $A^4 \times P^1$, where $(t_0 : t_1)$ are coordinates on P^1. Let, for example, $t_0 \neq 0$. This affine open set V is the spec-rum of some ring A, where $t_1 t_0^{-1} = L_i y^{-1} \in A$. Let the restriction $\mathfrak{G}$ on V correspond to the module $M = \mathfrak{A} \otimes_{\mathfrak{Z}} A$. The elements $f^k \otimes 1$ and $e^k \otimes 1$ generate M over A. Formulas (8) give:

$$e^n y = f^{p-n}\delta_n, \quad \text{if} \quad 0 < n < i, \quad e^n y = f^{p-n} L_i \delta_n, \quad \text{if} \quad i \leqslant n < j,$$
$$f^{p-n} x = e^n \varepsilon_n,$$

$j \leq n < p$, where δ_n, $\varepsilon_n \in \mathfrak{Z}$ are invertible in A. From here it is immediately evident that $e^k \otimes 1$, $\leqslant k < i$, $f^{p-k} \otimes 1$, $i \leqslant k < p$, are generators of M. Since their number is equal to p, then they are free enerators. This also means that the module M is free, and the sheaf $\mathfrak{G}$ is locally free.

Since the algebra $\mathfrak{A}$ operates on the sheaf $\mathfrak{F}$, then this is also valid for $\mathfrak{G}$, which by the same ›ken determines the set of p-dimensional representations. We introduce into this set a rigidity with the ıd of the section $1 \otimes 1$. We denote the obtained set by $U \to \mathfrak{B}$. A p-family is a family of rigid irreducible epresentations of the algebra A_1, in which the fibers over points of some open, everywhere dense set are rreducible and the fiber at each point x is generated by the vectors $\mathfrak{A} \cdot s(x)$, where s is the rigidity. In articular, the set $U \to \mathfrak{B}$ is a p-family.

Let X be a k-diagram, let M_X be a set of p-families with basis X. Let $E \to Y$ be a p-family and $Y_0 \in$ ' be an open, everywhere dense set, over the points of which the layers are irreducible. Let us denote by ıom^0(X, Y) the set of such mappings $f \in$ Hom (X, Y) for which $f^{-1}Y_0$ is everywhere dense in X. Then f^*E ı a p-family over X. Thus, the mapping α: Hom^0(X, Y) $\to M_X$ appears; the p-family $E \to Y$ is called niversal if α is one-to-one for any p-family X.

THEOREM 2. The p-set $U \to \mathfrak{B}$ is universal.

Let $E \to X$ be a p-family, s–the rigidity, $h \cdot s = a \cdot s$, $a \in \Gamma(X, \mathcal{O})$. Then $(h-a)^p - (h-a)$ commutes in nd E with all $u \in \mathfrak{A}$, is equal to 0 on s and by the same token, in view of the fact that E is a p-family, is qual to 0 on all of E. In the ring $F_p[T]$ we have the equality

$$T^p - T = \Pi_{k \in F_p} (T - k).$$

ifferentiating, we obtain

$$-1 = \sum_{l \in F_p} \Pi_{k \in F_p \setminus l} (T - k).$$

'e introduce the notation

$$P_l(T) = -\Pi_{k \in F_p \setminus l} (T - k).$$

Then

$$\sum_{l \in F_p} P_l(h-a) = 1, \quad P_l(h-a) P_k(h-a) = 0$$

on E, $l \neq k$, since $(h-a)^p = h-a$. Therefore,

$$E = \oplus P_l(h-a) E.$$

The subfiberings P_l (h−a) E are one-dimensional, since if a layer over the point x is irreducible, then $(P_l(h-a)E)_x$ are one dimensional.

It is easy to see that

$$e^k P_l(h-a) E = P_{l-2k}(h-a) E,$$
$$f^k P_l(h-a) E = P_{l+2k}(h-a) E.$$

In order to prove the theorem it is sufficient to construct the inverse mapping β: $M_X \to \mathrm{Hom}^0(X, \mathfrak{B})$. Let ξ: $E \to X$ be a p-family, let s be the rigidity. Since s is a weight vector, us and s are linearly dependent at each point for arbitrary $u \in \mathfrak{Z}$. We set $\underline{us} = \lambda_u s$. This determines the homomorphism $\mathfrak{Z} \to \Gamma(X, \mathcal{O})$ and by the same token the mapping $\bar{f}$: $X \to \bar{Z}$. It is necessary to construct f: $X \to \mathfrak{B}$, in order that $\pi f = \bar{f}$. We define f on affine, open subsets of X. Let W be an affine open subset of X. If $\bar{f}(W) \subset \bar{Z} \setminus \bar{S}$, then f on W at once is determined by the condition $\pi f = \bar{f}$. Let W be such that $\bar{f}(W) \subset U_\alpha$, $U_\alpha \ni s$. Let x be an arbitrary point of W, let E_x be a fiber of the family over x, $k_x = \mathcal{O}_x / \mathfrak{m}_x$, s is the rigidity in E_x

$$E_x = E_0 \oplus \ldots \oplus E_{p-1}$$

is the decomposition corresponding to the decomposition M. All E_k are one-dimensional, $E_0 = k_x s$. It is clear that the subspace, generated by $\mathfrak{U}s$, coincides with the subspace generated by $e^k s$, $f^l s$, and its intersection with E_{-2i} is $k_x e^i s + k_x f^{p-i} s$. According to the definition of a p-family, $E_{-2i} \neq 0$, from where it follows that either $e^i s(s)$ or $f^{p-i} s(x)$ is not equal to 0.

If $e \in E_{-2i}$, $e \neq 0$, then

$$e^i s(x) = \alpha(x) e, \quad f^{p-i} s(x) = \beta(x) e$$

and, thus, $x \to (\beta(x) : \alpha(x))$ determines the morphism f: $X \to P^1$. Let $f = (\bar{f}, \bar{\bar{f}})$: $W \to \bar{Z} \times P^1$. We may assume that f maps W into $A^4 \times P^1$ and for $P \in f(W)$ relations (10) are satisfied, which one can easily verify directly. Thus, we have the morphism $f = W \to \mathfrak{B}$. These morphisms for different open sets of W are consistent and determine the morphism β: $X \to \mathfrak{B}$. It remains to verify that $\alpha\beta = 1$ and $\beta\alpha = 1$. This is proved with the aid of arguments completely analogous to the one used in the corresponding place in the proof of Theorem 1. The fact that E_X is generated by the vector s(x) now plays the role of irreducibility of
1 the representation E_X.

LITERATURE CITED

1. C. Curtis, Representations of Lie algebras of classical type with applications to linear groups, J. Math. and Mech., 9, 307-326 (1960).
2. H. Zassenhaus, The representations of Lie algebras of prime characteristic, Proc. Glasgow Math. Assoc., 2, 1-36 (1954).
3. A. Grothendieck, Fondements de la géométrie algébrique, Paris (1962) (preprint).
4. N. Jacobson, Lie Algebras [Russian translation], Moscow (1964).
5. N. Jacobson, Theory of Rings [Russian translation], Moscow (1947).
6. J. P. Serre, Faisceaux algébriques cohérents. Coll. Papers I, n. 29.

Graded Lie algebras of finite characteristic
(with A. I. Kostrikin)

Izv. Akad. Nauk SSSR, Ser. Mat. **33**, No. 2, 251–322 (1969). Zbl. **211**, 53
[Math. USSR, Izv. **3**, No. 2, 237–304 (1970)]

An invariant description is given of finite-dimensional algebras of Cartan type and, with certain restrictions, they are identified with abstract graded Lie algebras defined over algebraically closed fields of characteristic $p > 7$.

Introduction

This paper is concerned with an attempt at new methods in the problem of classifying finite-dimensional simple Lie algebras over algebraically closed fields of characteristic $p > 0$. In the introduction we present the formulation of the question, the principal results, and the plan of our work.

1. Classical algebras. As is known, simple Lie algebras defined over fields of characteristic 0 have natural analogs over fields of characteristic $p > 0$. Their definition is based on a process of reduction modulo p. Specifically, by Chevalley [25], every simple Lie algebra $\mathfrak{G}$ over the field of complex numbers has a standard basis whose elements are uniquely determined up to a factor of ± 1 or up to an automorphism of $\mathfrak{G}$. The structural constants relative to this basis are integers. One can therefore consider the algebra with the same multiplication table defined over an arbitrary field k. For the algebras obtained in this way from the simple Lie algebras of type A_n, $B_n, \dots$, we shall use the corresponding notation $\widetilde{A}_n$, $\widetilde{B}_n$, etc.

It is possible to characterize these algebras by other properties. For example, they coincide with the Lie algebras of simple algebraic groups defined over an algebraically closed field k. On the other hand, if (for $p > 5$) we consider the class of algebras, indecomposable into a direct sum, which in some faithful representation have a nondegenerate bilinear symmetric form, then we obtain those same algebras, except for $\widetilde{A}_n$, $n + 1 \equiv 0\,(p)$, for which such a form does not exist [3].

Of all these algebras only the algebras $\widetilde{A}_n$, $n + 1 \equiv 0\,(p)$, are not simple; they have a one-dimensional center with simple quotient. The simple factor algebras of $\widetilde{A}_n$, $\widetilde{B}_n, \dots$, $\widetilde{E}_8$ we again denote by A_n, $B_n, \dots$, E_8, and they will be called the classical Lie algebras. Thus, $A_n = \widetilde{A}_n$ for $n + 1 \not\equiv 0\,(p)$, and $B_n = \widetilde{B}_n, \dots$, $E_8 = \widetilde{E}_8$. All classical algebras are Lie p-algebras [19].

2. Algebras of Cartan type. As opposed to the situation over fields of characteristic zero, the classical algebras do not exhaust all simple Lie algebras, or even simple Lie p-algebras, over fields of characteristic p. The first example of a nonclassical algebra was constructed by Witt in the 1930's, and subsequently many similar examples were found involving both p-algebras and simple Lie algebras that are not p-algebras.

In [21] we noted that all known examples of simple nonclassical Lie p-algebras have a striking resemblance to the simple infinite-dimensional Lie algebras arising in the theory of pseudo groups of É. Cartan [4]. Specifically, in the theory of Cartan a fundamental role is played by the four series of Lie algebras over a field k of characteristic 0. All these algebras consist of (continuous) derivatives of the ring $k[[x_1, \ldots, x_n]]$ of formal power series, i.e., of operators of the form

$$\mathfrak{D} = \sum_{i=1}^{n} f_i \frac{\partial}{\partial x_i}, \quad f_i \in k[[x_1, \ldots, x_n]].$$

The definitions are of the separate series are as follows:

1. All derivations of the ring $k[[x_1, \ldots, x_n]]$.

2. All derivations $\mathfrak{D}$ satisfying the condition $\mathfrak{D}\omega = 0$, $\omega = dx_1 \wedge \ldots \wedge dx_n$.

3. All derivations satisfying, for $n = 2n'$, the condition $\mathfrak{D}\omega = 0$, where ω is a nondegenerate differential 2-form with constant coefficients, i.e.,

$$\omega = \sum \alpha_{ij}\, dx_i \wedge dx_j, \quad \alpha_{ji} = -\alpha_{ij}, \quad \det(\alpha_{ij}) \neq 0.$$

4. All derivations satisfying, for $n = 2n' + 1$, the condition $\mathfrak{D}\omega = f\omega$, where f is a function in $k[[x_1, \ldots, x_n]]$, depending on $\mathfrak{D}$, and ω is a differential 1-form with polynomial coefficients of degree ≤ 1 and nondegenerate differential $d\omega$.

The definition of these simple (for char $k = 0$) Lie algebras continues to be meaningful for a field k of characteristic $p > 0$; however, the algebras are no longer simple, for in each of them the derivations $\mathfrak{D} = \sum f_i \frac{\partial}{\partial x_i}$ c $f_i \in (x_1^p, \ldots, x_n^p)$ form an ideal of finite codimension. In cases 1, 2, 3, and 4 we denote the corresponding quotient Lie algebra by W_n, $\widetilde{S}_{n-1}$, $\widetilde{H}_{n'}$, and $\widetilde{K}_{n'+1}$. Alternatively, these algebras can be described as consisting of those derivations of the ring $k[x_1, \ldots, x_n]/(x_1^p, \ldots, x_n^p)$ that satisfy one of the conditions 1, 2, 3, or 4. We introduce also the algebras S_{n-1}^* and $H_{n'}^*$, in which conditions 2 and 3 are replaced by the conditions $\mathfrak{D}\omega = c\omega$, $c \in k$. It is clear that $\widetilde{S}_{n-1}$ is an ideal in S_{n-1}^*, $\widetilde{H}_{n'}$ is an ideal in $H_{n'}^*$, and dimal in/$\widetilde{S}_{n-1}^*$ = $\dim H_n^*/\widetilde{H}_{n'} = 1$.

The Lie algebras constructed above are either simple or very close to simple. Namely, the algebra W_n is simple, and the algebras $\widetilde{S}_n$, $\widetilde{H}_n$, $\widetilde{K}_n$ contain ideals $S_n \subset \widetilde{S}_n$, $H_n \subset \widetilde{H}_n$, and $K_n \subset \widetilde{K}_n$, themselves simple algebras, for which the quotients are all abelian with the one exception of $\widetilde{S}_1$ and $\widetilde{H}_1$ (they are isomorphic), $\widetilde{S}_1/S_1$ being a three-dimensional nilpotent nonabelian algebra.

As in [21], we call the algebras W_n, S_n, H_n, and K_n general, special, Hamiltonian, and contact algebras, respectively. However, we call attention to the following. In contrast to [21], the subscript in the notation W_n, S_n, H_n, K_n now pertains to the realization of the elements of the algebra as derivations of the ring $k[x_1, \ldots, x_s]/(x_1^p, \ldots, x_s^p)$ in $s = n$, $n + 1$, $2n$, and $2n - 1$ variables, respectively, and has approximately the same intrinsic meaning as the subscript in the notation for the algebras of classical type. These are p-algebras, and they exhaust all examples we know of nonclassical simple Lie p-algebras (for $p > 5$).

In the construction it is possible to make a generalization that will give simple Lie algebras which are not Lie p-algebras. For this purpose take a fixed sequence of natural numbers $m_1, \ldots, m_n$. In place f the ring $k[x_1, \ldots x_n]$ consider the divided power algebra in n variables, and in place of the ring $[x_1, \ldots, x_n]/(x_1^p, \ldots, x_n^p)$ the subalgebra O of the divided power algebra which is generated by the $^{(p^j)}$, $0 \leq j < m_i$. In O consider the derivations $\partial_i : x_1^{(\alpha_1)} \ldots x_n^{(\alpha_n)} \to x_1^{(\alpha_1)} \ldots x_i^{(\alpha_i - 1)} \ldots x_n^{(\alpha_n)}$. The Lie lgebras of derivations $\mathfrak{D} = \sum f_i \partial_i$, $f_i \in O$, satisfying conditions 1, 2, 3, and 4 we denote as be-re by W, $\widetilde{S}$, $\widetilde{H}$, $\widetilde{K}$; omitting now their subscripts. The simple algebras W, S, H and K are likewise efined as before and will again be called algebras of Cartan type. Their invariant description, in-ependent of the choice of the generators $x_1, \ldots, x_n$, is given in §3 of Chap. I. They are Lie p-lgebras only for $m_1 = \ldots = m_n = 1$.

3. **Principal results.** In this work we consider only graded Lie algebras. An important role layed by the algebras of Cartan type in the study of abstract graded and, more generally, filtered Lie lgebras is connected with their possession of a natural filtration. Specifically, in an algebra $\mathfrak{L}$ of artan type let $\mathfrak{L}_i$ be the subalgebra consisting of all derivations $\mathfrak{D} = \Sigma f_k \partial_k$ such that f_k involves nly monomials $x_1^{(\alpha_1)} \ldots x_n^{(\alpha_n)}$ with $\alpha_1 + \ldots + \alpha_n \geq i+1$. The subalgebras $\mathfrak{L}_i$ define a filtration $\{\mathfrak{L}_i\}$ a $\mathfrak{L} = \mathfrak{L}_{-1}$ having the following important property: $\mathfrak{L}_{i+1}$, $i \geq 0$, consists of all elements $x \in \mathfrak{L}_i$ for hich $[\mathfrak{L}, x] \subseteq \mathfrak{L}_i$. Thus the filtration is wholly described by specifying the subalgebra $\mathfrak{L}_0$. We will aclude this property in the definition of filtered algebra, and for a graded algebra $L = \bigoplus_{i \geq -1} L_i$ will as-ume this property for the filtration $\{\mathfrak{L}_i; \mathfrak{L}_i = \bigoplus_{k \geq i} \mathfrak{L}_k\}$. With an arbitrary filtered algebra $\mathfrak{L}$ is associated graded algebra $\mathrm{Gr}\,\mathfrak{L}$ containing much information about $\mathfrak{L}$. In this connection we note that relative the natural filtration all algebras of Cartan type are in fact graded, except for algebras of the K eries. But these, too, have a gradation which, however, involves a term L_{-2} (cf. (8), §7, Chap. I).

Important invariants of the graded algebra

$$L = L_{-1} \oplus L_0 \oplus \ldots \oplus L_r$$

re the subalgebra L_0 and its representation Γ: $L_0 \to \mathrm{End}\, L_{-1}$ defined by commutation in L.

Our principal result is the enumeration for $p > 7$ *of all graded Lie algebras satisfying the con-itions*: 1) L_0 *is a p-algebra*; 2) Γ *is an irreducible p-representation of dimension* $\dim L_{-1} < p - 1$; 3) $_2 \neq 0$.

It is proved that *an arbitrary graded algebra with these properties is isomorphic either to an al-ebra* W *or to some subalgebra* L *of an algebra* S^* [*resp.* H^*] *such that* $S \subseteq L \subseteq S^*$ [$H \subseteq L \subseteq H^*$].†

Since the quotients S^*/S and H^*/H have a very simple structure, our result gives an effective escription of all algebras of the type in question. With reference to p-algebras, these theorems easily nply the following fact:

If $\dim L_{-1} < p - 1$, *in a simple graded Lie p-algebra* L, *then* L *is isomorphic either to a classical imple algebra or to a p-algebra of Cartan type.*

4. **Plan of the work.** In an arbitrary graded Lie algebra L the spaces L_i, $i \geq 1$, are "bounded bove" in terms of the algebra L_0 and its representation Γ in L_{-1}. Namely, we have always $L_i \subseteq L^{(i)}$, here $L^{(i)}$ (the ith Cartan prolongation) is defined inductively, beginning with $L^{(0)} = L_0$; $L^{(i+1)}$ $\mathrm{Hom}(L_{-1}, L^{(i)})$ consists of the homomorphisms f: $L_{-1} \to L^{(i)}$ for which

$$f(x)(y) = f(y)(x), \quad x, y \in L_{-1}$$

†See the footnote at the end of the paper.

(where we use the fact that $f(x) \in L^{(i)} \subset \mathrm{Hom}(L_{-1}, L^{(i-1)})$).

Chapter II is concerned with obtaining all p-algebras L_0 and their irreducible p-representations Γ in a space L_{-1}, $\dim L_{-1} < p-1$, for which $L^{(2)} \neq 0$. The result is very simple: L_0 can only be one of the algebras A_n, C_n, $A_n \oplus k$, $C_n \oplus k$, and Γ is its standard representation; i.e., the result is the same as over a field of characteristic 0 [13].

Unfortunately, the proof has not been accomplished without some sorting out. First of all, proceeding from the general properties of weights, we arrive at a fairly limited number of possibilities, which are all contained in the table in §4 of Chap. II. The class of algebras and their representations so obtained have a simple meaning: they are, firstly, the algebras and representations of the four types above, and secondly, those algebras and irreducible representations which are encountered in the graded classical Lie algebras [15]. For all algebras and representations of the latter type we must first calculate explicitly the first prolongation $L^{(1)}$, and then prove that $L^{(2)} = 0$. In this part, consequently, we operate rather by verification than by proof.

In Chap. III, for the possible types of L_0 and Γ such that $L^{(2)} \neq 0$, we construct all modules $L^{(i)}$ and prove that they reduce to the small number of above-described algebras of Cartan type. In case of simple p-algebras, the condition $L^{(2)} = 0$ distinguishes the classical algebras.

Chapter I has independent significance and contains all the necessary information about algebras of Cartan type, including the question of their simplicity, the conditions that they be p-algebras, etc. In particular, it is proved that *there exist only a finite number of nonisomorphic algebras of Cartan type for a given dimension.* This is not obvious, since the definition of algebras of types H and K depends on the choice of the form ω.

At the end of the paper is given an example which illustrates the necessity of the restriction $\dim L_{-1} < p-1$. Namely, there is constructed a graded Lie p-algebra with $\dim L_{-1} = p-1$, an irreducible representation Γ, and $L_2 \neq 0$, which is not isomorphic as a graded algebra to any algebra of Cartan type. However, this is relative only to its structure of graded (or filtered) algebra. Without taking into account this structure, it is the algebra H_1.

5. The role of the filtration $\{\mathfrak{L}_i\}$. Examples similar to the one mentioned above bear out the fact that in addition to the natural filtration $\{\mathfrak{L}_i\}$ defined in paragraph 3, the algebras of Cartan type may have other, so-called pathological, filtrations. In Chap. I we look at one of the possible ways of specifying the natural filtration $\{\mathfrak{L}_i\}$ in p-algebras of Cartan type that differs from the initial one in that the subalgebra $\mathfrak{L}_0$ from which is recovered the whole filtration is defined entirely by structural properties of the algebra. This definition is based on the simple fact that in any p-algebra of Cartan type there exist elements $x \neq 0$ for which $(\operatorname{ad} x)^2 = 0$. The subalgebra $\mathfrak{C}$ generated by these elements has as normalizer precisely the algebra $\mathfrak{L}_0$.

Thus, *all subalgebras $\mathfrak{L}_i$ are defined intrinsically and in particular are invariant relative to all automorphisms of the algebra $\mathfrak{L}$*, although of course they are not ideals. This interesting fact is connected with the presence of nilpotent elements in the automorphism schemes of algebras of Cartan type. We prove as another intrinsic characterization of the algebra $\mathfrak{L}_0$ that *it is the maximal subalgebra of $\mathfrak{L}$ invariant with respect to all automorphisms, i.e., it contains all invariant subalgebras.*

As a final remark, we note that in the present article we prove in one respect more, and in another ss, than in [21]; more, in that not only do we consider the case of p-algebras but we have dropped e requirement $\dim L_0 < p$; less, in that we restrict ourselves to graded algebras.

CHAPTER I. ALGEBRAS OF CARTAN TYPE

§1. Infinite-dimensional Lie algebras of finite characteristic

In the theory of Cartan, infinite-dimensional Lie algebras are realized as derivations of a ring of wer series. Over fields of finite characteristic the natural analog to the algebra of polynomials is e algebra of divided powers. We now describe this algebra (cf., for example, [5]).

1. The divided power algebra. From now on, k is a field of characteristic $p > 0$ and E is a finite-mensional vector space over k. We recall that the divided power algebra over E is given by genera-rs $x^{(h)}$, $x \in E$, $h \in Z$, $h \geq 0$, and the defining relations

$$(x+y)^{(h)} = \sum_{i=0}^{h} x^{(i)} y^{(h-i)}, \quad x, y \in E,$$
$$(\alpha x)^{(h)} = \alpha^h x^{(h)},$$
$$x^{(k)} x^{(l)} = \binom{k+l}{k} x^{(k+l)},$$
$$x^{(0)} = 1.$$

The divided power algebra over E will be denoted by $O(E)$. Its elements are "polynomials"

$$\sum \alpha_{h_1 \ldots h_n} x_1^{(h_1)} \ldots x_n^{(h_n)},$$

here $x_1, \ldots, x_n$ is a basis of E.

There is in $O(E)$ a gradation in which $\deg x^{(h)} = h$, $x \in E$:

$$O(E) = \sum_{i \geq 0} O(E)_i.$$

he corresponding filtration will be denoted by $\{\mathfrak{m}^{(i)}\}$:

$$\mathfrak{m}^{(i)} = \sum_{k \geq i} O(E)_k.$$

2. Special derivations of the algebra $O(E)$. Consider the derivations $\mathfrak{D}$ of $O(E)$ such that

$$\mathfrak{D}(x^{(h)}) = x^{(h-1)} \mathfrak{D}(x)$$

r all $x \in E$. We call them special derivations. It is simple to prove that the special derivations form Lie subalgebra and an $O(E)$-submodule in the algebra of all derivations of $O(E)$. The algebra of ecial derivations we denote by $W(E)$.

Theorem 1. *The algebra $W(E)$ is a free $O(E)$-module of rank $n = \dim_k E$.*

Proof. Let E^* be the dual space of E. For arbitrary $\xi \in E^*$ the map

$$\partial_\xi : x^{(h)} \to x^{(h-1)} \xi(x), \quad x \in E,$$

extends to a special derivation, which we denote again by ∂_ξ. Choose in E a basis $\{x_1, \ldots, x_n\}$, denote by $\{\xi_1, \ldots, \xi_n\}$ the dual basis in E^*, and let $\partial_i = \partial_{\xi_i}$. We check that the ∂_i form a free basis for $W(E)$ over $O(E)$. For $\mathfrak{D} \in W(E)$ set $\mathfrak{D}(x_i) = f_i \in O(E)$. Then the derivations $\mathfrak{D}$ and $\sum f_i \partial_i$ coincide on E. But a special derivation is obviously determined by its values on E, so that

$$\mathscr{D} = \sum_{i=1}^{n} f_i \partial_i. \tag{1}$$

The uniqueness of this representation follows from the fact that $f_i = \mathfrak{D}(x_i)$.

3. **Differential forms.** The module $\mathrm{Hom}_{O(E)}(W(E), O(E))$ will be denoted by $\Omega^1(E)$, and its r-th exterior power $\bigwedge^r \Omega^1(E)$ over $O(E)$ by $\Omega^r(E)$. For $f \in O(E)$ the element $df \in \Omega^1(E)$ is defined by the equation

$$(df)(\mathscr{D}) = \mathscr{D}f, \quad \mathscr{D} \in W(E).$$

It is obvious that

$$d(f+g) = df + dg, \quad dfg = fdg + gdf,$$

and if $\{x_1, \ldots, x_n\}$ is a basis of E, then the elements

$$dx_{i_1} \wedge \cdots \wedge dx_{i_r}, \quad i_1 < i_2 < \ldots < i_r,$$

form a free basis of $\Omega^r(E)$ over $O(E)$. The ring

$$\Omega(E) = \oplus\, \Omega^r(E)$$

is a graded algebra over $O(E)$, on which is defined the differential

$$d : \Omega(E) \to \Omega(E),$$

extending the differential d from $O(E)$ and having degree 1. Any derivation $\mathfrak{D}$ of the algebra $O(E)$ can be uniquely extended to $\Omega(E)$ as a derivation of this algebra commuting with the differential d.

In particular, for any special derivation $\mathfrak{D}$ and differential form $\omega \in \Omega^r(E)$ there is defined a form $\mathfrak{D}\omega \in \Omega^r(E)$.

4. **The series of simple Lie algebras.** Now we are able to introduce the four basic series of infinite-dimensional Lie algebras.

1) The algebra $W(E)$ has already been defined in paragraph 2.

2) The algebra $S(E)$ consists of all $\mathfrak{D} \in W(E)$ for which $\mathfrak{D}\omega = 0$. Here $\omega = dx_1 \wedge \ldots \wedge dx_n$, where $\{x_1, \ldots, x_n\}$ is a basis for E and $n > 2$.

3) Suppose $\dim E + n = 2n'$, $n' \geq 1$. The algebra $H(E)$ consists of all $\mathfrak{D} \in W(E)$ for which $\mathfrak{D}\omega = 0$. Here $\omega = \sum_{i=1}^{n'} dx_i \wedge dx_{n'+i}$, and $\{x_1, \ldots, x_n\}$ is a basis for E.

4) Suppose $n = \dim E$ is odd and $(n-1)/2 = n' \geq 1$. The algebra $K(E)$ consists of all $\mathfrak{D} \in W(E)$ for which $\mathfrak{D}\omega = u\omega$, $u \equiv u(\mathfrak{D}) \in O(E)$. Here

$$\omega = dx_n + \sum_{i=1}^{n'} (x_i\, dx_{n'+i} - x_{n'+i}\, dx_i),$$

nd $\{x_1, \ldots, x_n\}$ is a basis for E.

The algebras $W(E)$, $S(E)$, $H(E)$ in all four series are simple Lie algebras. The proof of this is ot difficult and will at once follow from the argument used later for the proof of more delicate facts.

§2. Invariant subalgebras of the algebra $O(E)$

The object of this section is to describe the graded subalgebras of the algebra $O(E)$ that are nvariant with respect to all special derivations of degree -1 of this algebra, i.e., derivations ∂_ξ, $\in E^*$. First we describe a construction that gives the result.

Definition. By a (generalized) flag $\mathfrak{F}$ we mean a decreasing system of subspaces of the space E:

$$E_0 \supseteq E_1 \supseteq \ldots \supseteq E_k \supset E_{k+1} = 0 \quad (E_k \neq 0). \tag{1}$$

epetitions are permitted in this sequence.

With each flag $\mathfrak{F}$ one can associate the subalgebra in $O(E)$ generated by all the elements

$$x^{(p^i)}, \quad x \in E_i. \tag{2}$$

e denote this algebra by the symbol $O(\mathfrak{F})$. It has a finite dimension over k, because the flag is nite, which we can easily find. If $\dim E_i = n_i$ and $\{x_{ij};\ j = 1, \ldots, n_i\}$ is a basis for the subspace E_i, en $x_{ij}^{(p^i)}$ is a system of generators for the algebra $O(\mathfrak{F})$. From the construction of the divided power gebra it follows that these generators satisfy no relations other than

$$(x_{ij}^{(p^i)})^p = 0.$$

herefore

$$\dim_k O(\mathfrak{F}) = p^{n_0 + \cdots + n_k}. \tag{3}$$

nce $O(\mathfrak{F})$ has a system of generators (2) consisting of homogeneous elements, it is a graded sub- gebra of $O(E)$. In view of the fact that

$$\partial_\xi x^{(p^i)} = x^{(p^i - 1)}\, \xi(x),$$

nd $x^{(p^i-1)}$ is a polynomial in the $x^{(p^j)}$, $j < i$ (or 1 for $i = 0$), the algebra $O(\mathfrak{F})$ is invariant with respect all derivations. In order to find a basis for the algebra $O(\mathfrak{F})$ we take in each subspace E_i vectors $_{ij}$ whose images form a basis in E_i/E_{i+1} (if $E_i = E_{i+1}$, then instead of vectors x_{ij} take the unit ement). Then the monomials

$$\prod_{i,j} (x_{ij}^{(\alpha_{ij})}, \quad 0 \leqslant \alpha_{ij} < p^{i+1}, \tag{4}$$

rm a basis for the algebra $O(F)$.

Theorem 1. *The subalgebra $O(\mathfrak{F})$ of the algebra $O(E)$ coincides with the set of all elements $\in O(E)$ satisfying the condition*

$$\partial_\xi^{p^i}(u) = 0, \quad \xi \in \operatorname{Ann} E_i \subset E^* \tag{5}$$

(where Ann E_i *is the annihilator in* E^* *of the subspace* E_i*).*

Proof. We note first that all elements $u \in O(\mathfrak{F})$ satisfy (5). Since $\partial_\xi^{p^i}$ is a derivation of the algebra $O(E)$, it suffices to prove (5) for generators of $O(\mathfrak{F})$, i.e., for $u = x^{(p^j)}$, $x \in E_j$. Then

$$\partial_\xi^{p^i} x^{(p^j)} = \begin{cases} 0, & \text{if } \quad i > j, \\ x^{(p^j - p^i)}\, \xi(x)^{p^i}, & \text{if } \quad i \leqslant j. \end{cases}$$

In the second case, $E_i \supseteq E_j \ni x$, i.e., Ann $E_j \supseteq$ Ann $E_i \ni \xi$, so that $\xi(x) = 0$.

We show now that any $u \in O(E)$ satisfying condition (5) is contained in $O(\mathfrak{F})$. For this we use the linearity of (5) and the fact that any $u \in O(E)$ can be represented as a linear combination of monomials (4) with no restrictions on the exponents α_{ij}. Indeed, the elements x_{ij} different from 1 form a basis for the space E. We must establish that if a monomial u satisfies (5), then $\alpha_{ij} < p^{i+1}$. Assume this is so for $i < l$, and suppose that $\alpha_{ls} \geq p^{l+1}$ for some index s. Pick an element $\xi \in$ Ann E_{l+1} for which $\xi(x_{ls}) = 1$ and $\xi(x_{lj}) = 0$ for $j \neq s$. Then

$$\partial_\xi^{p^{l+1}}(u) = x_{ls}^{(\alpha_{ls} - p^{l+1})} \prod_{(i,\, j) \neq (l,\, s)} x_{ij}^{(\alpha_{ij})} \neq 0,$$

contrary to (5).

The principal result of this section is

Theorem 2. *Any graded finite-dimensional subalgebra*

$$A = \sum_{s=0}^{\infty} A_s \quad (A_s = 0 \;\; \textit{for} \;\; s > s_0)$$

of the algebra $O(E)$ *which is invariant under all derivations* ∂_ξ, $\xi \in E^*$, *coincides with one of the algebras* $O(\mathfrak{F})$.

Proof. By hypothesis, $\partial_\xi A \subset A$ for all $\xi \in E^*$. We must prove the existence of a flag $\mathfrak{F}$ such tha

$$A_s = O(\mathfrak{F})_s \tag{6}$$

for all s. For $s = 1$ it obviously suffices to take the flag $\mathfrak{F}$: $A_1 = E_0 \supset E_1 = 0$ (it is easy to see that $A_0 \cong k$). We prove (6) by induction on s. Assume it holds for $s < m$, i.e., that we have a flag

$$\mathfrak{F}' : E_0 \supseteq E_1 \supseteq \cdots \supseteq E_k \supset E'_{k+1} = 0,$$

satisfying (6) for all $s < m$. From the definition of the flag we can assume that $p^k < m$.

First, observe that always

$$A_m \supseteq O(\mathfrak{F}')_m.$$

To see this, write the element $v \in O(\mathfrak{F}')_m$ in the form $v = \prod x_i^{(p^i)}$, $x_i \in E_i$, and remember that all $p^i < m$. In other words, $v = \prod v_s$, $v_s \in O(\mathfrak{F}')_s$, $s < m$; but since by our induction assumption $O(\mathfrak{F}')_s = A_s$, we have $v_s \in A$, and so $v \in A$.

We consider now two cases.

1) $m \neq p^l$. We prove that $A_m \subseteq O(\mathfrak{F}')_m$. Let $u \in A_m$, and η be an arbitrary element in E^*. Then $\eta u \in A_{m-1}$ because of the invariance of the algebra A. Therefore by the induction assumption and heorem 1, $\partial_\xi^{p^i} \partial_\eta u = 0$ for $\xi \in \operatorname{Ann} E_i$. It follows that $\partial_\eta \partial_\xi^{p^i} u = m - p^i \neq 0$ and η is an arbitrary ele-ent of E^*, we have $\partial_\xi^{p^i} u = 0$. This means that $u \in O(\mathfrak{F}')_m$.

2) $m = p^l$. The argument in case 1) proves that $\partial_\xi^{p^i} u = 0$ for $u \in A_m$ and any $\xi \in \operatorname{Ann} E_i$, $i < l$. riting the element $u \in A_m$ as a linear combination of monomials (4), we see, as in the proof of heorem 1, that this implies a condition on the exponents: $\alpha_{ij} < p^{i+1}$ for $i < l$. Thus A_m is contained n the space S generated by monomials (4) of degree m in which $\alpha_{ij} < p^{i+1}$ for $i < l$. The monomials n S but not in $O(\mathfrak{F}')_n$ are the $x_{l-1,j}^{(p^l)}$. Therefore $S/O(\mathfrak{F}')_m \cong E_m$, $k = l-1$. Under this isomorphism ne subspace A_m corresponds to a subspace in E_k which we denote by E_{k+1}. Obviously the flag

$$\mathcal{F} : E_0 \supseteq E_1 \supseteq \ldots \supseteq E_k \supseteq E_{k+1} \supset E_{k+2} = 0$$

atisfies condition (6) for $s \leq m$.

Remark. It is possible to obtain a more invariant description of the subspace E_{k+1}. For $u \in A_m$, $= p^l$, and $\xi \in E^*$, we obviously have $\partial_\xi^{p^l} u \in k$, so that the formula

$$(\xi, u) = \partial_\xi^{p^l} u \tag{7}$$

efines a scalar product between the elements of the spaces E^* and A_m. Denote by B the subspace n A_m which is the intersection of the kernels $\ker \partial_\xi^{p^i}$, $\xi \in \operatorname{Ann} E_i$, $i < l$. Then $B \supset O(\mathfrak{F}')_m$, and the calar product (7) is zero for $\xi \in \operatorname{Ann} E_k$ and $u \in O(\mathfrak{F}')_m$. Therefore it defines a pairing of the spaces $^*/\operatorname{Ann} E_k$ and $S/O(\mathfrak{F}')_m$, and so of the spaces $E^*/\operatorname{Ann} E_k$ and $B/O(\mathfrak{F}')_m$. Theorem 1 shows that this airing is not degenerate for the component $B/O(\mathfrak{F}')_m$, so that there is defined an imbedding

$$B / O(\mathcal{F}')_m \to (E^* / \operatorname{Ann} E_k)^* = E_k.$$

he image of this map is E_{k+1}.

§3. Lie algebras corresponding to flags

1. **Algebras of Cartan type.** From now on we consider only those flags in which $E_0 = E$. Let $\mathfrak{F}$ e such a flag. Note that $x^{(h)} \in O(\mathfrak{F})$ for arbitrary $x \in E_i$ and $h < p^{i+1}$. In accordance with paragraph of §1, a derivation $\mathfrak{D}$ of the algebra $O(\mathfrak{F})$ will be called special if

$$\mathcal{D}(x^{(h)}) = x^{(h-1)}\mathcal{D}(x)$$

r all $h < p^{i+1}$ and $x \in E_i$. As in §1, it is easy to verify that the set of all special derivations of ne algebra $O(\mathfrak{F})$ forms a Lie algebra, which is a free $O(\mathfrak{F})$-module of rank $n = \dim E$. This algebra is alled the general algebra relative to the flag $\mathfrak{F}$, and is denoted by $W(\mathfrak{F})$. As a free $O(\mathfrak{F})$-module,

$$W(\mathcal{F}) = \sum_{i=1}^{n} O(\mathcal{F})\, \partial_i, \tag{1}$$

here the ∂_i are characterized by the fact that $\partial_i(x) = \xi_i(x)$ for $x \in E$, $\{\xi_1, \ldots, \xi_n\}$ being a basis for * which can be taken compatible with the flag $\mathfrak{F}$. The latter means that $\{\xi_1, \ldots, \xi_n\}$ is dual to the asis $\{x_{ij}\}$ in E (cf. (4) §2).

From the decomposition (1) and formula (3) of §2 it follows that

$$\dim_k W(\mathcal{F}) = np^{n_0+\dots+n_k}, \tag{2}$$

where $n_i = \dim E_i$.

Theorem 1. *The algebra* $W(\mathfrak{F})$ *coincides with the subalgebra* U *of* $W(E)$ *consisting of derivations which leave invariant the algebra* $O(\mathfrak{F})$. *As a subalgebra of* $W(E)$, *the algebra* $W(\mathfrak{F})$ *is characterized by the conditions*:

$$(\mathrm{ad}\, \partial_\xi)^{p^i}\mathcal{D} = 0, \quad \xi \in \mathrm{Ann}\, E_i \subset E^*, \quad \mathcal{D} \in W(E).$$

Proof. Obviously any special derivation of $W(E)$ leaving invariant the ring $O(\mathfrak{F})$ defines a special derivation of this ring. Therefore we have the restriction homomorphism

$$U \to W(\mathcal{F}).$$

This homomorphism has no kernel, since any special derivation is uniquely determined by its action on E and according to our convention $E_0 = E$. Finally, the restriction map is an epimorphism, is seen from the decomposition (1) and the analogous decomposition for $W(E)$.

Let $\mathfrak{D} \in W(E)$ have the form

$$\mathcal{D} = \sum_{j=1}^{n} f_j \partial_j.$$

For any $\xi \in E^*$ we have

$$(\mathrm{ad}\, \partial_\xi)\, \mathcal{D} = \sum_{j=1}^{n} \partial_\xi (f_j)\, \partial_j$$

and

$$(\mathrm{ad}\, \partial_\xi)^{p^i} \mathcal{D} = \sum_{j=1}^{n} \partial_\xi^{p^i} (f_j)\, \partial_j,$$

so that the final conclusion of our theorem follows from Theorem 1 of §2. This completes the proof.

Now consider the algebras

$$S(E) \cap W(\mathcal{F}) = \tilde{S}(\mathcal{F}),$$
$$H(E) \cap W(\mathcal{F}) = \tilde{H}(\mathcal{F}),$$
$$K(E) \cap W(\mathcal{F}) = \tilde{K}(\mathcal{F}),$$

where S, H, and K are the infinite-dimensional algebras defined in paragraph 4 of §1. The second commutators $\mathfrak{L} = [[\tilde{\mathfrak{L}}, \tilde{\mathfrak{L}}], [\tilde{\mathfrak{L}}, \tilde{\mathfrak{L}}]]$ of the algebras $\tilde{\mathfrak{L}} = \tilde{S}, \tilde{H}, \tilde{K}$ will be denoted by $S(\mathfrak{F})$, $H(\mathfrak{F})$, $K(\mathfrak{F})$, respectively, and called the special, Hamiltonian, and contact algebras of the flag $\mathfrak{F}$. The algebras $W(\mathfrak{F})$, $S(\mathfrak{F})$, $H(\mathfrak{F})$, and $K(\mathfrak{F})$ will also be called algebras of Cartan type.

We now define in $L = W(\mathfrak{F})$ a gradation

$$W(\mathcal{F}) = \bigoplus_{i \geq -1} L_i,$$

by letting

$$L_i = \sum_{j=1}^{n} O(\mathcal{F})_{i+1}\partial_j = O(\mathcal{F})_{i+1}\cdot V,$$

where V is the space of all ∂_ξ, $\xi \in E^*$.

It is easy to see that $W(\mathcal{F})$ is a graded algebra, i.e.,

$$[L_i, L_j] \subseteq L_{i+j}.$$

We will sometimes consider it also as a filtered algebra with filtration $\{\mathfrak{L}_i\}$:

$$\mathfrak{L}_i = \sum_{j \geq i} L_j.$$

This filtration defines a filtration on each of the subalgebras $\widetilde{S}$, S, $\widetilde{H}$, H, $\widetilde{K}$, K.

The algebras $\widetilde{\mathfrak{L}}$ and $\mathfrak{L}$ and their filtrations will be given in explicit form in the sections following.

2. The variety of algebras corresponding to flags. We now determine how many nonisomorphic algebras of a given dimension can be obtained by the above construction. The main result is as follows:

Theorem 2. *Up to isomorphism there exist only a finite number of Lie algebras of Cartan type of a given dimension over k.*

The proof below contains additional information about the set of isomorphism classes of these algebras.

The definition of each of the algebras $W(\mathcal{F})$, $\widetilde{S}(\mathcal{F})$, $\widetilde{H}(\mathcal{F})$, and $\widetilde{K}(\mathcal{F})$ depends on a differential form ω and a flag $\mathcal{F}$. It is easy to see that an automorphism of the space E preserving the form ω up to scalar multiplication and taking the flag $\mathcal{F}$ into a flag $\mathcal{F}_1$ induces an isomorphism of the algebras determined by ω and the flags $\mathcal{F}$ and $\mathcal{F}_1$ respectively. Therefore it suffices to consider the set of orbits into which the manifold of flags is decomposed relative to the group of linear automorphisms $\mathrm{Aut}\,\omega$ of the form ω.

It will be more convenient to consider the action of the whole group $GL(E)$ both on the form ω and on the flag $\mathcal{F}$. We determine those forms into which ω can be carried.

In the case of the algebras $\widetilde{S}$, it is obvious that $\Psi\omega = (\det\Psi)\omega$, where $\Psi \in GL(E)$. In the case of $\widetilde{H}$, the form $\Psi\omega$ has the appearance $\sum \alpha_{ij} dx_i \wedge dx_j$, where (α_{ij}) is an arbitrary nondegenerate skew-symmetric matrix. We obtain, therefore, as the homogeneous space over $GL(E)$, a set isomorphic to the set of nondegenerate skew-symmetric matrices.

In the case of the algebra $\widetilde{K}$, the answer is somewhat more complicated. Using a theorem already proved by Lie and Frobenius (cf. [20], §55), one can show that under the action of the elements of $GL(E)$ there is obtained from ω the set of forms

$$\sum_{i=1}^{n} \beta_i \, dx_i + \sum_{i,j=1}^{n} \alpha_{ij} x_i \, dx_j,$$

where (α_{ij}) is a skew-symmetric matrix of rank $n-1$ and

$$\begin{vmatrix} \alpha_{11} \dots \alpha_{1n} & \beta_1 \\ \dots\dots & \\ \alpha_{n1} \dots \alpha_{nn} & \beta_n \\ -\beta_1 \dots -\beta_n & 0 \end{vmatrix} \neq 0. \tag{3}$$

If one interprets (α_{ij}) as the matrix of a skew-symmetric form α in E, and $\{\beta_1, \dots, \beta_n\}$ as the coefficients of a linear form β, then (3) is equivalent to the condition

$$\operatorname{rank} \alpha = n-1; \quad \beta \neq 0 \quad \text{on} \quad \operatorname{Ann}_\alpha E. \tag{4}$$

Thus we obtain as the homogeneous space over $GL(E)$ a set of forms isomorphic to the set of pairs (α, β) satisfying condition (4).

We see in sum that the totality of isomorphism classes of algebras of Cartan type is a surjective image of the set of orbits relative to $GL(E)$ of the following sets:

1) For W – all flags.

2) For $\widetilde{S}$ – all flags.

3) For $\widetilde{H}$ – the pairs $(\alpha, \mathfrak{F})$, where α is a nondegenerate skew-symmetric form on E and $\mathfrak{F}$ is a flag.

4) For $\widetilde{K}$ – the pairs $(r, \mathfrak{F})$ where r in turn is a pair (α, β) satisfying condition (4), and $\mathfrak{F}$ is a flag.

If in the sequence $(n_0, \dots, n_k)$, $n_i = \dim E_i \neq 0$, we retain only distinct integers, we get a sequence $(t_0, t_1, \dots, t_l)$, $t_0 > t_1 > \dots > t_l$, $0 \leqslant l \leqslant k$, which we shall call the flag type.

It is clear that flags are equivalent relative to $GL(E)$ if and only if their types coincide. Therefore, the sets of interest in cases 1)-4) can be described as follows:

1. The totality of different flag types.

2. The totality of different flag types.

3. For each flag type, the set of orbits of nondegenerate skew-symmetric forms relative to the automorphism group of any flag of this type.

4. The same as 3, with skew-symmetric forms replaced by pairs (α, β) satisfying condition (4).

We will show that we get finite sets in all cases. In cases 1 and 2 this is obvious. In case 3 the question comes down to the classification of nondegenerate skew-symmetric forms relative to the automorphism group of a fixed flag $\mathfrak{F}$. If $\mathfrak{F}$ has type $(t_0, t_1, \dots, t_l)$, its automorphism group is the group of generalized triangular matrices with squares of orders $t_{i+1} - t_i$ along the diagonal. It is clear that the group will be minimal if $(t_0, \dots, t_l) = (n, n-1, \dots, 2, 1)$. Therefore we can restrict ourselves to this case. But here the answer is given by Theorem 1 of §3, Chap. IX of [24]* according to which there are in all

$$k = \frac{n!}{2^{n/2} \cdot \left(\frac{n}{2}\right)!}$$

---

*The authors acknowledge their thanks to A. G. Èlašvili, who pointed out this theorem.

bits. Namely, as representatives of orbits we can take the skew-symmetric forms

$$\alpha_\pi = \sum_{i=1}^{n} \alpha_{i,\,\pi i}\, x_i y_{\pi i}, \quad \alpha_{\pi i,\, i} = -\alpha_{i,\,\pi i}, \tag{5}$$

ere π runs through the involutions of the symmetric group S_n acting on the set

$$M = \{1, 2, \ldots, n\},\ \pi i \neq i,\ i \in M,\ n = 2n'.$$

Finally, case 4 reduces essentially to case 3 if we note that the linear form $\beta = \beta_1 x_1 + \ldots + \beta_n x_n$ the pair (α, β) with $\alpha = \alpha_\pi$ (without loss of generality we assume $\alpha_{i,\,n} = \alpha_{n,\,i} = 0,\ 1 \leqslant i \leqslant n$ $= 2n' + 1$) can be reduced by an automorphism

$$\Psi : \begin{cases} x_i \to \gamma_i x_i + \delta_i x_{\pi i} \\ x_{\pi i} \to \qquad \gamma_i^{-1} x_{\pi i} \\ x_n \to \gamma x_n \end{cases}$$

a form β' with coefficients $\beta'_n = 1$, $\beta'_i = 0$ or 1, $\beta'_i \beta'_{\pi i} = 0$, $i < n$.

A "triangular" automorphism Ψ of the space E obviously preserves the form (5). This completes e proof of Theorem 2.

Remark. The enumeration of the different types of forms is reminiscent of the Bruhat decomposin. It may be, therefore, that there exists a more invariant classification of the algebras of Cartan pe.

§4. The general algebra $W(\mathfrak{F})$

1. Simplicity of the algebra $W(\mathfrak{F})$. It is convenient to choose a basis in E compatible with the g $\mathfrak{F}$. In contrast to (4) of §2, we eliminate double indices by excluding repetition of subspaces E_i the sequence (1) of §2. Thus $E = \langle x_1, \ldots, x_n \rangle$, and the basis of the algebra $O(\mathfrak{F})$ consists of nomials

$$x_1^{(h_1)} \ldots x_n^{(h_n)},$$
$$0 \leqslant h_i < p^{m_i}, \quad 1 \leqslant m_1 \leqslant m_2 \leqslant \cdots \leqslant m_n. \tag{1}$$

e number m_i will sometimes be called the height of the element x_i, and $m_n = \max m_i$ the height of e flag $\mathfrak{F}$. If $m_i = m_{i+1} = \ldots = m_{i+l_s}$ then $\{x_1, \ldots, x_{i+l_s}\}$ is the preimage of a basis of a factor ace E_s/E_{s+1} of dimension s_s. By definition,

$$\dim O(\mathscr{F}) = p^m, \quad m = m_1 + \ldots + m_n \geqslant n, \quad \dim W(\mathscr{F}) = np^m. \tag{2}$$

ese dimensions are written in another way in (3) of §2 and (2) of §3. The special derivations

$$\partial_i : O(\mathscr{F}) \to O(\mathscr{F})$$

e defined by the relation

$$x_j^{(h)} \to \delta_{ij} x_j^{(h-1)}, \quad h > 0.$$

If $\mathcal{D}_1 = \sum f_i\partial_i$, $\mathcal{D}_2 = \sum g_i\partial_i$, then $[\mathcal{D}_1, \mathcal{D}_2] = \sum u_i\partial_i$, where

$$u_i = \mathcal{D}_1 g_i - \mathcal{D}_2 f_i = \sum_{j=1}^{n} \{f_j(\partial_j g_i) - g_j(\partial_j f_i)\}. \tag{3}$$

As remarked in §3,

$$W(\mathcal{F}) = L_{-1} + L_0 + \ldots + L_r$$

is a graded algebra, where obviously

$$r + 1 = \sum_{i=1}^{n} (p^{m_i} - 1). \tag{4}$$

The representation

$$\Gamma : L_0 \to \operatorname{End} L_{-1}$$

coincides with the standard representation of the full matrix algebra $gl(n, k) \cong L_0$.

Proposition 1. *The general algebra $W(\mathfrak{F})$ is simple except for the case $p = 2$, $m = 1$. It is a Lie p-algebra if and only if $m = n$.*

Proof: For any polynomial $f \in O(\mathfrak{F})$ of degree > 0, we have $\partial_j f \neq 0$ for at least one j, $1 \leq j \leq n$. Since $(\operatorname{ad}\partial_j)\sum f_i\partial_i = \sum (\partial_j f_i)\partial_i$, there exists for all $0 \neq \mathfrak{D} \in L_k$, $k \geq 0$, an element $v \in V = L_{-1}$ such that $[v, \mathfrak{D}] \neq 0$. Suppose

$$\mathcal{D} = \mathcal{D}_{-1} + \mathcal{D}_0 + \ldots + \mathcal{D}_k, \quad \mathcal{D}_k \neq 0 \quad (\mathcal{D}_i \in L_i),$$

is an element of an ideal N of the general algebra $W(\mathfrak{F})$. Considering only the "highest" element $\mathfrak{D}_k$, we find, after $k+1$ commutations with suitable elements of V, a nonzero element $\mathfrak{D}'_{-1} \in N \cap V$. The representation Γ being irreducible, an L_0-submodule in V containing $\mathfrak{D}'_{-1}$ coincides with V, i. e., $V \subset N$. Further,

$$[V, L_k] = L_{k-1}, \quad 0 \leqslant k \leqslant r. \tag{5}$$

Indeed, if $h_j < p^{m_j} - 1$ for at least one j and $\sum h_s = k$, then

$$[\partial_j, x_1^{(h_1)} \ldots x_j^{(h_j+1)} \ldots x_n^{(h_n)} \partial_i] = x_1^{(h_1)} \ldots x_n^{(h_n)} \partial_i \in L_{k-1},$$

which gives (5). Thus

$$N \supset L_{-1} + L_0 + \ldots + L_{r-1}.$$

By (4), $r \geq 1$ if $p \neq 2$ or $m > 1$, i.e., $L_1 \neq 0$ and $L_0 \subset N$. This means that $L_r = [L_r, L_0] \subset N$, as is easily verified. This proves the simplicity of the algebra $W(\mathfrak{F})$.

For $m = n$, we have a flag $\mathfrak{F}$ of unit height, the algebra $O(\mathfrak{F})$ is obviously isomorphic with the truncated polynomial algebra $k[y_1, \ldots, y_n]$, $y_1^p = \ldots = y_n^p = 0$, and the Lie algebra $W(\mathfrak{F})$ is isomorphic with the algebra of all derivations of the algebra $k[y_1, \ldots, y_n]$. Therefore $W(\mathfrak{F})$ is a Lie p-algebra.

Now suppose $m > n$, so that x_n has height $m_n > 1$. If $W(\mathfrak{F})$ were a p-algebra, we should have

$$\partial_n^{[p]} = \sum_{i=1}^{n} a_i\partial_i, \quad a_i \in O(\mathfrak{F}).$$

But then

$$\partial_n = (\mathrm{ad}\,\partial_n)^p \cdot x_n^{(p)}\partial_n = [\partial_n^{[p]}, x_n^{(p)}\partial_n] = \sum_{i=1}^{n} [a_i\partial_i, x_n^{(p)}\partial_n] \in \mathfrak{L}_{p-2},$$

a contradiction, proving Proposition 1.

For the flag $\mathfrak{F}$: $E = E_0 \supset E_1 = 0$, i.e., for $m = n$, following the notation in [21] we write simply W_n for $W(\mathfrak{F})$. We identify W_n with the algebra of all derivations of the ring $O = k[y_1, \ldots, y_n]$, $y_1^p = \ldots = y_n^p = 0$.

We note that the algebra W_n was first studied by Jacobson [11]. The algebra W_1, the first example of a nonclassical simple Lie algebra, was found by Witt [6]. One can show that for $n = 1$ and arbitrary m the algebra $W(\mathfrak{F})$ coincides with the algebra of Zassenhaus (see, for example, [16]).

2. The subalgebra $\mathfrak{C}$ in W_n. Now assume that char $k = p > 2$.

Proposition 2. *In the algebra W_n we have the inclusion*

$$\mathfrak{L}_{\frac{p-1}{2}} \supset \mathfrak{C} \supseteq \mathfrak{L}_{\frac{p+1}{2}}.$$

For an element $C \in L_{(p-1)/2}$ the following are equivalent:

a) $C = \sum_{i=1}^{n} c_i \dfrac{\partial}{\partial x_i} \in \mathfrak{C}$;

b) $\operatorname{div} C = 0$.

Here div C *denotes the divergence of the operator C:*

$$\operatorname{div} C = \sum_{i=1}^{n} \frac{\partial c_i}{\partial x_i}.$$

The proof will be based on two lemmas. The first will be formulated in somewhat greater generality, needed for the argument in §5.

Lemma 1. *Let* $C = \sum_{i=1}^{n} c_i \dfrac{\partial}{\partial x_i} \in W_n$. *Then the conditions*

$$c_i c_j = 0, \quad \frac{\partial c_i}{\partial x_j} c_k = 0, \quad \sum_i \frac{\partial c_k}{\partial x_j}\frac{\partial c_j}{\partial x_i} = 0, \quad 1 \leqslant i, j, k \leqslant n, \tag{6}$$

are necessary and sufficient to have $(\mathrm{ad}\, C)^2 = 0$. *For $n = 2$, conditions (6) follow from the weaker requirement*

$$(\mathrm{ad}\, C|_{S_{n-1}})^2 = 0,$$

where S_{n-1} is the special algebra defined in §3 (see also §5).

Proof. Let $\mathcal{U} = \sum_{i=1}^{n} u_i \frac{\partial}{\partial x_i}$ be an arbitrary element of W_n, and

$$\mathcal{V} = \sum_{i=1}^{n} v_i \frac{\partial}{\partial x_i} = [[\mathcal{U}, C], C].$$

As is seen by straightforward calculation,

$$v_k = \sum_{i,j} \left\{ \frac{\partial^2 u_k}{\partial x_i \partial x_j} c_i c_j + \frac{\partial u_k}{\partial x_i} \frac{\partial c_i}{\partial x_j} c_j - 2 \frac{\partial u_i}{\partial x_j} \frac{\partial c_k}{\partial x_i} c_j + u_i \left(\frac{\partial c_k}{\partial x_j} \frac{\partial c_j}{\partial x_i} - \frac{\partial^2 c_k}{\partial x_i \partial x_j} c_j \right) \right\}.$$

Differentiating and adding the second group of identification in (6), it is easy to verify that they imply $v_k = 0$, $k = 1, \ldots, n$, i.e., $(\operatorname{ad} C)^2 = 0$. Now require that $v_k = 0$ for $k = 1, \ldots, n$ and for any choice of $\mathfrak{U}$ with the restriction indicated in the lemma. It is easily seen, and in the next section explicitly noted, that S_{n-1} consists of the operators $\mathfrak{U}$ with $\operatorname{div} \mathfrak{U} = 0$ and one further restriction on the homogeneous components of degree $\geq (n-1)p$. In the special cases below this restriction will automatically be fulfilled.

Setting $u_i = 1$ and $u_j = 0$ for all $j \neq i$, we have:

$$\sum_i \left(\frac{\partial c_k}{\partial x_j} \frac{\partial c_j}{\partial x_i} - \frac{\partial^2 c_k}{\partial x_i \partial x_j} c_j \right) = 0, \tag{7}$$

or equivalently,

$$\sum_j \frac{\partial c_k}{\partial x_j} \frac{\partial c_j}{\partial x_i} = \frac{1}{2} \sum_j \frac{\partial}{\partial x_i} \left(\frac{\partial c_k}{\partial x_j} c_j \right), \quad 1 \leqslant i, k \leqslant n. \tag{8}$$

If $\mathcal{U} = x_i \frac{\partial}{\partial x_j}$ (for $n > 2$ we take $i \neq j$), then, using (8),

$$v_k = -2 \frac{\partial c_k}{\partial x_j} c_i = 0, \quad k \neq j, \tag{9}$$

$$v_j = 2 \frac{\partial c_j}{\partial x_j} c_i - \sum_s \frac{\partial c_i}{\partial x_s} c_s = 0. \tag{10}$$

Setting $\mathcal{U} = x_i x_j \frac{\partial}{\partial x_k}$ (again $k \neq i, j$ for $n > 2$), we have

$$v_k = 2c_i c_j + x_i \left(\sum_s \frac{\partial c_j}{\partial x_s} c_s - 2 \frac{\partial c_k}{\partial x_k} c_j \right) + x_j \left(\sum_s \frac{\partial c_i}{\partial x_s} c_s - 2 \frac{\partial c_k}{\partial x_k} c_i \right) = 0,$$

from which, in combination with (10),

$$c_i c_j = 0, \tag{11}$$

here the indices i, j may be chosen arbitrarily for any n. In the case $n \leq 2$ the set of relations (7),)), and (11) coincides with conditions (6). In the case $n > 2$ we must add the result of differentiating lation (11). This proves the lemma.

Lemma 2. *If $f(x_1, \ldots, x_n)$ is a form of degree $m = (p+1)/2$ and $f^2 = 0$ in the ring O, then $f = l^m$, here l is a linear form.*

Proof. By means of a linear transformation we can assume that f contains a term in x_1^m and no erms with factor x_1^{m-1}. Suppose

$$f = x_1^m + x_1^k f_{m-k} + \ldots + f_m, \quad k \leqslant m-2,$$

here the forms $f_{m-k}, \ldots, f_m$ do not depend on x_1. The condition $f^2 = 0$ implies

$$x_1^{m+k} f_{m-k} = 0,$$

hich can hold only for $f_{m-k} = 0$. This proves the lemma.

We proceed now to the proof of Proposition 2. The inclusion $\mathfrak{C} \subset \mathfrak{L}_{(p-1)/2}$ follows from the conitions $c_i^2 = 0$ of Lemma 1, which show that $c_i \in (x_1, \ldots, x_n)^{(p+1)/2}$. To prove the inclusion $\mathfrak{L}_{(p+1)/2}$ $\mathfrak{C}$, consider the element

$$C = l^{(p+3)/2} f \frac{\partial}{\partial x_i},$$

here l is an arbitrary linear form and f an arbitrary polynomial. For this element conditions (6) are ulfilled, so that $C \in \mathfrak{C}$. On the other hand, an arbitrary element of $\mathfrak{L}_{(p+1)/2}$ may be represented as a inear combination of operators of this type. It remains only to prove the equivalence of (a) and (b). ote first of all that if, in the notation of Lemma 1, the polynomial c_i has the form d_i of lowest degree, nd $\mathfrak{D} = \sum_{i=1}^{n} d_i \frac{\partial}{\partial x_i}$, then $(\operatorname{ad} C)^2 = 0$ implies $(\operatorname{ad} \mathfrak{D})^2 = 0$. This comes from the bilinearity of the orresponding relations. Therefore we can restrict our consideration to elements C in which the c_i re forms of degree $m = (p+1)/2$.

A direct verification shows that for elements of the type

$$\mathcal{U} = l^m \sum_{i=1}^{n} a_i \frac{\partial}{\partial x_i}, \quad a_i \in k, \tag{12}$$

here l is a linear form, (a) and (b) are equivalent. Namely for each such element the first two relaions in (6) are fulfilled automically, and the last is equivalent to condition (b).

Suppose the element C satisfies condition (b). Then it is easily verified that it can be represented s a linear combination of elements of type (12) which also satisfy condition (b). This proves that (b) mplies (a).

Now suppose C satisfies condition (a). From the first relation in (6) for $i = j$ it follows that c_i^2 0, and by Lemma 2 this holds only for $c_i = l_i^m$. The relation $c_i c_j = 0$ for $i \neq j$ implies that all the orms l_i are proportional, and this means that C is of type (12). This proves the proposition.

Proposition 3. *The normalizer $N_{W_n}(\mathfrak{C})$ of the subalgebra $\mathfrak{C} \subset W_n$ coincides with $\mathfrak{L}_0$.*

Proof. The fact that $\mathfrak{L}_0 \subseteq N_{W_n}(\mathfrak{S})$ is a consequence of the inclusion $[\mathfrak{L}_{(p+1)/2}, \mathfrak{L}_0] \subseteq \mathfrak{L}_{(p+1)/2}$ and the easily verified identity

$$\operatorname{div}[\mathcal{D}_1, \mathcal{D}_2] = \mathcal{D}_1(\operatorname{div}\mathcal{D}_2) - \mathcal{D}_2(\operatorname{div}\mathcal{D}_1). \tag{13}$$

Suppose $\mathfrak{D} \notin \mathfrak{L}_0$ and $\mathfrak{D} \in N_{W_n}(\mathfrak{S})$. Then there exists an element $\mathfrak{D}$ with the same properties but contained in L_{-1}. As already noted, the algebra W_n is graded. We now use the irreducibility of the representation Γ of L_0 in the space L_{-1}. Since $L_0 \subset N_{W_n}(\mathfrak{S})$, it follows that $N_{W_n}(\mathfrak{S}) = W_n$, i.e., $\mathfrak{S}$ is an ideal in W_n. This contradicts the simplicity of the algebra W_n.

§5. The special algebra $S(\mathfrak{F})$

1. **Description of the algebra $\widetilde{S}(\mathfrak{F})$.** In paragraph 2 of §3 it was remarked that the algebra $\widetilde{S}(\mathfrak{F})$ is fully determined by the type of the flag $\mathfrak{F}$. We select in $O(\mathfrak{F})$ a basis consisting of monomials of the form (1) of §4. By assumption,

$$\mathcal{D}\omega = \Big(\sum_{i=1}^{n} \partial_i f_i\Big)\omega = 0,$$

whence

$$\sum_{i=1}^{n} \partial_i f_i = 0, \quad \text{or} \quad \operatorname{div}\mathcal{D} = 0$$

is a necessary and sufficient condition for an element $\mathcal{D} = \sum f_i \partial_i \in W(\mathcal{F})$ to belong to $\widetilde{S}(\mathfrak{F})$.

The closure relative to commutation of operators with zero divergence follows also from the identity (13) of §4, which holds in the general algebra $W(\mathfrak{F})$. We have

Proposition 1. *For $n > 2$ and arbitrary characteristic $p > 0$ of the base field k, the special algebra $S(\mathfrak{F})$ is simple. As a vector space over k, the algebra $S(\mathfrak{F})$ is spanned by the derivations*

$$\mathcal{D}_{i,j}\{u\} = (\partial_j u)\partial_i - (\partial_i u)\partial_j, \quad u \in O(\mathcal{F}), \tag{1}$$

and has dimension

$$\dim_k S(\mathcal{F}) = (n-1)(p^m - 1), \quad 2 < n \leqslant m,$$

where $m = m_1 + \ldots + m_n$ (m_i is the height of the element x_i).

$S(\mathfrak{F})$ is a restricted (Lie p- algebra) if and only if $m = n$. The basis elements of the factor space $\widetilde{S}(\mathfrak{F})/S(\mathfrak{F})$ have as inverse images the derivations

$$\mathcal{D}_i = \prod_{j \neq i} x_j^{(p^{m_j}-1)} \cdot \partial_i, \quad 1 \leqslant i \leqslant n. \tag{2}$$

The proof can be omitted since for $m = n$ it is contained in the work of Frank [8] and Block [2], and the general case is similar. The statement that for $m > n$ the algebra $S(\mathfrak{F})$ is not restrcited is proved in exactly the same way as the corresponding fact in Proposition 1 of §4 (replace the operator $x_n^{(p)}\partial_n$ by $D_{n,1}\{x_1^{(1)}x_n^{(p)}\}$; cf. (1)).

The gradation in $L = S(\mathfrak{F})$:

$$L = L_{-1} + L_0 + \ldots + L_r,$$

nduced by the gradation of the general algebra $W(\mathfrak{F})$, is obviously characterized by the fact that

$$L_k = \langle \mathcal{D}_{i,j}\{u\};\, u \in O(\mathcal{F})_{k+2}\rangle.$$

n particular, $L_{-1} = \langle\partial_1, \ldots, \partial_n\rangle$, while the subalgebra $L_0 \cong [gl(n, k), gl(n, k)] = A_{n-1}$ and for $\not\equiv 0(p)$ is isomorphic to A_{n-1}; L_0 acts on L_{-1} irreducibly. Note that the length r of the filtration $\{\mathfrak{L}_i\}$ s here one less than for $W(\mathfrak{F})$ (cf. (4) of §4).

For $m = n$ we denote the p-algebra $S(\mathfrak{F})$ by S_{n-1}.

2. The subalgebra $\mathfrak{C}$ in S_n.

Proposition 2. *The subalgebra* $\mathfrak{C}$ *of the algebra* S_n *coincides with the term* $\mathfrak{L}_{(p-1)/2}$ *of the filtra-on* $\{\mathfrak{L}_i\}$ *in* S_n.

Proof. By Lemma 1 of § 4, the subalgebra $\mathfrak{C}$ of S_{n-1} is given by the same conditions (6) of §4. roposition 2 therefore follows from Proposition 2 of §4.

Proposition 3. *The normalizer* $N_{S_n}(\mathfrak{C})$ *of the subalgebra* $\mathfrak{C} \subset S_n$ *coincides with* $\mathfrak{L}_0$.

The proof is the same as for Proposition 3 of §4.

§6. The hamiltonian algebra $H(\mathfrak{F})$

1. Description of the algebra $\widetilde{H}(\mathfrak{F})$, Char $k = p > 2$. In accordance with paragraph 2 of §3, for a xed flag $\mathfrak{F}$, there is involved in the definition of the algebra $\widetilde{H}(\mathfrak{F})$ in addition a skew-symmetric form . We shall sometimes write $\widetilde{H}(\mathfrak{F}, \omega)$, where

$$\omega = \sum_{i,j=1}^{n} \omega_{i,j}\, dx_i \wedge dx_j, \quad n = 2n',$$

$$M = (\omega_{i,j})_1^n, \quad M^* = -M, \quad \det M \neq 0. \tag{1}$$

e also set

$$M^{-1} = (\bar{\omega}_{i,j})_1^n. \tag{2}$$

derivation $\mathcal{D} = \sum f_i\partial_i \in W(\mathcal{F})$ is an element of the algebra $\widetilde{H}(\mathfrak{F}, \omega)$ if and only if $\mathfrak{D}\omega = 0$, i.e.,

$$\partial_s \bar{f}_t = \partial_t \bar{f}_s, \quad s, t = 1, \ldots, n, \tag{3}$$

here

$$\bar{f}_t = \sum_{i=1}^{n} \omega_{i,t} f_i.$$

he solution of the system (3) may obviously be written in the form

$$\bar{f}_t = \partial_t u + \alpha_t x_t^{(p^{m_t}-1)}, \quad \alpha_t \in k,$$

where m_i is the height of the variable x_i and u is an arbitrary element of $O(\mathfrak{F})$. Thus every derivation $\mathfrak{D} \in \widetilde{H}(\mathfrak{F}, \omega)$ is represented as a linear combination

$$\mathfrak{D} = \mathfrak{D}_u + \sum_{j=1}^{n} a_j Q_j, \quad u \in O(\mathfrak{F}), \quad a_j \in k,$$

where

$$\mathfrak{D}_u = \sum_{i=1}^{n} f_i \partial_i, \quad \sum_{i=1}^{n} \omega_{ij} f_i = \partial_j u, \quad 1 \leqslant j \leqslant n$$
$$\left(f_i = \sum_j \overline{\omega}_{j,i} \partial_j u\right);$$
$$Q_j = \sum_i a_i \partial_i, \quad \sum_i \omega_{ik} a_i = \delta_{j,k} x_j^{(p^{m_j}-1)}, \quad 1 \leqslant j \leqslant n.$$

When the field k has characteristic 0 (or when $O(\mathfrak{F}) = O(E)$), there are no operators Q_j and the expression $\mathfrak{D} = \mathfrak{D}_u$ is well known (see, for example, [14]).

Now let $\mathfrak{D}_v = \sum g_i \partial_i$, $\sum_i \omega_{ij} g_i = \partial_j v$. By the skew-symmetry of the matrix M (cf. [1]),

$$\sum_i (\partial_j g_i)(\partial_i u) = -\sum_i f_i (\partial_j \partial_i v),$$
$$\sum_i (\partial_j f_i)(\partial_i v) = -\sum_i g_i (\partial_j \partial_i u).$$

Therefore

$$[\mathfrak{D}_u, \mathfrak{D}_v] = \sum_i h_i \partial_i, \quad \sum_i \omega_{i,j} h_i = \partial_j w,$$

where

$$w = \sum_i f_i \partial_i v = -\sum_i g_i \partial_i u,$$

or, in the notation (2),

$$[\mathfrak{D}_u, \mathfrak{D}_v] = \mathfrak{D}_w, \quad w = \sum_{i,j} \overline{\omega}_{i,j} (\partial_i u)(\partial_j v). \tag{4}$$

In particular, when

$$M = \begin{pmatrix} 0 & -I_{n'} \\ I_{n'} & 0 \end{pmatrix},$$

we obtain a commutator law expressible by the ordinary Poisson bracket $\{\ ,\ \}$:

$$w = \{u, v\} = \sum_{i=1}^{n'} (\partial_i u \partial_{n'+i} v - \partial_{n'+i} u \partial_i v). \tag{4'}$$

Since $Q_j = \mathfrak{D}_{x_j^{(p^{m_j})}}$ in the algebra $H(E)$, we have

$$[Q_i, Q_j] = \bar{\omega}_{i,j}\mathfrak{D}_{x_i^{(p^{m_i}-1)}x_j^{(p^{m_j}-1)}}; \tag{5}$$

$$[\mathfrak{D}_u, Q_j] = \mathfrak{D}_w, \quad w = \sum_i \bar{\omega}_{i,j}(\partial_i u)\, x_j^{(p^{m_j}-1)}. \tag{6}$$

Summarizing, we have:

Proposition 1. *As a vector space over k, the algebra $\widetilde{H}(\mathfrak{F},\omega)$ is spanned by the derivations $\mathfrak{D}_u$, $u \in O(\mathfrak{F})$, and $u\,Q_j$, $1 \leqslant j \leqslant n$; $\dim \widetilde{H}(\mathfrak{F},\omega) = p^m + n - 1$. Multiplication in $\widetilde{H}(\mathfrak{F},\omega)$ is given by formulas (4)-(6). If $n > 2$ and the monomial $\prod_i x_i^{(p^{m_i}-1)}$ occurs in u with nonzero coefficient, then $\mathfrak{D}_u \notin [\widetilde{H}, \widetilde{H}]$.*

The last assertion is false for $n = 2$, as is seen from formula (5), while for $n > 2$ the monomial $\prod_i x_i^{(p^{m_i}-1)}$ could be obtained in formula (4) only when

$$u = \prod_i x_i^{(\alpha_i)}, \quad v = \prod_i x_i^{(\beta_i)}, \quad \alpha_i + \beta_i = p^{m_i} - 1$$

for $i \neq s, t$, $\alpha_s + \beta_s = p^{m_s}$, $\alpha_t + \beta_t = p^{m_t}$. But then $w = \gamma \cdot \prod_i x_i^{(p^{m_i}-1)}$, where

$$\gamma = \prod_{i\neq s,t}\binom{p^{m_i}-1}{\alpha_i}\cdot\left\{\binom{p^{m_s}-1}{\alpha_s-1}\binom{p^{m_t}-1}{\alpha_t} - \binom{p^{m_s}-1}{\alpha_s}\binom{p^{m_t}-1}{\alpha_t-1}\right\}\bar{\omega}_{s,t} = 0.$$

2. **The algebra $H(\mathfrak{F},\omega)$.** By definition, $H(\mathfrak{F},\omega) = [[\widetilde{H},\widetilde{H}],[\widetilde{H},\widetilde{H}]]$, but from Proposition 1 it follows almost immediately that $H = [\widetilde{H},\widetilde{H}]$ for $n > 2$. We have

Proposition 2. *The hamiltonian algebra $H(\mathfrak{F},\omega)$ is isomorphic to the algebra $O'(\mathfrak{F})/k$ with the multiplication law*

$$(u, v) \to \{u, v\}_\omega = \sum_{i,j}\bar{\omega}_{i,j}(\partial_i u)(\partial_j v). \tag{7}$$

Here $O'(\mathfrak{F})$ consists of all functions $u \in O(\mathfrak{F})$ not containing the term $\prod_i x_i^{(p^{m_i}-1)}$, and k is its subalgebra of contants; $\dim H(\mathfrak{F},\omega) = p^m - 2$, $m = m_1 + \ldots + m_n$. The algebra $H(\mathfrak{F},\omega)$ is simple for any $n = 2n' \geq 2$, char $k = p > 2$, and is restricted if and only if $m = n$. In this case, $\{\,,\}_\omega$ can be taken as the ordinary Poisson bracket (4').

The first part of Proposition 2 is in essence already proved. For $m = n$, when all nondegenerate skew-symmetric forms are equivalent, there is obtained a simple Lie p-algebra $H_{n'}$, considered earlier by Albert and Frank [1] and Block [2]. In this as well as in the general case, the simplicity is easily obtained if we note that the algebra $L = H(\mathfrak{F}, \omega)$ is graded: $L = L_{-1} + L_0 + \ldots + L_r$, $r = \sum_i (p^{m_i} - 1) - 2$; $L_i = O(\mathfrak{F})_{i+2}$. Further, $L_0 \cong C_{n'}$, and Γ: $L_0 \to \mathrm{Hom}(L_{-1}, L_{-1})$ is an n-dimensional irreducible representation of the symplectic algebra $C_{n'}$.

If $m > n$, suppose, say, $x_n^{(p+1)} \in H(\mathfrak{F}, \omega)$ (for simplicity we identify $H(\mathfrak{F}, \omega)$ with $O'(\mathfrak{F})/k$). Since the matrix M is nondegenerate, $\bar{\omega}_{j,n} \neq 0$ for some index j. Assume that $(x_j^{(1)})^{[p]} = u \in H(\mathfrak{F}, \omega)$ Then

$$(\bar{\omega}_{j,n})^p x_n = (\mathrm{ad}\, x_j^{(1)})^p x_n^{(p+1)} = \{u, x_n^{(p+1)}\}_\omega$$
$$= \sum_i \bar{\omega}_{i,n} (\partial_i u) x_n^{(p)} \in \mathfrak{L}_{p-2} = \sum_{s \geqslant p-2} L_s,$$

although in fact $x_n \notin \mathfrak{L}_0$. This proves that $H(\mathfrak{F}, \omega)$ is not a Lie p-algebra.

3. The subalgebra $\mathfrak{C}$ in $H_{n'}$.

Proposition 3. *The subalgebra $\mathfrak{C}$ of the algebra $H_{n'}$ coincides with the term $\mathfrak{L}_{(p-1)/2}$ of the filtration $\{\mathfrak{L}_i\}$ in $H_{n'}$.*

For the proof we need a lemma. Let $\mathfrak{C}'$ be the image of $\mathfrak{C}$ in the algebra O' (we omit the flag $\mathfrak{F}$: $E = E_0 \supset E_1 = 0$ corresponding to the truncated polynomial algebra). Obviously it is generated by elements $c \in O'$ for which $\{\{O', c\}, c\} \in k$.

Lemma. *If $c \in \mathfrak{C}'$, then* $\frac{\partial c}{\partial x_i}\frac{\partial c}{\partial x_j} = 0$, $1 \leqslant i, j \leqslant n$, *in* O'.

Proof. Consider a fixed element $c \in \mathfrak{C}'$, with $(\mathrm{ad}\, c)^2 = 0$. For convenience we use the symbol ϵ_i, equal to 1 for $i \leq n'$ and to -1 for $i > n'$. We also set $\pi i = i + \epsilon_i n'$. Observe that π is a permutation of order two on the set $1, 2, \ldots, n$. Obviously $\{x_i, c\} = \varepsilon_i \frac{\partial c}{\partial x_{\pi i}}$, and therefore the relation $\{\{x_i, c\}, c\} = 0$ in O'/k means that

$$a_i = \varepsilon_i \left\{\frac{\partial c}{\partial x_{\pi i}}, c\right\} \in k, \quad i = 1, \ldots, n.$$

It is clear that the map $u \to \{u, v\}$ for fixed v (and likewise $v \to \{u, v\}$ for fixed u) is a derivation of the ring O'. Therefore

$$\{\{x_i x_j, c\}, c\} = \left\{\varepsilon_j x_i \frac{\partial c}{\partial x_{\pi j}} + \varepsilon_i x_j \frac{\partial c}{\partial x_{\pi i}}, c\right\}$$
$$= 2\varepsilon_i \varepsilon_j \frac{\partial c}{\partial x_{\pi i}} \frac{\partial c}{\partial x_{\pi j}} + a_i x_j + a_j x_i = b_{ij} \in k. \tag{8}$$

Consider now for $i \neq j$ the identity

$$\left(\frac{\partial c}{\partial x_{\pi i}}\right)^2 \left(\frac{\partial c}{\partial x_{\pi j}}\right)^2 = \left(\frac{\partial c}{\partial x_{\pi i}} \cdot \frac{\partial c}{\partial x_{\pi j}}\right)^2,$$

which, in view of (8), gives:

$$(b_{ii} - 2a_ix_i)(b_{jj} - 2a_jx_j) = (b_{ij} - a_ix_j - a_jx_i)^2.$$

Since the right-hand side contains the term $a_i^2x_j^2$ but the left-hand side does not, we have

$$a_i = 0, \quad i = 1, \ldots, n. \tag{9}$$

We now use the fact that $\{\{(x_ix_j)^2, c\}, c\} = 0$ in O'/k. As already remarked, $u \to \{u, v\}$ is a derivation, so that

$$\{\{(x_ix_j)^2, c\}, c\} = 2x_ix_j\{\{x_ix_j, c\}, c\} + 2\{x_ix_j, c\}^2,$$

which in view of (8) is equal to

$$2x_ix_jb_{ij} + 2\left(\varepsilon_jx_i\frac{\partial c}{\partial x_{nj}} + \varepsilon_ix_j\frac{\partial c}{\partial x_{ni}}\right)^2,$$

or with the help of (9),

$$4x_ix_jb_{ij} + x_i^2b_{jj} + x_j^2b_{ii}.$$

By hypothesis this element must lie in k, from which it follows that $b_{ij} = 0$. This proves the lemma.

Proof of Proposition 3. We write $c \in \mathfrak{C}'$, with $(\operatorname{ad} c)^2 = 0$, in the form $c = c_s + c_{s+1} + \ldots$, where the c_i are homogeneous polynomials of degree i and $s > 0$. In view of the lemma, $(\partial c/\partial x_i)^2 = 0$, from which it follows that $s \geq (p+3)/2$.

Conversely, let $c = c_s + c_{s+1} + \ldots$, $s \geq (p+3)/2$. We will prove that $c \in \mathfrak{C}'$. It is sufficient to establish this for homogeneous elements c_i for $i \geq (p+3)/2$. Each such element is a linear combination of functions of the form $l^m v$, $m = (p+3)/2$, where $v = x_1^{\alpha_1} \ldots x_n^{\alpha_n}$, $\alpha_i \geq 0$, and l is a linear function. Using again the observation at the beginning of the proof of the lemma, we see first of all that

$$\{u, l^m v\} = l^{m-1}w, \quad w \in O',$$

for arbitrary $u \in O$, and then that

$$\{l^{m-1}w, l^m v\} = l^{2m-3}y, \quad y \in O.$$

Since $l^p = 0$ and $2m - 3 \geq p$, this expression is equal to zero. It remains to observe that to the functions in O' of the form $c_s + c_{s+1} + \ldots$, $s \geq (p+3)/2$, corresponds in the algebra $H_{n'}$ the subalgebra $\mathfrak{L}_{(p-1)/2}$.

Proposition 4. *The normalizer* $N_{H_{n'}}(\mathfrak{C})$ *of the subalgebra* $\mathfrak{C}$ *in* $H_{n'}$ *coincides with* $\mathfrak{L}_0$.

The proof is the same as for Proposition 3 of §5.

§7. The contact algebra $K(\mathfrak{F})$

1. Description of the algebra $\widetilde{K}(\mathfrak{F})$. Even though the results of this section are not required later, we will examine somewhat extensively the structure of the algebra $K(\mathfrak{F})$, since in the case of finite characteristic this has not been done up to now, even for the flag $\mathfrak{F}$: $E = E_0 \supset E_1 = 0$ for which $K(\mathfrak{F}) = K_{n'+1}$ is a Lie p-algebra. It would be very surprising if K_{n+1}, as the classical contact algebra extended to characteristic p, were found not to be isomorphic with the p-algebra studied five years ago

by Frank [9]. The authors recognize that they have verified the isomorphism only on the basis of an indirect argument.

Thus, let $n = 2n' + 1$, $p > 2$,

$$\omega = \sum_{i=1}^{n} \beta_i dx_i + \sum_{i=1}^{n-1} \alpha_{i,\pi i} x_i dx_{\pi i},$$

$$\beta_n = 1, \quad \beta_i \beta_{\pi i} = 0, \quad i < n; \quad \alpha_{i,\pi i} = -\alpha_{\pi i, i}.$$

As was already known to Lie [14], for characteristic 0 (of course when $\beta_i = 0$ for $i < n$), the coefficients f_i of a differential operator $\mathfrak{D} = \sum_{i=1}^{n} f_i \partial_i$, satisfying the condition $\mathfrak{D}\omega = u\omega$, $u \in O(\mathfrak{F})$, can be represented by a single function

$$f = \frac{1}{2}\left\{\sum_{i=1}^{n} \beta_i f_i + \sum_{i=1}^{n-1} \alpha_{i,\pi i} x_i f_{\pi i}\right\} \in O(\mathfrak{F}),$$

$$f_i = (x_i + \alpha_{i,\pi i}\beta_{\pi i})\partial_n f - \alpha_{i,\pi i}\partial_{\pi i} f, \quad i < n,$$

$$f_n = \Delta f,$$

where

$$\Delta f = 2f - \sum_{j=1}^{n-1} (x_j + \alpha_{j,\pi j}\beta_{\pi j})\partial_j f. \tag{1}$$

The operator $\mathfrak{D}$ corresponding to the function f will be denoted by $\mathfrak{D}_f$. Thus the algebra $\widetilde{K}(\mathfrak{F})$ defined in paragraph 1 of §3, or, more exactly, the algebra $\widetilde{K}(\mathfrak{F}, \omega)$, consists of all derivations $\mathfrak{D}_f$, where $f \in O(\mathfrak{F})$.

It is clear that $\dim \widetilde{K}(\mathfrak{F}) = p^m$, $m = m_1 + \ldots + m_n$. The operation of commutation [,] is represented directly in terms of the functions:

$$[\mathfrak{D}_u, \mathfrak{D}_v] = \mathfrak{D}_w,$$

$$w = \Delta u \cdot \partial_n v - \Delta v \cdot \partial_n u + \{u, v\}'_\omega, \tag{2}$$

where

$$\{u, v\}'_\omega = \sum_{i=1}^{n-1} \alpha_{i,\pi i} (\partial_i u)(\partial_{\pi i} v)$$

is the Poisson bracket only in the variables x_i, $i < n$; Δu and Δv are calculated by formula (1). For $m = n$ we identify $O(\mathfrak{F})$, as always, with the truncated polynomial algebra and reduce ω to the form

$$\omega = dx_n + \sum_{i=1}^{n'} (x_{n'+i} dx_i - x_i dx_{n'+i}),$$

so that the operation of commutation is given by the formula

$$\left.\begin{aligned} [\mathcal{D}_u, \mathcal{D}_v] &= \mathcal{D}_w, \\ w &= \frac{\partial u}{\partial x_n}\Delta v - \frac{\partial v}{\partial x_n}\Delta u + \{u, v\}', \\ \Delta u &= 2u - \sum_{i=1}^{n-1} x_i \frac{\partial u}{\partial x_i}. \end{aligned}\right\} \tag{2'}$$

he Poisson bracket $\{u, v\}'$ assumes the form (4′) of §6.

Proposition 1. *If $n+3 \neq 0\,(p)$, then $K(\mathfrak{F}) = \widetilde{K}(\mathfrak{F})$ is a simple Lie algebra of dimension p^m, $m = m_1 + \ldots + m_n$ (m_i the height of the variable x_i). If $n+3 \equiv 0\,(p)$, then the derived algebra $K(\mathfrak{F})$ of $\widetilde{K}(\mathfrak{F})$ is simple algebra of dimension $p^m - 1$ which consists of all operators $\mathfrak{D}_f$ with functions $f \in O(\mathfrak{F})$ not ontaining the term $\prod_i x_i^{(p^{m_i}-1)}$. For $m = n$, and only in this case, $K(\mathfrak{F}) = K_{n'+1}$ is a Lie p-algebra.*

The proof is divided into three parts.

1. We prove first that for $n+3 \equiv 0\,(p)$ the set of operators $\mathfrak{D}_f$, where $\deg f < \sum_{i=1}^{n} (p^{m_i} - 1)$, orms a subalgebra and in fact an ideal (namely the derived algebra) in $\widetilde{K}(\mathfrak{F})$. For this, in view of the ilinearity of (2), it is sufficient to verify that for arbitrary monomials

$$u = x^{(s)}, \quad (s) = (s_1, \ldots, s_n),$$
$$v = x^{(t)}, \quad (t) = (t_1, \ldots, t_n)$$

he function $w = w_1 + w_2 + w_3$, where

$$\begin{aligned} w_1 &= \left\{(2 + s_n - |s|)\binom{s+t-\varepsilon_n}{s} - (2 + t_n - |t|)\binom{s+t-\varepsilon_n}{t}\right\} x^{(s+t-\varepsilon_n)}, \\ w_2 &= \sum_{j=1}^{n-1} \alpha_{j,nj}\beta_{nj}\left\{\binom{s+t-\varepsilon_j-\varepsilon_n}{s-\varepsilon_n} - \binom{s+t-\varepsilon_j-\varepsilon_n}{t-\varepsilon_n}\right\} x^{(s+t-\varepsilon_j-\varepsilon_n)}, \\ w_3 &= \sum_{i=1}^{n-1} \alpha_{i,ni}\binom{s+t-\varepsilon_i-\varepsilon_{ni}}{s-\varepsilon_i} x^{(s+t-\varepsilon_i-\varepsilon_{ni})}, \end{aligned} \tag{3}$$

orresponding to the operators $[\mathfrak{D}_u, \mathfrak{D}_{\dot{u}}]$ does not contain the monomial $e = \prod_i x_i^{(\overline{m}_i)}$. Here we have set

$$\left.\begin{aligned} &\varepsilon_i = (0, \ldots, \underset{i}{1}, \ldots, 0), \quad |s| = s_1 + \ldots + s_n, \\ &(\overline{m}) = (\overline{m}_1, \ldots, \overline{m}_n), \quad \overline{m}_i = p^{m_i} - 1, \quad 1 \leqslant i \leqslant n, \\ &\binom{s}{t} = \prod_{i=1}^{n}\binom{s_i}{t_i}. \end{aligned}\right\} \tag{4}$$

ndeed, the monomial e could occur in only one of the functions w_1, w_2, w_3 (cf. (3)). If $(s + t - \epsilon_n) = (\overline{m})$, he coefficient of e in w_1, and consequently in w, is

$$(-1)^{|s|}\{(2 + s_n - |s|) + (2 + t_n - |t|\} = (-1)^{|s|}(n+3) = 0.$$

It is easy to verify likewise that for arbitrary n the coefficients of e in w_2 and w_3 are zero. This proves our assertion.

2. As in the statement of Proposition 1, we denote by $K(\mathfrak{F})$ the algebra $\widetilde{K}(\mathfrak{F})$ or, for $n+3 \equiv 0(p)$, the ideal of codimension 1 in $\widetilde{K}(\mathfrak{F})$ not containing the element $\mathfrak{D}_e$. Let N be an ideal of the algebra $K(\mathfrak{F})$, and $0 = \mathfrak{D}_f \in N$. Then also

$$[\mathcal{D}_1, \mathcal{D}_f] = 2\mathcal{D}_{\partial_n f} \in N. \tag{5}$$

Applying this operation repeatedly, we obtain an element $f \neq 0$ such that $\mathfrak{D}_f \in N$ and $\partial_n f = 0$. But then

$$[\mathcal{D}_{x_k}, \mathcal{D}_f] = \mathcal{D}_{\{x_k, f\}'_\omega} = \alpha_{k,\pi k}\mathcal{D}_{\partial_{\pi k} f} \in N \tag{6}$$

for any $k < n$. Making several such commutations, we conclude that $\mathfrak{D}_1 \in N$. If $u = u^{(s)}$ is an arbitrary monomial and $(s) \neq (\bar{m})$ (cf. (4)) for $n+3 \equiv 0(p)$, then formulas (5) and (6) show that $\mathfrak{D}\partial_{iv} \in N$, $1 \leq i \leq n$. Thus we get automatically in N all operators of the form $\mathfrak{D}_{x^{(s)}}$ with $|s| \leq |\bar{m}| - 1$ for $n+3 \not\equiv 0(p)$ and $|s| \leq |\bar{m}| - 2$ for $n+3 \equiv 0(p)$. It is easy to verify that linear combinations of these and of their products by pairs give all elements of the algebra $K(\mathfrak{F})$ (cf. (3)). Consequently, $N = K(\mathfrak{F})$.

3. For $m > n$ the contact algebra $K(\mathfrak{F})$ cannot be a Lie p-algebra. Indeed, if $m_n > 1$, then

$$\mathcal{D}_1^{[p]} = \mathcal{D}_a \in K(\mathcal{F}) \Rightarrow 2\mathcal{D}_1 = (\operatorname{ad} \mathcal{D}_1)^p \mathcal{D}_{x_n^{(p)}}$$
$$= [\mathcal{D}_1^{[p]}, \mathcal{D}_{x_n^{(p)}}] = [\mathcal{D}_a, \mathcal{D}_{x_n^{(p)}}] = \mathcal{D}_w$$

– an inconsistent equation, since formula (2) implies that $w \in \mathfrak{m}^{(p-2)}$ whenever $v \in \mathfrak{m}^{(p)}$. If $m_j > 1$ for some $j = \pi k < n$, then

$$\mathcal{D}_{x_k}^{[p]} = \mathcal{D}_a \in K(\mathcal{F}) \Rightarrow \alpha_{k,\pi k}\mathcal{D}_1 = (\operatorname{ad} \mathcal{D}_{x_k})^p \mathcal{D}_{x_{\pi k}^{(p)}}$$
$$= [\mathcal{D}_a, \mathcal{D}_{x_{\pi k}^{(p)}}] = \mathcal{D}_w, \quad w \in \mathfrak{m}^{(p-2)},$$

– again an inconsistent equation. Here we have made use of the relations

$$[\mathcal{D}_{x_k}, \mathcal{D}_{x_{\pi k}^{(s)}}] = \alpha_{k,\pi k}\mathcal{D}_{x_{\pi k}^{(s-1)}}$$

and

$$\alpha_{k,\pi k}^2 = 1.$$

Now let $m = n$. Since $K_{n'+1} = \widetilde{K}(\mathcal{F}),\ \mathcal{F}: E = E_0 \supset E_1 = 0$, is the algebra of all (not only special, as for $m > n$) derivations of the ring $O(\mathfrak{F})$ which multiply the form ω by a function in $O(\mathfrak{F})$, it is obvious that $(\mathfrak{D}_u)^{(p)} \in \widetilde{K}_{n'+1}$, and consequently $\widetilde{K}_{n'+1}$ is a Lie p-algebra. For $n+3 \not\equiv 0(p)$, $K_{n'+1}$ coincides with $\widetilde{K}_{n'+1}$ and is therefore a Lie p-algebra. It remains to prove that the algebra $K_{n'+1}$ is restricted for $n+3 \equiv 0(p)$. In other words, we must verify the invariance of the subalgebra $K_{n'+1} \subset \widetilde{K}_{n'+1}$ relative to the mapping $\mathfrak{D}_u \to \mathfrak{D}_u^p$. For this we calculate $\mathfrak{D}_u^p$ for $u = x^\alpha$, $\alpha = (\alpha_1, \ldots, \alpha_n)$. If $\alpha_n \geq 2$, then obviously $\mathfrak{D}_u^p = 0$. For $\alpha_n = 0$ or 1, let

$$\mathcal{D}_u^p = \mathcal{D}_v + \nu\mathcal{D}_{(x_1 \ldots x_n)^{p-1}}, \quad \nu \in k,$$

here v contains only terms of degree $< n(p-1)$. Commutating this equality with $\mathfrak{D}_1$ we get that

$$[\mathscr{D}_1, \mathscr{D}_u^p] = -2\mathscr{D}_{\frac{\partial v}{\partial x_n}} + 2\nu\mathscr{D}_{(x_1 \ldots x_{n-1})^{p-1}x_n^{p-2}} \tag{7}$$

sing the commutation law (2′)!). If $\alpha_n = 0$, the expression on the left is zero, i.e., in particular $= 0$. Let $\alpha_n = 1$. It is easy to verify by induction that if in general $u = gx_n$, where g is independent x_n, then $(\mathrm{ad}\,\mathfrak{D}_u)^k\mathfrak{D}_1 = \nu_k\mathfrak{D}_{g^k}$. Therefore (7) again gives $\nu = 0$. Thus for an arbitrary monomial v we ave $\mathfrak{D}_v^p \in K_{n'+1}$, and since $K_{n'+1}$ is a subalgebra of the p-algebra $\widetilde{K}_{n'+1}$, it is likewise a Lie p-lgebra. This completes the proof of Proposition 1.

2. **The filtration of the algebra $K_{n'+1}$.** We have

Lemma. *The subalgebra $\mathfrak{L}_i$ of the algebra $K_{n'+1}$ consists of operators $\mathfrak{D}_f$ with $f \in kx_n^{i+1} + \mathfrak{m}^{i+2}$, here $\mathfrak{m} = (x_1, \ldots, x_n)$.*

For $i = -1$ this is obvious; it says that $O = k + \mathfrak{m}$. Suppose it is proved for i. Take $\mathfrak{D}_f \in \mathfrak{L}_{i+1}$. hen $\mathfrak{D}_f \in \mathfrak{L}_i$ and $f = \alpha x_n^{i+1} + g$, $g \in \mathfrak{m}^{i+2}$. Since by hypothesis $f_j \in \mathfrak{m}^{i+2}$, the formula expressing f terms of f_j shows that $f \in \mathfrak{m}^{i+2}$, and those expressing f_j in terms of f show that $\partial f/\partial x_k \in \mathfrak{m}^{i+2}$, $\neq n$, i.e., has the form $\beta x_n^{i+2} + g'$, $g' \in \mathfrak{m}^{i+3}$.

As opposed to W_n, S_n, and H_n, the algebra K_n provided with its ordinary filtration $\{\mathfrak{L}_i\}$ is not aded, as is shown by the following:

Proposition 2. *Let $\widetilde{L}$ be the associated graded Lie algebra of $\widetilde{K}_{n'+1}$ (with its ordinary filtration). hen there is a natural isomorphism*

$$\widetilde{L} \cong O_1 \otimes H'_{n'} + W_1,$$

here $O_1 = k[x_n]/(x_n^p)$ and $H'_{n'} \cong O/k \cong H_{n'} + k(x_1 \ldots x_{n-1})^{p-1}$ (cf. paragraph 1 of §6). Under this somorphism,

$$\widetilde{L}_i \cong \sum_{k+l=i} \langle x_n^k \rangle \otimes (H'_{n'})_l + (W_1)_i$$

$(W_1)_i = 0$ for $i \geq p-1$).

Further, $O_1 \otimes H'_{n'}$ is an ideal in $\widetilde{L}$: if $\mathfrak{D}_{g_0} \in O_1 \otimes H'_{n'}$ and $D_{x_n^{s+1}} \in (W_1)_s$, then

$$[\mathscr{D}_{x_n^{s+1}}, \mathscr{D}_{g_0}] = \mathscr{D}_{h_0},$$

$$h_0 = (s+1)\, x_n^s \cdot \Delta g_0 - 2x_n^{s+1} \frac{\partial g_0}{\partial x_n}$$

where g_0 is a polynomial not containing pure powers of x_n).

The proof is obtained immediately from the lemma and the definition of the algebra $\widetilde{K}_{n'+1}$. Obiously $\widetilde{L}_i$ consists of operators $\mathfrak{D}_f$, where $f = \alpha x_n^{i+1} + f_0$, f_0 a homogeneous form of degree $i+2$ not ontaining the term x_n^{i+2}. Let $g = \beta x_n^{j+1} + g_0$, $\deg g_0 = j+2$. From formula (2′) applied to the lgebra $\widetilde{L}$, we get that

$$[\mathscr{D}_f, \mathscr{D}_g] = \mathscr{D}_h, \quad h = 2\alpha\beta(i-j)\, x_n^{i+j+1} + h_0,$$

where

$$h_0 = (\{f_0, g_0\}')_0 + \alpha(i+1)x_n^i\Delta g_0 - \beta(j+1)x_n^j\Delta f_0 + 2\beta x_n^{j+1}\frac{\partial f_0}{\partial x_n} - 2\alpha x_n^{i+1}\frac{\partial g_0}{\partial x_n}.$$

In particular,

$$[\mathcal{D}_{f_0}, \mathcal{D}_{g_0}] = \mathcal{D}_{\{f_0, g_0\}'}$$

is the operation of multiplication in the algebra $O_1 \otimes H'_{n'}$.

Remark. It is possible to introduce in K_n a gradation

$$K_{n'+1} = L = L'_{-2} + L'_{-1} + L'_0 + \ldots + L'_r, \quad [L'_s, L'_t] \subseteq L'_{s+t}, \tag{8}$$

by setting

$$L'_j = \langle \mathcal{D}_{x^{(s)}};\ |s| + s_n = j+2 \rangle$$

(using formulas (3) and the notation (4): compare the gradation of the algebra H_n in paragraph 2 of §6).

Obviously $L'_0 \cong C_{n'} \oplus k$ and Γ: $C_{n'} \to$ End L'_{-1} is the standard representation. If $\mathfrak{L}'_i = \sum_{j\geq i} L'_j$, then $\mathfrak{L}'_0 = \mathfrak{L}_0$, the subalgebra defining the ordinary filtration. All the results of this paragraph carry over at once to an arbitrary algebra $K(\mathfrak{F}, \omega)$ whenever the coefficients β_i, $i < n$, in ω are equal to zero.

3. The subalgebra $\mathfrak{C}$ in $K_{n'+1}$.

Proposition 3. *The subalgebra $\mathfrak{C}$ of the algebra $K_{n'+1}$ coincides with the algebra of elements $\mathfrak{D}_f$, $f \in \mathfrak{m}^{(p+3)/2}$.*

Proof. Let $[[\mathfrak{D}_u, \mathfrak{D}_c], \mathfrak{D}_c] = \mathfrak{D}_w$. It is obvious that w is contained in the ideal in O generated by $\frac{\partial c}{\partial x_i} \cdot \frac{\partial c}{\partial x_j}$ and $\frac{\partial^2 c}{\partial x_i \partial x_j}$. Now take $c = l^{(p+3)/2}v$, where l is a linear form. Then $(\text{ad } \mathfrak{D}_c)^2 = 0$. Since the elements of the form $l^{(p+3)/2}v$ generate $\mathfrak{m}^{(p+3)/2}$ as a linear space, it follows that $\mathfrak{D}_f \in \mathfrak{C}$ if $f \in \mathfrak{m}^{(p+3)/2}$.

Conversely, $\mathfrak{D}_c \in \mathfrak{C}$ and $c \notin \mathfrak{m}^{(p+1)/2}$. Then

$$c = c_s + c_{s+1} + \ldots, \quad s \leqslant \frac{p+1}{2}, \quad c_i \in \mathfrak{m}^i.$$

If u is a homogeneous polynomial of degree k, it is easily verified that the homogeneous component of lowest degree in w is $\{\{u, c_s\}', c_s\}'$, and the degree of this component is $k + 2s - 4$. Therefore when $w = 0$ we must have the relation

$$\{\{u, c_s\}', c_s\}' = 0.$$

Letting $u = x_i$ and $u = x_i^2$, $1 \leq i < n$, we obtain the relation

$$\left(\frac{\partial c_s}{\partial x_i}\right)^2 = 0,$$

$$(t-s)(t-1)x_n^{t+2s-2}=0.$$

$s \geq 1$, then for $t=0$ we arrive at the inequality $s \geq (p+3)/2$. Contradiction. For $s=0$, the same result is obtained by letting $t=2$.

Proposition 4. *The normalizer* $N_{K_{n'+1}}(\mathfrak{C})$ *of the algebra* $\mathfrak{C}$ *coincides with the subalgebra* $\mathfrak{L}_0$ *of* $_{n'+1}$.

We prove first that $N_{K_{n'+1}}(\mathfrak{C}) \supseteq \mathfrak{L}_0$. By the lemma, $\mathfrak{L}_0 = \langle \mathfrak{D}_g \rangle$, $g \in kx_n + \mathfrak{m}^2$. If $f \in \mathfrak{m}^{(p+3)/2}$, $g \in \mathfrak{m}^2$, and $\mathfrak{D}_h = [\mathfrak{D}_f, \mathfrak{D}_g]$, then $h \in \mathfrak{m}^{(p+3)/2}$, as follows at once from formula (2′). If $g = ax_n$, then $\{f, x_n\}' = 0$, and formula (2′) again shows that $h \in \mathfrak{m}^{(p+3)/2}$.

On the other hand, $N_{k_{n'+1}}(\mathfrak{C})$ cannot be bigger than the algebra $\mathfrak{L}_0$, since $\mathfrak{L}_0$ is maximal in $K_{n'+1}$ and the latter, as we established in Proposition 1, is simple.

§ 8. Invariant subalgebras of p-algebras

Theorem. *All subalgebras* $\mathfrak{L}_i$ *of the filtration of a p-algebra* $\mathfrak{L}$ *of Cartan type are invariant. Any proper invariant subalgebra is contained in* $\mathfrak{L}_0$.

Proof. Propositions 3 of § 4, 3 of § 5, 4 of § 6, and 4 of § 7 show that in an arbitrary algebra $\mathfrak{L}$ of Cartan type, $\mathfrak{L}_0 = N_{\mathfrak{L}}(\mathfrak{C})$. Since the subalgebra $\mathfrak{C}$ is by its definition invariant, so is its normalizer $\mathfrak{L}_0$. Finally, all terms of the filtration $\mathfrak{L}_i$ are defined by the subalgebra $\mathfrak{L}_0$, and this definition shows that they are invariant along with it. This proves the first part of the theorem. From this, incidentally, it follows that the automorphism group of the algebra $\mathfrak{L}$ induces groups of linear transformations of all the spaces L_i.

Suppose now that $\mathfrak{M}$ is an invariant subalgebra of $\mathfrak{L}$ not contained in $\mathfrak{L}_0$. We show first that the projection of $\mathfrak{M}$ on $V = \mathfrak{L}/\mathfrak{L}_0$ coincides with all of V. By hypothesis this projection is different from , i.e., it contains an element $e \in V$, $e \neq 0$, and consequently all elements of the form e^{ϕ}, $\phi \in \mathfrak{G}$, where $\mathfrak{G}$ is the group of linear transformations induced by the automorphism group of $\mathfrak{L}$ in V. We construct in explicit form a certain subgroup of the group $\mathfrak{G}$. To do so, note that each automorphism ϕ of the algebra O defines an automorphism (also denoted by ϕ) of the algebra W_n of derivations of O:

$$\mathcal{D}^{\varphi}(f) = (\mathcal{D}(f^{\varphi^{-1}}))^{\varphi}, \quad f \in O.$$

Similarly, if Ω is the algebra of exterior differential forms over O, then ϕ defines an automorphism in Ω, given for one-dimensional forms by the condition

$$(\mathcal{D}, \omega^{\varphi}) = (\mathcal{D}^{\varphi^{-1}}, \omega)^{\varphi}.$$

In particular, let ω be the differential form involved in the definition of an algebra $\mathfrak{L}$ of Cartan type (we can let $\omega = 0$ for $\mathfrak{L} = W_n$), and ϕ an automorphism of O such that $\omega^{\phi} = a\omega$, $a \in k$. Then ϕ stabilizes $\mathfrak{L} \subset W_n$ and defines an automorphism in $\mathfrak{L}$. We restrict ourselves to linear transformations of the variables $x_1, \ldots, x_n$ satisfying $\omega^{\phi} = a\omega$, $a \in k$. In the case of the algebra W_n, S_{n-1}, $H_{n'}$, and $K_{n'+1}$, the group $\mathfrak{G}_0$ induced by these automorphisms in V coincides, respectively, with $GL(n,k)$, $GL(n,k)$, $CSp(n,k) = Sp(n,k) \times k$, and G_n, where G_n is the group of automorphisms of the space V leaving fixed some hyperplane V_0 in V.

It is well known that the first three groups act irreducibly on V. Therefore, for $e \neq 0$, $e \in V$, the vectors e^{ϕ}, $\phi \in \mathfrak{G}_0$, in these cases generate the whole space V, and this means that the projection of $\mathfrak{M}$ on V coincides with V.

Let $\mathfrak{L} = K_{n'+1}$. Then $V_0 = L'_{-1}$ (cf. (8), § 7) is the unique proper invariant subspace relative to $\mathfrak{G}_0$ of the space V [17]. Therefore, the projection of $\mathfrak{M}$ on V coincides either with V or V_0. If it coincides with V_0, then the algebra generated by $\mathfrak{M}$ and $\mathfrak{L}_0$ also projects onto V_0, whereas it is known that $\mathfrak{L}_0$ is not contained in a bigger subalgebra of $K_{n'+1}$ (primitivity of the algebra $K_{n'+1}$ in the terminology of [18]). Thus, the projection of $\mathfrak{M}$ on V in all cases coincides with V.

Set $\mathfrak{M} \cap \mathfrak{L}_i = \mathfrak{M}_i$ and denote by Gr $\mathfrak{M} = M$ the graded algebra corresponding to this filtration. It is sufficient to prove that $M = L$. To do this we must use another automorphism of the algebra $\mathfrak{L}$. Let $\mathfrak{L}_r$ be the last nonzero term of the filtration. Then $(\operatorname{ad} x)^2 = 0$ for $x \in \mathfrak{L}_r$; therefore, the transformation $\phi_x = 1 + \operatorname{ad} x$ defines, as is well known, an automorphism of the algebra $\mathfrak{L}$, and so of the algebra L. It follows that $[\mathfrak{M}, x] \subset \mathfrak{M}$, and therefore $[M, L_r] \subset M$. Since we have proved that $V \subset M$, we have $[V, L_r] \subset M$. Similarly, $[V, x] \subset M$ if x is the image of any element of $\mathfrak{C}$.

Consider first the case that $\mathfrak{L}$ is an algebra of one of the first three types, i.e., $\mathfrak{L} \neq K_n$. Then, as we know, $\mathfrak{L}$ is graded: $\mathfrak{L} = L$. If $x \in L_r$, $x \neq 0$, then the ideal I_x generated by the element x in L coincides with L. On the other hand, I_x is spanned by elements of the form $[\ldots[x, y_1]\ldots y_k]$, $y_i \in L_{s_i}$. Using repeatedly the Jacobi identity, we can express all these elements in terms of those for which $s_i = 0$ or -1, and in fact (in view of the invariance of L_{-1} and L_r relative to L_0) in terms of products of the form $[\ldots[x', y_1]\ldots y_k]$, $x' \in L_r$, $y_i \in L_{-1}$. Thus $L_i = [L_r, V^{r-i}]$, which implies the inclusion $L_i \subset M$ for $i < r$. But since $[L_r, M] \subset M$, we have also $L_r \subset M$, or else the subalgebra L_r would be an ideal.

Turning now to the algebra $K_{n'+1}$, we see that as in the case of the first three algebras, $[\mathfrak{M}, \mathfrak{C}] \subseteq \mathfrak{M}$ (a consequence of the invariance of the subalgebra $\mathfrak{M}$). Since $\mathfrak{D}_{x_n^{p-1} x_1} \in \mathfrak{C}$ (cf. Proposition 3, § 7), we have in particular

$$[\mathscr{D}_{x_n^{p-1} x_1}, \mathscr{D}_{x_{n'+1}+u}] = \mathscr{D}_{x_n^{p-1}+f} \in \mathfrak{M},$$

$$\min \deg f \geqslant p \ (\text{if } \deg u > 1, \ \mathscr{D}_{x_{n'+1}+u} \in \mathfrak{M}). \tag{1}$$

Further $\mathfrak{L}_r \in \mathfrak{C}$, so that $[\mathfrak{M}, \mathfrak{L}_r] \subseteq \mathfrak{M}$; it follows, since $\pi_{-1}(\mathfrak{M}) = V$, that $\pi_{-1}(\mathfrak{M}) = L_{r-1}$ (where $\pi_i(\mathfrak{M})$ is the projection of $\mathfrak{M}$ on L_i). Since $L_i = \widetilde{L}_i \cong (O_1 \otimes H'_{n'})_i$, $i \geq p-1$ (cf. Proposition 2, § 7), then $\pi_i([\mathfrak{L}_r, \mathfrak{M}^{r-i}]) = L_i$, $i \geq p-1$, a reflection of a fact noted earlier about hamiltonian algebras. It remains only to observe that $(H'_{n'})_{-1} = \langle \mathfrak{D}_{x_1}, \ldots, \mathfrak{D}_{x_{n-1}} \rangle = V_0 \subset V$. On the other hand, $\pi_{p-2}([\mathfrak{L}_r, \mathfrak{M}^{r-(p-2)}]) \neq L_{p-2}$. More precisely,

$$\pi_{p-2}([\mathfrak{L}_r, \mathfrak{M}^{r-(p-2)}]) = \langle \mathscr{D}_{h_0}; \ \deg h_0 = p \rangle.$$

From (1) we see, however, that $\pi_{p-2}(\mathfrak{M}_{p-2}) = L_{p-2}$. Thus,

$$[\mathfrak{M}_{p-2}, \mathfrak{M}_{r-(p-2)}] \cong [L_{p-2}, L_{r-(p-2)}] = L_r = \mathfrak{L}_r = \mathfrak{M}_r],$$

or passing from the graded algebra to the original, $\mathfrak{M}_{p-2} = \mathfrak{L}_{p-2}$. The proof is now completed by formal operations within the algebra $K_{n'+1}$.

Remark. Computing the subalgebra $\mathfrak{C}$ for an algebra of Cartan type other than a p-algebra is appreciably more difficult. We conjecture, however, that the results of this section hold in the general case.

§ 9. Nonisomorphism of p-algebras of Cartan type

Theorem. *Two algebras of Cartan type are not isomorphic if they belong to different series, or to the same series and have different indices in that series.*

Proof. It is easy to show that the dimensions of two different algebras of Cartan type are equal only in the case of W_n, $n = p^s$, s even, and $K_{m'+1}$, $m = p^s + s = 2m' + 1$. That these algebras are nonisomorphic follows from the fact that their filtrations are defined invariantly. The algebra W_n is graded relative to this filtration, i.e., $\operatorname{Gr} W_n \cong W_n$, but the algebra $K_{m'+1}$ is not (cf. Proposition 2, § 7).

Remark. The analogous result is apparently true for arbitrary algebras of Cartan type. The proof rests on the question raised at the end of the preceding section.

§ 10. Strongly transitive subalgebras

We consider those subalgebras of graded p-algebras of Cartan type that contain $L_{-1} + L_0$. As a preliminary we describe the structure of the L_0-module L_s, mainly for $s = 1$.

We begin with the algebra W_n. Let

$$L_s' = \langle \mathcal{D} \in L_s;\ \operatorname{div} \mathcal{D} = 0 \rangle, \tag{1}$$

$$L_s'' = \left\langle \mathcal{T}_f;\ \mathcal{T}_f = f\mathcal{T}_1,\ \mathcal{T}_1 = \sum_{k=1}^{n} x_k \frac{\partial}{\partial x_k},\ \deg f = s \right\rangle. \tag{2}$$

Obviously L_s' and L_s'' are two proper subspaces in L_s ($L_s' = 0$ for $n = 1$). The invariance of L_s' relative to L_0 is a direct consequence of identity (13), § 4 . Since the algebra L_0 is spanned by operators of the form $x_i(\partial/\partial x_j)$, the invariance of L_s'' follows from the relations

$$\left[x_i \frac{\partial}{\partial x_j},\ \mathcal{T}_f\right] = \mathcal{T}_u, \quad u = x_i \frac{\partial f}{\partial x_j}. \tag{3}$$

By means of (3) it is easy to establish also the irreducibility of L_s''. It is well known that in the case of fields of characteristic zero we have the direct sum decomposition

$$L_s = L_s' \oplus L_s''.$$

This is also true in the general case provided $s + n \not\equiv 0\ (p)$. But

$$\operatorname{div} \mathcal{T}_f = (s+n)f, \quad s = \deg f,$$

and for $s + n \equiv 0\,(p)$ we get the inclusion $L_s'' \subset L_s'$.

1) We prove that the L_0-module L_s' is irreducible for $s + n \not\equiv 0\,(p)$, $s < p - 1$. Indeed, let N be a submodule of L_s' and

$$0 \neq \mathcal{A} = \sum_{k=1}^{n} a_k \frac{\partial}{\partial x_k} \in N.$$

Without loss of generality we can assume $a_1 = 0$. Since

$$\left[x_i \frac{\partial}{\partial x_j},\ \mathcal{A}\right] = \sum_{k=1}^{n} g_k \frac{\partial}{\partial x_k}, \quad g_k = \frac{\partial a_k}{\partial x_j} x_i - \delta_{j,k} a_i, \tag{4}$$

we obtain in the case $\partial a_1/\partial x_j \neq 0$, $j > 1$, after t-fold multiplication, $t \leq s$, by an operator of type $x_1(\partial/\partial x_j)$, the element

$$\mathcal{B} = x_1^{s+1}\frac{\partial}{\partial x_1} + \sum_{k=2}^{n} b_k \frac{\partial}{\partial x_k} \in N. \tag{5}$$

If $[\mathcal{B}, x_i(\partial/\partial x_j)] = 0$, $j \geq 2$, then according to (4), $\partial b_k/\partial x_j = 0$, $j \neq k$, since $s < p-1$. Thus

$$b_k = \sum_{l=0}^{s+1} \beta_{k,l}\, x_1^l x_k^{s+1-l}.$$

Now, setting $i = j = k$, we find that $\beta_{k,l} = 0$ for all $l \neq s$, i.e., $b_k = \beta_{k,s} x_1^s x_k$. If $j = k > 1$, $i = 1$, we get the equality $\beta_{k,s} = 1$, i.e., $\mathcal{B} = \mathcal{T}_{x_1^s}$, which contradicts the condition div $\mathcal{B} = 0$, $s + n \not\equiv 0(p)$. Thus

$$0 \neq \left[\mathcal{B}, x_{i_0}\frac{\partial}{\partial x_{j_0}}\right] = \mathcal{D} \in N$$

for some pair of indices $i_0 \geq 1$, $j_0 \geq 2$. Products of $\mathcal{D}$ with $x_1(\partial/\partial x_j)$, $j \geq 2$, lead to the element $x_1^{s+1} \sum_{k=2}^{n} \alpha_k(\partial/\partial x_k) \neq 0$ (possibly $\mathcal{D}$ already had this form). Multiplying by $x_i(\partial/\partial x_j)$, $\alpha_i \neq 0$, we can extract elements

$$\mathcal{E}_j = x_1^{s+1}\frac{\partial}{\partial x_j} \in N, \quad j > 1.$$

It is easy to see that the differential operators

$$\left[\ldots\left[\mathcal{E}_j, x_{k_1}\frac{\partial}{\partial x_1}\right], \ldots, x_{k_t}\frac{\partial}{\partial x_1}\right] \in N, \quad t \leqslant 2s+2, \quad k_i > 1,$$

generate the whole module L_s'.

2) Let $n = n_0 p - 1$. Then the L_0-module L_1' contains the unique submodule L_1'' (cf. (1) and (2)). Indeed, suppose a submodule $N \subset L_1'$ does not equal L_1''. As before, we find in N an operator $\mathcal{B} \neq 0$ of the form (5). This implies $N \supset L_1'$ except for the case $\mathcal{B} \in L_1''$, when we obtain only $N \supset L_1''$. Thus, we must consider the situation in which

$$\mathcal{A} \in N - N_1'', \text{ but } \left[x_1\frac{\partial}{\partial x_j}, \mathcal{A}\right] = \mathcal{T}_{f_j} \text{ for all } j > 1. \tag{6}$$

Condition (6) is equivalent to the system of relations

$$f_j x_k = \frac{\partial a_k}{\partial x_j} x_1 - \delta_{j,k} a_1, \quad k \geqslant 1, \quad j > 1,$$

which implies

$$a_1 = \left(\sum_{i=1}^{n} \alpha_i x_i\right) x_1, \quad a_k = \left(\sum_{i=1}^{n} \alpha_i x_i\right) x_k + \alpha_k' x_1^2, \quad k > 1,$$

o that

$$\mathfrak{A} = x_1^2 \sum_{k=2}^{n} \alpha'_k \frac{\partial}{\partial x_k} + \mathcal{T}_{\Sigma\alpha_i x_i}.$$

y assumption, $\alpha'_{k_0} \neq 0$ for some $k_0 > 1$ and $\mathcal{T}_{\Sigma\alpha_i x_i} \in L''_1 \subset N$. Consequently

$$0 \neq \mathfrak{D} = x_1^2 \sum_{k=2}^{n} \alpha'_k \frac{\partial}{\partial x_k} \in N \text{ and } N = L'_1.$$

3) We prove next that $L_1 \supset L'_1 \supset L''_1 \supset 0$ is the unique composition series of the module L_1 for $= n_0 p - 1$. Again, starting from an element $\mathfrak{A} \neq 0$ in a submodule N not contained in L'_1, we find an perator $\mathfrak{B} \neq 0$ of the form (5). Assume first that $\operatorname{div} \mathfrak{B} = \sum_{i=1}^{n} \beta_i x_i \neq 0$. An analysis of the condiions of type (6) reveals that they cannot be fulfilled by operators with nonzero divergence. In paricular, $\left[x_1 \frac{\partial}{\partial x_j}, \mathfrak{B}\right] = \mathfrak{D} = \sum_{i=2}^{n} d_i \frac{\partial}{\partial x_i} \neq 0$ for some $j > 1$. As in (1), $\mathfrak{D}$ leads to a nonzero operar $x_1^2 \sum_{k=2}^{n} \gamma_k \frac{\partial}{\partial x_k} \in N \cap L'_1$. In accordance with (2), we infer that $N \supset L'_1$; but then

$$\mathfrak{B} - \Big(\sum_{i \neq j} \beta_i x_i + \frac{\beta_j}{2} x_j\Big) x_j \frac{\partial}{\partial x_j} \in L'_1 \subset N$$

nd consequently

$$\Big(\sum_{i \neq j} \beta_i x_i + \frac{\beta_j}{2} x_j\Big) x_j \frac{\partial}{\partial x_j} \in N \quad \text{for } j = 1, \ldots, n.$$

, say, $\beta_1 \neq 0$, then

$$\beta_1 x_1 x_k \frac{\partial}{\partial x_1} = \left[x_k \frac{\partial}{\partial x_1}, \Big(\sum_{i=2}^{n} \beta_i x_i + \frac{\beta_1}{2} x_1\Big) x_1 \frac{\partial}{\partial x_1}\right] - \Big(\sum_{i=2}^{n} \beta_i x_i\Big) x_k \frac{\partial}{\partial x_1} \in N$$

r $k = 2, \ldots, n$, since $\operatorname{div}\left\{\Big(\sum_{i=2}^{n} \beta_i x_i\Big) x_k \frac{\partial}{\partial x_1}\right\} = 0$. Then also

$$\frac{\beta_1}{2} x_1^2 \frac{\partial}{\partial x_1} = \Big(\sum_{i=2}^{n} \beta_i x_i + \frac{\beta_1}{2} x_1\Big) x_1 \frac{\partial}{\partial x_1} - \sum_{i=2}^{n} \beta_i \Big(x_i x_1 \frac{\partial}{\partial x_1}\Big) \in N.$$

ut $\dim L_1/L'_1 = n$, so $N = L_1$. Now let $\operatorname{div} \mathfrak{B} = 0$. By virtue of an earlier observation, $\mathfrak{B} \notin L''_1$; conequently we again infer that $N \supset L'_1$, and then with the aid of an initial element $\mathfrak{A}$ we get the whole odule L_1.

We go back to the algebra $L = S_{n-1}$. By definition, $L_k = (S_{n-1})_k$ coincides for $k \geq 1$ with the module L'_k of the algebra W_n (cf. (1)); in addition, $(S_{n-1})_k$ behaves over $(S_{n-1})_0$ in the same way as over $(W_n)_0$. The preceding argument shows that the module $(S_{n-1})_1$ is irreducible for $n + 1 \neq 0(p)$ and contains a unique proper submodule L''_1 for $n + 1 \equiv 0(p)$.

Finally, consider the algebra $H_{n'}$, $n = 2n'$, which, in accordance with § 6, we can interpret as the factor ring O'/k with the multiplication law (4′) of § 7. The module L_s is generated by homogeneous polynomials of degree $s + 2$. For $s \leq p-3$ the irreducibility of L_s over $L_0 \cong C_{n'}$ is proved very simply. If $\partial u/\partial x_{n'+i} \neq 0$, then the condition $s \leq p-3$ implies that $\partial u/\partial x_{n'+i}/x_i \neq 0$, and consequently $0 \neq 2(\partial u/\partial x_{n'+i})/x_i = \{x_i^2, u\} \in N$. In this way we obtain a nonzero element $v \in N$ satisfying $\partial v/\partial x_{n'+i} = 0$, $i = 1, \ldots, n'$. Obviously $\{v, x_1 x_{n'+i}\} = (\partial v/\partial x_i)\, x_1$, and it is easy to see that $x_1^{s+2} \in N$. Since any monomial $x_1^{s_1} \ldots x_n^{s_n}$, $\sum s_i = s + 2$, can be written in the form

$$2^{-\mu}\{\ldots\{\{\ldots\{x_1^{s+2}, x_{i_1}x_{n'+1}\}, \ldots, x_{i_\nu}x_{n'+1}\}, x_{n'+j_1}^2\}, \ldots, x_{n'+j_\mu}^2\} \in N$$

with $1 \leq i_k, j_k \leq n'$, it follows that N coincides with L_s, and the irreducibility of L_s is proved.

Now we are able to prove the following:

Proposition. *Let $L = L_{-1} + L_0 + \ldots + L_r$ be a simple Lie p-algebra of type W_n, S_{n-1}, or $H_{n'}$, and let M be a subalgebra (not necessarily graded) containing $L_{-1} + L_0$. Then only the following cases are possible:*

1) $M = L_{-1} + L_0$;

2) $M = L$;

3) $L = W_n$, $M = L_{-1} + L_0 + L''_1$ ($\cong A_n$ *for* $n + 1 \not\equiv 0(p)$), *where* $L''_1 = \langle \mathfrak{I}_f;\ \deg f = 1\rangle$ (cf. (2));

4) $L = W_n$, $M = L_{-1} + L_0 + L'_1 + \ldots + L'_{r-1} \supset S_{n-1}$ (*cf.* (1) *for the definitions of* L'_s), $M/S_{n-1} \cong L_0/[L_0, L_0]$;

5) $L = S_{n-1}$, $n + 1 \equiv 0(p)$, $M = L_{-1} + L_0 + L''_1 \cong A_n$.

Proof. Let M contain the subalgebra $L_{-1} + L_0$ properly. If

$$a = a_{-1} + a_0 + \ldots + a_k \in M, \qquad k > 0, \quad a_k \neq 0,$$

then $b_1 = [\ldots[a_k, u_1], \ldots, u_{k-1}] \neq 0$ for some $u_i \in L_{-1}$, and we obtain an element $b_1 \in L_1 \cap M$. Thus, $L_1 \cap M \neq 0$ under our assumption. As earlier (cf. the proof of the theorem in § 8), consider the ideal I_c generated in L by an arbitrary nonzero element $c \in L_r$ (r the length of the filtration of L). Of course, I_c coincides with L; on the other hand, I_c is generated by elements of the form

$$[\ldots[[\ldots[c, x_1], \ldots, x_k], u_1], \ldots, u_l], \quad x_i \in L_0, \quad u_j \in L_{-1}.$$

This implies, among other things, the irreducibility of the L_0-module L_r in an arbitrary graded simple Lie algebra L. Suppose for the moment that $L_r \cap M \neq 0$. Then for $c \in L_r \cap M$ we get

$$M = I_c = L.$$

Thus case (2) of the proposition holds whenever $M \cap L_r \neq 0$.

The simplest situation is presented by the hamiltonian algebra H'_n. Having established the irreducibility of L_1, we know that $M \supset L_1$. In particular, $x_1^3 \in M$. If $x_1^i \in M$, then $\{x_1^i, x_1^2 x_{n'+1}\} = i x_1^{i+1} \in M$.

uppose we already know that $w = (x_1 \ldots x_k)^{p-1} x_{k+1}^i \in M$. Then

$$M \ni wx_{k+1} = \begin{cases} \{x_1 x_{k+1} x_{n'+1}, w\} & \text{for } k+1 \leqslant n', \\ \frac{1}{2+i}\{x_{k+1-n'} x_{k+1}^2, w\} & \text{for } k+1 > n' \quad i < p-2, \\ \{x_{k+2-n'} x_{k+1} x_{k+2}, w\} & \text{for } k+1 < n, \quad i = p-2. \end{cases}$$

Consequently, we obtain the monomial $(x_1 \ldots x_{n-1})^{p-1} x_n^{p-2} \in L_r \cap M$, thereby proving that M coincides with $L = H_{n'}$.

We turn now to the special algebras S_{n-1}. Suppose first that $L_1 \subset M$. We will not obtain in explicit form an element in $L_r \cap M$, but its existence follows from the following simple considerations: Assume we know that $0 \neq \mathfrak{A} = \sum_{i=1}^{n} a_i \frac{\partial}{\partial x_i} \in L_k \cap M$. Without loss of generality we take $a_1 \neq 0$. In addition we can ensure by transformations $\mathfrak{A} \to [\mathfrak{A}, x_1(\partial/\partial x_j)]$, $j > 1$, that $\partial^2 a_1/\partial x_1^2 \neq 0$, and if $k > p-3$, also $\partial^{p-1} a_1/\partial x_1^{p-1} \neq 0$. Since $\operatorname{div} \mathfrak{A} = 0$, the inequality $\partial a_l/\partial x_1 \neq 0$ (respectively, $\partial^{p-2} a_l/\partial x_1^{p-2} \neq 0$) must be satisfied as well for some $l > 1$. If $[\mathfrak{A}, x_s x_t(\partial/\partial x_1)] \neq 0$, $s, t \neq 1$, we obtain an element in $L_{k+1} \cap M$, which we deal with in the same way. If $[\mathfrak{A}, x_s x_t(\partial/\partial x_1)] = 0$ for all $s > 1$ and $t > 1$, then in particular $(\partial a_l/\partial x_1) x_s x_t = 0$, which under our choice of $\mathfrak{A}$ is possible only for $k = r = n(p-1) - 2$. Thus, $L_r \cap M \neq 0$ and $M = S_{n-1}$.

As we know, $L_1 \cap M \neq L_1$ is possible only for $n + 1 \equiv 0\,(p)$, and in this case $L_1 \cap M = L_1''$ (cf. (2) and 2)). It is easy to verify that $[L_1'', L_1''] = 0$; consequently $M = L_{-1} + L_0 + L_1''$ is a subalgebra in $S_{n_0 p-2}$, with $\dim M = 2n + (n^2 - 1) = (n_0 p)^2 - 2$. The Cartan subalgebra H_0 of L_0 remains a Cartan subalgebra in M, but the root system is enlarged by $2n$ roots $\pm\omega_1, \ldots, \pm\omega_n$. The corresponding root vectors e_{ω_i} lie in L_{-1}, the $e_{-\omega_i}$ in L_1''. The isomorphism of M with the algebra of all matrices of trace zero and order $n_0 p$ modulo its one-dimensional center is established either directly or with the aid of Theorems A and of [22].

Suppose now that M contains an element $\mathfrak{A} = \mathfrak{A}_1 + \ldots + \mathfrak{A}_k$, $\mathfrak{A}_k \neq 0$, which is not contained in $L_{-1} + L_0 + L_1''$. If $\mathfrak{A} = \mathfrak{A}_1$, then, in accordance with (2), the L_0-module generated by L_1'' and $\mathfrak{A}_1$ coincides with L_1. If $k > 2$, then after $k-2$ multiplications of $\mathfrak{A}$ by elements of L_{-1} we get $\mathfrak{A}' = \mathfrak{A}_1' + \mathfrak{A}_2' \in M$, $\mathfrak{A}_2' \neq 0$. Since $n + 2 \not\equiv 0\,(p)$, then according to (1) the images of the component $\mathfrak{A}_2'$ relative to operators from L_0 generate the whole module L_2. It remains only to use the observation that $[L_2, L_{-1}] = L_1$. Thus, a subalgebra $M \subset S_{n-1}$ properly containing $L_{-1} + L_0 + L_1''$ must coincide with S_{n-1}.

The definition of the general algebra W_n already implies that $M = W_n$ if $M \subset L_1$. Further, from the above remarks on subalgebras in S_n it follows that $M \supset S_{n-1} + \langle x_1(\partial/\partial x_1)\rangle$ in the case $M \cap L_1 = L_1'$. Let $\mathfrak{A} = \mathfrak{A}_1 + \ldots + \mathfrak{A}_k \in M$, $\mathfrak{A}_k \in L_k$, $\operatorname{div} \mathfrak{A}_k \neq 0$. For $k = 1$ we observe (cf. the beginning of this section and (3)) that the L_0-module generated by L_1' and $\mathfrak{A}$ coincides with L_1, so that $M = W_n$. For $k > 1$, using the condition

$$\frac{\partial}{\partial x_{i_{k-1}}}\left(\cdots\left(\frac{\partial}{\partial x_{i_1}}(\operatorname{div} \mathfrak{A}_k)\right)\cdots\right) \neq 0$$

for some sequence of indices $i_1, \ldots, i_{k-1}$), we get the element $\mathfrak{B}_1 = [\ldots[\mathfrak{A}_k, (\partial/\partial x_{i_1})], \ldots, \partial/\partial x_{i_{k-1}}]$

$\in M \cap L_1$ with div $\mathfrak{B}_1 \neq 0$. Again $M = L$. Finally, let $M \cap L_1 = L_1''$. To supplement the above remarks we should observe that $[L_k'', L_{-1}] \not\subseteq L_{k-1}''$, since

$$\left[\frac{\partial}{\partial x_i}, \mathcal{T}_f\right] = \mathcal{T}_{\frac{\partial f}{\partial x_i}} + f\frac{\partial}{\partial x_i}.$$

Hence, in the case $n + 1 \not\equiv 0(p)$, any element $\mathfrak{A} \in M$ not contained in $L_{-1} + L_0 + L_1''$, leads, when acted on by operators in L_{-1}, to an element $\mathfrak{B}_1 = \mathfrak{B}_1' + \mathfrak{B}_1'' \in M$ with $0 \neq \mathfrak{B}_1' \in L_1'$. The irreducibility of L_1' implies $M \cap L_1 = L_1$, contradicting the assumption. In the case $n + 1 \equiv 0(p)$ we proceed similarly. Here, the inclusion $L_1' \supset L_1''$ leads to the conclusion $M \cap L_1 = L_1'$, again contrary to our assumption. This completes the proof of the proposition.

CHAPTER II. THE ALGEBRA L_0

§ 1. Statement of results

Consider a finite-dimensional graded Lie algebra L over an algebraically closed field k of characteristic $p > 7$. Let

$$L = L_{-1} + L_0 + L_1 + \ldots + L_r$$

be its gradation, $L_{-1} = V$, and Γ: $L_0 \to \operatorname{End} V$ the representation defined by multiplication in L. From the definition of a graded algebra it follows that Γ is a faithful representation.

The main result of this chapter is the following

Theorem. *Under the conditions*

a) $L_2 \neq 0$,

b) dim $V < p-1$,

c) L_0 *is a Lie p-algebra,*

d) Γ *is an irreducible p-representation,*

only the following possibilities can occur:

1) $L_0 \cong A_n$ *and* Γ *is the standard representation;*

2) $L_0 \cong A_n \oplus k$ *and* $\Gamma|_{A_n}$ *is the standard representation;*

3) $L_0 \cong C_n$ *and* Γ *is the standard representation;*

4) $L_0 \cong C_n \oplus k$ *and* $\Gamma|_{C_n}$ *is the standard representation.*

The proof is divided into parts corresponding to Propositions 1-3 and Lemmas 1-15.

Proposition 1. *Under the hypothesis of the theorem, the algebra* L_0 *is semisimple and is a direct sum of simple classical algebras with nondegenerate trace form and possibly a one-dimensional center. The restriction of the representation* Γ *to each summand of* L_0 *is a p-representation.*

Proof. From conditions (b) and (d) of the theorem it follows that the algebra L_0 is reductive with center $\mathfrak{z}(L_0)$, dim $\mathfrak{z}(L_0) \leq 1$, since in this case the usual argument can be applied (cf. Chap. II of [19]). We can therefore use the theorem of [23], which states that

$$L_0 = M_1 \oplus \ldots \oplus M_k \oplus \mathfrak{z}(L_0),$$

where the M_i are classical simple algebras.

Without the condition dim $L_{-1} < p$, the representation $\Gamma_{|M_i}$ may not be a p-representation. There xists an example for $p = 3$. But in our case, for any three-dimensional simple subalgebra $A_1 \subseteq M_i$ the estriction $\Gamma_{|A_1}$ is a p-representation [12]. It follows, obviously, that $\Gamma_{|M_i}$ is a p-representation. his proves the proposition.

To study the representation of the algebra L_0 in V we can use the theory of irreducible p-repre- entations of simple classical algebras (cf. [7]). Indeed, in view of Proposition 1, the representation ' is known to be a tensor product of irreducible representations of the simple summands of L_0. By 7], an irreducible p-representation is given by its highest weight.

Proposition 2. *The difference of any two weights of the representation Γ is equal to the differ- nce of two roots of the algebra L_0.*

Proposition 3. *Up to its one-dimensional center, the algebra L_0 is simple. In the expression*

$$\varphi = \sum_{i=1}^{l} s_i \lambda_i$$

f the highest weight ϕ of the representation Γ in terms of the fundamental weights λ_i, the coefficients $_i$ are equal to 0.1 or 2, with $s_i = 2$ for at most one index i. There is also an exceptional pair $L_0 = G_2 \oplus$, $\phi = 3\lambda_2$. A similar statement holds for the coefficients t_i of the expression

$$\psi = -\sum_{i=1}^{l} t_i \lambda_i$$

f the lowest weight ψ of the representation Γ.

§ 2. Proof of Proposition 2

Take a fixed Cartan subalgebra H of the algebra L_0. We shall use the concept of weights of a epresentation of the algebra L_0, meaning by this the weights of the representation restricted to H. The epresentation Γ gives an embedding of L_0 into $V \otimes V^*$, where V^* is the dual of V. In the same way, e look on each of the L_0-modules L_i as an invariant subspace relative to L_0 in $V \otimes S^{i+1}(V^*)$. If $= \sum_i V_{\mu_i}$ and $V^* = \sum_i V^*_{-\mu_i}$ are the decompositions of V and V^* into weight spaces, there is corresponding decomposition of the space $V \otimes S^{i+1}(V^*)$, and so also a decomposition of its L_0-in- ariant subspace L_i into weight spaces with weights of the form $\mu_{k_0} - \mu_{k_1} - \cdots - \mu_{k_{i+1}}$. Denote by Λ_i ne set of weights of the representation Γ_i of L_0 in L_i. In particular, Λ_0 is the set of roots of the alge- ra L_0. We need the following

Lemma 1. *If $\mu \in \Lambda_{-1}$ and $\alpha \in \Lambda_0$ are such that $\mu + \alpha \in \Lambda_{-1}$, and e_α is a root vector, then there exists weight vector v_μ corresponding to μ such that $\Gamma(e_\alpha) v_\mu = [v_\mu, e_\alpha] \neq 0$.*

For the proof see [19], Russian p. 126. The argument is based on the complete reducibility of the estrictions of the representation Γ to the simple three-dimensional subalgebras, which in our case is nsured by the condition dim $V < p$.

We shall want also three simple lemmas from [17]. For the convenience of the reader we present ieir proofs, which do not depend upon the characteristic.

Lemma 2. $[L_1, V] \supset [L_0, L_0]$ (*Lemma* 5.1 *of* [17]).

Proof. If not, there is a decomposition of the algebra $L_0 = L_0' \oplus L_0''$ such that $\dim L_0'' > 1$ and $[L_1, V] \subset L_0'$. Correspondingly, $V = V' \otimes V''$ and $\Gamma = \Gamma' \otimes \Gamma''$. But then

$$\Gamma([x, v])(v' \otimes v'') = \Gamma'([x, v])v' \otimes v'' \tag{1}$$

for any x in L_1, v in V. Consider now the relation

$$[[x, v_1], v_2] = [[x, v_2], v_1].$$

We compute the two sides separately, setting $v_1 = v_1' \otimes v_1''$, $v_2 = v_2' \otimes v_2''$ and using condition (1). If v_1'' and v_2'' are taken not proportional (by assumption, $\dim V'' > 1$) we find that $\Gamma'([x, v_1]) = 0$ for all x and v_1, so that $[L_1, V] = 0$, i.e., $L_1 = 0$, a contradiction.

Lemma 3. *For each root* $\alpha \in \Lambda_0$ *there exists a weight* $\xi \in \Lambda_1$ *such that if* $\mu + \alpha$ *is a weight for some* $\mu \in \Lambda_{-1}$, *then* $\mu + \xi$ *is a root* (*Lemma* 5.3 *of* [17]).

Proof. By Lemma 2, a root vector e_α corresponding to α has the form

$$e_\alpha = \sum_i [u_\xi^i, v_\nu^i], \quad \xi + \nu = \alpha, \quad \xi \in \Lambda_1, \quad \nu \in \Lambda_{-1},$$

and since the root space is one-dimensional,

$$e_\alpha = [u_\xi, v_\nu].$$

From the condition $\mu + \alpha \in \Lambda_{-1}$ there follows, in accordance with Lemma 1, the existence of a weight vector v_μ for which $[v_\mu, e_\alpha] \neq 0$. Thus

$$0 \neq [v_\mu, e_\alpha] = -[[u_\xi, v_\nu], v_\mu] = -[[u_\xi, v_\mu], v_\nu],$$

whence $[u_\xi, v_\mu] \neq 0$, which means that $\mu + \xi$ is a root.

Lemma 4. *If* $\mu, \nu \in \Lambda_{-1}$, $\alpha \in \Lambda_0$ and $\mu + \alpha, \nu + \alpha \in \Lambda_{-1}$, *then* $\mu - \nu$ *is the difference of two roots* (*Lemma* 5.5 *of* [17]).

Indeed by the previous lemma there is a weight $\xi \in \Lambda_1$, for which $\mu + \xi$ and $\nu + \xi$ are roots. The difference $\mu - \nu = (\mu + \xi) - (\nu + \xi)$ gives the required representation.

We turn now to the proof of Proposition 2.

Let $\mu, \nu \in \Lambda_{-1}$. If there exists a root α such that $\mu + \alpha, \nu + \alpha \in \Lambda_{-1}$, then the proposition follows from Lemma 4. Consider now the case that for any $\alpha \in \Lambda_0$ either $\mu + \alpha \notin \Lambda_{-1}$ or $\nu + \alpha \notin \Lambda_{-1}$. Let v_μ and v_ν be arbitrary weight vectors for μ and ν. By hypothesis, $L_2 \neq 0$. We shall prove that $[[L_2, v_\mu], v_\nu] \neq 0$.

First note that whatever the weight vector $\lambda \in \Lambda_{-1}$, the space V is spanned by the vectors of the form $[[\dots[v_\lambda, e_{\beta_1}], \dots,], e_{\beta_k}]$, where the e_{β_i} are root vectors for which $\lambda + \beta_i \in \Lambda_{-1}$ (in other words, $[v_\lambda, e_{\beta_i}] \neq 0$). This follows from the fact that the space V, because of the irreducibility of Γ, is spanned by the products $[v_\lambda e_{\beta_1} \dots e_{\beta_k}]$ with arbitrary roots β_i, and if $[v_\lambda, e_{\beta_j}] = 0$ for some $j \leq k$, then

$$[v_\lambda e_{\beta_1} \dots e_{\beta_j}] = [v_\lambda e_{\beta_1} \dots [e_{\beta_{j-1}}, e_{\beta_j}]] + \dots + [v_\lambda [e_{\beta_1}, e_{\beta_j}], \dots, e_{\beta_{j-1}}],$$

and the product can be expressed in terms of nonzero commutators of smaller length.

We apply this remark to the case $\lambda = \mu$. We shall take only roots β_i, for which $\mu + \beta_i \in \Lambda_{-1}$, and this means, in view of the assumption about μ and ν, that $\nu + \beta_i \notin \Lambda_{-1}$, i.e., $[v_\nu, e_\beta] = 0$. If $[[L_2, v_\mu] v_\nu]$

0, then

$$[[L_2, [v_\mu, e_{\beta_1}]], v_\nu]$$
$$= [[[L_2, e_{\beta_1}], v_\mu], v_\nu] + [[L_2, [v_\mu, e_{\beta_1}]], v_\nu] + [[L_2, v_\mu], [v_\nu, e_{\beta_1}]]$$
$$= [[[L_2, v_\mu], v_\nu], e_{\beta_1}] = 0,$$

ince $[L_2, e_{\beta_1}] \subseteq L_2$ and $[v_\nu, e_{\beta_1}] = 0$. Applying successive transformations of this form, we get

$$[[L_2, [v_\mu e_{\beta_1} \dots e_{\beta_k}]], v_\nu] = 0.$$

n view of the remark above this implies the equality $[[L_2, V], u_\nu] = 0$ or equivalently, $[[L_2, u_\nu], V] = 0$. Ve are led to conclude that $[L_2, u_\nu] = 0$. But this is obviously impossible: commutating it with arbitrary elements $e_{\alpha_1}, \dots, e_{\alpha_s}$, we get relations of the form

$$[L_2, [v_\nu e_{\alpha_1} \dots e_{\alpha_s}]] = 0,$$

which imply that $[L_2, V] = 0$, i.e., $L_2 = 0$.

Thus, $[[L_2, v_\mu], v_\nu] \neq 0$. Consequently, there exists a weight vector $w_\eta \in L_2$, $\eta \in \Lambda_2$, such that $\neq [[w_\eta, v_\mu], v_\nu] = a \in L_0$, $a = e_\beta$, $\beta \in \Lambda_0$, or $a = h \in H$. The representation Γ being faithful, $[a, v_\lambda]$ 0 for some v_λ, $\lambda \in \Lambda_{-1}$, so that

$$[w_\eta v_\lambda v_\mu v_\nu] = [w_\eta v_\lambda v_\nu v_\mu] = [w_\eta v_\mu v_\nu v_\lambda] \neq 0.$$

hus $[w_\eta v_\lambda v_\mu] \neq 0$ and $[w_\eta v_\lambda v_\nu] \neq 0$, i.e.,

$$\eta + \lambda + \mu = \alpha_1 \in \Lambda_0,$$

$$\eta + \lambda + \nu = \alpha_2 \in \Lambda_0$$

nd so $\mu - \nu = \alpha_1 - \alpha_2$. This proves the proposition.

§ 3. Proof of Proposition 3

By Proposition 1,

$$L_0 = M_1 \oplus \dots \oplus M_k \oplus \mathfrak{z}, \tag{2}$$

where the M_i are classical simple algebras with invariant trace form and $\mathfrak{z}$ is the center of dimension at ost 1. We can understand by L_0, in general, an arbitrary abstract algebra with decomposition (2). It uffices to assume that there exists a faithful irreducible representation $\Gamma: L_0 \to \operatorname{End} V$ satisfying the onditions of Proposition 2.

Lemma 5. *In the decomposition* (2) *the number of summands* $k \leq 2$.

Proof. It is known that $\Gamma = \Gamma_0 \otimes \Gamma_1 \otimes \dots \otimes \Gamma_k$, where Γ_i for $i > 0$ is an irreducible representation of the algebra M_i, and Γ_0 is a one-dimensional representation of the center $\mathfrak{z}$. Consider the ighest weight ϕ of the representation Γ and the highest weights ϕ_i of the representations Γ_i. Then $\phi = \phi_0 + \phi_1 + \dots + \phi_k$. By the definition of the highest weight, there exist simple roots α_i, $i = 1, \dots k$, such that $\phi_i - \alpha_i$ are weights of the representation Γ_i. Then $\mu = \phi_0 + (\phi_1 - \alpha_1) + \dots + (\phi_k - \alpha_k)$ s also a weight of the representation Γ. By Proposition 2, the difference $\phi - \mu = \alpha_1 + \dots + \alpha_k$ must e the difference of some two roots of the algebra L_0. Since the α_i are roots of different simple algebras, this is possible only for $k \leq 2$.

Lemma 6. *If in the decomposition* (2) *the number of simple summands* $k = 2$, *then the differences* $\mu_i - \nu_i$ *between any two weights of the representation* Γ_i $(i = 1, 2)$ *is a root of the algebra* M_i.

Proof. If $\mu_i = \nu_i$, this is obvious. Let $\mu_i \neq \nu_i$. Consider any two distinct weights μ_j and ν_j of the representation Γ_j, $(j \neq i)$, and the two weights

$$\mu = \mu_0 + \mu_i + \mu_j, \qquad \nu = \mu_0 + \nu_i + \nu_j$$

of the representation Γ. By Proposition 2,

$$\mu - \nu = (\mu_i - \nu_i) + (\mu_j - \nu_j) = \beta_1 + \beta_2,$$

where β_1 and β_2 are roots of L_0. Each of these is a root of either M_1 or M_2. If they are both roots of the same (say M_i), then $\mu_j = \nu_j$, which contradicts the choice of the weights μ_j, ν_j. We can therefore suppose that β_1 is a root of M_i and β_2 a root of M_j; and then

$$\mu_i - \nu_i = \beta_1, \quad \mu_j - \nu_j = \beta_2.$$

Lemma 7. *In the decomposition* (2) *the number of summands* $k = 1$ *(or 0 when* $L_0 = \mathfrak{z}$ *is a one-dimensional algebra).*

Proof. If $k = 2$, then from the preceding lemma it follows almost immediately that the highest weights ϕ_1 and ϕ_2 of the representations Γ_1 and Γ_2 coincide with appropriate fundamental weights. Namely, suppose $\phi_1 = s_1\lambda_1 + \ldots + s_m\lambda_m$, where $s_1 \neq 0$ and $s_2 \neq 0$. Then $\phi_1 - \alpha_1 \in \Lambda_{-1}$ and $\phi_1 - \alpha_2 \in \Lambda_{-1}$ (α_i simple roots). But their difference $(\phi_1 - \alpha_1) - (\phi_1 - \alpha_2) = \alpha_2 - \alpha_1$ must be a root, which is impossible because the difference of simple roots is not a root. Further, if $\phi_1 = s_{i_1}\lambda_{i_1}$, $s_{i_1} \neq 0, 1$, then $\phi_1 - \alpha_{i_1}$, $\phi_1 - 2\alpha_{i_1} \in \Lambda_{-1}$ and (again by Lemma 6) $\phi_1 - (\phi_1 - 2\alpha_{i_1}) = 2\alpha_{i_1}$ is a root, which is impossible because $2\alpha_{i_1}$ is never a root. The same argument applies to the lowest weights ψ_1 and ψ_2. If $\psi_1 = -\lambda_{j_1}$, then $\phi_1 - \psi_1 = \lambda_{i_1} + \lambda_{j_1} = \theta_1$ is a root, and it is easy to see that θ_1 coincides with the highest root of the algebra M_1. The form of θ_1 for simple algebras shows (see, e.g., [13], pp. 699-700) that $\theta_1 = \lambda_{i_1} + \lambda_{j_1}$ only in the case $M_1 \cong A_n$ or $M_1 \cong C_n$. The corresponding representation Γ_1 is either the standard one or its dual. Without loss of generality we can suppose (cf. [13], Proposition 4.2) Γ_1 to be the standard representation. The same applies to the pair (M_2, Γ_2). Thus, with appropriate identification there remain three possibilities: 1) $L_0 = A_n \oplus A_m \oplus \mathfrak{z}$, 2) $L_0 = A_n \oplus C_m \oplus \mathfrak{z}$, 3) $L_0 = C_n \oplus C_m \oplus \mathfrak{z}$. That none of these is compatible with the condition $L_2 \neq 0$ can be seen by calculating the full prolongation $L_0^{(i)}$ (in the sense of Cartan) of the pair (L_0, Γ). Since $(A_n \oplus A_m \oplus \mathfrak{z})^{(2)} = 0$ (cf. [17], pp. 99-100) and $L_2 \subseteq L_0^{(2)}$, case (1) is eliminated. Further, there is an obvious natural imbedding $(A_n \oplus C_m \oplus \mathfrak{z})^{(i)} \subset (A_n \oplus A_{2m-1} \oplus \mathfrak{z})^{(i)}$, so that $L_0^{(2)} = 0$ in case (2) and for the same reason in case (3). This proves the lemma.

From now on we shall suppose that

$$L_0 = M \oplus \mathfrak{z}$$

and that

$$\varphi = \sum_{i=1}^{l} s_i\lambda_i, \quad \psi = -\sum_{i=1}^{l} t_i\lambda_i \tag{3}$$

e the expressions for the highest and lowest weights of the representation Γ of the simple algebra M terms of fundamental weights. If α_i is the simple root corresponding to the weight λ_i:

$$\frac{2(\lambda_i,\ \alpha_j)}{(\alpha_j,\ \alpha_j)} = \delta_{i,j},$$

en s_i is the length of the α_i-chain $\phi, \phi-\alpha_i, \ldots, \phi-s_i\alpha_i$ of the weight ϕ, which was used in the roof of Lemma 7. Note that in this statement s_i appears both as an element of the prime field of k (in rmula (3)) and as a natural number (the length of the chain). Since dim $V < p$, the one determines the ther.

To complete the proof of Proposition 3 it remains to look at the possible values of the coefficients $_i \geq 3$ for some i, then $\phi, \phi-\alpha_i, \phi-2\alpha_i, \phi-3\alpha_i \in \Lambda_{-1}$ and, by Proposition 2, $\phi-(\phi-3\alpha_i) = 3\alpha_i = \beta$ γ; $\gamma \in \Lambda_0$. Directly from the tables expressing all roots of an arbitrary classical Lie algebra in terms the fundamental weights λ_j, it is easy to see that this relation is possible only for $M = G_2$, $\phi = s_1\lambda_1$ $3\lambda_2$, when $3\alpha_2 = (\alpha_1+3\alpha_2)-\alpha_1$. If $s_1 \neq 0$, then $\phi-\alpha_1 \in \Lambda_{-1}$ and $\phi-3\alpha_2 \in \Lambda_{-1}$, whence $(\phi-\alpha_1)-(\phi$ $3\alpha_2) = 3\alpha_2-\alpha_1 = \beta-\gamma$; $(\beta, \gamma \in \Lambda_0)$. There are no such roots β, γ of the algebra G_2. Therefore $s_1 = 0$ nd $\phi = 3\lambda_2$.

Now suppose $s_i = s_j = 2$ for $i \neq j$. Then $\phi-2\alpha_i \in \Lambda_{-1}$ and $\phi-2\alpha_j \in \Lambda_{-1}$, so that $(\phi-2\alpha_i)-(\phi-2\alpha_j)$ $2\alpha_j-2\alpha_i = \beta-\gamma$ $(\beta, \gamma \in \Lambda_0)$, where we can take β and γ to be positive roots. From the tables exressing the roots in terms of fundamental weights one sees again that there are no such roots for any lassical simple algebra. The statement concerning the lowest weight ψ is proved similarly.

We remark that it is possible to obtain more precise information about the coefficients s_i, but we shall ot need it here.

§ 4. The table

From now on, l will be the rank of the simple component M of the algebra L_0, θ its highest root, nd $\{\alpha_1, \ldots, \alpha_l\}$ the system of simple roots. It is well known that in the expression

$$\theta = \sum_{i=1}^{l} m_i\alpha_i$$

ll coefficients m_i are different from zero. In addition, $(\alpha, \alpha) \neq 0$ in k for any root $\alpha \neq 0$; hence the ormula

$$(\varphi, \theta) = \frac{1}{2}\sum_{i=1}^{l} s_i m_i (\alpha_i, \alpha_i)$$

n the case of characteristic zero implies the inequality $(\phi, \theta) \neq 0$. The same is true in our case, as s easily seen by a simple verification. It is more convenient to write the expression for $\nabla = 2(\phi,$ $)/(\theta, \theta)$. The results are as follows:

A_l: $\nabla = s_1 + \ldots + s_l$, $\quad l+1$;

B_l: $\nabla = s_1 + 2(s_2 + \ldots + s_{l-1}) + s_l$, $\quad l > 2$, $\quad 2l \quad (\nabla = 2(s_1+s_2),\ l = 2)$;

C_l: $\nabla = s_1 + \ldots + s_l$, $\quad l+1$;

D_l: $\nabla = s_1 + 2(s_2 + \ldots + s_{l-2}) + s_{l-1} + s_l$, $\quad 2l-1$;

$G_2 : \nabla = 2s_1 + s_2, \quad 5;$
$F_4 : \nabla = 2s_1 + 3s_2 + 8s_3 + 4s_4, \quad 25;$
$E_6 : \nabla = s_1 + 2s_2 + 3s_3 + 2s_4 + s_5 + 2s_6, \quad 14;$
$E_7 : \nabla = s_1 + 2s_2 + 3s_3 + 4s_4 + 3s_5 + 2s_6 + 2s_7, \quad 21;$
$E_8 : \nabla = 2s_1 + 3s_2 + 4s_3 + 5s_4 + 6s_5 + 4s_6 + 2s_7 + 3s_8, \quad 35.$

The numbers on the right are the maximum values that ∇ may take on, as determined by Proposition 3. In the case of the algebras A_l, B_l, C_l, and D_l it is easy to verify that the maximal values given in the table for ∇ do not exceed dim V_0, where V_0 is the standard representation module, and since dim $V_0 \leq \dim V < p$ (cf. [19]), we have $\nabla \neq 0$. For the exceptional algebras it follows from the table that $\nabla \neq 0$ for $p > 31$. A more detailed analysis, which we omit, proves that this is true for $p > 7$. Consequently $(\phi, \theta) \neq 0$. The inequality $(\phi, \theta) \neq 0$ means that in all cases $\phi - \theta \in \Lambda_{-1}$, a fact necessary for the proof of the following lemma.

Lemma 8. *There exists a nonnegative root $\alpha \in \Lambda_0$ connected with ϕ, ψ, and θ by the relation*

$$\varphi - \psi = \theta + \alpha.$$

The proof is based on the same argument as the proof of Lemma 7.1, 7.2 in [13]. We need only note that the choice of the maximal weight $\xi \in \Lambda_1$ in Lemma 7.1 must be replaced because the concept of maximality is undefined in our case, by the condition of maximality of the root $\beta = \phi + \xi$. The proof of the equality $\beta = \theta$ goes through without change.

Remark. The construction in [13], which is very clever but holds only fot characteristic zero, shows that the statement of Lemma 8 can be changed to a stronger one: $\phi - \psi = \theta$. From this and from Proposition 3 the theorem would follow directly. Indeed, the condition s_i, $t_i = 0, 1, 2$ on the coefficients in (3) and the expressions for θ in terms of fundamental weights leave for (M, Γ) only the four possibilities indicated in the theorem (cf. the proof of Lemma 7).

We are able to make do with only Lemma 8 and Proposition 3, i.e., with the fact that

$$\theta + \alpha = \varphi - \psi = \sum_{i=1}^{l} (s_i + t_i)\lambda_i, \quad s_i, t_i = 0, 1, 2, \quad \alpha \geqslant 0. \tag{4}$$

Fortunately, there are not many nonnegative roots α satisfying (4). The results, obtained by directly checking the systems of roots, are described in the following table:

M	θ	2θ	$\theta + \alpha$, $0 < \alpha < \theta$
A_l	$\lambda_1 + \lambda_l$	$2\lambda_1 + 2\lambda_l$	$\lambda_2 + \lambda_{l-1}$, $\lambda_2 + 2\lambda_l$, $2\lambda_1 + \lambda_{l-1}$
B_l	λ_2	$2\lambda_2$	$2\lambda_1$, $\lambda_1 + \lambda_2$, $\lambda_1 + \lambda_3$, λ_3, λ_4
C_l	$2\lambda_1$	$4\lambda_1$	$2\lambda_2$, $2\lambda_1 + \lambda_2$
D_l, $l > 4$	λ_2	$2\lambda_2$	$2\lambda_1$, $\lambda_1 + \lambda_3$, λ_4
D_4	λ_2	$2\lambda_2$	$2\lambda_1$, $\lambda_1 + \lambda_3$, $2\lambda_3$, $2\lambda_4$, $\lambda_1 + \lambda_3 + \lambda_4$
G_2	λ_1	$2\lambda_1$	$2\lambda_2$, $3\lambda_2$, $\lambda_1 + \lambda_2$
F_4	λ_1	$2\lambda_1$	$2\lambda_4$, $\lambda_1 + \lambda_4$, λ_2, λ_3
E_6	λ_6	$2\lambda_6$	$\lambda_1 + \lambda_5$, λ_3
E_7	λ_6	$2\lambda_6$	λ_2, λ_3
E_8	λ_1	$2\lambda_1$	λ_2, λ_7

§ 5. Completion of the proof of the theorem

We now sort through all the representations in the above table and prove that they satisfy $L_0^{(2)}=0$. e first make some remarks to reduce the labor.

The roles of the highest and lowest weights are entirely equivalent. This means that it is un-cessary to describe representations Γ, say, both with weights $\phi = \lambda_2$, $\psi = -\lambda_{l-1}$ and also with eights $\phi = \lambda_{l-1}$, $\psi = -\lambda_2$. It suffices to take only the first. Further, the lowest [resp., highest] weight uniquely determined by the highest [lowest], and we will make use of this. But in large part the guments are quite trivial. For example, the equality $\phi - \psi = \theta + \alpha = \lambda_i$ is impossible, since $\phi \neq 0$ ıd $\psi \neq 0$.

If the representation Γ of L_0 leaves invariant (or invariant up to a scalar multiple) a nondegen-ate symmetric bilinear form, then $L_0^{(1)} = 0$ (or $L_0^{(2)} = 0$) (cf. [18]). In particular, this applies to the ljoint representation. This singles out the combination $\phi = \theta$, $\psi = -\theta$.

Finally, if $L_0 = M \oplus k$ and $L_0^{(2)} = 0$ for a certain representation, then $M^{(2)} = 0$ for the restriction f this representation to M. Therefore we will consider only the case $L_0 = M \oplus k$.

These remarks show that of the critical linear combinations $\theta + \alpha$ required for further analysis, ery few remain in the table. We shall consider the algebras of the several types separately.

1. The algebra $A_l \oplus k$. The combination $\phi = \lambda_1$, $\psi = -\lambda_l$ corresponds to the standard repre-entation of the algebra A_l (the first two cases of the theorem). If $\theta + \alpha = \lambda_2 + 2\lambda_l$, the possible ighest weights $\phi = \lambda_2$, $\phi = \lambda_l$ are excluded, because the corresponding lowest weights would be $\psi = -\lambda_{l-1}$ and $\psi = -\lambda_1$, not $-2\lambda_l$ and $-(\lambda_2 + \lambda_l)$ as required by the equality $\phi - \psi = \theta + \alpha$. For the ame reason we can eliminate the linear combination $\theta + \alpha = 2\lambda_1 + \lambda_{l-1}$, and of the possible decompo-itions of 2θ the only suitable one appears to be (up to symmetry between ϕ and ψ) $\phi = 2\lambda_1$, $\psi = -2\lambda_l$. ,emmas 9 and 10 below show, however, that in this case, and also in the case of a representation Γ vith $\phi = \lambda_2$, $\psi = -\lambda_{l-1}$, the condition $L_2 \neq 0$ is in contradiction with the equality $L_0^{(2)} = 0$. For purely echnical reasons we now rewrite l as $n-1$.

Lemma 9. *If $L_0 = \mathfrak{gl}(n, k)$ and Γ has highest weight $\phi = 2\lambda_1$, then $L_0^{(2)} = 0$.*

Proof. Let $L_0 = \mathrm{End}\, V$. Then Γ is the natural representation of the algebra L_0 in $S^2(V)$. Fixing a basis in V and identifying $S^2(V)$ with the space M_n^+ of symmetric matrices of order n, we can suppose he representation Γ to be given by the formula

$$\Gamma(A)S = AS + SA^*,$$

vhere $A \in L_0$, $S \in M_n^+$, and A^* is the transposed matrix. Let us find the first prolongation $L_0^{(1)}$ of the air (L_0, Γ). By definition, a linear transformation f belongs to $L_0^{(1)} \subset \mathrm{Hom}\,(M_n^+, L_0)$, if

$$f(S)T + Tf(S)^* = f(T)S + Sf(T)^* \tag{5}$$

for arbitrary matrices $S, T \in M_n^+$. Note that if A is an arbitrary symmetric matrix, then the function $f_A: S \to SA$ satisfies condition (5) and $f_A(E) = A$.

We prove first that $f(E)$ is a symmetric matrix for any $f \in L_0^{(1)}$. Set $T = E$ in (5). The relation obtained can be written in the form

$$(f(S) - f(E)S) + (f(S) - f(E)S)^* = 0,$$

so that the matrix

$$g(S)=f(S)-f(E)S.$$

Substituting into (5) the expression

$$f(S)=f(E)S+g(S) \tag{6}$$

and using the skew-symmetry of $g(S)$, we find that

$$f(E)[S,T]-[S,T]f(E)^*=[g(T),S]-[g(S),T],$$

for any $S,\ T\in M_n^+$. From this we get

$$\mathrm{Tr}(f(E)[S,T]-[S,T]f(E)^*)=0,$$

or

$$\mathrm{Tr}\, f(E)[S,T]=0,$$

since always $\mathrm{Tr}XY=\mathrm{Tr}\ Y^*X^*$. Any matrix B in the space M_n^- of skew-symmetric matrices can be represented as a linear combination of matrices $[S,\ T]$, where $S,\ T\in M_n^+$. Thus, $\mathrm{Tr}\, f(E)B=0$ for all $B\in M_n^-$. It follows at once that $f(E)\in M_n^+$.

Set

$$f(E)=A,\quad A\in M_n^+$$

and

$$f(S)=SA+h(S).$$

From the symmetry of the matrices S and T and the skew-symmetry of $g(S)$ in (6) it follows that the matrix $h(S)$ is skew-symmetric. In accordance with the observation at the beginning of the proof, $h(S)$ satisfies condition (5), which can be rewritten, since $h(S)\in M_n^-$, as

$$[h(S),T]=[h(T),S]. \tag{7}$$

Relation (7) obviously expresses the fact that the function h lies in the first prolongation of the algebra $(M_n^-;[\ ,\ ])$, represented in the space M_n^+ by the law

$$U(S)=[U,S],\qquad U\in(M_n^-;\ [\ ,\ ]).$$

This representation leaves invariant the nondegenerate symmetric form $\mathrm{Tr}\,ST$, and its first prolongation is therefore equal to zero. Thus, $h(S)=0$ and

$$f(S)=f_A(S)=SA,\quad A^*=A.$$

We prove now that the second prolongation is equal to zero. Since $L_0^{(1)}\cong S^2(V)$, it follows that $L_0^{(2)}\subset \mathrm{Hom}\,(S^2(V),\ S^2(V))$ and consists of those linear functions $S\to\Phi(S)$, $S^*=S$, such that

$$\Phi(S)^*=\Phi(S),\quad S\Phi(T)=T\Phi(S). \tag{8}$$

etting $T = E$, we get

$$\Phi(S) = S\Phi(E),$$

nd substituting this back into (8), we see that

$$[S, T]\Phi(E) = 0,$$

here $S, T \in M_n^+$. It easily follows that $\Phi(E) = 0$, and consequently $\Phi(S) = 0$. This proves the lemma.

Lemma 10. *If $L_0 = \mathfrak{gl}(n, k)$ and Γ has highest weight $\phi = \lambda_2$, then $L_0^{(2)} = 0$.*

Proof. Let $L_0 = \operatorname{End} V$, as in the preceding lemma. The space $V \wedge V$ of the natural representa-on Γ can be identified with the space M_n^- of skew-symmetric matrices of order n: $\Gamma(A)T = AT + TA^*$, $\in M_n^-$. Unfortunately, in M_n^- there is no good equivalent of the matrix $E \in M_n^+$, and an invariant argu-ent for the first, more essential, half of the proof of this lemma has not been found. Choose in M_n^- e standard basis $e_{i,j} = E_{i,j} - E_{j,i}$, $i < j$. Denote by f_{ij}^{st} the elements of the matrix $f_{ij} = f(e_{ij}) \in L_0$. he relation of type (5) can be rewritten in the form:

$$f_{ij}e_{kl} + e_{kl}f_{ij}^* = f_{kl}e_{ij} + e_{ij}f_{kl}^*, \quad i<j, \quad k<l. \tag{9}$$

From now on we assume $n \geq 5$; for the smallest possible value $n = 4$ the argument is similar. If e indices i, j, k, l are distinct, then obviously $f_{ij}^{st} = 0$, $f_{ij}^{si} = 0$, $f_{ij}^{sj} = 0$. In addition, $f_{ij}^{kk} + f_{ij}^{ll} = 0$, $< l$, $k \neq i, j$, $l \neq i, j$. Giving different values to k and l (here we use the assumption $n \geq 5$), we get $_j^{ss} = 0$, $s \neq i, j$. Thus, $f_{ij}^{st} = 0$ for all $s \neq i, j$, i.e., for the matrix $f_{i,j}$ only the i-th and j-th rows are ifferent from zero. Using this fact, we list the relations we need that follow from (9).

$i<j$, $k<l$ indices pairwise distinct	$k \neq i < j = l$	$k = i < j \neq l$	$i < j = k < l$
1) $f_{ij}^{il} + f_{kl}^{kj} = 0$ 2) $f_{ij}^{jl} - f_{kl}^{ki} = 0$ 3) $f_{ij}^{jk} + f_{il}^{li} = 0$	4) $f_{ij}^{ij} + f_{kj}^{kj} = 0$ 5) $f_{ij}^{jj} - f_{kj}^{ki} = 0$ 6) $f_{ij}^{ik} - f_{kj}^{jj} = 0$	7) $f_{ij}^{ji} + f_{il}^{li} = 0$ 8) $f_{ij}^{ii} - f_{il}^{lj} = 0$	9) $f_{ij}^{ij} - f_{jl}^{lj} = 0$ 10) $f_{ij}^{il} + f_{jl}^{jj} = 0$ 11) $f_{ij}^{jj} + f_{jl}^{li} = 0$

For fixed $j > 3$, we write $f_{ij}^{ij} = g_i$. By 4), $g_1 + g_k = 0$, $g_1 + g_i = 0$, $g_i + g_k = 0$, whence $g_i = 0$. If $= 2$ or 3, then 9) $\Rightarrow f_{jl}^{lj} = f_{ij}^{ij} = f_{jm}^{mj}$, $m > l > j$ (again using the assumption $n \geq 5$), and 7) $\Rightarrow$ (replac-ng i, j, l by j, m, l) $\Rightarrow f_{jl}^{lj} + f_{jm}^{mj} = 0$, so that $f_{jl}^{lj} = 0$, and consequently, $f_{ij}^{ij} = 0$. Now 9) (for $i = 1$) $\Rightarrow f_{jl}^{lj} = 0$ for all $j > 1$. Setting $h_l = f_{1l}^{l1}$, we get from 7) (for $i = 1$): $h_2 + h_l = h_3 + h_l = h_2 + h_3 = 0 \Rightarrow h_l$ h 0. For $j > 2$ we have: 8) $\Rightarrow f_{ij}^{ii} = f_{il}^{lj}$; 2) $(j \leftrightarrow l) \Rightarrow f_{il}^{lj} = f_{kj}^{ki}$; 5) $\Rightarrow f_{kj}^{ki} = f_{ij}^{jj}$, i.e., $f_{ij}^{ii} = f_{ij}^{jj}$. Finally, 1) $\Rightarrow f_{12}^{22} = -f_{2n}^{n1} = -f_{1l}^{l2}$; 5) $\Rightarrow f_{1l}^{11} = f_{12}^{11}$.

We list here the relations obtained:

$$f_{ij}^{ij} = 0 = f_{ij}^{ji}, \quad f_{ij}^{ii} = f_{ij}^{jj}, \quad i < j,$$

nd will make no further direct references to them.

Let $1 < i < j$. For $s \neq 1, i$, it follows from 1) that $f_{ij}^{is} = -f_{1s}^{ij}$. Further, $f_{ij}^{ii} = f_{ij}^{jj} = f_{1j}^{1i}$ (cf. 5)) and) $\Rightarrow f_{ij}^{i1} = f_{1j}^{ij} = f_{1j}^{11}$. Thus, $f_{ij}^{is} = \epsilon f_{1u}^{1v}$. In the same way, we find $f_{ij}^{js} = \epsilon f_{1u}^{1v}$. The elements f_{1u}^{1v} are not independent. Indeed, if $v > u > 1$, then 10) and 5) $\Rightarrow f_{1u}^{1v} = -f_{uv}^{uu} = -f_{uv}^{vv} = -f_{1v}^{1u}$. Using the equalities $f_{1u}^{1u} = 0$, we find that all elements f_{ij}^{st} are expressible linearly in terms of the f_{1u}^{1v}, where $n \geq u > v \geq 1$.

Thus, dim $L_0^{(1)} \leqslant \binom{n}{2}$. By altering the argument somewhat it is easy to show that this bound holds also for $n = 4$. But $\binom{n}{2}$ independent solutions of equation (9) can be exhibited immediately: namely, the function f_A (A any skew-symmetric matrix) with values $f_A(X) = XA$.

It is easy to prove now that the second prolongation is equal to zero. Since $L_0^{(1)} \cong M_n^-$, it follows that $L_0^{(2)} \subset \mathrm{Hom}\,(M_n^-, M_n^-)$ and consists of linear functions $X \to \Phi(X)$, $X^* = -X$, such that

$$\Phi(X)^* = -\Phi(X), \qquad Y\Phi(X) = X\Phi(Y). \tag{10}$$

Passing to transposed matrices, we get from (10) the relation

$$\Phi(X)Y = \Phi(Y)X.$$

Multiplying on the left by an arbitrary skew-symmetric matrix Z and using again relation (10), we find that

$$X\Phi(Z)Y = Y\Phi(Z)X.$$

In particular,

$$\mathrm{Tr}[X, Y]\Phi(Z) = 0.$$

From this it follows without difficulty that $\Phi(Z) = 0$ for any matrix $Z \in M_n^-$. This proves the lemma.

2. The algebra $B_l \oplus k$. From the table at the end of § 4 and the remarks at the beginning of this section it follows that in the relation $\theta + \alpha = \phi - \psi$ it suffices to consider the combinations $\phi = \lambda_1$, $\psi = -\lambda_1, -\lambda_2, -\lambda_3$. But $\psi \neq -\lambda_2$ or $-\lambda_3$, since the highest weight $\phi = \lambda_1$ corresponds to the standard representation Γ of the algebra B_l with lowest weight $\psi = -\lambda_1$.

The representation Γ of the algebra $L_0 \cong B_l \oplus k$ multiplies the nondegenerate symmetric bilinear form in the definition of the algebra B_l by a constant factor. It follows that $L_0^{(2)} = 0$, and this means that $L_2 = 0$.

3. The algebra $C_l \oplus k$. The combination $\phi = \lambda_1$, $\psi = -\lambda_1$ corresponds to the standard representation of the algebra C_l (the last two cases of the theorem). For $\theta + \alpha = 4\lambda_1$ and $\theta + \alpha = 2\lambda_1 + \lambda_2$ there are no suitable combinations for ϕ and ψ, since the adjoint representation with $\phi = 2\lambda_1$ is eliminated, while the highest weight $\phi = \lambda_2$ corresponds to lowest weight $\psi = -\lambda_2$. This last combination gives a decomposition for $\theta + \alpha = 2\lambda_2$. The answer to the question in the present case is given by

Lemma 11. *If $L_0 \cong C \oplus k$ and Γ has highest weight $\phi = \lambda_2$, then $L_0^{(2)} = 0$.*

Proof. By definition, C_l is the Lie algebra of all endomorphisms of the space U leaving invariant the nondegenerate bilinear form f. It is known that $\Gamma(C_l)$ acts in a natural manner on $\bigwedge^2(U)$: $\Gamma(A)(u \wedge v) = Au \wedge v + u \wedge Av$. But here $\Gamma(A)$, $A \in C_l$, leaves invariant the exterior power f^2 of the rm f:

$$f^2(x\wedge y,\ u\wedge v)=f(x,\ u)f(y,\ v)-f(x,\ v)f(y,\ u).$$

he form f^2 is nondegenerate and symmetric; therefore $L_0^{(2)}=0$. This proves the lemma.

4. The algebra $D_l\oplus k$, $l>4$. In this case only three representations are possible, correspond-g to $\theta+\alpha=2\lambda_2$, $2\lambda_1$, $\lambda_1+\lambda_3$. Of these, the first corresponds to the adjoint representation, the sec-nd to the standard, and consequently both leave invariant a symmetric form. Finally, the third case eliminated, since the representation with highest weight $\phi=\lambda_1$ has lowest weight $-\lambda_1$, not $-\lambda_3$.

5. The algebra $D_4\oplus k$. To the cases in paragraph 4 are to be added two eight-dimensional ir-ducible representations Γ_1 and Γ_2 with weights $\phi=\lambda_3$, $\psi=-\lambda_3$, and $\phi=\lambda_4$, $\psi=-\lambda_4$, respectively. he linear combination $\theta+\alpha=\lambda_1+\lambda_3+\lambda_4$, consequently, cannot be represented in the form $\phi-\psi$. is known (cf., e.g., [19], Chap. VII) that both representations Γ_1 and Γ_2 are realized in the space f even elements of a Clifford algebra. More precisely, let U be a vector space of dimension 8 over , $(\ ,\ )$ a nondegenerate symmetric form on $U\times U$; $U=U_1+U_2$, $U_1=\langle e_1,e_2,e_3,e_4\rangle$, $U_2=\langle f_1,f_2,f_3,$ $\rangle$; $(U_i,U_i)=0$, $i=1,2$; $(e_i,f_j)=\delta_{i,j}$. In the Clifford algebra $C=C(U,(\ ,\ ))$ the even elements e_j, f_if_j, $i>j$; $e_kf_l-\delta_{k,l}$, $1\le k,l\le 4$, generate a Lie algebra isomorphic to D_4, with the usual mul-plication $[u,v]=uv-vu$ (note that $e_if_j+f_je_i=2\delta_{ij}$). In C consider the algebra $C(U_1,(\ ,\ ))$, iden-fiable with the exterior algebra (of dimension 16) of the subspace U_1. Set

$$F=f_1f_2f_3f_4,\quad \mathfrak{S}=F\cdot C(U_1,\ (\ ,\ )),$$
$$\mathfrak{S}^+=\langle F;\ Fe_ie_j,\ i<j;\ Fe_1e_2e_3e_4\rangle,$$
$$\mathfrak{S}^-=\langle Fe_i;\ Fe_1\ldots\hat{e}_i\ldots e_4,\ i=1,2,3,4\rangle.$$

hen $\mathfrak{S}^+L_0\subseteq\mathfrak{S}^+$, $\mathfrak{S}^-L_0\subseteq\mathfrak{S}^-$, and the L_0-modules $\mathfrak{S}^+$, $\mathfrak{S}^-$ reduce precisely to the above irreducible re-esentations Γ_i of D_4.

Lemma 12. *Let Γ be the irreducible representation of D_4 corresponding to one of the modules $\mathfrak{S}^+$ $\mathfrak{S}^-$, $L_0\cong D_4\oplus k$. Then $L_0^{(2)}=0$.*

Proof. The argument below, as in the case of Lemma 10, is unfortunately not invariant.

1) $V=\mathfrak{S}^+$. Let g be an arbitrary element of $L_0^{(1)}\subset\operatorname{Hom}(\mathfrak{S}^+,L_0)$ with values

$$g(F)=\sum\alpha_{st}^0e_se_t+\sum\beta_{st}^0f_sf_t+\sum\gamma_{st}^0(e_sf_t-\delta_{st})+\lambda_0,$$
$$g(Fe_1e_2e_3e_4)=\sum\alpha_{st}^1e_se_t+\sum\beta_{st}^1f_sf_t+\sum\gamma_{st}^1(e_sf_t-\delta_{st})+\lambda_1,$$

$$g(Fe_ie_j)=\sum\alpha_{st}^{ij}e_se_t+\sum\beta_{st}^{ij}f_sf_t+\sum\gamma_{st}^{ij}(e_sf_t-\delta_{st})+\lambda_{ij}.$$

hroughout it is assumed that $s<t$ for the coefficients α_{st}, β_{st}, and $1\le s,t\le 4$ for γ_{st}. By the hy-othesis,

$$ug(v)=vg(u) \tag{11}$$

r any element $u,v\in\mathfrak{S}^+$.

Direct but rather laborious calculation shows that, using relation (11), one can express all coef-cients α, β, γ in terms of λ_0, λ_1, and λ_{ij}. So calculated, only the following coefficients can be different from zero:

$$\beta^0_{s,t} = -\frac{\lambda_{st}}{4}, \quad \gamma^0_{ss} = \frac{\lambda_0}{4},$$
$$\alpha^1_{st} = (-1)^{s+t+1}\frac{\lambda_{s't'}}{4}, \quad \gamma^1_{ss} = -\frac{\lambda_1}{4},$$
$$\alpha^{st}_{st} = \lambda_0; \quad \beta^{st}_{s't'} = (-1)^{s+t}\frac{\lambda_1}{2},$$
$$\gamma^{ij}_{ss} = \frac{\lambda_{ij}}{4}, \quad s \neq i, j, \quad \gamma^{ij}_{ii} = \gamma^{ij}_{jj} = -\frac{\lambda_{ij}}{4},$$
$$\gamma^{ij}_{it} = \frac{\lambda_{jt}}{2}, \quad j < t, \quad \gamma^{ij}_{it} = -\frac{\lambda_{tj}}{2}, \quad t < j,$$
$$\gamma^{ij}_{jt} = -\frac{\lambda_{it}}{2}, \quad i < t, \quad \gamma^{ij}_{jt} = \frac{\lambda_{ti}}{2}, \quad t < i.$$

Here, for the pair i, j $(1 \leq i < j \leq 4)$ we denote by i', j' the complementary pair in $\{1, 2, 3, 4\}$, with $i' < j'$.

This done, it is not difficult to establish that $L_0^{(2)} = 0$.

The case $V = \mathfrak{S}^-$ is similar.

6. **The algebra $G_2 \oplus k$.** It is well known that the seven-dimensional irreducible representation Γ of G_2 with highest weight $\phi = \lambda_2$ corresponds to an injective mapping $G_2 \to B_3$, where B_3 is an orthogonal algebra given by its standard representation. Since $(B_3 \oplus k)^{(2)} = 0$, we have $(G_2 \oplus k)^{(2)} = 0$. Note that the lowest weight is then $\psi = -\lambda_2$, so that the remaining combinations $\phi - \psi = 3\lambda_2$ and $\phi - \psi = \lambda_1 + \lambda_2$ cannot be realized.

7. **The algebra $F_4 \oplus k$.** We need consider only the representation Γ with $\phi = \lambda_4$, $\psi = -\lambda_4$, acting on the 26-dimensional space $J' \subset J$ of elements with zero trace in the 27-dimensional exceptional Jordan algebra J with operation $\cdot$ over k. The elements of $\Gamma(F_4)$ are derivations of the algebra J leaving invariant the nondegenerate (for $p > 3$) symmetric form $(a, b) = \mathrm{Tr}(a \cdot b)$ (by definition, the elements a, b of J are hermitian 3-by-3 Cayley matrices [19]); consequently, $L_0^{(2)} = 0$.

8. **The algebra $E_6 \oplus k$.** We need consider only the representation Γ with $\phi = \lambda_1$, $\psi = -\lambda_5$, acting on the exceptional Jordan algebra J. Consider first the restriction Γ_0 of this representation to the algebra E_6. It turns out to be possible to solve the problem of the second prolongation $E_6^{(1)}$ without calculating $E_6^{(2)}$. Namely, $\Gamma_0(E_6)$ leaves invariant the symmetric trilinear form of Freudenthal:

$$(x, y, z) \equiv (x \cdot y, z) - (\mathrm{Tr}\, x)(y, z) - (\mathrm{Tr}\, y)(x, z) - (\mathrm{Tr}\, z)(x, y) + 2(\mathrm{Tr}\, x)(\mathrm{Tr}\, y)(\mathrm{Tr}\, z).$$

Here, $(x, y) = \mathrm{Tr}(x \cdot y)$ (cf. [10]). The form $(\ ,\ ,\ )$ is such that for fixed $a \neq 0$ the bilinear form f_a: $f_a(y, z) = (a, y, z)$ is nondegenerate. A general answer is given by

Lemma 13. *Let L_0 be a Lie algebra of linear transformations acting on a vector space V over a field k of characteristic $p > 3$ and leaving invariant the nondegenerate symmetric trilinear form $(\ ,\ ,\ )$:*

$$(xA, y, z) + (x, yA, z) + (x, y, zA) = 0; \quad x, y, z \in V, \quad A \in L_0.$$

Then $L_0^{(2)} = 0$.

Proof. The invariance of the form (, ,) with respect to L_0 can be expressed briefly in the form

$$(xA, x, x) = 0. \tag{12}$$

Pick an arbitrary element $\Phi \in L_0^{(2)}$. By definition, Φ is a symmetric function on $V \times V$ with values in L_0 and

$$x\Phi(u, v) = u\Phi(x, v), \quad x, u, v \in V.$$

In particular,

$$\sum_{\pi \in S_3} x_{\pi 1}\Phi(x_{\pi 2}, x_{\pi 3}) = 6x_1\Phi(x_2, x_3) \tag{13}$$

(S_3 the symmetric group on three symbols).

Setting $A = \Phi(u, v)$ in (12) and using the invariance and symmetry of the form (, ,) we find that

$$\begin{aligned} 0 &= (x\Phi(u, v), x, x) = (u\Phi(x, v), x, x) \\ &= -2(u, x\Phi(x, v), x) = -2(u, v\Phi(x, x), x). \end{aligned}$$

Similarly, $(v, u\Phi(x, x), x) = 0$. Therefore

$$\begin{aligned} &(u, v, x\Phi(x, x)) \\ &= -(u, v\Phi(x, x), x) - (u\Phi(x, x), v, x) = 0. \end{aligned}$$

Polarizing this relation with the aid of (13) gives, finally:

$$(u, v, w\Phi(x, y)) = 0.$$

Here u, v, w, x, y are arbitrary elements of V. The nondegeneracy of the form implies that $w\Phi(x, y) = 0$, i.e., $\Phi = 0$. This proves the lemma.

For the representation Γ_0 of the algebra E_6 one can prove that in fact $E_6^{(1)} = 0$. This follows from Lemma 14.

Lemma 14. *Let an irreducible p-module L_{-1} over a classical simple Lie algebra L_0 satisfy the following conditions:*

1) $L_0^{(2)} = 0$;

2) $\dim L_{-1} < \min(p-1, \dim L_0)$.

Then $L_0^{(1)} = 0$.

Proof. Suppose $L_0^{(1)} \neq 0$. In the L_0-module $L_0^{(1)}$ consider the minimal submodule L_1. In view of the hypothesis $L_0^{(2)} = 0$, the space

$$L = L_{-1} + L_0 + L_1$$

forms a graded Lie algebra. We prove that L is a simple p-algebra.

Let N be an ideal in L, $N \neq 0$. Then $N \cap L_{-1} \neq 0$, so that $N \supset N_{-1}$ in view of the irreducibility of L_{-1}. It is easy to see that $[L_{-1}, L_1] = L_0$ because of the simplicity of the algebra L_0, and therefore $N \supset L_0$.

Finally, $[L_0, L_1] \neq 0$; otherwise L_1 would be one-dimensional and we should have, since $L_0 = [L_{-1}, L_1]$, that $\dim L_0 \leq \dim L_{-1}$, contradicting hypothesis 2). From the irreducibility of L_1 it follows that $L_1 \subset N$. Thus, $N = L$.

The p-algebra structure present, by hypothesis, on L_0 we now extend to L, setting $x_{-1}^{[p]} = x_1^{[p]} = 0$. That this extension is possible follows from Theorem 11, Chap. 5 of [19], and also from the fact that L_{-1} and L_1 are p-modules over L_0 and $(\operatorname{ad} x)^3 = 0$ for $x \in L_{-1}$ or L_1. From Lemma 2, § 4, Chap. III it follows that L is a classical simple algebra.

Now it is easy to conclude the proof of Lemma 14. First of all, note that $\dim L_1 = \dim L_{-1}$. Indeed, L_{-1} is a sum of one-dimensional root spaces (the roots correspond to the weights $\phi_i \in \Lambda_{-1}$). Since L is a classical simple algebra, with each root ϕ_i there occurs also $-\phi_i$, and the corresponding element $e_{-\phi_i}$ enters into L_1. Thus $L_1 \supseteq L_{-1}^*$, and since L_1 is by its choice irreducible, we have $L_1 = L_{-1}^*$. From this and the condition $\dim L_{-1} < p-1$ it follows that

$$\dim L \leqslant (p-2) + ((p-2)^2 - 1) + (p-2) < p^2 - 2p.$$

Therefore L cannot be the algebra A_{kp-1}. We need this to be able to assert the existence of a representation of L with nondegenerate form (cf. [3]), from which in turn it follows that every derivation of L is inner.

If we consider the derivation

$$d: x_{-1} + x_0 + x_1 \to -x_{-1} + x_1,$$

we arrive at an element e such that $d = \operatorname{ad} e$. This element must be contained in the algebra L_0 and lie in its center, which contradicts the fact that L_0 is a simple algebra. This proves Lemma 14.

We return now to our 27-dimensional module L_{-1} over the algebra $L_0 = E_6 \oplus k$. It is known [13] that the algebra E_7 can be represented in the form

$$E_7 = L_{-1} + (E_6 \oplus k) + L_1,$$

where L_{-1} is our representation and $L_1 = L_{-1}^*$.

The equality $(E_6 \oplus k)^{(2)} = 0$ that we need comes from the following general fact.

Lemma 15. *If the algebra L has a nondegenerate Killing form $(\,,)$ and*

$$L = L_{-1} + L_0 + L_1,$$
$$L_0 = L_0' \oplus \langle e \rangle,$$

where the representation Γ of L_0 in L_{-1} is irreducible and its restriction Γ' to L_0' such that $(L_0')^{(1)} = 0$, then $L_0^{(2)} = 0$.

The proof consists of two parts. We prove first that $L_0^{(1)} = L_1$. It is easy to see that the form $(\,,)$ establishes a duality between L_{-1} and L_1. We can normalize the element e so that

$$[x_i, e] = ix_i, \quad i = -1, 0, 1; \quad x_i \in L_i.$$

Let $f \in L_0^{(1)}$; in particular, $f \in \operatorname{Hom}(L_{-1}, L_0)$. Then $(f(\ ), e) \in L_{-1}^*$ and corresponds to an element $y \in L_1$ for which

$$(f(x), e) = (x, y), \quad x \in L_{-1}.$$

nce $y = [-y, e]$ and the Killing form is invariant, we have

$$(x, y) = ([x, -y], e)$$

nd so

$$(f(x) - [y, x], e) = 0,$$

e.,

$$g(x) \equiv f(x) - [y, x] \in L_0'.$$

Thus

$$g = f - [y,] \in (L_0')^{(1)}$$

nd by hypothesis $g = 0$. This means that $f \in L_1$.

It remains to prove that $L_0^{(2)} = 0$, i.e., that $L_1^{(1)} = 0$. Let $\Phi \in L_0^{(1)}$; in particular, $\Phi \in \mathrm{Hom}(L_{-1},$ $)$. For any element h of a Cartan subalgebra of L_0 and arbitrary root vectors $e_\phi, e_\psi \in L_{-1}$, ϕ, ψ Λ_{-1}, we have

$$([\Phi(e_\varphi), e_\psi], h) = (\Phi(e_\varphi), [e_\psi, h]) = \psi(h)(\Phi(e_\varphi), e_\psi).$$

nce $\Phi \in L_1^{(1)}$, it follows that

$$\psi(h)(\Phi(e_\varphi), e_\psi) = \varphi(h)(\Phi(e_\psi), e_\varphi).$$

ssuming that $\phi \neq \psi$, choose h so that $\psi(h) \neq 0$, $\phi(h) = 0$. We see then that

$$(\Phi(e_\varphi), e_\psi) = 0 \quad \text{for} \quad \varphi \neq \psi.$$

follows that

$$\Phi(e_\varphi) = c_\varphi e_{-\varphi} \in L_1, \; c_\varphi \in k.$$

utting these values into the formulas expressing the fact that $\Phi \in L_1^{(1)}$, we find that:

$$c_\varphi[e_\varphi, e_{-\psi}] = c_\psi[e_\psi, e_{-\varphi}]; \quad \varphi, \psi \in \Lambda_{-1}. \tag{14}$$

in (14) some $c_\phi \neq 0$, then for our weight ϕ and for weight ψ, $\psi \neq \phi$, $\psi \in \Lambda_{-1}$, the difference $\phi - \psi$ not a root. Otherwise in the left- and right-hand sides of (14) we should have root vectors belong g to opposite roots. But all weights of the module L_{-1} can be obtained from one of them by succes ve addition of simple roots. Therefore, there exists a weight $\psi = \phi \pm \alpha$, where α is a simple root. his proves Lemma 15.

9. The algebras $E_7 \oplus k$ and $E_8 \oplus k$. That the second prolongation is zero follows directly om the remarks at the beginning of § 5. This completes the proof of the theorem of Chap. II.

CHAPTER III. CLASSIFICATION OF THE IRREDUCIBLE GRADED LIE ALGEBRAS

§ 1. Statement of results

In this chapter we enumerate all irreducible graded Lie algebras of finite dimension over k for hich 1) $L_2 \neq 0$; 2) L_0 is a Lie p-algebra with faithful irreducible representation Γ on L_{-1}; 3) dim L_{-1}

$< p-1$. We also assume that char $k = p > 7$. We first recall, going back to the definitions and results of Chap. I, that the algebras $\tilde{S}(\mathcal{F})$ and $\tilde{H}(\mathcal{F})$ as vector spaces over k have the form

$$\tilde{S}(\mathcal{F}) = S(\mathcal{F}) + S_0,$$
$$\tilde{H}(\mathcal{F}) = H(\mathcal{F}) + H_0,$$

where $\mathcal{F}$ is a flag and S_0 and H_0 are homogeneous (relative to the given standard gradation) subspaces

$$S_0 = \left\langle \prod_{j\neq i} x_j^{(p^{m_j}-1)} \partial_i;\ i = 1, \ldots, n \right\rangle,$$
$$H_0 = \left\langle \mathcal{D}_e,\ e = \prod_{i=1}^{n} x_i^{(p^{m_i}-1)};\ Q_j,\ 1 \leqslant j \leqslant n \right\rangle.$$

For the definition of $\mathcal{Q}_j$ cf. § 6, Chap. I.

The study of the derivations $\mathfrak{D}$ for which $\mathfrak{D}\omega = \alpha\omega$, where $\alpha \in k$, $\omega = dx_1 \wedge \ldots \wedge dx_n$ or $\sum \omega_{ij} dx_i \wedge dx_j$, leads to the algebras $S^*(E)$, $H^*(E)$ and, correspondingly, the algebras

$$S^*(\mathcal{F}) = S^*(E) \cap W(\mathcal{F}), \quad H^*(\mathcal{F}) = H^*(E) \cap W(\mathcal{F}).$$

They differ from $\tilde{S}(\mathcal{F})$ and $\tilde{H}(\mathcal{F})$ only in the zero homogeneous component:

$$\tilde{S}(\mathcal{F})_0 \cong \tilde{A}_{n-1}, \quad n > 2; \quad \tilde{H}(\mathcal{F})_0 \cong C_l, \quad 2l = n,$$

whereas

$$S^*(\mathcal{F})_0 \cong \mathfrak{gl}(n, k); \quad H^*(\mathcal{F})_0 \cong C_l \oplus \mathfrak{z},$$

where $\mathfrak{z} = \left\langle \sum_{i=1}^{n} x_i \partial_i \right\rangle$ is the one-dimensional center.

For the definition of $\tilde{A}_{n-1}$ see the Introduction.

We first prove

Theorem 1. *Every graded Lie algebra L is isomorphic to a graded transitive subalgebra M of some algebra $W(E)$.*

As usual, transitivity is defined by the condition $M \supset W(E)_{-1}$.

A large part of Chap. III is concerned with the proof of the following fact, which is of independent interest.

Theorem 2. *Let L be a finite-dimensional graded transitive subalgebra of the algebra $W(E)$ over the field k, char $k = p > 3$, such that $L_2 \neq 0$ and V is the standard representation of the algebra $L_0 \cong \mathfrak{gl}(n, k)$, $\tilde{A}_{n-1}$, C_l or $C_l \oplus k$. Then there exists a flag $\mathcal{F}$ with respect to which only one of the following facts can hold:*

1) $L = W(\mathcal{F})$;
2) $S(\mathcal{F}) \subseteq L \subseteq S^*(\mathcal{F})$;
3) $H(\mathcal{F}) \subseteq L \subseteq H^*(\mathcal{F})$.

By the proposition of § 10, Chap. I, under the hypothesis of Theorem 2 there exist only five possibilities for L_0 and L_1:

	L_0	L_1
1)	$gl(n, k)$	$W(E)_1$
2)	$sl(n, k)$	$S(E)_1$
3)	$gl(n, k)$	$S(E)_1$
4)	$Sp(n, k)$	$H(E)_1$
5)	$Sp(n, k) \oplus k$	$H(E)_1$

We now state the main result of the paper.

Theorem 3. *Let L be a finite-dimensional graded Lie algebra over an algebraically closed field k of characteristic $p > 7$, satisfying the conditions:*

a) $\dim L_{-1} < p-1$;

b) Γ: $L_0 \to \operatorname{End} L_{-1}$ *is an irreducible (faithful) p-representation of the Lie p-algebra* L_0;

c) $L_2 \neq 0$.

*Then L is isomorphic to one of the algebras in the statement of Theorem 2. In particular, the number of such algebras of given dimension is finite.**

§ 2. Proof of Theorem 1

Let L be an arbitrary graded Lie algebra:

$$L = \bigoplus_{i \geqslant -1} L_i, \quad \dim_k L < \infty.$$

Recall that, by definition, $[L_i, L_j] \subseteq L_{i+j}$, with $[L_{-1}, x] \neq 0$ for any $0 \neq x \in L_i$, $i \geq 0$.

We construct on the space $E = L_{-1}^*$ the algebra of divided powers $O(E)$ and the corresponding Lie algebra $W(E)$ (cf. Chap. I). We then identify the subspaces L_{-1} and $V = W(E)_{-1}$.

The subalgebra $L_0 \subset L$ is contained in End V, and since $W(E)_0 = \operatorname{End} V$, there exists an isomorphic imbedding

$$\varphi_0 : L_0 \to W(E)_0.$$

We shall now extend ϕ_0 to an imbedding

$$\varphi = \sum_{i \geqslant -1} \varphi_i, \quad \varphi_{-1} = \mathrm{Id},$$

of the graded algebra L into $W(E)$. The mappings ϕ_i will be constructed so that for $x \in L_i$ and $v \in L_{-1}$ they satisfy the relations

$$[\varphi_i(x), \varphi_{-1}(v)] = \varphi_{i-1}([x, v]), \tag{1}$$

and we shall then verify that ϕ defines an isomorphic imbedding of L to $W(E)$.

*Cf. the footnote at the end of the paper.

Suppose we have proved the existence of $\phi_{-1}, \ldots, \phi_{m-1}$, where ϕ_k defines an imbedding of L_k into $W(E)_k$ for $k \leq m-1$ and

$$[\varphi_j(x_j), \varphi_k(x_k)] = \varphi_{j+k}([x_j, x_k]), \quad j+k \leqslant m-1. \tag{2}$$

For any x in L_m set

$$\psi_x(v) = \varphi_{m-1}([x, v]),$$

thereby defining the element $\psi_x \in \mathrm{Hom}\,(V, W(E)_{m-1})$ (recall that we have identified L_{-1} and V). We prove that in fact $\psi_x \in W(E)^{(1)}_{m-1}$ and so corresponds to an element of $W(E)_m$. To do this we must verify the relation

$$\psi_x(u)(v) = \psi_x(v)(u)$$

or, equivalently, the relation

$$[\varphi_{m-1}([x, u]), v] = [\varphi_{m-1}([x, v]), u]. \tag{3}$$

But in view of the induction assumption (2),

$$[\varphi_{m-1}([x, u]), v] = \varphi_{m-2}([[x, u], v]),$$

$$[\varphi_{m-1}([x, v]), u] = \varphi_{m-2}([[x, v], u]),$$

and now (3) follows from the Jacobi identity in L. Thus, we have proved the existence of a mapping $\phi_m : x \to \psi_x$, $x \in L_m$, satisfying relation (1) for all $v \in V$ and $i = m$.

That ϕ_m is linear and injective is obvious. It remains to verify that (2) holds for $j + k = m$:

$$[\varphi_j(x_j), \varphi_k(x_k)] = \varphi_m([x_j, x_k]).$$

For this it suffices to prove that

$$[[\varphi_j(x_j), \varphi_k(x_k)], v] = [\varphi_m([x_j, x_k]), v]$$

for all $v \in V$. But this follows directly from the induction assumption and the Jacobi identity in $W(E)$ and L:

$$\begin{aligned} &[[\varphi_j(x_j), \varphi_k(x_k)], v] \\ &= [[\varphi_j(x_j), v], \varphi_k(x_k)] + [\varphi_j(x_j), [\varphi_k(x_k), v]] \\ &= [\varphi_{j-1}([x_j, v]), \varphi_k(x_k)] + [\varphi_j(x_j), \varphi_{k-1}([x_k, v])] \\ &= \varphi_{j+k-1}([[x_j, v], x_k] + [x_j, [x_k, v]]) = \varphi_{m-1}([[x_j, x_k], v]) \\ &= [\varphi_m([x_j, x_k]), v]. \end{aligned}$$

This proves Theorem 1.

§ 3. Proof of Theorem 2

An essential role in the proof is played by the following

Proposition 1. *Let $L(\mathfrak{F})$ be one of the simple Lie algebras $W(\mathfrak{F})$, $S(\mathfrak{F})$, or $H(\mathfrak{F})$ over k, char $k = p > 2$, corresponding to the flag*

$$\mathcal{F} : E = E_0 \supseteq E_1 \supseteq \ldots \supseteq E_{q-1} \supset E_q = 0.$$

t

$$\bar{L}(\mathcal{F})_l = \sum_{i=1}^{l-1} [L(\mathcal{F})_i, L(\mathcal{F})_{l-i}], \quad l > 1; \quad L_{-1} = V.$$

hen we have the equality

$$\bar{L}(\mathcal{F})_l = L(\mathcal{F})_l, \quad l \neq p^t - 1,$$

cept in the case $l = p^t - 2$ and $L = H(\mathfrak{F})$, when

0) $H(\mathcal{F})_l/\bar{H}(\mathcal{F})_l \cong \langle \mathcal{D}_{x_j^{(p^t)}};\ x_j \in E_t \rangle \cong E_t$.

$l = p^t - 1$, then

$$L(\mathcal{F})_l / \bar{L}(\mathcal{F})_l \cong E_t \otimes V,$$

explicitly:

1) $W(\mathcal{F})_l/\bar{W}(\mathcal{F})_l \cong \langle x_j^{(p^t)} \partial_i;\ x_j \in E_t,\ 1 \leqslant i \leqslant n \rangle$,
2) $S(\mathcal{F})_l/\bar{S}(\mathcal{F})_l \cong \langle \mathcal{D}_{i,k}\{x_j^{(p^t)} x_i^{(1)}\};\ x_j \in E_t,\ 1 \leqslant i, k \leqslant n \rangle$,
3) $H(\mathcal{F})_l/\bar{H}(\mathcal{F})_l \cong \langle \mathcal{D}_{x_j^{(p^t)} x_k^{(1)}};\ x_j \in E_t,\ 1 \leqslant k \leqslant n \rangle$.

Corollary. *In the notation of Proposition 1, $\bar{L}(\mathfrak{F})_l = L(\mathfrak{F})_l$ for all $l \geq p^q$. In particular, the Lie algebra L of each of the three types is generated by its homogeneous components L_{-1}, L_0, L_1.*

Proof. Select in E a basis $\{x_1, \ldots, x_n\}$, compatible with the flag $\mathfrak{F}$. Any derivation can then be ·itten as a linear combination of "monomials" $x_1^{(h_1)} \ldots x_n^{(h_n)} \partial_i$, where $h_i < p^{s+1}$ if $x_i \in E_s$, $x_i \notin E_{s+1}$.

I) The algebra $W(\mathfrak{F})$. Note first that the subflag $\mathfrak{F}': E = E_0 \supseteq E_1 \supseteq \ldots \supseteq E_{t-1} \supset E'_t = 0$ of the flag determines subalgebras $O(\mathfrak{F}') \subset O(\mathfrak{F})$ and $W(\mathfrak{F}') \subset W(\mathfrak{F})$ such that, indeed $\bar{W}(\mathfrak{F})_{p^t-1} \subset W(\mathfrak{F}')$ and

$$\langle x_j^{(p^t)} \partial_i;\ x_j \in E_t,\ 1 \leqslant i \leqslant n \rangle \subseteq W(\mathcal{F})_{p^t-1}/\bar{W}(\mathcal{F})_{p^t-1}.$$

·elow we obtain the reverse inclusion.

Let $w_0 = \prod_{j \neq i} x_j^{(h_j)}$, $\mathcal{D} = w_0 x_i^{(h_i)} \partial_i \in W(\mathcal{F})_l$. Obviously

$$\mathcal{D} = [w_0 \partial_i,\ x_i^{(h_i+1)} \partial_i] \in \bar{W}(\mathcal{F})_l,$$

·cept in the cases a) $\deg w_0 \leq 1$, b) $h_i = p^{s+1} - 1$, $x_i \in E_s/E_{s+1}$, c) $h_i = 0$. We consider these cases ·parately.

a) If $w_0 = x_j^{(h_j)}$, $h_j \leq 1$, then

$$[x_j^{(h_j)} x_i^{(\alpha)} \partial_i,\ x_i^{(l+2-h_j-\alpha)} \partial_i] = \gamma \mathcal{D},$$

$$\gamma = \binom{l+1-h_j}{\alpha} - \binom{l+1-h_j}{\alpha-1},$$

$\alpha = 2-h_j, \ldots, l-h_j$. The coefficient γ is equal to zero only for $l+1-h_j = p^t$. The value $h_j = 0$, and consequently $l = p^t - 1$, corresponds to assertion 1) of the proposition. If $h_j = 1$, then $\mathfrak{D} \in \bar{W}(\mathfrak{F})_l$, since

$$\mathfrak{D} - x_j^{(2)} x_i^{(l-1)} \partial_j = [x_i^{(l)} \partial_j,\ x_j^{(2)} \partial_i] \in \overline{W}(\mathfrak{F})_l,$$

$$x_j^{(2)} x_i^{(l-1)} \partial_j = [x_i^{(l-1)} \partial_j,\ x_j^{(3)} \partial_j] \in \overline{W}(\mathfrak{F})_l.$$

b) Use the relation

$$2\mathfrak{D} = [x_i^{(2)} \partial_i,\ w_0 x_i^{(p^{s+1}-2)} \partial_i] \in \overline{W}(\mathfrak{F})_l.$$

c) Since

$$\Big(\prod_{j \neq i} \binom{h_j}{\alpha_j}\Big) \mathfrak{D}$$

$$= \Big[\Big(\prod_{j \neq i} x_j^{(\alpha_j)}\Big) \partial_i,\ \Big(\prod_{j \neq i} x_i^{(h_j-\alpha_j)}\Big) x_i^{(1)} \partial_i\Big] \in \overline{W}(\mathfrak{F})_l,$$

$2 \leqslant \sum_{j \neq i} \alpha_j \leqslant l+1$, while the coefficient on the left is identically equal to zero only for $\mathfrak{D} = x_j^{(l+1)} \partial_i$, $l+1 = p^t$, we again arrive at assertion 1).

II) The algebra $S(\mathfrak{F})$. For the reason indicated in paragraph I),

$$\langle \mathfrak{D}_{i,k} \{x_j^{(p^t)} x_i^{(1)}\};\ x_j \in E_l,\ 1 \leqslant i, k \leqslant n \rangle \subseteq S(\mathfrak{F})_{p^t-1} / \bar{S}(\mathfrak{F})_{p^t-1}.$$

By Proposition 1 of § 5, Chap. I, $S(\mathfrak{F})$ is spanned by the derivations $\mathfrak{D}_{i,j}\{u\}$, $u \in O(\mathfrak{F})$. We note the following relations:

$$[\mathfrak{D}_{i,j}\{u\},\ \mathfrak{D}_{i,j}\{v\}] = \mathfrak{D}_{i,j}\{\partial_j(u)\partial_i(v) - \partial_i(u)\partial_j(v)\}, \tag{1}$$

$$[\mathfrak{D}_{i,j}\{u\},\ \mathfrak{D}_{j,k}\{v\}] = \mathfrak{D}_{j,i}\{\partial_j(u)\partial_k(v)\}$$
$$+ \mathfrak{D}_{k,j}\{\partial_i(u)\partial_j(v)\} + \mathfrak{D}_{i,k}\{\partial_j(u)\partial_j(v)\}, \qquad k \neq i. \tag{2}$$

Set $w_0 = \prod_{s \neq i, j} x_s^{(h_s)}$, $w = w_0 x_i^{(h_i)} x_j^{(h_j)}$, where i, j is an arbitrary but fixed pair of indices, and deg $w_0 + h_i + h_j = l + 2 \geq 4$.

Suppose first that deg $w_0 \geq 2$ and $h_i + h_j \geq 2$. Then taking $u = w_0 x_i^{(1)}$, $v = x_i^{(h_i)} x_j^{(h_j+1)}$ or $u = w_0 x_j^{(1)}$, $v = x_i^{(h_i+1)} x_j^{(h_j)}$, we obtain by means of formula (1) the required inclusion $\mathfrak{D}_{i,j}\{w\} \in \bar{S}(\mathfrak{F})_l$ if $h_i < p^{m_i} - 1$ or, respectively, $h_j < p^{m_j} - 1$ (m_i the height of the variable x_i). If $h_i = p^{m_i} - 1$, $h_j = p^{m_j} - 1$, we ust instead use formula (2), taking, for example, $u = w_0 x_j^{(1)}$, $v = x_i^{(h_i)} x_j^{(h_j)} x_k^{(1)}$, $k \neq i, j$. This gives se, to be sure, to the element

$$\mathcal{D}_{i,k}\{w_0 x_k^{(1)} x_i^{(h_i)} x_j^{(h_j-1)}\}.$$

ut either it is equal to zero or, by the preceding (replace the pair i, j by i, k), it lies in $\bar{S}(\mathcal{F})_l$, except in the case $h_k = p^{m_k} - 2$.

We are thus concerned with the element $\mathcal{D}_{i,j}\{w\}$, where w is a monomial with exponents $h_i = p^{m_i}$ 1, $h_j = p^{m_j} - 1$, $h_k = p^{m_k} - 2$. Here we set

$$w_0 = w_0' x_k^{(h_k)}, \quad u = w_0' x_j^{(\alpha_j+1)} x_k^{(\alpha_k)}, \quad v = x_i^{(h_i)} x_j^{(h_j-\alpha_j)} x_k^{(h_k+1-\alpha_k)}$$

nd consider the linear system

$$[\mathcal{D}_{i,j}\{u\}, \mathcal{D}_{j,k}\{v\}] \equiv 0 \pmod{\bar{S}(\mathcal{F})_l}$$

r $\alpha_j = 0$, $\alpha_k = h_k$, and $\alpha_j = 1$, $\alpha_k = h_k$ with determinant 1.

It remains to consider the case that one of the conditions $\deg w_0 \geq 2$, $h_i + h_j \geq 2$, is not satisfied.

a) $h_i + h_j = 0$. Then $w = w_0$ and $\mathcal{D}_{i,j}\{w\} = 0$.

b) $h_i + h_j = 1$, say $h_i = 1$, $h_j = 0$; $\deg w_0 > 2$.

gain taking $w_0 = w_0' x_k^{(h_k)}$, $u = w_0' x_j^{(1)} x_k^{(\alpha)}$, $v = x_i^{(2)} x_k^{(h_k - \alpha)}$, we obtain by formula (1):

$$\binom{h_k}{\alpha} \mathcal{D}_{i,j}\{w\} \in \bar{S}(\mathcal{F})_l.$$

ere $\alpha = 0$ for $h_k = 1$, $\alpha = 1$ for $h_k = 2$. For $h_k > 2$ we can take $\alpha = 2, 3, \ldots, h_k - 1$, and $\binom{h_k}{\alpha} = 0$ nly when $h_k = p^t$. If $w = x_k^{(p^t)} x_i^{(1)}$, then $\mathcal{D}_{i,j}\{w\} = -x_k^{(p^t)} \partial_j$ is one of the elements named in assertion 2) of our proposition. If $w = w_0'' x_a^{(p^s)} x_b^{(p^t)} x_i^{(1)}$, then the inclusion $\mathcal{D}_{i,j}\{w\} \in \bar{S}(\mathcal{F})_l$ is again obained by formula (1), where we take $u = w_0'' x_a^{(p^s)} x_j^{(1)}$, $v = x_i^{(2)} x_b^{(p^t)}$.

c₁) $w = x_i^{(h_i)} x_j^{(h_j)} x_k^{(1)}$. If $h_i \geq 2$ and $h_j \geq 1$, set $u = x_j^{(1)} x_i^{(h_i)}$, $v = x_j^{(h_j)} x_k^{(2)}$ and use formula (2):

$$\begin{aligned} -\mathcal{D}_{i,j}\{w\} + h_j \mathcal{D}_{k,j}\{x_i^{(h_i-1)} x_j^{(h_j)} x_k^{(2)}\} + \mathcal{D}_{i,k}\{x_i^{(h_i)} x_j^{(h_j-1)} x_k^{(2)}\} \\ = [\mathcal{D}_{i,j}\{u\}, \mathcal{D}_{j,k}\{v\}] \in \bar{S}(\mathcal{F})_l. \end{aligned}$$

n view of the preceding argument, the terms $\mathcal{D}_{k,j}\{*\}$ and $\mathcal{D}_{i,k}\{*\}$ are contained in $\bar{S}(\mathcal{F})_l$, so that the equired inclusion is obtained. The same holds for $h_i \geq 1$, $h_j \geq 2$. If $w = x_i^{(h_i)} x_k^{(1)}$, we use the equality

$$\mathcal{D}_{i,j}\{w\} = \mathcal{D}_{k,j}\{x_i^{(h_i-1)} x_k^{(2)}\},$$

educing the problem to the previous case $\deg w_0 = h_i - 1 \geq 2$, $h_k + h_j = 2$.

c₂) $w = x_i^{(h_i)} x_j^{(h_j)}$. When $h_i \geq 2$, $h_j \geq 2$, we use formula (2) with $u = x_i^{(h_i)} x_j^{(1)}$, $v = x_j^{(h_j)} x_k^{(1)}$, which ives a reduction to c₁). In formula (1) set $u = x_i^{(1)} x_j^{(2)}$, $v = x_i^{(h_i)}$, when $p^t \neq h_i \geq 3$, $h_j = 1$. The monoial $w = x_i^{(p^t)} x_j^{(1)}$ gives an element specified in assertion 2) ($h_i = 1$, $h_i \geq 3$ give nothing new). It remains to consider the case $w = x_i^{(h_i)}$. Formula (1) with $u = x_i^{(\alpha)} x_j^{(1)}$, $v = x_i^{(h_i+1-\alpha)}$, $\alpha = 2, 3, \ldots, h_i$

-2, leads to the inclusion $\mathfrak{D}_{i,j}\{w\} \in \bar{\bar{S}}(\mathfrak{F})_l$, except in the cases $h_i = p^t + 1$ (an element in assertion 2)) and $h_i = p^t$. But

$$\mathfrak{D}_{i,j}\{x_j^{(p^t)}\} = \mathfrak{D}_{i,k}\{x_j^{(p^t-1)}x_k^{(1)}\} = [\mathfrak{D}_{i,k}\{x_j^{(1)}x_i^{(1)}x_k^{(1)}\},\ \mathfrak{D}_{i,k}\{x_j^{(p^t-2)}x_k^{(1)}\}] \in \bar{S}(\mathfrak{F})_l.$$

III) The algebra $H(\mathfrak{F})$. We use the isomorphism of $H(\mathfrak{F})$ with the algebra $O'(\mathfrak{F})/k$ provided by the composition law (7), § 6, Chap. I. As noted in paragraph 2, § 3, Chap. I, we can take the form ω to be reduced to the canonical form corresponding to the monomial matrix $M: M^{-1} = (\bar{\omega}_{ij})$, $\bar{\omega}_{i,j} = \pm 1, 0$. Suppose $\bar{\omega}_{i,j} \neq 0$ for some pair of indices i, j. Let

$$w = \prod_k x_k^{(h_k)}, \quad w_0 = \prod_{k\neq i,\, j} x_k^{(h_k)}, \quad \sum h_k = l + 2 \geqslant 4.$$

In view of the skew-symmetry of M, we can assume $h_i \geq h_j$. With $u = x_i^{(\alpha_i)}x_j^{(\alpha_j)}$, $v = w_0 x_i^{(h_i+1-\alpha_i)} \cdot x_j^{(h_j+1-\alpha_j)}$ we get:

$$\{u, v\}_\omega = \bar{\omega}_{i,j}\cdot\gamma\cdot w,$$

where

$$\gamma = \binom{h_i}{\alpha_i - 1}\binom{h_j}{\alpha_j} - \binom{h_i}{\alpha_i}\binom{h_j}{\alpha_j - 1}.$$

For elements not belonging to $\bar{H}(\mathfrak{F})_l$ we need look only among those w for which $\gamma = 0$ for any allowable α_i, α_j.

a) $h_i \geq 2$, $h_j \geq 2$. Take either $\alpha_i = 0$, $\alpha_j = h_j + 1$ or $\alpha_i = h_i + 1$, $\alpha_j = 0$. This is impossible only for $h_i = p^{m_i} - 1$, $h_j = p^{m_j} - 1$—values which for $n = 2$ give an element w not contained in $O'(\mathfrak{F})$ and consequently, in $H(\mathfrak{F})$. For $n > 2$ there exists for the same reason a pair of indices i', j' with $\bar{\omega}_{i',j'} \neq 0$, $h_{i'} < p^{m_{i'}} - 1$. In the case $h_{j'} = p^{m_{i'}} - 1$ we replace i, j by i', j', in our argument and get the necessary result. In the case $h_{j'} < p^{m_{j'}} - 1$, let

$$w_0' = \prod_{k\neq i,\, i',\, j,\, j'} x_k^{(h_k)}, \qquad u = x_i^{(p^{m_i}-1)}x_{i'}^{(h_{i'}+1)}, \qquad v = x_j^{(p^{m_j}-1)}x_{j'}^{(h_{j'}+1)}w_0'.$$

Then

$$\{u, v\}_\omega = \bar{\omega}_{i',j'}w + \bar{\omega}_{i,j}x_i^{(p^{m_i}-2)}x_j^{(p^{m_j}-2)}w_0''.$$

By what has already been proved, the element with coefficient $\bar{\omega}_{i,j}$ is contained in $\bar{H}(\mathfrak{F})_l$. Therefore also $w \in \bar{H}(\mathfrak{F})_l$.

b$_1$) $h_i = 2$, $h_j = 1$, $\deg w_0 \geq 1$. The values $\alpha_i = 3$, $\alpha_j = 0$ give $\gamma \neq 0$.

b$_2$) $h_i \geq 3$, $h_j = 1$. Obviously $\alpha_j = 0$, $\gamma = 0 \Rightarrow \binom{h_i}{\alpha_i - 1} = 0$ for $\alpha_i = 3, 4, \ldots, h_i$, i.e., $h_i = p^t$. If $n = 2$, then w is an element in assertion 3). For $n > 2$ and $\deg w_0 \geq 1$ (if $\deg w_0 = 0$, then w is the required element), again choosing a pair i', j' with $\bar{\omega}_{i',j'} \neq 0$, $h_{j'} \geq 1$, we let

$$u = x_i^{(p^t)}x_{i'}^{(h_{i'}+1)}, \quad v = x_j^{(1)}x_{j'}^{(h_{j'}+1)}w_0'$$

nd get

$$\{u, v\}_\omega = \bar{\omega}_{i',j'}w + \bar{\omega}_{i,j}x_i^{(p^t-1)}x_j^{(0)}w_0''.$$

follows that $w \in \bar{H}(\mathfrak{F})_l$, since this holds for the element with coefficient $\bar{\omega}_{i,j}$ (take $\alpha_i = 3$, $\alpha_j = 0$). ur argument fails to apply only for $h_{i'} + 1 = p^{m_{i'}}$ or $h_{j'} + 1 = p^{m_{j'}}$. But then deg $w_0 > 1$ and $\gamma \neq 0$, if e take $u = x_i^{(p^t+1)}$, $v = x_j^{(2)}w_0$.

c$_1$) $2 \leq h_i < p^{m_i} - 1$, $h_j = 0$, deg $w_0 \geq 2$. Take $u = x_i^{(h_i+1)}$, $v = w_0 x_j^{(1)}$.

c$_2$) $h_i = p^{m_i} - 1$, $h_j = 0$, deg $w_0 \geq 2$. Take $u = x_i^{(h_i)}$, $v = w_0 x_i^{(1)} x_j^{(1)}$.

c$_3$) $h_i \geq 3$, $h_j = 0$, $w_0 = x_k^{(1)}$. Taking $u = x_i^{(\alpha)}$, $v = x_k^{(1)} x_i^{(h_i+1-\alpha)} x_j^{(1)}$, $\alpha = 3, \ldots, h_i$, we see that $= 0 \Rightarrow h_i = p^t$ and $w = x_i^{(p^t)} x_k^{(1)}$, as desired.

c$_4$) $w = x_i^{(h_i)}$, $h_i \geq 4$. If $u = x_i^{(\alpha)}$, $v = x_i^{(h_i+1-\alpha)} x_j^{(1)}$, $\alpha = 3, \ldots, h_i - 1$, then $\gamma = \binom{h_i}{\alpha - 1} = 0 \Rightarrow h_i$ $= p^t + 1$ or $h_i = p^t$ – values specified in 3) and 0).

d) $h_s \leq 1$ for all s, which is possible, of course, only for $n > 2$. Let

$$\bar{\omega}_{i,j} \neq 0, \quad w = x_i^{(1)} x_j^{(h_j)} x_k^{(1)} w_0', \quad \deg w_0' \geqslant 2 - h_j.$$

If we take $u = x_i^{(2)} x_k^{(1)}$, $v = x_j^{(h_j+1)} w_0'$, then $\{u, v\} = \bar{\omega}_{i,j} w + x_i^{(2)} *$. By what has already been proved, $x_i^{(2)}* \in \bar{H}(\mathfrak{F})_l$. Consequently, also $w \in \bar{H}(\mathfrak{F})_l$. This completes the proof of Proposition 1.

Proposition 2. *In the notation of Proposition* 1,

1) $\tilde{L}(\mathfrak{F})_{l-1}^{(1)} = \tilde{L}(\mathfrak{F})_l$ *for* $l \neq p^t - 1$.

2) $\tilde{L}(\mathfrak{F})_{l-1}^{(1)} = \tilde{L}(\mathfrak{F}^0)_l$ *for* $l = p^t - 1$, *where the flag* $\mathfrak{F}^0$ *has the form* $\mathfrak{F}^0$: $E = E_0^0 \supseteq E_1^0 \supseteq \ldots \supseteq E_{t-1}^0 = E_t^0$, $E_{t+1}^0 = 0$ *and* $E_i^0 = E_i$ *for* $i = 0, 1, \ldots, t-1$.

In particular, $L(\mathfrak{F})_{l-1}^{(1)} = L(\mathfrak{F})_l$ *for* $l \geq p^q$.

Proof. I. The algebra $W(\mathfrak{F})$. First we prove 1. Let $\mathcal{D} = \sum u_i \partial_i$, $u_i \in O(E)_{l+1}$. By definition

$$\mathcal{D} \in W(\mathfrak{F})_{l-1}^{(1)} \Leftrightarrow \partial_\xi(u_i) \in O(\mathfrak{F})_l \quad \forall \xi \in E_0^*, \quad 1 \leqslant i \leqslant n.$$

onsequently

$$u_i \notin O(\mathfrak{F}) \Rightarrow u_i = \Sigma \alpha_s x_i^{(p^t)} + v_i, \quad v_i \in O(\mathfrak{F}), \tag{*}$$

hich contradicts the condition $l \neq p^t - 1$.

Now let $l = p^t - 1$. In this case the inclusion $\mathfrak{D} \in W(\mathfrak{F})_{l-1}^{(1)}$ again implies (*), where $x_s \in E_{t-1}$. lements of the form $x_s^{(p^t)}$ are not contained in $W(\mathfrak{F})$, but are contained in $W(\mathfrak{F}^0)$, and we get 2).

II. The algebra $\tilde{S}(\mathfrak{F})$. From case I it follows that if $\mathfrak{D} \in \tilde{S}(\mathfrak{F})_{l-1}^{(1)}$, then $\mathfrak{D} \in W(\mathfrak{F})_l$ and $l = p^t - 1$ nd $W(\mathfrak{F}^0)_l$ for $l = p^t - 1$. It remains to verify that div $\mathfrak{D} = 0$ or, equivalently, $\mathfrak{D}\omega = 0$. But this folows from the fact that $[\partial_\xi, \mathfrak{D}] \in \tilde{S}(\mathfrak{F})$. Indeed,

$$0 = [\partial_\xi, \mathcal{D}]\omega \Rightarrow \partial_\xi \cdot \mathcal{D}\omega = 0 \Rightarrow \mathcal{D}\omega = 0$$

$\partial_\xi \omega = 0$, since ω has constant coefficients).

III. The algebra $\tilde{H}(\mathfrak{F})$ is considered similarly.

We turn now to the proof of Theorem 2, considering separately the cases listed after the statement of the theorem.

1) $L_0 = \mathfrak{gl}(n, k)$, $L_1 = W(E)_1$. We prove that in this case $L = W(\mathfrak{F})$, i.e., that there exists a flag $\mathfrak{F}$ for which

$$L_i = W(\mathscr{F})_i \tag{3}$$

for all i. For $i = -1, 0, 1$, equation (3) is given, and we prove it from then on by induction. Suppose it holds for all $i < l$. More precisely,

$$L_i = W(\mathscr{F}')_i, \quad i < l, \tag{3'}$$

where

$$\mathscr{F}' : E = E_0 \supseteq E_1 \supseteq \ldots \supseteq E_{t-1} \supset E_t' = 0$$

is a certain subflag of the flag $\mathfrak{F}$. Under construction we can always choose $\mathfrak{F}'$ satisfying condition (3') such that $l \geq p^{t-1}$. Otherwise, we could choose a flag of height t_1 (where t_1 is the highest value of t satisfying $l \geq p^{t-1}$), for which all E_k with $0 \leq k \leq t_1 - 1$ are the same as in the flag $\mathfrak{F}'$. Therefore, we will always suppose $l \geq p^{t-1}$. If, further, $l + 1 \neq p^s$, then by Propositions 1 and 2 and in accordance with (3'),

$$\begin{aligned} L_l \supseteq \overline{L}_l &= \sum_{i=1}^{l-1} [L_i, L_{l-i}] = \overline{W}(\mathscr{F}')_l \\ &= W(\mathscr{F}')_l = W(\mathscr{F}')_{l-1}^{(1)} = L_{l-1}^{(1)}. \end{aligned}$$

But since, of course, we always have $L_l \subseteq L_{l-1}^{(1)}$, it follows that in the present case $L_l = L_{l-1}^{(1)} = W(\mathfrak{F}')$. From the same Propositions 1 and 2 it follows that a similar situation holds for $l + 1 = p^s$, $s > t$. Now let $l + 1 = p^t$. Consider, as in Proposition 2, the flag

$$\mathscr{F}^0 : E = E_0 \supseteq E_1 \supseteq \ldots \supseteq E_{t-1} = E_t^0 \supset E_{t+1}^0 = 0.$$

It is obvious (again using the induction equality (3')) that

$$\overline{W}(\mathscr{F}^0)_l = \overline{W}(\mathscr{F}')_l = \overline{L}_l \subseteq L_l \subseteq L_{l-1}^{(1)} = W(\mathscr{F}')_{l-1}^{(1)} = W(\mathscr{F}^0)_l,$$

where, by Proposition 1,

$$W(\mathscr{F}^0)_l / \overline{W}(\mathscr{F}^0)_l \simeq E_t^0 \otimes V \tag{4}$$

and all three spaces

$$\overline{W}(\mathscr{F}^0)_l, \quad L_l, \quad W(\mathscr{F}^0)_l$$

are invariant relative to L_0. The isomorphism (4) provides $E_t^0 \otimes V$ with the structure of an L_0-module:

$$\mathscr{D} : e \otimes v \to \mathscr{D}(e) \otimes v + e \otimes \mathscr{D}(v), \quad \mathscr{D} \in L_0,$$

isomorphic, as is easy to see, to the tensor product of the standard L_0-module V and a trivial (i.e., ne with zero action) module $E_0^{(t)}$. Indeed, this follows directly from the action of $\mathfrak{D} \in L_0$ on the inerse images $x_j^{(p^t)}\partial_i$ of the basis elements of the factor space $W_l/\overline{W}_l$:

$$x_j^{(p^t)}\partial_i \to [\mathfrak{D}, x_j^{(p^t)}\partial_i] = \mathfrak{D}(x_j^{(p^t)})\,\partial_i + x_j^{(p^t)}[\mathfrak{D}, \partial_i].$$

Obviously, $\mathfrak{D}(x_j^{(p^t)}) \in O(\mathfrak{F}')$ and the derivation $\mathfrak{D}(x_j^{(p^t)})\partial_i$ goes into zero under the mapping $(\mathfrak{F}^0)_l \to W(\mathfrak{F}^0)_l/\overline{W}(\mathfrak{F}^0)_l$ while $\partial_i \to [\mathfrak{D}, \partial_j]$ coincides precisely with the standard action of $\mathfrak{D}$ in V.

In accordance with the properties of semisimple modules, any invariant subspace in $E_t^0 \otimes V$ lative to L_0 has the form $E_t \otimes V$, where E_t is a uniquely defined subspace in E_t^0. Such a space is e factor space

$$L_l / \overline{L}_l \subseteq W(\mathcal{F}^0)_l / \overline{W}(\mathcal{F}^0)_l.$$

e have arrived, consequently, at a flag

$$\mathcal{F}'' : E = E_0 \supseteq E_1 \supseteq \ldots \supseteq E_{t-1} \supseteq E_t \supseteq E''_{t+1} = 0$$

ot excluding the possibility $E_t = E''_{t+1} = 0$, in which case our proof is automatically finished; posble also is the case $\mathfrak{F}'' = \mathfrak{F}^0$). We assert that

$$L_i = W(\mathcal{F}'')_i, \quad i < l+1 = p^t. \tag{5}$$

Indeed, the addition of the subspace E_t does not change the components $W(\mathfrak{F}')_i$ with $i < l$, when) coincides with (3′), while in dimension l we have added just elements not belonging to $W(\mathfrak{F}')_l$.

This completes the induction proof.

2) $L_0 = sl(n, k)$, $L_1 = S(E)_1$. Proceeding as in case 1), we assume that we have proved the reuired inclusion

$$S(\mathcal{F}')_i \subseteq L_i \subseteq \overline{S}(\mathcal{F}')_i, \quad i < l, \tag{6}$$

here $l > 1$, and

$$\mathcal{F}' : E = E_0 \supseteq E_1 \supseteq \cdots \supseteq E_{t-1} \supset E'_t = 0$$

some flag of height t. We must embed $\mathfrak{F}'$ in a new flag $\mathfrak{F}''$ with respect to which inclusion (6) holds r $i = l$.

From Proposition 1, § 5, Chap. I it follows that the algebra $\widetilde{S}(\mathfrak{F}')$ is obtained from the special gebra $S(\mathfrak{F}')$ by adding the n derivations

$$\mathfrak{D}_i = \Big(\prod_{j \neq i} x_j^{(p^{m_j}-1)}\Big)\partial_i$$

m_j is the height of the variable x_j in a basis compatible with the flag $\mathfrak{F}'$), with $[\widetilde{S}, \widetilde{S}] = S$. Further, it easy to see that $[\mathfrak{D}_i, \mathfrak{D}_j] = 0$ and

$$\overline{\overline{S}}(\mathcal{F}')_l = \sum_{i=1}^{l-1} [\widetilde{S}_i, \widetilde{S}_{l-i}] = \overline{S}(\mathcal{F}')_l \tag{7}$$

for all $l > 1$. It suffices simply to look at the form of the elements of $S_l/\overline{S}_l$ in assertion 2) of Proposition 1, taking into account the hypothesis $n > 2$.

First let $l \neq p^t - 1$. Then from Propositions 1 and 2 and the induction assumption (6) it follows that

$$L_l \supseteq \overline{L}_l = \overline{S}(\mathcal{F}')_l = S(\mathcal{F}')_l.$$

In addition,

$$L_l \subseteq L_{l-1}^{(1)} \subseteq \widetilde{S}(\mathcal{F}')_{l-1}^{(1)} = \widetilde{S}(\mathcal{F}')_l.$$

Now let $l = p^t - 1$. Observe that

$$\mathcal{D}_i \in \widetilde{S}(\mathcal{F})_{q_i}, \quad q_i + 1 = \sum_{j \neq i} (p^{m_j} - 1) = (p-1)\, p' \neq p^t = l + 1,$$

so that

$$\overline{S}(\mathcal{F})_l = S(\mathcal{F})_l,$$

whatever the flag $\mathfrak{F}$. As before, consider the flag $\mathfrak{F}^0$, setting $E_t^0 = E_{t-1}$. Then by the preceding observation and the induction inclusion (6),

$$\overline{S}(\mathcal{F}^0)_l = \overline{S}(\mathcal{F}')_l = \overline{L}_l \subseteq L_l \subseteq L_{l-1}^{(1)} \subseteq \widetilde{S}(\mathcal{F}')_{l-1}^{(1)} = \widetilde{S}(\mathcal{F}^0)_l = S(\mathcal{F}^0)_l,$$

so that, by Proposition 1,

$$L_l/\overline{L}_l \subseteq S(\mathcal{F}^0)_l/\overline{S}(\mathcal{F}^0)_l \cong E_t^0 \otimes V.$$

The preceding argument goes through word for word, and we arrive at the equality

$$L_l = S(\mathcal{F}'')_l$$

or, using (6),

$$S(\mathcal{F}'')_i \subseteq L_i \subseteq \widetilde{S}(\mathcal{F}'')_i, \quad i < l + 1,$$

where

$$\mathcal{F}'': E = E_0 \supseteq E_1 \supseteq \ldots \supseteq E_{t-1} \supseteq E_t \supseteq E''_{t+1} = 0$$

is the flag obtained from $\mathfrak{F}^0$ by replacing E_t^0 by a certain subspace E_t. This corresponds to the choice of a submodule in the L_0-module $E_t^0 \otimes V$. This completes the proof of part 2).

We remark that the addition to $\mathfrak{F}'$ of a nonvoid subspace E_t transfers some of the elements $\mathfrak{D}_i \in \widetilde{S}(\mathfrak{F}')$ (possibly after a linear change of variables) to the category of elements of $S(\mathfrak{F}'')$. At the same time, in the homogeneous components of higher degree there arise new elements of type $\mathfrak{D}_i$. It is also the case, incidentally, that the equality $L = S(\mathfrak{F})$ may hold, even though the proof involved the strict inclusion $S(\mathfrak{F}')_i \subset L_i$.

3) $L_0 = \mathfrak{gl}(n, k)$, $L_1 = S(E)_1$. Since L_0-modules in $E_t \otimes V$ coincide with $sl(n, k)$-submodules, the whole argument for part 2) also applies here (replace $\widetilde{S}$ by S^*!).

4) $L_0 = Sp(n, k)$, $L_1 = H(E)_1$. This is almost no different from the argument for part 2), but since in Proposition 1 the algebra $H(\mathfrak{F})$ occupies a special place, we give an outline of the proof.

As usual, we have the inclusion

$$\bar{H}(\mathcal{F}')_l \subseteq \bar{L}_l \subseteq L_l \subseteq L_{l-1}^{(1)} \subseteq \tilde{H}(\mathcal{F}')_{l-1}^{(1)}.$$

$l \geq p^t$, then from Propositions 1 and 2 it follows that

$$\bar{H}(\mathcal{F}')_l = H(\mathcal{F}')_l,$$
$$\tilde{H}(\mathcal{F}')_{l-1}^{(1)} = \tilde{H}(\mathcal{F}')_l,$$

ich gives a proof of the theorem in this case.

Now let $l < p^t$. Starting with the induction hypothesis

$$H(\mathcal{F}')_i \subseteq L_i \subseteq \bar{H}(\mathcal{F}')_i, \quad i < l, \tag{8}$$

e note that

$$\bar{H}(\mathcal{F}')_l \subseteq \bar{L}_l \subseteq \bar{\bar{H}}(\mathcal{F}')_l.$$

deed,

$$\bar{H}(\mathcal{F}')_l = \bar{L}_l = \bar{\bar{H}}(\mathcal{F}')_l. \tag{9}$$

o prove (9), we use formulas (5) and (6) of § 6, Chap. I, which imply that $\bar{H}(\mathcal{F}')_l = \bar{\bar{H}}(\mathcal{F}')_l$. The ly possible exception is the case $n = 2$, when for $e = x_1^{(p^{m_1}-1)} x_2^{(p^{m_2}-1)}$ the element $\mathfrak{D}_e$ is not con- ined in $H(\mathcal{F}')$, yet lies in $\bar{\tilde{H}}(\mathcal{F}')$ (cf. § 6, Chap. I). But this is impossible in our case by a dimen- on argument: $\mathfrak{D}_e \in \bar{\tilde{H}}(\mathcal{F}')_k$, where $k = p^{m_1} + p^{m_2} - 4$, $m_1 \geq 1$, $m_2 = t$, and $k > l$ since we have $l < p^t$; that $\mathfrak{D}_e \notin \bar{\tilde{H}}(\mathcal{F}')_l$.

As usual, we now consider two cases.

If $l \neq p^t - 1$, then

$$\bar{H}(\mathcal{F}')_l = H(\mathcal{F}')_l. \tag{10}$$

his follows directly from Proposition 1 if we notice that for $l = p^t - 2$ assertion 0) reduces to the uality $H(\mathcal{F}')_l / \bar{H}(\mathcal{F}')_l \cong E'_t = 0$. Now combining (9) and (10) with the obvious chain of inclusions $\subseteq L_l \subseteq L_{l-1}^{(1)} \subseteq \tilde{H}(\mathcal{F}')_{l-1}^{(1)}$ (induction!) and the equality $\tilde{H}(\mathcal{F}')_{l-1}^{(1)} = \hat{H}(\mathcal{F}')_l$ from Proposition 2, we ob- in (8) for $i = l$.

Finally, let $l = p^t - 1$. From $\mathcal{F}'$ we pass to the flag $\mathcal{F}^0$, as before. Obviously $\tilde{H}(\mathcal{F}^0)_l = \bar{H}(\mathcal{F}')_l$, that by (9) we find that

$$\bar{H}(\mathcal{F}^0)_l = \bar{L}_l \subseteq L_l \subseteq L_{l-1}^{(1)} \subseteq \tilde{H}(\mathcal{F}')_{l-1}^{(1)}.$$

Further, Proposition 2 implies that

$$\tilde{H}(\mathcal{F}')_{l-1}^{(1)} = \tilde{H}(\mathcal{F}^0)_l$$

d in turn

$$\bar{H}(\mathcal{F}^0)_l = H(\mathcal{F}^0)_l,$$

since $\widetilde{H}(\mathcal{F}^0)_k$ differs from $H(\mathcal{F}^0)_k$ only in the homogeneous components of degree $p^s - 2$ and $\sum (p^{m_i} - 1) - 2$, which cannot coincide with l.

Thus,

$$\overline{H}(\mathcal{F}^0)_l = \overline{L}_l \subseteq L_l \subseteq H(\mathcal{F}^0)_l,$$

whence

$$L_l / \overline{L}_l \subseteq H(\mathcal{F}^0)_l / \overline{H}(\mathcal{F}^0)_l,$$

or

$$L_l/\overline{L}_l \cong E_t \otimes V \subseteq E_t^0 \otimes V,$$

if we use Proposition 1. From the construction of the flag $\mathcal{F}''$: $E = E_0 \supseteq E_1 \supseteq \ldots \supseteq E_{t-1} \supseteq E_t \supseteq E''_{t+1} = 0$ it is now clear that the inclusions (8) hold, with $\mathcal{F}'$ replaced by $\mathcal{F}''$, for all $i < l+1$.

5) $L_0 = Sp(n, k) \oplus k$, $L_1 = H(E)_1$. The argument is a repetition of that for part 4). This completes the proof of Theorem 2.

§ 4. Proof of Theorem 3. An example

1. **Proof of Theorem 3.** From conditions a) and b) of Theorem 3 and from the theorem of Chap. II, it follows that L_0 is isomorphic to one of the algebras A_n, $A_n \oplus k$, C_l, $C_l \oplus k$, and Γ is the standard representation. In view of this, Theorem 3 follows from Theorem 2.

2. **Example.** We give an example to show that condition a) of Theorem 3 cannot be removed. Namely, there exist graded algebras with $\dim L_{-1} = p-1$ that are not isomorphic (as graded algebras to any of the algebras in Theorem 2. If we ignore gradation, however, the algebra we obtain is still isomorphic to one of the algebras of Theorem 2, namely, the algebra H_1. Thus it is simply a question of another gradation of H_1.

We recall that the algebra H_1 may be given as O'/k:

$$O' \subset k[x, y] / (x^p, y^p)$$

and consists of polynomials not containing the term $(xy)^{p-1}$. The commutator is defined by formula

$$[f, g] = \begin{vmatrix} f'_x & f'_y \\ g'_x & g'_y \end{vmatrix}.$$

We denote by L'_k the image in O'/k of the set of all polynomials of degree $k+1$ in x. Clearly the family $\{L'_k\}$ defines a gradation of length $p-2$ of the algebra H_1. Here, $\dim L_{-1} = p-1$. The algebra L'_0 consists of polynomials of the form $xf(y)$, and the mapping $xf(y) \to f(y)\,(d/dy)$ determines an isomorphism with the algebra W_1. Hence, the graded algebra $H_1 = \bigoplus L'_k$ is not isomorphic to any of the algebras in Theorem 2.

§ 5. Application to p-algebras

We begin with two general lemmas on p-algebras.

Lemma 1. *The structures of a graded algebra and of a Lie p-algebra without center are connected y the relation*

$$L_i^{[p]} \subset L_{ip}.$$

Proof. Let $x \in L_i$ and

$$x^{[p]} = \sum_{k \geqslant -1} x_k. \tag{1}$$

'e prove first of all that $x_{-1} = 0$. If $x_{-1} \neq 0$, then since the center is empty, there exists a homoge-eous element $y_s \in L_s$ such that $[x_{-1}, y_s] \neq 0$. From (1) we get that

$$[x^{[p]}, y_s] = [x_{-1}, y_s] + \ldots.$$

'he left-hand side is contained in L_{pi+s}, but the right has a nonzero component contained in L_{s-1}. 'e are led to a contradiction: $pi + s = s - 1$, $pi = -1$.

Thus, for any $x_k \neq 0$ in (1), we have $k \neq -1$, so that there exists an element y_{-1} in L_{-1} such that $_k, y_{-1}] \neq 0$. From (1) it follows that

$$[x^{[p]}, y_{-1}] = \sum [x_k, y_{-1}].$$

he left-hand side is contained in L_{pi-1}, and the right contains a nonzero component in L_{s-1}. There-re, $k = pi$. This proves the lemma.

Remark. From the lemma it follows that with the same hypothesis L_0 is a p-algebra and Γ: L_0 End L_{-1} is a p-representation.

Lemma 2. *Let*

$$L = L_{-1} \oplus L_0 \oplus L_1$$

e a graded simple Lie p-algebra in which L_0 is a direct sum of classical simple algebras and pos-ibly a one-dimensional center $\mathfrak{z}(L_0)$. Then L is isomorphic to a classical simple Lie algebra.

Proof. Denote by H_0 a Cartan subalgebra of L_0. Obviously

$$N_L(H_0) = C_L(H_0) = L'_{-1,0} \oplus H_0 \oplus L'_{1,0} = H,$$

here $N_L(H_0)$ is the normalizer of H_0 in L, C is its centralizer, $L'_{-1,0} = C_{L_{-1}}(H_0)$, and $L'_{1,0} = C_{L_1}(H_0)$. e prove that the $L'_{i,0}$ are the subspaces in L_{-1} and L_1, respectively, corresponding to zero weight f such exist).

Indeed, H_0 has a basis consisting of elements h for which $h^{[p]} - h = \lambda e$, $e \in \mathfrak{z}(L_0)$. If $L_{-1,0}$ is e subspace corresponding to zero weights in L_{-1}, and $u \in L_{-1}$, then $(\operatorname{ad} h)^{p^k} u = (\phi(\lambda)E + \operatorname{ad} h)u = 0$, om which it easily follows that $[h, u] = 0$, i.e., $L'_{-1,0} = L_{-1,0}$.

From the definition it follows that H_0 lies in the center of H and $[L_{-1,0}, L_{1,0}] \subseteq H_0$, so that H is a ilpotent subalgebra of class ≤ 2.

If $a = a_{-1} + a_0 + a_1 \in N_L(H)$, $a_i \in L_i$, then in particular $[H_0, a] \subseteq H$, whence $[H_0, a_{-1}] \subseteq L_{-1,0}$, $[H_0,$ $] \subseteq L_{1,0}$. Writing a_i as a linear combination of elements of the various weight spaces (or root spaces r $i = 0$) relative to H_0, we easily conclude that $a \in H$. Thus, $N_L(H) \subseteq H$ and H is a Cartan subalge-a of L. Since $\operatorname{cl}(H) \leq 2$ and $(\operatorname{ad} x)^3 = 0$ for $x \in L_i$, $i = \pm 1$, it follows that in the Cartan decomposition

$$L = H + \sum_{\alpha \neq 0} L^{(\alpha)}$$

the roots $\alpha \neq 0$ are linear functions on H, and each $L^{(\alpha)}$ is a weight space relative to H_0. Any element $a \neq 0$ in $H \cap L_{-1}$ or in $L^{(\alpha)} \cap L_{-1}$ satisfies the condition $(\operatorname{ad} a)^{p-1} = 0$ of Theorem 1 of [22]. By this theorem, L is an algebra of classical type so long as it contains no element $c \neq 0$ such that $(\operatorname{ad} c)^2 = 0$. If L contained such an element, then without loss of generality (cf. Theorem A, [22]) we could suppose that

$$\operatorname{ad} c (\operatorname{ad} x_1 \ldots \operatorname{ad} x_k) \operatorname{ad} c = 0, \quad k = 0, 1, \ldots, p-4, \tag{2}$$

for any elements $x_1, \ldots, x_k \in L$.

We have $c = c_{-1} + c_0 + c_1$. Obviously there always exist elements $a_1, \ldots, a_s \in L_{\pm 1}$, $s \leq 3$, for which

$$[ca_1 \ldots a_s] = c_0' \in L_0, \quad c_0' \neq 0.$$

Here

$$(\operatorname{ad} c_0')^2 = \sum \ldots \operatorname{ad} c (\operatorname{ad} a_{i_1} \ldots \operatorname{ad} a_{i_k}) \operatorname{ad} c \ldots = 0,$$

which follows directly from relations (2) and the inequalities $k \leq 2s \leq 6 < p-4$. Since the element $e \in \mathfrak{z}(L_0)$ acts as a scalar, we are led to the conclusion that $c_0' \in [L_0, L_0]$. But in the semisimple algebra $[L_0, L_0]$ of classical type there is no element $c_0' \neq 0$ with $(\operatorname{ad} c_0')^2 = 0$. This contradiction proves the lemma.

Remark. The proof of the lemma does not use any property of the representation $\Gamma: L_0 \to \operatorname{End} L_{-1}$.

Theorem. *A graded simple Lie p-algebra* ($p > 7$) *with* $\dim L_{-1} < p-1$ *is either a classical algebra or an algebra of Cartan type (other than a contact algebra).*

Proof. Let

$$L = L_{-1} + L_0 + L_1 + \ldots + L_r.$$

We prove that the L_0-module L_{-1} is irreducible. Indeed, since L is simple, it coincides with the ideal I_x generated by an arbitrary element $x \neq 0$ in L_{-1}. On the other hand, it is easy to verify that I_x is a sum of linear subspaces

$$[xL_0^{i_0}L_1^{i_1} \ldots L_r^{i_r}] \subseteq L_{-1+i_1+2i_2+\cdots+ri_r}.$$

Obviously, therefore,

$$L_{-1} \cap I_x = \sum_i [xL_0^i],$$

but this cannot coincide with L_{-1} if x is contained in a proper invariant subspace.

It is clear that $L_1 \neq 0$; otherwise L_{-1} would be an ideal in L. The case $L_2 = 0$ is considered in Lemma 2. Let $L_2 = 0$. Then by Theorem 3 either L coincides with $W(\mathfrak{F})$, or

$$S(\mathfrak{F}) \subseteq L \subseteq S^*(\mathfrak{F}),$$

$$H(\mathcal{F}) \subseteq L \subseteq H^*(\mathcal{F}),$$

where the flag $\mathfrak{F}$ has the form $E = E_0 \supset E_1 = 0$.

In the last two cases, since L is simple and $S(\mathfrak{F})$ and $H(\mathfrak{F})$ are ideals in $S^*(\mathfrak{F})$ and $H^*(\mathfrak{F})$, respectively, we conclude that L coincides with $S(\mathfrak{F})$ or $H(\mathfrak{F})$. This proves the theorem.†

V. A. Steklov Mathematical Institute
Academy of Sciences of the USSR

Received 10 OCT 68

BIBLIOGRAPHY

[1] A. A. Albert and M. S. Frank, *Simple Lie algebras of characteristic* p, Univ. e Politec. Torino. Rend. Sem. Mat. 14 (1954/55), 117-139. MR 18, 52.

[2] R. Block, *New simple Lie algebras of prime characteristic*, Trans. Amer. Math. Soc. 89 (1958) 421-449. MR 20 #6446.

[3] ———, *Trace forms on Lie algebras*, Canad. J. Math. 14 (1962), 553-564. MR 25 #3973.

[4] E. Cartan, *Les groupes de transformations continues, infinis, simples*, Ann. Sci. École Norm. Sup. (3) 26 (1909), 93-161.

[5] R. Cartier, *Séminaire "Sophus Lie" de la faculté des Sciences de Paris*, 1955/56, Secrétariat mathématique, Paris, 1957, pp. 3-7. MR 19, 431.

[6] H. J. Chang, *Über Wittsche Lie-Ringe*, Abh. Math. Sem. Hansischen Univ. 14 (1941), 151-184. MR 3, 101.

[7] C. Curtis, *Representations of Lie algebras of classical type with applications to linear groups*, J. Math. Mech. 9 (1960), 307-326. MR 22 #1634.

[8] M. S. Frank, *A new class of simple Lie algebras*, Proc. Nat. Acad. Sci. U.S.A. 40 (1954), 713-719. MR 16, 562.

[9] ———, *Two new classes of simple Lie algebras*, Trans. Amer. Math. Soc. 112 (1964), 456-482. MR 29 #133.

[10] H. Freudenthal, *Oktaven, Ausnahmegruppen und Oktavengeometrie*, Mathematisch Inst. Rijksuniversiteit, Utrecht, 1951. MR 13, 433.

[11] N. Jacobson, *Classes of restricted Lie algebras of characteristic* p, II, Duke Math. J. 10 (1943), 107-121. MR 4, 187.

[12] ———, *A note on three-dimensional simple Lie algebras*, J. Math. Mech. 7 (1958), 823-831. MR 20 #3901.

[13] S. Kobayashi and T. Nagano, *On filtered Lie algebras and geometric structures*, I, J. Math. Mech. 13 (1964), 875-908; III, J. Math. Mech. 14 (1965), 679-706. MR 29 #5961; MR 32 #5803.

† Added in Proof. M. Ju. Celoucov has recently established an isomorphism between the algebras of [9] and K_n (cf. the beginning of § 7, Chap. I). V. A. Kreknin has proved the theorem of § 8, Chap. I in the case of an arbitrary flag $\mathcal{F}$ (the question of whether $\mathfrak{E}_0$ and $N_{\mathfrak{g}}(\mathfrak{C})$ coincide, however, remains open). Finally, an analysis by V. A. Žerebcov of the proof of the results of [22] and [23] has shown that, as conjectured, the conditions that the algebras and their representations be restricted are superfluous. Hence, in the statements of the main results of the present paper (cf. the Introduction and the beginning of Chap. III) one can omit the assumption that L_0 and Γ are restricted.

[14] S. Lie, *Gesammelte Abhandlungen*, Band VI, Oslo, Leipzig, 1927, pp. 300-365.

[15] Takushiro Ochiai, *Classification of the finite nonlinear primitive Lie algebras*, Trans. Amer. Math. Soc. 124 (1966), 313-322. MR #4320.

[16] R. Ree, *On generalized Witt algebras*, Trans. Amer. Math. Soc. 83 (1956) 510-546. MR 18, 491.

[17] I. M. Singer and S. Sternberg, *The infinite groups of Lie and Cartan*, I, The transitive groups, J. Analyse Math. 15 (1965), 1-114. MR 36 #911.

[18] V. W. Guillemin and S. Sternberg, *An algebraic model of transitive differential geometry*, Bull. Amer. Math. Soc. 70 (1964), 16-47; Russian transl., Matematika 10 (1966), no. 4, 3-31. MR 30 #533.

[19] N. Jacobson, *Lie algebras*, Interscience Tracts in Pure and Appl. Math., no. 10, Interscience, New York, 1962; Russian transl., "Mir," Moscow, 1964. MR 26 #1345.

[20] F. Klein, *Vorlesungen über höhere Geometrie*, 3rd ed., Springer, Berlin, 1926; reprint Chelsea, New York, 1949; Russian transl., GONTI, Moscow, 1939.

[21] A. I. Kostrikin and I. R. Šafarevič, *Cartan's pseudogroups and Lie p-algebras*, Dokl. Akad. Nauk SSSR 168 (1966), 740-742 = Soviet Math. Dokl. 7 (1966), 715-718. MR 33 #7384.

[22] A. I. Kostrikin, *Squares of adjoint endomorphisms in simple Lie p-algebras*, Izv. Akad. Nauk SSSR Ser. Mat. 31 (1967), 445-487 = Math. USSR Izv. 1 (1967), 435-473. MR 36 #1501.

[23] ———, *A theorem on semisimple Lie p-algebras*, Mat. Zametki 2 (1967); 465-474 = Math. Notes 222 (1967), 773-778. MR 36 #1502.

[24] W. V. D. Hodge and D. Pedoe, *Methods of algebraic geometry*. Vol. I, Cambridge Univ. Press, New York, 1947; Russian transl., IL, Moscow, 1954. MR 10, 396.

[25] C. Chevalley, *Sur certains groupes simples*, Tohoku Math. J. (2) 7 (1955), 14-66; Russian transl., Matematika 2 (1958), no. 1, 3-53. MR 17, 467.

Le théorème de Torelli pour les surfaces algébriques de type K3

Actes Congrès Intern. Math., Nice 1970, 413–417 (1971). Zbl. **236,** 14016

Ce qui suit est l'exposition d'un travail effectué en commun avec I. I. Pjatetckii-Shapiro. Nous étudions les surfaces algébriques sur le corps des nombres complexes qui possèdent une classe canonique nulle. Cette condition équivaut au fait que la première classe de Chern est égale à 0 ou encore au fait que le groupe structural du fibré tangent peut être réduit au groupe spécial linéaire.

Il est bien connu qu'il existe deux classes de telles surfaces. L'une d'entre elles est constituée par les variétés abéliennes de dimension 2. Les surfaces de l'autre classe sont simplement connexes. On les appelle surfaces de type $K3$.

Une variété abélienne est uniquement déterminée par les périodes de ses formes différentielles holomorphes de degré 1. Il est facile d'en déduire qu'elle est également déterminée par les périodes de sa forme différentielle holomorphe de degré 2. Sur une surface de type $K3$, il n'y a pas de forme holomorphe de degré 1, mais, en revanche, il existe une forme holomorphe de degré 2, unique à un facteur constant près. Nous examinerons dans quelle mesure les périodes de cette forme différentielle déterminent la surface. La question ne se pose de façon naturelle que pour les surfaces polarisées, i. e. munies d'une classe d'homologie de dimension 2 correspondant à une section hyperplane. Pour énoncer le résultat, il est nécessaire d'introduire quelques notions:

Comme on le sait (cf. par exemple [1], chap. IX), pour une surface X de type $K3$, le groupe $H_2(X, \mathbb{Z})$ est un $\mathbb{Z}$-module libre de rang 22 et l'indice d'intersection y détermine une structure de réseau pair unimodulaire euclidien de signature (3, 19). On sait que les réseaux euclidiens qui possèdent ces propriétés sont tous isomorphes. Fixons l'un d'eux, soit L, et un certain vecteur l de L. Nous appellerons surface distinguée de type $K3$ un triplet $(X, \xi, \varphi) = \tilde{X}$ où X est une surface de type $K3$, $\xi \in H_2(X, \mathbb{Z})$ une classe de sections hyperplanes pour un certain plongement projectif et $\varphi: H_2(X, \mathbb{Z}) \to L$ un isomorphisme de réseaux euclidiens tel que $\varphi(\xi) = l$.

Posons $\tilde{\Omega} = \operatorname{Hom}(L, \mathbb{C})$ et définissons dans cet espace un produit scalaire bilinéaire sur $\mathbb{C}$ qui prolonge celui donné sur L. Désignons par Ω le domaine de l'espace projectif de dimension 21 correspondant à $\tilde{\Omega}$, dont les points correspondent aux vecteurs $\omega \in \tilde{\Omega}$ tels que

$$\omega^2 = 0$$
$$\omega . \overline{\omega} > 0,$$

et par $\Omega(l)$ l'ensemble des $\omega \in \Omega$ tels que $\omega . l = 0$. L'ensemble $\Omega(l)$ se présente comme la réunion disjointe de deux variétés complexes dont chacune est isomorphe à un domaine borné symétrique du type IV de la classification de E. Cartan.

Comme on le montre dans [1], chapitre IX, à toute surface distinguée de type $K3$, on peut associer un point $\tau(\tilde{X}) \in \Omega(l)$. Pour cela, considérons une forme différentielle régulière ω sur X et, pour $\gamma \in H_2(X, \mathbb{Z})$, posons

$$f(\gamma) = \int_\gamma \omega.$$

Il est clair que $f \in \operatorname{Hom}(H_2(X, \mathbb{Z}), \mathbb{C})$ et que φ transforme f en $g \in \operatorname{Hom}(L, \mathbb{C}) = \tilde{\Omega}$. Puisque la forme ω est déterminée de manière unique à un facteur constant près, le point correspondant de l'espace projectif, dont on vérifie facilement qu'il appartient à $\Omega(l)$, est donc parfaitement déterminé. Ce point est noté $\tau(\tilde{X})$. On l'appelle périodes de la surface X et on appelle τ l'application des périodes.

Le résultat fondamental (théorème de Torelli pour les surfaces de type $K3$) affirme que la surface distinguée $\tilde{X}$ de type $K3$ est déterminée de manière unique par ses périodes, i.e. par le point $\tau(\tilde{X}) \in \Omega(l)$.

Voici le schéma de la démonstration de ce théorème. Nous nous appuierons sur le « théorème de Torelli local » pour les surfaces de type $K3$ démontré par G. N. Turina dans [1], chapitre IX. On construit une famille analytique $\mathscr{X} \to S$ de surfaces de type $K3$ dont toutes les fibres sont des surfaces distinguées, toutes les surfaces distinguées figurant (à isomorphisme près) parmi les fibres ; cette famille est effectivement paramétrée par une base S de dimension 19. La construction repose sur la construction du schéma de Hilbert des surfaces de type $K3$ plongées dans un espace projectif et sur un passage au quotient selon le groupe projectif. Du résultat principal de [1], chapitre IX, résulte que l'application des périodes $\tau : S \to \Omega(l)$ est un isomorphisme local holomorphe de la base de notre famille dans $\Omega(l)$.

Nous construisons ci-dessous un ensemble partout dense $Z \subset \Omega(l)$ au-dessus duquel la représentation τ est bijective. Notre théorème découlera alors du résultat simple suivant :

Soit $f : \mathscr{U} \to V$ une application localement isomorphe de variétés analytiques. Supposons qu'il existe un ensemble $Z \subset V$, partout dense dans V, tel que $f^{-1}(z)$ soit réduit à un point pour tout point $z \in Z$. Alors f est un plongement.

Il reste à décrire la construction de l'ensemble $Z \subset \Omega(l)$ pour lequel on démontre d'abord le théorème de Torelli. Elle repose sur l'étude de certaines classes spéciales de surfaces du type $K3$.

Considérons une variété abélienne A de dimension 2 et son automorphisme θ : $\theta x = -x$, $x \in A$. Désignons par g le groupe composé de 1 et θ. L'espace quotient A/g est un espace complexe normal. Il possède 16 points doubles correspondant aux points d'ordre 2 de A. Chacun de ces points peut être désingularisé par une transformation quadratique ; la variété ainsi obtenue est une surface de type $K3$, appelée surface de Kummer.

Si la variété abélienne A est réductible, c'est-à-dire contient une courbe elliptique, alors la surface de Kummer correspondante est dite spéciale.

Il se trouve qu'il est possible d'exprimer qu'une surface de type $K3$ est une surface de Kummer spéciale en termes du réseau euclidien $H_2(X, Z)$ et de l'application des périodes τ.

De manière précise, la structure de surface algébrique sur X particularise dans le groupe d'homologie $H_2(X, \mathbb{Z})$ un sous-réseau de cycles algébriques que nous désignerons par S. Le théorème de Lefschetz caractérise le réseau S en termes de $H_2(X, \mathbb{Z})$ et de la représentation des périodes τ: si $\tilde{X} = (X, \varphi, \xi)$ est une surface distinguée, $\pi = \tau(\tilde{X})$ est une forme linéaire complexe sur L déterminée à une proportionnalité près, et $S = \operatorname{Ker} \varphi^*(\pi)$. Ainsi, le réseau S est défini par les périodes de la surface distinguée $\tilde{X}$.

Si A est une variété abélienne réductible et C une courbe elliptique contenue dans A, alors l'homomorphisme $A \to A/C$ définit une fibration en courbes elliptiques. Cette fibration définit un faisceau de courbes elliptiques sur la surface de Kummer spéciale X correspondante. Les classes des fibres d'un tel faisceau peuvent être décrites par leurs propriétés dans le réseau S. Cela donne un critère pour que X soit une surface de Kummer spéciale.

Le théorème de Torelli se démontre directement pour les surfaces de Kummer spéciales. Parmi les périodes de telles surfaces de Kummer figure l'ensemble partout dense dont nous avons besoin.

Indiquons quelques applications du théorème de Torelli. D'abord, il permet de décrire la « variété grossière des modules » des surfaces de type $K3$. Nous appellerons classe d'une telle surface le minimum des nombres $\frac{1}{2}\mathscr{D}^2$, où $\mathscr{D}$ est un diviseur effectif sur la surface et $\mathscr{D}^2 > 0$. Alors, la variété des modules M_k des surfaces de classe k est une variété algébrique $F - F_0$. Ici F est la compactification standard de l'espace $\Omega(l)/\Gamma$, quotient du domaine symétrique $\Omega(l)$ par son groupe discret d'automorphismes Γ, induit par les automorphismes du réseau L qui conservent un vecteur l, le vecteur l étant arbitraire de longueur $2k$. La sous-variété algébrique $F_0 \subset F$ se compose des images de la frontière de la compactification et du sous-ensemble défini par la condition $\omega . a = 0$, $a \in L$, $a . l = 0$, $a^2 = -2$.

Si $k = 2$, nous avons affaire aux surfaces de degré 4 dans $\mathbb{P}_3$. Dans ce cas, la variété M_2 est affine. Il est tout à fait vraisemblable qu'il en est toujours ainsi.

La variété M_k possède un modèle défini sur le corps $\mathbb{Q}$ et tel que, pour tout point $m \in M_k$, le corps $\mathbb{Q}(m)$ coïncide avec le « corps des modules » de la surface correspondante X au sens de Shimura [2]. En particulier, si X ne possède pas d'automorphisme $\neq 1$, alors $\mathbb{Q}(m)$ coïncide avec le corps de définition de cette surface.

Les autres applications sont liées à la structure du groupe des automorphismes de la surface de type $K3$. Soit S le groupe des classes de diviseurs d'une telle surface; on peut le considérer comme un réseau euclidien. Chaque vecteur $a \in S$ avec $a^2 = -2$ définit un automorphisme du réseau S:

$$x \mapsto x + (x . a) . a,$$

réflexion dans l'hyperplan orthogonal à a. Désignons par H le sous-groupe qu'ils engendrent dans le groupe de tous les automorphismes du réseau S; il est clair que c'est un sous-groupe distingué. Il résulte facilement du théorème de Torelli que le groupe $\operatorname{Aut} S/H$, « à des groupes finis près », est isomorphe au groupe des automorphismes de la surface. Ici deux groupes G et G' sont dits isomorphes à des groupes finis près s'il existe des sous-groupes $G_1 \subset G$ et $G'_1 \subset G'$ d'indices finis et des sous-groupes distingués finis $G_0 \subset G$, $G'_0 \subset G'_1$ tels que $G_1/G_0 \simeq G'_1/G'_0$.

Par exemple, si le réseau S est de rang 2, il est déterminé par une forme quadratique indéfinie à deux variables $F(x, y)$. Le groupe des automorphismes de la surface X est infini si et seulement si la forme F ne représente ni 0 ni -2. Pour toute forme quadratique paire à deux variables indéfinie F, il existe une surface correspondante. Dans le travail [3], Severi a étudié le cas des surfaces du quatrième degré qui contiennent une courbe de degré 6 et de genre 2. Pour ces surfaces, la forme F est du type

$$4x^2 + 12xy + 2y^2,$$

et, en accord avec nos résultats, le groupe des automorphismes est infini. Comme dernière application, considérons les surfaces pour lesquelles le rang du groupe S est de valeur maximum possible 20. Nous les appellerons singulières. Le complémentaire orthogonal de S dans $H_2(X, \mathbb{Z})$ est le réseau défini positif des « cycles transcendants ». Il est facile de déduire du théorème de Torelli qu'une surface singulière est entièrement définie par le réseau T. Le réseau T est pair, c'est-à-dire que $t^2 \equiv 0$ (mod. 2) pour tout $t \in T$, et tout réseau pair défini positif correspond à une certaine surface singulière. La surface singulière est de Kummer si et seulement si $t^2 \equiv 0$ (mod. 4) pour tout $t \in T$.

Passons à certains problèmes soulevés par le théorème de Torelli. Si l'on compare cette situation à celle des variétés abéliennes de dimension 2, il est clair que, dans la théorie des surfaces de type $K3$, il manque un analogue à la notion d'isogénie. Il est facile de le définir dans le langage des périodes. Sur le domaine symétrique $\Omega(l)$ agit un groupe discret Γ dont les transformations correspondent aux différents isomorphismes $H_2(X, \mathbb{Z}) \to L$ qui transforment ξ en l. Des points équivalents correspondent à une même surface. Le groupe Γ est composé des points entiers d'un certain groupe algébrique G, $\Gamma = G(\mathbb{Z})$. Il est naturel d'appeler isogènes des surfaces auxquelles correspondent des points dans $\Omega(l)$ qui se déduisent l'un de l'autre par des transformations du groupe $G(\mathbb{Q})$. Donc le problème d'un équivalent algébrique de cette notion se pose.

Plus précisément, soit $g \in G(\mathbb{Q})$, x et $y \in \Omega(l)$, $y = gx$, et X et Y des surfaces distinguées telles que $\tau(x) = X$, $\tau(y) = Y$. La transformation g détermine un homomorphisme $H_2(X, \mathbb{Z}) \to H_2(Y, \mathbb{Z})$. Cet homomorphisme est-il donné par une certaine corres-
1 pondance algébrique entre X et Y? Telle est la formulation précise de notre problème. Une réponse négative à cette question réfuterait la conjecture de Hodge. Une réponse positive démontrerait la méromorphie et l'équation fonctionnelle des fonctions ζ des surfaces singulières de type $K3$ (et de beaucoup d'autres classes intéressantes). La recherche des surfaces de type $K3$ sur un corps fini est d'un grand intérêt. Quelle peut être la valeur minimale du rang du groupe S des classes de diviseurs sur de telles surfaces? Si l'on admet la conjecture de Tate, il est facile de montrer qu'elle est $\geqslant 2$. Nous ne savons pas si cette valeur peut être atteinte. Quelles sont les surfaces qui
2 correspondent à la valeur maximale 22 de ce rang? Sont-elles toutes des surfaces de Kummer?

En conclusion mentionnons le problème extrêmement intéressant de la généralisation de ces considérations aux variétés algébriques de dimension quelconque et de
3 classe canonique nulle.

BIBLIOGRAPHIE

[1] « Алгебраические поверхности ». Труды Математического Института им. В. А. Стеклова, т. 75. Москва 1965.

[2] G. SHIMURA. — « Number fields and zeta functions associated with discontinuous groups and algebraic varieties ». Труды Международного Конгресса Математиков. Москва 1966, стр. 290-299.

[3] F. SEVERI. « Complementi alla teoria della base per la totalita delle curve di una superficie algebrica ». Rendiconti del circolo matematico di Palermo. T. 30, 1910, 265-288.

Steklov Mathematical Institute
Academy of Sciences of USSR
ul Vavilova 42,
Moscow V 333
(U. R. S. S.)

A Torelli theorem for algebraic surfaces of type K3

(with I. I. Piatetskij-Shapiro)

Izv. Akad. Nauk SSSR, Ser. Mat. **35**, 530–572 (1971). Zbl. **219**, 14021
[Math. USSR, Izv. **5**, No. 3, 547–588 (1972)]

In Memoriam: Galina Nikolaevna Tjurina

Abstract. In this paper it is proved that an algebraic surface of type $K3$ is uniquely determined by prescribing the integrals of its holomorphic differential forms with respect to a basis of cycles of the two-dimensional homology group, if the homology class of a hyperplane section is distinguished.

Introduction

This paper is devoted to algebraic surfaces with canonical class equal to zero. We shall only consider surfaces defined over the field of complex numbers. There this condition is equivalent to the first Chern class being equal to 0, or to the structure group of the tangent bundle being reducible to the special linear group.

It is well known that there are two classes of such surfaces. One of them consists of two-dimensional abelian varieties. The surfaces of the other class are simply connected. They are called $K3$ surfaces.

An abelian variety is uniquely determined by the periods of its holomorphic 1-forms. On $K3$ surfaces there are no holomorphic 1-forms, but, as for two-dimensional abelian varieties, there exists a unique (up to a constant multiple) holomorphic 2-form. The question of whether the periods of this differential form determine the surface bears the name of Torelli's theorem. It is natural to state it for polarized surfaces, i.e. to fix the class of a hyperplane section of the surface in the two-dimensional homology group. A discussion of this question and analogous ones for other types of algebraic varieties can be found in Griffiths' report [7].

In this article we give a positive answer to the question stated above. The proof is based on the "local Torelli theorem", which was proved by G. N. Tjurina ([1], Chapter IX) for $K3$ surfaces. The thought that the global Torelli theorem might follow from the local one was expressed to us by A. N. Tjurin.

§1. Statement of the problem. Basic results

The basic object to be studied in this paper is a $K3$ surface. This is a two-dimensional compact complex Kähler manifold X whose canonical class is trivial and whose one-dimensional Betti number equals zero. All the main results of the paper

AMS 1970 *subject classifications*. Primary 14C30, 14D20, 14J10; Secondary 10B10, 14G99.

relate to the case where X is an algebraic surface defined over the field of complex numbers.

As is shown in [1], pp. 212–213, for any $K3$ surface X the group $H_2(X,\mathbf{Z})$ has no torsion. The intersection index defines an integer-valued scalar product in it.

By a *Euclidean lattice* we shall mean a free module over $\mathbf{Z}$ of finite rank, in which there is defined a scalar product with values in $\mathbf{Z}$. It is obvious how to define an isomorphism of Euclidean lattices. A Euclidean lattice E will be called *even* if $x^2 \equiv 0$ (2) for all $x \in E$, *positive definite* if $x^2 > 0$ for $0 \neq x \in E$, and *unimodular* if its Gram determinant equals ± 1. If X is a $K3$ surface, the group $H_2(X, \mathbf{Z})$ is an even unimodular lattice. We shall denote it by H_X or H. The signature of the Euclidean space $H_X \otimes \mathbf{R}$ is (3,19) ([1], p. 214).

In order to state the main results of the paper, we introduce a new concept, analogous to that of the Teichmüller surface in the theory of Riemann surfaces. We consider a Euclidean lattice L which is even and unimodular and for which the signature of the space $L \otimes \mathbf{R}$ equals (3,19).

It is known ([16], p. 93) that all such lattices are isomorphic. We fix one of these and in it a vector $l \in L$.

Definition. A *marked* $K3$ surface is a triple (X, ϕ, ξ), where X is a $K3$ surface, $\phi: H_X \to L$ an isomorphism of Euclidean lattices, $\xi \in H_X$ the class corresponding to a very ample divisor on X and such that $\phi(\xi) = l$.

An isomorphism of marked surfaces (X, ϕ, ξ) and (X', ϕ', ξ') is an isomorphism of surfaces $f: X \to X'$ which takes ϕ into ϕ' and ξ into ξ'.

Thus the concept of a ruled surface depends on the choice of the vector l (it would be more precise to speak of a marked surface of "type l"). Since this choice will be fixed for the duration of the paper, we shall not distinguish this dependence.

The present paper is devoted to the study of the set of isomorphism classes of marked $K3$ surfaces. The results are based on the comparison (for such surfaces) of points of some symmetric space (the "periods of the surface"), to whose definition we now proceed.

Let $\tilde{\Omega}$ be the space of all complex linear functionals on L. On $\tilde{\Omega}$ we define as usual a scalar product; namely, we start by extending the scalar product on L to $L \otimes \mathbf{C}$, and then we use the natural isomorphism between $\tilde{\Omega}$ and $L \otimes \mathbf{C}$.

We now denote by Ω the set of all lines $\omega \in \tilde{\Omega}$ passing through the origin and such that

$$\omega_0^2 = 0, \tag{1}$$

$$\omega_0 \cdot \bar{\omega}_0 > 0 \tag{2}$$

for $\omega_0 \in \omega$.

It is not hard to prove that Ω is a complex manifold. Furthermore we let $\Omega(l)$ denote the set of all $\omega \in \Omega$ orthogonal to the vector l. $\Omega(l)$ is a disjoint union of two copies of a complex manifold, which is a symmetric domain in $\mathbf{C}^{19}$ of type IV according to É. Cartan's classification. For the proof of this assertion we consider the set G_l of all real transformations of the space $\tilde{\Omega}$ preserving the scalar product and taking the hyperplane $\omega \cdot l = 0$ into itself. It is not hard to see that G_l is isomorphic to the group of transformations preserving a quadratic form with two positive squares and 19 negative squares. From the definitions of G_l and $\Omega(l)$ it is obvious that G_l takes $\Omega(l)$ into itself. It is easy to check that G_l acts transitively on $\Omega(l)$. All our assertions on $\Omega(l)$ are proved via simple computations. (For more detail see [11], Russian p. 126.)

We will show how to associate a point in $\Omega(l)$ to a marked surface. Let $\tilde{X} = (X, \phi, \xi)$ be such a surface. It is known that on any $K3$ surface X there exists a unique (up to multiplication by a constant) form ω of type (2,0). The mapping

$$\gamma \to \int_\gamma \omega, \tag{3}$$

where $\gamma \in H_X$, defines a complex linear functional on H_X; via the isomorphism ϕ between H_X and L we define a linear functional on L. Since the form ω is determined up to a constant multiple, this linear functional is also determined up to multiplication by a constant. As is shown in [1], pp. 201–202, to construct a linear function in such a way, we need (1) and (2).

Since ξ is an algebraic homology class, $\int_\xi \omega = 0$. Hence the constructed point of Ω is contained in $\Omega(l)$.

The point of $\Omega(l)$ thus obtained will be denoted by $\tau(\tilde{X})$ and will be called a period of the marked surface $\tilde{X}$, and the mapping τ will be called the period mapping. The main result of this paper is:

1 **Torelli Theorem for $K3$ Surfaces.** *A marked $K3$ surface is uniquely determined by its periods.*

We outline the proof of this theorem. We construct an analytic family of $K3$ surfaces: $\mathfrak{X} \to S$, all of whose fibers are marked surfaces; among the fibers are all the marked surfaces (up to isomorphism), where the family is effectively parametrized and the dimension of the base is 19. The construction of such a family is based on the construction of the Hilbert scheme of $K3$ surfaces imbedded in projective space, and a factorization by the action of the projective group. It is explained in §2. From the main result of Chapter IX of [1] it follows that the period mapping $\tau: S \to \Omega(l)$ is a holomorphic isomorphism, locally isomorphic to a mapping of the base of our family into $\Omega(l)$.

In §7 it is shown that there exists an everywhere dense set $Z \subset \Omega(l)$ on which the mapping τ is one-to-one. After this our theorem follows from the following simple lemma.

Lemma 1. *Let $f: U \to V$ be a local isomorphism of analytic manifolds. Assume that there exists a set $Z \subset V$, everywhere dense in V, such that $f^{-1}(z)$ consists of* *exactly one point for all points $z \in Z$. Then f is an imbedding.*

Proof. We assume that for some point $v \in V$ there exist two distinct preimages: u_1 and u_2. By the condition there exist neighborhoods $U_1 \ni u_1$ and $U_2 \ni u_2$ where f is an isomorphism onto a neighborhood V' of the point v. The neighborhoods U_1 and U_2 can be chosen to be disjoint. Then any point $v' \in V'$ has at least two preimages. But by hypothesis V' should contain a point $z \in Z$, which gives a contradiction.

It remains to describe the construction of the set $Z \subset \Omega(l)$ for which we first prove a Torelli theorem. This construction is based on the construction of some special types of $K3$ surfaces.

We consider a two-dimensional abelian variety A and an automorphism θ of it: $\theta(x) = -x$, $x \in A$. We denote by g the group consisting of 1 and θ. The factor space A/g is a normal complex space. It has 16 double points, corresponding to the points of order two on A, each of which can be resolved by one σ-process. The variety obtained in this way is a $K3$ surface. It is called a Kummer surface.

If the abelian variety A is reducible, i.e. contains an elliptic curve, the corresponding Kummer surface will be called *special*.

It turns out that we can give a criterion for a $K3$ surface X to be a special Kummer surface, stated in terms of the Euclidean lattice H_X and the period mapping.

Namely, the structure of an algebraic surface on X distinguishes the sublattice of algebraic cycles in the homology group H_X, which we denote by S_X or S. Its orthogonal complement in H_X is the Euclidean lattice of "transcendental cycles", denoted by T_X or T. We know that $S \cap T = 0$, and hence $S \oplus T$ is a subgroup of finite index in H. A theorem of Lefschetz gives a characterization of the lattice S in terms of H and the period mapping τ: if $\tilde{X} = (X, \phi, \xi)$ is a marked surface and $\pi = \tau(\tilde{X})$ is a complex linear form on L, defined up to proportionality, then $S = \operatorname{Ker} \phi^*(\pi)$. Thus the lattices S and T are defined by the periods of the marked surface $\tilde{X}$.

If A is a reducible abelian variety and C is an elliptic curve contained in it, the homomorphism $A \to A/C$ defines a fibering by elliptic curves. This fibering defines a pencil of elliptic curves on the corresponding special Kummer surface X. The classes a of the generic fiber of such a pencil can be described by their properties in the lattice S_X. This also gives the criterion for X to be a special Kummer surface. A study of pencils of elliptic curves on a $K3$ surface is carried out in §3, and a proof of the criterion can be found in §4.

In §5 a Torelli theorem is proved for special Kummer surfaces. Among the periods of such surfaces we find our desired everywhere dense set $Z \subset \Omega(l)$, as is proved in §6, whereupon our proof of the Torelli theorem is complete.

The last sections are devoted to applications of the Torelli theorem. This theorem can be given in another, somewhat more convenient form. We shall call the classes $u \in H_X$ which are algebraic and correspond to effective cycles, *effective*.

Moreover, integration of the regular 2-form defines a linear function $\pi_X \in \mathrm{Hom}\,(H_X, \mathbf{C})$. The second form of Torelli's theorem reads:

Any isomorphism of lattices $H_X \to H_{X'}$ *which takes effective cycles into effective ones, and the form* $\pi_X \in \mathrm{Hom}\,(H_X, \mathbf{C})$ *into a form proportional to* $\pi_{X'} \in \mathrm{Hom}\,(H_{X'}, \mathbf{C})$, *is induced by the unique isomorphism of surfaces* $X \to X'$.

The condition on the forms π is equivalent to the preservation of the natural bigrading in $H^2(X, \mathbf{C}) = \mathrm{Hom}\,(H_X, \mathbf{C})$.

This form of the Torelli theorem allows us to study the group of automorphisms of $K3$ surfaces in §7. For this we use the following concepts. If E is a Euclidean lattice and $e \in E$ is a vector such that $(e^2) = -2$, a reflection through a hyperplane orthogonal to e is written in the form $s_e(x) = x + (x \cdot e)e$ and is an automorphism of E. The group generated by all such automorphisms will be denoted $\Gamma(E)$ and for short will be called the group generated by reflections. It is a normal subgroup of the group of all automorphisms of the lattice E, which we shall denote by $G(E)$. In §7 we prove that the group of automorphisms of a $K3$ surface X is isomorphic to the quotient group $G(S_X)/\Gamma(S_X)$ "up to finite groups". This result allows us to answer the question of the finiteness of the group of automorphisms of a surface in many cases.

The final section, §8, is devoted to surfaces X for which $\mathrm{rg}\, S_X = 20$. This is the maximum possible value of $\mathrm{rg}\, S_X$ for $K3$ surfaces X. Such surfaces will be called *exceptional*. Their classification is very simple: they are uniquely determined by the Euclidean lattice T_X, which in this case is an even, positive definite lattice. Conversely, each such lattice corresponds to an exceptional surface X. This surface X is a Kummer surface if and only if $t^2 \equiv 0\,(4)$ for all $t \in T_X$. As an example we find the lattice T_X for the surface X which is given by the equation $\Sigma_{i=0}^{3} x_i^4 = 0$ in $\mathbf{P}^3$, and prove that it is a Kummer surface.

The methods of this paper do not apply to $K3$ surfaces over an arbitrary field. It would be interesting to see whether or not the results of §7 on the structure of the group of automorphisms of such surfaces carry over to this case. On the other hand,
2 our methods also do not apply to nonalgebraic $K3$ surfaces. Nevertheless, it seems likely to us that the second form of the Torelli theorem should remain valid, since the statement at least makes no mention of algebraicity.

Finally, for the convenience of the reader we give a summary of the basic notation and properties of $K3$ surfaces, which we will assume to be known and use without further ado.

For a Euclidean lattice E: $G(E)$ is the group of automorphisms; $\Gamma(E)$ the subgroup generated by reflections; for $a \in E$, $a^2 = 0$, let $E(a) = E'/\mathbf{Z}a$, where E' is the sublattice spanned by all the vectors $b \in E$ such that $ab = 0$ and $b^2 = -2$; $E(a)$ has a natural structure of a Euclidean lattice.

For an algebraic surface X: $H_X = H_2(X, \mathbf{Z})$ with the structure of a Euclidean lattice; S_X the sublattice of algebraic classes; T_X the orthogonal complement to S_X in H_X.

For K3 surfaces X the Riemann-Roch inequality takes the form

$$l(D)+l(-D)\geqslant\frac{(D^2)}{2}+2, \tag{4}$$

and the formula for the genus of an irreducible curve $C \in X$ becomes

$$p_a(C)=\frac{(C^2)}{2}+1, \tag{5}$$

where $p_a(C) = g + \delta$, g being the genus of the normalization of the curve C, and δ the sum of the positive terms corresponding to the singular points of the curve C. In particular, $\delta = 0$ if and only if C is smooth.

From these formulas we get

Lemma 2. *On a K3 surface X:*

1. *For an irreducible curve $C \in X$, either $(C^2) \geq 0$ or $(C^2) = -2$; in the latter case the curve C is rational and smooth.*

2. *If $D \in S_X$ and $D^2 = -2$, then either D or $-D$ is equivalent to an effective divisor (perhaps reducible).*

3. *If $D \in S_X$ and $D^2 = 0$, then either D or $-D$ is equivalent to an effective divisor. In the latter case $l(D) \geq 2$.*

§2. Families of marked K3 surfaces

The goal of this section is to construct sufficiently universal analytic families of marked K3 surfaces. We will be interested in families $f: \mathfrak{X} \to S$ in which $\mathfrak{X}$ and the base S are smooth analytic varieties, the mapping f is nowhere degenerate, and all the fibers are K3 surfaces.

The family $f: \mathfrak{X} \to S$ will be called a family of marked surfaces if all of its fibers $X_s = f^{-1}(s)$, $s \in S$, have the structure of a marked K3 surface $\widetilde{X}_s = (X_s, \phi_s, \xi_s)$, where for sufficiently close points s and s' of the base the natural homomorphism of the fibers X_s and $X_{s'}$ takes ϕ_s into $\phi_{s'}$ and ξ_s into $\xi_{s'}$.

Our task is to construct a family such that any marked surface is isomorphic to one of its fibers, the family is effectively parametrized and the dimension of the base equals 19 (according to [1] this is the maximum possible value). The existence of such a family is established by Theorem 1. The idea of the proof is very simple. First we construct a family $\mathfrak{X}' \to S'$ consisting of K3 surfaces imbedded in some projective space $\mathbf{P}^n$, where S' is a subset of the corresponding Hilbert scheme. Then we construct a new family $\mathfrak{X}'' \to S''$, in which the fibers are obtained from the fibers of $\mathfrak{X}'$ at the expense of introducing all the possible marked surface structures among them. Its base S'' is an unramified covering of the base S'. Finally, the desired family is obtained by factoring $\mathfrak{X}''$ and S'' by the action of the group $PGL(n, \mathbf{C})$ of automorphisms of the projective space $\mathbf{P}^n$.

First we shall prove some auxiliary results. Let X be a $K3$ surface, $\mathcal{F}$ a very ample sheaf and $X \subset \mathbf{P}^n$ the corresponding imbedding into a projective space. We denote by C the corresponding divisor class, i.e. the class of hyperplane sections. Then

$$l(C) = n + 1. \tag{1}$$

According to a criterion of Kodaira ([13], Theorem 2.4), $H^1(X, \mathcal{F}) = 0$; and the Riemann-Roch theorem shows that

$$C^2 = 2n - 2. \tag{2}$$

Finally, by formula (5) of §1

$$g(C) = n, \tag{3}$$

where $g(C)$ is the genus of the curve C. From (2) and (3) it follows that C is a canonical curve.

The set of such surfaces X forms a family over an open set M of the Hilbert scheme. We recall that the tangent space $\Theta_{M,m}$ at a point $m \in M$ of the corresponding surface X is isomorphic to $H^0(X, \check{N}_{\mathbf{P}^n/X})$ and the point m is simple if $H^1(X, \check{N}_{\mathbf{P}^n/X}) = 0$. Here $N_{\mathbf{P}^n/X}$ is the normal bundle to X in $\mathbf{P}^n$, and $\check{E}$ is the sheaf of germs of sections of a bundle E (see [9], p. 23).

As usual, our family defines the characteristic homomorphism $\Theta_{M,m} \to H^1(X, \check{\Theta}_X)$, where Θ_X is the tangent bundle of X. In our case the homomorphism

$$\chi : H^0(X, \check{N}_{\mathbf{P}^n/X}) \to H^1(X, \check{\Theta}_X)$$

is the differential in the exact cohomology sequence of the exact sequence of bundles

$$0 \to \Theta_X \to \Theta_{\mathbf{P}^n}|_X \to N_{\mathbf{P}^n/X} \to 0, \tag{4}$$

where $\Theta_{\mathbf{P}^n}|_X$ is the restriction of the tangent bundle of $\mathbf{P}^n$ to X.

The action of the projective group $PGL(n)$ on $\mathbf{P}^n$ defines a homomorphism of the vector space of its Lie algebra $\psi : sl(n+1) \to \Theta_{M,m}$ which can be described in the following way. It is obvious that

$$sl(n+1) = H^0(\mathbf{P}^n, \check{\Theta}_{\mathbf{P}^n}).$$

Restriction defines a homomorphism

$$\rho : H^0(\mathbf{P}^n, \check{\Theta}_{\mathbf{P}^n}) \to H^0(X, (\check{\Theta_{\mathbf{P}^n}|_X})),$$

and the exact sequence (4) defines a homomorphism

$$\tau : H^0(X, (\check{\Theta_{\mathbf{P}^n}|_X})) \to H^0(X, \check{N}_{\mathbf{P}^n/X}).$$

Their composition then gives the desired homomorphism $\psi = \tau\rho$.

Proposition 1. *The scheme M is smooth, the sequence*

$$0 \to sl(n+1) \xrightarrow{\psi} \Theta_{M,m} \xrightarrow{\chi} H^1(X, \check{\Theta}_X) \tag{5}$$

is exact, and the cokernel of the homomorphism χ *has dimension one.*

The first assertion will be verified if we prove that $H^1(X, \check{N}_{P^n/X}) = 0$.

Since $H^0(X, \check{\Theta}_X) = H^2(X, \check{\Theta}_X) = 0$ (see [1], p. 214), the sequence (4) reduces to the exact sequence

$$\begin{aligned} 0 \to H^0(X, (\check{\Theta}_{P^n}|_X)) \to H^0(X, \check{N}_{P^n/X}) \to H^1(X, \check{\Theta}_X) \\ \to H^1(X, (\check{\Theta}_{P^n/X})) \to H^1(X, \check{N}_{P^n/X}) \to 0. \end{aligned} \tag{6}$$

We now use the standard exact sequence

$$0 \to O_X \to O_X(1)^{n+1} \to \check{\Theta}_{P^n}|_X \to 0$$

and the corresponding exact sequence of cohomology. Since $H^1(X, O_X) = 0$, $H^1(X, O_X(1)) = 0$ (by Kodaira's criterion) and $H^2(X, O_X(1)) = 0$ (by duality), we will obtain two exact sequences:

$$0 \to H^0(X, O_X) \to H^0(X, O_X(1))^{n+1} \to H^0(X, (\check{\Theta}_{P^n}|_X)) \to 0,$$

$$0 \to H^1(X, (\check{\Theta}_{P^n}|_X)) == H^2(X, O) \to 0.$$

An analogous sequence exists on the entire space $\mathbf{P}^n$. Restriction to X reduces to a commutative diagram

$$\begin{array}{ccccccccc} 0 \longrightarrow & H^0(\mathbf{P}^n, O_{P^n}) & \to & H^0(\mathbf{P}^n, O_{P^n}(1))^{n+1} & \to & H^0(\mathbf{P}^n, \check{\Theta}_{P^n}) & \longrightarrow 0 \\ & \downarrow & & \downarrow & & \downarrow & \\ 0 \to & H^0(X, O_X) & \longrightarrow & H^0(X, O_X(1))^{n+1} & \longrightarrow & H^0(X, (\check{\Theta}_{P^n}|_X)) & \to 0 \end{array}$$

Since the first two (from the left) vertical arrows are isomorphisms, the third one is also an isomorphism. This can be expressed in the equality

$$H^0(X, (\check{\Theta}_{P^n}|_X)) = sl(n+1).$$

Finally sequence (6) acquires the form

$$\begin{aligned} 0 \to sl(n+1) \xrightarrow{\psi} H^0(X, \check{N}_{P^n/X}) \xrightarrow{\chi} H^1(X, \check{\Theta}_X) \xrightarrow{\varphi} H^2(X, O_X) \\ \to H^1(X, \check{N}_{P^n/X}) \to 0. \end{aligned} \tag{7}$$

The left side of sequence (7) coincides with sequence (5).

As was proved in [1], Chapter IX, a polarization of the surface X gives a hyperplane in $H^1(X, \check{\Theta}_X)$ (which is isomorphic to the tangent space of the local moduli variety) which contains $\chi H^0(X, \check{N}_{P^n/X})$. Therefore the mapping ϕ is nonzero, and since the space $H^2(X, O_X)$ is one dimensional, it is an epimorphism. From this it follows that $H^1(X, \check{N}_{P^n/X}) = 0$ and the cokernel of χ has dimension one. This proves the proposition.

Proposition 2. *An automorphism of a $K3$ surface X which acts trivially on the group $H^2(X, \mathbf{Z})$ is the identity.*

Automorphisms which act trivially on $H^2(X, \mathbf{Z})$ in particular preserve the polarization of the surface X, and hence are induced by projective transformations under the corresponding imbedding of X into projective space. Therefore they form an algebraic group. The Lie algebra of this group coincides with $H^0(X, \check{\Theta}_X)$, and hence is zero. Thus the automorphisms which interest us form a finite group and, in particular, each such automorphism has finite order. We can restrict consideration to an automorphism of prime order p.

Let g be such an automorphism of the surface X with $g^p = 1$ and $g \neq 1$. Since g has finite order, it follows that for any fixed point $x \in X$ the automorphism $d_x g$ of the tangent space $\Theta_{X,x}$ is different from 1. This is easy to verify, considering the action of g on the power series ring in the local parameters in a neighborhood of the fixed point. Since g acts trivially on $H^2(X, \mathbf{Z})$, it in particular does not change the periods of the differential form $\omega \in H^0(X, \Omega^2)$, and hence the form itself as well. In view of this, for any fixed point $x \in X$ of the automorphism g the automorphism $d_x g$ in the tangent space $\Theta_{X,x}$ has determinant 1. In particular, from this it follows that g has only isolated fixed points: the curve of fixed points would determine a one-dimensional subspace of $\Theta_{X,x}$ which is $d_x g$-invariant for any point x of that curve. Since $\det d_x g = 1$ and $d_x g$ reduces to diagonal form, it would follow from this that $d_x g = 1$, and this, as we saw, is impossible.

Since the eigenvalues of $d_x g$ are different from 1, it follows that the index of each fixed point equals 1. Since g acts trivially on $H^2(X, \mathbf{Z})$, by the Lefschetz formula, the number of fixed points equals the Euler characteristic of the surface X, which is 24.

We denote by G the cyclic group generated by the automorphism g, and set $Y = X/G$. Since g preserves the regular differential form ω on X, ω can be considered as a rational form on Y. It will be regular at all the nonsingular points of this surface. The surface Y has 24 singular points. We explain their structure. In a neighborhood of each singular point the automorphism g can be given by a linear transformation with the matrix $\begin{pmatrix}\epsilon & 0\\ 0 & \epsilon^{-1}\end{pmatrix}$, where $\epsilon^p = 1$ and $\epsilon \neq 1$. If x and y are the local parameters of a fixed point, the set of invariants is generated by the functions $u = x^p$, $v = y^p$ and $w = xy$. Therefore a neighborhood of a singular point on X/G is locally given by the equation $u \cdot v = w^p$. This singularity is denoted in the theory of singular points on surfaces by A_{p+1}. It is a rational double point.

We denote by $f\colon X' \to Y$ the minimal resolution of the singular points of the surface Y. From the immediateness of their resolution by σ-processes or from the basic properties of rational double points it follows that the form ω is regular on X' and vanishes nowhere. From this it follows that the canonical class of the surface X'

equals 0. It is easy to see that it is simply connected and hence is a $K3$ surface. From the resolution of the 24 singular points of the surface Y there arise 24 algebraic cycles on X', which are pairwise nonintersecting and hence are linearly independent in $H^2(X', \mathbf{Z})$. But this contradicts the fact that the rank of the group $S_{X'}$ cannot be greater than 20. This contradiction proves the proposition.

In the proof of Theorem 1, stated below, we shall make use of the following simple construction. Let $f: \mathfrak{X}' \to S'$ be an arbitrary family of $K3$ surfaces, and in it distinguish a homology class ξ which on each fiber X'_s induces a very ample algebraic class $\xi_{s'} \in H_{X'_s}$, defining an imbedding into $\mathbf{P}^n$.

We consider the set S'_0 of the points $s' \in S'$ for which there is an isomorphism of the lattices $H_{X'_s}$, and L, taking the class $\xi_{s'}$ into l. As will be proved in the Appendix to §5, for this it suffices that the greatest common divisor of the coordinates be the same for $\xi_{s'}$ and l and that $\xi^2_{s'} = l^2$. Therefore S'_0 is a connected component of the base S'. Then we can construct a new family $\mathfrak{X}'' \to S''$. Its base S'' is an unramified covering $\nu: S'' \to S'_0$, where the fiber $\nu^{-1}(s')$ is identified with the set of isomorphisms of the Euclidean lattices $H_{X'_s}$, and L, taking the class $\xi_{s'}$ into l. This covering is a principal covering with structure group $\mathfrak{G}$, namely the group of automorphisms of the lattice L that leave the vector l unchanged. The covering ν is defined by the action of $\pi_1(S'_0)$ on $H_{X'_s}$, and consequently on the set of isomorphisms of $H_{X'_s}$ into L. The family $\mathfrak{X}'' \to S''$ is the inverse image of the family $\mathfrak{X}' \to S'$ by the covering $\nu: S'' \to S'_0$.

Theorem 1. *For any natural number n there exists a family of marked $K3$ surfaces $\mathfrak{X} \to S$, which possesses the following properties:*

a) *It is effectively parametrized.*

b) *For any marked $K3$ surface X for which the class ξ defines an imbedding into a projective space $\mathbf{P}^n$, there exists an isomorphism of it (as a marked surface) with a fiber X_s of the family $\mathfrak{X}$.*

c) *The base has dimension* 19.

We consider a family $f': \mathfrak{X}' \to S'$, where S' is the Hilbert scheme of the $K3$ surfaces $X \subset \mathbf{P}^n$ for which the hyperplane sections form a complete linear system. In its fibers we distinguish a cohomology class $\xi_{s'} \in H^2_{X'_s}$, corresponding to a hyperplane section. Let $f'': \mathfrak{X}'' \to S''$ be the corresponding family of marked surfaces whose construction is described immediately preceding the statement of the theorem. Via its action in $\mathbf{P}^n$ the projective group $G = PL(n, \mathbf{C})$ acts on $\mathfrak{X}'$ and S'_0, and hence also on $\mathfrak{X}''$ and S'', where this action commutes with f' and f''. We define $\mathfrak{X}$ and S as the quotients $\mathfrak{X}''/G$ and S''/G, and for this we should first of all prove that these quotients exist.

If an arbitrary complex Lie group G acts on an analytic manifold M, Palais [20] proved that the quotient M/G exists in the category of complex spaces, provided that the mapping

$$\psi : G \times M \to M \times M, \qquad \psi(g, m) = (gm, m) \tag{8}$$

is proper.

This test we also verify for our families. Then the properness of the mapping (8) for the manifolds $\mathfrak{X}''$ and S'' (and the group $G = PL(n)$) is an obvious formal consequence of the analogous fact for the varieties $\mathfrak{X}'$ and S'. In this case we have to dea with algebraic varieties and an analytic action of the group $PL(n)$. Therefore ψ is a morphism in the category of algebraic varieties. In order to prove that ψ is a proper morphism we can use the following criterion, connected with the behavior of discrete valuation rings ([8], Chapter II, §7). Let $f: X \to Y$ be a morphism of algebraic varieties over an algebraically closed field k, $U = \operatorname{Spec} O$, O a discrete valuation ring with residue class field k, $g: U \to X$ a rational morphism and $h: U \to Y$ a morphism, where $fg = h$. If for all such g and h there exists a morphism $g' : U \to X$, coinciding with g at a generic point of U, the morphism f is proper.

To realize this criterion in our situation we use Theorem 2 of the paper [15] of Matsusaka and Mumford. It asserts that if V and W are smooth polarized varieties ove Spec O, neither of which is ruled, and their specializations V_0 and W_0 over the close point of Spec O are also smooth polarized varieties, then the specialization of any isomorphism $\phi: V \to W$ is an isomorphism $\phi_0: V_0 \to W_0$.

It is easy to see that giving a morphism

$$h : \operatorname{Spec} O \to S' \times S'$$

and a rational morphism

$$g : \operatorname{Spec} O \to G \times S', \quad G = PL(n),$$

satisfying the criterion stated above, defines varieties V and W over Spec O and an isomorphism ϕ of them, to which we apply the theorem of Matsusaka-Mumford. On the other hand, g gives a rational morphism of Spec O into G and hence a morphism g' of it into the closure of the projective group G. By the theorem of Matsusaka and Mumford, the specialization of ϕ is an isomorphism, i.e. g' takes the closed point of Spec O into a point of G. This means that g' is a morphism of Spec O into G, i.e. the criterion of properness holds for the morphism ψ. The situation is analogous with the morphism $G \times X' \to X'$.

In this manner we prove that the quotients $\mathfrak{X}''/G = \mathfrak{X}$ and $S''/G = S$ exist as complex spaces. It is obvious that they define a family $f: \mathfrak{X} \to S$ of marked surfaces. According to Proposition 2, G acts on S'' and $\mathfrak{X}''$ without fixed points. Therefore $\mathfrak{X}$ and S are manifolds.

We shall prove that the family f is effectively parametrized. This means that for any point $s \in S$ the natural mapping $\Theta_{S,s} \to H^1(X_s, \check{\Theta}_{X_s})$ is an imbedding. It is obvious that

$$\Theta_{S,s} = \Theta_{S',s'}/sl(n+1),$$

and therefore our assertion follows from Proposition 1. Assertion b) of Theorem 1 holds by the very definition of the families $\mathfrak{X}'$, $\mathfrak{X}''$ and $\mathfrak{X}$.

Finally, the space $\Theta_{S,s}$ has dimension 19 according to the last assertion of Proposition 1, since the dimension of $H^1(X, \check{\Theta}_X)$ equals 20 ([1], p. 214). Theorem 1 is proved.

As was stated in §1, the association of a surface X_s to its periods, i.e. a point $z \in \Omega(l)$, defines a mapping $\tau: S \to \Omega(l)$.

Corollary of Theorem 1. *The mapping τ is holomorphic and is a local isomorphism of the analytic manifolds S and $\Omega(l)$.*

Since the family $\mathfrak{X}$ is effectively parametrized, Theorem 2 of [1], Chapter IX, guarantees that the mapping τ is holomorphic and is a local imbedding. Since the dimensions of the manifolds S and $\Omega(l)$ are the same (= 19), τ is a local isomorphism.

§3. Pencils of elliptic curves on K3 surfaces

This section and the following one are relevant to K3 surfaces over any algebraically closed field, provided the characteristic is different from 2 or 3. In this case we define a K3 surface by the vanishing of the canonical class and the group $H^1(X, O)$. This definition allows us to retain formulas (4) and (5) of §1 and also Lemma 2 of §1. The group S_X is defined as the group of divisor classes under algebraic equivalence. For K3 surfaces $S_X = \operatorname{Pic} X$. By the Néron-Severi theorem this group has a finite number of generators. In the case of K3 surfaces this group has no torsion. The proof of this is literally the same as the proof in [1], p. 212, of the fact that the group $H_2(X, \mathbf{Z})$ has no torsion for K3 surfaces X defined over $\mathbf{C}$. In this way S_X is a Euclidean lattice (relative to the intersection index).

By an elliptic curve we will understand a smooth curve of genus 1, and by a pencil of elliptic curves a morphism $f: X \to B$ onto a smooth curve B such that the fiber $f^{-1}(b)$ is an elliptic curve for almost all points $b \in B$.

Lemma 1. *If C is an elliptic curve on a K3 surface, then $C^2 = 0$ and $l(C) = 2$.*

The first assertion follows from formula (5) of §1. We shall prove the second. Let $\mathcal{F}_C$ be a sheaf corresponding to the curve C. In the exact sequence

$$0 \to O \to \mathcal{F}_C \to \mathcal{F}_C|_C \to 0 \tag{1}$$

we have $\mathcal{F}_C|_C = O_C$, since $C^2 = 0$. The exact cohomology sequence corresponding to (1) takes the form

$$0 \to k \to H^0(X, \mathcal{F}_C) \to k \to 0$$

by the definition of a K3 surface. From this it follows that $l(C) = 2$.

Theorem 1. *If an effective divisor $D > 0$ on a K3 surface X satisfies the conditions $D^2 = 0$ and $DE > 0$ for any effective divisor $E > 0$, then the linear system $|D|$ contains a divisor of the form mC, where $m \geq 0$ and C is an elliptic curve.*

Proof. We shall prove that the system $|D|$ contains a divisor with only one component. Let

$$D' = \sum_{i=1}^{r} k_i C_i, \quad k_i > 0, \quad r \geqslant 2, \tag{2}$$

where the curves C_i are different from each other and irreducible. Since $DC_i \geq 0$ and $D^2 = 0$, we have $DC_i = 0$, and since $C_i C_j \geq 0$ for $i \neq j$, we have $C_i^2 \leq 0$. We shall prove that $C_i^2 = 0$ for at least one divisor $D' \in |D|$ and one curve C_i in (2). For this we fix a projective imbedding of the surface X and let H denote the corresponding hyperplane section. Then $D \cdot H = \Sigma k_i C_i \cdot H$, and therefore the numbers k_i and $C_i \cdot H$ are bounded above for all expansions (2) of divisors $D' \in |D|$. The number $C_i \cdot H$ is equal to the degree of the curve C_i in the projective imbedding under consideration. There exist only a finite number of curves $C_i \subset X$ of given degree with $C_i^2 < 0$. Therefore there also exist only a finite number of divisors (2) with all $C_i^2 < 0$. But the system $|D|$ is infinite by Lemma 2 of §1.

We see that there exist divisors $D' \in |D|$ such that

$$D' = mC + \sum k_i C_i = mC + D_0, \tag{3}$$

$$C^2 = 0.$$

We observe that then $D_0^2 = 0$. In fact, since $D^2 = 0$ and $D \cdot D_0 = 0$, we have

$$2mC \cdot D_0 + D_0^2 = 0, \quad mC \cdot D_0 + D_0^2 \geqslant 0,$$

and since C and D_0 have no common components, also $C \cdot D_0 \geq 0$. From this it follows that $C \cdot D_0 = 0$, and hence also $D_0^2 = 0$. Among the divisors D' having an expansion like (3), we choose the one for which the number $D_0 \cdot H$ is smallest. We shall prove that for this $D_0 = 0$.

In fact, if $D_0 \neq 0$, then $l(D_0) \geq 2$. Therefore in the system $|D_0|$ there is a divisor D_0' intersecting the curve C. It must contain the curve as components:

$$D_0 \sim D_0' = l \cdot C + \overline{D}_0, \quad l > 0,$$

$$D' \sim D'' = (m + l) C + \overline{D}_0$$

and

$$H \cdot \overline{D}_0 = H \cdot D_0' - l \cdot C \cdot H < H \cdot D_0'.$$

Consequently we have proved that there exists a divisor $mC \in |D|$, where C is an irreducible curve.

As before $C^2 = 0$, and hence $l(C) \geq 2$. If the characteristic of the base field is zero, Bertini's theorem guarantees that the linear system $|C|$ contains smooth curves. If the characteristic is different from zero, this may not be the case, but in our case it follows from a formula of Tate [18] that such a pathology is only possible in characteristics 2 and 3. In all the remaining cases we find a smooth irreducible curve $C \in X$ with $C^2 = 0$. By formula (5) of §1 this is an elliptic curve.

Corollary 1. *If C is an elliptic curve, the class $c \in S_X$ containing C is primitive (i.e. $c \neq mc'$, $c' \in S_X$, $m \neq \pm 1$).*

In fact, let $c = mc'$, where we assume that $m > 0$. Then c' satisfies the conditions of Theorem 1. Hence there exists an elliptic curve C' such that $C \sim mm'C'$. By Lemma 1, $l(C) = l(C') = 2$. Therefore, if $mm' > 1$, all the curves of the pencil $|C|$ must be multiple. However, this is not the case for C itself.

Corollary 2. *A pencil of elliptic curves on a K3 surface has no multiple components.*

This corollary follows from Corollary 1 in an obvious manner.

We observe, although it will not be needed later, that the criterion of Theorem 1 can be stated only in terms of the Euclidean lattice S_X. Namely, from the corollary of Theorem 1 which will be proved in §6 we get

Corollary 3. *On a K3 surface X there exists a pencil of elliptic curves if and only if there exists an element $x \in S_X$, $x \neq 0$, for which $x^2 = 0$.*

By Theorem 1 of §6 there exists an automorphism $\phi \in \Gamma(S_X)$ such that $\phi(x)$ satisfies the conditions of the proposition. Therefore the class $\phi(x)$ contains a divisor mC, where C is an elliptic curve. By Lemma 1, $l(C) = 2$, and hence the system $|C|$ defines a rational map $f: X \to \mathbf{P}^1$. It will be a morphism, since $C^2 = 0$ and therefore the system has no base points. The morphism f also determines a pencil of elliptic curves containing C.

§4. Special Kummer surfaces

In what follows we shall use the following notation. In a four-dimensional lattice E with basis u_1, u_2, u_3, u_4, $u_i u_j = \delta_{ij}$, we consider a sublattice E' consisting of the vectors $\Sigma_{i=1}^4 x_i u_i$, $x_i \in \mathbf{Z}$, for which $\Sigma_{i=1}^4 x_i \equiv 0\ (2)$. It is easy to verify that this lattice is even. The lattice in which the scalar product differs from the product in E' by a sign will be denoted by G_4.

In this section, as in the preceding one, we shall assume that the ground field k is algebraically closed with arbitrary characteristic different from 2 and 3.

Theorem 1. *A K3 surface X is a special Kummer surface if and only if S_X contains an element $z \neq 0$ such that $z^2 = 0$, the class z contains an irreducible divisor and*

$$S_X(z) = G_4^4 \tag{1}$$

is a direct sum of four Euclidean lattices G_4.

See the end of §1 for the definition of $S_X(z)$.

Before beginning the proof of the theorem, we shall indicate a simple interpretation of the group $S_X(z)$ in our case. Suppose that the divisor class $z \in S_X$, $z \neq 0$, contains an elliptic curve C which is one of the fibers of the pencil $f\colon X \to \mathbf{P}^1$. If $y \in S_X$, $z \cdot y = 0$ and $y^2 = -2$, then, replacing y by $-y$ if necessary, we can find an effective divisor D in the class y such that $C \cdot D = 0$ and $D^2 = -2$. If the divisor D were not contained in the fibers of the pencil f, it would have isolated points of intersection with some fibers of it, which contradicts the condition $C \cdot D = 0$. Thus D is a linear combination of components of the fibers of the pencil f. Conversely, if a fiber D of the pencil f has more than one component, i.e. if $D = \sum_{i=1}^{r} n_i L_i$, $r \geq 2$, then $D \cdot L_i = 0$ (we can replace D by another fiber equivalent to it, which does not intersect L_i). From Zariski's connectedness theorem it follows that D is connected. Therefore $L_i \cdot L_j > 0$ for at least one $j \neq i$. From this it follows that $L_i^2 < 0$, and hence $L_i^2 = -2$.

Thus, in the case where the class z corresponds to a pencil of elliptic curves f, the Euclidean lattice $S_X(z)$ coincides with the quotient lattice generated by the components of the fibers of the pencil f modulo the cyclic sublattice generated by a fiber.

We shall prove the necessity of the condition of the theorem. Let A be a reducible two-dimensional abelian variety, $\phi\colon A \to B$ a homomorphism of it onto a curve B, g the group of automorphisms of A consisting of the identity automorphism and the reflection $x \to -x$, and X the Kummer surface which is obtained by a minimal resolution of the singularities of the surface A/g. The fibers $C_b = \phi^{-1}(b)$ are elliptic curves.

The group of automorphisms of B, which also consists of the identity and a reflection, will also be denoted by g.

We have a commutative diagram

$$\begin{array}{ccccc} A & \to & A/g & \leftarrow & X \\ \downarrow & & \downarrow & \swarrow{\scriptstyle f} & \\ B & \to & B/g & = & \mathbf{P}^1 \end{array}$$

We let a denote the generic fiber of the pencil $f\colon X \to \mathbf{P}^1$ and prove that relation (1) holds for it. According to the considerations stated after the statement of the theorem, we need to determine the degenerate fibers of the pencil f.

All the fibers C_b of the mapping ϕ are elliptic curves. If $b \in B$ and $2b \neq 0$, then C_b is mapped isomorphically into A/g and its image does not pass through the singular points. Therefore its image in X is also a nondegenerate fiber of the pencil f. If also $2b = 0$, then C_b is invariant relative to g and $C_b/g \simeq \mathbf{P}^1$, and the mapping $C_b \to C_b/g$ has two sheets. The image of C_b in A/g has the form $2L$, where L is an irreducible Weil divisor. The divisor L passes through 4 singular points of the surface A/g. If L_1, L_2, L_3, L_4 are its preimages in X, the fiber $f^{-1}(b)$ has, as is easily verified, the form $2L + L_1 + L_2 + L_3 + L_4$. In addition, L, L_1, L_2, L_3 and L_4 are isomorphic to $\mathbf{P}^1$, L intersects L_1, L_2, L_3, L_4 transversally, and they do not intersect each other. If we denote the divisor classes of L, L_1, L_2, L_3, L_4 by f, e_1, e_2, e_3, e_4, then in S_X

$$(f^2) = (e_i^2) = -2, \quad i = 1, 2, 3, 4,$$

$$(e_i e_j) = 0, \quad i \neq j,$$

$$e_i \cdot f = 1,$$

$$a = 2f + e_1 + e_2 + e_3 + e_4.$$

The quotient group of this lattice modulo $\mathbf{Z} \cdot a$ is isomorphic to the lattice G_4. The isomorphism is given by the formulas

$$f \to -u_1 - u_3,$$

$$e_1 \to u_1 + u_2,$$

$$e_2 \to u_1 - u_2,$$

$$e_3 \to u_3 + u_4,$$

$$e_4 \to u_3 - u_4.$$

Since the pencil f has 4 degenerate fibers, corresponding to the four points $b \in B$ with $2b = 0$, the lattice $S_X(a)$ is isomorphic to G_4^4.

We now prove the sufficiency of the condition of Theorem 1. By Theorem 1 of §3 the class z which satisfies the conditions of Theorem 1 contains an elliptic curve C which is a fiber of the pencil $f\colon X \to \mathbf{P}^1$.

Our next goal is to determine the degenerate fibers of the pencil f. We denote these fibers by $D_1, \cdots, D_r$. By a theorem of Zariski (see [1], Chapter VII), the irreducible components C_{ij} of the fiber $D_i = \Sigma_j n_{ij} C_{ij}$ are connected in the group $S_X/\mathbf{Z} \cdot z$ by the unique relation

$$D_i = \sum_j n_{ij} C_{ij} = 0.$$

The components of the different fibers are orthogonal to each other. As we saw, the group $S_X(z)$ coincides with the group generated in S_X by the elements C_{ij} factored by the subgroup $\mathbf{Z} \cdot z$. Therefore $S_X(z) = H_1 \oplus \cdots \oplus H_r$, where H_i is obtained from the Euclidean lattice generated by the components of the fiber D_i factored by $\mathbf{Z} \cdot D_i$. If the number of components of D_i is equal to n_i, the rank of H_i is $n_i - 1$. The condition of Theorem 1 means that

$$H_1 \oplus \ldots \oplus H_r = G_4 \oplus G_4 \oplus G_4 \oplus G_4, \tag{2}$$

and, in particular, $\Sigma_1^r (n_i - 1) = 16$.

Eichler proved (see a much simpler proof in [14]) that a positive definite lattice decomposes uniquely into indecomposable orthogonal summands. This is obviously also true for negative definite lattices. The lattice G_4 is indecomposable. In fact, it is even and has determinant 4, and there exist no even lattices of determinant 1 or 2 of ranks 2 and 3. Thus it follows from (2) that the lattices H_i corresponding to the reducible fibers are sums of the lattice G_4. In particular, $n_i = 5$, 9 or 17.

We now need to recall Kodaira's classification of reducible fibers (see [12] up to notation). In this connection we omit the fibers with $n_i < 5$. The remaining components are isomorphic to $\mathbf{P}^1$ and intersect transversally. The fibers are represented by diagrams in which the points represent the components, and if two components intersect, the corresponding points are joined by a segment. The number shows the multiplicity with which the corresponding component is contained in the fiber.

All the diagrams obtained are the so-called extended Dynkin diagrams ([4], p. 199) from which their notation is adopted.

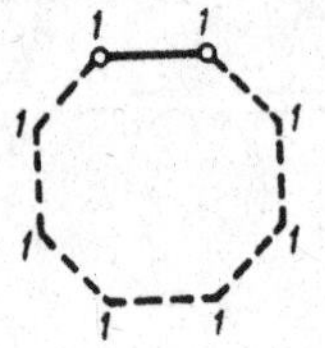

A_l: number of vertices $(l+1)$ D_l: number of vertices $(l+1)$

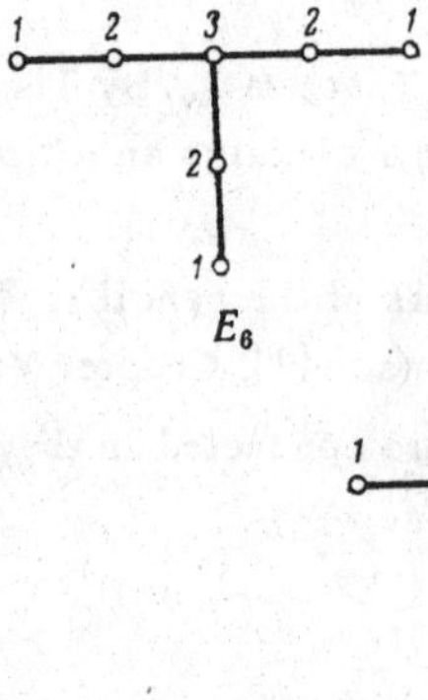

E_6

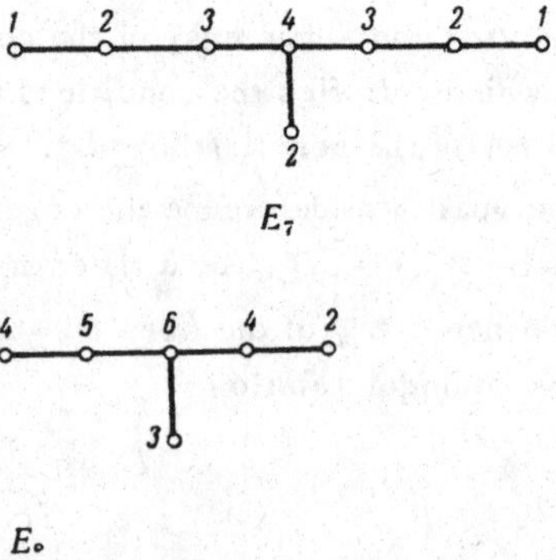

E_7

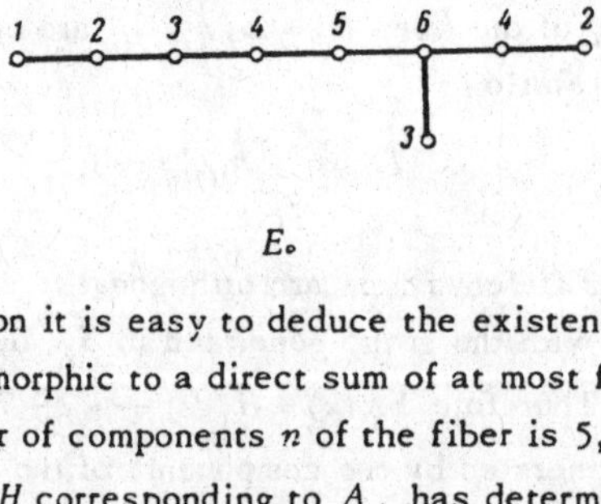

E_8

From this classification it is easy to deduce the existence of a fiber whose corresponding lattice H is isomorphic to a direct sum of at most four lattices G_4. This is the fiber D_4. If the number of components n of the fiber is 5, the fiber has the form A_4 or D_4. But the lattice H corresponding to A_4 has determinant 5 (easy to check) and hence is not isomorphic to G_4. If $n = 9$, the fiber has the form A_8, D_8 or E_8. For none of these is the lattice H isomorphic to $G_4 \oplus G_4$. This is easy to derive from counting the numbers of vectors of length -2. In G_4 this number is 24, and in $G_4 \oplus G_4$ it is 48. On the other hand, if the diagram corresponding to the fiber contains a chain  with components $C_1, \cdots, C_k$, the cycles $\pm(C_i + C_{i+1} + \cdots + C_{i+j})$ have square -2. The number of them is $k(k+1)/2$. From this it is easy to see that the number of vectors with square −2 is greater than 48 in the lattices corresponding to the fibers A_8, D_8 and E_8. The case $n = 17$ is considered analogously.

We see that the pencil f has precisely 4 reducible fibers of type D_4. We shall prove that it has no other degenerate fibers. For this we shall use the formula (see [10])

$$\chi = \sum_{i=1}^{r} \chi_i, \tag{3}$$

where χ is the Euler characteristic of the surface X and χ_i is that of the degenerate fiber D_i. From the Riemann-Roch formula

$$\frac{K^2 + \chi}{12} = p_a(X)$$

it follows that $\chi = 24$, since $K = 0$ and $p_a(X) = 2$. The fiber D_4 has Euler characteristic 6, and since there are four such fibers, it follows from (3) that $\chi_i = 0$ for the remaining fibers. This is possible only for a multiple elliptic fiber. We have already shown that there are no such fibers.

Thus we have constructed a pencil f on X with base $\mathbf{P}^1$ having 4 degenerate fibers of type D_4 and having no other degenerate fibers. From an analysis of the types of degenerate fibers (see [1], pp. 171–173) it follows that the fiber D_4 (in [1] it is denoted by B_6) corresponds to potential good reduction. More precisely, if $\pi: C \to \mathbf{P}^1$ is any two-sheeted covering ramified at a point x over which this fiber lies, then the normalization of the surface $X \times_{\mathbf{P}^1} C$ has a nondegerate fiber over the point $\pi^{-1}(x)$.

Let u_1, u_2, u_3, u_4 be the points on $\mathbf{P}^1$ over which the degenerate fibers lie, $\pi: C \to \mathbf{P}^1$ a two-sheeted covering ramified at these points. Then C is an elliptic curve and we can provide C with a group law such that π coincides with the quotient of C by the group of order two generated by the automorphism $c \to -c$. By what was stated above, the surface A is a smooth minimal model of the fibering $X \times_{\mathbf{P}^1} C \to C$ and defines a pencil of elliptic curves $\psi: A \to C$ with no degenerate fibers. The automorphism θ of the two-sheeted covering $C \to \mathbf{P}^1$ extends to A. We denote the group generated by it by g. Obviously X is birationally isomorphic to A/g. The regular 2-form on X gives a regular form on A which is different from 0, since the characteristic of the field is $\neq 2$.

Hence Theorem 1 reduces to the following assertion.

Lemma. *Let C be an elliptic curve, A a smooth surface with $p_g(A) \neq 0$, and ψ: $A \to C$ a morphism all of whose fibers are elliptic curves. Then A is an abelian variety.*

Moreover, suppose that an automorphism θ of order two acts on A such that $\psi\theta = -\psi$ and $p_g(X') \neq 0$ for the variety $X' = A/g$, $g = (1, \theta)$. Then there is a group law on A such that θ takes the form $\theta x = -x$.

It is proved in [19], p. 342 (transl. p. 110), that the pencil ψ becomes trivial after lifting to some unramified covering $h: C' \to C$, which is, of course, a homomorphism of elliptic curves. Therefore $A = (B \times C')/\Gamma$, where Γ is a finite group of automorphisms of the abelian variety $B \times C'$, where all the automorphisms $\gamma \in \Gamma$ have the form

$$\gamma(b, c') = (\varphi_\gamma(b), c' + e_\gamma),$$

where ϕ_γ is an automorphism of the curve B (which might not preserve the group structure) and $e_\gamma \in \operatorname{Ker} h$. The automorphism ϕ_γ has the form $\phi_\gamma(b) = \alpha_\gamma(b) + f_\gamma$, where α_γ is an automorphism of B as an abelian variety, and $f_\gamma \in B$. If $\alpha_\gamma \neq 1$, then $\phi_\gamma^*(\omega') = \lambda\omega'$, $\lambda \neq 1$, for a regular form ω' on B. Therefore if ω is the regular 2-form on $B \times C'$, then $\gamma^*\omega = \lambda\omega \neq \omega$. Since $H^0(A, \Omega^2) = H^0(B \times C', \Omega^2)^\Gamma$, if $\alpha_\gamma \neq 1$, then $p_g(A) = 0$, which contradicts the hypothesis of the lemma. Hence $\alpha_\gamma = 1$, $\phi_\gamma(b) = b + f_\gamma$ and A is the quotient of the abelian variety $B \times C'$ by a finite subgroup, and hence A is also an abelian variety.

It remains for us to explain how the automorphism θ of order two acts on A. We know that it agrees with the homomorphism $A \to C$ and that it induces the automorphism $c \to -c$ on C. On the abelian variety $B \times C'$ isogenous to A the analogous corresponding automorphism θ' has the form

$$\theta'(b, c') = (\alpha(b) + \xi(c') + b_0, -c'),$$

where α is an automorphism of the group B, ξ is a homomorphism of C' into B, and $b_0 \in B$. Since $(\theta')^2 = 1$, we have $\alpha^2 = 1$ and $(1 - \alpha)\xi = 0$. Since there exists a regular 2-form on A/g it follows that θ' does not change the regular 2-form on $B \times C'$. This is only possible for $\alpha = -1$. Then $2\xi = 0$, i.e. $\xi = 0$. Finally, at the expense of making another choice of the zero element in B we can get $b_0 = 0$. We find that $\theta' = -1$, and hence also $\theta = -1$. This proves the lemma and thus Theorem 1 as well.

We observe that the criterion of Theorem 1 can be formulated in terms of the lattice S_X alone: we do not require that the class z contain an irreducible curve. This follows from Corollary 3 of §3.

§5. The Torelli theorem for special Kummer surfaces

In this section we shall prove the Torelli theorem for special Kummer surfaces. Such a surface X is obtained from an abelian variety A, in which we give a fibering into elliptic curves $A \to C$, where C is an elliptic curve. By the same token, as we saw in §4, there is also a pencil of elliptic curves defined on X. Such a pencil will be called a Kummer pencil, and the class $a \in S_X$ defined by its fibers will be the Kummer vector.

Lemma 1. *Let X be a special Kummer surface, $f: X \to \mathbf{P}^1$ a Kummer pencil and $a \in S_X$ the Kummer vector. Then any effective cycle z such that* 1) $z^2 = -2$, 2) $z \cdot a = 0$, *and* 3) *z cannot be represented as a sum of effective cycles with the properties*

1) *and* 2), *is an irreducible component of some degenerate fiber of the pencil* f. *Conversely, the irreducible components of the degenerate fibers possess the properties* 1), 2) *and* 3).

Proof. We denote by $\mathfrak{F}_X(a)$ the sublattice of S_X generated by all the vectors with properties 1) and 2). As was shown in the proof of Theorem 1 of §4, $\mathfrak{F}_X(a)$ is generated by the irreducible components of the fibers. Thus any vector $z \in \mathfrak{F}_X(a)$ can be represented as a sum of vectors, each of which is a linear combination of the components of a singular fiber. Because the components of the different singular fibers are orthogonal, and because $x^2 \equiv 0$ (2) for any $x \in S_X$, it follows that if z satisfies 1) and 2), then $z = z' + \lambda a$, where z' is a linear combination of the irreducible components of a degenerate fiber. We recall that, as was shown during the course of the proof of Theorem 1 of §4, a degenerate fiber consists of 5 components f, e_1, e_2, e_3, e_4, among which the following relations hold:

$$\left.\begin{aligned} f^2 = e_i^2 = -2, \quad fe_i = 1, \quad i = 1, 2, 3, 4, \\ e_i \cdot e_j = 0, \quad i \neq j, \\ a = 2f + e_1 + e_2 + e_3 + e_4. \end{aligned}\right\} \tag{1}$$

Any vector z, $z^2 = -2$, which is an integral combination of the vectors f, e_1, e_2, e_3 and e_4 has the form

$$\pm (k_0 f + k_1 e_1 + k_2 e_2 + k_3 e_3 + k_4 e_4), \tag{2}$$

where all the $k_i \geq 0$ (it is easily verified that every vector z, $z^2 = -2$, has the form $e_i + \lambda a$, $f + \lambda a$, $f + e_i + \lambda a$ or $f + e_i + e_j + \lambda a$, $\lambda \in \mathbf{Z}$). The assertion of the lemma follows from (2).

For later use we need some facts about the structure of the homology group of Kummer surfaces. Let A be a two-dimensional abelian variety, θ an automorphism of A, $\theta(x) = -x$, $\mathfrak{F} = A/g$, $g = (1, \theta)$, and X the surface obtained from $\mathfrak{F}$ via a resolution of singularities. Thus we have a rational map

$$\pi : A \to X, \tag{3}$$

which is regular and one-to-one on $A-N$, where N is the set of points on A of order two. We define a mapping

$$\pi' : H_2(A, \mathbf{Z}) \to H_X = H_2(X, \mathbf{Z}), \quad \pi' = \pi_* \sigma_*^{-1}, \tag{4}$$

where σ is the inclusion $A-N \to A$. It is obvious that this map is an inclusion and that

$$\pi'\gamma_1 \cdot \pi'\gamma_2 = 2\gamma_1 \cdot \gamma_2. \tag{5}$$

Further, we let Π_X denote the submodule of S_X generated by the 16 homology classes e_i arising from the resolution of the singularities of $\mathfrak{F}$.

Lemma 2. *The orthogonal complement of* Π_X *in* H_X *is* $\pi' H_2(A, \mathbf{Z})$.

Lemma 3. *An automorphism of the lattice H_X, effecting a permutation of the vectors e_i, $i = 1, \cdots, 16$, leaving one of the vectors e_i fixed, and acting as the identity on the orthogonal complement of Π_X, is the identity on all of H_X.*

The proofs of both these lemmas will be found in the Appendix to this section.

3 **Lemma 4.** *Let A_1 and A_2 be two-dimensional complex tori. An isomorphism $\psi_1: H_2(A_1, \mathbf{Z}) \to H_2(A_2, \mathbf{Z})$, preserving the scalar product and the periods, is always induced by an isomorphism $\phi_1: A_1 \to A_2$. If the tori A_1 and A_2 are provided with a group structure, we may choose ϕ_1 so that it induces a group isomorphism.*

Proof. The assertion of the lemma follows easily from the fact that a matrix of rank 2 and size 2×4 is uniquely determined up to multiplication by a nondegenerate matrix of second order by its minors of second order, given up to multiplication by an arbitrary number.

Theorem 1. *Let X and X' be two $K3$ surfaces, where X is a special Kummer surface. If there exists an isomorphism ψ between the lattices H_X and $H_{X'}$ which preserves the periods and takes effective cycles into effective cycles, then: 1) X' is also a special Kummer surface, and 2) there exists an isomorphism $\phi: X \to X'$ inducing ψ.*

Proof. We consider a Kummer pencil on X. Let a be the corresponding Kummer vector. Then $a \cdot c \geq 0$ for any effective divisor c. Hence in view of the properties of the isomorphism ψ we have $\psi(a)^2 = 0$ and $\psi(a) \cdot c \geq 0$ for any effective divisor c on X'. Hence by Lemma 1 of §3 the class $\psi(a)$ contains an irreducible elliptic curve, and hence $\psi(a)$ defines a pencil of elliptic curves. From Theorem 1 of §4 it follows that this is a Kummer pencil, and hence X' is a special Kummer surface. This proves the first assertion of our theorem. We now pass to the proof of the second assertion. Let f' be a Kummer pencil on X', the homology class of whose generic fiber is $\psi(a)$. From Lemma 1 it follows that ψ takes the homology classes containing the irreducible components of the degenerate fibers into the homology classes containing the irreducible components of the degenerate fibers of f'. The classes l_i are characterized by the fact that they are included in the homology class a with coefficient 1. Hence

$$\psi(\Pi_X) = \Pi_{X'}. \tag{6}$$

We denote by A' the abelian variety from which X' is obtained, and by π' the rational map $A' \to X'$. From (6) and Lemma 2 it follows that ψ induces an isomorphism $\psi_1: H_2(A, \mathbf{Z}) \to H_2(A', \mathbf{Z})$, which obviously preserves periods, as does ψ. Hence by Lemma 4 it is induced by an isomorphism $\phi_1: A \to A'$. We now denote by l the homology class in S_X arising from the resolution of the singularity at the point $0 \in A$, and in A' we agree to choose a point 0 so that the cycle arising from it is $\psi(l)$. By Lemma 4 the isomorphism ϕ_1 can be chosen so that it induces a group isomorphism. We let ϕ denote the corresponding isomorphism of the surfaces, defined so that the diagram

$$\begin{array}{ccc} A & \xrightarrow{\varphi_1} & A' \\ \pi\downarrow & & \downarrow\pi' \\ X & \xrightarrow{\varphi} & X' \end{array}$$

is commutative. We now let ϕ_* denote the corresponding isomorphism in homology. Then it is clear that $\phi_*^{-1}\psi$ satisfies the conditions of Lemma 3, and hence is the identity. Thus we have proved the Torelli theorem for special Kummer surfaces.

We shall now prove that the two forms of the Torelli theorem mentioned in §1 are equivalent, and hence that the Torelli theorem in the first form also holds for special Kummer surfaces.

We shall call a vector $\xi \in S_X$ almost ample if $\xi^2 \geq 0$ and $\xi \cdot \alpha \geq 0$ for any class α containing an effective divisor.

Lemma 5. *Let X and X' be two K3 surfaces. If ϕ is an isomorphism between the Euclidean lattices S_X and $S_{X'}$, where for some very ample divisor class ξ, $\phi(\xi)$ is a very ample divisor class, then ϕ again takes an effective divisor class into an effective divisor class.*

Proof. It suffices to prove that if x is the homology class of an irreducible curve, $\phi(x)$ is an effective class. From the adjunction formula it follows that $x^2 \geq -2$. Consequently $\phi(x)^2 \geq -2$. Hence, by the Riemann-Roch inequality, $l(\phi(x)) + l(-\phi(x)) > 0$, and hence either $\phi(x)$ or $-\phi(x)$ is effective. We have $\phi(\xi)\cdot\phi(x) = \xi\cdot x > 0$, since ξ is an ample class. In view of the almost ampleness of the class $\phi(\xi)$, this excludes the case when the class $-\phi(x)$ is effective. Consequently $\phi(x)$ is an effective class.

From Lemma 5 it follows immediately that the second form of Torelli's theorem implies the first form. The converse is obvious.

Appendix to §5. Topology of Kummer surfaces[1)]

We consider a Kummer surface X obtained from a complex torus T. Thus there exists a rational map

$$\psi: T \to X \tag{1}$$

which is regular everywhere except at the points of T of order two. The set of points of order two on T is isomorphic to $(\mathbf{Z}/2\mathbf{Z})^4$. We choose this isomorphism once and for all, and let e_x, $x \in (\mathbf{Z}/2\mathbf{Z})^4$, denote the homology class of the cycle arising from the resolution of the singularity at the point $t_x \in T$.

We denote further by Π the set of all the integral homology classes expressed rationally by e_x, $x \in (\mathbf{Z}/2\mathbf{Z})^4$.

1) The description of the module Π on which the results of this appendix are based was obtained by D. B. Fuks. The authors heartily thank D. B. Fuks also for reading the text of this appendix and making a series of helpful comments.

Definition. Two complex structures in $\mathbf{R}^{2n}$ are said to be *equally oriented* if $J_2 = A^{-1} J_1 A$, $\det A > 0$, where J_2 and J_1 are the operators of multiplication by $\sqrt{-1}$ corresponding to the given structures.

Lemma 1. *In* $\mathbf{R}^{2n}$ *suppose we are given two equally oriented complex structures, and let* S *be the ball* $\Sigma_1^{2n} x_i^2 \leq 1$. *There exists a diffeomorphism of the ball* S *onto itself, which* 1) *inside a smaller ball is a linear transformation which takes one complex structure to the other,* 2) *is the identity on the boundary of* S *and* 3) *commutes with the mapping* $\theta: x \to -x$.

The proof follows from the connectedness of the group $SO(2n)$.

Lemma 2. *Let* T_1 *and* T_2 *be complex tori,* X_1 *and* X_2 *the corresponding Kummer surfaces, and* $\alpha: H_2(T_1, \mathbf{Z}) \to H_2(T_2, \mathbf{Z})$ *an isomorphism of Euclidean lattices. There exist diffeomorphisms* $\phi_1: T_1 \to T_2$ *and* $\phi_2: X_1 \to X_2$ *such that the diagram*

$$\begin{array}{ccc} T_1 & \xrightarrow{\varphi_1} & T_2 \\ \downarrow{\scriptstyle \psi_1} & & \downarrow{\scriptstyle \psi_2} \\ X_1 & \xrightarrow{\varphi_2} & X_2 \end{array} \qquad (2)$$

is commutative (*i.e.* $\phi_2 \circ \psi_1$ *and* $\psi_2 \circ \phi_1$ *are defined at the same points and* $\phi_2 \circ \psi_1 = \psi_2 \circ \phi_1$), *inducing a homomorphism in homology* ϕ_{1*} *which coincides with* α, $\phi_2(\Pi_1) = \Pi_2$.

Proof. We write T_1 and T_2 in the form $\mathbf{C}^2/\Gamma_1$ and $\mathbf{C}^2/\Gamma_2$, where Γ_1 and Γ_2 are lattices in $\mathbf{C}^2$. In these lattices we choose equally oriented bases, i.e. such that the real linear transformation taking one basis into the other (of course, preserving the numbering of the vectors) has positive determinant. This transformation gives an isomorphism $\beta: T_1 \to T_2$ (as groups, but not as complex manifolds), where two complex structures in the tangent space at any point of T_2, the natural one and the one transferred to T_1, have the same orientation. The canonical homomorphism of $SL(4, \mathbf{Z})$ into $G(\mathbf{Z}) = \operatorname{Aut} H_2(T, \mathbf{Z})$ is an epimorphism. In fact, the mapping $SL(4, \mathbf{R}) \to G(\mathbf{R})$ is an epimorphism by the connectedness of both groups. Let Γ be the full preimage of the group $G(\mathbf{Z})$; then Γ is a discrete subgroup of $SL(4, \mathbf{R})$ containing $SL(4, \mathbf{Z})$. But it is known [3] that the group $SL(4, \mathbf{Z})$ is a maximal discrete subgroup of $SL(4, \mathbf{R})$, and hence $\Gamma = SL(4, \mathbf{Z})$. Therefore β can always be chosen so that the induced mapping $\beta_*: H_2(T_1, \mathbf{Z}) \to H_2(T_2, \mathbf{Z})$ would coincide with α. From the fact that β is a group isomorphism it follows that it takes the points of order two on T_1 into points of order two on T_2. From Lemma 1 it is possible to correct our diffeomorphism β close to the points of order two so that it gives an isomorphism of the complex structures close to them. We denote this diffeomorphism by ϕ_1. By Lemma 1 we can guarantee that the mapping ϕ_1 still commutes with the mapping $x \to -x$, but it would already cease to be an isomorphism of groups. Since the correction took place on a small domain, the action of ϕ_1 on the homology remains as before, i.e. $\phi_{1*} = \alpha$. Since ϕ_1 commutes with the mapping $x \to -x$, it follows that ϕ_1 induces a mapping $\mathfrak{F}_1 \to \mathfrak{F}_2$, $\mathfrak{F}_i = T_i/(1, \theta)$, and since this map is a complex isomorphism close to the singular points of $\mathfrak{F}_1$ and $\mathfrak{F}_2$, it follows that it extends to a diffeomorphism $\phi_2: X_1 \to X_2$ which possesses all the necessary properties.

We now pass to the construction of a special system of homology classes on Kummer surfaces.

Lemma 3. *Let T be a torus, X the corresponding Kummer surface, and Π the sublattice of $H_2(X, \mathbf{Z})$ defined above. Each plane $v \in (\mathbf{Z}/2\mathbf{Z})^4$ can be associated to a homology class $f_v \in H_2(X, \mathbf{Z})$ so that*

$$f_v \cdot e_x = \begin{cases} 1 & \text{if } x \in v, \\ 0 & \text{if } x \notin v, \end{cases} \tag{3}$$

$$f_v \cdot f_{v'} = \begin{cases} -2 & \text{if } v = v', \\ -1 & \text{if } |v \cap v'| = 2, \\ 0 & \text{if } |v \cap v'| < 2, \end{cases} \tag{4}$$

where $|A|$ denotes the number of elements in the set A.

Proof. By Lemma 2 it suffices to prove the existence of the homology classes f_v for one Kummer surface X and a fixed mapping $\psi: T \to X$. For T we take $\mathbf{C}^2/D$, where D is the lattice consisting of all the vectors whose coordinates are numbers of the form $\alpha + \beta i$, $\alpha, \beta \in \mathbf{Z}$. In the torus T there are 24 two-dimensional (real) subtori which are invariant relative to $\theta: x \to -x$. There obviously exists a one-to-one correspondence between these subtori and the planes in $(\mathbf{Z}/2\mathbf{Z})^4$, under which a plane v corresponds to a subtorus T_v containing all the points t_x of order two with $x \in v$. We denote by F_v the fundamental domain in T_v relative to the mapping θ. We first consider those v for which T_v is a complex subtorus of T. It is easy to see that there are 8 of these. For such v, $\psi(F_v)$ represents a complex projective line in X. We denote its homology class by f_v. To define the remaining f_v we divide the remaining set of planes v into two subsets each of which consists of planes which either are pairwise parallel or pairwise intersect in a point. Let V be one of these subsets. Obviously, on T there exists a complex structure, equally oriented with the original one, in which all T_v, $v \in V$, are complex manifolds. Let T^V be the torus equipped with this new complex structure, and ϕ_1^V a diffeomorphism of T^V onto T which is the identity outside of sufficiently small neighborhoods of the points of order two and a complex isomorphism close to these points. The existence of such a ϕ_1^V, which commutes with θ, is ensured by Lemma 2. As above, we define homology classes f_v^V on X^V, and let $f_v = \phi_{2*}^V(f_v^V)$, where ϕ_2^V is the diffeomorphism between X^V and X corresponding to ϕ_1^V. Thus the classes f_v are defined. It remains for us to prove relations (3) and (4). We set $a_v = 2f_v + \Sigma_{x\in v} e_x$. Further, we denote by $\widetilde{T}_v$ some subtorus in T parellel to T_v and lying in general position, i.e. not containing points of order two of T; then it is clear that the homology class of the cycle $\psi(\widetilde{T}_v)$ equals a_v. Consequently the following relations hold:

$$a_v \cdot a_{v'} = \begin{cases} 2 \text{ if } |v \cap v'| = 1, \\ 0 \text{ if } |v \cap v'| \neq 1 \end{cases} \tag{5}$$

and a_v is homologous to $a_{v'}$ if v and v' are parallel. Since $\widetilde{T}_v$ does not contain points of order two, we have

$$a_v \cdot e_x = 0 \tag{6}$$

for any v and any $x \in (\mathbf{Z}/2\mathbf{Z})^4$. From (5), (6) and the obvious relations between the indexes of intersection of the e_x:

$$e_x \cdot e_y = \begin{cases} -2 \text{ for } x = y, \\ 0 \text{ for } x \neq y \end{cases} \tag{7}$$

we get (3) and (4).

Corollary. *For any three-dimensional subspace* $w \subset (\mathbf{Z}/2\mathbf{Z})^4$

$$g_w = \frac{1}{2} \sum_{x\in w} e_x \in \Pi \tag{8}$$

and consequently the discriminant $D(\Pi) \leq 2^6$.

Proof. Let $w = v_1 \cup v_2$, where v_1 and v_2 are parallel planes; then it is clear that

$$g_w = a_{v_1} - f_{v_1} - f_{v_2}. \tag{9}$$

From (9) it is clear that $g_w \in H_2(X, \mathbf{Z})$, and hence, by the definition of Π, (8) holds. There are 31 three-dimensional linear subspaces in $(\mathbf{Z}/2\mathbf{Z})^4$. The inequality $D(\Pi) \leq 2^6$ follows from this and from (7).

Lemma 4. 1) *The orthogonal complement of* Π *is* $\psi \cdot H_2(T, \mathbf{Z})$. 2) *The whole group* $H_2(X, \mathbf{Z})$ *is generated by the classes* f_v *and* e_x. 3) $D(\Pi) = D(\psi \cdot H_2(T, \mathbf{Z})) = 2^6$. 4) Π *is generated by* e_x *and* g_w *(see (8))*.

Proof. We let E be the lattice in $H_2(X, \mathbf{Z})$ generated by Π and all the f_v, and $\widetilde{E}$ the lattice generated by Π and all the a_v. If v_1 and v_2 are parallel, then $f_{v_1} - f_{v_2} \in \Pi$, since $a_{v_1} = a_{v_2}$ and $2f_v = a_v - \Sigma_{x\in v} e_x \in \widetilde{E}$. Consequently $(E : \widetilde{E}) \leq 2^6$. Let $f = \Sigma \lambda_v f_v \in \widetilde{E}$, where the sum extends over some system of 6 planes no two of which are parallel. If $f \in \widetilde{E}$, then $f = \Sigma \mu_v a_v + p$, where v runs through the same set as above and $p \in \Pi$. Multiplying f by $a_{v'}$, $|v \cap v'| = 1$, we will get that $\lambda_v = 2\mu_v$. Consequently $(E : \widetilde{E}) = 2^6$, whence

$$D(E) = 2^{-12} D(\widetilde{E}). \tag{10}$$

From the definition of $\widetilde{E}$ it follows that $D(\widetilde{E}) = 2^6 D(\Pi)$. Hence

$$D(E) = 2^{-6} D(\Pi). \tag{11}$$

From this and from the inequality proved in the corollary we get

$$D(\Pi) = 2^6, \quad D(E) = 1. \tag{12}$$

From the first of these equalities we get that Π is generated by e_x and g_w (see (8)), i.e. the fourth assertion of the lemma. From this and from $D(E) = 1$ we get the second assertion. The orthogonal complement to Π is obviously generated by a_v, and hence coincides with $\psi \cdot H_2(T, \mathbf{Z})$, whose discriminant equals 2^6. Thus Lemma 4 is proved.

Lemma 5. *Let an automorphism ϕ: $H_2(X, \mathbf{Z}) \to H_2(X, \mathbf{Z})$ have the following properties:*

1) $\phi(e_0) = e_0$.

2) $\phi(e_x) = e_{x'}$, $x, x' \in (\mathbf{Z}/2\mathbf{Z})^4$, *and consequently ϕ maps Π into itself.*

3) $\phi(g) = g$ *for all* g *in the orthogonal complement G of Π.*

Then ϕ is the identity transformation in $H_2(X, \mathbf{Z})$.

Proof. We recall that

$$a_v = 2f_v + \sum_{x \in v} e_x \in G. \tag{13}$$

Consequently

$$a_v = 2\varphi(f_v) + \sum_{x \in \varphi(v)} e_x.$$

We set

$$U(x) = \{v \cup \varphi(v)\} - \{v \cap \varphi(v)\}. \tag{14}$$

From (13) and (14) it follows that

$$\frac{1}{2} \sum_{x \in U(v)} e_x \in \Pi. \tag{15}$$

If the plane v contains a point o, then from condition 1) we get that $|U(v)| \leq 7$. From this and the fourth assertion of Lemma 4 we get that $U(v) = \emptyset$, and hence $\phi(v) = v$ for any plane v passing through the point o. Let $x \in (\mathbf{Z}/2\mathbf{Z})^4$, $x \neq o$, and let v_1 and v_2 be two planes whose intersection is $\{0, x\}$; then $\phi(x) \in \phi(v_1) \cap \phi(v_2) = v_1 \cap v_2$, and hence $\phi(x) = x$, i.e. $\phi(e_x) = e_x$ for any $x \in (\mathbf{Z}/2\mathbf{Z})^4$; hence $\phi(f_v) = f_v$ and therefore $\phi(l) = l$ for any $l \in H_2(X, \mathbf{Z})$.

In conclusion we shall give some facts on the structure of the lattice Π. Since these facts will not be used in the basic text of the paper, we merely state them without proof. The reconstruction of the proofs requires no fundamentally new concepts.

A) The group of automorphisms of the lattice Π is generated by automorphisms of the following forms:

1) reflections $S_x(p) = p + (p \cdot e_x)e_x$,

2) $\tilde{\phi}(e_x) = e_{\tilde{\phi}}(x)$, where $\tilde{\phi}$ is an arbitrary affine transformation in $(\mathbf{Z}/2\mathbf{Z})^4$ (not necessarily with $\tilde{\phi}(o) = o$).

B) Every automorphism of the lattice Π extends to an automorphism of all of $H_2(X, \mathbf{Z})$.

C) In the lattice L there exist primitive sublattices which are isomorphic to, but not isometric with, the lattice Π (see the Appendix to §6 for the definition of isomorphism and isometry of lattices).

D) Two primitive sublattices Π_1 and Π_2 of L, which are isomorphic to each other and for which there exist systems of homology classes f_v satisfying the conditions of Lemma 3, are isometric. The last assertion combined with Torelli's theorem (see §§1 and 7) gives the following necessary and sufficient condition for a $K3$ surface X to be Kummer: X is a Kummer surface if and only if there exist a primitive sublattice of S_X isomorphic to Π and a system of homology classes f_v in $H_2(X, \mathbf{Z})$ satisfying the conditions of Lemma 3.

§6. Completion of the proof of the Torelli theorem

The goal of this section is to construct the everywhere dense set $\mathrm{Z} \subset \Omega(l)$ on which Torelli's theorem is valid. As was explained in §1, the existence of such a set implies the validity of the Torelli theorem for all $\omega \in \Omega(l)$. As was proved in §5, the Torelli theorem holds for special Kummer surfaces. It therefore suffices to prove that the set of points $\omega \in \Omega(l)$ corresponding to the special Kummer surfaces is everywhere dense. In this section we shall actually prove somewhat more, namely, that the set of special Kummer surfaces with the maximum number of algebraic cycles is everywhere dense in $\Omega(l)$. We recall that we called such surfaces exceptional.

We need some general facts about $K3$ surfaces and automorphisms of the lattices S_X, which we now present. Let X be some $K3$ surface. We let $U \subset S_X$ denote the set of all effective divisor classes, $V \subset S_X$ the set of all divisor classes $a \in S_X$ such that $a^2 \geq 0$ and $au \geq 0$ for all $u \in U$. We let P_X denote the set of all $p \in U$, $p^2 = -2$. To each $p \in P_X$ we can associate a linear transformation (reflection) in H_X:

$$s_p(x) = x + (x \cdot p)\, p. \tag{1}$$

Let $\Gamma(X)$ be the group generated by the reflections s_p. Furthermore, let $\mathfrak{G}$ denote the group of all automorphisms of the lattice H_X which preserve periods, and $\mathfrak{G}^V$ the subgroup of it taking V into itself. We let δ denote the transformation $x \to -x$ and D the subgroup $(1, \delta)$.

Theorem 1. 1) *The group $\Gamma(X)$ is a normal subgroup of $\mathfrak{G}$.*

2) $\mathfrak{G} = \mathfrak{G}^V \cdot \Gamma(X) \cdot D$.

3) *For any $x \in S_X$ for which $x^2 = 0$, there exists a $\gamma \in \Gamma(X) \cdot D$ such that $\gamma \in V$.*

Proof. Let $a \in S_X$, $a^2 = -2$. Then the Riemann-Roch inequality yields that either a or $-a$ is effective, and hence $s_a \in \Gamma(X)$. In order to prove 1) we observe that $\Gamma(X) \subset \mathfrak{G}$, since s_p acts as the identity on the orthogonal complement to S_X in H_X and therefore preserves periods. Since $\Gamma(X)$ is obviously a normal subgroup of the group $G = \operatorname{Aut} S_X$ and $\mathfrak{G} \subset G$, we get 1). We prove 2). We let $\widetilde{V}$ denote the convex hull of V in $R = S_X \otimes \mathbf{R}$. From the general theory of groups generated by reflections (see [5], [6]) it follows that $\widetilde{V}$ is a fundamental domain for the group $\Gamma(X)$, i.e. the translates of $\widetilde{V}$ under the action of the group $\Gamma(X)$ fill up one half of the cone $x^2 > 0$ in R, where $\gamma_1 \widetilde{V}$ and $\gamma_2 \widetilde{V}$ can only intersect on a side. Let h be some element in general position in $\widetilde{V}$, i.e. such that $h \cdot p > 0$ for any $p \in P$ and $h^2 > 0$. Let $g \in \mathfrak{G}$. Since $\widetilde{V}$ is a fundamental domain, there exists a $\gamma \subset \Gamma(X) \cdot D$ such that $h_1 = \gamma g h \in \widetilde{V}$. We will show that then $g_1 = \gamma g \in \mathfrak{G}^V$. It clearly suffices to check that $g_1^{-1} P = P$. Let $p \in P$. Obviously either $g_1^{-1} p \in P$ or $-g_1^{-1} p \in P$. For $h \in U$ we have

$$(g_1^{-1}p) \cdot h = p \cdot (g_1 h) = p \cdot h_1 > 0, \quad h_1 \in U. \tag{2}$$

Consequently $g_1^{-1} p \in P$.

It remains to prove 3). Let h be an element in general position in V. Then $h \cdot x \in \mathbf{Z}$ and $h \cdot x \neq 0$. We assume that $h \cdot x > 0$, and we will show that there exists $\gamma \in \Gamma(X)$ such that $\gamma x \in V$. If $x \notin V$, there exists p such that $p \cdot x < 0$. We put $x_1 = s_p(x)$; then $x_1 \cdot h = x \cdot h + (p \cdot x)(p \cdot h) < x \cdot h$. Because $x_1 \cdot h \in \mathbf{Z}$ and $x \cdot h > 0$, since the reflection s_p maps the half-cone $x^2 > 0$ into itself, it follows that we obtain $x \in V$ by a finite number of steps. The proof is complete.

We observe that the same argument proves the decomposition $G = G^V \Gamma D$, where G is the group of automorphisms of the lattice S_X and G^V consists of the automorphisms preserving effective cycles.

Corollary. *If in the group S_X there is an element $x \neq 0$ such that $x^2 = 0$, then on X there exists a pencil of elliptic curves, the homology class of the fibers of which has the form γx, where $\gamma \in \Gamma(X) D$.*

In fact, 3) implies that there exists $\gamma \in \Gamma(X) D$ such that $\gamma x \in V$, and then, by §3, γx is the homology class of the generic fiber of a pencil of elliptic curves.

We now explain when an exceptional $K3$ surface is a Kummer surface. Then, of course, it will be a special Kummer surface.

Proposition 1. *Let B be a positive definite even lattice of rank 2. There exists a reducible two-dimensional abelian variety A for which $T_A \simeq B$.*

The definition of the lattice T_X is given in §1 for $K3$ surfaces. Here we use it without change for abelian varieties.

Proof. We consider a lattice E of rank 6 in which there exists a basis $e_1, e_2, e_3, e_4, e_5, e_6$ such that

$$e_i e_j = \begin{cases} 1 & \text{for } |i-j| = 3, \\ 0 & \text{for } |i-j| \neq 3. \end{cases}$$

This lattice is isomorphic to the lattice E_3 in the notation of the Appendix to §6. Let $\tilde{E}$ be the space of all complex linear functionals on E, equipped with a scalar product, carried over in the usual manner from E. As is well known, there exists a one-to-one correspondence between the complex linear functionals on E, considered up to multiplication by a constant and satisfying the conditions

$$\omega^2 = 0, \quad \omega \cdot \bar{\omega} > 0, \tag{3}$$

and two-dimensional complex tori.

By Theorem 1 of the Appendix to §6 there exists a sublattice of E which is isomorphic to B. We shall assume that $B \subset E$. We let B' denote the orthogonal complement to B in E. We construct a functional ω, starting with the condition $\omega(B') = 0$; then ω will be determined by its values on B. The proposition now follows from

Lemma 1. *For any positive definite two-dimensional lattice B there exists a complex-valued functional ω satisfying condition* (3). *These properties determine the functional uniquely up to multiplication by a complex number and a change of ω into $\bar{\omega}$.*

There exists a positive definite form $F(x_1, x_2)$ such that $F(x_1, x_2) = (x_1 e_1 + x_2 e_2)^2$ and $B = \mathbf{Z}e_1 + \mathbf{Z}e_2$. Then

$$F(x_1, x_2) = |x_1\mu_1 + x_2\mu_2|^2,$$

where $\mu_1, \mu_2 \in \mathbf{C}$. From this it follows that $\pi(x) = x_1\mu_1 + x_2\mu_2$, $x \in B$, is an isometric inclusion of B into $\mathbf{C}$.

A simple verification shows that $\pi(x) = \alpha \cdot x$, where $\alpha = (-\mu_2, \mu_1) \in B \otimes \mathbf{C}$, and $\alpha \cdot x$ denotes the scalar product in $B \otimes \mathbf{C}$, extended from B. Hence $\alpha^2 = 0$ and $\alpha \cdot \bar{\alpha} > 0$. If $\pi_1(x) = \alpha_1 \cdot x$ is another representation, then also $\alpha_1^2 = 0$, from which it follows easily that the vector α_1 is proportional to α or $\bar{\alpha}$.

Proposition 2. *Let B be a positive definite lattice of rank 2 such that $b^2 \equiv 0$ (4) for any $b \in B$. Then there exists a unique exceptional K3 surface X for which $T_X \simeq B$. This surface X is a special Kummer surface.*

We let B' denote the same group as B, but with the product of two vectors defined to be one-half their product in B. Let A be the abelian variety associated with B, whose existence was established in Proposition 1. We take for X the Kummer surface associated with A. It is easy to see that $T_X \simeq B'$. This proves the existence of the surface X.

Let X and Y be two exceptional $K3$ surfaces, where $T_X \simeq T_Y \simeq B$ and X is a special Kummer surface. We shall prove that X and Y are isomorphic. By Theorem 1 of the Appendix to §6 there exists an isomorphism $\phi': H_X \to H_Y$ inducing our given isomorphism $T_X \simeq T_Y$. It obviously maps S_X into S_Y. By Theorem 1 we can multiply ϕ' by an automorphism γ of the module H_Y, $\gamma \in \Gamma(Y) \cdot D$, such that $\phi = \gamma\phi'$ will take effective cycles into effective cycles. Since γ maps S_Y into itself and is the identity on T_Y, ϕ takes the period mapping π_X into π_Y. It remains to apply the Torelli theorem for special Kummer surfaces.

For a linear functional $\omega \in \Omega(l)$ we denote by L_ω the orthogonal complement to the kernel of ω in L. The following theorem represents the main assertion of this section.

Theorem 2. *Consider the set of all $\omega \in \Omega(l)$ for which*

1) $\operatorname{rg} L_\omega = 2$ *(hence the lattice L_ω is positive definite), and*

2) $a^2 \equiv 0\,(4)$ *for $a \in L_\omega$.*

This set is everywhere dense in $\Omega(l)$.

First we observe that there exist points ω satisfying the conditions of the theorem. For this we consider any lattice B satisfying the conditions of Proposition 2. By Theorem 1 of the Appendix to §6, in the orthogonal complement to the lattice $\mathbf{Z}l$ in L. there is a submodule isomorphic to B. To it we apply the arguments used in the proof of Proposition 1, i.e. we define ω so that $\omega(x) = 0$ for $x \in L$, $x \cdot B = 0$, and ω satisfies the conditions of Lemma 1 on B. By the same token we obtain the desired point $\omega \in \Omega(l)$.

Let ϕ be a transformation of $L \otimes \mathbf{C}$; we denote by $\tilde{\phi}$ the adjoint transformation in $\tilde{\Omega}$.

Further, we let Γ denote the set of all transformations of $L \otimes \mathbf{Q}$ which 1) take the given vector l into a proportional one with a positive coefficient of proportionality, 2) multiply the scalar product by a positive factor, and 3) together with their inverses can be written in a basis of L so that the denominators of the matrix elements will be relatively prime to 2. We denote by $\tilde{\Gamma}$ the induced group of automorphisms of $\Omega(l)$. The group of transformations of the Euclidean space $L \otimes \mathbf{R}$ satisfying conditions 1) and 2) will be denoted by G, and the corresponding group of automorphisms of $\Omega(l)$ by $\tilde{G}$. The group $\tilde{\Gamma}$ acts everywhere densely in $\Omega(l)$. In fact, the group $\tilde{G}$ acts transitively in $\Omega(l)$, and the group $\tilde{\Gamma}$, as we know from the theory of algebraic groups, is everywhere dense in the group $\tilde{G}$, considered as a group of transformations of $\Omega(l)$.

To complete the proof of the lemma it suffices to see that if ω satisfies the conditions of the theorem and $\tilde{\gamma} \in \tilde{\Gamma}$, then $\tilde{\gamma} \cdot \omega$ satisfies the same conditions. This follows easily from the following lemma.

Lemma 2. *Let ψ be a linear transformation of $L \otimes \mathbf{Q}$ preserving the scalar products up to multiplication by a constant and such that $\psi L \subset L$; then $L_{\tilde{\psi}\omega} \subset \psi^{-1} L_\omega$.*

Proof. Since $\psi L_{\widetilde{\psi}\omega} \subset L$, it suffices to prove that if $u \in L_{\widetilde{\psi}\omega}$ and $a \in L_{\omega}$, then $(\psi u)\cdot a = 0$, which obviously follows from the fact that

$$L_{\widetilde{\varphi}\omega} \otimes Q = \psi(L_\omega \otimes Q).$$

Appendix to §6: On even unimodular lattices[1]

Here we will consider only even unimodular Euclidean lattices. The simplest example is the lattice E_k of rank $2k$ having a basis $e_1, \cdots, e_k, e'_1, \cdots, e'_k$ such that

$$e_i e_j = 0, \quad e'_i e'_j = 0, \quad e_i e'_j = \delta_{ij}. \tag{1}$$

Two sublattices F_1 and F_2 which are mapped into each other by automorphisms of the lattice E will be called *isometric*. Obviously, isometric lattices are isomorphic, but the converse is in general false.

Lemma 1. *The lattice E_k forms a distinguished orthogonal direct summand in any lattice containing it.*

Proceeding by induction on k, we reduce the proof to the case $k = 1$. Let $E_1 \subset E$, let e, e' be the standard basis of E_1, and F the orthogonal complement to E_1 in E. Then for $x \in E$

$$x = (xe)e' + (x\cdot e')e + f, \quad f \in F,$$

which also proves the lemma.

Definition. Let $E = E_1 \oplus F$ be an orthogonal decomposition of a lattice E, let e, e' be the standard basis of E_1, and let $f_0 \in F$. The transformation

$$\left.\begin{aligned} &e \to e, \\ &e' \to e' + f_0 - \frac{1}{2}(f_0^2)e, \\ &f \to f - (f\cdot f_0)e, \quad f \in F, \end{aligned}\right\} \tag{2}$$

is called *the elementary transformation defined by e, e' and f_0.*

Lemma 2. *Let $E = E_1 \oplus F$. Any vector $x \in E$ of the form $x = \alpha e + e' + f$, $f \in F$, is isometric to a vector of the form $\beta e + e'$.*

For the proof it suffices to apply the elementary transformation corresponding to e, e' and $-f$.

Lemma 3. *Any primitive vector $f \in E_2$ is isometric to a vector of the form $\alpha e_1 + e'_1$.*

---

1) All the facts contained in this appendix are well known to specialists. It is written for the convenience of the reader, since in other accounts the analogous constructions are usually carried out in a more complicated situation. We thank L. N. Vaseršteĭn, who wrote this appendix at our request.

Let $f = \alpha_1 e_1 + \alpha_2 e_1' + \alpha_3 e_2 + \alpha_4 e_2'$. It is well known that there exist integral unimodular matrices A and B of second order such that

$$A\begin{pmatrix}\alpha_1 & -\alpha_3\\ \alpha_4 & \alpha_2\end{pmatrix}B=\begin{pmatrix}\alpha & 0\\ 0 & 1\end{pmatrix},$$

where $\alpha = \alpha_1\alpha_2 + \alpha_3\alpha_4$. The automorphism

$$\begin{pmatrix}e_1 & -e_2\\ e_2' & e_1'\end{pmatrix}\to B^{-1}\begin{pmatrix}e_1 & -e_2\\ e_2' & e_1'\end{pmatrix}A^{-1} \tag{3}$$

takes f into $\alpha e_1 + e_1'$.

Theorem 1. *If the lattice E contains a sublattice isomorphic to E_k, then* 1) *for any even (not necessarily unimodular) lattice of rank $\leq k$ there exists a primitive sublattice of E isomorphic to it, and* 2) *any isomorphic primitive sublattices of the same rank $l \leq k-1$ are isometric.*

We recall that a sublattice $F \subset E$ is called primitive if the group E/F has no torsion.

For the proof of the first assertion we can assume that $E = E_k$. Let $e_1, \cdots, e_k$, $e_1', \cdots, e_k'$ be a basis satisfying condition (1) and let $y_1, \cdots, y_l$ be a basis of the lattice F, $l \leq k$. We consider the vectors

$$x_i = e_i + \sum_{j=1}^{i} a_{ij} e_j', \quad i = 1, \ldots, l, \tag{4}$$

where $a_{ii} = \frac{1}{2} y_i^2$, $a_{ij} = y_i y_j$ for $i \neq j$, and we denote by F' the lattice spanned by them. Since e_i is included with coefficient 1 in these formulas, F' is primitive in E. Substitution shows that the mapping $y_i \to x_i$ is an isomorphism of F and F'.

We shall now prove the second assertion using induction on l. Start with $l = 1$. We must prove that in the lattice $E = E_2 \oplus V$ any two isomorphic primitive vectors are isometric. Let $x \in E$ and let

$$x = u + f, \quad u \in E_2, \quad f \in V, \qquad u = \alpha_1 e_1 + \alpha_2 e_1' + \alpha_3 e_2 + \alpha_4 e_2'. \tag{5}$$

Since the lattices E and E_2 are unimodular, so is V. Therefore there exist vectors $u' \in E_2$ and $f' \in V$ such that $x \cdot (u' + f') = u \cdot u' + f \cdot f' = 1$.

Via an elementary transformation we can make $\alpha_1 \neq 0$. Then via another elementary transformation corresponding to e_1', e_1 and some $f_0 \in V$ we can obtain that the vector u in (5) is primitive. Now by Lemma 3 we can replace by the vector $\alpha e + e'$, so that x is isometric to the vector $\alpha e + e' + f$. By Lemma 2 the vector x is isometric to the vector $\beta e + e'$, where $\beta = \frac{1}{2} x^2$.

Now assume the theorem is proved for the value l and let $y_1, \cdots, y_{l+1}$ be a basis of a primitive sublattice $F \subset E$. The basis $x_1, \cdots, x_{l+1} \in E_k \subset E$, given by formula

(4), defines a sublattice $F' \subset E_k$. We must prove that there exists an automorphism o the lattice E taking one basis into the other. For $y_1, \cdots, y_l$ and $x_1, \cdots, x_l$ we can assume this by induction. Therefore we shall assume that $y_i = x_i \in E_l$, $i = 1, \cdots, l$, $E_k = E_l \oplus E_{k-l}$, $E = E_l \oplus E_{k-l} \oplus V$, $k-l \geq 2$, and

$$y_{l+1} = u_1 + u_2 + v, \tag{6}$$

where $u_1 \in E_l$, $u_2 \in E_{k-l}$, $v \in V$.

We now want to map y_{l+1} to x_{l+1} by elementary transformations preserving the vectors $x_i = y_i$, $i \leq l$. We let H denote the orthogonal complement to the vectors x_i, $i \leq l$, in E (it looks like $H' \oplus E_{k-l} \oplus V$, where H' is the primitive sublattice generated by the vectors $z_i = e_i - \sum_{j=i}^{l} a_{ij} e_j'$, $i = 1, \cdots, l$). By the primitivity of F and the unimodularity of E, there exists a vector y in H such that $y \cdot y_{l+1} = 1$.

Making the elementary transformation (2) corresponding to e_{l+1}, e_{l+1}' and some $y \in H$ orthogonal to e_{l+1}, e_{l+1}', we can make the coefficient of e_{l+1} in u_2 nonzero. Then, making the elementary transformation corresponding to e_{l+1}', e_{l+1} and some y H orthogonal to e_{l+1}' and e_{l+1}, we can make the coefficients of e_{l+1} and e_{l+1}' in u_2 relatively prime, and hence the vector $u_2 + v$ is primitive. By what has been proved above we can reduce it to the form $e_{l+1} + \alpha e_{l+1}'$ by elementary transformations of the space $E_{k-l} \oplus V$. Thus we have reduced the vector y_{l+1} to the form

$$y_{l+1} = e_{l+1} + \alpha e_{l+1}' + \xi, \quad \xi \in E_l. \tag{7}$$

Since $y_{l+1} y_i = x_{l+1} y_i$ for $i = 1, \cdots, l$, it follows that $\xi y_i = 0$. Therefore the elementary transformation corresponding to the vectors e_{l+1}, e_{l+1}' and ξ does not change the vectors $y_1, \cdots, y_l$. A check shows that it kills the term ξ in (7). Finally, since $y_{l+1}^2 = x_{l+1}^2$, it follows that $\alpha = 0$, i.e. $y_{l+1} = x_{l+1}$. The theorem is proved.

§7. Automorphisms

The Torelli theorem allows us to study the structure of the group of automorphism of a K3 surface. The following description is an immediate consequence of the Torel theorem.

Proposition. *The group of automorphisms of a K3 surface X coincides with the group of automorphisms of the Euclidean lattice H_X which preserve the set of effective cycles and multiply the linear form π on H_X, defining the periods, by a complex number.*

We now give another description of the group of automorphisms. Actually, it give an answer "up to finite groups", but on the other hand this answer is formulated in terms of a purely algebraic invariant, the Euclidean lattice S_X.

We will say that two groups $\mathfrak{G}$ and $\mathfrak{G}'$ are isomorphic up to finite groups if there exist subgroups of finite index $\mathfrak{G}_1 \subset \mathfrak{G}$ and $\mathfrak{G}'_1 \subset \mathfrak{G}'$ and finite normal subgroups $\mathfrak{G}_0 \subset \mathfrak{G}_1$ and $\mathfrak{G}'_0 \subset \mathfrak{G}'_1$ such that the groups $\mathfrak{G}_1/\mathfrak{G}_0$ and $\mathfrak{G}'_1/\mathfrak{G}'_0$ are isomorphic.

Theorem 1. *The group of automorphisms of a K3 surface is isomorphic to the group* $G(S_X)/\Gamma(S_X)$ *up to finite groups.*

See the end of §1 for the definition of the groups $G(S_X)$ and $\Gamma(S_X)$.

Let $\mathfrak{G}$ be the group of automorphisms of the surface X. As $\mathfrak{G}_0$ we take the kernel of the natural homomorphism $\mathfrak{G} \to G(S)$. In particular, an automorphism $g_0 \in \mathfrak{G}_0$ preserves the polarization of the surface X and is therefore induced by a projective transformation. Thus $\mathfrak{G}_0$ is an algebraic group. Its Lie algebra is isomorphic to $H^0(X, \check{\Theta}_X)$. As we saw in §2, this space equals 0. Therefore the group $\mathfrak{G}_0$ is finite.

In the lattice H we now consider a sublattice T, the orthogonal complement to S. The lattice $H' = S \oplus T$ has finite index in H. Therefore its group of automorphisms contains a subgroup of finite index, consisting of automorphisms which extend to automorphisms of H. In particular, in the group $G(S)$ there exists a subgroup of finite index G_1 such that for $g_1 \in G_1$ the automorphism $(g_1, 1_T)$ is induced by an automorphism of H.

Each reflection $\gamma \in \Gamma(S)$ can be defined as a reflection relative to the same vector in the whole lattice H. Under this the vectors of T map into themselves. Therefore $\Gamma(S) \subset G_1$.

We denote by $\overline{\mathfrak{G}}_1$ the group of automorphisms of G_1 which are induced by automorphisms of the surface X, and by $\mathfrak{G}_1$ the group of the corresponding automorphisms of X. Then $\mathfrak{G}_1$ has finite index in $\mathfrak{G}$:

$$\overline{\mathfrak{G}}_1 \subset G_1, \quad \overline{\mathfrak{G}}_1 \simeq \mathfrak{G}_1/\mathfrak{G}_0.$$

On the other hand, by a basic theorem on groups generated by reflections (Theorem 1, §6), the group $G(S)$ is a semidirect product, $\Gamma(S) \cdot D \cdot G^V$, where G^V is a group consisting of the automorphisms which preserve the set of effective cycles, and $D = \{1, -1\}$. From this it follows that G_1 is a semidirect product of the normal subgroup $\Gamma(S)$ and the subgroup $\widetilde{\mathfrak{G}}_1$ consisting of the automorphisms $g_1 \in G_1$ which preserve effective cycles. In particular, $\widetilde{\mathfrak{G}}_1 \simeq G_1/\Gamma(S)$. We shall prove that $\widetilde{\mathfrak{G}}_1 = \overline{\mathfrak{G}}_1$, from which the theorem will also follow. By construction, $\widetilde{\mathfrak{G}}_1 \subset \overline{\mathfrak{G}}_1$. In order to prove the reverse inclusion, we observe that an automorphism $\tilde{g} \in \widetilde{\mathfrak{G}}_1$ can by definition be extended to an automorphism $g \in G(H)$, inducing 1 on T, and that the extension is unique. We check that the conditions of the proposition hold for g; this also proves the theorem. The first condition, that the set of effective cycles is preserved, holds by definition. The second condition follows from the fact that the form π equals 0 on S and is given by its restriction to T, but g does not change elements of T. This proves the theorem.

Corollary. *The group of automorphisms of a K3 surface X is finite if and only if the group* $G(S_X)/\Gamma(S_X)$ *is finite.*

Example. We consider the simplest nontrivial case, a surface X for which $\operatorname{rg} S_X = 2$. In this case S_X is a two-dimensional module corresponding to an indefinite quadratic form. Such a form has an infinite group of automorphisms if and only if it does not represent zero. If the group of automorphisms is infinite, it is generated by an element s of order two and an element a of infinite order, where $s^{-1}as = a^{-1}$.

If the form represents neither 0 nor -2, then the group $G(S)$ is infinite and $\Gamma(S) = 1$, and consequently the group of automorphisms of the surface is infinite. If the form represents -2, $\Gamma(S)$ contains just one element of order 2. Such an element has the form sa^k, and it is easy to see that the elements conjugate to it generate a subgroup of $G(S)$ of finite index. Therefore the group of automorphisms of the surface X is finite.

We see that the group of automorphisms of the surface X is infinite if and only if there are no vectors of length 0 or -2 in the module S_X.

In [17] Severi considered the simplest type of such surfaces—surfaces of degree four, containing a curve of genus 2 and degree 6. In this case the form corresponding to the module S_X has the form $2\lambda^2 + 12\lambda\mu + 4\mu^2$. It represents neither 0 nor -2, so that the group $G(S)$ is infinite.

It follows from the theorem in the Appendix to §6 that there exist surfaces of the type in question for which the module S_X is defined by any preassigned even, indefinite form.

We give two more examples, which were kindly communicated to us by E. B. Vinberg.

1. The group of automorphisms of the form $-2x^2-2y^2-2z^2+2xy+4yz$ of signature (1, 2) is isomorphic to $PGL(2, \mathbf{Z})$. From this it follows that if such a form corresponds to the lattice S_X for a $K3$ surface X, then X has a finite group of automorphisms.

2. We consider an even unimodular form of 18 variables of signature (1, 17). The form is uniquely determined by these properties ([16], p. 93). In its group of automorphisms the reflections generate a subgroup of finite index. Therefore a $K3$ surface X for which the lattice S_X corresponds to such a form also has a finite group of automorphisms.

From the results of the Appendix to §6 it easily follows that both these forms correspond to $K3$ surfaces.

§8. Exceptional surfaces

We recall that this was the name we gave to surfaces X for which $\operatorname{rg} S_X$ attains the maximum possible value of 20.

Proposition: *For an exceptional surface X the period mapping $\pi: H_X \to \mathbf{C}$ is uniquely determined by the structure of Euclidean lattice in H_X and the prescription of the submodule $S_X \subset H_X$.*

Since the periods of the algebraic cycles are 0, the form π is defined by its values on the group $T^* = H/S$. We furnish this group with a scalar product. For this we extend the product defined in H to $H \otimes \mathbf{R}$, and imbed T^* in the orthogonal complement to $S \otimes \mathbf{R}$. This imbedding induces the desired scalar product.

In view of Lemma 1 of §6 the inclusion $\pi: T^* \to \mathbf{C}$ is a similitude if the metric in $\mathbf{C}$ is defined as $|z|^2$. It is clear that this property defines π uniquely up to a multiple.

Remark. The Euclidean lattice T^* and the orthogonal complement T to S in H are reciprocal to each other. In fact, if $t \in T$ and $t^* \in T^*$, there exists $s^* \in S \otimes \mathbf{R}$ such that $s^* + t^* = h \in H$. Then $h \cdot t \in \mathbf{Z}$, and since $s^* \cdot t = 0$, then also $t^* \cdot t \in \mathbf{Z}$. If $\phi: T^* \to \mathbf{Z}$ is a homomorphism, it extends to a homomorphism $\phi': H \to \mathbf{Z}$ which is equal to 0 on S. Since the lattice H is unimodular, there exists a vector $h_0 \in H$ such that $\phi'(h) = h \cdot h_0$. Since $\phi'(S) = 0$, it follows that $h_0 \in T$, and hence $\phi(t^*) = t \cdot t^*$. Since the lattices T and T^* are two-dimensional, they are similar to each other.

Theorem. *An exceptional surface X is uniquely determined by the two-dimensional Euclidean lattice T_X.*

We must repeat the arguments used in the proof of Proposition 2 of §6. Suppose we are given two surfaces X and Y for which the lattices T_X and T_Y are isomorphic.

By the Appendix to §6, an isomorphism $T_X \to T_Y$ can be extended to an ismorphism $\phi': H_X \to H_Y$. It obviously maps S_X into S_Y. In view of Theorem 1 of §6 we can multiply ϕ' by an automorphism σ of the module H_Y, $\sigma \in \Gamma(H_Y)$, such that $\phi = \sigma\phi'$ will take effective divisors of H_X into effective ones of H_Y. Since elements of $\Gamma(H_X)$ take S_Y into itself and are the identity on T_Y, ϕ takes the period mapping π_X into π_Y. It remains to apply the Torelli theorem.

Thus an exceptional surface X is defined up to isomorphism by the class of the binary quadratic form $F(x_1, x_2)$, which gives a square in T_X. This form is positive definite and even. As is proved in the Appendix to §6, any lattice defined by such a form is isomorphic to a sublattice of the lattice L and therefore corresponds to a surface X.

The form F can be represented in the form $m \cdot F'$, where $m \in \mathbf{Z}$ and F' is a primitive even form. Consequently an exceptional surface has two invariants—the number m and the form F'. As was shown in §6, a surface X is Kummer if and only if $m \equiv 0\ (2)$. The form F is the primitive form corresponding to the lattice T, and by the same token also to the lattice T^*. In other words, the surface X is determined by the number m and the lattice $T^* \subset \mathbf{C}$, considered up to similitude.

Instead of the lattice T^* we can consider the elliptic curve $C = \mathbf{C}/T^*$ as an invariant of the surface. This curve obviously possesses complex multiplication.

The contrast between the surface X and the elliptic curve C can be described in the following way. Let X be an arbitrary Kähler manifold. The classical exact sequence of sheaves

$$0 \to \mathbf{Z} \to O \overset{\exp}{\to} O^* \to 0 \tag{1}$$

reduces to an exact sequence of cohomology groups, which, as was proved by Kodaira, gives the basic constructions of the theory of divisors if we restrict to the cohomology groups of dimensions 1 and 2. In arbitrary dimension p we analogously obtain the exact sequence

$$0 \to H^{0,p}(\mathbf{C})/\Pi^{0,p}H^p(X,\mathbf{Z}) \to H^p(O^*) \to H^{p+1}(\mathbf{Z}),$$

where $\Pi^{0,p}$ is the projection of $H^p(\mathbf{C})$ onto the subspace $H^{0,p}(X)$ of elements of type $(0, p)$.

In contrast to the case $p = 1$ we conclude that the group $H^{0,p}(\mathbf{C})/\Pi^{0,p}H^p(\mathbf{Z})$ does not have a natural complex structure—as a rule, the subgroup $\Pi^{0,p}H^p(X, \mathbf{Z})$ is not discrete in $H^{0,p}(X, \mathbf{C})$. Therefore the quotient group by it does not have a natural meaning. In fact, for example, in connection with the Brauer group only a periodic subgroup of it arises. However, it can occur that the subgroup $\Pi^{0,p}H^p(X, \mathbf{Z})$ is discrete in $H^{0,p}(X, \mathbf{C})$. This will happen if and only if the whole space $\bigoplus_{i=1}^{p-1} H^{p-i,i}(X, \mathbf{C})$ is generated by the integral cycles contained in it. In this case the group $H^{0,p}(X, \mathbf{C})/\Pi^{0,p}H^p(X, \mathbf{C})$ is a complex torus. In some cases it is connected with Weil's generalized Jacobians; however, this is far from being the case always—for instance, it is known not to occur if p is even.

The condition given above for the discreteness of the group $\Pi^{0,p} H^p (X, \mathbf{Z})$ in $H^{0,p} (X, \mathbf{C})$ for $p = 2$ means that all the cycles of type $(1, 1)$ are generated by the integral ones, i.e. by the algebraic cycles. In particular, it is valid for exceptional $K3$ surfaces. It is easy to see that the torus $H^{0,2} (X, \mathbf{C})/\Pi^{0,2} H^2 (X, \mathbf{Z})$ coincides with the curve C. Kuga observed that this criterion holds for Sato's varieties, which are families of abelian varieties.

In conclusion we consider a special exceptional surface X. It is given by the equation

$$x_0^4 - x_1^4 = x_2^4 - x_3^4 \tag{2}$$

in $\mathbf{P}^3$. Its exceptionalness is easily discerned from various considerations. It will also follow from the arguments given below. The curve C for it is found very simply, the point of the matter being that the surface X has automorphisms, for example,

$$\sigma(x_0 : x_1 : x_2 : x_3) = (ix_0 : x_1 : x_2 : x_3), \quad i^2 = -1,$$

which multiply its regular differential form by i. Therefore they must induce an automorphism of order 4 in the lattice T^*, and this lattice is uniquely determined up to a similitude: it is square. The curve C has the form $y^2 = x^3 - x$.

It is more difficult to find the factor m. The calculations we shall give only show that $m = 2$ or 4. However this is already sufficient to obtain the astonishing (at least to the authors) result: the surface X is Kummer!

The surface X has an obvious pencil of elliptic curves, depending on a parameter t and given by the equations

$$x_0^2 + x_1^2 = t(x_2^2 - x_3^2),$$
$$t(x_0^2 - x_1^2) = x_2^2 + x_3^2. \tag{3}$$

We calculate its degenerate fibers. There are automatically degenerate fibers corresponding to the values $t = 0, \infty, 1, -1, i, -i$. These fibers have type A_3 and their components are lines. The sum of their Euler characteristics is 24, i.e. the Euler characteristic of the entire surface. Therefore there are no other degenerate fibers. We denote by Λ_X the subgroup of S_X generated by the components of the fibers, and by Λ'_X the maximal primitive sublattice of S_X containing Λ_X. A quadratic form defined on Λ_X has a simple form. Namely, if from each degenerate fiber we choose any 3 components and complete this system of vectors by the fiber a, then we obtain a decomposition

$$\Lambda_X = M_1 \oplus M_2 \oplus M_3 \oplus M_4 \oplus M_5 \oplus M_6 \oplus \mathbf{Z}a, \tag{4}$$

where $\operatorname{rg} M_i = 3$, $a^2 = a \cdot M_i = 0$ and in a basis consisting of the three components of the fibers the matrix of the quadratic form corresponding to the module M_i has the form

$$A = \begin{pmatrix} -2 & 1 & 0 \\ 1 & -2 & 1 \\ 0 & 1 & -2 \end{pmatrix}. \tag{5}$$

We compute the power of 2 which the index $(\Lambda'_X : \Lambda_X)$ divides. For this we must interpret the pencil (3) as an elliptic curve $\mathfrak{X}$ over the field $k(t)$. Obviously the group Λ'_X/Λ_X is the group of divisor classes of finite order on $\mathfrak{X}$. The group of divisor classes of degree zero on an elliptic curve is isomorphic to the group of rational points of its Jacobian curve. The Jacobian curve for the elliptic curve

$$z^2 = au^2 + bv^2,$$
$$w^2 = cu^2 + dv^2$$

is known [2]. It is isomorphic to the curve

$$\mu^2 = \lambda(a + c\lambda)(b + d\lambda).$$

From this, after simple transformations, we find that the Jacobian curve $\mathfrak{A}$ of the curve $\mathfrak{X}$ is given by the equation

$$y^2 = x(x+1)(x+c^2), \quad c = \frac{t^2-1}{t^2+1}. \tag{6}$$

We need to find the points of order 2^k on $\mathfrak{A}$. We recall that the curve $y^2 = (x-e_1)(x-e_2)(x-e_3)$ has points of order two $(e_i, 0)$, $i = 1, 2, 3$, and the point (x, y) is divisible by 2 if and only if $x-e_i \in k(t)^{*2}$.

From this we see that on $\mathfrak{A}$ there are 3 points of order 2: $E_1 = (0, 0)$, $E_2 = (-1, 0)$, $E_3 = (-c^2, 0)$. A simple computation shows that they are all divisible by 2 and give the following points:

$$\begin{aligned} &\text{for a division of } E_1\text{: } (c \pm c(c+1)),\ (-c,\ \pm c(c-1)), \\ &\text{for a division of } E_2\text{: } (-d-1,\ \pm id(d+1)),\ (d-1,\ \pm id(d-1)), \\ &\text{for a division of } E_3\text{: } (c(id-c),\ \pm cd(id-c)),\ (-c(id+c),\ \pm cd(id+c)). \end{aligned} \tag{7}$$

Here $d = 2t/(t^2+1)$, so that $c^2 + d^2 = 1$.

Using the test given above for the divisibility of a point by 2, it is easy to perceive that none of the points (7) is divisible by 2. Thus we see that the 2-component of the $k(t)$-rational points of the curve $\mathfrak{A}$ is isomorphic to $\mathbf{Z}/4\mathbf{Z} \oplus \mathbf{Z}/4\mathbf{Z}$ and has order 16. By the same token $(\Lambda'_X : \Lambda_X) = 16k$, $k \equiv 1 \pmod 2$.

We now determine the group S_X. It is known that the group S_X/Λ_X is isomorphic to the group of divisor classes on the curve $\mathfrak{X}$. The structure of this group is the following. We denote by l the greatest common divisor of the degrees of the divisors on $\mathfrak{X}$. In terms of the surface X this is the greatest common divisor of the numbers $u \cdot a$, $u \in S$, where a is the fiber of the bundle defined by the curve (6). Let u be a curve such that $u \cdot a = l$. Then $S_X = \mathbf{Z} \cdot u + S_X^0$, where S_X^0 consists of the classes $v \in S_X$ for which $va = 0$. Then S_X^0/Λ_X is the group of divisor classes of degree 0 on $\mathfrak{X}$. The last group is isomorphic to the group of $k(t)$-rational points of the curve $\mathfrak{A}$.

For the determination of this group we need to apply arguments already used to a minimal nonsingular model of the surface (6). It has the same degenerate fibers as the curve (3). As before, $S_Y = \mathbf{Z}u' + S_Y^0$, and S_Y^0/Λ_Y is the group of $k(t)$-rational points on $\mathfrak{A}$. Therefore $\operatorname{rg} S_Y = \operatorname{rg} S_Y^0 + 1$. But $S_Y^0 \supset \Lambda_Y$ and $\operatorname{rg} \Lambda_Y = 19$. Since $\operatorname{rg} S_Y \leq 20$, we have $\operatorname{rg} S_Y^0 = 19$ and the index $(S_Y^0 : \Lambda_Y)$ is finite. In other words, the group of rational points on $\mathfrak{A}$ is finite. From this it follows that the group S_X^0/Λ_X isomorphic to it is finite, and hence $S_X^0 = \Lambda'_X$.

As a result we see that $S_X = \mathbf{Z} \cdot u + \Lambda'_X$. We set $\mathfrak{A} = \mathbf{Z}u + \Lambda_X$. Then

$$(S_X : \mathcal{A}) = (\Lambda'_X : \Lambda_X) = 16 \cdot k,$$

$$D(S_X) = D(\mathcal{A})/2^8k^2,$$

where $D(E)$ is the discriminant of the lattice E. Finally, $D(A)$ is easily computed. We join the bases of the modules M_i in which the matrix (5) was written, and add to them the vector u. We obtain a basis in which the Gram matrix acquires the form

$$\begin{pmatrix}
A & & & & & & & * \\
 & A & & & & & & \\
 & & A & 0 & & & & \vdots \\
 & & & A & & & & \\
0 & & & & A & 0 & & \\
 & & & & & A & & * \\
 & & & & & & 0 & l \\
* & \cdots & & & & * & l & 0
\end{pmatrix}.$$

We will obtain that $D(\mathfrak{A}) = |A|^6 l^2 = 2^{12} l^2$. Therefore $D(S_X) = 2^4 (l/k)^2$.

Among other things, we hence obtain that $k = 1$, i.e. that the group of points on the curve (6) is isomorphic to $Z/4Z + Z/4Z$. In fact, $l = 1$ or 2, since we know there exists an element $v \in S_X$ with $v \cdot a = 2$. This is the line $x_0 = x_1$, $x_2 = x_3$. From the fact that $D(S_X)$ is integral and k is an odd number, it follows that $k = 1$. Whether l equals one or two depends on whether or not the curve (3) has a $k(t)$-rational point. V. A. Dem'janenko has kindly informed us that he proved that the curve (3) has no rational points and hence $l = 2$.

As a result of all these considerations we see that

$$D(S_X) = 16l^2, \tag{8}$$

i.e. equals 64 in view of the result of Dem'janenko.

Now we use the fact that in a unimodular Euclidean lattice the discriminants of a primitive sublattice and its orthogonal complement coincide. We leave the elementary verification of this assertion to the reader. From this and from formula (8) it follows that also

$$D(T_X) = 16l^2 = 64. \tag{9}$$

We saw that the binary form F corresponding to the lattice T_X has the form $F = m \cdot F'$, where F' is an even primitive form corresponding to a square lattice, i.e.

$$F'(x_1, x_2) = 2x_1^2 + 2x_2^2.$$

Therefore $D(T_X) = D(F) = 4m^2$, and, comparing with (9), we obtain that $m = 2l$. Since m is even, the surface X is Kummer. From Dem'janenko's result it follows that $m = 4$. Therefore the corresponding form F has the form

$$F(x_1, x_2) = 8x_1^2 + 8x_2^2.$$

Received 26 JAN 70

BIBLIOGRAPHY

[1]* I. R. Šafarevič, et al., *Algebraic surfaces*, Trudy Mat. Inst. Steklov. 75 (1965) = Proc. Steklov Inst. Math. 75 (1965). MR 32 #7557; 35 #6685.

[2] M. I. Bašmakov and D. K. Faddeev, *Simultaneous representation of zero by a pair of quadratic forms*, Vestnik Leningrad. Univ. 14 (1959), no. 19, 43–46. (Russian) MR 22 #7976.

[3] A. Borel, *Density and maximality of arithmetic subgroups*, J. Reine Angew. Math. 224 (1966), 78–89. MR 34 #5824.

[4] N. Bourbaki, *Éléments de mathématique*. Fasc. XXXIV. *Groupes et algèbres de Lie*. Chap. IV: *Groupes de Coxeter et systèmes de Tits*. Chap. V: *Groupes engendrés par des réflexions*. Chap. VI: *Systèmes de racines*, Actualités Sci. Indust., no. 1337, Hermann, Paris, 1968. MR 39 #1590.

[5] H. F. M. Coxeter, *Discrete groups generated by reflections*, Ann. of Math. (2) 35 (1934), 588–621.

[6] È. B. Vinberg, *Geometric representations of Coxeter groups*, Uspehi Mat. Nauk 25 (1970), no. 2 (156), 267–268. (Russian) MR 41 #6985.

[7] P. A. Griffiths, *Periods of integrals on algebraic manifolds: Summary of main results and discussion of open problems*, Bull. Amer. Math. Soc. 76 (1970), 228–296; Russian transl., *Report on the variation of a Hodge structure*, Uspehi Mat. Nauk 25 (1970), no. 3, 175–234. MR 41 #3470.

[8] A. Grothendieck, *Éléments de géométrie algébrique*, Inst. Hautes Études Sci. Publ. Math. No. 8 (1961). MR 36 #177b.

[9] ——, *Techniques de descente et théorèmes d'existence en géométrie algébrique*. IV. *Les schémas de Hilbert*, Séminaire Bourbaki, 13ième année: 1960/61, fasc. 3, Exposé 221, Secrétariat mathématique, Paris, 1961. MR 27 #1339.

[10] I. V. Dolgačev and A. N. Paršin, *Different and disciminant of regular mappings*, Mat. Zametki 4 (1968), 519–524. (Russian) MR 38 #4489.

[11] C. L. Siegel, *Automorphic functions of several complex variables*, Mimeograph Notes, Göttingen, 1953; Russian transl., IL, Moscow, 1954.

[12] K. Kodaira, *On compact analytic surfaces*, Analytic Functions, Princeton Univ. Press, Princeton, N. J., 1960, pp. 121–135; Russian transl., Matematika 6 (1962), no. 6, 3–17. MR 25 #3939.

[13] ——, *On compact complex analytic surfaces*. I, Ann. of Math. (2) 71 (1960), 111–152; Russian transl., Matematika 6 (1962), no. 6, 19–56. MR 24 #A2396.

[14] M. Kneser, *Zur Theorie der Kristalgitter*, Math. Ann. 127 (1954), 105–106. MR 15,780.

2) *Editor's note*. All page references to [1] are to the English translation.

[15] T. Matsusaka and D. Mumford, *Two fundamental theorems on deformations of polarized varieties*, Amer. J. Math. 86 (1964), 668–684; correction, ibid. 91 (1969), 851. MR 30 #2005; 40 #1398.

[16] J.-P. Serre, *Cours d'arithmétique*, Collection SIP: "Le Mathématicien", 2, Presses Universitaires de France, Paris, 1970. MR 41 #138.

[17] F. Severi, *Complementi alla teoria della base per la totalità delle curve di una superficie algebrica*, Rend. Circ. Mat. Palermo 30 (1910), 265–288.

[18] J. Tate, *Genus change in inseparable extensions of function fields*, Proc. Amer. Math. Soc. 3 (1952), 400–406. MR 13,905.

[19] I. R. Šafarevič, *Principal homogeneous spaces defined over a function*, Trudy Mat. Inst. Steklov. 64 (1961), 316–346; English transl., Amer. Math. Soc. Transl. (2) 37 (1964), 85–114. MR 29 #110.

[20] R. Palais, *On the existence of slices for actions of non-compact Lie groups*, Ann. of Math. (2) 73 (1961), 295–323. MR 23 #A3802.

Translated by:
J. S. Joel

The arithmetic of K3 surfaces
(with I. I. Piatetskij-Shapiro)

Tr. Mat. Inst. Steklova **132**, 44 – 54 (1973). Zbl. **293**, 14010
[Proc. Steklov Inst. Math. **132**, 45 – 57 (1975)]

ABSTRACT. In this paper we study some properties of K3 surfaces and give arithmetic applications.

Bibliography: 8 items.

§1. The main results

A complete algebraic surface X is called a K3 surface if $H^1(X, O_X) = 0$ and if the canonical class K_X of X equals zero.

It is known (see [1], Chapter VIII) that, in addition to K3 surfaces, two-dimensional abelian varieties satisfy the condition $K_X = 0$. Over the field of complex numbers $\mathbf{C}$ a two-dimensional abelian variety is uniquely determined by the periods of its holomorphic one-forms. From this it follows that it is also determined by the periods of its holomorphic two-form. On a K3 surface there are no holomorphic one-forms, but there is a holomorphic 2-form, which is unique up to a constant multiple. The question naturally arises of "to what extent do its periods determine the surface?" To answer this we need to introduce some more concepts.

One sees that for all K3 surfaces X the homology groups $H_2(X, \mathbf{Z})$ together with the scalar product defined by the intersection index are all isomorphic. We fix one such Euclidean lattice L and a vector $l \in L$. A marked K3 surface is a triple $\widetilde{X} \approx (X, \xi, \varphi)$, where X is a K3 surface, $\xi \in H_2(X, \mathbf{Z})$ is the homology class of a hyperplane section under some projective imbedding of X and $\varphi\colon H_2(X, \mathbf{Z}) \to L$ is an isomorphism of Euclidean lattices such that $\varphi(\xi) = l$.

In the space $\widetilde{\Omega} = \mathrm{Hom}\,(L, \mathbf{C})$ we define a bilinear scalar product by extending the one given in L. The conditions

$$\omega^2 = 0, \qquad \omega\bar{\omega} > 0$$

AMS (MOS) subject classification (1970). Primary 14G13, 14J25; Secondary 14G25, 14J20.

**Editor's note.* The present translation incorporates suggestions made by the authors.

define a domain Ω in projective space corresponding to the vector space $\widetilde{\Omega}$. We let $\Omega(l)$ denote the connected component of a section, defined by the condition $\omega l = 0$. This is a homogeneous symmetric domain on which the group $G(R)$ acts, where G is the isotropy subgroup of the vector l in the orthogonal group of the lattice L. For a form $\eta \in \Gamma(X, \Omega_X^2)$ and $\gamma \in H_2(X, \mathbf{Z})$ we put

$$f(\gamma) = \int_\gamma \eta.$$

Obviously $f \in \operatorname{Hom}(H_2(X, \mathbf{Z}), \mathbf{C})$, and we see that φf defines a point in $\Omega(l)$. We denote this point by $\tau(\widetilde{X})$.

THEOREM 1 (*Torelli theorem for* K3 *surfaces*). *A marked* K3 *surface* $\widetilde{X}$ *is uniquely determined by a point* $\tau(\widetilde{X}) \in \Omega(l)$.

This theorem is proved in [2]. Here we illustrate some applications of it. We are basically interested in arithmetic applications, but since our method uses the analytic-geometric Theorem 1, it is impossible to consider these applications divorced from at least some geometric results.

K3 surfaces possess an integral invariant which is analogous to the genus of a curve; this is the smallest value of the number $\frac{1}{2}(D^2)$, where D is an irreducible curve on the surface with $(D^2) > 0$. We shall call it the class of the surface. It was proved in [1], Chapter VIII, that the divisor $2D$ is always very ample. Therefore all the surfaces with the same class define an open set S'_* in some Hilbert scheme.

In [2] it was proved that the scheme S' is smooth. On it we have a universal family $X' \to S'$ of K3 surfaces of a given class. Also in [2] we constructed a family $X'' \to S''$ of marked K3 surfaces of given class. Its base S'' is a complex manifold and an unramified covering of S', where the Galois group of the covering $f: S'' \to S'$ is isomorphic to the discrete group $G(\mathbf{Z})$, and a point of the fiber $f^{-1}(S')$ corresponds to all the ways of introducing the structure of a marked surface on the polarized K3 surface $X'_{S'}$. The period mapping defines a holomorphic mapping $\tau: S'' \to \Omega(l)$. In [2] we showed that the group $Pl(n)$ of projective transformations of the projective space $\mathbf{P}^n$ in which the surfaces $X'_{s'}, s' \in S'$, are imbedded acts in a natural way on S' and on S'', where τ is factored through the projection $p: S'' \to S''/Pl(n)$, and we have $\tau = hp$, and now Theorem 1 follows from the fact that $h: S''/Pl(n) \to \Omega(l)$ is an open immersion.

THEOREM 2. S'' *is connected.*

Consider a normal subgroup $\Gamma \subset G(\mathbf{Z})$ which is a congruence subgroup and contains no elements of finite order different from the identity. The variety $S = S''/\Gamma$ is a finitely sheeted covering of S' with Galois group $G(\mathbf{Z})/\Gamma$, and by a general theorem S is an algebraic variety. The period mapping defines a holomorphic map

$\pi: S \to \Omega(l)/\Gamma$. Since Γ acts without fixed points in $\Omega(l)$, we apply to π a theorem proved independently by Borel and Kobayshi [3], which says that π is a morphism of algebraic varieties. Together with Theorem 1 this yields that $\Omega(l) - \tau(S'')$ is contained in a proper complex submanifold, and hence $\tau(S'')$ is connected. From this it follows that the manifold $S''/Pl(n)$, homeomorphic to $\tau(S'')$, is connected, and hence S'' is also.

This result has two applications.

COROLLARY 1. *The manifold S' of all* K3 *surfaces of a given class is connected.*

This is obvious since S' is the image of S'' under f.

2 COROLLARY 2. *The image of the fundamental group $\pi(S', s'_0)$ in* $\operatorname{Aut} H^2(X'_{s'_0}, \mathbf{Z})$, *defined by the action on the cohomology of a fiber, coincides with* $G(\mathbf{Z})$.

In fact, if $H \subset G(\mathbf{Z})$ is the image of $\pi(S', s'_0)$ and s''_0 is a point lying over s'_0, then, covering all the mapping of the unit interval $f: I \to S'$, $f(0) = s'_0$, for covering mappings g with $g(0) = s''_0$ we obtain as $g(1)$ the points of some component of the manifold S'' in which only the points $H \cdot s''_0$ among those lying over s'_0 will be contained. Since the fiber over s'_0 has the form $G(\mathbf{Z})s'_0$, and S'' is connected, we have $G(\mathbf{Z}) = H$.

For example, arbitrarily deforming a nonsingular surface of degree 4 in $\mathbf{P}^3(\mathbf{C})$, as an automorphism of its cohomology we can take any automorphism of the corresponding integral quadratic form of signature (2.19).

We now proceed to a statement of the arithmetic applications.

3 THEOREM 3. *For a* K3 *surface over a finite field the Riemann hypothesis is true.*

The method, based on an application of Theorem 1, uses a lifting of a surface X defined over a finite field to a surface defined over a field of characteristic zero. This means that X is contained in the scheme $Y \to B$ as closed fiber, where the structure morphism is smooth, and the general fiber is a K3 surface over a field of characteristic 0. We shall assume that such a lifting exists, although it follows from the definition of a K3 surface that $H^2(X, \Theta) = 0$, and hence all the obstructions to lifting vanish. Let X be a smooth projective variety over $\mathbf{C}$, and let X^n denote the n-fold product of X with itself. The Hodge group of X will be an algebraic subgroup of the group of all automorphisms of $H^*(X, \mathbf{Q})$, distinguished by the condition of having trivial action on all the subspaces $H^{p,p} \cup H^{2p}(X^n, \mathbf{Q})$ (n and p are arbitrary). For abelian varieties the Hodge group was introduced and studied in [4]. We shall say that a smooth projective variety is a variety of CM-type if its Hodge group is commutative.

CONJECTURE. *Any variety of CM-type is defined over an algebraic number field, and its zeta-function is expressed by means of L-functions of one-dimensional characters of a finite extension of the field of definition of the variety.*

The excellent book of Shimura and Taniyama [5] contains a proof of this conjecture for abelian varieties. In this paper we prove it for K3 surfaces.

THEOREM 4. *A* K3 *surface of CM-type is defined over a number field, and its zeta-function is expressed by means of L-functions of one-dimensional characters of a finite extension of the field of definition.*

Theorem 3 was proved independently by P. Deligne, who kindly sent us a manuscript.** Since our proof is somewhat different from his (the use of the Torelli theorem introduces significant simplifications), we have resolved to introduce it here.

§2. Cohomology of fibers as Galois modules

Let X and S be irreducible algebraic varieties defined over a field k which is a finite extension of the field of rational numbers $\mathbf{Q}$, and let $f\colon X \to S$ be a smooth proper morphism over k. The Galois group $\mathrm{Gal}\,(\overline{k(S)}/k(S))$ of the algebraic closure $\overline{k(S)}$ of the field $k(S)$ acts on the l-adic étale cohomology $H^*(F, \mathbf{Q}_l)$ of the general fiber of the bundle $f\colon X \to S$ and so defines a homomorphism

$$\mathrm{Gal}\,(\overline{k(S)}/k(S)) \to \mathrm{Aut}\, H^*(F, \mathbf{Q}_l)$$

into the group of automorphisms of the cohomology algebra. We let $k(S)_l$ denote the subfield of $\overline{k(S)}$ corresponding to the kernel of this homomorphism. If we must emphasize its dependence on the bundle f, we will denote it by $k(S)_{f,l}$. Let k' be the extension of k obtained by adjoining all the l^nth roots of 1 to k. By the existence of a Verdier multiplication in the cohomology it follows that $k(S)_l$ contains a subfield over which k' has finite degree.

A bundle $f\colon X \to S$ is said to be *strictly nonconstant* if the algebraic closure of k in $k(S)_l$ has finite degree over its intersection with k'. Then we can find a finite extension of k such that k' will be algebraically closed in $(k(S))_l$.

LEMMA 1. *The bundle* $f\colon X \to S$ *is strictly nonconstant if and only if the group* $\mathrm{Gal}\,(K(S)_l/K'(S))$ *has finite index in* $\mathrm{Gal}\,(k(S)_l/k'(S))$ *for any extension* K *of the field* k.

Analogously to the field $k(S)_l$ we consider fields $k(S)_{l,n}$, for which the groups $\mathrm{Gal}\,(k(S)_{l,n}/k(S))$ are contained in $\mathrm{Aut}\, H^*(F, \mathbf{Z}/l^n\mathbf{Z})$. We have a commutative diagram

---

***Translator's note.* Deligne's paper has since appeared; see Invent. Math. 15 (1972), 206–226. Other important arithmetic results on K3 surfaces have been obtained by M. Artin and H. P. F. Swinnerton-Dyer; see Invent. Math. 20 (1973), 249–265.

$$\begin{array}{ccc} \mathrm{Gal}\,(K'(S)_{l,n}/K'(S)) & \xrightarrow{\ \varphi_n\ } & \mathrm{Gal}\,(k'(S)_{l,n}/k'(S)) \\ & \searrow \qquad \swarrow & \\ & \mathrm{Aut}\, H^*(F, \mathbf{Z}/(l^n)), & \end{array}$$

in which the slanting arrows are inclusions, and therefore the homomorphism φ_n will also be an inclusion. Its kernel corresponds to the subfield $k'(S)_{l,n} \cdot K'(S)$, and hence

$$K'(S)_{l,n} = k'(S)_{l,n} \cdot K'. \tag{1}$$

Suppose that the bundle $f\colon X \to S$ is strictly nonconstant and that k' is algebraically closed in $k'(S)_l$. Then $k'(S)_{l,n}$ is a regular extension of k'. Therefore, by the general properties of regular extensions,

$$[k'(S')_{l,n} \cdot K' : K'(S)] = [k'(S)_{l,n} : k'(S)].$$

Together with (1) this proves that φ_n is an isomorphism, and hence φ is also one. The converse is obtained by reversing these arguments.

In the applications we shall encounter, the field K will be the field of complex numbers $\mathbf{C}$. We consider an embedding of the field k over which S, X and f are defined into $\mathbf{C}$, and we shall assume that $k \subset \mathbf{C}$. We let $S_{\mathbf{C}}$ denote the variety of complex points of S. The fundamental group $\pi(S_{\mathbf{C}}, s_0)$ acts on the complex cohomology $H^*(F_{s_0}, \mathbf{Z})$ of a fiber of f. Here, of course, the image of elements of $\pi(S_{\mathbf{C}}, s_0)$ is contained in the group $\mathrm{Aut}\, H^*(F_{s_0}, \mathbf{Z})$.

§3. Kuga-Satake varieties

A construction of certain abelian varieties corresponding to K3 surfaces was made by M. Kuga and I. Satake. This construction goes as follows. Let X be a K3 surface and let $l \in H^2(X, \mathbf{Z})$ be the cohomology class corresponding to a hyperplane section.

We let $H^2(X, \mathbf{Z})^l = V \subset H^2(X, \mathbf{Z})$ denote the orthogonal complement to l. The signature of the intersection index form on V is $(2, 19)$.

Let $C^+ = C^+(V)$ be the even Clifford algebra associated with V. Now in $\mathrm{End}\,(C^+ \otimes \mathbf{R})$ we define a complex structure operator j.

Let $V_+ = (H^{2,0} + H^{0,2}) \cap V \otimes \mathbf{R}$ and let e_1, e_2 be any orthonormal basis of V. Put $e_+ = e_1 e_2$ and define a complex structure in V by

$$jx = e_+ x, \qquad x \in C^+. \tag{2}$$

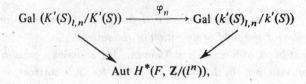

As is shown in [6], on $C^+ \otimes \mathbf{R}$ there exists a skew symmetric form $\alpha(x, y)$ which is integral on C^+ such that the form $\alpha(jx, y)$ is symmetric and positive definite. Hence $(C^+ \otimes \mathbf{R})/C^+$ is an abelian variety. Denote it by J_X. In [4] Mumford gave a construction for Hodge families of abelian varieties. From the results of [4], [7] it follows that on $\Omega(l)/\Gamma$ there is a Hodge family of abelian varieties corresponding to the spinor group; here Γ is any congruence subgroup of the spinor group. The canonical homomorphism of $\pi_1(\Omega(l)/\Gamma)$ into $\operatorname{Aut} H^1(J_X, \mathbf{Z})$ coincides with the natural inclusion of Γ into the spinor group. We know [4] that $\Omega(l)/\Gamma$ is defined over a finite extension of the field $\mathbf{Q}$, and the rational functions on $\Omega(l)/\Gamma$ possess the following property: at all the points corresponding to abelian varieties of CM-type their values belong to $\overline{\mathbf{Q}}$.

LEMMA 2. *The family of abelian varieties corresponding to* $\Omega(l)/\Gamma$ *is strictly nonconstant.*

PROOF. In [8] it was proved that for any abelian variety A a subgroup of finite index in the image of the Galois group in $\operatorname{Aut} H^1(A, \mathbf{Q}_l)$ is contained in the Hodge group of A. For the family defined above the Hodge group coincides with the spinor group. On the other hand, by construction of this family the fundamental group of the base is a congruence subgroup of the spinor group.

LEMMA 3. *If the Hodge group of the* (K3) *surface* X *is commutative, the corresponding abelian variety* J_X *is of CM-type.*

PROOF. From the construction of J_X we get the existence of an inclusion

$$V = H^2(X, \mathbf{Q})^l \to H^2(J_X, \mathbf{Q}) \tag{3}$$

which preserves the Hodge structure. From the definition of the Hodge group of any abelian variety A it follows that this group takes any subspace $V \subset H^k(A, \mathbf{Q})$ which satisfies the following property into itself:

$$V \otimes \mathbf{C} = \sum_{p+q=k} (V \otimes \mathbf{C}) \cap H^{p,q}. \tag{4}$$

From this and from (3) we get the existence of a homomorphism:

$$\tilde{\rho}:\ \operatorname{Hg}(J_X) \to \operatorname{Aut} H^2(X,\ \mathbf{Q})^l. \tag{5}$$

At a generic point this homomorphism coincides with the homomorphism of the spinor group into the orthogonal group. Hence its kernel is the group of order two. But this occurs for any X. Now suppose that the Hodge group of X is commutative. Hence the Hodge group of J_X is an extension of a commutative group by a group of order two. We know that the Hodge group of any abelian variety is reductive, from which it follows that it must be commutative if its quotient by a subgroup of order two is commutative. As is known from [4], the commutativity of the Hodge group of an abelian variety is equivalent to the fact that this variety is a variety of CM-type.

COROLLARY. *If* X *is a singular* K3 *surface, the abelian variety* J_X *is of CM-type.*

It suffices to verify that the Hodge group of X is commutative. This follows from the fact that Hg X must act trivially on $H^{1,1}(X, \mathbf{Q}) = S_X$. Hence it takes the orthogonal complement T_X of S_X into itself and the intersection index preserves the form. It remains to take into account that the group of automorphisms of a two-dimensional module equipped with a scalar product is commutative. We now give an example of a K3 surface of CM-type.

First we will prove the following property of a Hodge group.

LEMMA 4. *Let* X *be any smooth projective variety. The algebra* $\mathfrak{A}_X$ *of all endomorphisms of the additive group of* $H^*(X, \mathbf{Q})$ *which preserve the Hodge bigrading commutes with* Hg X.

PROOF. All such endomorphisms can be interpreted in a standard way as rational cohomology classes of type (n, n) on $X \times X$, where n is the dimension of X. By definition, the Hodge group preserves such classes, which gives the assertion of the lemma.

COROLLARY. *Let* X *be a* K3 *surface. If the algebra of endomorphisms of* $H^2(X, \mathbf{Q})$ *which preserve the Hodge bigrading has dimension* $\geqslant 22$ *and is semisimple, then* X *is a surface of* CM*-type.*

We give an example of K3 surfaces of CM-type. Let X be a CM-field, i.e. a field of the form $F(\sqrt{-\delta})$, where F is totally real and δ is totally positive, i.e. positive under any imbedding $F \to \mathbf{R}$. Now consider the triple (K, φ_1, λ), where φ_1 is an imbedding $F \to \mathbf{R}$ such that $\varphi_1(\lambda) > 0$ for $\lambda \in F$, and $\varphi(\lambda) < 0$ for the remaining imbeddings. For $\alpha, \beta \in K$ we define a scalar product in the following way:

$$(\alpha, \beta) = \operatorname{tr}(\alpha\lambda\beta^\sigma),$$

where σ is the automorphism of complex conjugation. We shall show that $(\alpha, \beta) = (\beta, \alpha)$ and that the quadratic form (α, α) has signature $(2, 2n-2)$, where $2n = [K:\mathbf{Q}]$. We have $\operatorname{tr} x = \operatorname{tr} x^\sigma$ for any $x \in K$. Hence $(\alpha, \beta) = \operatorname{tr} \alpha\lambda\beta^\sigma = \operatorname{tr} \beta\lambda\alpha^\sigma = (\beta, \alpha)$ ($\lambda \in F$ and hence $\lambda^\sigma = \lambda$).

To compute the signature we introduce suitable coordinates on $K \otimes \mathbf{R}$. Let $\psi_1, \cdots, \psi_{2n}$ denote a complete system of imbeddings $K \to \mathbf{C}$, numbered so that ψ_{2i-1} and ψ_{2i} will be conjugate. Put

$$x_{2i-1} = {}^1/_2\,(\psi_{2i-1}(\alpha) + \psi_{2i}(\alpha)),\quad x_{2i} = \frac{1}{2i}\,(\psi_{2i-1}(\alpha) - \psi_{2i}(\alpha)).$$

We obviously have

$$(\alpha, \alpha) = 2\sum \lambda_k\,(x_{2k-1}^2 + x_{2k}^2),$$

where $\lambda_k = \psi_{2k-1}(\lambda) = \psi_{2k}(\lambda)$. It remains to show that this form can be represented by a standard form–the intersection index on $H^2(X, \mathbf{Q})$. By Hasse's theorem [7] this is possible in every case if $2n \leqslant 16$, since the dimension of the standard module is 22.

We define a linear functional ω on the additive group of the field K as an imbedding of K in $\mathbf{C}$ that extends φ_1, and on the orthogonal complement of K in $H^2(X, \mathbf{Q})$ we set it equal to zero. It is obvious that in this way we determine a point in the domain $\Omega(l)$ for suitable l. Points thus obtained are dense in this domain. Therefore one of them corresponds to a surface of type K3 that will be a surface of CM-type.

§4. Zeta-functions of K3 surfaces

Our method of investigation is based on the construction of two bundles over the same base such that these bundles satisfy the conditions of Lemma 6 (below). One of these is the universal bundle $f\colon X \to S$ for K3 surfaces of a given class. The base was defined earlier. The other will be constructed beginning from the fibering into abelian varieties over the base $\Omega(l)/\Gamma$ introduced in the preceding section. An application of Lemma 6 allows us to carry the classical result about abelian varieties over to K3 surfaces.

On the variety $S = S''/\Gamma$ introduced in §1, we consider the family $f\colon X \to S$ obtained by pulling back the family $X' \to S'$ relative to the covering $S \to S'$. By Weil's criteiron the family f is defined over a finite extension of $\mathbf{Q}$.

LEMMA 5. *The bundle* $f\colon X \to S$ *is strictly nonconstant.*

We consider a homomorphism $\varphi\colon \pi(S_{\mathbf{C}}, s_0) \to \operatorname{Aut} H^2(F_{s_0}, \mathbf{Z})$, which is given by the action of the group $\pi(S_{\mathbf{C}}, s_0)$ on the cohomology of a fiber. The basis of the proof is the relation

$$\varphi \cdot \pi(S_{\mathbf{C}}, s_0) = \Gamma,$$

which immediately follows from Corollary 2 of §1.

We note that $H^*(F, \mathbf{Z}) = H^0(F, \mathbf{Z}) \oplus H^2(F, \mathbf{Z}) \oplus H^4(F, \mathbf{Z})$, and therefore the group $\operatorname{Aut} H^*(F, \mathbf{Z})$ contains $G(\mathbf{Z})$, where G is the orthogonal group of the quadratic form defined by multiplication of cohomology in the lattice $H^2(F, \mathbf{Z})^l$. We denote by G_l the completion of the group $\varphi \cdot \pi(S_{\mathbf{C}}, s_0)$ in the topology defined by congruence subgroups modulo l^n in the group $\operatorname{Aut} H^*(F_{s_0}, \mathbf{Z}) \supset G(\mathbf{Z})$. Since the approximation theorem holds for $G(\mathbf{Z})$, the completion of $G(\mathbf{Z})$ coincides with $G(\mathbf{Z}_l)$, and G_l is the closure of Γ in $G(\mathbf{Z}_l)$. By the hypothesis $(G(\mathbf{Z}_l)\colon G_l) < \infty$.

By the comparison theorem for étale cohomology the rings $H^*(F_{s_0}, \mathbf{Z}_l)$ and $H^*(F, \mathbf{Z}_l)$ are isomorphic, and the group G_l coincides with the image of the group $\operatorname{Gal}(\mathbf{C}(S)_l/\mathbf{C}(S))$. Since the image of the group $\operatorname{Gal} k(S)_l$ is obviously contained in $G(\mathbf{Z}_l)$, and since the image of $\operatorname{Gal}(\mathbf{C}(S)_l/\mathbf{C}(S))$ already has finite index in $G(\mathbf{Z}_l)$, we have

$$(\operatorname{Gal}(k(S)_l/k(S)) : \operatorname{Gal}(\mathbf{C}(S)_l\ \mathbf{C}(S))) < \infty.$$

Thus, a necessary condition for f to be strictly nonconstant has been obtained for $K = \mathbf{C}$. Since $\mathbf{C}$ can be taken as a universal domain in the sense of A. Weil, this condition holds for arbitrary K.

We now consider two bundles $f\colon X \to S$ and $f'\colon X' \to S$ over the same base. Having considered them over $\mathbf{C}$, we associate to each of them a "system of local coefficients" $H^*(F_S, \mathbf{Q})$, or, what amounts to the same, the sheaf $R^*f_*\mathbf{Q}$ on S. Suppose we are given a locally trivial subsheaf A of the sheaf $\operatorname{Aut} R^*f_*\mathbf{Q}$ and a homomorphism

$$\rho : A \to \operatorname{Aut} R^*f'_*\mathbf{Q}$$

of these sheaves, i.e. a system of homomorphisms

$$\rho_S : A_S \to \operatorname{Aut} H^*(F'_S, \mathbf{Q}),$$

compatible with the action of the fundamental group $\pi(S_\mathbf{C})$. By the same token, as we already saw, we have a homomorphism of groups

$$\rho_{\text{ét}} : A_\xi \to \operatorname{Aut} H^*(F', \mathbf{Q}_l),$$

where ξ is a generic point of the base S, and F' is the generic fiber of f'.

LEMMA 6. *If F and F' are strictly nonconstant bundles, then the homomorphism $\rho_{\text{ét}}$ commutes with the action of subgroups of finite index in the groups* $\operatorname{Gal}(k(S)_{f,l}/k'(S))$ *and* $\operatorname{Gal}(k(S)_{f',l}/k'(S))$.

In fact, by construction $\rho_{\text{ét}}$ commutes with the action of the group G_l, i.e. the group $\operatorname{Gal}(\mathbf{C}(S)_l/\mathbf{C}(S))$ (this group depends only on the base, and hence it is the same for both bundles f and f'). Since both bundles are strictly nonconstant, the group $\operatorname{Gal}(\mathbf{C}(S)_l/\mathbf{C}(S))$ is contained in the groups $\operatorname{Gal}(k(S)_{f,l}/k'(S))$ and $\operatorname{Gal}(k(S)_{f',l}/k'(S))$ as a subgroup of finite index. The assertion of the lemma now follows.

To construct the second bundle we make use of the period mapping. It is easy to see that this defines a holomorphic map $\pi\colon S \to \Omega/\Gamma$. On the other hand, recall that S and Ω/Γ have the structure of an algebraic variety defined over a finite extension of the field of rational numbers.

LEMMA 7. *The holomorphic mapping π coincides with the corresponding morphism of algebraic varieties, defined over a finite extension of the field of rational numbers.*

The fact that π is a morphism of varieties is true of any holomorphic mapping of the algebraic variety S into Ω/Γ. This is the content of the theorem of Borel and Kobayshi [3].

It remains to show that π is defined over a finite extension of $\mathbf{Q}$, or, what is the same, over $\overline{\mathbf{Q}}$. For this we recall that the group G of projective transformations of the space in which all the fibers of family f are imbedded acts on S. We consider an open set $U \subset S$ such that the quotient U/G exists. We must prove that the morphism π: $U \to \Omega/\Gamma$ is defined over $\overline{\mathbf{Q}}$, and that it can be represented as $\pi = h \cdot p$, where p: $U \to U/G$ is the natural projection, and h: $U/G \to \Omega/\Gamma$ is a morphism. It suffices to prove that h is defined over $\overline{\mathbf{Q}}$.

By the Torelli theorem proved in [2], h is an open immersion. We shall prove that $h^*(r) \in \overline{\mathbf{Q}}(U/G)$ for a function $r \in \overline{\mathbf{Q}}(\Omega/\Gamma)$. For this we use the fact that there are many points $x \in \Omega/\Gamma$ for which it is known that $r(x) \in \overline{\mathbf{Q}}$ for all $r \in \overline{\mathbf{Q}}(\Omega/\Gamma)$. These are the points corresponding to singular surfaces of CM-type (see the Corollary to Lemma 4 in §3). In particular, they are points corresponding to singular K3 surfaces. The point $h^{-1}(x)$ is defined over $\overline{\mathbf{Q}}$ if the point x corresponds to a singular K3 surface. This will be proved by a lemma. Thus the points $h^{-1}(x)$ corresponding to singular surfaces form an everywhere dense set in U/G, and for $r \in \overline{\mathbf{Q}}(\Omega/\Gamma)$ the function $h^*(r)$ takes algebraic values on this set. Then $h^*(r) \in \overline{\mathbf{Q}}(U/G)$ by Lemma 9.

LEMMA 8. *If a point $x \in \Omega/\Gamma$ corresponds to a singular* K3 *surface, then the point $h^{-1}(x) \in U/G$ is defined over a finite extension of the field* $\mathbf{Q}$.

It suffices to prove that the point $h^{-1}(x)$ is invariant relative to a subgroup of finite index in the group of all automorphisms of $\mathbf{C}$. Since $h^{-1}(x)$ is the image of a G-orbit on U, our assertion will be proved if we verify that this orbit is left stable by a subgroup of finite index of Aut $\mathbf{C}$. This assertion is equivalent to the fact that if X is the polarized K3 surface corresponding to the point x (i.e. X is singular), then among the polarized surfaces X^σ, $\sigma \in \text{Aut}\,\mathbf{C}$, there are only a finite number of nonisomorphic ones.

The last assertion follows from the classification of singular surfaces obtained in [2]. In fact, the Néron-Severi lattices S_X and S_{X^σ} of the surfaces X and X^σ are isomorphic. Therefore the orthogonal complements T_X and T_{X^σ} of these lattices in the homology group have the same discriminant, equal to the discriminant of S_X. But there exist only a finite number of nonisomorphic lattices with given discriminant, and the lattice T_X, as proved in [2] (the theorem in §8), uniquely determines the surface X up to isomorphism.

LEMMA 9. *Let M be an irreducible algebraic variety defined over a finite extension k of the field $\mathbf{Q}$. If $f \in \mathbf{C}(M)$ is such that for some everywhere dense set $Z \subset M_{\overline{\mathbf{Q}}}$ the values of f belong to $\overline{\mathbf{Q}}$, then $f \in \overline{\mathbf{Q}}(M)$.*

PROOF. Since M is defined over a finite extension k of $\mathbf{Q}$, it follows that there exist functions $f_0, f_1, \cdots, f_n$, $n = \dim M$, in $k(M)$ and an integer $m \geqslant 0$ such that any $f \in \mathbf{C}(M)$ can be represented uniquely in the form

$$f = \frac{1}{q} \sum_{k=0}^{m} p_k(f_1, \ldots, f_n) f_0^m,$$

where $p_k(f_1, \cdots, f_n)$, $k = 0, \cdots, m$, and $q(f_1, \cdots, f_n)$ are polynomials of $f_1, \cdots, f_n$ with complex coefficients; $(p_0, \cdots, p_m, q) = 1$.

Obviously, if $f = 0$ on an everywhere dense set of points, then $f \equiv 0$. By hypothesis, the infinite system of linear equations relative to the coefficients of q and p_k, $k = 0, \cdots, m$, of the form

$$q(f_1(x), \ldots, f_n(x)) f(x) = \sum p_k(f_1(x), \ldots, f_n(x)) f_0^k(x), \quad x \in M,$$

has a nonzero solution. This system is linear, and hence, if it has a nonzero solution in $\mathbf{C}$, it must have a nonzero solution in the field generated by the coefficients. From this it follows that it must have a solution in some finite extension of $\mathbf{Q}$. Thus there exists a function

$$\tilde{f} = \frac{1}{\tilde{q}} \sum \tilde{p}_k(f_1, \ldots, f_k) f_0^m$$

such that $\tilde{q}, \tilde{p}_k \in \overline{\mathbf{Q}}[T_1, \cdots, T_n]$ and $\tilde{f}(x) = f(x)$ for $x \in \mathbf{Z}$. Then $\tilde{f} = f$, which proves the lemma.

We now consider the family of abelian varieties A defined over $\Omega(l)/\Gamma$ in the introduction to the preceding section. We let X denote its pullback relative to the morphism $f\colon S \to \Omega(l)/\Gamma$. From what was proved earlier it follows that X' is defined over a finite extension of $\mathbf{Q}$ and it is strictly constant. The homomorphism $\tilde{\rho}$ introduced in the preceding section defines a homomorphism $\rho\colon \mathrm{Hg}\,(A_\xi)(\mathbf{Q}) \to \mathrm{Aut}\, H^*(F_\xi, \mathbf{Q})$, where A_ξ and F_ξ are fibers of the bundles A and F.

We verify that ρ satisfies the conditions of Lemma 2, i.e. that it commutes with the action of the group $\pi(S_{\mathbf{C}}, s_0)$. The morphism $S \to \Omega(l)/\Gamma$ defines an epimorphism $\pi(S, s_0) \to \Gamma$, and the group $\pi(S, s_0)$ acts on the cohomology of the fibers of the bundle X via this homomorphism. For the bundle X' this is true by definition. That the homomorphism ρ commutes with the action of the group Γ is obvious from the invariant character of its definition.

Thus we have verified that the bundles X and X' constructed in this way satisfy the conditions of Lemma 2, and we can apply this lemma. We obtain a subgroup G of finite index in $\mathrm{Gal}\,(k(S)_l/k'(S))$, whose actions on $\mathrm{Aut}\, H^*(F_S, \mathbf{Q}_l)$ and $\mathrm{Hg}\,(A_S)(\mathbf{Q}_l)$ are connected by the homomorphism ρ.

We shall prove that the group $\mathrm{Gal}\,(k(S)_l/k(S))$ contains a subgroup G of finite index possessing the same property.

The diagram

$$\begin{array}{ccc} & & \mathrm{Hg}\,(A_S)(\mathbf{Q}_l) \\ & \overset{\varphi}{\nearrow} & \\ \mathrm{Gal}\,(k(s)_l/k(s)) & & \downarrow \rho \\ & \underset{\Psi}{\searrow} & \\ & & \mathrm{Aut}\, H^*(F_S, \mathbf{Q}_l), \end{array}$$

where s is a generic point of the variety S, is commutative under a restriction on the subgroup Γ. On the other hand, passing to determinants, we obtain an analogous commutative diagram for the group $\mathrm{Gal}\,(k'(s)/k(s))$. From this it follows that the elements $\Psi(g)^{-1}(\rho\varphi)(g)$ commute with $\Psi(G)$. Since by hypothesis $\Psi(G)$ has finite index in the group of unimodular automorphisms of the algebra $H^*(F_S, Q_l)$, and the center of the latter group is finite, it follows that the function $\Psi(g)^{-1}(\rho\varphi)(g)$ is equal to 1 on some neighborhood of 1, i.e. on the group $\widetilde{G}$.

Let $s_0 \in S$ be a point defined over a finite extension of the field $\mathbf{Q}$. In any finite Galois extension $k(s)_l \supset L \supset k'(s)$, $[L: k'(S)] < \infty$, we consider the normalization S_L of the variety S. The inverse images of the point s_0 in S_L are represented by the Galois group of $L/k'(s)$ and define a decomposition group in the residue class field. Considering the union of all such fields J', we obtain a residue class field $k(s_0)_l$ and a Galois group isomorphic to the decomposition group

$$\mathrm{Gal}\,(k\,(s_0)_l/k'\,(s_0)) \subset \mathrm{Gal}\,(k\,(s)_l/k'\,(s)). \tag{6}$$

From the functorial character of the cohomology groups and from the inclusion (6) we get the existence of a commutative diagram

$$\begin{array}{ccc} & \nearrow & \mathrm{Hg}\,(A_{s_0})(Q_l) \\ \mathrm{Gal}\,(k(s_0)_l/k'(s_0)) & & \downarrow \rho \\ & \searrow & \mathrm{Aut}\, H^*(F_{s_0}, Q_l). \end{array}$$

In our case

$$\mathrm{Hg}\,(A_{s_0})\,(Q_l) \subset \mathrm{Aut}\, \Lambda^2 H^1\,(F'_{s_0}, Q_l)$$

and ρ preserves the eigenvalues of the Frobenius automorphism.

Since A_{s_0} is an abelian variety, Theorem 3 will follow from this in view of the well-known results of Weil.

If F_{s_0} is a surface of CM-type, then, by Lemma 3, A_{s_0} is an abelian variety of CM-type. This implies Theorem 4 in view of the results of Shimura and Taniyama [5].

BIBLIOGRAPHY

1. I. R. Šafarevič, et al., *Algebraic surfaces,* Trudy Mat. Inst. Steklov. 75 (1965) = Proc. Steklov Inst. Math. 75 (1965). MR 32 #7557; 35 #6685.

2. I. I. Pjateckiĭ-Šapiro and I. R. Šafarevič, *A Torelli theorem for algebraic surfaces of type* K_3, Izv. Akad. Nauk SSSR Ser. Mat. 35 (1971), 530–572 = Math. USSR Izv. 5 (1971), 547–588. MR 44 #1666.

3. S. Kobayashi, Preprint.

4. D. Mumford, *A note on Shimura's paper "Discontinuous groups and Abelian varieties",* Math. Ann. 181 (1969), 345–351. MR 40 #1400.

5. G. Shimura and Y. Taniyama, *Complex multiplication of abelian varieties and its applications to number theory,* Publ. Math. Soc. Japan, 6, Math. Soc. Japan, Tokyo, 1961. MR 23 #A2419.

6. M. Kuga and I. Satake, *Abelian varieties attached to polarized K_3-surfaces,* Math. Ann. 169 (1967), 239–242. MR 35 #1603.

7. Z. I. Borevič and I. R. Šafarevič, *Number theory,* "Nauka", Moscow, 1964; English transl., Pure and Appl. Math., vol. 20, Academic Press, New York, 1966. MR 30 #1080; 33 #4001.

8. I. I. Pjateckiĭ-Šapiro, *Interrelations between the Tate and Hodge conjectures for abelian varieties,* Mat. Sb. 85 (127) (1971), 610–620 = Math. USSR Sb. 14 (1971), 615–625. MR 45 #3422.

Translated by J. S. JOEL

Über einige Tendenzen in der Entwicklung der Mathematik

Jahrb. Akad. Wiss. Göttingen, 31–36 (1973)
[Math. Intell. 3, 182–184 (1981)]

Von I. R. Schafarevitsch

Träger des Dannie-Heineman-Preises 1973

Aus dem Englischen übersetzt von C. L. Siegel

Jedes Geschöpf ist geneigt, seine Umgebung als etwas zweifellos Vertrautes anzusehen, das keine Fragen hervorruft. Das ist die Haltung eines Mathematikers gegenüber seiner Wissenschaft. Nur wenn der Zufall ihm eine Gelegenheit bietet, seinen Gegenstand unter einem anderen Winkel zu betrachten, dann bemerkt er, wie unglaublich fremd das Phänomen ist, mit dem er sein Leben lang zu tun hatte. Ihre höchst dankenswerte Einladung gibt mir eine solche Gelegenheit, wobei ich ein paar Worte über Mathematik auch zu solchen meiner Kollegen sagen möchte, deren wissenschaftliches Arbeitsgebiet der Mathematik fern liegt.

Bei oberflächlichem Blick auf die Mathematik mag man den Eindruck haben, sie sei das Ergebnis der getrennten persönlichen Bemühungen von vielen Gelehrten, die über Länder und Zeiten verstreut waren. Jedoch die innere Logik der mathematischen Entwicklung erinnert einen viel mehr an das Werk eines einzigen Intellekts, der seinen Gedanken systematisch und beständig entwickelt und dabei die Verschiedenheit menschlicher Individualitäten nur als Mittel benutzt. Gleichwie bei der Aufführung einer Symphonie durch ein Orchester das Thema von einem Instrument zum anderen hinübergeht, und wenn einer der Mitwirkenden mit seinem Teil zu Ende ist, so folgt ein anderer mit fehlerfreier Genauigkeit.

Dies ist keinesfalls ein bloß rednerischer Vergleich. Die Geschichte der Mathematik weiß von vielen Fällen, in denen eine Entdeckung unbekannt bleibt und dann später ein zweites Mal von anderer Seite mit erstaunlicher Gleichartigkeit gemacht wird. Am Vorabend seines tödlichen Duells berichtete Galois in einem Briefe über einige grundlegende Sätze in der algebraischen Funktionentheorie. Mehr als 20 Jahre später wurden genau die gleichen Sätze von Riemann ohne Kenntnis dieses Briefes erneut entdeckt und bewiesen. Ein anderes Beispiel bildet die Begründung der nichteuklidischen Geometrie, die Lobatschewski und Bolyai unabhängig voneinander gelegt hatten, wonach sich dann herausstellte, daß schon 10 Jahre vorher Gauß und Schweikart ebenfalls unabhängig voneinander zu denselben Ergebnissen gekommen waren. Man wird von einem eigentümlichen Gefühl ergriffen, wenn man in den Arbeiten von 4 Gelehrten dieselben Figuren wie von einer Hand gezeichnet wiederfindet, obwohl die Verfasser ohne Verbindung untereinander waren.

Man wird von der Idee gepackt, daß solch eine wunderbar rätselhafte und geheimnisvolle menschliche Tätigkeit, die seit Jahrtausenden andauert, nicht auf bloßem Zufall beruhen kann, sondern irgendein Ziel haben muß. Wenn wir aber dies erkannt haben, so stehen wir vor der unvermeidlichen Frage: *Was ist dieses Ziel?*

Wie kann eine Wissenschaft als Ganzes, nicht bloß ein Teil davon, und nicht nur in einer Periode ihrer Entwicklung, ein einziges Ziel haben? Wir wollen versuchen, dies am Beispiel der Physik zu untersuchen, die mit der Mathematik stets so eng verbunden gewesen ist. Zu Newtons Zeiten wurde die Physik von einem Ziel angezogen, nämlich dem Aufbau eines sogenannten „Systems" des Universums, so daß man aus einigen einfachen Gesetzen die ganze Mannigfaltigkeit der physikalischen Welt logisch herleiten könnte. Lange schien es, daß Newton dieses Problem in der Hauptsache gelöst hätte und seine Nachfolger nur noch prüfen müßten, wie alle bekannten Naturerscheinungen durch sein System beschrieben würden. Nur am Rande der Physik waren die elektrischen Vorgänge, die nicht in jenes Schema paßten. Aber im 19. Jahrhundert rückten dann gerade die elektromagnetischen Erscheinungen in den Mittelpunkt der Physik. Obwohl dies in gewisser Weise die Newtonschen Begriffe erschütterte, so kam zu ihrer Unterstützung die Hoffnung auf, daß Newtons Mechanik in Verbindung mit Maxwells Theorie des elektromagnetischen Feldes zu einem vollständigen und endgültigen System des Universums führen könnte. Aber diese Erwartung erfüllte sich nicht, denn bald zerstörten Quantenmechanik und Relativitätstheorie alle alten Vorstellungen. Eine Zeitlang waren die Physiker von dem Wunsche angetrieben, aus der allgemeinen Relativitätstheorie oder der relativistischen Quantenmechanik eine umfassende Theorie der Elementarteilchen und ein neues System des Universums zu gewinnen. Das ist aber bis heute nicht gelungen, und es gibt wohl nicht mehr viele Physiker, die an die Realität solcher Hoffnungen glauben. Wenn überhaupt nach so vielen Änderungen wieder eine gewisse Ordnung im physikalischen Bilde des Weltalls eintreten würde, so wäre es schwierig, an die Endgültigkeit eines solchen Systems zu glauben.

Wenn wir wieder zur Mathematik zurückkommen, so müssen wir gestehen: Dieses allumfassende Ziel, das von der Physik in ihrem Ehrgeiz mehrfach aufgestellt wurde, allerdings ohne Erfolg, dieses Ziel ist in unserer Wissenschaft überhaupt noch nicht angestrebt worden. Welchen Einfluß hat das auf ihre Entwicklung?

Die Mathematik wächst rasch und ohne Unterbrechung, wobei sie keine Krisen oder Reformen wie die Physik kennt und uns mit neuen Ideen und konkreten Tatsachen bereichert. Ich bin fest überzeugt, daß die Ergebnisse der gegenwärtigen Mathematik nicht weniger vollkommen sind als die Leistungen der Klassiker des 19., 18., 17. Jahrhunderts, und daß sie sogar den Vergleich mit den Schöpfungen des hellenischen Geistes aushalten. Andererseits werden die klassischen Leistungen im Prinzip auch wiederum nicht von den besten heutigen übertroffen. Was ist nun der Wert einer unbegrenzten Anhäufung von Ideen, die im Grunde tatsächlich von gleicher Tiefe sind? Wird damit nicht die Mathematik eine besonders schöne Variante der „schlechten Unendlichkeit“ Hegels?

Wenn eine Tätigkeit kein Ziel hat, so verliert sie dadurch ihren Sinn. Vergleicht man die Mathematik mit einem Lebewesen, so erscheint sie nicht in irgendeiner sinnvollen und beabsichtigten Tätigkeit, sondern es sind eher instinktive Handlungen mit stereotyper Wiederholung, bis dann ein äußerer oder innerer Antrieb erfolgt.

Ohne ein bestimmtes Ziel kann die Mathematik keine Idee von ihrer eigenen Gestalt entwickeln. Was einzig und allein als Ideal übrigbliebe, wäre ein ungeregeltes Wachstum oder, besser gesagt, eine Ausdehnung nach allen Richtungen. Um einen anderen Vergleich zu bilden, so kann man sagen, daß die Entwicklung der Mathematik sich von dem Anwachsen eines Lebewesens unterscheidet, welches nämlich dabei die durch seine Begrenzung gegebene Form bewahrt. Jene Entwicklung ähnelt vielmehr dem Wachstum eines Kristalls oder der Diffusion eines Gases, das sich solange frei ausdehnt, bis ein äußeres Hindernis entgegentritt.

Eine derartige Entwicklung einer Wissenschaft widerspricht offensichtlich dem Empfinden sinnvoller Schönheit, wie es in uns unvermeidlich bei der Begegnung mit Mathematik entsteht; denn auch eine schöne Symphonie ist in ewiger Dauer unmöglich.

Tritt denn aber nur in unserer Wissenschaft ein solches Problem auf? Ich glaube nicht, daß die Mathematik von anderen Arten geistiger Tätigkeit wesentlich verschieden ist. Jedoch sind ihre Gegenstände abstrakter, und sie abstrahiert aus einer größeren Anzahl zufälliger Eigenschaften. Daher kann man in der Mathematik deutlich gewisse

Vorgänge allgemeiner Natur unterscheiden, die in anderen Gebieten nur unbestimmt erkennbar sind. Nun meine ich insbesondere, daß der bereits erwähnte Mangel an Zielen und Gestalt fast das gesamte Leben der heutigen Menschheit betrifft. So sehen wir neben der ziellos fortschreitenden Mathematik das Beispiel der Physik, die einem anscheinend unerreichbaren Ziel zustrebte und nun aber ebenfalls die Idee eines Ziels verloren hat.

Ein fieberartiger Drang nach Tätigkeit, ohne jede Gestalt, Ziel oder Sinn, abgesehen von grenzenloser Ausdehnung, hat das Menschengeschlecht in den letzten Jahrhunderten ergriffen. Das wurde „Fortschritt" genannt und ist seit einer gewissen Zeit eine Art Ersatz für Religion geworden. Die heutige industrielle Gesellschaft ist das letzte Ergebnis davon. Man hat wiederholt darauf hingewiesen, daß dieses Rennen als Lebensform einen inneren Widerspruch enthält, da es zu unheilvollen materiellen Folgen führt, zu einer für den Menschen unerträglichen Zunahme von Änderungen der Lebensweise, zur Übervölkerung und zur Zerstörung der Umwelt. Ich möchte am Beispiel der Mathematik auf die nicht weniger verderblichen Folgen für geistige Tätigkeit hinweisen, welche sinnlos wird, wenn ihr das endgültige Ziel fehlt.

Die Gefahr ist hier nicht bloß negativer Natur, sie besteht nicht nur darin, daß weder die unermüdliche Anstrengung der Menschheit noch das Leben ihrer begabtesten Vertreter durch das Verständnis einer inneren Bedeutung erleuchtet wird. Ferner besteht nur ein Teil der Gefahr in der Tatsache, daß wir ohne Kenntnis eines Zieles unserer Handlungen auch nicht ihre Ergebnisse vorhersagen können. Die geistige Beschaffenheit der Menschheit gestattet bei längerer Zeitdauer keine Verknüpfung mit einer Tätigkeit, deren Ziel und Bedeutung nicht angegeben wird. Dann tritt, wie bei vielen anderen Vorgängen, ein Mechanismus in Kraft, indem nämlich die Leute zu einem Ersatz greifen, wenn sie in dem ihnen Angebotenen nicht das Benötigte gefunden haben. Ein solches Beispiel ist jedem wohlbekannt: Als die Menschen mit dem Gott der Nächstenliebe gebrochen hatten, schufen sie sofort neue Götter, die dann Millionen an Menschenopfern verlangten. Nach derselben Regel wird die Menschheit versuchen, wenn ihre geistige Tätigkeit keinen Aufschluß über das Ziel liefert, diesen Aufschluß aus anderen Quellen zu erlangen. So kann z. B. ein Mathematiker den Sinn seiner Arbeit in der Erfüllung eines staatlichen Auftrages erblicken, wobei er mit der Berechnung einer Raketenbahn oder einem Abhörapparat zu tun hat, und wenn dies ein Gelehrter mit weitem Wirkungskreis wäre, so würde er eine ganze Gesellschaft aus Kreuzungen von Menschen und Rechenmaschinen planen. Solch eine Haltung ist nicht allein seelische Erniedrigung eines Gelehrten, sondern es geraten dadurch in die Mathematik auch Gebiete ohne jene göttliche Schönheit, welche die Kenner unserer Wissenschaft begeistert.

Mehr als zweitausend Jahre Geschichte haben uns davon überzeugt, daß die Mathematik aus sich allein kein eigentliches Ziel zur Steuerung ihres Fortschreitens angeben kann. Daher muß sie dieses Ziel von außen empfangen. Selbstverständlich bin ich weit von dem Versuche entfernt, eine Lösung für ein Problem aufzuweisen, das nicht nur das innere Problem der Mathematik ist, sondern überhaupt der ganzen Menschheit. Ich möchte nur die hauptsächlichen Richtungen bei einer Suche nach dieser Lösung angeben.

Es scheinen da zwei Richtungen möglich zu sein. In der ersten würde man versuchen, das Ziel der Mathematik aus ihren praktischen Anwendungen zu entnehmen. Aber es ist schwer zu glauben, daß eine höhere geistige Tätigkeit ihre Berechtigung in einer niederen materiellen finden könnte. In dem 1945 entdeckten apokryphen Thomas-Evangelium sagt Jesus ironisch:

Wenn das Fleisch wegen des Geistes entstanden ist, ist es ein Wunder; wenn aber der Geist wegen des Leibes — ein Wunder der Wunder.

Die gesamte Geschichte der Mathematik beweist überzeugend, daß ein solches Wunder von Wundern unmöglich ist. Blicken wir auf den entscheidenden Augenblick in der Entwicklung der Mathematik, den Augenblick, in dem sie ihren ersten für die Menschheit wichtigen Schritt tat und logische Beweisführung als Grundlage entstand, so sehen wir, daß der darin verwendete Stoff tatsächlich jede Möglichkeit praktischer Anwendung ausschloß.

Die ersten Sätze des Thales von Milet bewiesen Aussagen, die für jeden vernünftigen Menschen evident erschienen, daß z.B. der Kreis durch einen Durchmesser in zwei gleiche Teile zerlegt wird. Ein Genius war nötig, nicht um von der Richtigkeit solcher Aussagen überzeugt zu werden, sondern um die Notwendigkeit eines Beweises zu verstehen. Offenbar ist der praktische Wert solcher Entdeckungen gleich Null.

Ebenso sind in unserer Zeit trotz verschiedenartiger und tiefliegender Anwendungen der Mathematik hierdurch nicht ihre hervorragendsten Leistungen hervorgerufen worden. Wie kann man dann also von den Anwendungen erwarten, sie könnten der Mathematik ein Ziel aufweisen, das diese selbst nicht mit innerer Kraft finden konnte?

Wenn wir daher diesen Weg verwerfen, so bleibt meines Erachtens nur eine Möglichkeit: Das Ziel kann der Mathematik nicht durch eine niedriger stehende Art menschlicher Bestrebung gegeben werden, sondern eine höher stehende, durch die Religion.

Nun ist es natürlich sehr schwierig sich vorzustellen, wie das geschehen könnte. Aber es ist sogar noch schwieriger sich vorzustellen, wie die Mathematik imstande sein könnte, sich ins Unendliche zu entwickeln, ohne Wissen um einen tieferen Sinn. Weshalb? Bereits in der nächsten Generation wird sie im Strom der Veröffentlichungen untergehen, und das ist nur eine einfache äußere Gefahr.

Andererseits zeigt die Geschichte, daß im Prinzip jene Lösung möglich ist. Gehen wir noch einmal in die Zeit des Entstehens der Mathematik zurück, so sehen wir, daß sie damals ihr Ziel kannte und auf dem bereits genannten Wege gefunden hatte. Die Mathematik entstand als Wissenschaft während des 6. vorchristlichen Jahrhunderts in der religiösen Gemeinschaft der Pythagoreer und war ein Teil ihrer Religion. Sie hatte ein klares Ziel, nämlich die Vereinigung mit dem Göttlichen durch das Verständnis der kosmischen Harmonie, die sich in der Harmonie von Zahlen ausdrückt. Gerade dieses höchste Ziel war die Quelle der Kraft zu einer wissenschaftlichen Glanzleistung, die ganz allein steht: nicht die Entdeckung eines schönen Theorems, nicht die Schöpfung eines neuen Teiles der Mathematik, sondern die Schöpfung der Mathematik als solcher.

Damals, fast im Augenblick des Entstehens, zeigte sich die Besonderheit der Mathematik, daß sie die gemeinsamen Tendenzen im gesamten menschlichen Geistesleben deutlich machte. Gerade aus diesem Grunde diente dann die Mathematik als Modell bei der Ausbildung von Grundlagen für die deduktiven Wissenschaften.

Zum Schluß möchte ich der Hoffnung Ausdruck geben, daß aus demselben Grunde die Mathematik jetzt als Modell dienen möchte zur Lösung des Hauptproblems unserer Epoche:

Ein oberstes religiöses Ziel und den Sinn geistigen Schaffens der Menschheit zu erkennen.

Inseparable morphisms of algebraic surfaces
(with A. N. Rudakov)

Izv. Akad. Nauk SSSR, Ser. Mat. **40,** 1269–1307 (1976). Zbl. **365,** 14008
[Math. USSR, Izv. **10,** No. 6, 1205–1237 (1976)]

Abstract. We prove that no regular vector field exists on an algebraic $K3$ surface defined over an algebraically closed field of finite characteristic.
Bibliography: 19 titles.

In this paper, we consider algebraic varieties over an algebraically closed field k of finite characteristic p, and we consider the finite inseparable morphisms of degree p of such varieties. Such morphisms, from a variety X, correspond to rational vector fields D on X which satisfy the condition $D^p = fD$, where f is a rational function (such fields will be called *p-closed fields*). These morphisms can be described as the quotient morphisms by a p-closed vector field.

In §1, we introduce some general facts concerning vector fields: if the variety is smooth and the p-closed vector field has only divisorial singularities, then the quotient variety is also smooth (Theorem 1). In this case, the quotient morphism can be described locally as taking the pth root of some local parameter.

In §2, we derive a formula which connects the canonical class of a variety, the divisor of a p-closed vector field and the canonical class of the quotient by this vector field. Using this formula, we prove a converse of Theorem 1 for algebraic surfaces: if the quotient of a smooth surface by a p-closed vector field is smooth, then the vector field has only divisorial singularities (Theorem 4). If there is a nonzero regular vector field on a projective surface, there also exists a p-closed vector field. From the results we have mentioned, it is not difficult to show that the quotient of a $K3$ surface by a p-closed vector field must be a rational surface (Theorem 5). It follows that if there exists a nonzero regular vector field on a $K3$ surface, then the surface is unirational and has a pencil of elliptic or quasi-elliptic curves.

We then investigate the regular vector fields on surfaces with a pencil of curves. In particular, we give a complete list of all elliptic surfaces which have no multiple fibers and which admit regular vector fields (with one additional restriction for $p = 2$) (Theorem 6). Together with the results of §2, this implies that no regular vector fields exist on a $K3$

AMS (MOS) subject classifications (1970). Primary 14J10, 14A10, 14J15; Secondary 14C20, 14N10, 14B05, 14F05, 14L15, 14H20, 14G13, 14B20, 14F10, 14F15, 14H45, 14E30, 14E05, 14J25.

**Editor's note.* The present translation incorporates the corrections published in Izv. Akad. Nauk SSSR Ser. Mat. 41 (1977), 476, and also further corrections furnished by the authors, by B. Saint-Donat, and by M. Reid.

surface. From this it follows, of course, that for such surfaces there exists a formal moduli space which has, as in the characteristic-zero case, dimension equal to 20.

The authors wish to express their gratitude to I. V. Dolgačev for calling their attention to [8] and [9], and to V. Danilov, M. Reid, and A. N. Tjurin for helpful suggestions.

§1. Inseparable morphisms and vector fields

If K/L is an inseparable extension of degree p (where p is the characteristic), then the module of derivations of K/L is one dimensional; i.e., there exists a derivation D of K, such that $L = K^D = \{\alpha \in K; D(\alpha) = 0\}$, and D is uniquely determined up to a nonzero scalar multiple from K. Moreover, $D^p = \alpha D$, where $\alpha \in K$. Derivations having this property will be called *p-closed derivations*. Conversely, any p-closed derivation defines a subfield $L = K^D$ over which K is an inseparable extension of degree p.

Let X be an irreducible algebraic variety defined over an algebraically closed field K, and let D be a rational vector field on X, i.e. a derivation of the field $\mathrm{k}(X)/\mathrm{k}$. The usual construction defines a variety X^D and a morphism $\pi_D\colon X \longrightarrow X^D$. (If $X = \bigcup U_i$, where $U_i = \operatorname{spec}(A_i)$ is an affine open set, then $X^D = \bigcup \operatorname{spec}(A_i^D)$.) The following properties are well known:

1) π_D is a finite purely inseparable morphism.
2) If X is normal, so is X^D.
3) If D is p-closed, then $\deg \pi_D = p$.

Conversely, if $\pi\colon X \longrightarrow Y$ is a finite inseparable morphism of degree p, and if X and Y are normal, then there exists a rational p-closed vector field D such that $\pi = \pi_D$, $Y = X^D$, and D is uniquely defined up to a nonzero scalar multiple from $\mathrm{k}(X)$. Indeed, D is defined by the condition that $\mathrm{k}(Y) = \mathrm{k}(X)^D$. It follows that there exists a finite birational morphism $\varphi\colon X^D \longrightarrow Y$ which, in view of the fact that Y is normal, must be an isomorphism.

In what follows, we will usually denote a vector field on X by D if it is to be thought of as a derivation of $\mathrm{k}(X)$ and by θ if it is to be considered as a section of the tangent bundle. The vector corresponding to the point $x \in X$ will be denoted in the first case by D_x and in the second case by $\theta(x)$.

The existence of p-closed vector fields can sometimes be deduced from the following observation:

LEMMA 1. *If X is a complete variety on which a nonzero regular vector field exists, then a p-closed regular vector field exists on X, also.*

Indeed, letting Θ_X denote the tangent bundle, we have in this case that the space $H^0(X, \Theta_X)$ is a finite-dimensional Lie p-algebra over k. For any $\theta \in H^0(X, \Theta_X)$ such that $\theta \neq 0$, the elements θ^{p^i}, $i = 1, 2, \ldots$, generate a finite-dimensional abelian p-subalgebra A. The pth power operation in A is a p-linear transformation, which, as is well known, can be diagonalized. Therefore there exists a $D \in H^0(X, \Theta_X)$ with $D \neq 0$ such that $D^p = \lambda D$, where $\lambda \in \mathrm{k}$.

COROLLARY. *If a complete algebraic variety admits a nonzero regular vector field, then it also admits a nonzero regular vector field D such that $D^p = 0$ or $D^p = D$.*

Indeed, by the proof of the lemma, there exists a field $\bar{D}$ for which $\bar{D}^p = \lambda\bar{D}$, where $\lambda \in k$. If $\lambda \neq 0$, then for $D = \lambda^{1/(1-p)}\bar{D}$ we will have $D^p = D$.

The vector fields D for which $D^p = 0$ will be called *fields of additive type*, while those for which $D^p = D$ will be called *fields of multiplicative type.*

LEMMA 2. *On the projective line, every nonzero regular vector field can be written in terms of some system of coordinates either in the form d/dx or in the form $\alpha x d/dx$, where $\alpha \in k^*$. In the first case the field has one singular point and is a field of additive type, while in the second case it has two singular points and is proportional to a field of multiplicative type.*

Indeed, since the degree of the anticanonical class of the projective line is equal to 2, it follows that a nonzero regular vector field has either one or two singular points. If a regular vector field D has only one singular point at infinity, then $D = f(x)d/dx$, where f is a polynomial with no root. Therefore $D = \alpha d/dx$, and under the transformation $x \longrightarrow \alpha x$ it becomes d/dx. If, however, D has two singular points, 0 and ∞, then $D = \alpha x d/dx$. The remaining assertions of the lemma are obvious.

If X is smooth, then any rational vector field can be written in local coordinates at the point $x \in X$ in the form

$$D = \sum_1^n f_i \frac{\partial}{\partial x_i}, \quad \text{where} \quad f_i \in k(X).$$

There exists a function $h_x \in k(X)$, such that $f_i = h_x \bar{f}_i$, where $\bar{f}_i \in \mathcal{O}_x$ and $((\bar{f}_1), \ldots, (\bar{f}_n)) = (1)$. The functions h_x determine a divisor on X, which we will call a divisor of D and which we will denote by (D). If $D' = fD$, where $f \in k(X)$ and $f \neq 0$, then the fields D' and D will be called equivalent. To denote this, we will write $D' \sim D$. In this event, we have $(D') \sim (D)$.

If h is a local equation of a divisor of D in an open set U, then $h^{-1}D = \Sigma f_i \partial/\partial x_i$, where $f_i \in \mathcal{O}(U)$ and the ideal $(f_1, \ldots, f_n)$ determines a subvariety of codimension greater than 1 in U. When this subvariety is empty for arbitrary choice of the open set U, we will say that the field D *has only divisorial singularities.*

A finite inseparable morphism $\pi: X \longrightarrow Y$ of degree p establishes certain relations between the subvarieties of X and those of Y. Let D be a vector field of X which is regular at the general point v of a prime divisor $V \subset X$. We denote by D' a vector field which is equivalent to D and which is regular and not 0 at the point v. We will call V an *integral subvariety* for D if the vector D'_v is tangent to V. In this case, D defines a vector field on V; moreover, it is clear that V^D will be a subvariety of X^D.

PROPOSITION 1. *Let X and Y be varieties which are nonsingular in codimension* 1, *let $\pi: X \longrightarrow Y$ be a finite inseparable morphism of degree p, let D be the p-closed vector field which corresponds to π, let C be an irreducible divisor on X, and let C' be the image of C in Y.* If C *is an integral subvariety for D, then $\pi_* C = pC'$ and $\pi^* C' = C$. If, however, C is not an integral subvariety, then $\pi_* C = C'$ and $\pi^* C' = pC$.*

PROOF. Replacing D by an equivalent vector field, we may suppose that C does not belong to the divisor of D. Let u be a local equation of C in some open set. One can easily check that C is an integral subvariety for D if and only if $D(u) = 0$ on C. We denote by $\mathcal{O}$ the local ring of C. It is a one-dimensional regular local ring; moreover, the condition that $D(u) = 0$ on C is equivalent to the condition that $D(u) \equiv 0 \ (u)$ in $\mathcal{O}$.

The ring $\mathcal{O}^D$ is also one-dimensional and regular; moreover, $\mathcal{O}$ has rank p over $\mathcal{O}^D$. Let u and v be local parameters in $\mathcal{O}$ and in $\mathcal{O}^D$, respectively. It is known that two situations are possible: u can be chosen in such a way that either $u = v \in \mathcal{O}^D$, where $[\mathcal{O}/u: \mathcal{O}^D/u] = p$, $\pi_* C = pC'$, and $\pi^* C' = C$, or $v = u^p \epsilon$, where $\epsilon \in \mathcal{O}^*$, $\mathcal{O}/u = \mathcal{O}^D/u$, $\pi_* C = C'$, and $\pi^* C' = pC$. In the first situation, $u \in \mathcal{O}^D$ and $D(u) = 0$, so that C is an integral subvariety for D. Conversely, if C is an integral subvariety, then D determines a derivation of $\mathcal{O}/u$; moreover, $[\mathcal{O}/u: (\mathcal{O}/u)^D] = p$. Since $[\mathcal{O}/u: \mathcal{O}^D/v] \leqslant p$, the inclusions $\mathcal{O}^D/v \subset (\mathcal{O}/u)^D \subset \mathcal{O}/u$ show that we are in the first situation. The proposition is proved.

THEOREM 1. *Let X be a smooth variety, and let D be a p-closed vector field on X which has only divisorial singularities in a neighborhood of a point $x \in X$. Then in the completion $\hat{\mathcal{O}}_x$ of the local ring of the point x there exist local parameters $x_1, \ldots, x_n$ such that $D \sim \partial/\partial x_n$.*

COROLLARY. *Under the hypotheses of Theorem 1, the variety $Y = X^D$ is nonsingular at the point $\pi(x)$, and we may choose systems of local parameters $x_1, \ldots, x_n$ in $\hat{\mathcal{O}}_x$ and $y_1, \ldots, y_n$ in $\hat{\mathcal{O}}_{\pi(x)}$ such that $x_i = \pi^* y_i$ for $i < n$, and $x_n = (\pi^* y_n)^{1/p}$.*

PROOF. See [19], §3, especially Theorem 3, p. 110.

We will need the following result in §4.

THEOREM 2. *Let X be a smooth variety, and let D be a vector field of multiplicative type on X. If D assumes the value 0 at a point $x \in X$, then there exists a system of local parameters $x_1, \ldots, x_n$ in $\hat{\mathcal{O}}_x$ for which $D = \Sigma_1^n \alpha_i x_i \partial/\partial x_i$, where $\alpha_i \in \mathbf{F}_p$.*

PROOF. We recall that the Lie algebra of derivations on $\hat{\mathcal{O}}_x$ has a filtration $\mathfrak{L}_{-1} \supset \mathfrak{L}_0 \supset \cdots$, where $\mathfrak{L}_k = \bigoplus_i M_{k+1} \partial/\partial y_i$, and where $[\mathfrak{L}_r, \mathfrak{L}_s] \subset \mathfrak{L}_{r+s}$ and $\mathfrak{L}_r^p \subset \mathfrak{L}_{pr}$. From the condition $D^p = D$, it follows that $D \in \mathfrak{L}_0$ and $D \notin \mathfrak{L}_1$; therefore, for some local parameters $x_1, \ldots, x_n$, we have

$$D = \sum_{i=1}^n \alpha_i x_i \frac{\partial}{\partial x_i} + \sum_{i=1}^n P_i \frac{\partial}{\partial x_i}, \quad \text{where } P_i \in \mathfrak{m}_x^2 \text{ and } \alpha_i \in \mathbf{F}_p.$$

We proceed by induction. We assume that

$$D = \sum_{i=1}^n \alpha_i x_i \frac{\partial}{\partial x_i} + \sum_{i=1}^n P_i \frac{\partial}{\partial x_i} \quad \text{and} \quad P_i \in \mathfrak{m}_x^l.$$

We consider a change of the local parameters: $x_i' = x_i + \xi_i$. We require that

$$D(x_i + \xi_i) \equiv \alpha_i (x_i + \xi_i) \quad (\mathfrak{m}_x^{l+1}). \tag{3}$$

We introduce additional notation:

$$D_0 = \sum_{i=1}^n \alpha_i x_i \frac{\partial}{\partial x_i}, \quad D_1 = \sum_{i=1}^n P_i \frac{\partial}{\partial x_i}.$$

Then (3) shows that

$$\alpha_i x_i + P_i + D_0 \xi_i \equiv \alpha_i (x_i + \xi_i) \quad (\mathfrak{m}_x^{l+1}),$$

i.e., to determine the ξ_i, we must solve the congruences

$$\alpha_i \xi_i - D_0 \xi_i \equiv P_i \quad (\mathfrak{m}_x^{l+1}).$$

Using the well-known Jacobson identity (see [6], Chapter V, formula (63)), we obtain

$$D^p = (D_1 + D_0)^p = D_1^p + D_0^p + \sum_{j=1}^{p-1} s_j (D_1, D_0) = D_1 + D_0,$$

where $is_j(a, b)$ is the coefficient of λ^{i-1} in the expression $\mathrm{ad}(\lambda a + b)^{p-1} \cdot a$. Since $D_0 \in \mathfrak{L}_0$ and $D_1 \in \mathfrak{L}_{l-1}$, we have $s_j(D_1, D_0) \in \mathfrak{L}_{j(l-1)}$ and $D_1^p \in \mathfrak{L}_{p(l-1)}$. It follows that

$$(\mathrm{ad}\, D_0)^{p-1} D_1 \equiv D_1 \quad (\mathfrak{L}_l). \tag{4}$$

It is easy to see that

$$(\mathrm{ad}\, D_0) \left(\sum_{i=1}^{n} Q_i \frac{\partial}{\partial x_i} \right) = \sum_{i=1}^{n} (D_0 - \alpha_i E) Q_i \frac{\partial}{\partial x_i}$$

(where E denotes the identity transformation). Consequently (4) shows that

$$(D_0 - \alpha_i E)^{p-1} P_i \equiv P_i \quad (\mathfrak{m}_x^{l+1}).$$

This shows that to construct the new local parameters, it is enough for us to set

$$x_i' = x_i + \xi_i = x_i - (D_0 - \alpha_i E)^{p-2} P_i.$$

This completes the proof of the theorem.

To conclude this section, we introduce a formula for the number of singular points of a vector field of multiplicative type on an algebraic surface. This formula generalizes the well-known relation

$$\sum w(x_i) = e(C), \tag{5}$$

where $e(C)$ is the Euler characteristic of the smooth complete algebraic curve C and where $w(x_i)$ is the multiplicity of a singular point of a vector field on C. Namely, let D be a regular vector field of multiplicative type on a complete algebraic surface X, and let $M = \bigcup C_i$ be a reduced curve on X with components C_i which either are integral curves for D or are contained in the zero divisor of D. We denote by $w(M)$ the number of isolated singular points of the restriction of D to M, and we denote by $U(M)$ the part of the zero divisor U of D whose carrier is contained in M.

We assume that the curve M has only linear branches with transversal intersections at all of its singular points.

THEOREM 3. *For a vector field D of multiplicative type on a surface X and a reduced curve $M \subset X$ which satisfies the restrictions mentioned above,*

$$w(M) - U(M)^2 - U(M) \cdot K = e(M),$$

where K is the canonical class of X.

We first note that, by Theorem 2, all isolated singular points of the vector field D on M have multiplicity 1, and all components of the zero divisor of D on X also have multiplicity 1 and are smooth curves which are pairwise disjoint.

Along with the number $e(M)$, we consider the number

$$p(M)=w(M)+2\chi(\mathcal{O}_{U(M)}).$$

Theorem 3 asserts that $p(M) = e(M)$ (the "Euler characteristic" equals the "Poincaré characteristic").

We consider the normalization M^ν of M: $M^\nu = \bigcup C_i^\nu$, where $C_i^\nu \cap C_j^\nu = \varnothing$ for $i \neq j$.

We denote by r_k the multiplicity (number of different branches) of a singular point of M. Then obviously

$$e(M)=\sum e(C_i^\nu)-\sum(r_k-1).$$

We will prove an analogous formula for $p(M)$:

$$p(M)=\sum p(C_i^\nu)-\sum(r_k-1). \tag{6}$$

Since the divisor $U(M)$ consists of smooth disjoint curves, it follows that $\chi(\mathcal{O}_{U(M)}) = \Sigma\chi(\mathcal{O}_{C_i^\nu})$ for all of the C_i which are components of $U(M)$. Therefore (6) takes the form

$$w(M)=\sum w(C_j^\nu)-\sum(r_k-1), \tag{7}$$

where the first sum is over the components C_j which are not components of $U(M)$.

Let x be a point of M. Two cases are possible: only curves C_i which are not components of $U(M)$ pass through x, or some of those curves and exactly one component of $U(M)$ pass through x (in particular, x can be a nonsingular point on M). We denote the number of branches at x by r. Then the preimage of x in M^ν consists of r points.

In the first case, either x is not a singular point of D on M, in which event $r = 1$ and the contribution of this point to both sides of (7) is 0, or the contribution to the left-hand side of (7) is 1 and r preimages of x on C_i^ν are singular points with multiplicity 1, so that the contribution of x to the right-hand side is equal to $r - (r - 1) = 1$. In the second case, x is not an isolated singular point of M, and its contribution to $w(M)$ is equal to 0. Among the preimages of x, one point lies on the component where D is equal to 0, and the remaining $r - 1$ preimages are isolated singular points of D. Therefore, the contribution of x to the right-hand side of (7) is $(r - 1) - (r - 1) = 0$. This proves (7).

To prove that $p(M) = e(M)$, it remains to verify that $p(C^\nu) = e(C^\nu)$ for a smooth curve C^ν. If D does not vanish on C^ν, then this follows from (5); if $D = 0$ on C^ν, then it follows from the definition of $p(C)$ and from the fact that $e(C) = -2\chi(\mathcal{O}_C)$ for the smooth curve C. The theorem is proved.

§2. Ramification of an inseparable morphism

Let $\pi\colon X \longrightarrow Y$ be a finite inseparable morphism of degree p of smooth varieties, and let $k(X) = k(Y)(\alpha^{1/p})$, where $\alpha \in k(Y)$. The function α is defined to within a substitution

$$\alpha' = \sum_0^{p-1} c_i^p \alpha^i, \text{ where } c_i \in \mathrm{k}(Y). \tag{1}$$

Consequently, the differential forms $d\alpha'$ and $d\alpha$ are related by the equation

$$d\alpha' = \left(\frac{d\alpha'}{d\alpha}\right) d\alpha, \quad \text{where} \quad \frac{d\alpha'}{d\alpha} = \sum_0^{p-1} i c_i^p \alpha^{i-1}.$$

Therefore, the divisors of these forms are equivalent: $(d\alpha') \sim (d\alpha)$. (The divisor of a differential form $\Sigma f_i dx_i$ is defined in a way analogous to that in which the divisor of a vector field is defined.) The equivalence class of $(d\alpha)$, which is associated invariantly with the morphism π, is called the *ramification divisor class* of this morphism.

PROPOSITION 2. *If $\pi: X \longrightarrow Y$ is a finite inseparable morphism of degree p of smooth varieties, then the canonical classes of these varieties are related by the equation*

$$K_X = \pi^* K_Y - \frac{p-1}{p} \pi^*(V), \tag{2}$$

where V is the ramification of π.

For any α such that $\mathrm{k}(X) = \mathrm{k}(Y)(\alpha^{1/p})$, we define an operation $\varphi_\alpha: \Omega_Y^n \longrightarrow \Omega_X^n$. Namely, since $d\alpha \neq 0$, it follows that any form $\omega \in \Omega_Y^n$ can be written in the form $\omega = \eta \wedge d\alpha$, where $\eta \in \Omega_Y^{n-1}$. We set

$$\varphi_\alpha(\omega) = \pi^*(\eta) \wedge d\beta, \text{ where } \beta^p = \alpha. \tag{3}$$

This is well defined, since if $\eta \wedge d\alpha = 0$, then $\eta = \xi \wedge d\alpha$ and $\pi^*(\eta) = \pi^*(\xi) \wedge \pi^*(d\alpha) = 0$.

The dependence on the choice of α is expressed by the simple formula

$$\varphi_{\alpha'}(\omega) = \varphi_\alpha(\omega) \pi^* \left(\frac{d\alpha'}{d\alpha}\right)^{\left(\frac{1}{p}-1\right)}. \tag{4}$$

Indeed,

$$\varphi_{\alpha'}(\omega) = \pi^*\left(\eta \frac{d\alpha}{d\alpha'}\right) \wedge d\beta' = \pi^*\eta \left(\pi^* \frac{d\alpha'}{d\alpha}\right)^{-1} \wedge d\beta',$$

where, in notation as in (1) above,

$$d\beta' = \sum i\pi^* c_i \beta^{i-1} d\beta + \sum d(\pi^*(c_i)) \cdot \beta^i = \left(\pi^*\left(\frac{d\alpha'}{d\alpha}\right)\right)^{\frac{1}{p}} d\beta + \sum d(\pi^* c_i)\beta^i.$$

Since $\pi^*\Omega_Y^n = 0$, it follows that $\pi^*\eta \wedge d\pi^* c_i = 0$, from which (4) can be deduced.

Turning now to the proof of (2), we consider a p-closed vector field D on X such that $Y = X^D$. Since (2) is an equality of divisors, it is sufficient to check it on an open set whose complement has codimension greater than 1. Therefore, deleting from X the nondivisorial singularities of D, and deleting their images from Y, we reduce the proof to the case where D has only divisorial singularities.

It is enough to verify (2) in the completion $\hat{\mathcal{O}}_X$ of the local ring of an arbitrary point

$x \in X$. We apply the corollary to Theorem 1, and we use the notation which we introduced there. Let $\omega = f dy_1 \wedge \cdots \wedge dy_n$. Then $(\omega) = (f)$. By (3) and (4),

$$\varphi_\alpha(\omega) = \left(\pi^*\left(\frac{d\alpha}{dy_n}\right)\right)^{\frac{1}{p}-1} \varphi_{y_n}(\omega) = \left(\pi^*\left(\frac{d\alpha}{dy_n}\right)\right)^{\frac{1}{p}-1} (\pi^* f)\, dx_1 \wedge \ldots \wedge dx_n.$$

Since $d\alpha = (d\alpha/dy_n)dy_n$, it follows that $(d\alpha/dy_n) = (d\alpha) = V$, and we obtain (2).

REMARKS. 1. Formula (2) is not new: it follows from the results of [8] and [9]. In [9] the behavior of the canonical class was described in terms of the different of the morphism, while in [8] it is treated for projective varieties using global considerations. We preferred to provide a proof which seemed to us to be more direct.

2. It is not necessary to use Theorem 1 in order to prove Proposition 2. Since we are comparing two divisors, it is enough to show that any irreducible divisor C on X appears on both sides of (2) with the same coefficient. This can be done using the preceding proof where, however, we put the following conditions on the functions $x_1, \ldots, x_n$ and $y_1, \ldots, y_n$ (these conditions are weaker than those which guarantee Theorem 1):

$$x_i = \pi^* y_i, \quad \text{for} \quad i < n, \quad \text{and} \quad x_n^p = \pi^* y_n;$$

$$\nu_C(dx_1 \wedge \ldots \wedge dx_n) = \nu_{\pi C}(dy_1 \wedge \ldots \wedge dy_n) = \nu_{\pi C}(dy_n) = 0,$$

where ν_C is the valuation defined by the divisor C.

In order to construct such a system of functions, we need to look at two cases.

(1) $\pi_* C = \pi C$, so that π is a birational isomorphism $C \longrightarrow \pi C$. For x_n, we need to take a local equation of C, and for y_n we take x_n^p. For $y_1, \ldots, y_{n-1}$, we take functions such that their restrictions $\bar{y}_1, \ldots, \bar{y}_{n-1}$ to C form a separating basis, i.e. $d\bar{y}_1 \wedge \cdots \wedge d\bar{y}_{n-1} \neq 0$ on C.

(2) $\pi_* C = p\pi C$, so that π determines a morphism of degree p: $C \longrightarrow \pi C$. Then there exist functions $\bar{u}_1, \ldots, \bar{u}_{n-1}$ and $\bar{v}_1, \ldots, \bar{v}_{n-1}$ on C and πC, respectively, such that $\bar{u}_i = \pi^* \bar{v}_i$ for $i < n-1$, and $\bar{u}_{n-1}^p = \pi^* \bar{v}_{n-1}$. For $y_1, \ldots, y_{n-2}$, we must take functions whose restrictions to πC are $\bar{v}_1, \ldots, \bar{v}_{n-2}$, respectively; for y_{n-1}, we take a local equation of πC; for x_n, we take a function whose restriction to C is $\bar{u}_{n-1}$; and for y_n, we take x_n^p.

PROPOSITION 3. *If $\pi: X \longrightarrow Y = X^D$ is a morphism of smooth varieties which is determined by a p-closed vector field D, and if U is the divisor of D and V is the ramification of π, then*

$$U \approx -\frac{1}{p}\pi^*(V). \tag{5}$$

As in the proof of Proposition 2, we can assume that D has only divisorial singularities, and apply Theorem 1. Let $k(X) = k(Y)(\beta)$, where $\beta^p = \alpha \in k(Y)$. (5) will follow from the equation

$$(D) + \frac{1}{p}\pi^*((d\alpha)) = (D(\beta)). \tag{6}$$

It is sufficient to check this equation in the ring $\hat{\mathcal{O}}_X$ for arbitrary $x \in X$. By Theorem 1, let $D = f\partial/\partial x_n$. Then

$$D(\beta) = f\,\frac{\partial\beta}{\partial x_n} = f\left(\pi^*\,\frac{d\alpha}{dy_n}\right)^{1/p}$$

Since $(d\alpha) = (d\alpha/dy_n)$, it follows that

$$(D(\beta)) = (f) + \left(\frac{\partial\beta}{\partial x_n}\right) = (D) + \frac{1}{p}\pi^*(d(\alpha)).$$

As in the proof of Proposition 2, one can easily avoid using Theorem 1.

REMARK. Formula (5) shows that $-(1/p)\pi^*(V)$ can be interpreted as the divisor corresponding to a line bundle, the kernel of $d\pi\colon \Theta_X \longrightarrow \pi^*\Theta_Y$.

COROLLARY 1. *Under the hypotheses of Proposition* 3,

$$K_X = \pi^* K_Y + (p-1)U. \tag{7}$$

If the vector field D is regular on X, then $U \geqslant 0$, and (7) indicates a certain "monotonicity" of the canonical class. This can be made concrete; we have, for example,

COROLLARY 2. *If D is a regular p-closed vector field on a complete variety X which is nonsingular in codimension* 1, *and if $Y = X^D$ and the singularities of Y admit a resolution (for example, if* $\dim Y = 2$), *then $P_n(Y) \leqslant P_n(X)$ for $n \geqslant 1$ for the plurigenera.*

We recall that the plurigenus $P_n(Y)$ is the dimension of $H^0(Y_1, nK_{Y_1})$, where Y_1 is an arbitrary complete smooth model for Y and K_{Y_1} is the canonical class of Y_1.

PROOF. Let Y_1 be a complete smooth variety which dominates Y, and let $\varphi\colon Y_1 \longrightarrow Y$ be the corresponding morphism. We can remove from X any closed subvariety of codimension greater than 1 having smooth complement $\widetilde{X}$ and such that D has only divisorial singularities on $\widetilde{X}$. Set $\widetilde{Y} = \pi_D(\widetilde{X})$ and $\widetilde{Y}_1 = \varphi^{-1}(\widetilde{Y})$. Then $\widetilde{Y}_1$ and $\widetilde{Y}$ are isomorphic. Moreover,

$$H^0(Y_1, nK_{Y_1}) \subset H^0(\widetilde{Y}_1, nK_{\widetilde{Y}_1}) \approx H^0(\widetilde{Y}, nK_{\widetilde{Y}}). \tag{8}$$

On the other hand, the inclusion of fields $\pi_D^*\colon k(Y) \longrightarrow k(X)$ defines an inclusion

$$\pi_D^*\colon H^0(\widetilde{Y}, nK_{\widetilde{Y}}) \subset H^0(X, \pi^* nK_{\widetilde{Y}}) \subset H^0(\widetilde{X}, \pi^* nK_{\widetilde{Y}} + n(p-1)U). \tag{9}$$

We can apply (7) to $\widetilde{X}$, so that

$$H^0(\widetilde{X}, \pi^* nK_{\widetilde{Y}} + n(p-1)U) = H^0(\widetilde{X}, nK_{\widetilde{X}}). \tag{10}$$

Finally, $H^0(\widetilde{X}, nK_{\widetilde{X}}) = H^0(X, nK_X)$, since $X\backslash\widetilde{X}$ has codimension greater than 1 and nK_X is a line bundle. In view of this, the inclusions (8), (9), and (10) prove the corollary.

Further on, it will be important for us to know the extent of the class of morphisms of smooth varieties $X \longrightarrow Y$ to which the construction described in Theorem 1 can be applied, i.e. which morphisms can be represented as quotient morphisms by a vector field having only divisorial singularities. We will show that in dimension 2, these are all the finite inseparable morphisms of degree p.

THEOREM 4. *Any finite inseparable morphism of degree p of smooth surfaces* $\pi: X \longrightarrow Y$ *is a quotient morphism by a p-closed vector field which has no isolated singularities.*

Two simple results will be used in the proof:

LEMMA 1. *Let* $\varphi: Z \longrightarrow X$ *and* $\psi: Z \longrightarrow Y$ *be birational morphisms of surfaces; suppose that Z and Y are smooth and X is normal, and that* φ *contracts every curve which is contracted by* ψ. *Then the rational map* $\varphi\psi^{-1}$ *is a morphism.*

It follows from the hypotheses of the lemma that $\psi\varphi^{-1}$ does not contract any curve of X. Under these conditions the inverse transformation $\varphi\psi^{-1}$ is a morphism, as is proved in [18] (the lemma in Part I, Chapter IV, §3; actually, in the statement of that lemma, X is assumed to be smooth; however, this is not used in the proof).

LEMMA 2. *Let X be a smooth surface, let Y be normal, and let* $f: X \longrightarrow Y$ *be a birational morphism which does not contract any curve of X. Then f is an isomorphism.*

Indeed, under the hypotheses of the lemma, any point $x \in X$ is isolated in the fiber $f^{-1}(f(x))$. Therefore, the completed local ring $\hat{\mathcal{O}}_x$ is a module of finite type over $\hat{\mathcal{O}}_{f(x)}$ ([7], II, 6.25, p. 119). Since Y is normal, it follows that $\hat{\mathcal{O}}_{f(x)}$ is integrally closed; hence $\hat{\mathcal{O}}_x = \hat{\mathcal{O}}_{f(x)}$. Therefore, $f(x)$ is a simple point, so that the lemma follows from well-known properties of birational morphisms.

We turn to the proof of the theorem. We know that there exists a p-closed vector field D on X such that $Y = X^D$. We must show that D has no isolated singular points. This is a local question; therefore we can replace X and Y by the spectra of local rings, which are regular by hypothesis. Since in this case $\operatorname{Pic} X = \operatorname{Pic} Y = 0$, we may assume that the divisor of the vector field D equals 0, that D is regular on X, and that there is an isolated singular point of D at a unique closed point $x \in X$.

The plan of the proof is as follows. We consider a σ-process $X' \longrightarrow X$ with center at x. The vector field D determines a vector field on X'. We will assume that the theorem is false and that D vanishes at x. It follows that D is regular on X'. Indeed, we can describe the process in local coordinates by the formulas $x_1' = x_1$ and $x_2' = x_2/x_1$. Then $D(x_1') = D(x_1)$ and $D(x_2') = x_1^{-1}(D(x_2) - x_2'D(x_1))$. Since by hypothesis $D(x_1)(x) = D(x_2)(x) = 0$, this implies that $x_1^{-1}D(x_l) \in \mathcal{O}_{x'}$ for $l = 1, 2$.

Consider the variety $Y' = (X')^D$. One can easily verify that there exists a morphism τ such that the diagram

$$\begin{array}{ccc} X' & \xrightarrow{\pi'} & Y' \\ \sigma\downarrow & & \downarrow\tau \\ X & \xrightarrow{\pi} & Y \end{array}$$

commutes. The vector field D can have zeros on X', some of which can be isolated, so that, a priori, Y' can have singular points. Let $L = \sigma^{-1}(x)$, and set $M = \pi'(L)$. We will show that Y' is smooth and that $(\pi')^*M = L$. From these two statements, the theorem follows. Indeed, τ is then a σ-process and $M^2 = -1$ just as $L^2 = -1$. But this contradicts the general formula $((\pi')^*M)^2 = (\deg \pi')M^2$ and the fact that $\deg \pi' = p$. This contradiction proves the theorem.

In order to show that Y' is smooth, we first delete from X' the isolated singular points of D, and we denote the remaining scheme by $\widetilde{X}$. We set $\widetilde{Y} = \pi'(\widetilde{X})$, $\widetilde{L} = \widetilde{X} \cap L$ and $\widetilde{M} = \widetilde{Y} \cap M$. Then $\widetilde{Y}$ is smooth, and (7) yields

$$K_{\widetilde{X}} = (\pi')^* K_{\widetilde{Y}} + (p-1)\,\widetilde{U}. \tag{11}$$

Here obviously $K_{\widetilde{X}} = \widetilde{L}$ (since $K_X = 0$) and $\widetilde{U} = m\widetilde{L}$ with $m \geqslant 0$. We note that $(\pi')^* K_{\widetilde{Y}} > 0$. Indeed, if we let $\overline{Y}$ be a resolution of singularities of Y', then $\widetilde{Y}$ is isomorphic to an open subset $\widetilde{Y} \subset \overline{Y}$. But the composite of the morphisms $\overline{Y} \longrightarrow Y'$ and $\tau\colon Y' \longrightarrow Y$ is a birational morphism of smooth schemes $\overline{Y} \longrightarrow Y$ which is the product of σ-processes. Since $K_Y = 0$, it follows that $K_{\overline{Y}}$ consists of curves which arise from σ-processes with positive multiplicities; hence $K_{\overline{Y}} > 0$. Restricting to $\widetilde{Y}$, we see that $K_{\widetilde{Y}} > 0$, also. Moreover, $K_{\widetilde{Y}} \neq 0$, since τ shrinks $\widetilde{M}$ to a point and $\widetilde{M}$ must appear in $K_{\widetilde{Y}}$ with nonzero coefficient.

Since $\operatorname{Pic} \widetilde{X} = \operatorname{Pic} X = \mathbf{Z} \cdot L$, it follows that $(\pi')^* K_{\widetilde{Y}} = nL$ with $n > 0$. From (11) we obtain that $\widetilde{L} = n\widetilde{L} + (p-1)m\widetilde{L}$ with $n > 0$ and $m \geqslant 0$. Hence $n = 1$ and $m = 0$, i.e.

$$K_{\widetilde{Y}} = \widetilde{M}, \quad \widetilde{L} = (\pi')^* \widetilde{M}, \quad U = 0. \tag{12}$$

We now return to the scheme $\overline{Y}$, the resolution of the singularities of Y'. The morphism $\overline{Y} \longrightarrow Y$ can be decomposed into a product of σ-processes: $\sigma_1\colon Y_1 \longrightarrow Y$, $\sigma_2\colon Y_2 \longrightarrow Y_1$, etc. We have the commutative diagram

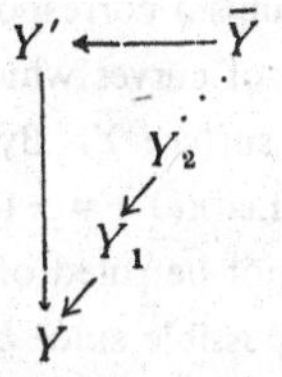

We saw that $\widetilde{Y} \subset \overline{Y}$, and it follows from (12) that the proper preimage M appears in the canonical class $\overline{Y}$ with coefficient 1:

$$K_{\overline{Y}} = M + \sum r_i C_i, \quad \text{where } C_i \neq M. \tag{13}$$

On the other hand, the only nonregular point of the rational transformation $Y \longrightarrow \overline{Y}$ is $y = \pi(x)$. Therefore $\overline{Y}$ can be obtained by a sequence of σ-processes at points which lie over this single point. From the formula for the behavior of the canonical class under a σ-process, it now follows that the unique curve F_1 which appears in the canonical class of $\overline{Y}$ with coefficient 1 is the curve F_1 contracted under the first morphism $Y_1 \longrightarrow Y$:

$$K_{\overline{Y}} = F_1 + \sum_{i \geqslant 2} k_i F_i, \quad k_i \geqslant 2. \tag{14}$$

Comparing (13) and (14), we see that $M = F_1$ (more precisely, their proper preimages in $\overline{Y}$ coincide). Lemma 1 now yields that the rational map $Y_1 \longrightarrow Y'$ is a morphism, and Lemma 2 yields that it is an isomorphism. Since the scheme Y_1 is smooth, it follows that Y' is also smooth. From this and (12), as we saw, the theorem follows.

COROLLARY. *Let $\mathfrak{O}$ be a two-dimensional regular complete local ring of characteristic*

$p > 0$ whose residue class field $\mathcal{O}/\mathfrak{m}$ is algebraically closed. If $\mathcal{O}' \subset \mathcal{O}$ is a regular local ring with the same residue class field $\mathcal{O}'/\mathfrak{m}' = k$, and if $\mathfrak{m}' \subset \mathfrak{m}$ and $\mathcal{O}' \supset \mathcal{O}^p$ and the rank of $\mathcal{O}$ as an $\mathcal{O}'$-module equals p, then $\mathcal{O} = \mathcal{O}'[x^{1/p}]$, where $x \in \mathfrak{m}$ and $x \notin (\mathfrak{m}')^2$.

This is a combination of Theorem 4 and the corollary to Theorem 1.

REMARK. Of course, the question arises whether this algebraic fact has an algebraic proof. Furthermore: does the analogous result hold for complete (or, more generally, for not necessarily complete) local rings of arbitrary dimension? This is true when $p = 2$. Indeed, the module $\mathcal{O}$ over the ring $\mathcal{O}'$ is free ([10], III, p. 88). Therefore, if $p = 2$, we have that $\mathcal{O} = \mathcal{O}'[\alpha]$, and is defined by the equation $\alpha^2 = x \in \mathcal{O}'$, where we may assume that $x \in \mathfrak{m}$. By the Jacobi criterion, $\mathcal{O}$ is regular if and only if $dx \neq 0$, i.e. $x \notin (\mathfrak{m})^2$.

The (affirmative) answer to the question posed above would follow from the results of [8] (Theorem 5). However, it appears to us that the proof of this theorem contains an error: Lemma 5, on which the proof essentially depends, is false.

To illustrate the concepts which we developed above, we apply them to the study of vector fields on $K3$ surfaces. According to Lemma 1 of §1, if $H^0(X, \Theta_X) \neq 0$, then there exists a nonzero regular p-closed vector field D on X.

THEOREM 5. *If X is a $K3$ surface, and if D is a nonzero p-closed regular vector field on X, then X^D is a rational surface.*

Let $\pi: X \longrightarrow Y = X^D$ be the morphism corresponding to D, let $\overline{Y} \longrightarrow Y$ be a resolution of singularities, and let E be the system of curves which are contracted under $\overline{Y} \longrightarrow Y$. Let us find the Kodaira dimension κ of the surface $\overline{Y}$. By Corollary 2 to Proposition 3, we have $P_m(\overline{Y}) \leqslant 1$ for the plurigenera $P_m(\overline{Y})$, i.e. $\kappa(\overline{Y}) = -1$ or 0. If $\kappa(\overline{Y}) = -1$, then $\overline{Y}$ is either rational or ruled ([3], part IV). It cannot be ruled of genus > 0, since then X would map onto a curve of genus > 0, which is impossible since $b_1(X) = 0$. Thus, to prove the theorem, we need only eliminate the case $\kappa(\overline{Y}) = 0$.

In this case, for a minimal model Y' of the surface $\overline{Y}$, we have $mK_{Y'} = 0$; where $m|12$ (see [2], page 2 of the preprint). We denote by F the system of curves which are shrunk by the morphism $\overline{Y} \longrightarrow Y'$. Then $mK_{\overline{Y}} \geqslant 0$ and supp $mK_{\overline{Y}} = F$; also, since $mK_{\overline{Y}} = 0$, it follows that $F \subset E$. In other words, we can apply Lemma 1 from the proof of Theorem 4 to the morphisms in the diagram

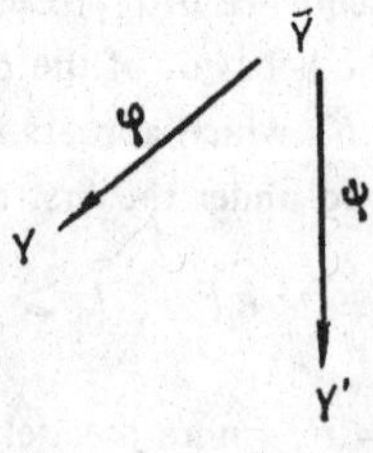

and get that $\xi = \varphi\psi^{-1}: Y' \longrightarrow Y$ is a morphism. Thus, Y is obtained by contracting curves on a model Y' which is already minimal.

For the Euler characteristic in etale cohomology we have

$$e(Y') = e(Y) + \sum (e(F_i) - 1), \tag{15}$$

where the F_i are the reduced connected curves which are contracted to distinct singular points on Y. Let us show that each F_i is a tree whose components are smooth rational curves. To do this, consider the rational map $\xi^{-1}\pi\colon X \longrightarrow Y'$, where $\xi\colon Y' \longrightarrow Y$ is a birational isomorphism. Let $\eta\colon X' \longrightarrow X$ be a birational morphism such that $\rho = \xi^{-1}\pi\eta\colon X' \longrightarrow Y'$ is a morphism (a resolution of the points of indeterminacy of $\xi^{-1}\pi$). Obviously η is a contraction of a system of connected nonintersecting trees Φ_i which consist of rational curves, and $F_i = \rho\Phi_i$. We denote by $\nu\colon Y'' \longrightarrow Y'$ the normalization of Y' in the field $k(X')$. Then ν is an inseparable morphism of degree p, and $X' \longrightarrow Y''$ is a birational morphism which consists of the contractions of certain components of the Φ_i. Therefore the F_i are also trees which consist of smooth rational curves.

We denote by n_i the number of components of F_i. Then $e(F_i) = n_i + 1$, and (15) assumes the form

$$e(Y') = e(Y) + \sum n_i. \tag{16}$$

On the other hand, since π is a radical morphism, it follows that the etale topologies of X and Y are isomorphic and $e(Y) = e(X) = 24$. Formula (16) shows that $e(Y') \geqslant 24$, and that $e(Y') = 24$ only if all the $n_i = 0$, i.e. only if Y is already a smooth surface. Finally, it can be seen from the enumeration of all the types of surfaces with $\kappa = 0$ ([2], preprint page 5) that $e(Y') \leqslant 24$ for such surfaces. Thus, we have the inequality $24 = e(Y) \leqslant e(Y') \leqslant 24$, i.e. $e(Y) = e(Y')$. Thus, Y is a smooth surface, so that, by Theorem 2, D has no isolated singular points. If we multiply by m, we can see from (7) that $U = 0$, i.e. D has no singular points at all. But no such vector field can exist on a surface X of type $K3$, since $e(X) \neq 0$.

COROLLARY 1. *If a nonzero regular vector field exists on a K3 surface, then it is unirational.*(1)

By Theorem 3, the field $k(X)$ is an inseparable extension of degree p of the field of rational functions. In other words, $k(X) = k(u, v, w)$, where $w^p = f(u, v)$. But then $k(X) \approx k(U^p, V^p, f^{1/p}(U, V)) \subset k(U, V)$, where $f^{1/p}$ is obtained from f by taking the pth root of all of the coefficients.

COROLLARY 2. *If a nonzero regular vector field exists on a K3 surface X, then X is an elliptic or quasi-elliptic surface.*

By Corollary 1, X is unirational. We may assume that there exists a morphism $f\colon Y \longrightarrow X$, where Y is a rational surface. Since every 2-cycle on Y is algebraic, it follows from the formula $f_*f^*x = (\deg f)x$ that all 2-cycles on X are also algebraic. Hence $\rho = b_2 = 22$ for X; thus Pic X has rank 22 and the intersection pairing defines an indeterminate quadratic form of rank 22 on Pic X. Such a form necessarily represents 0; that is, there exists a divisor $D \not\equiv 0$ with $D^2 = 0$. It then follows that an elliptic or quasi-elliptic pencil exists on X, as was shown in the proof of Theorem 1 in [14], §3 (in [14] the proof of the existence of an elliptic pencil is carried out for $p > 3$, but for $p = 2$ or 3 the same proof gives the existence of an elliptic or quasi-elliptic pencil).

(1) This result was announced without proof by M. Artin [1].

§3. Vector fields on surfaces having a pencil of curves

A pencil of curves on a smooth surface X over a curve B is a morphism

$$f: X \to B, \tag{1}$$

all the fibers of which are irreducible, except possibly for a finite number.

For a smooth morphism f, the exact sequence of sheaves

$$0 \to f^*\Omega^1_B \to \Omega^1_X \to \Omega^1_{X/B} \to 0$$

corresponds to an exact sequence of vector bundles and determines a dual exact sequence of vector bundles:

$$0 \to \Theta_{X/B} \to \Theta_X \to N_{X/B} \to 0, \tag{2}$$

where the restriction of $N_{X/B}$ to each fiber coincides with the normal bundle of this fiber. By definition, there exists an isomorphism

$$N_{X/B} \xrightarrow{\sim} f^*\Theta_B. \tag{3}$$

If $\theta \in H^0(X, \Theta_X)$, then the image of θ in $H^0(X, N_{X/B})$ will be denoted by θ_N and will be called the normal component of the vector field θ.

PROPOSITION 4. *If the morphism* (1) *is smooth and proper, then for any vector field θ on X, there exists a vector field $\tau \in H^0(B, \Theta_B)$ such that $\theta_N = f^*\tau$.*

PROOF. In view of the isomorphism (3), we may assume that $\theta_N \in H^0(X, f^*\Theta_B)$. If, for a covering U_i of the curve B, Θ_B is defined by the transition functions u^i_j, then $f^*\Theta_B$ is given in the covering $f^{-1}(U_i)$ by the functions $f^*(u^i_j)$. Let $(\theta_N)^i$ be the components of the section θ_N in this covering. Then

$$(\theta_N)^i = f^*(u^i_j)(\theta_N)^j.$$

Since f is proper, we have $f_*\mathcal{O}_X = \mathcal{O}_B$, i.e. there exist functions $\tau^i \in H^0(U_i, \mathcal{O}_B)$ such that $(\theta_N)^i = f^*\tau^i$. Then $\tau^i = u^i_j\tau^j$, i.e. the τ^i are compnents of a section $\tau \in H^0(B, \Theta_B)$, and $\theta_N = f^*\tau$.

PROPOSITION 5. *If the morphism* (1) *is smooth, and if for a vector field $\theta \in H^0(X, \Theta_X)$ there exists a vector field $\tau \in H^0(B, \Theta_B)$ such that $\theta_N = f^*(\tau)$, then, for any regular function u on B,*

$$f^*(\tau(u)) = \theta(f^*(u)). \tag{4}$$

PROOF. In view of the definition of $N_{X/B}$ in the exact sequence (2), we have that $\theta_N = \operatorname{res}\theta$ for $\theta \in H^0(X, \Theta_X)$, where res is the restriction of the linear form θ, given on the fibers of Ω^1_X, to the sub-bundle $f^*\Omega^1_B$, i.e.

$$\langle \theta, f^*\omega \rangle = \langle \theta_N, f^*\omega \rangle$$

where $\omega \in f^*\Omega^1_B$. Setting $\theta_N = f^*\tau$ and using the fact that $\langle f^*\tau, f^*\omega \rangle = f^*\langle \tau, \omega \rangle$, we get that $\langle \theta, f^*\omega \rangle = f^*\langle \tau, \omega \rangle$. In particular, if $\omega = du$, then $f^*(\tau(u)) = \theta(f^*(u))$.

We now suppose that the bundle X is proper but not necessarily smooth, and we denote by $\widetilde{B}$ the open set in B on which $f: \widetilde{X} \longrightarrow \widetilde{B}$ is smooth, where $\widetilde{X} = f^{-1}(B)$. Let $\theta \in H^0(X, \Theta_X)$. By Proposition 4, we have that $\theta_N = f^*(\tau)$ on the open set $\widetilde{X}$, where $\tau \in H^0(\widetilde{B}, \Theta_{\widetilde{B}})$. If $\widetilde{B}$ is not empty, then we can consider τ as a rational vector field on B. In what follows, we will also use τ to denote the normal component of the vector field θ.

PROPOSITION 6. *If not all of the fibers of the morphism f are degenerate, then the vector field τ is regular on B, and $\tau(b_0) = 0$ if the fiber $f^{-1}(b_0)$ is degenerate. If every fiber is degenerate, then $\theta \in H^0(X, \Theta_{X/B})$.*

Of course, the second case cannot arise if the characteristic of the base field is 0.

We assume first that $\widetilde{B}$ is not empty. Formula (4), which we proved for points of the open set $\widetilde{X}$, then holds also on all of X. Since θ is a regular vector field, this implies that if u is a function which is regular at a point $b \in B$, then $f^*(\tau(u))$ is regular at the points of $f^{-1}(b)$; hence $\tau(u)$ is also regular at b. This proves that τ is a regular vector field.

Now suppose that the fiber $f^{-1}(b_0)$ is degenerate, and let x_0 be one if its singular points. Let t be a local parameter at the point b_0. Then $f^*(t) \in \mathfrak{m}_{x_0}^2$, i.e. $f^*(t) = \Sigma u_i v_i$, where $u_i, v_i \in \mathfrak{m}_{x_0}$. Hence

$$\theta(f^*(t)) = \sum \theta(u_i) v_i + \sum u_i \theta(v_i) \in \mathfrak{m}_{x_0}.$$

Therefore $\theta(f^*(t))(x_0) = 0$, from which we can infer that $\tau(b_0) = 0$, also.

Addressing ourselves now to the case where all of the fibers are degenerate, we consider a single degenerate irreducible fiber $C = f^{-1}(b_0)$. Our assertion is equivalent to the statement that $\theta_N = 0$ at the nonsingular points of C, i.e. $\theta \in H^0(C, \Theta_C)$.

Let $\widetilde{C}$ be a normalization of the curve C, and let $\psi: \widetilde{C} \longrightarrow X$ be the composite of the normalization $\widetilde{C} \longrightarrow C$ with the inclusion $C \hookrightarrow X$. We set $\psi^*\Theta_X = E$ and $\xi = d\psi: \Theta_{\widetilde{C}} \longrightarrow E$. We will show that

$$\deg E/\xi(\Theta_{\widetilde{C}}) < 0. \tag{5}$$

Indeed, $\deg E = -\deg \psi^* K_X = -C \cdot K$. But by the adjunction formula we have $CK = 2g(\widetilde{C}) - 2 + 2\delta = C^2$, where $g(\widetilde{C})$ is the genus of $\widetilde{C}$ and δ is the "multiplicity measure" of the singular points of C. Since $\deg \Theta_{\widetilde{C}} = 2 - 2g(\widetilde{C})$, it follows that

$$\deg E/\xi(\Theta_{\widetilde{C}}) = C^2 - 2\delta.$$

Since $C^2 = 0$, it follows that $C^2 - 2\delta \leqslant 0$, and we have equality only when $\delta = 0$. But this implies that C is smooth, contradicting the fact that F_{b_0} is degenerate. This proves (5).

It follows from (5) that $H^0(\widetilde{C}, E/\xi(\Theta_{\widetilde{C}})) = 0$. Therefore, in the exact sequence

$$H^0(\widetilde{C}, \Theta_{\widetilde{C}}) \to H^0(\widetilde{C}, E) \to H^0(\widetilde{C}, E/\xi(\Theta_{\widetilde{C}}))$$

the last term is equal to 0; hence $\psi^*(\theta) \in H^0(\widetilde{C}, \Theta_{\widetilde{C}})$, which means that θ is tangent to C at its nonsingular points.

§4. Auxiliary lemmas

Proposition 6 allows us to investigate the pencils on which there exist nonzero regular vector fields. Our investigation will lead to a complete enumeration of all possible types of reduced elliptic pencils (with one additional restriction in characteristic 2). We recall that a pencil $f: X \longrightarrow B$ is called *elliptic* if almost all of the fibers are elliptic curves, and that f is called *reduced* if it has no multiple fibers. In this case, we call X a reduced elliptic surface.

In this section, we will prove several auxiliary results which will be needed when we study vector fields on reduced elliptic surfaces.

LEMMA 1. *A regular vector field θ_0 on a surface X_0 having a normal singularity x_0 can only define a regular vector field on a smooth surface X dominating X_0 if $\theta_0(x_0) = 0$.*

We will prove the lemma only in the specific case for which it will be needed, namely when the desingularized surface X has only smooth curves over x_0. (This is precisely what happens if the curves appearing in the resolution are components of a degenerate fiber of an elliptic pencil. Actually the general case can easily be dealt with using the arguments employed in the proof of Proposition 6.) Suppose that θ_0 defines a regular vector field θ on X, let $\pi: X \longrightarrow X_0$ be the natural mapping, let E be one of the curves in $\pi^{-1}(x_0)$, and let $x \in E$. It is well known that $(E^2) < 0$; hence the normal bundle of E has no sections. Therefore the normal component along E of θ equals 0, which means that θ is tangent to E. Since $\pi^*(f)$ is constant on E for any function f which is regular at the point x_0, and since θ is tangent to E, it follows that $\theta(\pi^*(f)) = 0$ on E, so that $\theta_0(f)(x_0) = \theta(\pi^*(f))(x) = 0$.

For the formulation of the next lemma, we recall the following concept. Let $f: X \longrightarrow B$ be a reduced elliptic pencil over a base which, in view of the local nature of the question, can be considered to be the spectrum of a regular one-dimensional local ring $\mathcal{O}$. Associated with this pencil is the invariant $\epsilon(X)$, which can be defined in two equivalent ways (see [12] and [5]): (1) $\epsilon(X)$ is the sum of the Euler characteristic of a closed fiber and the higher ramification index δ; and (2) $\epsilon(X)$ is that power of a local parameter which divides the discriminant of the Weierstrass model of the general fiber. If F_b is the closed fiber of X, then we will also denote $\epsilon(X)$ by $\epsilon(F_b)$. Let us consider those pencils whose general fibers are supersingular curves, and for which the invariant $\epsilon(X)$ assumes a given value n. Let $r(n)$ denote the maximal number r such that the normal component of any regular vector field on a pencil of this type has a zero of order $\geqslant r$ at the closed point of the base.

LEMMA 2. *If the base field has characteristic $p = 2$, then $\epsilon(X)$ is divisible by 4; if $p = 3$, then $\epsilon(X)$ is divisible by 3. For $p = 2$ we have $r(n) \geqslant n/4 - 3$, while for $p = 3$ we have $r(n) \geqslant n/3 - 4$.*

The proof is an immediate calculational verification which depends on the explicit form of the minimal Néron model for X (see [11]). It is carried out separately for $p = 2$ and $p = 3$. We will use the notation of [11]; in particular, we write the equation of the Weierstrass model of X in the form

$$y^2 + \lambda xy + \mu y = x^3 + \alpha x^2 + \beta x + \gamma. \tag{1}$$

Case $p = 2$. The supersingularity condition means that $\lambda = 0$; in addition, the discriminant of the model (1) is equal to μ^4 (see [17], page 62). Hence we have $\epsilon(X) = 4d$, where $d = \nu(\mu)$ and ν is the valuation of the local ring $\mathcal{O}$. It is easy to see that a vector field which is regular on the affine part of the model (1) can be written in the form

$$D=f\frac{\partial}{\partial x}+g\frac{\partial}{\partial t},$$

where $f, g \in \mathcal{O}$ and where t is a local parameter on the base. In this notation the normal component of D has the form $g\partial/\partial t$.

It follows from (1) that

$$D(y)=(x^2+\beta)f+(\alpha' x^2+\beta' x+\gamma'+\mu' y)g,$$

from which we can easily derive the relations

$$f+\alpha' g\equiv 0(t^d), \tag{2}$$

$$\beta' g\equiv 0(t^d), \tag{3}$$

$$\beta f+\gamma' g\equiv 0(t^d) \tag{4}$$

(the relation $\mu' g \equiv 0(t^d)$ holds automatically if the closed fiber is degenerate).

We consider all the types of degenerate fibers which are described in [11].

(b_m) This fiber is impossible for a supersingular curve, since $\nu(\lambda) = 0$ for such a fiber ([11], page 124), and supersingularity implies that $\lambda = 0$.

(c1) It follows from (2) that for $\nu(g) \leqslant d$ we have $\nu(f) \geqslant \nu(g)$; also, since $\nu(\gamma) = 1$ for this fiber ([11], page 124), it follows from (4) that $\nu(\beta f + \gamma' g) = \nu(\gamma' g) \geqslant d$. Therefore, $\nu(g) \geqslant d = \epsilon(X)/4$.

(c2) From (3) and the fact that $\nu(\beta) = 1$ for this fiber ([11], page 124) it follows that $\nu(g) \geqslant d = \epsilon(X)/4$.

(c3) Since the fiber must satisfy the condition $\nu(\bar{\gamma}) = 2$ ([11], page 124), it follows that $d = 1$, and the relations of the lemma hold automatically.

(c4) It follows from the conditions in [11], page 124, that $\nu(\gamma + \alpha\beta) = 3$. Then (2), (3) and (4) imply that $(\gamma + \alpha\beta)' g \equiv 0(t^d)$, so that $\nu(g) \geqslant d - 2 = \epsilon(X)/4 - 2$.

($c5_m$) In this case, consideration of the Weierstrass model gives no new restrictions on the vector field, and it is necessary to look at the Néron model. Its covering by charts, which are denoted by $A_1, \dots, A_{m+2}$, is described in [11].

(a) $m = 2n - 1$. In this case, it follows from the conditions in [11], page 125, that $d = n + 1$. The chart A_{m+2} is associated with the model (1) by means of the birational map $u = x/t^{n+1}$, $v = y/t^{n+1}$. Since

$$D(u)=D\left(\frac{x}{t^{n+1}}\right)=\frac{D(x)}{t^{n+1}}+(n+1)\frac{x}{t^{n+1}}\frac{D(t)}{t}=\frac{f}{t^{n+1}}+(n+1)u\frac{g}{t},$$

it follows from the regularity of D that $f \equiv 0(t^{n+1})$. In view of the condition $\nu(\alpha) = 1$ ([11], page 125), we get from (2) that $g \equiv 0(t^{n+1})$, also, i.e. $\nu(g) \geqslant d = \epsilon(X)/4$.

(b) $m = 2n$. The argument is similar. One must use the relation $\nu(\beta) = n + 2$ ([11], page 125) and the regularity of D on the chart A_{m+1} (use the function $v = y/t^{n+1}$). We get that $\nu(g) \geqslant d - 1 = \epsilon(X)/4 - 1$.

(c6) From the relation $\nu(\bar{\gamma}) = 4$ in [11], page 125, it follows that $d = 2$, and the relations in the lemma always hold.

(c7) In view of (3), the condition $\nu(\beta) = 3$ in [11], page 125, leads to the relation $\nu(g) \geqslant d - 2 = \epsilon(X)/4 - 2$.

(c8) It is necessary to use the fact that D is regular on the chart A_3, which is defined by the relations $u = x/t^2$ and $v = y/t^2$. Since the function $D(v)$ is regular on this chart, we get the relation

$$\beta f + \gamma' g \equiv 0\,(t^{d+2}).$$

As in the case of the fiber (c4), it follows from (2) and the condition $\nu(\beta) \geqslant 4$ in [11], page 125, that $\nu(\beta f + \gamma' g) = \nu(\gamma' g)$. And since, again by [11], page 125, we have $\nu(\gamma) = 5$, it follows that $\nu(g) \geqslant d - 2 = \epsilon(X)/4 - 2$.

Case $p = 3$. In order to get rid of the term involving xy, we carry out a transformation $y = z + \lambda x$ and get the relation

$$z^2 + a_3 z = x^3 + a_4 x + a_6, \tag{5}$$

where

$$a_3 = \mu, \quad a_4 = \beta - \lambda\mu, \quad a_6 = \gamma; \tag{6}$$

the term involving x^2 on the right-hand side fails to appear because of the supersingularity ([17], page 62), which evinces itself in the form

$$-\alpha = \lambda^2. \tag{7}$$

In addition, the discriminant is equal to $-a_4^3$ ([17], page 62). If we set $\nu(a_4) = d$, then we will have $\epsilon(X) = 3d$.

A vector field which is regular on the affine part of the model (5) must have the form

$$D = f\frac{\partial}{\partial z} + g\frac{\partial}{\partial t}, \quad f, g \in O.$$

Relation (5) yields

$$a_4 D(x) = (2z + a_3) f + (a_3' z - a_4' x - a_6') g;$$

hence

$$a_3 f - a_6' g \equiv 0\,(t^d), \qquad -f + a_3' g \equiv 0\,(t^d).$$

Eliminating f, we obtain

$$(\mu^2 + \gamma)' g \equiv 0\,(t^d). \tag{8}$$

In view of the conditions $\nu(\gamma) = 1$ and $\nu(\mu) \geqslant 1$ ([11], page 124), we get from (8) that $\nu(g) \geqslant d$.

(c2) The conditions $\nu(\beta) = 1$, $\nu(\lambda) \geqslant 1$ and $\nu(\mu) \geqslant 1$ ([11], page 124) show that $\nu(a_4) = 1$, i.e. $d = 1$, and the relations of the lemma hold automatically.

(c3) The relation $\nu(\mu^2 + \gamma) = 2$ ([11], page 124), together with (8), yields that $\nu(g) \geqslant d - 1$.

(c4) The conditions $\nu(\Delta) = 6$, $\nu(\beta) \geqslant 2$, $\nu(\gamma) \geqslant 3$ and $\nu(\lambda) \geqslant 1$ ([11], page 124), together with (7), show that $\nu(\beta) = 2$, and in view of the fact that $\nu(\mu) \geqslant 2$ ([11], page 124) we get that $\nu(a_4) = d = 2$, so that the conditions of the lemma hold automatically.

(c5) This case is impossible, since the condition $\nu(\alpha) = 1$ ([11], page 125) contradicts (7).

(c6) The condition $\nu(\mu^2 + \gamma) = 4$ ([11], page 125), together with (8), yields $\nu(g) \geqslant d - 3$.

(c7) The conditions $\nu(\beta) = 3$, $\nu(\mu) \geqslant 3$ and $\nu(\lambda) \geqslant 1$ show that $d = \nu(a_4) = 3$, and the conditions of the lemma hold automatically.

(c8) The conditions $\nu(\gamma) = 5$ and $\nu(\mu) \geqslant 3$ ([11], page 125) show that $\nu(\mu^2 + \gamma) = 5$, and it follows from (8) that $\nu(g) \geqslant d - 4$.

In the following lemmas, we examine the types of fibers of an elliptic pencil X which do not satisfy the transversality condition in Theorem 3. It is known that there are two such types: (c1) and (c2), in the notation of [11]. The fiber (c1) is a rational curve with a cusp, while (c2) consists of two smooth rational curves which have simple tangency at one point.

LEMMA 3. *Theorem* 3 *is also true for fibers of type* (c2).

PROOF. The vector field D cannot vanish on both components of the fiber: Since D is of multiplicative type, it follows that the components of the divisor of D do not intersect. Nor can D vanish on one of the components, since then it would have a singular point of multiplicity 2 on the other component, which would contradict Lemma 2 of §1. By that same lemma, D has two singular points on each component. If one of these points is the point at which the components intersect, then the number of points is equal to 3 and coincides with $e(F)$, where F is the fiber under consideration. Thus, it remains to show that the point at which the components intersect is a singular point of the vector field.

Let C_1 and C_2 be the components of F, and let x_0 be the point at which they intersect; we will assume that x_0 is not a singular point of D. Then the transformation $\pi: X \longrightarrow X^D$ defines a morphism of a neighborhood of x_0 onto a smooth surface. We set $C_i' = \pi C_i$ for $i = 1, 2$. Since C_1 and C_2 are integral curves for D, it follows by Proposition 1 that $C_i = \pi^* C_i'$. Hence $(C_1, C_2)_{x_0} = p(C_1', C_2')_{\pi(x_0)}$, which immediately leads to a contradiction for $p \neq 2$, since $(C_1, C_2)_{x_0} = 2$.

Suppose $p = 2$. We denote by x_1 and y_1 the singular points of D lying by assumption on C_1, and we denote those on C_2 by x_2 and y_2. An obvious verification shows that by carrying out a σ-process at any of these points, we obtain a surface on which no singular point of D lies on the new exceptional curve. We denote by $\overline{X}$ the surface which we get by carrying out σ-processes at all four points. On this surface, a fiber will consist of six rational curves: $\overline{C}_1$ and $\overline{C}_2$ (the preimages of C_1 and C_2), L_1 and M_1 (the preimages of x_1 and y_1), and L_2 and M_2 (the preimages of x_2 and y_2). The curves $\overline{C}_1$ and $\overline{C}_2$ are tangent at one point; furthermore, L_1 and M_1 intersect $\overline{C}_1$ transversally at two different points, and L_2 and M_2 intersect $\overline{C}_2$ similarly. The vector field D now has no singular points on the whole fiber. Under the transformation $\overline{\pi}: \overline{X} \longrightarrow \overline{X}^D$, the elliptic pencil of $\overline{X}$ goes to the elliptic pencil of $\overline{X}^D$, and our fiber of the pencil of $\overline{X}$ goes to a fiber of the pencil of $\overline{X}^D$ which consists of six curves which intersect transversally in the following manner:

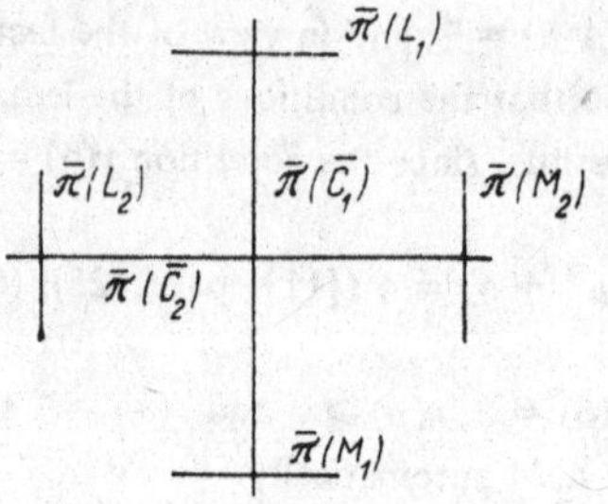

The curves $\bar{C}_1$ and $\bar{C}_2$ are integral curves for D on $\bar{X}$, whereas L_1, M_1, L_2 and M_2 are not. Therefore Proposition 1 yields that $\bar{\pi}(\bar{C}_i)^2 = \frac{1}{2}(\bar{C}_i)^2 = -2$ for $i = 1, 2$, and $\bar{\pi}(L_i)^2 = 2L_i^2 = -2$ and $\bar{\pi}(M_i)^2 = -2$ for $i = 1, 2$. Thus the fiber pictured above contains no exceptional curves and must be a fiber of a minimal model. However, in the list of possible fibers, such a type does not appear. This proves the lemma.

LEMMA 4. *For $p \neq 2$, no regular vector field whose normal component fails to be identically equal to zero can exist in a neighborhood of a fiber of type* (c1).

Since the statement is local in character, we can assume that the pencil has a section. Let us suppose that the Weierstrass form of the pencil has the form

$$y^2 + \lambda xy + \mu y = x^3 + \alpha x^2 + \beta x + \gamma. \tag{9}$$

If we make the transformation $\bar{y} = y + \frac{1}{2}(\lambda x + \mu)$, we reduce (9) to

$$\bar{y}^2 = x^3 + \bar{\alpha} x^2 + \bar{\beta} x + \bar{\gamma}, \text{ where } \bar{\gamma} = \gamma + \frac{1}{4}\mu^2. \tag{10}$$

It is easy to see that any regular vector field D on the model (10) can be given by the formula

$$D = f\bar{y}\frac{\partial}{\partial x} + g\frac{\partial}{\partial t},$$

where g and f are functions of a local parameter t on the base. It follows from (10) that

$$2\bar{y}D(\bar{y}) = (3x^2 + 2\bar{\alpha}x + \bar{\beta})D(x) + (\alpha' x^2 + \beta' x + \gamma')D(t).$$

Since $D(x) \equiv 0\ (\bar{y})$, it follows from this that

$$(\bar{\alpha}' x^2 + \bar{\beta}' x + \bar{\gamma}')g \equiv 0\ (\bar{y}),$$

i.e. $\bar{\alpha}' g = \bar{\beta}' g = \bar{\gamma}' g = 0$. By assumption, $g \neq 0$; hence $\bar{\gamma}' = 0$. Therefore $\gamma' + \frac{1}{2}\mu\mu' = 0$. For a minimal Weierstrass model of type (c1), every coefficient of the equation, and in particular μ, vanishes at the point over which the fiber is situated. Therefore $\nu(\mu) > 0$; hence $\nu(\gamma') > 0$, i.e $\nu(\gamma) > 1$. But the fiber (c1) is characterized by the condition $\nu(\gamma) = 1$ ([11], page 124). This contradiction proves the lemma.

The last two lemmas refine the structure of pencils with certain special fibers which will arise in the proof of Theorem 6.

LEMMA 5. *For $p = 2$, there exists no regular vector field D of multiplicative type in a neighborhood of a fiber corresponding to the extended root system D_n.*

PROOF. The fiber D_n contains a component $\overline{C}$ which intersects three components C_1, C_2 and C_3, where C_1 and C_2 intersect no other component (all the components are smooth rational curves). It follows from the properties of a vector field of multiplicative type and Lemma 2 of §1 that D must have a zero of multiplicity 1 on $\overline{C}$ and exactly one singular point on each of the components C_1 and C_2; call these points x_1 and x_2, respectively. As in the proof of Lemma 3, we denote by $\widetilde{X}$ the surface which we obtain from a given surface X by σ-processes at the point x_1. The morphism $\pi\colon \widetilde{X} \longrightarrow \widetilde{X}^D$ maps a neighborhood of C_1 onto a smooth surface. We denote by $\widetilde{C}_1$ the preimage of C_1 in $\widetilde{X}$. Then, as in the proof of Lemma 3, we have $\widetilde{C}_1^2 \equiv 0\ (2)$, which is impossible, since $C_1^2 = -2$ and hence $\widetilde{C}_1^2 = -3$.

For the formulation of the following lemma, we recall that for an elliptic pencil over $\mathbf{A}^1$ which has a section, we can find a model of type (9) which will be a minimal Weierstrass model for any point $a \in \mathbf{A}^1$. In other words, the coefficients λ, μ, α, β and γ in (9) will be polynomials and will satisfy the conditions for a minimal Weierstrass model at each point $a \in \mathbf{A}^1$. This follows from the argument used to prove Proposition 4 in Chapter III of [11], where the existence of a minimal Weierstrass model was established (see also the author's remark at the end of [11], §16, page 123). In particular, these conditions determine the coefficient λ uniquely up to a nonzero constant multiple. It is easy to see that λ coincides with the Hasse invariant of the curve given by equation (9).

LEMMA 6. *Let $p = 2$. For a reduced elliptic pencil, the invariant $\epsilon(X)$ (see the definition before Lemma 2) satisfies the condition $\epsilon \geqslant 4$ for any fiber of unstable type. If $X \longrightarrow \mathbf{A}^1$ is an elliptic pencil which has a section and a unique degenerate fiber of unstable type for which $\epsilon = 4$, then the Hasse invariant for this pencil has the form $\lambda = tu^2$, where t is a coordinate on $\mathbf{A}^1$.*

The first assertion follows from the explicit formula for the discriminant of the curve (9) ([11], page 94)

$$\Delta = \mu^4 + \lambda^3\mu^3 + \lambda^4\beta^2 + \lambda^4\mu^2\alpha + \lambda^5\mu\beta + \lambda^6\gamma. \tag{11}$$

Since for fibers of unstable type (i.e. type (c) in the terminology of [11]) all of the coefficients of (11) vanish at the point a over which the fiber lies, it follows that $\epsilon = \nu_a(\Delta) \geqslant 4$.

To prove the second statement of the lemma, we assume that the degenerate fiber of X is situated over the point of the base with coordinate $t = 0$. We can assume that (11) defines a Weierstrass model which is minimal on all of $\mathbf{A}^1$. Therefore the discriminant Δ can vanish only when $t = 0$; also, in view of the condition $\epsilon = 4$, we can assume (multiplying t by a constant, if necessary) that

$$\Delta = t^4. \tag{12}$$

Because the fiber over the point $t = 0$ is of unstable type, every coefficient of (11) vanishes when $t = 0$. We set $\lambda = \lambda_1 t$, $\mu = \mu_1 t$, $\alpha = \alpha_1 t$, $\beta = \beta_1 t$ and $\gamma = \gamma_1 t$. Substituting these values in (11) and using (12), we obtain, after simple transformations, that

$$\lambda_1^3 t^3 (\mu_1^3 + \lambda_1\beta_1^2 + \lambda_1\mu_1\alpha_1 t^2 + \lambda_1^2\mu_1\beta_1 t^2 + \lambda_1^3\gamma_1 t^2) = (1 + \mu_1)^4 \tag{13}$$

It follows that λ_1 and μ_1 are relatively prime. Therefore λ_1 is relatively prime to the sum in parentheses on the left-hand side of (13), so that any factor must appear in λ_1 with an even power, i.e. $\lambda_1 = u^2$.

§5. Vector fields on elliptic surfaces

THEOREM 6. *Let X be a reduced elliptic surface (which if $p = 2$ satisfies the additional condition $\chi(0_X) \equiv 0\ (2)$). If X has a regular nontrivial vector field, then it can be obtained from the following construction: X is the minimal elliptic surface birationally equivalent to Y/G, where $Y \longrightarrow C$ is an elliptic surface which is a principal fiber bundle with an elliptic curve A as structure group (any such surface is given by an element of $H^1(C, A)$), and G is a finite group of automorphisms of Y which takes fibers into fibers and commutes with an action of G on C, $\psi: G \longrightarrow \mathrm{Aut}\, C$. Three cases are then possible*:

a) *C has genus $g(C) > 1$ and $\psi(G)$ acts freely*;

b) *$g(C) = 1$, $\psi(G)$ acts freely and X is a hyperelliptic or Abelian surface*;

c) *$g(C) = 0$ and $X \approx A \times \mathbf{P}^1$ or is a rational surface.*

We remark that all of the constructed surfaces actually possess an elliptic pencil, given by the natural projection $Y/G \longrightarrow C/\psi(G)$, almost all fibers of which are equal to A. A more detailed description of the vector fields on these surfaces will be given after the proof of the theorem.

We begin with an enumeration of the reduced elliptic pencils $f: X \longrightarrow B$ on which there exist regular vector fields θ whose normal components θ_N differ from 0. By Proposition 6, in this case there exists a nonzero regular vector field τ on B which vanishes at the degenerate points of the morphism f. Therefore, $g(B) = 1$ or 0; in the case where $g(B) = 1$, the pencil f has no degenerate fibers, while in the case where $g(B) = 0$, the degenerate fibers lie over zeros of the vector field τ (they are consequently either 0, 1, or 2 in number).

We begin with some observations relating to the case where all the fibers have potentially good reduction; this implies that the absolute invariant j of the general fiber is a rational function on the base which nowhere takes the value ∞. Since the base is a complete curve, any such function is constant: $j = \alpha \in \mathrm{k}$. If A is an elliptic curve over k with absolute invariant α, then the general fiber of $X \longrightarrow B$ becomes isomorphic to A after some separable extension of $\mathrm{k}(B)$; the same holds for the Jacobian $\mathfrak{J} \longrightarrow B$ of $X \longrightarrow B$. Hence there exists a Galois cover $C \longrightarrow B$ such that

$$\mathfrak{J} \times_B C \backsim A \times C,$$

$\backsim$ denoting birational isomorphism compatible with the projection to C. If G is the Galois group of the cover $C \longrightarrow B$, then this implies that

$$\mathfrak{J} \backsim (A \times C)/G;$$

here G acts on C as a Galois group (which gives an inclusion $\psi: G \longrightarrow \mathrm{Aut}\, C$), and its action on A is defined by a homomorphism $\varphi: G \longrightarrow \mathrm{Aut}(A, 0)$ (the group of automorphisms of A preserving the group law). If φ were not a monomorphism, then we could replace C by a smaller cover corresponding to the kernel of φ; hence we can consider that φ is also an inclusion.

If $\mathfrak{I} \times_B C \backsimeq A \times C$ then $X \times_B C \backsimeq Y$, where Y is as in the statement of Theorem 6. Then

$$X \backsimeq Y/G. \tag{1}$$

The cover $C \longrightarrow B$ is only branched at points of B which correspond to degenerate fibers of the pencil $X \longrightarrow B$. Indeed, since this is a reduced pencil, it follows that the scheme $X \times_B \hat{\mathcal{O}}_b$, where $\hat{\mathcal{O}}_b$ is the completion of the local ring of a point $b \in B$, is isomorphic to its Jacobian. Let $c \in C$ be such that $c \longrightarrow b$, and suppose that c has stabilizer subgroup $G_c \subset G$. Then this Jacobian is defined by the action of G_c on A by means of automorphisms of A which preserve the composition law; therefore we can assume that $\varphi(G_c) \subset \mathrm{Aut}(A, 0)$, where $0 \in A$ is the zero element. If $\varphi(G_c) \neq 1$, then G_c acts nontrivially on A and hence on the one-dimensional l-adic cohomology of A. But by the Serre-Tate test [16] this contradicts the fact that the fiber $f^{-1}(b)$ is nondegenerate. Therefore $\overline{\varphi}(G_c) = 1$. Since φ is an injection, it follows that $G_c = 1$, also, if the fiber $f(b)$ is nondegenerate.

We can now set forth the plan for the proof of the theorem. We will consider two cases: I, when the genus of the basis is equal to 1; and II, when its genus is equal to 0. In case I all fibers are nondegenerate, so that we have a representation (1). We will show that the genus of the curve C is also equal to 1, and this case will lead to case b) of the theorem. In case II there will be separate considerations depending on whether there is at least one fiber without potentially good reduction or whether all fibers have potentially good reduction. We will show that the first case can never occur. In the second case, the representation (1) again leads to the assertion of the theorem.

The basic difficulties arise in the consideration of case II, where we successively consider types (1)–(6) below, which together embrace all the elliptic pencils on which there exist regular vector fields.

If all of the higher ramification indices are equal to 0, then we have types (1) and (2). This is the simplest part of the proof. The remaining parts consequently relate only to the case where the characteristic is 2 or 3. In the sequel, we will assume that the vector field D is of additive or multiplicative type (such a vector field always exists, in view of the corollary to Lemma 1 of §1). The normal component of D is of the same type as D itself. Therefore, in view of Lemma 2 of §1, we have one degenerate fiber in the first case and two in the second. The case where the vector field is of additive type divides into three types, (3), (4) and (5). If the unique degenerate fiber does not have potentially good reduction, we have type (3). Then there exists a covering of the base $B' \longrightarrow B$ such that the degenerate fiber is stable in $X' = X \times_B B'$. The proof in this case is based on the fact that a regular vector field on X determines a regular vector field on a minimal smooth model of X', which makes it possible to apply the results of cases (1) and (2). If the degenerate fiber has potentially good reduction (type (4)), then we again have a representation (1).

If A is supersingular (type (5)), then the surface X is rational, as follows easily from Lemma 2 of §4. Finally, type (6) corresponds to a vector field of multiplicative type. If the degenerate fibers are different from (c1) and (c2), then it is easy to infer from Theorem 3 that all the higher ramification indices are zero, so that we are back to types (1) and (2). The case where there are fibers of type (c1) or (c2) is analyzed on the basis of special considerations using the help of Lemmas 3, 4, 5, and 6 of §4.

We now present the proof.

I. *The genus of the base B is equal to* 1. In this case there are no degenerate fibers, C is an unramified covering of the elliptic curve B, $\psi(G)$ acts on C freely, and the birational isomorphism (1) is an isomorphism. Y is an abelian variety and corresponds to an element of $\mathrm{Ext}(C, A)$. If G acts on Y by translations, then X is an abelian variety. Otherwise, X is a hyperelliptic surface, and we have case b) of the theorem.

II. *The genus of the base B is equal to* 0.

(1) Every fiber is stable, i.e. every fiber either is nondegenerate or corresponds to the extended root system A_m (type (b_m) in Néron's classification). If there are no degenerate fibers, then in (1), C must be an unramified covering of $\mathbf{P}^1$, i.e. $C = \mathbf{P}^1$ and $J = A \times \mathbf{P}^1$. Thus $X \longrightarrow \mathbf{P}^1$ is a principle bundle with fiber A, and hence $X = A \times \mathbf{P}^1$ since $H^1(\mathbf{P}^1, A) = 0$. This type is included in case c) of the theorem. If degenerate fibers are present, we base our argument on the formula for the Euler characteristic $e(X)$ of the surface X (see [4] and [15]). If degenerate fibers are present, then $b_1 = 0$ (b_i is the i-dimensional Betti number), and the formula sssumes the form

$$e(X) = 2 + b_2 = \sum (e(F_i) + \delta_i) = \sum (m_i + \varepsilon_i + \delta_i), \tag{2}$$

where we sum over all the degenerate fibers. Here m_i is the number of components of the fiber; ϵ_i equals zero if the fiber is of type A_m, and one if it is simply connected; and δ_i is the higher ramification index.

In the case which we are considering, we have $\delta_i = 0$ (see [15]) and $\epsilon_i = 0$. If there is only one fiber, then (2) yields

$$2 + b_2 = m_1. \tag{3}$$

But the components of the fiber, together with any cycle which is transversal to the fiber, form an independent system of elements in the two-dimensional cohomology group; therefore $m_1 + 1 \leqslant b_2$, which contradicts (3).

If there are two degenerate fibers, then we get from (2) that

$$2 + b_2 = m_1 + m_2. \tag{4}$$

But the components of two different fibers are connected in the two-dimensional cohomology group by a single relation, so that they, together with a transversal cycle, form a system of $m_1 + m_2$ independent elements. Therefore $m_1 + m_2 \leqslant b_2$, which contradicts (4).

(2) There is no higher ramification. In this case, for every fiber there exists a covering of the base which has no wild ramification and on which the fiber becomes stable after lifting. Since there are not more than 2 branch points, it follows that there exists a global covering $C \longrightarrow \mathbf{P}^1$ with $C \simeq \mathbf{P}^1$ on which both fibers become stable after lifting (this covering has the simple form $t \longrightarrow t^m$). Using what we proved in case (1), we get that $X \times_B C \backsim A \times \mathbf{P}^1$, and $X \backsim (A \times \mathbf{P}^1)/G$, where G acts on $\mathbf{P}^1$ and A by means of $\psi\colon G \longrightarrow \mathrm{Aut}\,\mathbf{P}^1$ and $\varphi\colon G \longrightarrow \mathrm{Aut}\,A$. Each of the transformations $\psi(\sigma)$, where $\sigma \in G$ and $\sigma \neq 1$, has a fixed point $c \in \mathbf{P}^1$. If $\varphi(\sigma)$ had no fixed points (i.e. if it were a translation on A), then the fiber on X corresponding to the point c would be a multiple fiber, which contradicts the fact that the pencil X is reduced. If, however, $\varphi(\sigma)$ has fixed points, then $A/\varphi(G)$ is a rational curve.

The rational transformation $X \longrightarrow A/\varphi(G)$ has fiber $\mathbf{P}^1$. Hence X is rational: this case is included in type c) in the formulation of the theorem.

Types (3), (4), and (5) correspond to the case where the vector field D is of additive type. In this case, there is a unique degenerate fiber.

(3) The stable fiber corresponding to the unique degenerate fiber F_b is not smooth, and hence it corresponds to the extended root system A_m. We can assume that the fiber is not stable, since otherwise we would be dealing with type (1). Therefore, there exists an extension of degree 2, $\mathcal{O}' \supset \mathcal{O}$, of the local ring corresponding to the point b of the base, such that $X \otimes_B \mathcal{O}'$ has a stable fiber over the point b' corresponding to b; the fiber corresponds to the extended root system A_m. Here $m = -\nu_b(j)$, where j is the absolute invariant of the general fiber of the pencil on X. Since $\nu_{b'}(j) = 2\nu_b(j)$, it follows that m is even. We set $G = \mathrm{Gal}(\mathcal{O}'/\mathcal{O})$ and suppose that $G = \{g\}$; then $g(a) = -a$ on the group of smooth points of the fibering of $X \otimes \mathcal{O}'$ ([12], page 5).

We can assume that $p = 2$, as otherwise there would again be no higher ramification, and we would be back in type (2). Therefore there exists a separable covering of degree 2, $B' \longrightarrow B$, ramified in only one point b and for which $B' \otimes \mathcal{O} \approx \mathcal{O}'$, Indeed, since $B \approx \mathbf{P}^1$, we can assume that we have chosen a coordinate T on B with b the point at infinity. If $\mathcal{O}'$ is defined over $\mathcal{O}$ by the equation $Z^2 + Z = f$, and if $f_0(T^{-1})$ is the principal part of the function f (which is a polynomial in T^{-1}), then B' is given by the equation $Z^2 + Z = f_0(T^{-1})$.

We denote a minimal nonsingular model for the fibering of $X \otimes_B B'$ by $\widetilde{X}$. Our next goal will be to prove that a vector field D which is regular on X defines a vector field on $\widetilde{X}$ which is also regular. In order to do this, we consider the rational isomorphism $X \backsim \widetilde{X}/G$, where $G = \mathrm{Gal}(B'/B)$ (in our case, $|G| = 2$). Since G will in general have fixed points on $\widetilde{X}$, it follows that $\widetilde{X}/G$ will have singularities. We denote by $\overline{X}$ the surface which we get from a minimal resolution of the singularities of $\widetilde{X}/G$. Then X is the minimal elliptic surface corresponding to $\overline{X}$ (i.e. X is obtained by contracting the exceptional curves of the first kind on $\overline{X}$ which are contained in the fibers). We will show that $X = \overline{X}$; for this it is enough to show that no component of the image in $\overline{X}$ of a fiber of $\widetilde{X}$ is an exceptional curve of the first kind. Indeed, taking the quotient of the fiber by the group of automorphisms generated by $a \longrightarrow -a$, the components are identified in pairs, with the exception of the two corresponding to the identity and the element of order two in the quotient group of the group of nonsingular points by the identity component. Thus, the fiber $F_{b'}/G$ consists of $m/2$ components which form a cycle:

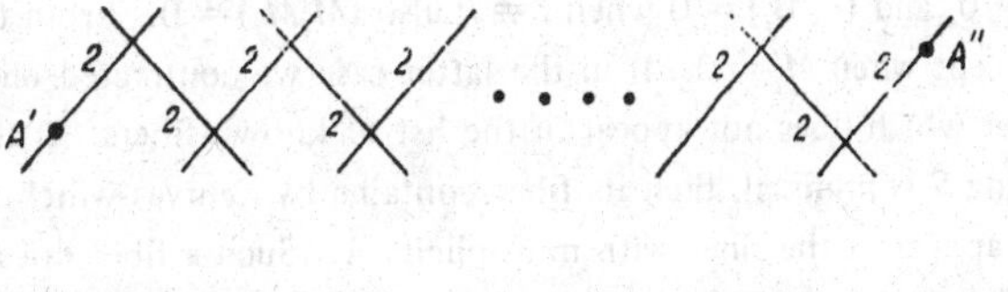

All of the components of this divisor have multiplicity 2, and on the first and last components there are unique points, A' and A'', respectively, which can be singular on $\widetilde{X}/G$. Moreover, these points are locally isomorphic, so that they are either both singular or both nonsingular. If they were both nonsingular, then the whole fiber of the smooth surface $\widetilde{X}/G$

could be contracted to a point, which is impossible. If both are singular, then after they have been resolved the divisor in the above figure will not contain any exceptional curve with self-intersection -1.

Let us denote by Y the surface which we obtain from X by deleting all of the cycles which arise in the resolution of the points A' and A'', and by $\widetilde{Y}$ the surface obtained from $\widetilde{X}$ by deleting the two fixed points of the nonidentity automorphism in G. Then the restriction of the covering $\widetilde{Y} \longrightarrow Y$ will be etale, and the regular vector field θ defined on X will determine a vector field on $\widetilde{Y}$ which is also regular. However, since $\widetilde{Y}$ is obtained from $\widetilde{X}$ by the deletion of two points, it follows that a vector field which is regular on $\widetilde{Y}$ is also regular on $\widetilde{X}$.

Thus a nonzero regular vector field on X would determine a regular vector field on $\widetilde{X}$, which would then correspond to the above case (1) with one degenerate fiber, which again gives a contradiction. Thus, this case can never occur.

(4) The unique degenerate fiber is of potentially good reduction. In this case, there exists a cover of the base $C \longrightarrow B$, for which we have a representation (1). Let us assume, in addition, that the curve A is not supersingular. We can restrict our attention to the cases $p = 2$ and $p = 3$, since otherwise the condition for case (2) above would hold. It is known that for $p = 2$ and $p = 3$ no supersingular curve has only two automorphisms (such curves are characterized by the condition $j = 0$). Therefore, we can take C to be a covering of degree two. It follows that in the case $p = 3$ there is no higher ramification, and we can confine our attention to the case $p = 2$.

We will now take a closer look at the structure of the birational isomorphism $X \backsim Y/G$, where $G = \{g\} = \operatorname{Aut} A$. Let $\overline{X}$ be the surface obtained from a minimal resolution of the singular points of the surface Y/G. The surface Y/G contains the degenerate fiber $S = A/G$ which has multiplicity two and which corresponds to the branch point of C/B. S contains two points, x_1 and x_2, which are the images of the points of order two on A and which are the only points which can be singular; moreover, the formal neighborhoods of these points are isomorphic. The points x_1 and x_2 are indeed singular, for otherwise the smooth surface Y/G would have a fiber $2S$ of multiplicity two. Resolution of x_1 and x_2 gives rise to identical cycles. The degenerate fiber on $\overline{X}$ will have the form $2S + T_1 + T_2$, where T_i is the divisor whose support intersects S at the point lying over x_i. If $(ST_i) > 1$, then $S^2 < -1$, and the curve S cannot be contracted. Suppose $(ST_i) = 1$; then $T_i = L_i + M_i$, where L_i is an irreducible curve with $(L_iS) = 1$, and the support of the divisor M_i does not intersect S. By contracting S, we obtain a fiber which consists of the two curves L_i and the divisors M_i. The L_i intersect in a single point through which neither M_i passes; moreover, $(L_iM_i) > 0$, and $(L_iM_j) = 0$ when $i \neq j$; also $(M_iM_j) = 0$. From this it follows that $(L_i^2) < -1$, except when $M_i = 0$. If in the latter case we contracted one of the L_i, we would obtain a fiber which does not appear in the list of known fibers. If the fibering obtained by contracting S is minimal, then its fiber contains two curves which intersect transversally and which appear in the fiber with multiplicity 1. Such a fiber does not appear in the list of known fibers, either. Thus the curve S cannot be contracted, and $\overline{X}$ is minimal, i.e. $\overline{X} = X$.

We denote by Y_1 the surface obtained from Y by deleting the fixed points of the automorphism g. The morphism $Y_1 \longrightarrow X$ is etale; therefore, the regular vector field θ on

X determines a regular vector field $\widetilde{\theta}$ on Y_1. Finally, since Y differs from Y_1 in only a finite set of points, it follows that $\widetilde{\theta}$ is regular on all of Y.

It is clear that the normal component of the vector field $\widetilde{\theta}$ also differs from 0. Therefore, we can apply to Y the argument given at the very beginning of the proof and conclude, in particular, that the genus of C is equal to 1 or 0. If $C = \mathbf{P}^1$, then the surface X is rational, as we noted in our analysis of case (2). If C is an elliptic curve, then the vector fields on such a surface were determined in our analysis of type I–they arise from the invariant vector fields on the abelian variety Y. By Lemma 1 of §4, the vector field θ, which is regular on X, determines a vector field which vanishes at the singular points of Y/G; hence on Y the vector field $\widetilde{\theta}$ vanishes at the fixed points of G. But if any vector field on Y vanishes at even one point, then the vector field itself is zero. Therefore G must act on Y freely, so that $\mathrm{Gal}(C/B)$ acts on C freely as well. This contradicts the fact that the genus of B is equal to 0.

(5) We make the same assumptions as in (4), except that here A is supersingular. In view of the fact that case (2) has already been analyzed, we can assume that $p = 2$ or $p = 3$ and apply Lemma 2 of §4. In our case, the normal component of the vector field has a zero of multiplicity 2 at the point corresponding to the degenerate fiber. Therefore, we obtain from Lemma 2 of §4 that $2 \geqslant n/4 - 3$, so that $n \leqslant 20$, if $p = 2$, and $2 \geqslant n/3 - 4$, so that $n \leqslant 18$, if $p = 3$. In the case where the degenerate fiber is unique, (2) shows that $n = \epsilon(X) = e(X)$; hence, in both cases, $e(X) \leqslant 20$. Since by the Riemann-Roch theorem we have $e(X) \equiv 0\ (12)$ (in the case where $K^2 = 0$), it follows that $e(X) \leqslant 12$; also, since there is a degenerate fiber, $e(X) > 0$. Therefore, $e(X) = 12$ and $\chi(X) = 1$, and it follows from the formula for the canonical class ([3], page 354) that the canonical class of the surface X is negative, i.e. X is rational.

(6) Finally, we assume that the vector field is of multiplicative type, so that the fibering of X has two degenerate fibers.

We will use the formula for the number of fixed points of a regular vector field on a complete surface:

$$w - U^2 - UK = e(X), \tag{5}$$

where U is the divisor of the vector field and w is the sum of the multiplicities of the isolated singular points.

In the case of an elliptic surface, the divisor of zeros of the vector field consists of components of the fibers, for otherwise the field would vanish at at least one point on each fiber, and its normal component would be identically equal to zero. Hence $UK = 0$. Therefore (5) assumes the form $w - U^2 = e(X)$. If we decompose both summands into terms corresponding to the degenerate fibers, we obtain, in the notation of Theorem 3, that

$$\sum p(F_b) = e(X). \tag{6}$$

We first consider the case where none of the fibers is of type (c1) or (c2). By Theorem 3, in this case $p(F_b) = e(F_b)$. Then (2) and (6) show that $\delta_b = 0$, i.e. we are back to case (2). According to Lemma 3 of §4, a fiber of type (c2) cannot occur, while for $p \neq 2$ Lemma 4 of §4 shows that no fiber of type (c1) is present. Thus, we can confine our attention to the case where $p = 2$, there are two degenerate fibers, F_1 and F_2, and F_1 is of type (c1).

Let us look more closely at the restriction of the vector field to the fiber F_1 of type (cl). We denote by Ω the completion of the local ring of the unique singular point of the fiber, and we denote by Ω^ν the normalization of Ω. We can assume that $\Omega^\nu = k[[t]]$, and that Ω consists of the $f \in \Omega^\nu$ such that $f \equiv \alpha(t^2)$, where $\alpha \in k$. A simple verification shows that any derivation of the ring Ω extends to a derivation D of the ring Ω^ν into its field of fractions such that $D = (f/t^2)d/dt$ for some $f \in \Omega$ (here we make use of the fact that $p = 2$). Therefore, a regular vector field D on the fiber F_1 determines a vector field D' on the normalization F_1^ν of F_1 which either is regular or has a pole of order 2 at the point corresponding to the singular point of the fiber. Since F_1^ν is a smooth rational curve, it follows that in the first case D' has 2 zeros on F_1^ν and in the second case D' has 4. In the second case we have $D(t^2)(0) \neq 0$, from which it follows that the singular point of the fiber is not a singular point of the vector field. Therefore in the first case there are no more than 3 singular points of the vector field on the fiber (there are in fact only 2, since the singular point of the fiber is not a singular point of the vector field), while in the second case there are four singular points of the vector field on the fiber. In other words,

$$p(F_1) \leqslant 4. \tag{7}$$

If the second degenerate fiber were also of type (cl), then (6) would imply that $e(X) \leqslant 8$. Since $e(X) \equiv 0 \pmod{12}$, this implies that $e(X) = 0$, so (2) implies that X does not have degenerate fibers at all. Thus we can apply Theorem 3 to the fiber F_2. Applying (2) and (6), we get that

$$p(F_1) + e(F_2) = 2 + \delta(F_1) + e(F_2) + \delta(F_2),$$

or, in view of (7),

$$\delta(F_1) + \delta(F_2) \leqslant 2.$$

But the contribution of F_1 to the discriminant of the fibering is equal to $\epsilon(F_1) = 2 + \delta(F_1)$, and by the first assertion of Lemma 5 of §4 this is $\geqslant 4$. Therefore $\delta(F_1) = 2$, $\epsilon(F_1) = 4$, and $\delta(F_2) = 0$. Thus (2) assumes the form

$$4 + e(F_2) = e(X). \tag{8}$$

If the fiber F_2 were of type other than A_n or D_n, then $e(F_2) \leqslant 10$. By (8) we would then have $e(X) \leqslant 14$, i.e. $e(X) = 12$, and the surface X would be rational. By Lemma 5 of §4, the fiber D_n cannot occur. It remains for us to consider the case where F_1 is of type (cl) with $\epsilon(F_1) = 4$ and F_2 is of type A_n.

We will show that such an elliptic surface can only exist (for $p = 2$) if $\chi(\mathcal{O}_X) \equiv 1$ (2); our argument does not assume the existence of a vector field on the surface. Moreover, by going to the Jacobian we can assume that X has a section.

Suppose that the fibering over $\mathbf{A}^1$ is given by (9) of §4, which defines a minimal Weierstrass model for any point $a \in \mathbf{A}^1$. Suppose that F_1 corresponds to the point $t = 0$, and that F_2 corresponds to the point $t = \infty$. We will determine $e(F_2)$. To do this, we note that for the fiber F_2 of type A_m we have $e(F_2) = m = -\nu_\infty(j)$. We will use the formula for j in characteristic 2: $j = \lambda^{12}/\Delta$ ([11], page 94). We will also use Lemma 6 of §4. As in that lemma, we have $\Delta = t^4$ and $\lambda = tu^2$. Therefore, $j = t^8u^{24}$ and $e(F_2) = 8 + 24k$, where $k = \deg u$.

Formula (2) yields $e(X) = 12 + 24k$; hence $\chi(\mathcal{O}_X) = 1 + 2k$. (We could have analyzed the case for which the fibering had types (c1) and A_m without using Lemma 6 of §4, but assuming the existence of a vector field and making a comparison of $e(X)$ and $e(X^D)$ based on the quotient morphism by this vector field.)

This concludes our investigation of vector fields which have nonzero normal components. It remains to consider vector fields which are tangent to the fiber of the morphism $f\colon X \longrightarrow B$, i.e. sections of the relative tangent bundle $\Theta_{X/B}$.

We first suppose that the pencil f has a section $s\colon B \longrightarrow X$. In this case the set X^0 of smooth points of f forms a group scheme over B. Since the general fiber of this scheme is an elliptic curve, it follows that the vector field $\theta \in H^0(\Theta_{X/B})$ is invariant with respect to the action of X^0/B, and that in order to determine it on X, it is enough to determine its restriction to $s(B)$. But the normal bundle to $s(B)$ has degree $(s(B))^2 = -\chi(\mathcal{O}_X) = -e(X)/12$. On the other hand, by (2) we have $e(X) \geqslant 0$, and $e(X) = 0$ only when f has no degenerate fiber. Therefore, except for this last case, the normal bundle of $s(B)$ cannot have a nonzero section, and θ must be tangent to $s(B)$. Since it is simultaneously tangent to the fiber, it follows that $\theta = 0$. If, however, f has no degenerate fiber, then, as we have already seen, there exists an unramified Galois covering $C \longrightarrow B$ with Galois group G, as well as an elliptic curve A and an inclusion $G \hookrightarrow \operatorname{Aut} A$, such that $X = (A \times C)/G$ and $H^0(\Theta_{X/B})$ coincides with the G-invariant vector fields on A. If $g(B) = 0$, this case is impossible; if $g(B) = 1$, it leads to case b) of the theorem; and if $g(B) > 1$, it leads to case a).

Suppose now that the pencil f has no section, and let $\tilde{f}\colon J \longrightarrow B$ be the corresponding Jacobian. We denote by X^0 and J^0, respectively, the sets of smooth points of the morphisms f and $\tilde{f}$. An obvious verification shows that for any smooth morphism $\bar{f}\colon V \longrightarrow B$ and any finite Galois covering $C \longrightarrow B$ with Galois group G, the regular vector fields on V which are tangent to the fibers of $\bar{f}$ are in one-to-one correspondence with the G-invariant regular vector fields on $V \times_B C$ which are tangent to the fibers of the projection onto C (G acts through C). Let us now consider a covering $C \longrightarrow B$ such that $X \times_B C$ and $J \times_B C$ are isomorphic. The isomorphism $\varphi\colon X \times_B C \simeq J \times_B C$ carries the action of G on $X \times_B C$, realized by means of the action on the factor C, into the following action on $J \times_B C$:

$$\sigma(x, c) = (x + \varphi_\sigma(c), \sigma(c)),$$

for $\sigma \in G$, where $\varphi_\sigma \in H^1(G, \operatorname{Hom}_B(C, J))$. Since θ is invariant under translations, its invariance with respect to this action is equivalent to invariance with respect to the trivial action on $J \times C$ through C. Therefore a vector field on X tangent to the fibers would determine a vector field on J^0, which, as we saw, is possible only if $J = (A \times C)/G$, where $C \longrightarrow B$ is an unramified covering with Galois group G. We have already found that in this case $X = Y/G$, where $Y \longrightarrow C$ is a principle fiber bundle with group A (although perhaps G and C may be different). Depending on whether $g(B) = 1$ or $g(B) > 1$, we arrive at cases b) or a) of the theorem. This completes the proof of Theorem 6.

REMARKS. 1) We do not know whether the condition $\chi(\mathcal{O}_X) \equiv 0$ (2) is essential for the truth of Theorem 6 in the case $p = 2$.

2) In case b) Y corresponds to an element of $\operatorname{Ext}(C, A)$, and in case a) to $\operatorname{Ext}(\mathfrak{J}(C), A)$,

where $\mathfrak{J}(C)$ is the Jacobian of C. In both these cases this element defines an isogeny of the abelian variety $A \times \mathfrak{J}(C)$, and in case a) Y is the inverse image of $A \times C$. From this remark it follows that we could even take $Y = A \times C'$ and $X = Y/G'$, but with C' and G' not necessarily reduced such that $C' \to B$ is a principal fiber bundle with structure group G'.

3) If the pencil $X \to B$ has a section, then $Y \to C$ is of the form $Y = A \times C$, and the action of G is determined by its action ψ on C and its action φ on A, with the action of $\varphi(G)$ preserving the group law on A. The surface $(A \times C)/G$ has a vector field θ_A tangent to the fibers if the differentials of every automorphism $\varphi(g)$ for $g \in G$ equal 1 (that is, if G is a p-group). Since all such vector fields can be transferred onto $X \to B$ from its Jacobian, under the same hypotheses they also exist on X.

In case a) there can never be any other vector fields. In case b), if X is an abelian surface then it has only invariant vector fields. Hyperelliptic surfaces are described in [2] and are of the form $(A \times C)/G$, with A and C elliptic curves, and G can be one of the following possibilities:

Group G	Action of G on $A \times C$	Condition on A	Characteristic
a1) $\mathbf{Z}/2\mathbf{Z}$	$(a, c) \to (-a, c+\gamma)$, $2\gamma = 0$	none	any
a2) $(\mathbf{Z}/2\mathbf{Z})^2$	$(a, c) \to (-a, c+\gamma)$, $(a+\alpha, c+\beta)$, $2\alpha = 2\beta = 2\gamma = 0$	none	$\neq 2$
a3) $\mathbf{Z}/2\mathbf{Z} \times \mu_2$	$(a, c) \to (-a, c+\alpha)$, $2\alpha = 0$; μ_2 acts on A and C	none	2
b1) $\mathbf{Z}/3\mathbf{Z}$	$(a, c) \to (\omega a, c+\gamma)$, $3\gamma = 0$, $\omega^3 = 1$, $\omega \in \operatorname{Aut} A$	$j(A) = 0$	any
b2) $(\mathbf{Z}/3\mathbf{Z})^2$	$(a, c) \to (\omega a, c+\gamma)$, $(a+\alpha, c+\beta)$, $3\alpha = 3\beta = 0$; γ and ω as in b1)	$j(A) = 0$	$\neq 3$
c1) $\mathbf{Z}/4\mathbf{Z}$	$(a, c) \to (ia, c+\gamma)$, $4\gamma = 0$, $i^4 = 1$, $i \in \operatorname{Aut} A$	$j(A) = 12^3$	any
c2) $\mathbf{Z}/2\mathbf{Z} \times \mathbf{Z}/4\mathbf{Z}$	$(a, c) \to (ia, c+\gamma)$, $(a+\alpha, c+\beta)$, $2\alpha = 2\beta = 0$; γ and i as in c1)	$j(A) = 12^3$	$\neq 2$
d) $\mathbf{Z}/6\mathbf{Z}$	$(a, c) \to (-\omega a, c+\gamma)$, $6\gamma = 0$, $\omega^3 = 1$, $\omega \in \operatorname{Aut} A$	$j(A) = 0$	any

The vector field θ_C tangent to C and 0 on A always gives a vector field on X. Furthermore, the field θ_A which is tangent to A and 0 on C gives a vector field on X if the automorphisms of A have differential equal to 1; that is, in the following cases: a1) with $p = 2$; a3) with $p = 2$; b1) with $p = 3$ and c1) with $p = 2$.

We have not compiled for case c) a complete list of the rational surfaces which possess vector fields. It would probably not be difficult to make such a list starting from the minimal models and keeping track of the effect of σ-processes on the dimension of the space of vector fields.

4) An interesting problem in the theory of algebraic surfaces is the question of when the group scheme of automorphisms of a smooth complete surface X is reduced. It is clear that this will be true if and only if $\dim H^0(X, \Theta_X) = \dim \operatorname{Aut} X$, where $\operatorname{Aut} X$ denotes the reduced group of automorphisms (or its connected component). The list of all possible cases given in Remark 1 allows us to determine easily which of them correspond to schemes of automorphisms which are not reduced.

In case a), if $d\varphi(G) = 0$ but $\varphi(G)$ does not consist of translations only ($\varphi(G) \not\subset A$), then, as we saw, there exists a vector field which is unique up to proportionality, so that $\dim H^0(X, \Theta_X) = 1$. However, there exists no algebraic group of automorphisms corresponding to this vector field. Such a group would be an elliptic curve Δ and would act fiberwise on X. But then X would be a principal fiber bundle with group Δ. Hence the one-dimensional Betti number would be given by $b_1(X) = b_1(B) + 2$. However, an independent method of computing the Betti numbers is in terms of the representation $X = Y/G$, and we get $H^1(X) = H^1(Y)^G = H^1(C)^G = H^1(B)$ and $b_1(X) = b_1(B)$. Thus in this case $\dim \operatorname{Aut} X = 0$.

In case b), the group corresponding to the vector field θ_C is always the group of translations on the points of the curve C; this group commutes with the automorphisms from G and is carried onto X. In the cases a1) ($p \neq 2$), a2) ($p \neq 3$), b2), c1) ($p \neq 2$), c2) and c3) there exist no other vector fields. Therefore, we have here that

$$\dim H^0(X, \Theta_X) = \dim \operatorname{Aut} X = 1.$$

However, in cases a1) ($p = 2$), b1) ($p = 3$) and c1) ($p = 2$) the vector field θ_A is G-invariant and, just as in case a), no corresponding group of automorphisms exists. Therefore in these cases $\dim H^0(X, \Theta_X) = 2$ and $\dim \operatorname{Aut} X = 1$, and the scheme of automorphisms is not reduced.

5) It would be interesting to extend Theorem 6 to the case of elliptic fiberings with multiple fibers. Here we would find new types of surfaces with nonreduced group schemes of automorphisms. For example, take $X = (A \times \mathbf{P}^1)/G$, where A is an elliptic curve and $G = \mathbf{Z}/2\mathbf{Z}$ and where the action is given (after a coordinate x has been chosen on $\mathbf{P}^1$) as $(a, x) \longrightarrow (a + \alpha, x + 1)$, where $2\alpha = 0$. Then besides the G-invariant vector fields θ_A and d/dx, to which the group $(a, x) \longrightarrow (a + a', x + c)$ corresponds, we have, when $p = 2$, the G-invariant field $(x^2 + x)d/dx$, to which no group of automorphisms corresponds. Therefore for this example we have $\dim H^0(X, \Theta_X) = 3$ and $\dim \operatorname{Aut} X = 2$.

§6. *K*3 surfaces

We give one application of Theorem 6.

THEOREM 7. *There exists no nonzero regular vector field on a K3 surface.*

1 By Corollary 2 to Theorem 5, on such a surface there exists either an elliptic or a quasi-elliptic pencil $f\colon X \longrightarrow \mathbf{P}^1$. We consider first the case where the fiber is elliptic. It follows from general properties of a $K3$ surface that the elliptic pencil is reduced ([14], §3, Corollary 2 of Theorem 1). Since $\chi(\mathcal{O}_X) = 2$ for a $K3$ surface, it follows that the condition $\chi(\mathcal{O}_X) \equiv 0$ (2) holds when $p = 2$. Consequently we can apply Theorem 6. But in cases a) and b) the surface is mapped onto a curve of genus > 0, which is impossible for a $K3$ surface, while in case c) it either is rational or has the form $A \times \mathbf{P}^1$, where $g(A) = 1$, i.e. again it cannot be a $K3$ surface.

The case of a quasi-elliptic pencil reduces to the elliptic case already considered, in view of the following result:

If there exists a quasi-elliptic pencil on a smooth K3 surface, then there is also an elliptic pencil on this surface.

The proof of this result will be published elsewhere.*

REMARKS. 1. Theorem 6 shows that smooth projective elliptic surfaces whose group schemes of automorphisms are not reduced are very rare exceptions. By Theorem 7, the group scheme of automorphisms of a $K3$ surface is reduced and discrete. It seems likely to us that for surfaces of general type, the scheme of automorphisms is always reduced. In other words, for surfaces of general type, $H^0(X, \Theta_X) = 0$.

2. The fundamental (although technically also the simplest) part of the proof of Theorem 6 was the examination of the fibers of type (1). It seems likely that an analogous proposition is also true for fiberings on curves of arbitrary genus, namely,

If $g(B) = 1$, then there exists no fibering $X \longrightarrow B$ which does not have degenerate fibers, while if $g(B) = 0$, none exists which has no more than two degenerate fibers.

This assertion is true over the field of complex numbers: it follows from the fact that the Teichmüller variety of curves of genus > 1 is a bounded domain, while a universal covering for B, if $g(B) = 1$, or for $\mathbf{P}^1 \setminus S$, if $|S| = 1$ or 2, cannot be mapped into a bounded domain. In the general case, this question is related to the so-called "finiteness conjectures" (see [13], Chapter 2, §3, Remark 3). If this proposition were true in the general case, it would be a significant step in the direction of a proof that $H^0(X, \Theta_X) = 0$ for surfaces of general type which possess morphisms onto algebraic curves, while this, in turn, would lend further support to the conjecture which was stated in Remark 1.

Added in translation. The authors are grateful to P. Deligne for a helpful discussion. In particular, he pointed out a gap in the proof of Theorem 7, which is corrected in this translation. He also found a simple (positive) answer to the question raised in §2 (cf. the Remark after the Corollary to Theorem 4).

Received 8/JUNE/1976

BIBLIOGRAPHY

1. M. Artin, *Supersingular K3 surfaces*, Ann. Sci. École Norm. Sup. (4) 7 (1974), 543–567. MR 51 #8116.

2. E. Bombieri and D. Mumford, *Enriques' classification of surfaces in* Char. *p*. II, Complex Analysis and Algebraic Geometry (Papers Dedicated to K. Kodaira), Iwanami Shoten, Tokyo; Cambridge Univ. Press, Cambridge, 1977, pp. 23–42.

3. Enrico Bombieri and Dale Husemoller, *Classification and embeddings of surfaces*, Proc. Sympos. Pure Math., Vol. 29, Amer. Math. Soc., Providence, R. I., 1975, pp. 329–420.

4. I. V. Dolgačev, *The Euler characteristic of a family of algebraic varieties*, Mat. Sb. 89 (131) (1972), 297–312 = Math. USSR Sb. 18 (1972), 303–319. MR 48 #6116.

5. I. V. Dolgačev and A. N. Paršin, *Different and discriminant of regular mappings*, Mat. Zametki 4 (1968), 519–524 = Math. Notes 4 (1968), 802–804.

6. Nathan Jacobson, *Lie algebras*, Interscience, New York, 1962. MR 26 #1345.

7. Alexandre Grothendieck, *Éléments de géométrie algébrique*. II, Inst. Hautes Études Sci. Publ. Math. No. 8 (1961). MR 36 #177b.

8. Reinhardt Kiehl and Ernst Kunz. *Vollständige Durchschnitte und p-Basen*, Arch. Math. (Basel) 16 (1965), 348–362. MR 32 #5659.

9. Ernst Kunz, *Über die kanonische Klasse eines vollständigen Modells eines algebraischen Funktionenkörpers*, J. Reine Angew. Math. 209 (1962), 17–28. MR 26 #4998.

10. Masayoshi Nagata, *Local rings*, Interscience, New York, 1962. MR 27 #5790.

11. André Néron, *Modèles minimaux des variétés abéliennes sur les corps locaux et globaux*, Inst. Hautes Études Sci. Publ. Math. No. 21 (1964). MR 31 #3423.

---

*Uspehi Mat. Nauk 33 (1978), no. 1 (199).

12. A. P. Ogg, *Elliptic curves and wild ramification*, Amer. J. Math. 89 (1967), 1–21. MR 34 #7509.

13. A. N. Paršin, *Algebraic curves over function fields*. I, Izv. Akad. Nauk SSSR Ser. Mat. 32 (1968), 1191–1219 = Math. USSR Izv. 2 (1968), 1145–1170. MR 41 #1740.

14. I. I. Pjateckiĭ-Šapiro and I. R. Šafarevič, *Torelli's theorem for algebraic surfaces of type K3*, Izv. Akad. Nauk SSSR Ser. Mat. 35 (1971), 530–572 = Math. USSR Izv. 5 (1971), 547–588. MR 44 #1666.

15. M. Raynaud, *Caractéristique d'Euler-Poincaré d'un faisceau et cohomologie des variétés abéliennes*, Seminaire Bourbaki 1964/65, Benjamin, New York, 1966, Exposé 286; reprinted in *Dix exposés sur la cohomologie des schémas*, North-Holland, Amsterdam; Masson, Paris, 1968. MR 33 #5420*l*; 39 #2777.

16. J.-P. Serre and J. Tate, *Good reduction of abelian varieties*, Ann. of Math. (2) 88 (1968), 492–517. MR 38 #4488.

17. P. Deligne, *Courbes elliptiques: formulaire d'après J. Tate*, Modular Functions of One Variable. IV (Proc. Internat. Summer School, Antwerp, 1972), Lecture Notes in Math., Vol. 476, Springer-Verlag, Berlin and New York, 1975, pp. 53–73. MR 52 #8135.

18. I. R. Šafarevič, *Basic algebraic geometry*, "Nauka", Moscow, 1972; English transl., Springer-Verlag, Berlin and New York, 1974. MR 51 #3162, 3163.

19. Conjeerveram Srirangachari Seshadri, *L'opération de Cartier. Applications*, Variétés de Picard (Séminaire C. Chevalley, 3ième Année: 1958/59), Exposé 6, École Norm. Sup., Paris, 1960, pp. 101–115. MR 28 #1094.

Translated by T. B. GREGORY
with the assistance of
B. SAINT-DONAT and M. REID

Quasi-elliptic surfaces of type K3

(with A. N. Rudakov)

Usp. Mat. Nauk **33**, No. 1, 227–228 (1978). Zbl. **383**, 14013
[Russ. Math. Surv. 33, No. 1, 215–216 (1978)]

We assume the ground field k to be algebraically closed.

THEOREM. *If on a smooth surface X of type* K3 *there is a quasi-elliptic bundle, then there is also an elliptic bundle on it.*

As was proved in [3], there is an elliptic or quasi-elliptic bundle on a surface of type K3 if and only if there is on it an algebraic cycle, not equivalent to zero, whose square is zero. Thus, the following criterion holds:

THEOREM. *For the existence of an elliptic bundle on a surface of type* K3 *it is necessary and sufficient that there should be on it an algebraic cycle D, $D \not\sim 0$, for which $(D)^2 = 0$.*

We mention that this result allows us to simplify the proof of Theorem 7 in [4] in the absence of regular vector fields on a smooth surface of type K3: there is no need to treat the case of a quasi-elliptic bundle separately.

PROOF. Let σ be the curve formed from the cusps of the fibres of a quasi-elliptic fibration. We claim that it is smooth. In a neighbourhood of a non-degenerate fibre this is proved in [1]. We recall that in a neighbourhood of some fibre the set X_0 of non-singular points of the fibres can be endowed with the structure of a family of groups, and a fibrewise action of X_0 on X can be defined that extends the group structure on X_0. The curve σ is invariant under this action, therefore, it intersects a degenerate fibre in a singular point.

On the other hand, as is shown in [1],

(a) $(\sigma, F) = p = \text{char}\, k$, where F is the fibre.

In our case $p = 2$ or 3, however, the intersection index of two curves having a singular point at their intersection is not less than 4, therefore, σ must be smooth. Since σ is birationally isomorphic to the base, it is rational, and $(\sigma)^2 = -2$.

In the usual way we can associate a graph with the system of non-singular rational curves on X. The action of the group of non-singular points of a degenerate fibre on the whole fibre induces an action of it on the corresponding graph, which determines a group Γ of automorphisms of the graph that is transitive on the set of vertices corresponding to components of multiplicity 1. We adjoin to the system of components the fibre of σ.

(b) The automorphisms in Γ extend to automorphisms of the graph of this enlarged system that leave the vertex σ in place.

We recall that from the formula for the Euler–Poincaré characteristic (see, for example, [2]) it follows that for a quasi-elliptic bundle

(c) $\Sigma\,(m_i - 1) = 20$, where m_i is the number of components of the ith degenerate fibre.

Arguing ex contrario, we assume that there is no elliptic bundle on X. We exclude successively the possible types of degenerate fibres on X.

Let $p = 2$. A fibre of type A_n cannot occur in a quasi-elliptic fibering, because the set of points of the base in the fibres over which a singular point has separate tangents must be open. From (a) and (b) it follows that fibres D_{2n+1}, E_6, or (c3) also cannot occur, and a fibre D_{2n} of σ intersects in a middle component. If there is a fibre E_7, then we can single out a subsystem E_6 in the system of components of the fibre and σ (Fig. 1)

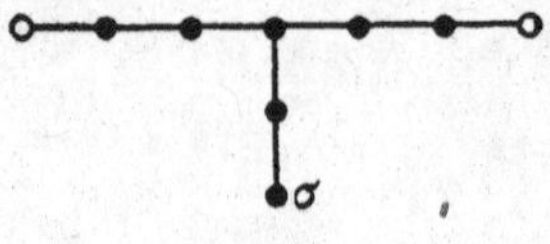

Fig. 1.

Taking the curves of this subsystem with the appropriate multiplicities, we obtain an effective divisor whose square is zero. By the Riemann–Roch theorem it is a fibre of an elliptic or quasi-elliptic bundle, which is impossible. Similarly a fibre D_{2n} with $n > 3$ leads to a system E_7. From (a) it follows that there are two possibilities for the graph of the system of components of a fibre E_8 and σ (Fig. 2). Clearly, in the first case we have a subsystem E_7 and in the second D_8.

From (b) it follows that a fibre D_6 cannot be unique. Adding σ and one component of any degenerate ibre to the system of curves of the fibre in question, we arrive at a system containing E_6. Two fibres D_4 ead to a system D_6 (Fig. 3). A fibre D_4 and two fibres ($c2$) give a D_5 (Fig. 4). By (c), the only remaining

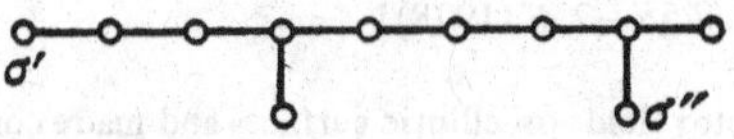

Fig. 2.

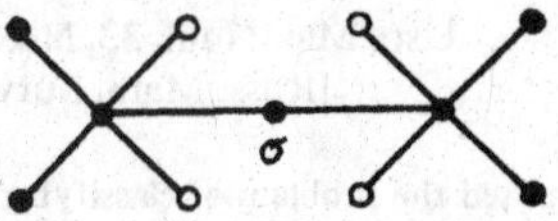

Fig. 3.

ossibility is 20 fibres ($c2$). However, four fibres ($c2$) lead to a D_4 if we take σ and one component from ach fibre.

Let $p = 3$. By analogy, a fibre A_n cannot be degenerate. Taking (a) and (b) into account, we see that the ibres ($c2$), D_n, and E_7 cannot occur. A fibre E_6 leads to a D_4. For a fibre E_8 there are three possibilities for

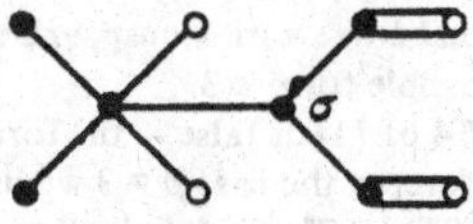

Fig. 4.

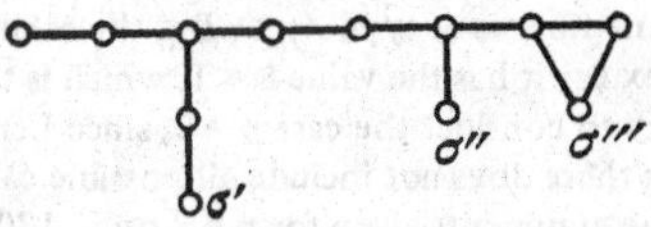

Fig. 5.

here are three possibilities for the position of σ, but the first leads to a subsystem E_6 and the second to D_7 (Fig. 5). In the system of components of two fibres E_8 and σ it is easy to single out a subsystem D_{14}. The only remaining possibilities are, by (c), a fibre E_8 and six fibres ($c3$) or ten fibres ($c3$). But four fibres ($c3$) and σ give a D_4. This proves the theorem.

References

[1] E. Bombieri and D. Mumford, Enquiries classification of surfaces in characteristic P. III, Invectiones Math. 35 (1976), 197–232.

[2] I. V. Dolgachev, The Euler characteristic of a family of algebraic varieties, Mat. Sb. 89 (1972), 297–312. MR 48 # 6116.
= Math. USSR-Sb. 18 (1972), 303–319.

[3] I. I. Pyatetskii–Shapiro and I. R. Shafarevich, A Torelli theorem for algebraic surfaces of type K3, Izv. Akad. Nauk SSSR Ser. Mat. 35 (1971), 530–572. MR 44 # 1666.
= Math. USSR-Izv. 5 (1971), 547–588.

[4] A. N. Rudakov and I. R. Shafarevich, Inseparable morphisms of algebraic surfaces, Izv. Akad. Nauk SSSR Ser. Mat. 40 (1976), 1269–1307.
= Math. USSR-Izv. 10 (1976), 1205–1237.

Received by the Editors 12 May 1977

Vector fields on elliptic surfaces

(with A. N. Rudakov)

Usp. Mat. Nauk **33**, No. 6, 231–232 (1978). Zbl. **411**, 14008
[Russ. Math. Surv. **33**, No. 6, 255–256 (1978)]

In [1] we posed the problem of classifying regular vector fields on elliptic surfaces and made considerable progress in studying it. Here we report the final solution of the problem.

Let $f\colon X \to B$ be a minimal reduced elliptic surface with base B, defined over an algebraically closed field k of characteristic $p > 0$. In [1] we showed that, with the exception of a few cases we did not consider, a non-zero regular vector field can exist on X only if X is a rational surface, an Abelian variety, a hyperelliptic surface, or if all the fibres are isomorphic to one another, and the vector field is tangent to the fibres. The singular cases not considered in [1] were $p = 2$ or 3, B is a rational curve, f has two degenerate fibres and, in addition:

a) for $p = 2$ one fibre is of type A_n, the other is a rational curve with a cusp, and the highest ramification index for it has the value $\delta = 2$, which is the least possible for $p = 2$;

b) for $p = 3$ one fibre is of type A_n or D_n, the other is a rational curve with a cusp, and the highest ramification index for it has the value $\delta = 1$, which is the least possible for $p = 3$.

(It is necessary to consider the case $p = 3$, since Lemma 4 in §4 of [1] is false – the form of the vector field given there does not include all possible cases. Therefore, in the case $p = 3$ we have to repeat word for word the arguments given for $p = 2$ on p. 1301–1302 of [1].) The determination of vector fields on arbitrary reduced elliptic surfaces reduces to the case when the surface has a section.

THEOREM. *All elliptic surfaces that have a section and satisfy the conditions* a) *or* b) *are given by the equations*:

(1) *for* $p = 2$: a) $y^2 + xy = x^3 + c\tau^2 x^2 + \tau^{2\cdot 4^n}$, $c \in k$, $n > 0$,

(2) *for* $p = 3$: б$_1$) $y^2 = x^3 + x^2 - \tau^{9^n}$,

(3) *or* б$_2$) $y^2 = x^3 + \tau x^2 - \tau^{3(9^n-1)}$, $n > 0$.

The equations (1), (2) and (3) give the Weierstrass minimal models of the open sets of these surfaces that are obtained by rejecting the degenerate fibre having a cusp. Regular vector fields on the Néron minimal models of these surfaces actually exist. They have the form $\alpha\tau \dfrac{\partial}{\partial\tau}$ in cases a) and b$_1$) and $\alpha\left(x\dfrac{\partial}{\partial x} + \tau\dfrac{\partial}{\partial\tau}\right)$ in case b$_2$). Here $\alpha \in k$. There are no other regular vector fields.

That these fields are regular in the charts described by (1), (2), and (3) (on the corresponding Néron minimal models) is verified very simply. To verify that they are regular in a neighbourhood of the remaining degenerate fibre we have to put $\tau = t^{-1}$. However, in this way we arrive at non-minimal models. The equations of the corresponding minimal models have the form

$$(4)\qquad \text{a)}\quad y^2 + t^{(2\cdot 4^n+1)/3} xy + \mu y = x^3 + (ct^{(4^{n+1}-4)/3} + \alpha)x^2 + x + \gamma,\quad c \in k,$$

$$\mu = t\Big(\sum_{i=n}^{1} t^{2(4^n - 4^i)/3}\Big),\quad \deg\alpha < (4^{n+1}-4)/3,$$

where α and γ are uniquely determined from the relation $\Delta = t^4$, which holds in case a),

$$(5)\qquad \text{б)}\quad y^2 = x^3 + t^{(3^m+1)/2} x^2 + \beta x + \gamma,$$

$m = 2n - 1$ in case b$_1$) and $2n$ in case b$_2$), $\beta = t\Big(-1 + \sum\limits_{i=m-1}^{1} t^{(3^{m-1} - 3^i)/2}\Big)$, and γ is determined from the relation $\Delta = t^3$, which holds in case b) (in both cases Δ is the discriminant of the curves (4) and (5)). After going over to this model we can easily verify that the vector field is regular.

The proof of the theorem is based on a study of the model in which the fibre A_n or D_n turns out to be infinitely distant (that is, the model for which, as a result, (4) and (5) are satisfied). The conditions a) and b) are expressed by the equations $\Delta = t^4$ and $\Delta = t^3$, respectively, which are then solved. For this solution the basic fact is that in case a) the coefficient of xy, and in case b) the coefficient of x^2, is the degree of t. Since in both cases this coefficient is the Hasse invariant, the interpretation of this relation is that the Hasse invariant vanishes only in degenerate fibres, and all non-degenerate fibres are non-supersingular.

If it is known that the corresponding coefficient has the form t^k, then the equations $\Delta = t^4$ and $\Delta = t^3$ can be solved inductively, by expressing the solution corresponding to k in terms of the solution corresponding to $k-1$, from which we obtain (4) and (5), and then (1), (2), and (3). This indicates, apparently, that there are interesting correspondences between surfaces given by (4) or (5) for different values of n.

Reference

[1] A. N. Rudakov and I. R. Shafarevich, Non-separable morphisms of algebraic surfaces, Izv. Akad. Nauk SSSR Ser. Mat. **40** (1976), 1269–1307.
= Math. USSR-Izv. **10** (1976), 1205–1237.

Received by the Editors 24 April 1978

Supersingular K3 surfaces over fields of characteristic 2

(with A. N. Rudakov)

Izv. Akad. Nauk SSSR, Ser. Mat. **42,** 848–869 (1978). Zbl. **404,** 14010
[Math. USSR, Izv. **13,** No. 1, 147–165 (1979)]

Abstract. In this paper the unirationality of supersingular $K3$ surfaces over a field of characteristic 2 is proved and a classification of such surfaces is given.

Bibliography: 14 titles.

Introduction

An algebraic $K3$ surface is called ***supersingular*** if all of its cycles are algebraic–in other words, if the rank of the Picard group equals 22. This paper is devoted to studying supersingular $K3$ surfaces over fields of characteristic 2. We show that every such surface contains a quasielliptic pencil; from this we easily deduce the unirationality of these surfaces.

In [1], Artin introduced an important invariant of a supersingular $K3$ surface: the discriminant of the quadratic form defined by the intersection number in the Picard group. For a surface defined over a field of characteristic p, this discriminant is of the form $p^{2\sigma}$, where σ can have the values $1, 2, \ldots, 10$. As Artin showed in the same paper, in the versal family of deformations of a surface, the points corresponding to surfaces with a given value of σ form a subscheme of the base of dimension $\geqslant \sigma - 1$. We show the irreducibility of the 'variety of moduli' of surfaces with a given value of σ and we show that the 'number of moduli' of surfaces with a given value of σ is equal to $\sigma - 1$. The precise meaning of this assertion is the following. For every value of σ, $1 \leqslant \sigma \leqslant 10$, we build an irreducible family $\pi\colon \mathscr{X} \longrightarrow S$, whose fibers $\pi^{-1}(s)$, $s \in S$, are $K3$ surfaces with the given σ. This family has the property that an algebraic group G acts on its base S so that the stabilizer of the general point is zero-dimensional. Further, the fibers $\pi^{-1}(s_1)$ and $\pi^{-1}(s_2)$ over two equivalent (with respect to the group G) points s_1 and s_2 are isomorphic, and $\dim S - \dim G = \sigma - 1$. Since Artin showed in [1] that in the versal family of deformations of a surface with a given value of σ the points of the base corresponding to surfaces with the same σ form a subscheme of dimension $\geqslant \sigma - 1$, it follows from this that the number of moduli is exactly equal to $\sigma - 1$.

The above results are obtained by combining geometric considerations with a detailed study of the integral quadratic form defined in the Picard group by the intersection number.

1980 *Mathematics Subject Classification.* Primary 14J25, 14J10; Secondary 14C22.

The last problem is related to the theory of integral quadratic forms. As an application of the results obtained here, we propose a significantly simpler proof than the one given in our paper [11] of the fact that on a $K3$ surface there do not exist nonzero, regular vector fields.

In this paper, a big role is played by irreducible fibers of elliptic and quasielliptic fibrations and their connection with simple root systems. We decided to abandon the previously proposed notation for irreducible fibers used by Kodaira [6] and Néron [8], and instead to denote an irreducible fiber by the same symbol as the corresponding root system: for example, a fiber whose graph coincides with the extended Dynkin diagram A_n will be called a fiber of type A_n. This will not cause ambiguity for fibers of type A_n, $n > 2$, D_n, E_6, E_7 and E_8. It is only necessary to be more precise for the fiber consisting of two tangents whose graph coincides with the Dynkin diagram A_1 and for the fiber consisting of three components intersecting in one point whose graph coincides with the Dynkin diagram A_2. We will call these fibers *fibers of type* $^*A_1$ *and* $^*A_2$. For the remaining two irreducible degenerate fibers, we will write A_0 for the fiber with a singular point and distinct tangents, and $^*A_0$ for the fiber with a cusp.

§1. Unirationality

Let a bilinear scalar product $\mathbf{Q}^n \otimes \mathbf{Q}^n \longrightarrow \mathbf{Q}$, which we will denote by (x, y), be defined in n-dimensional space over the rationals. A submodule N of rank n over $\mathbf{Z}$, $N \subseteq \mathbf{Q}^n$, will be called a *Euclidean lattice.* We denote by N^* the dual lattice $\{\alpha \in \mathbf{Q}^n \mid (\alpha, N) \in \mathbf{Z}\}$. In what follows, we will assume that N is integral, i.e. that $N \subset N^*$. The lattice $N = \mathbf{Z}e$ of rank 1 with the condition $e^2 = 1$ (or -1) we denote by $I_{(+)}$ (respectively $I_{(-)}$); we denote by $N^{(m)}$ the direct sum of m copies of N. The lattice whose vectors are the same as in N and with scalar product defined to be $k \cdot (x, y)$ will be denoted by $N(k)$.

We will assume that the following conditions are satisfied.

a) $n = 6 \bmod 8$.

b) The quadratic form defined in $N \otimes \mathbf{R}$ has positive inertia index 1 and negative inertia index $n - 1$.

c) N is even, i.e. $x^2 \equiv 0 \bmod 2$ for $x \in N$.

d) $N \supset 2N^*$.

Under these conditions, (x, y) defines on the space $N/2N^*$ a nondegenerate product with values in the two-element field $\mathbf{F}_2$ for which $x^2 = 0$. Therefore, our product can be considered simultaneously also as antisymmetric, so that $\operatorname{rg}_{\mathbf{F}_2} N/2N^* = 0 \bmod 2$. We put

$$\operatorname{rg}_{\mathbf{F}_2} N/2N^* = 2\tau.$$

Then $n = 2\tau + 2\sigma$, where $2\sigma = \operatorname{rg}_{\mathbf{F}_2} N^*/N$. Since $|d(N)| = |N^*/N|$, it follows that $d(N) = -2^{2\sigma}$, where $d(N)$ is the discriminant of N.

We remark for future use that condition d) is "hereditary" in the sense that if $N \supset M$ and it is satisfied for M, then it is also true for N. Indeed, $M^* \supset N^*$, and if $N \supset M \supset 2M^*$ then $N \supset M \supset 2M^* \supset 2N^*$.

LEMMA 1. *If conditions* a), b) *and* d) *hold, then* $N \subset S$, *where* $S = I_{(+)} \oplus I_{(-)}^{(n-1)}$.

PROOF. If $|d(N)| \geqslant 4$, then there exists a lattice $N' \supset N$, $(N' : N) = 2$, having the same properties as N. We will give a proof of this property due to Kneser ([5], p. 247). The product (x, y) defines in N^*/N the product $\langle x, y\rangle$ with values in $\frac{1}{2}\mathbf{Z}/\mathbf{Z} = \mathbf{F}_2$. Since the form $\langle x, y\rangle$ is bilinear, $u(x) = \langle x, x\rangle$ is a linear form. Since by assumption $\dim_{\mathbf{F}_2} N^*/N > 1$, it has a nonzero kernel. Let $\bar{e} \neq 0$, $u(\bar{e}) = 0$, and let $\bar{e}$ be the image of a vector $e \in N^*$. Then the lattice $N' = \{e, N\}$ has all the necessary properties, as can be immediately checked

It is obvious that $d(N') = \frac{1}{4}d(N)$. Continuing the same process we arrive in the end at a lattice $S \supset N$ with $|d(S)| = 1$. According to general properties of unimodular lattices (see [13], pp. 92–93), and because of condition b), if S is even then $S \simeq U_2 \oplus E_8^{(k)}$, where U_2 is a lattice of rank 2 corresponding to the quadratic form $2x_1x_2$ and E_8 is a negative definite even lattice of rank 8. If, however, S is odd, then $S \simeq I_{(+)} \oplus I_{(-)}^{(n-1)}$. Because of condition a), $n \not\equiv 2 \bmod 8$, and therefore the first case is impossible. The second case leads to the assertion of the lemma.

LEMMA 2. *Under conditions* a)–d) $N \supset M$, $M \simeq S(2)$, *where* S *is as in Lemma* 1.

PROOF. According to Lemma 1, $S \supset N$, and hence $N^* \supset S^* = S \supset N \supset 2N^*$. Since $|d(N)| = 2^{2\sigma}$, therefore $\mathrm{rg}_{\mathbf{F}_2} S/N = \sigma$; and, because of the assumption, $\mathrm{rg}_{\mathbf{F}_2} N/2N^* = 2\tau$. In $N/2N^*$, as we have seen, a skew-symmetric bilinear product with values in $\mathbf{F}_2$ is defined. Let us choose some maximal isotropic subspace of dimension τ and denote its inverse image in N by M. Then $S \supset N \supset M \supset 2N^*$. Also, $\mathrm{rg}_{\mathbf{F}_2} N/M = \tau$, so that $\mathrm{rg}_{\mathbf{F}_2} S/M = \sigma + \tau = n/2$, i.e. $|d(M)| = 2^n$. By assumption we have $(x, y) \in 2\mathbf{Z}$ for $x, y \in M$, i.e. $M \simeq M_1(2)$ for some integral lattice M_1, and $d(M) = 2^n d(M_1)$. Hence M_1 is unimodular, and then, as we have seen, $M_1 \simeq S$.

We now pass to applications to $K3$ surfaces. If for an integral lattice N we have $f \in N$ and $f^2 = 0$, then we put $N(f) = f^\perp/\mathbf{Z}f$, where $f^\perp \subset N$ is the submodule orthogonal to f. Obviously under condition b) (x, y) defines in $N(f)$ a nondegenerate and negative definite product, which we will also denote by (x, y). We denote by $R(f)$ the root system of $N(f)$, i.e. the set of vectors $x \in N(f)$ such that $(x, x) = -2$, and by $H(f)$ the sublattice generated by it. If $R(f) = \bigoplus R_i(f)$ is a decomposition of $R(f)$ into indecomposables, then $H(f) = \bigoplus H_i(f)$, where the $H_i(f)$ are generated by the $R_i(f)$.

Now let X be a $K3$ surface defined over an algebraically closed field k, $N = \operatorname{Pic} X$, $f \in N$, $f^2 = 0$, and let f correspond to the fiber of an elliptic or a quasielliptic fibration $\varphi\colon X \longrightarrow \mathbf{P}^1$. The root of the system $R(f)$ defines a class $D \in \operatorname{Pic} X$ such that $D^2 = -2$ and $DF = 0$. From $D^2 = -2$ according to the Riemann-Roch theorem it follows that $D > 0$ or $-D > 0$, and the condition $D \cdot F = 0$ shows that D consists of components of fibers of the fibration φ. Hence $R(f) = Q/\mathbf{Z}f$, where Q is the set of divisors with square -2, consisting of components of fibers of φ.

If φ has degenerate fibers F_i, then $R(f) = \bigoplus R_i(f)$, $R_i(f) = Q_i/\mathbf{Z}f$, where Q_i is the set of divisors with square -2 consisting of components of F_i.

Hence in particular it is clear that the module N and the element f, corresponding to the fibration φ, determine the types of the reducible degenerate fibers of the fibration.

Clearly, $H(f) = V(f)/\mathbf{Z}f$, where $V(f)$ is the subgroup generated by components of fibers and $H_i(f) = V_i(f)/\mathbf{Z}f$, where $V_i(f)$ is generated by the components of F_i.

THEOREM 1. *On a supersingular K3 surface over a field of characteristic 2 there exists a quasielliptic pencil.*

PROOF. Let X be such a surface. As Artin showed [1], for $N = \operatorname{Pic} X$ condition d) is satisfied (in any characteristic p, $N \supset pN^*$). Conditions a), b) and c) are well known; also, $n = 22$. Therefore we can apply Lemma 2 to N. According to that lemma $N \supset M$, where $M = \bigoplus_{-1}^{20} f_i\mathbf{Z}$, $f_{-1}^2 = 2$, $f_i^2 = -2$ for $i \geqslant 0$, and $f_if_j = 0$ if $i \neq j$. We set $f = f_{-1} + f_0$. Then $f^2 = 0$, and according to [10] we may consider that f corresponds to an elliptic or a quasielliptic pencil φ (if necessary after performing an automorphism of the lattice N). Our aim is to prove that φ cannot be elliptic and so must be quasielliptic. We will assume, therefore, that φ is elliptic. By construction $f_i \in f^\perp$, $i = 1, 2, \ldots, 20$, and (identifying them with their images mod $\mathbf{Z}f$) we have $M(f) = \bigoplus_1^{20} \mathbf{Z}f_i$, $f_i^2 = -2$. Obviously, $N(f) \supset H(f) \supset M(f)$ (since the f_i are roots). Since $M(f) \supset 2M(f)^*$, therefore $H(f) \supset 2H(f)^*$ and consequently $H_i(f) \supset 2H_i(f)^*$ for all degenerate fibers of φ. In addition, $\Sigma \operatorname{rg} H_i(f) \leqslant 20$, since by taking from each fiber all the components except one and by adjoining to them a fiber and an arbitrary divisor $D \notin f^\perp$ we will obtain an independent set of elements of N. But in our case we have 20 elements f_i contained in $H(f)$, and therefore $\Sigma \operatorname{rg} H_i(f) = 20$. Theorem 1 will therefore be proved if we establish the following result, which is true over fields of characteristic $p = 2$ and 3:

PROPOSITION. *A pencil with fiber F of arithmetic genus 1 on a K3 surface is quasielliptic if and only if the following conditions are fulfilled:*
1) $H_i(f) \supset pH_i(f)^*$ *for every reducible fiber* F_i.
2) $\Sigma \operatorname{rg} H_i(f) = 20$.

Suppose a pencil with fiber F is quasielliptic. For every degenerate fiber the group $H_i^*(f)/H_i(f)$ is isomorphic to the group of components of multiplicity one of that fiber. This is easy to see by considering the possible types of degenerate fibers; probably it follows from general properties of Dynkin diagrams. But in the case of a quasielliptic fibration, the group of all sections of the fibration $X \times_B \hat{\mathcal{O}}$, where $\hat{\mathcal{O}}$ is the completion of the local ring of a point of the base, has period p (the characteristic of the ground field), since it embeds into the additive group. Therefore, in our case $p \cdot H^*(f)/H(f) = 0$, i.e. condition 1) is satisfied. In order to prove 2) we observe that $N(f)/H(f)$ is isomorphic to the group of rational points of the generic fiber of the Jacobian fibration corresponding to φ. Since, for the same reasons as above, this group has period p and is finitely generated, it is therefore finite. On the other hand, $\operatorname{rg} N(f) = \operatorname{rg} N - 2 = 20$. Therefore $\operatorname{rg} H(f) = 20$, and 2) follows.

Suppose now conditions 1) and 2) are satisfied. Then the relation between the Euler characteristics of the fibers gives

$$\sum (e(F_i) + \delta_i) \leqslant 24, \tag{1}$$

where the sum is taken over the reducible fibers, $e(F_i)$ is the Euler characteristic of the fiber and δ_i is the higher ramification index ([11], proof of Theorem 6). If the number of components of F_i is $n_i + 1$ (so that $n_i = \operatorname{rg} R_i(f)$), then $e(F_i) = n_i + 2 - \epsilon_i$, where $\epsilon_i = 0$ with the single exception of fibers of type A_n, where $\epsilon_i = 1$. Therefore, (1) can be written as

$$\sum (n_i + 2 + \delta_i - \varepsilon_i) \leqslant 24. \tag{2}$$

Condition 2) can be rewritten as

$$\sum n_i = 20. \tag{3}$$

Subtracting (3) from (2), we obtain

$$\sum (2 + \delta_i - \varepsilon_i) \leqslant 4. \tag{4}$$

We first consider the case $p = 3$. Condition 1) is satisfied by irreducible root systems A_2, E_6, and E_8. Therefore, irreducible degenerate fibers can only be of types A_2, ${}^*A_2$, E_6 and E_8. Obviously, if for such fibers (4) is satisfied then the sum on the left in (3) is always less than 20.

Now let $p = 2$. Condition 1) is satisfied by the irreducible root systems A_1, D_{2n}, E_7 and E_8. Therefore reducible degenerate fibers can only be of the types A_1, ${}^*A_1$, D_{2n}, E_7 and E_8. In addition, from formulas describing all the types of degenerate fibers ([8], pp. 124–125) it follows that for a fiber different from A_n we have $\delta_i \geqslant 1$, and therefore $2 + \delta_i - \epsilon_i \geqslant 3$; and for the fiber D_{2n} we have $\delta_i \geqslant 2$. It follows from this that (3) and (4) can be satisfied only if φ has a single degenerate fiber of type D_{20} with $\delta = 0$. We will show that such an elliptic pencil does not exist.

Let j denote the absolute invariant of the fibration. We first consider the case where j is nonconstant.

Let the only degenerate fiber be over the point $\infty \in \mathbf{P}^1$. Then $j(\alpha) \neq \infty$ if $\alpha \neq \infty$, and that means $j(\infty) = \infty$. Hence the corresponding stable fiber is of type A_n, and the degenerate fiber takes on that form after a quadratic covering of the base ramified only at ∞ and with $\delta = 1$ (see for example [9], p. 5). Such a covering of $\mathbf{P}^1$ is rational. After the covering we obtain a pencil $\varphi'\colon X' \longrightarrow \mathbf{P}^1$ with a single degenerate fiber of type A_n. Such a pencil is impossible (see [11], proof of Theorem 6). (As Deligne has shown, an elliptic pencil with base $\mathbf{P}^1$ and a single degenerate fiber always has constant absolute invariant.)

If j is constant and $j = 0$, then the general fiber of φ is supersingular. In such a case the equation of the general fiber has the form $y^2 + \mu y = x^3 + \alpha x^2 + \beta x + \gamma$, and for the discriminant there is the simple formula $D = \mu^4$ ([14], p. 62). The presence at ∞ of the fiber D_{20} gives the condition $\nu_\infty(\mu) \geqslant 10$ ([8], p. 125), which contradicts the condition $\nu_\infty(D) = 24$.

Finally, in the case $j \neq 0$ the general fiber is a nonsupersingular curve C. We may assume that the pencil has a section by passing to its Jacobian fibration. Since in this case $|\mathrm{Aut}\, C| = 2$, therefore X is birationally equivalent to $(C \times B)/G$, where $|G| = 2$ and B is a two-sheeted covering of the base $\mathbf{P}^1$, ramified at ∞ with higher ramification index $\delta = 1$ (since $\delta = 2$ for φ). Therefore B is again rational, and then $(C \times B)/G$ has a projection onto the rational curve C/G with a rational fiber B and therefore is itself rational. This contradicts the birational isomorphism of $(C \times B)/G$ with X.

COROLLARY. *A supersingular $K3$ surface over a field of characteristic 2 is unirational.*

Let $X \longrightarrow B$ be a quasielliptic pencil on a supersingular $K3$ surface, and let σ be the curve formed by the cusps of the fibers. As follows from the results of [2], the mapping $\sigma \longrightarrow B$ is purely inseparable, and consequently the curve σ is rational. We set $X' = X \times_B \sigma$.

On X' we have a pencil whose general fiber is a rational curve. Therefore X' is rational, which means that X is unirational.

§2. Classification of lattices

We give here the full classification of lattices satisfying conditions a)–d) of §1.

DEFINITION. Even lattices N for which $(x, x) \in 4\mathbf{Z}$ for $x \in 2N^*$ are called *lattices of type* I; otherwise they are called *lattices of type* II.

It is easy to see that a lattice is of type I if and only if the product $\langle x, y\rangle$ on N^*/N with values in $\mathbf{F}_2$ defined in §1 is skew symmetric, i.e. $\langle x, x\rangle = 0$.

THEOREM 2. *A lattice satisfying conditions* a)–d) *from* §1 *is uniquely determined by the rank n, the invariant σ and the type; also,* $0 \leqslant \sigma \leqslant n/2$. *If* $\sigma = 0$, *there are no such lattices; if* $\sigma = 1$, *there exists only a lattice of the first type, and if* $\sigma = n/2$, *only of the second type.*

For the remaining values of σ there exist lattices of the first type and of the second type.

The proof is based on the fact that for lattices of the type considered the class and the genus are the same. As is known, the genus equals the spinor genus for indefinite lattices of rank $\geqslant 3$ ([7], p. 319). Since $n \equiv 6 \bmod 8$, therefore $n > 3$. The equality of the spinor genus with the class is a purely local condition. It is satisfied if, for every prime number p, $N \otimes \mathbf{Z}_p$ contains as a direct summand a lattice of the form $M(k)$, where M is a unimodular lattice of rank $\geqslant 2$ for $p \neq 2$ and of rank $\geqslant 3$ for $p = 2$ ([7], p. 304). In our case $N \otimes \mathbf{Z}_p$ is unimodular for $p \neq 2$, and the necessary condition is satisfied. When $p = 2$, let us consider an arbitrary basis $\bar{e}_1, \ldots, \bar{e}_{2\tau}$ of the space $N/2N^*$ and any preimages e_i of the vectors $\bar{e}_i$ in N. Let us set $N_0 = \bigoplus \mathbf{Z}_2 e_i$. Since the scalar product in $N/2N^*$ is nondegenerate (because of the definition of N^*), the lattice N_0 is unimodular. Therefore it splits off as a direct summand: $N \otimes \mathbf{Z}_2 = N_0 \oplus N_1$. Since $\operatorname{rg}_{\mathbf{F}_2} N/2N^* = 2\tau$, $N_0 = N_0^*$ and $\operatorname{rg}_{\mathbf{F}_2} N_0/2N_0^* = 2\tau$, it follows that $N_1 = 2N_1^*$. That means that $N_1 = M_1(2)$, where M_1 is an integral lattice over $\mathbf{Z}_2$. Finally, $\operatorname{rg} M_1 = 2\sigma$ and $\nu_2(d(N_1)) = \nu_2(d(N)) = 2\sigma$, whence it follows that M_1 is unimodular. Since $2\sigma + 2\tau = n \geqslant 6$, we see that either $2\sigma \geqslant 4$ or $2\tau \geqslant 4$, and the needed condition is satisfied.

Thus, an isomorphism of two lattices N of the type considered reduces to an isomorphism of the lattices $N \otimes \mathbf{Z}_p$ for all p. But the lattices $N \otimes \mathbf{Z}_p$ are unimodular for $p \neq 2$, and their discriminant is determined by giving σ: $d(N) = -2^{2\sigma}$; therefore they are isomorphic. Therefore the lattices N of the type considered by us are uniquely determined by giving a lattice $N \otimes \mathbf{Z}_2$. Let us observe that, because of Lemma 1, $N \subset S$, where S is a fixed lattice described in the statement of the lemma. Therefore $N \otimes \mathbf{Q} = S \otimes \mathbf{Q}$, which means that $N \otimes \mathbf{Z}_2 \otimes \mathbf{Q}_2 = S \otimes \mathbf{Q}_2$. We now describe all the even lattices $\overline{N}$ over $\mathbf{Z}_2$ of rank n and such that $\overline{N} = N_0 \oplus M_1(2)$, where N_0 and M_1 are unimodular and $\overline{N} \otimes \mathbf{Q}_2 = S \otimes \mathbf{Q}_2$.

A. Suppose the lattice M_1 is even. As is easy to see, an even unimodular lattice is of the form $U_2^{(k)}$, where U_2 is a lattice corresponding to the form $2x_1x_2$, or of the form $U_2^{(k-1)} \oplus V_2$, where V_2 is the unique two-dimensional even unimodular lattice over $\mathbf{Z}$ which does not represent 0 (it corresponds to the form $2x_1^2 + 2x_1x_2 + 2x_2^2$). Therefore in our case four cases are possible:

1. $N_0 = U_2^{(\tau)}$, $M_1 = U_2^{(\sigma)}$.
2. $N_0 = U_2^{(\tau-1)} \oplus V_2$, $M_1 = U_2^{(\sigma)}$.
3. $N_0 = U_2^{\tau}$, $M_1 = U_2^{(\sigma-1)} \oplus V_2$.
4. $N_0 = U_2^{(\tau-1)} \oplus V_2$, $M_1 = U_2^{(\sigma-1)} \oplus V_2$.

We will explain which of the above agree with the condition $\overline{N} \otimes \mathbf{Q}_2 = S \otimes \mathbf{Q}_2$.

In cases 2 and 3

$$d(\overline{N}) = (-1)^{\sigma+\tau-1} \cdot 3 \cdot 2^{2\sigma} = (-1)^{\frac{n}{2}-1} \cdot 3 \cdot 2^{2\sigma} = 3 \cdot 2^{2\sigma}$$

(because $n \equiv 6 \bmod 8$). Therefore the condition $d(\overline{N}) = d(S) \cdot c^2$, $c \in (\mathbf{Q}_2^*)^2$, is not satisfied here.

In case 1 it is necessary to compute the Hasse symbol $\epsilon(\overline{N})$ (see [13], p. 64). Obviously $\epsilon(S) = (-1)^{(n-1)(n-2)/2} = 1$. We observe that $U_2(2) \otimes \mathbf{Q}_2 \simeq U_2 \otimes \mathbf{Q}_2$; therefore, in case 1, $\overline{N} \otimes \mathbf{Q}_2 \simeq U^{(n/2)} \otimes \mathbf{Q}_2$. Hence

$$\varepsilon(\overline{N}) = (-1)^{\frac{1}{2}\frac{n}{2}\left(\frac{n}{2}-1\right)} = -1.$$

This case is also impossible.

Therefore for the type A there remains only one possibility:

$$\overline{N} \simeq U_2^{(\tau-1)} \oplus V_2 \oplus U_2^{(\sigma-1)}(2) \oplus V_2(2). \tag{5}$$

B. Suppose the lattice M_1 is odd. If $N_0 = U_2^{(\tau-1)} \oplus V_2$, then $y_2 \not\equiv 0\ (4)$ if $y \in V_2$ but $y \notin 2V_2$, i.e. $y^2 \equiv 2\ (4)$. Adding to the vector y from V_2 a vector $x \in M_1(2)$ such that $x^2 \notin 4\mathbf{Z}_2$, we obtain a vector y_1 with $y_1^2 \equiv 0 \bmod 4$, and that means a new decomposition with $N_0 = U_2^{(\tau)}$. If rg $M_1 > 4$, then it contains vectors e_1 and e_2 with $e_1^2 = 1$, $e_2^2 = -1$ and $e_1 e_2 = 0$. Indeed, since rg $M_1 \geqslant 4$, M_1 represents all the numbers in $\mathbf{Z}_2^*$. In particular, there exists an $e_1 \in M_1$ with $e_1^2 = 1$. The lattice $e_1^\perp$ is unimodular. If it is odd, then analogously we find in it a vector e_2. If $e_1^\perp$ is even, then the lattice $U_2 = \mathbf{Z}_2 a \oplus \mathbf{Z}_2 b$, $a^2 = b^2 = 0$, $ab = 1$, splits off from it as a direct summand. And for $e_1' = e_1 + a$ and $e_2' = -e_1 + b \in (e_1')^\perp$ we have $(e_1')^2 = 1$ and $(e_1')^2 = 1$; therefore now $(e_1')^\perp$ is already odd.

By applying this procedure several times we obtain vectors $e_i \in M$, $i = 1, \ldots, 2\sigma - 4$, $e_i^2 = 1$ for $i \equiv 1 \bmod 2$, $e_i^2 = -1$ for $i \equiv 0 \bmod 2$, $e_i e_j = 0$ for $i \neq j$. Then $M_1 = \bigoplus_{i=1}^{2\sigma-4} \mathbf{Z} e_i \oplus M_1'$, rg $M_1' = 4$ and M_1' is odd. Repeating the same considerations, we find in M_1' a basis $e_{2\sigma-3}, e_{2\sigma-2}, e_{2\sigma-1}, e_{2\sigma}$ with $e_i e_j = 0$ for $i \neq j$ and $e_{2\sigma-i}^2 = \epsilon_i \in \mathbf{Z}_2^*$, $i = 0, 1, 2, 3$.

Using the condition $d(N) = -2^{2\sigma}$ we obtain that $\epsilon_0 \epsilon_1 \epsilon_2 \epsilon_3 = 1$. Finally we use the fact that $\epsilon(\overline{N}) = \epsilon(S) = 1$. Arguing as in A, we obtain that $\epsilon(M') = 1$. It is easy to see that these conditions on M' define M' uniquely: $M' \simeq I_+^{(4)}$. Hence

$$\overline{N} \simeq U_2^{(\tau)} \oplus I_{(+)}(2)^{(\sigma+2)} \oplus I_{(-)}(2)^{(\sigma-2)}. \tag{6}$$

Thus we have found two possibilities for $\overline{N}$, given by (5) and (6). It is easy to verify that (5) gives a lattice of type I and (6) a lattice of type II. Clearly for both types $\sigma > 0$; for type I $\tau > 0$ and for type II $\sigma > 1$.

It remains to show that all these lattices are indeed realized by lattices over $\mathbf{Z}$ satisfying conditions a)–d) of §1.

We consider three cases.

a) $\sigma = n/2$. Here there exists only a lattice of type II, for which we can take the lattice $S(2) = \bigoplus_1^n \mathbf{Z}e_i$, where $e_1^2 = 2$, $e_i^2 = -2$ for $i > 1$, and $e_ie_j = 0$ for $i \neq j$.

b) $\sigma = n/2 - 1$. The lattice of type I will be

$$N = \{S(2), f\},\ f = \frac{1}{2}\sum_1^n e_i.$$

It is even, since $n \equiv 2 \bmod 4$, and

$$2N^* = \left\{\sum x_ie_i,\ \sum x_i \equiv 0 \bmod 2\right\},$$

from which we see that it belongs to type I. A lattice of type II will be $U_2 \oplus I_{(-)}(2)^{(n-2)}$.

c) $\sigma \leqslant n/2 - 2$. We will use two constructions:

A.

$$N = U_2 \oplus N',\ N' \subset I_{(-)}^{(n-2)},\ I_{(-)}^{(n-2)} = \bigoplus_{i=1}^{n-2} \mathbf{Z}\, e_i,\ e_i^2 = -1,$$

and N' consists of elements of the form $\Sigma_1^{n-2} x_ie_i$, where

$$x_1 + \ldots + x_{k_1} \equiv x_{k_1+1} + \ldots + x_{k_1+k_2} \equiv \ldots$$
$$\equiv x_{k_1+\ldots+k_{r-1}+1} + \ldots + x_{k_1+\ldots+k_r} \equiv 0 \bmod 2. \tag{7}$$

The evenness of the lattice N' (and hence also of N) is equivalent to the condition

$$k_1 + \ldots + k_r = n-2. \tag{8}$$

It is easy to check that

$$2(N')^* = \{f_1, \ldots, f_r, 2N'\},\ f_i = e_{k_1+\ldots+k_{i-1}+1} + \ldots + e_{k_1+\ldots+k_i},$$

and the inclusion $N' \supset 2(N')^*$ is equivalent to the condition

$$k_i \equiv 0 \bmod 2,\ i = 1, \ldots, r.$$

A lattice is of type I if and only if $f_i^2 \equiv 0 \bmod 4$, i.e. $k_i \equiv 0 \bmod 4$, $i = 1, \ldots, r$. Finally $(S: N) = 2^r$, so that $\sigma = r$.

Setting $k_1 = k_2 = \cdots = k_{r-1} = 2$ and $k_r = n - 2 - 2(r-1)$, we obtain for $0 \leqslant r \leqslant n/2 - 2$ lattices of type II with all the possible values of σ.

Setting $k_1 = k_2 = \cdots = k_{r-1} = 4$ and $k_r = n - 2 - 4(r-1)$, we obtain, on account of the condition $k_r \geqslant 4$, lattices of type I with values $1 \leqslant \sigma \leqslant (n-2)/4$.

B. We set

$$N = U_2 \oplus N',\quad N' = \{I_{(-)}(2)^{(n-2)}, f_1, \ldots, f_r\},$$

where

$$I_{(-)}(2)^{(n-2)} = \bigoplus \mathbf{Z}\, e_i,\quad e_i^2 = -2,\quad e_ie_j = 0 \text{ for } i \neq j,$$

$$f_s = \frac{1}{2}(e_{k_1+\ldots+k_{s-1}+1} + \ldots + e_{k_1+\ldots+k_s}).$$

Integrality and evenness of N' (and hence also of N) is assured by the condition $k_s \equiv 0 \bmod 4$, $s = 1, \ldots, r$. The relation $N' \supset 2(N')^*$ is always satisfied, since $N' \supset I_{(-)}(2)^{(n-2)}$. A simple computation shows that $2(N')^*$ consists of the elements $\Sigma x_i e_i \in I_{(-)}(2)^{(n-2)}$ that satisfy (7). Therefore N' is of type I if and only if conditions (8) are satisfied. Finally,

$$(N' : I_{(-)}(2)^{(n-2)}) = 2^r$$

and therefore $\sigma = n/2 - r - 1$. Setting $k_1 = \cdots = k_{r-1} = 4$ and $k_r = n - 2 - 4(r-1)$, as before we obtain lattices of type I with $(n-2)/4 \leqslant \sigma \leqslant n/2$.

REMARK. The results obtained here are not tied to the prime number 2. Let us call a lattice N *p-elementary* if $N \supset pN^*$ for some prime p. We set $n = \operatorname{rg} N$ and $\operatorname{rg}_{\mathbf{F}_p} N^*/N = s$, and assume that the following conditions hold:

a) $n \equiv 6 \pmod 8$.

b) The quadratic form defined in $N \otimes \mathbf{R}$ has positive inertia index 1 and negative inertia index $n - 1$.

c) N is even.

d) N is p-elementary, $p \neq 2$.

e) $s \equiv 0\ (2)$, $s = 2\sigma$.

Then a simpler result holds than for $p = 2$; namely,

THEOREM 2′. *A lattice satisfying conditions* a)–e) *is uniquely determined by its rank and the invariant* σ. *Always* $0 < \sigma < n/2$, *and for any such value of* σ *there exists a lattice with such an invariant.*

The proof is analogous to the proof of Theorem 2. Again we see that class and genus are the same. From conditions a), b), d) and e) it follows that $d(N) = -p^{2\sigma}$. If q is a prime number different from 2 and p, then $N \otimes \mathbf{Z}_q$ is a unimodular lattice and consequently isomorphic to $U_2^{(n/2)}$ or $U_2^{(n/2-1)} \oplus V_2$, where V_2 is a unimodular anistropic lattice of rank 2 over $\mathbf{Z}_q$. Comparison of discriminants shows that the second case is impossible, and therefore

$$N \otimes \mathbf{Z}_q = U_2^{\left(\frac{n}{2}\right)}, \quad q \neq 2, p.$$

Completely in the same way one proves that $N \otimes \mathbf{Z}_2 = U_2^{(n/2)}$.

For a form $N \otimes \mathbf{Z}_p$ there are 4 possibilities:

1) $U_2^{(n/2-\sigma)} \oplus U_2(p)^{(\sigma)}$;

2) $U_2^{(n/2-\sigma-1)} \oplus V_2 \oplus U_2(p)^{(\sigma)}$;

3) $U_2^{(n/2-\sigma)} \oplus U_2(p)^{(\sigma-1)} \oplus V_2(p)$;

4) $U_2^{(n/2-\sigma-1)} \oplus V_2 \oplus U_2(p)^{(\sigma-1)} \oplus V_2(p)$,

where V_2 is again a unimodular anisotropic lattice of rank 2 over $\mathbf{Z}_p$. The required discriminant corresponds only to cases 1) and 4). In order to choose between them we compute the invariant ϵ_q for the candidates that we have for $N \otimes \mathbf{Z}_q$ (for the definition of ϵ_q, see [13], p. 64). A simple computation shows that

$$\varepsilon_q\left(U_2^{\left(\frac{n}{2}\right)}\right) = 1 \text{ for } q \neq 2, p; \quad \varepsilon_2\left(U_2^{\left(\frac{n}{2}\right)}\right) = -1,$$

$$\varepsilon_p\left(U_2^{\left(\frac{n}{2}-\sigma\right)} \oplus U_2(p)^{(\sigma)}\right) = 1.$$

Therefore this collection of lattices cannot be obtained in the form $N \otimes \mathbf{Z}_q$ from some integral lattice N: it does not satisfy the relation

$$\prod_q \varepsilon_q (N \otimes \mathbf{Z}_q) = 1$$

(see [13], p. 78). Therefore $N \otimes \mathbf{Z}_p$ has to correspond to case 4). This determines $N \otimes \mathbf{Z}_q$ uniquely for all prime numbers q; and this means that N is also uniquely determined.

The invariant σ cannot take the values 0 or $n/2$ (see [1]). In fact, in the first case the lattice N would be even and unimodular and its rank would satisfy $n \equiv 2 \pmod 8$. On the other hand, σ cannot equal $n/2$, since then N would be of the form $M(p)$ and the lattice M would be even and unimodular, which is impossible for the same reason.

It remains only to exhibit a lattice N for a given value of σ. We will use the following auxiliary lattices over $\mathbf{Z}$. H_p is an even lattice of rank 4 and with discriminant $-p^2$ (it can be realized as a maximal order in a quaternion algebra ramified only at p and ∞ with product $\mathrm{tr}(xy^*)$, where tr is the reduced trace and * denotes the involution). Let $E = \bigoplus \mathbf{Z}e_i$, $(e_ie_j) = 0$ for $i \neq j$, and $e_i^2 = a_i$, $a_i \equiv 1$ (2). By E_* we denote the lattice $E_0 + \epsilon\mathbf{Z}$, where $E_0 \subset E$ and it consists of all vectors $\Sigma x_i e_i$ with $\Sigma x_i \equiv 0$ (2), and $\epsilon = \frac{1}{2}\Sigma e_i$. E_* is an even integral lattice if the condition $\Sigma a_i \equiv 0$ (8) is satisfied. (This construction is an elementary special case of the operation of "making even" an odd lattice introduced by B. B. Venkov.) Let $n = 8m + 6$. If $\sigma = 2\bar{\sigma} + 1$ is odd, we set

$$N_\sigma = U_2 \oplus H_p \oplus (I(p)^{(4\bar{\sigma})} \oplus I^{(8m-4\bar{\sigma})})_*.$$

If $\sigma = 2\bar{\sigma} + 2$, we set

$$N_\sigma = U_2(p) \oplus H_p \oplus (I(p)^{(4\bar{\sigma})} \oplus I^{(8m-4\bar{\sigma})})_*.$$

It is easy to check that $\Sigma a_i \equiv 0$ (8) for such lattices E to which we apply the operation E_*, and further that N_σ has the needed value σ and that all the conditions of Theorem 2′ are satisfied.

Geometric application. We give a simpler proof of the result of [11] that on a $K3$ surface there does not exist a nonzero regular vector field. In [11] it is shown (quite simply) that on a surface X which has a nonzero vector field θ there exists a pencil $\varphi: X \longrightarrow \mathbf{P}^1$ of arithmetic genus 1.

Let us suppose that φ is elliptic. The field θ is either tangent to the fibers of the pencil (this case easily leads to a contradiction) or it has a nonzero "normal component" θ_N which is a regular vector field on $\mathbf{P}^1$, where θ_N vanishes only at those points over which the degenerate fibers lie. Therefore the number of degenerate fibers does not exceed two, and if on the surface X we construct an elliptic pencil with three degenerate fibers, the result will be proved. But even in the case of two degenerate fibers, the field θ belongs to the "multiplicative type", and in [11] it was shown (also quite simply) that if both of the degenerate fibers are reducible then there is no higher ramification. In case there is no higher ramification, the theorem easily follows, as was shown in [11]. Because of this we can restrict ourselves to the case when $p = 2$ or 3. Thus it is enough to construct on X an elliptic pencil that has at least two reducible fibers. A pencil of arithmetic genus 1 corresponds to an element $f \in N$, $f^2 = 0$, and its reducible fibers correspond to irreducible components of the root system of the lattice $N(f)$. Finally, the case of a quasielliptic pencil is distinguished from the case of an elliptic pencil by means of the proposition in §1. As was shown in [11], a surface X on which there exists a nonzero vector field is unirational, hence supersingular, so that the results obtained above are applicable to it. In particular, because of Theorems 2 and 2′ it is enough for $p = 2$ and $p = 3$ and for every value of σ, $0 < \sigma < 11$, to construct an arbitrary lattice N satisfying the conditions of these theorems and such that there exists in it an element f, $f^2 = 0$, such that the root system $R(f)$ of the lattice $N(f)$ decomposes into at least two irreducible components but not satisfying conditions 1) and 2) of the proposition. We will give N in the form $U_2 \oplus N'$ or $U_2(p) \oplus N'$, where $f \in U_2$ or $f \in U_2(p)$.

The case $p = 3$. We denote by V_2 the negative definite even lattice of rank 2 and discriminant 3. As is easily seen, $H_3 = V_2 \oplus V_2$, and the root system V_2 is of type A_2. Therefore for $p = 3$ the lattice N_σ, constructed above for any p, satisfies the necessary conditions (it is very easy to check that for it conditions 1) and 2) of the proposition fail).

The case $p = 2$. Let $E_{16} = \bigoplus \mathbf{Z}e_i$, the e_i being orthonormal; $E'_{16} \subset E_{16}$ consists of vectors $\Sigma x_i e_i$, $\Sigma x_i \equiv 0$ (4), and $F_{16} = E'_{16} + e\mathbf{Z}$, where $e = \frac{1}{4}\Sigma e_i$. F_{16} is an odd unimodular lattice. For any odd lattice S we denote by S_{even} the sublattice of elements whose square is even. We write down the lattices that we need:

σ	N	$R(f)$, $f \in U_2$, $U_2(2)$
1	$U_2 \oplus (F_{16} \oplus I^{(4)})_{\text{ч}}$	$D_5 \oplus A_{15}$
2	$U_2(2) \oplus (F_{16} \oplus I^{(4)})_{\text{ч}}$	$D_5 \oplus A_{15}$
3	$U_2 \oplus D_4(2) \oplus E_8^{(2)}$	$E_8^{(2)}$
4	$U_2(2) \oplus D_4(2) \oplus E_8^{(2)}$	$E_8^{(2)}$
5	$U_2 \oplus D_4 \oplus E_8 \oplus E_8(2)$	$D_4 \oplus E_8$
6	$U_2(2) \oplus D_4 \oplus E_8 \oplus E_8(2)$	$D_4 \oplus E_8$
7	$U_2 \oplus D_4^{(3)} \oplus E_8(2)$	$D_4^{(3)}$
8	$U_2(2) \oplus D_4^{(3)} \oplus E_8(2)$	$D_4^{(3)}$
9	$U_2 \oplus M_{12} \oplus E_8(2)$	$A_1^{(12)}$
10	$U_2(2) \oplus M_{12} \oplus E_8(2)$	$A_1^{(12)}$

where

$$M_{12} = \left(\bigoplus_{i=1}^{12} \mathbf{Z}e_i,\ e_i^2 = -2,\ e_ie_j = 0\right) + \sum_{i=1}^{12} \frac{1}{2} e_i.$$

§3. Classification of surfaces

THEOREM 3. *The lattice* $N = \operatorname{Pic} X$ *for a supersingular* $K3$ *surface* X *over a field of characteristic* 2 *belongs to type* I.

PROOF. Since $\operatorname{rg} N = 22$, according to Theorem 2 the invariant σ can only take values from 1 to 11. We consider three cases:

a) Let $\sigma = 11$. In this case, according to Theorem 2, there exists only a lattice of type II, and it is isomorphic to

$$S(2)=\oplus \mathbf{Z}e_i,\ e_1^2=2,\ e_i^2=-2,\ 2\leqslant i\leqslant 22.$$

If N were isomorphic to this lattice, then the element $f = e_1 + e_2$ would correspond to an elliptic or a quasielliptic fibration, and since $N(f)$ is generated by images of the elements e_i, $i = 2, \ldots, 22$, therefore, according to the property formulated in §1, the corresponding fibration should have 20 fibers of type $^*\mathrm{A}_1$, and according to Lemma 3 it must be quasielliptic. In any quasielliptic fibration there is defined a curve ξ formed by the cusps of the fibers and $\xi \cdot F = 2$ (see [2], where this curve is denoted by σ). From this it follows that if the nondegenerate fiber has components C_i and C_i', then $(\xi, C_i) = (\xi, C_i') = 1$. This, however, contradicts the fact that in the lattice $N \simeq S(2)$ the product of any two elements is even. Therefore, this case is impossible.

b) Let $\sigma = 10$, and suppose that N is of type II. As we saw in proving Theorem 2, such a lattice is given by $U_2 \oplus I_{(-)}(2)^{(20)}$. Therefore, according to Theorem 2, in this case we would have an isomorphism $N \simeq U_2 \oplus I_{(-)}(2)^{(20)}$. If $U_2 = \{f, g\}$, where $f^2 = g^2 = 0$ and $fg = 1$, then f corresponds as in case a) to a quasielliptic fibration with 20 degenerate fibers F_i of type $^*\mathrm{A}_1$, where the fibration has a section since there exists a divisor g with $fg = 1$. Let s be such a section and ξ the curve formed by the cusps of the fibers, and $F_i = C_i + C_i'$. We can assume that $sC_i = 1$ and therefore $sC_i' = 0$. Since f, s and C_i, $i = 1, \ldots, 20$, form a basis of the lattice N, we have

$$\xi = 2s + \sum k_i C_i + lf,\quad k_i,\ l \in \mathbf{Z}.$$

Considering the intersection with C_i, we obtain that $1 = 2 - 2k_i$. This contradiction shows that N cannot belong to type II.

c) Let $\sigma < 10$. In this case, as we saw in the proof of Theorem 2, the lattice N has the form $U_2 \oplus N'$, where N' contains $n - 2$ independent elements with square -2.

Let $U_2 = \{f, g\}$, where $f^2 = g^2 = 0$ and $fg = 1$. Since U_2 is unimodular, it follows that $N \simeq U_2 \oplus N(f)$, and as in the analysis of case b) we can compute that f corresponds to a quasielliptic pencil having a section s.

We set $N'' = U_2 \oplus H(f)$ and $N' = (N'', \xi)$; we will show that the lattice N' belongs to type I. Since the module $(N')^*/N'$ is "bigger" than the module N^*/N, the property of the scalar product $\langle x, x\rangle = 0$ carries over from the first to the second; hence the theorem will follow.

We have the inclusion $(N')^*/N'' \subset (N'')^*/N''$, in which evidently $(N')^*/N'' = \xi^\perp$, where ξ is considered as an element of the group $(N'')^*/N''$. We need to show that $\langle x, x\rangle = 0$ for $x \in \xi^\perp$. We set $\langle x, x\rangle = u(x)$ for $x \in (N'')^*/N''$. Obviously u is a linear function, and we need to show that

$$\xi^\perp \in \operatorname{Ker} u. \tag{9}$$

Since U_2 is unimodular, we have $(N'')^*/N'' = H(f)^*/H(f)$, and if $H(f) = \bigoplus H_i(f)$, it is enough to check (9) for all the groups $H_i^*(f)/H_i(f)$. This check has to be carried out for all degenerate fibers of types E_8, E_7, $^*\mathrm{A}_1$ and D_{2n}. The lattice E_8 is unimodular, and therefore $\mathrm{E}_8^* = \mathrm{E}_8$, so for it our claim is trivial. For the remaining types, the action of ξ as a linear form on $H(f)$ is defined based on the fact that it is invariant with respect to the action of $H_i^*(f)/H_i(f)$ on the extended Dynkin diagram corresponding to this fiber (see [12]). In addition, ξ turns out to be an element of the basis dual to the basis consisting of simple roots, where this basis element corresponds to a "symmetric" simple root. In other words, ξ is one of the "basic weights" in the terminology of [3]. Since for all simple root systems the "basic weights" ω_i are written out in [3], we obtain an explicit representation for ξ. In particular, for fibers of types ${}^*\mathrm{A}_1$ and E_7 we get that H_i^*/H_i is one-dimensional and $\xi \neq 0$; therefore $\xi^\perp = 0$ and (9) is obvious. For the fiber D_{2n}, using formulas from [3], pp. 256–257, we obtain that $\xi = \overline{\omega}_n$ and $\xi(x) = u(x)$, which also proves (9). This ends the proof of Theorem 3.

§4. Construction of universal families

The construction of families which we will now explain resembles somewhat the construction of lattices over $\mathbf{Z}$ of type I at the end of the proof of Theorem 2; in both cases the set of possible values of σ splits into two parts, for each of which a separate construction is proposed.

a) *The case* $10 \geqslant \sigma \geqslant 6$.

1) $\sigma = 10$. The lattice Pic X in this case is isomorphic to the lattice obtained by adjoining to the lattice with basis $C_0, \ldots, C_{22}$, where the C_i are orthogonal, $C_0^2 = +2$ and $C_i^2 = -2$ for $i > 0$, the element $\epsilon = \frac{1}{2}\Sigma_0^{22} C_i$. The element $F = C_0 - C_1$ has square 0, and we may consider that it is realized by the fiber of a pencil whose generic curve has arithmetic genus 1. The classes $C_2, \ldots, C_{22}$ are orthogonal to F, and we can assume that they are realized by components of fibers. We obtain 20 fibers of type ${}^*\mathrm{A}_1$. The pencil will be quasielliptic. Components of the fiber are the curves C_i and $F - C_i$, $i \geqslant 2$. The class of C_1 has square -2; therefore it contains a unique effective divisor s for which $(s, F) = 2$. Since $(\epsilon, F) = 2$ and (C_i, F) is even, the intersection with the fiber of an arbitrary cycle is even. We will show that the divisor s is irreducible. This follows from the following assertion.

LEMMA 3. *Let F be a fiber of a pencil of curves of arithmetic genus* 1 *on a* $K3$ *surface, and let s be an effective divisor satisfying the conditions $s^2 = -2$, $(s, C_i) \geqslant 0$ for all components C_i of fibers of the pencil, and $0 < (s, F) \leqslant (D, F)$ for an arbitrary divisor D for which $(D, F) > 0$. Then s is an irreducible curve.*

If s were reducible, the only possible decomposition would be $s = \overline{s} + x$, where $\overline{s}$ is a multiple section and x is a divisor consisting of components of fibers. We have $\overline{s}^2 = s^2 - 2sx + x^2$. Also $(x)^2 \leqslant 0$ since the intersection form is nonpositive definite on the sublattice generated by components of fibers, and $(s, x) \geqslant 0$ since $(s, C_i) \geqslant 0$. Consequently $\overline{s}^2 \leqslant s^2 = -2$. Because $\overline{s}$ is an irreducible curve, we thus have $\overline{s}^2 = -2$ and $x = 0$. It is easy to check, using standard exact sequences, that $H^1(rs + mF) = 0$ for $m \geqslant r$. Hence, by the Riemann-Roch theorem,

$$l(rs+mF) = -r^2+2rm+2 \quad \text{for} \quad m \geqslant r.$$

Acting by analogy with the construction of the Weierstrass normal form of an elliptic curve we may choose functions $x \in L(s + F)$ ($x \notin L(F)$) and $y \in L(2s + 3F)$, with y linearly independent of $1, x, x^2, t, t^2, t^3, xt, xt^2, x^2t$, where $t \in L(F)$ is a function on the base. Then the monomials of the form $x^i y^j t^k$ which belong to $L(4s + 6F)$ are linearly dependent. This leads to the relation

$$y^2 + (b_1x^2 + b_2x + b_3)y = a_2x^4 + a_3x^3 + a_4x^2 + a_5x + a_6, \qquad (*)$$

where a_i and b_i are polynomials in t of degree i. Let us show that in our case $b_1 = b_2 = b_3 = 0$. We put

$$\varphi(x) = b_1x^2 + b_2x + b_3,\ f(x) = a_2x^4 + a_3x^3 + \ldots + a_6$$

and consider the cases where $\deg \varphi$ is 2, 1, or 0. Suppose $\deg \varphi = 2$. Then $\varphi'_x = 0$. In fact, if φ is irreducible and $\varphi'_x \neq 0$, then on computing the singular points of the curve (*) we find that the two roots of φ give two distinct singular points of it, contradicting the fact that the arithmetic genus is 1. If, however, φ has two roots in the field $k(t)$, the same computations show that one of these roots gives a singular point of the curve (*) rational over $k(t)$, which would imply the rationality of the surface X. The equations for the singular points of the surface are now of the form

$$f'_x = 0, \quad \varphi = 0, \quad \varphi'_t y = f'_t, \quad y^2 + \varphi y = f.$$

Writing f in the form $g(x^2) + f'_x \cdot x$, we reduce the last equation to the form $y^2 = g(x^2)$. Substituting $y = f'_t/\varphi'_t$, we obtain an equation depending only on x^2. The value of x^2 is determined by the condition $\varphi = 0$ (since $\deg \varphi = 2$ and $\varphi'_x = 0$). We obtain a relation on the coefficients of (*) which, as computations show, is of degree 12 in t. Since the values of t corresponding to singular points of the surface should give the value 0 for this expression, this contradicts the fact that our pencil has 20 degenerate fibers.

If $\deg \varphi = 1$, the same considerations show that a singular point of the curve (*) is rational over $k(t)$, which leads to rationality of the surface X. Finally, in the case when $\deg \varphi = 0$ but $\varphi \neq 0$ we have an elliptic pencil.

Therefore our equation is reduced to the form

$$y^2 = a_2(t)x^4 + a_3(t)x^3 + a_4(t)x^2 + a_5(t)x + a_6(t).$$

In the general case such an equation defines a quasielliptic $K3$ surface with 20 fibers of type $^*A_1$.

Permissible transformations $y \longrightarrow y + b_1x^2 + b_2x$ show that a point of the base S of this family is given by the polynomials a'_2, a_3, a'_4, a_5 and a'_6. The total number of parameters entering into them is 16. The transformations $x \longrightarrow \alpha x + c_1$, $\deg c_1 \leqslant 1$, $y \longrightarrow \beta y$ and the transformations $\varphi \in SL(2, k)$ of the parameter t define the group G of automorphisms of the base of the family. Obviously $\dim G = 7$, so that $\dim S - \dim G = 9 = \sigma - 1$. Let us check that the stabilizer of the generic point of the base is finite. It is easy to see that

$$a_3 \rightarrow \frac{\alpha^3}{\beta^2} a_3^\varphi, \text{ where } a_i^\varphi(t) = a_i\left(\frac{at+b}{ct+d}\right)(ct+d)^i.$$

If the roots of a_3 are distinct, the transformations φ preserving a_3 form a finite group. The condition of invariance of the coefficients of a_2 and a_4 easily gives that $c_1 = 0$ for $\varphi = 1$ and $\alpha = \beta = 1$.

2) $\sigma = 9$. The lattice Pic X in this case is obtained by adjoining to the lattice with basis $C_0, \ldots, C_{22}$, where the C_i are orthogonal, $C_0^2 = 2$, $C_i^2 = -2$ for $i \geqslant 1$, the elements $\frac{1}{2}(C_0 + C_1)$ and $\epsilon = \frac{1}{2}\Sigma_1^{22} C_i$. We consider the elements $F = \frac{1}{2}(C_0 - C_1)$ and C_1. Then $(F)^2 = 0$, $(C_1)^2 = -2$, and $(F, C_1) = 1$. The curves in the class F form a pencil of curves of arithmetic genus 1. The class C_1 contains a unique effective curve s.

According to Lemma 3, to prove the irreducibility of s it is enough to check the nonnegativity of intersections with components of fibers. In our case, as is easy to see, there are 20 fibers of type $^*\mathrm{A}_1$, and consequently the pencil is quasielliptic and their components are C_i and $F - C_i$ for $i \geqslant 2$. We have $(s, C_i) = 0$ and $s \cdot (F - C_i) = 1$, which means that the curve s is irreducible.

As in the previous case it is easy to show that $H^1(rs + mF) = 0$ for $m \geqslant 2r$ and to find $l(rs + mF)$ according to the Riemann-Roch theorem. If we choose $x \in L(2s + 4F)$, $x \notin L(s + 4F)$, and $y \in L(3s + 6F)$, y linearly independent of $1, x, t, t^2, t^3, t^4, t^5, t^6, tx, t^2x$, we see that there is a relation between the elements $x^i y^j t^k$ belonging to $L(6s + 12F)$, which can be reduced to the form

$$y^2 = x^3 + a_4(t)x^2 + a_8(t)x + a_{12}(t). \tag{10}$$

As in case 1), a point of the base S of this family is defined by the polynomials a_4', a_8 and a_{12}', so that $\dim S = 17$. The transformations $x \longrightarrow \lambda^2 x + c_6$, $y \longrightarrow \lambda^3 y$ and $\varphi \in \mathrm{SL}(2, \mathrm{k})$ define an automorphism group G, with $\dim G = 9$, so that $\dim S - \dim G = 8 = \sigma - 1$. In order to show the finiteness of the stabilizer of the generic point it is enough to consider the transformation of the coefficient a_8: $a_8 \longrightarrow (a_8^\varphi + c_4^2)\lambda^{-4}$; hence $a_8' \longrightarrow (a_8')^\varphi \cdot \lambda^{-4}$. Since $a_8' = u_3^2$, where $\deg u_3 = 3$, we conclude as in case 1) that φ belongs to a finite group and $\lambda = 1$. Considering the coefficient a_4 gives us that $c_4 = 0$.

3) $\sigma = 8$. The lattice Pic X for $\sigma = 8$ can be obtained from the lattice with basis s, $F, C_1, \ldots, C_{20}$, where $s^2 = -2$, $F^2 = 0$, $s \cdot F = 1$, the C_i are orthogonal, and $C_i^2 = -2$, by adding the elements $D_1 = \frac{1}{2}\Sigma_1^8 C_i$ and $\epsilon = \frac{1}{2}\Sigma_9^{20} C_i$. We may consider that F consists of curves of a pencil with arithmetic genus 1. It is then easy to check that the components of fibers are C_i and $F - C_i$. In this way we obtain 20 fibers of type $^*\mathrm{A}_1$, and consequently the pencil is quasielliptic. The class of s contains an irreducible curve, a section of the pencil, which we will also denote by s.

We can choose the equation of the surface in the form (10) so that the section s will be at infinity. Let us consider now the class $s' = s - D_1 + 2F$. Then $(s')^2 = -2$, $s' \cdot F = 1$ and the intersections of s' with components of fibers are nonnegative. According to Lemma 4 this class contains an irreducible curve, which we will also denote by s'. Since $ss' = 0$, the section s' is given by the equations $x = u(t)$ and $y = v(t)$, where u and v are polynomials in t. Also,

$$\deg u \leqslant (s', (x)_\infty) = (s'\ 2s + 4F) = 4.$$

Carrying out the substitution $x \longrightarrow x + u(t)$, $y \longrightarrow y + v(t)$, we transform the equation of the surface into the form

$$y^2 = x(x^2 + a_4(t)x + a_8(t)). \tag{11}$$

A point of the base S is determined by the polynomials a_4' and a_8; $\dim S = 11$.

Admissible transformations have the form $x \longrightarrow \lambda^2 x$, $y \longrightarrow \lambda^3 y$ and $\varphi \in SL(2, k)$, so that $\dim G = 4$ and $\dim S - \dim G = 7 = \sigma - 1$. The finiteness of the stabilizer of the general point obviously follows from considerations used in cases 1) and 2).

4) $\sigma = 7$. The lattice Pic X is obtained by adding to the lattice considered in the previous case the element

$$D_2 = \frac{1}{2}\sum_{i=1}^{4} C_i + \frac{1}{2}\sum_{i=9}^{12} C_i.$$

Again we have a quasielliptic pencil with 20 fibers of type $^*A_1$ with components C_i and $F - C_i$. Repeating the considerations of the previous case, we conclude that the equation of the surface has the form (11) and a section $s' = s - D_1 + 2F$ is given by equations $x = 0$ and $y = 0$. Let us consider the section $s'' = s - D_2 + 2F$. Since $s'' \cdot s = 0$, this section is given by the equations $x = \bar{u}(t)$ and $y = \bar{v}(t)$, and, analogously, $\deg \bar{u}(t) \leqslant 4$ and $\deg \bar{v}(t) \leqslant 6$.

Let t_1, t_2, t_3, t_4 be the values of t corresponding to fibers that contain C_1, C_2, C_3, C_4 respectively. These values are all distinct and we may consider them to be finite. We note that $(s'; C_i) = (s''; C_i) = 1$ for $i = 1, 2, 3, 4$. Since $s \cdot C_i = 0$, it follows that s does not pass through the components C_i and the Weierstrass model of the surface X is obtained from the minimal model by means of contracting the curves C_i into singular points of the surface and of the corresponding fibers. Thus the sections s' and s'' on the Weierstrass model have four points in common, the singular points of the fibers $t = t_i$ for $i = 1, 2, 3, 4$. Therefore $\bar{u}(t_i) = 0$ and $\bar{v}(t_i) = 0$ for $i = 1, 2, 3, 4$. Hence $\deg \bar{u} = 4$ and $\bar{v}$ is divisible by $\bar{u}$. We set $\bar{v} = w \cdot \bar{u}$ and make the change of variable $y \longrightarrow y + wx$; then the equation of the surface takes the form

$$y^2 = x(x + a_4(t))(x + b_4(t)), \tag{12}$$

where $a_4 = \bar{u}_4$. A point of the base S is determined by the polynomials a_4 and b_4; and $\dim S = 10$. The group G is given by the same transformations as in the previous case, so that $\dim S - \dim G = 6 = \sigma - 1$.

Finiteness of the stabilizer of the general point is checked in the same way.

5) $\sigma = 6$. In order to obtain the lattice Pic X we add to the elements F, s, $C_1, \ldots, C_{20}$, D_1, ϵ and D_2 constructed in the previous cases also the element

$$D_3 = \frac{1}{2}\sum_{i=1}^{4} C_i + \frac{1}{2}\sum_{i=13}^{16} C_i.$$

The class F gives us a pencil of curves of arithmetic genus 1. It is easy to check that its fibers are 16 fibers of type $^*A_1$ with components C_i and $F - C_i$, $i = 1, \ldots, 16$, and a fiber of type D_4 with components

$$\varepsilon - D_3 - D_2 + C_1 + C_2 + C_3 + C_4 = \frac{1}{2}\sum_{i=17}^{20} C_i,$$
$$F - C_{20}, \quad (-C_{19}), \quad (-C_{18}), \quad (-C_{17}),$$

the first of which is double. Such a pencil is quasielliptic. Again we consider the sections

$$s' = s - D_1 + 2F,$$
$$s'' = s - D_2 + 2F,$$
$$s''' = s - D_3 + 2F.$$

It is not hard to check that these are irreducible rational curves, and $(s')^2 = (s'')^2 = (s''')^2 = -2$. Also the elements s, s', s'' and s''' are pairwise orthogonal. Using our previous considerations, we can assume that the equation of the surface has the form (12) and the section s is at infinite distance, s' is given by the equations $x = 0$ and $y = 0$, s'' is given by the equations $x = \bar{u}$ and $y = \bar{v}$, and s''' by the equations $x = \bar{\bar{u}}$ and $y = \bar{\bar{v}}$. The degrees of $\bar{u}$ and $\bar{\bar{u}}$ are equal to 4, and the degrees of $\bar{v}$ and $\bar{\bar{v}}$ are equal to 6. If to fibers with components C_1, C_2, C_3, C_4 correspond values of t equal to t_1, t_2, t_3, t_4, then, applying the same considerations as in case 4) to s'' and s''', we obtain that $\bar{u}(t_i) = \bar{\bar{u}}(t_i) = 0$. Then $\bar{\bar{u}} = \alpha\bar{u}$. We recall that $a_4 = \bar{u}$ and the equation of the surface can be written in the form

$$y^2 = x(x + \bar{u})(x + b_4).$$

Substituting $x = \alpha\bar{u}$ and $y = \bar{\bar{v}}$, we obtain

$$\bar{\bar{v}}^2 = \alpha\bar{u}(\alpha + 1)\bar{u}(\alpha\bar{u} + b_4).$$

The derivative of the left-hand side is equal to 0; therefore $\alpha\bar{u}' + b_4' = 0$, i.e. $\alpha a_4' + b_4' = 0$, so that the derivatives a_4' and b_4' are proportional.

Therefore the normal form has the same form (12) as in case 4):

$$y^2 = x(x + a_4(t))(x + b_4(t)),$$

but with the supplementary condition that a_4' and b_4' are proportional. Since a_4' and b_4' are squares of linear polynomials, this represents one condition; and $\dim S - \dim G = 5 = \sigma - 1$.

b) *The case* $5 \geqslant \sigma \geqslant 1$. We know that any supersingular surface contains a quasielliptic pencil $X \longrightarrow \mathbf{P}^1$. As we saw in proving Theorem 3, if $\sigma \neq 10$ we can assume that this pencil has a section. The equation of the pencil can in this case be written in the form

$$y^2 = x^3 + ax + b, \quad a, b \in \mathrm{k}(\mathbf{P}^1),$$

and, as was shown when considering case 2), if $\mathrm{k}(\mathbf{P}^1) = \mathrm{k}(t)$, then $\deg a \leqslant 8$ and $\deg b \leqslant 12$

The points over which the nondegenerate fibers lie are defined by the equation $D = 0$, where $D = (a')^2 a + (b')^2$.

By considering all the possible types of fibers and the corresponding normal forms of minimal models ([8], pp. 124–125) it is not hard to verify that for a degenerate fiber with $n + 1$ components over the point $t = \alpha$ we have $v_\alpha(D) = n$. It is probable that one can use the ideas of [4] to get a proof of this fact not connected with going through separate types of fibers.

A particularly simple case is when $a \in (\mathrm{k}(t)^*)^2$, i.e. $a' = 0$. Then $D = (b')^2$, whence it follows in particular that all the numbers $n_i = \mathrm{rg}\, H_i(f)$ have to be even.

Conversely, if all the numbers n_i are even, then $D = \Delta^2$, $\Delta \in \mathrm{k}(t)$, and if $a' \neq 0$ we would obtain that $a = ((b' + \Delta)/a')^2$, i.e. again $a' = 0$. Therefore $a' = 0$ if and only if all the n_i are even.

This is the case if $1 \leqslant \sigma \leqslant 5$ and also all $n_i = 0 \bmod 4$, so that $D = \Delta_1^4$. Indeed, we can exhibit lattices $N = U_2 \oplus N(f), f \in U_2, f^2 = 0$, belonging to type I and having the required values of σ:

σ	$N(f)$
5	$D_4^{(5)}$
4	$D_4^{(3)} \oplus D_8$
3	$D_4 \oplus D_8^{(2)}$
2	$D_4 \oplus D_8 \oplus E_8$
1	$D_4 \oplus E_8^{(2)}$

Since $\deg a \leqslant 8$, we have $a = a_1^2$, $\deg a_1 \leqslant 4$. Also, a_1 and b are defined up to the addition of arbitrary squares ([14], p. 57). Therefore we may assume that $a_1 = \lambda t + \mu t^3$ and $b = t\Delta_1^2$.

We still have to learn how to distinguish degenerate fibers of types E_8 and D_8 with $n = 8$. Suppose that such a fiber lies over the point $t = 0$ and that the equation of the surface has the form

$$y^2 = x^3 + (\lambda t + \mu t^3)^2 x + t^5 \cdot c, \quad c(0) = \nu \neq 0.$$

It is not hard to show that the fiber over $t = 0$ is of type E_8 if and only if $\lambda = 0$. If $\lambda = 0$ we immediately obtain the normal form of the fiber E_8 ([8], p. 125). If $\lambda \neq 0$, one needs to make the transformation $x \longrightarrow x + r$, where $r = \lambda t + (\lambda^{-1}\nu)^{1/2} t^2$, in order to kill the t^2 term in the coefficient of x and the t^5 term in the constant term, and to obtain the normal form of a fiber of type D_8 ([9], p. 125). Now we can easily write down the equations of universal families.

6) $\sigma = 5$. The equation is

$$y^2 = x^3 + (\lambda t + \mu t^3)^2 x + t^3 (t+1)^2 (t+\alpha)^2 (t+\beta)^2.$$

There are degenerate fibers of type D_4 at the points ∞, 0, 1, α, and β. "Moduli" are α, β, λ, and μ.

7) $\sigma = 4$. The equation is

$$y^2 = x^3 + (\lambda t + \mu t^3)^2 + t^3 (t+1)^2 (t+\alpha)^4.$$

There are degenerate fibers of type D_4 at ∞, 0, and 1, and of type D_8 at α. "Moduli" are α, λ and μ.

8) $\sigma = 3$. The equation is

$$y^2 = x^3 + (\lambda t + \mu t^3)^2 x + t^5 (t+1)^4.$$

There are degenerate fibers of type D_4 at ∞, and of type D_8 at 0 and 1. "Moduli" are λ and μ.

9) $\sigma = 2$. The equation is

$$y^2 = x^3 + \mu t^6 \cdot x + t^5 (t+1)^4.$$

There are degenerate fibers of type D_4 at ∞, of type E_8 at 0, and of type D_8 at 1. The modulus is μ.

10) $\sigma = 1$. The equation is

$$y^2 = x^3 + t^5(t+1)^4.$$

There are degenerate fibers of type D_4 at ∞, and of type E_8 at 0 and 1. There are no "moduli".

Received 25/DEC/1977

BIBLIOGRAPHY

1. M. Artin, *Supersingular K3 surfaces*, Ann. Sci. École Norm. Sup. (4) 7 (1974), 543–568.

2. E. Bombieri and D. Mumford, *Enriques' classification of surfaces in* char. *p*. III, Invent. Math. 35 (1976), 197–232.

3. N. Bourbaki, *Groupes et algèbres de Lie*. Chaps. 4, 5, 6, Actualités Sci. Indust., no. 1337, Hermann, Paris, 1968.

4. V. I. Dolgačev and A. N. Paršin, *The different and discriminant of regular mappings*, Mat. Zametki 4 (1968), 519–524; English transl. in Math. Notes 4 (1968).

5. Martin Kneser, *Klassenzahlen definiter quadratischer Formen*, Arch. Math. (Basel) 8 (1957), 241–249.

6. K. Kodaira, *On compact complex analytic surfaces*. I, Ann. of Math. (2) 71 (1960), 111–152.

7. O. T. O'Meara, *Introduction to quadratic forms*, Springer-Verlag, Berlin; Academic Press, New York, 1963.

8. André Néron, *Modèles minimaux des variétés abéliennes sur les corps locaux et globaux*, Inst. Hautes Études Sci. Publ. Math. No. 21 (1964).

9. A. P. Ogg, *Elliptic curves and wild ramification*, Amer. J. Math. 89 (1967), 1–21.

10. I. I. Pjateckiĭ-Šapiro and I. R. Šafarevič, *A Torelli theorem for algebraic surfaces of type K3*, Izv. Akad. Nauk SSSR Ser. Mat. 35 (1971), 530–572; English transl. in Math. USSR Izv. 5 (1971).

11. A. N. Rudakov and I. R. Šafarevič, *Inseparable morphisms of algebraic surfaces*, Izv. Akad. Nauk SSSR Ser. Mat. 40 (1976), 1269–1307; English transl. in Math. USSR Izv. 10 (1976).

12. ———, *On quasi-elliptical surfaces of type K3*, Uspehi Mat. Nauk 33 (1978), no. 1 (199), 227–228. (Russian)

13. Jean-Pierre Serre, *Cours d'arithmétique*, Presses Univ. de France, Paris, 1970.

14. P. Deligne, *Courbes elliptiques: formulaire d'après J. Tate*, Modular Functions of One Variable. IV, Lecture Notes in Math., Vol. 476, Springer-Verlag, Berlin and New York, 1975, pp. 53–73.

Translated by PIOTR BLASS

On the degeneration of K3 surfaces over fields of finite characteristic

(with A. N. Rudakov)

Izv. Akad. Nauk SSSR, Ser. Mat. **45**, 646–661 (1981). Zbl. **465**, 14014
[Math. USSR, Izv. **18**, No. 3, 561–574 (1982)]

ABSTRACT. The authors describe degenerations of strongly elliptic surfaces of type $K3$, with Hasse invariant equal to zero, over fields of finite characteristic.
Bibliography: 9 titles.

The starting point of this paper is the question of to what extent a supersingular $K3$ surface is determined by its periods (for the definition of the period mapping for supersingular $K3$ surfaces, see [7]). As is well known, this question reduces to the question of the possible degenerations of similar surfaces. We resolve the question under the assumption that the surface is strongly elliptic, i.e. contains an elliptic pencil with a section. As is well known, this imposes one condition on the moduli of the surface. Under these assumptions we can solve a much more general problem: to describe all the degenerations of the surfaces whose Hasse invariant is 0; in other words, for the height introduced by Artin [1] we have $h > 1$ (for a supersingular surface $h = \infty$).

The authors thank P. Deligne, V. S. Kulikov, and T. Zink for useful criticisms of the results of this paper.

§1. The Hasse invariant of an elliptic $K3$ surface

We shall consider algebraic $K3$ surfaces over an algebraically closed field k of finite characteristic p different from 2 or 3. For such a surface X the space $H^2(X, \mathcal{O})$ is one-dimensional, so the action of the Frobenius morphism on it is given (after choosing a basis) by a single element $\lambda(X) \in k$, which is called the *Hasse invariant* of the surface X. If ω is a differential form generating the dual basis of the space $H^0(X, \Omega^2)$, then the Hasse invariant can be determined using the Cartier operation C: if $C\omega = \alpha\omega$, then $\lambda(X) = \alpha^p$.

Now assume that the surface X is strongly elliptic; that is, it has a morphism $f\colon X \to \mathbf{P}^1$ whose general fiber is an elliptic curve and whose corresponding pencil has a section $s\colon \mathbf{P}^1 \to X$. In this case the pencil f can be given in Weierstrass normal form:

$$y^2 = x^3 + a(t)x + b(t), \tag{1}$$

where t is a coordinate on $\mathbf{P}^1$, and $a(t)$ and $b(t)$ are polynomials of degrees 8 and 12. More precisely, the meaning of this assertion is as follows. Consider $a(t)$ and $b(t)$ as sections of the sheaves $\mathcal{O}(8)$ and $\mathcal{O}(12)$ over $\mathbf{P}^1$. Let E be the sheaf $\mathcal{O} \oplus \mathcal{O}(4) \oplus \mathcal{O}(6)$; in the

1980 *Mathematics Subject Classification*. Primary 14J25; Secondary 14M99.

projective bundle $\mathbf{P}(E)$ the subvariety X^* is given by the equation

$$y^2z = x^3 + a(t)xz^2 + b(t)z^3,$$

where x, y, and z are sections of the sheaves $\mathcal{O}(4)$, $\mathcal{O}(6)$, and $\mathcal{O}$. If the discriminant $4a^3 + 27b^2$ is not identically 0 and if there is no point $\xi \in \mathbf{P}^1$ at which a and b have zeros of multiplicities $\geqslant 4$ and $\geqslant 6$ respectively, then X^* has only double rational singular points, and their minimal resolution leads to a smooth $K3$ surface X.

In the normal form (1) the Hasse invariant of the surface X is computed as follows. The form that is a basis of $H^0(X, \Omega^2)$ can be taken to be in the form $\omega = y^{-1}\,dx \wedge dt$ (its regularity is evident for $t \neq \infty$, and for $t = \infty$ it becomes evident after the standard substitution $t = \tau^{-1}$). Since

$$\omega = \frac{y^{p-1}}{y^p}\,dx \wedge dt = \frac{(x^3 + a(t)x + b(t))^{(p-1)/2}}{y^p}\,dx \wedge dt,$$

to compute $C\omega$ we must write the polynomial $(x^3 + a(t)x + b(t))^{(p-1)/2}$ in the form $\sum_{0 \leqslant i,j \leqslant p-1} A_{ij}x^it^j$. Then

$$C\omega = \frac{(A_{p-1,p-1})^{1/p}}{y}\,dx \wedge dt$$

and $\lambda(X) = A_{p-1,p-1}$. Computing degrees makes it evident that

$$(x^3 + a(t)x + b(t))^{(p-1)/2}$$

contains only one term of the form $\Lambda(t)^{kp-1}$, namely for $k = 1$. The coefficient $\Lambda(t)$ of x^{p-1} is by definition the Hasse invariant on the general fiber of the elliptic pencil f; that is, of the curve (1) over the field $k(t)$. Recall that if we give a and b the weights 4 and 6, then the Hasse invariant $\Lambda(a, b)$ of the curve $y^2 = x^3 + ax + b$ will be isobaric of weight $p - 1$ (this follows from the fact that $\Lambda(a, b)$ is the coefficient of x^{p-1} in

$$(x^3 + ax + b)^{(p-1)/2},$$

and the polynomial $x^3 + ax + b$ is isobaric of weight 6 if x is assigned weight 2). In particular,

$$\Lambda(c^4a, c^6b) = c^{p-1}\Lambda(a, b). \tag{2}$$

If a and b are polynomials in t of degrees 8 and 12, then from what has been proved it follows that $\Lambda(a(t), b(t))$ has degree $2p - 2$. Therefore, it contains only one term λt^{kp-1}, namely the term for $k = 1$. Hence we see that, for a surface X given by equation (1), $\lambda(X) = \lambda$; that is, the Hasse invariant of an elliptic $K3$ surface equals the middle coefficient (that of t^{p-1}) in the Hasse invariant of the general fiber of the elliptic bundle. (It is easy to check that in a polynomial of degree $\leqslant 2p - 2$ the coefficient of t^{p-1} gets multiplied by $(\det \varphi)^{p-1}$ under a linear-fractional transformation φ.)

§2. Moderate degenerations

An elliptic bundle of type $K3$ given by equation (1) is determined by the coefficients of the polynomials a and b. It is natural to assign weight 4 to the coefficients of the polynomial a, and weight 6 to those of b. Therefore, we shall represent this bundle by a point in the weighted-projective space $\mathfrak{P}$ of type $(4^9, 6^{13})$. We shall also consider the points of $\mathfrak{P}$ that do not correspond to $K3$ surfaces. The equations (1) corresponding to them will be called *degenerations of $K3$ surfaces*.

The group of linear-fractional transformations of the projective line $\mathbf{P}^1$ with coordinate t defines a group of automorphisms of the weighted-projective space $\mathfrak{P}$. We can introduce the concepts of stable and semistable points with respect to this group in the sense of invariant theory [6]. To determine what points are stable or semistable one can use Hilbert's criterion ([6], Theorem 2.1). Since this problem is solved separately for the coefficient a and the coefficient b in the equation, the problem is solved completely as in the case of the action of SL(2) on $S^n(\mathbf{P}^1)$, i.e. as in the theory of binary forms ([6], Proposition 4.1). We give the answer:

The point of the space $\mathfrak{P}$ defined by equation (1) *is not stable if there exists a point $\xi \in \mathbf{P}^1$ at which a and b have zeros of multiplicities $\geqslant 4$ and $\geqslant 6$. This point is not semistable if at some point $\xi \in \mathbf{P}^1$ a and b have zeros of multiplicities > 4 and > 6.*

Later on we shall be interested only in semistable points and the surfaces corresponding to them.

As we mentioned in §1, the surface given by equation (1) may not be a $K3$ surface in two cases.

I. The discriminant of the right side of the equation is identically 0. In this case the right side has a multiple root $c(t) \in k[t]$, and therefore

$$a = -3c^2, \qquad b = 2c^3, \qquad \deg c = 4.$$

We shall call the degeneration *moderate* if the polynomial $c(t)$ has no multiple roots. Note that from the semistability condition it follows that c has no root of multiplicity > 2.

II. The discriminant of the right side of (1) is not equal to 0 identically, but a and b have zeros of multiplicities $\geqslant 4$ and $\geqslant 6$ at some point $\xi \in \mathbf{P}^1$. It is evident that there can be one or two such points. The condition that in the neighborhood of such a point ξ equation (1) defines a nonminimal Weierstrass model. Therefore, in this case we shall call the fiber over the point ξ *nonminimal*. If $t(\xi) = t_0$, then a minimal model in the neighborhood of ξ is given by

$$y_1^2 = x_1^3 + a_1x_1 + b_1, \qquad a_1 = \frac{a}{(t-t_0)^4}, \qquad b_1 = \frac{b}{(t-t_0)^6},$$

$$x_1 = x(t-t_0)^2, \qquad y_1 = y(t-t_0)^3. \tag{3}$$

The minimality of the model (3) in the neighborhood of ξ follows from the semistability condition.

In this case we shall call the degeneration *moderate* if for each nonminimal fiber the corresponding fiber of the minimal model (3) is smooth. Note that from the semistability condition it follows that $a_1(t)$ and $b_1(t)$ in (3) do not both vanish at ξ. Therefore, in the semistable case we have either a moderate degeneration or the fiber over the point 0 must be of multiplicative type.

A more invariant meaning of these conditions in both cases is the following. In case I the surface is not normal: it has a curve of singularities $x = c(t)$, $y = 0$. After normalization, the preimage of the curve of singularities becomes the curve $s^2 = 3c(t)$. Our condition is then that its genus is 1. In case II the surface has an isolated singular point $t = t_0$, $x = 0$, $y = 0$. Equation (3) determines a minimal resolution of this singular point (i.e. a resolution and contraction of all exceptional curves of genus 1). The preimage of the singular point will be the fiber $t = t_0$. Our condition is that its genus is 1. Thus in both cases we are dealing with "elliptic singularities", whose resolution gives rise to curves of arithmetic genus 1. The moderation condition is that the geometric genus of this curve also equals 1; that is, that we have a "nondegenerate elliptic singularity".

Before formulating our first theorem we note that the Hasse invariant of the surface with equation (1) is determined by polynomial formulas containing the coefficients of the polynomials $a(t)$ and $b(t)$. Therefore, it is defined for degenerate surfaces too. In this way we shall understand the Hasse invariant of degenerate curves of genus 1 written in Weierstrass form.

LEMMA. *The Hasse invariant of a degenerate elliptic curve of multiplicative type is different from* 0.

In this case the open set of nonsingular points of the curve can be isomorphically mapped onto the multiplicative group G_m; under this the regular differential form corresponds to the invariant form $\omega = t^{-1}\,dt$ on G_m. The equality $C\omega = \omega$ is checked immediately.

The lemma can be proved differently if the curve is written in the form $y^2 = x^2(x-\alpha)$, where $\alpha \neq 0$ because the curve is of multiplicative type. The Hasse invariant equals the coefficient of x^{p-1} in $(x^2(x-\alpha))^{(p-1)/2} = x^{p-1}(x-\alpha)^{(p-1)/2}$; that is, it equals $(-\alpha)^{(p-1)/2}$.

THEOREM 1. *A semistable degeneration of a strongly elliptic* $K3$ *surface with Hasse invariant* 0 *is moderate*.

We consider separately the degenerations of types I and II.

I. As we have already said, because the degeneration is semistable, the polynomial $c(t)$ cannot have a root of multiplicity >2. If the degeneration were not moderate, then c would have a double root. We can take this root to be 0, so that $c = t^2 d$. Then, because of (2),

$$\Lambda(a, b) = t^{p-1} d^{(p-1)/2} \Lambda(-3, 2).$$

Since the curve $y^2 = (x-1)^2(x+2)$ is of multiplicative type, $\Lambda(-3, 2) \neq 0$. Therefore, the vanishing of the Hasse invariant means that $d(0) = 0$; that is, that c has a triple root.

II. In this case the bundle (1) has one or two nonminimal fibers. Suppose one of them lies over the point $t = 0$; that is, $a = t^4 a_1$ and $b = t^6 b_1$. Then $\Lambda(a, b) = t^{p-1}\Lambda(a_1, b_1)$, and by definition $\lambda(X) = \Lambda(a_1, b_1)(0)$, i.e. it equals the Hasse invariant of the corresponding fiber of the minimal model (3). By the condition, $\Lambda(a_1, b_1)(0) = 0$. Hence, if in the model (3) the fiber over the point 0 is degenerate, then its Hasse invariant is 0. By the lemma it is then of additive type. But this is possible only if $a_1(0) = b_1(0)$, whence it follows that a and b have roots of multiplicity >4 and >6 at the point $t = 0$, which contradicts the semistability of the degeneration.

§3. Degenerations of families

We shall now consider families of strongly elliptic $K3$ surfaces over the base $S = \operatorname{Spec}(k[[\varepsilon]])$ with general point η and closed point s. We shall assume that the general fiber X_η of the family is a $K3$ surface and is written in the form (1), and the preimage X_s of the closed point s can degenerate. By a *reconstruction* of a family $\mathfrak{X} \to S$ we shall mean a family $\mathfrak{X}' \to S$ whose general fiber is isomorphic to the general fiber of the initial family. We shall say that a degeneration of the family $\mathfrak{X} \to S$ is *of chain type* if for the preimage

X_s of the closed point we have $X_s = \Sigma_0^n X_i$, $X_i \cap X_j = \varnothing$ for $|i-j| > 1$, X_0 and X_n are rational surfaces, the X_i $(0 < i < n)$ are smooth vector bundles over the elliptic curve C, and $X_i \cap X_{i+1}$ is a section of this bundle over X_i for $0 < i < n$ and over X_{i+1} for $0 \leq i < n-1$. Moreover, the three-dimensional scheme $\mathfrak{X}$ is smooth at the points of the curve $X_i \cap X_{i+1}$, X_i and X_{i+1} intersect transversally, and the only possible singular points on the fibers (the general and the closed fiber) are double rational points (on the surfaces X_0 and X_n).

THEOREM 2. *Any family of strongly elliptic $K3$ surfaces with a moderate degeneration can be reconstructed, after a change of base, into a family which either is nondegenerate or has a degeneration of chain type.*

We again treat separately the two possible types of degenerations of elliptic $K3$ surfaces.
Degeneration of type I.

LEMMA. *Let $f, g \in k[[\varepsilon]][t]$, $f \not\equiv 0$ (ε), and suppose $f^3 - g^2$ is divisible by ε^{2k}, $k > 0$ (which, of course, implies the existence of a polynomial $h_0 \in k[t]$ such that $f \equiv h_0^2$ (ε) and $g \equiv h_0^3$ (ε)). Assume that h_0 has no multiple roots. Then we can write*

$$f \equiv h^2 + \varepsilon^k U \quad (\varepsilon^{2k}), \qquad g \equiv h^3 + \tfrac{3}{2}\varepsilon^k U h \quad (\varepsilon^{2k}).$$

Assume that $f \equiv h^2$ (ε^r) and $g \equiv h^3$ (ε^r), $r < 2k$. Writing $f = h^2 + \varepsilon^r U + \varepsilon^{2r} V$ and $g = h^3 + \varepsilon^r U_1 + \varepsilon^{2r} W$, from the relation $f^3 \equiv g^2$ (ε^2), by comparing the terms in ε^r we find that $U_1 = \frac{3}{2} U h$; and from the relation $f^3 \equiv g^2$ (ε^{2r+1}) by comparing the terms in ε^{2r} we find that $U \equiv 0$ (h_0, ε) (here we use the fact that h_0 has no multiple roots). The assertion of the lemma follows by induction.

COROLLARY. *Let $a, b \in k[[\varepsilon]][t]$. Assume that the discriminant of the polynomial $X^3 + aX + b$ is not divisible by ε^{2k}, $k > 0$, whence follows a representation $X^3 + aX + b \equiv (X - c_0)^2(X + 2c_0)$ (ε). Assume that $c_0(t)$ has no multiple roots. Then, by substituting for X, we can transform the polynomial $X^3 + aX + b$ to $X^3 + CX^2 + \varepsilon^k AX + \varepsilon^{2k} B$, where $A^2 - 4BC \not\equiv 0$ (ε).*

Since $4a^3 + 27b^2 \equiv 0$ (ε^{2k}), by the lemma (putting $a = -3f$ and $b = 2g$) we can write $a \equiv -3h^2 - 3U\varepsilon^k$ (ε^{2k}), $b \equiv 2h^3 + 3Uh\varepsilon^k$ (ε^{2k}) or $a = -3h^2 + A\varepsilon^k$, $b = 2h^3 - Ah\varepsilon^k + B\varepsilon^{2k}$, where $A = -3U$. Hence

$$X^3 + aX + b = (X - h)^2(X + 2h) + \varepsilon^k AX - \varepsilon^k hA + \varepsilon^{2k} B.$$

Putting $3h = C$, after substituting $X - h \to X$, we get the required representation. Substituting the expressions $a = -3h^2 + A\varepsilon^k$ and $b = 2h^3 - Ah\varepsilon^k + B\varepsilon^{2k}$ into the formula for the discriminant D, we get $D \equiv C^2(4BC - A^2)$ (ε^{2k}), where $C = 3h$. The assertion of the corollary now follows from the assumption on the discriminant.

We can now go on to investigate degenerations of type I. We may assume that the family $\mathfrak{X}$ is given by the equation

$$y^2 = x^3 + a(t, \varepsilon)x + b(t, \varepsilon), \qquad a(t, \varepsilon), b(t, \varepsilon) \in k[[\varepsilon]][t], \tag{4}$$

in $\mathbf{P}(E)$, where $E = \mathcal{O} \oplus \mathcal{O}(4) \oplus \mathcal{O}(6)$ is a sheaf over $\mathbf{P}_S^1$, $S = \operatorname{Spec} k[[\varepsilon]]$. Passing to a covering of the base S if necessary, we may assume that the discriminant of the left side of

(4) is divisible by an even power of ε; if this power is $2k$, on the basis of the lemma we can write the equation in the form

$$y^2 = x^3 + C(t)x^2 + \varepsilon^h A(t, \varepsilon)x + \varepsilon^{2k} B(t, \varepsilon),$$
$$A(t, \varepsilon), B(t, \varepsilon), C(t, \varepsilon) \in k[[\varepsilon]][t],$$
$$\deg C = 4, \quad \deg A = 8, \quad \deg B = 12, \quad A^2 - 4BC \not\equiv 0\ (\varepsilon). \tag{5}$$

The reconstruction of the family (4) we need will be gotten by using a σ-process in the variety $\mathbf{P}(E)$ along the curve defined by the equations $x = 0$, $y = 0$ and $\varepsilon = 0$. This curve lies entirely in the finite part of the variety $\mathbf{P}(E)$ (with respect to x and y) in which $\mathfrak{X}$ is given by (4). We have to consider two charts, which are the preimages of the sets $t \neq \infty$ and $t \neq 0$ on $\mathbf{P}^1_S$. Since they are completely similar, we restrict ourselves to the first. We denote by $\mathfrak{X}_1$ the proper preimage of the variety $\mathfrak{X}$ under the σ-process. $\mathfrak{X}_1$ is covered by three charts:

(1) $y = ux$, $\varepsilon = \eta x$,

$$u^2 = x + C(t, \eta x) + \eta^k A(t, \eta x)x^{k-1} + \eta^{2k} B(t, \eta x)x^{2k-2}.$$

The preimage X_s of the closed point is given by the equation $\varepsilon = 0$, and we put $X_s = X_0 \cup X_1$,

$$X_0\colon \eta = 0, \qquad u^2 = x + C(t,0),$$
$$X_1\colon x = 0, \qquad u^2 = C(t,0), \quad \text{for } k > 1,$$
$$x = 0, \quad u^2 = C(t,0) + \eta A(t,0) + \eta^2 B(t,0) \quad \text{for } k = 1.$$

(2) $x = vy$, $\varepsilon = \theta y$,

$$1 = v^3 y + C(t, \theta y)v^2 + \theta^k y^{k-1} A(t, \theta y)v + \theta^{2k} y^{2k-2} B(t, \theta y),$$
$$X_s = X_0 \cup X_1, \qquad X_0\colon \theta = 0, \qquad 1 = v^2(vy + C(t,0)),$$
$$X_1\colon y = 0, \qquad 1 = C(t,0)v^2 \quad \text{for } k > 1,$$
$$y = 0, \qquad 1 = C(t,0)v^2 + A(t,0)v\theta + B(t,0)\theta^2 \quad \text{for } k = 1.$$

(3) $x = x_1\varepsilon$, $y = y_1\varepsilon$,

$$y_1^2 = x_1^3\varepsilon + C(t, \varepsilon)x_1^2 + \varepsilon^{k-1} A(t, \varepsilon)x_1 + \varepsilon^{2k-2} B(t, \varepsilon).$$

The preimage of the closed point is irreducible:

$$X_s = X_1\colon \varepsilon = 0, \qquad y_1^2 = C(t,0)x_1^2 \quad \text{for } k > 1,$$
$$\varepsilon = 0, \qquad y_1^2 = C(t,0)x_1^2 + A(t,0)x_1 + B(t,0) \quad \text{for } k = 1.$$

The surfaces X_1 in all three charts are glued into a single surface X_1. The surfaces X_0 in charts (1) and (2) are glued into a surface $x = 0$, $\eta = 0$, $u^2 = C(t, 0)$. By the condition our degeneration is moderate, and therefore the curve is smooth. It is easy to see that at its points the variety $\mathfrak{X}_1$ is smooth.

The surface X_0 is rational and smooth. For $k > 1$ the surface X_1 has, in the chart (3), the double curve Y_1: $y_1 = x_1 = 0$. The normalization of X_1 is a ruled surface with elliptic base. The curves Y_0 and Y_1 do not intersect.

If $k > 1$, then we must carry out a σ-process along Y_1. Under this the component X_0 is mapped isomorphically; the preimage of X_1, as is easily checked, coincides with its normalization and is a smooth elliptic surface; and in the degenerate fiber there arises one more component X_2, which for $k > 2$ is an elliptic ruled surface with a double curve Y_2. This process continues for k steps. After a σ-process with a number less than k we add one more elliptic surface to our chain. After the kth σ-process the last surface will have equation

$$y_k^2 = C(t,0)x_k^2 + A(t,0)x_k + B(t,0) \tag{6}$$

in the chart (3). Since, according to the lemma, $A(t,0)^2 - 4B(t,0)C(T,0) \neq 0$, this surface will contain a pencil of rational curves; therefore, it is rational.

It remains for us to investigate the singular points of the surface (6). We put $A(t,0) = a(t)$, $B(t,0) = b(t)$, $C(t,0) = c(t)$ and write (6) in the form

$$y^2 = cx^2 + ax + b. \tag{7}$$

The singular points correspond to the values t_0 for which $(a^2 - 4bc)(t_0) = 0$. We can take $t_0 = 0$. If $c(0) \neq 0$, then, putting $c^{1/2}x = x_1$ and substituting, we bring (7) into the form $y_1^2 = x_1^2 + t^n u(t)$. This is a singularity of type A_{n-1}. Since our degeneration is moderate, the remaining possibility is that $c(0) = 0$ and $c'(0) \neq 0$. Then $a(0) = 0$ and, by a similar transformation, (7) is brought into the form $y_1^2 = tx_1^2 + t^n$. This singularity is of type D_{n+1}. By the same token Theorem 2 is proved in the case of a degeneration of type I.

Degeneration of type II. We consider a somewhat more general situation: the family $\mathfrak{X}$ of elliptic surfaces given by the equation

$$\begin{gathered} y^2 = x^3 + a(t, \varepsilon)x + b(t, \varepsilon), \\ a(t, \varepsilon), b(t, \varepsilon) \in k[[\varepsilon]][t], \\ a(t, \varepsilon) \in (t, \varepsilon)^4, \qquad b(t, \varepsilon) \in (t, \varepsilon)^6. \end{gathered} \tag{8}$$

We shall be interested in the degrees (with respect to t) not of $a(t, \varepsilon)$ and $b(t, \varepsilon)$ but of $a(t,0)$ and $b(t,0)$, and we shall assume that

$$\deg a(t,0) \leqslant 4r, \qquad \deg b(t,0) \leqslant 6r. \tag{9}$$

In this case the preimage X_s of the closed point is given by an equation

$$y^2 = x^3 + t^4\alpha(t)x + t^6\beta(t),$$

i.e. is not minimal. We define an "elementary reconstruction" whose purpose is to replace this fiber by the corresponding minimal (for $t = 0$) fiber

$$y_1^2 = x_1^3 + \alpha(t)x_1 + \beta(t).$$

This can be done by putting $x = t^2x_1$ and $y = t^3y_1$. However, in any three-dimensional family $\mathfrak{X}$ such a reconstruction can be gotten only by adjoining one more component to the closed fiber.

This new family $\mathfrak{X}_1$ is defined over the base Σ gotten from $\mathbf{P}_S^1$ by a σ-process at the point $t = 0$, $\varepsilon = 0$. Recall that Σ has a morphism σ: $\Sigma \to \mathbf{P}_S^1$, and if $\mathbf{P}_S^1 = V_1 \cup V_0$, where $V_1 = \{t \neq \infty\}$ and $V_0 = \{t \neq 0\}$, then σ^{-1}: $V_0 \to V$ is an isomorphism and $\sigma^{-1}(V_1) = U_1 \cup U_2$,

$$\begin{gathered} U_1 = \operatorname{Spec} k[[\varepsilon]][t, \varepsilon_1]/(t\varepsilon_1 - \varepsilon), \\ U_2 = \operatorname{Spec} k[[\varepsilon]][t_2, \varepsilon]/(t_2\varepsilon - t), \end{gathered}$$

and the subsets $U_{12} \subset U_1$ and $U_{21} \subset U_2$ are defined by the conditions $\varepsilon_1 \neq 0$ and $t_2 \neq 0$ and are isomorphically mapped onto one another because of the relation $\varepsilon_1 t_2 = 1$.

The family $\mathfrak{X}_1$ is defined over U_1 by the equation

$$y_1^2 = x_1^3 + a_1x_1 + b_1,$$
$$a_1 = a(t, t\varepsilon_1)/t^4, \qquad b_1 = b(t, t\varepsilon_1)/t^6,$$

and over U_2 by the equation

$$y_2^2 = x_2^3 + a_2x_2 + b_2,$$
$$a_2 = a(t_2\varepsilon, \varepsilon)/\varepsilon^4, \qquad b_2 = b(t_2\varepsilon, \varepsilon)/\varepsilon^6.$$

Over $U_{12} \simeq U_{21}$ these families are glued together by the formulas

$$x_1 = x_2\varepsilon_1^2, \qquad y_1 = y_2\varepsilon_1^3 \quad \text{on } U_{12},$$

$$x_2 = x_1t_2^2, \qquad y_2 = y_1t_2^3 \quad \text{on } U_{21}.$$

Over U_1 for $t \neq 0$ the formulas $x_1 = x/t^2$ and $y_1 = y/t^3$ give an isomorphism of open sets of the families $\mathfrak{X}_1$ and $\mathfrak{X}_2$. Over U_2 for $\varepsilon \neq 0$ the same isomorphism is given by the formulas $x_2 = x/\varepsilon^2$ and $y_2 = y/\varepsilon^3$. Thus $\mathfrak{X}_1$ and $\mathfrak{X}_2$ are isomorphic on V_1 outside the preimage of the point ($t = 0$, $\varepsilon = 0$). Thanks to this construction, over V_1 the family $\mathfrak{X}_1$ is glued to the preimage of $\mathfrak{X}$ over the set V_0 and gives a single family $\mathfrak{X}_1$ over Σ, called an *elementary reconstruction* of the family $\mathfrak{X}$ in the fiber $t = 0$.

In $\mathfrak{X}_1$ the preimage X_s of the closed point decomposes into 2 components: $X_s = X_0 \cup X_1$, where X_0 is defined over U_1 by the equations $\varepsilon_1 = 0$ and $y_1^2 = x_1^3 + a_1(t,0)x_1 + b_1(t,0)$; that is, $y_1^2 = x_1^3 + \alpha(t)x_1 + \beta(t)$. Now $\deg \alpha(t) \leqslant 4(r-1)$ and $\deg \beta(t) \leqslant 6(r-1)$. This component does not intersect the preimage of U_2. To investigate the component X_1, put

$$a(t, \varepsilon) = \sum_{i+j \geqslant 4} \alpha_{ij}t^i\varepsilon^j, \qquad b(t, \varepsilon) = \sum_{i+j \geqslant 6} \beta_{ij}t^i\varepsilon^j.$$

Then X_1 is given over U_1 by the equations

$$t = 0, \quad y_1^2 = x_1^3 + \left(\sum_{i+j=4} \alpha_{ij}\varepsilon_1^j\right)x_1 + \sum_{i+j=6} \beta_{ij}\varepsilon_1^j,$$

and over U_2 by

$$\varepsilon = 0, \quad y_2^2 = x_2^3 + \left(\sum_{i+j=4} \alpha_{ij}t_2^i\right)x_2 + \sum_{i+j=6} \beta_{ij}t_2^i. \tag{10}$$

The curve $Y_0 = X_0 \cap X_1$ is given by the equations $\varepsilon_1 = 0$, $t = 0$ and $y_1^2 = x_1^3 + a_1(0,0)x_1 + b_1(0,0)$ or $y_1^2 = x_1^3 + \alpha(0)x_1 + \beta(0)$. If this curve is smooth, then, as is easily checked, the family $\mathfrak{X}_1$ is smooth at an arbitrary point of Y_0: X_0 and X_1 are also smooth at this point and intersect transversally along Y_0.

Now we go over to the description of a sequence of elementary reconstructions whose result is the reconstruction whose existence is claimed by the theorem. We begin with a family $\mathfrak{X}$ of surfaces of type $K3$ having a moderate degeneration of type II for $\varepsilon = 0$. In it let the preimage X_s of the closed point have a nonminimal fiber for $t = 0$; that is, it satisfies (8) with $r = 2$. Making an elementary reconstruction in this fiber, we then get a family $\mathfrak{X}^{(1)}$ in which the preimage of the closed point consists of two components: $X_s^{(1)} = X_0^{(1)} \cup X_1^{(1)}$, where $X_0^{(1)} \cap X_1^{(1)} = Y_1$ is a smooth elliptic curve; the component $X_0^{(1)}$ is birationally isomorphic to X_0, but has a minimal fiber for $t = 0$, and the component $X_1^{(1)}$ is given by (10) with $r = 1$ and so either is a rational surface (if all its fibers are minimal) or has one more nonminimal fiber (and then its minimal model is isomorphic to

$\mathbf{P}^1 \times Y_0$). Note that in this last case the discriminant of the elliptic pencil (10) is not zero for $t_2 = 0$, i.e. is not identically zero. The same is also true for the component $X_0^{(1)}$: it is either rational or has one more nonminimal fiber Y_1, and then its minimal model is isomorphic to $\mathbf{P}^1 \times Y_1$, whence $Y_1 \simeq Y_0$. If both components $X_0^{(1)}$ and $X_1^{(1)}$ are rational surfaces, then our process stops and the theorem is proved. If this does not happen, then one or both of them have a nonminimal fiber that does not intersect Y_0, and we make an elementary reconstruction along this fiber. As a result of each reconstruction the component in whose nonminimal fiber we performed this reconstruction is replaced by the component $X_i^{(2)} \simeq \mathbf{P}^1 \times Y_0$ ($i = 0$ or 1); in the corresponding minimal fiber this component is intersected by a new component $X_j^{(2)}$, which again is a rational surface (if it is minimal) or a nonminimal model of the surface $\mathbf{P}^1 \times Y_0$ if it has a nonminimal fiber. The component intersecting $X_1^{(2)}$ will be denoted by $X_2^{(2)}$ (if it arises), and the component intersecting $X_0^{(2)}$ by $X_{-1}^{(2)}$ (if it arises). After several steps we get a family whose degenerate fiber is a chain $\bigcup_{i=-r}^{q} X_i$, $X_i \cap X_{i+1} \simeq Y_0$, X_i $(-r < i < q)$ is isomorphic to $\mathbf{P}_0^1 \times Y_0$, and X_{-r} and X_q are minimal rational elliptic surfaces or have one nonminimal fiber. In the first case our process is complete, and the theorem is proved; in the second case we must continue it, making elementary reconstructions in the nonminimal fibers. It thus remains for us to prove that sooner or later our process stops. This is proved in like ways for each end of the chain ($i > 0$ or $i < 0$), so we explain it for one end ($i > 0$).

Assume that the process of reconstructing the families $\mathfrak{X}^{(i)}$ continues to infinity; from this assumption we shall derive a contradiction to the fact that the general fiber is a $K3$ surface (more precisely, to the fact that its Weierstrass model is minimal). By construction we start from the case where the degenerate fiber has a nonminimal fiber over $t = 0$. It can turn out that after several elementary reconstructions this situation still occurs. We shall prove, however, that this cannot be so for *all* elementary reconstructions; that is, sooner or later we arrive at a chain $X_0 \cup \cdots \cup X_q$ such that the surface X_q is an elliptic bundle with nonminimal fiber over the point $t = \lambda \neq 0$. In fact, if for all elementary reconstructions there were a nonminimal fiber over the point $t = 0$, then a sequence of s reconstructions would be given by the formulas $t = t_s \varepsilon^s$, and the bundle gotten would have in Weierstrass form an equation (8) with coefficients

$$a_s(t_s, \varepsilon) = a(t_s\varepsilon^s, \varepsilon)/\varepsilon^{4s}, \qquad b(t_s, \varepsilon) = b(t_s\varepsilon^s, \varepsilon)/\varepsilon^{6s}. \tag{11}$$

But the general fiber of the family has a minimal fiber over $t = 0$; therefore, either $a \not\equiv 0$ (t^4) or $b \not\equiv 0$ (t^6). For example, in the first case $a = \sum \alpha_{ij} t^i \varepsilon^j$ contains a term $\alpha_{ij} t^i \varepsilon^j$ with $\alpha_{ij} \neq 0$, $0 \leqslant i \leqslant 4$. After a change of base $\varepsilon = \varepsilon_1^k$ and a transformation by (11) it will yield a term $\alpha_{ij} t^i \varepsilon_1^{kj+(i-4)s}$. We can choose s so that $kj + (i - 4)s = 0$ (since $i < 4$); that is, we put

$$s = \frac{kj}{4 - i} \tag{12}$$

(because of the choice of k, we find that this number is integral). As a result $a_s(t_s, \varepsilon)$ will contain a term $\alpha_{ij} t^i$ with $\alpha_{ij} \neq 0$; and this now shows that $a_s(t_s, 0) \not\equiv 0$ (t^4); that is, the new bundle has a minimal fiber over $t = 0$.

Thus, after some, say s, steps the nonminimal fiber in the bundle $\mathfrak{X}^{(s)}$ turns out to lie over the point $t = \lambda \neq 0$. We shall prove that we did not need a change of base for the first s reconstructions. Indeed, our assumption means that

$$a_s(t_s, 0) = \alpha(t_s - \lambda)^4, \qquad b_s(t_s, 0) = \beta(t_s - \lambda)^6,$$

where either $\alpha \neq 0$ or $\beta \neq 0$. If, for example, $\alpha \neq 0$, then $a_s(t_s, 0)$ contains a term γt_s^3 with $\gamma \neq 0$. Assume that in the process we made a change of base $\varepsilon \to \varepsilon^k$ and that the term γt_s^3 came from a term $\delta t^3 \varepsilon^j$ in $a(t, \varepsilon)$. Then according to (12) we have $s = kj$; that is, $s \equiv 0\,(k)$ and our transformation $t = t_s \varepsilon_1^s$ is written in the form $t = t_s \varepsilon^{s/k}$ without a change of base.

Consequently, if our reconstruction process is continued infinitely, then (according to the first remark) it is necessary infinitely often to change the point over which there is a nonminimal fiber; therefore, according to the second remark, the whole process goes forward without a change of base. If during the first s steps there is a nonminimal fiber over the point $t = 0$, and on the $(s + 1)$th step it lies over a point $t = \lambda \neq 0$, then

$$a_s(t_s, \varepsilon) \equiv \alpha(t_s - \lambda)^4(\varepsilon), \qquad b_s(t_s, \varepsilon) \equiv \beta(t_s - \lambda)^6(\varepsilon).$$

From the fact that $t = t_s \varepsilon^s$ and (11) we find that

$$a(t_s\varepsilon^s, \varepsilon) \equiv \alpha\varepsilon^{4s}(t_s - \lambda)^4(\varepsilon^{4s+1}), \qquad b(t_s\varepsilon^s, \varepsilon) \equiv \beta\varepsilon^{6s}(t_s - \lambda)^6(\varepsilon^{6s+1}),$$
$$a(t, \varepsilon) \equiv \alpha(t - \lambda\varepsilon^s)^4(\varepsilon^{4s+1}), \qquad b(t, \varepsilon) \equiv \beta(t - \lambda\varepsilon^s)^6(\varepsilon^{6s+1}).$$

Putting $t - \lambda\varepsilon^s = t_1$, we find that

$$a(t_1, \varepsilon) \in (t_1^4, \varepsilon^{4s+1}), \qquad b(t_1, \varepsilon) \in (t_1^6, \varepsilon^{6s+1}).$$

Continuing such substitutions we find a power series

$$\varphi(\varepsilon) = \lambda\varepsilon^s + \lambda_1\varepsilon^{s_1} + \lambda_2\varepsilon^{s_2} + \cdots, \qquad s < s_1 < s_2 \cdots,$$

with the property that for $\tau = t - \varphi(\varepsilon)$ we have $a(\tau, \varepsilon) \equiv 0\,(\tau^4)$ and $b(\tau, \varepsilon) \equiv 0\,(\tau^6)$.

This means that for $t = \varphi(\varepsilon)$ the general fiber has a nonminimal fiber, i.e. is not a $K3$ surface. This contradiction proves the theorem.

REMARKS. 1. The last argument in the proof of the theorem can be clarified by the following illustration. We assign to each term $\alpha_{ij}t^i\varepsilon^j$ with $\alpha_{ij} \neq 0$ in the expansion $a(t, \varepsilon) = \sum \alpha_{ij}t^i\varepsilon^j$ the point in the plane with coordinates $(3i, 3j)$, and to a term $\beta_{ij}t^i\varepsilon^j$ with $\beta_{ij} \neq 0$ in the expansion $b(t, \varepsilon) = \sum \beta_{ij}t^i\varepsilon^j$ the point with coordinates $(2i, 2j)$. We get a set of points $\mathfrak{M}$ on the surface, where $\mathfrak{M}$ is contained in the "half-band" $y \geqslant 0$, $0 \leqslant x \leqslant 24$, the segment of the x-axis with $0 \leqslant x < 12$ does not contain points of $\mathfrak{M}$, and the point $(12, 0)$ is contained in $\mathfrak{M}$. An elementary reconstruction is equivalent to a change of coordinates

$$x' = x, \qquad y' = y + x - 12,$$

under which the y-axis does not change, while the x-axis turns around the point $(12, 0)$. A sequence of l transformations of such a form together with a change of basis $\varepsilon = \varepsilon_1^k$ is equivalent to the change of coordinate systems

$$x' = x, \qquad y' = y + \gamma(x - 4)$$

with an arbitrary rational $\gamma = l/k > 0$. As a result of such a transformation we can take the x-axis to be the "base line" of the set $\mathfrak{M}$ (see the figure). Then the new x'-axis will contain at least 2 points of the set $\mathfrak{M}$. This means that a nonminimal fiber of the corresponding component lies over a point $t = \lambda \neq 0$. As has been proved, in this case the angular coefficient γ must be an integer. The substitution $t_1 = t - \lambda$ is equivalent to some reconstruction of the set $\mathfrak{M}$ under which all the points of this set, except for $(12, 0)$, lie strictly above the new x-axis. If this process continues to infinity, then the x-axis, turning around the point $(12, 0)$, rises to an unbounded height, and in the limit we get an equation for the family $\mathfrak{X}$ such that for it the set $\mathfrak{M}$ contains no points in the band $0 \leqslant x < 12$. This means that $a(t, \varepsilon) \equiv 0\ (t^4)$ and $b(t, \varepsilon) \equiv 0\ (t^6)$, i.e. the general fiber of the family is not a $K3$ surface.

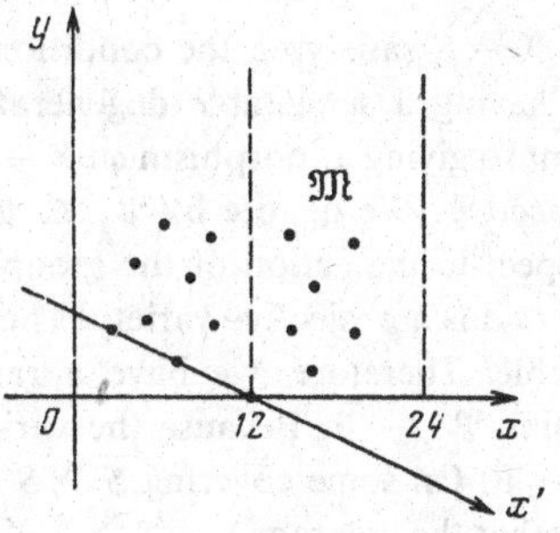

FIGURE 1

2. If the degeneration is not moderate, then the same process leads to some reconstruction of it in which the image of the closed point also has the form of a chain, but its components have somewhat more complicated singularities. It seems likely that further reconstructions of this chain may lead to families with degenerations of the same type as those found in characteristic 0 by Kulikov [5].

3. It would be interesting to try to apply analogous arguments to investigate $K3$ surfaces that are double planes. Useful considerations in this direction are contained in Horikawa's paper [4].

4. We give one example of Theorem 2. Let A_ε be a family of elliptic curves that does not degenerate for $\varepsilon = 0$, and let B_ε be a family such that B_0 is of multiplicative type. Take X_ε to be the Kummer surface $(A_\varepsilon \times B_\varepsilon)/G$, $G = \{1, \sigma\}$, $\sigma(x, y) = (-x, -y)$. The projections $A_\varepsilon \times B_\varepsilon \to A_\varepsilon$ and $A_\varepsilon \times B_\varepsilon \to B_\varepsilon$ define pencils $X_\varepsilon \to A_\varepsilon/G$ and $X_\varepsilon \to B_\varepsilon/G$ which give X_ε two structures of strongly elliptic surface. If A_ε and B_ε are given by the equations

$$A_\varepsilon\colon y^2 = x^3 + a(\varepsilon)x + b(\varepsilon), \qquad B_\varepsilon\colon v^2 = u^3 + \alpha(\varepsilon)u + \beta(\varepsilon),$$

then the corresponding equations for X_ε have the form

$$v^2 = u^3 + \alpha(\varepsilon)\big(t^3 + a(\varepsilon)t + b(\varepsilon)\big)^2 u + \beta(\varepsilon)\big(t^3 + a(\varepsilon)t + b(\varepsilon)\big)^3,$$

$$y^2 = x^3 + a(\varepsilon)\big(t^3 + \alpha(\varepsilon)t + \beta(\varepsilon)\big)^2 x + b(\varepsilon)\big(t^3 + \alpha(\varepsilon)t + \beta(\varepsilon)\big)^3.$$

The degenerations are of types I and II respectively. If a minimal model of the family B_ε has a degenerate fiber C of type A_n, then the degenerate fiber of the family X_ε can be taken in the form $(A_0 \times C)/(1, \sigma)$, where σ is an automorphism of order 2 on A_0 and C. If $C = \Sigma_{0 \leqslant i \leqslant n} C_i$ is the connected component of the identity, then

$$C/G = C_0/G + C_{(n+1)/2}G + \sum_{0<i<(n+1)/2} C_i$$

(by a change of base we may consider $n + 1$ even). Then $(A_0 \times G)/G$ has the form of a chain $X_0 \cup \cdots \cup X_{(n+1)/2}$, in which $X_0 = (A_0 \times C_0)/G$, $X_{(n+1)/2} = (A_0 \times C_{(n+1)/2})/G$ is a rational surface, and $X_i = A_0 \times C_i$ for $0 < i < (n+1)/2$.

We return to the investigation of elliptic $K3$ surfaces over a field of finite characteristic.

THEOREM 3. *Any family of strongly elliptic $K3$ surfaces the Hasse invariant of whose general fiber equals* 0 *can be reconstructed, after a change of base, into a family which either does not degenerate or has a degeneration of chain type whose degenerate fiber has smooth components.*

Let there be given a family $\mathfrak{X} \to S$ satisfying the condition $\lambda(X_\eta) = 0$. First of all we construct a reconstruction $\overline{\mathfrak{X}}$ having a semistable degeneration. For this we note that giving the family $\mathfrak{X}$ is equivalent to giving a morphism $\varphi\colon S \to \mathfrak{P}$ of the base of this family into the weighted-projective space $\mathfrak{P}$. We denote by $\mathfrak{P}_{ss} \subset \mathfrak{P}$ the open set of semistable points of the space $\mathfrak{P}$ with respect to the action of the group $G = \mathrm{PGL}(2)$. As we know, the quotient $\overline{\mathfrak{P}} = \mathfrak{P}_{ss}/G$ exists and is a projective variety. Then $\varphi(\eta)$ corresponds to some $K3$ surface and so is semistable. Therefore, we have a rational mapping $f\colon S \rightsquigarrow \overline{\mathfrak{P}}$, $f = \psi\varphi$, where ψ is the projection $\mathfrak{P}_{ss} \to \overline{\mathfrak{P}}$. Because the variety $\overline{\mathfrak{P}}$ is complete, f can be extended to a morphism $f\colon S \to \overline{\mathfrak{P}}$. On some covering $\overline{S} \to S$ this morphism can be lifted to a morphism $\bar{f}\colon \overline{S} \to \mathfrak{P}_{ss}$ such that the diagram

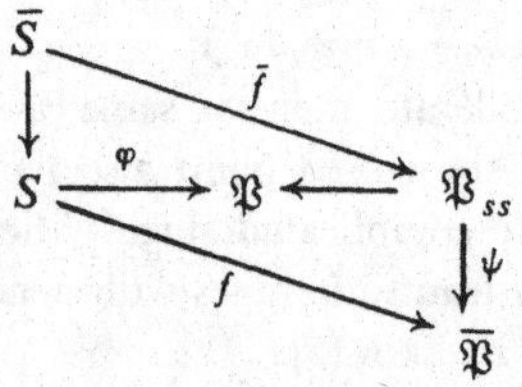

is commutative. The inverse image of the family of bundles defined over $\mathfrak{P}$ and $\mathfrak{P}_{ss}$ gives the family $\overline{\mathfrak{X}}$ over $\overline{S}$ that we require.

We now apply Theorem 1 to $\overline{\mathfrak{X}}$: it guarantees that the degeneration of the family $\overline{\mathfrak{X}}$ is moderate. Thus, according to Theorem 2, there exists a reconstruction $\mathfrak{X}'$ of this family which either does not degenerate or has a degeneration of chain type.

Suppose that in our reconstruction the degeneration has the form of a chain $X_s = X_0 \cup X_1 \cup \cdots \cup X_m$, where $X_1, \ldots, X_{m-1}$ are isomorphic to $\mathbf{P}^1 \times C$ (C is an elliptic curve) and X_0 and X_m are rational surfaces with pencils of elliptic curves all of whose fibers are minimal. The only thing distinguishing our situation from what we must prove is the possible presence of singular points on X_0 and X_m. But minimal models of elliptic surfaces have only double rational singular points. Therefore, we can apply the theorem proved by Tjurina [8] and Brieskorn [3] over fields of characteristic 0 and transferred by Artin [2] to fields of finite characteristic. According to this theorem, double rational singular points can be resolved using some covering of the base $S' \to S$. It remains for us to observe what happens to the other surfaces of our chain under this covering. If the degree of the covering is n, then locally $\mathfrak{X} \times_S S'$ is given by the equation $xy = \tau^n$. This equation coincides with the equation defining the singularity A_{n-1}. Therefore, using the same σ-process as in resolving the singularity A_{n-1}, we resolve our singularity; and as a result, between X_i and X_{i+1} there is pasted in another $n - 1$ surface isomorphic to $\mathbf{P}^1 \times C$. As a result we get a degeneration whose existence was asserted by the theorem.

THEOREM 4. *A family of strongly elliptic $K3$ surfaces with zero Hasse invariant can be reconstructed* (*after a change of base*) *into a nondegenerating family if and only if its monodromy, acting on the group $H^2(X_\eta)$, is finite.*

In view of Theorem 3 it suffices for us to show that if the family has a degeneration of chain type, then its monodromy is infinite. For varieties over the field of complex numbers the theory of mixed Hodge structures establishes a connection between the monodromies of the family and the geometry of a degenerate fiber. This is done using the Clemens-Schmidt exact sequence or Steenbrink's spectral sequence. These results, in the form useful for us, are introduced in the first section of Kulikov's paper [5]. Recently Rapoport and Zink [9] have shown that these results also hold over fields of finite characteristic. In particular, they establish a necessary and sufficient condition for finite monodromy, which previously had been proved only over the field of complex numbers. The following consequence will suffice for us. Suppose a family $\mathfrak{X} \to S$ is regular, and the degenerate fiber X_s decomposes into smooth surfaces X_i having normal intersection. Then the monodromy is infinite if the restriction homomorphism

$$\bigoplus_i H^1(X_i) \to \bigoplus_{i<j} H^1(X_i \cap X_j) \qquad (*)$$

is not epimorphic.

It remains for us to apply this criterion to a degeneration of chain type: $X_s = X_0 \cup \cdots \cup X_n$. We put $X_i \cap X_{i+1} = Y_i$, $i = 0, 1, \ldots, n-1$. Since Y_i is an elliptic curve, it follows that $\operatorname{rk} H^1(Y_i) = 2$ and $\operatorname{rk}(\bigoplus H^1(X_i \cap X_j)) = 2n$. Among the surfaces X_i the surfaces X_0 and X_n are rational, and therefore $H^1(X_0) = H^1(X_n) = 0$; and $X_1, \ldots, X_{n-1}$ are elliptic ruled surfaces, so that $\operatorname{rk} H^1(X_i) = 2$ for $i = 1, \ldots, n-1$. Therefore, $\operatorname{rk}(\bigoplus(X_i)) = 2(n-1)$, so that the homomorphism $(*)$ cannot be epimorphic.

THEOREM 5. *Supersingular strongly elliptic $K3$ surfaces are not degenerate.*([1])

In other words, any family $\mathfrak{X} \to S$ consisting of such surfaces can be reconstructed into a smooth family. Actually, for a supersingular surface X we have

$$H^2(X, \mathbf{Q}_l) = (\operatorname{Pic} X) \otimes \mathbf{Q}_l,$$

and, since $\operatorname{Pic} X_\eta$ is defined over a finite covering of the base, the monodromy is finite. On the other hand, the one-dimensional formal group of a $K3$ surface [1] is additive for a supersingular surface, so the Hasse invariant of such a surface is 0 (the tangent space to this group coincides with $H^2(X, \mathcal{O})$, so that the Hasse invariant of the surface is different from 0 if and only if this formal group is isomorphic to G_m). Thus the assertion of the theorem follows from Theorem 4.

As is well known, Theorem 5 implies an analog of Torelli's theorem for strongly elliptic supersingular $K3$ surfaces and for the period mapping defined by Ogus [7]. From this it also follows that in this case the period mapping is epimorphic.

Received 20/JAN/81

BIBLIOGRAPHY

1. M. Artin, *Supersingular $K3$ surfaces*, Ann. Sci. École Norm. Sup. (4) **7** (1974), 543–567.
2. ______, *Algebraic construction of Brieskorn's resolutions*, J. Algebra **29** (1974), 330–348.
3. E. Brieskorn, *Singular elements of semi-simple algebraic groups*, Proc. Internat. Congr. Math. (Nice, 1970), vol. 2, Gauthier-Villars, Paris, 1971, pp. 279–284.
4. Eiji Horikawa, *Surjectivity of the period map of $K3$ surfaces of degree* 2, Math. Ann. **228** (1977), 113–146.

([1]) The authors have recently proved Theorem 5 without assuming strong ellipticity.

5. Viktor S. Kulikov, *Degenerations of K3 surfaces and Enriques surfaces*, Izv. Akad. Nauk SSSR **41** (1977), 1008–1042; English transl. in Math. USSR Izv. **11** (1977).

6. David Mumford, *Geometric invariant theory*, Academic Press, New York, and Springer-Verlag, Berlin, 1965.

7. Arthur Ogus, *Supersingular K3 crystals*, Journées de Géométrie Algébrique de Rennes (1978), Vol. II, Astérisque, Vol. 64, Soc. Math. France, Paris, 1979, pp. 3–86.

8. G. N. Tjurina, *Resolution of singularities of flat deformations of double rational points*, Funkcional. Anal. i Priložen. **4** (1970), no. 1, 77–83; English transl. in Functional Anal. Appl. **4** (1970).

9. B. Rapoport and T. Zink, *Über die lokalen Faktoren der ζ-Funktion einiger Shimura-mannigfältigkeiten*, Preprint, 1980.

Translated by M. ACKERMAN

On some infinite-dimensional groups. II

Izv. Akad. Nauk SSSR, Ser. Mat. **45,** No. 1, 214–226 (1981). Zbl. **475,** 14036
[Math. USSR, Izv. **18,** No. 1, 185–194 (1982)]

ABSTRACT. This paper contains proofs of some previously announced theorems on infinite-dimensional algebraic groups, and, in particular, on the structure of the group of automorphisms of an affine space as an infinite-dimensional algebraic group.
Bibliography: 9 titles.

This paper touches on a lecture [9] I gave 15 years ago at a conference devoted to the 100th anniversary of Castelnuovo's birth. The immediate stimulus to its appearance was a recent review of that lecture [7].

In the lecture [9] I wanted to direct attention to the fact that some naturally occurring groups (for example, the group of automorphisms of a polynomial ring in several variables) can be naturally considered as infinite-dimensional analogues of algebraic groups. For them we can introduce the concept of a Lie algebra, and some well-known difficult conjectures turn out to be completely verifiable on the Lie algebra level. Hence, for these groups one can prove some weakened forms of these conjectures.

I had hoped that hints contained in [9] would suffice for a skilled mathematician to develop the course of the argument. But from Professor Kambayashi's review [7] it is evident that he ran up against difficulties in attempting to establish the proofs. Therefore, I here offer a detailed exposition of some material in the lecture [9]. Some insignificant changes in the formulations have been introduced to simplify the exposition.

§1

DEFINITION. By an *infinite-dimensional algebraic variety* over a field k we shall mean the inductive limit X of a directed system (X_i, f_{ij}) of algebraic varieties over the field k, where the morphisms $f_{ij}\colon X_i \to X_j$ $(i < j)$ are closed embeddings.

Further, we shall consider only the case where the set of indices is the set of natural numbers $\mathbf{N}$ or has such a cofinal subset. We shall always assume that the field k is algebraically closed.

Each of the X_i will be considered with its Zariski topology, and we endow X with the topology of the inductive limit, in which a set $Z \subset X$ is closed if and only if its preimage in each X_i is closed. In particular, each X_i is closed in X.

1980 *Mathematics Subject Classification.* Primary 14L99; Secondary 20G99.

A continuous mapping $f\colon X \to Y$ of two infinite-dimensional algebraic varieties will be called a *morphism* if for any X_i in the system (X_i) defining X there is a Y_j in the system (Y_j) defining Y such that $f(X_i) \subset Y_j$ and the restriction $f\colon X_i \to Y_j$ is a morphism of (finite-dimensional) algebraic varieties.

Irreducibility and connectedness of an infinite-dimensional algebraic variety are defined as irreducibility and connectedness of the corresponding topological space.

PROPOSITION 1. *An infinite-dimensional algebraic variety $X = \varinjlim X_i$ is irreducible if and only if the set of irreducible components of the varieties X_i, ordered by inclusion, is directed.*

PROOF. Let us call irreducible components X_i^λ and X_j^μ of the varieties X_i and X_j *comparable* if there exists an irreducible component X_k^ν of a variety X_k such that $X_i^\lambda \subset X_k^\nu$ and $X_j^\mu \subset X_j^\nu$. The condition of being directed means that any two components X_i and X_j are comparable. Suppose this condition is satisfied, and X is reducible: $X = X' \cup X''$, X' and X'' closed, $X' \neq X$, $X'' \neq X$. Suppose $a' \notin X'$ and $a'' \notin X''$. Then $a', a'' \in X_i$ for some i. Suppose $a' \in X_i^\lambda$, $a'' \in X_i^\mu$, where X_i^λ and X_i^μ are irreducible components of X_i. By the comparability condition, there is an irreducible component X_j^ν of some X_j such that $X_i^\lambda \subset X_j^\nu$ and $X_i^\mu \subset X_j^\nu$. Then X_j^ν is contained either in X' or in X''. For example, if $X_j^\nu \subset X'$, then $a'' \in X_i^\mu \subset X_j^\nu \subset X'$, which contradicts the assumption.

Assume that X is irreducible, and that X_i^λ and X_j^μ are two incomparable components of the varieties X_i and X_j. Denote by S' the set of irreducible components X_k^ν such that $X_k^\nu \not\supset X_i^\lambda$, by S'' the set of components X_l^σ for which $X_l^\sigma \not\supset X_j^\mu$, by X' the union of all the components in S', and by X'' the union of all the components in S''. If X_k^ν is a component in S' and X_n^ρ is a component of X_n such that $X_k^\nu \supset X_n^\rho$, then X_n^ρ is also contained in S'. Therefore, S' (and likewise S'') is a closed set, $X' \cup X'' = X$, because of the incomparability of X_i^λ and X_j^μ, $X' \neq X$, since X_i^λ is not contained in S', and $X'' \neq X$ for a like reason. Therefore, X is reducible.

There is an analogous characterization of connected varieties. Two components X_i^λ and X_j^μ of the varieties X_i and X_j will be called *connected* if for some variety X_k there are components $X_k^1, \dots, X_k^r$ such that $X_i^\lambda \subset X_k^1$, $X_j^\mu \subset X_k^r$ and $X_k^\nu \cap X_k^{\nu+1} \neq \varnothing$.

PROPOSITION 2. *An infinite-dimensional algebraic variety X is connected if and only if any two components X_i^λ and X_j^μ of any of the varieties X_i and X_j are connected.*

The proof is completely analogous to that of Proposition 1. The only change relates to proving that if the components X_i^λ and X_j^μ are not connected, then X is not connected. Here one must take S' to be the set of all components connected with X_i^λ, and S'' to be the set of components not connected with X_i^λ.

We omit the obvious definition of the product of infinite-dimensional algebraic varieties.

A set G that is a group and an infinite-dimensional algebraic variety will be called an *infinite-dimensional algebraic group* if the inversion mapping $G \to G$ and the multiplication $G \times G \to G$ are morphisms.

PROPOSITION 3. *A connected infinite-dimensional algebraic group is irreducible.*

Let G be a connected infinite-dimensional algebraic group, $G = \varinjlim X_i$. Let X_i^λ and X_j^μ be irreducible components of the varieties X_i and X_j. Because the variety G is connected, by Proposition 2 there is a variety X_k with components $X_k^1, \dots, X_k^r$ such that $X_i^\lambda \subset X_k^1$,

$X_j^\mu \subset X_k^r$ and $X_k^\nu \cap X_k^{\nu+1} \neq \varnothing$. Suppose $X_k^\nu \cap X_k^{\nu+1} \ni g_\nu$. The set $X_k^1 g_1^{-1} X_k^2 g_2^{-1} \cdots X_k^{r-1} g_{r-1}^{-1} X_k^r$ is contained (by the definition of an algebraic group) in some variety X_l. This set is an irreducible finite-dimensional variety, since it is the image of the irreducible variety $X_k^1 \times \cdots \times X_k^r$ under the natural morphism. Therefore, it is contained in some irreducible component X_l^ρ of the variety X_l. Then

$$X_i^\lambda = X_i^\lambda g_1^{-1} g_1 g_2^{-1} g_2 \cdots g_{r-1}^{-1} g_{r-1} \subset X_l^\rho, \qquad X_j^\mu = g_1 g_1^{-1} \cdots g_{r-1} g_{r-1}^{-1} X_j^\mu \subset X_l^\rho.$$

Therefore, X_i^λ and X_j^μ are comparable, and this, by Proposition 1, implies the irreducibility of X.

If all the X_i are affine varieties, then X will be called an *affine* infinite-dimensional variety. If $k[X_i]$ is the ring of regular functions on X_i, then $k[X] = \varprojlim k[X_i]$ will be called the *ring of regular functions* on X. The topology of the projective limit turns $k[X]$ into a topological ring. A morphism $f\colon X \to Y$ defines a continuous homomorphism $f^*\colon k[Y] \to k[X]$.

If $x \in X$, then the correspondence $f \to f(x)$ is a continuous homomorphism of the ring $k[X]$ into the field k. Thus a one-to-one correspondence is established between the points of the infinite-dimensional affine variety X and the continuous homomorphisms of the topological ring $k[X]$ into the field k.

Let X and Y be two infinite-dimensional affine varieties. A morphism $f\colon X \to Y$ is called a *closed embedding* if the corresponding homomorphism of rings $f^*\colon k[X] \to k[Y]$ is a strict epimorphism, i.e. if f^* is an epimorphism and the natural homomorphism $k[X]/\operatorname{Ker} f^* \to k[Y]$ is an isomorphism of topological rings. Identifying points with their corresponding continuous homomorphisms into k, we thus get an embedding of Y into X, and Y becomes a closed subset of X. In this case we shall call Y a *closed subvariety*.

Proceeding from these definitions, we define in an obvious way an *affine infinite-dimensional group* and a *closed subgroup* of it (which is, by definition, a closed subvariety).

In the sequel we shall restrict ourselves to considering affine infinite-dimensional algebraic groups, since the author knows no nontrivial examples of another type (examples will be studied in §2).

If x is a point of an infinite-dimensional affine variety X, then the functions $f \in k[X]$ for which $f(x) = 0$ form a closed maximal ideal $\mathfrak{m}_x$ of the ring $k[X]$. We shall denote by $\mathfrak{m}_x^{(n)}$ the closure of the nth power of the ideal $\mathfrak{m}_x$ in $k[X]$. If $\mathfrak{m}_{x,i}$ is the maximal ideal of the point x in the ring $k[X_i]$ for $X_i \ni x$, then $\mathfrak{m}_x^{(n)} = \varprojlim \mathfrak{m}_{x,i}^{(n)}$ and $\mathfrak{m}_x^{(n)}/\mathfrak{m}_x^{(i)} = \varprojlim \mathfrak{m}_{x,i}^n/\mathfrak{m}_{x,i}^l$ for $n < i$. The topological vector space $\mathfrak{m}_x/\mathfrak{m}_x^{(2)}$ (with the topology of the projective limit) is called the *cotangent space* to X at the point x, and the dual space $T_{x,X} = \operatorname{Hom}_C(\mathfrak{m}_x/\mathfrak{m}_x^{(2)}, k)$ (of continuous homomorphisms) is called the *tangent space*. A closed embedding $f\colon Y \to X$ determines a homomorphism $(df)_y\colon T_{y,Y} \to T_{f(y),X}$.

Our goal is to prove the following result.

THEOREM 1. *Let G be a connected affine infinite-dimensional algebraic group defined over a field k of characteristic* 0, *H a closed subgroup of it, and $f\colon H \to G$ an embedding. If $(df)_e\colon T_{e,H} \to T_{e,G}$ is an isomorphism, then f is an isomorphism* (*i.e.* $H = G$).

Before going on to the proof, we introduce another definition. Denote by $\hat{S}^n(\mathfrak{m}_x/\mathfrak{m}_x^{(2)})$ the completed nth symmetric product (see [3], p. 75) of the topological vector space $\mathfrak{m}_x/\mathfrak{m}_x^{(2)}$. By continuity, there is defined a homomorphism

$$\varphi_n\colon \hat{S}^n(\mathfrak{m}_x/\mathfrak{m}_n^{(2)}) \to \mathfrak{m}_x^{(n)}/\mathfrak{m}_x^{(n+1)},$$

which is an epimorphism since it is the projective limit of the epimorphisms

$$S^n\left(\mathfrak{m}_{x,X_i}/\mathfrak{m}^2_{x,X_i}\right) \to \mathfrak{m}^n_{x,X_i}/\mathfrak{m}^{n+1}_{x,X_i}$$

and since the space $\hat{S}^n(\mathfrak{m}_x/\mathfrak{m}_x^{(2)})$ is linearly compact. The variety X is called *smooth* at a point x (and the point x is called *simple* on X) if all the homomorphisms φ_n are isomorphisms. Otherwise, the point x is called *singular*.

Theorem 1 is a consequence of two other results.

THEOREM 2. *If $f: Y \to X$ is a closed embedding of infinite-dimensional affine varieties, X is irreducible, Y is smooth at a point $y \in Y$, and the homomorphism $(df)_y: T_{y,Y} \to T_{f(y),X}$ is an isomorphism, then f is also an isomorphism.*

THEOREM 3. *An infinite-dimensional algebraic group defined over a field of characteristic* 0 *is a smooth variety.*

Let us prove Theorem 2. It suffices to show that the homomorphism f^* is a monomorphism; then, from the definition of a closed embedding it will follow that f is an isomorphism of the topological rings $k[Y]$ and $k[X]$, and hence $Y = X$.

Let $u \in k[X]$ and $f^*(u) = 0$. Denote by Ω_y the topological vector space $\mathfrak{m}_y/\mathfrak{m}_y^{(2)}$. By hypothesis $\Omega_y \simeq \Omega_{f(y)}$. We have a commutative diagram

$$\begin{array}{ccc} \hat{S}^n(\Omega_{f(y)}) & \stackrel{\sim}{\to} & \hat{S}^n(\Omega_y) \\ \downarrow & & \downarrow\wr \\ \mathfrak{m}^{(n)}_{f(y)}/\mathfrak{m}^{(n+1)}_{f(y)} & \to & \mathfrak{m}^{(n)}_y/\mathfrak{m}^{(n+1)}_y \end{array}$$

in which the vertical homomorphisms are epimorphisms, and the upper horizontal and the right-hand vertical are, by hypothesis, isomorphisms. Hence it follows that the lower horizontal homomorphism is a monomorphism. Thus, if $u \in \mathfrak{m}^{(n)}_{f(y)}$ and $f^*(u) = 0$, then $u \in \mathfrak{m}^{(n+1)}_{f(y)}$, and so $u \in \bigcap \mathfrak{m}^{(n+1)}_{f(y)}$. Let X_i^λ be some irreducible component of the variety X_i passing through $f(y)$. Denote by u_i^λ the restriction of u to X_i^λ and by $\tilde{\mathfrak{m}}$ the maximal ideal of the point $f(y)$ in $k[X_i^\lambda]$. From the relation proved above it follows that $u_i^\lambda \in \bigcap \tilde{\mathfrak{m}}^n$, i.e. $u_i^\lambda = 0$. Let X_i^μ be another component of X_i. As we saw in the proof of Proposition 1, these components are comparable, i.e. are contained in some irreducible component X_j^ν of some subvariety X_j. Arguing as above, we find that $u = 0$ on X_j^ν, and this means that $u = 0$ on X_i^μ. Thus $u = 0$ on X_i, and this is true for all i, so that $u = 0$.

For finite-dimensional algebraic groups Theorem 3 is well known. Usually the proof of the corresponding fact is based on the fact that the variety G is homogeneous, and the set of its simple points is open and nonempty. But such reasoning is not applicable to infinite-dimensional varieties, since such a variety may have no simple points at all. For example, if C is a finite-dimensional algebraic variety with a singular point c_0, $X_i = C^i$, an embedding $X_i \to X_{i+1}$ is defined by $(c_1, c_2, \ldots, c_i) \to (c_0, c_1, \ldots, c_i)$ and $X = \bigcup X_i$, then no point of X is simple. But for a base field of characteristic 0 there is another proof, due to Cartier, that a group variety is smooth. We shall explain this too, omitting some details; it is completely analogous to the finite-dimensional case (see [5] or [8], §§11 and 12).

We shall consider only affine varieties and shall use the language of rings instead of the dual geometrical language. If X and Y are affine infinite-dimensional algebraic varieties, then, as is easily seen, $k[X \times Y] = k[X] \hat{\otimes} k[Y]$ (the completed tensor product). Let G be an affine infinite-dimensional algebraic group defined over a field k, $A = k[G]$. The multiplication morphism $\mu: G \times G \to G$ defines a continuous homomorphism $\mu^*: A \to A \hat{\otimes} A$, and the homomorphism "value at the unit element e" is a continuous homomorphism $\varepsilon: A \to k$. We shall denote by $\mathfrak{m}$ the ideal $\mathfrak{m}_e = \operatorname{Ker} \varepsilon$. μ^* and ε are connected by the well-known relations

$$(\mu^* \hat{\otimes} 1)\mu^* = (1 \hat{\otimes} \mu^*)\mu^*, \qquad (\varepsilon \hat{\otimes} 1)\mu^* = (1 \hat{\otimes} \varepsilon)\mu^* = \mathrm{Id}.$$

From these relations it follows that for $a \in \mathfrak{m}$

$$\mu^* a = a \hat{\otimes} 1 + 1 \hat{\otimes} a + u, \qquad u \in \mathfrak{m} \hat{\otimes} \mathfrak{m}.$$

We shall consider, not further specifying this, only continuous differential topological k-algebras A.

A derivation D is called *invariant* if $\mu^* D = (1 \hat{\otimes} D)\mu^*$. Multiplying this equality by $1 \hat{\otimes} \varepsilon$, we get $D = (1 \hat{\otimes} \varepsilon D)\mu^*$. Put $\varepsilon D = f$. This is a homomorphism $A \to k$; and from the fact that D is a derivation it follows that $f(\mathfrak{m}^{(2)}) = 0$. Moreover, $f(1) = 0$, so that f is determined by a linear form, which we shall denote by the same letter: $f \in \operatorname{Hom}_C(\mathfrak{m}/\mathfrak{m}^{(2)}, k)$. Conversely, from a form $f \in \operatorname{Hom}_C(\mathfrak{m}/\mathfrak{m}^{(2)}, k)$ one constructs the invariant derivation $D_f = (1 \hat{\otimes} f)\mu^*$, where we consider f as a homomorphism on $\mathfrak{m}$ that equals 0 on $\mathfrak{m}^{(2)}$ and we extend it to all of A by putting $f(1) = 0$. Here $1 \hat{\otimes} f$ is considered as the homomorphism $A \hat{\otimes} A \to A \hat{\otimes} k$ mapping $a \hat{\otimes} b \to af(b)$. The invariance of D_f easily follows from the diagram

$$\begin{array}{ccccc} A & \xrightarrow{\mu^*} & A \hat{\otimes} A & \xrightarrow{1 \otimes f} & A \\ \mu^* \downarrow & & \mu^* \otimes 1 \downarrow & & \mu^* \downarrow \\ A \hat{\otimes} A & \xrightarrow{1 \otimes \mu^*} & A \hat{\otimes} A \hat{\otimes} A & \xrightarrow{1 \otimes (1 \otimes f)} & A \hat{\otimes} A \end{array}$$

The verification that D_f is a derivation is also elementary. It suffices to check the equality $D_f(ab) = D_f(a)b + aD_f(b)$ for $a, b \in \mathfrak{m}$. We must write $\mu^* a$ and $\mu^* b$ in the forms $\mu^* a = 1 \hat{\otimes} a + a \hat{\otimes} 1 + u$ and $\mu^* b = 1 \hat{\otimes} b + b \hat{\otimes} 1 + v$, where $u, v \in \mathfrak{m} \hat{\otimes} \mathfrak{m}$, and substitute in our equality. After obvious transformations we get

$$D_f(ab) = af(b) + bf(a) + (1 \hat{\otimes} f)((a \hat{\otimes} 1)v) + (1 \hat{\otimes} f)(u(b \hat{\otimes} 1)),$$
$$D_f(a)b + aD_f(b) = af(b) + bf(a) + a(1 \hat{\otimes} f)(v) + b(1 \hat{\otimes} f)(v).$$

It remains to verify that $(1 \hat{\otimes} f)((x \hat{\otimes} 1)z) = x(1 \hat{\otimes} f)(z)$ for $z \in \mathfrak{m} \hat{\otimes} \mathfrak{m}$. It suffices to consider the case $z = z_1 \hat{\otimes} z_2$, $z_1, z_2 \in \mathfrak{m}$, when this equality is obvious.

We have thus established an isomorphism between the space of invariant derivations of the ring A and the space $\operatorname{Hom}_C(\mathfrak{m}/\mathfrak{m}^{(2)}, k) \simeq T_e$. Since the space of invariant derivations is a Lie algebra, by the same token the space T_e is endowed with a structure of Lie algebra. This algebra is called the Lie algebra of the group G and is denoted by $\mathfrak{L}(G)$.

After these preparations, Theorem 3 is easily proved. Put $\mathfrak{m}/\mathfrak{m}^{(2)} = \Omega$, and let φ_k: $\hat{S}^k(\Omega) \to \mathfrak{m}^{(k)}/\mathfrak{m}^{(k+1)}$ be the homomorphisms introduced earlier (they are obviously surjective). For $k < n$ suppose that φ_n is injective (φ_1 is tautologically an isomorphism). We assume that $\varphi_n(x) = 0$ for $x \in \hat{S}^n(\Omega)$. For any $f \in \operatorname{Hom}_C(\Omega, k)$ we have $D_f \mathfrak{m}^{(k)} \subset \mathfrak{m}^{(k-1)}$. Denote by $\overline{D}_f$ the uniquely determined derivation on $\hat{S}(\Omega) = \oplus \hat{S}^k(\Omega)$ equal to f on Ω. Then $D_f \varphi_n = \varphi_{n-1} \overline{D}_f$, and $\varphi_{n-1} \overline{D}_f(x) = 0$. Since φ_{n-1} is injective by assumption, we have $\overline{D}_f(x) = 0$ for any $f \in \operatorname{Hom}_C(\Omega, k)$. It remains to note that if for a linearly compact space Ω and for some $x \in \hat{S}^n(\Omega)$ we have $\overline{D}_f(x) = 0$ for an arbitrary derivation $\overline{D}_f$, then

$x = 0$ (under the assumption of characteristic 0). This can be inferred, for example, from an analogue of Euler's formula. By assumption the space Ω has a basis $\{e_i\}$, $e_i \to 0$. If f_i is the dual basis in $\mathrm{Hom}_C(\Omega, k)$, and $x \in \hat{S}^n(\Omega)$, then

$$nx = \sum \bar{D}_{f_i}(x)e_i.$$

It suffices to check this equality for $x = e_{i_1} \hat{\otimes} \cdots \hat{\otimes} e_{i_n}$, when it reduces to Euler's classical identity.

We have thus established that the point e is simple. The simplicity of the other points follows from the homogeneity of a group variety.

REMARK. The arguments explained above can be given a more local character by considering the local ring of the point e instead of the ring $k[G]$. It is then necessary to choose between two variants of the definition of the local ring of a point x of an infinite-dimensional algebraic variety $X = \bigcup X_i$. This ring can be defined as the projective limit of the local rings $\mathcal{O}_{x,X_i}$ of the point x on all the varieties X_i containing it. This ring is complete with respect to the topology of the projective limit, but it is not the fiber of some natural sheaf. The other variant is to consider functions regular in some neighborhood of x. This ring is the fiber of the "structure sheaf $\mathcal{O}_X$" of the variety X. Both definitions are suitable for a proof of Theorem 3.

§2

Let $\mathbf{A}^n$ be the n-dimensional affine space over an algebraically closed field k of characteristic 0. By $G = \mathrm{Aut}(\mathbf{A}^n)$ we denote the group of all automorphisms of $\mathbf{A}^n$. G is the group of automorphisms of the algebra $k[\mathbf{A}^n] = k[T_1, \ldots, T_n]$. In the sequel we shall denote this algebra by R. Having chosen a coordinate system x in $\mathbf{A}^n$, we assign to an element $g \in G$ n elements $(g(x_1), \ldots, g(x_n)) \in R^n$. We shall assign to this element the list of $2n$ elements $(g(x_1), \ldots, g(x_n); g^{-1}(x_1), \ldots, g^{-1}(x_n)) \in R^{2n}$. The polynomial ring R can be considered as an infinite-dimensional affine space in which the coordinates are the coefficients of a polynomial. The spaces R^n and $R^{2n} = R^n \times R^n$ are also isomorphic to infinite-dimensional affine spaces; in this role we shall denote R^n by $\mathbf{A}'$, the other copy of R^n by $\mathbf{A}''$, and $R^{2n} = R^n \times R^n$ by $\mathbf{A}$. Thus $G \subset \mathbf{A} = \mathbf{A}' \times \mathbf{A}''$, and we have the "equation" $g \circ h = e$, $g \in \mathbf{A}'$, $h \in \mathbf{A}''$. This equation is indeed an identity with respect to $x_1, \ldots, x_n$, from which, by equating coefficients, we get an infinite number of equations defining G in $\mathbf{A}$.

A structure of infinite-dimensional variety in A is defined by the subspaces $\mathbf{A}_i$, where $\mathbf{A}_i$ consists of the lists of $2n$ polynomials each of whose degrees is $\leqslant i$. The induced structure on G gives, as one easily sees, a structure of affine infinite-dimensional algebraic group on G. Our goal is to define the Lie algebra $\mathfrak{L}(G)$ of this group.

The tangent space $T_{e,\mathbf{A}}$ of the infinite-dimensional affine space $\mathbf{A}$ at the point e corresponding to the unit element of the group G may be identified with the space itself, $T_{e,\mathbf{A}} \simeq \mathbf{A}$. Thus, an element $\xi \in T_{e,\mathbf{A}}$ can be given by a list $(P_1, \ldots, P_n; Q_1, \ldots, Q_n)$, $P_i, Q_i \in R$. Elements of $T_{e,G} \subset T_{e,\mathbf{A}} = \mathbf{A}$ are given in the same way. If $\xi \in T_{e,G}$, then $Q_i = -P_i$. In fact, for $f \in T_{e,G}$ and $u \in k[\mathbf{A}]$, from $u|_G = 0$ it follows that $\langle f, u \rangle = 0$. Recall that $\langle f, uv \rangle = \langle f, u \rangle v(e) + u(e)\langle f, v \rangle$. We apply this to the "equation" $g \circ h = e$. Let $g(x_i) = \sum a_i^{(\nu)} x^{(\nu)}$ and $h(x_j) = \sum b_j^{(\nu)} x^{(\nu)}$, where $a_i^{(\nu)}, b_j^{(\nu)} \in k[\mathbf{A}]$. Here $x^{(\nu)} = x_1^{\nu_1} \cdots x_n^{\nu_n}$, $\alpha_i^{(\nu)} = \alpha_i^{(\nu_1, \ldots, \nu_n)}$, etc. Then $\langle f, g(x_i) \rangle = P_i$ and $\langle f, h(x_j) \rangle = Q_j$. If $\langle f, a_i^{(\nu)} \rangle = \alpha_j^{(\nu)} \in k$ and $\langle f, b_j^{(\nu)} \rangle = \beta_j^{(\nu)}$, then our "equation" assumes the form $\sum a_i^{(\nu)} h^{(\nu)} = x_i$, whence

$\langle f, \Sigma a_i^{(\nu)} h^{(\nu)} \rangle = 0$, where we put $h = (h(x_1), \ldots, h(x_n))$. But

$$\left\langle f, \sum a_i^{(\nu)} h^{(\nu)} \right\rangle = \left(\sum \alpha_i^{(\nu)} h^{(\nu)}\right)\Big|_{h=e} + \sum \left(\frac{\partial g(x_i)}{\partial x_j}(h) \langle f, h_j \rangle \right)\Bigg|_{g=h=e} = P_i + Q_i.$$

Hence $Q_i = -P_i$. Therefore, in the sequel we shall consider $\mathfrak{L}(G)$ as a subspace of $\mathbf{A}' = R^n$, since an element of $\mathfrak{L}(G)$ is uniquely determined by this projection.

In the same way we shall consider the embedding $(F_1, \ldots, F_n)$: $G \to \mathbf{A}'$, and thus $\mathfrak{L}(G)$ consists of those lists $(P_1, \ldots, P_n)$ such that $P_i = \langle f, F_i \rangle$ for some $f \in T_{e,G} \subset T_{e,\mathbf{A}'}$.

We consider the algebra $\mathfrak{L}(G)$ as a subalgebra of derivations of the ring R. Namely, if the affine algebraic group G acts on an affine variety X (both may be infinite-dimensional), then any element $f \in T_{e,G}$ determines a derivation D_f of the ring $k[X]$. If the action of G on X is given by a morphism φ: $G \times X \to X$ corresponding to a continuous homomorphism φ^*: $k[X] \to k[X] \hat{\otimes} k[G]$, then $D_f = (1 \hat{\otimes} f)\varphi^*$. In general, the product $D_{f_2} \circ D_{f_1}$ is not a derivation. It is the differential operator D_f defined by the same formula $D_f = (1 \hat{\otimes} f)\varphi^*$ using the convolution $f = f_1 * f_2$ of the mappings f_1 and f_2. Namely, f_1 and f_2 will be considered as homomorphisms $k[G] \to k$ by defining them on $\mathfrak{m}_e$ as the composition $\mathfrak{m}_e \to \mathfrak{m}_e/\mathfrak{m}_e^{(2)} \xrightarrow{f} k$ and defining them on $k[G]$ by the condition $f_i(1) = 0$, and setting f equal to $(f_1 \hat{\otimes} f_2)\mu^*$. The equality $D_{f_2} \circ D_{f_1} = D_f$ immediately follows from considering a commutative diagram, in which we put $k[G] = A$ and $k[X] = R$:

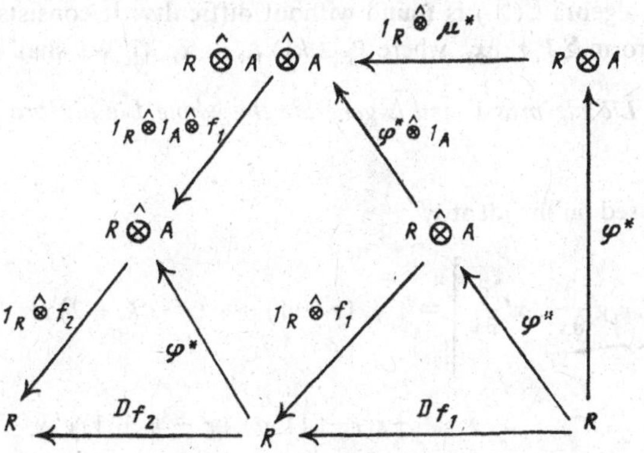

Setting $X = G$ and $\varphi = \mu$, we see that the Lie algebra $\mathfrak{L}(G)$ is isomorphic to the Lie algebra $\mathrm{Hom}_C(\mathfrak{m}/\mathfrak{m}^{(2)}, k)$ with respect to convolution. For arbitrary X we have a homomorphism of the latter algebra into the Lie algebra of derivations Diff $k[X]$ of the algebra $k[X]$. Hence it follows that, by composing an invariant derivation on G with the derivation of the ring $k[X]$ corresponding to the same function f, we get a homomorphism of $\mathfrak{L}(G)$ into Diff $k[X]$. In our case $X = \mathbf{A}^n$, $G = \mathrm{Aut}\, \mathbf{A}^n$, this homomorphism obviously is a monomorphism. We shall also investigate the Lie algebra $\mathfrak{L}(G)$ as a Lie subalgebra of Diff $k[\mathbf{A}^n]$.

We denote by $J_g(x)$ the Jacobian of the endomorphism g of the space $\mathbf{A}^n$. From the relation $J_{g_1 g_2}(x) = J_{g_1}(x) J_{g_2}(g_1(x))$ it follows that if g has an inverse, then $J_g(x) = c$, which gives us another "equation" of the group $G \subset \mathbf{A}'$, which we shall also investigate. If $f \in T_{e,G}$, g is an endomorphism of $\mathbf{A}^n$ and $g(x_i) = F_i(x_1, \ldots, x_n)$, then

$$\langle f, J_g \rangle = \langle f, J(F_1, \ldots, F_n) \rangle$$
$$= J(\langle f, F_1 \rangle, F_2, \ldots, F_n)|_{g=e} + \cdots + J(F_1, F_2, \ldots, \langle f, F_n \rangle)|_{g=e}.$$

Hence

$$\langle f, J_g \rangle = \frac{\partial \langle f, F_1 \rangle}{\partial x_1} + \cdots + \frac{\partial \langle f, F_n \rangle}{\partial x_n}.$$

Thus $\partial P_1/\partial x_1 + \cdots + \partial P_n/\partial x_n = c \in k$ for $(P_1, \ldots, P_n) \in \mathfrak{L}(G)$.

We have proved the following result.

LEMMA 1. *The Lie algebra $\mathfrak{L}(G)$ of the infinite-dimensional algebraic group $G = \mathrm{Aut}(\mathbf{A}^n)$ is contained in the algebra $\Lambda \subset \mathrm{Diff}\, k[\mathbf{A}^n]$ consisting of the differential operators of the form $D = \Sigma P_i \partial/\partial x_i$ with $\mathrm{Div}\, D = 0$. That Λ is a Lie algebra easily follows from the proved identity*

$$\mathrm{Div}[D_1, D_2] = D_1(\mathrm{Div}\, D_2) - D_2(\mathrm{Div}\, D_1).$$

In the group G we distinguish two important subgroups. One of them is $\mathrm{Aff}(n)$, the group of affine transformations; its Lie algebra consists of the differential operators $\Sigma P_i \partial/\partial x_i$ in which P_i is linear. We shall denote this group by Γ. The other, $\mathfrak{T}$, is the group of triangular transformations of the form

$$g(x_i) = x_i + f_i(x_1, \ldots, x_{i-1}), \qquad i = 1, \ldots, n.$$

Every transformation of this form is an automorphism, and therefore the equations of $\mathfrak{T}$ in $\mathbf{A}'$ are simply expressed as reflections in 0 of some coefficients of the polynomials $g(x_i)$. In view of this, the algebra $\mathfrak{L}(\mathfrak{T})$ is found without difficulty: it consists of the differential operators of the form $\Sigma P_i \partial/\partial x_i$, where $P_i \in k[x_1, \ldots, x_{i-1}]$. We shall denote it by Δ.

LEMMA 2. *The Lie algebras Γ and Δ generate the whole Lie algebra Λ over a field k of characteristic* 0.

The proof is based on the identity

$$\Big[\underbrace{y\frac{\partial}{\partial x}, \ldots, y\frac{\partial}{\partial x}}_{k}, x^r\frac{\partial}{\partial y}\Big] = -kr(r-1)\cdots(r-k+2)y^{k-1}x^{r-k+1}\frac{\partial}{\partial x}$$

$$+ r(r-1)\cdots(r-k+1)y^k x^{r-k}\frac{\partial}{\partial y},$$

which is easily verified by induction on r.

We write an element $x_1^{\alpha_1}\cdots x_n^{\alpha_n}\partial/\partial x_i$ with $i < n$ in the form

$$Cx_i^{\alpha_i}x_n^{\alpha_n}\frac{\partial}{\partial x_i}, \quad C = x_1^{\alpha_1}\cdots x_{i-1}^{\alpha_{i-1}}x_{i+1}^{\alpha_{i+1}}\cdots x_{n-1}^{\alpha_{n-1}}.$$

By the identity referred to above, the algebra generated by the elements $x_n\partial/\partial x_i$ and $x_i^r\partial/\partial x_n$ contains an element of the form

$$Cx_i^{\alpha_i}x_n^{\alpha_n}\frac{\partial}{\partial x_i} + C \cdot F(x_i, x_n)\frac{\partial}{\partial x_n}.$$

Hence for any $P_1, \ldots, P_{n-1}$ there is an element of the form

$$\eta = \sum_{i=1}^{n-1} P_i\frac{\partial}{\partial x_i} + Q\frac{\partial}{\partial x_n}$$

belonging to the algebra generated by the subalgebras Γ and Δ. Let

$$\xi = \sum_{i=1}^{n} P_i \frac{\partial}{\partial x_i} \in \Lambda.$$

Of course, the element η we have constructed belongs to Λ. Therefore,

$$\xi - \eta = (P_n - Q)\frac{\partial}{\partial x_n} \in \Lambda.$$

This means that $\partial(P_n - Q)/\partial x_n = c$, i.e. $P_n - Q = cx_n + U$, $U \in k[x_1, \ldots, x_{n-1}]$. Thus

$$\xi = \eta + cx_n \frac{\partial}{\partial x_n} + U\frac{\partial}{\partial x_n},$$

i.e., it belongs to the subalgebra generated by Γ and Δ.

COROLLARY. *The Lie algebra of the group G is Λ.*

The algebra Λ is not simple: it contains the ideal Λ_0 consisting of the differential operators D with $\operatorname{Div} D = 0$.

LEMMA 3. *Over a field of characteristic* 0 *the Lie algebra Λ_0 is simple.*

In fact, if a nonzero element $\Sigma P_i \partial/\partial x_i$ belongs to an ideal of Λ_0, then that ideal also contains

$$\left[\frac{\partial}{\partial x_j}, \sum P_i \frac{\partial}{\partial x_i}\right] = \sum \frac{\partial P_i}{\partial x_j}\frac{\partial}{\partial x_i}.$$

Applying this operation several times, we find in our ideal a nonzero element with constant coefficients $\Sigma c_i \partial/\partial x_i$, $c_i \in k$. After changing variables, we may consider that this is $\partial/\partial x_1$. Then the same ideal also contains the elements

$$\left[\frac{\partial}{\partial x_1}, \sum X_i \frac{\partial}{\partial x_i}\right] = \sum \frac{\partial X_i}{\partial x_1}\frac{\partial}{\partial x_i},$$

where $\eta = \Sigma X_i \partial/\partial x_i$ is an arbitrary element of the algebra Λ_0. If we are given an element $\xi = \Sigma Q_i \partial/\partial x_i$ of Λ_0, then we can represent it in such a form, taking X_i from the condition $\partial X_i/\partial x_1 = Q_i$. This is all that need be done so that the element η may belong to Λ_0. Since $\operatorname{Div} \xi = 0$, we have $\partial \operatorname{Div} \eta/\partial x_1 = 0$, i.e. $\operatorname{Div} \eta = F(x_2, \ldots, x_n)$. Without disturbing the condition $\partial X_i/\partial x_1 = Q_i$ we can change X_i to $X_i + \Phi_i(x_2, \ldots, x_n)$. This changes $\operatorname{Div} \eta$ to $\Sigma \partial \Phi_i/\partial x_i$, and it remains to choose Φ_i so that $F + \Sigma \partial \Phi_i/\partial x_i = 0$.

To infer some facts about algebraic groups from these assertions about Lie algebras, we need one more simple result.

LEMMA 4. *The group $G = \operatorname{Aut} \mathbf{A}^n$ is irreducible.*

By Proposition 3 it suffices to show that G is connected. For this the method Alexander used to prove the connectedness of the group of homeomorphisms of a ball [1] is appropriate. As an algebraic group, G decomposes into the direct product of two subgroups: the group of affine transformations $\operatorname{Aff}(n)$ and the group H whose transformations are given by the formulas

$$h(x_i) = x_i + f_i^{(2)}(x_1, \ldots, x_n) + f_i^{(3)}(x_1, \ldots, x_n) + \cdots + f_i^{(k)}(x_1, \ldots, x_n),$$

where $f_i^{(k)}$ is a form of degree k. Therefore, it suffices to prove the connectedness of H.

Denote by C_λ the homothety $C_\lambda(x_i) = \lambda x_i$. For $h \in H$ and $h_\lambda = C_\lambda^{-1} h C_\lambda \in H$ we have

$$h_\lambda(x_i) = x_i + \lambda f_i^{(2)} + \lambda^2 f_i^{(3)} + \cdots + \lambda^{k-1} f_i^{(k)}.$$

Hence it is evident that $C_\lambda^{-1} h C_\lambda$ is an algebraic curve joining the element h (when $\lambda = 1$) to the unit element (when $\lambda = 0$).

From Lemmas 2 and 4 and Theorem 1 follows

THEOREM 4. *As an algebraic group, the group G is generated by the subgroups* Aff(n) *and* $\mathfrak{T}$; *i.e., every closed algebraic subgroup $G' \subseteq G$ containing* Aff(n) *and* $\mathfrak{T}$ *coincides with G.*

Since $J_g(x) \in k^*$ for $g \in G$, the mapping $g \to J_g$ is a homomorphism $G \to G_m$. Its kernel is a subgroup G_0 whose Lie algebra is Λ_0. In the same way, Lemma 4 implies

THEOREM 5. *As an algebraic group, the group G is simple.*

For $n = 2$ it is well known that G is generated by the subgroups Aff(2) and $\mathfrak{T}$ as an abstract group as well. Of course, Theorem 4 is only a weak approximation of this result. However, for Theorem 5, a full strengthening is impossible; as V. I. Danilov has proved, as an abstract group the group G is not simple [4], even for $n = 2$.

Theorems 1–4 cover the contents of §1 of the lecture [9]. In conclusion we make some bibliographical remarks on §2 of the lecture. Theorem 6 establishes the property just mentioned: for $n = 2$ the group G is generated by the subgroups Aff(2) and $\mathfrak{T}$. It did not pretend to novelty, but, as pointed out in [9], it was derived by a new method—by embedding $\mathbf{A}^2$ into $\mathbf{P}^2$ and by decomposing each automorphism $g \in G$, as a birational automorphism of $\mathbf{P}^2$, into a product of σ-processes. Theorem 7 states that G is even an amalgamated product of these subgroups; and Theorem 8 infers from this the general form of finite-dimensional algebraic subgroups of G. Gizatullin and Danilov [2] generalized these results to a very wide class of homogeneous (and quasihomogeneous) algebraic surfaces. Finally, Theorem 9 states that the affine plane $\mathbf{A}^2$ has no forms. This was also generalized to the same class of surfaces in the same paper [2]. My earlier, more elementary, proof was also published. On Professor Kambayashi's request I communicated to him a similar means of proving Theorems 8 and 9. He took on himself the task of publishing these arguments [6]; evidently, the more general results of [2] escaped his attention.

Received 23/SEPT/80

BIBLIOGRAPHY

1. J. W. Alexander, *On the deformation of an n-cell*, Proc. Nat. Acad. Sci. U.S.A. **9** (1923), 406–407.

2. M. H. Gizatullin and V. I. Danilov, *Automorphisms of affine surfaces*. I, Izv. Akad. Nauk SSSR Ser. Mat. **39** (1975), 523–565; English transl. in Math. USSR Izv. **9** (1975).

3. A. Grothendieck, *Éléments de géométrie algébrique*. I, Inst. Hautes Études Sci. Publ. Math. No. 4 (1960).

4. V. I. Danilov, *Nonsimplicity of the group of unimodular automorphisms of an affine plane*, Mat. Zametki **15** (1974), 289–293; English transl. in Math. Notes **15** (1974).

5. P. Gabriel, *Étude infinitésimale des schémas en groupes*, Schémas en Groupes (Séminaire de Géometrie Algébrique du Bois-Marie, 1962–1964, SGA-3), Exposés 7A, 7B, 2nd ed., Lecture Notes in Math., vol. 151, Springer-Verlag, 1969, pp. 411–475, 476–562.

6. T. Kambayashi, *On the absence of nontrivial separable forms of the affine plane*, J. Algebra **35** (1975), 449–456.

7. ______, Review of [9], Math. Reviews **58** (1979), #5697.

8. David Mumford, *Abelian varieties*, Tata Inst. Fund. Res., Bombay, and Oxford Univ. Press, 1970.

9. I. R. Shafarevich [Šafarevič], *On some infinite-dimensional groups*, Rend. Mat. e Appl. (5) **25** (1966), 208–212; also published in Atti Simpos. Internaz. Geom. Algebrica (Rome, 1965), Cremonese, Rome, 1967, pp. 208–212.

Translated by M. ACKERMAN

Surfaces of type K3 over fields of finite characteristic

(with A. N. Rudakov)

Itogi Nauki Tekh., Ser. Sovrem. Probl. Mat. **18**, 115–207 (1981). Zbl. **518**, 14015
[J. Sov. Math. **22**, No. 4, 1476–1533 (1983)]

INTRODUCTION

The present work is a survey of one direction in the geometry of algebraic surfaces over fields of finite characteristic. We are speaking of surfaces of type K3 distinguished by the property that for them $K_X=0$, $H^1(X, \mathcal{O}_X)=0$.

Over the field of complex numbers these surfaces are studied according to an idea expressed already by Severi with the help of the periods of two-dimensional differential forms. Namely, it follows from the definition that on a surface of type K3 there exists a unique (up to a multiple) regular differential form ω and that the two-dimensional Betti number of this surface is equal to 22. Integrals of the form over the basis cycles of the group $H_2(X, \mathbf{Z})$ satisfy relations analogous to the Riemann relations on an algebraic curve: they express the fact that $\omega\cup\omega=0$, $\omega\cup\bar{\omega}>0$ [where ω is understood as an element of $H^2(X, \mathbf{C})$]. If we consider polarized surfaces with polarization $l\in H_2(X, \mathbf{Z})$, then there is added a linear relation on the periods: $\int_l \omega=0$. A point in $P^{21}(\mathbf{C})$ with homogeneous coordinates $\int_{c_i}\omega$, where $c_1,\dots,c_{22}$ is a basis of $H_2(X, \mathbf{C})$, belongs by virtue of these relations to an 19-dimensional manifold Ω which is found to be a symmetric space. It has been proved [6] that surfaces of type K3 over the complex number field are uniquely determined by the corresponding point of the manifold (an analogue of Torelli's theorem) and that any point of this manifold corresponds to some surface [4].

Those works which are surveyed here show that there is hope of finding an analogue of this theory at least for certain surfaces of type K3 over fields of finite characteristic. At the present time it is possible to speak of this hope for so-called supersingular surfaces, i.e., surfaces for which all cycles in the two-dimensional (étale) homology group are algebraic – the existence of such surfaces is a phenomenon peculiar to geometry over fields of finite characteristic. For these surfaces it is possible to define a "space of periods" Ω and a "period mapping" from the manifold of moduli into Ω, and there is reason to hope that an analogue of Torelli's theorem holds. However, this has been proved so far only in a very special case (fields of characteristic 2). The construction of periods for supersingular surfaces of type K3 was first proposed by Artin [10] who used flat homology, but later Ogus [33] showed that the theory can naturally be constructed in the framework of crystalline homology, and this is the approach we adopt. Another method intensively applied here is the arithmetic of quadratic forms (in our case the quadratic form is defined by the index of intersection on the group of algebraic cycles of the surface). Applications of this same method to the theory of surfaces of type K3 over the field of complex numbers are contained in the preceding paper of V. V. Nikulin in this collection. It may be said that after so many applications of algebraic geometry to number theory we finally have an application of number theory to algebraic geometry.

We also present other results related to this circle of questions: the absence of vector fields on surfaces of type K3 and the lift of a surface of type K3 defined over a field of finite characteristic to characteristic 0.

We have attempted to write this survey so that it would give a coherent picture of the entire area. It has not been possible to present complete proofs, since this would require an entire book. In those cases where we have been able to contribute something new to the exposition we have presented all details, while in the remaining cases only a summary of the proof is given, and the reader is referred to the literature for details.

Proofs of some results of interest to us and a number of useful remarks were communicated by P. Deligne. We have also made use of suggestions of V. V. Nikulin concerning various

Translated from Itogi Nauki i Tekhniki, Sovremennye Problemy Matematiki, Vol. 18, pp. 115-207, 1981.

questions of the arithmetic of quadratic forms. We are glad to express our gratitude to them both.

1. Even, p-Elementary, Hyperbolic Lattices

A lattice over an integral domain A is an A-module of finite rank without torsion on which there is defined a symmetric bilinear function (a, b) with values in A; (a, b) is called the product of a and b, and (a, a) is henceforth denoted by a^2. If K is the field of fractions of A, then the *dual* lattice to a lattice S is the submodule $S^*\subset S\otimes_A K$, consisting of all elements x such that $(x, a)\in A$ for all $a\in A$ [we assume that the product (a, b) is extended to $S\otimes_A K$ by linearity]. It is obvious that $S\subset S^*$.

A lattice over **Z** is called an integral lattice. An integral lattice is called hyperbolic if the quadratic form x^2 on $S\otimes\mathbf{R}$ has index of inertia $(1, n-1)$, where $n = \operatorname{rk} S$, i.e., in some basis it can be written in the form $x_1^2 - x_2^2 - \ldots - x_n^2$. An integral lattice S is called p-elementary for some prime number p if $pS^* \subset S$. Thus, S^*/S is an elementary p-group. If $\operatorname{rk} S = n$ and $|S^*/S| = p^s$, then for a hyperbolic lattice S $d(S) = (-1)^{n-1}\cdot p^s$, where $d(S) = \pm|S^*/S|$ is the discriminant of the lattice S.

A lattice S is called even if $x^2 \equiv 0 \pmod 2$ for all $x\in S$.

We henceforth use the following notation. If S is a lattice with product (x, y), then the new lattice defined on the same set S with product $\langle x, y\rangle = n(x, y)$ is denoted by $S(n)$. We denote by I the one-dimensional lattice with basis vector e for which $e^2 = 1$.

If S is a 2-elementary lattice and if $x^2 \equiv 0 \pmod 4$ for $x\in S$, then S is called a lattice of type I; otherwise it is called a lattice of type II. This concept can be clarified as follows. For any p-elementary lattice S the product extended to $S^*\subset S\otimes\mathbf{Q}$, has range in $(1/p)\mathbf{Z}$ and defines a product in S^*/S with range in $(1/p)\mathbf{Z}/\mathbf{Z}$. For $p = 2$ a lattice belongs to type I if and only if this product in S^*/S with range in $(1/2)\mathbf{Z}/\mathbf{Z}$ is skew-symmetric, i.e., $x^2 = 0$ for $x\in S^*/S$.

A lattice generated by a root system of type A_n, D_n or E_n we denote by $Q(A_n)$, $Q(D_n)$, $Q(E_n)$ (see [17], p. 166). We here assume that these lattices are negative definite, so that for the roots there is the equality $x^2 = -2$. It is easy to see that among these lattices A_{p-1} are p-elementary for any p; $Q(A_2)$, $Q(E_6)$, $Q(E_8)$ are p-elementary for $p = 3$; $Q(A_1)$, $Q\times(D_{2n})$, $Q(E_7)$, $Q(E_8)$ are p-elementary for $p = 2$. The lattices $Q(E_7)$, $Q(E_8)$, and $Q(D_{2n})$ or $n \equiv 0 \pmod 2$ belong to type I, and $Q(A_1)$ and $Q(D_{2n})$ for $n \equiv 1 \pmod 2$ belong to type II. The direct sum of m lattices isomorphic to N is denoted by N^m.

As previously, we define the number s by the condition $d(S) = (-1)^{n-1}p^s$.

For a p-elementary lattice we always have $s \leqslant n$, since $(S^*: S) = p^s$, $S^*\supset S\supset pS^*$ and $(S^*: pS^*) = p^n$.

THEOREM. An even, hyperbolic, p-elementary lattice of rank n for $p \neq 2$ and $n > 2$ is uniquely determined by its discriminant (i.e., the number s), while for $p = 2$ and $n > 4$ it is determined by the discriminant and the property that it belongs to type I or II.

For $p \neq 2$ a lattice corresponding to given values of n and $s \leqslant n$ exists if and only if the following conditions are satisfied:

$$n\equiv 0 \pmod 2 \text{ and}$$
$$\text{for } s\equiv 0 \pmod 2,\ n\equiv 2 \bmod (4),$$
$$\text{for } s\equiv 1 \pmod 2,\ p\equiv (-1)^{n/2-1} \pmod 4.$$
$$\text{and moreover } n>s>0,\ \text{ if } n\equiv 2 \pmod 8.$$

For $p = 2$ a lattice of type I corresponding to given values of n and s exists only for $s \equiv 0 \pmod 2$, $n \equiv 2 \pmod 4$ and, moreover, $n > s > 0$ if $n \not\equiv 2 \pmod 8$. A lattice of type II exists only for $s > 0$ and $n \equiv s \pmod 2$, while for $s = 2$ we must additionally have $n \not\equiv 6 \pmod 8$ and for $s = 1$, $n \equiv 1 \pmod 8$ or $n \equiv 3 \pmod 8$.

The theorem is a special case of more general results obtained by Nikulin [5]. We shall present a more elementary proof following the scheme set forth in [8]; moreover, we hereby give an explicit construction of the lattices with given invariants which we shall need later

The proof is based on the fact that for indeterminate lattices of rank $n \geq 3$ the genus coincides with the spinor genus [34, p. 319]. The coincidence of the spinor genus with the class is a purely local condition, and it is satisfied if from any p, $S \oplus \mathbf{Z}_p$ contains a $\mathbf{Z}_p$-lattice of the form M(k) as a direct factor where M is a unimodular lattice of rank ≥ 2 for $p \neq 2$ and ≥ 3 for $p = 2$ [34, p. 304]. Any q-elementary lattice T over $\mathbf{Z}_q$ can be represented in the form

$$T = M \oplus N(q), \ \mathrm{rg} N = s, \tag{1}$$

where M and N are unimodular. A proof is easily obtained if we take for M a unimodular sublattice in T such that $M/qM = T/qT^*$. From the decomposition (1) it follows immediately that under the hypotheses of the theorem the criterion presented above for coincidence of the spinor genus and the class is satisfied. Thus, in our case the class and the genus coincide, so that uniqueness of a lattice S with given invariants will be proved if we show that the lattices $S \otimes \mathbf{Z}_q$ are uniquely determined for all prime numbers q.

For $q \neq p$ the lattice $S \otimes \mathbf{Z}_q$ is unimodular, and for $q = 2$ it is also even. Such a lattice is uniquely determined for $q \neq 2$ by the lattice $S \otimes \mathbf{Z}/q\mathbf{Z}$, and for $q = 2$ by the "quadratic form" $^1/_2 x^2 \pmod 2$ on $S \otimes \mathbf{Z}/2\mathbf{Z}$. A lattice over $F_q = \mathbf{Z}/q\mathbf{Z}$ for $q \neq 2$ is uniquely determined by its discriminant, and hence this is also true for the lattice $S \otimes \mathbf{Z}_q$ for $q \neq p$, $q \neq 2$. A unimodular, even lattice over $\mathbf{Z}_2$ reduces to the form U_2^m or $U_2^{m-1} \oplus V_2$, where U_2 and V_2 are even, unimodular lattices of rank 2 characterized by the fact that U_2 represents 0 while V_2 does not. In particular, its rank is even.

We first consider the case $p \neq 2$. Then $S \otimes \mathbf{Z}_2$ is unimodular and even, whence it follows that $n \equiv 0 \pmod 2$. As we have seen, the lattice $S \otimes \mathbf{Z}_q$ for $q \neq 2$ is uniquely determined by prescribing the discriminant. As shown above, $S \otimes \mathbf{Z}_2 = U_2^{n/2}$ or $U_2^{n/2-1} \otimes V_2$ and since $d(U_2) = -1$, $d(V_2) = 3$, $d(S) = (-1)^{n-1} p^s$ we find first of all that the type of the lattice is determined by its discriminant, and, secondly, we obtain the conditions presented in the formulation of the theorem. Finally, for $S \otimes \mathbf{Z}_p$ in view of (1) we obtain 4 possibilities, since for M as well as for N there are 2 possibilities depending on their discriminants: $M = I^{n-s}$ or $I^{n-s-1} \oplus I(\varepsilon)$, $N = I^s$ or $I^{s-1} \oplus I(\varepsilon)$, where $(\varepsilon/p) = -1$. We use the relation $\varepsilon_q(S \otimes \mathbf{Z}_q) = 1$, where ε_q is the Hasse invariant [37, p. 64]. Since $\varepsilon_q(S \otimes \mathbf{Z}_q) = 1$ for $q \neq 2$ and we know $\varepsilon_2(S \otimes \mathbf{Z}_2)$ and $\varepsilon_\infty(S \otimes \mathbf{R})$, from this we find $\varepsilon_p(S \otimes \mathbf{Z}_p)$, which gives the $\mathbf{Q}_p$-lattice $S \otimes \mathbf{Q}_p$ up to isomorphism. It remains to verify that among the four possibilities for $S \otimes \mathbf{Z}_p$ this (i.e., the type of $S \otimes \mathbf{Q}_p$) selects only one; in other words, these 4 lattices over $\mathbf{Z}_p$ remain nonisomorphic after multiplication by $\mathbf{Q}_p$. According to Witt's theorem, it suffices to verify that the 4 lattices $I \oplus I(p)$, $I(\varepsilon) \oplus I(p)$, $I \oplus I(\varepsilon p)$, and $I(\varepsilon) \oplus I(\varepsilon p)$ remain nonisomorphic after multiplication by $\mathbf{Q}_p$. This follows in an obvious manner by considering their discriminants and Hasse invariants.

Suppose now that $p = 2$. The lattice $S \otimes \mathbf{Z}_q$, $q \neq 2$, is again uniquely determined by its discriminant. To investigate the lattice $S \otimes \mathbf{Z}_2$ we use the two identities

$$U_2 \oplus I(\varepsilon) \simeq I(\varepsilon)^2 \oplus I(-\varepsilon) \simeq I(\varepsilon) \oplus I \oplus I(-1), \tag{2}$$

$$V_2 \oplus I(2\varepsilon) \simeq U_2 \oplus I(-6\varepsilon), \tag{3}$$

where $\varepsilon \in \mathbf{Z}_2^*$.

The first is obtained if from bases e, f in U_2 ($e^2 = f^2 = 0$, $(e, f) = 1$) and h in $I(\varepsilon)$ we form the vectors $a = h + e$, $b = h - \varepsilon f$ which generate the sublattice $I(\varepsilon)^2$. Being unimodular, it separates as a direct factor; the remaining factor is determined by the discriminant. The identity $I \oplus I(-1) \simeq I(\varepsilon) \oplus I(\varepsilon)$ follows from the fact that the form $x^2 - y^2$ represents any $\varepsilon \in \mathbf{Z}_2^*$. The relation (3) follows similarly from the fact that the lattice on the left contains the sublattice U_2. Indeed, if f is a basis in V_2 and h a basis in $I(2\varepsilon)$, $e^2 = f^2 = 2$, $(e, f) = 1$, then $e + h$ and f generate an even sublattice W, and $d(W) \equiv -1 \pmod 8$ whence it follows that $W \simeq U_2$.

It is easy to see that a lattice S belongs to type I or II depending on whether in the decomposition (1) of $S \otimes \mathbf{Z}_2 = M \oplus N(2)$ the lattice N is even or odd.

Type I. Since M is unimodular and even, it follows that $n \equiv s \pmod 2$; since N is even and unimodular, it follows that $n \equiv s \equiv 0 \pmod 2$. We again have for each of the lattices M and N 2 possibilities: U_2^k or $U_2^{k-1} \oplus V_2$; this gives a total of 4 possibilities. The condition $d(S) = (-1)^{n-1} 2^s$ leaves two possibilities: $M = U_2^{\frac{n-s}{2}}$, $N = U_2^{s/2}$ and $M = U_2^{\frac{n-s}{2}-1} \oplus V_2$,

$N=U_2^{s/2-1}\oplus V_2$; these match up only for n/2 ≡ 1 (mod 2), i.e., n ≡ 2 (mod 4), which gives the necessary condition in the formulation of the theorem. Under this condition it remains for u to check, as for p ≠ 2, that the two remaining types are distinguished by the value of the Hasse invariant $\varepsilon_2\,(S\otimes \mathbf{Z}_2)$. Again, for this it suffices to verify that $U_2\oplus U_2(2)$ and $V_2\oplus V_2(2)$ have distinct Hasse invariants. Since over $\mathbf{Q}_2$, $U_2\simeq I\oplus I(-1)$, $V_2\simeq I(2)\oplus I(6)$, this verification is trivial.

Type II. Since N is odd, from it we can factor out the term I(ε). Using (3), we see that the two possibilities for M reduce to a single possibility: M = $U_2^{(n-s)/2}$. If N represents zero, then there exists a vector $e\in N$, $e\notin 2N$, $e^2=0$ and $f\in N$ with (e, f) = 1. Replacing f by $f+\alpha e$, $\alpha\in \mathbf{Z}_2$, it is possible to arrange that f^2 = 0 or 1. If f^2 = 1, then e and f generate $I\oplus I(-1)$, while if f^2 = 2, then they generate U_2. In the latter case $N=U_2\oplus N'$, N' is odd, and hence it contains the term I(ε). Applying (2), we see that in any case N contain the term $I\oplus I(-1)$,, i.e., $N=I\oplus N'$, where N' is odd. Repeating this process we arrive at the representation N = $I^{s-k}\oplus F$, F is odd, k = rk F = 4 if s ⩾ 4 and F = N if s ⩽ 4.

Arguing as in the case p ≠ 2, we see that uniqueness of a lattice of type II with the given invariants follows from the next assertion.

<u>LEMMA.</u> A unimodular, odd lattice F of rank k ⩽ 4 over $\mathbf{Z}_2$ is uniquely determined by the lattice $F\otimes\mathbf{Q}_2$ over $\mathbf{Q}_2$. For k ⩾ 3 there exists such a lattice with any values $d(F)\in\mathbf{Z}_2^*$ and $\varepsilon_2(F)$.

The proof of the lemma consists of a number of elementary verifications, and we omit it.

It follows from the lemma that consideration of the discriminant and Hasse invariant $\varepsilon_2(S\otimes\mathbf{Z}_2)$ cannot give us any restrictions on the existence of lattices for s ⩾ 3. We shall further consider certain necessary conditions for the existence of the required lattices whic follow from the preceding argument for s ⩽ 2, s = 2. We set n = 2k + 2. $S\otimes\mathbf{Z}_2=U_2^k\otimes N(2)$, $d(N) = (-1)^{k+1}$ since $d(S) = -2^2$. Therefore, $N\simeq I(\eta)\oplus I((-1)^{k+1}\eta)$, $\eta\in\mathbf{Z}_2^*$. Since $\varepsilon_\infty(S\otimes\mathbf{R})=(-1)^k$, it follows that also $\varepsilon_2(S\otimes\mathbf{Z}_2)=(-1)^k$.

To compute ε_2 we can set $U_2\simeq I\oplus I(-1)$ over $\mathbf{Q}_2$, i.e.,

$$S\otimes\mathbf{Q}_2\simeq I^k\oplus I\,(-1)^k\oplus I\,(2\eta)\oplus I\,((-1)^{k+1}2\eta).$$

A trivial computation gives $\varepsilon_2(S\otimes\mathbf{Z}_2)=(-1)^{\frac{k(k-1)}{2}+k\frac{\eta-1}{2}}$. The relation $\varepsilon_2 = (-1)^k$ can be written in the form

$$\frac{k\,(k+1)}{2}+k\frac{\eta-1}{2}\equiv 0 \pmod 2.$$

The condition for the solvability of this equation is k ≡ 1 (mod 2), i.e., n ≡ 0 (mod 4) or $k\equiv\frac{k\,(k+1)}{2}\equiv 0 \pmod 2$, i.e., k ≡ 0 (mod 4), n ≡ 2 (mod 8). They together are equivalent to the condition $n\not\equiv 6 \pmod 8$. Thus, $s=1$, $n=2m+1$. $S\otimes\mathbf{Z}_2=U_2^m\oplus I\,(2\eta)$, $\eta\in\mathbf{Z}_2^*$, $d\,(S)=2$; hence $\eta = (-1)^m$. Since $\varepsilon_\infty(S\otimes\mathbf{R})=(-1)^m$, it follows that $\varepsilon_2(S\otimes\mathbf{Z}_2)=(-1)^m$. To compute ε_2 we again write U_2 as $I\oplus I(-1)$ over $\mathbf{Q}_2$:

$$S\otimes\mathbf{Q}_2=I^m\oplus I\,(-1)^m\oplus I\,(2\,(-1)^m),$$

whence

$$\varepsilon_2\,(S\otimes\mathbf{Z}_2)=(-1)^{\frac{m(m+1)}{2}}.$$

and the equality $\varepsilon_2 = (-1)^m$ gives m ≡ 0 or 1 (mod 4), i.e., n ≡ 1 (mod 8) or n ≡ 3 (mod 8).

With this we have proved the assertion of the theorem concerning the uniqueness of a lattice with given invariants, and we have derived necessary conditions for its existence. When these necessary conditions are satisfied it remains to explicitly construct the corresponding lattices.

p ≠ 2. We shall use the following construction which is a special case of the operation of "evenization" of an odd lattice due to Venkov [2]. In the lattice $E=\oplus I(a_i)$ with basis $\{e_i\}$, $(e_i, e_j) = 0$ for i ≠ j, $e^2 = a_i$, $a_i \equiv 1 \pmod 2$, we consider the sublattice $E_0 = \{\Sigma x_i e_i;\ \Sigma x_i \equiv 0 \pmod 2\}$, and we set $E_* = E_0 + \varepsilon\mathbf{Z}$, $\varepsilon = {}^1/{}_2\Sigma e_i$; E_* is an integral, even lattice if the condition $\Sigma a_i \equiv 0 \pmod 8$ is satisfied.

We consider the lattice S which is the direct sum of the following two or three terms:) U_2 or $U_2(p)$; 2) $(I(-p)^k \oplus I(-1)^l)_*$ for $kp + l \equiv 0 \pmod 8$; 3) there may further be the term $_p(-1)$, where H_p is an even, positive definite lattice of rank 4 and discriminant p^2: it can e defined as the maximal order in the quaternion algebra ramified only at p and ∞ with prodct (x, y) = tr(xy*), where x ↦ x* is the standard involution.

The lattice

$$U_2 \oplus (I(-p)^k \oplus I(-1)^l)_* \tag{4}$$

ives an even lattice with invariants n and s if k = s, k + l = n − 2, pk + l ≡ 0 (mod 8), .e., (p − 1)s + n − 2 ≡ 0 (mod 8) or [(p − 1)/2]s + n/2 ≡ 1 (mod 4). Similarly, the lattice

$$U_2 \oplus H_p \oplus (I(-p)^k \oplus I(-1)^l)_*$$

ives the required answer under the condition

$$\frac{p-1}{2}(s-2)+\frac{n}{2}\equiv 3 \pmod 4.$$

eplacing U_2 by $U_2(p)$ in (4), we must set k = s − 2 instead of k = s, and we can thus obtain ith the help of our stock of lattices the required lattice if one of the following conruences is satisfied:

$$\frac{p-1}{2}(s-2)+\frac{n}{2}\equiv 1,\ \ 3 \pmod 4,$$

.e., the congruence

$$\frac{p-1}{1}(s-2)+\frac{n}{2}\equiv 1 \pmod 2.$$

f s ≡ 0 (mod 2), then it is equivalent to the condition n ≡ 2 (mod 4). If s ≡ 1 (mod 2), hen the congruence takes the form

$$\frac{p-1}{2}+\frac{n}{2}\equiv 1 \pmod 2,$$

nd this is easily seen to be equivalent to $p\equiv(-1)^{\frac{n}{2}-1} \pmod 4$. An exception here is the case = 1 when the replacement of s by s − 2 is not admissible. In this case it is necessary to onstruct a lattice T with s = 3 by the method indicated above. The three-dimensional lattice T^*/T over F_p must represent zero. If $T^* \supset S \supset T$ and S is such that S/T gives a one-dimensional, sotropic subspace in T^*/T, then it is easy to see that S will satisfy the required conditions ith s = 1.

p = 2. Type I: n ≡ s ≡ 0 (mod 2), n = 4m + 2. Here we use the following construction: f E is a 2-elementary lattice, then we set $\tilde{E} = E^*(2)$. This is always an integral lattice. f E is even and of type I, then E* is also. It is obvious that $d(\tilde{E}) = 2^{\mathrm{rk}E}/d(E)$. For s ⩽ m

$$S=U_2\oplus\Big(\bigoplus_1^{s/2} Q(D_{4n_i})\Big) \quad \text{for} \quad \sum_1^{s/2} n_i=m$$

ive the required lattices. For 4m > s ⩾ 2m the required lattices are given by

$$S=U_2\oplus\Big(\bigoplus_1^{2m-\frac{s}{2}} \widetilde{Q(D_{4n_i})}\Big) \quad \text{for} \quad \sum_1^{2m-\frac{s}{2}} n_i=m,$$

nd for s = 4m by

$$S=U_2(2)\oplus\widetilde{Q(D_{4m})}.$$

Type II: n = s + 2m. For s > 2 the required lattice is given by

$$S=U_2\oplus I(2)^{s-2}+Q(D_{2m}).$$

or s = 2 this formula gives a lattice of type II only for m ≡ 1 (mod 2). The necessary conition formulated in the theorem is given by m ≢ 2 (mod 4), i.e., m ≡ 0 (mod 2) already imlies that m ≡ 0 (mod 4). If m = 4k, then the required lattice is given by

$$S=I(2)\oplus I(-2)\oplus Q(E_8)^k.$$

For s = 1, n = 8k + 1 the required lattice is

$$S=I(-2)\oplus Q(E_8)^k,$$

while for n = 8k + 3 it is

$$S=U_2\oplus I(-2)+Q(E_8)^k.$$

For s = 0, n = 8k + 2 the required lattice is

$$S=U_2\oplus I(-2)+Q(E_8)^k.$$

The proof of the theorem is complete.

Remarks. 1. Of course, the explicit decompositions presented are not the only possible ones. For example, for m > 4 we can everywhere replace $Q(D_{2m})$ by $Q(D_{2m-8})\oplus Q(E_8)$. We shall make use of this below.

2. Using a more precise criterion for the coincidence of the spinor genus and the genus, V. V. Nikulin showed that it is possible to eliminate the condition n > 4 for p = 2.

In conclusion we present a criterion that a lattice be p-elementary.

Proposition. Suppose that a lattice T is related to a unimodular lattice S by a pair of homomorphisms

$$T\underset{\psi}{\overset{\varphi}{\rightleftarrows}}S,$$

satisfying the relations

$$(\varphi(x), \varphi(y))=p(x, y), \ x, y\in T, \tag{5}$$

$$(\varphi(x), u)=(x, \psi(u)), \ x\in T, \ u\in S, \tag{6}$$

$$\psi\varphi=p1_T. \tag{7}$$

Then T is p-elementary.

Suppose first that $U\subset S$ is any sublattice which is complete in S, i.e., $(U\otimes\mathbf{Q})\cap S=U$, and let θ be orthogonal projection onto the subspace $U\otimes\mathbf{R}$. We shall show that θ(S) = U*, where U* is the dual lattice to U in the subspace $U\otimes\mathbf{Q}$.

Indeed, for $x\in S$, $u\in U$, $(\theta x, u)=(x, u)\in\mathbf{Z}$, and hence $\theta(S)\subset U^*$. Equality is proved by computing the indices. Obviously, $U\subset\theta(S)\subset U^*$ and (U*:U) = |d(U)|. On the other hand,

$$\theta(S)/U\simeq S/(U\otimes U^\perp).$$

It is easy to see (due to the fact that S is unimodular) that $d(U^\perp)$ = d(U), and hence $d(U\oplus U^\perp)=d(U)^2$; therefore, (θ(S):U) = |d(U)|.

We note now that in view of (5) and (7) θ = (1/p) φψ is the orthogonal projection of S onto $\varphi(T)\otimes\mathbf{R}$. Applying the remark just made, we find that $\theta(S)=(\overline{\varphi(T)})^*$, where $\overline{\varphi(T)}=(\varphi(T)\otimes\mathbf{Q})\cap S$. On the other hand, $p\theta=\varphi\psi$, and hence $p\theta(S)\subset\varphi(T)$. As a result, we see that

$$p(\overline{\varphi(T)})^*\subset\varphi(T). \tag{8}$$

We consider the sequence of sublattices

$$\varphi(T)\subset\overline{\varphi(T)}\subset(\overline{\varphi(T)})^*\subset\varphi(T)^*.$$

From (8) it follows that

$$p\varphi(T)^*\subset\overline{\varphi(T)}. \tag{9}$$

Together (8) and (9) show that

$$p^2\varphi(T)^*\subset\varphi(T). \tag{10}$$

From the definitions it follows immediately that $\varphi(T^*)\subset p\varphi(T)^*$. Therefore, $p^2\varphi(T)^*\subset p\varphi(T^*)$, and (10) implies that $\varphi(T)\supset p\varphi(T^*)$, whence it follows that $T\supset pT^*$.

2. Curves of Genus 0 and 1 on Surfaces of Type K3

The basic numerical relations of the theory of algebraic surfaces assume a simpler form for a surface X of type K3. For an irreducible curve $C \subset X$ Noether's formula takes the form

$$C^2 = 2p_a - 2, \tag{1}$$

where p_a is the arithmetic genus of the curve C, $p_a = g + \delta$, g is the genus of the normalization $C^\vee$ of the curve C, and δ is defined by the singularities of this curve. More precisely, if $\nu: C^\vee \to C$ is the normalization morphism, then

$$\delta = \sum_{x \in C} l(\tilde{\mathcal{O}}_x / \mathcal{O}_x), \quad \tilde{\mathcal{O}}_x = \bigcap_{\substack{y \in C^\vee \\ \nu(y) = x}} \mathcal{O}_y, \tag{2}$$

where the sum extends over all points x of the curve C.

From formula (1) it follows that $D^2 \equiv 0 \pmod 2$ for any divisor D on X, i.e., that the quadratic form defined by the intersection index on the group $S_X = \operatorname{Pic} X$ is even.

The Riemann–Roch theorem for the case of a curve $C \subset X$ becomes the relation

$$l(C) = \frac{C^2}{2} + 2. \tag{3}$$

This follows from the fact that $H^1(X, \mathcal{O}) = 0$, $H_2(X, \mathcal{O}) = 1$ by definition, $l(K - C) = l(-C) = 0$, and $H^1(X, \mathcal{L}(C)) = 0$; the last relation follows from consideration of the exact sequence

$$0 \to \mathcal{L}(-C) \to \mathcal{O} \to \mathcal{O}_C \to 0,$$

which gives $H^1(X, \mathcal{L}(-C)) = 0$, and from the duality theorem.

In view of (1) curves C of arithmetic genus 0 are characterized by the condition $C^2 = -2$. For them obviously $q = \delta = 0$, i.e., they are isomorphic to $\mathbf{P}^1$. From (3) we obtain $l(C) = 1$ for them, i.e., the linear system $|C|$ consists of a single curve.

The theory of curves of arithmetic genus 1 is much more substantial. If C is such a curve, then, according to formula (1), $C^2 = 0$. From formula (3) we see that $l(C) = 2$, i.e., the system $|C|$ is one-dimensional, and since $C^2 = 0$, it has no basis points, i.e., they define a morphism $\varphi: X \to \mathbf{P}^1$, with fibers which are divisors of this system. We shall henceforth call φ a pencil of curves of genus 1.

The formula $p_a = g + \delta$ admits two possibilities: a) $g = 1$, $\delta = 0$ and b) $g = 0$, $\delta = 1$. In case a) the curve C is a smooth elliptic curve, and hence all but a finite number of curves of the system $|C|$ are such curves. In this case φ is called an elliptic pencil. In case b) we apply formula (2). It is obvious that $\tilde{\mathcal{O}}_x \supset k + R_x \supset \mathcal{O}_x$, where R is the radical of the semilocal ring $\tilde{\mathcal{O}}_x$, i.e.,

$$R_x = \bigcap m_y, \quad y \in C^\vee, \quad \nu(y) = x$$

and m_y is a maximal ideal of the local ring of the point y on $C^\vee$. There are thus two possibilities: b_1) $\mathcal{O}_x = k + R_x$ and b_2) $\tilde{\mathcal{O}}_x = k + R_x$. We consider them following the work [9].

We shall treat the case b_1). Since $l(\tilde{\mathcal{O}}_x / k + R_x) = n - 1$, where n is the number of points in the set $\nu^{-1}(x)$, it follows that in our case $n = 2$, i.e., $\nu^{-1}(x) = \{y_1, y_2\}$ and $m_x = m_{y_1} \cap m_{y_2}$. It is obvious that $(m_x/m_x^2) = 2$ and the homomorphisms $m_x/m_x^2 \to m_y/m_{y_i}^2$ define two nonproportional linear forms. This implies that in the tangent space T_x at the point x on X (or, equivalently, on C) the images of the tangents at the points y_i are nonzero and distinct. In other words, the local index of intersection of the curve C and any other smooth curve E at the point x is equal to 2 if E is not tangent to one of these two lines and is greater than 2 if the tangent to E coincides with one of them. Obviously, this is possible only if in some local coordinates u, v the equation of the curve C has the form $uv + f$, $f \in M_x^3$, where M_x is the maximal ideal of the point x on the surface X.

A special feature of the case b_1) is that it is not stable, i.e., it can occur only for a finite number of curves of the pencil $|C|$. Indeed, suppose that it occurs for some curve C and that t is the local parameter of points $\varphi(C) \in \mathbf{P}_1$. According to the foregoing, in some neighborhood of a singular point x of the curve C and in some local coordinates $t = uv + f$.

The singular points of neighboring fibers are given by the conditions $\partial t/\partial v = 0$, $\partial t/\partial v = 0$. Since $\partial t/\partial u = v + g$, $\partial t/\partial v = u + h$, $g, h \in M_x^2$, these functions are relatively prime in a neighborhood of the point x, and hence in a sufficiently small neighborhood there are no other singular points of curves of the pencil $|C|$.

Of course, in the case of a field of characteristic 0 the property we have obtained follows from Bertini's theorem according to which a linear system without fixed points on a smooth surface has only a finite number of singular curves, and in the context of differential geometry it agrees the Sard's lemma which is analogous to it. It is thus the more interesting that in the case of b_2) and fields of characteristic 2 and 3 we arrive at a different result. We henceforth follow the exposition in the work [16].

In the case b_2) $\tilde{\mathcal{O}}_x = k + R_x$, and hence $n = 1$, since $\tilde{\mathcal{O}}_x$ is the local ring of the nonsingular point $y \in C^\nu$, $\tilde{\mathcal{O}}_x = \mathcal{O}_{y,C^\nu}$. By hypothesis $l(\mathcal{O}_{y,C^\nu}/\mathcal{O}_{x,C}) = 1$, i.e., $l(m_y/m_x) = 1$, where m_y is the maximal ideal of the ring $\mathcal{O}_{y,C^\nu}$. Since the point y is simple, it follows that $m_y = (\tau)$. It is obvious that $m_x \subseteq m_y^2$, for otherwise $m_x = m_y$. Now from the condition $l(m_y/m_x) = 1$ it follows that $m_x = m_y^2$, and hence $\mathcal{O}_{x,C} = k + (\tau^2)$. The epimorphism $\rho: \mathcal{O}_{x,X} \to \mathcal{O}_{x,C}$ establishes the isomorphism $m_{x,X}/m_{x,X}^2 \simeq m_{x,C}/m_{x,C}^2$. Therefore, there exist functions $u, v \in m_{x,X}$ which are local parameters at the point x such that $\rho(u) = \tau^3$, $\rho(v) = \tau^2$. Hence $u^2 = v^2$ on C, i.e., the local equation of C in some neighborhood of the point x can be written in the form $u^2 - v^3 = 0$, and the mapping φ can be written in the form

$$t = \varepsilon(u^2 - v^3), \quad \varepsilon \in \mathcal{O}_{x,X}^*.$$

We consider an arbitrary algebraic surface X and on it a pencil $\varphi: X \to B$ of curves of arithmetic genus 1. We suppose that the set of singular points of this pencil is infinite. Then obviously each fiber must have a singular point and, moreover, only one. All these singular points form a curve ξ. In a neighborhood of a point x on ξ $\frac{\partial t}{\partial u} = \frac{\partial t}{\partial v} = 0$. If $p \neq 2$, then $\frac{\partial t}{\partial u} = \frac{\partial \varepsilon}{\partial u}(u^2 - v^3) + 2\varepsilon u \notin m_{x,X}^2$. Therefore, $dt/du = 0$ must already be the local equation of the curve ξ. In particular, the curve ξ is smooth. If, moreover, $p \neq 3$, then $\frac{\partial t}{\partial v} = \frac{\partial \varepsilon}{\partial v}(u^2 - v^3) - 3\varepsilon v^2$ must be divisible by $\partial t/\partial u$ which is impossible, since even the quadratic terms of $\partial t/\partial v$ are not divisible by the linear terms of $\partial t/\partial u$. Hence, for $p \neq 2$ and 3 the set of singular curves of our pencil is again finite. Let $p = 3$. As we have seen, the curve ξ in this case is smooth. We shall find the index of intersection of this curve with a fiber C of the mapping φ. Since $C \cap \xi = \{x\}$, it follows that

$$(C, \xi) = (C, \xi)_x = l\left(\mathcal{O}_{x,X} \Big/ \left(t, \frac{\partial t}{\partial u}\right)\right) = l(\mathcal{O}_x/(u^2 - v^3, u)) = 3.$$

On the other hand, since ξ intersects each fiber in one point, the mapping φ restricted to ξ is a purely nonseparable morphism of degree 3. This implies the curve ξ is isomorphism to B.

It remains to consider the case $p = 2$. In this case

$$\frac{\partial t}{\partial v} = \frac{\partial \varepsilon}{\partial v}(u^2 - v^3) + \varepsilon v^2 \equiv (\alpha u + \beta v)^2 \, (m_x^3), \quad \beta \neq 0.$$

Thus, we have a curve ξ and a function $h = \partial t/\partial v$ vanishing at 0 on ξ, while at any point y of some open set on ξ $h \equiv \lambda^2 (m_y^3)$, $\lambda \in m_y$ $\lambda \notin m_y^2$. It is easy to see that this is possible only in the case where in a neighborhood of the point $x(h) = 2\xi$, i.e., $h = \eta\psi^2$, where $\eta \in \mathcal{O}_x^*$, and ψ is the local equation of the curve ξ. In particular, from this it follows that the curve ξ is smooth. To find its index of intersection with a fiber of the pencil φ, we note that

$$(\xi, C) = (\xi, C)_x = \frac{1}{2}(h, C)_x = \frac{1}{2}(u^2 - v^3, v^2) = \frac{4}{2} = 2.$$

As in the case $p = 3$, we see that the curve ξ is isomorphic to B, and the mapping $\varphi: \xi \to B$ is a purely nonseparable morphism of degree 2.

Pencils of curves of arithmetic genus 1 in which all fibers have a singular point are called quasielliptic. The arguments presented prove two assertions.

Proposition 1. An irreducible curve C on a surface of type K3 has arithmetic genus 1 if and only if $C^2 = 0$. It determines an elliptic or quasielliptic pencil.

THEOREM. There can exist a quasielliptic pencil on an algebraic surface only if the characteristic of the base field is equal to 2 or 3. All curves of the pencil except a finite number have a unique singular point which is a cusp. These singular points are arranged on a smooth curve the projection of which onto the base of the pencil is a purely nonseparable morphism of degree equal to the characteristic.

The question of the existence of quasielliptic pencils over a field of characteristic 2 or 3 is solved by explicitly constructing them. We set $K = k(\mathbf{P}^1)$; then the common fiber of the pencil $\varphi: X \to B$ is a curve C over the field K. Its genus (if we use, for example, the definition of the genus contained in the book of Chevalley [18]) is equal to 1. In the case when the pencil is elliptic this curve is smooth and elliptic, while in the case where it is quasielliptic it is not smooth although it is normal. It is then called quasielliptic. Its unique singular point is determined by a general point s of the curve ξ, and $[K(s):K] = p$. If a curve of genus 1 over an arbitrary field has over this field a rational point, then the Riemann–Roch theorem gives for it the existence of the "Weierstrass normal form"

$$Y^2+\lambda XY+\mu Y=X^3+\alpha X^2+\beta X+\gamma. \tag{4}$$

For $p \neq 2$ we may assume that $\lambda = \mu = 0$. A singular point then corresponds to a multiple root of the right side. For $p \neq 3$ this root always belongs to K, and the curve is not normal. This again demonstrates that quasielliptic pencils exist only for p = 2 and 3. For p = 3 there is the possible case

$$Y^2=X^3+\gamma, \tag{5}$$

which if $\gamma \notin K^3$ gives a normal curve having the singular point $X = -\gamma^{1/3}$, $Y = 0$.

If p = 2 for the singular point $\lambda X + \mu = 0$, $\lambda Y = X^2 + \beta$, and hence this point is rational except in the case $\lambda = 0$, $\mu = 0$. In the latter case by the usual transformation we can reduce the equation to the form

$$Y^2=X^3+aX+b,$$

and if $a \notin K$ or $b \notin K^2$, we obtain a normal curve with singular point $X = a^{1/2}$, $Y = b^{1/2}$.

If C is a quasielliptic curve over a field K which possesses a rational point $\mathcal{O}$, then on the open set $C \setminus s$ (s is the singular point of the curve C) it is possible to define a group law which converts $C \setminus s$ into an algebraic group with $\mathcal{O}$ the zero point. The proof is the same as in the case of elliptic curves. Over the extension K(s) the curve $C \setminus s$ is isomorphic to $\mathbf{P}^1 \setminus s \simeq \mathbf{A}^1$; since the unique algebraic group law on $\mathbf{A}^1$ is the law of the additive group G_a, it follows that $C \setminus s$ is a form of the additive group. All forms of the additive group are classified in the work of Russel [35]. As he showed, they are subgroups of the group $G_a \times G_a$ and are given there by equations

$$y^{p^n}=x+a_1x^p+\ldots+a_mx^{p^m}. \tag{6}$$

It is easy to show how equations of quasielliptic curves can be reduced to this form. For p = 3, setting $x = 1/Y$, $y = X/Y$ in Eq. (5), we reduce it to the form

$$y^3=x-\gamma x^3.$$

For p = 2 it is necessary to write Eq. (4) in the form $Y^2 = (X^2 + a)X + b$, square it, divide both sides by $(X^2 + a)^4$, and set $1/(X^2 + a) = x$, $Y = (X^2 + a) = y$. It then takes the form

$$y^4=x+ax^2+b^2x^4.$$

In both cases the meaning of these transformations is that the singular point s is taken to infinity. Of course, the fact that the law of addition hereby goes over into the additive law for x and y requires additional (but not difficult) verification.

As in the case of elliptic curves, a quasielliptic curve not having a rational point over K is a smooth homogeneous space over some quasielliptic curve possessing a rational point. Here, however, the theory of principal homogeneous spaces can be given a more explicit form:

as Russel showed, all principal homogeneous spaces over the group (6) have the form

$$v^{p^n}=a+u+a_1u^p+\ldots+a_mu^{p^m},$$

and the action of a point (x, y) on the point (u, v) is defined to be (x + u, y + v). The group of the principal homogeneous spaces is isomorphic to $G_a(K)/M$, where M is the subgroup consisting of elements of the form $v^{p^n} - u - a_1u^p - \ldots - a_mu^{p^m}$, u, v ∈ K.

Proposition 2. If on a surface X there exists a quasielliptic pencil $\varphi: X\to \mathbf{P}^1$, then the surface is unirational.

Suppose that ξ is the curve formed by cusps of the fibers. Since the morphism $\varphi:\xi\to\mathbf{P}^1$ is purely nonseparable, ξ is a rational curve. We set $X' = X \times_{\mathbf{P}^1} \xi$. On X' we have a pencil the general fiber of which is a rational curve. Therefore, X' is rational, and X is unirational.

COROLLARY. On a surface X possessing a quasielliptic pencil $\varphi: X\to\mathbf{P}^1$, all two-dimensional cycles are algebraic, i.e., $b_2(X) = \rho(X)$, where $\rho(X)$ = rk S(X) is the Picard number.

This follows from Proposition 2 and the obvious observation that on a unirational surface all two-dimensional cycles are algebraic. Indeed, if $\pi:\overline{X}\to X$ is a morphism of a rational surface and $\alpha\in H^2(X)$, then $(\deg \pi)\alpha = \pi_*\pi^*\alpha$. On $\overline{X}$ all cycles are algebraic including $\pi^*\alpha$ which gives the assertion.

Finally, we return to the case of a pencil of curves of genus 1 $\varphi: X\to\mathbf{P}^1$ on a surface of type K3. This pencil contains no multiple fibers. Indeed, if C_0 is one of its fibers and $C_0 = mE$, then $E^2 = 0$, and hence $l(E) \geqslant 2$. This would imply that the set of divisors mE', E' > 0, E' ∼ E is a least one-dimensional, i.e., all the fibers of the pencil φ must be multiple, while the general fiber is irreducible. Therefore, if φ also has no section, then it has one locally at each point (in the étalé topology). If $\psi: Y\to\mathbf{P}^1$ is a bundle with a section (a Jacobian bundle) over which φ is a principal homogeneous bundle, then φ and ψ are locally isomorphic (in the étalé topology). The canonical class of a surface with a pencil of curves of genus one is determined by the local (in the étalé topology) properties of the degenerate fibers (see [15, p. 354]). Therefore, X is a surface of type K3 only if Y is such a surface, and X is a locally trivial, principal homogeneous space.

3. Roots and Elliptic Pencils

We shall recall some properties of root systems which we need. We shall consider three types of root systems which we call hyperbolic, parabolic, and elliptic.

1. Let S be an integral, hyperbolic lattice. A root of the lattice S is an element $r\in S$, such that $r^2 = -2$. The collection R(S) of all roots is called a hyperbolic root system.

We consider the cone K in the space $S\otimes\mathbf{R}$, consisting of elements x for which $x^2 > 0$. This cone decomposes into two connected components called skirts. If the equation of the cone in coordinates has the form $x_1^2 - x_2^2 - \ldots - x_n^2 > 0$, then one skirt is defined by the additional condition $x_1 > 0$ and the other by the condition $x_1 < 0$. From this it is easy to see that for two points x and y in the same skirt (x, y) > 0. We choose one of the two skirts and denote it by K_+. The set of points $x\in K_+$, for which (x, r) ≠ 0 for all $r\in R(S)$, decomposes into connected components called cells. Each cell is an open, convex set. Since a cell is the intersection of K_+ and the subspaces containing it are defined by the hyperplanes (x, r) = 0, $r\in R(S)$, each cell is determined by the choice for each pair of roots (r, −r) of one of them and is given by the conditions

$$x\in K_+,\ (x, r)>0$$

for the chosen roots r. These roots determined by the choice of cell U are called positive. The set of positive roots is denoted by $R_+(S)$. Obviously, $R(S) = R_+(S) \cup (-R_+(S))$. For positivity of a root r the inequality (h, r) > 0 already suffices, where h is any element of the cell U. If L is one of the hyperplanes with equation (x, r) = 0, $r\in\mathbf{R}(S)$, U is contained in the half space (x, r) > 0, and L ∩ U is open in L, then L is called a wall of the cell U. Roots corresponding to walls of the cell U are called simple.

The reflection corresponding to a root r is the transformation

$$s_r: x\to x+(x, r)r.$$

It takes r into $-r$ and leaves the elements of the hyperplane $(x, r) = 0$ unchanged. It is obvious that s_r is a lattice automorphism (in particular, it does not change the product).

Proposition 1. Each point $x \in \bar{K}_+ \cap S$ by means of a finite number of reflections corresponding to simple roots can be taken into a point of $\bar{U}$.

If $x \notin \bar{U}$, then $(x, r) < 0$ for some simple root r. Then for some $h \in U$ and $x_1 = s_r(x)$, $(x_1, h) < (x, h)$. Since x, x_1 and $h \in \bar{K}_+$, it follows that $(x_1, h) > 0$. In view of the fact that $(x_1, h) \in \mathbf{Z}$, we must in a finite number of such transformations arrive at a point of $\bar{U}$.

Remark. It can be proved that $\bar{U}$ is a fundamental domain in $S \otimes \mathbf{R}$ for the discrete group generated by the reflections s_r, where r are simple roots of the lattice S.

2. Let $f \in S$, $f \neq 0$, $f^2 = 0$. The roots of the lattice S orthogonal to f form a parabolic root system.

3. Let $f^\perp$ be the orthogonal complement to f. It is obvious that the product (x, y), $x, y \in f^\perp$, remains unchanged under the substitutions $x \to x + kf$, $y \to y + lf$ and hence defines a product in the lattice $S_f = f^\perp/f\mathbf{Z}$. The lattice S_f is negative definite. Its elements with square -2 are called roots. They form an elliptic root system which we denote by R_f. This system decomposes into the direct sum of pairwise orthogonal and further indecomposable root systems

$$R_f = \oplus R_i,$$

which are called irreducible systems. The lattice generated by the system of vectors R_i is called the lattice of roots and is denoted by $Q(R_i)$, while the lattice dual to it is called the lattice of weights and is denoted by $P(R_i)$. There is the following rule for calculating the lattice $P(R_i)/Q(R_i)$ for an irreducible root system R_i. In the lattice $Q(R_i)$ it is possible to find a basis $r_1, \ldots, r_m$ consisting of roots such that any root can be expressed in terms of it as a linear combination with integer coefficients of the same sign. There exists a maximal root $\rho = \sum^m k_j r_j$, possessing the property that for any root $r = \sum_1 l_j r_j$, $k_1 \geq l_1, \ldots, k_m \geq l_m$. It is obvious that the lattice $Q(R_i)$ has the basis $r_1, \ldots, r_m$, while the lattice $P(R_i)$ has the corresponding basis $\omega_1, \ldots, \omega_m$. The elements ω_j for j such that in the representation $\rho = \sum k_j r_j$, $k_j = 1$, are then representative of all nonzero cosets of $P(R_i)$ with respect to $Q(R_i)$ (see [7, p. 177]).

We shall now present the application of these concepts to the theory of surfaces of type K3. Let X be such a surface. We set $S = \operatorname{Pic} X$. In the cone K consisting of elements D with $D^2 > 0$ we choose for K_+ that skirt containing ample divisors. Then all divisors of $K_+ \cap S$ are effective, since for $D^2 > 0$ either D or $-D$ is effective and from the fact that $D \in K_+$, it follows that $DH > 0$, where H is ample, and hence D rather than $-D$ is effective. If C is a divisor with $C^2 = -2$, then either C or $-C$ is effective. Choosing from the pair $(C, -C)$ the effective divisor, we obtain a cell U consisting of divisors D for which $D^2 > 0$ and $DC > 0$ for any $C > 0$, $C^2 = -2$. Then $DD_1 > 0$ for any $D_1 > 0$. Indeed, if $D_1 = \sum_1^N m_i C_i$ is the decomposition into irreducible components where for $i \leq n$, $C_i^2 = -2$, while for $i > n$, $C_i^2 \geq 0$, then $DC_i > 0$ for $i \leq n$ by hypothesis and $DC_i \geq 0$ for $i > n$, since C_i has nonnegative index of intersection with any component of the divisor D. Thus, we always have $DD_1 \geq 0$. Moreover, for the component C_i for which $C_i^2 > 0$, $DC_i > 0$, since D and C_i lie in the same skirt K_+. This implies that the equality $DD_1 = 0$ is possible only if all $C_i^2 = 0$ and $DC_i = 0$. Now the orthogonal complement to a vector D with $D^2 > 0$ is a negative definite lattice, and therefore $C_i^2 = 0$, $DC_i = 0$ imply that $C_i = 0$, and hence also $D_1 = 0$. Thus, for $D_1 > 0$, $DD_1 > 0$ as well, i.e., the divisor D contained in the cell U is ample. Since in deriving this property we have used only the condition $DC > 0$ for irreducible curves with $C^2 = -2$, the simple roots corresponding to the cell U are contained among such curves. Conversely, any irreducible curve C with $C^2 = -2$ is a simple root. For the proof it is necessary to demonstrate the existence of a divisor $E > 0$ for which $E^2 > 0$, $EC = 0$ and $EC' > 0$ for any other irreducible curve C' with $(C')^2 = -2$. We use the fact that the curve C can be contracted to a singular point by means of a morphism $\varphi: X \to X'$ onto a projective surface X'. If H is a hyperplane section of X', then $E = \varphi^* H$ satisfies the required conditions. Thus, simple roots coincide precisely with smooth, rational curves on X.

Classes $D\in\bar{U}$ possess the following property: $D^2 \geqslant 0$, $DC \geqslant 0$ for any curve C with $C^2 = -2$ and hence also for any effective divisor C. Such classes are called quasiample. Among them classes with $D^2 = 0$ will be especially interesting to us in view of the following property of them.

Proposition 2. A class of divisors D on a surface of type K3 contains an irreducible curve of arithmetic genus 1 if and only if it is quasiample, primitive, and $D^2 = 0$. Here an element x of a lattice S is called primitive if it cannot be represented in the form $x = my$, $m\in\mathbf{Z}$, $y\in S$, $m>1$.

If a class D contains an irreducible curve C of arithmetic genus 1, then $C^2 = 0$, and hence $D^2 = 0$; the pencil $|C|$ does not contain multiple fibers, and hence D is primitive. Finally, it suffices to verify the condition $DE \geqslant 0$ for an irreducible curve E, and for such a curve it is obvious, since the pencil $|C|$ contains a curve C_1 not intersecting E. If E is not a component of C_1, then $EC_1 > 0$, while if it is a component, then for another curve C_2 of the same pencil $EC_2 = 0$.

Thus, only sufficiency of the condition is in need of proof. From the Riemann–Roch theorem it follows that either $l(D) \geqslant 2$ or $l(-D) \geqslant 2$. Let H be some polarization of the surface; then by hypothesis $DH \geqslant 0$, whence it follows that the class $-D$ cannot contain an effective divisor, for this would imply that $DH < 0$. Thus, $l(D) \geqslant 2$, and D contains effective divisors. If $D = \Sigma k_iC_i$ is the decomposition of one of them into irreducible components, then $D^2 = 0$ and $DC_i \geqslant 0$ give $DC_i = 0$, while since $C_iC_j \geqslant 0$ for $i \neq j$ it follows that $C_i^2 \leqslant 0$. As we have seen, this is possible only for $C_i^2 = 0$ or $C_i^2 = -2$. We shall prove that at least one effective divisor D of our class contains at least one irreducible component with $C_i^2 = 0$. Indeed, for all curves C_i contained in decompositions of the divisors D of our class $k_iC_iH \leqslant DH$, and hence both k_i and C_iH are bounded. There are only a finite number of classes C with given square and given CH. This follows from the fact that the lattice $S = \operatorname{Pic} X$ is hyperbolic. Therefore, in $S\otimes\mathbf{R}$ we can represent $C = \alpha H + Y$, where Y lies in the space $H^{\perp}$ with a negative definite form. The condition $\alpha H^2 = CH$ gives a restriction on α, and $\alpha^2H^2 + Y^2 = C^2$ gives a restriction on Y^2. Thus, C lies in a compact domain. Now each class with $C_i^2 = -2$ contains only one effective divisor, and we have found that the class D contains only a finite number of effective divisors which contradicts the condition $l(D) \geqslant 2$.

Let C be a component with $C^2 = 0$, and let $D = mC + D_0$ be a representation with $m > 0$, $D_0 > 0$. Replacing in the condition $DD = 0$ one of the factors D by its decomposition, we find that $CD + D_0D = 0$. Since by the quasiample condition $CD \geqslant 0$, $D_0D \geqslant 0$, it follows that $CD = 0$, $D_0D = 0$. Substituting here again the representation for D, we find that $CD_0 = 0$, $D_0^2 = 0$.

If D_0 and C were linearly independent, then they would generate a two-dimensional subspace in the cone with equation $x^2 = 0$ in $S\otimes\mathbf{R}$. Such a subspace cannot exist, since on the sphere over which the cone is constructed there are no lines. Therefore, C and D_0 must be proportional, and since the class C is primitive, it follows that $D_0 = m_0C$, $m_0\in\mathbf{Z}$. Hence $D = (m + m_0)C$ which is possible in view of the primitivity of D only for $m + m_0 = 1$. The proof of the proposition is complete.

Proposition 1 can now be interpreted in the following manner.

Proposition 3. Any effective class D with $D^2 > 0$ can by a finite number of reflections corresponding to smooth rational curves be transformed into an ample class. Any effective primitive class f with $f^2 = 0$ can analogously be transformed into a class defining a sheaf of curves of arithmetic genus 1.

We have, in particular, the following criterion.

THEOREM. On a surface X of type K3 there exists a pencil of curves of arithmetic genus 1 if and only if there exists a class $D\in\operatorname{Pic} X$, $D\neq 0$, with $D^2 = 0$.

If f is a class corresponding to a pencil of curves of arithmetic genus 1, then the roots orthogonal to f are linear combinations of components of the pencil $|f|$: indeed, if $C^2 = -2$ and $C > 0$, then $Cf = 0$ is possible only when C is a component of f. In the expansion of the root system R_f into irreducible systems, $R_f = \otimes R$, the individual terms correspond to reducible fibers F_i of the pencil f. Indeed, the components of different fibers do not intersect and hence are orthogonal in S; the components of the same fiber cannot be partitioned into two groups with representatives which do not intersect due to the connectedness of the fiber.

Finally, we clarify how the lattice $P(R_F)$ and the group $P(R_F)/Q(R_F)$ are interpreted. Let $\varphi: X \to \mathbf{P}^1$ be our pencil. We denote by $\mathcal{O}$ the local ring of the point over which there is the fiber F corresponding to our irreducible root system and by $\hat{\mathcal{O}}$ its completion. We set $\hat{X} = X \otimes_{\mathbf{P}^1} \hat{\mathcal{O}}$. Since $X \to \mathbf{P}^1$ is a locally trivial bundle, it follows that $\hat{X}$ has a section, and therefore in the decomposition

$$F = \sum_0^m k_i C_i \qquad (1)$$

of the closed fiber into irreducible components at least one multiplicity is equal to 1 — the one for which the component C_i intersects the section. Suppose this is k_0. Then the representatives $C_1, \ldots, C_m$ in $\overline{F}/ZF$ (which we denote by the same letters) form a basis of the root system R_F. Indeed, if some root, i.e., a divisor with $D^2 = -2$, can be represented as a linear combination of curves $C_1, \ldots, C_m$ with coefficients of different signs, then $D = D_1 - D_2$, $D_1 > 0$, $D_2 > 0$, and D_1 and D_2 are relatively prime. Hence

$$-2 = D^2 = D_1^2 - 2D_1D_2 + D_2^2,$$

which is impossible, since $D_1^2 \leqslant -2$, $D_2^2 \leqslant -2$, $D_1D_2 \geqslant 0$. We shall prove that $\rho = \sum_1^m k_i C_i$ is a maximal root. If this is not so, then the maximal root $r = \sum_1^m l_i C_i$ can be represented in the form $r = \rho + E$, $E = \sum_1^m h_i C_i > 0$, $E \neq 0$. Now $\rho \sim -C_0$, since $D \sim 0$ on X. Therefore, we obtain

$$-2 = r^2 = (-C_0 + E)^2 = (-C_0)^2 - 2C_0E + E^2 = -2 - 2C_0E + E^2 = 0, \quad -2C_0E + E^2 = 0,$$

while $-C_0E \leqslant 0$, $E^2 < 0$.

By the criterion presented the nonzero elements of the group $P(R_F)/Q(R_F)$ are determined by those elements ω_i of the basis dual to the basis C_i, $i \geqslant 1$ for which $k_i = 1$ in the expansion (1).

We consider the group Pic $(\hat{X})$ and its mapping into $P(R_F) = Q(R_F)^*$ determined by the index of intersection with the components C_i, $i = 1, \ldots, m$. The image $\Pi \subset P(R_F)$ of the "Picard group up to numerical equivalence." It is obvious that $Q(R_F) \subset \Pi$. On the other hand, the nonzero elements of $P(R_F)/Q(R_F)$ have as representatives ω_i for which $k_i = 1$. Now precisely these elements are given by sections intersecting the components C_i for which $K_i = 1$. Therefore, $\Pi = P(R_F)$ and the classes of $P(R_F)/Q(R_F) = \Pi/Q(R_F)$ can be given by sections intersecting the components of the fiber with multiplicity 1. Sections s_1 and s_2 are added as divisors in Π. Let $s_1 \oplus s_2$ be their sum as sections, i.e., for $x \in \mathbf{P}^1$, $(s_1 \oplus s_2)(x)$ is the sum in the fiber F_x of the points $s_1(x)$ and $s_2(x)$. Then on a general fiber F_η, $(s_1 \oplus s_2)(\eta) \sim s_1(\eta) + s_2(\eta) - \mathcal{O}(\eta)$, where $\mathcal{O}$ is the null section. From this it follows that on $\hat{X}$, $s_1 \oplus s_2 \sim s_1 + s_2 - \mathcal{O} + D$, where D consists of components of the closed fiber. Therefore, $\Pi/Q(R_F)$ is also isomorphic to the group of sections modulo $Q(R_F)$ added fiberwise. The latter group is obviously isomorphic to the "group of components of multiplicity 1," i.e., to the factor group of the group of nonsingular points of the fiber with respect to its connected component of unity. This gives the interpretation of the group $P(R_F)/Q(R_F)$.

We present the results of our comparison in Table 1. The latter part of this table remains in force for any algebraic surface X with a pencil of curves all of arithmetic genus 1, $\varphi: X \to B$. If $f \in \operatorname{Pic} X$ is a fiber of the pencil then in the group $f^\perp$ we call the elements D with $D^2 = -2$ roots; in exactly the same way they are linear combinations of the components of fibers of the pencil. Roots D with $D > 0$ we call positive roots, while simple roots are irreducible components of the fibers. The last three rows of the tables are now preserved in the general case as well.

The comparison affords the possibility of applying to the theory of pencils of arithmetic genus 1 the classification of root systems (in which all vectors have the same length). Thus, the reducible fibers can be classified according to the known series A_n, D_n, E_6, E_7, and E_8. From the known classification of degenerate fibers it follows that the type of the root system is uniquely determined by the fiber with the exception of types A_2 and A_1, and with the exception of these two cases we say that the fiber has type $A_n, \ldots, E_8$. In the case of type A_2 the fiber may consist of three lines which intersect pairwise and of three lines which intersect at a single point. In the first case we say that the fiber has type A_2 and in the

TABLE 1

Hyperbolic lattices s	Surfaces X of type K3
Hyperbolic lattice S	Pic X
Cell U	Ample classes of divisors
Positive roots	Effective divisors C with $C^2=-2$
Simple roots	Smooth rational curves
Elements $h\in\bar{U}$ with $f^2=0$	Pencils of curves of arithmetic genus 1
Roots of the lattice $f\perp$	Divisors C with $C^2=-2$, consisting of components of the pencil
Simple roots of the lattice $f\perp$	Irreducible components of the fibers
Simple roots of the lattice $f^{\perp}$	Reducible fiber F_i of the pencil f
Irreducible terms R_{F_i} of the root system R_f	
The expression $f-C_0$ in terms of irreducible components of the fiber F (C_0 is a component of the fiber F of multiplicity 1)	The maximal root of the root system R_F, corresponding to the fiber F
The group $P(R_F)/Q(R_F)$	"The group of components of multiplicity 1" of a fiber, i.e., the factor group of the group of nonsingular points of the fiber F with respect to the connected component of unity

second case the type $^*A_2$. In exactly the same way, a fiber consisting of two lines intersecting transversally at two points we assign to type A_1, while a fiber consisting of two intersecting lines we assign to type $^*A_1$. Finally, an irreducible, degenerate fiber we denote by A_0 if it has a single point with separated tangents and by $^*A_0$ if this point is a cusp.

4. A Criterion for Quasiellipticity of a Pencil

Let $\varphi: X\to \mathbf{P}^1$ be a fibration of a smooth surface into curves of arithmetic genus 1. We shall assume that it is minimal (i.e., without exceptional curves of genus 1 in the fibers) and reduced (i.e., without multiple fibers).

We recall the basic formulas concerning such pencils. The Riemann–Roch theorem gives

$$e(X)=12\chi(\mathcal{O}_X), \tag{1}$$

where e(X) is the Euler characteristic in the étalé cohomology. On the other hand,

$$e(X)=\sum_F (e(F)-e(F_0)+d(F))+e(\mathbf{P}^1)e(F_0), \tag{2}$$

where F_0 is the general fiber of the bundle φ, the sum extends over all fibers F of this bundle (but the terms are nonzero only for degenerate fibers), and d(F) is the invariant of wild ramification. By definition,

$$d(F)=\sum_{i=1}^{\infty}\frac{e_i}{e_0}\dim_{\mathbf{F}_l} M/M^{G_i}, \tag{3}$$

where $M=\mathfrak{U}_l$, $\mathfrak{U}$, is the general fiber of the bundle as an elliptic curve over the field $k(\mathbf{P}_1)$, and G_i are the ramification groups corresponding to the point $x\in\mathbf{P}^1$, over which F lies and which are contained in the Galois group of the field $K(\mathfrak{U}_l)/K$, where K is the field of fractions of the completed local ring of the point of the base over which there lies the fiber F. Regarding formula (2), see [3].

Since $\dim_{\mathbf{F}_l} M = 2$, it follows that $\dim_{\mathbf{F}_l} M^{G_i} = 2$ if G_i acts trivially on M. We note that the case $\dim_{\mathbf{F}_l} M^{G_i} = 1$ is impossible. Indeed, then M has a line on which G_i acts trivially, and hence the action of G_i on M reduces to triangular form. As is known, the action of G on $\Lambda^2 M$ is determined by the action on roots of unity of degree l, and since the field K is algebraically closed, it follows from this that the elements of the group G_i defines unimodular transformations of M. Because of this, the action of G_i reduces to the strict triangular form

$$T_g=\begin{pmatrix}1 & \chi(g)\\ 0 & 1\end{pmatrix}, \tag{4}$$

where $\chi \in \mathrm{Hom}(G_i, \mathbf{Z}/l)$. Since G_i for $i \geqslant 1$ is a p-group and $l \neq p$, it follows that $\chi = 0$ and $\mathfrak{r}_g = 1$. Thus, $\dim M/M^{G_i} = 0$ if $G_i = \{1\}$ and is equal to 2 otherwise. Therefore, formula (3) can be rewritten in the form

$$d(F) = \frac{2}{e_0} \sum_{i=1}^{m} e_i, \qquad (5)$$

where the sum extends over all i for which $e_i > 1$. The same argument shows that if a fiber F has multiplicative type, i.e., the connected component of the group of its nonsingular points is isomorphic to G_m, then $d(F) = 0$. Indeed, since $(G_m)_l = \mathbf{Z}/l$, it follows that $\mathfrak{A}_l$ contains a vector invariant relative to G, and the action of G reduces to the form (3), whence $|G| = 1$ or l (see [31, p. 5]).

For fibrations into curves of any genus it is known that $\mathbf{e}(F) \geqslant \mathbf{e}(F_0)$, and equality holds only for nondegenerate fibers [1, p. 55].

For the case of an elliptic fibration $\mathbf{e}(F_0) = 0$, while for a quasielliptic fibration $\mathbf{e}(F_0) = 2$; therefore, it follows from (2) that $\mathbf{e}(X) \geqslant 0$, and, according to (1), $\chi(\mathcal{O}_X) > 0$. The condition $\mathbf{e}(X) = \chi(\mathcal{O}_X) = 0$ shows that the fibration has no degenerate fibers, and this, as is known, is possible only in the case $X = \mathbf{P}^1 \times F$, while the condition $\chi(\mathcal{O}_X) = 1$, $\mathbf{e}(X) = 12$ characterizes rational surfaces X. We henceforth assume that

$$\chi(\mathcal{O}_X) \geqslant 2. \qquad (6)$$

It is known that $\mathcal{R}^1\varphi_* \mathcal{O}_X$ is a one-dimensional pencil corresponding to the divisor D for which $\deg D = -\chi(\mathcal{O}_X)$ [15, p. 354]. From $\chi(\mathcal{O}_X) > 0$ it follows that $H^1(X, \mathcal{O}_X) = 0$, and hence $b_1(X) = 0$. The canonical class K_X is equal to some multiplicity of a fiber of the pencil.

The structure of root systems corresponding to degenerate fibers makes it possible to distinguish elliptic and quasielliptic pencils. As before, we assume that the pencil $\varphi: X \to \mathbf{P}^1$ is minimal, reduced, and $\chi(\mathcal{O}_X) \geqslant 2$.

Below we simplify notation by using for the lattices denoted earlier by $Q(R_F)$ and $D(R_F)$ the symbols $Q(F)$ and $P(F)$.

<u>THEOREM.</u> A pencil $\varphi: X \to \mathbf{P}^1$ of curves of arithmetic genus 1 is quasielliptic if and only if the following conditions are satisfied:

1) $p = 2$ or 3;

2) for any fiber F the lattice $Q(F)$ is p-elementary;

3) $\sum_F n(F) = b_2(X) - 2$, where $n(F)$ is the rank of the root system corresponding to the fiber F [i.e., F consists of $n(F) + 1$ components], and the sum extends over all fibers F of the pencil φ.

It obviously suffices to prove the assertion for a Jacobian fibration of the pencil φ, so that we may henceforth assume that φ has a section.

We have already seen that quasielliptic pencils exist only over fields of characteristic 2 and 3. For a quasielliptic pencil the group of sections of the fibration $X \times_B \hat{\mathcal{O}}$, where $\hat{\mathcal{O}}$ is the completed local ring of a point of the base, can be imbedded in the additive group, and it therefore has period p. As we have seen, for any fiber F the group $P(F)/Q(F)$ is isomorphic to the group of multiplicity 1 components of this fiber, i.e., to some factor group of the group $X \times_B \hat{\mathcal{O}}$. Therefore, it also has period p, i.e., condition 2) is satisfied. Finally, the group $S_f/Q(R_f)$ is isomorphic to the group of sections of the fibration φ. By the same considerations as above, this group has period p. Since it is finitely generated, it is finite, and $\operatorname{rk} S_f = \operatorname{rk} Q(R_f) = \Sigma \operatorname{rk} Q(F) = \Sigma n(F)$. On the other hand, because of the corollary of Proposition 2, Section 2, $\rho(X) = b_2(X)$, and $\operatorname{rk} S_f = \rho - 2$. Condition 3) follows from this.

Suppose now that conditions 1), 2), and 3) are satisfied. We assume that φ is elliptic, and we shall arrive at a contradiction. In this case relation (2) can be rewritten in the form

$$\sum (n(F) + 2 + d(F) - \varepsilon(F)) = b_2 + 2, \qquad (7)$$

where the sum extends over all fibers, $\varepsilon(F) = b_1(F) = 1$ for fibers of type A_n, and are equal to 0 in the remaining cases. [We observe that condition 3) for a quasielliptic pencil follows from relations (2), since in these cases there are no fibers of type A_n, because a singular point with separated tangents is stable.] Condition 3) and relation (7) give

$$\sum_F (2+d(F)-\varepsilon(F))=4. \tag{8}$$

We consider the case p = 3. Only the root systems A_2, E_6, and E_8 satisfy condition 2). If the number of fibers of type A_2 and $*A_2$ is equal to k and the number of those of type E_6 and E_8 is equal to l, then (8) gives $k + 2l \leqslant 4$, while condition 3) gives $b_2 - 2 \leqslant 2k + 8l$. Since $\chi(\mathcal{O}_X)\geqslant 2$, it follows that $e(X) \geqslant 24$ and $b_2 \geqslant 22$. We therefore obtain $k + 2l \leqslant 4$, $2k + 8l \geqslant 20$ which is impossible.

Suppose now that p = 2. It is obvious that in this case there can only be the root systems A_1, D_{2n}, E_7 and E_8. For fibers of types D_{2n}, E_7 and E_8 $d(F) \geqslant 1$. Indeed, $d(F) = 0$ would imply that the field $K(\mathfrak{U}_l)/K$ has odd degree, where K is the field of fractions of the completed local ring of the point x over which the fiber F lies. Now fibers of type D_{2n}, E_7, and E_8 have components of even multiplicity which cannot have multiplicity one after covering of a base of odd degree. Hence, after the covering of the base corresponding to the extension $K(\mathfrak{U}_l)/K$, we again obtain one of the fibers of this type. Now the set of nonsingular points of these fibers is a 2-group, and the group $\mathfrak{U}_l$ must reduce to 0. This contradicts the fact that since $l \neq 2$ it reduces monomorphically. Thus, each fiber having type distinct from A_0, $*A_0$, A_1 or $^*A_1$ gives in the sum (8) a term $\geqslant 3$. Therefore, there can only be one such fiber, and in order to satisfy condition 3) for $\chi(\mathcal{O}_X)\geqslant 2$, i.e., $b_2 \geqslant 22$, this fiber must have type D_{2n}. Thus, only the following two cases are possible: one fiber of type D_{2n} with $d(F) = 2$ or one fiber of type D_{2n} with $d(F) = 1$ and one fiber of type A_0.

We shall deduce the theorem from the following result.

LEMMA. There exists a covering of the base $B \to \mathbf{P}^1$ such that the fibration $X \times_{\mathbf{P}^1} B \to B$ has only stable, degenerate fibers (i.e., of type A_n), and $\deg B/\mathbf{P}^1 = 2$, $B/\mathbf{P}^1$ is ramified only at that point over which there is a fiber of type D_{2n}, and the index of the ramification at this point is equal to 1 (i.e., $G = G_1$, $G_2 = 1$).

We shall show how the theorem follows from the lemma. From the formula for the genus of a covering it follows that the genus of B is equal to 0, i.e., $B\simeq\mathbf{P}^1$. The fibration $X \times_{\mathbf{P}^1} B \to B$ (more precisely, a smooth, minimal model of it) has no more than three fibers of type A_n. We shall prove that then it has no such fiber. Indeed, if the degenerate fibers have the form A_{n_i}, $i = 1,\dots,r \leqslant 3$, then according to formula (2), $\sum n_i+r=b_2+2$, and this together with (8) gives r = 4. If there are no degenerate fibers, then $X\times_{\mathbf{P}^1}B\simeq A\times B$, where A is an elliptic curve. Then $p_g(X\times_{\mathbf{P}^1}B)=0$, and hence also $p_g(X) = 0$, i.e., $\chi(\mathcal{O}_X)\leqslant 1$.

We proceed to the proof of the lemma. We note that the assertion of the lemma is of local character: if $\mathcal{O}$ is the completion of the local ring of that point $P\in\mathbf{P}^1$, over which there lies a fiber of type D_{2n} and $\bar{X}=X\times_{\mathbf{P}^1}\mathcal{O}$, then it suffices to prove the existence of an extension $\bar{\mathcal{O}}/\mathcal{O}$ with the required properties. It is easy to see that there exists a covering $B/\mathbf{P}^1$ with completion which at the point lying over P coincides with $\bar{\mathcal{O}}$. It satisfies the conditions of the lemma.

There are two possible cases: a) $j(P) = \infty$ and b) $j(P) \neq \infty$, where j is the absolute invariant of the fibration of X. As is known, in case a) a stable fiber corresponding to the fiber F over the point P has type A_m, while in case b) it is nondegenerate. We consider case a). A curve with $j(P) = \infty$ has over P either a fiber of type A_n or it acquires such a fiber over an extension $\mathcal{O}/\mathcal{O}$ of degree 2. Indeed, in Eq. (4), Sec. 2 $\lambda \neq 0$, since for $\lambda = 0$ by the formula for j [28, p. 303] we obtain j = 0. Taking $\lambda x + \mu$ for the new variable u, we reduce the equation to the form

$$y^2+uy=u^3+au^2+bu+c,$$

and by the translation $u \to u + a$ we make $a = 0$. By the transformation $y \to y + \xi$ we replace b by $b + \xi^2 + \xi$. We can therefore arrange that b = 0, possibly, after an extension of degree 2. In the equation $y^2 + uy = u^3 + c$, $j = 1/c$ (by the formula for j), and $j(P) = \infty$ implies that $c(P) = 0$. We see that a degenerate fiber F has a double point x with separated tangents whence it follows easily that the group law defined on $F \setminus x$ gives a group isomorphic to G_m, and this corresponds to a fiber of type A_n. Let $\bar{\mathcal{O}}/\mathcal{O}$ be the corresponding extension, and let K and $\bar{K}$ be the fields of fractions of the rings $\mathcal{O}$ and $\bar{\mathcal{O}}$. As we have already seen, the Galois group of the extension $\bar{K}(\mathfrak{A}_l)/\bar{K}$ has order 1 or l. Moreover, $[K(\mathfrak{A}_l):K]$ is even, since the fiber F has wild ramification. This implies that $[K(\mathfrak{A}_l):K]=2$ or $2l$. If this degree is equal to 2, then $\bar{K}=K(\mathfrak{A}_l)$ both from Eq. (5) and from the fact that d(F) = 1 or 2, hence d(F) = 2, $e_1 = 2$, $e_2 = 1$, **so that the lemma is proved. If the degree is equal to** $2l$, **then** $K(\mathfrak{A}_l)\supset\bar{K}$ and hence the Galois group G of the extension $K(\mathfrak{A}_l)/K$ has a normal divisor H of order l. Hence $G = H \times G_1$, where G_1 is the group of the ramification and $|G_1| = 2$. From formula (5) and from the assumption d(F) = 1 we find that

$$e_1 = \ldots = e_r = 2, \quad e_{r+1} = 1, \quad \frac{1}{2l}(2r) = \frac{r}{l} = \frac{1}{2},$$

i.e., $r = l/2$. Since l is odd, this is a contradiction. Similarly, from d(F) = 2 we find that $r = l$. For the field $\bar{K}$ suppose that $e_1 = \ldots = e_s = 2$, $e_{s+1} = 1$. We must find a relation between r and s. A simple argument shows that $r = ls$ [this can be obtained directly from the definition or from computation of the discriminant of the field $K(\mathfrak{A}_l)=L\otimes_K\bar{K}$ $[L:K]=l$].

From this we find that s = 1, i.e., the assertion of the lemma in case a).

In case b) there exists a covering $\bar{\mathcal{O}}/\mathcal{O}$, a curve $\bar{X}$ with a degenerate fiber over $\bar{\mathcal{O}}$, and an extension of the Galois group G of the covering $\bar{\mathcal{O}}/\mathcal{O}$ to $\bar{X}$ such that X is birationally isomorphic to $\bar{X}/G$. We denote by A a closed fiber of $\bar{X}/\bar{\mathcal{O}}$. The group G is isomorphic to a subgroup of the group Aut A.

Case b_1): $\deg\bar{\mathcal{O}}/\mathcal{O}=2$. In particular, this is the case if the curve A is not supersingular, since then |Aut A| = 2 (see, for example, [28, p. 304]). Since |G| = 2, it follows that $e_0 = e_1 = 2$, and from formula (5) it follows that d(F) ⩾ 2, while the condition d(F) = 2 is possible only for $e_r = 1$ which proves the lemma in this case.

Case b_2): $\deg\bar{\mathcal{O}}/\mathcal{O}>2$. In particular, the curve A is supersingular. Then |Aut A| = 24 and [28, p. 304] the group Aut A is an extension of the group of quaternions by means of a group of order 3. We shall prove that this case cannot occur in the situation considered in the lemma. We first assume that d(F) = 1. If $e_0 = e_1$, then the first term in formula (5) would already be equal to 2. Therefore, in our case $e_0 \neq e_1$, and this is possible only for $e_0/e_1 = 3$. In the group Aut A there exist (up to isomorphism) only two subgroups the order of which is even and divisible by 3: these are the entire group Aut A and an Abelian group of order 6. In the latter case $e_0 = 6$, and if $e_1 = \ldots = e_k = 2$, $e_{k+1} = 1$, then d(F) = 4k/6 = 2k/3 which is impossible for d(F) = 1. Therefore [for d(F) = 1], there remains the sole possibility $G\simeq\mathrm{Aut}\,A$.

Thus, $[\bar{K}:K] = 24$, where $\bar{K}$ and K are the fields of fractions of the rings $\bar{\mathcal{O}}$ and $\mathcal{O}$. Here $\bar{K}\supset L\supset L_0\supset K$, $[L_0:K]=3$, $[L:L_0]=4$ and $\mathrm{Gal}(L/L_0)\simeq(\mathbf{Z}/2)^2$, while $[\bar{K}:L]=2$. $L_0=K(\tau)$, $\tau^3=t$, where t is a local uniformizing parameter in K, $L=L_0(u_1, u_2)$, $u_1^2+u_1=\alpha_1$, $u_2^2+u_2=\alpha_2$, $\alpha_1, \alpha_2\in L$, $\bar{K}=L(\nu)$, $\nu^2+\nu=\mu+L$. The relation d(F) = 1 is possible only for

$$e_0=24, \quad e_1=8, \quad e_2=e_3=2, \quad e_4=1.$$

We note that the type of a degenerate fiber in $\bar{X}/G$ depends only on the field $\bar{K}/K$ and the isomorphism $\mathrm{Gal}(\bar{K}/K)\simeq\mathrm{Aut}\,A$. Indeed, at any point $x\in A$ we can take for local parameters some local parameter s of the point x on A and a uniformizing parameter t in K, after which the action of the stationary subgroup of the point x in $k[[s, t]]$ is determined. Using this fact, we restrict ourselves to finding a degenerate fiber in $\bar{X}/G$ for the case $X=A\times\bar{\mathcal{O}}$. For this we use the following global argument. We construct an extension $\mathcal{K}/k(\mathbf{P}^1)$, which after completion at the point P coincides with K and which is "minimally branched," i.e., it has a subfield $\Lambda_0\supset k(\mathbf{P}^1)$ $[\Lambda_0:k(\mathbf{P}^1)]=3$, ramified only at two points, while $\bar{K}/\Lambda_0$ is ramified at a single point — the point over which P lies. We choose t as a coordinate on $\mathbf{P}^1$, so that $k(\mathbf{P}^1) = k(t)$, and we consider the covering $B_0 \to \mathbf{P}^1$, $k(B_0) = \Lambda_0 = k(\tau)$, $\tau^3 = t$. We denote by p(f) the principal part of the function $f\in L$ at the point ∞. For a_1 and $a_2\in\Lambda_0$ we choose functions regular away from ∞ for which $p(a_i) = p(\alpha_i)$ — they are uniquely determined by this property. We set $\Lambda = \Lambda_0(U_1, U_2)$, $U_i^2 + U_i = a_i$, and we denote by B the normalization of B_0 in Λ. Then B is ramified over B_0 only at the point τ = ∞, and it is completely ramified at this point; we denote by b the preimage of this point. The completion of Λ_0 at the point ∞ is isomorphic to L_0, while Λ is in b — L. From the values we know of the numbers e_i for the field $\bar{K}$ it is easy to find that for Λ/Λ_0, $\bar{e}_0 = \bar{e}_1 = 4$, $\bar{e}_2 = 1$. The Hurwitz formula for the genus then tells us that the genus of the curve B is equal to 0. We denote by π a local uniformizing parameter at the point b such that Λ = k(π), we find $m\in\Lambda$, which is regular away from b and such that $p(m) = p(\mu)$, and we set $\mathcal{K}=\Lambda(V)$, $V^2+V=m$. The field $\mathcal{K}/\Lambda$ is ramified only at the point b, and at this point it is completely ramified. We denote the normalization of B in $\mathcal{K}$ by C and the preimage of the point b in G by c. It is obvious that the completion of $\mathcal{K}$ at the point c is isomorphic to $\bar{K}$. This implies that $\mathcal{K}/k(t)$, just as $\bar{K}/K$, is the Galois extension $\mathrm{Gal}(\mathcal{K}/k(t))=G\simeq \mathrm{Aut}\,A$. In analogy to the foregoing, we find that for the field $\mathcal{K}/\Lambda$ and the point b, $\bar{e}_0 = \bar{e}_1 = \bar{e}_2 = \bar{e}_3 = 2$, i.e., $m=\alpha_0\pi^3+\alpha_1\pi^2+\alpha_2\pi+\alpha_3$, $\alpha_i\in k$. We denote by $\mathfrak{G}$ the Galois group C/B, and we consider $C\times A/\mathfrak{G}$. Writing the equation of A in the form $y^2 + y = x^3$ and of C in the form $V^2+V=\alpha_0\pi^3+\alpha_1\pi^2+\alpha_2\pi+\alpha_3$, we can represent the action of the generator of the group $\mathfrak{G}$ as $\sigma y = y + 1$, $\sigma V = V + 1$, $\sigma(x) = x$, $\sigma(\pi) = \pi$. Hence the invariants of the group $\mathfrak{G}$ are generated by the functions x, π and $z = y + V$. They are connected by the equation $z^2+z=x^3+\alpha_0\pi^3+\alpha_1\pi^2+\alpha_2\pi+\alpha_3$ of third degree, whence it follows that the surface $C\times A/\mathfrak{G}$ is rational. Hence the surface $C \times A/G$ is rational as the image of a rational surface under a separable morphism. But then for the degenerate fiber F over the point ∞, $\mathbf{e}(F) \leqslant 12$ which contradicts the relation (1) with $\chi(\mathcal{O}_X)\geqslant 2$. This completes the treatment of the case $d(F) = 1$, $|G| > 2$.

It remains to consider the case $d(F) = 2$. Then F is the unique degenerate fiber, and, as is known, X becomes isomorphic to $B \times A$ after lifting to some minimal covering $B \to \mathbf{P}^1$; for B it is possible to take the normalization of $\mathbf{P}^1$ in the field $K(\mathfrak{A}_3)$, $K=k(\mathbf{P}^1)$. As we have seen, $\deg(B/\mathbf{P}^1)\,|\,24$. We establish first of all that $\det(B/\mathbf{P}^1)$ is not divisible by 3, i.e., it is a divisor of 8. Indeed, in this case the covering $\bar{B}/\mathbf{P}^1$ would have to factor through a covering of $\bar{B}/\mathbf{P}^1$ of degree 3 which has at least 2 branch points, so that $B/\mathbf{P}^1$ also has at least 2 branch points. Now each branch point of the covering $B/\mathbf{P}^1$ gives a degenerate fiber of the fibration of X. Indeed, if $H\subset \mathrm{Gal}(B/\mathbf{P}^1)$ is the stationary subgroup of a point $b\in B$, then for a prime number $l \neq 2$ the group $\mathfrak{A}(\mathcal{O})_l=\mathfrak{A}(\mathcal{O})_l^H$, where $\mathcal{O}$ is the completed local ring of the point b and $\mathcal{O}$ is the point corresponding to it on $\mathbf{P}^1$. Moreover, $\mathfrak{A}(\mathcal{O})_l=A_l$, and if H acts nontrivially on A_l, then $|\mathfrak{A}(\mathcal{O})_l|<l^2$; this shows that the fiber is degenerate. We have thus arrived at a contradiction to the fact that X has a unique degenerate fiber.

Finally, the last case is when $\deg B/\mathbf{P}^1 = 8$ or 4, i.e., the Galois group of $B/\mathbf{P}^1$ is the group of quaternions or $\mathbf{Z}/4$. Here $e_0 = e_1$, and the first term in formula (5) is equal to 2. Therefore, the condition $d(F) = 2$ gives $e_2 = 1$. This is impossible, since $G/G_2 = G_1/G_2$ by the general theory has period 2. The proof of the lemma and hence of the theorem is complete

Remarks. 1. The theorem is also true for the case of a fibration with a base of genus $g > 0$ in a trivial manner. Condition 3) alone is sufficient, since the relations analogous to 2) and 3) for $g > 1$ also lead to a contradiction; while for $g = 1$ they show that the fibration has no degenerate fibers.

2. The theorem is obviously not true for a fibration of $\mathbf{P}^1 \times A$ when all conditions 1)-3) are tautologically satisfied. It is also not true for rational surfaces. Moreover, in this case it is not possible to distinguish a quasielliptic fibration from an elliptic fibration either on the basis of the root system R_f or on the basis of the lattice S_f. For example, the elliptic fibration defined by the equation $y^2 + y = x^3 + t$ and the quasielliptic fibration defined by with equation $y^2 = x^3 + t$ both have a unique degenerate fiber for $t = \infty$ of type E_8. Therefore, for both R_f is E_8 and $S_f = Q(E_8)$.

3. It would be nice to "generalize" the proposition to the case $p > 3$, i.e., to prove that conditions 2) and 3) are not satisfied for $p > 3$. However, this is not the case. For example, from what will be proved below it follows that there exist the following surfaces of type K3: a) X over a field of characteristic 11 and an elliptic pencil f on it for which $S_f = A_{10}^2$ (there are further 2 fibers of type A_0 or 1 of type $*A_0$); b) X over a field of characteristic 7 and a pencil f with $S_{f_1}=E_8\oplus A_6^2$. It is possible to show that these are the only types of surfaces of type K3 on which there is an elliptic pencil satisfying conditions 2) and 3) of the theorem. It is not hard to enumerate the types of degenerate fibers which can occur a priori on an arbitrary elliptic surface with this property, but we do not know which of them are actually realized.

As an illustration of the theorem, we shall prove the following assertion which refines the theorem of Sec. 3.

Proposition. If on a surface X of type K3 there exists a quasielliptic pencil, then on it there also exists an elliptic pencil.

We prove the proposition by constructing on X a system of -2 curves R such that the corresponding diagram belongs to one of the types of the extended diagrams $*A_1$, $*A_2$, D_{2n}, E_6, E_7

or E_8 and does not satisfy condition 2) of the theorem. A linear combination of the curves in R corresponding to the maximal root them possesses the property that $D > 0$, $DC_i = 0$ for all $C_i \in R$, whence $D^2 = 0$. According to Proposition 2, Sec. 3, D defines a pencil of curves of arithmetic genus 1 which, according to the theorem (necessity of conditions 2), is elliptic. To construct the system R we begin with a system R_0 corresponding to a degenerate fiber (or fibers) of a given quasielliptic pencil, and we adjoin to it a curve ξ consisting of cusps of this pencil. According to the theorem of Sec. 2, ξ is a smooth, rational curve, and hence $\xi^2 = -2$. According to the same theorem, $\xi D = p$. Moreover, it is obvious that ξ is invariant relative to translations by local sections in a neighborhood of the fiber. Therefore, ξ intersects D along a component invariant relative to the action of the group $P(R_0)/Q(R_0)$ corresponding to the root system R_0. From this it is possible for all fibers [except E_8 where the group $P(R_0)/Q(R_0)$ is equal to 1] to uniquely determine how ξ intersects D: along a "central component" contained in D with multiplicity p, and the intersection is transversal For example, for p = 2 the system $E_7 \cup \xi$ is determined by the graph

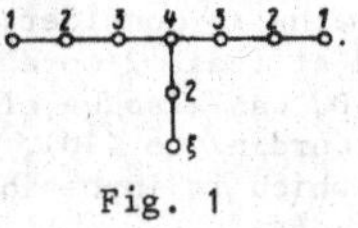

Fig. 1

In the system $R_0 \cup \xi$ we choose a subsystem R_1 which again determines a pencil of curves of genus 1 if it also determines a quasielliptic pencil, we go over from it to a system R_2, etc. until we arrive at the required system R. We write this process $R_0 \Rightarrow R_1$, $R_1 \Rightarrow R_2$, etc.

We note that in any characteristic (2 or 3) for any fibers

$$\sum_1^4 F_i \Rightarrow D_4$$

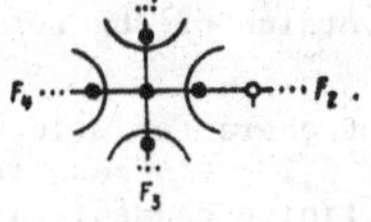

Fig. 2

$$\sum_1^3 F_i \Rightarrow E_6, \quad \text{if} \quad F_i \neq {}^*A_1, {}^*A_2 \tag{10}$$

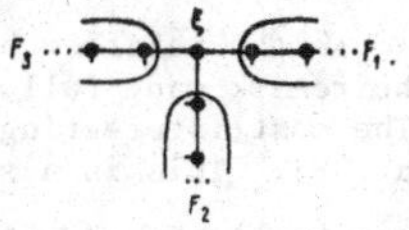

Fig. 3

The new system R_1 is denoted by dark dots.

We now consider separately the cases p = 2 and 3.

$p = 3$
$R = {}^*A_2$, E_6 or E_8
$E_6 \Rightarrow D_4$

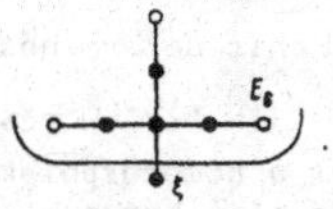

Fig. 4

The number of fibers cannot be <3 because of the condition $\Sigma n_i = 20$, $n_i \leqslant 8$. According to (9), the number of fibers is 3, and according to (10), one of them is $^*A_2$; otherwise we arrive at E_6 and D_4. For the remaining two fibers $n_1 + n_2 = 18$ which is impossible, since $n_i \leqslant 8$.

$p=2$

$R = {}^*A_1, D_{2n}, E_7$ or E_8

$E_7 \Rightarrow E_6$

Fig. 5

Similarly, $D_{2n} \Rightarrow E_7$ for $n \geqslant 4$. It remains to consider the possibilities $R = {}^*A_1, D_4, D_6, E_8$. Aside from the fiber D_4, there must be at least 2 more fibers (because $\Sigma n_i = 20$, $n_i = 8$). Now $D_4 \oplus F_1 \oplus F_2 \Rightarrow D_5$, and therefore the fiber D_4 can also be eliminated. Then, according to (9), the number of fibers is equal to 3, and according to (10), one of them must have type $^*A_1$. For the remaining two fibers $n_1 + n_2 = 19$ which is impossible, since $n_i \leqslant 8$. The proof of the proposition is complete.

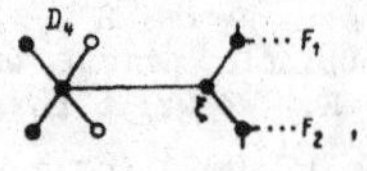

Fig. 6

5. Supersingular and Elementary Surfaces of Type K3

A surface X of type K3 is called *supersingular* if rk Pic X = 22. Since $b_2(X) = 22$, it follows that the equality rk Pic X = 22 implies that $S_X \otimes \mathbf{Q}_l = H_{et}^2(X, \mathbf{Q}_l)$, where S_X = Pic X, i.e., "all cycles are algebraic." Another version of the definition of the concept of supersingularity will be discussed in Sec. 9.

It is obvious that over a field of characteristic 0 there exist no supersingular surfaces of type K3: in this case rk Pic X $\leqslant$ 20. In many respects these surfaces are a typical phenomenon of geometry over fields of finite characteristic.

The simplest example of a supersingular surface of type K3 is the Kummer surface X corresponding to an Abelian variety A = C × C, where C is a supersingular elliptic curve and the characteristic $p \neq 2$. Indeed, $S_A = (C \times c)\mathbf{Z} \oplus (c \times C)\mathbf{Z} \oplus \mathrm{End}\, C$, and since in the present case rk × End C = 4, it follows that rk S_A = 6. On the other hand rk S_X = rk A + 16 (the 16 lines corresponding to fixed points of the automorphism $a \to -a$ on A are glued in).

Another type of examples (we discuss the relation between them below) are unirational surfaces of type K3. According to the remark made following Proposition 2 in Sec. 2, a unirational surface is supersingular. The most interesting example of a unirational surface that is not rational is due to Shioda [39]. This is a smooth surface in $\mathbf{P}^3$ with the equation

$$x_0{}^p x_1 + x_1{}^p x_0 = x_2{}^p x_3 + x_3{}^p x_2. \tag{1}$$

By the transformation $x_0 = u_0 + u_1$, $x_1 = u_0 - u_1$, $x_2 = u_2 + u_3$, $x_3 = u_2 - u_3$ it can be reduced to the "Fermat surface"

$$u_0^{p+1} - u_1^{p+1} = u_2^{p+1} - u_3^{p+1}. \tag{2}$$

The unirationality of the surface (1) becomes obvious if it is written in the inhomogeneous form

$$x^p y + y^p x = z^p + z.$$

It is obvious that over the field k(y) this determines a group manifold G with law

$$(x_1, z_1) + (x_2, z_2) = (x_1 + x_2, z_1 + z_2).$$

Since the mapping $(x, z) \to z$ determines a homomorphism $G \to G_a$, G is the form of an additive group. In view of the fact that an additive group has no separable forms, G becomes isomorphic to G_a over the field $k(y^{1/p^m})$ for some m. Thus,

$$k(x, y, z) \subset k(x, y^{1/p^m}, z) \simeq k(y^{1/p^m})(G_a).$$

The corresponding transformation can also easily be written in explicit form. Setting $y = t^p$, we write the equation in the form

$$(xt - z)^p = z - xt^{p^2}.$$

The transformation $xt - z = u$, $z - xt^{p^2} = v$ takes this equation into $u^p = v$, so that $k(x, t, z) = k(u, t)$.

Of course, Eq. (1) does not always define a surface of type K3 (only for $p = 3$). However, if $p \equiv 3 \pmod 4$, then setting in (2) $u_i^{(p+1)/4} = v_i$, we obtain a smooth surface of fourth degree, i.e., a surface of type K3. As Shioda showed [40], the surface of fourth degree obtained coincides with the Kummer surface considered above [for $p \equiv 3 \pmod 4$]. He also showed that for $p \equiv 1 \pmod 4$ the supersingular Kummer surface of type K3 is unirational [41].

These and some other examples make the following question interesting: is any supersingular surface of type K3 unirational? In Sec. 5 we shall prove that this is so over a field of characteristic 2.

A supersingular surface X of type K3 over a field of characteristic p is called elementary if its lattice $S_X = \operatorname{Pic} X$ is p-elementary. In Sec. 8 we shall see that elementarity already follows from supersingularity, so that this is a very tentative definition.

An example of an elementary surface is a unirational surface X of type K3 such that there exists a rational mapping $\varphi: \mathbf{P}^2 \to X$, which is nonseparable and of degree p. We have already seen that it is supersingular. The fact that the lattice S_X is p-elementary follows from the proposition in Sec. 1. Indeed, suppose that $\varphi: \mathbf{P}^2 \to X$ is the rational mapping which exists by hypothesis. Let Y be such a surface, and let $f: Y \to \mathbf{P}^1$ be a rational morphism such that $\pi: Y \to X$ is a morphism. Since f can be decomposed into a composition of σ-processes, it follows that $T_Y = \operatorname{Pic} Y$ is a unimodular lattice: more precisely, $T_Y = I \oplus I \times (-1)^m$ if f can be decomposed into m-processes. The operations of direct and inverse images give the homomorphisms $\pi^*: S_X \to T_Y$ and $\pi: T_Y \to S_X$ satisfying the conditions of the proposition.

The classification of p-elementary lattices of Sec. 1 can be applied to elementary surfaces of type K3; this shows that for such surfaces the lattice S_X is uniquely determined for $p \neq 2$ by the invariant $s(d(S_X) = p^{-s})$, while for $p = 2$ it is further required to know whether it belongs to type I or II. We shall now show that from this there follow certain corollaries regarding the geometry of elementary surfaces of type K3 over a field of characteristic 2.

THEOREM. On any elementary surface of type K3 over a field of characteristic 2 there exists a quasielliptic pencil.

It suffices for us to construct an element $f \in S = \operatorname{Pic} X$, such that $f^2 = 0$, and the lattice S_f satisfies the hypotheses of the theorem in Sec. 4. Since according to the theorem of Sec. 1 the lattice S is uniquely determined by the fact that it is hyperbolic, 2-elementary, belongs to type I or II, and its discriminant is equal to $-2^{2\sigma}$, it suffices for any value of σ, $1 \leqslant \sigma \leqslant 10$, to display a lattice with these properties and with such a vector f. For lattices of type II it is possible to set $S = U_2 \oplus I(-2)^{2\sigma-2} \oplus Q(D_{22-2\sigma})$ [for $\sigma \leqslant 7$ it is possible to replace $Q(D_{22-2\sigma})$ by $Q(D_{14-2\sigma}) \oplus Q(E_8)$, while for $\sigma \leqslant 3$ it can be replaced by $Q(D_{6-2\sigma}) \oplus Q(E_8^2)$]. The corresponding list for lattices of type I appears below (see top of next page). It is assumed that f is contained in the factor U_1 or U_2 (2) (we take those lattices which were presented in the proof of the theorem of Sec. 1).

Remarks. 1. The rather involved arguments in the proof of the theorem in Sec. 4 are not necessary if we only wish to verify that the pencils f presented in this table are not elliptic: it is obvious that for them relation (2) of Sec. 4 is already not satisfied.

2. A simpler proof can be suggested (see [8, p. 851]). On the vector space $S/2S^*$ there is defined a nondegenerate, bilinear form with values in $\mathbf{F}_2$. Because of the parity of S, this form is skew-symmetric, and hence $S/2S^*$ contains an isotropic subspace of half the dimension. We denote by M the preimage of this subspace: $S \supset M \supset 2S$. By construction $M = N(2)$, where N is some integral lattice, and a simple computation of discriminants shows that N is unimodular. By the general theory of unimodular lattices $N = I \oplus I(-1)^{10}$, i.e., $M = I(2) \oplus I \times (-2)^{19}$. Suppose that the basis corresponding to this decomposition is $e_1, \ldots, e_{20}$, $e_1^2 = 2$, $e_i^2 = -2$, $i > 1$. The vector $f = e_1 + e_2$ defines a pencil of curves of genus 1 for which it is very easy to verify conditions 2) and 3) of the proposition. Thus, this is a quasielliptic pencil. However, this proof appeals to the theorem of Sec. 4 in all generality.

σ	1	2	3	4	5
S	$U_2\oplus Q(D_{20})$	$U_2\oplus Q(D_4\oplus D_{16}$	$U_2\oplus Q$ $(D_4\oplus D_8{}^2)$	$U_2\oplus Q$ $(D_4{}^3\oplus D_8)$	$U_2\oplus Q(D_4{}^5)$
R_f	D_{20}	$D_4\oplus D_{16}$	$D_4\oplus D^2{}_8$	$D_4{}^3\oplus D_8$	$D_4{}^5$

σ	6	7	8	9	10
S	$U_2\oplus Q(D_4{}^3\oplus D_8)$	$U_2\oplus Q(D_4\oplus D_8{}^2)$	$U_2\oplus Q$ $(D_4\oplus D_{16})$	$U_2\oplus \widetilde{Q}(D_{20})$	$U_2(2)\oplus$ $\widetilde{Q(D_{20})}$
R_f	$D_4{}^3\oplus A_1{}^8$	$D_4\oplus A_1{}^{16}$	$D_4\oplus A_1{}^{16}$	$A_1{}^{20}$	$A_1{}^{20}$

3. A weaker analogue of this theorem also holds for characteristic 3: on any elementary surface X of type K3 with $\sigma(X) \leqslant 6$ for p = 3 there exists a quasielliptic pencil.

We write out immediately the corresponding lattices:

σ	1	2	3
S	$U_2\oplus Q(A_2{}^2\oplus E_8{}^2)$	$U_2\oplus Q(A_2{}^3\oplus$ $\oplus E_6\oplus E_8)$	$U_2\oplus Q(A_2{}^6\oplus$ $\oplus E_8)$
σ	4	5	6
S	$U_2\oplus Q(A_2{}^7\oplus E_6)$	$U_2\oplus Q(A_2{}^{10})$	$U_2(3)\oplus Q(A_2{}^{10})$

It is easy to show that for $\sigma > 6$, p = 3 a quasielliptic pencil does not exist.

4. As is evident from the tables presented, for p = 2, $\sigma < 10$ and for p = 3, $\sigma < 6$ there exists a quasielliptic pencil with a section. It is easy to show that in the remaining cases there exists no such pencil.

COROLLARY. An elementary surface of type K3 over a field of characteristic 2 is unirational.

Suppose that $\varphi: X \to \mathbf{P}^1$ is a quasielliptic pencil the existence of which is established in the proposition, and let ξ be the curve formed by the cusps of its fibers. As shown in Sec. 2, the morphism $\varphi: \xi \to \mathbf{P}^1$ is nonseparable and has degree 2. We set $Y = X \times_{\mathbf{P}^1} \xi$. The projection $Y \to X$ is also a nonseparable mapping of degree 2. On the other hand, on Y there is defined a pencil $Y \to \xi$ for which the fibers are rational and have a singular point which is now a section. Resolving this singular point (for example, on a general fiber), we obtain a surface $\tilde{Y}$ with a pencil of smooth, rational curves which is thus rational. The rational morphism $\tilde{Y} \to X$ bears witness to the fact that the surface X is unirational.

Proposition. For an elementary surface X of type K3 over a field of characteristic 2 the lattice S = Pic X belongs to type I.

We suppose that the lattice S belongs to type II. If $d(S) = -2^{2\sigma}$, then according to the theorem of Sec. 1, S is isomorphic to the lattice $U_2 \oplus I(-2)^{2\sigma-2} \oplus Q(D_{22-2\sigma})$ (according to this same theorem, $\sigma > 1$ always). An element $f \in U_2$ with $f^2 = 0$ determines a pencil of curves of genus 1 which satisfies the hypotheses of the theorem of Sec. 4 and must therefore be quasielliptic. This pencil has $2\sigma - 2$ reducible fibers of the type $*A_1$ and one fiber of type $D_{22-2\sigma}$ (for $\sigma = 11$ there is none). Let ξ be the curve formed by the singular points of the fibers. As we know, $\xi \cdot f = 2$. In correspondence with the decomposition of the lattice, suppose that $\xi = \xi_0 + \sum \xi_i + \bar{\xi}, \xi_0 \in U_2, \xi_i \in I(-2), \bar{\xi} \in Q(D_{22-2\sigma})$. If C, C' are components of the i-th fiber $*A_1$, then $\xi C = \xi C' = 1$, and hence $\xi_i C = 1$. Now in the lattice $I(-2) = \mathbf{Z}C$, $C^2 = -2$ there is no element with this property. This contradiction proves the proposition.

6. Vector Fields

One of the central questions of the theory of surfaces of type K3 is the study of their deformations and, in particular, the calculation of the "number of moduli." This problem is related to the computation of the dimensions of the groups $H^1(X, \Theta)$ and $H^2(X, \Theta)$, where Θ is the tangent bundle of the surface X. According to the Riemann–Roch theorem, $\dim H^0(X, \Theta) - \dim H^1(X, \Theta) + \dim H^2(X, \Theta) = c_2(\Theta) = -20$ (c_2 is the second characteristic class). On the other hand, by duality $\dim H^2(X, \Theta) = \dim H^0(X, \Theta)$, and hence the question reduces to the computation of the dimension of $H^0(X, \Theta)$, i.e., of the space of vector fields on the surface X. In the case of characteristic 0, $H^0(X, \Theta) = 0$ [1, p. 179]. We shall show that the same is true for fields of any characteristic. Hence $\dim H^1(X, \Theta) = 20$, $\dim H^2(X, \Theta) = 0$, whence it follows that there exists a smooth, formal manifold of moduli of dimension 20. The numbers $h^{i,j} = \dim \times H^j(X, \Omega^i)$ will thus be defined. For this it is necessary to use the fact that for a surface of type K3, $\Theta \simeq \Omega$, and $H^1(X, \mathcal{O}) = k$ by obvious considerations of duality and the relation $\Sigma(-1)^{i+j}h^{i,j} = c_2 = 24$. As a result, we obtain the following table:

j

1	0	1	
0	20	0	
1	0	1	i

The basis for the proof of the relation $H^0(X, \Theta) = 0$ for a surface X of type K3 is the fact that a surface of type K3 on which there exists a nonzero, regular vector field is elementary. This result is proved in [7]. Here we present a sketch of the proof, presenting in detail only those arguments which are simpler than those in [7]. The details omitted are all contained in [7].

The basis is the operation of factorization of an algebraic variety X with respect to a rational vector field D. A vector field is called p-closed if $D^p = \alpha D$, where $\alpha \in k(X)$. On each open, affine set $U \subset X$ we consider the ring of invariants $k[U]^D = \{g \in k[U], Dg = 0\}$ and the affine variety U^D = Speck $[U]^D$. These varieties are glued together into a Euclidean variety which we denote by X^D. There is defined a finite, nonseparable morphism $\pi_D: X \to X^D$ of degree p. Conversely, each such morphism $\varphi: X \to Y$ for a normal variety Y can be represented in the form $\varphi = \pi_D$, where D is some p-closed, rational vector field on X.

If X is a complete variety, then the Lie p-algebra $H^0(X, \Theta)$ is finite-dimensional, and therefore it contains a one-dimensional Lie p-subalgebra, i.e., an element $D \in H^0(X, \Theta)$, for which $D^p = 0$ or $D^p = D$. In the first case the vector field is called a field of additive type, while in the second case it is said to be of multiplicative type. In both cases it is p-closed. Therefore, in studying a variety X with $H^0(X, \Theta) \neq 0$, we can always assume that on it there exists a vector field of additive or multiplicative type which is thus p-closed.

If X is a variety which is smooth at the point x, then we define the germ of divisors C_x = l.c.d. $(D(f), f \in \mathcal{O}_x)$. For different simple points these germs are glued together into a single divisor which is called the divisor of the field D and is denoted by (D). If u is the local equation of (D), then locally $D = uD_1$, where D is regular and vanishes on a subset of codimension $\geqslant 2$.

If a vector field D is regular at a point x and does not vanish there (i.e., $D\mathcal{O}_x \not\subset m_x$), then locally it can be represented in a very simple form [38, part II]. Namely, because of the condition $D\mathcal{O}_x \not\subset m_x$, there exists $u \in m_x$ with $D(u) \notin m_x$. It is easy to see that $u \notin m_x^2$. Multiplying, if necessary, D by a function $h \in \mathcal{O}_x^*$, it may be assumed that D(u) = 1, whence, of course, it follows that $D^p = 0$. We set $A = \mathcal{O}_x^D + \mathcal{O}_x^D u + \ldots + \mathcal{O}_x^D u^{p-1}$. For $g(T) \in \mathcal{O}_x^D[T]$, $D(g(u)) = g'(u)$. From this it follows easily that $A = \bigoplus_{i=0}^{p-1} \mathcal{O}_x^D u^i$. We shall prove that $A = \mathcal{O}_x$. Let $v \in \mathcal{O}_x$. Since $D^p = 0$, there exists a minimal $k \geqslant 0$ for which $D^k v \in A$. Let $D^k v = g(u)$. Then $0 = D^{p-k}(g(u)) = g^{(p-k)}(u)$, whence $g(T) = h^{(k)}(T)$, $h \in \mathcal{O}_x^D[T]$. Thus, $D^k v = D^k(h(u))$, i.e., $DD^{k-1}(v - h(u)) = 0$

$$D^{k-1}(v - h(u)) = w \in \mathcal{O}_x^D, \quad D^{k-1}v = h^{(k-1)}(u) + w \in A,$$

in contradiction to the minimality of k. Choosing now a system of local parameters $u_1,\ldots,u_n\in\mathcal{O}_x$ so that $u_1 = u$ and setting $u_i=a_0^{(i)}+a_1^{(i)}u+\ldots+a_{p-1}^{(i)}u^{p-1}$, $a_k^{(i)}\in\mathcal{O}_x^D$, we find that $u_i\equiv a_0^{(i)}+a_1^{(i)}(x)u\,(m_x^2)$, whence it follows that $u, a_0^{(2)},\ldots,a_0^{(n)}$ form a system of local parameters. In other words, we have found a system of local parameters $t_1,\ldots,t_n$ such that $D(t_1) = 1$, $D(t_i) = 0$ for $i > 1$. If $\hat{\mathcal{O}}_x=k[[t_1,\ldots,t_n]]$, then it is obvious that $D = \partial/\partial t_1$ and $\hat{\mathcal{O}}_x^D=k[[t_1^p, t_2,\ldots,t_n]]$. In particular, we obtain a proof of the following result.

LEMMA 1. If a vector field D is regular and does not vanish at a simple point $x\in X$, then the point $\pi_D(x)\in X^D$ is also simple.

As Deligne has noted, the converse of this assertion can be proved by similar arguments.

LEMMA 2. If the points $x\in X$ and $\pi_D(x)\in X^D$ are simple, then the vector field D can be multiplied by a function such that it is regular and does not vanish at the point x.

We can assume that the vector field D is regular at the point x and does not vanish outside some subset Z of codimension $\geqslant 2$. By hypothesis, $\mathcal{O}_x$ and $\mathcal{O}_x^D$ are regular local rings. This implies [30, p. 88] that $\mathcal{O}_x$ is a free module over $\mathcal{O}_x^D{:}\,\mathcal{O}_x=\bigoplus_{i=0}^{p-1}\mathcal{O}_x^D a_i$. Then for any point $y\in\operatorname{Spec}\mathcal{O}_x$, $\mathcal{O}_y=\oplus\mathcal{O}_y^D a_i$. For $y\notin Z$ we constructed another such basis in the proof of Lemma 1.1, $u,\ldots,u^{p-1}$. It is obvious that $\det(D^i(u^j))\notin m_y$, since the matrix $D^i(u^j)$ is triangular with nonzero constant elements on the diagonal. From this it follows easily that this is also true for any other basis; in particular, $\det(D^i(a_j))\notin m_y$. Now the element $\det(D^i(a_j))$, which does not vanish off Z, $\operatorname{codim} Z \geqslant 2$, also does not vanish at x. This implies that D also does not vanish at x: otherwise the matrix $(D^i(a_j)(x))$ would contain an entire row of zeros.

If the vector field D at each point can be multiplied by a function so that it becomes regular and nonvanishing, then it is said to have divisor singularities. In this case it defines a line bundle $L\subset\Theta$, corresponding to the divisor D. As we have seen, the variety X^D is smooth. We have the sequence of morphisms

$$X\xrightarrow{\pi_D}X^D\to X^{(p)}\xrightarrow{\pi_D^{(p)}}(X^D)^{(p)},$$

where $X \to X^{(p)}$ is the absolute Frobenius mapping. For each point $x\in X$ we obtain the exact sequence

$$0\to L_x\to\Theta_x\to\pi_D^*\Theta_{\pi(x)}\to L_{\pi(x)}^{(p)}\to 0,$$

whence by passing to determinants we obtain the formula

$$K_X=\pi_D^*K_{X^D}+(p-1)(D). \tag{1}$$

If a vector field is regular on X, then $(D) \geqslant 0$. From (1) it then follows easily that if for X^D there is a possible resolution of singularities (for example, $\dim X = 2$), then

$$P_m(X^D)\leqslant P_m(X), \tag{2}$$

where $P_m(X)$ are multiple genera (see [7, p. 1280]).

If D vanishes at a point $x\in X$, then the point $\pi_D(x)\in X^D$ is, in general, singular. In the case of surfaces these singular points are somewhat reminiscent of "factorial singularities" in characteristic 0 because of the following result.

LEMMA 3. If $\dim X = 2$ and X is smooth at the point x, then the singular point $\pi_D(x)\in X^D$ can be resolved into a tree of rational, simply connected curves. (For the proof see [7, p. 1284].)

Proposition 1. If X is a surface of type K3 and D is a regular vector field on X, then X^D is a rational surface.

The proof uses the following simple assertion.

LEMMA 4. Let Y be a complete, normal surface, and let S be the set of its singular points. If $K_{Y-S} = 0$ and the Kodaira dimension $\varkappa$ of the surface Y is equal to 0, then the minimal resolution of the singular points of S is a minimal surface. (The proof is contained in [7, p. 1284].)

Proof of Proposition 1. Let $Y \to X^D$ be a minimal resolution of the singular points of the surface X^D. We apply to Y the classification of surfaces according to their Kodaira dimension

$\varkappa$ (see, for example, [15]). From the inequalities (2) it follows that $\varkappa=-1$ or 0. If $\varkappa=-1$, then Y is a rational or ruled surface. It cannot be a nonrational ruled surface, since for it $q = b_1/2 = 0$ because of the fact that $b_1 = 0$ for X, and $\pi_D: X \to X^D$ is a homeomorphism of étalé topologies. Thus, for $\varkappa=-1$ the theorem is true, and it remains for us to verify that the case $\varkappa=0$ is impossible. If $\varkappa=0$, then by Lemma 4, Y is a minimal surface. According to the classification [15, p. 373] in this case

$$\mathbf{e}(Y)<24, \tag{3}$$

where $\mathbf{e}$ is the Euler–Poincaré characteristic. On the other hand, $\mathbf{e}(Y) = \mathbf{e}(X^D) + \Sigma(\mathbf{e}(F_i) - 1)$, where F_i are cycles contractible by the morphism $\varphi: Y \to X^D$. Finally, $\mathbf{e}(X^D) = \mathbf{e}(X) = 24$. According to Lemma 3, F_i is a tree, and if the number of its irreducible components is n_i, then $\mathbf{e}(F_i) - 1 = n_i$. As a result,

$$\mathbf{e}(Y) = 24 + \Sigma n_i,$$

whence in combination with (3) we find that all $n_i = 0$, i.e., X^D is smooth. According to Lemma 2, from this it follows that D has only divisor singularities. Formula (1) together with $K_X = 0$, $mK_Y = 0$ for some $m > 0$ gives $(D) = 0$, i.e., D has no singularities at all. Such a vector field cannot exist on X, because $\mathbf{e}(X) \neq 0$.

COROLLARY. If on a surface X of type K3 there exists a nonzero, regular vector field D, then X is elementary.

From the existence of the morphisms

$$X \xrightarrow{\pi_D} X^D \to X^{(p)}$$

it follows that there exists a nonseparable morphism of degree p,

$$(X^D)^{(p^{-1})} \to X,$$

where $(X^D)^{(p^{-1})}$ is obtained from X^D by the mapping inverse to the absolute Frobenius mapping. Since the surface X^D is rational, the same is true for $(X^D)^{(p^{-1})}$ as well, and hence X is elementary, as we have seen in Sec. 5.

Before the proof of the main result of this section – the relation $H^0(X, \Theta)=0$ for a surface X of type K3 – we make several remarks.

All these remarks are based on the fact that on a supersingular surface X of type X3 there are many pencils of arithmetic genus 1. Indeed, according to the theorem of Sec. 3, an element $x \in S_X$ with $x^2 = 0$ is such a pencil. Since $\operatorname{rk} S_X = 22$ and the lattice $S_X \otimes \mathbf{R}$ is indefinite, by the general theory of quadratic forms such a vector must exist. In addition, the theorem in Sec. 4 provides us with the possibility of determining whether the corresponding fibration is elliptic or quasielliptic.

1. If $f: X \to B$ is a smooth fibration into curves of any genus, then there is the exact sequence of vector bundles

$$0 \to \Theta_{X/B} \to \Theta_X \to f^*(\Theta_B) \to 0.$$

For $D \in H^0(X, \Theta_X)$, its image in $H^0(X, f^*(\Theta_B))$ always has the form $f^*(\Delta)$, where $\Delta \in H^0(B, \Theta_B)$; this follows from the fact that f is a proper morphism with connected fibers (see [7, p. 1286]). The vector field Δ on B is called the normal component of the vector field D and is denoted by D_N. A simple verification shows [7, p. 1286] that D_N extends as a regular vector field to those points $b \in B$, for which the fiber $f^{-1}(b)$ is degenerate, and in this case $D_N(b) = 0$. This gives strong restrictions: if $B = \mathbf{P}^1$, then a nonzero vector field on B can vanish at no more than two points; therefore, if a fibration has 3 degenerate fibers, then the normal component of a regular vector field on X is equal to 0, i.e., the field is tangent to the fibers.

2. Let $f: X \to B$ be a fibration into elliptic curves (with base B which is not necessarily complete), let D be a regular vector field on X, and let s_0 be a section which defines a group law on the set of nonsingular points of any fiber F_b, so that $s_0(b)$ is the zero of this group. If s is some other section, then translation by s defines an automorphism t_s of the surface X. Then $t_s D = D$. Indeed, on restriction to each nondegenerate fiber F, D defines a section $\overline{D}$ invariant relative to translations of the bundle $E = \Theta_X|_F$, while the translation t_s becomes translation t_α at a point $\alpha \in F$. It is obvious that the translation does not change

the normal component, and therefore $t_\alpha\overline{D} = D + V(\alpha)$, where $V(\alpha)$ is tangent to the fiber F. We have thus defined a mapping $\alpha \to V(\alpha) \in H^0(\Theta_{X/B})$. Considering a formal neighborhood (or a Henzelization) of the fiber F, we may assume that through each point of the fiber there passes a section, and we obtain a morphism $F \to H^0(\Theta_{X/B})$, which must obviously be zero. Since t_sD and D coincide on restriction to any nondegenerate fiber, it follows that $t_sD = D$.

This remark has some useful applications. First of all, suppose C is a component of a degenerate fiber F_b. Since the normal component D_N of a regular vector field D on X vanishes at the point b, it follows that D is tangent to C. If $U \subset F_b$ is the set of nonsingular points of the fiber F_b, then arguing as above we find that $D|_U$ is invariant relative to the group structure on U.

The second application is that a regular vector field D defined on an elliptic fibration $X \to B$ carries over to its Jacobian fibration. This follows from the fact that the Jacobian fibration is constructed by factorization by means of translations relative to which D is invariant, as we have seen. If $X \to \mathbf{P}^1$ is an elliptic pencil on a surface of type K3, then its Jacobian fibration is also a surface of type K3. Therefore, it suffices to verify the assertion $H^0(X, \Theta_X) = 0$ for surfaces of type K3 with an elliptic pencil for their Jacobian fibrations.

In particular, we can resolve the question of vector fields on surfaces of type K3 having an elliptic pencil $f: X \to \mathbf{P}^1$ under the condition that the normal component of this field is equal to 0. As we have seen, it may be assumed that the pencil has a section s. Since $c(\mathbf{P}^1) = C \simeq \mathbf{P}^1$, it follows that $C^2 = -2$. Therefore, the normal bundle to C is negative, and hence has no sections. Therefore, the restriction of D to C must be tangent to C. On the other hand, since $D_N = 0$, it follows that D is tangent to the fibers F of the fibration f. Therefore, D must vanish at the point of intersection of C and F. Now a vector field on an elliptic curve F which is equal to 0 at the single point $C \cdot F$ is equal to 0 everywhere. Thus, $D = 0$. Thus, we can henceforth restrict consideration to vector fields D with $D_N \neq 0$.

3. By direct computation in the ring of formal power series it is possible to prove [7, p. 1274] that in some coordinate system a field of multiplicative type which vanishes can be reduced to the form $\sum \alpha_i x_i \frac{\partial}{\partial x_i}$, $\alpha_i \in \mathbf{F}_p$. This implies the important corollary: a manifold on which a vector field of multiplicative type vanishes is smooth.

If $f: X \to \mathbf{P}^1$ is a fibration of the surface X and D is a regular vector field on X with normal component D_N which is nonzero while D is of additive or multiplicative type, then D_N has the same type. On the other hand, it is easy to see that a vector field on $\mathbf{P}^1$ has additive type if it vanishes at a single point and has multiplicative type if it vanishes at two points. In view of the foregoing, we find the following "exclusion rules" for the extension on a surface of type K3 with an elliptic pencil $f: X \to \mathbf{P}^1$ of a regular vector field D of additive or multiplicative type with $D_N \neq 0$. A vector field with $D_N \neq 0$ cannot exist if the pencil f has more than two degenerate fibers. If it has 2 degenerate fibers, then this field must be of multiplicative type.

For fields of multiplicative type there are special "exclusion rules." Namely, for all degenerate fibers besides those of type A_n the connected component of the group of nonsingular points is isomorphic to G_a. As we have seen, a vector field must be tangent to this component and be invariant relative to G_a. Therefore, if the field is of multiplicative type, then on this component it must be equal to 0. If two such components intersect, as occurs in fibers of type $^*A_1$ and $^*A_2$, then we obtain a contradiction to the property of fields of multiplicative type formulated above. On the other hand, the vector field is tangent to each component of a degenerate fiber and therefore vanishes at a point of intersection of two components. If one component intersects 3 others, then on it the field vanishes at three points, and hence it is identically zero. The first consideration is applicable to components of multiplicity 1 of a fiber D_n, and the second to a component intersecting them. Figure 7 shows the components on which the field vanishes.

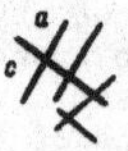

Fig. 7

The vector field is equal to zero on components a, b, and c. This contradicts the property already applied of fields of multiplicative type. As a result, we obtain still another "exclusion rule."

A vector field D of multiplicative type cannot exist if the pencil f has a fiber of type $*A_1$, $*A_2$ or D_n.

Combining these "exclusion rules," we arrive at the following result.

Proposition 2. If on a surface X of type K3 there exists an elliptic pencil with three degenerate fibers or with two such fibers one of which has type $*A_1$, $*A_2$ or D_n, then $H^0(X, \Theta) = 0$.

It is now easy to prove the basic result.

THEOREM. On a surface X of type K3 there exists no regular vector field except the zero field.

We suppose that $H^0(X, \Theta) \neq 0$. According to the corollary of Proposition 1, the surface X is elementary. In view of Proposition 2, the theorem will be proved if we show that on any elementary surface of type K3 there exists an elliptic pencil having at least 3 degenerate fibers or 2 degenerate fibers one of which has type $*A_1$, $*A_2$ or D_n. The proof breaks into three parts depending on the value of p.

1) $p > 3$. In this case there is no higher ramification, and formula (2) of Sec. 4 acquires the form

$$\sum_F \mathbf{e}(F) = 24.$$

From this it follows easily that any elliptic pencil possesses the required property. In other words, a pencil is impossible which has 1 degenerate fiber or 2 fibers of which none has the type $*A_1$, $*A_2$, D_n. Indeed, in the case of one degenerate fiber F, $\mathbf{e}(F) = 24$. If $\operatorname{rk} F = n$, i.e., F consists of $n + 1$ components, then $\mathbf{e}(F) = n + 2$ or $n + 1$ and $n \geqslant 22$. Now the components of the fiber and the section are independent in Pic X, and hence $n + 2 \leqslant 22$, $n \leqslant 20$. In the case of two fibers F_1 and F_2 of ranks n_1 and n_2 a similar contradiction is obtained by noting that no fiber has type D_n, and since $n_1 + n_2 \geqslant 20$, one of the fibers, for example, F_1, has type A^n and form it $\mathbf{e}(F_1) = n_1 + 1$. Together with the condition $\mathbf{e}(F_2) = n_2 + 1$ or $n_2 + 2$ we find that $n_1 + n_2 \geqslant 21$. Since the components of the fiber F_1, n_2 components of the fiber F_2, and the section are independent in Pic X, we again obtain the contradiction: $n_1 + n_2 \leqslant 20$.

For $p = 3$ and 2 we consider the pencil which obviously exists if we use the form of the lattice S = Pic X given by the theorem of Sec. 1. Namely, for $S = U_2 \oplus N$ or $S = U_2(p) \oplus N$ we take $f \in U_2$ or $f \in U_2(p)$. Then $Sf = N$.

2) $p = 3$. It is easy to see that $H_3 = Q(A_2)^2$ [by the way, we could immediately use $Q(A_2)^2$ in place of H_3 in the construction of the lattices in the theorem of Sec. 1]. We use the term $(I(-3)^k \oplus I(-1)^l)_* -$ in our case we always have $l \equiv 0 \pmod 4$. If $l > 0$, then elements $x \in I(-1)^l = \oplus \mathbf{Z} e_i$, $e_i^2 = -1$ of the form $e_i \pm e_j$ are roots orthogonal to the two root systems already constructed, and therefore the pencil determined by the class of f has at least 3 reducible fibers. For $l = 0$ we have only reducible fibers corresponding to the two terms $Q(A_2)$. They have type A_2 or $*A_2$. If at least one of them has type $*A_2$, then the theorem has been proved. If they both have type A_2, then since for a fiber of type A_n, $d(A_n) = 0$, they give a contribution of 4 to the left side of formula (2) in Sec. 4. The right side is equal to 24, and hence there must exist other terms, i.e., at least 3 degenerate fibers. Our pencil is not quasielliptic, since the root system of the lattice $(I(-3)^k \oplus I(-1)^l)_*$, as is easily seen, has type D_{16-k}. Therefore, it is not 3-elementary for $k < 16$, while for $k = 16$ its rank is less than the rank of the lattice.

3) $p = 2$. In this case s is even: $s = 2\sigma$. We proceed from the proof of Theorem 1, Sec. 5 and the list of quasielliptic pencils and supersingular surfaces presented there. We apply to them the proposition of Sec. 4, the proof of which contains a method of constructing elliptic pencils on the basis of quasielliptic pencils. Namely, the proof contains the relation

$$D_4 \oplus F_1 \oplus F_2 \Rightarrow D_5, \tag{4}$$

which on the basis of fibers D_4, F_1, F_2 of a quasielliptic pencil constructs a new pencil with fiber D_5. It is possible to strengthen this relation by noting that if the number of components of some F_i is greater than two, then among those components which together with the curve ξ form the fiber D_5 there are still components not intersecting them and which hence lie in another fiber. Therefore, (4) is true in the stronger form $D_4 \oplus F_1 \oplus F_2 \Rightarrow D_5 \oplus F$, if rk $F_1 > 1$ or rk $F_2 > 1$. A pencil having fibers D_5 and F satisfies all our conditions.

Fibers D_4 and 2 further fibers of ranks greater than 1 are found in all the lattices of the table in the theorem of Sec. 5 corresponding to the values $\sigma = 1,\ldots,6$. Using the remark to the theorem of Sec. 1, it is only necessary to replace everywhere the root system D_n, $n > 8$ by $D_{n-8} \oplus E_8$, and D_m, $n > 16$ by $D_{m-16} \oplus E_8^2$. We then obtain the following lattices:

σ	1	2	3	4
S	$U_2 \oplus Q(D_4 \oplus E_8^2)$	$U_2 \oplus Q(D_4 \oplus D_8 \oplus E_8$	$U_2 \oplus Q(D_4 \oplus D_8^2)$	$U_2 \oplus Q(D_4^2 \oplus \oplus D_8)$

σ	5	6
S	$U_2 \oplus Q(D_4^5)$	$U_2 \oplus Q\widetilde{(D_4^3 \oplus D_8)}$

In the cases σ = 7, 8, 9, 10 it is easy to indicate directly the lattices we require:

σ	7	8	9	10
S	$U_2 \oplus Q(D_4^3 \oplus \oplus Q(E_8)(2)$	$U_2(2) \oplus Q(D_4^3) \oplus \oplus Q(E_8)(2)$	$U_2 \oplus Q(\widetilde{D}_{12}) \oplus \oplus Q(E_8)(2)$	$U_2(2) \oplus \oplus Q(\widetilde{D}_{12}) \oplus \oplus Q(E_8)(2)$
$R(f)$	D_4^3	D_4^3	A_1^{12}	A_1^{12}

In all cases we have no less than three degenerate fibers. The pencil f is not quasielliptic, since rk $R(f) < 20$. The proof of the theorem is complete.

Remark. The role played in this proof by the presence of an elliptic pencil on the surface X suggests the idea of trying to find vector fields on elliptic surfaces. This is done in the works [7, 9]. If the characteristic of the field $p > 3$, then the answer is as was to be expected: vector fields can exist only in a very simple situation: either all fibers of the pencil are isomorphic to one and the same elliptic curve A (and there are no degenerate fibers) and A acts fiberwise on X (the existence of a vector field is then obvious) or X is an Abelian variety or a hyperelliptic surface (a factor of an Abelian variety by a finite graph), or X is rational. By the way, it is possible to conclude already from these simple examples that the scheme of automorphisms of an algebraic surface over a field of finite characteristic may be nonreduced.

Actually, nontrivial examples of elliptic surfaces having nonzero, regular vector fields exist only over fields of characteristic $p = 2$ or 3. They have all been enumerated. As we have seen, the general case reduces to the case where the fibration has a section. Under this assumption all possible surfaces are given by the equations (t is the parameter on the base $\mathbf{P}^1$):

$$p=3.\quad y^2 = x^3 + x^2 - t^{3^{2n}},$$
$$y^2 = x^3 + tx^2 - t^{3^{2n}-3},\ n>0,$$
$$p=2.\quad y^2 + xy = x^3 + ct^2x^2 + t^{2^{2n+1}},\ c \in k,\ n>0.$$

The enumeration of rational surfaces possessing vector fields is apparently a rather complicated problem.

So far no example is known of a surface of general type on which there exists a vector field, and it would be interesting to determine if such surfaces exist.

7. Deformations

The relation $H^0(X, \Theta) = 0$ proved in the preceding section for a surface X of type K3 makes it possible to investigate deformations of these surfaces. The computation of the numbers $h^{i,j}$ presented in Sec. 6 shows that $\dim_k H^1(X, \Theta)=20$, $H^2(X, \Theta)=0$. Because of this, we can apply Schlessinger's theory of formal deformations [36]. From this theory we obtain [in view of the values presented for $\dim H^i(X, \Theta)$] the prorepresentability of the functor Def X which is defined on the category of Artinian local k-algebras and assigns to an algebra A the set of classes (up to isomorphism) of flat deformations of a surface X over A. Moreover, the universal family is formally smooth, and the base has dimension 20, i.e., is isomorphic to $S = \operatorname{Spf} k[[T_1,\dots,T_{20}]]$. There is thus a plane family $\mathscr{X}\to S$, with closed fiber isomorphic to X which possesses the universal property.

Fibers of the family $\mathscr{X}$, lying over nonclosed points are, in general, formal schemes. From the geometric point of view it is more interesting, of course, to investigate deformations with fibers which are algebraic or even projective surfaces of type K3. For this it suffices to take some very ample invertible pencil L on X and to consider a subscheme $\Sigma(L)\subset S$, such that L extends to the restriction of $\mathscr{X}$ to $\Sigma(L)$.

LEMMA 1. Let S be the spectrum of a local ring, possibly formal, let $\mathscr{X}\to S$ be a formal family of surfaces of type K3, and let L be an invertible pencil on the fiber $X=\mathscr{X}_s$ over a point $s\in S$. There exists a closed subscheme $T\subset S$, such that L extends to $\mathscr{X}|_T=\mathscr{X}_T$, $\mathscr{X}_T\to T$ is a universal family with this property, and T is defined in S by one equation.

The existence of T is proved by a standard application of deformation theory as is the existence of a universal deformation (verification of the required conditions can be found in [19]). We shall verify that T is defined by one equation. Let S = Spec A, T = Spec B, let $\mathfrak{m}$ be a maximal ideal of A, let B = A/J, B' = A/J$\mathfrak{m}$, and let T' = Spec B'. Since for any ideal M with $M^2 = 0$, $(1 + M)^* \simeq M$, we have the exact sequence of homomorphisms of pencils on $\mathscr{X}$:

$$0\to \mathcal{O}_{\mathscr{X}_T}\otimes J/J\mathfrak{m}\to \mathcal{O}^*_{\mathscr{X}_{T'}}\to \mathcal{O}^*_{\mathscr{X}_T}\to 0.$$

From the exact sequence of homologies we obtain the sequence of homomorphisms

$$H^1\left(\mathscr{X}, \mathcal{O}^*_{\mathscr{X}_{T'}}\right)\to H^1\left(\mathscr{X}, \mathcal{O}^*_{\mathscr{X}_T}\right)\to H^2(\mathscr{X}, \mathcal{O}_{\mathscr{X}_T}\otimes J/J\mathfrak{m}),$$

which shows that the obstruction to the extension of the **invertible** pencil $\mathscr{L}$ from $\mathscr{X}_T$ to $\mathscr{X}_{T'}$ lies in $H^2(\mathscr{X}, \mathcal{O}_{\mathscr{X}_T}\otimes J/J\mathfrak{m})$. Since we are dealing with a family of surfaces of type K3, it follows that

$$H^2(\mathscr{X}, \mathcal{O}_{\mathscr{X}_T}\otimes J/J\mathfrak{m})\simeq H^2(\mathscr{X}, \mathcal{O}_{\mathscr{X}_T})\otimes J/J\mathfrak{m}\simeq J/J\mathfrak{m}.$$

[we choose a basis in $H^2(\mathscr{X}, \mathcal{O}_{\mathscr{X}_T})$ to establish this isomorphism]. Let $f\in J$ be the element of J defining this obstruction, and let $B''=B/(J\mathfrak{m}+(f))$, $T''=\operatorname{Spec} B''$. Then the obstruction to extending the pencil $\mathscr{L}$ to $\mathscr{X}_{T''}$ vanishes, i.e., $\mathscr{L}$ extends to $\mathscr{X}_{T''}$. In view of the universality of T this is possible only for T" = T, i.e., $J = J\mathfrak{m}+(f)$. By Nakayama's lemma this implies that J = (f).

Returning to the universal family $\mathscr{X}\to S$, we obtain for any invertible pencil L on X a subscheme $\Sigma(L)\subset S$, with dimension equal to 19 or 20. The question is which of these two possibilities actually occurs. Deligne showed [19] that the first occurs if, of course, the sheaf is nontrivial. We shall present a sketch of the proof in a somewhat more special case in order to present his arguments in the simplest form.

LEMMA 2. Let X be an algebraic surface such that $H^1(X, \mathcal{O})=H^0(X, \Omega^1)=0$, and let S be the Neron–Severi group of X. Then the mapping assigning to an invertible pencil the Chern class defines an imbedding

$$c_1: S/pS\hookrightarrow H^2_{\mathrm{DR}}(X).$$

By definition, $H^2_{\mathrm{DR}}(X)$ are the hyperhomologies of the double complex

$$\begin{array}{ccccc} C_2(\Omega^0) & \xrightarrow{d} & C_2(\Omega^1) & \xrightarrow{d} & C_2(\Omega^2)\\ \partial\uparrow & & \partial\uparrow & & \partial\uparrow\\ C_1(\Omega^0) & \xrightarrow{d} & C_1(\Omega^1) & \xrightarrow{d} & C_1(\Omega^2),\\ \partial\uparrow & & \partial\uparrow & & \partial\uparrow\\ C_0(\Omega^0) & \xrightarrow{d} & C_0(\Omega^1) & \xrightarrow{d} & C_0(\Omega^2)\end{array}$$

where C_i are the i-dimensional Čech chains, the horizontal differentials are defined in terms of taking the differential of differential forms, and the vertical differentials are taken in the Čech sense. Therefore, an element of $H^2_{DR}(X)$ is given by a triple $(\alpha_0, \alpha_1, \alpha_2) \in C_2(\Omega^0) \oplus C_1(\Omega^1) \oplus C_0(\Omega^2)$, and the Chern class of a pencil L with transition functions $f_{ij} \in \mathcal{O}_X^*(U_i \cap U_j)$ is defined as (0, d log f_{ij}, 0). If this cycle is a coboundary of a chain $(h_{ij}, \omega_i) \in C_1(\Omega^0) \oplus C_0(\Omega^1)$, then

$$f_{ij}^{-1} df_{ij} = \omega_i - \omega_j + dh_{ij}, \tag{1}$$

and $d\omega_i = 0$, $dh_{ij} = 0$. In view of the condition $H^1(X, \mathcal{O}) = 0$, we may assume that $h_{ij} = 0$ [by changing, if necessary, ω_i in (1)]. Applying the Cartier operator to both sides of Eq. (1) and subtracting from (1), we obtain (since log f = d log f), $C\omega_i - C\omega_j = \omega_i - \omega_j$, i.e., $C\omega_i - \omega_i = C\omega_j - \omega_j$. Therefore, the forms $C\omega_i - \omega_i$ define a single form $\eta \in H^0(X, \Omega^1)$. By hypothesis $\eta = 0$. Thus, $C\omega_i = \omega_i$, and by the well-known property of the Cartier operator from this it follows that $\omega_i = d\log\varphi_i$, $\varphi_i \in \mathcal{O}(U_i)$, possibly with smaller sets U_i. As a result, (1) can be written in the form $d\log f_{ij} = d\log\varphi_i\varphi_j^{-1}$, i.e., $f_{ij} = \varphi_i\varphi_j\psi_{ij}^p$, and this means that the pencil is divisible by p in S.

Proposition 1. If an invertible pencil L is not divisible by p in S_X, then the scheme $\Sigma(L)$ has dimension 19.

Otherwise L could be extended as an invertible pencil $\mathcal{L}$ to the entire universal family $\mathcal{X}$. This assumption must lead to a contradiction. The proof is based on applying standard properties of the de Rham cohomology of the family $\mathcal{X}/S$. However, the special nature of our situation is that now these properties must be applied in a "formal" version. However, the difference here is minor: if we use, for example, fully the algebraic exposition at the beginning of the work [26], then the proofs are preserved almost literally. We shall list those properties we require.

The de Rham cohomology is defined by means of the complex of formal differential forms where by definition $\Omega^i_{\mathcal{X}/S}$ is the projective limit of ordinary differential forms $\Omega^i_{\mathcal{X}'}$, $\mathcal{X} = \mathcal{X} \times_S S'$, and S runs through a system of infinitesimal neighborhoods of the point s in S. By definition, for the family $f: \mathcal{X} \to S$, $H^i_{DR}(\mathcal{X}/S) = R^i f_*(\dot{\Omega}_{\mathcal{X}/S})$. The cohomology groups $H^i_{DR}(\mathcal{X}/S)$ are modules of finite type over $\mathcal{O}_S$. There exists the spectral sequence

$$E_1^{ij} = H^j(\mathcal{X}, \Omega^i_{\mathcal{X}/S}) \Rightarrow H^*_{DR}(\mathcal{X}/S). \tag{2}$$

In the usual way this spectral sequence generates in $H^i_{DR}(\mathcal{X}/S)$ a filtration F. In our case, in analogy to what we showed in Sec. 6, the spectral sequence (2) degenerates in the number E_1. Simple considerations of commutative algebra show that the $\mathcal{O}_S$-modules and $H^j(\mathcal{X}, \Omega^i_{\mathcal{X}/S})$ and $H^i_{DR}(\mathcal{X}/S)$ are free and of finite type, and the restrictions $H^j(\mathcal{X}, \Omega^i_{\mathcal{X}/S}) \otimes k \to H^j(X, \Omega^i_{X/k})$, $H^i_{DR}(\mathcal{X}/S) \otimes k \to H^i_{DR}(X/k)$ are isomorphisms. From this, in particular, it follows that the filtration F in $H^2_{DR}(\mathcal{X}/S)$ consists of free $\mathcal{O}_S$-modules and has the form

$$H^2_{DR}(\mathcal{X}/S) = F^0 \supset F^1 \supset F^2 \supset F^3 = 0,$$
$$\operatorname{rk} F^0 = 22, \quad \operatorname{rk} F^1 = 21, \quad \operatorname{rk} F^2 = 1.$$

Differentiation of the de Rham cohomology classes with respect to a parameter defines operations ∇, the so-called Gauss–Minin connections:

$$\nabla: H^i_{DR}(\mathcal{X}/S) \to \Omega^1_{S/k} \otimes H^i_{DR}(\mathcal{X}/S).$$

Elements $x \in H^2_{DR}(\mathcal{X}/S)$, corresponding to invertible sheaves, i.e., having the form $c_1(\mathcal{L})$, are horizontal, i.e.,

$$\nabla x = 0. \tag{3}$$

The relations

$$\nabla F^i H^2_{DR}(\mathcal{X}/S) \subset \Omega^1_{S/k} \otimes F^{i-1} H^2_{DR}(\mathcal{X}/S)$$

(the so-called Griffiths transversality relations) are satisfied which define mappings for the graded components of the filtration F $\operatorname{gr}^i H^2_{DR}(\mathcal{X}/S) = H^{2-i}(\mathcal{X}, \Omega^i_{\mathcal{X}/S})$,

$$\operatorname{gr}^i \nabla: H^{2-i}(\mathcal{X}, \Omega^i_{\mathcal{X}/S}) \to \Omega^1_{S/k} \otimes H^{3-i}(\mathcal{X}, \Omega^{i-1}_{\mathcal{X}/S}), \tag{4}$$

which in view of the product rule for the connection ∇ are $\mathcal{O}_s$-homomorphisms. These mappings have the following interpretation. By duality, the homomorphisms $\operatorname{gr}^i\nabla$ define the mappings

$$\nabla_i:\Theta_S\to \mathrm{Hom}\,(H^{2-i}(\mathscr{X}, \Omega^i_{\mathscr{X}/S}), H^{3-i}(\mathscr{X}, \Omega^{i-1}_{\mathscr{X}/S})). \tag{5}$$

Let

$$\mathrm{Kod}\,(\mathscr{X}/S):\Theta_S\to H^1(\mathscr{X}, \Theta_{\mathscr{X}/S})$$

be the Kodaira mapping corresponding to the family $\mathscr{X}/S$. Then there is the commutative diagram

$$\begin{array}{ccc} & \overset{\mathrm{Kod}(\mathscr{X}/S)}{\nearrow} & H^1(\mathscr{X}, \Theta_{\mathscr{X}/S}) \\ \Theta_S & & \downarrow{\scriptstyle \varphi_i} \\ & \underset{\nabla_i}{\searrow} & \mathrm{Hom}\,(H^{2-i}(\mathscr{X}, \Omega^i_{\mathscr{X}/S}), H^{3-i}(\mathscr{X}, \Omega^{i-1}_{\mathscr{X}/S})), \end{array} \tag{6}$$

where φ_i is defined by the multiplication of cohomologies and the pairing $\Theta_{\mathscr{X}/S}\otimes\Omega^i_{\mathscr{X}/S}\to\Omega^{i-1}_{\mathscr{X}/S}$. After these preliminary facts, the proofs of which all exactly repeat the arguments of parts 1.1, 1.2, and 1.4 of the work [26], it is possible to proceed to the proof of the proposition.

We assume that the sheaf L extends to a pencil $\mathscr{L}$ on the scheme $\mathscr{X}$. Then its Chern class $c_1(\mathscr{L})$ is an element $x\in H_{DR}^2(\mathscr{X}/S)$, whose restriction at the point $s\in S$ is the class $c_1(L)$. Since by hypothesis L as an element of the group S_X is not divisible by p, according to the lemma $c_1(L) \neq 0$. Hence $x \neq 0$. By construction $x\in F^1H_{DR}^2(\mathscr{X}/S)$, and according to (3) $\nabla x = 0$. Now these properties contradict one another: the mapping ∇ is a monomorphism on $F^1H_{DR}^2(\mathscr{X}/S)$. For the verification it suffices for us to show that $\mathrm{gr}^1\nabla$ and $\mathrm{gr}^2\nabla$ are monomorphic. This will be proved if we verify that ∇_1 and ∇_2 are even isomorphisms. We use the interpretation given by the diagram (6). The mapping $\mathrm{kod}(\mathscr{X}/S)$ is an isomorphism. Since it is a question of a homomorphism of free modules, it suffices to verify that it is an isomorphism after tensor multiplication by k, i.e., $\Theta_{S,s}\to H^1(X, \Theta_{X/k})$ is an isomorphism. This follows immediately from the fact that S is the base of a universal family. Thus, it suffices to show that φ_1 and φ_2 are isomorphisms. By the same argument we see that for this it suffices that $\varphi_i\otimes k$ be isomorphisms. These are homomorphisms of the cohomologies of the surface X:

$$\varphi_1\oplus k: H^1(X, \Theta_X)\otimes H^1(X, \Omega^1)\to H^2(X, \mathcal{O}_X),$$
$$\varphi_2\oplus k: H^1(X, \Theta_X)\otimes H^0(X, \Omega^2)\to H^1(X, \Omega^1).$$

The fact that they are isomorphisms follows from the universal duality theorems if we use the fact that on a surface of type K3 $\Theta_X\simeq\Omega_X^1$, $\mathcal{O}_X\simeq\Omega_X^2$.

Remarks. 1. The condition of nondivisibility by p of the image of the pencil in the group S_X was necessary, since we use only the de Rham cohomology. In the work [19] the theorem is proved without this assumption, but for this the crystalline cohomology is used in place of the de Rham cohomology.

2. We shall apply the argument of the proposition proved to the base $\Sigma(L)$ in place of S. Let $x=c_1(\mathscr{L})$, where $\mathscr{L}$ is the pencil extending L to the universal family over $\Sigma(L)$. Then $\mathrm{gr}^1x\in\mathrm{Ker}\,\mathrm{gr}^1\nabla$ [see (3)]. Hence $(\nabla_1\Theta_{\Sigma(L)})x=0$. Since φ_1 is an isomorphism (the proof made no use of the fact that S was the base of a universal family), it follows that $\langle \mathrm{Kod}\,\Theta_{\Sigma(L)}, \mathrm{gr}^1x \rangle = 0$. Hence,

$$\langle \mathrm{Kod}\,\Theta_{\Sigma(L),s}, \mathrm{gr}^1x \rangle = 0. \tag{7}$$

On the other hand, we know that $\dim\Sigma(L) = 19$. Therefore, if $\mathrm{gr}^1c_1(L)\neq 0$ [i.e., $c_1(L)\notin F^2H_{DR}^2\times(X)$], then (7) gives and equation for $\mathrm{Kod}\,\Theta_{\Sigma(L),s}$ and thus shows that $\Sigma(L)$ is smooth.

Suppose now that U is a set of inversible pencils on X. We denote by $\Sigma(U)$ the subscheme in S to which all pencils of U extend. The linear hull of a set T of elements of a vector space we denote by <T>.

Proposition 2. Let $c[U]=\langle \mathrm{gr}^1c_1(L), L\in U \rangle\subset H^1(X, \Omega^1)$. Then $\dim\Sigma(U)\leqslant 20-\dim c[U]$.

Proof. Suppose that $L\in U$ and the image of L in S_X/pS_X is not equal to zero. Then $\Sigma(U)\subset\Sigma(L)$ and $\mathrm{Kod}\,\Theta_{\Sigma(U),s}\subset\mathrm{Kod}\,\Theta_{\Sigma(L),s}$. Hence by formula (7)

$$\langle \mathrm{Kod}\,\Theta_{\Sigma(U),s}, \mathrm{gr}^1c_1(L) \rangle = 0. \tag{8}$$

In the image, if L in S_X/pS_X is equal to zero, formula (7) also remains true, since $gr^1 c_1(L) = 0$. As a result we find that

$$\langle \mathrm{Kod}\, \Theta_{\Sigma(U),s}, c[U] \rangle = 0. \tag{9}$$

This shows that

$$\dim \Sigma(U) \leqslant \dim \mathrm{Kod}\, \Theta_{\Sigma(U),s} \leqslant 20 - \dim c[U].$$

From Proposition 1 there follows one important consequence: the possibility of lifting a surface of type K3 defined over a field of characteristic p to characteristic 0.

THEOREM 1 [19]. For any surface of type K3 over a field k of characteristic $p > 0$ there exists a normalization ring V which is a finite extension of the ring W(k) and a flat, smooth proper scheme $\mathscr{X}/V$, for which $\mathscr{X} \otimes_V k = X$.

We consider the "broader" deformation functor Def X defined on the category of Artinian local algebras over W with the same definition of the sets Def X as at the beginning of this section. As in the case considered above, it can be proved that this functor is prorepresentable, that the universal family has a base $\overline{\Sigma} = \mathrm{spf}\, W[[T_1, \ldots, T_{20}]]$, and that for any sheaf $L \in \mathrm{Pic}\, X$ there exists a subscheme $\overline{\Sigma}(L)$ defined by one equation $f \in W[[T_1, \ldots, T_{20}]]$ which classifies the deformations to which the pencil L extends. Those schemes S and $\Sigma(L)$ considered in the theorem are obtained from $\overline{\Sigma}$ and $\Sigma(L)$ by reduction modulo p. Since according to the proposition $\Sigma(L) \neq S$, it follows that $f \not\equiv 0 \pmod p$. Suppose that $\overline{\Sigma}(L) = \mathrm{Spf}\, \overline{R}$; then p is not a divisor of zero in $\overline{R}$. From this it follows that in $\overline{R}$ there exists a system of parameters $p, x_1, \ldots, x_{19}$ (i.e., a system of elements with length equal to the dimension of the ring $\overline{R}$ which generates a primary ideal corresponding to the maximal ideal). Then the ring $R = \overline{R}/(x_1, \ldots, x_{19})$ is a module of finite type over W. For the algebra $R \otimes_W K$ (where K is the field of fractions of W) there exists a homomorphism $R \otimes_W K \to K_1$, where K_1 is a finite extension of K.

Let V be the ring of integral elements of K_1; we obtain a homomorphism $R \to V$. The composition $\overline{R} \to R$ gives a homomorphism $\overline{R} \to V$ or a morphism $\mathrm{Spf}\, V \to \overline{\Sigma}(L)$. Let $\mathscr{X} \to \Sigma(L)$ be the family to which, by assumption, the pencil L extends. Then $\mathscr{Y} = \mathscr{X} \otimes_{\overline{R}} V$ gives a family of deformations $\mathscr{Y} \to \mathrm{Spf}\, V$, to which the pencil L extends. We now choose L so that it is very ample and is not divisible by p in the group S_X. This choice is possible: if $L_\perp$ is an ample, primitive element of S, then a sufficiently large multiple $L = nL_\perp$ [in particular, with $(n, p) = 1$] satisfies these conditions. The Grothendieck theorem of algebraization is now applicable to the formal scheme $\mathscr{Y}/\mathrm{Spf}\, V$ ([22], 5.4.5), whence the assertion of the theorem follows.

Remark. So far it is not known whether any surface can be lifted to a proper, smooth, flat scheme over W. Ogus in the work [33] obtained results very closely related to this assertion.

8. Crystalline Cohomologies

Crystalline cohomologies are defined for varieties over a field of finite characteristic p and for them are a replacement of the l-adic étalé cohomologies for $l = p$. Here we present the proofs of those properties of crystalline cohomologies which we shall use below. For a systematic survey of this theory see [14, 13, and 24]. We shall consider a smooth, complete variety X over an algebraically closed field k of characteristic p. Let W(k), or simply W, be the ring of Witt vectors over k. Crystalline cohomologies are denoted by $H^i_{\mathrm{cris}}(X)$ or $H^i(X/W)$. This is a module over W possessing the following properties:

1. Functoriality: $H_i(X/W)$ is a functor from the category of complete, smooth algebraic varieties to the category of W-modules.

2. Finiteness: $\mathrm{rk}_W H^i(X/W) < \infty$, $H^i(X/W) = 0$ for $i > 2 \dim X$.

3. Multiplicative properties: the Künneth formula holds, an anticommutative multiplication is defined, and the Poincaré duality holds if the modules $H^1(X/W)$ are free of torsion.

4. The "first Chern class" is defined, i.e., a homomorphism

$$c_1 : \mathrm{Pic} X \to H^2(X/W).$$

In the case of algebraic surfaces the intersection index in Pic X and the multiplication defined by the Poincaré duality $H^2 \otimes H^2 \to W$ are consistent.

5. If $H^i_{DR}(X)$ are the de Rham cohomologies of the variety X, then

$$\dim_k H_{DR}^i(X) = \mathrm{rk}_W H^i(X/W) + \tau_i + \tau_{i+1},$$

where τ_i is the number of torsion generators in $H^i(X/W)$.

In particular, if all the modules $H^i(X/W)$ are free, then

$$H^i_{DR}(X) = H^i(X/W)_W \otimes k.$$

6. Comparison theorems:

$$\mathrm{rk}_W H^i(X/W) = b_i,$$

where b_i are the Betti numbers defined by means of l-adic cohomologies.

7. By functoriality the absolute Frobenius operation $(x \to x^p)$ defines a homomorphism $F: H^i(X/W) \to H(X/W)$ which is a φ-linear mapping, where φ is the Frobenius automorphism of the ring W. For an n-dimensional variety X the homomorphism F acts on the fundamental class in the homologies of highest dimension $H^{2n}(X/W)$ as multiplication by p^n.

Properties 5 and 6 afford the possibility of finding conditions under which the cohomology theory is especially close to that which holds for projective varieties over the field of complex numbers. We set $h^{p,q} = \dim_k (X, \Omega^p)$, where Ω^p is the pencil of p-dimensional differential forms. We suppose that for $0 \leqslant m \leqslant 2\dim X$

$$\sum_{p+q=m} h^{p,q} = b_m. \tag{1}$$

On the other hand, we recall that for the de Rham cohomology there exists the spectral sequence

$$E_1^{p,q} = H^q(X, \Omega^p) \Rightarrow H^{p+q}_{DR}(X), \tag{2}$$

whence $\dim_k H^m_{DR}(X) \leqslant \sum_{p+q=m} h^{p,q}$. In combination with properties 5 and 6 this shows that $\sum_{p+q=m} \times h^{p,q} \geqslant b_i + \tau_i + \tau_{i+1}$, so that in view of (1) we find the equalities $\tau_i = 0$, i.e., the group $H^i(X/W)$ is a free W-module, and hence $\dim_k H^m_{DR}(X) = \sum_{p+q=m} h^{p,q}$, whence it follows that the spectral sequence degenerates in the first term.

As stated at the beginning of Sec. 5, the basic theorem of this section makes it possible to find the numbers $h^{p,q}$ for a surface of type K3. The values presented there show that relation (1) is satisfied, and hence the W-module $H^2(X/W)$ is free (the remaining cohomology groups are not interesting):

$$b_0 = b_4 = 1, \ b_1 = b_3 = 0.$$

The object arising in this situation has its own name. A *crystal* is a free W-module H of finite rank and a φ-linear transformation $F: H \to H$. A crystal is denoted by the symbol (H, F). If $\mathscr{X} \to B$ is a smooth, proper family of algebraic varieties, then their crystalline cohomology groups form an object called a *crystal over the base B*. We do not give its definition here (referring the reader to the works [24, 27, or 32]) although we shall use it occasionally.

We shall always have in mind also the lattice structure on $H = H^2(X/W)$ given by the Poincaré duality. Thus, in H there is given a scalar product compatible with the scalar product in S_X on the image of S_X under the mapping c_1, and

$$(Fx, Fy) = p^2(x, y), \tag{3}$$

by Property 7. We observe that the lattice H is unimodular.

In the remainder of this section we present some direct applications of the properties of crystalline cohomology listed above to the theory of supersingular surfaces of type K3. Henceforth we understand X to be such a surface. The crystal $H^2(X/W)$ we denote by H.

<u>Proposition 1.</u> The discriminant of the lattice $S = \mathrm{Pic}\, X$ has the form $-p^{2\sigma}$.

It is obvious that the discriminant is negative, since the lattice is hyperbolic and of even rank 22. We shall show that up to sign it is a power of p, i.e., it is not divisible by any prime number $l \neq p$. For this we note that the l-component $\mathrm{Br}X(l)$ of the Brauer group of the surface X is equal to 0. This follows from the fact that the group $B = \mathrm{Br}\,X(l)$ has a subgroup B^0 infinitely divisible by l which is isomorphic to $(\mathbf{Q}_l/\mathbf{Z})^\tau$, where τ is the "number of transcendental cycles," i.e., $b_2 - \rho$, $\rho = \mathrm{rk}\ S$, and the group B/B^0 is finite and dual to the torsion group of the group S (see [23, pp. 82 and 147]). In our case $\rho = b_2 = 22$, and S has no torsion.

From the exact Kummer sequence $0 \to \mu_l \to G_m \to G_m \to 0$, because of the fact that $\mathrm{Br}\,X(l) = H^2(X, G_m)(l) = 0$, we now obtain the isomorphism $S/lS \simeq H^2(X, \mu_l)$. This isomorphism obviously commutes with multiplication. In view of the Poincaré duality in étalé cohomology groups the multiplication on $H^2(X, \mu_l)$ is nondegenerate, and so it is nondegenerate in S/lS and hence in $S\otimes\mathbf{Z}_l$, i.e., the lattice $S\otimes\mathbf{Z}_l$ is unimodular.

Thus, $d(S) = -p^s$. It remains to show that s is even. According to Property 4 we have the mapping $S \to H$ which is an imbedding since it is consistent with multiplication. It extends to an imbedding $\iota: S\otimes W \to H$. Since both W-modules have the same rank 22, the module $H/\iota(S\otimes W)$ has finite length, i.e., it is isomorphic to $\oplus W/p^{n_i}$. If $\sum n_i = \sigma$, then $d(S\otimes W) = d(\iota(S\otimes W)) = d(H)\,p^{2\sigma}$. Now the lattice H is unimodular in view of the fact that the product satisfies the Poincaré duality theorem. Therefore, $d(H) = (1)$, and we obtain the required assertion. It is instructive to compare this argument with the first proof of Artin based on the application of plane cohomology. The imbedding $S \to H$ extends to a homomorphism $S\otimes \mathbf{Z}_p \to H$ which is also an imbedding, since the product is preserved.

<u>Proposition 2 (see [33]).</u> The image of the group $S\otimes\mathbf{Z}_p$ in H coincides with the set of those elements $x\in H$, for which $Fx = px$.

By Lemma 2 of Sec. 7 the homomorphism $S/pS \to H/pH = H^2_{DR}(X)$ is an imbedding, whence it follows that $H/S\otimes\mathbf{Z}_p$, has no torsion. It is obvious that elements $x\in S\otimes\mathbf{Z}_p$, considered as elements of H possess the property $Fx = px$. We consider the submodule $T \subset H$ consisting of all such elements. The rank of the module T_p over $\mathbf{Z}_p$ does not exceed 22. This follows from the fact that its elements which are linearly independent over $\mathbf{Z}_p$ are also linearly independent over W. Indeed, the relation $\alpha_1 x_1 + \dots \alpha_n x_n = 0$, $\alpha_i \in W$ can be divided by α_i divided by the least power of p to obtain a relation in which, say, $\alpha_1 = 1$. Applying F to it and subtracting from the original relation, we obtain a relation between, x ,...,x_n and thus in a finite number of steps we obtain $x_n = 0$. On the other hand, $T \supset S$, and therefore $\mathrm{rk}\ z_p\ T \geqslant 22$, i.e., $\mathrm{rk}\ z_p\ T = 22$. If $S\otimes\mathbf{Z}_p \neq T$, then the group $T/S\otimes\mathbf{Z}_p$ would be finite, i.e., would have torsion contradicting the fact that $H/S\otimes\mathbf{Z}_p$ has no torsion.

<u>LEMMA 1.</u> If in a crystal (U, Φ) of rank n over W there exists a system of linearly independent elements x ,...,x_n for which $\Phi x_i = x_i$, then there is also a basis possessing this property.

If C is the matrix of the endomorphism Φ is some basis of U and A is the matrix expressing the system $x_1,\dots,x_n$ in terms of this basis, then $C = A^\varphi A^{-1}$. Therefore, $\det C \in W^*$, i.e., $C_0 = C\otimes k$ is a nondegenerate matrix over k. This implies that C can be represented in the form $C = B^\varphi B^{-1}$, where B is a matrix over W such that $\det B \in W^*$. Such a representation is obtained successively modulo increasing powers of p. Modulo p the existence of a matrix B_0 over the field k with the condition $B_0^\varphi B_0^{-1} = C_0$ is a well-known fact [25, p. 213]. The passage to higher powers of p is still more obvious. The equality $C = B^\varphi B^{-1}$ implies that in U there exists a basis $y_1,\dots,y_n$ for which $\Phi y_i = y_i$.

<u>LEMMA 2.</u> The mapping induced by the Frobenius mapping in $H\otimes k$,

$$F\otimes k: H\otimes k \to H\otimes k$$

has rank 1.

According to Property 5, $H\otimes k \simeq H^2_{DR}(X)$, and $F\otimes k$ is the mapping $\overline{F}$ which F induces in the de Rham cohomologies. We use the filtration which the spectral sequence $E^{i,j} = H^j(X, \Omega^i) \Rightarrow H^{i+j}_{DR}(X)$ defines in $H^2_{DR}(X)$. The action of the mapping F on the de Rham complex takes all differential forms into 0 and hence annihilates the first term $F^1H^2_{DR}(X)$ of the filtration Since $\mathrm{gr}^0 H^2_{DR}(X) \simeq H^2(X, \mathcal{O})$, the rank of the mapping $\overline{F}$ is equal to 1 or $\overline{F} = 0$, and it remains for us to show that the latter case is impossible.

The assumption $\bar{F} = 0$ is equivalent to $FH \subset pH$, i.e., $F = p\Phi$, where Φ is a p-linear endomorphism of the module H. Applying Lemma 1 to the crystal (H, Φ), we find that in H there exists a basis $x_1, \ldots, x_{22}$ for which $\Phi x_i = x_i$, i.e., $Fx_i = px_i$. According to Proposition 2, this means that $S \otimes W = H$, and hence $d(S \otimes W) = d(S) \in W^*$. In view of the proposition it follows from this that the lattice S is unimodular, and this is impossible, since according to the general theory [37, p. 92] for a unimodular, even, hyperbolic lattice of rank n, $n \equiv 2 \cdot (\mathrm{mod}\ 8)$.

LEMMA 3. $F^n H \subset p^{n-1} H$.

The proof is by induction on n. Suppose that the assertion has been proved for some value of n. We set $\Psi = p^{-n+1}F$. For those independent elements $x_1, \ldots, x_{22}$ for which $Fx_i = px_i$ we also have $\Psi x_i = px_i$. Hence $\Psi^N \to 0$ as $N \to \infty$. In particular, $\bar{\Psi} = \Psi \otimes k$ is nilpotent over $H \otimes k$. Moreover, it commutes with $\bar{F} = F \otimes k$, since $\bar{F}$ has rank 1; from this it follows that $\bar{\Psi}\bar{F} = 0$. This means that ΨE is divisible by p, i.e., the assertion of the lemma is true for n + 1.

THEOREM. For a supersingular surface X of type K3 the lattice S = Pic X is p-elementary. (For a proof based on flat cohomology see [10].)

We note first that the submodule $S \otimes W$ in H is characterized as the set of elements $x \in H$, for which

$$F^n x \in p^n H \text{ for all n.} \tag{4}$$

Since for $x \in S$, $Fx = px$, for $x \in S \otimes W$ condition (4) is satisfied. We denote by K the set of all elements $x \in H$, satisfying condition (4). Since $H \supset K \supset S \otimes W$, it follows that K is a module of rank 22. By definition, $F|_K$ is divisible by p. We set $\Phi = p^{-1}F$. According to Lemma 1, K has a basis $t_1, \ldots, t_{22}$ for which $\Phi t_i = t_i$. By Proposition 2, $t_i \in S \otimes \mathbf{Z}_p$, and hence $K = S \otimes W$. It is obvious that the submodule K (i.e., as we have shown, $S \otimes W$) is the maximal submodule in H on which F is divisible by p, i.e., which is invariant relative to $\Psi = p^{-1}F$.

According to Lemma 3, the elements $x \in pH$ satisfy conditions (4), i.e., $pH \subset S \otimes W$. This implies that $S^* \otimes W \supset p^{-1}H$ (in view of the fact that H is unimodular), and hence $p(S^* \otimes W) \subset H$. It is obvious that $p(S^* \otimes W)$ is invariant relative to Ψ and by the remark made above $pS^* \otimes W \subset S \otimes W$. From this it follows that $pS^* \subset S$.

We thus see that in the definition of an elementary surface of type K3 the requirement of p-elementarity of the lattice S was superfluous. It follows from the fact that the surface is supersingular. With this the term "elementary surface" ceases to exist.

Proposition 1 and Theorem 1 together with the results of Sec. 1 shows that for a supersingular surface of type K3 its lattice S = Pic X is determined by a single invariant — its discriminant, and we always have $d(S) = -p^{2\sigma}$. We call the number σ the Artin invariant of the surface X and write $\sigma(X)$ or $\sigma(S)$.

As an application of the results obtained we prove the following assertion.

Proposition 3. An automorphism ψ of a supersingular surface X of type K3 which acts trivially on the group S_X is the identity. [The proof is somewhat analogous to the proof of the theorem in Sec. (6)].

We may restrict ourselves to the case where ψ is an automorphism of finite order. Indeed, automorphisms acting trivially on the group S_X form an algebraic group, and in each algebraic group over a field of characteristic $p > 0$ there exist elements of finite order distinct from 1. We shall assume that $\psi^l = 1$, where l is a prime number.

If f is the class of a curve of arithmetic genus 1, then the condition $\psi^* f = f$ implies that ψ preserves the fibering $X \to \mathbf{P}^1$ with fibers belonging to the class f, i.e., there exists an automorphism $\psi_0 : \mathbf{P}^1 \to \mathbf{P}^1$ for which the following diagram is commutative:

$$\begin{array}{ccc} X & \xrightarrow{\psi} & X \\ \downarrow & & \downarrow \\ \mathbf{P}^1 & \xrightarrow{\psi_0} & \mathbf{P}^1 \end{array}.$$

We choose the pencil f in the form suggested by the theorem of Sec. 1, i.e., we assume that $f \in U_2$ or $f \in U_2(p)$ in the representation of the lattice S_X presented there. We break the entire argument into cases: A. $\psi_0 \neq 1$ and B. $\psi_0 = 1$; each of these cases we break into subcases: 1. $l \neq p$ and 2. $l = p$.

A.1. For any reducible curve M we set $w(M) = \nu + \Sigma \mathbf{e}(C_i)$, where ν is the number of fixed points of ψ on M, C_i are the components of M which are fixed curves, and $\mathbf{e}(C_i)$ is the Euler characteristic. We use the relations

$$w(M) = \mathbf{e}(M),\quad \Sigma w(F) = 24, \tag{5}$$

where the sum extends over all fibers of the pencil f in which there lie fixed points or curves. It is obvious that these fibers are situated over two fixed points of ψ_0 on $\mathbf{P}^1$, so that in the sum there are two terms. For $p = 2$ or $p = 3$, $\sigma < 9$ the pencils we consider have no fewer than three reducible fibers, and since ψ takes components of reducible fibers into themselves, this case is impossible. For $p = 3$, $\sigma = 9$, 10 and for $p > 3$ we have no more than two reducible fibers F_i, $i = 1,\ldots,r$. If n_i are the ranks of these fibers, then from the form of the lattices presented in the theorem of Sec. 1 it follows that $\sum_1^r n_i < 20$. Since $\mathbf{e} \times (F_i) \le n_i + 2$ and $r \le 2$, it follows that $\Sigma \mathbf{e}(F_i) < 24$ in contradiction to the relations (5).

A.2. By the same considerations we must have only one reducible fiber. Now for $p = 2$, 3, and for $p > 3$, $\sigma < 9$ there are 2 reducible fibers, so that it remains for us to consider the case $p > 3$, $\sigma = 9$, 10. The action of the automorphism ψ carries over from the fibering $X \to \mathbf{P}^1$ to its Jacobian fibering $J \to \mathbf{P}^1$. For $p > 3$, $\sigma = 9$, 10 it (as well as X) has reducible fibers the sum of the ranks of which is $\leqslant 4$ (they correspond to the root system of the lattice H_p). Since the group of sections of the fibering $J \to \mathbf{P}^1$ is isomorphic to S_f/R_f in the notation of Sec. 3, the rank of this group is $\geqslant 16$. If $\mathfrak{A}$ is a general fiber of the fibering, then $\operatorname{rk} \mathfrak{U}(k(\mathbf{P}^1)) \geqslant 16$.

Writing the equation of the curve in the form $y^2 = x^3 + ax + b$ and choosing the coordinate t on $\mathbf{P}^1$ so that $\psi_0{}^*t = t + 1$, we find from this that $a(t + 1) = a(t)$, $b(t + 1) = b(t)$, i.e., $a = a_1(t^p - t)$, $b = b_1(t^p - t)$ (which leads to a contradiction for large p). Since ψ acts trivially on the group $\mathfrak{U}(k(\mathbf{P}^1))$ (ψ takes each section into itself), it follows that $\mathfrak{U}(k(t)) = \mathfrak{U}_1(k(\tau))$, where $\mathfrak{U}_1$ is the curve defined by the equation $y_1^2 = x_1^3 + a_1(\tau)x_1 + b_1(\tau)$. The surface corresponding to this curve over the field $k(\tau)$ must be rational, since $\deg a \leqslant 8$, $\deg b \leqslant 12$, and hence $\deg a_1 \leqslant 4$, $\deg b_1 \leqslant 6$. Therefore, $\operatorname{rg} \mathfrak{U}_1(k(\tau)) \leqslant 10$, and this contradicts the fact that $\operatorname{rk} \mathfrak{U}(k(t)) \geqslant 16$.

B. In this case ψ defines an automorphism of a general fiber of $X \to \mathbf{P}^1$ and $J \to \mathbf{P}^1$ as an elliptic curve over $k(\mathbf{P}^1)$, For $p = 2$ the pencil we have chosen is quasielliptic, and it is easily seen that a quasielliptic curve having a rational point possesses only the identity automorphism. Therefore, ψ acts trivially on J. For $p = 3$, $\sigma = 1$, $2 \operatorname{Aut} J/k(\mathbf{P}^1) \equiv \mathbf{Z}/2$, and the group is faithfully represented by permutations of the components of the fibers $^*A_2$. Therefore, in our case the action of ψ on J is again trivial. Finally, for $p = 3$, $\sigma > 2$ and for $p > 3$ $\operatorname{rk} J(k(\mathbf{P}^1)) > 0$, and from $\psi s = s$, where $s \in J(k(\mathbf{P}^1))$, it follows that $(\psi - 1)s = 0$, i.e., $n(\psi - 1)s = 0$, where n is the norm of the endomorphism and $s \in \operatorname{tors} J(k(\mathbf{P}^1))$. Therefore, in all cases the action of ψ on J is trivial. From this it follows that on X, $\psi = t_s$ is translation by a section s of the fibering of J. In the principal homogeneous space $X \to \mathbf{P}^1$ there is a p-fold section θ, and $t_s\theta = \theta$ in S_X provided that $ps = 0$. Thus, the case B.1 is impossible. Finally, for $p \neq 2$, $|\operatorname{tors} J(k(\mathbf{P}^1))|$, being always of degree 2, is not divisible by p. Therefore, it remains to treat the case $l = p = 2$.

For $p = 2$, $\sigma < 10$ the pencil we have chosen has a section, and therefore $X = J$. If $\sigma = 10$, then $\operatorname{tors} J(k(\mathbf{P}^1)) = 2$, and this group consists of two elements: the zero section $\mathcal{O}$ and the section s. The fibers of the fiberings $X/\mathbf{P}^1$ and $J/\mathbf{P}^1$ have the type $^*A_1$, and our problem reduces to determining whether the transformation t_s preserves or interchanges them, while the answer will be the same in $X/\mathbf{P}^1$ and $J/\mathbf{P}^1$. In the latter fibering our question is equivalent to the following: do s and $\mathcal{O}$ intersect the same component of a degenerate fiber or different components? An answer is given directly by considering the lattice S. In our case $S = U_2 \oplus \widetilde{Q}(D_{20})$. Let $U_2 = f\mathbf{Z} + \mathcal{O}\mathbf{Z}$, $f^2 = 0$, $\mathcal{O}^2 = -2$, $f\mathcal{O} = 1$, $\widetilde{Q}(D_{20}) = \bigoplus_1^{20} \mathbf{Z}e_i + \varepsilon\mathbf{Z}$, $e_i^2 = -2$, $e_ie_j = 0$ for $i \neq j$, $\varepsilon = \frac{1}{2}\sum_1^{20} e_i$. The components of the i-th degenerate fiber correspond to e_i and $f - e_i$.

It is obvious that $\mathcal{O}$ does not intersect e_i and intersects $f - e_i$. An expression for s is obtained uniquely from the conditions $sf = 1$, $se_i \geqslant 0$, $s(f - e_i) \geqslant 0$, $s^2 = -2$. We find that $s = 5f + \mathcal{O} - \varepsilon$ and $se_i = 1$, $s(f - e_i) = 0$, i.e., s intersects e_i and does not intersect $f - e_i$. Thus, $t_s \neq 1$ on S_X, and hence $\psi = 1$.

Remark. For $p > 2$ our assertion follows from a more general result of Ogus: an automorphism of any surface X of type K3 is equal to 1 if it acts trivially on the group $H^2(X/W)$ [33, p. 15]. Below, however, we shall need precisely the case $p = 2$. We have presented the proof for any p, since it uses more elementary arguments than the proof of Ogus. If we wish to restrict attention to the case $p = 2$, then some parts of it can be omitted.

. The Formal Brauer Group. The Height of a Surface of Type K3

Still another method of studying deformations of a surface X is related to the concept of the formal Brauer group (see [10, 12]). Let G be an algebraic group. We first define a functor of deformation cohomologies $\Phi^r(X, G)$ on the category of Artinian local k-algebras with residue field k by setting

$$\Phi^r(X, G)(A) = \ker(H^r(X \times A, G) \to H^r(X, G)). \tag{1}$$

Using Schlessinger's theory [36] it is possible to prove the representability of the functor $\Phi^2(X, G_m)$ if X is a smooth, complete surface of type K3. This makes it possible to define a smooth, formal group $\widehat{\mathrm{Br}}\, X$ so that there is the isomorphism of functors

$$(\widehat{\mathrm{Br}}\, X)(A) \simeq \Phi^2(X, G_m)(A).$$

$\widehat{\mathrm{Br}}\, X$ is called the formal Brauer group of the surface X.

For what follows it is convenient for us to rewrite formula (1) somewhat. If G is an algebraic group and A is an Artinian local k-algebra, then we can define the sheaf G_A on the topology X_{et} by setting

$$G_A(U) = G(U \times A).$$

Then $H^r(X \times A, G) = H^r(X, G_A)$, and

$$\Phi^r(X, G)(A) = \ker(H^r(X, G_A) \to H^r(X, G)).$$

The natural projection $U \times A \to U$ gives us the homomorphism of sheaves

$$G_A \to G.$$

We denote by IG_A the kernel of this homomorphism. For a commutative group G it is easy to verify that there is the isomorphism

$$G_A \simeq IG_A \times G.$$

From this we obtain the formula

$$\Phi^r(X, G)(A) = H^r(X, IG_A).$$

In some cases the sheaf IG_A can easily be computed. For example, if $A = k[\varepsilon]$, $\varepsilon^2 = 0$, then IG_A is nothing other than the tangent space to the group G at the point e:

$$IG_{k[\varepsilon]} = \ker(G_{k[\varepsilon]} \to G) \approx TG.$$

In the case where the functor $\Phi^r(X, G)$ is a formal group we can also compute the tangent space $T\Phi^r(X, G)$ to this group. Since $\Phi^r(X, G)(k) = 0$, it follows that

$$T\Phi^r(X, G) = \Phi^r(X, G)(k[\varepsilon]) = H^r(X, TG).$$

For the case of the formal Brauer group we obtain

$$T\,\widehat{\mathrm{Br}}\, X = H^2(X, G_a) = H^2(X, \mathcal{O}_X). \tag{2}$$

Thus, the formal Brauer group of a K3 surface is one-dimensional.

The height is defined for one-dimensional formal groups. We recall the definition. Let $[p](t)$ be the formal series expressing multiplication by p in the group. If it is identically equal to 0, then the height of the group is said to be equal to ∞. In this case the group is isomorphic to an additive group. Otherwise it is always the case that for some $h \geqslant 1$

$$[p](t) = a_h t^{p^h} + \sum_{k > p^h} c_k t^k, \quad a_h \neq 0. \tag{3}$$

The number h is called the height.

It is known that over an algebraically closed field of characteristic $p > 0$ a one-dimensional formal group is uniquely determined by its height. In particular, a group of height

1 is a multiplicative group. It is also characterized by the fact that on its tangent space the Frobenius mapping (and on the cotangent space the Cartier operator C) is nonzero. We shall also call the height of the group $\hat{\mathrm{Br}}\, X$ the height of the surface X and denote it by h(X).

A surface X of type K3 is an ordinary surface if h(X) = 1. In view of (2) and the remarks made above a surface of type K3 is ordinary if and only if $C\omega \neq 0$, where $\omega \in H^0(X, \Omega^2)$, $\omega \neq 0$, and C is the Cartier operator.

The basis for the classification of formal groups is the assignment to each such group G of a certain W-module **D**G called the Dieudonné module and equipped with a φ-linear endomorphism F. To verify the properties which are used below, a definition in terms of "formal curves" is more convenient.

From the functoriality of the definition of the group $\hat{\mathrm{Br}}\, X$ it follows easily that

$$\mathbf{D}\hat{\mathrm{Br}}\, X = H^2(X, \mathbf{D}G_m). \tag{4}$$

Since the Dieudonné module of the group G_m is the ring of Witt vectors, it follows that

$$H^2(X, \mathbf{D}G_m) = H^2(X, W(\mathcal{O}_X)). \tag{5}$$

The sheaf cohomology of Witt vectors has been considered already by Serre. In our case this cohomology can be computed in terms of the crystalline cohomology of X.

Proposition 1. The module $\mathbf{D}\hat{\mathrm{Br}}\, X \otimes \mathbf{Q}$ as a module over W with an operator F is isomorphi to

$$H/H_1 \otimes \mathbf{Q},$$

where $H = H^2_{\mathrm{cris}}(X/W)$, $H_1 = \{x \in H,\ F^{n+rg}\,{}^H x \in p^n H$ for all n $\geqslant$ 0}.

A proof is given in the work [12].

It is known that the height of a one-dimensional formal group Z can be computed in terms of its Dieudonné module:

$$h(Z) = \dim \mathbf{D}Z \otimes \mathbf{Q}, \quad \text{if} \quad h(Z) < \infty,$$
$$h(Z) = \infty \Leftrightarrow \dim \mathbf{D}Z \otimes \mathbf{Q} = 0.$$

Thus,

$$h(X) = \infty \Leftrightarrow H/H_1 \text{ is of finite length.}$$

We note that the submodule $c_1(S_X) \otimes W$ is contained in H_1; hence if rk S_X = rk H, then h(X) = ∞ We have thus obtained the proposition.

Proposition 2. If a surface X is supersingular, then h(X) = ∞.

So far it has not been proved that the converse is true, although this is probable. This was established by Artin for elliptic surfaces X [10].

It is possible to prove the following assertion.

Proposition 3. If $\mathscr{X} \to T$ is a smooth family of K3 surfaces over an irreducible base T and for any point $t \in T$ the fiber $\mathscr{X}_t$ has height ∞ while for some point t_0 the fiber $\mathscr{X}_{t_0}$ is supersingular, then for any point $t \in T$, $\mathscr{X}_t$ is a supersingular surface.

It suffices to prove the proposition when T is the spectrum of a local ring. The genera case follows from this by specialization. The consideration of the case where the base is th spectrum of a local ring is based on a crystalline version of the theory of obstructions to the extension of algebraic cycles [33].

The concept of deformation cohomologies can be extended when X is given over Spec R, wher R is a local ring. We first define a functor on the category of local Artinian R-algebras with residue field k by the formula

$$\Phi^r(X, G)(A) = \ker(H^r(X \times A, G) \to H^r(X \times k, G)).$$

Using the Schlessinger condition it can be shown that $\Phi^2(X, G_m)$ is representable by a smooth, formal group if X is a surface of type K3. This group $\hat{\mathrm{Br}}\, X$ possesses the property that if $\mathrm{Spec}\, R_1 \hookleftarrow \mathrm{Spec}\, R$, then $(\hat{\mathrm{Br}}\, X)_{\mathrm{Spec}\, R_1} = \hat{\mathrm{Br}}\, (X|_{\mathrm{Spec}\, R_1})$. It is obvious that $\hat{\mathrm{Br}}\, X$ is a one-dimensional formal group over the ring R.

If $\mathscr{X}\to T$ is a family of K3 surfaces over a local base, then we can consider subschemes T_i defined by the condition that $h(\hat{B}r\, X|_{T_i}) \geqslant i$. In view of formula (3), the subscheme T_i is distinguished in T_{i-1} by one equation; raising the height of a one-dimensional formal group by 1 is presented by the vanishing of one coefficient a_h.

Proposition 4. For a smooth, complete surface X of type K3 of finite height

$$\operatorname{rk} \operatorname{Pic} X \leqslant 22-2h(X).$$

This is a special case of the Artin–Mazur inequality

$$\operatorname{rk} \operatorname{Pic} X \leqslant b_2-2h(X),$$

proved in [12] for a much larger class of surfaces. For the proof we consider a filtration on the module H:

$$H_0=H,$$
$$H_i=\{x\in H,\ F^{n+\operatorname{rk} H}x\in p^{in}H \text{ for all } n>0\}.$$

It is clear that the submodules H_i are invariant relative to F. For $N \subset H$ we denote by $N^{\perp}$ the set $\{x\in H,\ \forall y\in N,\ (x,y)\in pW\}$. Using formula (10) of Sec. 8, it is then easy to show that $H^1\subset H_2^{\perp},\ H_3\subset H_0^{\perp}=0$. A finer analysis of the structure of the crystal H based on the concept of slope [12] shows that $H_1 = H_2^{\perp}$. Then $\operatorname{rk} H_0/H_1 = \operatorname{rk} H_2 = h(X)$. On the other hand, $\operatorname{rk} S_X \leqslant \operatorname{rk} H_1/H_2$ and $\operatorname{rk} H = 22$. Hence

$$22\geqslant 2h(X)+\operatorname{rk} S_X.$$

Thus, if X is a K3 surface, then h(X) can be equal to $1,\ldots,10,\infty$. If $\mathscr{X}\to T$ is a family of algebraic surfaces of type K3, then the filtration T breaks off at T_{10}:

$$(T_{11})_{\mathrm{red}}=(T_{\infty})_{\mathrm{red}}. \tag{6}$$

THEOREM. If $\mathscr{X}\to T$ is a complete family of formal deformations of a supersingular surface X and there is given a p-elementary sublattice $N \hookrightarrow S_X$ of rank 22 and discriminant $p^{-2\sigma}$, then the scheme $\Sigma(N)$ defined in Sec. 7 is a smooth subscheme of dimension $\sigma - 1$.

Proof. On the one hand, by formula (9) of Sec. 7 for the dimension of the tangent space $t = \dim T\Sigma(N)$ the following condition is satisfied:

$$t\leqslant 20-\dim c[N].$$

Now c[N] is the linear hull of the images $\operatorname{gr}^1 c_1(u),\ u\in N$ in $H^1(X,\Omega^1)$, and hence

$$20-\dim c[N]=\dim H^1(X,\Omega^1)/c[N]\leqslant\dim(F^1H^2_{\mathrm{DR}}(X)/\langle c_1(N)\rangle)\leqslant\dim(H^2_{\mathrm{DR}}(X)/\langle c_1(N)\rangle)-1.$$

Now for a p-elementary lattice $H_{\mathrm{DR}}^2(X)/\langle c_1(N)\rangle k\simeq H^2(X/W)/\langle c_1(N)\rangle_W$, i.e., the factor of a unimodular lattice with respect to a sublattice of discriminant $p^{-2\sigma}$, and it is clear that

$$\dim_k H^2(X/W)/\langle c_1(N)\rangle_W=\sigma.$$

We have thus proved that $\dim T\Sigma(N)\leqslant\sigma-1$. On the other hand, we choose $n_0\in N$ such that image $c_1(n_0)$ in $H^2_{\mathrm{DR}}(X)$ is not equal to 0. Then $\dim\Sigma(n_0) = 19$. Such an element n_0 can be chosen to be very ample. Then in the family $\mathscr{X}_{(\Sigma n_0)}\to\Sigma(n_0)$ it is possible to consider the formal Brauer group and to construct in $\Sigma(n_0)$ a filtration with respect to height. We have

$$\dim\Sigma(n_0)_{11}\geqslant\dim\Sigma(n_0)-10=9, \tag{7}$$

and such an inequality holds for each irreducible component in $\Sigma(n_0)_{11}$, but $(\Sigma(n_0)_{11})_{\mathrm{red}} = (\Sigma(n_0)_{\infty})_{\mathrm{red}}$ in view of (6).

Let U be an irreducible component of $\Sigma(n_0)_{\infty}$, and let η be a general point in U. By Proposition 3, $\operatorname{rk}\operatorname{Pic}\mathscr{X}_\eta=22$, and passing possibly to an étalé covering, we may assume that there exists a lattice M and an isomorphism

$$M\tilde{\to}\operatorname{Pic}\mathscr{X}_U\tilde{\to}\operatorname{Pic}\mathscr{X}_\eta.$$

Then M is p-elementary, and $U\subset\Sigma(M)$. By the foregoing

$$\dim U\leqslant\dim T\Sigma(M)\leqslant\sigma(M)-1.$$

Comparing this with (7), we see that $\dim U = 0$, and U is a smooth subscheme in T.

If U contains a general point of $\Sigma(N)$, then $M \subset N$ and it is then easy to see that

$$\dim c[N] = \dim c[M] + \dim N/M.$$

Now since

$$\dim U = \dim TU = \dim (H^1(X, \Omega^1)/c[M])$$

and the condition for the extendability of one algebraic class is given by one equation, it follows that

$$\dim \Sigma(N) = \dim (H^1(X, \Omega^1)/c[M]) - \dim N/M = \dim T\Sigma(N).$$

The proof of the theorem is complete.

In conclusion we present a result concerning crystalline cohomologies of surfaces of type K3 defined over a field of characteristic 2.

We call an algebraic surface strongly elliptic if on it there is a pencil of elliptic curves with a section.

Proposition 5. If X is a strongly elliptic surface of type K3 over a field of characteristic 2, then the Poincaré duality converts the module $H^2(X/W)$ into an even lattice over W.

Thus, it is necessary to prove that for $x \in H^2(X/W)$, $x^2 \equiv 0 \pmod 2$. This is equivalent to the property that for $x \in H_{DR}^2(X)$, $x^2 = 0$. The idea of the proof is that the assertion is verified for a specific strongly elliptic surface of type K3 and is then carried over "by continuity" to all remaining surfaces. In the proof we shall use the properties of ordinary crystals expounded in detail, for example, in [20].

We shall first prove that the assertion is true for an ordinary surface rk $S_X = 20$. Indeed, from the general properties of crystals it follows (see [20], 1.3.3) that for such a surface for any p the crystal $H^2(X/W)$ can be decomposed into a direct sum of three crystals:

$$H^2(X/W) = H_0^2(X/W) \oplus H_1^2(X/W) \oplus H_2^2(X/W),$$

where rk H_0^2 = rk $H_2^2 = 1$, rk $H_1^2 = 20$, and in $H_1^2(X/W)$ there is a basis consisting of elements x for which $Fx = p^i x$. Let T be the set of $x \in H_1^2(X/W)$, for which $Fx = px$. It is obvious that we have the imbedding $S_X = S \subset T$, while, according to Lemma 2 in Sec. 7, $S \otimes \mathbf{Z}_p$ is distinguished as a direct factor in T and hence so is $S \otimes W$ in $H_1^2(X/W)$. If rk S = rk $H_1^2 = 20$, then $S \otimes W = H_1^2(X/W)$. From the properties of the product in $H^2(X/W)$ it follows that $H_1^2(X/W)$ is orthogonal to $H_0^2(X/W)$ and $H_2^2(X/W)$, while each of the last two spaces is isotropic. Thus, for p = 2 it suffices for us to prove that $x^2 \equiv 0 \pmod 2$ for $x \in H_1^2(X/W)$. Now this is obvious due to the parity of the lattice S and the fact that $H_1^2(X/W) = S \otimes W$.

As an example of such a surface we can take the Kummer surface corresponding to an Abelian variety A × A, where A is a nonsupersingular elliptic curve with complex multiplication (for example, given by the equation $y^2 + xy = x^3$). For it rk $S_{A\times A} = 4$, and hence rk S_X = rk $S_{A\times A} + 16 = 20$. Because of the criterion presented in this section, the surface X is ordinary, since the curve A is not supersingular and for $\omega \in H^0(A \times A, \Omega^2)$, $C\omega \neq 0$, while ω carries over from A × A to X and gives us there a form for which also $C\omega \neq 0$. Finally, the surface X is strongly elliptic, since the projection A × A → A (onto the second factor) after factorization defines an elliptic pencil X → P^1 in which the image of the curve A × 0 is a section.

All strongly elliptic surfaces of type K3 "form" a quasiprojective variety: they are given by the equation

$$y^2 + a_1(t)xy + a_3(t)y = x^3 + a_2(t)x^2 + a_4(t)x + a_6(t), \tag{8}$$

where $a_i(t)$ is a polynomial of degree 2i, and they are therefore defined by a point of the weighted projective space M in which the coefficients of the polynomials $a_i(t)$ serve as coordinates, and the coefficients of $a_i(t)$ have weight i. Points to which there correspond surfaces which are not of type K3 form a subvariety M_0. In $M \setminus M_0$ any two points can be joined by an irreducible curve. In particular, we can join by a curve B' that strictly elliptic surface X for which we have already proved the assertion with a surface X_1 for which we wish to prove it. Over B' we have a family $\mathscr{X}' \to B'$, all fibers of which are strictly elliptic surfaces of type K3. Since surfaces written in the Weierstrass form (8) have only double rational singular points, according to the work [11] for the family $\mathscr{X}' \to B'$ there exists a resolution of singularities, i.e., a family $\mathscr{X} \to B$ and morphisms $f: \mathscr{X} \to \mathscr{X}'$, $g: B \to B'$, such that the square

$$\begin{array}{ccc} \mathscr{X} & \xrightarrow{f} & \mathscr{X}' \\ \downarrow & g & \downarrow \\ B & \rightarrow & B' \end{array}$$

is commutative, the base B is smooth and complete, and for g(b) = b', $f:\mathscr{X}_b\to\mathscr{X}'_{b'}$ is a resolution of the singularities of the fiber $\mathscr{X}'_{b'}$. Thus, the smooth surfaces X and X' are included in the smooth family $\mathscr{X}\to B$.

It remains for us to prove that if $\mathscr{X}\to B$ is a connected, smooth family of surfaces of type K3 one of the fibers of which $X=\mathscr{X}_b$ is an ordinary surface and for $x\in H^2_{DR}(X)$ $x^2 = 0$, then the same is true for all fibers. According to an idea set forth by Deligne, this can be done as follows. From the general properties of sheaves over families of varieties it is easily found that in the case of surfaces of type K3 a sheaf relative to the de Rham cohomology $H^2(X/B)$ is locally free. It therefore suffices for us to prove that for some affine open set $U\subset B$, $U=\operatorname{Spec} A$, the module $H = \Gamma(U, H^2(X/B))$ is an even lattice over A. By the usual reduction this reduces to the case where A is a local ring $\mathcal{O}_b$ and $A=\hat{\mathcal{O}}_b$ is its completion, i.e., $A \simeq k[[t]]$. We set $A = W[[t]]$. The crystal corresponding to the family $\mathscr{X}$ defines a free module $\overline{H}$ over $\overline{A}$, while $H=\overline{H}\otimes_{\overline{A}} A$ and $\overline{H}$ is a lattice over $\overline{A}$. From the fact that the surface X is ordinary it follows that there exists a filtration in $\overline{H}$: $\overline{U}_0\subset\overline{U}_1\subset\overline{U}_2=\overline{H}$, $\operatorname{rk}\overline{U}_0 = 1$, $\operatorname{rk}\overline{U}_1/\overline{U}_0 = 20$, $\operatorname{rk}\overline{U}_2/\overline{U}_1 = 1$, $\overline{U}_1 = \overline{U}_0^{\perp}$. The crystal $\overline{U}_1/\overline{U}_0$ is a twisted unit crystal, whence $\overline{U}_1/\overline{U}_0\simeq Z_p^{20}\otimes\overline{A}$, where Z_p^{20} is a lattice of rank 20 over $\mathbf{Z}_p$. In H there thus arises the filtration $U_0\subset U_1\subset U_2=H$ with analogous properties, in particular, $U_1/U_0=(\mathbf{Z}/p)^{20}\otimes A$. If $(\mathbf{Z}/p)^r\oplus\mathbf{Z}/pe_i$, then $(e_i, e_j)\in\mathbf{Z}/p$, and hence for p = 2, $e_i^2 = 0$ this property holds in the module $U_1/U_0\otimes k$ (i.e., at the point b). Thus, the lattice U_1/U_0 is even, and hence the lattice U_1 is also even (since $U_0 = U_1^{\perp}$). From the properties of general crystals we find that $H=U_1\oplus L$, where $L = H^0(U, R^0f_*\Omega^2)$ corresponds to the smallest term of the Hodge filtration. It is obvious that (L, L) = 0 (for any p), whence it follows that the lattice H is even.

Remark. The proof remains in force for any surface of type K3 which is in the same "connected component" of the manifold of modules as the surface X constructed above. However, it is not known whether any surface of type K3 possesses this property. Over the field of complex numbers the parity of the lattice $H^2(X, Z)$ for a surface X of type K3 is deduced from Wu's theorem on the connection between the Stiefel classes and the Steenrod operation. It would be interesting to know if an analogue of this theorem (under some conditions) is true for the de Rham cohomology or for crystalline cohomology modulo torsion of algebraic varieties over a field of characteristic 2. For example, is a lattice $H^2(X/W)$ (modulo torsion) even if and only if the canonical class of the surface X is even? It is also of interest to investigate this question for 2-adic étalé cohomologies of algebraic varieties defined over a field of characteristic $p \neq 2$.

10. Periods of Supersingular Surfaces

Following [27] we shall study in more detail the structures arising on two-dimensional K3 cohomologies. We recall the facts already obtained.

1. The module $H = H^2(X/W)$ is a free module over the ring W with a φ-linear mapping F.

2. Multiplication of cohomologies defines in the module H a unimodular scalar product with values in W for which $(Fx, Fy)=p^2(x, y)^{\varphi}$.

We denote $H\otimes_W k$ by $\overline{H}$ and $F\otimes_W k$ by $\overline{F}$.

3. The rank of the mapping $\overline{F}:\overline{H}\to\overline{H}$ is equal to 1.

We denote henceforth by $T = T(H)$ the $\mathbf{Z}_p$-submodule $\{x\in H,\ Fx=px\}$ in H which is called the Tate module of the crystal H. The mapping assigning to an invertible sheaf its first Chern class defines an imbedding of the lattice S_X in H, and we henceforth identify S_X with its image in H.

4. The lattice imbeddings $S_X\subset T\subset H$, are defined, where $S_X\otimes\mathbf{Z}_p=T$, $T\otimes W\subset H\subset T^*\otimes W$. The lattice T is p-elementary, i.e., the $\mathbf{Z}_p$-module T^*/T is annihilated by multiplication by p.

It will be convenient for us to fix a sublattice $N \subset S_X$ of full rank.

Definition. The imbedding φ of the lattice $N \hookrightarrow S_X \hookrightarrow H$ we call an N-structure on the surface X if N is a lattice of full rank in S_X and

$$N^* \otimes W \supset H. \tag{1}$$

The lattice N will usually be p-elementary. An N-structure on X will sometimes be written as (X, φ).

A Euclidean K3-crystal is a crystal H with a scalar product for which properties 2 and 3 hold. An N-structure in a Euclidean K3-crystal H is an imbedding $N \hookrightarrow T(H) \subset H$, for which $H \subset N^* \otimes W$.

Suppose that the lattice N is p-elementary, and $\dim_k N^*/N = 2\sigma$. Let $D = (N^* \otimes W)/(N \otimes W) = N^*/N \otimes k$ be a vector space over the field k. Using the representation $D = N^*/N \otimes k$, we define in D the operator $1 \otimes \varphi$ of the Frobenius endomorphism which we denote simply by φ. As usual, in N^*/N there is defined a scalar product obtained by multiplication of the scalar product in N^* by p. We extend it to all of D and denote it by [,]. Since $N \otimes W \subset H \subset N^* \otimes W$, to H there corresponds a subspace $\bar{H} \subset D$.

Proposition 1. Let p > 2. We assign to the Euclidean K3-crystal H with an N-structure its image $\bar{H}$ in the space D. We thus establish a one-to-one correspondence between Euclidean K3-crystals with an N-structure and subspaces in D with the properties

a) $\dim V = \sigma$,

b) $\dim(V \cap \varphi V) = \sigma - 1$,

c) V is isotropic relative to the scalar product [,] in D.

We observe that the scalar product in N^*/N and hence also in D is nondegenerate; condition a) therefore implies that V is a maximal isotropic subspace.

To consider the case p = 2 we must augment these definitions somewhat. We shall assume that the lattice N is not only 2-elementary but also even. Then the quadratic form (x, x) takes integral values on N^*, is even on N, and defines a quadratic form Q on N^*/N and hence on D which is compatible with the scalar product. For $x, y \in D$

$$Q(x+y) = Q(x) + Q(y) + [x, y].$$

We call a Euclidean crystal H even if the scalar product defines an even lattice in H.

Proposition 1'. Let p = 2. We assign to an even Euclidean K3-crystal H with an N-structure its image $\bar{H}$ in D. This establishes a one-to-one correspondence between the set of subspaces $V \subset D$ with properties a), b), and c)': V is isotropic relative to the quadratic form Q.

We shall prove both propositions simultaneously, and for p ≠ 2 we denote by Q the quadratic form (1/2)[x, x] on D.

Let H be a Euclidean K3-crystal satisfying the hypotheses of the proposition. The lattice H contains the lattice $N \otimes W$; therefore, $d(N \otimes W) = d(H)\, p^{2l}$, where $l = \dim_k H/N \otimes W$. Then $k = \dim \bar{H}$, and from the unimodularity of H it follows that $k = \sigma$. The lattice H is defined on W, and hence if an element x in $N^* \otimes W$ belongs to H, then $(x, x) \in W$ is equivalent to the property that for the class $\bar{x}$ in $N^*/N \otimes k$, $[x, x] = 0$. This proves condition c) for p > 2. In the case p = 2 the evenness of the lattice H implies that $(x, x) \in 2W$, and this is equivalent to $Q(\bar{x}) = 0$.

We now consider the mapping F on $N^* \otimes W$. It coincides with $p \otimes \varphi$. We consider $H_1 = \{x \in H, Fx \in pH\}$. The image of this set in $\bar{H}$ coincides with $\operatorname{Ker} \bar{F}$. Condition 3 implies that $\dim_k H/H_1 = 1$. Since $H_1 \supset N \otimes W$, this is equivalent to the property that the image of H_1 in D has dimension $\sigma - 1$. However,

$$(p \otimes \varphi)\, x \in pH \Leftrightarrow (1 \otimes \varphi)\, x \in H.$$

Therefore, the image of H_1 in D coincides with $\bar{H} \cap \varphi \bar{H}$, which proves condition b).

If now V is a subspace in D with properties a), b), and c), then we consider the complete preimage H in $N^* \otimes W$ of the subspace V. This is a free W-module. Condition c) shows that the values of the scalar product on H belongs to W, and c') implies that moreover $(x, x) \in 2W$. Since $F = p \otimes \varphi$, it follows that $F(N^* \otimes W) \subset N \otimes W$, and hence H is F-invariant. We have thus obtained the structure of a K3-crystal on H. From a) it follows that the lattice H is unimodular, and property 3 follows from b). It is easy to see that property 2 is also satisfied for H, and the imbedding $N \subset H$ gives an N-structure in H. The proposition has thus been proved.

Suppose now that D is a vector space of dimension 2σ over a field k and on D there is given a nondegenerate quadratic form Q and a Frobenius mapping φ. We call a characteristic subspace in D a subspace V of dimension σ which is isotropic relative to the form Q and such that

$$\dim(V \cap \varphi V) = \sigma - 1. \tag{2}$$

Proposition 2. If the set of characteristic subspaces is nonempty, then in the Grassmannian manifold G of subspaces of dimension σ of the space D the characteristic subspaces correspond to points of a smooth, complete submanifold $P = \hat{P}[Q]$ of dimension $\sigma - 1$.

It is known that the maximal isotropic subspaces of a nondegenerate quadratic form can be parametrized by a smooth, complete subscheme in G. We denote it by R, $\dim R = \sigma(\sigma - 1)/2$. In $R \times R$ we now consider the subscheme Π consisting of pairs of points (v_1, v_2) such that for the corresponding subspaces (V_1, V_2) the condition $\dim(V_1 \cap V_2) \geqslant \sigma - 1$ is satisfied. This condition is also equivalent to $\dim(V_1 + V_2) \leqslant \sigma + 1$. We shall henceforth use the same notation for a point in G and the subspace corresponding to it.

It is clear that Π is a closed subscheme. We shall show that it is smooth. We recall that subspaces of dimension σ form two families, the intersection of two subspaces of the same family have even codimension in each of them, and the intersection of subspaces of different families is odd. If V_1 and V_2 are of the same family and $\dim(V_1 \cap V_2) \geqslant \sigma - 1$, then $V_1 = V_2$; if they are from different families, then $\dim(V_1 \cap V_2) = \sigma - 1$. Thus, the subscheme Π consists of the union of the diagonal Δ in $R \times R$ and the subscheme Π' consists of pairs of subspaces of different families, and $\Delta \cap \Pi' = \varnothing$. We must prove that Π' is smooth. For this it suffices to prove the smoothness of the fibers of the projection $\pi_2 : \Pi \to R$ onto the second factor. We recall that subspaces in a neighborhood of a subspace L in the Grassmannian manifold G can be represented in the form

$$(E + A)L, \text{ where } A \in \mathrm{Hom}(L, D).$$

A parametrization of this neighborhood is hereby obtained if we consider the projection τ: $\mathrm{Hom}(L, D) \to \mathrm{Hom}(L, D/L)$. The subspace $(E + A)L$ depends only on the image τA, and, choosing some section of the mapping τ, we obtain a parametrization of an affine neighborhood of the point L in G. If L is a characteristic subspace, then the characteristic condition for $(E + A)L$ is that for all $l_1, l_2 \in L$

$$[l_1, Al_2] + [Al_1, l_2] = 0. \tag{3}$$

If $(L_1, L_2) \in \Pi'$ and $((E + A)L_1, L_2) \in \Pi'$, then, in addition to condition (3), the following condition must be satisfied

$$\dim(E + A)L_1 \cap L_2 = \sigma - 1. \tag{4}$$

For convenience of computations we choose a basis in the space D. We first choose a basis $e_{\sigma+2}, \ldots, e_{2\sigma}$ in $L_1 \cap L_2$, and then augment it by a vector $e_{\sigma+1}$ to a basis in L_1 and by a vector e_1 to a basis in L_2 so that $[e_1, e_{\sigma+1}] = 1$. It is easy to see that these vectors can be augmented to a basis in D in which the matrix of the form Q has the form $\begin{pmatrix} 0 & E \\ E & 0 \end{pmatrix}$. Using this basis, we can establish the isomorphism

$$D \simeq k^{2\sigma} \simeq k^{\sigma} \times k^{\sigma}.$$

Then $L_1 = \{(0, x), x \in k^{\sigma}\}$, and the characteristic subspace in some neighborhood of L_1 can be written in the form $L = \{(xB, x), x \in k^{\sigma}\}$, where B is a skew-symmetric, square matrix of order σ. We remark that $L_2 = \{(y_1, 0, \ldots, 0, y_2, \ldots, y_\sigma), y_i \in k\}$, and hence condition (4) is equivalent to

$$(0, y_2, \ldots, y_\sigma) B \in \{(x, 0, \ldots, 0), x \in k\}. \tag{5}$$

This implies that the matrix B has the form

$$B = \begin{pmatrix} 0 & b_1 \ldots b_{\sigma-1} \\ -b_1 & \\ \vdots & 0 \\ -b_{\sigma-1} & \end{pmatrix}.$$

We thus find that the fiber $\pi_2^{-1}(L_2)$ is a smooth subscheme of dimension $\sigma - 1$. Hence, Π' is a smooth, complete manifold of dimension $\sigma(\sigma - 1)/2 + (\sigma - 1)$.

We now consider the graph Γ of the Frobenius mapping on R. By definition, $P=\pi_1(\Gamma\cap\Pi')$. However, the mapping $\pi_i|\Gamma$ establishes an isomorphism of Γ and R, and therefore $P\simeq\Gamma\cap\Pi'$. The completeness of P follows immediately from the completeness of Γ and Π'. We shall show that Γ and Π' intersect transversally at each point. We represent the tangent space of R × R as the direct sum $\Theta_R\oplus\Theta_R$. Then $\Theta_\Gamma=\{(v, (d\varphi)v)\}$. Now $d\varphi=0$, whence it follows immediately that $\Theta_{\Gamma\cap\Pi'}=\mathrm{Ker}\, d\pi_2$. According to what was proved above $\dim \mathrm{Ker}\, d\pi_2 = \sigma - 1$. Then

$$\dim(\Theta_\Gamma+\Theta_\Pi)=\dim\Theta_\Gamma+\dim\Theta_\Pi-\dim(\Theta_{\Gamma\cap\Pi})=\frac{\sigma(\sigma-1)}{2}+\left(\frac{\sigma(\sigma-1)}{2}+\sigma-1\right)-(\sigma-1)=\dim(R\times R).$$

For the transversal intersection of two smooth submanifolds of a smooth manifold the intersection is also smooth, and its dimension coincides with the dimension of the tangent space. The proposition has thus been proved.

We note that the manifold $\Pi'\subset R\times R$ consists of pairs of characteristic subspaces lying in different families, and hence in order that the manifold P[Q] be nonempty it is necessary and sufficient that the mapping φ permute the families of characteristic subspaces of the quadratic form Q. In this case the set of geometric points of the scheme P also consists of two components which are permuted by the mapping φ. It is possible to equip the scheme P with the structure of a scheme over $\mathbf{F}_{p^2}$, and it will be absolutely irreducible (see [33]).

Proposition 3. We retain the notation of Proposition 2. If the set of characteristic subspaces is nonempty, then the scheme P[Q] is unirational.

In a vector space of dimension 2σ over $\mathbf{F}_p$ there are two classes of quadratic forms representatives of which can be taken to be

$$Q=x_1x_{\sigma+1}+\ldots+x_\sigma x_{2\sigma}, \tag{6}$$

$$Q=u(x_1, x_{\sigma+1})+x_2x_{\sigma+2}+\ldots+x_\sigma x_{2\sigma}, \tag{7}$$

where $u(x_1, x_{\sigma+1})$ is a binary quadratic form which does not represent 0 over $\mathbf{F}_p$. In the first case φ preserves the connected components of the scheme R; therefore, there are no characteristic subspaces, and we may assume that our form Q has the form (7) in some coordinate system.

We shall assume that the form Q has this form in the basis $e_1',\ldots,e_{2\sigma}'$, and we arrive at another basis $e_1,\ldots,e_{2\sigma}$ by setting $e_1=\alpha e_1'+\beta e_{\sigma+1}'$, $e_{\sigma+1}=\alpha^\varphi e_1'+\beta^\varphi e_{\sigma+1}'$, $e_i=e_i'$ when $i \neq 1, \sigma + 1$, where $\alpha, \beta\in\mathbf{F}_{p^2}$ and are such that in this basis the form Q takes the form (6). The mapping φ is defined as follows:

$$\varphi e_1=e_{\sigma+1},\quad \varphi e_{\sigma+1}=e_1,\quad \varphi e_i=e_i \text{ for } i\neq 1, \sigma+1.$$

If we set $K_1 = \langle e_{\sigma+1},\ldots,e_{2\sigma}\rangle$, $K_2 = \langle e_1, e_{\sigma+2},\ldots,e_{2\sigma}\rangle$, then we find ourselves in a situation analogous to that considered in the proof of Proposition 2 with $\varphi K_1=K_2$, $\dim(K_1+\varphi K_1)=\sigma+1$. We again write the vectors in D in terms of their coordinates in the basis $e_1,\ldots,e_{2\sigma}$, separating these coordinates into two groups with σ coordinates in each. Suppose that for $x, y\in k^{2\sigma}$, $\pi(x, y) = y$ and $l(x, y) = x_1$ is the first element of the row x.

Let C be a skew-symmetric matrix, and let $M_C = \{(xC, x)\}$. We compute more explicitly the condition $\dim M_C\cap M_C^\varphi=\sigma-1$. We note that $\dim\pi(M_C)=\sigma$ and $\dim\pi(M_C^\varphi)=\sigma-1$; hence $\pi|_{M_C}$ is an isomorphism, and $\dim(M_C\cap M_C^\varphi)=\sigma-1$ if and only if $M_C\cap M_C^\varphi=M_C\cap\pi^{-1}(\pi M_C^\varphi)$. We denote the intersection on the right by T. We always have the inclusion $M_C\cap M_C^\varphi\subset T$. For equality it is necessary and sufficient that $M_C^\varphi\supset T$. We write out the basis of T:

$$f_i=(\varepsilon_iC, \varepsilon_i)+c_{1i}^\varphi(\varepsilon_1C, \varepsilon_1) \text{ for } i>1,$$

where $\varepsilon_i = (0,\ldots 0\overset{i}{1}0\ldots 0)$. We note that $\pi(f_i)=\pi((\varepsilon_iC, \varepsilon_i)^\varphi)$, i.e., $f_i'=f_i-(\varepsilon_iC, \varepsilon_i)^\varphi\in\ker\pi$. On the other hand, $(\ker\pi)\cap M_C^\varphi$ is a one-dimensional space, and l is a nontrivial linear function on it. The element $(\varepsilon_1C, \varepsilon_1)^\varphi$ may be considered a basis element in this space and $l((\varepsilon_1C, \varepsilon_1)^\varphi)=1$. Hence in order that f_i' belong to $(\ker\pi)\cap M_C^\varphi$, it is necessary and sufficient that $f_i'=l(f_i')(\varepsilon_1C, \varepsilon_1)^\varphi$. Thus, we arrive at the system of equations

$$f_i-(\varepsilon_iC, \varepsilon_i)^\varphi=c_{1i}(\varepsilon_1C, \varepsilon_1)^\varphi.$$

or

$$(e_l C, e_l)+c_{1l}^p(e_1 C, e_1)=(e_l C, e_l)^{\varphi}+c_{1l}(e_1 C, e_1)^{\varphi}. \tag{8}$$

We set $x_i = c_{1i}$; the system (8) can then be rewritten in the form

$$c_{jl}^p - c_{jl} = x_l x_j^p - x_l^p x_j, \quad 2 \leqslant l < j \leqslant \sigma. \tag{9}$$

It now remains for us to prove the following lemma.

LEMMA 1. The affine, algebraic variety defined by the system of equations

$$y_{ij}^p - y_{ij} = \varphi_{ij}(x_i, x_j), \ i<j, \ i, j=1,\ldots,n,$$

where φ_{ij} is a p-polynomial in each variable, is unirational.

Proof of the Lemma. We shall indicate a means of successively constructing a unirational parametrization. We first consider a single equation $y_{in}^p - y_{in} = \varphi_{in}(x_i, x_n)$. It defines a one-dimensional subgroup U_i in $G_a \times G_a$ over the field $k(x_n)$ — the form of an additive group. Over some extension $k(x_n^{p^{-N}})$ this form becomes trivial. This means that over $k(x_n^{p^{-N}})$ the isomorphism $G_a \to U_i$ is defined. It is given by p-polynomials $y_{in} = \psi_i(z_i)$, $x_i = \varphi_i(z_i)$. Repeating this argument for all $i = 1,\ldots,n-1$, we find for sufficiently large N a system of p-polynomials ψ_i, φ_i, $i=1,\ldots,n-1$ with coefficients in $k(x_n^{p^{-N}})$. Setting $t_n = x_n^{p^{-N}}$, over the field $k(t_n)$ we consider the system of equations

$$y_{ij}^p - y_{ij} = \varphi_{ij}(\varphi_i(z_i), \varphi_j(z_j)) \ i, j=1,\ldots,n-1, \ i<j.$$

We obtain a system of the same form with a lower number of indices. Assuming by induction that it defines a unirational variety, we immediately obtain the unirationality of the variety of the original system. The proof of the lemma is complete.

The computations made in the proof of Proposition 3 assume an especially clear form for $\sigma = 3$, $p > 2$. Then the quadric $x_1x_2 + x_3x_4 + x_5x_6 = 0$ in the projective space $\mathbf{P}^5$ coincides with the Grassmannian of lines in $\mathbf{P}^3$ in its canonical imbedding in $\mathbf{P}^5$. The two families of isotropic subspaces are realized as lines passing through a point in $\mathbf{P}^3$ and as lines lying in one plane. The mapping φ is a mapping $\mathbf{P}^3 \to \check{\mathbf{P}}^3$ which can easily be found explicitly:

$$(\alpha_0:\alpha_1:\alpha_2:\alpha_3)^{\varphi} = (\alpha_1^p : -\alpha_0^p : -\alpha_3^p : \alpha_2^p).$$

The condition on the dimension of the intersection means that the point x belongs to the plane φx.

Thus, the manifold P[Q] is isomorphic over $\mathfrak{k}$ to the surface $x_0x_1^p - x_1x_0^p + x_2x_3^p - x_3x_2^p = 0$ in $\mathbf{P}^3$ which after a change of coordinates becomes the Fermat surface $z_0^{p+1} + z_1^{p+1} + z_2^{p+1} + z_3^{p+1} = 0$. Unirationality of this surface was proved in Sec. 5. For p = 2 this is a rational surface, for p = 3 it is a surface of type K3, and for p > 3 it is a surface of general type.

We shall further show how on the basis of the computations carried out above to determine the tangent space to the manifold P = P[Q]. For this we consider $L \in P$ and the ring $k[\varepsilon]$, $\varepsilon^2 = 0$. The tangent space is composed of characteristic subspaces of the form $(E + \varepsilon A)L$, where $A \in \mathrm{Hom}(L, D)$, and such a subspace depends only on the image of A under the projection τ: Hom (L, D) → Hom (L, D/L). We note that $\varphi(E+\varepsilon A)L = \varphi L$, since $\varepsilon^p = 0$; therefore condition (2) can be written in the form

$$\dim((E+\varepsilon A)L \cap \varphi L) = \dim(L \cap \varphi L) = \sigma - 1.$$

This means that

$$A(L \cap \varphi L) \subset \varphi L. \tag{10}$$

Now the image τA under condition (10) is uniquely determined by its restriction to $L \cap \varphi L$, which can be an arbitrary mapping on φL. Indeed, $(L \cap \varphi L)^{\perp} = L + \varphi L$, and hence condition (3) imposes no restrictions on $A|_{L \cap \varphi L}$. If l_0 is a complementary vector to $L \cap \varphi L$ in L, then the value of Al_0 is determined by the conditions $[Al_0, l_0] = 0$, $[Al_0, l] = -[l_0, Al]$ for $l \in L \cap \varphi L$ up to addition of a vector in $L \cap \varphi L$. From these considerations it is easy to find that τA is uniquely determined by the image of $A|_{L \cap \varphi L}$ under the projection

$$\mathrm{Hom}(L \cap \varphi L;\ \varphi L) \to \mathrm{Hom}(L \cap \varphi L,\ \varphi L/(L \cap \varphi L)).$$

Thus, summarizing what has been said above, we obtain

$$\Theta_{P,L} \simeq \mathrm{Hom}(L \cap \varphi L,\ \varphi L/(L \cap \varphi L)). \tag{11}$$

We consider an even, hyperbolic, p-elementary (of type 1 for $p = 2$) lattice N of rank 22 with $\sigma(N) = \sigma$. From the enumeration of them given in the theorem of Sec. 1 it follows easily that for them formula (7) holds. The corresponding space P[Q] is called the space of periods and is denoted by $P[\sigma]$. Propositions 2 and 3 in this case can be combined into a theorem.

THEOREM 1. The period manifold $P[\sigma]$ for $0 < \sigma < 11$ is smooth, complete, has dimension $\sigma - 1$, and is unirational.

In Proposition 1 it was shown how to construct a point in the period manifold on the basis of a crystal H over a field. This construction can be generalized to the case of crystals over an arbitrary base. We fix a lattice T with Artin invariant σ.

Proposition 4. There exists a K3-crystal $\mathscr{H}$ with a T-structure over $P[\sigma]$ with the following universal property: any K3-crystal with a T-structure (for $p = 2$ any even K3-crystal with a T-structure) over a smooth scheme R is isomorphic to the inverse image $f^*(\mathscr{H})$ for some uniquely determined morphism $f: R \to P[\sigma]$. In the case $R = \mathrm{Spec}\, k$ and a crystal H over R the point $f(R)$ in $P[\sigma]$ is determined by the characteristic subspace $\varphi^{-1}[H]$.

This result was proved by Ogus [33] for $p > 2$; we shall not present the proof. The case $p = 2$ is considered in exactly the same way due to the added condition of evenness.

On the basis of this proposition it is possible to construct a period mapping for supersingular K3 surfaces. Let $\mathscr{X}/R$ be a family of supersingular K3 surfaces over a smooth base R equipped with a T-structure. For $p = 2$ it is necessary to assume that all the fibers are strictly elliptic. This is the case if $\sigma < 10$ [32]. Then $R^2_{\pi_*}\mathcal{O}_{\mathscr{X}}$ – a supersingular K3-crystal over R with a T-structure for $p = 2$ – is even because of Proposition 5 of Sec. 9. The mapping $f: R \to P[\sigma]$ for which $R^2\pi_*\mathcal{O}_{\mathscr{X}} = f^*(\mathscr{H})$, is uniquely determined from Proposition 4 and is called the period mapping for the family $\mathscr{X}/R$.

It is possible to compute the differential of the period mapping. If X is a supersingular K3 surface with a T-structure, then we assign to X the subspace $L = L(X)$ defined by the condition

$$\varphi L = \mathrm{Im}(H \to T^*/T \otimes k), \tag{12}$$

where $H = H^2(X/W)$.

Using the inclusion $T \otimes W \subset H \subset T^* \otimes W$ and the isomorphism $(T^* \otimes W)/(T \otimes W) \cong D$, we can interpret an element $x \in L \cap \varphi L$ as the image in D of an element $x \in T^* \otimes W$, satisfying the conditions $x \in H$, $\varphi x \in H$. Since the action of F on H and $\varphi = 1 \otimes \varphi$ on $T^* \otimes W$ are related by $F = p\varphi$, it follows that $Fx \in pH$. Therefore, $Fx = 0$ in $H^2_{DR}(X)$. On the other hand, from the definition of the filtration in H^2_{DR}, from the action of F on $E_2^{i,j}$, and from Lemma 2 of Sec. 8 it follows that $\mathrm{Ker}\, F = F^1$. Therefore, $L \cap \varphi L = \mathrm{Im} F^1 H$, whence in view of (11)

$$\Theta_{P,L} \simeq \mathrm{Hom}(L \cap \varphi L,\ \varphi L/(L \cap \varphi L)) \simeq \mathrm{Hom}(\mathrm{Im}\, F^1 H,\ \mathrm{Im}\, H/\mathrm{Im}\, F^1 H).$$

We now observe that $\mathrm{Im}\, F^1 H$ in the space $D = T^*/T \otimes k$ coincides with $gr^1 H/c[T]$ (in the notation of Sec. 7), and $\mathrm{Im}\, H/\mathrm{Im}\, F^1 H$ coincides with $gr^0 H$. Thus, we find that

$$\Theta_{P,L(X)} \simeq \mathrm{Hom}(gr^1 H/c[T],\ gr^0 H). \tag{13}$$

The essential factor, however, is that the isomorphism of formula (13) exists not simply at each point but also possesses the following property.

Proposition 5. If $\mathscr{X} \to R$ is a smooth family of supersingular K3 surfaces with a T-structure and $\pi: R \to P$ is the period mapping, then its differential $d\pi$ at the point r, considering the isomorphism (13), becomes the mapping ∇^1 given for $H = H^2(X_r/W)$ by the Gauss–Manin connection. In other words, the following diagram is commutative:

$$\begin{array}{ccc} \Theta_{R,r} & \xrightarrow{\text{Kod}} & H^1(X_r, \Omega^1) \\ \downarrow & \searrow & \downarrow \wr \\ \mathrm{Hom}(gr^1 H/c[T],\ gr^0 H) & \longrightarrow & \mathrm{Hom}(gr^1 H,\ gr^0 H) \end{array},$$

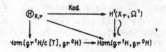

where kod is the Kodaira–Spencer mapping. (For a proof see the work of Ogus [33].)

Recalling the considerations of Sec. 7, as a corollary we obtain a local Torelli theorem.

THEOREM 2. If $\mathscr{X} \to \Sigma_T$ is a universal family of deformations of a singular K3 surface X with a T-structure for which $\sigma(T) < 10$ for $p = 2$, then:

a) Σ_T is a smooth scheme of dimension $\sigma(T) - 1$.

b) the period mapping $\Sigma_T \to P$ is étalé.

Proof. We may assume that $\Sigma_T = \operatorname{Spec} R$, where R is a complete local ring. It is then clear that $\Sigma_T = \Sigma(T)$ is a smooth scheme of dimension $\sigma(T) - 1$ by Theorem 1 of Sec. 9, whence a) follows. Part b) follows from the fact that in view of a) the family $\mathscr{X} \to \Sigma$ coincides with a family of deformations of X preserving algebraic cycles in T, $\mathscr{X} \to \Sigma(T)$. In view of the preceding proposition and the fact that kod is a monomorphism, we find that df is a monomorphism, and the equality of dimensions shows that it is even an isomorphism.

11. The Manifold of Moduli

We choose some even, p-elementary, hyperbolic lattice N of rank 22 (of type 1 for $p = 2$). As we know, for each σ, $1 \leqslant \sigma \leqslant 10$ there exists only one such lattice N for which $\sigma(N) = \sigma$.

We note that any supersingular surface X of type K3 for which $\sigma(X) \leqslant \sigma(N)$ has at least one N-structure. In order to see this it suffices to prove that the lattice S_X has a sublattice isomorphic to N and, conversely, that the lattice N can be imbedded in a lattice N_1 isomorphic to S_X. Imbeddings of N in an integral lattice are described by isotropic subspaces in the space N^*/N of dimension 2σ over $\mathbf{F}_p$.

Since any quadratic form of rank 3 over $\mathbf{F}_p$ represent 0, it follows that $N^*/N \simeq U_2^{\sigma-1} \oplus V_2$ or $\simeq U_2^{\sigma}$, where V_2 is a two-dimensional, anisotropic space (it can be shown that the first case occurs). From this it follows that there is an isotropic subspace N_1/N of any dimension $\leqslant \sigma - 1$. The corresponding lattice N_1 possesses all the required properties and is uniquely characterized by them; hence

In the case of a family $f: X \to T$ an N-structure need not exist. Let T be the spectrum of a complete, local integral domain. We denote by L an invertible sheaf on X having in its general fiber a positive square d and by c a constant such that in any p-elementary hyperbolic lattice E of rank 22 for any vector $l \in E$, $l^2 = d$, vectors $x \in E(l)$ with $|x^2| < c$ generate the entire lattice $E(l)$. The existence of such a constant follows easily from the Minkowski lemma. We consider elements $x \in \operatorname{Pic} X/T$, $(L, x) = 0$, $|x^2| < c$. From the general properties of Picard schemes it follows that these elements determine a scheme which is proper and unramified over T. They therefore determine a reducible scheme which decomposes into a finite number of copies isomorphic to T. Thus, $\operatorname{Pic} X/T \simeq N_1 \times T$ for some lattice N_1. Choosing an N-structure in N_1, we obtain an N-structure in X. Then for any point $t \in T$ in the fiber over t an N-structure is also defined.

If X/T is a family over an arbitrary irreducible and locally irreducible base, then the same arguments show that Pic X/T defines an unramified Galois covering $T' \to T$, and the Galois group Γ of this covering acts on $\operatorname{Pic} X_{\bar\eta}$, where $\bar\eta$ is a general geometric point of the base T. The family $X' = X \times_T T'$ is already such that $\operatorname{Pic} X'/T' = N_1 \times T'$, and it therefore has an N-structure. Here the surfaces $X_{t_i'}$ lying over points t_i' covering the same point t are all isomorphic (they are isomorphic to X_t), but they define on X_t different N-structure which go over into one another by means of automorphisms of the group Γ.

LEMMA 1. Let (X, L) be a deformation of a supersingular surface X_0 of type K3 and an ample sheaf L_0, while the base T of this deformation is the spectrum of a complete, regular, local ring and L_0^2 is not divisible by p. Let $\bar{T}$ be a supersingular surfaces in T, and let $\bar{T} = \bigcup \Sigma_i$ be the decomposition of $\bar{T}$ into irreducible components. Let Θ be the tangent space to T at a closed point t_0, and let T_i be the tangent spaces to Σ_i, $T_i \subset T$. We denote by δ, δ_i the mappings of the Gauss–Manin connection $\operatorname{gr}^1 \nabla$ which arise in the families over T and over Σ_i. We suppose that δ maps Θ onto $\operatorname{Hom}(\operatorname{gr}^1 H/c_1(L_0), \operatorname{gr}^0 H)$ epimorphically, where $H = H^2(X_0/W)$, and that $\operatorname{Ker} \delta = \operatorname{Ker} \delta_i$ for all i. Then the schemes Σ_i are smooth and of the same dimension, and for each i there is defined a natural N-structure in the crystal H_{Σ_i}, where N is the lattice defined by the condition $\sigma(N) = 10$. For different i the sublattices $N \to \operatorname{Pic} X/\Sigma_i \to S_{X_0}$ in S_{X_0} are distinct.

Proof. We denote by v the dimension of $\operatorname{Ker}\delta = \operatorname{Ker}\delta_i$. Then $\dim T = \dim\Theta = 19 + v$. The subscheme T is determined by the condition $h(X_{\bar{T}}) = \infty$, and therefore for each irreducible component Σ_i, $\dim\Sigma_i \geqslant 9 + v$.

On the other hand, if η_i is a general point of Σ_i, then the lattice $\operatorname{Pic} X_{\eta_i}$ has rank 22; therefore, carrying out the same arguments as in Theorem 1 of Sec. 9 and using the diagram (14) of Sec. 10, we find that $\dim\Sigma_i \leqslant v + \sigma(\operatorname{Pic} X_{\eta_i}) - 1$.

From this we immediately find that $\sigma(\operatorname{Pic} X_{\eta_i}) = 10$, $\dim\Theta_i = \dim\Sigma_i = v + 9$, and hence Σ_i is a smooth subscheme. Moreover, Σ_i is defined as a subscheme in T to which it is possible to extend cycles from the sublattice $\operatorname{Pic} X_{\eta_i} \hookleftarrow \operatorname{Pic} X_0$; therefore, for different i these sublattices are distinct. Setting $N = \operatorname{Pic} X_{\eta_i}$, we obtain the required N-structure.

We choose one of the cells U relative to the hyperbolic system of roots in N, and we consider all possible imbeddings of the lattice N in p-elementary lattices of the same rank $N \subset N_i$. Since they correspond to isotropic subspaces of the space N^*/N over $\mathbf{F}_p$, there are a finite number of them. In $N_i \otimes \mathbf{R}$ the set U may not be a cell, but it decomposes into a finite number of cells. In view of the finiteness of these decompositions, there is a vector $l \in U \cap N$, which lies inside each cell for any imbedding $N \to N_i$. If $N_i = \operatorname{Pic} X_i$ and the cell in which l lies corresponds to ample classes, then l corresponds to an ample class. Multiplying, if necessary, l by $k > 2$, we may assume that l corresponds to a very ample class. We call such a vector $l \in N$ a universal polarization. From the foregoing it follows easily that there always exists a universal polarization l with $l^2 \not\equiv 0 \pmod p$. A certain universal polarization $l \in N$ with $l^2 \not\equiv 0 \pmod p$ will be fixed below. If the N-structure (X, φ) is such that $\varphi(l)$ is a polarization of the surface X, then we speak of a universal polarization of the N-structure.

The isomorphism of an N-structure with a universal polarization is defined in the obvious manner.

It is obvious that any supersingular surface of type K3 possesses an N-structure with a universal polarization. Indeed, we have seen that some N-structure $\varphi_1 : N \to \operatorname{Pic} X$ exists. By hypothesis, the vector $\varphi_1(l)$ is contained inside one of the cells U_i of the lattice Pic X. There exists an automorphism σ of this lattice composed of reflections such that σU_i is a cell of ample classes. Then the class $\sigma\varphi_1(l)$ is very ample (by construction), and $\varphi = \sigma\varphi_1$ defines an N-structure with a universal polarization.

The next result uses an assertion which is "surely" true, but we have been unable to find the proof of it in the literature. The assertion is that a smooth surface X of type K3 imbedded in the projective space $\mathbf{P}^N$ determines a stable point of the corresponding Hilbert scheme relative to the action of the group PGL(N). The assertion is evidently "easier" than the corresponding result of Giescker for surfaces of general type [21]. We hope that a proof will soon be published by one of the specialists. The following theorem must for the time being be considered a provisional assertion.

THEOREM. In any positive characteristic p there exists a family of surfaces of type K3 $f : \mathscr{X} \to M$, which satisfies the following properties:

1) The family possesses an N-structure with a universal polarization 1.
2) Any supersingular surface of type K3 is isomorphic to one of the fibers of the family
3) If two fibers $f^{-1}(m_1)$ and $f^{-1}(m_2)$ are isomorphic together with their N-structures and polarizations, then $m_1 = m_2$.
4) M and F are smooth.
5) The period mapping $\pi : M \to P$ corresponding to the family f is étalé.

Proof. Let N be the lattice and $l \in N$ be the vector which were fixed previously, and suppose that $l^2 = c$. We consider the Hilbert scheme H_n of surfaces of type K3 imbedded in the projective space $\mathbf{P}^n$ by means of an invertible sheaf with square c.

Let $\bar{H}_n \subset H_n$ be a subscheme the points of which correspond to smooth, supersingular surfaces, and let $\bar{H}_n = \bigcup \Sigma_i$ be its decomposition into irreducible components in a formal neighborhood of the point $s \in \bar{H}_n$.

We shall verify that Lemma 1 is applicable to H_n, $\bar{H}_n$ and Σ_i. In the work [6] (pp. 537-538) from a consideration of standard exact sequences it is deduced that the manifold H_n (there it is denoted by M) is smooth, and there is the exact sequence

$$0 \to \mathrm{sl}(n+1) \to T_{H_n,s} \to H^1(X_s, \Theta)^{L_0} \to 0,$$

where sl(n + 1) is the Lie algebra of the group G = PGL(n) realized as a subspace of the tangent space $T_{H_n,s}$ at the point $s \in H_n$. From this it follows that Ker δ = sl(n + 1), and the mapping δ is epimorphic. Since the manifold Σ_i is invariant relative to G, it follows that $\mathrm{sl}(n+1) \subset T_i$, i.e., Ker δ = Ker δ_i, as required in Lemma 1. We can apply this lemma, and we see that each branch Σ_i of the manifold $\overline{H}_n$ is smooth.

We denote by V_α the irreducible components of the manifold $\overline{H}_n$. On each of them by restriction from H_n there is defined a family $\mathcal{Y}_\alpha \to V_\alpha$ of supersingular surfaces of type K3. The group G acts on each of the V_α as well as on the family $\mathcal{Y}_\alpha$, and these actions are compatible.

Let V_α^ν be the normalization of the variety V_α. According to Lemma 1, all the V_α^ν are smooth. The normalization morphism $V_\alpha^\nu \to V_\alpha$ defines on each of them a family of surfaces $\overline{\mathcal{Y}}_\alpha \to V_\alpha^\nu$, while the group G acts in compatible fashion both on V_α^ν and on $\overline{\mathcal{Y}}_\alpha$. Let $V_\alpha' \to V_\alpha^\nu$ be the minimal, unramified covering such that $\mathcal{Y}_\alpha'/V_\alpha'$ has the N-structure (N, φ_α). The existence of such a covering has been demonstrated earlier. We select those families for which for some choice of the isomorphism φ_α an element $l \in N$ corresponds to the polarization defined by the original imbedding in P^n: only for these do we retain the notation $\mathcal{Y}_\alpha'$.

Our immediate goal is to consider the factor V_α'/G, and we must first convince ourselves that it exists. This follows from the criterion presented in the book of Mumford [29, p. 38]. Since all points of the scheme H_n and hence also V_α are stable (see the remark preceding the formulation of the theorem), the result we need follows from Proposition 1.18 of [29, p. 44].

Since the scheme V_1^ν, and hence also V_α', is smooth because of Lemma 1, it suffices for us to show that G acts on V_α' without fixed points. Now on V_α' there is defined a family of surfaces $\mathcal{Y}_\alpha'$, to which the action of G extends. This action is consistent with the action on the Picard groups of surfaces with the N-structure defined on the family $\mathcal{Y}_\alpha'/V_\alpha'$. Therefore, a fixed point would determine a surface X, and an automorphism of it would be trivial on PicX. In view of the proposition in Sec. 8, such an automorphism is trivial.

The disjoint union of schemes $M_\alpha = V_\alpha'/G$ is a base of the same family that we construct. The family $\mathcal{X}$ over the base M is constructed in an entirely similar way proceeding from the family defined over the base H_n. The family $\mathcal{X}_\alpha$ over M_α is defined as the factor $\mathcal{Y}_\alpha'/G$. The existence of the factor follows from [29], Proposition 2.18 (p. 65).

We shall verify that the family $\mathcal{X}/M$ constructed possesses properties 1)-5) presented in the theorem. Properties 1) and 4) are obvious from the construction. Property 2) is also obvious: any supersingular surface of type K3 possesses an N-structure, and the element l involved in its choice defines on it a polarization of degree c. Thus, a polarized surface X corresponds to some point of the scheme and hence to at least one point of at least one scheme M_α.

We shall verify property 3). If the fibers X_{m_1} and X_{m_2} are isomorphic as polarized surfaces, then to them there correspond G-equivalent points of the scheme $\overline{H}_n$. It may be assumed that this is one and the same point $\overline{m} \in \overline{H}_n$. According to Lemma 1, different branches Σ_i at this point differ by their images $\varphi_i(N)$ in Pic $X_{\overline{m}}$. Since $\varphi_1(N) = \varphi_2(N)$, to the points m_1 and m_2 there corresponds one and the same point on some scheme V_α^ν. Finally, as explained at the beginning of the section, from the fact that the morphisms φ_1 and φ_2 taking $N \to \mathrm{Pic}\, X_{\overline{m}}$ coincide it follows that to the points m_1 and m_2 there corresponds one and the same point on V_α'.

It remains to check that the period mapping is étalé on each M_α. If $\Theta_x(V_\alpha')$ is the tangent space at the point $x \in V_\alpha'$, then the action of the group G on V_α' determines an imbedding $\mathrm{sl}(n+1) \subset \Theta_x(V_\alpha')$ and $\Theta_y(M_\alpha) = \Theta_x(V_\alpha')/\mathrm{sl}(n+1)$, where $y \in M_\alpha$ is the point corresponding to $x \in V_\alpha'$ under factorization with respect to G. On the other hand, $\Theta_x(V_\alpha') \simeq \Theta_x(V_\alpha^\nu)$, and the last space is isomorphic to $\Theta_s(\Sigma_i)$ for $s \in H_n$ and some i. As proved in Lemma 1, Ker δ_i = sl(n + 1), dim Im δ_i = 9. From this it follows that for the Kodaira–Spencer mapping δ_α corresponding to the family $\mathcal{Y}_\alpha/M_\alpha$, $\mathrm{Ker}\,\delta_\alpha = 0$, $\dim \mathrm{Im}\,\delta_\alpha = 9$. In view of Proposition 4 of Sec. 10 this implies the étalé property of the period mapping.

The theorem proved establishes such a close relation between the manifold of moduli M and the manifold of periods P that there naturally arises the question of the extent to which the properties of one carry over to the other. Since the manifold P is complete, the first question is whether M is complete. The assertion of its completeness is equivalent to the following property of families of supersingular surfaces of type K3.

For the morphism $\mathscr{X}\to \operatorname{Spec} R$, where R is a one-dimensional, regular, complete local ring with general fiber which is a supersingular surface of type K3, there exists a finite covering R'/R and a morphism $\mathscr{X}'\to \operatorname{Spec} R'$, such that the general fiber $\mathscr{X}'$ is isomorphic to the general fiber of $\mathscr{X}\otimes_R R'$, while a closed fiber is also a smooth surface of type K3.

This assertion can be briefly described as the property that supersingular surfaces of type K3 "do not degenerate." There are several arguments in favor of this conjecture. First of all there is the analogy with Abelian varieties where an analogous fact holds: Abelian varieties for which the Lie p-algebra is unipotent "do not degenerate." This follows from the theory of stable reduction of Mumford and Grothendieck according to which for each family of Abelian varieties $\mathscr{X}\to \operatorname{Spec} R$ there exists a family $\mathscr{X}'\to \operatorname{Spec} R'$ with the properties described above for which a closed fiber is an extension of an Abelian variety by means of a product of groups G_m. Still more convincing is the comparison with what we have in characteristic 0. From the results of Kulikov [4] it follows that over the field of complex numbers the desired family $\mathscr{X}'$ exists if and only if the monodromy of the original family $\mathscr{X}$ is finite. If it is assumed that such a result is true over fields of arbitrary characteristic, then this would imply the property formulated above, since it is obvious that in the family of supersingular surfaces of type K3 the monodromy is finite. Finally, in Sec. 12 we shall prove this property over fields of characteristic 2.

Let us suppose that the conjecture that supersingular surtaces of type K3 "do not degenerate" is true. Then we would have an étalé morphism of complete manifolds M → P, and the number of preimages of each point would have to be the same. If there were to exist a point $x_0\in P$, for which it were possible to prove that it had a unique preimage, then this would imply that the period mapping is an isomorphism. In other words, the global Torelli theorem would be true in the following formulation.

If (X, φ) and (X', φ') are two supersingular surfaces of type K3 with an N-structure, then any isomorphism of N-structures f

$$N \begin{matrix} \overset{\varphi}{\nearrow} & H^2(X/W) \\ & \downarrow f \\ \underset{\varphi'}{\searrow} & H^2(X'/W) \end{matrix},$$

taking a cell of effective divisors in $\operatorname{Pic} X \hookrightarrow H^2(X/W)$ into a cell of effective divisors in Pic X' induces a unique isomorphism of the surfaces X and X'.

In particular, from this it follows that the group of automorphisms of a supersingular surface X of type K3 is isomorphic to the factor group Aut S_X/G, where G is a normal divisor generated by reflections relative to the root system in S_X.

For p > 2 Ogus found in [33] a supersingular surface X such that at the point $\pi_0\in P$ corresponding to it the period mapping is one-to-one: this is the Kummer surface corresponding to a supersingular Abelian variety.

In the next section we shall show that such a surface also exists for p = 2.

Example. We shall consider the conjecture that a family of surfaces of type K3 with finite monodromy does not degenerate for an example which Deligne brought to our attention.

As Shioda showed [42], the Kummer surface A × A/G, G = {1, g}, gx = –x, where A is a supersingular elliptic curve over a field of characteristic 2, is not a surface of type K3, but it is rational. We consider the family $A_\varepsilon^{(1)} \times A_\varepsilon^{(2)}/G$ defined over k[[ε]], where at a general point either $A_\varepsilon^{(1)}$ or $A_\varepsilon^{(2)}$ is a nonsupersingular elliptic curve, and $A_0^{(1)} = A_0^{(2)} = A$. According to what has been said above, this family degenerates. If the conjecture advanced above is true, then there must exist a rearrangement of this family which does not degenerate. We shall show that this is in fact the case.

Let R be a ring of characteristic, and let 2, $\overline{R}=R[y, y_2]$, $y_i^2+P_iy_i+Q_i=0$, P_i, $Q_i\in R$, $i=1, 2$, let $g\in \operatorname{Aut}\overline{R}/R$, $gy_i=y_i+P_i$, $G=\{1, g\}$. It is easy to see that $\overline{R}^G=R[z]$, $z=P_2y_1+P_1y_2$, $z^2+P_1P_2z+P_1^2Q_2+P_2^2Q_1=0$. For $u=z/P_1P_2$, $u^2+u=Q_1/P_1^2+Q_2/P_2^2$. We set $R=k[[\varepsilon]][x_1, x_2]$, $P_i=\lambda_ix+\mu_i$, $Q_i=f_i(x_i)=x_i^3+\alpha_ix_i^2+\beta_ix_i+\gamma_i$, λ_i, μ_i, α_i, β_i, $\gamma_i\in k[[\varepsilon]]$. Then Spec $\overline{R}^G$ is an affine open set of

$A_\varepsilon^{(1)}\times A_\varepsilon^{(2)}/G$, where $A_\varepsilon^{(i)}$ is given by the equation $y_i^2+(\lambda_i x_i+\mu_i)y_i=f_i(x_i)$. In particular, if $A_\varepsilon^{(1)}+A_\varepsilon^{(2)}=A$, where A is given by the equation $y^2 + y = f(x)$, then the equation $z^2 + z + f_1(x_1) = f_2(x_2) = 0$ determines a cubic, i.e., it is a rational surface; this is Shioda's result.

In the general case for $A_\varepsilon^{(1)}\times A_\varepsilon^{(2)}/G$ we obtain the equation $u^2+u=\frac{f_1(x_1)}{(\lambda_1 x_1+\mu_1)^2}+\frac{f_2(x_2)}{(\lambda_2 x_2+\mu_2)^2}$. The conditions $A_1^{(i)} = A$ can be written in the form $\lambda_i(0) = 0$, $\mu_i(0) \neq 0$, $i = 1, 2$. We suppose that λ_1/λ_2, $\lambda_2/\lambda_1 = \lambda$, and we set $\lambda_1 x_i = s_i$. After multiplication by λ_1^3 we obtain the equation $\lambda_1^3(u^2+u)=\frac{s_1^3+\lambda_1\varphi_1}{(s_1+\mu_1)^2}+\frac{s_2^3+\lambda_1\varphi_2}{(\lambda s_2+\mu_2)}$, where $\varphi_i=\alpha_i s_i^2+\beta_i s_i\lambda_1+\gamma_i\lambda_1^2$. Setting $\lambda_1=\nu^2$, $\nu\in k[[\varepsilon]]$, $v^3 u=v$, we obtain for $\varepsilon = 0$ the equation $v^2=\frac{s_1^3}{(s_1+m_1)^2}+\frac{s_2^3}{(ls_2+m_2)^2}$, where $m_i=\mu_i(0)$, $l=\lambda(0)$. If $v(s_1+m_1)(s_2+m_2)=w$, then $w^3=s_1^3(ls_2+m_2)^2+s_2^3(s_1+m_1)^2=s_2^3(s_1+m_1)^2+l^2s_1^3s_2^2+m_2^2s_1^3$. We set $s_1 = t$, $s_2(s_1+m_1)^2 = \xi$, $w(s_1 + m_1)^2 = \eta$. Multiplying the equation by $(s_1 + m_1)^4$, we obtain $\eta^2 = \xi^3 + l^2t^3\times\xi^2 + (m_1 + t)^4m_2^2t^3$. This equation can be reduced to the form $\eta^2 = \xi_1^3 + l^4t^6\xi_1 + (m_1 + t)^4\times m_2^2t^3$. This is a quasielliptic pencil in the discriminant of which there are the points $t = 0$, $t = m_1$, and $t = \infty$ with multiplicities 4, 8, and 8; from this it follows that we obtain a surface of type K3. The surfaces of type K3 to which our families specialize are supersingular, since they possess a quasielliptic pencil. It is easy to determine just what surfaces are obtained in this manner. For $l = 0$ if we set $t = \tau^{-1}$, then we obtain the standard form of a pencil with fibers D_4, E_8, E_8 indicated in [8] (p. 868). For $l \neq 0$, applying the criteria in this same work, it is easy to find that the pencil has a fiber of type D_4 for $t = 0$ and fibers of type D_8 for $t = m_1$ and $t = \infty$. This implies that the lattice S_X for this surface X contains the sublattice $U_2\oplus Q(D_4)\oplus Q(D_8)^2$. However, this is not the entire lattice S_X, since the pencil has the nonzero section $\xi=l^{-1}m_2(m_1+t)^2$, $\eta=(l^{-1}m_2)^{3/2}(m_1+t)^3$. Therefore, for the lattice S_X $\sigma \leqslant 2$. It is not hard to see that for general l, $\sigma = 2$ (perhaps for all $l \neq 0$?). Thus, the limits of nonsupersingular Kummer surfaces when the Abelian variety itself becomes supersingular are for $p = 2$ supersingular surfaces X with $\sigma(X) = 1, 2$. In other words, for $p = 2$ just these surfaces are the natural analogues of supersingular Kummer surfaces which for $p > 2$ are also characterized by the condition $\sigma(X) = 1, 2$.

12. Period Mapping over a Field of Characteristic 2

Quasielliptic Pencils. As in Sec. 2, we say that on a surface X there is given a quasielliptic pencil if there is given a morphism of the surface X onto P^1 with a general fiber hich is a rational curve of arithmetic genus 1 over the field $k(P^1)$. Such pencils can exist only for $p = 2$ or 3. A quasielliptic pencil with a section can be given in Weierstrass normal form. The affine part of the surface in Weierstrass normal form is given as a subscheme in the body of the fibering of $\mathcal{O}(2n)\oplus\mathcal{O}(3n)$ over P^1 defined

$$\begin{aligned}&\text{for } p=2 \text{ by the equation } y^2=x^3+ax+b, && a\in H^0(\mathcal{O}(4n)),\\ &&& b\in H^0(\mathcal{O}(6n)),\\ &\text{for } p=3 \text{ by the equation } y^2=x^3+b, && b\in H^0(\mathcal{O}(6n)),\end{aligned} \tag{1}$$

where x and y are coordinates in the first and second terms. The morphism defining the pencil is induced by a projection. To obtain the complete surface it is necessary to complete each fiber with the "infinitely distant" point. We denote this surface by $X_{(a,b)}$ for $p = 2$ and by $X_{(b)}$ for $p = 3$.

Before formulating some properties of quasielliptic surfaces, we define the operation of differentiation of sections of sheaves $\mathcal{O}(px)$ over P^1. We note that if $(t_0:t_1)$ is a coordinate system on P^1 and the element $s\in H^0(\mathcal{O}(pk))$ can be represented in this system in the form $F(t_0, t_1)$, then $t_0\frac{\partial F}{\partial t_0}+t_1\frac{\partial F}{\partial t_1}=0$. Hence there is defined a form $\partial_{(t_0:t_1)}F$ of degree $pk - 2$ such that $\frac{\partial F}{\partial t_1}=t_0\cdot\partial_{(t_0:t_1)}F$.

LEMMA. If $(t_0, t_1) = (u_0, u_1)A$, then

$$\partial_{(t_0:t_1)}F=\partial_{(u_0:u_1)}F\cdot\det A^{-1}.$$

This is easily verified by direct computation. Thus, up to a constant multiple the derivative s' of a section s does not depend on the coordinate system.

We define the discriminant of the Weierstrass normal form of a quasielliptic pencil by setting

$$D=(a')^2a+(b')^2 \text{ for } p=2 \text{ and } D=(b')^2 \text{ for } p=3.$$

The discriminant is defined up to a constant multiple and belongs to $H^0(\mathbf{P}^1, \mathcal{O}(12n-4))$.

Proposition 1. 1) Let p = 2. An isomorphism of the surfaces $X_{(a,b)}$ and $X_{(a_1,b_1)}$ which preserves the structure of a quasielliptic pencil exists if and only if the pair (a_1, b_1) can be obtained from (a, b) by successive application of the transformations

$$\begin{aligned}
&\text{(a) } a\to a+c^4\\
&\quad\ \ b\to b+ac^2+d^2, \text{ where } c\in H^0(\mathcal{O}(n)),\ d\in H^0(\mathcal{O}(3n));\\
&\text{(b) } a\to a^\sigma\\
&\quad\ \ b\to b^\sigma, \text{ where } \sigma\in \mathrm{Aut}(\mathbf{P}^1);\\
&\text{(c) } a\to \varepsilon^4 a\\
&\quad\ \ b\to \varepsilon^6 b, \text{ where } \varepsilon\in k.
\end{aligned}$$

All such transformations generate a group G. Transformations of type (a) form a normal subgroup H in G. Transformations of types (b) and (c) commute and $G/H\simeq \mathrm{PSL}(2,k)\times G_m$.

2) Let p = 3. An isomorphism of surfaces $X_{(b)}$ and $X_{(b_1)}$ which preserves the structure of a quasielliptic pencil exists if and only if the element b_1 can be obtained from b by successive application of the transformations

$$\begin{aligned}
&\text{(a) } b\to b+d^3, \text{ where } d\in H^0(\mathcal{O}(2n));\\
&\text{(b) } b\to b^\sigma, \text{ where } \sigma\in \mathrm{Aut}(\mathbf{P}^1);\\
&\text{(c) } b\to \varepsilon^{6}b, \text{ where } \varepsilon\in k.
\end{aligned}$$

These transformations generate a group G, while transformations of type (a) form a normal subgroup H in it and

$$G/H\simeq \mathrm{PSL}(2,k)\times G_m.$$

The proof follows from formulas presented in [28, p. 301].

We note that for p = 3 the subgroup H is isomorphic to the additive group of the vector space $H^0(\mathcal{O}(2n))$, while for p = 2 it is isomorphic to the group of matrices

$$\left\{\begin{pmatrix}1 & c & d\\ 0 & 1 & c^2\\ 0 & 0 & 1\end{pmatrix},\ c\in H^0(\mathcal{O}(n)),\ d\in H^0(\mathcal{O}(3n))\right\}.$$

Proposition 2. Under transformations of the type (a) the discriminant of the surface does not change. If the discriminant is identically equal to 0, then the surface $X_{(a,b)}$ is not normal and has a curve of singular points. If $D \neq 0$, then the surface $X_{(a,b)}$ has a finite number of singular points situated in fibers corresponding to the zeros of a section of D.

This is easily verified by standard computations.

As before, let $a\in H^0(\mathcal{O}(4n))$, and suppose that $(t_0:t_1)$ is a coordinate system on $\mathbf{P}^1$. We denote by $Sa(t_0, t_1)$ the form in the coordinates obtained as follows:

$$Sa(t_0, t_1)=a(t_0, t_1)-a(1, 0)t_0^{4n},$$

that is, we subtract from a section of fourth degree, so that the difference vanishes at the origin. We shall find the following propoposition useful below.

Proposition 3. Let p = 2. For a surface $X = X_{(a,b)}$ given by Eq. (1) $\chi(\mathcal{O}_X)=n$ if and only if for any coordinate system $(t_0:t_1)$ on $\mathbf{P}^1$ either $Sa(t_0, t_1)$ is not divisible by t_1^4 or $D = (a')^2a + (b')^2$ is not divisible by t_1^{12}.

Let p = 3. For a surface $X = X_b$, $\chi(\mathcal{O}_X)=n$, if and only if for any coordinate system $(t_0:t_1)$ on $\mathbf{P}^1$, $D = (b')^2$ is not divisible by t_1^{12}.

Proof. It is well known that $\chi(\mathcal{O}_X)\leq n$ always and $\chi(\mathcal{O}_X)<n$ if and only if by transformations of type (a) it is possible to reduce the equations to the form

$$y^2=x^3+(u^4a_1)x+u^6b_1,$$

where $\deg u > 0$. It is then possible to make a birational transformation $y_1 = yu^{-3}$, $x_1 = xu^{-2}$ which establishes a birational isomorphism between the original surface and a new surface th equation

$$y_1^2 = x_1^3 + a_1 x_1 + b_1,$$

which the degrees of the sections a_1 and b_1 are less. In this case $\chi(\mathcal{O}_X) < n$. However, the conditions of the proposition are violated for some coordinate system $(t_0:t_1)$ we can ansform the equation with transformations of the type (a), so that a and b acquire factors and t_1^6. This proves the proposition in one direction. The converse assertion follows in elementary way from the form of transformations of type (a).

We note that in the hypotheses of the proposition after resolution of the singular points the fibers we obtain a relatively minimal model of the pencil which for n > 1 is a smooth nimal model of the surface.

Family of Quasielliptic Pencils. The construction of surfaces $X(a,b)$ and $X(b)$ affords the possibility of constructing a family of quasielliptic pencils over a base T, where $=H^0(\mathcal{O}(4n)) \times H^0(\mathcal{O}(6n))$ for p = 2 and $T = H^0(\mathcal{O}(6n))$ for p = 3. In both cases T is an affine ace.

We shall study the action on T of the group G of isomorphisms of quasielliptic pencils. first consider the action of the subgroup H of transformations of type (a), and we conruct the ring of H-invariant functions on T.

We fix a coordinate system $(t_0:t_1)$ on $\mathbf{P}^1$. We write the sections a and b as forms in t_1:

$$a = \sum_{i=0}^{4n} a_i t_0^{4n-i} t_1^i, \quad b = \sum_{i=0}^{6n} b_i t_0^{6n-i} t_1^i.$$

We may consider the coefficients a_i, b_i as functions on the space T. For p = 3 the functions $\{b_i,\ i \not\equiv 0(3)\}$ are invariant under the action of H. For p = 2 the functions $\{a_i,\ i \not\equiv$ $(4)\}$ are also invariant. Moreover, in this case we consider polynomials in a_i, b_i which e coefficients of the discriminant D. We denote them by d_i. They are also invariant, since e discriminant does not change under the action of H. Moreover, their expressions in terms a_i, b_i have the form

$$d_k = f_k(a_i,\ i \not\equiv 0(4)) \text{ for } k \not\equiv 0 \pmod 4,$$
$$d_{4j} = b_{2j+1}^2 + f_{4j}(a_i).$$

We shall indicate the explicit form of several functions:

$$d_0 = b_1^2 + a_1^2 a_0;\ d_1 = a_1^3;\ d_2 = a_1^2 a_2;$$
$$d_3 = a_1^2 a_3;\ d_4 = b_3^2 + a_3^2 a_0 + a_1^2 a_4.$$

Proposition 4. The ring of H-invariant functions on T is a polynomial ring. As genertors of this ring it is possible to take

$$\{a_i,\ d_j,\ i \not\equiv 0\ (4),\ j \equiv 0\ (4)\} \text{ for } p=2,\ \{b_i,\ i \not\equiv 0\ (3)\} \text{ for } p=3.$$

Proof. The Case p = 2. From the explicit formulas for transformations belonging to H t follows that any geometric point $t \in T$ can be taken by a transformation in H into a point such that $a_i(\bar{t}) = 0$ for $i \equiv 0 \pmod 4$, $b_i(\bar{t}) = 0$ for $i \equiv 0\ (2)$. From this it follows that $_j(\bar{t}) = b_{2j+1}^2(\bar{t})$. If f is an H-invariant function on S, then

$$f(t) = f(\bar{t}) = F(a_i, b_{2j+1},\ i \not\equiv 0(4))(\bar{t}) = F(a_i, d_{4j}^{1/2},\ i \not\equiv 0(4))(\bar{t}) = F(a_i, d_{4j}^{1/2})(\bar{t})$$

or some polynomial F. However, by restricting the function $F(a_i, d_{4j}^{1/2})$ to linear manifolds points in T of the form b = 0, a' fixed, it is easily found that such a function belongs the ring of regular functions on T only if it can be expressed in terms of a_i and d_{4j} in ntegral powers.

The Case p = 3. Here the assertion of the proposition is obvious.

Proposition 5. We denote by $\bar{T}$ the spectrum of the ring of H-invariant functions. Then e scheme T is a geometric factor for the action of H on T.

Proof. We must verify four conditions (see [29, Def. 0.6]). Conditions (i) and (iv) re immediately obvious. To prove condition (ii) it is necessary to verify that two geometric oints t_1 and t_2 with the same values of the invariant functions lie on the same orbit. Ineed, for p = 2, using the transformations in H, we may assume that $a_i(t_1) = a_i(t_2)$ for all i.

Then $d_{4j}(t_1) = d_{4j}(t_2)$ imply $b_{2j+1}(t_1) = b_{2j+1}(t_2)$, and, using obvious transformations in H, we can arrange that $b_j(t_1) = b_j(t_2)$ for all j. For p = 3 the argument is still simpler. Condition (iii) is the universal submersibility of the morphism $\varphi: T \to \overline{T}$. For p = 3 φ is simply the factorization of a linear space with respect to a subspace, and universal submersibility of such a morphism is easily verified. For p = 2 φ can be represented as the composition of a purely nonseparable finite morphism of affine spaces, an isomorphism, and a factorization of a linear space with respect to a subspace, and the universal submersibility for p = 2 follows from the universal submersibility of each factor.

The action of the group G on T defines an action of G/H on $\overline{T}$. The action of the factor G_m on $\overline{T}$ is described very simply: $\overline{T}$ is a linear space, and G_m acts coordinatewise. A geometric factor of the subscheme $T \setminus \{0\}$ exists and is a weighted projective space P. For p = 3 this is a space of type (6^{3n}) isomorphic to a simply projective space. [A weighted projective space has type (a^n, b^m) if n coordinates have weight a and m have weight b.] For p = 2 it has type $(4^{3n}, 12^{3n})$ and is isomorphic to a weighted projective space of type $(1^{3n}, 3^{3n})$. To establish the isomorphism we define the degree of coordinate functions on T by setting

$$\deg a_i = 1,\ \deg d_{4j} = 3 \text{ for } p=2,\ \deg b_i = 1 \text{ for } p=3.$$

We note that this definition of degree agrees with the change of coordinate functions which arise in changing coordinates on $\mathbf{P}^1$. It remains for us to consider the action of the group PSL(2, k) on the weighted projective space P. This action admits a linearization L: for p = 3 we set $k = \mathcal{O}_P(1)$, while for p = 2 $L = \mathcal{O}_P(3)$ [$\mathcal{O}_P(n)$ denotes the sheaf defined by rational functions for the weight n].

We denote by P_* and P_{**} the subsets of stable and semistable points, respectively, and by $\overline{T}_*$, $\overline{T}_{**}$ and T_*, T_{**} their full preimages in $\overline{T}$ and T.

Proposition 6. An open subset of points $t \in T$, such that for the corresponding surface $X = X_{(t)}$, $\chi(\mathcal{O}_X) = n$ is contained in T_* for all $n \geq 2$, $T_* = T_{**}$ for p = 3 and p = 2, $n \equiv 0 \times$ (mod 2). For n = 2 the subset $T_* = T_{**}$ coincides precisely with the subset of those points s for which $\chi(\mathcal{O}_X) = 2$ and hence those for which the surface X is a K3 surface.

Proof. We apply the criterion of stability of Hilbert–Mumford [29, Theorem 2.1]). For this we must compute the action of one-parameter subgroups of the group PSL(2, k) on sections of the linearization L. The one-parameter subgroups correspond to coordinate systems on $\mathbf{P}^1$. Let $\lambda \subset PSL(2, k)$ be a subgroup. Then there exists a coordinate system $(t_0:t_1)$ such that $\lambda(\alpha)t_0 = \alpha^{-1}t_0$, $\lambda(\alpha)t_1 = \alpha t_1$. Fixing this coordinate system, we consider the functions

$$a_i,\ i \not\equiv 0(4),\ d_{4j} \text{ for } p=2,$$
$$b_i,\ i \not\equiv 0(3) \text{ for } p=3$$

on $\overline{T}$. Then for p = 3 they form a basis of the sections of the pencil, while for p = 2 such a basis is obtained by taking all possible products of the three functions a_i and the functions d_{4j}. The action of the subgroup λ on these functions is defined by the formulas

$$\lambda(\alpha)a_i = \alpha^{2i-4n}a_i,$$
$$\lambda(\alpha)d_i = \alpha^{2i-12n+4}d_i,$$
$$\lambda(\alpha)b_i = \alpha^{2i-6n}b_i.$$

Thus, for p = 2 a point $x \in P$ is not stable if for some choice of coordinate system

$$a_i(x) = 0 \text{ for } 2i-4n<0,\ i \not\equiv 0(4),$$
$$d_{4j}(x) = 0 \text{ for } 8j-12n+4<0,$$

and it is not semistable if the same is true for

$$2i-4n \leqslant 0,\ i \not\equiv 0(4),$$
$$8j-12n+4 \leqslant 0.$$

For even n these conditions are equivalent, since $2i - 4n = 0$ implies $i = 2n \equiv 0(4)$ and $8j - 12n + 4 = 0$ implies $j = \frac{3n-1}{2} \notin \mathbf{Z}$. It has thus been shown that $P_{**} = P_*$. The remaining assertions are obtained by simply comparing the stability criterion and the hypothesis of Proposition 2. The case p = 3 is considered similarly. The condition of instability in this case is the following: for some coordinate system

$$b_i(x)=0 \text{ for } 2i-6n<0,\ i\not\equiv 0(3),$$

the equality $2i - 6n = 0$ implies $i \equiv 0(3)$ for all n.

We thus find that for the open subset $T_* \subset T$ there exists a geometric factor with respect to the action of G $T_*/G \simeq P_*/\mathrm{PSL}(2, k)$ and in the case where $T_* = T_{**}$ this factor is a complete variety (see [29, p. 187]).

<u>Completeness of the Base of a Universal Family.</u> We recall the criterion for completeness of algebraic varieties. We shall recall a well-known criterion for completeness (we have already appealed to it in Sec. 11).

Let T be a quasiprojective algebraic variety, and let R, R' be schemes each of which is the spectrum of a one-dimensional, complete, regular local ring.

If for some rational mapping $\varphi: R \dashrightarrow T$ there exists a regular mapping $\psi/R' \to T$ and a finite morphism $\pi: R' \to R$ such that the diagram

$$\begin{array}{ccc} R' & & \\ \downarrow & \searrow & \\ R & \dashrightarrow & T \end{array}$$

is commutative, then the variety T is complete.

We now consider a universal family $\mathscr{X} \to M$ of supersingular K3 surfaces with Artin invariant which satisfies the inequality

$$\sigma(X) \leqslant 9 \quad \text{for} \quad p=2,$$
$$\sigma(X) \leqslant 5 \quad \text{for} \quad p=3.$$

<u>THEOREM.</u> The base M of the universal family $\mathscr{X} \to M$ indicated above is a complete variety.

<u>Proof.</u> We shall use the criterion of completeness formulated above. Let R again be the spectrum of a complete, one-dimensional, regular local ring. Prescription of a rational mapping $\varphi: R \dashrightarrow M$ is equivalent to prescribing a smooth family of supersingular K3 surfaces $V_r \to \{r\}$ over $\{r\} = R \setminus r_0$. By considering a covering $R' \to R$, we may assume that the family is defined over $\{r'\}$ and has an N-structure where $\sigma(N) = 9$ or 5, respectively, if p = 2 or 3.

According to Remark 4 to the theorem of Sec. 5, in the cases considered in a general fiber of our family there exists a quasielliptic pencil with a section, so that we can use Proposition 6. To prove the theorem it suffices for us to show that after a further possible lift of the base $R'' \to R'$ the family $V_{r''} \to \{r''\}$ can be augmented to a smooth family of supersingular K3 surfaces $V_{R''} \to R''$.

In order to simplify notation, we assume that we are given a smooth family $V_r \to \{r\}$ of Jacobian quasielliptic surfaces of type K3 over a general point of the scheme R. It is clear that such a family can be obtained as an inverse image of the family of Jacobian quasielliptic surfaces $U \to T$ (where n = 2) relative to some morphism $\sigma: \{r\} \to T_*$ which was constructed earlier. Here two distinct morphisms σ and σ_1 which are equivalent relative to the action of the group G on T give identical inverse images of the family. In our case because of Proposition 6 the geometric factor T_*/G is a complete variety, and hence the mapping $\bar{\sigma}: \{r\} \to T_*/G$ obtained from σ can be extended to a morphism $\bar{\sigma}: R \to T_*/G$. For some finite covering π': $R' \to R$ we can lift the mapping $\bar{\sigma}$ to a mapping $\tau: R' \to T_*$ so that the diagram

$$\begin{array}{ccc} R' & \xrightarrow{\tau} & T_* \\ {\scriptstyle \pi'}\downarrow & & \downarrow \\ R & \xrightarrow{\bar{\sigma}} & T_*/G \end{array}$$

is commutative. On the basis of the properties of the geometric factor, for some finite covering $\pi'': R'' \to R'$ the points $\pi''\tau(r')$ and $\pi''\cdot\pi'\sigma(r)$ are equivalent relative to the action of the group G, and hence the inverse images of the families $V_r \to \{r\}$ and $U \to T$ relative to the morphisms $\pi''\cdot\pi': \{r''\} \to \{r\}$ and $\pi''\tau: \{r''\} \to T$ are isomorphic. In other words, this means that the lifted family $V_{r''} \to \{r''\}$ is augmented to a smooth family of quasielliptic Jacobian surfaces $U_{R''} \to R''$. Since the image of a closed point r'' belongs to S_*, the family $U_{R''}$ is a family of K3 surfaces and, of course, the value of the Artin invariant for a closed fiber is subject to the same inequality as a general fiber. Thus, the smooth family we require has been constructed, and the theorem is proved.

We thus have a mapping of complete, smooth varieties $\pi: M \to P[9]$, where π is the period mapping and M is the manifold of moduli. We shall prove that it is one-to-one at some point $m \in M$. Namely, we consider a surface X with $\sigma(X) = 1$. For it is possible to take (in view of the theorem of Sec. 1) S_X in the form

$$S_X = U_2 \oplus Q(D_4 \oplus E_8{}^2). \tag{2}$$

Considering a primitive element $f \in U_2$ with $f^2 = 0$ corresponding to the effective divisor F on X, we find on X a quasielliptic pencil with three degenerate fibers. It is easy to see (see 8, p. 868]) that this pencil reduces uniquely (up to automorphism of X) to the form

$$y^2 = x^3 + t^5(t+1)^4 \tag{3}$$

(a degenerate fiber of D_4 is located at ∞, while the types E_8 are at 0 and 1).

On the surface X we introduce an N-structure with polarization $\varphi: N \to S_X$, and we denote by m the corresponding point on M. If $m' \in M$ is another point with $\pi(m) = \pi(m')$, then to m' there must also correspond a surface X' with $\sigma(X') = 1$ and a polarization $\varphi': N \to S_X$. Since $\pi(m) = \pi(m')$, the diagram

$$N \begin{array}{c} \varphi \\ \searrow \\ \varphi' \end{array} \begin{array}{ccc} S_X & \longrightarrow & H^2(X/W) \\ \downarrow \psi & & \downarrow \psi \\ S_{X'} & \longrightarrow & H^2(X'/W) \end{array}$$

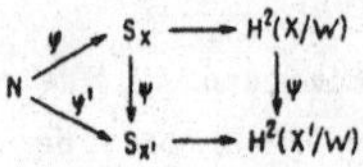

is commutative for some isomorphism $\psi: H^2(X/W) \to H^2(X'/W)$ which, as is easily seen, defines an isomorphism $\psi: S_X \to S_{X'}$. We set $\psi(f) = f'$. This element must define a pencil with the same properties as f which for the same reasons can be written in the form (3). There is thus defined an isomorphism $u: X \to X$ which generates ψ in the cohomology groups. This means that m = m'.

In view of the remarks made at the end of the next section, this implies the global Torelli theorem for surfaces over a field of characteristic 2 with $\sigma(X) < 10$.

Similar arguments are also applicable for $p = 3$, $\sigma(X) < 6$. In place of the lattice (2) it is necessary to take only the lattice

$$U_2 \oplus Q(E_6{}^2 \oplus E_8).$$

UNSOLVED QUESTIONS

1. Determine finite group schemes acting on elliptic surfaces. (Their Lie algebras are determined in the works [31, 33].)

2. On what quasielliptic surfaces do there exist nonzero, regular vector fields?

3. On what surfaces of general type do there exist nonzero, regular vector fields?

4. Prove the stability of a smooth surface of type K3.

5. (Artin). Prove that the two definitions of supersingularity for surfaces of type K3 coincide, i.e., that the condition $h = \infty$ implies that $\rho = 22$. Artin proved this for elliptic surfaces of type K3.

6. (Artin, Shioda). Is a supersingular surface of type K3 unirational?

7. For each supersingular surface X of type K3 does there exist a purely nonseparable rational mapping $f: P^2 \dashrightarrow X$? A positive answer would imply a positive answer to question 6. It is more likely, however, that the answer is negative.

8. For any two supersingular surfaces X and Y of type K3 does there exist a purely nonseparable rational mapping $f: X \dashrightarrow Y$? In view of the fact that Kummer surfaces are unirational, a positive answer would imply a positive answer to question 6. This question is a special case of the general question concerning "isogenies" of a surface of type K3. If, in general, the answer is negative, it would be interesting to determine when such a mapping exists, for example, in terms of the periods of the surfaces X and Y.

9. Is it true that supersingular surfaces of type K3 do not degenerate (in the sense indicated in Sec. 11)?*

*The authors have recently found that the answer to this question is positive.

10. Is it true that a family of surfaces of type K3 with finite monodromy do not degenerate? A positive answer would, of course, imply a positive answer to question 9.

11. (Ogus). Prove a global Torelli theorem for supersingular surfaces of type K3.

12. Find an analogue of formulas of Wu for crystalline cohomologies (or for de Rham cohomologies) of algebraic varieties over a field of characteristic 2. For example, is a quadratic form in $H^2_{DR}(X)$ for a surface X even if and only if the canonical class of X is even? (At least in the case where the de Rham cohomologies give the "true" Betti numbers.) For surfaces of type K3 this question, as we have seen in Sec. 9, is related to the question of whether any surface of type K3 is "connected" with the continuous curve constructed in the proof of Proposition 5, Sec. 9. Therefore, the question of the irreducibility of the "manifold of moduli" of surfaces of type K3 having a given least degree of polarization is of interest. More precisely, can any two such surfaces be included in an irreducible family? (Over a field of characteristic $p > 0$. Over a field of characteristic 0 this follows from the global Torelli theorem.)

LITERATURE CITED

1. Algebraic Surfaces, Tr. Mat. Inst. im. V. A. Steklova, 75 (1965).
2. B. B. Venkov, "On the classification of integral, even, unimodular, 24-dimensional quadratic forms," Tr. Mat. Inst. im. V. S. Steklova, 148, 65-76 (1978).
3. I. V. Dolgachev, "The Euler characteristic of algebraic varieties," Mat. Sb., 89, No. 2, 297-312 (1972).
4. V. S. Kulikov, "Degenerations of K3 surfaces," Izv. Akad. Nauk SSSR, Ser. Mat., 41, No. 5, 1008-1042 (1977).
5. V. V. Nikulin, "Integral symmetric bilinear forms and some of their geometric applications," Izv. Akad. Nauk SSSR, Ser. Mat., 43, No. 1, 111-177 (1979).
6. I. I. Pyatetskii-Shapiro and I. R. Shafarevich, "The Torelli theorem for algebraic surfaces of type K3," Izv. Akad. Nauk SSSR, Ser. Mat., 35, 530-572 (1971).
7. A. N. Rudakov and I. R. Shafarevich, "Nonseparable morphisms of algebraic surfaces," Izv. Akad. Nauk SSSR, Ser. Mat., 40, No. 6, 1269-1307 (1976).
8. A. N. Rudakov and I. R. Shafarevich, "Supersingular surfaces of type K3 over fields of characteristic 2," Izv. Akad. Nauk SSSR, Ser. Mat., 42, No. 4, 848-869 (1978).
9. A. N. Rudakov and I. R. Shafarevich, "Vector fields on elliptic surfaces," Usp. Mat. Nauk, 33, No. 6, 231-232 (1978).
10. M. Artin, "Supersingular K3 surfaces," Ann. Sc. Ec. Norm. Sup. 4-e Serie, 7, fase 4, 543-567 (1974).
11. M. Artin, "Algebraic construction of Breskorn's resolution," J. Algebra, 29, 330-348 (1974).
12. M. Artin and B. Mazur, "Formal groups arising from algebraic varieties," Ann. Sc. Ec. Norm. Sup. 4-e serie, 10, 87-132 (1977).
13. P. Berthelot, "Cohomologie cristalline des schemas de characteristique p > 0," Lect. Notes Math., 407 (1974).
14. P. Berthelot and A. Ogus, Notes on Crystalline Cohomology, Princeton Univ. Press (1978).
15. E. Bombieri and D. Husemoller, "The classification and embeddings of surfaces," Proc. of Symposia in Pure Mathematics, 29, 329-421 (1974).
16. E. Bombieri and D. Mumford, "Enriques' classification of surfaces in char. p. III," Inventiones Mathem., 35, 197-232 (1976).
17. N. Bourbaki, Groupes et Algêbras de Lie, Chaps. 4, 5, et 6, Eléments de Mathématique, Fasc. XXXIV, Paris (1968).
18. C. Chevalley, Introduction to the Theory of Algebraic Functions of One Variable, Baltimore (1951).
19. P. Deligne, "Relèvements des surfaces K3 en characteristique nulle (rédigé par L. Illusie)," Preprint (1979).
20. P. Deligne (avec la collaboration de L. Illusie), "Cristaux ordinaires et coordonnées canoniques," Preprint (1979).
21. D. Giescker, "Global moduli for surfaces of general type," Preprint (1979).
22. A. Grothendieck, "Eléments de géométrie algébrique. III," Publications Mathématiques, No. 11, 349-511 (1961).
23. A. Grothendieck, "Le groupe de Brauer. III," Dix Exposés sur la Cohomologie des Schémas, Paris—Amsterdam (1968), pp. 88-188.
24. L. Illusie, "Report on crystalline cohomology," Proceedings of Symposia in Pure Mathematics, 29, 459-479 (1974).

25. N. Jacobson, Lie Algebras, New York–London (1962).
26. N. Katz, "Algebraic solutions of differential equations, p-curvature and the Hodge filtration," Inventiones Mathem., 18, 1-118 (1972).
27. N. Katz, "Travaux de Dwork," Séminaire Bourbaki, exp. 409, Lecture Notes in Math., 317 (1973), pp. 167-200.
28. S. Lang, Elliptic Functions, Addison-Wesley, Reading, Mass. (1974).
29. D. Mumford, Geometric Invariant Theory, Springer-Verlag, Berlin (1965).
30. M. Nagata, Local Rings, Wiley-Interscience, New York (1962).
31. A. Ogg, "Elliptic curves and wild ramification," Am. J. Math., 89, No. 1, 1-21 (1967).
32. A. Ogus, "F-crystals and Griffiths transversality," Int. Symp. on Alg. Geometry, Kyoto (1977), pp. 15-44.
33. A. Ogus, "Supersingular K3-crystals," Astèrisque, 64, 3-86 (1979).
34. T. O'Meara, The Arithmetic Theory of Quadratic Forms, New York (1950).
35. P. Russel, "Forms of the affine line and its additive group," Pac. J. Math., 79, No. 3, 411-449 (1964).
36. M. Schlessinger, "Functors of Artin rings," Trans. Am. Math. Soc., 130, 205-222 (1968).
37. J. P. Serre, Cours d'Arithmétique, Presses Univ. de France, Paris (1970).
38. C. S. Seshadri, L'Operation de Cartier. Applications. Variétés de Picard, Seminaire C. Chevalley, Paris (1960).
39. T. Shioda, "An example of unirational surfaces in characteristic p," Math. Ann., 211, 233-236 (1974).
40. T. Shioda, "Algebraic cycles on certain K3 surfaces in characteristic p," Proceedings of the International Conference on Manifolds and Related Topics in Topology, Univ. of Tokyo Press (1975), pp. 357-364.
41. T. Shioda, "Some results on unirationality of algebraic surfaces," Math. Ann., 230, 153-168 (1977).
42. T. Shioda, "Kummer surfaces in characteristic 2," Proc. Jpn. Acad., 50, No. 9, 718-722 (1974).

The influence of height on degenerations of algebraic surfaces of type K3

(with A. N. Rudakov and T. Zink)

Izv. Akad. Nauk SSSR, Ser. Mat. **46**, 117–134 (1982). Zbl. **492**, 14024
[Math. USSR, Izv. **20**, No. 1, 119–135 (1983)]

Dedicated to the Memory of O. N. Vvedenskiĭ

ABSTRACT. The authors announce the conjecture that a family of $K3$ surfaces the Artin height of whose generic fiber is greater than 2 does not degenerate; they prove this conjecture for surfaces of degree 2. As a corollary it is shown that a family of supersingular $K3$ surfaces does not degenerate; i.e., its variety of moduli is complete.
Bibliography: 18 titles.

Introduction

For an algebraic variety X defined over an algebraically closed field $\mathbf{k}$ of finite characteristic, its crystalline cohomology groups $H^i_{\text{crys}}(X)$ contain information on the geometry of X analogous to that which, over the field of complex numbers, can be obtained by endowing X with a Kähler structure. The elucidation of this information in various special cases is an intriguing problem but has been little investigated. The groups H^i_{crys} are modules over the ring $W(\mathbf{k})$ of Witt vectors, and the Frobenius endomorphism acts on them. Similar objects (crystals) are described by not very complicated invariants (at least, after multiplying by the field of fractions of the ring $W(\mathbf{k})$), and it is a matter of understanding the geometric meaning of these invariants. For example, the crystal corresponding to the group H^2_{crys} for an algebraic surface with geometric genus 1 may be taken to be an integer h satisfying $1 \leqslant h \leqslant b_2/2$. A $K3$ surface provides a more interesting example. In this case (and in a series of others) the geometric meaning of the invariant h was clarified by Artin [1]. He showed that the surface can be compared with a formal group, $\widehat{\mathrm{Br}}(X)$, the so-called formal Brauer group; in our case this group is one-dimensional and h is its height (in this case $h \geqslant 11$ means that $h = \infty$; i.e., the formal group is unipotent).

In this paper we want to turn our attention to the fact that the invariant arising in this way—the height of a $K3$ surface—plays a role in an interesting geometrical question about the possible degenerations of $K3$ surfaces. Let X/S be a family of surfaces whose base S is the spectrum of a one-dimensional regular local ring and whose general fiber is a $K3$ surface. By a *reconstruction* of this family we mean a family X'/S', where $S' \to S$ is a finite morphism and the general fibers of the families $X \times_S S'$ and X' are isomorphic. The

1980 *Mathematics Subject Classification*. Primary 14J25; Secondary 14L05.

problem is: For a given family X/S, to find a reconstruction X' with the simplest degeneration (i.e. the closed fiber). In particular, if the family has a reconstruction for which the morphism $X' \to S'$ is smooth, it is called *nondegenerating*.

Our basic observation is that in some cases one can construct a reconstruction X'/S' such that either it does not degenerate or its closed fiber has height 1 or 2 (the definition of height in this case continues to hold for the degenerate fiber). We prove this in two cases: a) for families with "standard degenerations" (according to Kulikov [7] any families in characteristic 0 can be reduced to these), and b) for surfaces of degree 2, i.e. double planes and elliptic surfaces. These two cases are not very far from each other, since in case b) there is a reconstruction whose degenerate fiber is different from type a) only by stiffenings of smooth rational components [16]. The problem remains open of whether such an assertion holds in the general case.

Under very broad assumptions the group $\widehat{\mathrm{Br}}$ can be defined for a family X/S as a formal group over S. (We give a simple exposition of this question in §1.) Then the height of $\widehat{\mathrm{Br}}$ can only jump under specialization. In particular, in cases a) and b) we find that if the height of the general fiber of the family X/S is greater than 2, then the family does not degenerate. Supersingular surfaces (i.e. those on which all two-dimensional cycles are algebraic) have $h = \infty$ and are realized as surfaces of degree 2, and so come under case b). Therefore, supersingular surfaces do not degenerate; i.e., the variety of their moduli is complete. Hence, as we know, the period mapping introduced by Ogus [11] for such
1 surfaces is epimorphic, and there is an analog of Torelli's theorem for it.

As another application of our result we show that the two definitions of supersingular $K3$ surfaces ("all cycles are algebraic" and $h = \infty$) agree for surfaces of degree 2.

§1. Specialization of the formal Brauer group

In [1] and [3] the concept of the formal Brauer group of an algebraic variety was introduced and its application to the geometry of $K3$ surfaces over a field of finite characteristic was pointed out. We shall be interested in a family X/S whose general fiber X_η is a $K3$ surface and whose base S is the spectrum of a one-dimensional regular local ring with generic point η. Then the formal Brauer group of the generic fiber X_η is defined and is a one-dimensional formal group over the field $\mathbf{k}(\eta)$. The structure of this group (over an algebraically closed field $\mathbf{k}(\eta)$) is determined by one natural number, the height [8]. One can often define the Brauer group for the closed fiber X_0 too and prove that the height does not decrease under specialization. This is the case when there is a formal Brauer group of the scheme X/S as a formal group over S. The conditions under which such a situation occurs are given by Artin and Mazur [3] and Raynaud [13]. For the reader's convenience we present a more elementary exposition of this question, following, with some simplifications, ideas of Schlessinger [17].

Let X be a locally noetherian scheme. Denote by $\mathrm{Nil}\,\mathcal{O}_X$ the category of all commutative quasiprojective $\mathcal{O}_X$-algebras N for which there exists a natural number m such that $N^m = 0$. The category of sheaves of sets over X will be denoted by $\widetilde{\mathrm{Ens}}\,X$. We shall consider functors $F\colon \mathrm{Nil}\,\mathcal{O}_X \to \widetilde{\mathrm{Ens}}\,X$ having the following property of "localness":

(a) *If $j\colon U \to X$ is an open embedding, then*

$$j_*(F(N) \mid U) = F\big(j_*(N \mid U)\big).$$

To each quasicoherent $\mathcal{O}_X$-module M one can assign a functor h_M with property (a):

$$h_M(N) = \widetilde{\operatorname{Hom}}_{\mathcal{O}_X}(M, N),$$

where $\widetilde{\operatorname{Hom}}$ is the sheaf of homomorphisms.

If M is a quasicoherent $\mathcal{O}_X$-module, then, putting $M^2 = 0$, we get an $\mathcal{O}_X$-algebra as well as an embedding $\operatorname{Mod} \mathcal{O}_X \to \operatorname{Nil} \mathcal{O}_X$ of the category of quasicoherent $\mathcal{O}_X$-modules into the category $\operatorname{Nil} \mathcal{O}_X$. The restriction of the functor F to $\operatorname{Mod} \mathcal{O}_X$ will be denoted by t_F. The sheaf $t_F(M)$ can be endowed with a structure of $\mathcal{O}_X$-module:

$$t_F(M) \oplus t_F(M) \tilde{\to} t_F(M \oplus M) \overset{t_F(\Sigma)}{\to} t_F(M),$$

$$t_F(M)(U) \overset{t_F(s)}{\to} t_F(M)(U), \qquad s \in \mathcal{O}_X(U).$$

If the functor F is permutable with inductive limits, one thus gets a quasicoherent $\mathcal{O}_X$-module. Indeed, it suffices to consider the case where X is affine. Let $U = D(f)$, $f \in \Gamma(X, \mathcal{O}_X)$. Then

$$(t_F(M)(X))_f = \varinjlim \left(t_F(M)(X) \overset{f^n}{\to} t_F(M)(X) \right)$$

$$= t_F\left(\varinjlim M \overset{f^n}{\to} M \right)(X) = t_F\big(j_*(M \mid U)\big)(X) = t_F(M)(U).$$

The quasicoherence now follows from [5], (1.4.1).

If $\varphi\colon Y \to X$ is a morphism, putting

$$(\varphi^* F)(N) = \varphi^*\big(F(\varphi_* N)\big),$$

gives us a functor on Y analogous to the functor F on X.

Definition. A functor $F\colon \operatorname{Nil} \mathcal{O}_X \to \widetilde{\operatorname{Ens}}\, X$ is called a *formal variety* over X if it is isomorphic to h_p, where P is a locally free $\mathcal{O}_X$-module of finite type. A functor $\operatorname{Nil} \mathcal{O}_X \to \operatorname{Ab} X$ into the category of sheaves of abelian groups on X is called a *formal group* over X if, as a functor into the category $\widetilde{\operatorname{Ens}}\, X$, it is a formal variety.

Proposition 1. *A functor F: $\operatorname{Nil} \mathcal{O}_X \to \widetilde{\operatorname{Ab}}\, X$ is a formal group if and only if condition* (a) *and the following conditions hold*:

(b) *F is exact.*

(c) *F is permutable with inductive limits.*

(d) *$t_F(\mathcal{O}_X)$ is a coherent $\mathcal{O}_X$-module.*

The proof is a simple degeneration of an argument of Schlessinger's [17]. By (a) we may assume $X = \operatorname{Spec} A$ is affine. Consider the canonical morphism

$$t_F(A) \otimes M \to t_F(M). \tag{1}$$

By (c) this is an isomorphism for any free modules. Since both functors in (1) are right-exact, using a free resolution we get an isomorphism for any M. Since by (b) the functor F is exact and $t_F(A)$ is a coherent sheaf, it follows that $t_F(A)$ is flat and therefore is a locally free sheaf.

We shall denote the A-module corresponding to the sheaf $t_F(A)$ by the same letter, and put $P = \operatorname{Hom}_A(t_F(A), A)$. Then the morphism (1) can be written as

$$h_P(M) = \operatorname{Hom}_A(P, M) \to t_F(M). \tag{2}$$

Denote by J the augmentation ideal of the symmetric algebra $S(P)$. Then for $N \in \operatorname{Nil} A$ a homomorphism of modules $P \to N$ defines a homomorphism $J/J^s \to N$ for all $s \gg 0$, and

$$h_P(N) = \varprojlim \operatorname{Hom}_{\mathrm{Alg}}(J/J^s, N).$$

Therefore

$$\operatorname{Hom}(h_P, F) = \varprojlim_s F(J/J^s). \tag{3}$$

Denote by $\xi_0 \in t_F(P)$ the image of the unit endomorphism of the module P under the homomorphism (2). Since F is right-exact, the element $\xi_0 \in F(J/J^2)$ can be lifted to an element $\xi \in \varprojlim_s F(J/J^s)$. By (3) we may assume that $\xi \in \operatorname{Hom}(h_P, F)$. It is easy to see that ξ is an isomorphism. To do this, with the aid of some short exact sequences, we must reduce the assertion to the case $N^2 = 0$. From this the proposition follows.

Let $f\colon X \to S$ be a morphism of locally noetherian schemes. For a formal group Φ over X we define its higher direct images as functors on $\operatorname{Nil} \mathcal{O}_S$:

$$\left(R^i f_* \Phi\right)(N) = R^i f_*(\Phi(f^*N)).$$

Proposition 2. *Suppose that a morphism $f\colon X \to S$ is proper and flat, and Φ is a formal group on X such that for some i and all $s \in S$*

$$H^{i-1}\left(X_s, t_\Phi(\mathcal{O}_X) \otimes \mathcal{O}_{X_s}\right) = H^{i+1}\left(X_s, t_\Phi(\mathcal{O}_X) \otimes \mathcal{O}_{X_s}\right) = 0.$$

Then $R^i f_* \Phi$ is a formal group over S.

Proof. We first check that

$$R^{i-1} f_*(\Phi(f^*N)) = R^{i+1} f_*(\Phi(f^*N)) = 0. \tag{4}$$

Since the functor f^* is exact, we can reduce everything to the case $N^2 = 0$. In proving the preceding proposition we saw that the canonical morphism

$$\Phi(f^*N) \xleftarrow{\sim} t_\Phi(\mathcal{O}_X) \otimes f^*N, \qquad N^2 = 0,$$

is an isomorphism. As we know, for any coherent sheaf $\mathcal{F}$ on X

$$\left(R^m f_* \mathcal{F}\right)^\wedge_s \xrightarrow{\sim} \varprojlim H^m\left(X_s, \mathcal{F} \otimes_{\mathcal{O}_S} \mathcal{O}_s/\mathfrak{m}_s^n \mathcal{O}_s\right).$$

Therefore, it suffices to consider only the case where the scheme S is artinian. Then everything easily reduces to the case $N = \mathcal{O}_s/\mathfrak{m}_s\mathcal{O}_s$, for which the assertion is contained in the condition.

From (4) it follows that the functor $F = R^i f_* \Phi$ is exact; this is condition (b) of Proposition 1. The remaining conditions are easily verified. From Proposition 1 it follows that $R^i f_* \Phi$ is a formal group.

Let $h\colon T \to S$ be a closed embedding. Then the conditions of Proposition 2 hold for $f_T\colon X \times_S T \to T$, and one easily checks that

$$h^*\left(R^i f_* \Phi\right) = \left(R^i f_T\right)_*(h_1^* \Phi),$$

where h_1 is the projection $X \times_S T \to X$. Hence

$$\left(R^i f_* \Phi\right) | T = R^i f_{T_*}(\Phi \,|\, X \times_S T), \tag{5}$$

i.e., the functor $R^if_*\Phi$ is permutable, in a natural sense, with restriction to a closed subscheme.

We consider the case where $\Phi = \hat{G}_m$, so that $t_\Phi(\mathcal{O}_X) = \hat{G}_a$. The scheme X/S will have relative dimension 2, and its fibers X_s will satisfy the condition $H^1(X_s, \mathcal{O}_{X_s}) = 0$. By Proposition 2, the functor $R^2f_*\hat{G}_m$ is then a formal group. It is called the *formal Brauer group* and denoted by $\widehat{\mathrm{Br}}(X/S)$. It is easy to check that if $F = \widehat{\mathrm{Br}}(X/S)$, then $t_F(\mathcal{O}_S) = R^2f_*\hat{G}_a$.

Turning to the case where S is the spectrum of a one-dimensional regular local ring and the general fiber of the scheme X/S is a $K3$ surface, we see that if the formal group $\widehat{\mathrm{Br}}(X/S)$ exists, it is one-dimensional. Proposition 2 and (5) give us

PROPOSITION 3. *Let S be the spectrum of a one-dimensional regular local ring and X/S a scheme whose general fiber is a $K3$ surface. If $H^1(X_0, \mathcal{O}_{X_0}) = 0$ for the closed fiber X_0, then the formal Brauer group $\widehat{\mathrm{Br}}(X/S)$ exists, is one-dimensional, and its height does not decrease when passing from the general to the closed fiber.*

§2. Standard degenerations

A family X/S, where S is the spectrum of a one-dimensional regular local ring with generic point η, and X_η is a $K3$ surface, by definition has a standard degeneration if one of the following conditions holds:

A. X/S is smooth; that is, the closed fiber X_0 is a smooth $K3$ surface.

In cases B and C the fiber X_0 is reducible, $X_0 = \bigcup V_\alpha$ is its decomposition into irreducible components, and $\Pi(X_0)$ is the polyhedron of this covering. We assume that the components V_i have multiplicity 1 and have normal intersection.

B. The polyhedron $\Pi(X_0)$ has the form

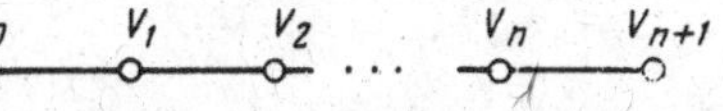

Here V_0 and V_{n+1} are smooth rational surfaces, and V_α, $\alpha = 1,\dots,n$, are ruled surfaces over an elliptic curve C; $V_\alpha \cap V_{\alpha+1}$ is a section of V_α for $\alpha = 1,\dots,n$ and of $V_{\alpha+1}$ for $\alpha = 0,\dots,n-1$.

C. $\Pi(X_0)$ is isomorphic to a triangulation of the sphere S^2. The components V_i are rational surfaces, and $V_\alpha \cap V_\beta$ are rational curves.

This terminology is justified by the fact that, as Kulikov has shown [7], over the field $\mathbf{C}$ any family of $K3$ surfaces has a reconstruction whose degeneration is of standard type.

THEOREM 1. *Let X/S be a family of $K3$ surfaces with a degeneration of standard type. Then the formal group $\widehat{\mathrm{Br}}(X/S)$ exists. The group $\widehat{\mathrm{Br}}(X_0)$ is isomorphic in case* B *to the formal group corresponding to the elliptic curve C, and in case* C *to the multiplicative group.*

Case A is evident because of Proposition 2. In cases B and C the condition $H^1(X_0, \mathcal{O}) = 0$ of Proposition 2 is checked by starting from the spectral sequence of the covering $X_0 = \bigcup V_\alpha$. We omit this verification, since it is analogous to, though simpler than, the considerations below. Thus the formal group $\widehat{\mathrm{Br}}(X/S)$ exists. We shall find $\widehat{\mathrm{Br}}(X_0)$. By definition,

$$\widehat{\mathrm{Br}}(X_0)(N) = H^2\big(X_0, (1 + \mathcal{O}_{X_0} \otimes N)^*\big), \qquad N \in \mathrm{Nil}\,\mathbf{k}.$$

To compute the group on the right we use the spectral sequence of the covering $X_0 = \bigcup_0^{n+1} V_\alpha$. Put

$$V^{[i]} = \coprod_{\alpha_1 < \cdots < \alpha_i} V_{\alpha_1} \cap \cdots \cap V_{\alpha_i} = \coprod_{\alpha_1 < \cdots < \alpha_i} V_{\alpha_1, \ldots, \alpha_i}.$$

Then

$$E_1^{ij} = H^j(V^{[i]}, (1 + \mathcal{O}_{V^{[i]}} \otimes N)^*) \Rightarrow H^{i+j}(X_0, (1 + \mathcal{O}_{X_0} \otimes N)^*).$$

In cases B and C, $H^2(V_\alpha, \mathcal{O}_{V_\alpha}) = 0$. Therefore the E_1 term has the form

$$\begin{aligned} \oplus H^1(V_\alpha, (1 + \mathcal{O}_{V_\alpha} \otimes N)^*) \xrightarrow{d^1} \oplus H^1(V_{\alpha\beta}, (1 + \mathcal{O}_{V_{\alpha\beta}} \otimes N)^*) \\ \oplus H^0(V_\alpha, (1 + \mathcal{O}_{V_\alpha} \otimes N)^*) \\ \to \oplus H^0(V_{\alpha\beta}, (1 + \mathcal{O}_{V_{\alpha\beta}} \otimes N)^*) \\ \to \oplus H^0(V_{\alpha\beta\gamma}, (1 + \mathcal{O}_{V_{\alpha\beta\gamma}} \otimes N)^*). \end{aligned}$$

We consider case B. Then $V_{\alpha\beta\gamma} = \varnothing$. Evidently, the homomorphism in the lower row is an epimorphism. Hence

$$H^1(X_0, (1 + \mathcal{O}_{X_0} \otimes N)^*) = \operatorname{Ker} d^1,$$

$$H^2(X_0, (1 + \mathcal{O}_{X_0} \otimes N)^*) = \operatorname{Coker} d^1.$$

In general, let $V \to C$ be a ruled surface and $D \subset V$ a section. Then

$$H^1(V, (1 + \mathcal{O}_V \otimes N)^*) \to H^1(D, (1 + \mathcal{O}_D \otimes N)^*)$$

is an isomorphism. To see this one must use the fact that both functors are exact; this enables us to restrict our attention to the case $N = \mathbf{k}$, where the assertion is evident. Likewise,

$$H^1(V_0, (1 + \mathcal{O}_{V_0} \otimes N)^*) = H^1(V_{n+1}, (1 + \mathcal{O}_{V_{n+1}} \otimes N)^*) = 0,$$

since V_0 and V_{n+1} are rational surfaces. Hence d^1 is an injective homomorphism with cokernel $H^1(C, (1 + \mathcal{O}_C \otimes N)^*)$. This gives the assertion of the theorem in case B.

In case C the lower row vanishes and the spectral sequence degenerates to the sequence

$$\bigoplus_\alpha \hat{G}_m(N) \to \bigoplus_{\alpha\beta} \hat{G}_m(N) \to \bigoplus_{\alpha\beta\gamma} \hat{G}(N).$$

This is the simplicial complex of some triangulation of the sphere S^2, tensored by $\hat{G}_m(N)$. Therefore, this second cohomology group is isomorphic to $\hat{G}_m(N)$. The theorem is proved.

COROLLARY. *Let X/S be a family of $K3$ surfaces over a field of characteristic $p > 0$ having a standard degeneration. If the height of a generic fiber X_η (i.e. of the formal group $\widehat{\mathrm{Br}}(X_\eta)$) is greater than 2, then the family does not degenerate.*

Indeed, if it did degenerate (i.e. if case B or C occurred), then the height of the fiber X_0 would not be greater than 2. By Proposition 3, this contradicts the fact that the height of the fiber X_η is greater than 2 by assumption.

§3. Degenerations of double planes

We shall now consider the case where the generic fiber of the family X/S is a double plane of type $K3$. In this case it is given by an equation $z^2 = f(x, y)$, where f is a polynomial of degree 6. Here we shall assume that the characteristic p of the base field is different from 2 or 3. Our goal is to make a list of surfaces such that for any family of double planes of type $K3$ there is a reconstruction whose closed fiber belongs to our list.

In the projective space $\mathfrak{P}$ of plane curves of degree 6 we consider the natural action of the group $G = \mathrm{PGL}(2)$ and the set $\mathfrak{P}_{ss}$ of points that are semistable with respect to this action. From the fact that the quotient $\mathfrak{P}_{ss}/G$ is a projective variety, it easily follows that the desired list may be taken to be the aggregate of all double planes with semistable ramification curve (the appropriate argument is given, for example, by Mumford [10]). However, this list may be somewhat shortened. Namely, it is easy to verify that we may restrict ourselves to those semistable curves whose G-orbit is closed in the set $\mathfrak{P}_{ss}$. Indeed, let $Y \to S$ be a family of semistable curves with generic fiber Y_η and closed fiber Y_0, and let y_η and y_0 be the corresponding points of $\mathfrak{P}_{ss}$. We assume that the orbit Gy_0 is not closed in $\mathfrak{P}_{ss}$, and let $g_t(y_0)$ be a one-parameter family of points in whose closure lies the point z_0 from the closure of the orbit Gy_0. We can assume that the base of this family is a scheme S' isomorphic to S. Choosing an isomorphism φ of these schemes, we consider the family corresponding to $g_t y_{\varphi(t)}$. Evidently it is a reconstruction of the desired family, and its closed fiber corresponds to z_0.

The enumeration of all semistable curves of degree 6 whose orbits are closed in $\mathfrak{P}_{ss}$ is gotten with no special difficulty with the aid of a classical device of invariant theory, the "Hilbert triangle" (see, for example, [9]). This has been done by many authors (Reid, Horikawa, Todorov, Shah). We give the answer (see, for example, [18], Theorem 2.4).

A. *Isolated singular points of multiplicity* 2 *or* 3 *that decompose into singular points of multiplicity* 2 *after one* σ*-process.* If a curve C of degree 6, $f = 0$, has such a singular point, then the double plane X: $z^2 = f$ has a double rational singular point. The correspondence between the types of singular points is given in the table below.

Thus, if the surface X belongs to this type, then its nonsingular minimal model is a $K3$ surface.

Point on C	Point on X
nonsingular	nonsingular
double point with distinct tangents	A_1
cusp	A_n
triple point with three distinct tangents	D_4
triple point with two tangents, one of which is double	D_n
triple point with one triple tangent, which becomes a point of multiplicity 2 after one σ-process	E_6, E_7, E_8

B. *Two isolated triple points that remain triple points after one* σ*-process* (the case of one such point gives a nonclosed orbit). If there are singular points at the origin of the coordinates and at the infinitely distant point on the x-axis, then the equation of the curve has the form $x^3 + \alpha xy^4 + \beta y^6 = 0$, and the double plane has the equation $z^2 = x^3 + \alpha xy^4 + \beta y^6$, where $4\alpha^3 + 27\beta^2 \neq 0$. The mapping $(x, y, z) \to z$ defines a pencil of elliptic curves on this surface. Putting $z = z_1 y^3$ and $x = x_1 y^2$, we get a nonsingular surface $\mathbf{P}^1 \times C$, where C is the elliptic curve $z_1^2 = x_1^3 + \alpha x_1 + \beta$. The mapping $(x_1, y_1, z_1) \to (x_1 y^2, y_1, z_1 y^3)$ is not a morphism at the infinitely distant points of the fiber s over $y = 0$ and $y = \infty$. After σ-processes at these points we get a smooth surface $\overline{X}$ and a morphism $f\colon \overline{X} \to X$, where the fibers of $\overline{X}$ over the points $y = 0$ and $y = \infty$ have the form $C + L$, where $L \simeq \mathbf{P}^1$, and the morphism f contracts the curves C to points.

C. *A unique double component that intersects the others transversally*. The equations of a curve of degree 6 can have the following forms:

$C_1\colon f = g^2$, where $g = 0$ is a smooth cubic.

$C_2\colon f = h^2 k$, where h and k are transversally intersecting conics, and the conic $h = 0$ is smooth.

$C_3\colon f = l_0^2 l_1 l_2 l_3 l_4$, where the $l_i = 0$ are distinct lines, and the lines $l_i = 0$, $i = 1, 2, 3, 4$, pass through one point not lying on the line $l_0 = 0$.

The corresponding surfaces $z^2 = f$ are not normal. In case C_1 the surface X is reducible and consists of two rational components that intersect along an elliptic curve. In case C_2 the normalization $\overline{X}$ of the surface X is smooth or has an ordinary double point and the morphism $\nu\colon \overline{X} \to X$ maps the elliptic curve $C \subset \overline{X}$ onto a line $L \subset X$ as a two-sheeted covering with 4 branch points, and outside C is an isomorphism.

In case C_3 the normalization $\overline{X} \to X$ is constructed in the same way, but is not smooth; it is given by an equation $z^2 = l_1 l_2 l_3 l_4$. It has an isolated singular point at the point of intersection of the lines $l_i = 0$. When the singularities are resolved, in exactly the same way as in case B there arises an elliptic curve that contracts to a singular point. Thus the resolution of the singularities has the form $f\colon \tilde{X} \to X$, where $\tilde{X}$ is a nonminimal model of the elliptic surface $\mathbf{P}^1 \times C$, f is an isomorphism outside two nonintersecting elliptic curves
2 C and $C' \subset \tilde{X}$, it contracts C to a point, and it maps C' onto a line as a two-sheeted covering with 4 branch points.

D. *A double component that does not transversally intersect two components.*

$D_1\colon f = gh^2$, where g and h are conics with two tangent points, and h is smooth.

$D_2\colon f = l_1^2 l_2^2 l_3^2$, where the $l_i = 0$ are three lines not passing through one point.

E. *A triple component.* $f = g^3$, where $g = 0$ is a smooth conic.

The degenerations of type E were investigated by Horikawa in [6]. He showed that they can be reduced to degenerations analogous to A–D. Namely, for any family of double planes of degree 6 whose degenerate fiber has type E, there is a reconstruction whose degenerate fiber is a strongly elliptic surface that is semistable in the sense of the classification of elliptic $K3$ surfaces or is gotten from such surfaces by contracting a rational curve with square -2 to an ordinary double point. This means that the equation of the surface can be written in the form $y^2 = x^3 + a(t)x + b(t)$, where a is a polynomial of degree 8, b is a polynomial of degree 12, and there is no point at which both a and b vanish with multiplicity > 4 and > 6 respectively.

Another proof of Horikawa's theorem was given by Shah [18]. Formally speaking, both proofs assume that the base field is the field of complex numbers. But this assumption is hardly used. For the reader's convenience we give below a simplified variant of the Horikawa-Shah proof that holds for characteristic > 11. For the case of characteristic > 3 we refer the reader to the original proof of Horikawa.

A strongly elliptic $K3$ surface X is given by an equation $y^2 = x^3 + a(t)x + b(t)$, where $a(t)$ and $b(t)$ are polynomials of degree $\leqslant 8$ and $\leqslant 12$ respectively (sections of $\mathcal{O}(8)$ and of

$\mathcal{O}(12)$ on $\mathbf{P}^1$). Therefore, such surfaces are parametrized by the points of a weighted projective space $\mathfrak{P}$ in which the coordinates of the polynomial $a(t)$ have weight 4 and those of $b(t)$ have weight 6. The space $\mathfrak{P}$ is acted on by the group $\mathrm{PGL}(1) = \mathrm{Aut}(\mathbf{P}^1)$, where $\mathbf{P}^1$ is the base of the bundle.

Horikawa's theorem allows us to replace the degeneration type E in our list by semistable degenerations of elliptic $K3$ surfaces whose orbits are closed in the set of semistable ones. The degenerations thus obtained divide into 4 types:

A′. *The surface X has only rational double points and is a Weierstrass model of an elliptic $K3$ surface.*

B′. *The surface X has 2 isolated triple points.* In this case $a = \alpha c^4$ and $b = \beta c^6$, where $c(t)$ is a polynomial of degree 2 without multiple roots and $4\alpha^3 + 27\beta^2 \neq 0$. The resolution of the singularities of the surface X has the same form as in case B of the degeneration of double planes.

C′. *The surface X has a curve of singular points whose preimage under the normalization is smooth.* In this case $a = 3c^2$ and $b = -2c^3$, where $c(t)$ is a polynomial of degree 4 without multiple roots. The resolution of the singularities has the same form as in case C_2 of the degeneration of double planes.

D′. *The surface X has a curve of singular points whose preimage under the normalization is not smooth.* The formulas are the same as in case C′, but the polynomial $c(t)$ has one root of multiplicity 2 and two simple roots.

§4. Horikawa's theorem

We denote by S^n the space of homogeneous polynomials in 3 variables x, y, and z over a field $\mathbf{k}$ of characteristic p, by $Q \in S^2$ a nondegenerate quadratic form, and by $G = \mathrm{SO(Q)}$ its group of automorphisms with determinant 1. Evidently, G acts on S^n, and QS^{n-2} is an invariant subspace.

LEMMA. *For $p > 2n$ the space QS^{n-2} has a G-invariant supplement in S^n.*

PROOF. We put $Q = xy - z^2$ and set up an isomorphism of the conic $Q = 0$ with $\mathbf{P}^1$ by the formulas $x = u^2$, $y = v^2$ and $z = uv$. Under this isomorphism G is identified with the image of SL(2) in its standard action on u and v. In SL(2) we consider the Cartan subgroup $u \to \lambda u$, $v \to \lambda^{-1}v$ and the corresponding Borel subgroup B that preserves u up to multiplication. Then B preserves the line $\mathbf{k}x^n$ in S^n, and the subspace L spanned by Bx^n coincides with the linear hull of the $g_\lambda x^n$, where $g_\lambda(u) = u + \lambda v$ and $g_\lambda(v) = v$. Evidently, L is G-invariant, and $\dim L \leqslant 2n + 1$. On the other hand, $\dim S^n/QS^{n-2} = 2n + 1$; therefore, we shall have proved that L is a G-invariant supplement of QS^{n-2} once we show that $\dim L = 2n + 1$ and $L \cap QS^{n-2} = 0$. For this we consider the operator ρ of restriction to Q and show that $\dim_\rho L = 2n + 1$. Indeed, under the given isomorphism of the conic with $\mathbf{P}^1$ we have $\rho x = u^2$, $\rho x^n = u^{2n}$, and $\rho g_\lambda(x^n) = (u + \lambda v)^{2n}$. Because of the condition $p > 2n$, all the binomial coefficients in the expansion of $(u + \lambda v)^{2n}$ are different from 0; therefore, the subspace spanned by all the forms $(u + \lambda v)^{2n}$ has dimension $2n + 1$.

In particular, if $n \leqslant 6$, the lemma is applicable for $p > 11$, which explains this restriction on the characteristic, which we shall henceforth impose.

We pass immediately to the proof of Horikawa's theorem. Let $S = \operatorname{Spec} \Lambda$, where Λ is a local ring with maximal ideal (π). Applying the preceding lemma three times, we see that any polynomial of degree 6 in 3 variables can be uniquely represented in the form

$F = \alpha Q^3 + F_2Q^2 + F_4Q + F_6$, where $F_i \in S^i$, and is contained in the G-invariant supplement of QS^{i-2} constructed in the lemma. Since the coefficients F belong to Λ, and $F(x, y, z) \equiv Q^3 \pmod{\pi}$, we have $\alpha \equiv 1\ (\pi)$, and we may assume that $\alpha = 1$ and all F_i are divisible by π. By an ordinary transformation $Q \to Q - \frac{1}{3}F_2$ we can get rid of the coefficient of Q^2. In this Q does not change (mod π); making a linear identity transformation (mod π), we can reduce Q to the preceding form. We now have $F = Q^3 + F_4Q + F_6$, where F_4 and F_6 are divisible by π. After a change of basis we may assume that $F_4 = \pi^{4k}A$ and $F_6 = \pi^{6k}B$, where A or B is not divisible by π. Thus the equation of the double plane has the form

$$w^2 = Q^3 + \pi^{4k}QA + \pi^{6k}B.$$

Divide both sides by π^{6k} and put

$$w = \pi^{3k}v, \qquad Q = \pi^{2k}u. \tag{1}$$

Then the equations of our surface assume the form

$$v^2 = u^3 + Au + B. \tag{2}$$

For $\pi = 0$ we hence get $Q = 0$ and

$$v^2 = u^3 + au + b, \tag{3}$$

where a and b are the restrictions of the polynomials A and B to the conic $Q = 0$. Since this conic is isomorphic to $\mathbf{P}^1$, we thus get an elliptic surface in which a and b are polynomials of degrees $\leqslant 8$ and $\leqslant 12$, and one of them is not 0.

We shall show that there is a transformation $g \in G(K)$, where K is the field of fractions of Λ such that for the polynomial $g \cdot F$ in (3) we get a semistable elliptic surface. For this we write F in the form $Q^3 + AQ + B$, where $A \in L$ and $B \in M$ lie in corresponding G-invariant subspaces. We consider the action of the group G on the multiply projective space $\mathbf{P}$ corresponding to the G-module $L \oplus M$, where L has weight 4 and M has weight 6. On the other hand, we consider the set of surfaces of the form (3) which are given by points (a, b) in a multiply projective space $\mathbf{P}'$. $\mathbf{P}'$ is acted on by the group PL(1), a homomorphic image of which is the group G. The homomorphism ρ of restriction to the conic $Q = 0$ defines an isomorphism of the spaces $\mathbf{P}$ and $\mathbf{P}'$ that commutes with the actions of PL(1) and G. By a standard argument already used above, there is a transformation $g' \in \mathrm{PL}(1)(K)$ that maps our family of elliptic surfaces into a family with semistable degeneration of the fiber. With the aid of the isomorphism we find that the corresponding transformation $g \in G(K)$ has the required property.

To complete the proof of the theorem it remains for us to show that the formulas (2) describe a family of projective surfaces whose closed fiber is isomorphic to a surface with equation (3). For this we consider a second chart, in which the coordinates $\xi = u/v$, $\eta = 1/v$ are defined and which, together with the chart (2), define a projective variety over S. In the new chart

$$\eta = \xi^3 + A\xi\eta^2 + B\eta^3, \qquad \eta Q = \pi^{2k}\xi. \tag{4}$$

The closed fiber $\pi = 0$ decomposes into two components, $X_0 = X_1 \cup X_2$, where X_1 is given by the equation $Q = 0$ and X_2 by $\eta = 0$. Obviously $X_2 \simeq \mathbf{P}^2$.

From (4) we see that $\eta\varphi = \xi^3$, where $\varphi = 1 - A\xi\eta - B\eta^2$ is an invertible element. Hence $\xi^3 Q = \pi^{2k}\xi\varphi$; that is, $\xi^2 Q = \pi^{2k}\varphi$. We see that our variety is not normal, and its normalization is given by the equations

$$\zeta\xi = \pi^k, \qquad \zeta^2\varphi = Q. \tag{5}$$

Let X_0', X_1', and X_2' be the preimages of the closed fiber X_0 and its components X_1 and X_2. Then the Weil divisor X_2' is not locally principal, but $X_2'' = kX_2'$ is locally principal and is given in the chart (5) by the equation $\xi = 0$. Since $\varphi = 1$ on X_2', X_2' is a two-sheeted covering of the plane with ramification curve of degree 2, i.e. a quadric. We shall show that this quadric can be contracted to a point. Let Y be the conic $X_1' \cap X_2' \subset X_2'$. Evidently, the restriction of the divisor X_2'' (or its sheaf) to X_2' is a negative multiple of Y. Hence, if H is a very ample divisor for which $H \cdot X_2'$ is a multiple of Y, then there are $a > 0$ and $b > 0$ such that $aH + bX_2''$ upon restriction to X_2' yields a divisor equivalent to 0. Likewise, the linear system $|aH + bX_2''|$ contracts X_2' to a point. It remains to clarify what happens to X_1' during this. In the chart (2), X_1' is mapped isomorphically. In the chart (4) the conic $Y = X_1' \cap X_2'$ contracts to a point. It is easy to see that $(Y^2) = -2$ on X_1', since $(Y^2) = 2$ on X_2'. Multiplying $aH + bX_2''$ by a sufficiently large number, we guarantee that the point to which Y contracts is normal. Since $(Y^2) = -2$, this point is an ordinary double point. By a theorem of Tyurina, Brieskorn and Artin [2] (in our case, just a simple special case of this theorem), there is a reconstruction of our family in which this point is resolved.

§5. Degenerations of $K3$ surfaces of degree 2 and height greater than 2

THEOREM 2. *Let X/S be a family defined over a field of characteristic $p = 2$ or 3, whose general fiber is a $K3$ surface of degree 2. There is a reconstruction X'/S' of this family for which the Brauer group $\widehat{\mathrm{Br}}(X'/S')$ is defined and either the family X'/S' is nondegenerate or the group $\widehat{\mathrm{Br}}(X_0')$ of a closed fiber has height $\leqslant 2$.*

A $K3$ surface of degree 2 is either a strongly elliptic surface (i.e. is given by an equation $y^2 = x^3 + a(t)x + b(t)$, where $\deg a(t) \leqslant 8$ and $\deg b(t) \leqslant 12$) or a double plane (i.e. is given by an equation $z^2 = f(x, y)$, where f is a polynomial of degree 6).

By §3, the family X/S has a reconstruction whose degenerate fiber belongs to one of the types A, A′; B, B′; C_1, C_2, C_3, C′; D_1, D_2, D′. We shall examine them in order.

A and A′. In this case all the fibers have only rational double points. By the Tyurina-Brieskorn-Artin theorem already cited [2], the family has a reconstruction all of whose fibers are smooth, i.e. the family itself is smooth.

In cases B, B′, C_1, C_2, C_3, C′, D_1, D_2, and D′ we shall show that the hypothesis of Theorem 1 is satisfied; that is, $\widehat{\mathrm{Br}}(X/S)$ is given by a one-dimensional formal group over S. Moreover, we shall verify that for the closed fiber X_0 the formal group $\widehat{\mathrm{Br}}(X_0)$ has height $\leqslant 2$. Hence this case cannot occur if the height of the formal group $\widehat{\mathrm{Br}}(X_\eta)$ is greater than 2.

We verify that the hypothesis of Theorem 1 is satisfied; that is, that $H^1(X_s, \mathcal{O}_{X_s}) = 0$ for any fiber X_s of the family X. This is a general property of a family of strongly elliptic surfaces and double planes. Let an elliptic surface be given by an equation

$$y^2 = x^3 + A(t_0 : t_1)x + B(t_0 : t_1), \tag{1}$$

where A is a form of degree 8, and B is a form of degree 12. We ascribe to the variables t_0 and t_1 the weight 1; to x, 4; and to y, 6—considering the relation in the weighted projective space of the appropriate type. Equation (1) is the set of zeros of a section of the sheaf $\mathcal{O}(12)$ in this space. Likewise, the equation of a double plane will be written as $z^2 = F(x: y: u)$ and interpreted at the set of zeros of a section of the sheaf $\mathcal{O}(6)$ in the weighted projective space where x, y, and u have weight 1, and z has weight 2. If X is our surface and $\mathbf{P}$ is the weighted projective space containing it, then from the exact sequence

$$0 \to \mathcal{O}_{\mathbf{P}}(-n) \to \mathcal{O}_{\mathbf{P}} \to \mathcal{O}_X \to 0$$

(where $n = 12$ or 6), and from the relation $H^i(\mathbf{P}, \mathcal{O}_{\mathbf{P}}(n)) = 0$ for $i = 1$ and 2 (see [4], p.8), it follows that $H^1(X, \mathcal{O}_X) = 0$.

It now remains to compute the group $\widehat{\mathrm{Br}}(X_0)$ for all surfaces X of type B, B′, C_1, C_2, C_3, C_1', D_1, D_2, or D′.

B and B′. The resolution of singularities has the form $f\colon \overline{X} \to X$, where $\overline{X}$ is the surface $C \times \mathbf{P}^1$ with two points blown up, and f contracts two nonintersecting elliptic curves C_1 and C_2 to points x_1 and x_2. Denote by $\mathfrak{F}_X$ the sheaf $(1 + \mathcal{O}_X \otimes N)^*$, where N is some nilpotent algebra. To compute the group $H^2(X, \mathfrak{F}_X)$ we use the spectral sequence

$$E_2^{i,j} = H^j\big(X, R^i f_* \mathfrak{F}_{\overline{X}}\big) \Rightarrow H^{i+j}\big(\overline{X}, \mathfrak{F}_{\overline{X}}\big). \tag{2}$$

We have $R^2 f_* \mathfrak{F}_{\overline{X}} = 0$, since the dimension of the fibers of the morphism f is $\leqslant 1$, and the sheaf $\mathfrak{F}_{\overline{X}}$ is "almost coherent", i.e. has a filtration with coherent quotients and in fact its quotients are isomorphic to $\mathcal{O}_{\overline{X}}$.

Also, $R^0 f_* \mathfrak{F}_{\overline{X}} = \mathfrak{F}_X$, since the fibers are complete and connected, and $R^1 f^* \mathfrak{F}_{\overline{X}}$ is concentrated at the points x_1 and x_2.

We shall prove that for each of the points $x = x_i$, $i = 1, 2$, the fiber of the sheaf $R^1 f_* \mathfrak{F}_{\overline{X}}$ is isomorphic to $\hat{C}(N)$, where $C = C_i$. We must prove that the restriction homomorphism

$$\big(R^1 f_*(1 + \mathcal{O}_{\overline{X}} \otimes N)^*\big)_x \to H^1(C, (1 + \mathcal{O}_C \otimes N)^*) \tag{3}$$

is an isomorphism. Using short exact sequences, this assertion is easily reduced to the case $N = k$, where the homomorphism (3) has the form

$$\big(R^1 f_* \mathcal{O}_{\overline{X}}\big)_x \to H^1(C, \mathcal{O}_C).$$

We use the relation

$$\big(R^1 f_* \mathcal{O}_{\overline{X}}\big)^{\wedge}_x \simeq \varprojlim H^1(C, \mathcal{O}_{\overline{X}}/\mathfrak{m}_x^n) = \varprojlim H^1(C, \mathcal{O}_{\overline{X}}/I^n),$$

where I is the sheaf of ideals defining C. We shall show that $H^1(C, I^n/I^{n+1}) = 0$, whence our claim easily follows. For this we note that I/I^2 defines the conormal bundle on C, and I^n/I^{n+1} defines its nth power. Since $(C^2) = -1$, these bundles are positive; and since C is an elliptic curve, $H^1(C, I^n/I^{n+1}) = 0$.

Thus the second term of the spectral sequence (2) is given by the diagram

$$\begin{matrix} 0 & 0 & 0 \\ \widehat{\mathrm{Pic}}_C(N)^2 & 0 & 0 \\ H^0(X, \mathfrak{F}_X) & H^1(X, \mathfrak{F}_X) & H^2(X, \mathfrak{F}_X) \end{matrix}$$

and the differential d_2 reduces to $d\colon H^0(X, \widehat{\mathrm{Pic}}_C(N)^2) \to H^2(X, \mathfrak{F}_X)$.

From the structure of the sheaf $\mathfrak{F}_X$ and the fact that $H^1(X, \mathcal{O}_X) = 0$ it follows that $H^1(X, \mathfrak{F}_X) = 0$. In the same way, from $H^2(\overline{X}, \mathcal{O}_{\overline{X}}) = 0$ it follows that $H^2(\overline{X}, \mathfrak{F}_{\overline{X}}) = 0$.

Finally,

$$H^1(\bar{X}, \mathcal{F}_{\bar{X}}) = \widehat{\mathrm{Pic}}(N).$$

Thus

$$\mathrm{Ker}\, d = \widehat{\mathrm{Pic}}(N), \qquad \mathrm{Coker}\, d = 0.$$

In other words, we have the exact sequence

$$0 \to \widehat{\mathrm{Pic}}_C(N) \to \widehat{\mathrm{Pic}}_C(N)^2 \to H^2(X, \mathcal{F}_X) \to 0,$$

form which it follows that $H^2(X, \mathcal{F}_X) = \widehat{\mathrm{Pic}}(N)$.

C_1 is a special case of the situation investigated in §2.

C_2 and C_2' lead to a resolution of singularities $f: \bar{X} \to X$, where $\bar{X}$ is a rational surface, $C \subset \bar{X}$ is an elliptic curve, $f(C) = L$ is a rational curve, f defines a two-sheeted covering $C \to L$ with 4 branch points, and $f: \bar{X} - C \to X - L$ is an isomorphism. In this case the fibers of the morphism f are zero-dimensional, and therefore $R^1 f_* \mathcal{F}_{\bar{X}} = R^2 f_* \mathcal{F}_{\bar{X}} = 0$.

The second term of the spectral sequence has the form

$$\begin{matrix} 0 & 0 & 0 \\ 0 & 0 & 0 \\ H^0(X, \mathcal{G}) & H^1(X, \mathcal{G}) & H^2(X, \mathcal{G}) \end{matrix}$$

where $\mathcal{G} = R^0 f_* F_{\bar{X}}$.

Since the surface $\bar{X}$ is rational, $H^2(\bar{X}, \mathcal{F}_{\bar{X}}) = H^1(\bar{X}, \mathcal{F}_{\bar{X}}) = 0$. Using the spectral sequence, we now see that $H^2(X, \mathcal{G}) = H^1(X, \mathcal{G}) = 0$.

Consider the exact sequence

$$0 \to \mathcal{F}_X \to \mathcal{G} \to \mathcal{H} \to 0. \tag{4}$$

The sheaf $\mathcal{H}$ is concentrated on L. We shall prove that its restriction to L is gotten from an analogous exact sequence

$$0 \to \mathcal{F}_L \to \mathcal{G}_L \to \mathcal{H}|L \to 0, \tag{5}$$

where $\mathcal{G}_L = R^0 f_* \mathcal{F}_C$. Arguing as in the preceding case, we reduce this assertion to the case $N = \mathbf{k}$, where it is equivalent to the claim that the conductor of the covering $\bar{X} \to X$ coincides with the sheaf of ideals defining L. This is evident, since locally X is defined by an equation $z^2 = x^2 y$, where $x = 0$ is the equation of L, and $\mathbf{k}[\bar{X}] = \mathbf{k}[X] + \mathbf{k}[X] z/x$.

From the exact sequence (4) and the fact that $H^1(X, \mathcal{G}) = H^2(X, \mathcal{G}) = 0$, we find that $H^1(X, \mathcal{H}) \simeq H^2(X, \mathcal{F}_X)$. On the other hand, $H^1(L, \mathcal{F}_L) = H^2(L, \mathcal{F}_L) = 0$, since the curve L is rational and, from the exact sequence, $H^1(L, \mathcal{G}_L) \simeq H^1(C, \mathcal{F}_C)$. Combining these equations, we see that the sequence (5) requires that $H^1(L, \mathcal{G}_L) \simeq H^1(L, \mathcal{H}|L)$. By definition,

$$H^2(X, \mathcal{F}_X) \simeq H^1(X, \mathcal{H}) \simeq H^1(C, \mathcal{F}_C) = \widehat{\mathrm{Pic}}_C(N).$$

Thus

$$\widehat{\mathrm{Br}}_X(N) = \mathrm{Pic}_C(N).$$

C_3 leads to a resolution of singularities $f: \tilde{X} \to X$, $\tilde{X}$ smooth and birationally isomorphic to $\mathbf{P}^1 \times C$, where C is an elliptic curve, $C, C' \subset \tilde{X}$ are two nonintersecting elliptic curves, $f(C) = x$, $f(C') = L$ is a rational curve, $f: C' \to L$ is a covering of degree 2 with 4 branch points, and $f: \tilde{X} - C - C' \to X - x - L$ is an isomorphism.

Put $R^0 f_* \mathcal{F}_{\tilde{X}} = \mathcal{G}$. The second term of the spectral sequence will have the form

$$\begin{matrix} 0 & 0 & 0 \\ \widehat{\mathrm{Pic}}_{C'}(N) & 0 & 0 \\ H^0(X, \mathcal{G}) & H^1(X, \mathcal{G}) & H^2(X, \mathcal{G}) \end{matrix}$$

and the differential d_2 reduces to the homomorphism $d\colon \widehat{\mathrm{Pic}}_{C'}(N) \to H^2(X, \mathcal{G})$. As in the preceding case we have exact sequences

$$\begin{gathered} 0 \to \mathcal{F}_X \to \mathcal{G} \to \mathcal{H} \to 0, \\ 0 \to \mathcal{F}_L \to \mathcal{G}_L \to \mathcal{H}\,|\,L \to 0, \end{gathered} \tag{6}$$

where $\mathcal{H}$ is concentrated on L and $\mathcal{G}_L = R^0 f_* \mathcal{F}_{C'}$. As before,

$$H^1(X, \mathcal{H}) = \widehat{\mathrm{Pic}}_{C'}(N).$$

Moreover,

$$H^1(\tilde{X}, \mathcal{F}_{\tilde{X}}) = \widehat{\mathrm{Pic}}_C(N), \qquad H^2(\tilde{X}, \mathcal{F}_{\tilde{X}}) = 0.$$

We shall prove that $H^1(X, \mathcal{G}) = 0$. For this we consider the normalization $\nu\colon X^\nu \to X$. Evidently, $\mathcal{G} = R^0 \nu_* \mathcal{F}_{X^\nu}$ and

$$H^1(X, \mathcal{G}) \simeq H^1(X^\nu, \mathcal{F}_{X^\nu}) \simeq \widehat{\mathrm{Pic}}\, X^\nu(N).$$

But $\widehat{\mathrm{Pic}}\, X^\nu = 0$, since X^ν is given by the equation $z^2 = l_1 l_2 l_3 l_4$, and we have verified that for such surfaces $H^1(X^\nu, \mathcal{O}_{X^\nu}) = 0$. The spectral sequence now shows that, because of the equation $H^1(\tilde{X}, \mathcal{F}_{\tilde{X}}) = 0$, the homomorphism d is 0. By $H^2(\tilde{X}, \mathcal{F}_{\tilde{X}}) = 0$, from this it follows in the same way that $H^2(X, \mathcal{G}) = 0$. From the exact sequence (6) it then follows that $H^2(X, \mathcal{F}_X) \simeq H^1(X, \mathcal{H}) \simeq \widehat{\mathrm{Pic}}_{C'}(N)$; that is, $\widehat{\mathrm{Br}}(X) = \widehat{\mathrm{Pic}}(C')$.

D_1, D_2, and D'. In these cases it would also be possible to compute the group $\widehat{\mathrm{Br}}(X)$ for a degenerate fiber and show that it has height 1. But it is easier to compute the Hasse invariant of the surface X and show that it is nonzero. The Hasse invariant $\lambda(X)$ is defined by $\lambda(X) = \alpha^p$ if $C\omega = \alpha\omega$, where ω is a basis of the one-dimensional space $H^0(X, \Omega^2_X)$ and C is the Cartier operator. For a double plane given by the equation $z^2 = f(x, y)$, where f is a polynomial of degree 6, we have $\omega = z^{-1}\, dx \wedge dy$. From the equation

$$\omega = z^{p-1}/z^p \cdot dx \wedge dy = f(x, y)^{(p-1)/2}/z^p \cdot dx \wedge dy$$

and the definition of the operator C it easily follows that (if the degree of f is 6) $\lambda(X)$ is the coefficient of $x^{p-1}y^{p-1}$ in $f(x, y)^{(p-1)/2}$. If we write f in homogeneous form as $F(x_0\colon x_1\colon x_2)$, then $\lambda(X)$ will be the coefficient of $x_0^{p-1}x_1^{p-1}x_2^{p-1}$ in $F^{(p-1)/2}$.

In case D_1 we can put $h = x_0x_2 + x_1^2$ and $g = x_0x_2 + ax_1^2 = h + (a-1)x_1^2$ for $a \neq 1$. Then

$$F^{(p-1)/2} = h^{p-1}\left(h + (a-1)x_1^2\right)^{(p-1)/2}.$$

The terms containing h in the second factor give an expression divisible by h^p, and so cannot contain the term we need. Therefore, such a term can only be contained in

$$(a-1)^{(r-1)/2} h^{p-1} x_1^{p-1} = \left(x_0x_2 + x_1^2\right)^{p-1} x_1^{p-1} \cdot (a-1)^{(p-1)/2}.$$

Again, the terms containing x_1 in the first factor will be divisible by x_1^p and cannot contain the needed term. As a result we find that $\lambda(X) = (a-1)^{(p-1)/2} \neq 0$, since $a \neq 1$.

In case D_2 we can take the lines l_i on the coordinate axes, and we then get

$$F^{(p-1)/2} = x_0^{p-1}x_1^{p-1}x_2^{p-1};$$

that is, $\lambda(X) = 1$.

In case D′ the computation is similar. One need only use the expression of the Hasse invariant for an elliptic surface. This expression is presented in [15]. The theorem is proved.

As in §2, we obtain the following assertion.

Corollary. *Let X/S be a family of $K3$ surfaces of degree 2 over a field of characteristic $p > 3$. If the general fiber of the family has height > 2, the family does not degenerate.*

In this way we have another proof of the main result of [16].

§6. Supersingular surfaces

A $K3$ surface is called *supersingular* if all its two-dimensional cycles are algebraic; that is, if its Picard number ρ is 22. Of course, such surfaces exist only over fields of characteristic $p > 0$. As Artin showed [1], a supersingular $K3$ surface has infinite height; i.e., its Brauer group $\widehat{\mathrm{Br}}(X)$ is unipotent. (In [1] Artin calls a surface of infinite height supersingular; but for purely technical reasons we shall find it more convenient to use another terminology.)

Theorem 3. *A family X/S of $K3$ surfaces defined over a field of characteristic $p > 3$ whose general fiber is supersingular does not degenerate.*

To reduce this assertion to the theorem it suffices to show that a supersingular $K3$ surface has degree 2. This means that its Picard lattice $S(X)$ represents 2. All possible types of Picard lattice for supersingular $K3$ surfaces are enumerated in [14]. From our list it is evident that the lattice $S(X)$ either represents a lattice U corresponding to the quadratic form $2x_1x_2$ (in which case $S(X)$ evidently represents 2, and the surface X is elliptic) or represents a lattice $H^p \oplus U(p)$, where H_p is an even lattice of rank 4 and discriminant $-p^2$, and $U(p)$ corresponds to the quadratic form $2px_1x_2$. In the latter case it suffices for us to verify that H_p represents all even integers modulo p. This is obvious, since the rank of H_p is 2 modulo p, and a form of rank 2 over a field $\mathbf{F}_p$ represents all elements of $\mathbf{F}_p^*$.

As is well known, from Theorem 3 it follows that for the period mapping of supersingular $K3$ surfaces introduced by Ogus [11] there is an analog of Torelli's theorem, and each point of the space of periods corresponds to some surface.

In conclusion we present one application of Theorem 2 that connects the concept of supersingularity with the height of a $K3$ surface. In [1] Artin showed that the unipotence of the group $\widehat{\mathrm{Br}}(X)$ implies the supersingularity of the surface X if we assume that X is elliptic. He turned his attention to the natural question of whether this assertion is true in the general case. We can answer it in the case of a surface of degree 2 (and, of course, characteristic $p > 3$).

Theorem 4. *If the height of a $K3$ surface X of degree 2 defined over a field of characteristic $p > 3$ is infinite, then X is supersingular.*

We shall use two of Artin's results proved in [1]. The first is the already formulated assertion that an elliptic surface of type X with infinite height is supersingular; the second is that, for a smooth family X/S of $K3$ surfaces of infinite height with connected base S, the Picard number of the fiber is constant. In particular, if one of the fibers is supersingular, they all are.

Let X_0 be a $K3$ surface of degree 2 and infinite height. If X_0 is elliptic, then it is supersingular by the result cited above. Thus we may assume that X_0 is a double plane. If it has singular points (double rational points), it is also elliptic. Indeed, in this case among the curves gotten by resolving the singular points there is a curve C with $(C^2) = -2$. On the other hand, if D is the preimage of a line on the plane that X_0 covers twice, then $(D^2) = 2$ and $(C \cdot D) = 0$. Therefore, the Picard lattice of X_0 represents the quadratic form $2x_1^2 - 2x_2^2$, and so it represents 0. Hence X_0 is elliptic [12].

Finally, we may assume that X_0 is a smooth double plane. As in §3 we consider the projective space $\mathfrak{P}$ of curves of degree 6 and in it the affine open set U corresponding to the smooth curves. Let $x_0 \in U$ be the point corresponding to X_0. Since U is affine, the quotient $\Sigma = U/\mathrm{PGL}(2)$ is also affine. Denote by y_0 the image of x_0 in Σ. The condition that the height of the surface be $\geqslant h$ cuts out on $\mathfrak{P}$, as on Σ, a closed subvariety of codimension 1. Since the height is infinite if it is greater than 10, and $\dim \Sigma = 19$, through x_0 there passes a smooth closed curve D all of whose points correspond to surfaces of infinite height, and its image in Σ is one-dimensional. We now apply Theorem 3, from which it follows that the curve D also has a covering that is an open subset of a complete curve C such that the preimage of the natural family defined on D (because of the embedding $D \hookrightarrow \mathfrak{P}$) is on C a nondegenerating family of surfaces of degree 2. If the fiber over at least one point $c \in C$ is an elliptic surface (in particular, a singular double plane), then, by Artin's results cited above, X_0 is supersingular. But if all these fibers are smooth double planes, then a morphism $C \to \Sigma$ is defined whose image contains the image of D. Since C is complete, and Σ is affine, this image must be a point, contradicting the one-dimensionality of the image of D. This contradiction proves the theorem.

Received 3/AUG/81

BIBLIOGRAPHY

1. M. Artin, *Supersingular K3 surfaces*, Ann. Sci. École Norm. Sup. (4) **7** (1974), 543–567.

2. ______, *Algebraic construction of Brieskorn's resolutions*, J. Algebra **29** (1974), 330–348.

3. M. Artin and B. Mazur, *Formal groups arising from algebraic varieties*, Ann. Sci. École Norm. Sup. (4) **10** (1977), 87–131.

4. I. Dolgachev, *Weighted projective varieties*, Preprint, 1978.

5. A. Grothendieck, *Éléments de géométrie algébrique*. I, Inst. Hautes Études Sci. Publ. Math. No. 4 (1960).

6. Eiji Horikawa, *Surjectivity of the period map of K3 surfaces of degree 2*, Math. Ann. **228** (1977), 113–146.

7. Viktor S. Kulikov, *Degenerations of K3 surfaces and Enriques surfaces*, Izv. Akad. Nauk SSSR Ser. Mat. **41** (1977), 1008–1042; English transl. in Math. USSR Izv. **11** (1977).

8. Vu. I. Manin, *Theory of commutative formal groups over fields of finite characteristic*, Uspekhi Mat. Nauk **18** (1963), no. 6 (114), 3–90; English transl. in Russian Math. Surveys **18** (1963).

9. David Mumford, *Geometric invariant theory*, Springer-Verlag, Berlin, and Academic Press, New York, 1965.

10. ______, *Stability of projective varieties*, Enseignement Math. (2) **23** (1977), 39–110; reprint, Monographic de l'Enseignement Math., No. 24, L'Enseignement Math., Geneva, 1977.

11. Arthur Ogus, *Supersingular K3 crystals*, Journées de Géométrie Algébrique de Rennes (1978), Vol. II, Astérisque No. 64, Soc. Math. France, Paris, 1979, pp. 3–86.

12. I. I. Pyatetskiĭ-Shapiro and I. R. Shafarevich, *A Torelli theorem for algebraic surfaces of type K3*, Izv. Akad. Nauk SSSR Ser. Math. **35** (1971), 530–572; English transl. in Math. USSR Izv. **5** (1971).

13. Michel Raynaud, "*p-torsion*" *du schéma de Picard*, Journées de Géométrie Algébrique de Rennes (1978), Vol. II, Astérisque No. 64, Soc. Math. France, Paris, 1979, pp. 87–148.

14. A. N. Rudakov and I. R. Shafarevich, *Supersingular $K3$ surfaces over fields of characteristic* 2, Izv. Akad. Nauk SSSR Ser. Mat. **42** (1978), 848–869; English transl. in Math. USSR Izv. **13** (1979).

15. ______, *On the degeneration of $K3$ surfaces over fields of finite characteristic*, Izv. Akad. Nauk SSSR Ser. Mat. **45** (1981), 646–661; English transl. in Math. USSR Izv. **18** (1982).

16. ______, *On the degeneration of surfaces of type $K3$*, Dokl. Akad. Nauk SSSR **259** (1981), 1050–1052; English transl. in Soviet Math. Dokl. **24** (1981).

17. Michael Schlessinger, *Functors of Artin rings*, Trans. Amer. Math. Soc. **130** (1968), 208–222.

18. Jayant Shah, *A complete moduli space for $K3$ surfaces of degree* 2, Ann. of Math. (2) **112** (1980), 485–510.

Translated by M. ACKERMAN

Zum 150. Geburtstag von Alfred Clebsch

Math. Ann. **266**, 135–140 (1983). Zbl. **519**, 01013

Der Gründer dieser Zeitschrift, Alfred Clebsch, war eine markante und originelle mathematische Persönlichkeit – sogar gemessen an seinem Jahrhundert, das so reich an mathematischen Talenten war.

Clebsch war ein sehr vielseitiger Mathematiker, und seinen Einfluß erfuhren – und erfahren in gewissem Maße bis heute – verschiedene Gebiete unserer Wissenschaft. Aber das Hauptwerk seines Lebens war sein Beitrag zur Entwicklung der algebraischen Geometrie. Dank der Arbeiten von Clebsch entstand die algebraische Geometrie als Wissenschaft, mit denselben Fragestellungen und Vorstellungen von den Gegenständen ihrer Forschung, die sie in großen Zügen bis zum heutigen Tage bewahrt hat.

Der Anfang der geometrischen Betätigung von Clebsch fällt in das Jahr 1860. Um diese Zeit wurde das Gesicht jenes Gebietes, das später algebraische Geometrie genannt werden sollte, geformt durch die Arbeiten von Poncelet, Chasles, Cayley, Sylvester, Salmon, Möbius, Hesse und Plücker. Allgemein üblich waren damals die projektive Sichtweise und die Betrachtung von Punkten mit komplexen Koordinaten. Detailliert ausgearbeitet war die projektive Theorie der Gebilde 2. Ordnung. Entdeckt waren viele Eigenschaften der ebenen Kurven 3. Ordnung (z. B. die Konfiguration ihrer 9 Wendepunkte), 4. Ordnung (28 Doppeltangenten) und der Flächen 3. Ordnung (mit ihren 27 Geraden). Resultate allgemeineren Charakters, die ebene algebraische Kurven beliebiger Ordnung betrafen, hatte Plücker erhalten: die Formeln, die seinen Namen tragen, verbinden die Ordnung der Kurve, ihre Klasse und die Anzahl ihrer Doppelpunkte und Doppeltangenten. Es lag eine Menge von prägnanten konkreten Fakten vor, die das Vorhandensein von allgemeinen Resultaten und Konzeptionen zu vermuten erlaubte.

Die tiefe und kühne Idee von Clebsch bestand darin, daß man dieses umfassende Konzept nicht in der Geometrie selbst suchen sollte, sondern in den kürzlich (1856) erschienen Arbeiten Riemanns, die damals aufgefaßt wurden als

---

* Aus dem Russischen übersetzt von Meinhard Peters, Mathematisches Institut der Universität, D-4400 Münster, Bundesrepublik Deutschland

Teil der Funktionentheorie, und die sich den Integralen und θ-Funktionen widmeten und auf dem analytischen Extremalprinzip von Dirichlet basierten. Im Grunde genommen war das eine Umdeutung der Ideen von Abel und Riemann, ihre Übersetzung in eine andere Sprache, aber eine höchst nichttriviale Übersetzung. Wenn für Abel die Gleichung $F(w,z)=0$, F ein Polynom, eine „Irrationalität" w und für Riemann eine algebraische Funktion w (gegeben auf einer Riemannschen Fläche) bestimmte, so war das für Clebsch die Gleichung einer algebraischen Kurve, und jede neue Tatsache über diese Gleichung spiegelte eine gewisse geometrische Eigenschaft der Kurve wider. Alle drei Betrachtungsweisen existieren noch heute und haben sich beispielsweise widergespiegelt in der Unterhaltung zweier heute lebender bekannter Spezialisten in der algebraischen Geometrie. Als einer von ihnen ausrief: „Verbindet sich mit den Worten ‚algebraische Kurve' bei Ihnen denn wirklich überhaupt kein anschauliches Bild?", da antwortete der andere: „Doch, natürlich, hier ist es". Und er schrieb: $F(X, Y)=0$.

Die neue Konzeption wendet Clebsch erstmals 1863 an in der Abhandlung „Ueber die Anwendung der Abel'schen Functionen in der Geometrie", veröffentlicht in Crelles Journal. Diese Abhandlung kann meines Erachtens auch als Zeugnis der Geburt der algebraischen Geometrie angesehen werden, als erster Schrei des Neugeborenen.

In der Abhandlung „Ueber einen Satz von Steiner und einige Punkte der Theorie der Curven dritter Ordnung", die nur einen Monat vor der obengenannten Arbeit abgefaßt wurde und im gleichen Band von Crelles Journal erschien, betrachtet Clebsch noch Kurven nur 3. Ordnung. Er beweist, daß sie parametrisiert werden durch elliptische Funktionen (vielleicht wird diese Tatsache hier erstmalig festgestellt), und gibt eine geometrische Interpretation des Additionstheorems elliptischer Funktionen: Punkte P_1, P_2, P_3 einer Kurve 3. Ordnung liegen dann und nur dann auf einer Geraden, wenn für die ihnen entsprechenden Argumentwerte z_1, z_2, z_3 der elliptischen Funktionen $z_1+z_2+z_3$ eine Periode ist. Hieraus folgen dann alle früher bekannten Eigenschaften der Wendepunkte und andere Resultate von Steiner, einfach als Behauptungen über die Lage im Periodenparallelogramm der Punkte, deren n-te Vielfachheit eine Periode ist.

In der Abhandlung über die Anwendung Abelscher Funktionen werden analoge Untersuchungen angestellt für Kurven beliebiger Ordnung mit beliebigen singulären Punkten. Der Satz von Abel und das von Riemann gelöste Umkehrproblem Abelscher Funktionen geben eine sehr genaue Information über diejenigen meromorphen Funktionen auf der Riemannschen Fläche zur Gleichung $F(w,z)=0$, welche vorgegebene Nullstellen und Pole haben. Übersetzt in die geometrische Sprache gibt dies eine geradezu erschöpfende Beschreibung der Kurven, die die oben gegebene Kurve in vorgegebenen Punkten mit vorgegebenen Ordnungen berührt. Diese Methode erzeugt einen Wasserfall von hervorragenden geometrischen Resultaten: der Satz von Plücker, daß eine Kurve 4. Ordnung 28 Doppeltangenten (d. h. Geraden, die sie in zwei Punkten berühren) hat, ist das elementarste Beispiel.

Die folgenden Jahre widmet Clebsch mit seinen Mitarbeitern und Schülern der systematischen Ausarbeitung der neuen Richtung. In abgeschlossener Gestalt erscheint sie in den „Vorlesungen über Geometrie", die 1871–1872 von Clebsch gehalten und von Lindemann herausgegeben wurden (Band I). Erstaunlich, wie sehr die Theorie in 8 Jahren vorangekommen ist! Wir treffen hier auf den Begriff der eindeutigen (jetzt: birationalen) Transformation algebraischer Kurven und auf die klare Einsicht, daß die grundlegende Aufgabe das Studium der Kurven bis auf solche Transformationen ist. Es wird das Geschlecht definiert (diese Bezeichnung stammt auch von Clebsch), und zwar rein geometrisch durch die Ordnung der Kurve und die Vielfachheit ihrer Doppelpunkte. Es wird die Invarianz des Geschlechts bei „eindeutigen Transformationen" bewiesen. Auch der Satz von Riemann-Roch wird geometrisch bewiesen, und es stellt sich heraus, daß die Kurven vom Geschlecht 0 parametrisiert werden durch rationale Funktionen und die vom Geschlecht 1 durch elliptische. Es werden Modelle minimaler Ordnung für Kurven gegebenen Geschlechts gefunden und gezeigt, daß die Anzahl der „Moduln", von denen Kurven vom Geschlecht p bis auf „eindeutige Transformationen" abhängen, gleich $3p-3$ ist. Es werden Integrale erster, zweiter und dritter Gattung konstruiert (rein algebraisch, d. h. es werden Integranden konstruiert). Es wird der Satz von Abel bewiesen. Es wird (in heutiger Sprechweise) die Jacobische Mannigfaltigkeit der Kurve konstruiert. Alles das wird rein synthetisch dargelegt (wir würden sagen „über beliebigem Grundkörper"), aber es stellt sich auch die Verbindung mit der topologischen und der analytischen Sichtweise Riemanns heraus. Es ist fast unnötig hinzuzufügen, daß die Theorie auf eine sehr große Zahl konkreter geometrischer Probleme angewandt wird.

Der nächste Schritt war der Übergang zur Betrachtung algebraischer Mannigfaltigkeiten höherer Dimension, vor allem algebraischer Flächen. Auch hier führt Clebsch den Begriff einer eindeutigen (d. h. birationalen) Transformation ein und betrachtet zuerst Flächen, die man „eindeutig" auf eine Ebene abbilden kann, also rationale Flächen. Er stellt die Rationalität der Flächen 3. Ordnung fest, verschiedener singulärer Flächen 4. und 5. Ordnung und der „Doppelebene", die definiert wird durch die Gleichung $f_4(x,y)=z^2$, wobei f_4 ein Polynom 4. Grades ist. Schließlich, in einer kurzen Note in den Comptes Rendus von 1868, führt er Begriffe ein, die danach eine grundlegende Rolle in der Geometrie der algebraischen Flächen spielten. Er definiert ein doppeltes Integral erster Gattung, das mit der algebraischen Fläche zusammenhängt, und das „Geschlecht der Fläche", die Anzahl der linear unabhängigen Integrale 1. Gattung. Ohne Beweis behauptet er die Invarianz des Geschlechts bei „eindeutigen" (birationalen) Transformationen. Als abschließende Aufgabe stellte Clebsch die Konstruktion einer allgemeinen Theorie der „eindeutigen Transformationen", bei der die Theorie der Invarianten (gegenüber projektiven Abbildungen) nur die Einführung sein sollte; es ist offensichtlich, daß vor seinen Augen die allgemeinen Konturen einer „birationalen algebraischen Geometrie" erschienen.

Einige Untersuchungen von Clebsch wurden erst Jahrzehnte später fortgesetzt. So studierte er (in heutiger Terminologie) rationale Flächen über einem nicht algebraisch-abgeschlossenen Körper, betrachtete die verallgemeinerte Jacobische Mannigfaltigkeit einer singulären rationalen Kurve, fand die bireguläre Klassifikation der rationalen Regelflächen.

Wie so oft fand diese ganze Umwälzung – die Begründung eines neuen Gebietes der Wissenschaft – beinahe sprunghaft in weniger als 10 Jahren statt, zwischen Clebschs erster Arbeit 1863 bis zu seinem Tode im Jahre 1872.

Den zweiten Platz nach der algebraischen Geometrie im Schaffen von Clebsch nahm die Invariantentheorie ein. Um den Beitrag von Clebsch zu diesem Gebiet zu charakterisieren, berufe ich mich auf den Mathematiker, der hierzu den größten Beitrag geleistet hat, nämlich Hilbert. In seinem Vortrag auf dem Internationalen Mathematiker-Kongreß 1893 teilt Hilbert die Entwicklung jedes Gebietes in drei Perioden: die naive, die formale und die kritische. In der Invariantentheorie, so meint er, stellen seine eigenen Arbeiten die dritte Periode dar, die zweite dagegen wird charakterisiert durch die Arbeiten von Clebsch und Gordan.

Die grundlegende Frage, auf die sich die Anstrengungen der Mathematiker konzentrierten, war hier die Suche nach einem vollständigen System von Invarianten für Formen gegebenen Grades n einer gegebenen Anzahl m von Variablen, d. h. die Aufstellung eines solchen endlichen Systems von Invarianten, durch welches sich alle übrigen polynomial ausdrücken lassen. Für binäre Formen (d. h. für $m=2$) wurde die Existenz eines endlichen vollständigen Systems von Invarianten bewiesen von Gordan, dem langjährigen Mitarbeiter Clebschs. Für Formen einer beliebigen Anzahl von Variablen wurde dieses Resultat von Hilbert erhalten. Der Beweis von Hilbert zerfällt in zwei heterogene Teile. Der eine von ihnen gehört – aus der heutigen Sichtweise – zu den Grundlagen der kommutativen Algebra (Hilbertscher Basissatz) oder der homologischen Algebra (Satz über die Syzygien). Der andere dagegen benutzt die Existenz eines Differentialoperators, der invariant gegenüber der Gruppe $SL(m)$ ist, des Projektors auf den Ring der Invarianten.

Die Idee der letzteren Überlegung schreibt Hilbert (Ges. Abh. Bd. II, S. 248) Clebsch zu (obwohl Differentialoperatoren in der Invariantentheorie auch vorher angewandt wurden von Sylvester und Cayley). Da der erste Teil des Beweises von Hilbert in keiner Weise davon abhängt, daß wir Invarianten gerade der Gruppe $SL(m)$ suchen, stellte Hilbert die Frage („14. Hilbertsches Problem"), ob ein analoger Endlichkeitssatz bestehe für die Invarianten einer beliebigen Untergruppe $G \subset SL(m)$. 1958 gab Nagata eine negative Antwort auf diese Frage. Auf diese Weise liegt die tiefe Ursache der Existenz eines endlichen vollständigen Systems von Invarianten in der Existenz eines Differentialoperators, der von Clebsch gefunden wurde. Heute wissen wir, daß die Existenz eines analogen Projektors bei einer beliebigen (algebraischen) Gruppe G verknüpft ist mit der Reduktivität der Gruppe G (s. Nagata "Lectures on the Fourteenth Problem of Hilbert", Bombay 1965).

Das Niveau, das in der Invariantentheorie zur Zeit von Clebsch erreicht war, spiegelt sich in seinem Buch „Theorie der binären algebraischen Formen" wider. In ihm spielt eine große Rolle die Methode der Darstellung der Form $f(x_1, x_2; y_1, y_2)$, die von zwei Paaren von Variablen abhängt (mit möglicherweise verschiedenen Homogenitätsgraden), durch die „Polaren" der Formen, die nur von einem Paar abhängen. Wenn man den Raum S_n der Formen vom Grade n als Darstellungsmodul der Gruppe $SL(2)$ ansieht (hierdurch werden bekanntlich alle endlichdimensionalen irreduziblen Darstellungen dieser Gruppe erfaßt), so ist der Raum der Formen $f(x_1, x_2; y_1, y_2)$ mit Grad n in x_1, x_2 und Grad m in y_1, y_2 ein Darstellungsmodul $S_n \otimes S_m$. Eine Formel aus dem Buch von Clebsch (die sogenannte Formel von Clebsch-Gordan) gibt eine explizite Zerlegung dieser Darstellung in irreduzible, d. h. sie bestimmt die Struktur des Ringes der irreduziblen

Darstellungen der Gruppe SL(2). Diese Formel hat wegen des Zusammenhangs zwischen den Gruppen SL(2), $O(3)$ und der Lorentz-Gruppe eine große Fülle von Anwendungen, besonders in der Quantenmechanik, siehe zum Beispiel H. Weyl „Gruppentheorie und Quantenmechanik".

In der Analysis verdankt man Clebsch eine fundamentale Bedingung für vollständige Integrierbarkeit eines Systems linearer partieller Differentialgleichungen, die heute Satz von Frobenius heißt (ein Artikel von Clebsch in Crelles Journal, Bd. 65). Clebsch erweist sich auch als Autor einer Reihe von Arbeiten über Physik, nämlich über Optik, Elastizitätslehre und Hydrodynamik.

Ich werde seine Untersuchungen über die Bewegung eines festen Körpers ohne äußere Kräfte in idealer Flüssigkeit zur Sprache bringen, wobei er zwei Fälle vollständiger Integrierbarkeit der zugehörigen Gleichungen fand (Math. Ann. Bd. 3, 1877). Diese Frage wurde in allerletzter Zeit von neuem beleuchtet dank einer gruppentheoretischen Interpretation. Zugrunde liegt die Beobachtung, daß die Bahn einer beliebigen Liegruppe G im Raume $\mathscr{L}$ ihrer koadjungierten Darstellung eine natürliche symplektische Struktur besitzt („Struktur von Kirillov"). Wenn wir auf $\mathscr{L}$ eine Hamiltonsche Funktion vorgeben (gewöhnlich wird eine quadratische Form genommen), kommen wir zu einem Hamiltonschen (auf jeder Bahn allgemeiner Lage) dynamischen System. Wenn wir beispielsweise $G = \mathrm{SO}(3)$ nehmen, erhalten wir auf diese Weise die Gleichungen der Bewegung eines festen Körpers ohne äußere Kräfte. Auf dem gleichen Wege kann man auch die Bewegungsgleichungen eines festen Körpers in einer idealen Flüssigkeit erhalten: sie hängen zusammen mit der Gruppe $G = E(3) = \mathrm{SO}(3) \cdot \mathbb{R}^3$ aller Bewegungen des dreidimensionalen Raumes. Viele der auf diese Weise entstehenden Systeme von Differentialgleichungen erweisen sich als vollständig integrierbar. Zum Beispiel wurden neue Fälle von vollständiger Integrierbarkeit bei Gleichungen erhalten, die mit der Gruppe SO(4) zusammenhängen („Bewegung eines vierdimensionalen festen Körpers mit einem Fixpunkt"). Die Liealgebra der Gruppe SO(4) kann man erhalten als „Deformation" aus der Liealgebra der Gruppe $E(3)$. Dabei erweist es sich, daß die erneut erhaltenen Fälle vollständiger Integrierbarkeit als „Deformationen" der Fälle erscheinen, die von Clebsch betrachtet wurden. Man fand auch andere Fälle vollständiger Integrierbarkeit von Gleichungen, die mit $E(3)$ zusammenhängen, analog dem Fall von Clebsch. Daher wurde diese Arbeit von Clebsch in allerletzter Zeit zum Ausgangspunkt bei der Suche nach neuen vollständig integrierbaren Gleichungen, die mit verschiedenen Liegruppen zusammenhängen. [Siehe V. I. Arnold "Mathematical Methods of Classical Mechanics", Springer 1978, S. P. Novikov, Uspechi Mat. Nauk t. 37 (1982), No. 5, V. V. Trofimov und A. T. Fomenko, Trudy Sem. po vektornomu i tensornomu analisu, vypusk 21, MGU.]

Clebsch gehörte zu einem neuen, erst ins Dasein tretenden Typ von Mathematikern: für ihn war angestrengte pädagogische Tätigkeit charakteristisch, lebendige Gemeinschaft mit jüngeren Mathematikern, mit einem Wort: die Gründung einer Schule. Dabei bricht er mit den klassischen Vorbildern, deren markantester Vertreter im 19. Jahrhundert Gauß war. Seine Vorgänger sind Jacobi und Dirichlet, seine Nachfolger Klein, Hilbert und viele andere. Heute, mehr als ein Jahrhundert später, können wir die Aufmerksamkeit auch auf die Schattenseiten dieses Typs von Organisation wissenschaftlicher Tätigkeit lenken:

eine gewisse Standardisierung mathematischen Denkens, eine Mechanisierung des schöpferischen Prozesses, die Gefahr der Erstarrung und Entartung der Schule. Aber damals war das der sich gerade erst bildende Stil mathematischen Lebens, der für die Zukunft kolossale Erfolge versprach.

Um Clebsch sammelte sich ein Kreis enthusiastischer Mitarbeiter und Schüler: Gordan, Brill, Lüroth, Zeuthen, und die berühmtesten: Noether und Klein. Sie empfanden stark das Neue ihrer Richtung: in ihren Arbeiten finden sich ständig die Bezeichnungen „Neuere Geometrie" (algebraische Geometrie), „Neuere Algebra" (Invariantentheorie und vielleicht Galoistheorie). Anscheinend schloß sich auch Jordan dieser Richtung an. Klein schreibt (in den „Vorlesungen über die Entwicklung der Mathematik im 19. Jahrhundert"): „Als Clebsch 1872 so plötzlich starb, waren wir, die wir uns von ihm hatten führen lassen, naturgemäß in starke persönliche Gegensätze mit einem großen Teil der übrigen Mathematiker verstrickt." Offensichtlich als Sprachrohr der neuen Richtung war die von Clebsch (gemeinsam mit C. Neumann) gegründete Zeitschrift Mathematische Annalen gedacht. Diese festgefügte Richtung, die sehr bewußt ihre „Avantgarde"-Rolle erlebte, kann man wahrscheinlich vergleichen mit dem Bourbakismus im 20. Jahrhundert. Clebsch war ihr Gründer und ihre Seele. Klein schreibt, daß man bei der Bewertung von Clebschs Beitrag auch die Arbeiten aller Mathematiker seiner Schule rechnen muß. Tatsächlich kann man oft sehen, wie in ihren Arbeiten die Ideen von Clebsch realisiert werden. Das glänzendste Beispiel scheint das grandiose Gebäude der birationalen algebraischen Geometrie zu sein, das von Noether nach dem Vorhaben von Clebsch errichtet wurde. Diese besondere Rolle von Clebsch erklärt den großen Einfluß, den er auf seine Zeitgenossen ausübte (dieser ist beispielsweise sichtbar in dem Nachruf, der von seinen Schülern im 7. Band der Mathematischen Annalen veröffentlicht wurde, oder in den Clebsch gewidmeten Seiten des Buches von Klein). Diese Rolle erklärt auch, warum man ihn jetzt vielleicht nicht so hoch schätzt, wie er es verdient.

Eingegangen am 8. November 1983

On the degeneration of K3 surfaces

(with A. N. Rudakov)

Tr. Mat. Inst. Steklova **166,** 222–234 (1984). Zbl. **577,** 14027
[Proc. Steklov Inst. Math. **166,** 247–259 (1986)]

ABSTRACT. Some special degenerations of algebraic surfaces are studied, and an explicit form of resolution of singularities of families with such degenerations is given. In particular, this result is applicable to K3 surfaces of degree 2. This implies that supersingular K3 surfaces do not degenerate.

Figures: 5. Bibliography: 10 titles.

This paper contains proofs of the results announced in [1]. We are talking about algebraic K3 surfaces, i.e. the surfaces X whose canonical class is trivial ($K_X = 0$) and $H^1(X, \mathcal{O}_X) = 0$. We consider surfaces over an algebraically closed field of characteristic p. In characteristic 0 results more general than ours follow from the classification of degenerations of K3 surfaces given by Kulikov [2]. Therefore we are interested in surfaces over fields of characteristic $p > 0$. We assume that $p > 3$. It would be interesting to carry over the results to the cases $p = 3$ and $p = 2$. In the former case the situation seems to be quite similar. On the contrary, the latter case, $p = 2$, will probably lead to new considerations.

This paper, or rather Theorem 1, has some points in common with a paper of Shah [3]. Although Shah considers varieties only over the field of complex numbers and he, unlike us, is interested in mixed Hodge structures connected with singular varieties, some of his arguments do not depend on these assumptions and deal with analysis of the same (and more general) degenerations as in Theorem 1 of this paper. At one point we use an argument suggested by Shah. It looks like we could have used more of his results. But his arguments in our situation probably would have been more cumbersome than the direct consideration based on the specific nature of the case. (All this relates to Theorem 1, and not Theorems 2 and 3 and the applications to supersingular surfaces.)

DEFINITION. Singularities of an algebraic surface X are called *moderate* if they belong to one of the following types:

a) A rational double point.

b) Isolated singular points of imbedded dimension 3 obtained by contracting an elliptic curve of square -1.

Locally (and formally) they are given by the equation

$$z^2 = x^3 + \alpha xy^4 + \beta y^6, \qquad 4\alpha^3 + 27\beta^2 \neq 0. \tag{1}$$

b′) Isolated singular points of imbedded dimension 3 obtained by contracting an elliptic curve of square -2.

1980 *Mathematics Subject Classification* (1985 *Revision*). Primary 14J28, 14J17.

Locally (and formally) they are given by equation

$$z^2 = x^4 + \alpha x^2y^2 + \beta xy^3 + \gamma y^4, \tag{2}$$

where the form on the right-hand side has no multiple factors.([1]) For more on singularities of types b) and b′) see, for example, [4].

c) A double curve C of singularities with a finite number of vertices $c_1, \dots, c_n$. In a formal neighborhood of a point $c \in C$, where $c \neq c_i$, the surface is given by equation $xy = 0$, and in a neighborhood of some c_i the equation is $x^2 - zy^2 = 0$ (n can be equal to 0).

Let $\mathscr{X} \to S$ be a flat family of algebraic surfaces with base $S = \operatorname{Spec} k[[\varepsilon]]$. A family $\overline{X} \to S$ whose generic fiber is isomorphic to the generic fiber of $\mathscr{X}$ is called a *modification* of the family $\mathscr{X}$.

A sequence $X_1, \dots, X_n$ of components of the closed fiber of a family $\mathscr{X}$ is called a *chain* if all the X_i are smooth ruled surfaces with base C, $X_i \cap X_{i+1}$ is a section on both X_i and X_{i+1}, and $X_i \cap X_j = \varnothing$ if $j \neq i-1, i, i+1$. By a *chain beginning with component* X_0 we mean a sequence $X_0, X_1, \dots, X_n$, where $X_1, \dots, X_{n-1}$ is a chain, $X_0 \cap X_1$ is a section on X_1, $X_0 \cap X_i = \varnothing$ when $i > 1$, $X_{n-1} \cap X_n$ is a section on X_{n-1} and X_n is a rational surface. A *chain connecting components* X_0 *and* X_{n+1} is a sequence $X_0, \dots, X_{n+1}$, where $X_1, \dots, X_n$ is a chain, $X_0 \cap X_1$ is a section on X_1, $X_n \cap X_{n+1}$ is a section on X_n, $X_0 \cap X_i = \varnothing$ if $i > 1$ and $X_i \cap X_{n+1} = \varnothing$ if $i < n$. A *ring fixed at component* X_0 is a sequence $X_0, \dots, X_n$, where $X_1, \dots, X_n$ is a chain, $X_0 \cap X_1$ is a section on X_1, $X_n \cap X_0$ is a section on X_n and $X_0 \cap X_i = \varnothing$ if $1 < i < n$.

THEOREM 1. *Let $\mathscr{X} \to S$ be a flat family of surfaces whose generic fiber is smooth or has rational double points and whose closed fiber has only moderate singularities. There exist a finite covering of the base $S' \to S$ and a modification $\overline{\mathscr{X}} \to S'$ of the family $\mathscr{X} \times_S S'$ with the following properties. The components of the closed fiber of $\overline{\mathscr{X}}$ are smooth or have rational double points and meet transversely. These components belong to one of the following types:* a) *models Z_i of the components of the closed fiber of the family $\mathscr{X}$,* b) *ruled surfaces,* c) *rational surfaces. Ruled and rational surfaces form chains beginning with components Z_i; ruled surfaces form chains connecting components Z_i and Z_j or rings fixed at components Z_i. All ruled surfaces have as bases either elliptic curves or curves C_i like those in the definition of moderate singularities of the closed fiber of the family $\mathscr{X}$ or their double covers ramified at points $c_1, \dots, c_n$. Different chains and rings do not have common points and components, except the components Z_i. Different components Z_i do not meet.*

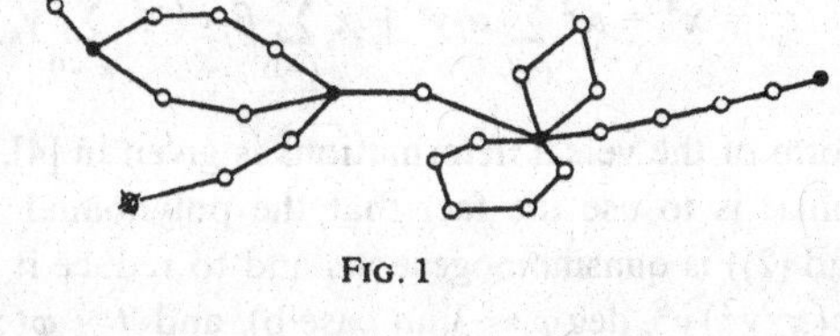

FIG. 1

([1]) For some incomprehensible reason, in paper [1] we did not include the singularities of type b′) in the list. This was pointed out to us by A. Ogus. As the reader will see, types b) and b′) are treated in exactly the same way.

Thus the components form a graph of the type shown on Figure 1, where black points denote Z_i, white points denote X_i which are ruled, and crosses denote rational X_i.

We start off with a remark one should have in mind while proving the theorem. We shall resolve singular points and the curves of the closed fiber one after another. This will not affect the singularities considered in the previous steps because the modifications we construct are isomorphisms off a given point or a curve and hence are strictly local. A lifting of the base will not affect the types of the singular points (of the hypothesis of the theorem) since the closed fiber does not change. Let us check that the properties of already constructed chains are also preserved. If X' and X'' are smooth surfaces belonging to a chain and meeting transversally along a curve C, then the local equation of the family $\mathscr{X}$ is $xy = \varepsilon$, where $x = 0$ and $y = 0$ are local equations of X' and X''. After lifting the base this equation (maybe in different local coordinates) will be $xy = \varepsilon^k$. Hence locally our family looks like $U \times T$, where $U \subset C$ is an open set and T is the germ of the singularity A_{k-1}. Under a resolution, the singularity A_{k-1} resolves into a chain of $[k/2]$ rational curves. Therefore, a new chain of $[k/2]$ ruled surfaces with base C is pasted in the original chain at this place.

After these remarks we can start proving the theorem. Notice that we can consider singular points and curves of types a), b), b′) and c) separately and restrict ourselves to formal neighborhoods of the singular points.

In case of singularities of type a) our assertion is trivial. In cases b) and b′) we can reduce our family $\mathscr{X}$ to a certain normal form. For this we construct a versal family of deformations for each of those singularities. According to [5] they can be constructed as follows. If an isolated singularity has equation $F(x, y, z) = 0$ then the space $\mathcal{O}_{\mathscr{P}}/(F, F'_x, F'_y, F'_z)$, where $\mathcal{O}_{\mathscr{P}}$ is the local ring of point $\mathscr{P}$ in the 3-space, is of finite dimension. Choose any finite subset of functions $\Phi_i \in \mathcal{O}_{\mathscr{P}}$ generating this space. Then the family $F + \sum \lambda_i \Phi_i$ is versal. In our case $F = z^2 - f(x, y)$, where f is a quasihomogeneous polynomial [in case b) x should be assigned weight 2 and y weight 1; in case b′) both x and y have weight 1]. Because of that we have $f \in (f'_x, f'_y)$, and since $F'_x = -f'_x$, $F'_y = -f'_y$ and $F'_z = 2z$ our quotient space becomes $\mathcal{O}/(f'_x, f'_y)$, where $\bar{\mathcal{O}}$ is the local ring of the point $(0, 0)$ on the plane.

Elementary calculations show that in our case the versal deformations can be written in the following form:

$$\text{b)} \quad z^2 = x^3 + x \sum_{i=0}^{4} \alpha_i y^i + \sum_{j=0}^{6} \beta_j y^j,$$

$$\text{b}') \quad z^2 = x^4 + x^2 \sum_{i=0}^{2} \alpha_i y^i + x \sum_{j=0}^{3} \beta_j y^j + \sum_{k=0}^{4} \gamma_k y^k.$$

(A more compact form of the versal deformations is given in [4].) The simplest way to obtain our formulas is to use the fact that the polynomial f in the equations $z^2 = f(x, y)$ ((1) and (2)) is quasihomogeneous and to reduce it to a polynomial in one variable: $f = \varphi(x/y^2)y^6$, $\deg \varphi = 3$, in case b), and $f = \varphi(x/y)y^4$, $\deg \varphi = 4$, in case b′); then, because of the assumptions, the polynomials φ and φ' are relatively prime and we can get all monomials as linear combinations of them.

Thus any family of surfaces with the degenerate fiber of types b) and b′) can be given in a neighborhood of a singular point of the fiber by the equations

$$\text{b)}\quad z^2 = x^3 + a_4(y,\varepsilon) + a_6(y,\varepsilon), \tag{3}$$

$$\text{b}')\quad z^2 = x^4 + a_2(y,\varepsilon)x^2 + a_3(y,\varepsilon)x + a_4(y,\varepsilon), \tag{4}$$

with the following conditions satisfied: $\deg_y a_i(y,\varepsilon) \leqslant i$, and the polynomial $x^3 + a_4(y,0)x + a_6(y,0)$ in case b) and the polynomial $x^4 + a_2(y,0)x^2 + a_3(y,0)x + a_4(y,0)$ in case b′) do not have multiple roots. Since the generic fiber has only rational double points, the conditions $a_i(y,\varepsilon) \equiv 0\ (y^i)$ ($i = 4, 6$ in case b) and $i = 2, 3, 4$ in case b′)) cannot be satisfied simultaneously. Using the substitution $\varepsilon \to \varepsilon^k$, if necessary, we shall always assume

$$a_i \in (y,\varepsilon)^i. \tag{5}$$

From here on the proof in cases b) and b′) will exactly follow [6] (the proof of Theorem 2, case of degenerations of the second kind). We reproduce it here for three reasons: 1) for the convenience of the reader; 2) because only type b) singular points are studied in [6]; and 3) a method used by Shah in [3] allows to shorten the proof.

Three-dimensional varieties given by equations (3) and (4) define bundles over $\mathbf{A}^1 \times S$, where $S = \operatorname{Spec} k[[\varepsilon]]$, whose fibers are affine curves (elliptic curves over the generic point of S). We complete every fiber by a nonsingular point at infinity. In other words our variety will consist of two pieces, one of which will be given by (3) or (4) and the other by

b) $u = v^3 + a_4 v^2 u + a_6 u^3$,

b′) $u^2 = 1 + a_2 v^2 + a_3 v^3 + a_4 v^4$

with gluing functions

b) $u = 1/z$, $v = x/z$,

b′) $u = (z/x)^2$, $v = 1/x$.

Thus we obtain a bundle $\mathscr{X} \to \mathbf{A}^1 \times S$ whose projection is a proper morphism.

Let us construct a new bundle, $\mathscr{X}' \to \Sigma$, which we call an *elementary transformation* of the original one. Σ is obtained from $\mathbf{A}^1 \times S$ by blowing up at $(0,0)$ and is covered by two charts:

$$U_1 = \operatorname{Spec}(k[[\varepsilon]][y,\varepsilon_1]/(y\varepsilon_1 - \varepsilon)), \qquad U_2 = \operatorname{Spec}(k[[\varepsilon]][y_2,\varepsilon]/(y_2\varepsilon - y)),$$

which contain subsets $U_{12} \subset U_1$ and $U_{21} \subset U_2$ defined by the conditions $\varepsilon_1 \neq 0$ and $y_2 \neq 0$, isomorphic to each other because $\varepsilon_1 y_2 = 1$.

Over U_1 the family $\mathscr{X}_1'$ is defined in cases b) and b′) by the following equations:

b) $z_1^2 = x_1^4 + a_4^{(1)}x_1 + a_6^{(1)}$,

b′) $z_1^2 = x_1^4 + a_2^{(1)}x_1^2 + a_3^{(1)}x_1 + a_4^{(1)}$, $a_i^{(1)} = a_i(y, y\varepsilon_1)y^{-i}$.

Over U_2 the family $\mathscr{X}_2'$ is defined in cases b) and b′) by

$$\text{b)}\quad z_2^2 = x_2^3 + a_4^{(2)}x_2 + a_6^{(2)}, \tag{6}$$

$$\text{b}')\quad z_2^2 = x_2^4 + a_2^{(2)}x_2^2 + a_3^{(2)}x_2 + a_4^{(2)}, \qquad a_i^{(2)} = a_i(y_2\varepsilon,\varepsilon)\varepsilon^{-i}. \tag{7}$$

Over $U_{12} \simeq U_{21}$ these families are glued together by formulas in which we assume that the numbers λ and μ are not equal to 2 and 3 in case b) and 1 and 2 in case b′). Then the formulas are

$$x_1 = x_2\varepsilon_1^\lambda, \quad z_1 = z_2\varepsilon_1^\mu \quad \text{on } U_{12},$$

$$x_2 = x_1 y_2^\lambda, \quad z_2 = z_1 y_2^\mu \quad \text{on } U_{21}.$$

Over U_1 when $y \neq 0$ the preimages of $\mathscr{X}$ and $\mathscr{X}'$ are isomorphic:

$$x_1 = xy^{-\lambda}, \qquad z_1 = zy^{-\mu}.$$

Over U_2 when $\varepsilon \neq 0$ the preimages of $\mathscr{X}$ and $\mathscr{X}'$ are isomorphic:

$$x_2 = x\varepsilon^{-\lambda}, \qquad z_2 = z\varepsilon^{-\mu}.$$

Similarly to what we did with families (3) and (4), we add points at infinity to the fibers of $\mathscr{X}_1'$ and $\mathscr{X}_2'$. It is easy to see that the gluing automorphisms extend to these completed families. Thus we obtain a family $\mathscr{X}'$ with proper morphism $\mathscr{X}' \to \Sigma$ isomorphic to the original family $\mathscr{X}$ away from the point $y = 0$, $\varepsilon = 0$. We call this family an *elementary transformation* of family $\mathscr{X}$.

In $\mathscr{X}'$ the preimage $\mathscr{X}_s'$ of the closed point $s \in S$ splits into two components: $X_s' = X_0 \cup X_1$. The component X_0 is defined over U_1 by the equation $\varepsilon_1 = 0$, and, in cases b) and b′),

b) $z_1^2 = x_1^3 + a_4(,0)x_1 + a_6(y,0)$,

b′) $z_1^2 = x_1^4 + a_2(y,0)x_1^2 + a_3(y,0)x_1 + a_4(y,0)$.

This component does not meet the preimage of U_2. X_0 is birationally isomorphic to the closed fiber X_s of the bundle $\mathscr{X} \to S$.

To describe the component X_1 we set

$$a_i(y,\varepsilon) = \sum_{j+k>i} \alpha_{ijk} y^j \varepsilon^k, \qquad A_i(y,\varepsilon) = \sum_{j+k=i} \alpha_{ijk} y^j \varepsilon^k.$$

Then X_1 is given over U, in cases b) and b′), by the equations

b) $y = 0$, $z_1^2 = x_1^3 + A_4(1,\varepsilon_1)x_1 + A_6(1,\varepsilon_1)$,

b′) $y = 0$, $z_1^2 = x_1^4 + A_2(1,\varepsilon_1)x_1^2 + A_3(1,\varepsilon_1)x_1 + A_4(1,\varepsilon_1)$

and over U_2 by equations

b) $\varepsilon = 0$, $z_2^2 = x_2^3 + A_4(y_2,1)x_2 + A_6(y_2,1)$,

b′) $\varepsilon = 0$, $z_2^2 = x_2^4 + A_2(y_2,1)x_2^2 + A_3(y_2,1)x_2 + A_4(y_2,1)$.

The curve $Y = X_0 \cap X_1$ is given by $\varepsilon_1 = t = 0$ and

b) $z_1^2 = x_1^3 + a_4^{(1)}(0,0)x_1 + a_6^{(1)}(0,0)$,

b′) $z_1^2 = x_1^4 + a_2^{(1)}(0,0)x_1^2 + a_3^{(1)}(0,0)x_1 + a_4^{(1)}(0,0)$.

Because of our assumptions, Y is a smooth curve and the component X·, which is smooth at all points of Y, maps onto the original closed fiber $X_1 \to X_s$ so that Y contracts to a singular point $\varepsilon = x = y = z = 0$ of this fiber. It is isomorphic to $Y \times \mathbf{A}^1$. It is easy to see that at all points of Y all the families $\mathscr{X}'$ are smooth and X_0 and X_1 meet transversally along Y.

If the polynomials $A_i(y_2,1)$ with $i = 4$ or 6 in case b) and $i = 2$, 3 or 4, in case b′) are not the ith power of the same linear polynomial $y_2 - \lambda$, then the component X_1 is a rational surface having only rational double points and the process of constructing the chain for the given singular point ends.

If the indicated representation in the form of powers of linear polynomials obtains, then after a change of variable $y_2 - \lambda \to y_2$ we almost get into the previous situation. Component X_1 has a singular point of type b) or b′) in the chart U_2. The only difference is that our surface X_1 is a bundle not over $\mathbf{A}^1$ but over $\mathbf{P}^1$. Nevertheless our local considerations are completely applicable to the open subset of X_1 lying over $\mathbf{A}^1$ (in chart U_2).

Thus successive consideration of elementary transformations gives us a sequence of families $\mathscr{X}_r$ such that the degenerate fiber of each family is a chain $X_0, \ldots, X_r$ beginning with X_0:* the degenerate fiber of the original family with the resolved singular point, $X_1, \ldots, X_{r-1}$ are isomorphic to $H \times \mathbf{P}^1$ and intersect transversely along a curve isomorphic to Y, X_r is either of the same type or is a rational surface with rational double points in which case the construction of the chain ends.

Let us prove that the process of constructing a chain cannot be infinite. To this end we recall that the polynomial $a_i(y, 0)$ [$i = 4$ or 6 in case b) and $i = 2$, 3 or 4 in case b′)] has leading term of degree i. The corresponding polynomial $a_i(y, \varepsilon)$ will have a root $\lambda(\varepsilon') \in k[[\varepsilon']]$, $\lambda(0) = 0$, where $k[[\varepsilon']]/k[[\varepsilon]]$ is a finite extension. Passing to the covering $S' \to S$, $S' = \operatorname{Spec} k[[\varepsilon']]$ combined with the shift $y = y - \lambda$, we can assume that one of the polynomials a_i (let it be a_m) is divisible by y (this is Shah's trick). Although we do it only in the beginning of the process, this property will carry on through the whole chain. Indeed the next term of the chain is given by (6) and (7), which obviously preserve the desired property. Then if the component X_1 in the family $\mathscr{X}'$ has a singularity of type b) or b′), then the polynomials $a_i(y_2, 0)$ must be the ith power of the monomial $y_2 - \lambda$ (when $i = 4$ or 6 in case b) and $i = 2$, 3 or 4 in case b′)). But by assumption $a_m(y_2, 0)$ is divisible by y_2; hence $\lambda = 0$. Thus the new singularity will always be at the origin of the coordinate system in which equations (6) and (7) are written.

A sequence of s transformations is given by the formulas $y = y_s\varepsilon^s$ and equations (6) and (7) with coefficients

$$a_i^{(s)}(y_s, \varepsilon) = a_i(y_s\varepsilon^s, \varepsilon)\varepsilon^{-is}. \tag{8}$$

Now let us use the condition $a_i \not\equiv 0\,(y^i)$ (when $i = 4$ or 6 in case b) and $i = 2$, 3 or 4 in case b′)). Let $a_l \not\equiv 0\,(y^l)$. This means that a_l contains a term $\alpha_{ljk}y^j\varepsilon^k$ with $\alpha_{ljk} \neq 0$ for some j, where $0 \leq j < 1$. After lifting the base $\varepsilon = \varepsilon_1^q$ and performing the transformations (8) it will become $\alpha_{ljk}y_s^j\varepsilon_1^{qk+(j-l)s}$. Set $s = qk/(l - j)$, choosing q so that it will be integral (note that $s > 0$ since $j < 1$). Then our term gives the term $\alpha_{ljk}y_s^j$, where $0 \leq j < 1$, i.e. after the sth step the surface X will be rational and the process will stop.

Now consider the remaining case: singularities of type c). We shall gradually simplify the singularity along the curve C by blowing up along this curve and similar curves arising in the course of blowings-up. We shall keep track of what happens in neighborhoods of those points on the curve where (3) and (4) hold and at the generic point of the curve.

The completed local ring of the generic point of the curve C can be given as a quotient of a regular local ring with parameters x and y, i.e. "defined by the equation" $x^2 - ay^2 = 0$, where $a \notin (x, y)$, and hence gives a nonzero function $\alpha \in k(C)$. Because of this the family $\mathscr{X}$ in a neighborhood of the generic point of C can be defined by

$$x^2 - ay^2 + \varepsilon^{2k}b = 0 \tag{9}$$

(the exponent of ε can be made even by lifting the base). Notice that $b \not\equiv 0\,(\varepsilon)$. Then $b \in (x, y)$, since otherwise we could achieve $b = 0$ in (9) by the change of variables

$x \to x + P$, $y \to y + Q$, and this would contradict the fact that the generic fiber of the family $\mathscr{X}$ has only rational double points. Similarly in a neighborhood of the point c of the curve C where the surface X is given by equation $xy = 0$, the deformation can be written as

$$xy + \varepsilon^{2k}b = 0, \qquad b \notin (x, y). \tag{10}$$

If in some neighborhood of a point c_i of the curve C the surface X is given by the equation $x^2 - zy^2 = 0$, then the deformation can be given by

$$x^2 - zy^2 + \varepsilon^r \cdot b \cdot y + \varepsilon^{2s}c = 0, \tag{11}$$

where either $b \in (x, y, z)$ or $c \in (x, y)$. We also note, using the substitution $y \to y - \varepsilon^r/2z \cdot b$, that if (11) holds at c_i then (10) holds at a point c from some neighborhood of c_i and in this case $k = \min(r, s)$. If $r < s$ then $b \in (x, y, z)$, and if $s < r$ then $c \in (x, y)$.

Now use a blowing-up on the curve C. In one of the charts it will be given by the formulas $x = u\varepsilon$ and $y = v\varepsilon$, and the family $\mathscr{X}$ will have equation

$$u^2 - av^2 + \varepsilon^{2k-2}b = 0$$

at the generic point of C,

$$uv + \varepsilon^{2k-2}b = 0$$

in neighborhoods of points $c \neq c_i$ and

$$u^2 - 2v^2 + \varepsilon^{r-1} \cdot b \cdot v + \varepsilon^{2s-2} \cdot c = 0$$

in neighborhoods of points c_i $(i = 1, \ldots, n)$. If $k > 1$ then the pasted-in variety Y is given by the equation $\varepsilon = 0$ and the equations

$$u^2 - av^2 = 0, \qquad uv = 0, \qquad u^2 - zv^2 = 0.$$

It follows that if α is a perfect square (which is equivalent to the function $\alpha \in k(C)$, the restriction of a to C, being a perfect square), then Y splits into two components Y' and Y'', each of which is a ruled surface over C, and they intersect transversely along a curve which is a section of both components. If a (and hence α) is not a square then Y is a ruled surface with base C, having a double curve with n vertices.

Absolutely standard calculations in the remaining charts show that the strict transform X_1 of the degenerate fiber X_0 of the original family is smooth in some neighborhood of the preimage of C, i.e. coincides with the normalization of X_0 in this neighborhood. The preimage C' of C on X_1 consists of nonsingular points of the family. The surfaces Y and X_1 are also smooth at these points and intersect transversely.

Now we can have three cases.

1. The degenerate fiber X of the original family is reducible and its two components X' and X'' intersect transversely along the curve C. Obviously the function $\alpha \in k(C)$ is then a perfect square. In this case the passage from a component of the closed fiber X_0 in some neighborhood of C to the degenerate fiber of the new family can be described as in Figure 2.

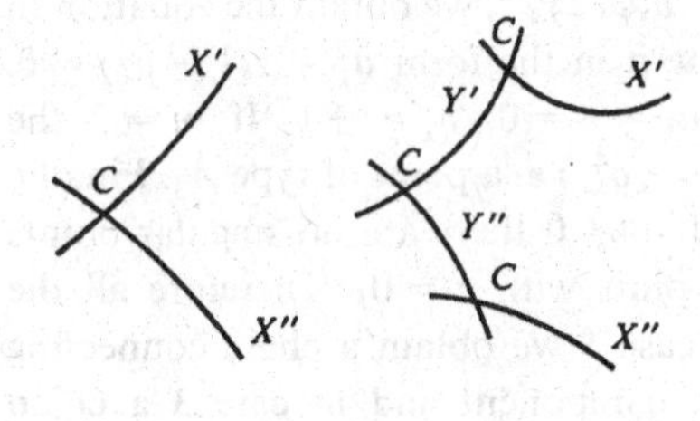

FIG. 2

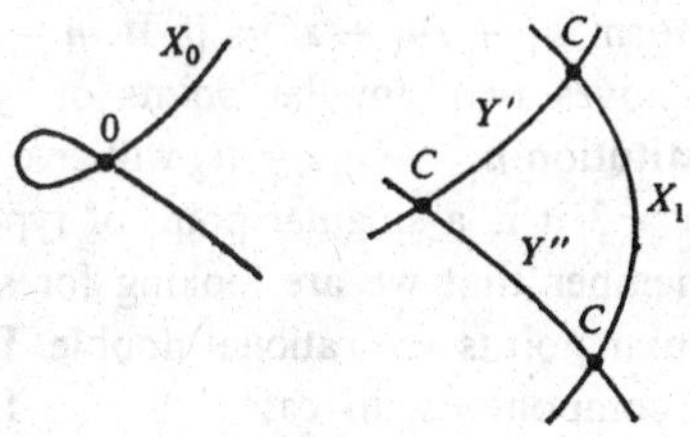

FIG. 3

2. The degenerate fiber X_0 of the original family is irreducible in some (Zariski) neighborhood of the curve C but the function α is a perfect square. This is shown on Figure 3.

3. The function α is not a square and the fiber X_0 is irreducible in some neighborhood of C. In this case the variety Y is irreducible and meets X_1 along a curve C' which is a double cover of C with ramification points $c_1, \ldots, c_n$ (see Figure 4).

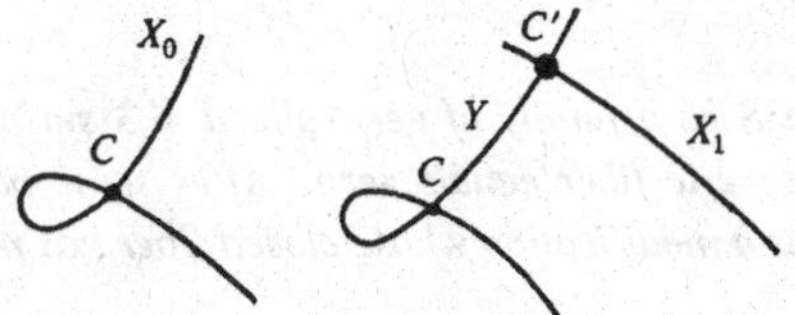

FIG. 4

In this figure the curve C' is shown as a thick point to emphasize that it covers C with multiplicity 2.

After blowing up k times on C we obtain a pasted-in variety Y which is given in a neighborhood of the generic point of C, of a point $c \neq c_i$ and one of the points c_i by equations

$$u^2 - \bar{a}v^2 = \bar{b}, \tag{12}$$

$$uv + \bar{b} = 0, \tag{13}$$

$$u^2 - zv^2 + \bar{b}v + \bar{c} = 0, \tag{14}$$

where $\bar{a}$, $\bar{b}$ and $\bar{c}$ denote the restrictions to $\varepsilon = 0$ or (in case (14)) -0. We saw that in (12) $\bar{a} \neq 0$ and $\bar{b} \neq 0$, so that Y_k is birationally a bundle of quadrics, i.e. a ruled surface over C. It remains to study its singularities, and for this we turn to equations (13) and (14). If z is a local coordinate inducing a local coordinate on the curve C, then in (13) we can set $\bar{b} = z^m\varphi(z)$, where $\varphi(z)$ is an invertible series. Indeed, the terms containing u or v can be killed by the substitution $u \to u + P$, $v \to v + Q$. Setting $u = u_1\varphi(z)$ we obtain the equation in the form $u_1v + z^m = 0$, i.e. a singular point of type A_{m-1}.

In (14) we can again assume $\bar{c} = z^m\varphi(z)$ to be a power series in z and $\bar{b} = \beta$ a constant, where either $\beta \neq 0$ or $\varphi(z) \neq 0$.

If $\beta = 0$ then, setting $u = u_1\varphi(z)^{1/2}$ and $v = v_1\varphi(z)^{1/2}$, we obtain the equation in the form $u_1^2 + zv_1^2 + z^m = 0$. If $m = 1$ we write it in the form $u_1^2 + z(1 - v_1^2) = 0$. This gives two singular points of type A_1: $u_1 = z = 0$, $v_1 = \pm 1$. If $m = 2$ the substitution $u_1 - \frac{1}{2}v_1^2 z = u_2$ will give $u_2^2 + z^2 + \frac{1}{4}v_1^4$, i.e. a point of type A_3. Finally, if $m \geqslant 3$ it is a singular point of type D_{m+1}. If $\beta \neq 0$ there are no singular points (remember that we are looking for singular points with $z = 0$). Therefore all the singular points are rational double. Hence in case 1 we obtain a chain connecting two components, in case 2 a ring fixed at a component and in case 3 a chain beginning with a component.

Henceforth by a *generalized* K3 *surface* we understand a surface having only rational double points as singularities and whose minimal resolution is a K3 surface.

Let X be a generalized K3 surface over a field of characteristic $p > 3$. Its Hasse invariant $\lambda(X)$ is given by condition $C(\omega) = \lambda(X) \cdot \omega$, where ω is a nonzero regular differential form on X (i.e. a nonzero section of the dualizing sheaf ω_X) and C the Cartier operator. Since the form ω is defined up to a constant factor $\gamma \neq 0$, $\lambda(X)$ is defined up to the factor γ^{1-p}. But the condition $\lambda(X) = 0$ is invariant.

By definition, a K3 surface X has degree 2 if on its nonsingular model there exists a divisor D with $D^2 = 2$.

THEOREM 2. *Let $\mathscr{X} \to S$ be a family of generalized* K3 *surfaces of degree* 2 *whose Hasse invariant of the generic fiber equals zero. After some base lifting $S' \to S$ the family $\mathscr{X}' = \mathscr{X} \times_S S'$ has a modification whose closed fiber has moderate singularities.*

It is known (see, for example, [7]) that a K3 surface of degree 2 either is a double plane or has a bundle of elliptic curves with a section (i.e. it is a strongly elliptic surface, in the terminology of [6]). In the first case the surface is given by the equation

$$z^2 = f(x, y), \tag{15}$$

where f is a polynomial of degree 6, and in the second case by

$$z^2 = x^3 + a(y)x + b(y), \tag{16}$$

where a is a polynomial of degree 8 and b is a polynomial of degree 12. We shall consider these two cases with regard to the type of the generic fiber of the family.

In both cases we study the set $\mathfrak{P}$ of all surfaces of the given type. In the first case $\mathfrak{P}$ is a projective space of all plane curves of degree 6; in the second the weighted projective space of pairs of polynomials a, b of the given degrees with coefficients of a having degree 4 and those of b having degree 6. In both cases there is an action of a semisimple algebraic group G on $\mathfrak{P}$: in the first case $G = PGL(2)$, the group of projective transformations of the coordinates x and y; in the second $G = PGL(1)$, the group of projective transformations of the coordinate y. The general principles of the invariant theory allow us to compile a list of singular surfaces so that for any family of surfaces of the given type there exists a modification whose closed fiber belongs to our list. This list is the set of the semistable points of $\mathfrak{P}$ whose orbits are closed in the set of all semistable points (see, for example, [8]). Thus the problem is reduced to a classification of the semistable points (having a closed orbit) in the above G-varieties. The solution to this problem is well known (see, for example, [8]) and we only state the result for the convenience of the reader.

In the case of strongly elliptic surfaces the corresponding degenerations can be of the following four types (in the notation of [8]).

A′. The surface X has only rational double points as singularities.

B′. X has two isolated triple points. In this case in (16) $a = \alpha c^4$ and $b = \beta c^6$, where c is a polynomial of the second degree with distinct roots and $4\alpha^3 + 27\beta^2 \neq 0$.

C′. X has a curve of singular points whose preimage under the normalization is smooth. In this case in (16) $a = -3c^2$ and $b = -2c^3$, where c is a polynomial of degree 4 without multiple roots.

D′. X has a curve of singularities whose preimage under the normalization is not smooth. The formulas for a and b from C′ still hold, but the polynomial $c(y)$ has one double root and two simple roots.

In the case of double planes there are five types. We assume that the surface is given by (15).

A. The curve $f(x, y) = 0$ has isolated singular points of order 2 or 3, but they split into singular points of order 2 after a single blowing-up. Then X has only rational double points.

B. The curve $f(x, y) = 0$ has two triple points which remain triple points after a blowing-up. Then X has two isolated singular points and can be given by the equation

$$z^2 = x^3 + \alpha y^4 x + \beta y^6, \qquad 4\alpha^3 + 27\beta^2 \neq 0.$$

C. The curve $f(x, y) = 0$ has a unique double component which meets the others transversely. The following types are possible for the equation of X:

C_1: $z^2 = g^2$, where $g = 0$ is a smooth cubic.

C_2: $z^2 = h^2 k$, where $h = 0$ and $k = 0$ are conics meeting transversely and the conic $h = 0$ is smooth.

C_3: $z^2 = l_0^2 l_1 l_2 l_3 l_4$, where the $l_i = 0$ are distinct curves; the curves $l_i = 0$, $i = 1, 2, 3, 4$, pass through a point not lying on $l_0 = 0$.(²)

D. The curve $f(x, y) = 0$ has a double component not transversal to others. The equation can be of the following forms:

D_1: $z^2 = gh^2$, where $g = 0$ and $h = 0$ are conics osculating at two points and the conic $h = 0$ is smooth.

D_2: $z^2 = l_1^2 l_2^2 l_3^2$, where the $l_i = 0$ are lines not passing through one point.

E. The curve $f(x, y) = 0$ has a triple component. The equation of X is $z^2 = g^3$, where $g = 0$ is a smooth conic.

As Horikawa [7] showed, any family of double planes having degenerate fiber of type E has a modification whose degenerate fiber belongs to one of the types A′–D′. (His proof was given for fields of characteristic 0 but it is correct for any characteristic $p > 3$.) Thus the degenerations of type E can be removed.

It remains to check that the degenerations of types A′–D′ and A–D either do not come from the families whose Hasse invariant is zero or, if they do, have moderate singularities.

(²) At this point we would like to correct a wrong statement in [8], Russian p. 124, English p. 126, concerning singular surfaces of type C_3. It was said there that the resolution of such a surface is its nonminimal model of the form $C \times \mathbf{P}^1$, where C is an elliptic curve. The correct statement is that it is a minimal model of a ruled surface with base C. The argument which followed does not depend on this.

First of all we shall prove that degenerations D′, D_1 and D_2 do not come from families whose Hasse invariant of the generic fiber is zero. To do this we fix a model of our family. We consider the strongly elliptic surface given by (16) as a subvariety of the bundle $\mathbf{P}(E)$ on $\mathbf{P}^1$ given by the equation

$$z^2u = x^3 + a(y)xu^2 + b(y)u^3$$

where $E = \mathcal{O} \oplus \mathcal{O}(4) \oplus \mathcal{O}(6)$; $\mathbf{P}(E)$ is the projectivization of E, y is the coordinate on $\mathbf{P}^1$, x, z and u are sections of $\mathcal{O}(4)$, $\mathcal{O}(6)$ and $\mathcal{O}$, and, finally, $a(y)$ and $b(y)$ are considered as sections of $\mathcal{O}(8)$ and $\mathcal{O}(12)$. We consider the double plane given by (15) as a subvariety of $\mathbf{P}(E)$ given by the same equation, where E is the bundle $\mathcal{O}(3) \oplus \mathcal{O}(1)$ on $\mathbf{P}^2$, x and y are sections of $\mathcal{O}(1)$, z is a section of $\mathcal{O}(3)$, and $f(x, y)$ and z^2 are sections of $\mathcal{O}(6)$. In both cases the family $\mathcal{X}$ is considered as a hypersurface in $\mathbf{P}(E) \times S$. According to our assumptions, the generic fiber of $\mathcal{X}$ is a generalized K3 surface.

Let ω_X and $\omega_{X/S}$ be the dualizing sheaves on the surface X of given type and on the family $\mathcal{X}$, respectively. Since all our surfaces X are linearly equivalent in $\mathbf{P}(E)$ and $H^i(\mathbf{P}(E), \mathcal{O}) = 0$ when $i > 0$, $H^2(X, \mathcal{O}_X)$ is the same for all surfaces, i.e. equals 1. Therefore, $R^0f_*\omega_{X/S}$ is a locally constant invertible sheaf (here $f\colon X \to S$ is the projection). The Cartier operator C acts on it fiberwise as a p-linear transformation. Therefore if $C = 0$ (i.e. the Hasse invariant of the generic fiber is zero) the same is true for the closed fiber of $\mathcal{X}$. Finally, to compute the Hasse invariant of a specific surface X we should have a nonzero section of ω_X. It can be obtained by standard construction of the Koszul resolution of the sheaf $\omega_{\mathbf{P}(E)}$ because X is a hypersurface in $\mathbf{P}(E)$ (see, for example, [9], p. 245). In our case this section is given by the form $y^{-1}dx \wedge dt$ for the strongly elliptic surface given by (16), and by the form $z^{-1}dx \wedge dy$ for the double plane (15). Now we can apply the Cartier operator to this form as usual. The corresponding (very simple) calculations are contained in [6] and [8]. For the strongly elliptic surface (16) the Hasse invariant equals the coefficient of $x^{p-1}y^{p-1}$ in the polynomial $(x^3 + a(y)x + b(y))^{(p-1)/2}$, and for the double plane (15) the same coefficient in the polynomial $f(x, y)^{(p-1)/2}$. It was shown in [6] and [8] (by equally simple calculations) that for the degenerate surfaces of types D′, D_1 and D_2 this coefficient is different from zero. Hence the degenerations of types D′, D_1 and D_2 cannot come from families whose Hasse invariant of the generic fiber is zero.

In conclusion we shall check that the degenerate surfaces of types A′–C′ and A–C have moderate singularities. This follows directly from the defining equations, so we only state which singularities exactly these degenerations have.

A′–a)

B′–b). Singularities at two points where $c(y) = x = z = 0$.

B′–c). The curve C is given by equation $z = 0$ and $x = c(y)$, where c_1, c_2, c_3 and c_4 are the roots of the polynomial $c(y)$.

A–a)

B–b). Singularities at two points $y = 0$ and $y = \infty$.

C_1–c). The curve C is given by $g = z = 0$. The set $\{c_1, \ldots, c_n\}$ is empty.

C_2–c). C is given by $h = z = 0$.

Let c_1, c_2, c_3, c_4 be the points of intersection of the conics $h = 0$ and $k = 0$.

C_3–c) + b′). The curve $l_0 = z = 0$ has a singularity of type c), where c_1, c_2, c_3, c_4 are the points of intersection of l_0 with the lines $l_i = 0$, $i = 1, 2, 3, 4$. The common point of l_1, l_2, l_3, l_4 giyes a singularity of type b). The theorem is proved.

In Figure 5 we show the graphs of the degenerate fibers which appear if a family $\mathscr{X}$ with singularities of the degenerate fiber of types A′–C′ or A–C is replaced by the modification $\mathscr{X}$ constructed in Theorem 1.

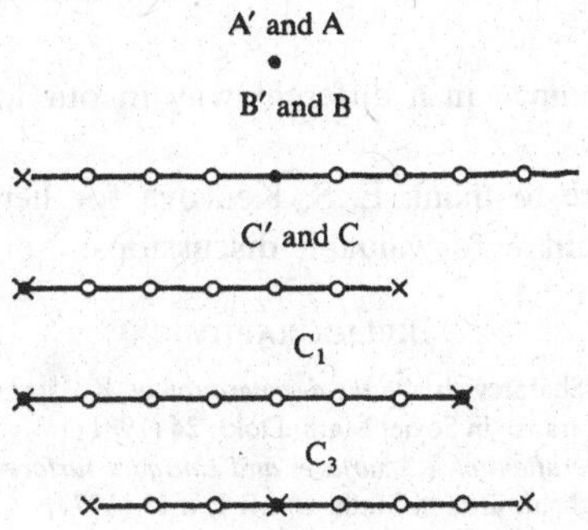

FIG. 5

Note that in our case, as directly follows from the description of singularities A′–C′ and A–C, the components of the degenerate fiber of the family $\mathscr{X}$ (denoted by dark points) are generalized K3 surfaces in cases A′ and A, elliptic ruled surfaces in cases B′ and B, and rational surfaces in case C. The ruled surfaces denoted by circles are elliptic.

Now we can literally carry over the result of [6] concerning families of strongly elliptic surfaces (Theorem 4) to arbitrary families of generalized K3 surfaces of degree 2. We call such a family *nondegenerating* if its closed fiber is also a generalized K3 surface.

THEOREM 3. *A family of generalized* K3 *surfaces of degree* 2 *with Hasse invariant zero has a nondegenerating modification* (*possibly, after base lifting*) *if and only if its monodromy acting on the cohomology group* $H^2_{\text{et}}(X_\eta, \mathbf{Q}_l)$ *of the generic fiber is finite.*

We can use Theorems 1 and 2. It remains to check that in a family $\mathscr{X} \to S$, whose degenerate fiber is one of the types B′, B, C′ or C shown on Figure 5, the monodromy is infinite. First of all, by the Brieskorn-Tjurina-Artin theorem, we can find a transformation $\mathscr{X}' \to S'$ (possibly after base covering) whose fibers do not have rational double points. As we noted at the beginning of the proof of Theorem 1, the only change in the degenerate fiber will be some elongation of chains. Hence its graph will still be among those in Figure 5. To study the monodromy of the family $\mathscr{X}' \to S'$ we use the following necessary condition for finiteness of monodromy (as was shown in [6], it follows from the results of [10]): the homomorphism

$$\bigoplus_i H^1(X_i, \mathbf{Q}_l) \to \bigoplus_{i<j} H^1(X_i \cap X_j, \mathbf{Q}_l) \tag{17}$$

is an epimorphism (here X_i stands for all the components of the degenerate fiber). But, as in [6], the rank of the source is always less than the rank of the target; hence this homomorphism cannot be epic. More precisely, if the number of the elliptic ruled surfaces in the degenerate fiber is n then the group in the left side of (17) has rank $2n$. The rank of the group in the right side (as we see from Figure 5) is $2n + 2$ in cases B′, B, C′, C_1 and C_2, and $2n + 4$ in case C_3.

As in [6], we get

COROLLARY. *A family of supersingular* K3 *surfaces in characteristic $p > 3$ has a nondegenerating modification* (*i.e. a modification whose closed fiber is also a* K3 *surface*).

This corollary was obtained in a different way in our joint work with Thomas Zink.

The authors would like to thank E. S. Kedrova for her help in preparing the manuscript, and V. S. Kulikov for valuable discussions.

BIBLIOGRAPHY

1. A. N. Rudakov and I. R. Shafarevich, *On the degeneration of* K3 *surfaces*, Dokl. Akad. Nauk SSSR **259** (1981), 1050–1052; English transl. in Soviet Math. Dokl. **24** (1981).

2. Viktor S. Kulikov, *Degenerations of* K3 *surfaces and Enriques surfaces*, Izv. Akad. Nauk SSSR Ser. Mat. **11** (1977), 1008–1042; English transl. in Math. USSR Izv. **11** (1977).

3. Jayant Shah, *Insignificant limit singularities of surfaces and their mixed Hodge structure*, Ann. of Math. (2) **109** (1979), 497–536.

4. Kyoji Saito, *Einfach-elliptische Singularitäten*, Invent. Math. **23** (1974), 289–325.

5. G. N. Tyurina, *Locally semiuniversal flat deformations of isolated singularities of complex spaces*, Izv. Akad. Nauk SSSR Ser. Mat. **33** (1969), 1026–1058; English transl. in Math. USSR Izv. **3** (1969).

6. A. N. Rudakov and I. R. Shafarevich, *On the degeneration of* K3 *surfaces over fields of finite characteristic*, Izv. Akad. Nauk SSSR Ser. Mat. **45** (1981), 646–661; English transl. in Math. USSR Izv. **18** (1982).

7. Eiji Horikawa, *Surjectivity of the period map of* K3 *surfaces of degree* 2, Invent. Math. **228** (1977), 113–146.

8. A. N. Rudakov, T. Tsink [Thomas Zink] and I. R. Shafarevich, *The influence of height on degenerations of algebraic* K3 *surfaces*, Izv. Akad. Nauk SSSR Ser. Mat. **46** (1982), 117-134; English transl. in Math. USSR Izv. **20** (1983).

9. Robin Hartshorne, *Algebraic geometry*, Springer-Verlag, 1977.

10. M. Rapoport and Th. Zink, *Über die lokale Zetafunktion von Shimuravariatäten. Monodromiefiltration und verschwindende Zyklen in ungleicher Charakteristik*, Invent. Math. **68** (1982), 21–101.

Translated by A. MARTSINKOVSKY

Notes

This section consists of comments on particular passages of some of the papers in this volume. The passages in question are identified by means of a number in the margin. In the comments below, the designation X.y refers to the passage numbered y on page X of this volume.

On the normalizability of topological fields

1.1 For a better exposition of the proof (with extension to noncommutative fields) see A. G. Kurosh, "Lectures on general algebra", Ch. 1, § 10, n. 5, Chelsea Publishing Company, N.J. 1965.

1.2 As far as I remember, I learned about the notion of a bounded set (in a linear topological space) in a conversation with I. M. Gelfand.

On Galois groups of $\mathfrak{p}$-adic fields

4.1 At the same time I considered also the case of a global field. The result is given in my doctoral thesis "Investigations on finite extensions" (Steklov Mathematical Institute, 1946). Proofs are completely parallel to the local case. For a description of the fundamental class I used the notion of "ideal algebra" introduced by E. Artin and discussed in a paper of S. MacLane and O. F. G. Schilling, Trans. Am. Math. Soc. **50,** 295–384 (1941). After the notion of the fundamental class was developed by A. Weil (J. Math. Soc. Japan **3,** 1–35 (1951)), E. Artin and J. Tate showed that the consideration of both local and global cases is incorporated in their general theory of Class Formations ("Class Field Theory", Princeton (1961)).

In the local case an analogous problem but taking into account also ramification groups was solved in: Shankar Sen and J. Tate, J. Indian Math. Soc. **27,** n. 3–4, 197–202 (1963).

On p-extensions

6.1 In a more modern terminology this means that the Galois group of the maximal p-extension of a local field k is a free pro-p-group if k does not contain any of the p-th roots of unity. If these roots are contained in k the Galois group of the maximal p-extension was determined in the papers: S. P. Demushkin, Izv. Akad. Nauk SSSR, Ser. Mat. **25,** 329–346

(1961) (if $p \neq 2$ or $\sqrt{-1} \in k$), J-P. Serre, Collected Papers II, n. 58 ($p=2$, $[k:\mathbb{Q}] = 2m+1$) and S. P. Demushkin, Sib. Mat. J. **4**, 951–955 (1963) (in the remaining case).

11.2 The groups $\mathbf{P}^{(i)}$ are in the modern terminology free groups in a certain variety of groups. The results obtained here were carried over to this more general setting by I. A. Maltsev: Mat. Sb. **26**, 19–33 (1950). As a result the proofs became much more transparent.

14.3 Recently P. Crew has generalized this formula to étale coverings of degree p of arbitrary algebraic varieties in characteristic p as a relation between certain Euler-Poincaré characteristics in cristalline cohomology: Compos. Math. **52**, I, 31–45 (1984).

A general reciprocity law

31.1 A more elegant exposition is due to H. Hasse: Math. Nachr. **5**, 322–327 (1957).

37.2 For a generalization to an arbitrary power of p, see A. I. Lapin, Izv. Akad. Nauk SSSR, Ser. Mat. **17**, 31–50 (1951).

51.3 In this paper I tried to pursue as far as possible the analogy between the norm residue symbol and the residue of a differential on an algebraic curve. A simpler expression for the norm residue symbol was given later by S. V. Vostokov: Izv. Akad. Nauk SSSR, Ser. Mat. **42**, 1288–1321 (1978). Cf. also lectures of H. Brückner at the University of Essen, about which I found a reference in the book of J. Neukirch "Class Field Theory", Springer-Verlag, 1986.

For generalizations to formal groups see S. V. Vostokov, Izv. Akad. Nauk SSSR, Ser. Mat. **43**, 765–792 (1978) and V. A. Kolivagin, same volume, 1054–1121. Cf. also A. Wiles, Ann. of Math. **107**, 235–254 (1978).

On the construction of fields with a given Galois group of order l^α

88.1 H. Koch noticed that in the paper of A. I. Skopin only the case $p \neq 2$ is considered. However it is easy to verify that for $p=2$ all the properties that we need are also valid. For this, consider a filtration $F^\nu = (N_c \cap N^\nu) N_{c+1}$ of the group $Z = N_c/N_{c+1}$, where $\{N^\nu\}$ is the descending central series. Part of the arguments of the paper [3] of the bibliography to this paper shows that $F^\nu/F^{\nu+1}$ is isomorphic to the module of Lie polynomials (for any P). Inverse images of the F in S define a central series – a refinement of $\{N_c\}$. We have then to replace the series $\{N_c\}$ by this refinement (this is what is actually done in § 3.5 of this paper). Cf. also paper 92 of the Bibliography.

89.2, 90.3 In both cases N_{c+1} should be replaced by N_{c+1}, N^1, where N^1 is the commutant of S. It is shown here that $x \mapsto x^{l^c}$ defines an isomorphism $\mathfrak{G}_d^{(c+1)}/\Phi(\mathfrak{G}_d^{(c+1)}) \to F^1$. (Cf. Note 1.)

Construction of fields of algebraic numbers with given solvable Galois group

139.1 The results of the papers 16–19 can now be deduced much more simply, although the main ideas of the proofs remain unchanged. The main reason why such a simplification became possible is the utilization of the duality theorems of Poitou-Tate for Galois cohomology of finite modules which were proved several years later. For a new exposition along these lines of the results of paper 19 see V. V. Ishanov, Izv. Akad. Nauk SSSR, Ser. Mat. **40,** 3–25 (1976). Concerning the paper 18 (as well as 34 and 39) see M. I. Bashmakov, Mat. Zametki **4,** 137–140 (1968) and J. Neukirch, Invent. Math. **21,** 59–116 (1973). A simplified and unified exposition can be found in the book "The Embedding Problem in Galois Theory" by V. V. Ishanov, B. B. Lurie and D. K. Faddeev, to be published by "Nauka", Moscow. Cf. also notes to papers 34 and 39.

Cohomology groups of nilpotent algebras

200.1 The analogy between the cohomology groups of algebras and loop-spaces of finite simply-connected *CW*-complexes (see note 2) suggests that an analogue of Theorem I is true for the loop-space of a bouquet of *CW*-complexes. Indeed, a generalization to coproducts in certain categories has been found which includes both these cases. See: P. J. Hilton a. G. Steer. Topology **8,** N. 3, 243–251 (1969).

202.2 In the case of a group ring of a finite group, rationality of the Poincarè series follows from the fact that cohomology ring is finitely generated, as proved by E. S. Golod, Dokl. Akad. Nauk SSSR **125,** 4, 703–706 (1959), and B. B. Venkov, Dokl. Akad. Nauk SSSR **127,** 943–944 (1959).

The question about the rationality of the Poincaré series connected with a nilpotent algebra or a noetherian local ring was independently raised by J-P. Serre: Lect. Notes in Math. II, IV-52 (1965). He also raised an analogous question for the loop space of a finite *CW*-complex. It turned out that both questions are connected: J. E. Roos, Lect. Notes in Math. **740,** 285–322 (1979). Unfortunately both questions have negative answers: D. Anick, C. R. Acad. Sci. Paris **290 A,** 729–732 (1980), J. E. Roos and C. Lofwall, same volume, 733–736.

The imbedding problem for split extensions

205.1 For a complete proof see: V. V. Ishanov, Izv. Akad. Nauk SSSR, Ser. Mat. **40,** 3–25 (1976).

The imbedding problem for local fields

231.1 Taking into account this Remark it is easy to see that Theorem 4 follows from the local duality theorem of Poitou-Tate (J. Tate, Intern. Congress of

Math., 288–295, Stockholm (1962), G. Poitou, Cohomologie galoisienne des modules finis. Dunod, 1967). More precisely, Theorem 4 is equivalent to the duality between 0-dimensional and 2-dimensional cohomology groups. Duality between 1-dimensional cohomology groups easily follows from this by a purely formal argument: cf. J-P. Serre, Lect. Notes in Math. **5,** 11–22 (1964). (Of course all this is "easy to see" only after the Poitou-Tate theorem has been formulated and proved.)

Principal homogeneous spaces defined over a function field

238.1 Most of the results of § 1–3 were independently obtained by A. Ogg: Ann. Math., II. Ser. **76,** 185–212 (1962).

248.2 This duality was investigated by O. Vedenski, Dokl. Akad. Nauk SSSR **219,** 6, 1291–1293, Izv. Akad. Nauk SSSR, Ser. Mat. **40,** 969–992 (1976) and L. Bégueri, Mémoires de la Soc. Math. de France, n. 4, 1980.

258.3 A. Grothendieck observed that Theorem 3 can be interpreted as a formula for the Euler-Poincaré characteristic of a sheaf on an algebraic curve in the étale topology. See J. S. Milne "Etale Cohomology", p. 190, Princeton (1980).

259.4 The proof of the theorem contains an error and it is not clear whether the theorem is true. It can be verified in the case of an elliptic curve where the group Z is equal to 0. The theorem is not used in the subsequent part of the paper.

260.5 The proof of the theorem is inaccurate. A minor change of the argument gives the correct answer: under the conditions of the theorem the variety is obtained from a constant one by means of a purely inseparable isogeny: cf. A. Grothendieck, Invent. Math. **2,** 59–78 (1966).

The second obstruction for the imbedding problem of algebraic number fields

272.1 The root numbers introduced in § 1 can be interpreted as elements of the group $H^1(\mathbf{G}, (A/A^1)')$ which as elements of the bigger group $H^1(\mathbf{G}, A')$ become trivial in all localizations (here **G** is the Galois group of the algebraic closure of the field Ω and A' is the character group of the group A). Using this it is easy to see that Theorem 1 is equivalent to the Poitou-Tate duality between the groups of locally trivial 1-dimensional and 2-dimensional cohomology classes (if one proves this duality by the usual method of "dévissage"). Cf. the note to paper 34.

Algebraic number fields

285.1 Only recently appeared a paper of Y. Ihara where this question was related with such arithmetical invariants as Jacobi sums etc.: "Profinite braid groups, Galois representations and complex multiplications", Ann. Math. **123,** 43–106 (1986).

287.2 The Galois group of the algebraic closure of the local field was described by means of generators and relations by A. V. Jakovlev: Izv. Akad. Nauk SSSR, Ser. Mat. **32,** 2131–2169 (1968). See also corrections in the same journal, **43,** 212–213 (1978). It was investigated further in the papers of U. Jannsen and U. Jannsen a. K. Wingberg: Invent. Math. **70,** 53–69 and 71–98 (1982).

288.3 This result and the inequality (3) are proved in paper 42.

289.4 This conjecture is proved in paper 43.

290.5 Here two statements made in the lecture are mixed into one: formulation of a result and of a conjecture. The result was restricted to the case of hyperelliptic curves while the conjecture concerned general curves. For a correct formulation of both the result and the conjecture see the talk of A. N. Parshin: Congrès Intern. de Math. Nice I, 467–471 (1970).

The proof of the result about hyperelliptic curves was never published, but the proof in the case of elliptic curves was reproduced in: J-P. Serre, "Abelian *l*-adic Representations and Elliptic Curves" Benjamin, IV-7 (1968) and S. Lang, "Elliptic Functions", Addison-Wesley, p. 231 (1973). The proof in the hyperelliptic case is practically identical.

The conjecture became much more attractive after A. N. Parshin proved that it implies the Mordell conjecture: cf. his talk cited above. In 1983 it was proved by G. Faltings (Invent. Math. **73,** 439–366 (1983)), with a proof of the Mordell conjecture as a consequence.

291.6 This question was answered positively even in a more general form: there are no abelian schemes over $\mathbb{Z}$. This was proved first in dimensions 2 and 3 by V. A. Abrashkin (Izv. Akad. Nauk SSSR, Ser. Mat. **41,** 937–956 (1977)), who connected it with the theory of finite group schemes over $\mathbb{Z}$, and then in the general case by J-M. Fontaine (Invent. Math. **81,** 515–538 (1985)), who also used finite group schemes over $\mathbb{Z}$, and later but independently by V. A. Abrashkin (Dokl. Akad. Nauk SSSR **283,** 6, 1289–1293 (1985)).

292.7 Cf. 290.5.

293.8 A positive answer to this question was obtained through the joint efforts of A. N. Parshin (Izv. Akad. Nauk SSSR, Ser. Mat. **32,** 1191–1219 (1968)) and S. Yu. Arakelov (Izv. Akad. Nauk SSSR, Ser. Mat. **35,** 1269–1293 (1971)).

Extensions with given ramification points

307.1 For other cases where in a similar situation one can determine all relations see: H. Koch "Galoissche Theorie der *p*-Erweiterungen", Berlin 1970.

312.2 Cf. paper 43.

313.3 On the basis of the result of paper 43 it was possible to prove that the minimal number of relations of a closed *p*-group is $\leqq 3$ and to clarify to a considerable extent their structure, although their complete description is still lacking. Cf. e.g. I. V. Andojski and V. M. Zvetkov, Izv. Akad. Nauk, Ser. Mat. **49,** 1322–1328 (1985).

On class field towers

317.1 It has been proved that this estimate is exact, i.e. $z(d)/d^2 \to \frac{1}{4}$ if $d \to \infty$. See J. Wislicency, Math. Nachr. **102,** 57–78 (1981).

320.2 By improving this method used in the proof of this theorem E. S. Golod solved several well-known problems of algebra: the Burnside problem (for an unbounded exponent), the problem of Kurosh about algebraic algebras, etc.; see Izv. Akad. Nauk SSSR, Ser. Mat. **28,** 273–276 (1964).

327.3 Now simpler examples are known, e.g. $\mathbb{Q}(\sqrt{-3\cdot 7\cdot 17\cdot 19})$. See J. Martinet, Invent. Math. **44,** 65–73 (1978). One also knows quadratic fields with prime discriminant for which the class field tower is infinite, e.g. $\mathbb{Q}(\sqrt{-p})$, $p = 3321607$ or 4724490703. See Diaz y Diaz et al. Math. Comp. **93,** 146, 836–840 (1979). It was proved also that the class field tower is infinite for every field normal over $\mathbb{Q}$ whose discriminant has sufficiently many prime divisors compared to its degree over $\mathbb{Q}$: A. Brumer and M. Rosen, Nagoya Math. Journ. **23,** 97–101 (1963).

328.4 Now these estimates have been improved considerably, e.g. for imaginary fields $23 \leqq \liminf \sqrt[n]{D_n} \leqq 93$ and assuming the Generalized Riemann Hypothesis one can replace 23 by 45. See the paper of Martinet cited above.

Algebraic surfaces

332.1 This theory was extended to algebraic surfaces over fields of arbitrary characteristic in a series of papers of E. Bombieri and D. Mumford. See a survey of E. Bombieri and D. Husemoller in "Algebraic Geometry", Arcata, 329–340 (1974).

Galois theory of transcendental extensions and uniformization

388.1 It seems that the idea of considering projective limits of algebraic varieties on which locally compact groups are acting "almost transitively" occurred independently to several mathematicians: cf. D. Mumford, Congr. Intern. de Math. Nice I, 457–465 (1970) and the book of G. Shimura "Introduction to the Arithmetic Theory of Automorphic Functions", Princeton (1971).

389.2 This conjecture was proved for algebraic curves uniformized by a certain class of discrete groups acting on an upper half plane in a paper of D. Každan: Funct. Anal. Appl. **2,** 36–39 (1968).

391.3 Cf. note 6.

404.4 All proalgebraic varieties are assumed irreducible and normal.

406.5 Omit this sentence.

417.6 No proof of the result announced in the paper 28 was published and the result, as the work of Igusa shows, is incorrect. So this case of Proposition 4 remained unproved as well as proposition 1 case 3) that is based on it. It was later proved by T. Miyake, Ann. Math. **95,** 243–252 (1972).

Cartan pseudogroups and Lie *p*-algebras

423.1 This conjecture has been proved: R. Block and R. Wilson, Proc. Nat. Acad. Sci. USA **81,** 5271–5274 (1984).

423.2 For the proof see paper 62.

On some infinitedimensional groups

431.1 Here I used the so-called "Jacobian conjecture". At the time when this paper was written I was not aware that all published "proofs" of this conjecture are incorrect. However for our purpose it is unnecessary: see paper 81.

432.2 Theorems 1–5 were generalized by M. H. Gisatullin and V. I. Danilov to a very wide class of affine algebraic surfaces. See their paper: Izv. Akad. Nauk SSSR, Ser. Mat. **39,** 523–565 (1975).

434.3 For an answer see the papers of M. H. Gisatullin, Izv. Akad. Nauk, Ser. Mat. **35,** 1047–1071 (1971) and M. H. Gisatullin and V. I. Danilov, Izv. Akad. Nauk SSSR, Ser. Mat. **38,** 42–118 (1974).

Irreducible representations of a simple three-dimensional Lie algebra over a field of finite characteristic

442.1 For a generalization of a part of the results to arbitrary Lie algebras, see F. D. Veldkamp, Journ. Pure Appl. Algebra **2,** 231–248 (1972).

Results of our paper are clarified by a paper of A. N. Rudakov: Izv. Akad. Nauk SSSR, Ser. Mat. **34,** 231–248 (1972). Namely, the universal family constructed in our paper has dimension 3 while irreducible complex representations of the group A_1 (the Lorenz group) depend on 1 parameter. It may seem that this contradicts the analogy of the two phenomena asserted in the Introduction. However, in the paper of Rudakov cited above it is proved that, in contrast to the classical case, inner automorphisms of the algebra act in a nontrivial way on the set of its irreducible representations. In our case the quotient by this action is one-dimensional – just as it should be!

Le théorème de Torelli pour les surfaces algébriques de type *K*3

514.1 A complete theory of isogenies for singular *K*3-surfaces (i.e. those for which Picard number = 20) was developed by H. Inose and T. Shioda: "Complex analysis and Algebraic Geometry", Cambridge Univ. Press, 119–136 (1977). S. Mukai proved that the cohomology class defining an isogeny is algebraic if the Picard number is ≧11: "On the moduli space of bundles on $K3$ surfaces". Preprint, Nagoya Univ. **40,** 1983. V. V. Nikulin generalized it to $K3$-surfaces containing an elliptic curve: Usp. Mat. Nauk, **40,** n. 5, 212–213 (1985).

514.2 These are the so-called supersingular $K3$-surfaces. Their theory was initiated by M. Artin: Ann. Sci. Ec. Norm. Sup. Ser. **4,** 7, 543–567 (1974). Cf. also papers 77, 79, 82, 87.

514.3 In this connection see for example A. Beauville, Journ. of Diff. Geom. **18,** 755–782 (1983).

A Torelli theorem for algebraic surfaces of type $K3$

518.1 V. S. Kulikov proved the existence of an algebraic $K3$-surface with arbitrary preassigned periods satisfying conditions (1) and (2): Usp. Mat. Nauk **32,** 257–258 (1977).

520.2 In this formulation the Torelli theorem for Kählerian $K3$-surfaces was proved by D. Burns and M. Rapoport (following an idea of P. Deligne) Ann. Sci. Ec. Norm. Sup. Ser. **4,**8, 235–274 (1975). As every complex analytic $K3$-surface is Kählerian (Y.-T. Siu, Invent. Math. **73,** 139–165 (1983)), this conjecture is now established in full generality.

536.3 In this formulation Lemma 4 is incorrect. Fortunately, isomorphisms arising from Kummer surfaces enjoy the property stated in the Lemma. See V. V. Nikulin, Izv. Akad. Nauk SSSR, Ser. Mat. **39,** 261–275 (1975).

547.4 This theorem is due to D. G. James: Pacific Journ. of Math. **26,** 261–275 (1975). The proof is quite elementary.

The arithmetic of $K3$-surfaces

559.1 This is obviously incorrect: the domain $\Omega(l)$ has 2 components. The reasoning is correct but proves that S has 2 components. The same applies to Corollaries 1 and 2. In particular the image of the monodromy in the group of automorphisms of the lattice of primitive cycles has index 2: these are automorphisms with spinor-norm 1. From this it is easy to prove that in the group of automorphisms of the full cohomology group, all elements with spinor-norm 1 are induced by diffeormorphisms. Recently S. K. Donaldson proved that the remaining automorphisms are not induced by diffeomorphisms: see Proceedings of the Georgia Topology Conference 1985. Preprint.

560.2 Here only the subgroup of primitive cycles is intended instead of the full cohomology group.

560.3 In the meantime P. Deligne has proved the Riemann Hypothesis for arbitrary projective varieties. However, the semisimplicity of the Frobenius endomorphism, which for $K3$ surfaces follows from our proof, is still not proved in the general case.

Inseparable morphisms of algebraic surfaces

607.1 The main application of the theorem: the existence of a lifting of a $K3$-surface to characteristic 0. See: P. Deligne and L. Illusie, Lect. Notes in Math. 868, 58–72. Simultaneously it was proved that the number of moduli of an algebraic $K3$ surface is equal to 19. For a simpler proof of Theorem 7, see paper 78. For a proof based on another principle, see N. O. Nygaard, Ann. of Math. **110,** 515–528 (1979).

Surfaces of type $K3$ over fields of finite characteristic

712.1 In the case of $K3$-surfaces that are representable as double planes this is proved in the paper 82.

712.2 This is proved in papers 82 and 87.

713.3 Cf. papers 82, 87 and the note 1 to paper 82.

713.4 This question was also raised by P. Deligne. For an answer see A. Ogus, Arithmetic and Geometry, Birkhäuser, II, 361–394 (1983).

The influence of height on degeneration of algebraic surfaces of type $K3$

716.1 To do this it is necessary to construct a moduli space of supersingular $K3$-surfaces. This can be done in an elementary way using the fact that all such surfaces are representable as double planes with a ramification curve of degree 6. However, small characteristics require a separate discussion. A totally different construction, valid for all characteristics, is given in: A. Ogus, Arithmetic and Geometry, Birkhäuser, II, 361–394 (1983).

722.2 Cf. note on p. 747.

725.3 A commentary to the translation: "of degree 2" means "representable as a double plane".

Bibliography

The papers and monographs are listed in chronological order. The page numbers in the right-hand column indicate where an article can be found in these Collected Papers. Articles marked with an asterisk (*) were translated especially for this publication. References to the reviews published in Zentralblatt für Mathematik (Zbl.) have been given as far as could be ascertained.

1943

1. On the normalizability of topological fields. Dokl. Akad. Nauk SSSR **40,** 149–151 (1943). In English: C. R. Acad. Sci. (Dokl.) Paris **40,** 133–135 (1943) 1

1945

2. On absolute Galois-groups of relative-abelian 𝔭-adic fields. In: Referaty nauchno-issledovatel'skikh rabot za 1943–44. Otd. Fiz.-matem. Nauk Akad. Nauk SSSR, Moscow Leningrad (1945) **67** (in Russian)

1946

3. On 𝔭-extensions. In: Referaty nauchno-issledovatel'skikh rabot za 1945. Otd. Fiz.-matem. Nauk Akad. Nauk SSSR, Moscow Leningrad (1946) **55** (in Russian)
4. On Galois groups of 𝔭-adic fields. Dokl. Akad. Nauk SSSR **53,** No. 1, 15–16 (1946). In English: C. R. Acad. Sci. (Dokl.) Paris **53,** 15–16 (1946) 4
5. Investigations on finite extensions of fields. Résumé of a doctoral thesis. Usp. Mat. Nauk **2,** No. 2, 223–226 (1946) (in Russian)

1947

6. On 𝔭-extensions. Mat. Sb., Nov. Ser. **20** (62), 351–363 (1947). Zbl. **41,** 171. In English: Transl., II. Ser., Am. Math. Soc. **4,** 59–72. Zbl. **71,** 33 6

1948

7. On a general law of reciprocity. Usp. Mat. Nauk **3,** No. 3 (25), 165 (1948) (in Russian)

1949

8. On a general law of reciprocity. Dokl. Akad. Nauk SSSR **64,** 25–28 (1949) (in Russian) Zbl. **32,** 392

1950

9. A general reciprocity law. Mat. Sb., Nov. Ser. **26** (68), 113–146 (1950). Zbl. **36,** 159. In English: Transl., II. Ser., Am. Math. Soc. **4,** 73–106 (1956). Zbl. **71,** 33 20
10. Algebraic geometry. Bolsh. Sov. Ehnts., 2nd ed., vol. 2, 62–63 (1950) (in Russian)

1951

*11. A new proof of the Kronecker-Weber theorem. Tr. Mat. Inst. Steklova **38,** 382–387 (1951). Zbl. **53,** 355 54

1952

12. On a general law of reciprocity and its applications in the theory of algebraic numbers. C. R. 1. Congr. Math. Hungr. 1950, 291–298 (1952) (in Russian and Hungarian). Zbl. **49,** 28
13. A conference on algebra and number theory. Usp. Mat. Nauk **7,** No. 3, 151–154 (1952) (in Russian)

1953

14. A commentary to the note: On the number of solutions of a congruence of degree 3. In: Voronoj, G. F.: Collected Papers, Vol. 3, 205. Kiev: Izd. Akad. Nauk Ukrainskoj SSR (1953) (in Russian)
15. A commentary to the paper: Remarks on Fermat's last theorem concerning the unsolvability of the equation $x^p + y^p = z^p$ in integers x, y and z with an odd prime p. In: Voronoj, G. F. Collected Papers, Vol. 3, 247. Kiev: Izd. Akad. Nauk Ukrainskoj SSR (1953) (in Russian)

1954

16. On the construction of fields with a given Galois group of order l^a. Izv. Akad. Nauk SSSR, Ser. Mat. **18,** 261–296 (1954). Zbl. **56,** 33. In English: Transl., II. Ser., Am. Math. Soc. **4,** 107–142 (1956). Zbl. **71,** 33 62
17. On an existence theorem in the theory of algebraic numbers. Izv. Akad. Nauk SSSR, Ser. Mat. **18,** 327–334 (1954). Zbl. **56,** 34. In English: Transl., II. Ser., Am. Math. Soc. **4,** 143–150 (1956). Zbl. **71,** 33 98
18. On the problem of imbedding fields. Izv. Akad. Nauk SSSR, Ser. Mat. **18,** 389–418 (1954). Zbl. **57,** 33. In English: Transl., II. Ser., Am. Math. Soc. **4,** 151–183 (1956). Zbl. **71,** 33 106

19. Construction of fields of algebraic numbers with given solvable Galois group. Izv. Akad. Nauk SSSR, Ser. Mat. **18,** 525–578 (1954). Zbl. **57,** 274. In English: Transl., II. Ser., Am. Math. Soc. **4,** 185–237 (1956). Zbl. **71,** 33 139

20. On extensions of fields of algebraic numbers solvable in radicals. Dokl. Akad. Nauk SSSR **95,** No. 2, 225–227 (1954) (in Russian) Zbl. **55,** 31

*21. On the imbedding problem for fields. Dokl. Akad. Nauk SSSR **95,** No. 3, 459–461 (1954). Zbl. **55,** 31 59

22. On the solution of equations of higher degree (Sturm method). Moscow: Gostekhteorizdat (1954) 24 p. Zbl. **56,** 254. In German: Berlin: Deutscher Verlag der Wissenschaften (1956) 29 p. zbl. **72,** 138

23. Preface to "automorphic functions of several complex variables" by C. L. Siegel. Moscow-Leningrad. Izd. Innostrannoj Literatury (1954) 3–4 (in Russian)

24. (with D. E. Menshov et al.) 16th Moscow Mathematics Olympiad. Usp. Mat. Nauk **9,** No. 3, 257–262 (1954) (in Russian)

1956

25. Galois theory and arithmetic of number fields. Tr. 3-go Vsesoyuznogo Mat. S'ezda **8.** Moscow (1956) (in Russian)

1957

*26. On birational equivalence of elliptic curves. Dokl. Akad. Nauk SSSR **114,** No. 2, 267–270 (1957). Zbl. **81,** 153 192

*27. Exponents of elliptic curves. Dokl. Akad. Nauk SSSR **114,** No. 4, 714–716 (1957). Zbl. **81,** 154 197

*28. (with A. I. Kostrikin) Cohomology groups of nilpotent algebras. Dokl. Akad. Nauk SSSR **115,** No. 6, 1066–1069 (1957). Zbl. **83,** 30 200

1958

*29. The imbedding problem for split extensions. Dokl. Akad. Nauk SSSR **120,** No. 6, 1217–1219 (1958). Zbl. **85,** 26 205

30. Dmitrij Konstantinovich Faddeev (on his 50th birthday). Usp. Mat. Nauk **13,** No. 1 (79), 233–236 (1958). Zbl. **78,** 3

31. Analytic manifolds and algebraic geometry (Survey article). Usp. Mat. Nauk **13,** No. 2, 233 (1958) (in Russian)

*32. Impressions from the International Mathematical Congress in Edinburgh. Usp. Mat. Nauk **14,** No. 2, 243–246 (1958) 233

1959

*33. The group of algebraic principal homogeneous varieties. Dokl. Akad. Nauk SSSR **124,** No. 1, 42–43 (1959). Zbl. **115,** 389 208

34. (with S. P. Demushkin) The imbedding problem for local fields. Izv. Akad. Nauk SSSR, Ser. Mat. **23,** No. 6, 823–840 (1959). Zbl. **93,** 44. In English: Transl., II. Ser., Am. Math. Soc. **27,** 267–288 (1963). Zbl. **127,** 265 211

1961

35. Principal homogeneous spaces defined over a function field. Tr. Mat. Inst. Steklova **64,** 316–346 (1961). Zbl. **129,** 128. In English: Transl., II. Ser., Am. Math. Soc. **37,** 85–114 (1964). Zbl. **142,** 184 237
36. Boris Nikolaevich Delone (on his 70th birthday). Usp. Mat. Nauk **16,** No. 3, 239–241 (1961). Zbl. **98,** 9. In English: Russ. Math. Surv. **16,** No. 3, 151–156 (1961)

1962

37. In memory of Francesco Severi. Vestn. Akad. Nauk SSSR **2,** 99–100 (1962) (in Russian)
38. Contributii Sovietice la theoria lui Galois. Bucuresti, 1–167 (1962). Acad. Republici Populare Romine. Inst. de Studii Romino-Sovietica, Biblioteca analelor Romino-Sovetice. Seria teknica No. 4. Asupra problemei scufundarii corpurilor, An. Rom.-Soviet., Ser. Mat.-Fiz. **16,** No. 3, 3–36 (1962). Construirea corparilor de numere algebrice cu grup al lui Galois rezolubil. An. Rom.-Soviet., Ser. Math.-Fiz. **16,** No. 3, 37–93 (1962)
39. (with S. P. Demushkin) The second obstruction for the imbedding problem of algebraic number fields. Izv. Akad. Nauk SSSR, Ser. Mat. **26,** No. 6, 911–924 (1962). Zbl. **115,** 37. In English: Transl., II. Ser., Am. Math. Soc. **58,** 245–260 (1966). Zbl. **199,** 98 267

1963

40. Algebraic number fields. Proc. Int. Congr. Math., Stockholm 1962, Inst. Mittag-Leffler, Djursholm, 163–176 (1963). Zbl. **126,** 69. In English: Transl., II. Ser., Am. Math. Soc. **31,** 25–39 (1963). Zbl. **133,** 293 283
41. Einige Anwendungen der Galoisschen Theorie und Diophantischen Gleichungen. Ber. Dirichlet-Tagung, (Schr. Inst. Math. Berlin 13) **81, 82** (1963). Zbl. **114,** 264
42. Extensions with given ramification points. Publ. Math., Inst. Haut. Etud. Sci. **18,** 295–319 (1964). Zbl. **118,** 275. In English: Transl., II. Ser., Am. Math. Soc. **59,** 128–149 (1966). Zbl. **199,** 97 295

1964

43. (with E. S. Golod) On class field towers. Izv. Akad. Nauk SSSR, Ser. Mat. **28,** No. 2, 261–272 (1964). Zbl. **136,** 26. In English: Transl., II. Ser., Am. Math. Soc. **48,** 91–102 (1965). Zbl. **148,** 281 317
44. Yurij Manin. Molodoj Kommunist **3,** 61 (1964) (in Russian)
45. (with S. P. Novikov and I. I. Piatetskij-Shapiro) Fundamental directions in the development of algebraic topology and algebraic geometry. Usp. Mat. Nauk **19,** No. 6, 75–82 (1964). Zbl. **168,** 439. In English: Russ. Math. Surv. **19,** No. 6, 67–73 (1964)

46. (with Z. I. Borevich) Number theory. Moscow: Nauka (1964). Zbl. **121,** 42. 2nd ed. 1972. 3rd compl. ed. 1985, 504 p. Zbl. **592,** 12001. In German: Basel-Stuttgart: Birkhäuser Verlag (1966). Zbl. **134,** 273. In English: New York-London: Academic Press (1966). Zbl. **145,** 49. In French: Paris: Gauthier-Villars (1967). Zbl. **145,** 49. In Japanese: Series of Math. Sc. **14.** Tokyo: Yoshioka Shoten (1971)
47. Conference on number theory. Oberwolfach (FRG), 6–12 Sept. 1964. Vestn. Akad. Nauk SSSR **12,** 63–64 (1964)
48. Textbook on higher algebra. Moscow: MGU (1963) 1–38 (in Russian)

1965

49. (with B. G. Averbukh et al.) Algebraic surfaces. Tr. Mat. Inst. Steklova **75,** 3–215 (1965). Zbl. **154,** 210. In English: Proc. Steklov Inst. Math. **75** (1965), 281 p. (1967). In German: Leipzig: Akademische Verlagsgesellschaft 303 p. (1968). Zbl. **154,** 210.
Preface, Ruled Surfaces, Surfaces with a pencil of elliptic curves 329
50. Preface to the multi-author volume Complex Spaces. Moscow: MIR (1965) 5–10 (in Russian)

1966

51. (with A. I. Kostrikin) Cartan pseudogroups and Lie p-algebras. Dokl. Akad. Nauk SSSR **168,** No. 4, 740–742 (1966). Zbl. **158,** 38. In English: Sov. Math., Dokl. **7,** 715–718 (1966) 422
52. (with A. I. Kostrikin) Cartan pseudogroups and Lie p-algebras. Tezisy Tr. Nauchn. Soobshch. Mezhdunarod. Kongressa Matematikov, Sekt. 2, Moscow: MIR (1966) 44 (in Russian)
53. (with I. I. Piatetskij-Shapiro) Galois theory of transcendental extensions and uniformization. Izv. Akad. Nauk SSSR, Ser. Math. **30,** No. 3, 671–704 (1966). Zbl. **218,** 14024. In English: Transl., II. Ser., Am. Math. Soc. **69,** 111–145 (1968) 387
54. (with I. I. Piatetskij-Shapiro) Galois theory of transcendental extensions and uniformization. In: Contemporary Problems in the Theory of Analytic Functions, Erevan 1965, 262–264. Moscow: Nauka (1966) (in Russian). Zbl. **196,** 532
55. Lectures on minimal models and birational transformations of two-dimensional schemes. Tata Institute of Fundamental Research, Bombay (1966) 175 p. Zbl. **164,** 517
56. (with A. A. Kirillov) The second summer school in topology. Usp. Mat. Nauk **21,** No. 2, 257–258 (1966) (in (Russian)
57. Über das Klassenkörperturmproblem. Berichte, Math. Forschungsinst. Oberwolfach, Heft 2, 265 (1966)

1967

58. On some infinitedimensional groups. In: Simposio Internazionale di Geometria Algebrica. Roma: Edizioni Cremonese, 208–212 (1967) = Rend. Mat. Appl., V. Ser. **25,** 208–212 (1966). Zbl. **149,** 390 430
59. (with A. N. Rudakov) Irreducible representations of a simple 3-dimensional Lie algebra over a field of finite characteristic. Mat. Zametki **2,** No. 5, 439–454 (1967). Zbl. **184,** 60. In English: Math. Notes **2,** Nos, 1, 2, 760–767 (1968) 435
60. (with J. T. Tate) The rank of elliptic curves. Dokl. Akad. Nauk SSSR **175,** No. 4, 770–773 (1967). Zbl. **168,** 422. In English: Sov. Math., Dokl. **8,** No. 4, 917–920 (1967) 426

1968

61. Algebraic geometry. Moscow: MGU (1968) 250 p. (in Russian) (preliminary version of 64)

1969

62. (with A. I. Kostrikin) Graded Lie algebras of finite characteristic. Izv. Akad. Nauk SSSR, Ser. Mat. **33,** No. 2, 251–322 (1969). Zbl. **211,** 53. In English: Math. USSR, Izv. **3,** No. 2, 237–304 (1970) 443
63. Zeta functions. Moscow: MGU (1969) 148 p.
64. Foundations of algebraic geometry. Usp. Mat. Nauk **24,** No. 6, 3–184 (1969). Zbl. **204,** 213. In English: Russ. Math. Surv. **24,** No. 6, 1–178 (1969). In German: Friedrich Vieweg & Sohn (1972). Zbl. **248,** 14001. In Hungarian: (in 2 Parts) Magyar Tud. Akad., mat. fiz. Tud. Oszt., Koezl. **22,** 79–184 (1974) and **22,** 283–360 (1975)

1970

65. Einleitung zu H. Koch: Galoissche Theorie der *p*-Erweiterungen. Berlin: Deutscher Verlag der Wissenschaften. Mathematische Monographien. Bd. 1 (1970) 3–4. Zbl. **216,** 47
66. Le Théorème de Torelli pour les surfaces algébriques de type K3. Actes Congrès Intern. Math., Nice 1970, **1,** 413–417 (1971). Zbl. **236,** 14016 511
67. (with I. I. Piatetskij-Shapiro) A Torelli theorem for algebraic surfaces of type K3. Izv. Akad. Nauk SSSR, Ser. Mat. **35,** 530–572 (1971). Zbl. **219,** 14021. In English: Math. USSR, Izv. **5,** No. 3, 547–588 (1972). Zbl. **253,** 14006 516
68. Lectures on higher algebra. Moscow: MGU (1971) 40 p. (in Russian)
69. (with V. I. Arnold, I. M. Gelfand, Yu. I. Manin, B. G. Mojshezon, S. P. Novikov) Galina Nikolaevna Tyurina (orbituary). Usp. Mat. Nauk **26,** No. 1 (157), 207–211 (1971). Zbl. **205,** 299. In English: Russ. Math. Surv. **26,** No. 1, 193–198 (1972). Zbl. **232,** 01016

70. Basic algebraic geometry. Moscow: Nauka (1972) 567 p. Zbl. **258,** 14001. In German: Berlin: Deutscher Verlag der Wissenschaften (1972). Zbl. **318,** 14001. In English: Grundlehren der math. Wiss. 213. Berlin Heidelberg New York: Springer-Verlag (1974). Zbl. **284,** 14001. Springer Study Edition of Grundlehren der math. Wiss. 213. Berlin Heidelberg New York: Springer-Verlag (1977). Zbl. **362,** 14001. In Rumanian: Buchresti: Editura Stiintifica si enciclopedia (1976)

1973

71. (with I. I. Piatetskij-Shapiro) The arithmetic of K3 surfaces. Tr. Mat. Inst. Steklova **132,** 44–54 (1973). Zbl. **293,** 14010. In English: Proc. Steklov Inst. Math. **132,** 45–57 (1975). Zbl. **305,** 14014. 558
72. Über einige Tendenzen in der Entwicklung der Mathematik. Jahrb. Akad. Wiss. Göttingen, 31–36 (1973). In English: Math. Intell. **3,** 182 –184 (1981) 571

1976

73. (with A. N. Rudakov) Inseparable morphisms of algebraic surfaces. Izv. Akad. Nauk SSSR, Ser. Mat. **40,** 1269–1307 (1976). Zbl. **365,** 14008. In English: Math. USSR, Izv. **10,** No. 6, 1205–1237 (1976). Zbl. **379,** 14006 577

1977

74. Remarks on the paper "Inseparable morphisms of algebraic surfaces". Izv. Akad. Nauk SSSR, Ser. Mat. **41,** 476 (1977) (in Russian). Zbl. **365,** 14009

1978

75. (with A. N. Rudakov) Quasi-elliptic surfaces of type K3. Usp. Mat. Nauk **33,** No. 1, 227–228 (1978). Zbl. **383,** 14013. In English: Russ. Math. Surv. **33,** No. 1, 215–216 (1978). Zbl. **404,** 14012 610
76. (with A. N. Rudakov) Vector fields on elliptic surfaces. Usp. Mat. Nauk **33,** No. 6, 231–232 (1978). Zbl. **411,** 14008. In English: Russ. Math. Surv. **33,** No. 6, 255–256 (1978). Zbl. **424,** 14010 612
77. (with A. N. Rudakov) Supersingular K3 surfaces over fields of characteristic 2. Izv. Akad. Nauk SSSR, Ser. Mat. **42,** 848–869 (1978). Zbl. **404,** 14010. In English: Math. USSR, Izv. **13,** No. 1, 147–165 (1979). Zbl. **424,** 14010 614

1981

78. (with A. N. Rudakov) Surfaces of type K3 over fields of finite characteristic. In: Itogi Nauki Tekh., Ser. Sovrem. Probl. Mat. **18,** 115–207 (1981). Zbl. **518,** 14015. In English: J. Sov. Math. **22,** No. 4, 1476–1533 (1983) 657

79. (with A. N. Rudakov) On the degeneration of K3 surfaces over fields of finite characteristic. Izv. Akad. Nauk SSSR, Ser. Mat. **45,** 646–661 (1981). Zbl. **465,** 14014. In English: Math. USSR, Izv. **18,** No. 3, 561–574 (1982). Zbl. **491,** 14025 633
80. (with A. N. Rudakov) On the degeneration of surfaces of type K3. Dokl. Akad. Nauk SSSR **259,** 1050–1052 (1981). Zbl. **524,** 14031. In English: Sov. Math., Dokl. **24,** No. 1, 163–165 (1981)
81. On some infinite-dimensional groups. II. Izv. Akad. Nauk SSSR, Ser. Mat. **45,** No. 1, 214–226 (1981). Zbl. **475,** 14036. In English: Math. USSR, Izv. **18,** No. 1, 185–194 (1982). Zbl. **491,** 14025 647

1982

82. (with A. N. Rudakov and T. Zink) The influence of height on degenerations of algebraic surfaces of type K3. Izv. Akad. Nauk SSSR, Ser. Mat. **46,** 117–134 (1982). Zbl. **492,** 14024. In English: Math. USSR, Izv. **20,** No. 1, 119–135 (1983). Zbl. **509,** 14036 715

1983

83. (with V. V. Nikulin) Geometries and groups. Moscow: Nauka (1983) 240 p. Zbl. **528,** 51001. In English: Berlin Heidelberg New York Tokyo: Springer-Verlag (1987)
84. Zum 150. Geburtstag von Alfred Clebsch. Math. Ann. **266,** 135–140 (1983). Zbl. **519,** 01013 732
85. Preface to "Etale cohomologies" by J. S. Milne. Moscow: MIR (1983) 5–6. Zbl. **544,** 14015

1984

86. (with A. N. Parshin) On the arithmetic of algebraic manifolds. Tr. Mat. Inst. Steklova **168,** 72–97 (1984). Zbl. **605,** 14001. In English: Proc. Steklov Inst. Math. **168,** 75–99 (1986)
87. (with A. N. Rudakov) On the degeneration of K3 surfaces. Tr. Mat. Inst. Steklova **166,** 222–234 (1984). Zbl. **577,** 14027. In English: Proc. Steklov Inst. Math. **166,** 247–259 (1986). Zbl. **592,** 14035 738
88. Henri Poincaré. Thoughts on science (review). Tekhnika i Nauka **2,** 42–43 (1984) (in Russian)
89. Basic Notions of Algebra. In: Itogi Nauki i Tekh. Ser. Sovrem. Probl. Mat. Fundam. Napr. v. **11,** 1–288 (1986). In English: Encyclopaedia of Mathematical Sciences, Vol. 11. Berlin Heidelberg New York (in preparation)
90. On Luroth's problem. Tr. Mat. Inst. Steklova **183** (1988)

To appear

91. (with V. A. Iskovski) Algebraic Surfaces. In: Itogi Nauki i Tekh. Ser. Sovrem. Probl. Mat. Fundam. Napr.
92. On factors of a descending central series. Mat. Zametki

Acknowledgements

We would like to thank the original publishers of I. R. Shafarevich's papers for granting permission to reprint them here.

The numbers following each source correspond to the numbering of the bibliography.

Reprinted from *Actes Congrès Intern. Math.* © Gauthier-Villars: 66
Reprinted from *C. R. Acad. Sci. (Doklady)* © Acad. Sci.: 1, 4
Reprinted from *J. Sov. Math.* © Plenum Publishing Corporation: 78
Reprinted from *Jahrbuch der Akademie der Wissenschaften, Göttingen* © Akademie der Wissenschaften: 72
Reprinted from *Math. Notes* © Plenum Publishing Corporation: 59
Reprinted from *Math. USSR, Izv.* © American Mathematical Society: 62, 67, 73, 77, 79, 81, 82
Reprinted from *Proc. Steklov Inst. Math.* © American Mathematical Society: 49, 71, 87
Reprinted from *Russ. Math. Surv.* © London Mathematical Society: 75, 76
Reprinted from *Sov. Math., Dokl.* © American Mathematical Society: 51, 60
Reprinted from *Transl., II. Ser., Am. Math. Soc.* © American Mathematical Society: 6, 9, 16, 17, 18, 19, 34, 35, 39, 40, 42, 43, 53